PROGRESS IN BRAIN RESEARCH
VOLUME 50
REFLEX CONTROL OF POSTURE AND MOVEMENT

Recent volumes in PROGRESS IN BRAIN RESEARCH

Volume 49: The Cholinergic Synapse, S. Tuček (Ed.) – 1979

Volume 48: Maturation of the Nervous System, M.A. Corner, R.E. Baker, N.E. van de Pol, D.F. Swaab and H.B.M. Uylings (Eds.) – 1978

Volume 47: Hypertension and Brain Mechanisms, W. De Jong, A.P. Provoost and A.P. Shapiro (Eds.) – 1977

Volume 46: Membrane Morphology of the Vertebrate Nervous System. A Study with Freeze-etch Technique, C. Sandri, J.M. van Buren and K. Akert – 1977

Volume 45: Perspectives in Brain Research, M.A. Corner and D.F. Swaab (Eds.) – 1976

Volume 44: Understanding the Stretch Reflex, H. Homma (Ed.) – 1976.

PROGRESS IN BRAIN RESEARCH

VOLUME 50

REFLEX CONTROL OF POSTURE AND MOVEMENT

Proceedings of an IBRO Symposium held in Pisa, Italy, on September 11-14, 1978

EDITED BY

R. GRANIT

Nobel Institute for Neurophysiology, Karolinska Institutet, Stockholm (Sweden)

AND

O. POMPEIANO

Institute of Physiology II, University of Pisa, Pisa (Italy)

ELSEVIER/NORTH-HOLLAND BIOMEDICAL PRESS
AMSTERDAM/NEW YORK/OXFORD
1979

PUBLISHED BY:
ELSEVIER/NORTH HOLLAND BIOMEDICAL PRESS
335 JAN VAN GALENSTRAAT, P.O. BOX 211
AMSTERDAM, THE NETHERLANDS

SOLE DISTRIBUTORS FOR THE U.S.A. AND CANADA:
ELSEVIER NORTH-HOLLAND INC.
52 VANDERBILT AVENUE
NEW YORK, N.Y. 10017, U.S.A.

ISBN 0-444-80099-9

Library of Congress Cataloging in Publication Data

Main entry under title:

Reflex control of posture and movement.

(Progress in brain research; v. 50)
Includes index.
1. Kinesiology—Congresses. 2. Reflexes—Congresses. 3. Posture—Congresses. 4. Proprioception—Congresses. 5. Sensory-motor integration—Congresses. I. Granit, Ragnar, 1900–
II. Pompeiano, O. III. International Brain Research Organization. IV. Series.
QP376.P7 vol. 50   [QP303]   612'.82'08s
ISBN 0-444-80099-9   [599'.01'852]   79-10006

WITH 338 ILLUSTRATIONS AND 6 TABLES

© ELSEVIER/NORTH-HOLLAND BIOMEDICAL PRESS, 1979

ALL RIGHTS RESERVED.
NO PART OF THIS PUBLICATION MAY BE REPRODUCED, STORED IN A RETRIEVAL SYSTEM, OR TRANSMITTED IN ANY FORM OR BY ANY MEANS, ELECTRONIC, MECHANICAL, PHOTOCOPYING, RECORDING OR OTHERWISE, WITHOUT THE PRIOR PERMISSION OF THE COPYRIGHT OWNER.

PRINTED IN THE NETHERLANDS

Preface

The last decades have seen a great many papers devoted to the proprioceptive mechanisms engaged in posture and movement.

Following the pioneering studies of the physiological behaviour of mammalian spindles and Golgi tendon organs, responding to stretch and contraction, a generation of competent experimenters has elucidated in detail the role of the proprioceptive input in the reflex regulation of muscle tone and servoassistance of movement. Similarly, the reactions of motoneurons and interneurons have been subjected to study, in relation to both proprioceptive and supraspinal descending volleys, with a view to understanding interactions between central programmes and peripheral feedback control of movement.

In recent years, more consideration has been given to the mechanisms that are subjected to proprioceptive influences from the neck and the extrinsic eye muscles; new experimental facts have served to throw light upon the effects of neck afferents on the motoneurons of their own muscles, as well as on supraspinal structures involved in the cervical control of posture and movement.

The proprioceptive input from the somatic musculature is supplemented by other afferent systems intervening in the regulation of tonic and phasic motor activity. Detailed studies have greatly clarified the significance of labyrinthine regulation of oculomotor and spinal motor activity, especially that of the neck motoneurons. Likewise the influence of neck and visual inputs on the vestibulo-ocular and vestibulo-spinal reflex arcs has been studied in several investigations.

Specialization may now be said to have advanced to a level where it has become hard for a single worker to embrace adequately both the structural and functional aspects within one circumscribed field of study. This is the point at which the present Symposium volume will become useful. One has, in current work, become increasingly aware of the urgent need for understanding how the proprioceptive, labyrinthine, and visual inputs enter the motor control systems, and how they interact at segmental and suprasegmental levels to ensure appropriate adjustment of the motoneuronal output. It is hoped that this volume will stimulate investigators to bridge the enormous gap between the actual complexity of each individual contributing system, and the emerging principles of their interaction in the service of an integrated final act.

To this end, the general theme of the Symposium was treated in several sessions, each one devoted to one of its aspects. Clarifying discussions were devoted to the role of the proprioceptive, labyrinthine and visual inputs in the control of reflex posture and movements. Some sections were concerned with the integration of different sensory inputs at segmental and supraspinal levels and to their significance for the coordination of body, head, and eye movements.

The plan of the Symposium and the selection of participants were completed after consultations with members of the Organizing Committee. For valuable suggestions thanks are due to Drs. E. Bizzi, E.V. Evarts, V. Henn, M. Ito, A. Lundberg, W.D. Willis and V.J. Wilson.

The meetings were held in the Scuola Normale Superiore, the old institution harbouring European cultural and scientific traditions, sited at Pisa's beautiful mediaeval Piazza dei Cavalieri. This privilege was highly appreciated by the participants and we, at this time, want to express our thanks to the Rector, Professor E. Vesentini and the Dean, Professor L.A. Radicati, of the Scuola for the permission to gather for lectures and discussions at this historical site. Their presence at the inauguration was gratefully noted. Similarly, the presence of the Vice Rector of Pisa University, Professor G. Falcone, the Dean of its Medical School, Professor G. Pellegrino, and the head of its Physiological Institute, Professor G. Moruzzi, was greatly appreciated.

The Fondazione Giovanni Lorenzini, the Italian National Research Council, the International Brain Research Organization, and the National Institute of Neurological and Communicative Disorders and Stroke at the National Institutes of Health in Bethesda, U.S.A. kindly agreed to sponsor the Symposium. Organizing it was made possible by a generous grant from the Fondazione Giovanni Lorenzini. For this we want to express our gratitude to its President, Professor R. Paoletti, and its Secretary, Dr. E. Folco. Grants from the Italian National Research Council and the companies of Montedison and Formenti are gratefully acknowledged. Finally, we owe a quite special debt of gratitude to the assistants, secretaries and technicians of the Institute of Physiology for their care and cooperation in the service of the Symposium. Help was willingly given and efficiently put into action.

Ragnar Granit
Ottavio Pompeiano

Pisa, September 1978

List of Contributors

V.C. ABRAHAMS – Department of Physiology, Queen's University, Kingston, Ontario K7L 3N6, Canada.
J.H.J. ALLUM – Institut für Hirnforschung, Universität Zürich, 8029 Zürich, Switzerland.
J.H. ANDERSON – Laboratory of Neurophysiology, University of Minnesota Medical School, Minneapolis, MN 55455, U.S.A.
H. ASANUMA – The Rockefeller University, New York, NY 10021, U.S.A.
G.B. AZZENA – Istituto di Fisiologia Umana, Università di Sassari, 07100 Sassari, Italy.
R. BAKER – Department of Physiology and Biophysics, New York University Medical Center, New York, NY 10016, U.S.A.
C. BATINI – Laboratoire de Psychophysiologie Sensorielle, Université Pierre et Marie Curie, 75230 Paris, France.
A. BERTHOZ – Laboratoire de Physiologie du Travail du C.N.R.S., Département de Physiologie Neurosensorielle, 75005 Paris, France.
E. BIZZI – Department of Psychology, Massachusetts Institute of Technology, Cambridge, MA 02139, U.S.A.
R.H.I. BLANKS – Max-Planck-Institut für Hirnforschung, Neurobiologische Abteilung, 6000 Frankfurt/M.-Niederrad, F.R.G.
R. BORTOLAMI – Istituto di Anatomia degli Animali Domestici, Università di Bologna, 40126 Bologna, Italy.
R.M. BOWKER – Marine Biomedical Institute, Departments of Psychiatry and Behavioral Sciences and Physiology and Biophysics, University of Texas Medical Branch, Galveston, TX 77550, U.S.A.
P. BUISSERET – Laboratoire de Neurophysiologie, Collège de France, 75231 Paris, France.
R.E. BURKE – Laboratory of Neural Control, National Institute of Neurological and Communicative Disorders and Stroke, National Institutes of Health, Bethesda, MD 20014, U.S.A.
H.J. BÜDINGER – Department of Neurology, St. Elisabethan Hospital, Ravensburg, F.R.G.
U. BÜTTNER – Neurologische Klinik, Universität Zürich, 8091 Zürich, Switzerland.
J.A. BÜTTNER-ENNEVER – Institut für Hirnforschung, Universität Zürich, 8029 Zürich, Switzerland.
A.J. CASTIGLIONI – Marine Biomedical Institute, Departments of Psychiatry and Behavioral Sciences and Physiology and Biophysics, University of Texas Medical Branch, Galveston, TX 77550, U.S.A.
P.D. CHENEY – Department of Physiology and Biophysics, and Regional Primate Research Center, University of Washington, Seattle, WA 98195, U.S.A.
H. COLLEWIJN – Department of Physiology, Faculty of Medicine, Erasmus University, Rotterdam, The Netherlands.
N. CORVAJA – Istituto di Fisiologia Umana, Cattedra II, Università di Pisa, 56100 Pisa, Italy.
J.D. COULTER – Marine Biomedical Institute, Departments of Psychiatry and Behavioral Sciences and Physiology and Biophysics, University of Texas Medical Branch, Galveston, TX 77550, U.S.A.
J.H. COURJON – Laboratoire de Neurophysiologie Expérimentale, I.N.S.E.R.M. Unité 94, 69500 Bron, France.
M. CROMMELINCK – Laboratoire de Neurophysiologie, Université Catholique de Louvain, 1200 Bruxelles, Belgium.
C. DARLOT – Laboratoire de Physiologie du Travail du C.N.R.S., Département de Physiologie Neurosensorielle, 75005 Paris, France.
J. DELGADO-GARCIA – Departamento de Fisiologia, Facultad de Medicina, Sevilla, Spain.
F. DENOTH – Istituto di Elaborazione dell'Informazione, C.N.R., 56100 Pisa, Italy.
J. DICHGANS – Neurologiache Klinik, Eberhard-Karls-Universität Tübingen, 7400 Tübingen-1, F.R.G.
N. DIERINGER — Max-Planck-Institut für Hirnforschung, Neurobiologische Abteilung, 6000 Frankfurt/M.-Niederrad, F.R.G.
M. DOERR – Neurologische Klinik, Abteilung für Neurophysiologie, Universität Freiburg i. Br., 7800 Freiburg i. Br., F.R.G.

J.R. DUFRESNE – Laboratory of Neurophysiology, University of Minnesota Medical School, Minneapolis, MN 55455, U.S.A.
C.-F. EKEROT – Institute of Physiology, University of Lund, S–223 62 Lund, Sweden.
K. EZURE – Department of Physiology, Institute of Brain Research School of Medicine, University of Tokyo, Tokyo, Japan.
M. FAVILLA – Istituto di Fisiologia Umana, Cattedra II, Università di Pisa, 56100 Pisa, Italy.
E.E. FETZ – Department of Physiology and Biophysics, and Regional Primate Research Center, University of Washington, Seattle, WA 98195, U.S.A.
K. FUKUSHIMA – The Rockefeller University, New York, NY 10021, U.S.A.
Y. GAHERY – C.N.R.S., Institut de Neurophysiologie et Psychophysiologie, Département de Neurophysiologie Générale, 13274 Marseille, France.
B. GHELARDUCCI – Istituto di Fisiologia Umana, Cattedra II, Università di Pisa, 56100 Pisa, Italy.
J.M. GOLDBERG – Department of Pharmacological and Physiological Sciences, University of Chicago, Chicago, IL 60637, U.S.A.
R. GRANIT – Nobel Institute for Neurophysiology, Karolinska Institutet, S–104 01 Stockholm, Sweden.
S. GRILLNER – Fysiologiska Institutionen III, Karolinska Institutet, S–114 33 Stockholm, Sweden.
A.F. GROOTENDORST – Department of Physiology, Faculty of Medicine, Erasmus University, Rotterdam, The Netherlands.
O.-J. GRÜSSER – Physiologisches Institut, Freie Universität Berlin, 1000 Berlin 33, F.R.G.
A.E. HENDRICKSON – Department of Ophthalmology, University of Washington, Seattle, WA 98195, U.S.A.
V. HENN – Neurologische Klinik, Universität Zürich, 8091 Zürich, Switzerland.
K. HEPP – Theoretische Physik, Eidgenössische Technische Hochschule, 8093 Zürich, Switzerland.
R. HESS – Mass-Planck-Institut für biophysikalische Chemie, Neurobiologische Abteilung, 3400 Göttingen-Nikolausberg, F.R.G.
S.M. HIGHSTEIN – Department of Neuroscience, Rose F. Kennedy Center, Albert Einstein College of Medicine, Bronx, NY 10461, U.S.A.
N. HIRAI – Laboratory of Physiology, Institute of Basic Medical Sciences, University of Tsukuba, Niihari-gun, Ibaraki-ken 300–31, Japan.
S. HOMMA – Department of Physiology, School of Medicine, Chiba University, Chiba, Japan.
T. HONGO–Laboratory of Physiology, Institute of Basic Medical Sciences, University of Tsukuba, Niihari-gun, Ibaraki-ken 300-31, Japan.
J. HORCHOLLE – BOSSAVIT, Laboratoire de Physiologie Nerveuse du C.N.R.S., 91190 Gif-sur-Yvette, France.
M. ITO – Department of Physiology, Faculty of Medicine, University of Tokyo, 7-3-1 Hongo, Bunkyo-ku, Tokyo, Japan.
E. JANKOWSKA – Department of Physiology, University of Göteborg, S-400 33 Göteborg, Sweden.
M. JEANNEROD – Laboratoire de Neuropsychologie Expérimentale, I.N.S.E.R.M. Unité 94, 69500 Bron, France.
R. JUNG – Neurologische Klinik, Abteilung für Neurophysiologie, Universität Freiburg i. Br., 7800 Freiburg i. Br., F.R.G.
E.L. KELLER – Department of Electrical Engineering and Computer Sciences and Electronics Research Laboratory, University of California, Berkeley, CA 94720, U.S.A.
J. KIMM – Departments of Otolaryngology and Physiology and Biophysics, University of Washington, Seattle, WA 98195, U.S.A.
J. KRÖLLER – Physiologisches Institut, Freie Universität Berlin, 1000 Berlin 33, F.R.G.
M. LACOUR – Laboratoire de Psychophysiologie, Faculté des Sciences, Marseille, France.
W. LANG – Institut für Hirnforschung, Universität Zürich, 8029 Zürich, Switzerland.
Y. LAPORTE – Laboratoire de Neurophysiologie, Collège de France, 75231 Paris, France.
B. LARSON – Institute of Physiology, University of Lund, S-223 62 Lund, Sweden.
M.McD. LEWIS – Sherrington School of Physiology, St. Thomas's Hospital Medical School, London SE1 7EH, U.K.
R. LLINÁS – Department of Physiology and Biophysics, New York University Medical Center, New York, NY 10016, U.S.A.
J. LOPEZ-BARNEO – Laboratoire de Physiologie du Travail du C.N.R.S., Département de Physiologie Neuro-sensorielle, 75005 Paris, France.
A. LUNDBERG – Department of Physiology, University of Göteborg, S-400 33 Göteborg, Sweden.

M. MAEDA – Department of Neurosurgery, School of Medicine, Juntendo University, 3-1-3 Hongo, Bunkyo-ku, Tokyo, Japan.
P.C. MAGHERINI – Istituto di Fisiologia Umana, Cattedra II, Università di Pisa, 56100 Pisa, Italy.
O. MAMELI – Istituto di Fisiologia Umana, Università di Sassari, 07100 Sassari, Italy.
E. MANNI – Istituto di Fisiologia Umana, Università Cattolica del Sacro Cuore, 00100 Roma, Italy.
J. MASSION – Institut de Neurophysiologie et Psychophysiologie, C.N.R.S., Département de Neurophysiologie Générale, 13274 Marseille Cedex 2, France.
R.A. McCREA – Department of Physiology and Biophysics, New York University Medical Center, New York, NY 10016, U.S.A.
T. MERGNER – Neurologische Klinik, Universität Ulm, Ulm-Donau, F.R.G.
M. MEULDERS – Laboratoire de Neurophysiologie, Université Catholique de Louvain, 1200 Bruxelles, Belgium.
D.L. MEYER – Neurobiology Unit, Department of Psychiatry, University of Göttingen, 3400 Göttingen, F.R.G.
Y. MIYASHITA – Department of Physiology, Faculty of Medicine, University of Tokyo, 7-3-1 Hongo, Bunkyo-ku, Tokyo, Japan.
E.A. MURRAY – Marine Biomedical Institute, Departments of Psychiatry and Behavioral Sciences and Physiology and Biophysics, University of Texas Medical Branch, Galveston, TX 77550, U.S.A.
Y. NAKAJIMA – Department of Physiology, School of Medicine, Chiba University, Chiba, Japan.
L.M. NASHNER – Neurological Sciences Institute, Good Samaritan Hospital and Medical Center, Portland, OR 97209, U.S.A.
O. OSCARSSON – Institute of Physiology, University of Lund, S-223 62 Lund, Sweden.
B.W. PETERSON – The Rockefeller University, New York, NY 10021, U.S.A.
O. POMPEIANO – Istituto di Fisiologia Umana, Cattedra II, Università di Pisa, 56100 Pisa, Italy.
W. PRECHT – Max-Planck-Institut für Hirnforschung, Neurobiologische Abteilung, 6000 Frankfurt/M. – Niederrad, F.R.G.
A. PROCHAZKA – Sherrington School of Physiology, St. Thomas's Hospital Medical School, London SE1 7EH, U.K.
H. REISINE – Department of Neuroscience, Rose F. Kennedy Center, Albert Einstein College of Medicine, Bronx, NY 10461, U.S.A.
F.J.R. RICHMOND – Department of Physiology, Queen's University, Kingston, Ontario K7L 3N6, Canada.
T.D.M. ROBERTS – Institute of Physiology, University of Glasgow, Glasgow G12 8QQ, Scotland, U.K.
P.K. ROSE – Department of Physiology, Queen's University, Kingston, Ontario K7L 3N6, Canada.
A. ROUCOUX – Laboratoire de Neurophysiologie, Université Catholique de Louvain, 1200 Bruxelles, Belgium.
S. SASAKI – Laboratory of Physiology, Institute of Basic Medical Sciences, University of Tsukuba, Niihari-gun, Ibaraki-ken 300–31, Japan.
Th. SAVIDES – Neurologische Klinik, Abteilung für Neurophysiologie, Universität Freiburg i.Br., 7800 Freiburg i.Br., F.R.G.
K.-P. SCHAEFER – Neurobiology Unit, Department of Psychiatry, University of Göttingen, 3400 Göttingen, F.R.G.
R. SCHMID – Istituto di Elettronica, Università di Pavia, 27100 Pavia, Italy.
P. SCHMIDT – Neurologische Klinik, Abteilung für Neurophysiologie, Universität Freiburg i.Br., 7800 Freiburg i.Br., F.R.G.
T. SHIBAZAKI – Department of Neurophysiology, Institute of Brain Research, School of Medicine, Tokyo University, 7-3-1 Hongo, Bunkyo-ku, Tokyo, Japan.
H. SHIMAZU – Department of Neurophysiology, Institute of Brain Research, School of Medicine, University of Tokyo, 7-3-1 Hongo, Bunkyo-ku, Tokyo, Japan.
J.I. SIMPSON – Department of Physiology and Biophysics, New York University Medical Center, New York, NY 10016, U.S.A.
J.F. SOECHTING – Laboratory of Neurophysiology, University of Minnesota Medical School, Minneapolis, MN 55455, U.S.A.
K.-H. SONTAG – Max-Planck-Institut für Experimentelle Medizin, Abteilung Biochemische Pharmakologie, 3700 Göttingen, F.R.G.
R.E. SOODAK – Department of Physiology and Biophysics, New York University Medical Center, New York, NY 10016, U.S.A.

N. SPROTT – Department of Physiology, Queen's University, Kingston, Ontario K7L 3N6, Canada.
M. STANOJEVIĆ – Istituto di Fisiologia Umana, Cattedra II, Università di Pisa, 56100 Pisa, Italy.
A. STARITA – Istituto di Elaborazione dell'Informazione, C.N.R., 56100 Pisa, Italy.
G. STENHOUSE – Institute of Physiology, University of Glasgow, Glasgow G12 8QQ, Scotland, U.K.
C.A. TERZUOLO – Laboratory of Neurophysiology, University of Minnesota Medical School, Minneapolis, MN 55455, U.S.A.
U. THODEN – Neurologische Klinik, Abteilung für Neurophysiologie, Universität Freiburg i.Br., 7800 Freiburg i.Br., F.R.G.
E. TOLU – Istituto di Fisiologia Umana, Università di Sassari, 07100 Sassari, Italy.
Y. UCHINO – Department of Physiology, Kyorin University, School of Medicine, Mitaka, Tokyo, Japan.
P.P. VIDAL – Laboratoire de Physiologie du Travail du C.N.R.S., Départment de Physiologie Neurosensorielle, 75005 Paris, France.
W. WAESPE – Neurologische Klinik, Universität Zürich, 8091 Zürich, Switzerland.
P. WAND – Max-Planck-Institut für Experimentelle Medizin, Abteilung Biochemische Pharmakologie, 3400 Göttingen, F.R.G.
D. WENZEL – Neurologische Klinik, Abteilung für Neurophysiologie, Universität Freiburg i.Br., 7800 Freiburg i.Br., F.R.G.
K.N. WESTLUNG – Marine Biomedical Institute, Departments of Psychiatry and Behavioral Sciences and Physiology and Biophysics, University of Texas Medical Branch, Galveston, TX 77550, U.S.A.
G. WILHELMS – Neurobiology Unit, Department of Neurosciences, University of California, San Diego, La Jolla, CA 92093, U.S.A.
W.D. WILLIS – Marine Biomedical Institute, University of Texas Medical Branch, Galveston, TX 77550, U.S.A.
V.J. WILSON – The Rockefeller University, New York, NY 10021, U.S.A.
J.A. WINFIELD – Departments of Otolaryngology and Physiology and Biophysics, University of Washington, Seattle, WA 98195, U.S.A.
S.P. WISE – Marine Biomedical Institute, Departments of Psychiatry and Behavioral Sciences and Physiology and Biophysics, University of Texas Medical Branch, Galveston, TX 77550, U.S.A.
M. YAMAMOTO – Department of Physiology, Faculty of Medicine, University of Tokyo, 7-3-1 Hongo, Bunkyo-ku, Tokyo, Japan.
K. YOSHIDA – Department of Physiology, Institute of Basic Medical Sciences, University of Tsukuba, Niihari-gun, Ibaraki-ken 300–31, Japan.
P. ZARZECKI – Department of Physiology, Queen's University, Kingston, Ontario K7L 3N6, Canada.
H. ZIERAU – Neurobiology Unit, Department of Psychiatry, University of Göttingen, 3400 Göttingen, F.R.G.

Contents

Preface v

List of Contributors vii

Opening Lecture – Some comments on "tone", R. Granit (Stockholm, Sweden) xvii

SECTION I – PROPRIOCEPTIVE INFLUENCES FROM LIMB RECEPTORS

A – Proprioceptive Control on Spinal Motoneurons

1. On the intrafusal distribution of dynamic and static fusimotor axons in cat muscle spindles, Y. Laporte (Paris, France) 3

2. Multisensory control of spinal reflex pathways, A. Lundberg (Göteborg, Sweden) 11

3. New observations on neuronal organization of reflexes from tendon organ afferents and their relation to reflexes evoked from muscle spindle afferents, E. Jankowska (Göteborg, Sweden) . 29

4. Input-output relationship in spinal motoneurons in the stretch reflex, S. Homma and Y. Nakajima (Chiba, Japan) 37

5. Contribution of different size motoneurons to Renshaw cell discharge during stretch and vibration reflexes, P. Wand and O. Pompeiano (Pisa, Italy) 45

6. The role of synaptic organization in the control of motor unit activity during movement, R.E. Burke (Bethesda, MD, U.S.A.) 61

7. Adaptive properties of the myotatic feedback, C.A. Terzuolo, J.F. Soechting and J.R. Dufresne (Minneapolis, MN, U.S.A.) 69

B – Proprioceptive Control of Supraspinal Structures and Supraspinal Influences on the Spinal Cord

1. Information carried by the spinocerebellar paths, C.-F. Ekerot, B. Larson and O. Oscarsson (Lund, Sweden) 79

2. Responses of dorsal spinocerebellar tract neurons to signals from muscle spindle afferents, O.-J. Grüsser and J. Kröller (Berlin, F.R.G.) 91

3. Effects of high threshold muscle afferent volleys on ascending pathways, W.D. Willis (Galveston, TX, U.S.A.) 105

4. Proprioceptive influences on somatosensory and motor cortex, P. Zarzecki and H. Asanuma (New York, NY, U.S.A.) 113

5. Vestibulospinal, reticulospinal and interstitiospinal pathways in the cat, K. Fukushima, B.W. Peterson and V.J. Wilson (New York, NY, U.S.A.) 121

6. Muscle fields and response properties of primate corticomotoneuronal cells, E.E. Fetz and P.D. Cheney (Seattle, WA, U.S.A.) 137

7. Interpretation of supraspinal effects on the gamma system, R. Granit (Stockholm, Sweden) 147

8. Discharge rates of muscle afferents during voluntary movements of different speeds, M.McD. Lewis, A. Prochazka, K.-H. Sontag and P. Wand (London, U.K. and Göttingen, F.R.G.) 155

9. Supraspinal control of ascending pathways, W.D. Willis (Galveston, TX, U.S.A.) 163

C – Reflex Control of Posture and Movement

1. Organization and programming of motor activity during posture control, L.M. Nashner (Portland, OR, U.S.A.) 177

2. Coupled stretch reflexes in ankle muscles: an evaluation of the contributions of active muscle mechanisms to human posture stability, J.H.J. Allum and H.J. Büdingen (Zürich, Switzerland and Ravensburg, F.R.G.) 185

3. The role of vision in the control of posture during linear motion, A. Berthoz, M. Lacour, J.F. Soechting and P.P. Vidal (Paris and Marseille, France) 197

4. Direction-specific vestibular and visual modulation of fore- and hindlimb reflexes in cats, U. Thoden, J. Dichgans, M. Doerr and Th. Savides (Freiburg i.Br., F.R.G.) 211

5. Diagonal stance in quadrupeds: a postural support for movement, J. Massion and Y. Gahery (Marseille, France) 219

6. Interaction between central and peripheral mechanisms in the control of locomotion. S. Grillner (Stockholm, Sweden) 227

7. Two functions of reflexes in human movement: interaction with preprograms and gain of force, R. Jung (Freiburg, i.Br., F.R.G.) 237

SECTION II – PROPRIOCEPTIVE INFLUENCES FROM NECK RECEPTORS

1. What are the proprioceptors of the neck? F.J.R. Richmond and V.C. Abrahams (Kingston, Ontario, Canada) 245

2. Proprioceptive and somatosensory influences on neck muscle motoneurons, P.K. Rose and N. Sprott (Kingston, Ontario, Canada) 255

3. Cortical, tectal and medullary descending pathways to the cervical spinal cord, J.D. Coulter, R.M. Bowker, S.P. Wise, E.A. Murray, A.J. Castiglioni and K.N. Westlund (Galveston, TX, U.S.A.) 263

4. Tonic cervical influences on forelimb and hindlimb monosynaptic reflexes, U. Thoden and D. Wenzel (Freiburg i.Br., F.R.G.) 281

SECTION III – PROPRIOCEPTIVE INFLUENCES FROM EYE MUSCLE RECEPTORS

1. Peripheral and central organization of the extraocular muscle proprioception in the ungulata, E. Manni and R. Bortolami (Roma and Bologna, Italy) 291

2. Properties of the receptors of the extraocular muscles, C. Batini (Paris, France) 301

3. Extraocular muscle input to the cerebellar cortex, C. Batini (Paris, France) 315

4. Proprioceptive influences from eye muscle receptors on cells of the superior colliculus, V.C. Abrahams (Kingston, Ontario, Canada) 325

5. Extraocular muscle afferents and visual input interactions in the superior colliculus of the cat, C. Batini and G. Horcholle-Bossavit (Paris, France) 335

6. Does extraocular proprioception influence the development of visual processes and the oculomotor system? P. Buisseret (Paris, France) 345

SECTION IV – LABYRINTHINE INFLUENCES ON THE MOTOR SYSTEM

A – Labyrinthine Receptors and their Influences on the Vestibular Nuclei

1. Vestibular receptors in mammals: afferent discharge characteristics and efferent control, J.M. Goldberg (Chicago, IL, U.S.A.) 355

2. Labyrinthine influences on the vestibular nuclei, W. Precht (Frankfurt/M.-Niederrad, F.R.G.) ... 369

3. Response of vestibular and cerebellar neurons to rotational stimulation, J. Kimm and J.A. Winfield (Seattle, WA, U.S.A.) 383

B – Labyrinthine Influences on Spinal Motoneurons

1. Reactions to overbalancing, T.D.M. Roberts and G. Stenhouse (Glasgow, U.K.) 397

2. Semicircular canal and macular influences on neck motoneurons, M. Maeda (Tokyo, Japan) ... 405

3. Role of vestibular inputs in the organization of motor output to forelimb extensors, J.H. Anderson, J.F. Soechting and C.A. Terzuolo (Minneapolis, MN, U.S.A.) 413

4. Efferent and afferent responses during falling and landing in cats, M.McD. Lewis, A. Prochazka, K.-H. Sontag and P. Wand (London, U.K. and Göttingen, F.R.G.) 423

C – Labyrinthine Influences on Oculomotor Neurons

1. Synaptic and functional organization of vestibulo-ocular reflex pathways, S.M. Highstein and H. Reisine (Bronx, NY, U.S.A.) 431

2. Labyrinthine influences on motoneurons responsible for vertical eye movements in the rabbit, B. Ghelarducci, M. Favilla and A. Starita (Pisa, Italy) 443

3. Vestibulo-ocular reflex pathways of rabbits and their representation in the cerebellar flocculus, M. Yamamoto (Tokyo, Japan) 451

4. Canal-otolith convergence on cat ocular motoneurons, W. Precht, J.H. Anderson and R.H.I. Blanks (Frankfurt/M.-Niederrad, F.R.G.) 459

5. Vestibular unit activity during nystagmus, H. Shimazu (Tokyo, Japan) 469

6. Organization and control of the vestibulo-ocular reflex, R. Schmid and M. Jeannerod (Pavia, Italy and Bron, France) 477

D – Interaction of Labyrinthine and Proprioceptive Neck Inputs

1. Otoliths and uprightness, T.D.M. Roberts (Glasgow, U.K.) 493

2. Neck and macular labyrinthine influences on the cervical spino-reticulocerebellar pathway, O. Pompeiano (Pisa, Italy) 501

3. Neck and macular labyrinthine influences on the Purkinje cells of the cerebellar vermis, F. Denoth, P.C. Magherini, O. Pompeiano and M. Stanojević (Pisa, Italy) 515

4. The neck and labyrinthine influences on cervical spinocerebellar tract neurones of the central cervical nucleus in the cat, N. Hirai, T. Hongo, S. Sasaki and K. Yoshida (Ibaraki-ken, Japan) .. 529

5. A role of neck afferents on vestibulocollic reflex elicited by dynamic labyrinthine stimulation, K. Ezure, S. Sasaki, Y. Uchino and V.J. Wilson (Ibaraki-ken, Japan) 537

6. Neck influences on the vestibulo-ocular reflex arc and the vestibulocerebellum, M. Maeda (Tokyo, Japan) 551

7. Vestibular-neck interaction in abducens neurons, U. Thoden and P. Schmidt (Freiburg i.Br., F.R.G.) 561

8. Vestibular influences on the cat's cerebral cortex. T. Mergner (Ulm-Donau, G.F.R.) 567

9. The vestibulocortical pathway: neurophysiological and anatomical studies in the monkey, U. Büttner and W. Lang (Zürich, Switzerland) 581

E – Compensation of Labyrinthine Functions

1. Somatosensory and cerebellar influences on compensation of labyrinthine lesions, K.-P. Schaefer D.L. Meyer and G. Wilhelms (Göttingen, F.R.G. and San Diego, CA, U.S.A.) 591

2. Cerebellar contribution in compensating the vestibular function, G.B. Azzena, O. Mameli and E. Tolu (Sassari, Italy) 599

3. Synaptic mechanisms involved in compensation of vestibular function following hemilabyrinthectomy, N. Dieringer and W. Precht (Frankfurt/M.-Niederrad, F.R.G.) 607

SECTION V – VISUAL INFLUENCES ON THE MOTOR SYSTEM

A – Reticular Control of Eye Movements

1. Organization of reticular projections onto oculomotor neurones, J.A. Büttner-Ennever (Zürich, Switzerland) 619

2. Organization of reticular projections to the vestibular nuclei in the cat, N. Corvaja, T. Mergner and O. Pompeiano (Pisa, Italy) 631

3. Neuronal activity preceding rapid eye movements in the brain stem of the alert monkey, K. Hepp and V. Henn (Zürich, Switzerland) 645

4. Afferent and efferent organization of the prepositus hypoglossi nucleus, R.A. McCrea, R. Baker and J. Delgado-Garcia (New York, NY, U.S.A. and Seville, Spain) 653

5. Functional role of the prepositus hypoglossi nucleus in the control of gaze, J. Lopez-Barneo, C. Darlot and A. Berthoz (Paris, France) 667

B — Visual Control of Eye Movements Interaction of Visual and Labyrinthine Inputs

1. Motion information in the vestibular nuclei of alert monkeys: visual and vestibular input vs. optomotor output, W. Waespe and V. Henn (Zürich, Switzerland) 683

2. Interaction of visual and canal inputs on the oculomotor system via the cerebellar flocculus, Y. Miyashita (Tokyo, Japan) 695

3. Visual-vestibular interactions and the role of the flocculus in the vestibulo-ocular reflex, J. Kimm, J.A. Winfield and A.E. Hendrickson (Seattle, WA, U.S.A.) 703

4. The accessory optic system and its relation to the vestibulocerebellum, J.I. Simpson, R.E. Soodak and R. Hess (New York, NY, U.S.A. and Göttingen-Nikolausberg, F.R.G.) 715

5. Colliculoreticular organization in the oculomotor system, E.L. Keller (Berkeley, CA, U.S.A.) ... 725

6. Labyrinthine and visual inputs to the superior colliculus neurons, M. Maeda, T. Shibazaki and K. Yoshida (Tokyo and Ibaraki-ken, Japan) 735

7. Visual fixation: a collicular reflex? A. Roucoux, M. Crommelinck and M. Meulders (Bruxelles, Belgium) 745

C – Compensation and Adaptation of Labyrinthine Functions by Visual Input

1. Adaptive modification of the vestibulo-ocular reflex in rabbits affected by visual inputs and its possible neuronal mechanisms, M. Ito (Tokyo, Japan) 757

2. Modification of central vestibular neuron response by conflicting visual-vestibular stimulation, E.L. Keller and W. Precht (Berkeley, CA, U.S.A. and Frankfurt/M.-Niederrad, F.R.G.) 763

3. Adaptation of optokinetic and vestibulo-ocular reflexes to modified visual input in the rabbit, H. Collewijn and A. F. Grootendorst (Rotterdam, The Netherlands) 771

4. Visual substitution of labyrinthine defects, J.H. Courjon and M. Jeannerod (Bron, France) 783

SECTION VI – COORDINATION OF EYE-HEAD MOVEMENTS

1. Strategies of eye-head coordination, E. Bizzi (Cambridge, MA, U.S.A.) 795

2. Neural activity pattern in different brain stem structures during eye-head movements, K.-P. Schaefer, D.L. Meyer and H. Zierau (Göttingen, F.R.G.) 805

SUBJECT INDEX 813

Opening Lecture

Some Comments on "Tone"

RAGNAR GRANIT

Nobel Institute for Neurophysiology, Karolinska Institutet, S-104 01 Stockholm (Sweden)

The relation of posture to movement was a subject of great actuality to physiologists working within the first half of this century. The outcome of it all was that tone became defined as postural reflexes adjusting body to ground and parts of the body to one another. Familiar to all are the postural adjustments so thoroughly elucidated by Magnus and De Kleijn. Their findings tended to clench the notion of tone as postural reflexes, one to which Sherrington also adhered.

From the declining interest in reflexes such tonus problems have suffered without being revived and systematically pursued as studies of motor control and its organization. The work on gamma-spindle problems in the early fifties drew attention to another and, indeed, much older way of thinking of "tone", not necessarily as reflex but as a kind of state of light excitation, the opposite to "slackness". Thus, for instance, the transition from sleep to arousal is reflected on the motor side by a mobilization of the gamma-spindle apparatus. This, in the cat, is another state, one of preparedness and proprioceptive awakening with subliminal or liminal activation of the motoneurons. We shall probably hear of recent contributions to this concept of "tone" by Grillner, present at this Symposium; I mean the monoaminergic mobilization of a new state of motor preparedness in the spinal cord of the spinalized cat.

This particular aspect of "tone" has been much studied in the cat with its prominent "red" muscles loaded with spindles and run by small alpha motoneurons which are easily excited to long-lasting discharges outlasting the stimulus. If I have understood Henneman correctly, he would think of this faculty merely as a matter of neuron size. My own view has always been, and still is, that behind such observations there is a specific neural organization. This may well be represented at different levels, not only in the spinal cord. However, it is not my intention today to place before you the arguments for a specific tonic neural organization in the cat. I merely wanted to refer briefly to the kind of work that in the present time has sustained a concept of "tone" as a state, naturally also reflected in reflex excitability.

I have taken up this subject in order, firstly, to draw attention to man and some clinical work of interest in this context and, secondly to make a final suggestion. In trying to understand how "the body has, as it were, an uncanny knowledge of all the relevant laws of mechanics" Martin (1967, 1977), has been led to the concept of a specific tonic organization. From his clinical experience he mentions a number of cases in which the voluntary activity is intact while tone is lost; for instance, a girl whose head fell down if she closed her eyes but who was capable of raising her head promptly when asked to do so, still with closed eyes. Though vestibularis may be regarded as a specifically postural receptor, it can be functionally normal in a patient incapable of standing, rising from a chair or holding up his head without a voluntary effort. The upright position of man seems to have made proprioceptive information relatively more important. A patient incapable of walking may be induced to do so when forced to imitate the rocking sideways movements that the centre of gravity executes in walking. These follow from the shift of the weight from one leg to the other. These examples are given by Martin.

Apparently, Denny-Brown (1962) does not agree with this idea of a tonic governor or organization, single or hierarchically organized. He states: "In disagreement with earlier views, we find that the motor disorder of extrapyramidal disease points clearly to identity of the mechanisms of posture and movement, for the more completely the projected type of movement disintegrates the more obviously there is abnormality of posture. Movement in its most simple form is change of posture" (p. 124).

In spite of these opposite views the two authorities are in essential agreement about the leading role in tone of that great output organization, the globus pallidus. One therefore wonders whether their difference of opinion does not chiefly concern the degree of independence that is allotted to the globus pallidus. After bilateral destruction of this structure in the monkey, Denny-Brown found the animal unable to stand, its hand and feet quite useless, no righting reflexes. Brooks (1975) has recently applied his technique of local cooling to the globus pallidus of the monkey. He found that, provided vision was excluded, there was a breakdown of alternating movements, suggesting, indeed, a regulatory role in posture for this structure in agreement with the ideas of Martin.

In thinking, as I do, about hierarchically organized tonic centres the question arises as to how the two aspects of tone, that of a "state" discussed above, and the control of postural reflexes are related. To be at all relevant for its purpose a postural response must enter into any movement from its very beginning and participate in it all the time, unless interrupted or suppressed by a ballistic motor act in which the contractile energy, released by a brief burst of spikes, is transferred into the mass of the moving parts. Therefore, the required postural organization should be specially directed towards neurons capable of an easy and early mobilization and capable also of supporting maintained activity. Different channels must have access to the same postural governor.

When Hagbarth and Eklund (1966) demonstrated the slow rise of the motor response elicited by vibration at the tendon of tibialis anterior in man, they seemed to me to have described activity of a tonic organization, now known to be spinal (Gillies et al., 1971; Burke et al., 1976). Their access to it was largely if not wholly proprioceptive, unphysiological though it be to mobilize spindles bypassing the link between alpha and gamma motoneurons. This organization would seem to be spinal in the same way as stepping is spinally organized. Supraspinal control is known to be present in both cases. It seems likely that the same spinal organization is mobilized by the basal ganglia to produce *both* manifestations of "tone", the state of preparedness and the postural reflexes.

My comments on "tone" have served the purpose of preparing the ground for their final aim which is to suggest that neurophysiology should return to the types of experiment carried out by Magnus and De Kleijn and cease to regard them as a finished chapter. There are available in them precise tests that can be related to the issues that I have discussed here. And it is a curious fact that the Dutch workers never took to the stereotactic technique of Horsley and Clarke, invented about fifteen years before their time.

REFERENCES

Brooks, V.B. (1975) Roles of cerebellum and basal ganglia in initiation and control of movement. *J. Canad. Sci. Neurol.*, 2: 265–277.
Burke, D., Hagbarth, K.-E., Löfstedt, L. and Wallin, B.G. (1976) The responses of human muscle spindle endings to vibration of non-contracting muscles. *J. Physiol. (Lond.)*, 261: 673–693.
Denny-Brown, D. (1962) *The Basal Ganglia and Their Relation to Disorders of Movement*. Oxford Univ. Press, London.
Gillies, J.D., Burke, D.J. and Lance, J.W. (1971) Tonic vibration reflex in the cat. *J. Neurophysiol.*, 34: 252–262.
Hagbarth, K.-E. and Eklund, G. (1962) Motor effects of vibratory muscle stimuli in man. In *Muscular Afferents and Motor Control. Nobel Symposium I,* R. Granit (Ed.). Almqvist and Wiksell, Stockholm, pp. 177–186.
Horsley, V. and Clarke, R.H. (1908) The structure and functions of the cerebellum examined by a new method. *Brain*, 31: 45–124.
Martin, J.P. (1967) *The Basal Ganglia and Posture*. Medical Publ. Co., London.
Martin, J.P. (1977) A short essay of posture and movement. *J. Neurol.*, 40: 25–29.

SECTION I

PROPRIOCEPTIVE INFLUENCES FROM LIMB RECEPTORS

A

Proprioceptive Control of Spinal Motoneurons

On The Intrafusal Distribution of Dynamic and Static Fusimotor Axons in Cat Muscle Spindles

Y. LAPORTE

Laboratoire de Neurophysiologie, Collège de France, 75231 Paris (France)

Three kinds of muscle fibre are found in cat spindles: the type 1 nuclear-bag fibres (bag$_1$ fibres), the type 2 nuclear-bag fibres (bag$_2$ fibres) and the nuclear-chain fibres (Ovalle and Smith, 1972; Banks et al., 1975, 1977; Barker et al., 1976a). Typically, each intrafusal bundle consists of one bag$_1$ fibre, one bag$_2$ fibre and several (4–6) chain fibres. The bag$_2$ fibres are much longer and wider than chain fibres, but their ultrastructure and histochemical profile resemble those of the chain fibres. On the other hand the bag$_1$ fibres, which are only slightly shorter and thinner than the bag$_2$ fibres, differ markedly from these especially in the intracapsular region where the sarcomeres lack an M line or have a faint double M line. In the equatorial region of the spindle bundle, the bag$_2$ fibre and the chain fibres are closely associated, whereas the bag$_1$ fibre is situated at some distance from this group of fibres, a disposition that is useful for identifying bag fibres in serial transverse sections.

The dynamic fusimotor axons exert their actions through the bag$_1$ fibres and the static axons through both chain and bag$_2$ fibres. This statement is based on a series of experiments (for details, see the recent review by Laporte, 1978) which started by the demonstration that the motor endings of single static axons (trail endings) lie on bag as well as on chain fibres (Barker et al., 1973). Cinematographical observations of living spindles (mostly tenuissimus spindles) have shown that the stimulation of dynamic γ axons elicits a weak and slow contraction in one of the bag fibres, whereas the stimulation of static γ axons elicits the contraction of the chain fibres and/or of one of the bag fibres (Bessou and Pagès, 1975; Boyd et al., 1977). As originally shown by Bessou and Pagès (1975), the contraction of the bag fibre activated by static axons although weaker than that of the chain fibres, is much stronger and faster than that of the bag fibre activated by dynamic axons. The two bag fibres have respectively been named by Boyd et al. (1977) the "dynamic nuclear bag fibre" and the "static nuclear bag fibre".

The distribution of fusimotor axons to intrafusal muscle fibres has also been studied with the glycogen-depletion technique of Edström and Kugelberg (1968), which consists of mapping, from serial transverse sections stained for glycogen, the muscle fibres that have been depleted of their glycogen content, following prolonged stimulation of their motor supply. This technique, originally developed to study the motor units of α axons, was first applied to the fusimotor system by Brown and Butler (1973, 1975). More recently, in experiments in which bag$_1$ and bag$_2$ fibres were identified, it was found that dynamic γ axons consistently induced glycogen depletion

– indicating neural activation – in bag₁ fibres whereas static axons elicited depletion not only in chain fibres but also in bag₂ fibres (Barker et al., 1976b). As will be discussed later, depletion in bag₁ fibres was also observed. By itself the demonstration that static γ axons supply bag₂ fibres does not prove that these fibres have a static action of their own. Conceivably, bag₂ fibres might exert a dynamic action which could interact with the static action of chain fibres. However, by taking advantage of the variability in intrafusal distribution of individual static axons, it has been possible to rule out this possibility. In some spindles a given static axon may supply only bag₂ fibres whereas in others it supplies either chain fibres alone or both chain and bag₂ fibres. Boyd et al. (1977) have reported the static action exerted on primary endings of spindles in which the only activated fibre was a fast-contracting bag fibre. In agreement with this observation Jami and Petit (see Fig. 1) have observed the static action of an axon, which, as shown by the glycogen-depletion technique, only supplied the bag₂ fibre of a tenuissimus spindle. All this converging evidence indicates that the bag₁ fibre may be equated with the dynamic bag fibre and the bag₂ with the static bag fibre.

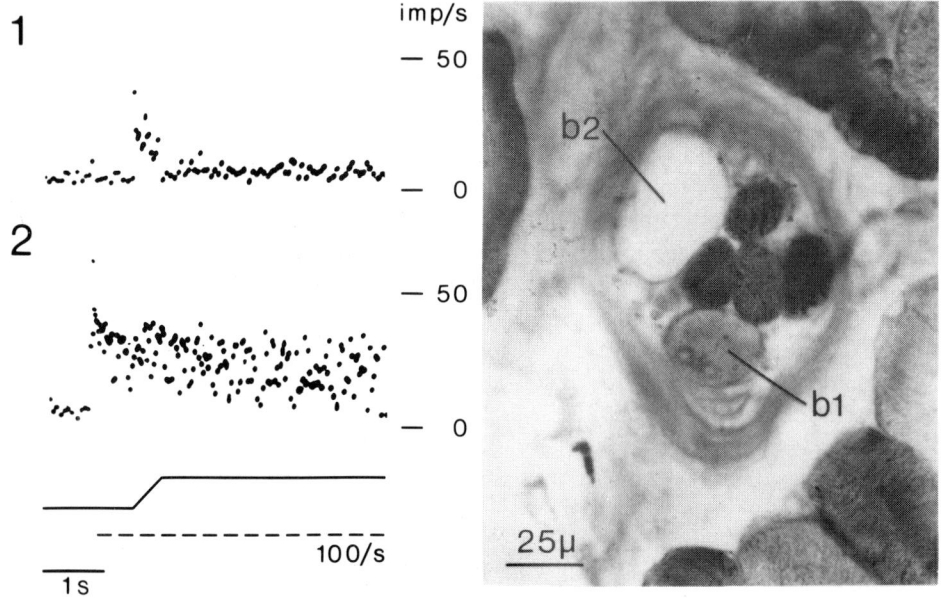

Fig. 1. Static action exerted by a single γ axon which only supplied the bag₂ fibre of a tenuissimus spindle. The left side of the figure shows discharges from the primary ending of the spindle, led from a dorsal root filament and recorded with an instantaneous frequency meter. 1) Passive response to a ramp-and-hold stretch. 2) Response during stimulation of the γ axon at 100 sec^{-1}. Note the typical static alteration of the response and the irregularity of the firing. The spindle to which the primary ending belonged was located by very localized pressure (technique of Bessou and Laporte, 1962). Its position in the muscle was marked by a thin thread knotted on the side of the muscle. The portion of the muscle containing the spindle was then treated for glycogen detection after the γ axon was stimulated in a way known to elicit glycogen depletion in intrafusal muscle fibres (see Harker et al., 1977). On the right, photomicrograph of a transverse section of the spindle showing in the upper left side a totally blanched large-diameter fibre. Examination on serial sections of the whole spindle showed that no other fibre was depleted. The depletion in the bag₂ fibre occurred in both poles, over a stretch of 250 μm in the distal pole and of 850 μm in the proximal pole. The centres of the depleted zones were in the capsular sleeves, resp. at 800 μm and 1250 μm from the equator. The depleted fibre was identified as bag₂ fibre on account of its diameter, length and close association with chain fibres in the equator. (From Jami and Petit, unpublished observations.)

Two points will be considered in the present review: the distribution of the dynamic γ axons that exert a category II action; and the innervation of the bag$_1$ fibres by static γ axons.

DISTRIBUTION OF DYNAMIC γ AXONS EXERTING A CATEGORY II ACTION

Emonet-Dénand et al. (1977) have recently subdivided the *actions* exerted by fusimotor axons on primary endings in six categories ranging from apparently "pure" dynamic action to apparently "pure" static action (see Fig. 4). Categories I and II are clearly dynamic since in both cases the responses of primary endings to ramp-and-hold stretches show a marked increase in the dynamic index associated with a slow decay of firing after the dynamic phase of the stretch. As illustrated by Fig. 2, the two categories differ by the magnitude of the acceleration of the discharge observed at constant muscle length, and by the variability of the discharge; these are both distinctively larger in category II responses. In the peroneus brevis muscle, category I responses are about three times as common as category II whereas in the tenuissimus muscle nearly all the responses appear to belong to category I. Most dynamic responses observed in the soleus muscle belong to category II.

Frequency grams of primary endings (see Bessou et al., 1968a) elicited by stimulation of dynamic γ axons giving category II responses have recently been studied (Emonet-Dénand and Laporte, 1978) in order to obtain some information on the contraction of intrafusal muscle fibres responsible for these responses.

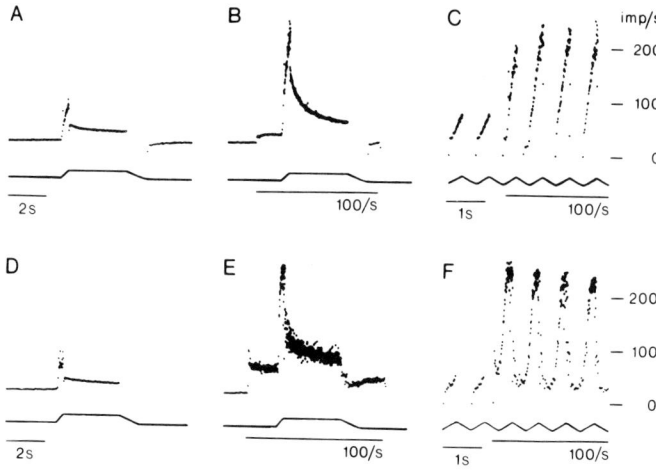

Fig. 2. Comparison of category I and category II dynamic responses. A and D) Passive responses of two primary endings (cat peroneus brevis muscle) to a 2 mm ramp-and-hold stretch followed by a slower release. B and E) Alteration of the responses elicited by the stimulation at 100 sec^{-1} of two single dynamic axons (each one acting on a different spindle). B and E respectively illustrate category I and category II responses. Note in E the stronger excitatory effect of the fusimotor stimulation at the initial length, the irregularity of the discharge and its persistence during muscle release. C and F) Responses to triangular stretching before and during stimulation at 100 sec^{-1} of the axons. Note in F that the ending continues to fire even when stretching passes through its minimal value. (From Emonet-Dénand et al., 1977.)

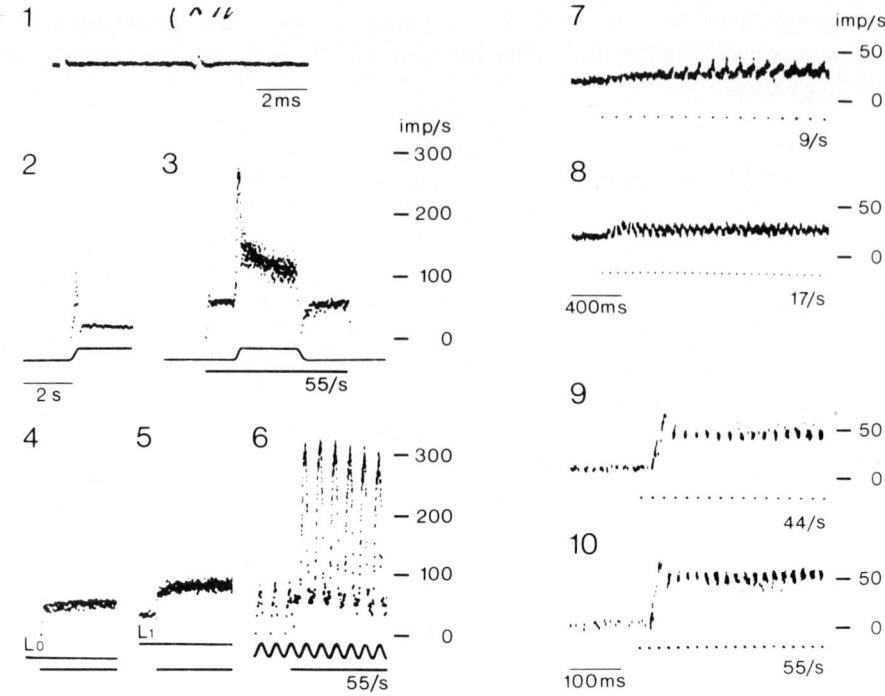

Fig. 3. Frequency grams of a primary ending due to the stimulation of a dynamic γ axon giving a category II response. Cat flexor hallucis longus muscle. On the left side of the figure records showing the action of the γ axon. 1) Action potential of the axon led from the muscle nerve after stimulation of a ventral root filament; conduction velocity: 40 m/sec. 2) Passive response of the ending to a ramp stretch of 0.8 mm. 3–6). The lower bar in each record indicates the stimulation of the axon at 55 sec^{-1}. 3) Same ramp stretch as in 2; a typical category II response is observed. 4, 5) Acceleration of the firing observed for two constant lengths L_0 and L_1. 6) Sinusoidal stretch (1.5 Hz, 1 mm amplitude); a marked increase in the peak-to-peak modulation is observed during the stimulation. On the right side of the figure, frequency grams obtained by superimposition of about 20 records of instantaneous frequency (see Bessou et al., 1968a) while the muscle length remains constant. The points situated at the lower part of each record indicate the stimulation. Note the slower sweep speed used for records 7 and 8 (see text). (From Emonet-Dénand and Laporte, 1978.)

These graphs, as illustrated by Fig. 3, show distinct increments in frequency whose periodicity is equal to that of the stimulation (records 7 and 8) and even "driving" (records 9–10) for rates of stimulation ranging from 9 to 44/sec (indicating a relatively strong and fast intrafusal contraction). These frequency grams differ markedly from those elicited in tenuissimus muscles by dynamic γ stimulation which display a smooth contour for rates of stimulation as low as 50/sec (Bessou et al., 1968b). Emonet-Dénand et al. (1977) suggested that category II responses resulted from the concomitant activation of a bag$_1$ fibre (i.e., the fibre responsible for dynamic action) and of a different functional type of intrafusal muscle fibre, whether a bag$_2$ or a chain fibre. This suggestion was supported by the observation that a category I response can be converted into a category II response by stimulating a static axon together with the dynamic axon responsible for the category I response, especially when the static axon was stimulated at a lower frequency than the dynamic one (see Figs. 5 and 11 of their paper).

The frequency grams shown in Fig. 3 are probably not incompatible with this assumption, but another possibility should be considered: the contraction responsible for category II action is engendered only in the bag_1 fibre but it is relatively strong and/or close to the sensory terminal. This would agree with the observation that strong dynamic effects can be elicited by axons giving category II responses for rates of stimulation as low as 20–30/sec, which is not the case for axons giving category I responses.

In favour of the latter assumption is also the fact that all grades between category I and II responses can be observed. If category I responses were due to bag_1 fibres alone and category II responses to fast contracting intrafusal fibres together with bag_1 fibres, all responses should be expected to fall into one of two distinct classes.

It would be desirable to determine the actual distribution of the axons responsible for category II actions by the glycogen-depletion method. However, the precise localization of spindles that is necessary for exact histophysiological correlation can be achieved only in the tenuissimus muscle but unfortunately, in this muscle it is difficult (probably for mechanical reasons) to obtain dynamic and static responses easily classifiable in one of the six categories which have been defined in the peroneus brevis muscle.

Barker et al. (1977) have reported the distribution of a peroneus brevis dynamic β axon whose stimulation activated three primary endings. Glycogen depletion was observed in three spindles of this muscle; in two of them the bag_1 was the only fibre depleted, whereas in the third spindle a chain fibre was depleted in addition to the bag_1 fibre. In two spindles the stimulation of the β axon exerted a typical category I action, but the response given by the third spindle (presumably the one in which a chain fibre was depleted in addition to the bag_1 fibre) was not a category II response. It had features of evenly balanced mixed static and dynamic actions and fell in category III of Emonet-Dénand et al. (1977) (unclassifiable action).

INNERVATION OF BAG_1 FIBRES BY γ STATIC AXONS

As recalled in the introduction, cinematographical analysis of living spindles and mapping of intrafusal distribution of fusimotor axons by the glycogen-depletion technique have given complementary information which leaves little doubt that dynamic axons supply bag_1 fibres and static axons chain and bag_2 fibres. However, one point remains on which the two methods have not given complementary results, namely, the innervation of some bag_1 fibres by static axons.

In tenuissimus spindles, Barker et al. (1976b) found that stimulation of single static axons induced glycogen depletion as often in bag_1 as in bag_2 fibres. This result was in line with prior studies in which the two kinds of bag fibre were not distinguished, but in which both bag fibres were found to be depleted in some spindles (Brown and Butler, 1973, 1975). That some static axons, in addition to their statogenic effectors, the chain and bag_2 fibres, may also supply bag_1 fibres is supported by the physiological observations reported by Emonet-Dénand et al. (1977). In their study of the fusimotor innervation they collected all types of response of primary endings during fusimotor stimulation without selecting the most typical ones. They found that the majority of static responses observed during ramp-and-hold stretches could be ascribed to an apparently "pure" static action (category VI responses), but that some of them were suggestive of an admixture of a dynamic action because, on completion of the dynamic phase of stretching, they showed a slow decay of firing comparable to

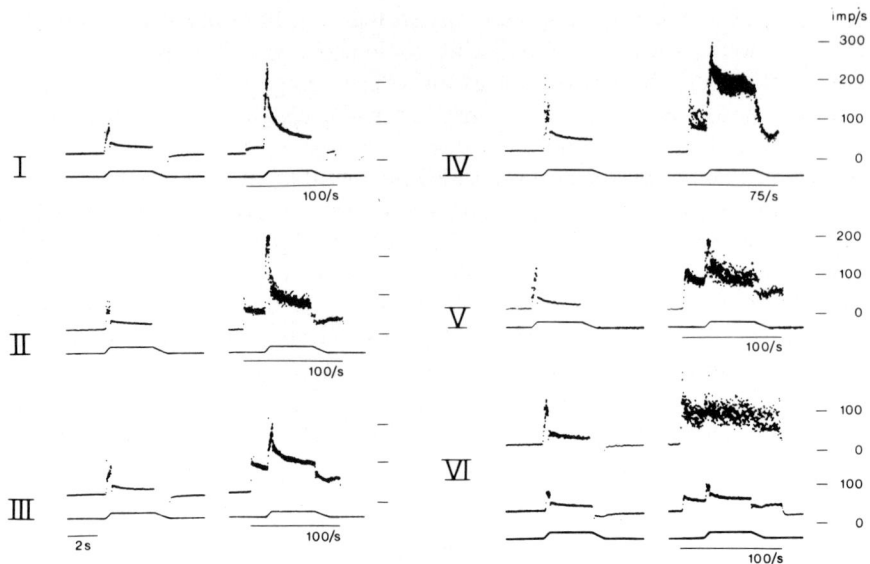

Fig. 4. Categories of fusimotor actions. Each pair of records shows: on the left, the passive response of a primary ending (cat peroneus brevis muscle) to a 2 mm ramp-and-hold stretch followed a few seconds later by a slower release; on the right, the response during repetitive stimulation of a single γ axon, indicated by the bars. I) Category I response. Purely dynamic action, characterized by marked increase in dynamic index with slow decay of firing after the dynamic phase of stretch and regular discharge. II) Category II response. Same dynamic features as in I, but greater excitation at initial length and appreciable variability of discharge. III) Category III response. Unclassifiable. Dynamic and static features equally balanced. IV) Category IV response. Static action modified by dynamic action. Strong static features (considerable excitation at initial length, firing during release, irregularity of the discharge) but slow decay after ramp stretch. V) Category V response. Static action with conceivable dynamic participation. Strong static features with slight sign of dynamic action. VI) Category VI response. Purely static action. Considerable excitation with usually a decrease of the dynamic index; no slow decay after ramp stretch, nearly always firing on release; sometimes gross variability in discharge (upper record) sometimes regular discharge (lower record). (From Emonet-Dénand et al., 1977.)

that given by the stimulation of dynamic axons (see Fig. 4). These responses were classified in categories IV and V, respectively called "static action modified by dynamic action" and "static action with conceivable dynamic participation". Furthermore they readily obtained responses of categories IV and V by combined stimulation of a dynamic axon giving a category I response and of a static axon giving a category VI response.

Since these observations were made on the peroneus brevis muscle Emonet-Dénand et al. (1978) studied the distribution of static axons in the spindles of this muscle with the glycogen-depletion method. In each experiment a small number of static axons (3–7), previously identified by their actions on the responses of at least one primary ending to ramp-and-hold stretches, were stimulated. The whole muscle was sectioned in order to search for glycogen-depletion in as many spindles as possible. Of 167 spindles examined, 53 contained depleted intrafusal muscle fibres. In 33 spindles (62%) the depletion included only chain and bag_2 fibres (chain only in 14 spindles, chain and bag_2 fibres in 18 spindles, bag_2 alone in 1 spindle). However in 20 spindles (38%) the bag_1 fibre was depleted in addition to other fibres (with chain fibres in 4 spindles, with chain and bag_2 fibres in 13 spindles and with bag_2 only in 3 spindles). Of the 102 static actions classified by Emonet-Dénand et al. (1977), 68 (67%) belonged to category VI (apparently pure static action) and 34 (33%) to categories IV (11%) and V (22%) suggestive of a dynamic participation. These findings are apparently in conflict with the cinematographical observations of Bessou and Pagès (1975) and of Boyd et al. (1977), who have never seen the same bag fibre (or rather the same pole of a bag fibre) contracting after stimulating either a static axon or a dynamic axon. This negative finding, as already pointed out (Barker et al., 1976b; Emonet-Dénand et al., 1977), may be due to the difficulty of observing a weak contraction in a bag_1 fibre while a much stronger contraction is taking place in nearly intrafusal fibres such as chain fibres. Another possibility should also be considered, namely, that a bag_1 pole which is supplied by a collateral of a static axon is not supplied by a dynamic axon.

The innervation in about one-third of the spindles of bag_1 fibres by static axons seems too large to be dismissed as an ontogenic imperfection, but its functional consequence is not yet understood. It has been suggested (Emonet-Dénand et al., 1977) that it could preserve, during strong activation of the static fusimotor system, the responsiveness of the bag_1 terminals by preventing slackness of the bag_1 fibres when the dynamic system is not active. Further studies of this innervation are obviously needed.

SUMMARY

In the first part of this paper the evidence showing that dynamic axons exert their effect through nuclear bag_1 fibres and static axons through both nuclear chain and nuclear bag_2 fibres is briefly reviewed. The second part deals with dynamic γ axons giving category II responses. Frequency grams of primary endings observed during stimulation of these axons are described. Two interpretations of category II responses are discussed. The innervation of some bag_1 fibres by static axons is studied in the last part. Recent experiments carried out on cat peroneus brevis muscles with the glycogen-depletion method show that after stimulation of static axons, bag_1 fibres, in addition to other fibres, are depleted (indicating neural activation) in about one third of the spindles.

ACKNOWLEDGEMENT

Recent investigations presented in this review were supported by grants from INSERM (ATP 76–61) and from the Foundation for French Medical Research.

REFERENCES

Banks, R., Harker, D. and Stacey, M. (1977) A study of mammalian intrafusal muscle fibres using a combined histochemical and ultrastructural technique. *J. Anat. (Lond.)* 123: 783–796.

Banks, R., Barker, D., Harker, D. and Stacey, M. (1975) Correlation between ultrastructure and histochemistry of mammalian intrafusal muscle fibres, *J. Physiol. (Lond.)*, 252: 16–17P.

Barker, D., Emonet-Dénand. F., Laporte, Y., Proske, U. and Stacey, M. (1973) Morphological identification and intrafusal distribution of the endings of static fusimotor axons in the cat, *J. Physiol. (Lond.)*, 230: 405–427.

Barker, D., Banks, R., Harker, D., Milburn, A. and Stacey, M. (1976a) Studies on the histochemistry, ultrastructure, motor innervation and regeneration of mammalian intrafusal muscle fibres. In *Progress in Brain Research, Vol. 44, Understanding the Stretch Reflex*, S. Homma (Ed.) Elsevier, Amsterdam, pp. 67–88.

Barker, D., Emonet-Dénand, F., Harker, D., Jami, L. and Laporte, Y. (1976b) Distribution of fusimotor axons to intrafusal muscle fibres in cat tenuissimus spindles as determined by the glycogen depletion method, *J. Physiol. (Lond.)*, 261: 49–70.

Barker, D., Emonet-Dénand, F., Harker, D., Jami, L. and Laporte, Y. (1977) Types of intra- and extrafusal muscle fibre innervated by dynamic skeleto-fusimotor axons in cat peroneus brevis and tenuissimus muscles, as determined by the glycogen-depletion method, *J. Physiol. (Lond.)*, 266: 713–726.

Bessou, P. and Laporte, Y. (1965) Technique de préparation d'une fibre afférente I et d'une fibre afférente II innervant le même fuseau neuro-musculaire chez le Chat, *J. Physiol. (Paris)*, 57: 511–520.

Bessou, P. and Pages, B. (1975) Cinematographic analysis of contractile events produced in intrafusal muscle fibres by stimulation of static and dynamic fusimotor axons, *J. Physiol. (Lond.)*, 252: 397–427.

Bessou, P., Laporte, Y. and Pages, B. (1968a) A method of analysing the responses of spindle primary endings to fusimotor stimulation, *J. Physiol. (Lond.)*, 196: 37–45.

Bessou, P., Laporte, Y. and Pages, B. (1968b) Frequency grams of spindle primary endings elicited by stimulation of static and dynamic fusimotor fibres, *J. Physiol. (Lond.)*, 196: 47–63.

Boyd, I., Gladden, M., McWilliam, P. and Ward, J. (1977) Control of dynamic and static nuclear bag fibres and nuclear chain fibres by γ and β axons in isolated cat muscle spindles, *J. Physiol. (Lond.)* 265: 133–162.

Brown, M. and Butler, R. (1973) Studies on the site of termination of static and dynamic fusimotor fibres within muscle spindles of the tenuissimus muscle of the cat, *J. Physiol. (Lond.)*, 233: 553–573.

Brown, M. and Butler, R. (1975) An investigation into the site of termination of static gamma fibres within muscle spindles of the cat peroneus longus muscle, *J. Physiol. (Lond.)*, 247: 131–143.

Edström, L. and Kugelberg, E. (1968) Histochemical composition, distribution of fibres and fatiguability of single motor units, *J. Neurol. Neurosurg. Psychiat.*, 31: 424–433.

Emonet-Dénand, F. and Lapore, Y. (1978) Frequencegrammes dûs à la stimulation d'axones γ dynamiques exerçant des effets du type II, *C.R. Acad. Sci.*, 287D: 531–534.

Emonet-Dénand, F., Laporte, Y., Matthews, P. and Petit, J. (1977) On the subdivision of static and dynamic fusimotor actions on the primary ending of the cat muscle spindle, *J. Physiol. (Lond.)*, 268: 827–861.

Emonet-Dénand, F., Jami, L., Laporte, Y. and Tankov, N. (1978) Glycogen-depletion elicited in peroneus brevis spindles by static γ axons, *Neurosci. Lett., Suppl. 1*: S 93.

Harker, D., Jami, L., Laporte, Y. and Petit, J. (1977) Fast conducting skeletofusimotor axons supplying intrafusal chain fibres in the cat peroneus tertius muscle, *J. Neurophysiol.*, 40: 791–799.

Laporte, Y. (1978) The motor innervation of the mammalian muscle spindle In *Studies in Neurophysiology*, R. Porter (Ed.), University Press, Cambridge.

Ovalle, W. and Smith, R. (1972) Histochemical identification of three types of intrafusal muscle fibres in the cat and monkey based on the myosin ATPase reaction, *Canad. J. Physiol. Pharmacol.*, 50: 195–202.

Multisensory Control of Spinal Reflex Pathways

A. LUNDBERG

Department of Physiology, University of Göteborg, S-400 33 Göteborg (Sweden)

A new era in research on spinal reflexes was started when Lloyd (1943, 1946) used the monosynaptic test technique to investigate the time course and linkage of the excitatory and inhibitory synaptic actions evoked from different afferent fibre groups. These studies were then followed up with intracellular recording of the synaptic potentials in motoneurones (Eccles, 1953). The subsequent work to define the receptor origin of the different afferent fibre groups has led to remarkable advances which now allow the study of central actions by selective activation of afferents from different receptor systems (cf. Jack, 1978). The focusing on the reflex actions evoked by individual receptor systems has tended to generate the tacit assumption that reflex lines from primary afferents are private and not shared by the different afferent systems, even when interneurones are interposed. Yet, already some time ago recording from spinal interneurones revealed that convergence from different afferents was not only common but even the rule (Hongo et al., 1966). Subsequent systematic investigations revealed convergence from primary afferents at an interneuronal level in different reflex pathways. I will discuss the possible role of wide convergence of afferents with different receptor origin on interneurones in reflex pathways to motoneurones (cf. also Jankowska, 1979).

REFLEX PATHWAYS FROM GROUP I MUSCLE AFFERENTS

The interneurone in the reciprocal Ia inhibitory pathway is one of the most extensively investigated neurones in the CNS. Convergence has been shown indirectly by spatial facilitation of Ia IPSPs recorded in motoneurones or by direct recording from interneurones which Jankowska and Roberts (1972a, b) identified as belonging to the reciprocal Ia inhibitory pathway. The summarising circuit diagram (Fig. 1) published by Lindström (1973) is now incomplete but suffices to emphasise how extensive connexions are to these interneurones, not only from descending motor pathways but also from primary afferents (for reference, see Hultborn, 1972, 1976). Convergence on the reciprocal Ia inhibitory interneurones closely resembles that of agonist alpha motoneurones and it has been postulated that these parallel connexions are required to link reciprocal inhibition to activation of alpha motoneurones. These interneurones appear to be target cells for other neuronal pathways in much the same way as motoneurones (Hultborn, 1976). It may therefore be argued that these interneurones

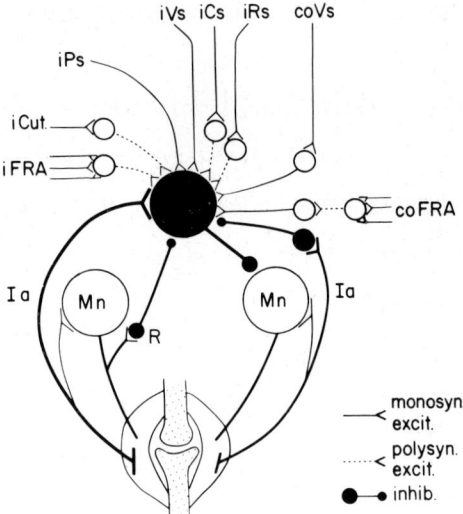

Fig. 1. Circuit diagram of some connexions to the interneurone in the reciprocal Ia inhibitory pathway. i, ipsilateral; co, contralateral; Vs, Cs, Rs, Ps, vestibulo-, cortico-, rubro- and propriospinal tracts. Cut, cutaneous afferents; FRA, flexor reflex afferents; Mn, motoneurones; R, Renshaw cells. (From Lindström, 1973; for reference, see also Hultborn, 1972, 1976).

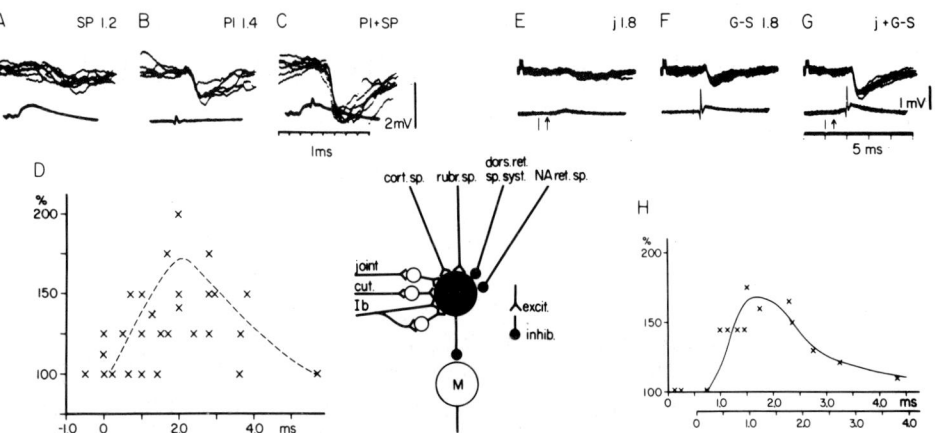

Fig. 2. Convergence on interneurones in the Ib inhibitory pathway to motoneurones. Upper traces, intracellular recordings from motoneurones to gastrocnemius-soleus (A–C) and to flexor digitorum longus (E–G). Lower traces, incoming volleys recorded from the L7 dorsal root entry zone. A–C show facilitatory interaction in inhibitory transmission between a volley in the cutaneous superficial peroneal (SP) nerve and a group I volley in the nerve to plantaris (Pl). E–G, a similar facilitation between a volley in the posterior nerve to the knee joint (j) and a group I volley in the nerve to gastrocnemius-soleus (G–S). Stimulus strength is given in each record in multiple of thresholds for the nerve. Corresponding graphs D and H show the time course of facilitation obtained by varying the conditioning testing interval. The arrival of impulses in the fastest joint nerve afferents is shown by arrow in E and G. Zero in lower abscissa in H gives time of arrival of effective impulses in the joint nerve, evoked when the stimulus strength was increased, 1.4–1.5 × threshold. Circuit diagram gives neuronal connexions to Ib inhibitory interneurones (for reference to the descending effects see Lundberg et al. 1978). This diagram is incomplete since Fetz et al. (1979) have evidence for other very interesting connexions. (From Lundberg et al. 1977a, 1978.)

represent a very special case that cannot be taken to reflect a general characteristic of reflex pathways. However, Fig. 2 shows convergence from different afferents also on interneurones in the reflex pathway traditionally described as the Ib inhibitory pathway (cf., however, Fetz et al., 1979; Jankowska, this volume, part IA, chapter 3). The group I test IPSPs in Fig. 2 are markedly facilitated by conditioning volleys in cutaneous afferents (A–C) and joint afferents (E–G) with a time course that in both cases suggests a disynaptic linkage to the inhibitory interneurones (D, H). The circuit diagram of connexions to those interneurones from higher centres and primary afferents in Fig. 2 is only one year old but already outdated since there is now evidence that also Ia afferents excite these interneurones and that both Ia and Ib afferents may inhibit the same interneurones (cf. Jankowska, chapter IA3).

In summary, there is evidence that different afferents converge at an interneuronal level to form multisensory reflex feedback systems. A relatively large number of such systems may exist in which afferents of different receptor origin act together in a variety of combinations. It is conceivable that afferents from a particular receptor system in a given receptive field (say Ib afferents from one muscle) have diverging connexions to interneurones of several parallel reflex feedback systems which control motoneurones to a certain muscle. Higher centres may select one of them (by interneuronal facilitation) to regulate a certain movement but activity in the given afferents nevertheless reaches the interneurones of the other feedback systems. In this connexion it is interesting to note that IPSPs from primary afferents are so common in interneurones (Hongo et al., 1966, 1972; Jankowska, chapter IA3). Accidental coactivation of parallel systems may indeed be prevented by inhibition from the selected interneuronal pathway.

REFLEX PATHWAYS FROM THE FRA

If different afferent systems act together they may be classified according to their common central actions. This was done long ago for the FRA (flexor reflex afferents), comprising group II and III muscle afferents, joint afferents and cutaneous afferents (Eccles and Lundberg, 1959a, Holmqvist and Lundberg, 1961). The reason for grouping them together was not only that similar reflex effects, excitation in flexors and inhibition in extensors, were evoked in spinal cats from all these afferents but also that they acted together on a variety of ascending pathways; in all cases the receptive field was very wide, for muscle afferents comprising both extensors and flexors. Early on it was suggested that the ascending pathways influenced from the FRA signalled the activity in interneurones of reflex pathways from the FRA (Lundberg, 1959; Lundberg and Oscarsson, 1962). This hypothesis received a new dimension when it appeared that descending tracts excite the interneurones of reflex pathways from the FRA so that they can mediate effects from higher motor centres to motoneurones (cf. below). With Oscarsson's terminology the interneurones of reflex pathways from the FRA function as "lower motor centres" and ascending FRA pathways signal the activity in these lower motor centres (Oscarsson, 1973; Ekerot et al., this volume, part IB, chapter 1). The ascending FRA information is by far the dominating action in the spino-reticulocerebellar and spino-olivocerebellar pathways; it is found in one of the channels in the dorsal spinocerebellar tract and is also signalled (mainly as inhibition) in the majority of the ventral spinocerebellar tract neurones (Oscarsson, 1973). Why does cerebellum require all this information about activity in the interneurones of reflex pathways from the FRA? Clearly it must be related to their role as lower motor centres but that does not explain *the role of the primary afferent activity in the control of these lower motor centres.* This is the special problem dealt with in this review.

The FRA concept has been much disliked but may now be less unpalatable when it is clear that convergence of afferents with different receptor function occurs in different reflex pathways. Published criticism is not abundant but Matthews (1972) has voiced it forcefully. His main concern is the secondary spindle afferents which were included among the FRA by Eccles and Lundberg (1959a). I admit there is still uncertainty in this case but it is not a crucial question for the assessment of the FRA hypothesis. In the criticism of the FRA concept two misunderstandings are common. It is often assumed: 1) that the contributory afferents (other than secondary spindle afferents) are activated only by strong nociceptive-like stimuli, and 2) that this activity evokes a fixed flexion reflex pattern, i.e., excitation of flexors and inhibition of extensors. With these assumptions it is admittedly difficult to give the FRA a role in normal motor control. In my opinion both of them are wrong (cf. below). The second one is clearly related to the terminology – the FRA is really a misnomer! I wish to recall that in the title of the original paper the afferents were defined as those which *may* evoke flexion reflex; already at that time there was evidence for alternative pathways (Eccles and Lundberg, 1959a).

Prior to the general discussion of the function of the FRA it is necessary to mention two different classes of reflexes from the FRA, the short-latency reflexes in the acute spinal cat and (at some length) the late longlasting reflexes which appear in acute spinal cats after DOPA i. v. Fig. 3C–F, shows short-latency EPSPs in a flexor motoneurone before DOPA. These EPSPs are effectively depressed after DOPA (G–I) but reappear as the effect of DOPA wears off (Andén et al., 1966a). After DOPA a train of volleys instead evokes a late longlasting discharge in ipsilateral flexor

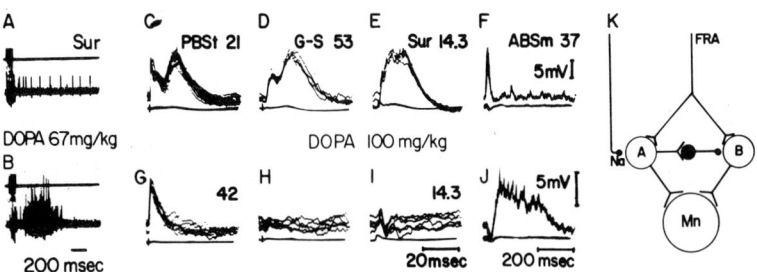

Fig. 3. Changes in excitatory transmission to flexor motoneurones from the FRA after DOPA i.v. Acute spinal cats. A and B) Recordings from the nerve to tenuissimus and triphasically from the sciatic nerve before and after DOPA. The same train of volleys given in the cutaneous sural (Sur) nerve before DOPA evokes a shortlasting intense discharge with some after-discharge after DOPA, a characteristic late longlasting discharge appears (note time scale). The intracellular records in C–J are from motoneurones to the knee flexors posterior biceps semitendinosus (PBSt) sampled before and after DOPA i.v. Records C–E and corresponding lower G–I show the effect of single volleys in the PBSt, G–S and Sur nerves at strengths indicated. Note the disappearance of short-latency FRA EPSPs after DOPA. A short train of volleys in the nerve to ABSm (anterior biceps-semimembranosus) gave a shortlasting early EPSP before (F) and a late very longlasting EPSP after (J) DOPA (note the slow time scale). See text for explanation of schematic circuit diagram K in which A and B each represent a chain of interneurones transmitting before (A) and after (B) DOPA. (From Andén et al. 1966a and Jankowska et al. 1967a.)

efferents (B) and intracellularly a late longlasting EPSP (J) not found before DOPA (slow time base in A, B, F, J). Since these late EPSPs were evoked from cutaneous afferents, joint afferents and high threshold muscle afferents from flexors and extensors they were attributed to the FRA. The hypothesis forwarded to account for this reflex change is shown in the circuit diagram K. DOPA gives transmitter release from a descending noradrenergic pathway which inhibits the short-latency reflex pathway A, thereby releasing transmission in pathway B from an inhibitory control normally exerted from pathway A. The finding that prolongation of repetitive stimulation of the FRA delays the onset of the effects mediated by pathway B (Jankowska et al., 1967a, b) was taken as evidence for an inhibition from pathway A to B.

Corresponding to the excitation of ipsilateral flexors from the FRA there was also late crossed longlasting activation of extensors with very strong reciprocity in activation of flexors and extensors. The further analysis revealed *mutual inhibitory connexions between interneurones transmitting late excitation to flexors and extensors*. The intracellular records in Fig. 4 show very effective depression of the late EPSP from the ipsilateral FRA in a flexor motoneurone by conditioning volleys in the contralateral FRA (A–C) and, vice versa, the obliteration of the late EPSP evoked from the contralateral FRA in an extensor motoneurone by conditioning volleys in the ipsilateral FRA (G–I). These findings were extended by recording from interneurones (location in Q, Fig. 4) in which a late longlasting train of impulses was evoked from either the ipsilateral (E) or contralateral FRA(K). Observe in D–F that the discharge from ipsilateral FRA is completely inhibited by preceding stimulation of the contralateral FRA, and in J–L the equally effective inhibition of the discharge from the contralateral FRA by conditioning stimulation of the ipsilateral FRA.

Since the conditioning FRA volleys after DOPA did not evoke primary afferent depolarisation in their own terminals (Andén et al., 1966b; Jankowska et al., 1966) it was concluded that the inhibition was exerted at an interneuronal level as shown in P, Fig. 4. Each single intercalated excitatory neurone in this diagram represents a chain of interneurones but the close correlation of the interneuronal discharges with the time course of the EPSPs in motoneurones (Jankowska et al., 1967b) strongly suggests the interneuronal recording was close to the motoneurones, possibly from last-order interneurones. FRA volleys from the limb opposite to that exciting the interneurones gave a very longlasting complete inhibition of glutamate evoked discharges in these interneurones (M–O, Fig. 4) which strongly suggests that inhibition is postsynaptic.

The half-centre organization (strong mutual inhibitory connexions between neurones exciting muscles with opposite function) of this interneuronal network was an exciting finding because here was the first experimental evidence for a neuronal system which could give alternating activation of extensors and flexors as suggested by Brown (1911, 1913, 1914). An excitatory drive to both interneuronal pools may give alternating activation. When DOPA was given after pretreatment with the monoamine oxidase inhibitor, Nialamide, spontaneous bursts occurred in the interneurones which gave bursts of motoneuronal discharges (Fig. 5A). Stimulation of the FRA during these bursts gave alternating flexor and extensor activation – described as spinal stepping by Jankowska et al. (1967a). Similar alternating efferent discharges were evoked also in the absence of spontaneous bursts when suitably timed short trains of volleys were given in the ipsilateral and contralateral FRA (D, Fig. 5). There can be no reasonable doubt that alternating discharges as in Fig. 5 depend on the half-centre organization in the network released after DOPA. This

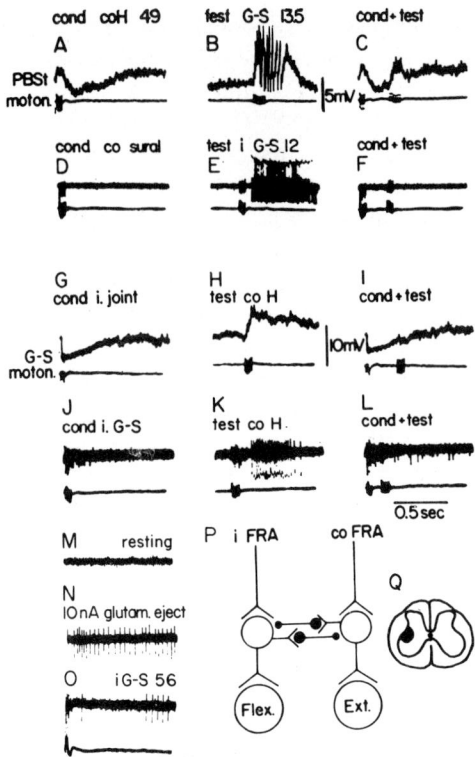

Fig. 4. The half-centre organisation of the interneuronal network released after DOPA. Acute spinal cats after DOPA 100 mg/kg i.v. A–C and G–I are intracellular recordings from motoneurones. In the knee flexor motoneurone (A–C) a train of volleys in high threshold muscle afferents evoked the characteristic longlasting EPSP (B) which was effectively inhibited (C) by a preceding train of volleys in contralateral high threshold afferents (H, hamstring). Conversely in the extensor motoneurone (G–I) the longlasting EPSP was evoked from contralateral high threshold muscle afferents (H) and inhibited from the ipsilateral joint nerve (I). The corresponding extracellular recordings D–F and J–L are from interneurones (location within black region, Q) which were excited from ipsilateral and contralateral high threshold muscle afferents resp. Records F and L show the complete inhibition of these discharges by conditioning stimulation of FRA in the opposite limb. M–O are from an interneurone responding as in J–L with a late longlasting discharge from the co FRA which was completely inhibited from the ipsilateral FRA. The interneurone had no resting discharge (M). Ejection of glutamate (10 nA) from the recording electrode (2 M glutamate solution) produced the discharge in N which was effectively silenced for more than 0.5 sec by a short train of volleys in the ipsilateral FRA (O). The time course of inhibition of transmission to motoneurones was equally longlasting (fig. 9, Jankowska et al., 1967a). See text for explanation of circuit diagram in P in which excitatory interneurones to flexor and extensor motoneurones represent a chain.
(M–O from Bergmans et al., 1967; cf. also Lundberg, 1969. A–C from Jankowska et al. 1967a, b.)

conclusion should be kept in mind when assessing further work in this field. These alternating discharges can be viewed as precursors of fictive locomotion, i.e., the maintained locomotor activity which after Nialamid and DOPA may occur spontaneously or be evoked by continuous bilateral stimulation (Grillner and Zangger, 1974). It seems safe to postulate that there is no qualitative difference between the shortlasting alternate discharges in Fig. 5 and the maintained alternate activity in fictive locomotion. It then follows that also the alternating activity in fictive locomotion depends on the half-centre organization of the interneuronal network released after DOPA.

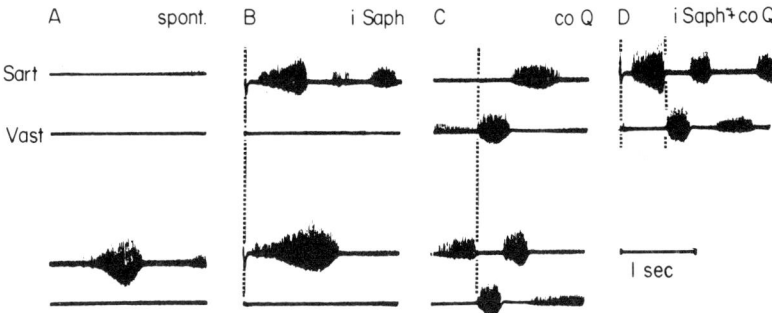

Fig. 5. Alternating discharges in flexor and extensor efferents after Nialamid and DOPA. Acute spinal cat pretreated with Nialamid (10 mg) and given DOPA 100 mg/kg. Recordings from the nerves to a flexor (medial sartorius, Sart) and to an extensor (medial vastus of quadriceps, Vast). A) Resting activity absent in upper records but a characteristic flexor burst in the lower records. B) Flexor discharges evoked by stimulation (short train) of ipsilateral cutaneous nerve (saphenus, i Saph) at time given by the vertical line. C) Discharges evoked by stimulation of contralateral high threshold muscle afferents (short train) from quadriceps (co Q) at time given by the vertical line. Note in the upper records that the extensor discharge is followed by a flexor burst. In the lower records in C, stimulation of co Q is given during a spontaneous burst which is then interrupted and resumed after cessation of the extensor discharge; a second extensor discharge then occurs after this flexor burst. D, combined ipsilateral and contralateral stimulation also gives alternating activation (From Jankowska et al., 1967a, record D added.)

> Edgerton et al. (1976) correlated efferent activity and interneuronal activity during fictive locomotion in spinal cats. They concluded that the interneuronal network giving alternate activity in fictive locomotion is identical with the network transmitting the late reflexes from the FRA after DOPA, but do not seem to accept the findings which revealed the half-centre organization of this network (cf. also Grillner, 1975). In their discussion they mention three proposed spinal generator models, among them Brown's half-centres, but conclude that "in the vertebrate locomotor system there is no direct evidence for either type of model". I maintain that the findings by Jankowska et al. (1967a, b) provide strong support for the half-centre model. This statement should not be taken to imply that the central step generator at large can give only strict alternating activity to flexor and extensor motoneurones because there is some evidence for a centrally generated more differentiated output (Grillner and Zangger, 1975). Nevertheless, strict alternating activation is such a characteristic feature of locomotor-like activity in many experimental situations that it seems premature to abandon the hypothesis of a simple generator (for which there is experimental evidence) merely because it does not account for *all* findings regarding the locomotor output. The central generator at large may well consist of a network giving alternating activity to which are coupled other central mechanisms providing a more differentiated output.

A hypothesis for the function of the FRA was indicated with the suggestion that activation of the "DOPA-network" could occur reflexly as a result of FRA activity produced by the movement (Jankowska et al., 1967a). Experimental evidence now available is compatible with the hypothesis that the FRA play a role in stepping. Acute spinal cats injected with DOPA or Clonidine showed no spontaneous activity but under certain conditions performed partial or more complete walking movements when placed on a moving treadmill belt (Budakova, 1971, 1973; Forssberg and Grillner, 1973). The passive and active movements thus appear to evoke afferent activity producing and maintaining spinal locomotor movements. I suggest that these stepping movements are generated by FRA-activity evoked by the passive and active movements and that this activity drives the half-centre network. Alternating activation of flexors and extensors may be produced by this mechanism but not walking movements that are timed to the speed of a treadmill belt. However, the reflex effects after DOPA do not consist of a fixed pattern of ipsilateral flexion and contralateral extension because it has now been shown that the *destination of the reflex response depends on the limb position*. Grillner (1973) first observed that ipsilateral flexor activation could be changed to ipsilateral extensor activation by placing the hip in flexion. Grillner and Rossignol (1978) found a similar crossed reflex reversal governed by the hip position of the recording side. Since the late reflexes from the FRA are mediated by the interneuronal network which may generate locomotor movement, Grillner and Rossignol (1978) suggested as one possible explanation of this reflex reversal "that the central network for locomotion is effectively driven by a rather unspecific input, and that this input is gated to the 'extensor' and then to the 'flexor' part of the network, depending at least partially on the position of the limb". I am strongly in favour of this hypothesis. A gating mechanism depending on limb position makes it much more easy to understand how activity in the FRA may contribute to locomotion since the required switch between the two half-centres would then be appropriately timed by the locomotor movement itself.

I have emphasised the FRA reflexes released after DOPA since this is the only case in which there are some experimental findings suggesting that their activity may contribute in movement regulation but the effect after DOPA is only one of many FRA effects. The short-latency FRA reflexes in the acute spinal cat are very pronounced, the interneurones of these pathways are excited from higher motor centres and the activity in many spinocerebellar pathways relates to these short-latency FRA reflex paths – all these findings have to be accounted for. I will now discuss a more general FRA hypothesis which was presented at the Biophysical Congress in Moscow (Lundberg, 1972). I will briefly consider 5 assumptions quoted from the proceedings of this congress.

1. The afferents comprising the FRA converge on common interneurones

Excitatory convergence from high threshold muscle afferents, joint afferents and cutaneous afferents is found in many interneurones (Hongo et al., 1966). The spatial facilitation technique has been used to show that it occurs in interneuronal pathways to motoneurones but only few results are published (cf. Jankowska et al., 1967a, 1973). Spatial facilitation in crossed pathways to extensor motoneurones is shown in Fig. 6. A–F refer to short-latency pathways open in the acute spinal cat and G–I to the pathway released after DOPA. Note the facilitatory interaction between joint afferents and cutaneous afferents (A–C), between muscle afferents and cutaneous afferents (D–F) and between joint afferents and muscle afferents (G–I). The convergence on common interneurones suggests that the different afferent systems combine in the activation of interneurones (cf. also assumption 5).

2. The FRA have access to a number of alternative spinal reflex pathways

Eccles and Lundberg (1959a) reported that volleys in the FRA in acute spinal cats occasionally evoked inhibition in flexor motoneurones instead of the more characteristic excitation. This inhibitory pathway could be analysed in decerebrate cats with low brain stem lesions. In the decerebrate state there is a very effective tonic inhibition of transmission from the FRA (Eccles and Lundberg, 1959b). A low pontine lesion releases the inhibitory but not the excitatory pathways from the FRA from this control and thus allows systematic investigation of the inhibitory FRA pathway to flexors (Holmqvist and Lundberg, 1961; Fedina and Hultborn, 1972). The possibility

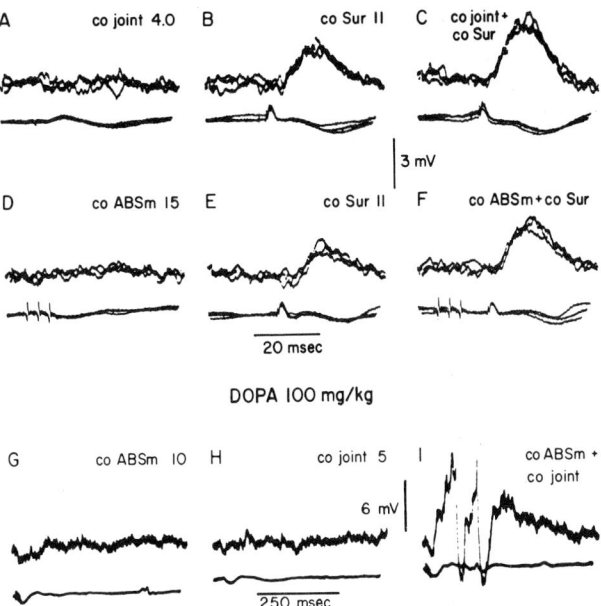

Fig. 6. Convergence at interneuronal level in the crossed extensor reflex pathway from the FRA. Acute spinal cats. A–F were recorded before DOPA from an adductor femoris motoneurone and G–I after DOPA from a motoneurone to quadriceps. Spatial facilitation was investigated as in Fig. 2. C, F and I show the effect of combined stimulation of the different nerves indicated. Note that these EPSPs are larger than the algebraic sum of the individual responses in the two left columns. (From Fu, Jankowska and Lundberg unpublished observations.)

that the excitatory and inhibitory paths were supplied by afferents of different receptor function was rejected since it would require "two sets of receptor systems from skin, muscle and joint with afferents in the same diameter ranges and both of them supplying action from the same receptive fields" (Holmqvist and Lundberg, 1961). In addition, these authors found that excitation or inhibition of flexors could be evoked by the same adequate stimuli. It was therefore postulated that the same afferent systems have alternative excitatory and inhibitory pathways to flexor motoneurones. Also extensor motoneurones receive alternative short-latency excitatory and inhibitory reflex pathways from the FRA. In this case no particular CNS state has been defined, which favours this excitatory pathway at the expense of the inhibitory but in some spinal cats FRA volleys evoke EPSPs in extensor motoneurones instead of the usual IPSPs (Holmqvist and Lundberg, 1961; Lundberg et al., 1977b). These EPSPs, which may be as large as those usually evoked only in flexor motoneurones, indicate the existence of a potent excitatory FRA pathway to extensors which is usually closed in the acute spinal cat. Jankowska et al. (1973) tentatively suggested the existence of three short-latency excitatory FRA pathways to extensors as shown in the circuit diagram J, Fig. 7; A is activated from the ipsilateral FRA, B is bilaterally supplied and C the most commonly seen pathway supplied by the contralateral FRA alone. Fig. 7 illustrates evidence for the excitatory pathway activated from both hindlimbs; note the spatial facilitation in G–I.

The definition of pathway B was aided by the finding that its last order interneurones are monosynaptically activated from long descending propriospinal fibres (Jankowska et al., 1973) and also

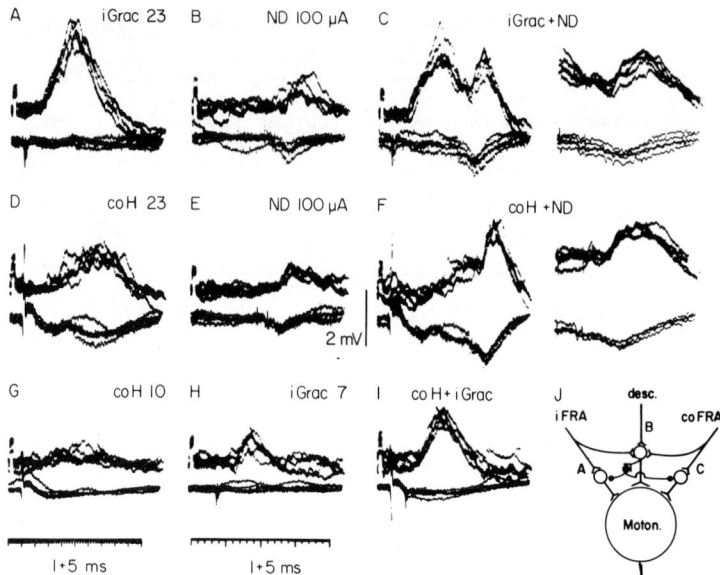

Fig. 7. Evidence for an excitatory reflex pathway to extensors bilaterally activated from the FRA with monosynaptic excitation of last order neurones from vestibulospinal fibres. The cat was anaesthetized with chloralose 60 mg/kg and hemisected (right side) in the lower thoracic region. The tip of a tungsten electrode was located in the left Deiters' nucleus (ND) in the region of the maximal antidromic field potential. The EPSPs in A and D were evoked from high threshold muscle afferents from both limbs as indicated. G–I show spatial facilitation between these nerves at weaker strengths. Double stimuli in Deiters' nucleus were required to evoke the disynaptic test EPSP in B and E. Note facilitation of the test EPSP in C and F by the conditioning volleys from either hindlimbs. Pair of records in C and F were obtained simultaneously at different speeds. (From Bruggencate and Lundberg, unpublished observations.)

from vestibulospinal fibres as shown in B, C and E, F, Fig. 7 (cf. also assumption 4). The recognition of the bilateral FRA reflex pathway is of considerable interest since many ascending neurones, which were assumed to signal activity in the reflex pathway from the FRA, are activated from both hindlimbs (the bVFRT neurones of Lundberg and Oscarsson, 1962). Like the last-order interneurones of the bilateral reflex pathway from the FRA, the bVFRT neurones are monosynaptically excited from the vestibulospinal tract and from long propriospinal neurones (Holmqvist et al., 1960; Grillner et al., 1968).

The most dramatic example of alternative reflexes from the FRA is provided by the late longlasting reflexes evoked after DOPA which were discussed in some detail above. In this case it is convenient to view the entire network mediating excitation both to extensors and flexors (B in diagram K, Fig. 3) as a pathway alternative to the entire set of short-latency pathways from the FRA (A in the same diagram).

3. *There are mutual inhibitory interactive connexions at an interneuronal level between the alternative reflex pathways from the FRA*

Volleys in primary afferents evoke IPSPs in the majority of interneurones; inhibition from the FRA is particularly common. Excitation or inhibition may be evoked from different or from the same afferents (Hongo et al., 1966, 1972; Jankowska, 1979). As already discussed in the first section these findings may suggest inhibition of one reflex pathway by activity in another.

The alternative short-latency EPSPs and IPSPs which can be evoked from the FRA in the same motoneurones (cf. above) appear to be almost mutually exclusive (Eccles and Lundberg, 1959a; Lundberg et al., 1977b) which might suggest mutual inhibitory connexions between these alternative pathways with opposite effect on the same motoneurones. In Fig. 3 I have indicated an interactive inhibition from the short-latency FRA reflex pathway to the pathways released after DOPA, the reason being that the activation of the latter pathway from the FRA could be delayed by prolongation of the respective stimulation.

It is tempting to speculate over the possibility that activity in the network released after DOPA may also inhibit transmission in the short-latency pathways (cf. diagram K. Fig. 3). The reason for this highly tentative suggestion is the following. Activity in the FRA is assumed to drive the half-centre network and give alternating activation of flexors and extensors in fictive locomotion. It is difficult to understand how this driving can be executed with any security if activity in the same afferents via pathway A (Fig. 3, K) can inhibit transmission to the half-centre network. This difficulty is resolved by hypothesising an inhibition also in the opposite direction so that once pathway B (Fig. 3, K) is activated it cannnot any longer be inhibited via the short-latency pathway A.

For the sake of the discussion below we will assume mutual inhibitory connexions between the short-latency excitatory reflex pathways to extensors but only one of these hypothetical inhibitory connexions is indicated in circuit diagram L, Fig. 7.

4. The interneurones of different alternative reflex pathways from the FRA can be activated by descending pathways from higher levels of the neuraxis

Excitatory action from higher motor centres in interneurones of FRA reflex pathways was shown by facilitatory interaction between volleys in descending tracts and the FRA in transmission to motoneurones. Lundberg and Voorhoeve (1962) found that conditioning volleys in corticospinal fibres markedly facilitated transmission of EPSPs and IPSPs from the FRA to ipsilateral motoneurones. Hongo et al. (1969) made similar findings for the rubrospinal tract and Bruggencate and Lundberg (1974) found that vestibulospinal volleys facilitated transmission from the FRA to contralateral motoneurones.

Reversal of the conditioning testing order revealed that conditioning volleys in the contralateral FRA facilitated transmission in the disynaptic excitatory vestibulomotoneuronal pathway to extensors (Bruggencate and Lundberg, 1974). These results show that excitatory interneurones which project directly to extensor motoneurones and are monosynaptically excited from vestibulospinal fibres also receive excitation from the contralateral FRA. The same technique was used to show that long descending propriospinal fibres and vestibulospinal fibres project directly to last order neurones of an excitatory pathway to extensor motoneurones which is activated from the FRA in both hindlimbs (Fig. 7). There is also evidence for direct connexions of rubrospinal fibres to last order interneurones in excitatory and inhibitory pathways from the FRA to ipsilateral motoneurones (Baldissera et al., 1971).

Investigations of excitatory actions from higher motor centres on the interneuronal pathways released after DOPA are highly desirable because such information may help us to understand how the spinal locomotor generator is commanded into action.

5. Some of the afferents belonging to the FRA are excited during an active limb movement

As mentioned above it is sometimes assumed that the FRA are activated only by strong mechanical nociceptive-like stimuli. There is some evidence that afferents belonging to this system are activated in normal movements. It was reported long ago that neurones belonging to the direct spinocerebellar tracts which were excited or inhibited from the FRA could be influenced in the same way by muscle stretch or contraction (Holmqvist et al., 1956; Oscarsson, 1957). In a similar way ascending neurones which are activated from the FRA in both hindlimbs (bVFRT, cf. Lundberg and Oscarsson, 1962) can be activated by passive limb movements in acute spinal cats (unpublished observations).

With regard to reflex pathways the results are scanty since it is difficult to obtain conditions which allow unambiguous interpretations. The best evidence is related to the late FRA reflexes after DOPA; the locomotor movements found in DOPA treated acute spinal cats placed on a moving treadmill belt were thought to be due to FRA activity produced by the passive and active movements (see above). The previous experiment of Fig. 8 indeed shows that late reflex discharges can be produced after DOPA by a muscle contraction caused by stimulation of its α-efferents (Fig. 8, cf. legend).

> A reflex response as in Fig. 8 was found only in one-third of the experiments but in another experiment a late reflex response, which remained after transection of the nerve to gastrocnemius-soleus, was produced by contraction of a small muscle in the base of the tail and disappeared when this muscle was denervated. Similar findings were made by Fu, Jankowska and Lundberg (unpublished results) who stimulated (6 stimuli at 300/Hz) α-afferents in the transected L7 ventral root to evoke contraction in

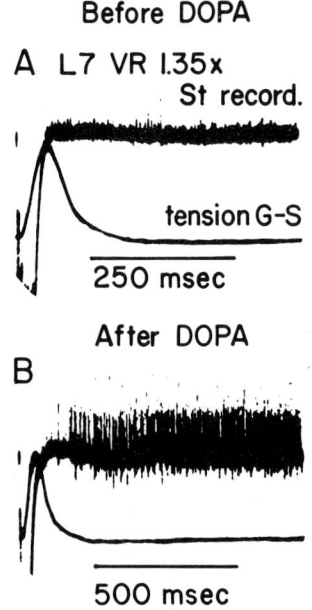

Fig. 8. Late reflex discharge evoked by muscle contraction after DOPA. Acute spinal cat. The L7 ventral root was transected and the peripheral end stimulated with a train at 1.35 × threshold which was subthreshold for γ-efferents. All muscles supplied by the L7 ventral root were denervated except gastrocnemius-soleus; the lower traces show the isometric contraction in this muscle recorded with a strain gauge. Upper traces, efferent discharge in the nerve to the knee flexor semitendinosus (St) which was conducted in the intact S1 ventral root. Note that reflex response appearing in B after DOPA closely resembles the reflex discharge that can be evoked from the FRA after DOPA (Figs. 3, 5). No discharge was evoked after transection of the nerve to G-S. Observe the different time calibrations in A and B. (From Bergmans, Grillner and Lundberg, unpublished observations.)

undissected muscles below the knee; the muscles above the knee were denervated and the surgical wounds in the thigh were carefully infiltrated with a local anaesthetic. The characteristic late longlasting discharge was evoked in afferents to semitendinosus (via the intact SI root). The reflex discharge remained after a further denervation of all muscles except the pretibial flexors and extensor digitorum brevis when active ankle flexion was tested against a 1 kg load. The reflex response decreased but did not disappear when ankle flexion was prevented.

Reflexes evoked by muscle contraction are probably mediated by group III muscle afferents (cf., however, Paintal, 1960; Bessou and Laporte, 1961). Recent findings (Lundberg et al., 1978) show a contribution from joint afferents to short-latency FRA pathways activated in a lower threshold range than was originally reported by Eccles and Lundberg (1959a). Threshold measurements for FRA effects evoked by electrical stimulation of the posterior nerve to the knee joint suggest that afferents in the diameter range 10–5 μ contribute effectively. Since fibres in this diameter range respond to joint movement and only very few to noxious stimuli (Burgess and Clark, 1969) it is likely that afferents from receptors which respond to joint movement contribute to the FRA.

The extensive convergence of different afferents covering a very large receptive field which suggests that the FRA pathways depend on simultaneous activity in fibres from many different receptors for their effective activation. That makes it difficult to elucidate the effect of a specific receptor system or of receptors in a small part of the receptive field. Feedback activation of FRA pathways during a movement presumably depends on spatial summation of activity from many sources which may give effective activation of the interneurones even if the movement produces only moderate activity in each individual afferent system.

Assumption 5 does not exclude that also nociceptive afferents belong to the FRA. Ascending FRA pathways which are activated by limb movements and light mechanical skin stimuli are stongly activated also by pinching of the skin, probably due to activation of nociceptive afferents (Lundberg and Oscarsson, 1962). With regard to reflex pathways there is a parallelism in the control of and in distribution of reflex actions evoked by volleys in the FRA and by adequate stimulation of nociceptors. In the decerebrate cat the descending tonic inhibition influences transmission in reflex pathways from the FRA and the flexor reflex from nociceptors in the same way. When the tonic inhibition operates effectively on the former pathway then even very strong pinching of the skin fails to evoke a flexor reflex (Holmqvist and Lundberg, 1961). It is noteworthy that when FRA volleys inhibit flexors (decerebrate cat with a low pontine lesion) inhibition of flexors is also evoked by pinching of the skin (Holmqvist and Lundberg, 1961). Correspondingly, it was shown recently that intra-arterial injection of algesic agents into a muscle produces the same reflex actions as volleys in the FRA; when the latter evokes excitation in extensor motoneurones then the algesic agents have the same effects (Kniffki, Schomburg and Steffens, to be published). The most simple explanation of all these findings is that nociceptive afferents also converge onto the reflex pathways from the FRA and thus belong to them. It is not unlikely that nociceptive afferents also have access to other reflex pathways which may mediate a purer nociceptive reflex.

Summarising FRA hypothesis

There is clearly strong experimental evidence for some of these "assumptions", but I prefer to discuss all of them as such since it is not known if they apply to all FRA pathways. The main difficulty when trying to assign a functional role to the FRA is the great number of central pathways on which they may act. A working hypothesis regarding the function of the reflex pathways from the FRA in motor control may, however, be based on the descending excitation of these interneuronal pathways ("lower motor centres" cf. Oscarsson, 1973). Let us assume that a descending command gives selective activation of interneurones belonging to one of the alternative FRA pathways to a given set of motoneurones (pathway B in diagram J, Fig. 7). Through the interactive inhibitory connexions from the FRA there is inhibition of transmission in the other excitatory pathways to the same motoneurones (A and C in the diagram) and also in the inhibitory FRA pathway to the same motoneurones. When the activity in the originally selected interneuronal pathway discharges motoneurones and produces a limb movement, then this movement produces activity in the FRA which, since the inhibition of the alternative pathways prevents their activation, is channelled back only to the already activated interneuronal pathway. The same scheme applies to each of these alternative excitatory pathways in Fig. 7 since each of them is assumed to have inhibitory connexions to the others. The FRA may thus function as a *segmental multisensory activating system common for many interneuronal pathways but nevertheless in usage be reserved for one pathway at a time.*

The hypothesis does not necessitate that the descending command for the movement is mediated exclusively by the interneurones of the FRA pathway. The main activation may be via another neuronal system but the positive feedback may be called into operation if the descending command includes subsidiary facilitation of the interneurones in an FRA pathway. The assumption of inhibitory interactive connexions makes it easy to understand how the FRA activity may exert a selective central action. But the existence of these connexions (the evidence is indeed indirect) is not prerequisite, since inhibition may be supplied by descending pathways (cf. Lundberg, 1966) or by other mechanisms like the limb position (Grillner and Rossignol, 1978). The possible role of the inhibitory reflex pathways from the FRA within the framework of this hypothesis also needs a comment. The descending command may comprise facilitation of inhibitory FRA pathways to motor nuclei whose activation is not required in the intended movement. If so, the FRA activity, which reinforces activation in some motor nuclei, may inhibit others.

> The positive feedback provided by the FRA might get out of hand if operating without control. The descending inhibition of interneurones from the lower brain stem (Lundberg, 1966) probably has an important regulatory function but segmental presynaptic inhibition may also play a role in the control of transmission from the FRA. The FRA evoke primary afferent depolarisation in their own terminals (Eccles et al., 1962), which suggests a negative feedback control of transmission from the FRA so that excess activity automatically curtails transmission. It is likely that the brain can regulate the gain of this negative feedback by facilitation or inhibition of interneuronal transmission to the terminals of FRA (Lundberg, 1966). In this connexion, it is of interest that one of the spino-olivocerebellar channels with its own longitudinal zone in the anterior vermis appears to signal activity in the pathway to FRA terminals, while other spino-olivary cerebellar channels are assumed to signal activity in FRA pathways to motoneurones (Andersson and Sjölund, 1978). It seems likely that the brain independently can regulate transmission in FRA pathways to primary afferent terminals and to motoneurones. Failure of either regulation might give motor disorder; it has been suggested that certain phenomena in spasticity might be due to decreased inhibitory control of the FRA pathways from the dorsal reticulospinal system (Burke and Lance, 1973). Dysregulation of a positive feedback mechanism can have severe consequence and it might be of interest to consider an unexplained phenomenon like muscle cramp (cf. Layzer and Rowland, 1971) in the light of the FRA hypothesis.

COMMENTS

The FRA concept has been a useful tool in the study of the functional organisation of segmental reflex pathways and ascending spinal tracts, and of the actions the latter exert on higher centres. The disclosure of longitudinal microzones in the cerebellar cortex is based largely on the results of detailed investigations of the projection of FRA activated ascending pathways (Oscarsson, 1973; Andersson and Oscarsson, 1978). The variety of central actions evoked from the FRA is bewildering and there is clearly a need for some unifying functional hypothesis. It is possible to evade the problem by arguing that interneurones of the reflex pathway from the FRA can be utilized by higher centres relatively independent of primary afferent activity; descending inhibition operating at an early stage in the interneuronal pathways (Lundberg, 1966) might leave last order interneurones at the exclusive disposal of the higher centres. However, in the hypothesis discussed above I have preferred to try to give a functional meaning to primary afferent activity and to the convergence of excitation from primary afferents and descending pathways.

Convergence from different primary afferent systems on common interneurones is a characteristic feature not only of reflex pathways from the FRA but also of other reflex pathways (cf. also Jankowska, this volume, chapter IA3) and may well be a general rule at least for reflex pathways from proprioceptors. It is relevant to raise the question why nature allows information from different receptors to become mixed up at an interneuronal level. I believe that the answer is quite simple. Reflexes serve to give feedback control of movements commanded from higher centres. It can be assumed that all active movements lead to activation of receptors in muscles, joints and skin. If so, why should a feedback system controlling a movement utilise only a single afferent system to control the movement. From a teleological point of view it seems much more sensible that different receptors, which might give useful information, combine in this regulation and this is best achieved if their afferents converge on the same interneurones so that they can form a common reflex pathway. Relative lack of place specificity is probably unique for the FRA pathways. In other reflex pathways afferents of different receptor origin may be drawn from very specific fields which are related to the particular movements subserved. Convergence of afferents from different receptors should not be taken to indicate loss of information from individual receptor systems but integration of it at a premotoneuronal level.

The road which may lead to an understanding of the organisation also of other interneuronal systems than the reciprocal Ia inhibitory pathway and the recurrent inhibitory pathway has been taken by Jankowska and her collaborators with the studies of the axonal projections of functionally identified interneurones (Czarkowska et al., 1976; Jankowska, this volume, chapter IA3). At present it is anybody's guess how many different interneuronal feedback systems the spinal cord possesses. The problem now at hand is to define them and investigate them in isolation from each other. Only then can the problem be approached how they interact and how higher centres which control them can handle a broad spectrum of regulatory mechanisms; in other words how they function together in normal movements.

SUMMARY

This review deals with convergence from different primary afferents on interneurones in reflex pathways.

The first section summarises results showing that interneurones of the reciprocal Ia inhibitory pathway receive a wide convergence resembling that found in agonist motoneurones. Interneurones in reflex pathways from Ib afferents are excited from cutaneous and joint afferents and also from Ia afferents (cf. Jankowska, chapter IA3).

The second, major section describes reflexes from the FRA, i.e. reflex actions from high threshold muscle afferents, joint afferents and cutaneous afferents which are mediated by common interneurones. A distinction is made between short-latency reflexes in the acute spinal and long-latency reflexes in acute spinal cats after DOPA i.v.; the half-centre organisation of the interneuronal network mediating the latter reflexes is recapitulated and discussed in relation to spinal generation of alternating activity in flexors and extensors. An important role of reflex pathways from the FRA in motor control is indicated by findings showing that their interneurones are excited from different higher motor centres and that many ascending pathways, spinocerebellar in particular, signal information regarding their interneuronal activity. It is emphasised that volleys in the FRA do not evoke a fixed pattern of ipsilateral flexion and crossed extension but that the FRA have alternative excitatory and inhibitory reflex lines to both flexors and extensors. It is postulated that afferents belonging to the FRA may be activated during an active movement and suggested that the FRA function as a multisensory activating system common for many interneuronal pathways but in actual usage is reserved for the particular pathway(s) selected by higher centres.

It thus appears that interneurones in all reflex pathways from muscle afferents investigated are activated from extramuscular afferents. It is pointed out that active movements excite receptors in muscle, skin and joints and postulated that reflex feedback systems controlling movements, rather than depending on a single afferent system, combine activity from different receptors which may give useful information.

ACKNOWLEDGEMENT

I am indebted to Dr. Hans Hultborn for valuable comments on the manuscript.

REFERENCES

Andén, N.-E., Jukes, M.G.M., Lundberg, A. and Vyklický, L. (1966a) The effect of DOPA on the spinal cord. 1. Influence on transmission from primary afferents. *Acta physiol. scand.*, 67: 373–386.

Andén, N.-E., Jukes, M.G.M., Lundberg, A. and Vyklický, L. (1966b) The effect of DOPA on the spinal cord. 3. Depolarisation evoked in the central terminals of ipsilateral Ia afferents by volleys in the flexor reflex afferents. *Acta physiol. scand.*, 68: 322–336.

Andersson, G. and Oscarsson, O. (1978) Climbing fibre microzones in cerebellar vermis and their projection to different groups of cells in the lateral vestibular nucleus. *Exp. Brain Res.*, 32: 565–578.

Andersson, G. and Sjölund, B. (1978) The ventral spino-olivocerebellar system in the cat. IV. Spinal transmission after administration of Clonidine and l-Dopa. *Exp. Brain Res.*, 33: 227–240.

Baldissera, F., Ten Bruggencate, G. and Lundberg, A. (1971) Rubrospinal monosynaptic connexion with last order interneurones of polysynaptic reflex paths. *Brain Res.*, 27: 390–392.

Bergmans, J., Fedina, L. and Lundberg, A. (1967) The mechanism of long lasting inhibitory actions in interneurones. *Arch. Int. Physiol.*, 75: 864–867.

Bessou, P. and Laporte, Y. (1961) Étude des récepteurs musculaires innervés par les fibres afférentes du groupe III (fibres myelinisées fines), chez le chat. *Arch. Ital. Biol.*, 99: 293–321.

Brown, T.G. (1911) The intrinsic factors in the act of progression in the mammal. *Proc. roy. Soc. B*, 84: 308–319.

Brown, T.G. (1913) The phenomenon of "narcosis progression" in mammals. *Proc. roy. Soc. B,* 86: 140–164.
Brown, T.G. (1914) On the nature of the fundamental activity of the nervous centres; together with an analysis of the conditioning of rhythmic activity in progression, and a theory of the evolution of function in the nervous system. *J. Physiol. Lond.,* 48: 18–46.
Bruggencate, G. ten and Lundberg, A. (1974) Facilitatory interaction in transmission to motoneurones from vestibulospinal fibres and contralateral primary afferents. *Exp. Brain Res.,* 19: 248–270.
Budakova, N.N. (1971) Stepping movements evoked by a rhythmic stimulation of a dorsal root in mesencephalic cat. *Sechenov Physiol. J. USSR,* 57: 1632–1640 (in Russian).
Budakova, N.N. (1973) Stepping movements in the spinal cat due to DOPA administration. *Sechenov Physiol. J. USSR,* 59: 1190–1198 (in Russian).
Burgess, P.R. and Clark, J.F. (1969) Characteristics of knee joint receptors in the cat. *J. Physiol. Lond.,* 203: 317–335.
Burke, D. and Lance, J.W. (1973) Studies of the reflex effects of primary and secondary spindle endings on spasticity. In *New Developments in Electromyography and Clinical Neurophysiology, Vol. 3*, J.E. Desmedt (Ed.). Karger, Basel, pp. 475–495.
Czarkowska, J., Jankowska, E. and Sybirska, E. (1976) Axonal projections of spinal interneurones excited by group I afferents in the cat, revealed by intracellular staining with horseradish peroxidase. *Brain Res.,* 118: 115–118.
Eccles, J.C. (1953) *The Neurophysiological Basis of Mind: The Principles of Neurophysiology*. Clarendon Press, Oxford.
Eccles, R.M. and Lundberg, A. (1959a) Synaptic actions in motoneurones by afferents which may evoke the flexion reflex. *Arch. Ital. Biol.,* 97: 199–221.
Eccles, R.M. and Lundberg, A. (1959b) Supraspinal control of interneurones mediating spinal reflexes. *J. Physiol. Lond.,* 147: 565–584
Eccles, J.C., Kostyuk, P.G. and Schmidt, R.F. (1962) Presynaptic inhibition of the central actions of flexor reflex afferents. *J. Physiol. Lond.* 161: 258–281.
Edgerton, V.R., Grillner, S., Sjöström, A. and Zangger, P. (1976) Central generation of locomotion in vertebrates. In *Neural Control of Locomotion*. R.M. Herman, S. Grillner, P.S. Stein and D.G. Stuart (Eds.) Plenum Press, New York and London, pp. 439–464.
Fedina, L. and Hultborn, H. (1972) Facilitation from ipsilateral primary afferent of interneuronal transmission in the Ia inhibitory pathway to motoneurones. *Acta physiol. scand.,* 86: 59–81.
Fetz, E.E., Jankowska, E., Johannisson, T. and Lipski, J. (1979) Autogenetic inhibition of motoneurones by impulses in group Ia muscle spindle afferents. *J. Physiol. Lond.,* in press.
Forssberg, H. and Grillner, S. (1973) The locomotion of the acute spinal cat injected with Clonidine i.v. *Brain Res.,* 50: 184–186.
Grillner, S. (1973) Locomotion in the spinal cat. In *Control of Posture and Locomotion*. R.B. Stein, K.B. Pearson, R.S. Smith and J.B. Redford (Eds). Plenum Publ. Corp., New York, pp. 515–535.
Grillner, S. (1975) Locomotion in vertebrates: central mechanisms and reflex interaction. *Physiol. Rev.,* 55: 247–304.
Grillner, S. and Zangger, P. (1974) Locomotor movements generated by the deafferented spinal cord. *Acta physiol. scand.,* 91: 38A–39A.
Grillner, S. and Zangger, P. (1975) How detailed is the central pattern generation for locomotion? *Brain Res.,* 88: 367–371.
Grillner, S. and Rossignol, S. (1978) Contralateral reflex reversal controlled by limb position in the acute spinal cat injected with clonidine i.v. *Brain Res.,* 144: 411–414.
Grillner, S., Hongo, T. and Lund, S. (1968) The origin of descending fibres monosynaptically activating spino-reticular neurones. *Brain Res.,* 10: 259–262.
Holmqvist, B. and Lundberg, A. (1961) Differential supraspinal control of synaptic actions evoked by volleys in the flexion reflex afferents in alpha motoneurones. *Acta physiol. scand.,* 54: Suppl. 186, 1–51.
Holmqvist, B., Lundberg, A. and Oscarsson, O. (1956) Functional organisation of the dorsal spinocerebellar tract in the cat. V. Further experiments on convergence of excitatory and inhibitory actions. *Acta physiol. scand.,* 38: 76–90.
Holmqvist, B., Lundberg, A. and Oscarsson, O, (1960) A supraspinal control system monosynaptically connected with an ascending spinal pathway. *Arch. ital. Biol.,* 98: 402–422.
Hongo, T., Jankowska, E. and Lundberg, A. (1966) Convergence of excitatory and inhibitory action on interneurones in the lumbosacral cord. *Exp. Brain Res.,* 1: 338–358.

Hongo, T., Jankowska, E. and Lundberg, A. (1969) The rubrospinal tract. II. Facilitation of interneuronal transmission in reflex paths to motoneurones. *Exp. Brain Res.*, 7: 365–391.

Hongo, T., Jankowska, E. and Lundberg, A. (1972) The rubrospinal tract. IV. Effects on interneurones. *Exp. Brain Res.*, 15: 54–78.

Hultborn, H. (1972) Convergence on interneurones in the reciprocal Ia inhibitory pathway to motoneurones. *Acta physiol. scand.*, 85: Suppl. 375, 1–42.

Hultborn, H. (1976) Transmission in the pathway of reciprocal Ia inhibition to motoneurones and its control during the tonic stretch reflex. In *Progress in Brain Research, Vol. 44, Understanding the Stretch Reflex*, S. Homma (Ed.) Elsevier, Amsterdam, pp. 235–255.

Jack, J.J.B. (1978) Some methods for selective activation of muscle afferent fibres. In *Studies in Neurophysiology presented to A.K. McIntyre*. R. Porter (Ed.). Cambridge Univ. Press.

Jankowska, E., and Roberts, W. (1972a) Synaptic actions of single interneurones mediating reciprocal Ia inhibition of motoneurones. *J. Physiol., Lond.*, 222: 623–642.

Jankowska, E. and Roberts, W.J. (1972b) An electrophysiological demonstration of the axonal projections of single spinal interneurones in the cat. *J. Physiol. Lond.*, 222: 597–622.

Jankowska, E., Lund, S. and Lundberg, A. (1966) The effect of DOPA on the spinal cord. 4. Depolarization evoked in the central terminals of contralateral Ia afferent terminals by volleys in the flexor reflex afferents. *Acta physiol. scand.*, 68: 337–341.

Jankowska, E., Lundberg, A. and Stuart, D. (1973) Propriospinal control of last order interneurones of spinal reflex pathways in the cat. *Brain Res.*, 53: 227–231.

Jankowska, E., Jukes, M.G.M., Lund, S. and Lundberg, A. (1967a) The effect of DOPA on the spinal cord. 5. Reciprocal organization of pathways transmitting excitatory action to alpha motoneurones of flexors and extensors. *Acta physiol. scand.*, 70: 369–388.

Jankowska, E., Jukes, M.G.M., Lund, S. and Lundberg, A. (1967b) The effect of DOPA on the spinal cord. 6. Half-centre organization of interneurones transmitting effects from the flexor reflex afferents. *Acta physiol. scand.*, 70: 389–402.

Layzer, R.B. and Rowland, L.P. (1971) Cramps. *N. Engl. J. Med.*, 285: 31–40.

Lindström, S. (1973) Recurrent control from motor axon collaterals of Ia inhibitory pathways in the spinal cord of the cat. *Acta physiol. scand.*, Suppl. 392, 1–43.

Lloyd, D.P.C. (1943) Neuron patterns controlling transmission of ipsilateral hind limb reflexes in cat. *J. Neurophysiol.*, 6: 293–315.

Lloyd, D.P.C. (1946) Facilitation and inhibition of spinal motoneurones. *J. Neurophysiol.*, 9: 421–438.

Lundberg, A. (1959) Integrative significance of patterns of connections made by muscle afferents in the spinal cord. *Symp. XXI Int. Physiol. Congr. Buenos Aires 1959*, pp. 1–5.

Lundberg, A. (1966) Integration in the reflex pathway. In *Muscular Afferents and Motor Control. Nobel Symposium 1*, R. Granit (Ed.) Almqvist and Wiksell, Stockholm, pp. 275–305.

Lundberg, A. (1969) Convergence of excitatory and inhibitory action on interneurones in the spinal cord. In *The Interneuron. UCLA Forum Med. Sci. No. 11*, M.A.B. Brazier (Ed.) University of California Press, Los Angeles, CA, pp. 231–265.

Lundberg, A. (1972) The significance of segmental spinal mechanisms in motor control. *Symposial paper 4th International Biophysics Congress, Moscow 1972.*

Lundberg, A., and Oscarsson, O. (1962) Two ascending spinal pathways in the ventral part of the cord. *Acta physiol. scand.*, 54: 270–286.

Lundberg, A. and Voorhoeve, P. (1962) Effects from the pyramidal tract on spinal reflex arcs. *Acta physiol. scand.*, 56: 201–219.

Lundberg, A., Malmgren, K. and Schomberg, E.D. (1977a) Cutaneous facilitation of transmission in reflex pathways from Ib afferents to motoneurones. *J. Physiol. Lond.*, 265: 763–780.

Lundberg, A., Malmgren, K. and Schomberg, E.D. (1977b) Comments on reflex actions evoked by electrical stimulation of group II muscle afferents. *Brain Res.*, 122: 551–555.

Lundberg, A., Malmgren, K. and Schomberg, E.D. (1978) Role of joint afferents in motor control exemplified by effects on reflex pathways from Ib afferents. *J. Physiol. Lond.*, 284: 327–343.

Matthews, P.B.C. (1972) *Mammalian Muscle Receptors and their Central Actions*. Edward Arnold Ltd. London.

Oscarsson, O. (1957) Functional organization of the ventral spino-cerebellar tract in the cat. II. Connections with muscle, joint, and skin nerve afferents and effects on adequate stimulation of various receptors. *Acta physiol. scand.*, 42: Suppl. 146, 1–107.

Oscarsson, O. (1973) Functional organization of spino-cerebellar paths. In *Handbook of Sensory Physiology. Vol. II. Somatosensory System*, A. Iggo (Ed.) Springer, Berlin-Heidelberg-New York, pp. 339–380.

Paintal, A.S. (1960) Functional analysis of group III afferent fibres of mammalian muscles. *J. Physiol Lond.*, 152: 250–270.

New Observations on Neuronal Organization of Reflexes from Tendon Organ Afferents and their Relation to Reflexes Evoked from Muscle Spindle Afferents

E. JANKOWSKA

Department of Physiology, University of Göteborg, S-400 33 Göteborg
(Sweden)

Spinal reflexes from group Ia muscle spindle afferents and from group Ib tendon organ afferents have usually been analysed as if they were mediated by separate neuronal channels. Such an approach turned out to be justified in the case of the Ia reciprocal inhibition because interneurones which mediate it (Hultborn et al., 1971; Jankowska and Roberts, 1972) lack any input from Ib afferents. The neuronal circuitry of polysynaptic Ia excitatory actions (for references, see Hultborn and Wigström 1979) is still unknown and not much can be said about properties of the interposed neurones. With regard to reflexes evoked from Ib afferents there is, on the other hand, accumulating evidence that they may be mediated largely by interneurones used in common by Ia and Ib reflex pathways. Two groups of observations leading to this conclusion will be summarized in this report.

LAMINAE V–VI INTERNEURONES AS TARGET CELLS OF BOTH Ia AND Ib AFFERENTS

The first systematic intracellular study of interneurones with group I input by Eccles et al. (1960) suggested a selective monosynaptic input from either Ia or Ib afferents to two subgroups of these interneurones denoted A and B. Subsequent investigations by Hongo et al. (1966, 1972) showed, however, that Ia and Ib afferents converge onto a number of laminae V–VI interneurones; monosynaptic EPSPs were found to be evoked in individual interneurones by lowest threshold afferents in one hindlimb nerve and by higher threshold group I afferents in another, or the same nerve. For technical reasons the analysis had to be limited primarily to synaptic actions from knee flexor posterior biceps and semintendinosus (PBSt) and knee extensor quadriceps (Q), whose Ia and Ib afferents show clearest differences in threshold to electrical stimuli and in conduction velocity (Bradley and Eccles, 1953; Laporte and Bessou, 1957; Lundberg and Eccles, 1957a; Coppin et al. 1969). There were nevertheless indications for convergence of Ia and Ib afferents from other nerves as well. In addition, disynaptic excitation by Ib afferents was found to be combined with monosynaptic excitation by Ia afferents (Eccles et al., 1960) and disynaptic inhibition from either Ia or Ib afferents with excitation.

We have recently reinvestigated contribution of muscle spindle and tendon organ afferents to excitation and inhibition of laminae V–VI interneurones using intracellular recording (Czarkowska et al., 1979; Jankowska et al., 1979). One of our aims was to analyse contribution of Ia and Ib afferents from a number of different muscles. Those from knee flexors and extensors were stimulated electrically while Ia muscle spindle afferents from ankle extensors (medial gastrocnemius, MG; lateral gastrocnemius and soleus, LGS; and plantaris, Pl) were activated by adequate stimuli (brief muscle stretches with 1.5–2.0 msec rise time and 30–35 μm amplitude at initial tension 5 N). Additional effects of larger stretches were considered to be due to either Ib or higher threshold Ia afferents, because under our experimental conditions a certain proportion of Ib afferents was excited by stretches of 40–50 μm, while activation of some 20–30% of Ia afferents required stretches of up to about 60 μm. We therefore attributed to Ib afferents only the difference between postsynaptic potentials evoked by electrical stimuli maximal for group I afferents in a given nerve, and the 60 μm stretches.

Convergence of Ia and Ib afferents of PBSt and Q was found in 22% of 33 interneurones excited from these muscles (Czarkowska et al., 1979) and in a majority of interneurones excitation from one or another subgroup of group I afferents from

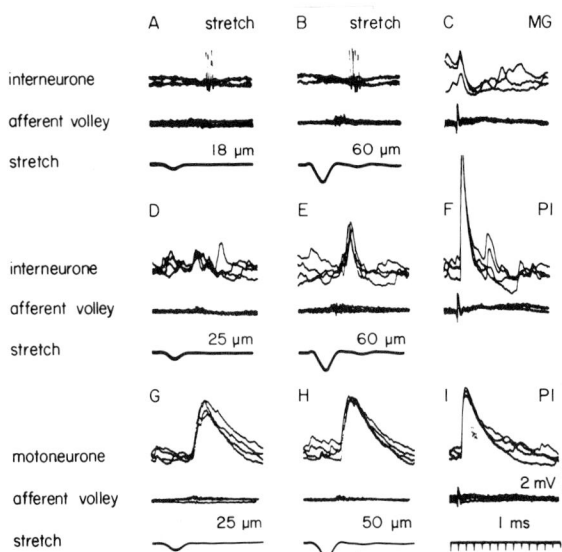

Fig. 1. Excitation by Ia afferents and excitation combined with inhibition from Ib afferents in a laminae V–VI interneurone. Extracellular (A–B, upper traces) and intracellular (C–F, upper traces) records from the same interneurone to be compared with records from a plantaris motoneurone (G–I) penetrated during the same experiment. Middle traces, records of afferent volleys from the surface of the spinal cord near L7 dorsal root entry zone. Lower traces, records of changes in muscle length (increase in length downwards). Left column (A, D, G), effect of muscle stretches submaximal for Ia afferents and subthreshold for Ib afferents. Middle column (B, E, H), effects of larger muscle stretches, near maximal for Ia afferents and subthreshold for Ib afferents. Right column (C, F, I), effects of electrical stimulation of medial gastrocnemius (MG) and plantaris (Pl) nerves. Note excitation of the interneurone by small muscle stretches (A, B), an increase of excitation and appearance of IPSPs cutting short the EPSPs with larger stretches (cf. amplitude of the EPSPs in D and E and time course of the EPSPs in E with that of EPSPs in D and H), and much larger EPSPs and IPSPs evoked from the nerves. Note also that 25 μm stretches were practically maximal for Ia afferents in this cat as judged from EPSPs in H and I; only minimal heteronymous EPSPs were evoked in this motoneurone. (From Jankowska, Johannisson and Lipski, unpublished observations.)

knee flexors and extensors was combined with excitatory actions from group I afferents from other nerves.

Convergence of Ia and Ib afferents of MG, LGS and Pl appeared in a much higher proportion of 70 interneurones with input from these muscles (Jankowska et al., 1979). More than 40% were co-excited by Ia afferents (Fig. 1A, D) and by Ib afferents (see difference between effects of stimulation of all group I afferents and of stimulation maximal for group Ia afferents, Fig. 1E, F). For interneurones with only excitatory input from ankle extensors the proportion was even higher (Fig. 2). Since in many interneurones excitation was combined with inhibition from Ia, Ib or both Ia and Ib afferents (Fig. 2) only about one-third of interneurones with excitatory input from Ib afferents was found not to be affected in one way or another (excited or inhibited) by Ia afferents (Fig. 2). Thus intracellular records gave a different picture of the input to laminae V–VI interneurones than that based on extracellular records (Lucas and Willis, 1974). We found no evidence for laminae V–VI interneurones selectively excited by Ia afferents from GS and Pl.

For nearly 50 interneurones with the above described input we were able to define their axonal projections, by classifying them to one of the 6 previously differentiated types of projections (Czarkowska et al., 1976). The comparison of the synaptic input with axonal projections showed that co-excitation, co-inhibition or excitation by one subgroup and inhibition by another subgroup of group I afferents occurs in

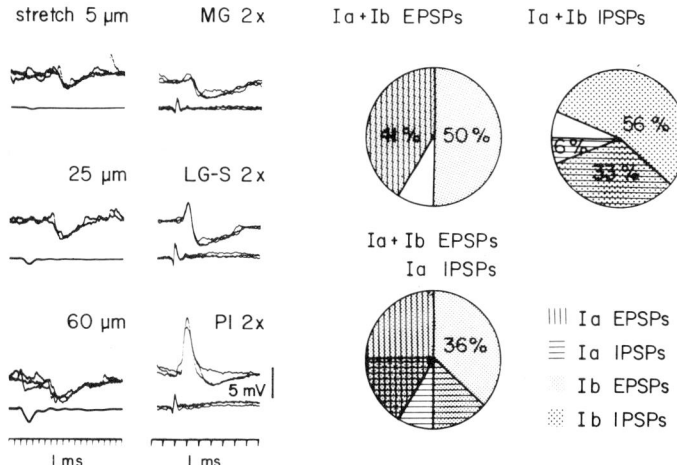

Fig. 2. Excitation from Ib afferents combined with inhibition from Ia afferents in a laminae V–VI interneurone. Upper traces, intracellular records. Lower traces, records of change in muscle length in the left column and records of afferent volleys in the right column. Effects of muscle stretches with increasing amplitudes to the left and of electrical stimulation of medial gastrocnemius (MG), lateral gastrocnemius (LGS) and plantaris (Pl) nerves to the right. Note that IPSPs of practically the same amplitude were evoked by all muscle stretches and that EPSPs appeared only to electrical stimulation of two of the three stimulated nerves, evidencing Ia origin of IPSPs and Ib origin of EPSPs. Somewhat longer IPSPs following MG and LGS stimuli might indicate a certain contribution of Ib afferents to the inhibition of the interneurone. Upper diagrams show relative proportions of interneurones: excited by only Ib afferents (dotted) and co-excited by Ia and Ib afferents (hatched and dotted), and inhibited by only Ia afferents (hatched) or Ib afferents (dotted) and co-inhibited by Ia and Ib afferents (hatched and dotted). Lower diagram shows proportions of interneurones with input from only Ib (dotted) afferents and with Ib excitation combined with Ia excitation and/or inhibition (hatched and dotted). (From Jankowska, Johannisson and Lipski, unpubl. obs.)

interneurones projecting to motor nuclei as well as in those which apparently terminate only on other interneurones. It was a feature of interneurones with ipsilateral as well as of those with crossed projections.

Generally one may, therefore, conclude that a variety of spinal reflexes evoked from group I afferents should depend on activation of both muscle spindle and tendon organ afferents, and be mediated by common rather than separate interneurones. As a particular consequence of co-excitation of the same interneurones by Ia and Ib afferents, one may further expect some similar reflex actions to be evoked by activation of these afferents and a mutual facilitation of their effects. We have now evidence that this is the case for autogenetic inhibition of motoneurones, or more exactly, for inhibition of motoneurones from homonymous and synergistic muscles.

ORIGIN OF AUTOGENETIC INHIBITION OF MOTONEURONES FROM Ia AS WELL AS FROM Ib AFFERENTS

Recent series of experiments (Fetz et al., 1979) revealed that inhibition of motoneurones may be evoked from Ia muscle spindle afferents as well as from Ib tendon organ afferents of the homonymous and synergistic muscles. The Ia inhibition of such an origin was demonstrated primarily for motoneurones of medial and lateral gastrocnemius and plantaris on brief stretches of these muscles. The minimal effective stretches were 10–20 μm in amplitude and much below threshold for Ib afferents (above 40 μm). The inhibition was demonstrated under several different experimental conditions as exemplified in Figs. 3 and 4. In motoneurones of Fig. 3 it was evoked while the nerves to their homonymous muscles were intact. The IPSPs were therefore visualized as a hump on the decay phase of the monosynaptic EPSPs after their reversal by hyperpolarization of the motoneurones and chloride injection and as cutting short the decay phase of the EPSPs after membrane depolarization.

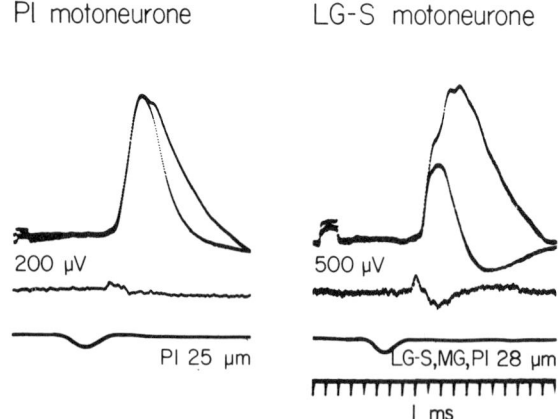

Fig. 3. IPSPs evoked by small stretches of homonymous muscles and by homonymous plus synergistic muscles. Averaged records from plantaris (Pl) and lateral gastrocnemius (LGS) motoneurones (upper traces), and from the surface of the spinal cord (middle traces, with higher amplification to the right) and records of changes in muscle length (lower traces). Photographically superimposed intracellular records taken during hyperpolarization (40 nA) and depolarization (20 nA). (Modified from Fig. 3 of Fetz et al., 1979.)

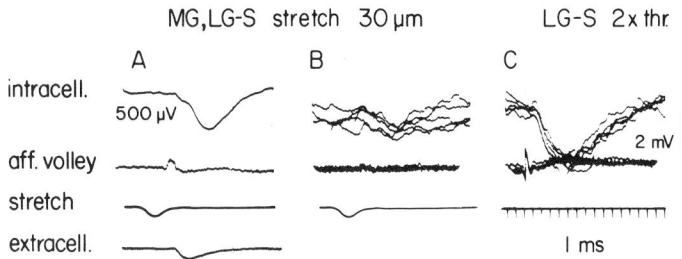

Fig. 4. Comparison of IPSPs evoked from Ia, and from all group I afferents of synergistic muscles. Intracellular records from a Pl motoneurone. From top to bottom: intracellular records taken during depolarization of the motoneurones (50 nA), afferent volleys recorded from L7 dorsal root entry zone, changes in muscle length and extracellular field potential to the same stretch recorded just outside the motoneurone. A and B) Averaged and single superimposed records of IPSPs evoked by stretches of triceps surae estimated to activate about 80% of Ia afferents from these muscles. C) IPSPs evoked by electrical stimuli supramaximal for group I afferents in lateral gastrocnemius and soleus. (From Fetz, Jankowska and Lipski, unpublished observations.)

In the plantaris motoneurone of Fig. 4A the Ia IPSP was evoked only from synergistic muscles from triceps surae, and therefore without any preceding EPSP (cf. Eccles et al., 1957b).

Control experiments confirmed that the observed inhibition disappeared after cutting the nerves of the stretched muscles, and was therefore not evoked from some other, unintentionally activated receptors. Moreover, stretches used to evoke the inhibition were below threshold for discharging motoneurones; it was thus not evoked secondarily to activation of Renshaw cells and could be differentiated from the recurrent inhibition of motoneurones.

When compared to IPSPs evoked by selective electrical stimulation of Ib afferents using method of Coppin et al. (1970) the stretch-evoked Ia IPSPs appeared with similar latencies (di- and trisynaptically) but had much smaller amplitudes. Ia IPSPs were 16–35% of the amplitudes of Ib IPSPs, although the latter were evoked by a smaller proportion, probably less than 50% as compared to about 80% of Ia afferents of the same muscles. The Ia and Ib IPSPs constituted about 10% and 25–66% of IPSPs evoked from all group I afferents, respectively.

Autogenetic inhibition could also be evoked in a few posterior biceps-semitendinosus motoneurones by near threshold (1.1–1.2 times threshold) electrical stimulation of the nerves (Fetz, Jankowska and Lipski, unpublished observations) (Fig. 5). In view of observations of Laporte and Bessou (1957) and Coppin et al. (1969) on thresholds of group I afferents in biceps and semitendinosus, this effect could be likewise attributed to the muscle spindle afferents.

These results are in general agreement with some previous observations which left open the possibility that Ia afferents may contribute to the inhibition evoked from group I afferents from homonymous and synergistic muscles. Inhibition from low threshold group I afferents stimulated electrically, was in fact already reported in one of the first studies on group I inhibitory actions on motoneurones (Eccles et al., 1957c) but in view of a possible overlap between the thresholds of Ia and Ib afferents to electrical stimuli, it was considered as due to Ib rather than to Ia afferents. Lundberg et al. (1977) found similarly low, or even lower thresholds for evoking IPSPs from quadriceps in ankle and toe extensors. These were sometimes (during facilitation by

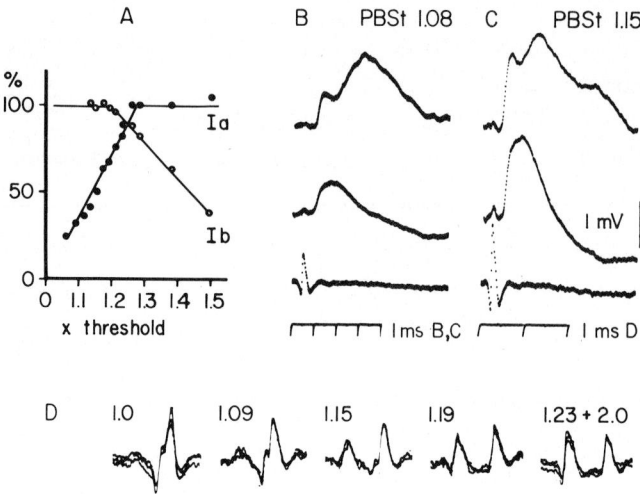

Fig. 5. Autogenetic inhibition from Ia afferents of posterior biceps and semitendinosus. A) Amplitudes of the first (Ia) and the second (Ib) components of the afferent volleys, evoked by the first and the second of two stimuli applied to the PBSt nerve, as a function of stimulus intensities; sample records are shown in E. Intervals between the two stimuli were shorter than the refractory period of the tested afferents. The second stimulus was maximal for group I afferents and intensity of the first was increased as indicated by multiples of threshold above the records. Decrease of the second component of the afferent volley evoked by the second stimulus defined threshold for Ib afferents. B, C) Postsynaptic potentials evoked in a PBSt motoneurone by stimulation of the PBSt nerve with intensities 1.08 and 1.15 × threshold. From top to bottom: potentials recorded in the motoneurone during hyperpolarization (50 μA), potentials recorded during depolarization (20 μA) and afferent volleys. (From Fetz, Jankowska and Lipski, unpublished observations.)

cutaneous afferents) lowered to 1.1–1.15 times threshold, and much below threshold for the Ib component of the afferent incoming volley. A facilitation of IPSPs evoked from Ib afferents from G–S by conditioning volleys in Ia afferents from Q was occasionally observed by these authors (Lundberg et al., 1977) and found also between Ia and Ib actions from triceps surae and plantaris (Fetz, Jankowska and Lipski, unpublished observations).

It might be recalled in this context that the autogenetic inhibition does not represent the only known combined reflex di- or polysynaptic action of muscle spindle and tendon organ afferents. Both these groups of afferents contribute to the crossed reflexes from group I afferents (Perl, 1959; Holmqvist, 1961; Baxendale and Rosenberg, 1976, 1977). Both contribute to the presynaptic depolarization of group I afferents from flexors as well as from extensors (for reference, see Schmidt 1973), and the information from both is jointly forwarded by some of the ventral spinocerebellar tract cells (Lundberg and Weight, 1971).

CO-EXCITATION OF MUSCLE SPINDLE AND TENDON ORGAN AFFERENTS

In view of the conclusion that reflexes from tendon organ afferents and some reflexes from group I muscle spindle afferents may be mediated by common neuronal pathways, the co-excitation of these afferents by various stimuli becomes of particular interest. It has long been known (cf. Granit, 1955; Matthews, 1972) that due to the γ-system the Ia and Ib afferents may be excited in parallel even during muscle contractions, which should otherwise unload muscle spindles. Recent studies on skeletofusimotor β-fibres (Emonet-Dénand and Laporte, 1975; Laporte and Emonet–Dénand, 1976) revive the problem by showing that β-fibres may as effectively activate Ia as Ib afferents, that they innervate a considerable number of muscle spindles and constitute a high proportion of all the extrafusal motor fibres. There are thus good peripheral conditions for combined reflex actions of Ia muscle spindle and Ib tendon organ afferents.

There is no doubt that some of the reflex actions of Ia afferents would utilize separate neuronal pathways (e.g., via monosynaptic connexions with motoneurones and via interneurones mediating Ia reciprocal inhibition of antagonists). To what extent other Ia and Ib actions are subserved by common or separate neurones and are jointly or independently controlled by various segmental and supraspinal neuronal systems remains to be established.

SUMMARY

Input from group Ia muscle spindle and group Ib tendon organ afferents to laminae V–VI interneurones has been reinvestigated with both electrical and adequate stimulation of these afferents. The reported observations show that a great proportion (about two-thirds) of interneurones excited by Ib afferents are co-excited and/or inhibited by Ia afferents from the same, synergistic or other muscles. These interneurones are thus shared by Ia and Ib reflex pathways. Ia autogenetic inhibition was revealed as a particular case of similar reflex actions of Ia and Ib afferents from a given group of muscles. It was demonstrated in triceps surae and plantaris motoneurones on stretches of their homonymous or synergistic muscles, subthreshold for Ib afferents.

ACKNOWLEDGEMENTS

I wish to express my thanks to Drs. Czarkowska, Fetz, Johannisson, Lipski and Sybirska for their permission to present some of our unpublished materials. The reported studies were supported by the Swedish Medical Research Council (project No. 94).

REFERENCES

Baxendale, R.H. and Rosenberg, J.R. (1976) Crossed reflexes evoked by selective activation of muscle spindle primary endings in the decerebrate cat. *Brain Res.*, 115: 324–327.

Baxendale, R.H. and Rosenberg, J.R. (1977) Crossed reflexes evoked by selective activation of tendon organ afferent axons in the decerebrate cat. *Brain Res.*, 127: 323–326.

Bradley, K. and Eccles, J.C. (1953) Analysis of the fast afferent impulses from thigh muscles. *J. Physiol. (Lond.)*, 122: 462–473.

Coppin, C.M.L., Jack, J.J.B. and McIntyre, A.K. (1969) Properties of group I afferent fibres from semitendinosus muscle in the cat. *J. Physiol. (Lond.)*, 203: 45–46P.

Coppin, C.M.L., Jack, J.J.B. and MacLennan, C.R. (1970) A method for the selective activation of tendon organ afferent fibres from the cat soleus muscle. *J. Physiol. (Lond.)*, 210: 18–20.

Czarkowska, J., Jankowska, E. and Sybirska, E. (1976) Axonal projections of spinal interneurones excited by group I afferents in the cat, revealed by intracellular staining with horseradish peroxidase. *Brain Res.*, 118: 115–118.

Czarkowska, J., Jankowska, E. and Sybirska, E. (1979) Common interneurones in reflex pathways from group Ia muscle spindle and group Ib tendon organ afferents. I. Interneurones with input from knee flexors and extensors. In preparation.

Eccles, J.C., Eccles, R.M. and Lundberg, A. (1957a) Synaptic actions on motoneurones in relation to the two components of the group I muscle afferent volley. *J. Physiol. (Lond.)*, 136: 527–546.

Eccles, J.C., Eccles, R.M. and Lundberg, A. (1957b) The convergence of monosynaptic excitatory afferents on to many different species of alpha motoneurones. *J. Physiol. (Lond.)*, 137: 22–50.

Eccles, J.C., Eccles, R.M. and Lundberg, A. (1957c) Synaptic actions motoneurones caused by impulses in Golgi tendon organ afferents. *J. Physiol. (Lond.)*, 138: 227–252.

Eccles, J.C., Eccles, R.M. and Lundberg, A. (1960) Types of neurone in and around the intermediate nucleus of the lumbosacral cord. *J. Physiol. (Lond.)*, 154: 89–114.

Emonet-Dénand, F. and Laporte, Y. (1975) Proportion of muscle spindles supplied by skeletofusimotor axons (β-axons) in peroneus brevis muscle of the cat. *J. Neurophysiol.*, 38: 1390–1394.

Fetz, E., Jankowska, E., Johannisson, T. and Lipski, J. (1979) Autogenetic inhibition of motoneurones by impulses in group Ia muscle spindle afferents. *J. Physiol. (Lond.)*, in press.

Granit, R. (1955) *Receptors and Sensory Perception*. Yale University Press, New Haven.

Holmqvist, B. (1971) Crossed spinal reflex actions evoked by volleys in somatic afferents. *Acta physiol. scand.*, 52, Suppl. 181.

Hongo, T., Jankowska, E. and Lundberg, A. (1966) Convergence of excitatory and inhibitory action on interneurones in the lumbosacral cord. *Exp. Brain Res.*, 1: 338–358.

Hongo, T., Jankowska, E. and Lundberg, A. (1969) The rubrospinal tract. II. Facilitation of interneuronal transmission in reflex paths to motoneurones. *Exp. Brain Res.*, 7: 365–391.

Hultborn, H. and Wigström, H. (1978) Motor response with long latency and maintained duration evoked by activity in Ia afferents. *Progr. in Clin. Neurophysiol., Vol. 8*, J. Desmedt (Ed.) Karger, Basel.

Hultborn, H., Jankowska, E. and Lindström, S. (1971) Recurrent inhibition of interneurones monosynaptically activated from group Ia afferents. *J. Physiol. (Lond.)*, 215: 613–636.

Jankowska, E., Johannisson, T. and Lipski, J. (1979) Common interneurones in reflex pathways from group Ia muscle spindle and group Ib tendon organ afferents. II. Interneurones with input from ankle extensors. In preparation.

Laporte, Y. and Bessou, P. (1957) Distribution dans les sous-groupes rapide et lent du groupe I des fibres Ia d'origine fusoriale et des fibres Ib d'origine golgienne. *J. Physiol. (Paris)*, 49: 252–253.

Laporte, Y. and Emonent-Dénand, J. (1976) The skeleto-fusimotor innervation of cat muscle spindle. In *Progress in Brain Research, Vol. 44, Understanding the Stretch Reflex*, S. Homma (Ed.) Elsevier, Amsterdam, pp. 99–106.

Lucas, M.E. and Willis, W.D. (1974) Identification of muscle afferents which activate interneurons in the intermediate nucleus. *J. Neurophysiol.*, 37: 282–293.

Lundberg, A., Malmgren, K. and Schomburg, E.D. (1977) Cutaneous facilitation of transmission in reflex pathways from Ib afferents to motoneurones. *J. Physiol. (Lond.)*, 265: 763–780.

Lundberg, A. and Weight, F. (1971) Functional organization of connexions to the ventral spinocerebellar tract. *Exp. Brain Res.*, 12: 295–316.

Matthews, P.B.C. (1972) *Mammalian Muscle Receptors and their Central Actions*. Edward Arnold, London.

Perl, E.R. (1959) Effects of muscle stretch on excitability of contralateral motoneurones. *J. Physiol. (Lond.)*, 145: 193–203.

Schmidt, R.F. (1973) Control of the access of afferent activity to somatosensory pathways. In: *Handbook of Sensory Physiology. Vol. II, Somatosensory System*. A. Iggo (Ed.), Springer-Verlag, Berlin, pp. 151–206.

Input-Output Relationship in Spinal Motoneurons in the Stretch Reflex

S. HOMMA and Y. NAKAJIMA

*Department of Physiology, Chiba University
School of Medicine,
Chiba (Japan)*

Brief stretching of a muscle with triangular pulses can excite primary endings of the muscle spindle at low amplitude (Bianconi and Van der Meulen, 1963; Homma, 1966; Brown et al., 1967; Homma et al., 1971a; Matthews, 1972). The stretching of an agonistic muscle elicits ripples of excitatory postsynaptic potential (EPSP) in an alpha-motoneuron, whereas the stretching of an antagonistic muscle produces ripples of inhibitory postsynaptic potential (IPSP) in the same alpha-motoneuron. Our investigations aim at calculation of the rising phase of the EPSP or the rising phase of the IPSP by means of cross-correlation analysis of random triangular stretches of agonistic or antagonistic muscle and alpha-motoneuronal activities elicited reflexively by the stretches of these muscles.

RANDOM TRIANGULAR STRETCHING

Cats were anesthetized by intraperitoneal injection of 5 ml/kg of 10% urethane and 1% chloralose. Tendons of the gastrocnemius, soleus and tibialis anterior muscle were cut and these muscles were separated from the surrounding tissues. The tendons were linked tightly to vibrators with steel hooks. The muscles were stretched by triangular pulses with a rising and falling time of 4 msec. Intervals of the triangular pulses changed randomly; the minimum and maximum intervals were 20 or 30 msec and 80 msec, respectively. Activity of alpha-motoneurons in the lumbar spinal cord was recorded either intracellularly or extracellularly. In the latter case, the central cut end of the L7 ventral root was split until a functionally single fiber responding to a brief manual stretch of the muscle was obtained. These motoneuronal action potentials are called motor unit spikes in this paper. Cross-correlograms between onsets of the triangular pulses and the motor unit spikes were calculated by computer.

EXCITATORY PRIMARY CORRELATION KERNEL

When a muscle was stretched with triangular pulses, intracellular recording of an alpha-motoneuron innervating the muscle revealed ripples of EPSP corresponding to the triangular stretches (Homma, 1976; Homma et al., 1970). Continuous application

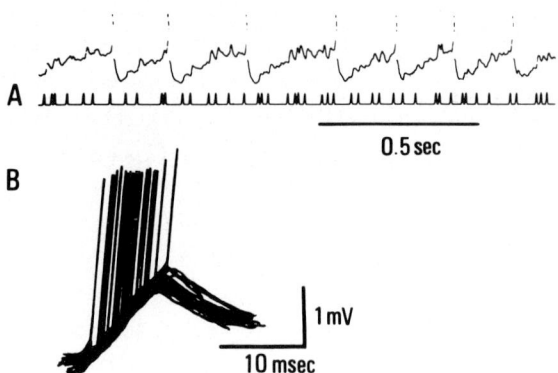

Fig. 1. A) Membrane potential change (upper trace) recorded intracellularly from the gastrocnemius motoneuron during random triangular stretches (lower trace) of the gastrocnemius muscle. B) Superimposed spike potentials of A. Only components which deflected toward the overshoot potential were superimposed. Timing was taken at the beginning of the stretches.

of the triangular stretches caused temporal summation of the EPSP ripples and when the summated membrane potential attained the critical firing level, the alpha-motoneurone fired as shown in Fig. 1A (Homma and Kanda, 1973).

Fig. 1B shows a superposition of the EPSP ripples which elicited spike potentials. Obviously the spike potentials take place during the rising phase of the EPSPs. Therefore we can conclude that motor unit spikes "break out" within the time-to-peak of the EPSPs. Furthermore, since these spikes occur most frequently on the steepest rising slope of the EPSPs and less frequently on the slower slopes both at the start and near the summit of the EPSPs, we can calculate the time course of an EPSP from a probability density distribution of the spikes (Knox and Poppele, 1977; Homma and Nakajima, 1978).

Fig. 2A shows motor unit spikes reflexively elicited by random triangular stretches of the gastrocnemius muscle. Fig. 2B shows the cross-correlogram of the motor unit spikes and the onsets of the random triangular pulses. The prominent kernel in Fig. 2B corresponds with the probability density distribution of the motor unit spikes

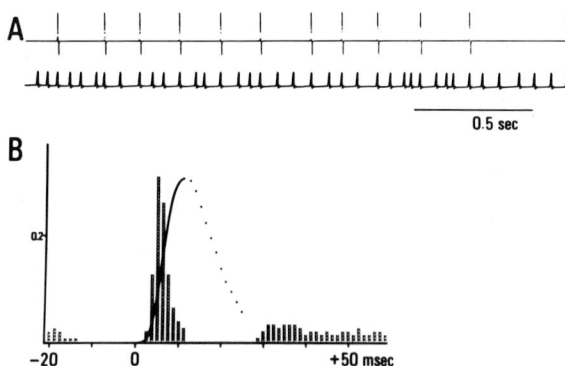

Fig. 2. A) Motor unit spikes (upper trace) of the gastrocnemius motoneuron activated by random triangular stretches (lower trace) of the gastrocnemius muscle. B) Cross-correlogram of the motor unit spikes and the stretches. The solid line was obtained by integrating the primary correlation kernel. Ordinate and abscissa indicate probability of spike occurrence and recurrence time, resp.

which responded to the triangular pulses with a suitable conduction time and synaptic delay (Homma et al., 1971b; Hagbarth, 1973; Godaux et al., 1975). This kernel is called the primary correlation kernel (Knox, 1974). Though spikes accompany secondary correlation kernels in the cross-correlograms, these kernels in Fig. 2B compose a plateau because of the random intervals of the triangular pulses. Since the minimum interval of the random stretches was 20 msec in this case, the secondary kernels do not take shape around the primary correlation kernel within 20 msec of either the negative or the positive recurrence time. The time lag of the primary correlation kernel indicates the time from the onset of a stretch to a resultant motor unit spike, the so-called reponse time (Homma and Nakajima, 1978). The mean value of the minimum response time was 3.2 ± 0.6 msec for the 28 gastrocnemius motor unit spikes. Apparently the mean value indicates that the motor unit spikes are elicited by a mono-synaptic transmission mechanism in the stretch reflex. On the other hand, the distribution width of the primary correlation kernel has been supposed to indicate a probability density function of motor unit spikes which occur during the rising phase of EPSPs as mentioned above and shown in Fig. 1B. Using a neuron model, Knox (1974) showed that the width of the kernel, which is called the correlation time, is related to the derivatives of postsynaptic potentials. Our experimental results (Fig. 1B) strongly support his theoretical point of view. The primary correlation kernel was integrated and fitted to the following equation by means of the least mean square.

$$Y = t^p \cdot \exp\left(\frac{-p \cdot t}{CT}\right) \quad (1)$$

t: time; CT: correlation time; p: power.

The integrated kernel (Y) is shown in Fig. 2B by the solid line, which rises slowly after the onset, then becomes very steep, and slows down again near the summit. Therefore, the line probably illustrates the time course of the rising phase of an EPSP. The falling phase of the EPSP is shown by dots because it is only based on calculation with equation (1) above (Homma and Nakajima, 1978).

Ten examples were integrated and after normalization the results were superimposed as shown in Fig. 5A. Fig. 5A shows that curves attain their peaks with an initially slow, then steep, and finally slow time course. Thus the primary correlation kernel makes it possible to calculate the time course and the time-to-peak of an EPSP. The mean width of primary correlation kernels in the 28 gastrocnemius motor units was 4.7 ± 1.1 msec.

With these statistical analyses it becomes possible to calculate the time course of EPSPs elicited on the gastrocnemius motoneuron by the proprioceptive afferent impulses from the homonymous muscle stretched with triangular pulses.

FACILITATORY PRIMARY CORRELATION KERNEL

Spindle primary afferents from the soleus muscle have a facilitatory effect on the alpha-motoneurone which innervates the gastrocnemius muscle. Since the facilitatory effect is exerted by EPSPs elicited by the spindle primary afferent impulses of the

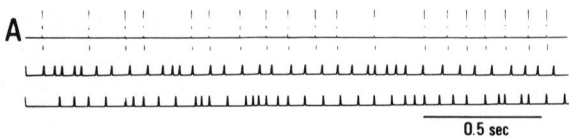
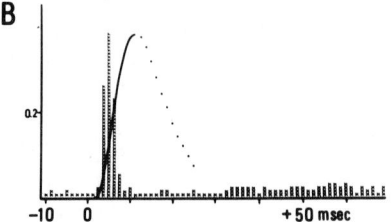

Fig. 3. A) Motor unit spikes (upper trace) of the gastrocnemius motoneuron activated by random triangular stretches of the gastrocnemius muscle (middle trace) and of the soleus muscle (lower trace). B) Cross-correlogram of the spikes and the stretches of the soleus muscle. The facilitatory primary correlation kernel was integrated and is shown by the solid line.

agonistic muscle, the time course of the EPSP can be calculated by the same analytical methods.

Fig. 3A shows motor unit spikes of the gastrocnemius motoneurone which responded to the random stretches of both the gastrocnemius and the soleus muscle. Fig. 3B shows the cross-correlogram of the gastrocnemius motor unit spikes and the random triangular stretches of the soleus muscle. The prominent kernel in Fig. 3B is a primary correlation kernel of the unit activities which was correlated with the stretches of the soleus muscle. This kernel is thought to indicate the rising phase of an EPSP elicited on the gastrocnemius motoneuron by the primary spindle afferent impulses originating from the heteronymous soleus muscle. Since in this case the triangular stretches of the soleus muscle alone can not activate the gastrocnemius motoneuron, it was supposed that the triangular stretches of the soleus muscle have a facilitatory effect on the gastrocnemius motoneurons. Therefore, the kernel in Fig. 3B is called a facilitatory primary correlation kernel. This correlation kernel was integrated and the result is indicated in Fig. 3B by the solid line. This curve is presumed to indicate the rising phase of an EPSP elicited on the gastrocnemius motoneuron by the stretches of the soleus muscle. Curves obtained from ten examples were superimposed and are shown after normalization in Fig. 5B. The mean value of the minimum response time from the onsets of the triangular stretches to the onsets of the EPSP was 2.4 ± 0.9 msec, which time corresponds with a latency of monosynaptic transmission. The mean width of the kernel was 9.0 ± 1.0 msec.

INHIBITORY PRIMARY CORRELATION TROUGH

Spindle primary afferent impulses from the tibialis anterior muscle, which is antagonistic to the gastrocnemius muscle, have an inhibitory effect on the gastrocnemius motoneurons. Since the effect is exerted by IPSPs, the time course of the IPSP can be calculated by the same analysis as above. Fig. 4A shows motor unit spikes of the gastrocnemius motoneuron together with random triangular stretches of the gastrocnemius and tibialis anterior muscle. Responding to the random triangular

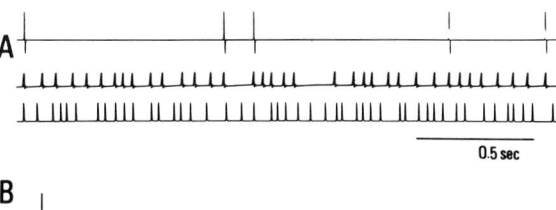
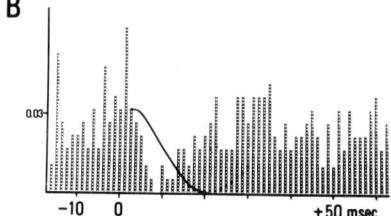

Fig. 4. A) Motor unit spikes (upper trace) of the gastrocnemius motoneuron produced by random triangular stretches of the gastrocnemius muscle (middle trace) and of the tibialis anterior muscle (lower trace). B) Cross-correlogram of the spikes and the stretches of the tibialis anterior muscle. The inhibitory primary correlation trough was integrated and is shown by the solid line.

stretches of the tibialis anterior muscle, the gastrocnemius motoneron decreased its discharge frequency. The inhibited motor unit spikes should be well correlated with the stretches of the tibialis anterior muscle. Fig. 4B shows the cross-correlogram of the motor unit spikes of the gastrocnemius motoneuron and the onsets of the triangular stretches of the tibialis anterior muscle. Fig. 4B clearly shows that the spike activities are depressed in corresponence with the stretch phase of the tibialis anterior muscle. This decreased probability of spikes as shown in the cross-correlogram is called the inhibitory primary correlation trough. The mean time from the onset of the stretch to the initiation of the trough is 4.4 ± 0.9 msec, which is equivalent to the latency of a polysynaptic Ia inhibitory pathway. Since this trough is closely related to IPSPs, the integrated trough would indicate a rising phase of the IPSP. This integrated trough is shown in Fig. 4B by the solid line. The dotted line represents the returning phase of the IPSP, which is based on calculation with equation (1). Ten examples of the rising phase were superimposed and are shown after normalization in Fig. 5C. The mean width of the inhibitory primary correlation trough was 21.0 ± 7.0 msec, which is longer than that of EPSP.

TIME COURSES OF THE EPSP AND IPSP

After obtaining the cross-correlogram of the random triangular stretches of the gastrocnemius muscle and the gastrocnemius motor unit spikes which responded to the stretches, the primary correlation kernel was integrated. The integrated curve shows the rising phase of the EPSP which was elicited on the gastrocnemius motoneuron. Ten examples of the curve were superimposed and are shown in Fig. 5A. These curves indicate the rising phase of the EPSP caused by the proprioceptive homonymous input.

Simultaneously with the stretches of the gastrocnemius muscle, the soleus muscle, which is agonist to the gastrocnemius muscle, was stretched with triangular pulses. From the cross-correlogram of the gastrocnemius motor unit spikes and the stretches

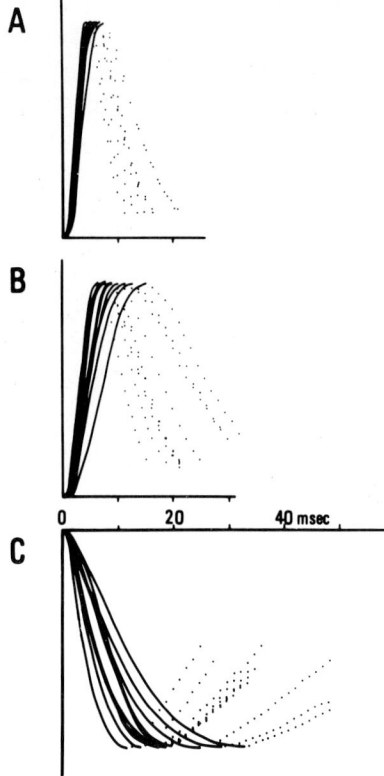

Fig. 5. Ten examples of the integrated primary correlation kernel, the facilitatory primary correlation kernel, and the inhibitory primary correlation trough are shown in A, B and C, respectively. Their amplitudes are normalized.

of the soleus muscle, a facilitatory primary correlation kernel was obtained and integrated. The integrated curve shows the rising phase of the EPSP elicited on the gastrocnemius motoneuron by spindle primary afferent impulses from the soleus muscle. Ten examples of the curve were superimposed and are shown in Fig. 5B. They indicate the rising phase of an EPSP caused by Ia impulses from the heteronymous muscle. The time course of the EPSP caused by this heteronymous input is longer than that of the EPSP caused by the homonymous input.

Simultaneously with the stretches of the gastrocnemius muscle, the tibialis anterior muscle, which is antagonist to the gastrocnemius muscle, was stretched with triangular pulses. The cross-correlogram of the gastrocnemius motor unit spikes and the stretches of the tibialis anterior muscle showed an inhibitory primary correlation trough. Integration of the trough indicates the rising phase of an IPSP. Ten examples of the integration were superimposed and are shown in Fig. 5C. They show the rising phase of the IPSP caused by the input from the antagonistic muscle. The time-to-peak of the IPSP is much longer than that of the EPSP.

SUMMARY

Random triangular stretches of a muscle can activate the stretch reflex center and elicit random motor unit spikes. A primary correlation kernel is revealed by cross-correlation analysis of the motor unit spikes and the muscle stretches. Integration of the primary correlation kernel was thought to indicate the rising phase of the EPSP elicited on an alpha-motoneuron by spindle primary afferent impulses from the stretched muscle.

Simultaneous random triangular stretches of an agonistic and an antagonistic muscle increased the motor unit spikes or depressed them in correspondence with the stretch phase. Cross-correlograms of motor unit spikes with the agonistic muscle or the antagonistic muscle stretches revealed a facilitatory primary correlation kernel or an inhibitory primary correlation trough, respectively. Integration of the kernel or trough made it possible to calculate the rising phase of the EPSP due to input from the agonist or the rising phase of the IPSP due to input from the antagonist. That is, statistical analysis in our investigation made it possible to calculate the time course of postsynaptic potentials, i.e., the EPSP due to the proprioceptive input from the homonymous muscle, the facilitatory potential due to the input from the heteronymous muscle, and the inhibitory potential due to the input from the antagonist.

Thus the coding process between input and output carried out by postsynaptic potentials was statistically quantified. This kind of analysis seems to have application to the input-output relation of a general neuronal circuit.

REFERENCES

Bianconi, R. and Van der Meulen, J.P. (1963) The response to vibration of the end-organs of mammalian muscle spindles. *J. Neurophysiol.*, 26: 177–190.

Brown, M.C., Engberg, I. and Matthews, P.B.C. (1967) The relative sensitivity to vibration of muscle receptors of the cat. *J. Physiol. (Lond.)*, 192: 773–800.

Godaux, E., Desmedt, J.E. and Demart, P. (1975) Vibration of human limb muscles: the alleged phase-locking of motor unit spikes. *Brain Res.*, 100: 175–177.

Hagbarth, K.-E. (1973) The effect of muscle vibration in normal man and in patients with motor disorders. In *New Developments in Electromyography and Clinical Neurophysiology, Vol. 3*. J. E. Desmedt (Ed.) Karger, Basel, pp. 428–443.

Homma, S. (1966) Firing of the cat motoneurone and summation of the excitatory postsynaptic potential. In *Muscular Afferent and Motor Control*, R. Granit (Ed.) Almqvist and Wiksell, Stockholm, pp. 235–244.

Homma, S. (1976) Frequency characteristics of the impulse decoding ratio between the spinal afferents and efferents in the stretch reflex. In *Progress in Brain Research, Vol. 44, Understanding the Stretch Reflex*, S. Homma (Ed.). Elsevier, Amsterdam, pp. 132–140.

Homma, S. and Kanda, K. (1973) Impulse decoding process in stretch reflex. In *Motor Control*. A.A. Gydikov, N.T. Tankov and D.S. Kosarov, (Eds.). Plenum Press, New York, pp. 45–64.

Homma, S., Ishikawa, K. and Stuart, D.G. (1970) Motoneuron responses to linearly rising muscle stretch. *Amer. J. phys. Med.*, 49: 290–306.

Homma, S. and Nakajima, Y. (1979) Coding process in human stretch reflex analyzed by phase-locked spikes. *Neurosci. Lett.*, 11: 19–22.

Homma, S., Kanda, K. and Watanabe, S. (1971a) Monosynaptic coding of group Ia afferent discharges during vibratory stimulation of muscles. *Jap. J. Physiol.*, 21: 405–417.

Homma, S., Kanda, K. and Watanabe, S. (1971b) Tonic vibration reflex in human and monkey subjects. *Jap. J. Physiol.*, 21: 419–430.

Knox, C.K. (1974) Cross-correlation functions for a neuronal model. *Biophys. J.*, 14: 567–582.

Knox, C.K. and Poppele, R.E. (1977) Correlation analysis of stimulus-evoked changes in excitability of spontaneously firing neurons. *J. Neurophysiol.*, 40: 616–625.

Matthews, P.B.C. (1972) *Mammalian Muscle Receptors and their Central Actions*. Edward Arnold, London.

Contribution of Different Size Motoneurons to Renshaw Cell Discharge During Stretch and Vibration Reflexes

P. WAND* and O. POMPEIANO

Istituto di Fisiologia Umana, Cattedra II, Università di Pisa, 56100 Pisa (Italy)

INTRODUCTION

In the decerebrate animal, the reflex contraction of a stretched muscle (Liddell and Sherrington, 1924) was generally attributed to autogenetic excitation of the homonymous motoneurons due to stimulation of group Ia afferents (cf. Matthews, 1972). It was recently suggested that not only the muscle spindle primary endings but also the secondary endings play an important role in the stretch reflex (Matthews, 1969). In particular, by comparing the relative strength and the mode of interaction of the myographically recorded reflex responses to static stretch and to high frequency vibration of the soleus muscle in the decerebrate cat, it appeared that the response to static stretch was relatively much greater than that elicited by muscle vibration for comparable frequencies of discharge of the primary endings of the muscle spindles. From this finding it was postulated that the secondary endings of the muscle spindles, which are stimulated during static muscle stretch but not during vibration, contributed excitation to the stretch reflex rather than the classically believed inhibition.

These conclusions have been criticized on the basis of the fact that the reflex response to a stretch may have been obscured by the length-force relationship of the contracting muscle (Grillner, 1970; Grillner and Udo, 1970, 1971a). Further arguments have been developed to support (Matthews, 1970; McGrath and Matthews, 1973; Fromm et al., 1977; Kanda and Rymer, 1977) or disprove (Grillner, 1973) the original hypothesis.

Recent experiments have indicated that the muscle spindle secondary endings may indeed monosynaptically and polysynaptically excite extensor motoneurons (Kirkwood and Sears, 1974, 1975; Lundberg et al., 1975, 1977; Stauffer et al., 1976). However, in a recent study it was shown that in decerebrate cats, the only significant excitatory action contributing to stretch and vibration reflexes of the soleus muscle originated from Ia afferents (Jack and Roberts, 1978). In addition to these findings it has been proposed that the differences in reflex tension to static stretch and vibration may depend not only upon various amounts of central excitation elicited by

---

*"Habilitationsschrift" of P. Wand submitted to the Faculty of Medicine, University of Göttingen (1979). P. Wand's present address: Max-Planck-Institut für Exp. Medizin, Hermann-Rein-Str. 3, 3400 Göttingen, G.F.R.

the group Ia and group II afferent inputs, but also upon different amounts of central inhibition triggered by the Ia afferent pathway in the two experimental conditions (Thoden et al., 1972).

Two basic mechanisms of autogenetic inhibition tend to reduce excitation of extensor motoneurons elicited by the corresponding Ia pathway. The first is a mechanism of presynaptic inhibition due to primary afferent depolarization (PAD) of both the homonymous and heteronymous Ia afferents (Barnes and Pompeiano, 1970; Magherini et al., 1972; Thoden et al., 1972). Muscle vibration was particularly effective in producing PAD of the Ia pathway, while static stretch was apparently inoperative in this respect. The second is a mechanism of postsynaptic inhibition due to recurrent excitation of Renshaw cells (Renshaw, 1941, 1946). It was shown that Renshaw cells were mainly sensitive to dynamic changes in muscle length, as produced by muscle vibration, but less sensitive to static muscle stretch (Hellweg et al., 1974; Pompeiano et al., 1974b, 1975b, c; Pompeiano and Wand, 1976). It has been hypothesized that static stretch and vibration of the GS muscle activate small and large motoneurons in different proportions in the two experimental conditions, and that axon collaterals of large phasic motoneurons, on the average, make more powerful excitatory connections onto Renshaw cells than do those orginating from small tonic motoneurons. The results presented here provide evidence supporting this hypothesis (Wand et al., 1977a, 1979a).

METHODS

Cats of either sex were decerebrated at a precollicular level under ether anesthesia. The nerve supply to the isolated gastrocnemius-soleus (GS) muscle of one leg was left intact, while the rest of the leg was completely denervated. To avoid transmission of vibration through bone, the left femur was osteotomized and 1 cm of bone removed.

The lumbo-sacral segments L5-S2 of the spinal cord were exposed by laminectomy and the ventral roots L6-S1 of the left side of the animal were cut. Conventional techniques were employed for recording monosynaptic reflexes from ventral roots and discharges of single motor axons originating from the GS pool in functionally isolated ventral root filaments. Motoneurons were classified according to their 'critical firing level' (CFL), following a method described by Clamann et al. (1974b).

The animals were paralyzed with gallamine triethiodide (Flaxedil, 2–4 mg/kg i.v.) and artificially ventilated. The condition of the animal was monitored continuously by recording blood pressure, body temperature, and end-tidal CO_2.

All motoneurons tested were submitted both to longitudinal muscle vibration of prolonged duration (cf. Morelli et al., 1970) and static muscle stretch, resting length being considered that muscle length which produced a deflection of about 0.2 N on the myograph.

Action potentials were recorded on film and simultaneously shaped into standard pulses by a window-discriminator. These pulses were then fed into a digital signal averager (Correlatron 1024, Laben) to analyze the occurrence of pulses relative in time to the presentation of a stimulus. Repeated sweeps build up sequential pulse density histograms. The resulting data were displayed on an X–Y plotter and printed out on a typewriter for further evaluation of the results (Programma P101, Olivetti).

RESULTS

Motoneuronal discharge during muscle vibration

In the present experiments, the activity of 94 GS motoneurons whose CFL varied from 1 to 92% was recorded from isolated ventral root filaments of L7 or S1. It has been postulated that motoneurons of different size have different CFL, the smaller the size, the lower the CFL (Clamann et al., 1974b). All these units responded to prolonged periods (1 sec) of longitudinal vibration of the homonymous muscle, which had been stretched by about 20% of its resting length (mean 10 mm, range 8–14 mm).

If the vibration of the GS muscle was carried out at 200/sec and at amplitudes such as to excite all the primary endings of the muscle spindles (Bianconi and Van der Meulen, 1963; Brown et al., 1967; Stuart et al., 1970), most of the motoneurons excited by muscle vibration discharged throughout the stimulation period. However, some units showed a very regular discharge with fixed decoding ratios as shown by the regularity at which the interspike-intervals occurred (Fig. 1a–f), while other units showed an extremely variable decoding ratio due to grouped spikes separated by variable long intervals (Fig. 1g–l).

As to the development of the response, the more irregularly firing motoneurons showed a sudden increase in the discharge frequency at the beginning of muscle vibration, whereas this increase was less pronounced in motoneurons firing with fixed decoding ratios (see Fig. 2a and b). This early or *phasic* response to muscle vibration gradually decreased during the first 100 msec of stimulation to a steady, albeit lower level, which was then maintained throughout the vibratory period (*tonic* response).

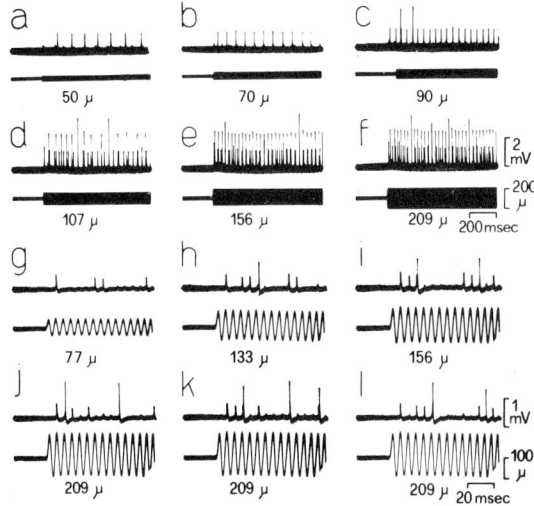

Fig. 1. Discharge patterns of different size motoneurons during progressively increasing amplitudes of vibration. Precollicular decerebrate cats with ventral roots L6-S1 cut, paralyzed with Flaxedil. GS muscle fixed at 10 mm of initial extension. a–f) The smaller unit (CFL 14%) was recruited at smaller amplitude during vibration of the GS muscle at 180/sec than the larger unit (CFL 34%). Both units exhibited a regular discharge during vibration. g–l) A pair of units from another cat. The small (CFL 8%) and the large unit (CFL 59%) showed again the order of recruitment during vibration according to the size principle, but both units discharged more irregularly. Moreover, the small unit seemed to be often silenced following a spike of the larger one. Note three spike amplitudes in recording a–f and g–l, the highest of which is due to superposition of the action potentials of the two units. (From Wand et al., 1979a.)

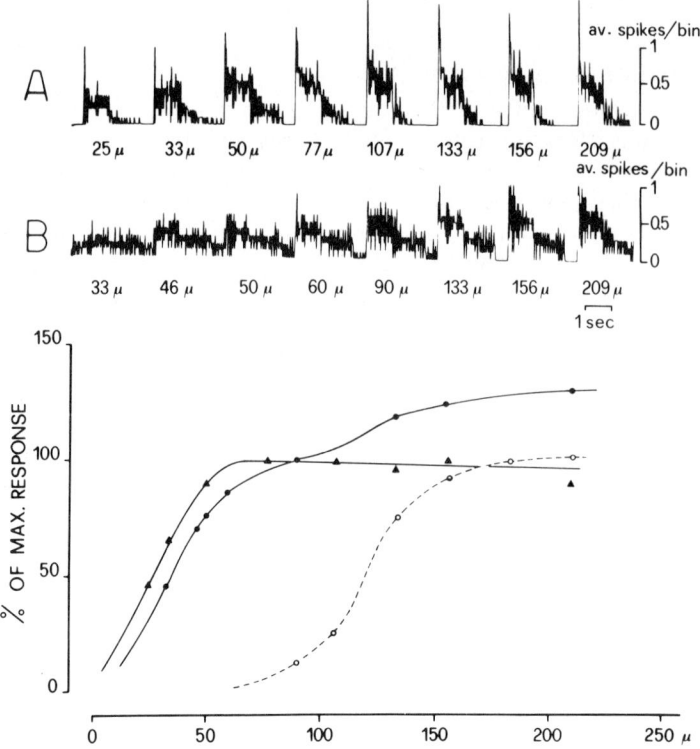

Fig. 2. Relation between discharge frequency of GS motoneurons and amplitude of vibration. Same experimental conditions as in Fig. 1. A) Excitation of a GS motoneuron (CFL 2%) during vibration of the homonymous muscle at 200/sec for 1 sec and at different amplitudes as indicated at the bottom. The GS muscle was fixed at 10 mm of initial extension. Computer records were built up by using 128 bins with 20 msec dwell time/bin averaging 10 consecutive sweeps. The percentage changes in discharge frequency obtained throughout the vibration periods for increasing amplitudes have been plotted in the diagram below (triangles). The maximal response (mean frequency = 28.6 imp/sec) was obtained with 77 μm peak-to-peak amplitude, i.e., when all the group Ia muscle afferents had been recruited by the stimulus. B) Response of another GS motoneuron (CFL 16%) to vibration of the homonymous muscle. Experimental conditions as in A. The development of the response of the motoneuron to increasing amplitudes of vibration has been plotted in the diagram marked with dots. The average response, when all the group Ia afferents had been recruited by the stimulus, corresponded to 21.1 imp/sec at an amplitude of 90 μm. The contribution of the high threshold (probably group II) muscle afferents to the unit's response is shown by the further increase in discharge frequency for vibration amplitudes above 100 μm (mean frequency = 27.7 imp/sec at an amplitude of 209 μm). The third curve in the diagram (circles) refers to a GS motoneuron (CFL 3%) recorded in another experiment in which vibration at 200/sec for 1 sec at different amplitudes was applied to the GS muscle fixed at 8 mm of initial extension only. (From Wand et al., 1979a.)

The transition from the *phasic* to the *tonic* response indicated that a balance was reached at the motoneuronal level between the autogenetic excitation produced by mechanically induced Ia volleys, and both the presynaptic (Barnes and Pompeiano, 1970; Thoden et al., 1972) and postsynaptic inhibitory effects (Pompeiano et al., 1974a, 1975a) due to autogenetic depolarization of the Ia primary afferents, and recurrent excitation of Renshaw cells respectively.

The different-size motoneurons were then studied during muscle vibration of different amplitudes, the vibration frequency and the muscle length being held constant. In these cases, the recruitment of small-size motoneurons occurred at amplitudes of vibration up to 100 μm, whereas larger-size motoneurones were activated by amplitudes exceeding 100 μm (Anastasijević et al., 1968, 1971, 1976; Westbury, 1971, 1972). Fig. 1 shows the order of recruitment according to cell size (Henneman, 1957; Henneman et al., 1965a, b, 1974; Clamann et al., 1974a) of pairs of units during progressively increasing amplitudes of vibration.

Small-size motoneurons usually showed a wide scattering in their threshold amplitude. Those which started to discharge tonically at threshold amplitudes of vibration of 10–20 μm showed an increase in magnitude of their response up to a maximum value for amplitudes of 50–90 μm, thus showing an S-shaped stimulus-response relationship (see diagram in Fig. 2, triangles). Only a few motoneurons exhibited a further increase in firing rate after having reached an initial plateau, when the vibration amplitude was raised from 100 up to 300 μm (see diagram in Fig. 2, dots), i.e., when the amplitude was sufficient to recruit the muscle spindle secondary endings (Stuart et al., 1970). The threshold of the motoneuronal discharge strongly depended on an appropriate prestretch of the muscle on which vibration was superimposed, as shown by the curve marked with circles in Fig. 2; in this case the stimulus-response curve had approximately the same shape as reported above, but was shifted to the right.

Motoneurons which discharged throughout the stimulation period were also observed during muscle vibration of an appropriate range of frequencies, in this case the vibration amplitude and muscle length being held constant. The mean frequency of discharge of these motoneurons at each frequency of vibration was calculated for 1 sec periods. Since the vibration used was of sufficient amplitude to produce 'driving' of all the primary endings of the muscle spindles (Bianconi and Van der Meulen, 1963; Brown et al., 1967; Stuart, 1970), the relationship between the mean afferent and efferent frequencies could be expressed in terms of an average increase in the discharge rate of motoneurons (imp/sec$_\alpha$) per average increase in the discharge rate of the Ia afferents (imp/sec$_{Ia}$).

The frequency transfer curve obtained for each motoneuron specifies the stationary transfer properties of the corresponding monosynaptic reflex pathway activated by a fixed number of stimulated afferent fibers. However, a more complete description of the behavior of the whole motoneuronal pool can be obtained by considering a family of frequency transfer curves corresponding to different size motoneurons. Therefore, the range and gain constant of the linear part of the input-output relation of 94 motoneurons of different size were calculated, and the results were plotted in Fig. 3. Large-size motoneurons tended to be recruited with high frequencies of vibration according to the size principle (Henneman, 1957; Henneman et al., 1965a, b, 1974; Clamann et al., 1974a), while small-size motoneurons showed a wide distribution in threshold frequencies of recruitment. Moreover, the average firing rate of small motoneurons of similar size (1–20% CFL) differed largely (6–34 imp/sec) for fixed parameters of vibration (1 sec at 150/sec, 156 μm amplitude), while the gain constant of the frequency transfer curves relating mean afferent and efferent frequencies varied from 0.04 to 0.22 imp/sec$_\alpha$ per imp/sec$_{Ia}$ (mean 0.11, n = 49). The average gain constant for all motoneurons was 0.14 imp/sec$_\alpha$ per imp/sec$_{Ia}$ (n = 94), whereas the gain constants for the *phasic* and the *tonic* responses corresponded on the average to 0.21 and 0.12 imp/sec$_\alpha$ per imp/sec$_{Ia}$ respectively (n = 77).

The wide scattering of firing thresholds of motoneurons of different size and the large differences in firing rates and gain constants of motoneurons of comparable size,

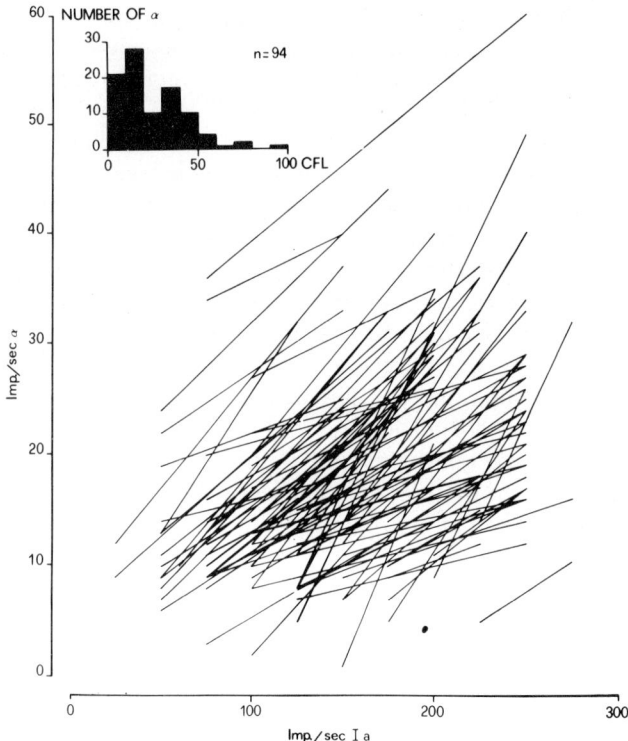

Fig. 3. Frequency transfer curves of GS motoneurons of different size during vibration of the homonymous muscle. All experiments were performed in decerebrate, deefferented cats with the GS muscle fixed at an average length of 10 mm (range 8–14 mm). Each line defines gain constant and linear range of the frequency transfer curve of 94 individual GS motoneurons during 1 sec vibration periods at 156 μm peak-to-peak amplitude (increasing series of vibration frequencies from 25/sec up to 275/sec; step 25/sec). The mean discharge rate of the motoneurons (imp/sec$_\alpha$) calculated throughout the period of vibration is plotted against the frequencies of vibration used, i.e., the discharge frequencies in the Ia afferents (imp/sec$_{Ia}$). Gain constants were calculated with the method of the least squares. The insert histogram shows the distribution of all the recorded motoneurons relative to size (CFL range 1 to 92%; class interval 10%). (From Wand et al., 1979a.)

as evaluated during muscle vibration, indicated that there is no strict separation of motoneurons into small tonic, and large phasic ones, and that the properties of a motoneuron are not entirely dependent on its size (Henneman and Harris, 1976; Harris and Henneman, 1977; Wand et al., 1979a), but also on synaptic connections leading to different strengths of synaptic excitation.

Rate modulation and recruitment of motoneurons with stretch

Among the 94 motoneurons of different size responding to longitudinal muscle vibration, 30 units showed a dynamic response to ramp-and-hold stretches, and of these, 21 (2–49% CFL) also showed a tonic discharge during maintained stretch. Stretch-evoked responses of motoneurons (see Fig. 4A) always occurred in strict order of size, i.e., the units with smallest CFL had the lowest thresholds to stretch, while the cells with largest CFL had the highest thresholds (Henneman et al., 1965a, b).

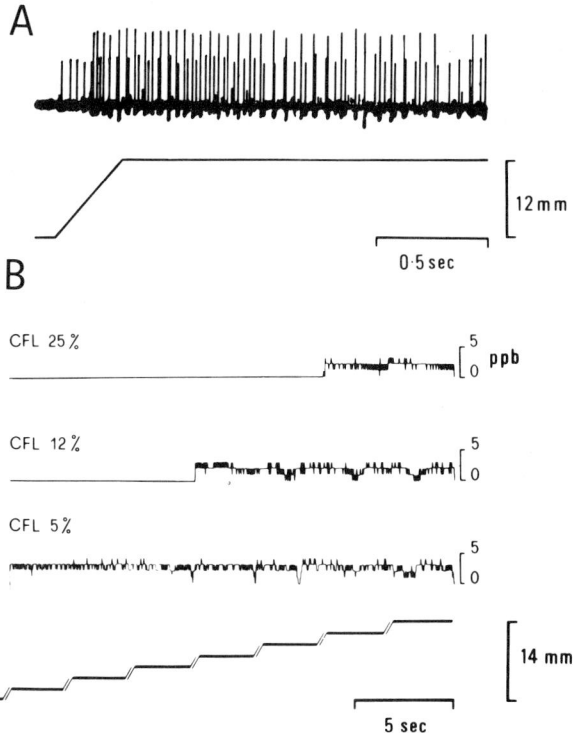

Fig. 4. Excitation of individual GS motoneurons of different size caused by stretch of the homonymous muscle. Same experimental conditions as in Fig. 1. A) Discharge activity and order of recruitment of a small and a larger size GS motoneuron (CFL 8 and 21% respectively) recorded from a S1 ventral root filament during ramp-and-hold stretch of the GS muscle from zero up to 12 mm extension (stretch velocity 25 mm/sec; the lower trace indicates muscle length). B) Effect of progressively increasing static stretch of the GS muscle from zero up to 14 mm (step 2 mm) on three GS motoneurons of different size. The computer records were built up by using 128 bins with 200 msec dwell time/bin. The analysis of the sweep started 2 sec after the end of the dynamic part of the stretch. The three motoneurons (CFL 5, 12, and 25%) started to discharge tonically during maintained muscle elongation at 2, 8 and 12 mm extension of the muscle respectively. Note the lack of increase in firing rates with further increase in muscle length. The lowermost trace again indicates muscle length. (From Wand et al., 1979b.)

In order to estimate the relationship between afferent and efferent frequencies of the motoneuron pool during the stretch reflex, units were subject to progressively increasing static stretch of the GS muscle. Kernell (1976) reported increases in firing rates of soleus motoneurons as other motoneurons were recruited. In our study, however, when the muscle was extended and held at increasing lengths, the motoneurons fired in a quite narrow frequency range (10.0 imp/sec ± 2.4, SD), and except for the initial recruitment of the individual units, virtually no change in frequency occurred for changes in static muscle length (see Fig. 4B; Grillner and Udo, 1971a, b; Grillner, 1973). Under the chosen experimental parameters, therefore, the estimate of the input-output relationship was 0.15 imp/sec/mm based on the assumptions firstly, that the total output frequency of the motoneuron pool was the result of recruitment, and not frequency-coding, and secondly, that only a fraction of the pool contributed to the motor output during static stretch (cumulative frequency distribution).

Since the average discharge rate of the primary spindle receptors recorded from the deefferented GS muscle, in our preparation corresponded to 2.62 imp/sec/mm static stretch (Pompeiano et al., 1974b, 1975b; cf. also Granit, 1958; Matthews and Stein, 1969), we may convert the increase in firing rate of the motoneurons during static stretch into an average 0.06 imp/sec of the motoneurons for each imp/sec in the Ia afferents (Wand et al., 1977b, 1979b). This value is 2.4 times smaller than that evaluated during 1 sec periods of muscle vibration.

The differences in gain factors observed with vibration and static stretch may be explained by the fact that in the former condition the synchronous high frequency volleys in the Ia pathway are the dominant excitatory input, while in the latter condition the discharge in the excitatory Ia pathway will be asynchronous, and the contribution of the group Ib inhibitory pathway (Granit, 1970; Watt et al., 1976) and the group II muscle afferent pathway will be relatively much more substantial, resulting in a mixed excitatory-inhibitory input.

However, regardless of the central connections of the group Ib and group II muscle afferents, their excitation should be identical under both conditions, as muscle vibration was always superimposed on a background of static stretch. Therefore, the possible explanation for the differences in gain factors observed with static stretch and vibration, might be the synchronicity and the larger amount of Ia excitation during vibration recruiting more motoneurons of larger size.

DISCUSSION

The contribution of motoneurons of different size to Renshaw cell discharge can be understood if we first consider the organization of autogenetic reflex pathways. Assuming that activity in the Ia pathway monosynaptically excites the motoneuron pool, then a fraction of the whole pool will reflexly discharge in response to the incoming excitation. Small tonic motoneurons with small axons will discharge at lower thresholds than larger phasic motoneurons with larger axons (Granit et al., 1957a; Kuno, 1959; Henneman et al., 1965a, b; Tan, 1971; Tan et al., 1972). The action potentials in the motor axons will be transmitted in recurrent collaterals (Cajal, 1909; Prestige, 1966; Scheibel and Scheibel, 1966, 1971; Szentágothai, 1967). These recurrent collaterals terminate on Renshaw cells (Renshaw, 1941, 1946; Eccles et al., 1954) which in turn terminate on motoneurons, building up a negative feedback mechanism. Direct excitatory connections from recurrent collaterals to motoneurons may also exist (Cullheim et al., 1977). Therefore, each efferent impulse sent from the motoneuron pool to the periphery will be the result of integration of Ia excitation, recurrent inhibition, and possible recurrent excitation (the relative importance of recurrent excitation remains to be investigated).

The results of the present study, combined with the data obtained in previous investigations on Renshaw cells (Pompeiano et al., 1974b, 1975b), are schematically summarized in Fig. 5. It appears in particular, that in order to fire the motoneurons of the GS pool with an average 1 imp/sec by static stretch of the homonymous muscle, these must receive on the average 17.0 imp/sec from the Ia afferents and 5.8 imp/sec from the Renshaw cells. During 1 sec periods of muscle vibration, the same output was achieved with an average of only 7.1 imp/sec from the Ia pathway, but an average of 10.4 imp/sec from the Renshaw cells. The disproportion between the Ia excitatory

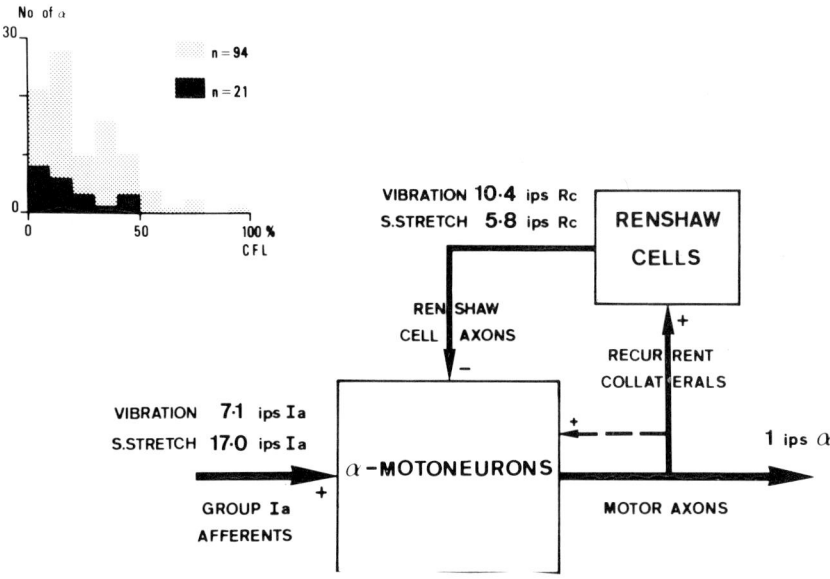

Fig. 5. Transfer and integration of Ia excitation and Renshaw inhibition by the GS motoneuron pool during static stretch and vibration of the homonymous muscle. The fraction of the motoneuron pool responding to static stretch of the GS muscle covers only small-size motoneurons (range 1–49% CFL), while that responding to muscle vibration is composed of both small- and large-size motoneurons (range 1–92% CFL). The data concerning Renshaw cells are from Pompeiano et al. (1975b). It appears that in order to produce an average of 1 imp/sec during static stretch of the homonymous muscle, the GS motoneurons must receive on the average 17.0 imp/sec from the Ia afferents and 5.8 imp/sec from the Renshaw cells. During 1 sec periods of muscle vibration the same average output is achieved with an average of only 7.1 imp/sec from the Ia afferents but an average of 10.4 imp/sec from the Renshaw cells. This difference in the decoding ratios of motoneurons of different size excited either by static stretch or vibration is even more pronounced for the phasic component of the motoneuronal response elicited during muscle vibration. The insert histogram shows the fraction of GS motoneurons discharging tonically during maintained muscle stretch (n = 21) relative to our total sample of GS motoneurons responding to muscle vibration (n = 94). (From Wand et al., 1979b.)

and the recurrent inhibitory volleys during muscle vibration as compared to static stretch, became even stronger when calculated for the early *phasic* response of motoneurons and Renshaw cells during vibration. In this early response, only an average 4.8 imp/sec from the Ia afferents, but an average 13.8 imp/sec from the Renshaw cells generated an average of 1 output pulse from the motoneurons. For the late *tonic* response of motoneurons and Renshaw cells, these values corresponded to an average 8.3 imp/sec from the Ia pathway and an average 9.0 imp/sec from the Renshaw cells (Pompeiano et al., 1974b, 1975b; Wand et al., 1979a, b).

From these discrepancies in the amounts of Ia excitation and Renshaw inhibition necessary to generate an average of 1 output pulse from the motoneuron pool, the following considerations can be made. Each primary spindle ending in a muscle under tension discharges in a quasi-periodic manner, and the different stretch thresholds (Hunt, 1954) cause the average discharge frequency of each ending to differ. The times of occurrence of impulses in one ending will then be statistically independent of the times of occurrence of impulses in other endings. The convergence of these asynchronous impulse patterns onto single motoneurons represents a situation very different from that of synchronous stimulation. In particular the asynchronous input during static stretch is mainly, if not exclusively effective on small-size tonic motoneurons as shown in the present study (cf. also Granit et al., 1957a, b; Henneman et al., 1965a, b; Kernell, 1966; Burke, 1967, 1968a, b; Granit, 1970; Burke and Ten Bruggenkate, 1971; Wand et al., 1979b), while that produced synchronously during muscle vibration is effective on both small-size and large-size phasic motoneurons (cf. also Anastasijević et al., 1968, 1971, 1976; Westbury, 1971, 1972; Wand et al., 1979a).

One finding consistently reported in the literature is that the amount of recurrent inhibition is proportional to the amount of motor activity (Haase and Vogel, 1971; Ross et al., 1972, 1975, 1976; Ryall et al., 1972; Benecke et al., 1974; Pompeiano et al., 1974a, 1975a; Pompeiano and Wand, 1976). In this context, it should be kept in mind that both the number of motoneurons recruited, and their individual firing rates, determine the amount of motor activity present (the firing rates may in turn be dependent on motoneuronal size). Our results support these findings, as during static stretch only the fraction of small-size motoneurons of the pool was active, firing in a narrow range at low frequency (Grillner and Udo, 1971a, b; Grillner, 1973; Wand et al., 1979b). Under this condition, Renshaw inhibition was relatively weak (Pompeiano et al., 1974b, 1975b; Pompeiano and Wand, 1976; Cleveland and Ross, 1977). During muscle vibration, when all the recorded motoneurons, regardless of size, participated in the motoneuronal response, firing at higher average frequencies in a wider range (Anastasijević et al., 1968; Brown et al., 1968; Homma et al., 1967, 1970a, b, 1971, 1972; cf. Eccles et al., 1958; Granit, 1970, 1972), Renshaw inhibition was much more powerful (Pompeiano et al., 1974b, 1975b; Pompeiano and Wand, 1976; Cleveland and Ross, 1977).

The hypothesis that the recurrent collaterals of the large phasic motoneurons make more powerful synaptic connections with the Renshaw cells than do axon collaterals of the small tonic motoneurons (Ryall et al., 1972; cf. Granit, et al., 1957a; Eccles et al., 1961), while small tonic motoneurons are in their turn subject to greater recurrent inhibition than are the large phasic ones (Granit et al., 1957a; Kuno, 1959; Eccles et al., 1961; Tan, 1972, 1975; Tan et al., 1972; cf. Holmgren and Merton, 1954), was also supported by the observation that the Renshaw cells were mainly sensitive to the velocity of muscle stretch, with little sensitivity to static muscle length (Hellweg et al., 1974; Pompeiano et al., 1975c).

The results summarized in Fig. 5 provide further evidence in favor of this hypothesis. Firstly, the larger the size of the motoneurons recruited by a vibratory stimulus, the higher was the amount of Renshaw inhibition exerted onto the pool. In other words, large phasic motoneurons greatly contributed to Renshaw cell excitation. Secondly, the smaller the size of the motoneurons recruited by a stretch stimulus, the higher was the amount of Ia excitation necessary to generate an average 1 output pulse in motoneurons when Renshaw inhibition was simultaneously operative. In other words, the small tonic motoneurons were more sensitive to recurrent inhibition than the larger phasic ones.

As to the possible function of Renshaw inhibition, it should be pointed out that Renshaw cells, like other interneurons, are subject to biasing from both afferent and supraspinal sources (Granit, 1970; Haase et al., 1975). Hence, it is to be expected that the recurrent feedback mechanism will display a high degree of flexibility, essential for a number of tasks attributed to Renshaw inhibition in regulating the organized discharge of motoneurons. However, some of the motoneurons are lacking recurrent effects. In anatomical studies Scheibel and Scheibel (1966) found 20–30% of ankle extensor motoneurons lacking in recurrent collaterals, and recurrent effects are entirely absent in motoneurons innervating the eye muscles of the cat (Sasaki, 1963) and in the motoneurons of the phrenicus (Gill and Kuno, 1963a, b).

Recurrent inhibition is thought to be one of the mechanisms by which the activity of motoneurons is limited and stabilized (Granit et al., 1960; Granit, 1970; Haase et al., 1975). However, experiments by Redman and Lampard (1967) in which the frequency transfer characteristics in stochastically stimulated motoneurons were measured before and after the infusion of dehydro-β-erythroidine indicated that recurrent inhibition did not change the frequency at which limiting occurred. According to these authors, repolarizing current during "after-hyperpolarization" and the depolarization limitations of the Ia afferent pathways were the principal factors in limiting the discharge frequency of motoneurons.

The finding that small tonic motoneurons are preferentially inhibited by Renshaw cells, led Eccles et al. (1961) to suggest that the recurrent process serves as a mechanism of "motor contrast", focussing motor performance during rapid movements onto fast-contracting muscles by suppressing small tonic motoneurons, thus preventing interference with slow-acting muscles.

The observation that Renshaw cells are mainly sensitive to stretch velocity (Hellweg et al., 1974; Pompeiano et al., 1975c) and the crucial finding that they act in parallel onto corresponding motoneurons and Ia inhibitory interneurons (Hultborn et al., 1968, 1971a, b; Benecke et al., 1975; cf. Hultborn, 1976) as well as onto static fusimotor motoneurons (Fromm and Noth, 1976; Ellaway and Trott, 1978), suggests that the mechanism of recurrent inhibition may operate as a switching device. This mechanism could represent, at a spinal cord level, an essential contribution to the performance of fast alternating movements, as well as to the execution of locomotor commands.

SUMMARY

1. Selective and intense activation of muscle spindle primary endings by vibration of the GS muscle excited 94 homonymous motoneurons of different sizes (1–92% CFL). Motoneurons were recruited with increasing amplitude and/or frequency of vibration according to the size principle. However, small-size motoneurons (1–20% CFL) showed a wide scattering in thresholds and great differences in firing rates (6–34 imp/sec) during vibration with fixed parameters. The gain constant of the input-output relationship of these motoneurons varied from 0.04–0.22 imp/sec$_\alpha$ per imp/sec$_{Ia}$ (mean 0.11, n = 49), but being on the average smaller than that of all motoneurons (mean 0.14, n = 94). These findings indicate that the population of small motoneurons was not a homogeneous one and that cell properties were not entirely dependent on cell size.

2. Among the 94 motoneurons activated with vibration, 30 units showed a dynamic response to muscle stretch, 21 of these units showing an additional tonic response to maintained stretch (2–49% CFL). The gain constant of the frequency transfer curve relating the mean afferent and efferent frequencies for the latter fraction of the motoneuron pool during static stretch, corresponded to 0.06 imp/sec$_\alpha$ per imp/sec$_{Ia}$ and was 2.4 times lower than that calculated for muscle vibration.

3. A comparison of the frequency transfer functions of the GS motoneuron pool

during static stretch and vibration was made in the light of the amount of Renshaw inhibition produced under the two conditions as evaluated in a previous study. It appeared that the larger the size of the motoneurons recruited by a vibratory stimulus, the greater was the amount of Renshaw inhibition produced by these motoneurons, while the small-size motoneurons, participating in the motoneuronal response during static stretch, were relatively weak in exciting Renshaw cells.

4. It is concluded that i) the afferent input produced asynchronously during static stretch is mainly effective on small-size motoneurons firing at low average frequencies, while that produced synchronously during vibration is effective not only on small, but also on larger-size motoneurons, giving rise to higher average firing rates for comparable discharge frequencies in the Ia afferents, and ii) the discharge in the recurrent collaterals of the larger phasic motoneurons is the main source of excitation of Renshaw cells.

ACKNOWLEDGEMENTS

The authors are greatly indebted to Dr. N.A. Fayein for her collaboration in part of the experiments. The investigations summarized in the present paper were supported by the European Training Program in Brain and Behaviour Research, by a Public Health Service Research Grant NS 07685–11 from the National Institute of Neurological Diseases and Stroke, N.I.H., U.S.A., by a Research Grant from the Consiglio Nazionale delle Ricerche, Italy, and by the "Sonderforschungsbereich 33" of the Deutsche Forschungsgemeinschaft.

REFERENCES

Anastasijević, R., Cvetković, M. and Vučo, J. (1971) The effect of short-lasting repetitive vibration of the triceps muscle and concomitant fusimotor stimulation on the reflex response of spinal alpha motoneurons in decerebrated cats. *Pflügers Arch.*, 325: 220–234.

Anastasijević, R., Stanojević, M. and Vučo, J. (1976) Patterns of motoneuronal units discharge during naturally evoked afferent input. In *Progress in Brain Research, Vol. 44, Understanding the Stretch Reflex*, S. Homma (Ed.), Amsterdam, Elsevier, pp. 267–278.

Anastasijević, R., Anojčić, M., Todorović, B. and Vučo, J. (1968) The differential reflex excitability of alpha motoneurons of decerebrate cats caused by vibration applied to the tendon of the gastrocnemius medialis muscle. *Brain Res.*, 11: 336–346.

Barnes, C.D. and Pompeiano, O. (1970) Presynaptic and postsynaptic effects in the monosynaptic reflex pathway to extensor motoneurons following vibration of synergic muscles. *Arch. Ital. Biol.*, 108: 259–294.

Benecke, R., Hellweg, C. and Meyer-Lohmann, J. (1974) Activity and excitability of Renshaw cells in non-decerebrate and decerebrate cats. *Exp. Brain Res.*, 21: 113–124.

Benecke, R., Böttcher, U., Henatsch, H.-D., Meyer-Lohmann, J. and Schmidt, J. (1975) Recurrent inhibition of individual Ia inhibitory interneurones and disinhibition of their target α-motoneurones during muscle stretches. *Exp. Brain Res.*, 23: 13–28.

Bianconi, R. and Van der Meulen, J.P. (1963) The response to vibration of the end organs of mammalian muscle spindles. *J. Neurophysiol.*, 26: 177–190.

Brown, M.C., Engberg, I. and Matthews, P.B.C. (1967) The relative sensitivity to vibration of muscle receptors of the cat. *J. Physiol. (Lond.)*, 192: 773–780.

Brown, M.C., Lawrence, D.G. and Matthews, P.B.C. (1968) Reflex inhibition by Ia afferent input of spontaneously discharging motoneurons in the decerebrate cat. *J. Physiol. (Lond.)*, 198: 5–7P.

Burke, R.E. (1967) Motor unit types of cat triceps surae muscle. *J. Physiol. (Lond.)*, 193: 141–160.

Burke, R.E. (1968a) Group Ia synaptic input to fast and slow twitch motor units of cat triceps surae. *J. Physiol. (Lond.)*, 196: 605–630.

Burke, R.E. (1968b) Firing patterns of gastrocnemius motor units in the decerebrate cat. *J. Physiol. (Lond.)*, 196: 631–654.

Burke, R.E. and Ten Bruggenkate, G. (1971) Electrotonic characteristics of alpha motoneurones of varying size. *J. Physiol. (Lond.)*, 212: 1–20.

Cajal, S.R. y (1909) *Histologie du Système Nerveux de l'Homme et des Vertébrés, Vol. 1*, Maloine, Paris, pp. 361–368.

Clamann, H,P., Gillies, J.D. and Henneman, E. (1974a) Effects of inhibitory inputs on critical firing level and rank order of motoneurons. *J. Neurophysiol.*, 37: 1350–1360.

Clamann, H.P., Gillies, J.D., Skinner, R.D. and Henneman, E. (1974b) Quantitative measures of output of a motoneuron pool during monosynaptic reflexes. *J. Neurophysiol.*, 37: 1328–1337.

Cleveland, S. and Ross, H.-G. (1977) Dynamic properties of Renshaw cells: frequency response characteristics. *Biol. Cybernetics*, 27: 175–184.

Cullheim, S., Kellerth, J.-O. and Conradi, S. (1977) Evidence for direct synaptic interconnections between cat spinal α-motoneurons via the recurrent axon collaterals: a morphological study using intracellular injection of horseradish peroxidase. *Brain Res.*, 132: 1–10.

Eccles, J.C., Fatt, P. and Koketsu, K. (1954) Cholinergic and inhibitory synapses in a pathway from motor-axon collaterals to motoneurons. *J. Physiol. (Lond.)*, 126: 524–562.

Eccles, J.C., Eccles, R.M. and Lundberg, A. (1958) The action potentials of the alpha motoneurones supplying fast and slow muscles, *J. Physiol. (Lond.)*, 142: 275–291.

Eccles. J.C., Eccles, R.M., Iggo, A. and Ito, M. (1961) Distribution of recurrent inhibition among motoneurones. *J. Physiol. (Lond.)*, 159: 479–499.

Ellaway, P.H. and Trott, J.R. (1978) Autogenetic reflex action on to gamma motoneurones by stretch of triceps surae in the decerebrated cat. *J. Physiol. (Lond.)*, 276: 49–66.

Fromm, C. and Noth, J. (1976) Reflex responses of gamma motoneurones to vibration of the muscle they innervate. *J. Physiol. (Lond.)*, 256: 117–136.

Fromm, C., Haase, J. and Wolf, E. (1977) Depression of the recurrent inhibition of extensor motoneurons by the action of group II afferents. *Brain Res.*, 120: 459–468.

Gill, P.K. and Kuno, M. (1963a) Properties of phrenic motoneurones. *J. Physiol. (Lond.)*, 168: 258–273.

Gill, P.K. and Kuno, M. (1963b) Excitatory and inhibitory actions on phrenic motoneurons. *J. Physiol. (Lond.)*, 168: 274–289.

Granit, R. (1958) Neuromuscular interaction in postural tone of the cat's isometric soleus muscle. *J. Physiol. (Lond.)*, 143: 387–402.

Granit, R. (1970) *The Basis of Motor Control*. Academic Press, New York.

Granit, R. (1972) *Mechanisms Regulating the Discharge of Motoneurons*. Liverpool University Press, Liverpool.

Granit, R., Haase, J. and Rutledge, L.T. (1960) Recurrent inhibition in relation to frequency of firing and limitation of discharge rate of extensor motoneurones. *J. Physiol. (Lond.)*, 154: 308–328.

Granit, R. Pascoe, J.E. and Steg, G. (1957a) Behaviour of tonic α and γ motoneurones during stimulation of recurrent collaterals. *J. Physiol. (Lond.)*, 138: 381–400.

Granit, R., Phillips, C.G., Skoglund, S. and Steg, G. (1957b) Differentiation of tonic from phasic alpha ventral horn cells by stretch, pinna and crossed extensor reflexes. *J. Neurophysiol.*, 20: 470–481.

Grillner, S. (1970) Is the tonic stretch reflex dependent upon group II excitation? *Acta physiol. scand.*, 78: 431–432.

Grillner, S. (1973) A consideration of stretch and vibration data in relation to the tonic stretch reflex. In: *Control of Posture and Locomotion*. R.B. Stein, K.B. Pearson, R.S. Smith and J.B. Redford (Eds). Plenum Press, New York, pp. 397–405.

Grillner, S. and Udo, M. (1970) Is the tonic stretch reflex dependent on suppression of autogenetic inhibitory reflexes? *Acta physiol. scand.*, 79: 13–14A.

Grillner, S. and Udo, M. (1971a) Motor unit activity and stiffness of the contracting muscle fibres in the tonic stretch reflex. *Acta physiol. scand.*, 81: 422–424.

Grillner, S. and Udo. M. (1971b) Recruitment in the tonic stretch reflex. *Acta physiol. scand.*, 81: 571–573.

Haase, J. and Vogel, B. (1971) Die Erregung der Renshaw-Zellen durch reflektorische Entladungen der α-Motoneurone. *Pflügers Arch.*, 325: 14–27.

Haase, J., Cleveland, S. and Ross, H.-G. (1975) Problems of post-synaptic autogenous and recurrent inhibition in the mammalian spinal cord. *Rev. Physiol. Biochem. Pharmacol.*, 73: 73–129.

Harris, D.A. and Henneman, E. (1977) Identification of two species of alpha motoneurons in cat's plantaris pool. *J. Neurophysiol.*, 40: 16–25.

Hellweg, C., Meyer-Lohmann, J., Benecke, R. and Windhorst, U. (1974) Responses of Renshaw cells to muscle ramp stretch. *Exp. Brain Res.*, 21: 353–360.

Henneman, E. (1957) Relation between size of neurons and their susceptibility to discharge. *Science*, 126: 1345–1347.

Henneman, E. and Harris, D.A. (1976) Identification of fast and slow firing types of motoneurons in the same pool. In *Progress in Brain Research, Vol. 44, Understanding the Stretch Reflex*, S. Homma (Ed.) Elsevier, Amsterdam, pp. 377–380.

Henneman, E., Somjen, G. and Carpenter, D.O. (1965a) Functional significance of cell size in spinal motoneurons. *J. Neurophysiol.*, 28: 560–580.

Henneman, E., Somjen, G. and Carpenter, D.O. (1965b) Excitability and inhibitability of motoneurons of different size. *J. Neurophysiol.*, 28: 599–620.

Henneman, E., Clamann, H.P. Gillies, J.D. and Skinner, R.D. (1974) Rank order of motoneurons within a pool: law of combination. *J. Neurophysiol.*, 37: 1338–1349.

Holmgren, B. and Merton, P.A. (1954) Local feedback control of motoneurons, *J. Physiol. (Lond.)*, 123: 47P.

Homma, S., Ishikawa, K. and Watanabe, S. (1967) Optimal frequency of muscle vibration for motoneuron firing. *J. Chiba Med. Soc.*, 43: 190–196.

Homma, S., Ishikawa, K. and Stuart, D.G. (1970a) Motoneuron responses to linearly rising muscle stretch. *Amer. J. phys. Med.*, 49: 290–306.

Homma, S., Kobayashi, H. and Watanabe, S. (1970b) Vibratory stimulation of muscles and stretch reflex. *Jap. J. Physiol.*, 20: 309–319.

Homma, S., Kanda, K. and Watanabe, S. (1971) Monosynaptic coding of group Ia afferent discharges during vibratory stimulation of muscle. *Jap. J. Physiol.*, 21: 405–417.

Homma, S., Kanda, K. and Watanabe, S. (1972) Preferred spike intervals in the vibration reflex. *Jap. J. Physiol.*, 22: 421–432.

Hultborn, H. (1976) Transmission in the pathway of reciprocal Ia inhibition to motoneurons and its control during the tonic stretch reflex. In *Progress in Brain Research, Vol. 44, Understanding the Stretch Reflex*, S. Homma (Ed.). Elsevier, Amsterdam.

Hultborn, H., Jankowska, E. and Lindström, S. (1968) Recurrent inhibition from motor axon collaterals in interneurones monosynaptically activated from Ia afferents. *Brain Res.*, 9: 367–369.

Hultborn, H., Jankowska, E. and Lindström, S. (1971a) Recurrent inhibition from motor axon collaterals of transmission in the Ia inhibitory pathway to motoneurones. *J. Physiol. (Lond.)*, 215: 591–612.

Hultborn, H., Jankowska, E. and Lindström, S. (1971b) Recurrent inhibition of interneurones monosynaptically activated from group Ia afferents. *J. Physiol. (Lond.)*, 215: 613–636.

Hunt, C.C. (1954) Relation of function to diameter in afferent fibers of muscle nerves. *J. gen. Physiol.*, 38: 117–131.

Jack, J.J.B. and Roberts, R.C. (1978) The role of muscle spindle afferents in stretch and vibration reflexes of the soleus muscle of the decerebrate cat. *Brain Res.*, 146: 366–372.

Kanda, R. and Rymer, W.Z. (1977) An estimate of the secondary spindle receptor afferent contribution to the stretch reflex in extensor muscles of the decerebrate cat. *J. Physiol. (Lond.)*, 264: 63–87.

Kernell, D. (1966) Input resistance, electrical excitability, and size of ventral horn cells in cat spinal cord. *Science* 152: 1637–1640.

Kernell. D. (1976) Recruitment, rate modulation and the tonic stretch reflex. In *Progress in Brain Research, Vol. 44, Understanding the Stretch Reflex*, S. Homma (Ed.). Elsevier, Amsterdam, pp. 257–266.

Kirkwood, P.A. and Sears, T.A. (1974) Monosynaptic excitation of motoneurones from secondary endings of muscle spindles. *Nature (Lond.)*, 252: 243–244.

Kirkwood, P.A. and Sears, T.A. (1975) Monosynaptic excitation of motoneurones from muscle spindle secondary endings of intercostal and triceps surae muscle in the cat. *J. Physiol. (Lond.)*, 245: 64–66P.

Kuno, M. (1959) Excitability following antidromic activation in spinal motoneurones supplying red muscles. *J. Physiol. (Lond.)*, 149: 374–393.

Liddell, E.G.T. and Sherrington, C. (1924) Reflexes in response to stretch (myotatic reflexes). *Proc. roy. Soc., B* 96: 212–242.

Lundberg, A., Malmgren, K. and Schomburg, E.D. (1975) Characteristics of the excitatory pathway from group II muscle afferents to alpha motoneurones. *Brain Res.*, 88: 538–542.

Lundberg, A., Malmgren, K. and Schomburg, E.D. (1977) Comments on reflex actions evoked by electrical stimulation of group II muscle afferents. *Brain Res.*, 122: 551–555.

Magherini, P.C., Pompeiano, O. and Thoden, U. (1972) The relative significance of presynaptic and postsynaptic effects on monosynaptic extensor reflexes during vibration of synergic muscles. *Arch. ital. Biol.*, 110: 70–89.

Matthews, P.B.C. (1969) Evidence that the secondary as well as the primary endings of muscle spindles may be responsible for the tonic stretch reflex of the decerebrate cat. *J. Physiol. (Lond)*, 204: 365–393.

Matthews, P.B.C. (1970) A reply to the criticism of the hypothesis that the group II afferents contribute excitation to the stretch reflex. *Acta physiol. scand.*, 79: 431–433.

Matthews, P.B.C. (1972) *Mammalian Muscle Receptors and their Central Actions.* E. Arnold, London.

Matthews, P.B.C. (1973) A critique of the hypothesis that the spindle secondary endings contribute excitation to the stretch reflex. In *Control of Posture and Locomotion.* R.B. Stein, K.B. Pearson, R.S. Smith and J.B. Redford (Eds). Plenum Press, New York, pp. 227–243.

Matthews, P.B.C. and Stein, R.B. (1969) The sensitivity of muscle spindle afferents to small sinusoidal changes of length. *J. Physiol. (Lond.)*, 200: 723–743.

McGrath, G.J. and Matthews, P.B.C. (1973) Evidence from the use of vibration during procaine nerve block that the spindle group II fibres contribute excitation to the tonic stretch reflex of the decerebrate cat. *J. Physiol. (Lond.)*, 235: 371–408.

Morelli, M., Nicotra, L., Barnes, C.D., Cangiano, A., Cook Jr., W.A. and Pompeiano, O. (1970) An apparatus for producing small-amplitude high-frequency sinusoidal stretching of the muscle. *Arch. ital. Biol.*, 108: 222–232.

Pompeiano, O. and Wand, P. (1976) The relative sensitivity of Renshaw cells to static and dynamic changes in muscle length. In *Progress in Brain Research, Vol. 44, Understanding the Stretch Reflex*, S. Homma (Ed.). Elsevier, Amsterdam, pp. 199–222.

Pompeiano, O., Wand, P. and Sontag, K.-H. (1974a) Excitation of Renshaw cells by orthodromic group Ia volleys following vibration of extensor muscles. *Pflügers Arch.*, 347: 137–144.

Pompeiano, O., Wand, P. and Sontag, K.-H. (1974b) A quantitative analysis of Renshaw cell discharges caused by stretch and vibration reflexes. *Brain Res.*, 66: 519–524.

Pompeiano, O., Wand, P. and Sontag, K.-H. (1975a) Response of Renshaw cells to sinusoidal stretch of hindlimb extensor muscles. *Arch. ital. Biol.*, 113: 205–237.

Pompeiano, O., Wand, P. and Sontag, K.-H. (1975b) The relative sensitivity of Renshaw cells to orthodromic group Ia volleys caused by static stretch and vibration of extensor muscles. *Arch. Ital. Biol.*, 113: 238–279.

Pompeiano, O., Wand, P. and Sontag, K.-H. (1975c) The sensitivity of Renshaw cells to velocity of sinusoidal stretches of the triceps surae muscle. *Arch. ital. Biol.*, 113: 280–294.

Prestige, M.C. (1966) Initial collaterals of motor axons within the spinal cord of the cat. *J. comp. Neurol.*, 126: 123–126.

Redman, S.J. and Lampard, D.G. (1967) Monosynaptic stochastic stimulation of spinal motoneurones in the cat. *Nature (Lond.)*, 216: 921–922.

Renshaw, B. (1941) Influence of discharge of motoneurons upon excitation of neighboring motoneurons. *J. Neurophysiol.*, 4: 167–183.

Renshaw, B. (1946) Central effects of centripetal impulses in axons of spinal ventral roots. *J. Neurophysiol.*, 9: 191–204.

Ross, H.-G., Cleveland, S. and Haase, J. (1972) Quantitative relation of Renshaw cell discharges to monosynaptic reflex height. *Pflügers Arch.*, 332: 73–79.

Ross, H.-G., Cleveland, S. and Haase, J. (1975) Contribution of single motoneurons to Renshaw cell activity. *Neurosci. Lett.*, 1: 105–108.

Ross, H.-G., Cleveland, S. and Haase, J. (1976) Quantitative relation between discharge frequencies of a Renshaw cell and an intracellular depolarized motoneuron. *Neurosci. Lett.*, 3: 129–132.

Ryall, R.W., Piercey, M.F., Polosa, C. and Goldfarb, J. (1972) Excitation of Renshaw cells in relation to orthodromic and antidromic excitation of motoneurons. *J. Neurophysiol.*, 35: 137–148.

Sasaki, K. (1963) Electrophysiological studies on oculomotor neurons of the cat. *Jap. J. Physiol.*, 13: 287–302.

Scheibel, M.E. and Scheibel, A.B. (1966) Spinal motoneurons, interneurons and Renshaw cells. A Golgi study. *Arch. ital. Biol.*, 104: 328–353.

Scheibel, M.E. and Scheibel, A.B. (1971) Inhibition and the Renshaw cell. A structural critique. *Brain Behav. Evol.*, 4: 53–93.

Stauffer, E.K., Watt, D.G.D., Taylor, A., Reinking, R.M. and Stuart, D.G. (1976) Analysis of muscle receptor connections by spike triggered averaging. 2. Spindle group II afferents, *J. Neurophysiol.*, 39: 1393–1402.

Stuart, D.G., Mosher, C.G., Gerlach, R.L. and Reinking, R.M. (1970) Selective activation of Ia afferents by transient muscle stretch. *Exp. Brain Res.*, 10: 477–487.

Szentágothai, J. (1967) Synaptic architecture of the spinal motoneuron. *Electroenceph. Clin. Neurophysiol.*, Suppl. 25: 4–19.

Tan, Ü. (1971) Changes in firing rates of extensor motoneurones caused by electrically increased spinal inputs. *Pflügers Arch.*, 326: 35–47.

Tan, Ü. (1972) The role of recurrent and presynaptic inhibition in the depression of tonic motoneuronal activity. *Pflügers Arch.*, 337: 229–239.

Tan, Ü. (1975) Post-tetanic changes in the discharge pattern of the extensor alpha motoneurones. *Pflügers Arch.*, 353: 43–57.

Tan, Ü., Yörükan, S. and Ridvanağaoğlu, A.Y. (1972) A quantitative analysis of the motoneuronal depression produced by increasing the stimulus parameters of afferent tetanization. *Pflügers Arch.*, 333: 240–257.

Thoden, U., Magherini, P.C. and Pompeiano, O. (1972) Evidence that presynaptic inhibition may decrease the autogenetic excitation caused by vibration of extensor muscles. *Arch. ital. Biol.*, 110: 90–116.

Wand, P., Pompeiano, O. and Fayein, N.A. (1977a) Response of different types of α-motoneurons to muscle vibration. *Proc. XXVII Int. Congr. Physiol. Sci., Paris*, 13: 800, no. 2380.

Wand, P., Pompeiano, O. and Fayein, N.A. (1977b) The relative sensitivity of α-motoneurons of different size to static stretch and vibration of the homonymous muscle. *Pflügers Arch.*, Suppl. 368: R35.

Wand, P., Pompeiano, O. and Fayein, N.A. (1979a) Impulse decoding process in extensor motoneurons during the vibration reflex. *Arch. ital. Biol.*, in press.

Wand, P., Pompeiano, O. and Fayein, N.A. (1979b) The relative sensitivity of different size motoneurons to orthodromic group Ia volleys caused by static stretch and vibration of extensor muscles. *Arch. ital. Biol.*, in press.

Watt, D.G.D., Stauffer, E.K., Taylor, A., Reinking, R.M. and Stuart, D.G. (1976) Analysis of muscle receptor connections. 1. Spindle primary and tendon organ afferents. *J. Neurophysiol.*, 39: 1375–1392.

Westbury, D.R. (1971) The response of α-motoneurones of the cat to sinusoidal movements of the muscles they innervate. *Brain Res.*, 25: 75–86.

Westbury, D.R. (1972) A study of stretch and vibration reflexes of the cat by intracellular recording from motoneurones. *J. Physiol. (Lond.)*, 226: 37–56.

The Role of Synaptic Organization in the Control of Motor Unit Activity during Movement

R. E. BURKE

Laboratory of Neural Control, National Institute of Neurological and Communicative Disorders and Stroke, National Institutes of Health, Bethesda, MD 20014 (U.S.A.)

Sherrington introduced many ideas now thoroughly incorporated into our models of CNS action, for example, the concept of neural control through the recruitment of neurons belonging to functionally defined populations. Sherrington also recognized that the action of a motoneuron is indivisibly linked with that of its innervated muscle fibers and coined the term "motor unit" to express this linkage. Thus, we usually think of movement control in terms of the recruitment and derecruitment of motor units belonging to various muscles, or "motor pools".

While working with Sherrington at Oxford, Denny-Brown (1929) described the process of motor unit recruitment in terms of relative "threshold grades" for motoneuron activation. He showed that the spectrum of threshold grades is related to the characteristics of the innervated muscle fibers and noted that the motoneurons with the lowest threshold grades usually innervate "red, slow" muscle fibers. He also showed, however, that different input systems can produce alterations in the sequence of threshold grades.

Over the past 50 years of continued study, Denny-Brown's description of the recruitment process has been confirmed repeatedly, with no essential modification. To summarize the present state of descriptive results, the spectrum of motoneuron threshold grades (from low to high) and the consequent order of motoneuron recruitment (from first recruited to last) are usually directly correlated with: 1) the amplitude of the muscle unit EMG potential (Denny-Brown and Pennybacker, 1938; Norris and Gasteiger, 1955; Olson et al., 1968); 2) the force output of the muscle unit (Burke, 1968b; Milner-Brown et al., 1973; Monster and Chan, 1977; Stephens and Usherwood, 1977); 3) the conduction velocity of the motoneuron axon (Henneman et al., 1965, 1974; Burke, 1968b; Clamann and Henneman, 1976; Freund et al., 1975); and 4) to the observed (Davis, 1971; in lobster ganglia) or inferred size of the motoneurons (Henneman et al., 1965, 1974; Burke, 1968b). Threshold grades are related inversely to: 1) muscle unit twitch contraction times (Burke, 1968b; Milner-Brown et al., 1973; Stephens and Usherwood, 1977); 2) muscle unit fatigue resistance (Stephens and Usherwood, 1977); and 3) the amplitude of functionally relevant synaptic potentials (Burke, 1968b).

Clues to an understanding of the factors underlying the process of motoneuron recruitment can presumably be found in these descriptions and two rather different hypotheses are of current interest. The first hypothesis assumes that the critical factor that produces variation in threshold grades is the presynaptic organization of synaptic input, i.e., the qualitative and quantitative distribution of synaptic effects among pool motoneurons. This hypothesis was implicit in Denny-Brown's 1929 paper and was made explicit in later studies of motoneuron firing probabilities (Lloyd and McIntyre, 1955; Rall and Hunt, 1956) and of synaptic potential amplitudes (Eccles et al., 1957; Burke, 1968a, b). In contrast, the second hypothesis assumes that threshold gradations result from properties intrinsic to the motoneurons themselves. This idea was implicit in the suggestion of Granit and colleagues that motoneurons could be divided into small 'tonic' and larger 'phasic' categories, but it received major emphasis in interpretations drawn from studies of Henneman and coworkers (Henneman et al., 1965, 1974), which suggested that threshold grades are fixed in strict relation to motoneuron size. The second hypothesis is thus often referred to as the "size principle".

Both hypotheses are compatible with most of the available descriptions of motor unit recruitment in animal and human muscles. There is one point, however, where the two lead to opposite predictions. If threshold grades are intrinsically specified by motoneuron properties, then they should be invariant despite differences in input drive (see Henneman et al., 1965, 1974). On the other hand, if threshold gradation depends mainly on synaptic organization, alternative recruitment patterns or differential control of particular motoneuron groups could result provided that the relevant input systems distribute to pool motoneurons with differing patterns.

Clear and unequivocal demonstration of differential bias within a given motor pool, or of major shifts in recruitment sequences, is technically difficult and may require rather special conditions but such evidence is available. For example, threshold scaling in cat hindlimb motoneuron pools can differ when comparing activation by supraspinal as opposed to primary afferent inputs (Denny-Brown, 1929; Kernell and Sjöholm, 1975; Clamann and Kulkulka, 1977), or in comparisons of responses to homonymous and heteronymous monosynaptic input (Lloyd and McIntyre, 1955), or when input from distal skin regions is superimposed on proprioceptive reflexes (Kanda et al., 1977). Results analogous to this last study have recently been reported in human subjects (Stephens et al., 1978). Whether or not certain usually high threshold motor units can respond preferentially during rapid voluntary movements in humans is a topic of some controversy (see Desmedt and Godaux, 1977), but a recent study by Grimby and Hannerz (1977) suggests that this is possible under some conditions (see also Borg et al., 1978). Finally, Rall and Hunt (1956) have demonstrated that the firing probabilities of individual motoneurons (i.e., their threshold grades) vary during monosynaptic reflexes in a manner independent of the fluctuations in overall pool responses, implying shifting threshold sequences.

These results, taken together, strongly suggest that threshold grades are not immutably identified with individual motoneurons, as would be the case if the controlling factors were entirely postsynaptic (i.e., cell size, input resistance, membrane properties, etc.). Rather, it can be argued that the presynaptic organization of inputs to the motor pool is the major factor that determines the ordering of thresholds and recruitment (see Burke, 1973).

One experimental index of synaptic organization is the efficacy of synaptic input to motoneurons from a given functionally-defined afferent system, as measured, for example, by the amplitude of the monosynaptic EPSPs produced by electrical stimulation of all the group Ia afferents from particular muscle nerves (Eccles et al., 1957; Burke, 1968a, b; Burke et al., 1976). The responsiveness of cat medial gastrocnemius (MG) motor units during decerebrate stretch reflexes is correlated with group Ia synaptic efficacy defined in this way (Burke, 1968b). There is evidence that the ordering of threshold grades during stretch reflexes is similar to that during many other types of motor actions (Denny-Brown, 1929; Henneman et al., 1965, 1974; Burke and Edgerton, 1975). Thus, we have found it of interest to examine two points: 1) the interrelation between group Ia EPSP amplitudes and other characteristics of motor units; and 2) the input-output relations of the same motor unit pool given that recruitment may occur in strict accord with Ia EPSP amplitudes (Burke et al., 1976).

The 3-dimensional graph in Fig. 1 shows the interrelation between homonymous (MG) group Ia EPSP amplitudes (right horizontal abscissa), muscle unit tetanic tension (vertical ordinate), motor axon conduction velocity (left horizontal ordinate), and motor unit type (symbols; defined by the mechanical responses of the muscle unit portions as discussed in Burke et al., 1973), in a population of 93 MG motor units (data from Burke et al., 1976). There is a linear negative correlation between EPSP amplitude and tetanic force (see also Fig. 3 in Burke et al., 1976). The correlation between EPSP amplitude and conduction velocity of the motor axons is more complex ("L" shaped) and no correlation can be seen among the fast twitch units (types FF, F(int) and FR considered together), despite a 10-fold range in EPSP size. Even

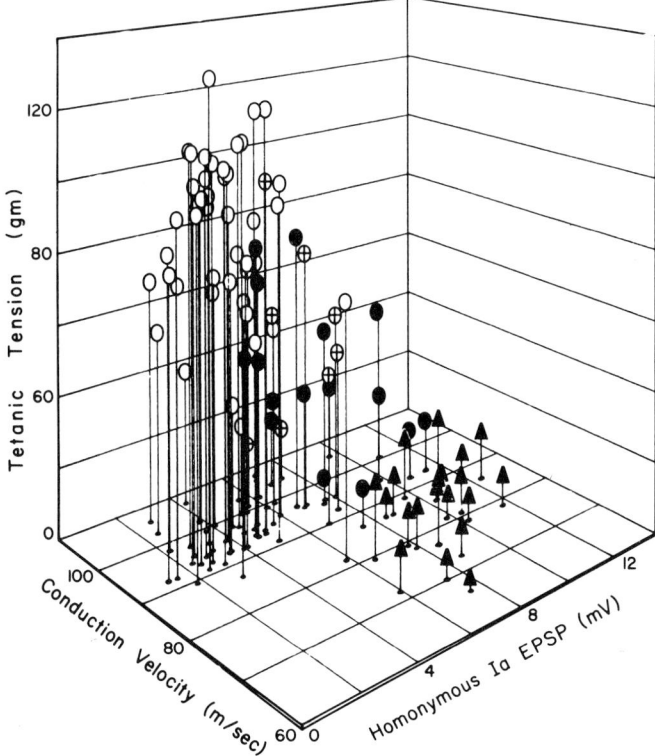

Fig. 1. 3-dimensional diagram showing the interrelation between peak amplitude of homonymous group Ia EPSPs (right horizontal abscissa), motor axon conduction velocities (left horizontal ordinate) and tetanic tension outputs (vertical ordinate) in a sample of 93 MG motor units pooled from several experiments. Each point represents a different unit, with unit type denoted by different symbols: ○ = FF; ⊕ = F(int); ● = FR; ▲ = S. See text for discussion. (From Burke et al., 1976.)

though axonal conduction velocity is a relatively inaccurate predictor of motoneuron size (Barrett and Crill, 1974) and input resistance (Burke, 1968), these data suggest that variations in motoneuron size cannot explain the range of synaptic amplitudes observed. The resistance of unit muscle fibers to fatigue is ordered as follows: type S > type FR > type F(int) > type FF (Burke et al., 1973). The ordering of Ia EPSP amplitudes is the same, so that there is a clear positive correlation between EPSP size and muscle unit fatigue resistance. If MG unit recruitment occurs in accordance with Ia EPSP amplitudes, then type S units would, on the average, have the lowest threshold grades and be recruited first while the type FF units should have the highest threshold grades and be recruited last. In order to examine this generalization in more detail, the data in Fig. 1 can be rearranged into a recruitment model.

If we assume that: 1) the MG unit sample in Fig. 1 is representative of the MG pool as a whole; 2) that MG units are recruited strictly in order of decreasing Ia EPSP amplitude; and 3) that each unit recruited adds its full tetanic tension to the cumulative total force produced by all previously recruited units, then a 'recruitment diagram' can be constructed (Fig. 2). These assumptions, although much oversimplified, can nevertheless be defended as reasonable approximations to the actual situation (see discussion in Burke et al., 1976 and Walmsley et al., 1978). The 3-dimensional diagram in Fig. 2 shows the cumulative MG output force (vertical ordinate) that would be developed as units are sequentially recruited (right horizontal

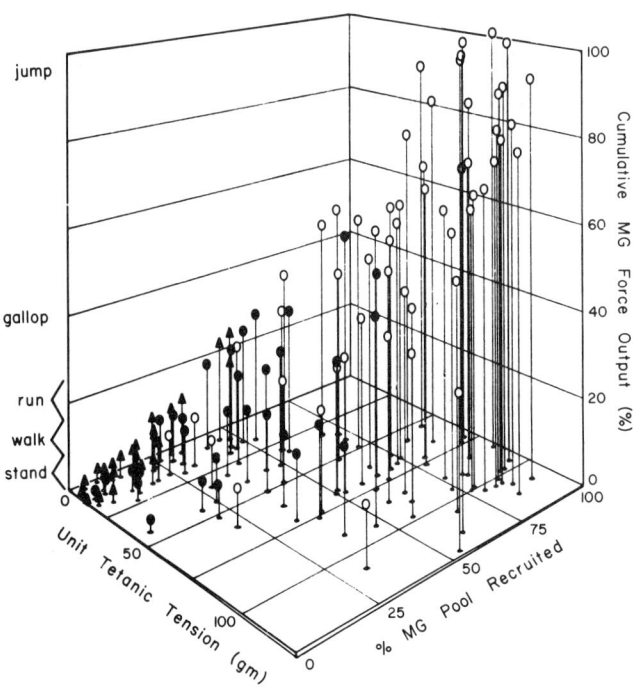

Fig. 2. 3-dimensional diagram, based mainly on data in Fig. 1, showing the relation between MG pool recruitment (right horizontal abscissa; scaled incrementally in %), individual motor unit tetanic force output (left horizontal ordinate) and cumulative MG force output expected as each unit is recruited in sequence (vertical ordinate). See text for further description and discussion of underlying assumptions.

abscissa) according to decreasing Ia EPSP size plotted incrementally. The left horizontal ordinate shows the tetanic tension developed by each individual motor unit. Motor unit types are denoted by the same symbols used in Fig. 1 except that, for simplicity, both FR and F(int) units are indicated by filled ovals.

If the underlying assumptions are accepted, Fig. 2 indicates that MG recruitment should begin largely with activation of type S units, with some admixture of type FR. Both groups of units produce individually small tetanic tensions (left ordinate), and thus the cumulative output force (vertical ordinate) produced by the first-recruited 25% of the MG pool is less than 10% of the maximum force possible from the entire MG. Units in the second 25% of the recruitment sequence are predominantly type FR units with a wider range of individual force outputs, so that, at 50% recruitment, the cumulative output force is about 25% of maximum. Note that the first-recruited half of the MG unit pool consists of all the type S units and most (87%) of the fatigue-resistant FR and F(int) units in the data sample, with a few, mostly small force, fatigable type FF units. In contrast, the higher threshold half of the pool consists mainly of FF units that produce, on the average, rather large individual tetanic forces.

We have recently recorded the force generated by the whole MG muscle in intact, freely moving cats (Walmsley et al., 1978). In these animals (of about the same body weight as the cats used to produce the data of Figs. 1 and 2), quiet quadrupedal standing required only about 6–7% of maximum MG force (400–500 g, from MG muscles that produced fused tetanic forces of 8–9 kg). This 'postural' output force could be supplied by the first-recruited 25% of the MG pool. Treadmill locomotion was associated with greater peak MG forces, up to about 25% of cumulative maximum during fast running (3 m/sec). The approximate ranges of peak forces during standing, walking and running are indicated on the vertical ordinate at the left. It is of interest that the entire range of locomotion may require recruitment of only about half of the MG unit population, consisting almost entirely of fatigue resistant motor units of both slow and fast twitch types. Only during gallop and vertical jumping were MG forces greater than 25% maximum observed, but vertical jumps of over 1 m produced total MG forces of 8–9 kg, i.e., the level predicted for 100% recruitment. Such output levels must involve nearly full recruitment of FF units in the second half of the recruitment sequence and it is of interest that this half can account for about 75% of the total MG force available.

The recruitment sequence suggested in Fig. 2 is entirely consistent with all of the available evidence about normal recruitment in both animal and human muscles (see Burke and Edgerton, 1975 for references). It is also essentially the same as that envisioned by the 'size principle' (see Henneman and Olson, 1965), although strict application of the size principle as the recruitment rule for Fig. 2 would give indiscriminant recruitment of FR and FF units. The use of a recruitment rule based on synaptic organization, in contrast, produces a much more intuitively satisfying pattern. It is obviously advantageous to use slowly contracting, very fatigue-resistant motor units for postural actions such as quiet standing. It seems also advantageous, both mechanically and energetically, to use fatigue resistant but larger and more rapidly contracting motor units (i.e., type FR) for locomotion. Predators such as the cat must also be able to gallop and jump vigorously but need to do so infrequently, and these requirements seem ideally met by the type FF units. These functional interrelations lend weight to the view that synaptic organization is the key to understanding the control of motor unit recruitment.

SUMMARY

Currently available evidence suggests that the qualitative and quantitative organization of synaptic inputs to motoneuron pools is the critical factor that controls motor unit recruitment patterns. A recruitment model that incorporates this hypothesis, based on data from cat medial gastrocnemius (MG) motor units, is compared with actual MG forces generated during free movement in intact animals.

REFERENCES

Barrett, J.N. and Crill, W.E. (1974) Specific membrane properties of cat motoneurones. *J. Physiol. (Lond.)*, 239: 301–324.

Borg, J., Grimby, L. and Hannerz, J. (1978) Axonal conduction velocity and voluntary discharge properties of individual short toe extensor motor units in man. *J. Physiol (Lond.)*, 277; 143–152.

Burke, R.E. (1968a) Group Ia synaptic input to fast and slow twitch motor units of cat triceps surae. *J. Physiol. (Lond.)*, 196: 605–630.

Burke, R.E. (1968b) Firing patterns of gastrocnemius motor units in the decerebrate cat. *J. Physiol. (Lond.)*, 196: 631–654.

Burke, R.E. (1973) On the central nervous system control of fast and slow twitch motor units. In *New Developments in Electromyography and Clinical Neurophysiology* J.E. Desmedt (Ed.) Karger, Basel, pp. 69–94.

Burke, R.E. and Edgerton, V.R. (1975) Motor unit properties and selective involvement in movement. In *Exercise and Sport Sciences Reviews, Vol. 3*, J.H. Wilmore and J.F. Keogh (Eds.), Academic Press, New York, pp. 31–81.

Burke, R.E., Levine, D.N., Tsairis, P. and Zajac, F.E. (1973) Physiological types and histochemical profiles in motor units of the cat gastrocnemius. *J. Physiol. (Lond.)*, 234: 723–748.

Burke, R.E., Rymer, W.Z., Walsh Jr. J.V., (1976) Relative strength of synaptic input from short-latency pathways to motor units of defined type in cat medial gastrocnemius. *J. Neurophysiol.*, 39: 447–458.

Clamann, H.P. and Henneman, E. (1976) Electrical measurement of axon diameter and its use in relating motoneuron size to critical firing level. *J. Neurophysiol.*, 39: 844–851.

Clamann, H.P. and Kulkulka, C.G. (1977) Reversals of recruitment order in medial gastrocnemius produced by stimulation of Deiters nucleus. *Neurosci. Abstr.*, 3: 269, No. 853.

Davis, W.J. (1971) Functional significance of motoneuron size and soma position in swimmeret system of the lobster. *J. Neurophysiol.*, 34: 274–288.

Denny-Brown, D. (1929) On the nature of postural reflexes. *Proc. roy. Soc. B*, 104: 252–301.

Denny-Brown, D. and Pennybacker, J.B. (1938) Fibrillation and fasciculation in voluntary muscle. *Brain*, 61: 311–344.

Desmedt, J.E. and Godaux, E. (1977) Ballistic contractions in man: characteristic recruitment pattern of single motor units of the tibialis anterior muscle. *J. Physiol. (Lond.)*, 264: 673–694.

Eccles, J.C., Eccles, R.M. and Lundberg, A. (1957) The convergence of monosynaptic excitatory afferents on to many different species of alpha motoneurones. *J. Physiol. (Lond.)*, 137: 22–50.

Freund, H.J., Büdingen, H.J. and Dietz, V. (1975) Activity of single motor units from human forearm muscles during voluntary isometric contractions. *J. Neurophysiol.*, 38: 933–946.

Granit, R., Henatsch, H.-D. and Steg, G. (1956) Tonic and phasic ventral horn cells differentiated by post-tetanic potentiation in cat extensors. *Acta Physiol. Scand.*, 37: 114–126.

Grimby, L. and Hannerz, J. (1977) Firing rate and recruitment order of toe extensor motor units in different modes of voluntary contraction. *J. Physiol. (Lond.)*, 264: 865–879.

Henneman, E. and Olson, C.B. (1956) Relations between structure and function in the design of skeletal muscles. *J. Neurophysiol.*, 28: 581–598.

Henneman, E., Somjen, G. and Carpenter, D.O. (1965) Functional significance of cell size in spinal motoneurons. *J. Neurophysiol.*, 28: 560–580.

Henneman, E., Clamann, H.P., Gillies, J.D. and Skinner, R.D. (1974) Rank order of motoneurons within a pool: law of combination. *J. Neurophysiol.*, 34: 1338–1349.

Kanda, K., Burke, R.E. and Walmsley, B. (1977) Differential control of fast and slow twitch motor units in the decerebrate cat. *Exp. Brain Res.*, 29: 57–74.

Kernell, D. and Sjöholm, H. (1975) Recruitment and firing rate modulation of motor unit tension in a small muscle of the cat's foot. *Brain Res.*, 98: 57–72.

Lloyd, D.P.C. and McIntyre, A.K. (1955) Monosynaptic reflex responses of individual motoneurons. *J. gen. Physiol.* 38: 771–787.

Milner-Brown, H.S., Stein, R.B. and Yemm, R. (1973) The orderly recruitment of human motor units during voluntary isometric contractions. *J. Physiol. (Lond.)*, 230: 359–370.

Monster, A.W. and Chan, H. (1977) Isometric force production by motor units of extensor digitorum communis muscle in man. *J. Neurophysiol.*, 40: 1432–1443.

Norris, F.H. and Gasteiger, E.L. (1955) Action potentials of single motor units in normal muscle. *Electroenceph. clin. Neurophysiol.*, 7: 115–126.

Olson, C.B., Carpenter, D.O. and Henneman, E. (1968) Orderly recruitment of muscle action potentials. *Arch. Neurol.*, 19: 591–597.

Rall, W. and Hunt, C.C. (1956) Analysis of reflex variability in terms of partially correlated excitability fluctuations in a population of motoneurons. *J. gen. Physiol.*, 39: 397–422.

Stephens, J.A., Garnett, R. and Buller, N.P. (1978) Reversal of recruitment order of single motor units produced by cutaneous stimulation during voluntary muscle contraction in man. *Nature (Lond.)*, 272: 362–364.

Stephens, J.A. and Usherwood, T.P. (1977) The mechanical properties of human motor units with special reference to their fatiguability and recruitment threshold. *Brain Res.*, 125: 91–97.

Walmsley, B., Hodgson, J.A. and Burke, R.E. (1978) The forces produced by medial gastrocnemius and soleus muscles during locomotion in freely moving cats. *J. Neurophysiol.*, 41: 1203–1216.

Adaptive Properties of the Myotatic Feedback

C.A. TERZUOLO, J.F. SOECHTING and J.R. DUFRESNE

Laboratory of Neurophysiology, University of Minnesota Medical School, Minneapolis, MN 55455 (U.S.A.)

Confronted with the problem of "understanding" how the stretch reflex is utilized in the course of different goal-oriented motor tasks, chronic intact preparations and human subjects have been frequently used in recent years by many investigators. Our contribution to this effort has been to provide quantitative data in a form adequate to permit deductions about the operational characteristics of the system during different tasks.

In this paper we shall begin by reviewing briefly the premises upon which the approach is predicated and defining the terminology introduced. We will then summarize some of the most recent data we have obtained on the stated problem.

WHY LINEAR SYSTEMS ANALYSIS AND MYOTATIC FEEDBACK

In man, and under most experimental conditions, direct measurements are restricted to the EMG activity and the muscle length, the latter in terms of angular position. In attempting to relate these two variables under a variety of conditions, one implicitly assumes that the dynamic relationship between muscle length and receptor output is relatively invariant of the exact form and magnitude of the forces which are responsible for changing the muscle length. The question then is: is this approximation justifiable?

In this context, it is well known that the primary endings of the muscle spindles behave linearly only within a restricted range of input amplitudes and frequencies (cf. Chen and Poppele, 1978; Hulliger et al., 1977; Matthews, 1972). However, the following considerations are pertinent. First of all, by avoiding abrupt transitions in muscle length one can reasonably expect to prevent in large measure the engagement of some non-linear properties of muscle spindle receptors. This is not too large a restriction since under most physiological conditions changes of muscle length are quite slow, because of inertia. For instance, the spectral analysis of the angular displacement of the human forearm, during even the fastest possible movements and their intentional arrest, shows that the amplitude at frequencies above 6 Hz is rather small, and most of the power resides at frequencies below 5 Hz. Only in the case of externally applied perturbations, such as when the reflex is elicited by tapping directly the tendon of the muscle, the input contains a substantial amount of power at higher frequencies. In this condition the operation of the reflex can obviously be dominated by the engagement of non-linear properties of the primary endings, which in turn bring about non-linear behavior at subsequent stages (cf. Burke et al., 1976; Granit, 1970; Robles and Soechting, 1979). This situation may also prevail when single torque pulses (and, more generally, input functions which produce sharp transitions in muscle length) are applied to the human forearm. All one can say a priori, however, is that a judicious choice of imposed variables may presumably permit the description of the behavior of the system within the limits of its quasi-linear behavior. Thus, it goes without saying that only by making such an attempt, can the question stated above be answered.

Secondly, the point needs to be stressed that the assumption of quasi-linear behavior actually is not as restrictive as it may first seem. It requires only that the reflex change in motor output behave reasonably linearly when external variables, such as the amplitude or duration of an applied perturbation, are changed. It does not require that the reflex output be independent of such intrinsic variables as the particular motor task and mean muscle length or tension. In fact, the approach permits the identification of the manner in which the reflex response depends on such factors. These internal variables are here defined as "operating points" of the system.

We shall now turn to identify the system we are dealing with. Given the approach outlined above, it is not possible in intact preparations to deduce the relative contribution by each type of muscle afferent to the measured change in EMG activity. Consequently an operational terminology should be adopted to describe the input-output relation between the overall input from muscle receptors – expressed in terms of angular position and its lower order derivatives (i.e., the kinematic variables) – and the motor output changes. The term "myotatic feedback" or "loop" will be used here to denote this relationship. It is meant to include also the central processing of the sensory input data.

One last point. Given the limitations of the approach, the mechanisms whereby the gain and dynamic properties of the loop are modified can only be inferred indirectly. The actions by static and dynamic γ-motoneurons upon the muscle spindle receptors (Chen and Poppele, 1978; Hulliger et al., 1977; Matthews, 1972) are among the possible mechanisms. Alone they are sufficient to point out the presence of adaptive properties whereby the operational characteristics of the loop can vary in relation to the constraints associated with different operating points. Therefore our approach has been to utilize operating points associated with some identifiable constraints.

POSITION CONTROL

Measurements of loop properties can be most simply made at a stable operating point. Two such points are specified by the instructions: "resist" and "do not resist" an externally applied perturbation. In other words, we wish to establish how the internal state associated with the intentional attempt to maintain a constant position influences the operational characteristics of the myotatic loop, this with respect to a state which excludes any action directed to control intentionally the position of the body segment to which the perturbation is applied. Note that since the work of Hammond (1960) it is already known that a large difference exists in the magnitude of the reflex motor output between these two conditions. The question, when posed in terms of a control system, is the following: how do the parameters of the myotatic feedback, namely the dependency of the motor output on each of the kinematic variables, change?

Pseudo-random torque sequences, that is, trains of flexion and extension torque pulses with random-like properties were applied to the forearm of normal human subjects by means of a torque motor. This choice of input was dictated mainly by the following reason: we wanted to provide varying levels of initial conditions, including the background activity of both muscle receptors and α-motoneurons. In this way one can reasonably expect the receptor and motor outputs to be modulated more smoothly. Since the approach allows the use of averaging techniques, one can eventually extract – on a statistical basis – the parameters of the feedback under a range of conditions which establish their general validity. The reader is referred to the original paper (Dufresne et al., 1978) for all technical details. Only the main findings and conclusions will be mentioned here.

The upper row of Fig. 1 shows the results obtained by cross-correlating the current applied to the torque motor (taken as the input) and each of the three output variables measured, namely angular position and EMG activities of biceps and triceps muscles.

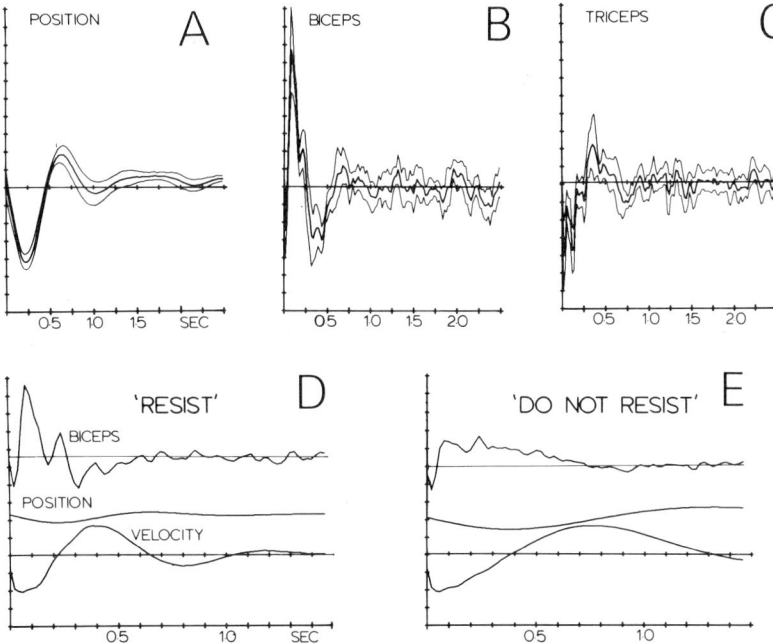

Fig. 1. A, B, and C are the impulse response functions for the angular displacement (A), biceps EMG activity (B) and triceps EMG activity (C). The heavy lines are the means; the lightest ones denote ± 1 SD. These impulse responses were obtained from ensemble averages of three different torque sequences at three different amplitudes (8, 16 and 24 Nm), when the task was to resist the perturbation. In D and E the impulse response for the biceps for the "resist" and "do not resist" tasks are compared to the impulse responses of angular position and its first derivative. Note the damping of the movement when the task is to resist. Also, the peak of the EMG impulse response coincides with the peak velocity in D while in E it is shifted in time toward the peak position. Scaling of the impulse response functions (per vertical division): A) 7×10^{-4} rad/Nm; B) $0.12\, \mu$V/Nm; C) $0.094\, \mu$V/Nm; D) biceps: $0.25\, \mu$V/Nm, position: 6×10^{-3} rad/Nm, velocity: 1.2×10^{-2} rad/Nm/sec; E) biceps: $0.25\, \mu$V/Nm, position: 1.2×10^{-2} rad/Nm, velocity: 2.5×10^{-2} rad/Nm/sec.

The three "impulse responses" represent the average output, plus and minus one standard deviation, to an average torque pulse. This mean is independent of pulse amplitude and the specific structure of the sequence, that is, the initial conditions. Since the variances are rather small one can already suggest that each of the outputs are reasonably independent of the variables just stated. The results of a Fourier analysis are in agreement with this conclusion over the frequency range (0.5 – 8 Hz) which is functionally significant for naturally occurring movements (see above). Thus it becomes appropriate to inquire about the dependence of the motor output on each of the kinematic variables using a linear model.

In brief, by comparing in Fig. 1D the changes in EMG activity with the kinematic variables one can see that they follow rather closely the velocity parameter – this when the subject attempts to resist the applied perturbation. Instead, the maximum of the EMG activity is delayed and occurs closer to the peak displacement when the

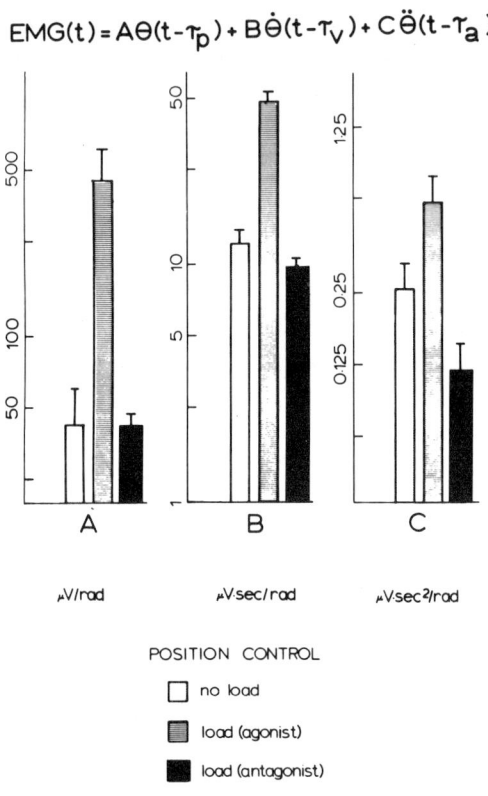

Fig. 2. Parametric model used to relate EMG density to kinematic variables and values of each coefficient, for different tasks. Data are for the biceps muscle of one subject. The scales are logarithmic and have been scaled to approximate the relative amplitude of the kinematic variables for the three conditions considered. Therefore the relative weight of the contribution by each coefficient (A, B and C) can be inferred. Bars denote 1 SD. Note the coefficient B is larger than coefficients A and C for all conditions. Also, all coefficients are larger in the loaded agonist muscle. As for a comparison between coefficients A and B, the percent increase of A when the muscle is loaded is much greater than that of B. Note also the difference in value of the coefficient A between the loaded and unloaded conditions, and the fact that the ratio between A and B for the loaded condition versus the unloaded condition is such that the velocity parameter is much more prominent in the antagonist than in the agonist. The behavior of the coefficients shown is representative of all data available.

subject does not resist the perturbation (Fig. 1E). This difference in the parameters of the feedback between the two operating points was quantitated by using a linear parametric model (Fig. 2) to relate position (θ), velocity ($\dot{\theta}$), and acceleration ($\ddot{\theta}$) to the EMG density. A time delay for each parameter is included. The coefficients A, B, and C express the amount by which the EMG density depends on each kinematic variable. Fig. 2 shows the values of each coefficient for those conditions to be considered here and in which the relative magnitudes of the kinematic variables are approximately the same. For the "do not resist" task instead, the position variable becomes much larger than the velocity variable. In this case the feedback operates mostly as a position feedback (Fig. 1E). Instead the velocity and acceleration parameters (B and C) are clearly dominant in all the conditions shown in Fig. 2, in which the subject intentionally resists the applied perturbation.

Note that it was previously shown that during the intentional arrest of an ongoing fast movement the myotatic feedback also operates as a delayed velocity feedback. This is not surprising, the aim being the same whether the subject is asked to oppose, i.e., to arrest, a displacement produced by an externally applied force or to arrest a previously initiated movement. The operational characteristics are instead quite different when the subject is asked either to accelerate maximally a movement or to oppose a steady torque. In the latter case the data of Fig. 2 shows that while there is an increase in gain for the muscle which is being loaded, there is also a relative decrease of the dependence on the velocity parameter. As for the muscle which is being unloaded (antagonist), the absolute gain of the loop is much less but the dependence of the motor output on the velocity is much higher than its dependence on the position parameter. Consequently the combined action by agonist and antagonist under asymmetrical loading is such that slow drifts (due to the applied load) are most effectively controlled by the agonist.

Fig. 3 illustrates for two different sequences, the extent to which the model duplicates the experimental data. It is quite apparent that the major features of the EMG outputs are predicted by the model. This lends support to the estimates for the parameters of the feedback, as given in Fig. 2.

We come now to consider the anatomo-physiological substrates of myotatic feedback, as defined above, and its adaptive properties. The first point which should

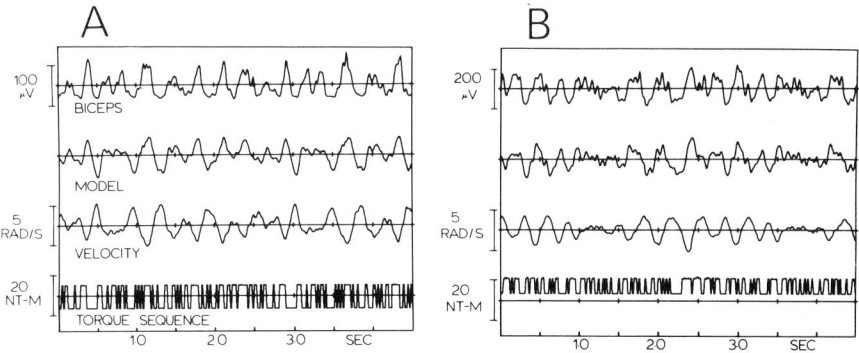

Fig. 3. Comparison of experimental data to model predictions. Two examples (one for each of two torque sequences) of the predictions of the linear parametric model (see Fig. 2) are shown together with the experimental data (ensemble averages). In both A and B the task was to "resist" the perturbation, but in B the muscle (biceps) was loaded by a steady torque. In each case the model fits the data to within 20–25% (computed as the squared residual error/the variance of the experimental data).

be stressed is the following. The velocity dependency of the motor output when the task is to resist the perturbation can be accounted for in terms of a monosynaptic reflex since the delay is on the average 20 ± 5 msec. The acceleration and position components instead have much longer delays. Even so the coupling between velocity and motor output has been shown to require the presence of cerebellar activities, both in human subjects (Soechting, 1973; Terzuolo and Viviani, 1973; Viviani and Terzuolo, 1973) and experimental animals (Soechting et al., 1976). Data are also available which demonstrate that the velocity and acceleration parameters are represented in the activity of interpositus and rubral neurons during physiological movements (Burton and Onoda, 1978; Soechting et al., 1978). Moreover, the acceleration component of the motor output during some type of movements was shown to be affected by lesions of the thalamic nuclei which convey somato-sensory input data and cerebellar outflow to the cortex (Ranish and Soechting, 1976).

Because of the known actions of rubral neurons upon static and dynamic γ-motoneurons (Appelberg and Jeneskog, 1972; Hongo et al., 1969), speculations are plausible about the involvement of γ-motoneurons in bringing about some of the adaptive changes of the myotatic loop we have encountered. In all instances the simplest mechanism consistent with the data is that of selective biasing actions upon either the static or dynamic γ-motoneurons of both agonist and antagonist muscles. Modification of segmental reflex mechanisms by actions exerted upon interneuronal pools are also likely (Lundberg, 1975). More specifically, given the above conclusion about the velocity parameter, the following mechanisms can be suggested to be responsible for the changes in operational characteristics described above:

1. When the task is to arrest a movement or to maintain the position, a bias is placed upon the dynamic γ-motoneurons whereby the velocity sensitivity of the primary endings is increased (cf. Matthews, 1972), thus shifting the operation of the loop toward the velocity parameter. The functional significance of this shift is too obvious to be discussed (cf. Matthews, 1972).

2. Intentional actions directed either to accelerate a motion or to oppose a steady force, are accompanied by a relative decrease of the dependence on the velocity parameter in the agonist loop. This is most likely due to increase in γ-static activity which Vallbö (1970) has observed, in human subjects, during isometric contraction. The increase in gain as a function of the mean EMG activity would instead result from the fact that the activity of a larger number of motor units can be modulated when more of them are active.

Short of considering in detail the data pertinent to other motor tasks, only the more general conclusions can be presented here. These are the following:

i) The constraints associated with different operating points are met by selectively adapting the operational characteristics of the loop in both agonist and antagonist.

ii) These changes in operational characteristics (gain and dynamics) are controlled independently of the level of α-motoneuron activity.

iii) They are determined centrally, being part of the operating point as defined by the task.

SUMMARY

Relationships between the overall input from muscle receptors, expressed in terms of angular position and its low order derivatives, and motor output changes during externally applied perturbations were quantitated using linear systems analysis techniques. These relationships are denoted by the term "myotatic feedback" and were defined at specified operating points of the system. The operating points were characterized, for example, by the instruction given the subject and the mean level of force produced by the perturbation. It was found that the myotatic feedback could be adequately characterized by a simple linear relation between angular position and its derivatives, the importance of the parameters depending on the operating point. These findings are discussed in relation to other motor tasks and the possible anatomo-physiological substrates of the observed behavior.

ACKNOWLEDGEMENTS

This work was supported by U.S. Public Health Service Grant NS-02567. Computer facilities were made available by the Air Force Office of Scientific Research, AFSC (Grant AFOSR-1221).

REFERENCES

Appelberg, B. and Jeneskog, T. (1972) Mesencephalic fusimotor control. *Exp. Brain Res.*, 15: 97–112.
Burke, R.E., Rudomin, P. and Zajac, F.E. (1976) The effect of activation history on tension production by individual muscle units. *Brain Res.*, 109: 515–530.
Burton, J.E. and Onoda, N. (1978) Comparison of unit activity in interpositus and red nuclei during intentional movement. *Brain Res.*, 152: 41–63.
Chen, W.J. and Poppele, R.E. (1978) Small signal analysis of the response of mammalian muscle spindles with fusimotor stimulation and a comparison with large signal responses. *J. Neurophysiol.*, 41: 15–27.
Dufresne, J.R., Soechting, J.F. and Terzuolo, C.A. (1978) EMG response to pseudo-random torque disturbances of human forearm position. *Neuroscience*, 3: 1213–1226.
Granit, R. (1970) *The Basis of Motor Control*. Academic Press, London.
Hammond, P.H. (1960) An experimental study of servo action in human muscular control. *Proc., 3rd Int. Confer. med. Electron.*, 190–199.
Hongo, T., Jankowska, E. and Lundberg, A. (1969) The rubrospinal tract. 2. Facilitation of interneuronal transmission in reflex paths to motoneurones. *Exp. Brain Res.*, 7: 365–391.
Hulliger, M., Matthews, P.B.C. and Noth, J. (1977) Static and dynamic fusimotor action on the response of Ia fibers to low frequency sinusoidal stretching of widely ranging amplitudes. *J. Physiol. (Lond.)*, 267: 811–838.
Lundberg, A. (1975) Control of spinal mechanisms from the brain. In *The Nervous System, Vol. 1.* D.B. Tower (Ed.) Raven Press, New York, pp. 253–265.
Matthews, P.B.C. (1972) *Mammalian Muscle Receptors and their Central Actions*. Camelot Press, London.
Ranish, N.A. and Soechting, J.F. (1976) Studies on the control of some simple motor tasks. Effects of thalamic and red nuclei lesions. *Brain Res.*, 102: 339–345.
Robles, S.S. and Soechting, J.F. (1979) Dynamic properties of cat tenuissimus muscle. *Biol. cybern.*
Soechting, J.F. (1973) Modeling of a simple motor task in man. Motor output dependence on sensory inputs. *Kybernetik*, 14: 25–34.
Soechting, J.F., Burton, J.E. and Onoda, N. (1978) Relationships between sensory input, motor output and unit activity in interpositus and red nuclei during intentional movement. *Brain Res.*, 152: 65–79.
Soechting, J.F., Ranish, N.A., Palminteri, R. and Terzuolo, C.A. (1976) Changes in a motor pattern following cerebellar and olivary lesions in the squirrel monkey. *Brain Res.*, 105: 21–44.
Terzuolo, C.A. and Viviani, P. (1973) Parameters of motion and EMG activities during some simple motor tasks in normal subjects and cerebellar patients. In *The Cerebellum, Epilepsy and Behavior*. I.S. Cooper, M. Riklan and R.S. Snider (Eds.) Plenum Press, New York, pp. 173–215.
Vallbö, A.B. (1970) Discharge patterns in human muscle spindle afferents during isometric voluntary contractions. *Acta physiol. scand.*, 80: 552–560.
Viviani, P. and Terzuolo, C.A. (1973) Modeling of a simple motor task in man: intentional arrest of an ongoing movement. *Kybernetik*, 14: 35–62.

B

Proprioceptive Control of Supraspinal Structures and
Supraspinal Influences on the Spinal Cord

Information Carried by the Spinocerebellar Paths

C.-F. EKEROT, B. LARSON and O. OSCARSSON

Institute of Physiology, University of Lund, S-223 62 Lund (Sweden)

About twenty spinocerebellar paths terminating as mossy fibres or climbing fibres in the anterior lobe have been defined. Among these paths a minority carry information primarily about peripheral events. The best known and only quite convincing examples are the dorsal spinocerebellar tract (DSCT) and its forelimb equivalent, the cuneocerebellar tract (CCT) (Oscarsson, 1973). These two tracts contain components activated monosynaptically from afferents of muscle spindles, tendon organs, joint receptors and different types of cutaneous receptors (Fig. 1A). The receptive fields are restricted and often small. The modality and space specific information carried by these tracts is well suited for signalling peripheral events.

The other spinocerebellar paths have a complex organization and the majority of them appear to carry information about interneuronal activity signalling the "internal state" of lower motor centres. These centres would be reflex arcs, links in descending motor paths, or collections of neurones responsible for motor patterns like stepping and scratching. More likely, these centres would perform a combination of these functions. The segmental connections of the ascending paths might be organized as shown in the alternative diagrams of Fig. 1B. In diagram I the ascending path is activated or inhibited by the interneurones forming the motor centre, whereas in diagram II the ascending path is formed by axon collaterals from these interneurones.

The mossy fibre paths provide instructive examples of both kinds of segmental organization shown in Fig. 1B. An example of the former kind (diagram I) is supplied by the ventral spinocerebellar tract (VSCT), which has been discussed in recent papers (Lundberg, 1971; Lindström, 1973; Fu et al., 1977). The VSCT carries information about the activity evoked in segmental reflex arcs by different descending paths like the corticospinal, rubrospinal and vestibulospinal tracts and by different segmental paths like the muscle spindle and tendon organ afferents and the Renshaw cells. During locomotion and scratching the VSCT neurones are rhythmically active. The rhythmic activity is largely independent of the peripheral input and descending paths indicating that it signals activity in the spinal mechanisms generating the motor patterns (Shik and Orlovsky, 1976; Arshavsky et al., 1978b). It is unknown how the pattern generator is related to the reflex arcs mentioned above.

A good example of the organization in diagram II of Fig. 1B is provided by a path recently investigated by Lundberg and his collaborators (Illert et al., 1976a, b, 1977; Illert and Tanaka, 1978; Illert and Lundberg, 1978). The main connections are shown in Fig. 2. Excitation of forelimb motoneurones from the brain is relayed in the 3rd and

INFORMATION FORWARDED

A. Receptor activity signalling peripheral events

DSCT (CCT)	Components
Proprioceptive	1. Muscle spindle
	2. Tendon organ
	3. Ruffini
Exteroceptive	1. FA hair rec.
	2. SA touch rec.
	3. Hair + touch
	4. SA pad pressure

B. Interneuronal activity signalling "internal state"

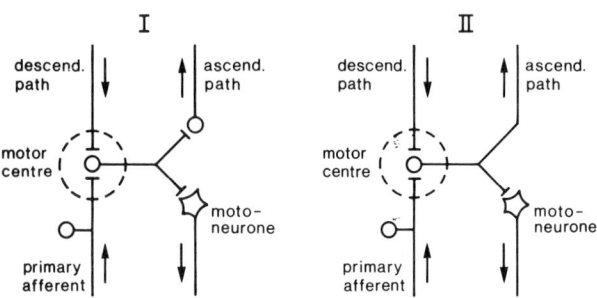

Fig. 1. Categories of information forwarded by the spinocerebellar paths. See text. Abbreviations: FA, fast adapting SA, slowly adapting; rec., receptor.

4th cervical segments by propriospinal neurones (PN) which are monosynaptically activated from several higher motor centres and project directly to the motoneurones. The PN are also monosynaptically activated from cutaneous and group I muscle afferents in forelimb nerves. Branches of the PN axons ascend to the lateral reticular nucleus (LRN), a main source of mossy fibres to the cerebellar anterior lobe. Through these branches the LRN would receive information identical with that reaching the forelimb motoneurones through the descending PN branches. It is interesting that the higher centres projecting to the PN, i.e., the sensorimotor cortex, the mesencephalic tectum and the red nucleus, also project to the LRN (Corvaja et al., 1977). This might permit the LRN to compare commands issued by the higher centres with the effects these commands evoke in the PN where signals from various motor centres and from spinal afferents are integrated (cf. Miller and Oscarsson, 1970; Zangger and Wiesendanger, 1973).

The bilateral ventral flexor reflex tract (bVFRT) is the main afferent path from the hindlimb segments of the cord to the LRN (Oscarsson, 1973; Clendenin et al., 1974). The segmental connections of this tract seem to combine the two kinds of organization shown in Fig. 1B. The behaviour of the LRN neurones is different in different preparations. In the lightly anaesthetized animal or in the decerebrate preparation with certain descending paths interrupted, there are strong effects from the periphery. The typical LRN neurone is then activated and inhibited from nerves in all four limbs. The effects are evoked by stimulation of cutaneous and high threshold muscle afferents, i. e., the flexor reflex afferents (FRA). Stimulation of cutaneous receptors permits

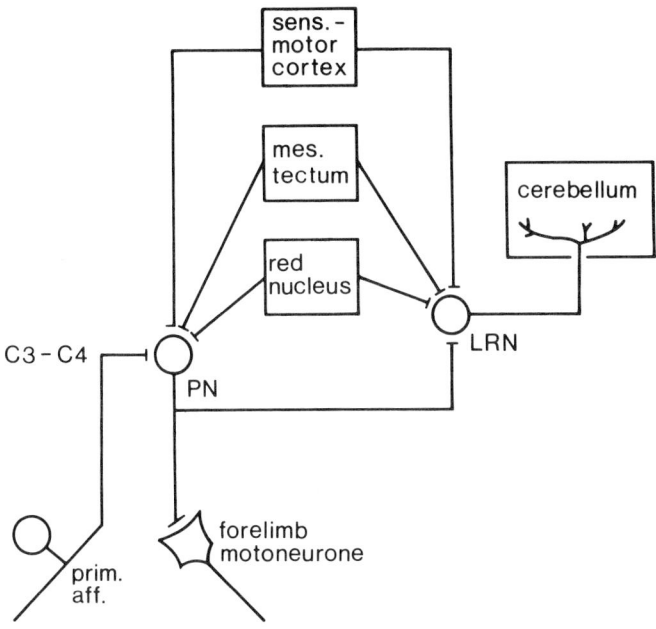

Fig. 2. Organization of propiospinal motor path with ascending axon collaterals to the lateral reticular nucleus (LRN) as suggested by Lundberg and his collaborators. Propriospinal neurones (PN) in the C3–C4 segments of the spinal cord activate monosynaptically forelimb motoneurones and project with axon collaterals to the LRN. The PN are monosynaptically activated by primary afferents in forelimb nerves and by descending paths from the sensorimotor cortex, mesencephalic tectum and red nucleus. These structures have also direct connections with the LRN. The axons of the LRN terminate as mossy fibres in the cerebellar anterior lobe. See text.

identification of excitatory and inhibitory areas covering most of the body surface. On the other hand, the effects from the periphery are depressed in the unanaesthetized decerebrate preparation with intact descending paths. The behaviour of the LRN neurones is again different in preparations performing stepping or scratching movements induced from motor pattern generators in the spinal cord (Shik and Orlovsky, 1976; Arshavsky et al., 1978a). In these preparations the LRN neurones display a modulation of their discharge in relation to the rhythmic movements, whereas effects from peripheral nerves and receptors are depressed.

Fig. 3 shows some of the connections of the bVFRT and LRN. The bVFRT neurones consist of equal proportions of excitatory and inhibitory neurones (in Fig. 3 indicated by combining the symbols for excitatory and inhibitory neurones) (Ekerot and Oscarsson, 1975). Many of the bVFRT neurones have collaterals descending for several segments in the ventral funiculus before terminating on unknown neurones in the ventral horn (Andersson and Ekerot, unpublished observations). The ascending axons reaching the LRN inform about the excitatory and inhibitory actions exerted by the bVFRT at the segmental level (cf. Fig. 1B: II). The rhythmic activity of the LRN neurones in preparations performing stepping or scratching movements suggests that the bVFRT carries information about activity in the motor pattern generators. Furthermore, it suggests that the bVFRT neurones are an integral part of these generators.

The bVFRT neurones are monosynaptically activated from a component of the

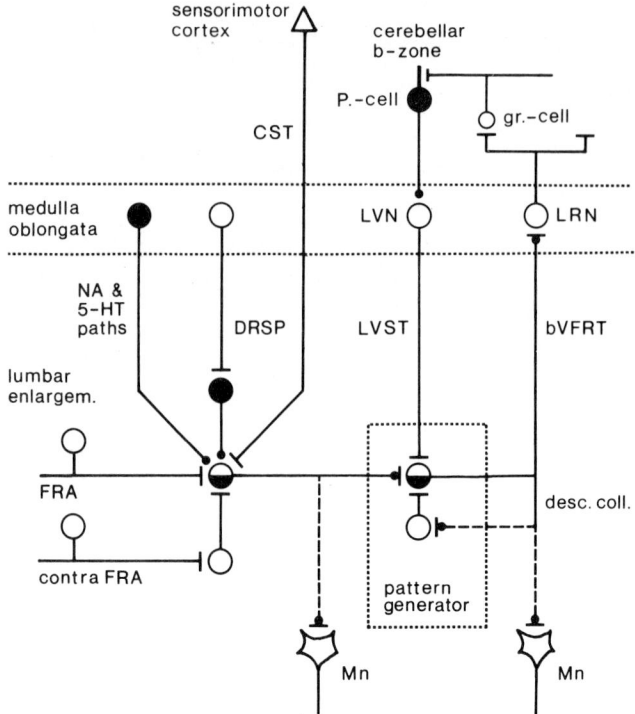

Fig. 3. Connections of the spino-reticulocerebellar path formed by the bilateral ventral flexor reflex tract (bVFRT) and the lateral reticular nucleus (LRN), as explained in the text. Abbreviations: P.-cell, Purkinje cell; gr.-cell, granule cell; CST, corticospinal tract; LVN, lateral vestibular nucleus; NA & 5-HT paths, noradrenergic and serotonergic paths; DRSP, dorsal reticulospinal path; LVST, lateral vestibulospinal tract; FRA, flexor reflex afferents; desc. coll., descending collateral; Mn, motoneurone. Excitatory neurones are indicated by white cell bodies and T-shaped endings, inhibitory neurones by black cell bodies and synaptic knobs. Neurone groups consisting of excitatory and inhibitory cells are indicated by combining the symbols for excitatory and inhibitory neurones. Uncertain connections indicated by interrupted lines. See text.

lateral vestibulospinal tract (LVST) (see Oscarsson, 1973). These LVST neurones are inhibited from the Purkinje cells in the b-zone of the cerebellum but apparently not directly activated from the vestibular nerve (Ekerot and Oscarsson, unpublished observations). The connections between the LVST and bVFRT neurones might be responsible for the control that the cerebellum exerts on locomotion through the LVST (Shik and Orlovsky, 1976).

Under some conditions the bVFRT is strongly influenced by the FRA in all four limbs (see above). The effects are evoked through excitatory and inhibitory interneurones which can be facilitated by the corticospinal tract and inhibited by several bulbospinal paths (see Oscarsson, 1973). The dorsal reticulospinal path (DRSP) is tonically active in the decerebrate preparation and responsible for the depression of peripheral effects in this preparation. The depression of peripheral effects during locomotion and scratching (Arshavsky et al., 1978a) is presumably due to descending noradrenergic and serotonergic paths (Grillner, 1976). The bVFRT might carry information about the transmittability through the FRA activated interneurones and their effects on segmental interneurones and motoneurones (cf. Fig. 1B: I).

After these examples from the mossy fibre system we will turn to the climbing fibre

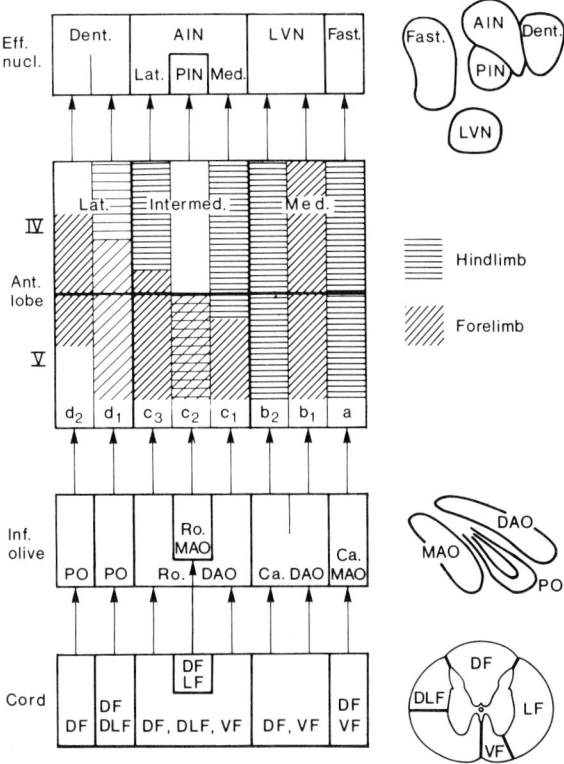

Fig. 4. Connections between spino-olivocerebellar paths (SOCPs) ascending through different funiculi of the spinal cord, regions of inferior olive, sagittal zones in cortex of cerebellar anterior lobe (lobules IV and V), and efferent cerebellar nuclei. Explanations: DF, DLF, LF and VF, dorsal, dorsolateral, lateral and ventral funiculi. PO, principal olive; Ro. and Ca., rostral and caudal parts of MAO and DAO, medial and dorsal accessory olives. Anterior lobe divided into sagittal zones labelled with letters and indices using a nomenclature modified from Voogd (1969). Somatotopical organisation of SOCP input to different zones indicated. Dent., dentate nucleus. Lat. and Med., lateral and medial parts of AIN, anterior interpositus nucleus. PIN, posterior interpositus nucleus. LVN, lateral vestibular nucleus. Fast., fastigial nucleus. (From Oscarsson, 1969, 1973; Voogd, 1969; Oscarsson and Sjölund, 1977a; Groenewegen and Voogd, 1977; Groenewegen et al., 1979; Ekerot and Larson, 1979a.)

paths. A large number of spino-olivocerebellar paths (SOCPs) have been described. They ascend through different funiculi of the spinal cord, relay in different regions of the inferior olive, and project as climbing fibres to different sagittal zones in the anterior lobe, as shown in Fig. 4 (Oscarsson, 1973, 1976; Groenewegen and Voogd, 1977; Groenewegen et al., 1978). Each zone has a private efferent path through one of the cerebellar nuclei (Voogd, 1969). It has been argued that each zone forms a functional unit controlling a certain motor mechanism (Oscarsson, 1973, 1976). Only some of the SOCPs will be discussed below.

Fig. 5 shows the organization of the five SOCPs ascending through the ventral funiculus (VF-SOCPs) (Oscarsson and Sjölund, 1977a, b). They are activated by the FRA from wide receptive fields. Three paths, terminating in the a, c_1 and c_3 zones (and denoted a-VF-SOCP, etc.), are activated exclusively from nerves in the ipsilateral hindlimb, whereas two paths, terminating in the b_1 and b_2 zones, are bilaterally activated from the forelimbs and hindlimbs, respectively (Fig. 5A, B). The segmental

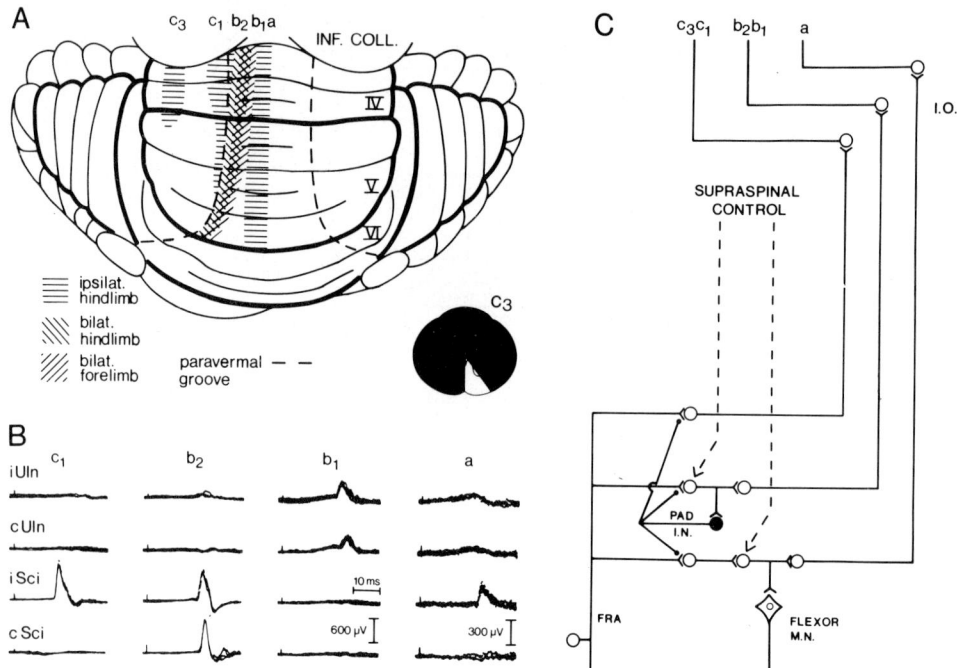

Fig. 5. Spino-olivocerebellar paths ascending through ventral funiculus (VF-SOCPs). A) Termination zones (c_3, c_1, b_2, b_1 and a) of the five VF-SOCPs in lobules IV and V of cerebellar anterior lobe. Receptive fields of different paths indicated (see key). Inf. coll., inferior colliculus. B) Responses recorded from surface of c_1, b_2, b_1 and a zones on stimulation of ipsilateral and contralateral (i, c) ulnar and sciatic nerves (Uln, Sci) in preparation with spinal cord interrupted at C3 except for contralateral ventral funiculus and adjacent part of lateral funiculus (inset in A). Superposed traces, positivity upwards. Higher amplification in records from the a zone. C) Tentative organization of the paths. See explanation in text. I.O., inferior olive; PAD I.N., interneurone responsible for primary afferent depolarization; FRA, flexor reflex afferents; M.N., motoneurone. (From Oscarsson and Sjölund, 1977a, b; Andersson and Sjölund, 1978; Sjölund, 1978.)

connections of the paths are tentatively shown in C. The a- and b-VF-SOCPs are activated through interneurones which are under effective supraspinal control. The interneurones of the a-paths behave, under a variety of tests, as those mediating the segmental flexion reflex, whereas those of the b-paths behave as if activated from the interneurones responsible for the primary afferent depolarization (PAD) evoked in the FRA (Andersson and Sjölund, 1978; Sjölund, 1978). These SOCPs may carry information about the activity in the reflex arcs discussed, but another possibility is that they relay information about other segmental mechanisms with a similar organization. The c_1- and c_3-VF-SOCPs are monosynaptically activated by the primary afferents and the transmission is little influenced by supraspinal control systems. These characteristics suggest that these paths carry information about peripheral events. However, the input from the FRA and the wide receptive fields make it unlikely that they relay modality and space specific information. They might possibly forward information about the primary afferent input to segmental motor mechanisms (Oscarsson and Sjölund, 1977b).

Fig. 6 shows the organization of the SOCPs ascending through the dorsal funiculus (DF-SOCPs) which project to all the identified zones of the anterior lobe (Ekerot and

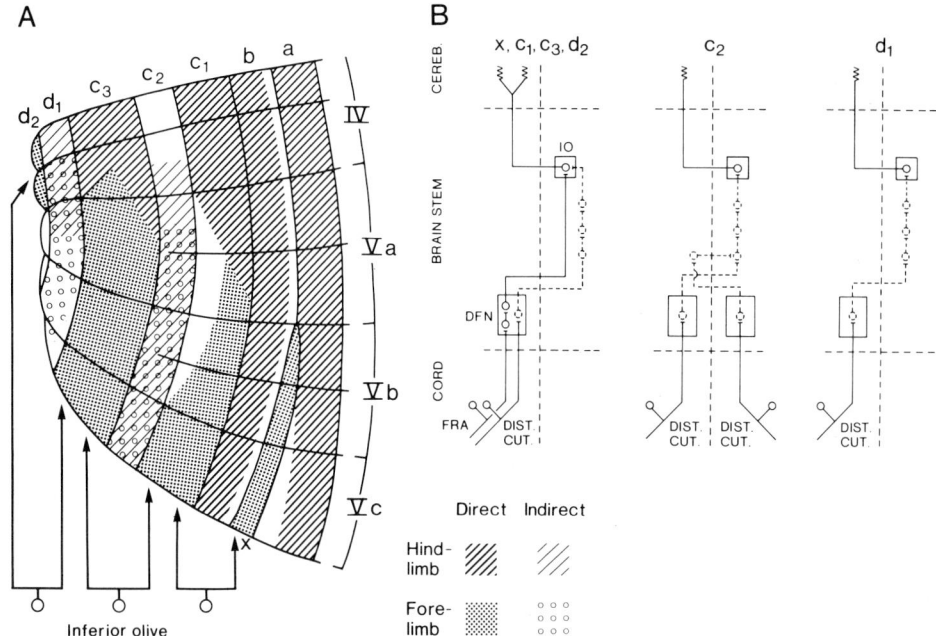

Fig. 6. Spino-olivocerebellar paths ascending through the dorsal funiculi (DF-SOCPs). A) Termination zones of DF-SOCPs in lobules IV and V of cerebellar anterior lobe. The zones are labelled as in Fig. 4 (the x zone can at present not be classified as belonging to either the a or b zone). Four of the zones, x, c_1, c_3 and d_2, are innervated in pairs by climbing fibres which branch as shown in the lower part of the diagram. The key explains the markings indicating the forelimb-hindlimb somatotopy in the termination zones of the direct and indirect paths (i.e., the paths without and with neurones intercalated between the dorsal funiculus nuclei and inferior olive). B) Tentative organization of some of the DF-SOCPs as explained in the text. IO, inferior olive; DFN, dorsal funiculus nuclei; FRA, flexor reflex afferents; Dist. cut., distal cutaneous afferents. Uncertain connections indicated by interrupted lines. Midline indicated by vertical interrupted line. (From Ekerot and Larson, 1977, 1979a.)

Larson, 1979a, b). Four of the zones, x, c_1, c_3 and d_2 are each innervated by two DF-SOCPs which converge at the olivary level (Fig. 6B). One set of DF-SOCPs is activated by the FRA from the ipsilateral limbs through paths with disynaptic relays in the dorsal funiculus nuclei (DFN). The other set of DF-SOCPs is activated by distal cutaneous afferents from the ipsilateral limbs and is interrupted by interneurones located between the DFN and inferior olive. Recent observations show that the x, c_1, c_3 and d_2 zones are innervated in pairs by climbing fibres which branch as shown in Fig. 6A (Ekerot and Larson, 1977, and unpublished). (These findings indicate that the simple scheme in Fig. 4 has to be revised with respect to the olivocerebellar projections.) Three of the zones, c_1, c_3 and d_2 have a detailed somatotopical organization. The DF-SOCPs to these zones carry information about peripheral events with a high degree of spatial discrimination and might be related to the control from the pars intermedia of fine movements in the ipsilateral limbs (Chambers and Sprague, 1955). However, the FRA input and the disynaptic relay in the DFN in the one set of DF-SOCPs and the intercalated interneurones in the other set are not satisfactorily explained if monitoring of peripheral events were the sole function of these paths. They might forward additional information about activity in the preolivary relays which might belong to motor paths.

The four zones discussed above interdigitate with four other zones receiving private and independent DF-SOCPs (Ekerot and Larson, 1979a). Those projecting to the c_2 and d_1 zones have a complex organization (Fig. 6B). Both are activated by distal cutaneous afferents and relayed through chains of interneurones before reaching the inferior olive. The c_2 path is bilaterally activated and has no somatotopical organization, whereas the d_1 path is activated from the ipsilateral limbs and has a crude somatotopical organization. These two paths might carry information about activity in the brain stem interneurones which possibly represent lower motor centres. The DF-SOCPs projecting to the a and b zones are activated from the FRA in hindlimb nerves but have not been investigated in detail.

These accounts have not taken into consideration that the climbing fibres to each cerebellar zone are usually activated from two or three SOCPs ascending through different funiculi of the cord (Fig. 4), from the contralateral sensorimotor cortex, and in at least one case, from the contralateral red nucleus (Oscarsson, 1973). The convergence occurs at the olivary or possibly in some cases, at a preolivary level. One would expect that the information carried by the ascending and descending paths to a certain zone is specifically related to the function of the motor mechanism assumed to be controlled by this zone. This contention is indeed strongly supported by the observations of Jeneskog and Johansson (1977) on the paths projecting to the hindlimb part of the d_1 zone. The following description is largely based on their observations and suggestions and will be discussed in connection with the hypothetical diagram in Fig. 7

In 1969 Miller et al. reported that stimulation close to the contralateral red nucleus evoked short-latency (monosynaptic) climbing fibre responses in the hindlimb part of the d_1 zone. This part is also reached by an SOCP ascending through the dorsolateral funiculus (DLF-SOCP) which is monosynaptically activated by distal cutaneous afferents and relays through a chain of interneurones in the brain stem (Larson et al., 1969). Jeneskog (1974a,b) showed that the rostral part of the red nucleus contains the neurones which give rise to the rubro-olivary path projecting to the d_1 zone and, after a relay in the brain stem, to a path descending in the contralateral dorsolateral funiculus to the lumbar enlargement. This path was called the rubrobulbospinal path (RBSP) and tentatively identified with the previously described dorsal reticulo-spinal system (Engberg et al., 1968). The RBSP would have complex segmental actions: inhibition of effects from the FRA to motoneurones and primary afferents and facilitation of dynamic, but not static, gamma-motoneurones (Appelberg, 1967; Engberg et al., 1968). Interestingly, distal cutaneous afferents appear to have similar segmental effects which are exerted through interneurones shared with the RBSP (Jeneskog and Johansson, 1977).

The observations of Jeneskog and Johansson (1977) suggest a close functional relation between the ascending and descending climbing fibre paths to the hindlimb area of the d_1 zone: stimulation of the RBSP and stimulation of distal cutaneous afferents evoke the same segmental effects and activate the same olivary neurones. It has been suggested that each sagittal zone in the anterior lobe controls a particular motor mechanism (Oscarsson, 1973, 1976). The d_1 zone would control a mechanism involving segmental interneurones inhibiting FRA paths and facilitating dynamic gamma-motoneurones. Furthermore, it has been suggested that the olivary relays to the anterior lobe act as comparators of command signals from higher motor centres and of the activity these signals evoke at lower levels (Miller and Oscarsson, 1970). In

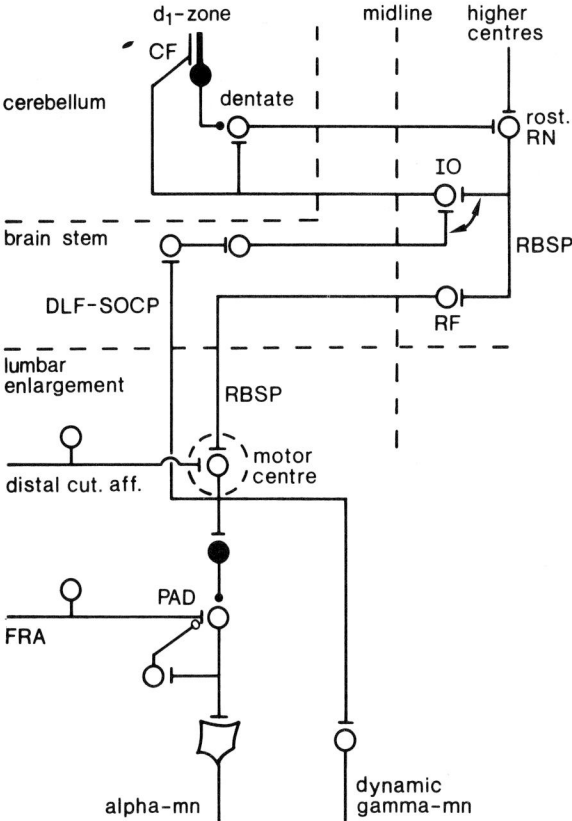

Fig. 7. Possible functional relation between sagittal zone in cerebellar cortex, motor centre controlled by this zone, and ascending and descending climbing fibre paths to this zone. Based largely on the analysis of the climbing fibre paths to the hindlimb part of the d_1 zone made by Jeneskog and Johansson (1977). In the diagram, the d_1 zone is assumed to control a motor centre in the lumbar enlargement, which facilitates dynamic gamma-motoneurones and inhibits effects from the flexor reflex afferents (FRA) to alpha-motoneurones and primary afferents (primary afferent depolarization, PAD). This centre is controlled by the rubrobulbospinal path (RBSP) which originates from the rostral part of the red nucleus (rost. RN) and which is under control from higher motor centres and from the d_1 zone and dentate nucleus. After a relay in the reticular formation (RF) the RBSP descends in the contralateral dorsolateral funiculus to the lumbar enlargement. The motor centre is also controlled by distal cutaneous afferents. The activity in the motor centre is monitored by a spino-olivocerebellar path ascending through the dorsolateral funiculus (DLF-SOCP), which is formed by ascending axon collaterals from the neurones in this centre. The part of the inferior olive (IO) projecting to the d_1 zone might possibly compare (arrow) the descending command signals from the RN with the activity these signals evoke in the motor centre which is also under segmental afferent control. On the basis of the comparison the d_1 zone might help to integrate motor activity evoked from higher centres and reflex activity. See text.

this case the olive might act as a comparator of command signals from the red nucleus and the effects these signals evoke in the segmental mechanism which is also influenced by activity in cutaneous afferents. On the basis of the information from the olive the d_1 zone might integrate motor activity elicited from a higher centre with segmental reflex activity. Necessary corrections might be executed through the dentate nucleus and its connections with the red nucleus (Allen and Tsukahara, 1974; Armstrong, 1974).

SUMMARY

Information from the spinal cord reaches the anterior lobe of the cerebellum through about twenty paths which terminate as mossy fibres or climbing fibres. Only a few of these paths (DSCT and CCT) carry information primarily about peripheral events. The majority of the other paths provide the cerebellum with information about the activity in groups of segmental or suprasegmental neurones presumably representing lower motor centres. The ascending paths are either activated or inhibited from the neurones forming these centres or formed by ascending axon collaterals of these neurones. The connections to the ascending paths are complex and different for each path indicating that they carry information from different motor centres.

The SOCPs are particularly interesting since they project to narrow sagittal zones in the cerebellar cortex. Each zone presumably controls a particular motor mechanism and the olivary path reaching it would be expected to carry information about the function of this mechanism. Recent observations support this contention for the path to the hindlimb part of the d_1 zone.

ACKNOWLEDGEMENT

This work was supported by grants from the Swedish Medical Research Council (Project No. 01013) and the Medical Faculty, University of Lund.

REFERENCES

Allen, G.I. and Tsukahara, N. (1974) Cerebrocerebellar communication systems. *Physiol. Rev.*, 54: 957–1006.

Andersson, G. and Sjölund, B. (1978) The ventral spino-olivocerebellar system in the cat. IV. Spinal transmission after administration of clonidine and l-dopa. *Exp. Brain Res.*, 33: 227–240.

Appelberg, B. (1967) A rubro-olivary pathway. II. Simultaneous action on dynamic fusimotor neurones and the activity of the posterior lobe of the cerebellar cortex. *Exp. Brain Res.*, 3: 382–390.

Armstrong, D.M. (1974) Functional significance of connections of the inferior olive. *Physiol. Rev.*, 54: 358–417.

Arshavsky, Yu.I., Gelfand, I.M., Orlovsky, G.N. and Pavlova, G.A. (1978a) Messages conveyed by spinocerebellar pathways during scratching in the cat. I. Activity of neurons of the lateral reticular nucleus. *Brain Res.*, 151: 479–491.

Arshavsky, Yu.I., Gelfand, I.M. Orlovsky, G.N. and Pavlova, G.A. (1978b) Messages conveyed by spinocerebellar pathways during scratching in the cat. II. Activity of neurons of the ventral spinocerebellar tract. *Brain Res.*, 151: 493–506.

Chambers, W.W. and Sprague, J.M. (1955) Functional localization in the cerebellum. I. Organization in longitudinal corticonuclear zones and their contribution to the control of posture, both extrapyramidal and pyramidal. *J. comp. Neurol.*, 103: 105–129.

Clendenin, M., Ekerot, C.-F., Oscarsson, O. and Rosen, I. (1974) The lateral reticular nucleus in the cat. II. Organization of component activated from bilateral ventral flexor reflex tract (bVFRT). *Exp. Brain Res.*, 21: 487–500.

Corvaja, N., Grofová, I., Pompeiano, O. and Walberg, F. (1977) The lateral reticular nucleus in the cat. I. An experimental anatomical study of its spinal and supraspinal afferent connections. *Neuroscience*, 2: 537–553.

Ekerot, C.-F. and Oscarsson, O. (1975) Inhibitory spinal paths to the lateral reticular nucleus. *Brain Res.*, 99: 157–161.

Ekerot, C.-F. and Larson, B. (1977) Three sagittal zones in the cerebellar anterior lobe innervated by a common group of climbing fibres. *Proc., XXVII Int. Congr. Physiol. Sci., Paris*, XIII: 208, no. 604.

Ekerot, C.-F. and Larson, B. (1979a) The dorsal spino-olivocerebellar system in the cat. I. Functional organization and termination in eight sagittal zones of the anterior lobe. *Exp. Brain Res.*, in press.

Ekerot, C.-F. and Larson, B. (1979b) The dorsal spino-olivocerebellar system in the cat. II. Somatotopical organization. *Exp. Brain Res.*, in press.

Engberg, I., Lundberg, A. and Ryall, R.W. (1968) Reticulospinal inhibition of transmission in reflex pathways. *J. Physiol. (Lond.)*, 194: 201–223.

Fu, T.-C., Jankowska, E. and Tanaka, R. (1977) Effects of volleys in cortico-spinal tract fibres on ventral spino-cerebellar tract cells in the cat. *Acta physiol. scand.*, 100: 1–13.

Grillner, S. (1976) Some aspects on the descending control of the spinal circuits generating locomotor movements. In *Neural Control of Locomotion*. R.M. Herman, S. Grillner, S.G. Stein and D.G. Stuart (Eds.). Plenum Publ. Corp., New York, pp. 351–375.

Groenewegen. H.J. and Voogd, J. (1977) The parasagittal zonation within the olivocerebellar projection. I. Climbing fiber distribution in the vermis of the cerebellum. *J. comp. Neurol.*, 174: 417–488.

Groenewegen, H.J., Voogd, J. and Freedman, S.L. (1979) The parasagittal zonation within the olivocerebellar projection. II. Climbing fiber distribution in the intermediate and hemispheric parts of the cat cerebellum. *J. comp. Neurol.*, 183: 551–602.

Illert, M. and Lundberg, A. (1978) Collateral connexions to the lateral reticular nucleus from cervical propiospinal neurones projecting to forelimb motoneurones in the cat. *Neurosci. Lett.*, 3: 167–172.

Illert, M. and Tanaka, R. (1978) Integration in descending motor pathways controlling the forelimb in the cat. 4. Corticospinal inhibition of forelimb motoneurones mediated by short propriospinal neurones. *Exp. Brain Res.*, 31: 131–141.

Illert, M., Lundberg, A. and Tanaka, R. (1976a) Integration in descending motor pathways controlling the forelimb in the cat. 1. Pyramidal effects on motoneurones. *Exp. Brain Res.*, 26: 509–519.

Illert, M., Lundberg, A. and Tanaka, R. (1976b) Integration in descending motor pathways controlling the forelimb in the cat. 2. Convergence on neurones mediating disynaptic cortico-motoneuronal excitation. *Exp. Brain Res.*, 26: 521–540.

Illert, M., Lundberg, A. and Tanaka, R. (1977) Integration in descending motor pathways controlling the forelimb in the cat. 3. Convergence on propriospinal neurones transmitting disynaptic excitation from the corticospinal tract and other descending tracts. *Exp. Brain Res.*, 29: 323–346.

Jeneskog, T. (1974a) Parallel activation of dynamic fusimotor neurones and a climbing fibre system from the cat brain stem. I. Effects from the rubral region. *Acta physiol. scand.*, 91: 223–242.

Jeneskog, T. (1974b) Parallel activation of dynamic fusimotor neurones and a climbing fibre system from the cat brain stem. II. Effects from the inferior olivary region. *Acta physiol. scand.*, 92: 66–83.

Jeneskog, T. and Johansson, H. (1977) The rubro-bulbospinal path. A descending system known to influence dynamic fusimotor neurones and its interaction with distal cutaneous afferents in the control of flexor reflex afferent pathways. *Exp. Brain Res.*, 27: 161–179.

Larson, B., Miller, S. and Oscarsson, O. (1969) Termination and functional organization of the dorsolateral spino-olivocerebellar path. *J. Physiol. (Lond.)*, 203: 611–640.

Lindström, S. (1973) Recurrent control from motor axon collaterals of Ia inhibitory pathways in the spinal cord of the cat. *Acta physiol. scand.*, Suppl. 392.

Lundberg, A. (1971) Function of the ventral spinocerebellar tract. A new hypothesis. *Exp. Brain Res.*, 12: 317–330.

Miller, S., Nezlina, N. and Oscarsson, O. (1969) Climbing fibre projection to cerebellar anterior lobe activated from structures in midbrain and from spinal cord. *Brain Res.* 14: 234–236.

Miller, S. and Oscarsson, O. (1970) Termination and functional organisation of spino-olivocerebellar paths. In *The Cerebellum in Health and Disease*. W.S. Fields and W.D. Willis (Eds.). Warren H. Green, St. Louis, pp. 172–200.

Oscarsson, O. (1973) Functional organization of spinocerebellar paths. In *Handbook of Sensory Physiology, Vol. II, Somatosensory System*. A. Iggo (Ed.). Springer-Verlag, Berlin-Heidelberg-New York, pp. 339–380.

Oscarsson, O. (1976) Spatial distribution of climbing and mossy fibre inputs into the cerebellar cortex. *Exp. Brain Res.*, Suppl. 1, 36–42.

Oscarsson, O. and Sjölund, B. (1977a) The ventral spino-olivocerebellar system in the cat. I. Identification of five paths and their termination in the cerebellar anterior lobe. *Exp. Brain Res.*, 28: 469–486.

Oscarsson, O. and Sjölund, B. (1977b) The ventral spino-olivocerebellar system in the cat. III. Functional characteristics of the five paths. *Exp. Brain Res.*, 28: 505–520.

Shik, M.L. and Orlovsky, G.N (1976) Neurophysiology of locomotor automatism. *Physiol. Rev.*, 56: 465–501.

Sjölund, B. (1978) The ventral spino-olivocerebellar system in the cat. V. Supraspinal control of spinal transmission. *Exp. Brain Res.*, 33: 509–522.

Voogd, J. (1969) The importance of fibre connections in the comparative anatomy of the mammalian cerebellum. In *Neurobiology of Cerebellar Evolution and Development*. R. Llinás (Ed.). American Medical Association, Chicago, pp. 493–514.

Zangger, P. and Wiesendanger, M. (1973) Excitation of lateral reticular nucleus neurones by collaterals of the pyramidal tract. *Exp. Brain Res.*, 17: 144–151.

Responses of Dorsal Spinocerebellar Tract Neurons to Signals from Muscle Spindle Afferents

O.-J. GRÜSSER and J. KRÖLLER

Physiologisches Institut, Freie Universität, 1000 Berlin 33, (F.R.G.)

INTRODUCTION

The proprioceptive neurons of Clarke's column transmit and integrate the signals from muscle spindle afferents under the control of propriospinal and corticofugal signals (cf. Lundberg, 1964; Oscarsson, 1965, 1973). Over the last six years, we studied three classes of dorsal spinocerebellar tract (DSCT)-neurons activated or inhibited by ipsilateral proprioceptive signals from the gastrocnemius muscles:

i) "Monomuscular" DSCT-neurons, which were activated by stretching the gastrocnemius muscle, exhibited a short latency (3.0–4.1 msec) of their first action potential evoked by electrical stimulation of the gastrocnemius nerve (Oscarsson, 1973). During muscle twitch, a typical discharge pause was observed in contrast to the less frequently encountered DSCT-neurons activated by gastrocnemius Golgi-receptors. Monomuscular DSCT-neurons were either inhibited by or did not respond to electrical stimulation of other nerves of the lower leg (*N. peroneus com., N. femoralis*, distal *N. tibialis*).

ii) "Stretch-inhibited" DSCT-neurons responded to stretch of the ipsilateral gastrocnemius muscle with a transient inhibition of their predominantly low spontaneous impulse rates ($\leq$ 12 impulses.sec^{-1}) and were activated by electrical stimulation of one or more hindlimb nerves.

iii) "Convergent" DSCT-neurons (Kuno et al., 1973a) were less activated by stretch of the gastrocnemius muscle than the monomuscular DSCT-neurons, by comparable mechanical stimulus parameters. They were also activated by electrical stimulation of one or more of the nerves mentioned above.

In the following we will describe the responses of monomuscular DSCT-neurons and discuss the following questions:

1. What are the signal transmission properties of monomuscular proprioceptive DSCT-neurons (henceforth called simply DSCT-neurons) as compared to the presynaptic Ia-fibers, when constant stretch, periodic (sinewave) or aperiodic (white noise) mechanical stimuli are applied to the gastrocnemius muscle?

2. How many muscle spindle afferents from the gastrocnemius muscle converge on a single DSCT-neuron? What is the contribution of each of these inputs to the output discharge pattern of Clarke's column cells, sending their axon into the DSCT?

METHODS

The data described in this report are part of the results obtained in 72 anesthetized (30–35 mg/kg pentobarbital i.p.) and spinalized (Th8) cats; laminectomy and tungsten microelectrode recordings from the DSCT at Th10 or Th11, another laminectomy at and below L5; ventral roots L7 and S1 cut. The gastrocnemius muscle and the gastrocnemius nerve were prepared and the soleus muscle was separated; the tendon was cut and fixed to the stimulation apparatus (Fig. 1a). The animals were fixed

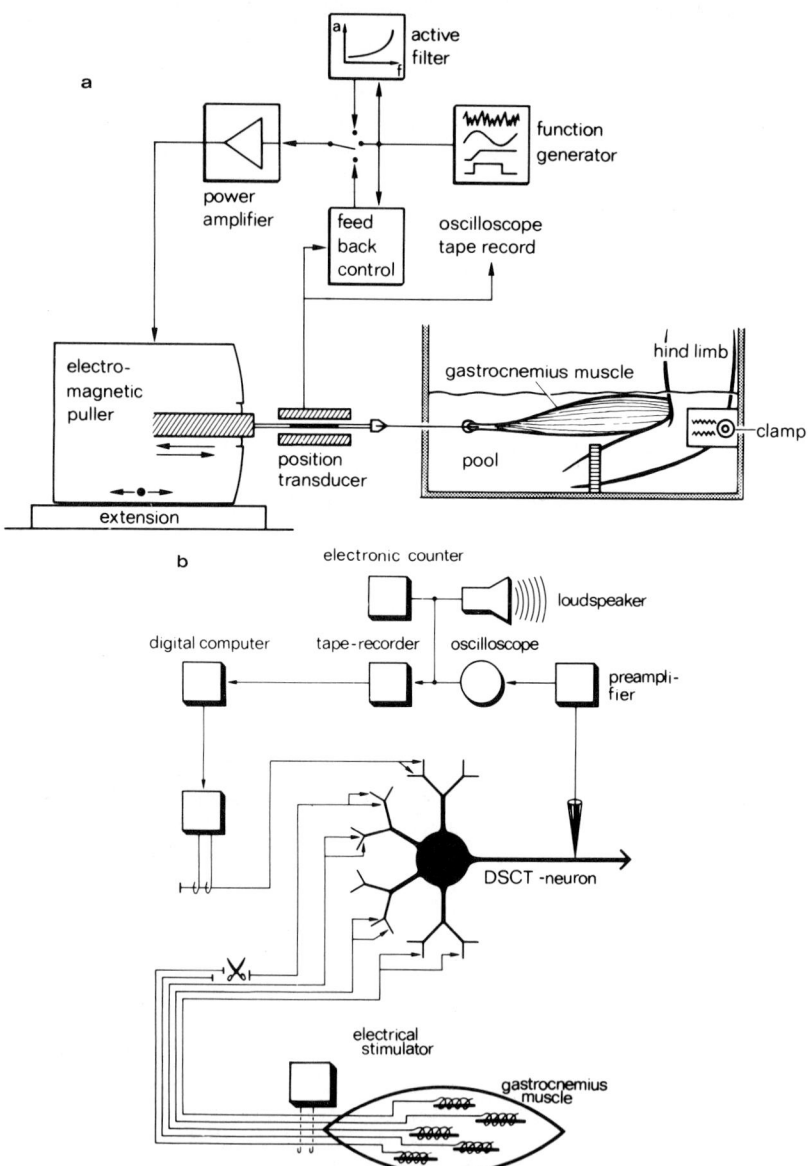

Fig. 1. a) Schematic diagram of the mechanical stimulation apparatus. The gastrocnemius muscle was stimulated by three types of stimuli: steady stretch, sinewave and wide band white noise stimuli. b) Schematic diagram of the recording technique of action potentials of a DSCT-neuron during progressive deafferentation. The cut filaments containing one exciting afferent Ia-fiber were electrically stimulated.

horizontally in a specially constructed animal holder. The dura was removed above the exposed parts of the spinal cord, the latter being covered with warm paraffin oil (36–38°C). The bones of the prepared lower leg were fixed with rigid clamps and the leg covered by a second paraffin oil pool (36–38°C). Bipolar electrical stimulation electrodes were placed on the gastrocnemius nerve of the *N. tibialis* above the *N. gastrocnemius*, the *N. tibialis* below the *N. gastrocnemius*, the *N. peroneus com.* and the *N. femoralis*. The bioelectrical signals and the stimulation signals (muscle length and tension, Fig. 1a; electrical stimuli, Fig. 1b) were taperecorded and later processed by a Linc 8 or HP 21 MX digital computer. In some of the experiments, *simultaneous* recordings of Ia action potentials were performed with microelectrodes positioned 1–3 mm cranial to the entrance of the respective dorsal rootlets into the spinal cord at the segment S1 or L7. Action potentials of muscle spindle afferents were also recorded by means of the more conventional method applying microdissection of single axons of primary or secondary afferent fibers.

Stimulation

Mechanical stimuli were applied to the gastrocnemius muscle of an electronically controlled electromagnetic puller with function generator signals as input (sinewave frequency 0.1–100 Hz, amplitude $\leq$ 4 mm at $\leq$ 30 Hz and $\leq$ 1.5 mm at $\leq$ 100 Hz; steady stretch of 0–12 mm). In some experiments mechanical noise was used (Gaussian amplitude distribution, variable upper frequency cutoff between 20 and 100 Hz).

Electrical stimuli

0.2 msec constant current shocks of 0.1–3.0 V were applied to the nerves mentioned above or to isolated small dorsal root filaments containing one muscle spindle afferent axon.

RESULTS

Comparison between DSCT-neurons and muscle spindle afferent discharge patterns

Constant muscle length. At steady stretch of the gastrocnemius muscle, the average impulse rate of DSCT-neurons was equal to or up to 30% higher than that of their presynaptic Ia afferents (Fig. 2a). A considerable difference, however, was found in the *regularity* of successive impulse intervals: high regularity in primary and secondary muscle spindle afferents and a considerable temporal fluctuation of successive impulse intervals in DSCT-neurons. Most DSCT-neurons displayed a monomodal interval histogram. The average coefficient of variation (c.v. = standard deviation/arithmetic mean) of Ia interval histograms (steady stretch) was in the range of 0.015–0.075 (Matthews and Stein, 1969; Eysel and Grüsser, 1970; Matthews, 1972). The c.v.-values for DSCT-neurons were between 0.10 and 0.40 (mean 0.265 ± 0.075), excluding the few neurons with a bimodal interval histogram (cf. also Jansen et al., 1966). In various neurons the c.v. did not depend on the impulse rate (Fig. 2b). As in Ia fiber discharge patterns, a serial correlation of successive intervals was present in DSCT-neurons, but the linear serial correlation coefficient $r_{1,2}$ between successive intervals only deviated significantly from 0 at impulse rates above 25 impulses.sec^{-1}. Above these values, a linear correlation between $r_{1,2}$ and the average impulse interval (i. e., a hyperbolic relationship with the average impulse frequency) was found

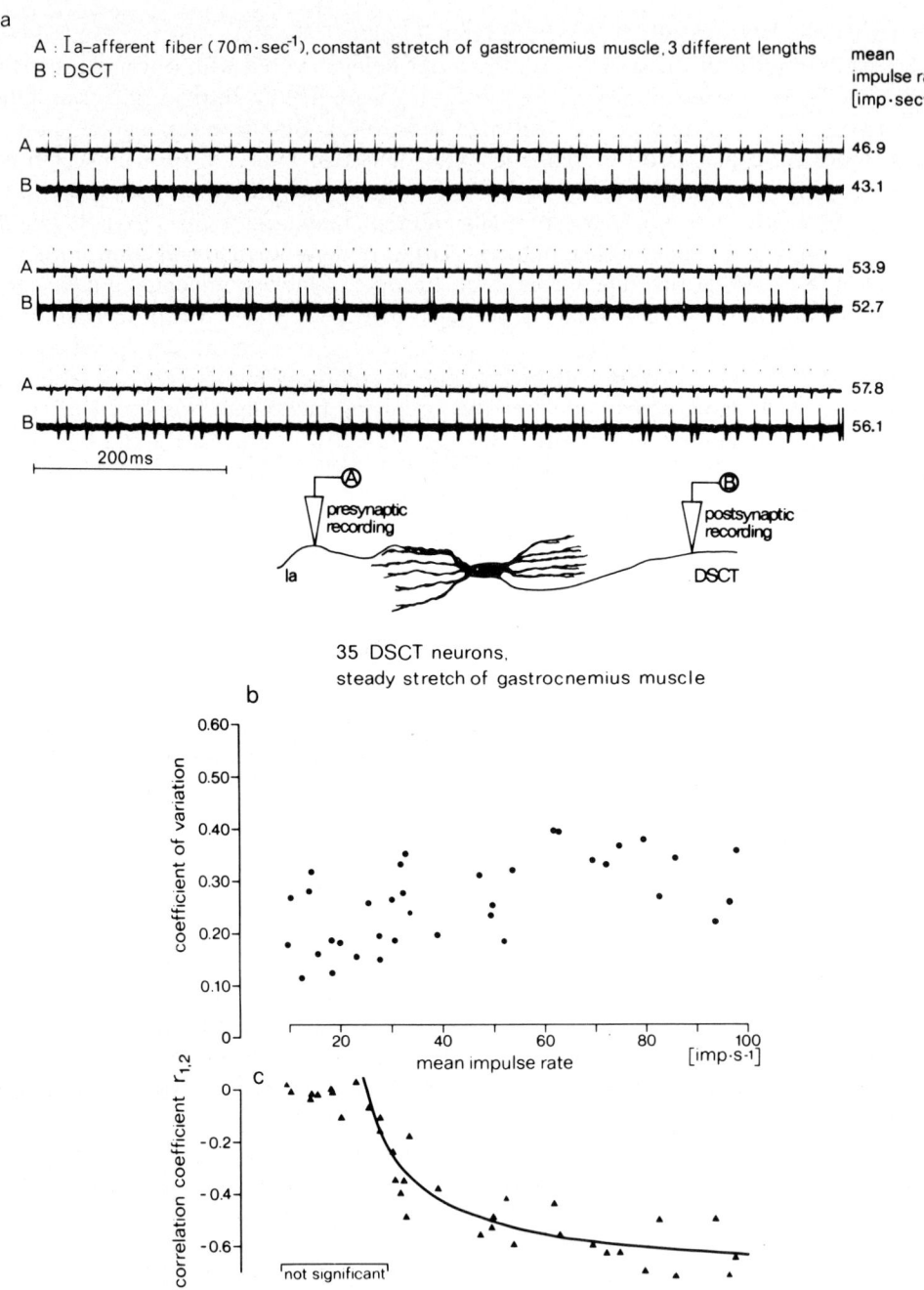

Fig. 2. a) Simultaneous recordings of the action potentials of a presynaptic Ia-fiber and of a DSCT-neuron driven by this fiber as revealed by the cross-correlogram. Three different stretches (2, 4, and 5 mm over relaxed muscle length) lead to three different average impulse rates. b) Relationship of the coefficient of variation of the first order impulse *interval histogram* of 35 DSCT-neurons (ordinate) recorded at different muscle lengths and of the average impulse rate (abscissa). c) Dependency of the first order correlation coefficient $r_{1,2}$ (interval sequence of a steady state impulse rate) on the average impulse rate. 35 DSCT-neurons discharging at different impulse rates. The correlation coefficient $r_{1,2}$ deviated significantly from 0 above an impulse rate of 25 impulses.sec^{-1}.

(Fig. 2c). Higher order serial correlation coefficients ($r_{1,3}$, $r_{1,4}$ ... $r_{1,n}$) did not deviate significantly from 0 above the fourth order. Thus, the higher order correlation coefficient and the Markov chain properties found in muscle spindle afferent impulse patterns (Eysel and Grüsser, 1970) are, as anticipated, not transmitted to the next postsynaptic level. The serial correlation coefficients in DSCT-neurons are probably caused mainly by the *relative refractory period* of these neurons and the *afterhyperpolarization* following each action potential (Gustafsson and Zangger, 1978; Jansen et al., 1966). A linear relationship between the average impulse rate $\bar{R}$ and the increase in muscle length S ($0 \leq S \leq 12$ mm) was found to be valid:

$$\bar{R} = aS + R_o \text{ [impulses.sec}^{-1}] \tag{1}$$

with $4 < a < 8$; R_o is the impulse rate obtained with a relaxed muscle (cf. Jansen and Rudjord, 1965).

Sinewave mechanical stimulation (0.05–100 Hz). With comparable sinewave stimuli, the neuronal response modulation was always higher (10–120%) in DSCT-neurons than in their presynaptic Ia fibers (Fig. 3a–c); otherwise, similar amplitude frequency

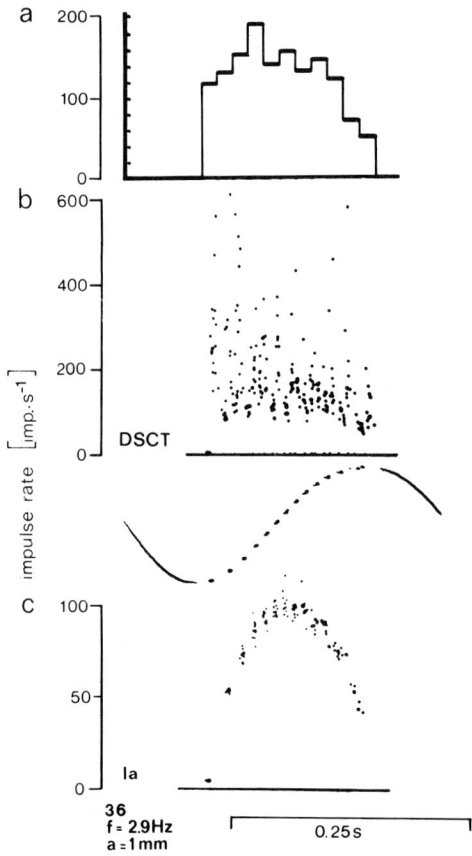

Fig. 3. a) PST-histogram of the responses of a DSCT-neuron; b) Dot-display of the instantaneous impulse rate of the same response, and c) The response of a presynaptic Ia-fiber elicited by identical sinewave stimuli (2.9 Hz, 1 mm amplitude). Note the considerably larger scatter in (b) as compared with (c).

characteristics of DSCT-neurons were found as compared to the presynaptic Ia fiber responses. Above 1 Hz, the slope of the gain was, as in muscle spindle afferents, larger than below 1 Hz (cf. also Matthews, 1972; Jansen et al.,1967):

$$\log R_{max} = b \log f + c \; [\text{impulses.sec}^{-1}] \tag{2}$$

whereby f is the sinewave frequency, b and c = constants. b was in the stimulus range ≤ 1 Hz ≈ 0.1, in the stimulus range > 1 Hz ≈ 0.6. R_{max} is the value of the maximum bin of the respective PST-histograms. At stimulus amplitudes > 0.2 mm, the *non-linearities* described for Ia and II fiber responses (Grüsser and Thiele, 1968) were also found for the DSCT-neuron responses: a non-sinusoidal time course of the impulse rate and discharge pauses during decrease in muscle length (Fig. 3). During these pauses, negative impulse rates would characterize a linear extension of the impulse rate time course. Under natural stimulus conditions, this "compensatory" information during the discharge pauses is probably conveyed to the central nervous system by the DSCT-neurons connected with muscle spindle afferents from the respective antagonistic muscles.

With increasing stimulus amplitude and frequency, the response fluctuation of DSCT-neurons decreased. This was probably due to the higher *stimulus-induced synchronization* of the different Ia fiber discharges activating the respective DSCT-neurons. Since 10–18 muscle spindle afferents converge at a single Clarke's column cell (cf. p. 100) and 2–3 synchronously elicited EPSPs seem to be sufficient to discharge a postsynaptic action potential (Walløe, 1968), this increasing regularity of DSCT impulse intervals was not surprising.

The maximum neuronal impulse rate R_{max} of the PST-histograms were related to the stimulus sinewave amplitude A by a power function (Kröller, 1978):

$$\log R_{max} = a \log A + c \; [\text{impulses.sec}^{-1}] \tag{3}$$

Mechanical noise stimuli. Periodic sinewave stimuli of a constant frequency are only partially useful in testing the frequency response characteristics of a non-linear neuronal system. We have therefore used mechanical noise stimulation with Gaussian amplitude distribution. To facilitate the data analysis, the stimuli were composed of periods with identical noise sequences lasting 0.6–20 sec. These periods were considerably longer than the "memory" of the system investigated. We will present the system-theoretical analyses of our data in a later report. At present, only two findings shall be emphasized:

i) *The deterministic structure* of the neuronal responses was higher in Ia fibers than in DSCT-neurons and increased in both systems as the amplitude and the upper frequency limit of the noise stimuli increased (Fig. 4a, b).

ii) The effective *average* stimulus can be determined by measuring the *average pre-event stimulus* (APES; Johannesma in Grashuis, 1977). With this technique a *backward* averaging of the *stimuli* preceding each action potential is performed. The APES corresponds to the first (linear) kernel of the Wiener non-linear system analysis method applied to the responses to white noise stimuli. As for Ia fibers, the APES for DSCT-neurons is also characterized by a negative part (reduced muscle length) before the rapid increase in muscle length preceding the neuronal impulses appeared (Fig. 4c). The APES of Ia fibers and DSCT-neurons changed in a similar way as the upper frequency limit of the noise stimulation was reduced.

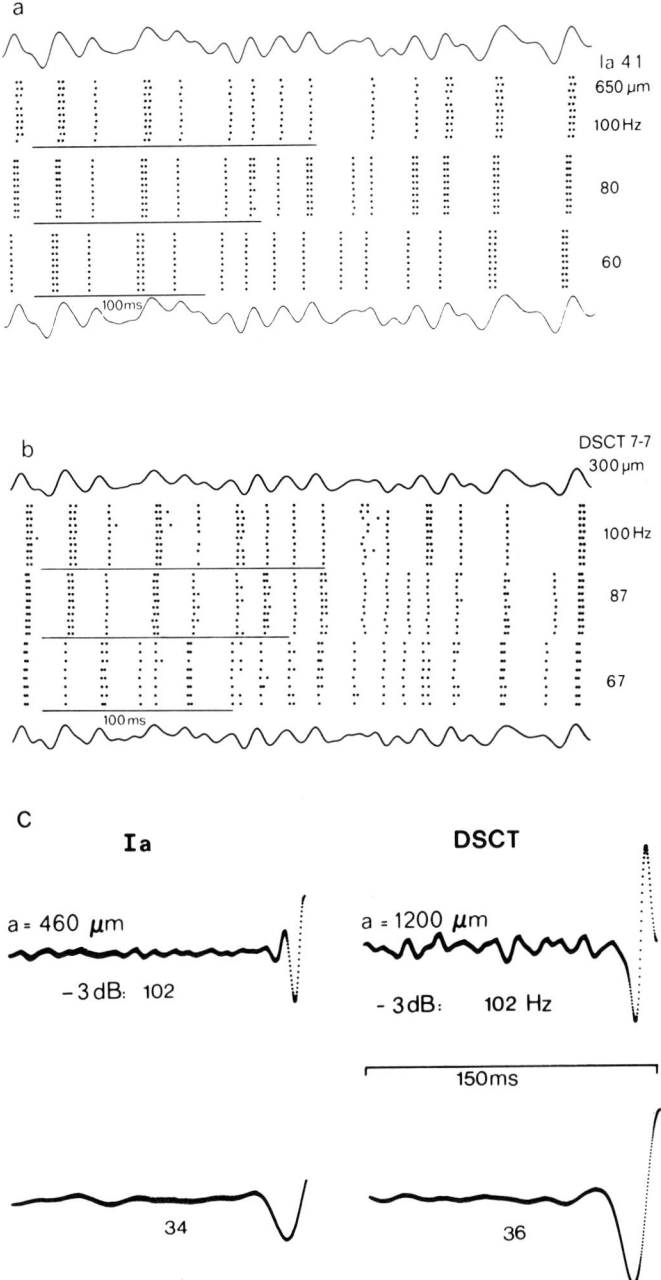

Fig. 4. a) Dot-display of a Ia-muscle spindle afferent impulse pattern activated by white noise mechanical stimuli at different upper frequency limits (100, 80, 60 Hz; max. stretch 650 μm). b) Responses of a DSCT-neuron to white noise mechanical stimuli of the gastrocnemius muscle (max. stretch 300 μm) at three different upper frequency limits. Note the change in time scale (horizontal bar: 100 msec). Comparing (a) and (b), one immediately recognizes the less deterministic structure of the DSCT-impulse pattern. c) Average pre-event stimulus (APES) of a presynaptic Ia-muscle spindle afferent (left side) and a DSCT-neuron, obtained with mechanical noise stimulation and two different upper frequency limits. Note the high similarity in the APES's, except for the time shift.

Due to the larger latency (conduction time, synaptic delay, delay between EPSP and action potential) of the DSCT-neuron discharges, the APES of DSCT-neurons is, of course, shifted 4–5 msec to the left, as opposed to the APES of Ia fibers. Comparison of the APES's and the sinewave data indicates that in the frequency range investigated (≤ 100 Hz), with respect to the frequency transfer properties, the DSCT-neurons act as a wide band filter which does not seriously affect the information about frequency of muscle length variation transmitted to them by the Ia fibers (Kröller, 1978).

Ten to eighteen muscle spindle afferents have excitatory inputs at one DSCT-neuron.

From results of intracellular studies (Eide et al., 1967, 1969a, b; Kuno et al., 1973b), it seems that the amplitude of the individual EPSPs evoked by different afferent axons at Clarke's column neurons varies over a 1 : 10 range. This finding indicated that the individual contribution of a single Ia afferent axon to the DSCT-neuron excitation level might vary considerably from axon to axon. This conclusion, however, was not confirmed in our present study. By the following method, we first tried to estimate the number of excitatory muscle spindle afferents converging at a DSCT-neuron (Fig. 1b): by electrical microstimulation of intact small rootlets, the part of the S1- or L7-roots was determined in which effective afferent fibers to the respective DSCT-neuron recorded were located. Thereafter, very fine fiber bundles were prepared under the microscope. Identical mechanical stimuli (sinewave stimuli 3–4 Hz, 1.5–3 mm amplitude) were used in all following measurements. When a small fiber bundle containing at least one afferent "effective" axon driving the recorded DSCT-neuron was cut, the neuronal impulse rate immediately decreased by about 3–4 impulses. Surprisingly, the contribution of each single afferent muscle fiber to the output impulse rate varied very little from one effective axon to the next (about 1 : 3 at maximum). Therefore, an approximate *linear* relationship between the number of axons cut and the neuronal impulse rate was found (Fig. 5a, b). Additional micro-dissection and electrical stimulation (20–80 stimuli.sec^{-1}) were used after each transection to verify that only *one* effective fiber driving the recorded DSCT-cell was included in the small bundle of axons cut. When the neuronal impulse rate decreased considerably by more than 3–4 impulses.sec^{-1} after the cutting of a fiber bundle, as a rule, two fibers were found to have been cut simultaneously. By further micro-dissection, these fibers could be separated into two different bundles.

As the neuronal impulse rate reached a value near 0 impulses.sec^{-1}, at least two effective axons were still connected with the respective DSCT-neuron. This was proved

Fig. 5. a) Recordings of a DSCT-neuron impulse pattern elicited by sinewave stimuli (3.4 Hz, 1.5 mm amplitude). In steps, afferent fibers were cut. Numbers of effective afferent fibers cut as indicated. On the right hand side, the responses to sinewave mechanical stimuli of the gastrocnemius muscle during electrical stimulation of a filament containing two effective afferent fibers is shown. Note the periodic activation to sinewave mechanical stimuli visible in this case (11 effective afferent fibers were cut). After the interruption of the 12th axon, no periodic responses were obtained during mechanical stimulation and simultaneous electrical stimulation of 2, 3 or 5 active afferent fibers. b) Dot-display of the impulse pattern of a DSCT-neuron responding to sinewave mechanical stimuli of the gastrocnemius muscle (1.5 mm amplitude, 3.4 Hz). The changes in the impulse pattern with successive interruption of the afferent input to the DSCT-neuron are shown. The numbers indicate the number of effective afferent axons cut. c) Relationship of the average impulse rate ($\bar{R}$, ordinate) and the numbers of transected "effective" filaments containing an effective muscle spindle axon. The vertical arrows give the increase in neuronal activation induced by electrical stimulation of the cut effective filament (100 stimuli.sec^{-1}).

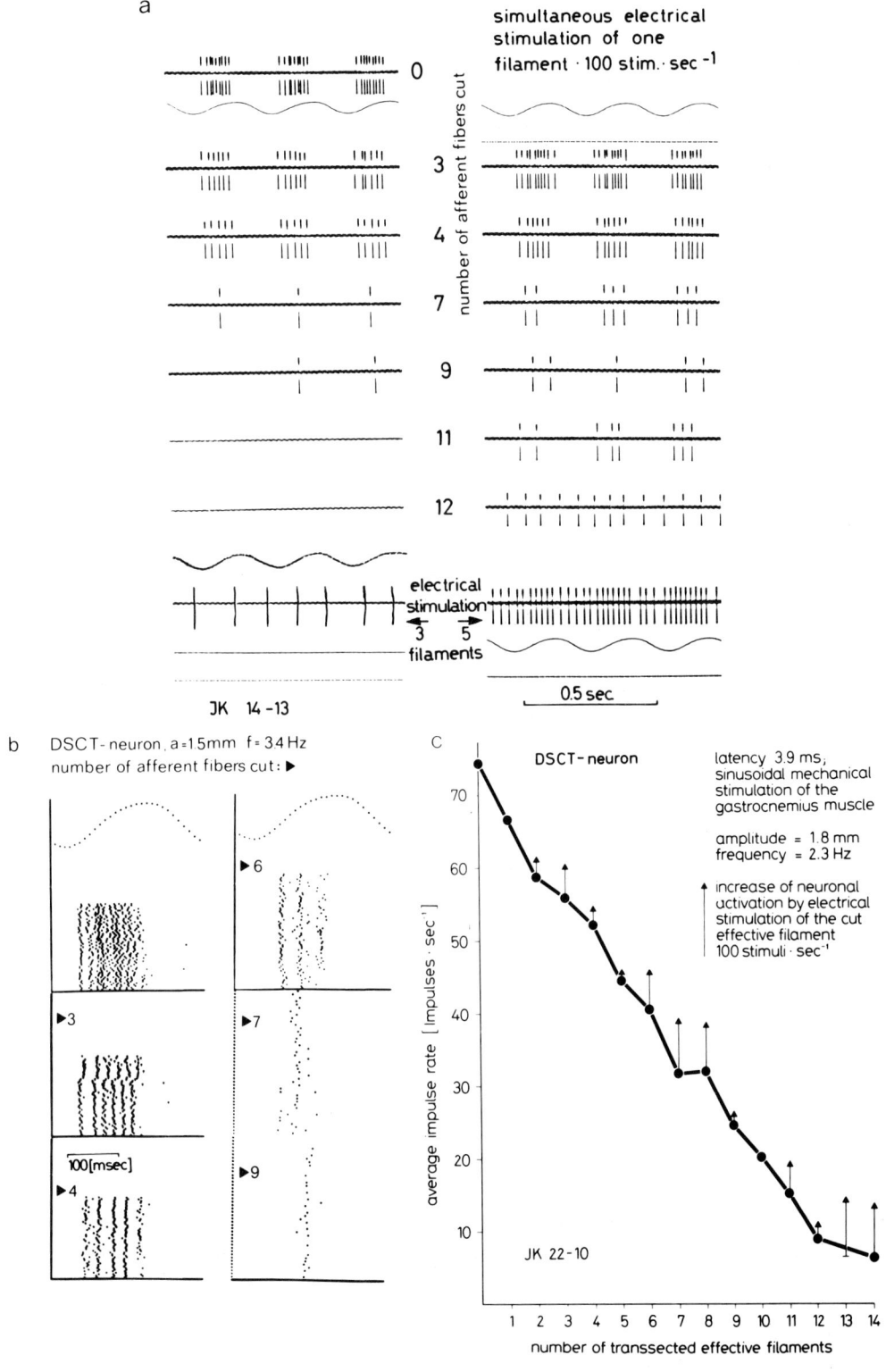

by the following technique: several bundles of cut dorsal root filaments containing 2–4 effective axons were stimulated together by electrical shocks at 60–100 stimuli.sec^{-1}. This led to a postsynaptic impulse rate of the DSCT-neuron in a range between 20 and 40 impulses.sec^{-1}. This electrically induced postsynaptic impulse pattern was periodically modulated in time when effective afferent axons from gastrocnemius muscle spindles were still connected with the respective DSCT-cell and the mechanical sinewave stimulation of the gastrocnemius muscle was turned on (Fig. 5a). After the interruption of 2–3 additional filaments, this sinewave modulation of the electrically induced discharge pattern disappeared.

From these findings we could conclude:

i) About 10–18 Ia fibers drive each monomuscular DSCT-neuron receiving its proprioceptive input from the gastrocnemius muscles only (cf. Eide et al., 1969).

ii) The relative contribution of these effective afferent axons originating at the muscle spindles in one hindlimb muscle to the output impulse pattern induced by mechanical stimulation of the muscle varies very little from one effective input to the next.

The signal transmission from a single Ia fiber to the output impulse pattern of a DSCT-neuron

Recording the discharge pattern of a DSCT-neuron and a presynaptic Ia fiber (Fig. 2) simultaneously, we tested by means of cross-correlograms whether the output signals were driven by the recorded presynaptic fiber. In the positive case, the cross-correlogram between the two impulse sequences might exhibit a periodic oscillation (Fig. 6, steady stretch). A flat null cross-correlogram was obtained when the presynaptic fiber had no connection with the respective DSCT-neuron. At a relatively high stretch level, the contribution of one afferent Ia fiber to the output impulse pattern consisted of an increase in the impulse probability of 0.05–0.08 for the time 3–4 msec after the Ia impulses (distance of the recording electrodes about 5 cm). Thereafter, the discharge probability decreased within 2–3 msec to the spontaneous value. The autocorrelogram of the DSCT-neuron impulse rate exhibited an absolute refractory period of about 1 msec and a recovery time constant of about 15–25 msec (Fig. 6a).

Another way to measure the contribution of the impulse pattern carried along a single excitatory axon forming terminals at a DSCT-neuron was by *electrical* stimulation of a small fiber bundle containing *one* effective Ia axon disconnected from its muscle spindle receptor. Such fiber bundles were prepared during the experiments described in the preceding Section (Fig. 1b). We stimulated these microfilaments with electrical stimulus trains of different frequencies or with electrical stimulus trains of a constant *average* stimulus rate using a computer-generated random distribution of successive stimulus intervals. The interval histogram of the latter stimulus sequence had a squarewave shape between a minimal and maximal interval; both values could be varied by the computer program. Figs. 6b and c exhibit auto- and cross-correlograms obtained with such presynaptic random impulse sequences. From these figures, one can recognize how the average output rate of DSCT-neurons depended on the input impulse rate of one excitatory Ia fiber. This relationship was a non-linear function of the input impulse rate:

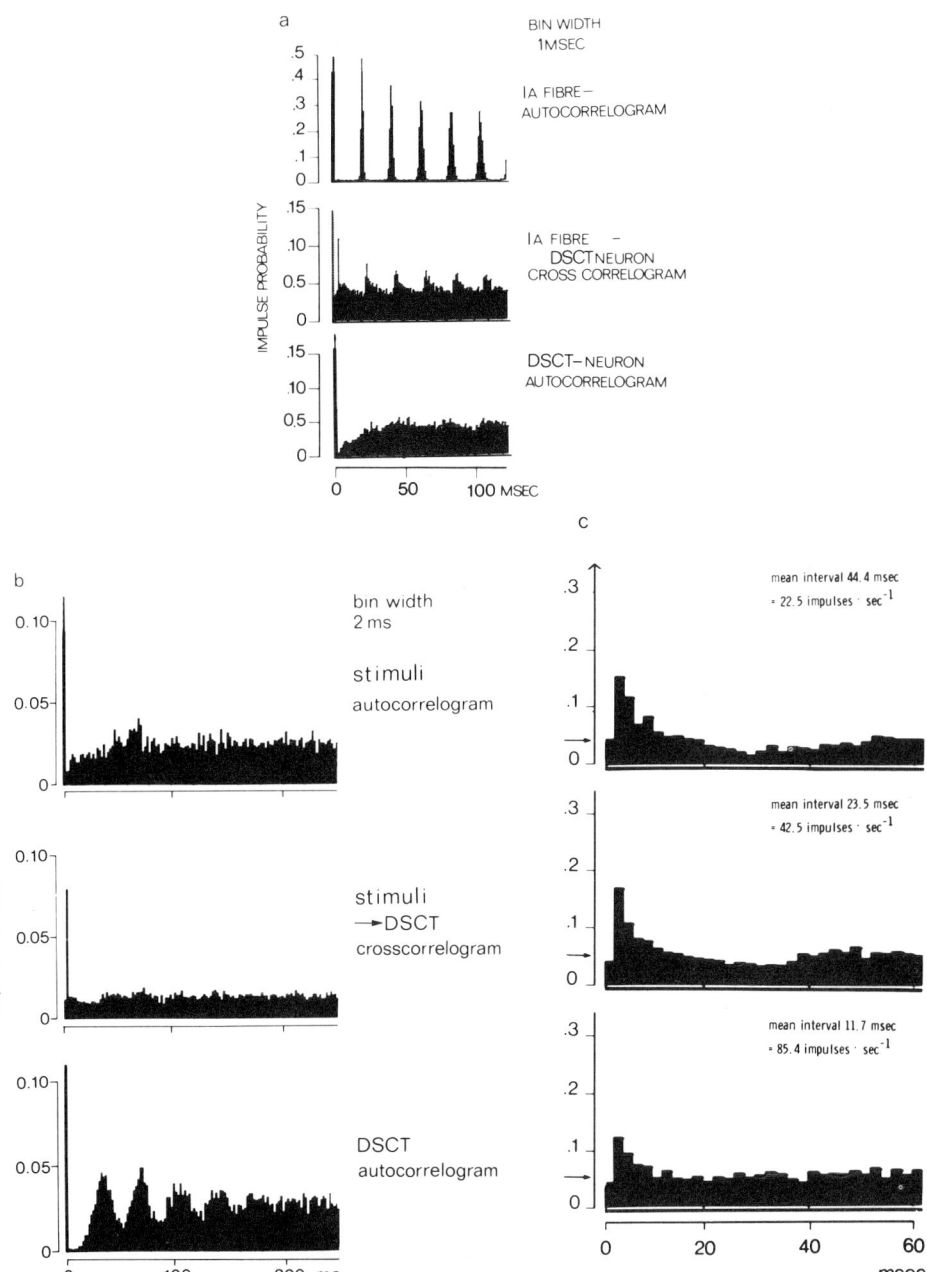

Fig. 6. a) Auto-correlogram of a Ia-gastrocnemius muscle spindle afferent, cross-correlogram between this muscle spindle afferent and a DSCT-neuron activated by this afferent fiber and auto-correlogram of the DSCT-neuron discharge pattern. b) Auto-correlogram of the electrical stimulus train applied to a dorsal root filament containing one effective Ia-fiber, cross-correlogram between the stimulus pattern and the DSCT-neuronal discharge, and DSCT-neuronal discharge auto-correlogram. Note the oscillatory behaviour of the latter without corresponding oscillations in the presynaptic interval distribution. c) Cross-correlogram between aperiodic electrical stimulus sequence applied to Ia-afferent fiber and the response pattern of a DSCT-neuron excited by this afferent fiber. Values obtained at three different mean stimulus rates (22.5, 42.5 and 85.4 impulses.sec^{-1}) applied to the presynaptic axon.

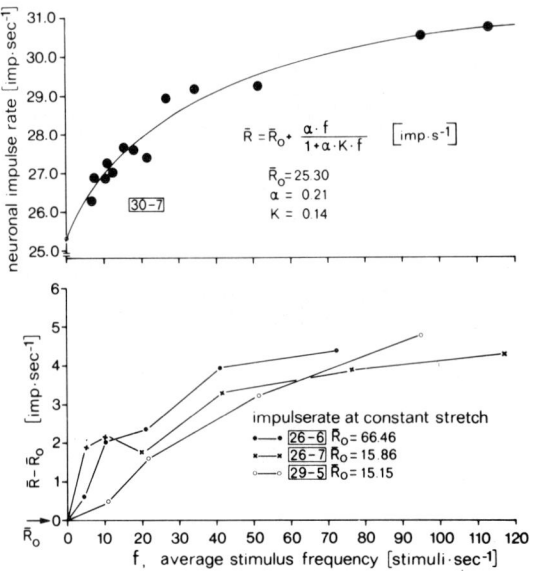

Fig. 7. Relationship between the average neuronal impulse rate (ordinate) and the average stimulus frequency obtained in four different DSCT-neurons. Stimulation of a small dorsal root filament containing 1 exciting Ia-fiber. In the upper part of the figure, the curve computed according to the hyperbolic self-inhibition function, fitting the experimental data optimally, is drawn.

$$\bar{R} = R_o + \frac{S_r}{1 + k_i S_r} \quad [\text{impulses.sec}^{-1}] \qquad (4)$$

whereby R_o is the impulse rate without stimulation of the afferent axon, S_r the average stimulus rate and k_i a constant (Fig. 7).

The maximum coupling between presynaptic and postsynaptic discharges was reached 2–3 msec after the stimuli (increase in discharge probability at max. 0.1) and decreased thereafter within 4 msec to a stimulus-independent spontaneous value. The auto-correlograms of some DSCT-neurons exhibited some oscillations despite the aperiodic presynaptic stimulus train. This indicates an intraspinal or intraneuronal feedback mechanism.

DISCUSSION

The temporal and spatial summing properties of DSCT-neurons were restricted essentially to the first 4 msec after a presynaptic action potential arrived at Ia terminals and elicited a postsynaptic EPSP after a synaptic delay ≤ 0.6 msec. The study of the quantitative interaction of different Ia fiber inputs contributing to the output discharge pattern of one DSCT-neuron indicated a rather similar efficiency in the synaptic contacts formed by many Ia fibers at the neuron. In addition, the excitatory effects elicited by *different* Ia axons were summed fairly linearly as long as the discharge patterns of the different presynaptic fibers were not correlated to each other. In contrast, a strong non-linear component was present when the influence of the *presynaptic impulse rate of one presynaptic Ia axon* on the postsynaptic discharge pattern was tested. Because this non-linearity (Fig. 7) was not dependent on the actual discharge rate of the DSCT-neurons, a presynaptic inhibitory interaction of the signals mediated via interneurons should be discussed. Another possible explanation for this self-inhibiting effect regulating the efficiency of one presynaptic axon terminal would be a frequency filtering by the synaptic processes or the local subsynaptic membrane.

The number of Ia fibers converging on a single DSCT-neuron (12–18) corresponds approximately to the number of large dendrites of Clarke's column cells (Loewy, 1970; and Fig. 2). It is, therefore, very tempting to assume that by the serial synapses described in anatomical studies (Albert and Szentágothai, 1955; Réthelyi and Szentágothai, 1973), each afferent fiber occupies only one dendrite of a nerve cell in Clarke's column.

Finally, we think that our data give clear evidence that the contribution of the temporal discharge pattern in individual Ia fibers exciting a DSCT-neuron to the output pattern of that neuron is almost negligible. Presynaptic input patterns appear at a postsynaptic level only when the impulse pattern of presynaptic exciting axons shows a stimulus-induced synchronization. Therefore the simple data analysis of PST-histograms, rather than a sophisticated impulse pattern analysis, probably leads to physiologically meaningful interpretations (cf. Eckhorn et al., 1976), when the responses of DSCT-neurons to mechanical stimulation are analyzed.

SUMMARY

By means of tungsten microelectrodes, the action potentials of muscle spindle afferents and of DSCT-neurons were recorded. For quantitative studies we selected DSCT-neurons activated predominantly by Ia fibers originating in the gastrocnemius muscle.

1. Each DSCT-neuron receives the input from 12–18 muscle spindle afferents, predominantly of the Ia-type.

2. The contribution of each afferent muscle spindle with an excitatory input to a DSCT-neuron is very similar for all 12–18 afferents. Therefore, the DSCT-neuron activity decreased approximately linearly with the number of effective afferent Ia axons cut.

3. The number of large dendrites of a Clarke column cell corresponds well with the number of muscle spindle afferent fibers with excitatory terminals at a single DSCT-neuron.

4. Each effective muscle spindle afferent contributed 3–4 impulses/sec to the impulse pattern elicited by sinewave stretching of the gastrocnemius muscle (1.5–2 mm amplitude, 3–4 Hz frequency).

5. Simultaneous recordings of the action potentials of DSCT-neurons and Ia fibers were performed. Auto-correlogram and cross-correlogram techniques were applied to estimate the contribution of a single presynaptic Ia fiber to the postsynaptic DSCT discharge pattern.

6. With sinewave stimuli or mechanical noise stimulation (Gaussian amplitude distribution), the impulse pattern of muscle spindle afferent fibers clearly exhibits non-linear components with respect to the stimulus. This non-linearity was present in the discharge pattern of the DSCT-neurons, but in the latter the deterministic component of the impulse pattern decreased, as compared to the presynaptic Ia fibers. Within the frequency range 0.1–100 Hz, the DSCT-neurons constitute a wide band filter with little distortion of the signals about muscle length, velocity and acceleration transmitted to these neurons by the afferent Ia fibers.

ACKNOWLEDGEMENTS

The work was supported by grants of the Deutsche Forschungsgemeinschaft (Gr 161). We thank Ing. L. Weiss for the computer programs, Mrs. M. Klingbeil and Mrs. J. Dames for careful technical assistance and Mrs. Hauschild for typing the manuscript. The mechanical stimulation apparatus was built in the precision shop of our Institute by Mr. A. Ewald. We thank Mr. H. Dannenberg for valuable technical advice in the construction of this stimulus apparatus

REFERENCES

Eckhorn, R., Grüsser, O.-J., Kröller, J., Pellnitz, K. and Pöpel, B. (1976) Efficiency of different neuronal Codes: information transfer calculations for three different neuronal systems. *Biol. Cybernetics.*, 22: 49–60.

Eide, E., Fedina, L., Jansen, J., Lundberg, A. and Vyklicky, L. (1967) Unitary excitatory postsynaptic potentials in Clarke's column neurons. *Nature (Lond.)*, 215: 1176–1177.

Eide, E., Fedina, L., Jansen, J., Lundberg, A. and Vyklicky, L. (1969a) Properties of Clarke's column neurones. *Acta physiol. scand.*, 77: 125–144.

Eide, E., Fedina, L., Jansen, J., Lundberg, A. and Vyklicky, L. (1969b) Unitary components in the activation of Clarke's column neurones. *Acta physiol. scand.*, 77: 145–158.

Eysel, U.Th. and Grüsser, O.-J. (1970) The impulse pattern of muscle spindle afferents. A statistical analysis of the response to static and sinusoidal stimulation. *Pflügers Arch.*, 315: 1–26.

Grashuis, I.L. (1977) The Pre-event Stimulus Ensemble; an Analysis of the Stimulus Response Relation for Complex Stimuli. Ph.D. dissertation. Univ. of Nijmegen, pp. 155.

Grüsser, O.-J. and Thiele, B. (1968) Reaktionen primärer und sekundärer Muskelspindelafferenzen auf sinusförmige mechanische Reizung, I. Variation der Sinusfrequenz. *Pflügers Arch.*, 300: 161–184.

Gustaffson, B. and Zangger, P. (1978) Effect of repetitive activation on the afterhyperpolarization in dorsal spinocerebellar tract neurones. *J. Physiol. (Lond.)*, 275: 303–319.

Jansen, J.K.S. and Rudjord, T. (1965) Dorsal spinocerebellar tract. Response pattern of nerve fibres to muscle stretch. *Science*, 149: 1109–1111.

Jansen, J.K.S., Nicolaysen, K. and Rudjord, T. (1966) Discharge pattern of neurons of the dorsal spinocerebellar tract activated by static extension of primary endings of muscle spindles. *J. Neurophysiol.*, 29: 1061–1086.

Jansen, J.K.S., Poppele, R.E. and Terzuolo, C.A. (1967) Transmission of proprioceptive information via the dorsal spinocerebellar tract. *Brain Res.*, 6: 382–384.

Kröller, J. (1978) Recordings from dorsal spinocerebellar tract fibers during sinusoidal stretching of the cat's gastrocnemius muscle. *Neurosci. Lett.*, Suppl. 1: S99,

Kuno, M., Muños-Martinez, E.J. and Randić, M. (1973a) Sensory inputs to neurones in Clarke's column from muscle, cutaneous and joint receptors. *J. Physiol. (Lond.)*, 228: 327–342.

Kuno, M., Muñoz-Martinez, E.J. and Randić, M. (1973b) Synaptic action on Clarke's column neurones in relation to afferent terminal size. *J. Physiol. (Lond.)*, 228: 343–360.

Loewy, A.D. (1970) A study of neuronal types in Clarke's column in the adult cat. *J. Comp. Neurol.*, 139: 53–80.

Lundberg, A. (1964) Ascending spinal hindlimb pathways in the cat. In *Progress in Brain Research, Vol. 12, Physiology of Spinal Neurons*, J.C. Eccles and J.P. Schadé (Eds.). Elsevier, Amsterdam.

Matthews, P.B.C. (1972) *Mammalian Muscle Receptors and their Central Actions*. Arnold, London.

Matthews, P.B.C. and Stein, R.B. (1969) The regularity of primary and secondary muscle spindle afferent discharges. *J. Physiol. (Lond.)*, 202: 59–82.

Oscarsson, O. (1965) Functional organisation of the spino- and cuneocerebellar tracts. *Physiol. Rev.*, 45: 495–522.

Oscarsson, O. (1973) Functional organisation of spinocerebellar paths. In *Handbook of Sensory Physiology, Vol. II, Somatosensory System*. A. Iggo (Ed.). Springer, Berlin-Heidelberg-New York, pp. 339–380.

Réthelyi, M. and Szentágothai, J. (1973) Distribution and connections of afferent fibers in the spinal cord. In *Handbook of Sensory Physiology, Vol. II, Somatosensory System*. A. Iggo (Ed.). Springer, Berlin-Heidelberg-New York, pp. 207–252.

Szentágothai, J. and Albert, A. (1955) The synaptology of Clarke's column. *Acta Morph. Hung.*, 5: 43–51.

Walløe, L. (1968) *The Transfer of Signals Through a Second Order Sensory Neuron*. Thesis, Oslo University.

Effects of High Threshold Muscle Afferent Volleys on Ascending Pathways

W.D. WILLIS

Marine Biomedical Institute, University of Texas Medical Branch at Galveston, TX 77550 (U.S.A.)

INTRODUCTION

High threshold muscle afferents can be taken to include fibers belonging to groups II, III and IV of the Lloyd and Chang Classification (1948; cf., Eccles and Lundberg, 1959; Mense and Schmidt, 1974). These afferents are known to excite neurons contributing to several ascending pathways, including the spinoreticular, spinocerebellar and spinocervical tracts (Lundberg and Oscarsson, 1961, 1964; Oscarsson, 1977; Kniffke et al., 1977). Recently, our laboratory has been able to demonstrate that high threshold muscle afferents also activate neurons which belong to the spinothalamic tract in the monkey (Foreman et al., 1977, 1979a, b). This work will be reviewed here.

METHODS

The animals were cynomolgous monkeys *(Macaca fascicularis)*. Anesthesia was induced with halothane and nitrous oxide and maintained with α-chloralose and a continuous infusion of sodium pentobarbital. The lumbosacral spinal cord was exposed by laminectomy, and a craniotomy allowed the introduction of a concentric bipolar stimulating electrode into the thalamus. Spinothalamic tract cells were identified in extracellular recordings using glass microelectrodes (4 M NaCl) by antidromic activation from the contralateral thalamus (cf., Trevino et al., 1973). The criteria for antidromic invasion included collision with orthodromic spikes. Muscle afferents were activated either by electrical stimulation or by intra-arterial injection of algesic chemicals into the circulation of the triceps surae muscles (cf. Franz and Mense, 1975). In the latter case, the limb was denervated except for the nerves to the triceps surae muscles.

RESULTS

Electrical stimulation of muscle afferents

The responses of 92 spinothalamic tract cells to electrical stimulation of one or more muscle nerves were tested. Using graded strengths of stimulation, it was possible to determine which fiber groups had an excitatory action. Only 19 of the spinothalamic tract cells studied could be activated by group I fibers. However, in 4 of these cases the excitation was apparently monosynaptic. For example, in Fig. 1A, a spinothalamic tract cell discharged when the hamstring nerve was stimulated at a strength of 1.4 times the threshold for the largest fibers. The latency of the spike was reduced to less than 1 msec as the group I volley was increased to maximum size (Fig. 1D).

Spinothalamic tract cells were much more responsive to high threshold muscle afferents than to group I afferents. The cell in Fig. 1 was only discharged once with a maximum group I volley, yet it discharged repetitively when group II (Fig. 1B) or group III (Fig. 1C) afferents were stimulated. Fig. 2 summarizes the stimulus strengths at which graded muscle afferent volleys first caused a discharge of spinothalamic tract cell. Most of the cells were activated by group II fibers. However, some required the inclusion of group III fibers in the afferent volley. Some spinothalamic tract cells could not be excited by stimulation of the muscle nerves employed.

Electrical stimulation at a strength sufficient to activate group IV muscle afferents produced a late discharge in the three experiments in which this was tried. Fig. 3 is a poststimulus time histogram showing the discharges produced by such a stimulus. The initial discharge occurred in two peaks which were due to the actions of group II and III fibers, respectively. The late discharge was attributed to group IV fibers.

Chemical stimulation of high threshold muscle afferents

The intra-arterial injection of algesic chemicals (bradykinin, serotonin, potassium ions) has been found in cats to excite selectively the group III and IV, but not group I and II, muscle afferents (Mense and Schmidt, 1974; Franz and Mense, 1975; Fock

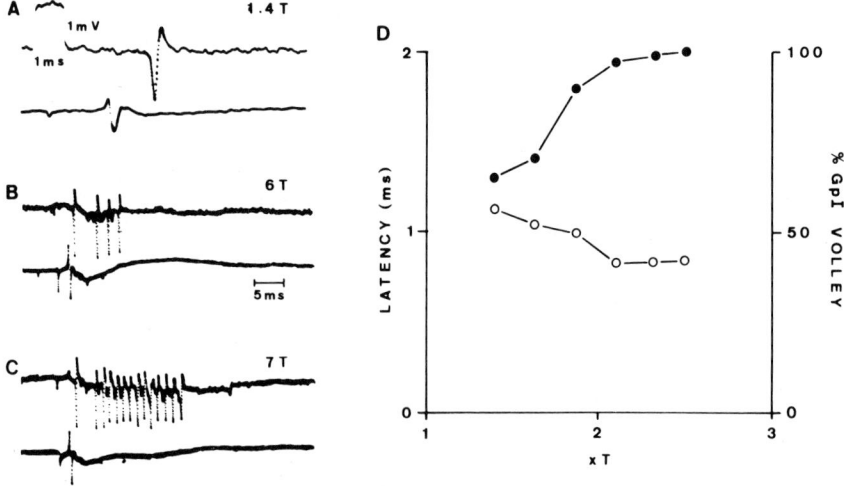

Fig. 1. Responses of a spinothalamic tract cell to graded stimulation of the hamstring nerve. The upper records in A–C are microelectrode recordings and the lower records are from the cord dorsum. The stimulus strengths applied to the hamstring nerve are indicated in multiples of threshold for the most excitable afferents: 1.4, 6 and 7T. The graph in D shows the size of the group I afferent volley as a percentage of maximum (●, right ordinate) and the latency of the spike evoked by the group I volley (○, left ordinate). (From Foreman et al., 1978a.)

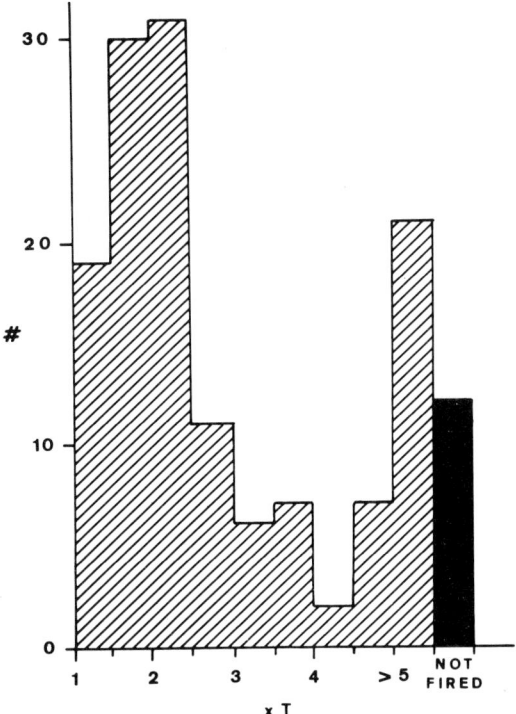

Fig. 2. Histogram showing the threshold strengths for eliciting initial discharges in spinothalamic tract cells using electrical stimulation of muscle nerves. The abscissa indicates the strength as a multiple of threshold for the most excitable fibers of the nerves, and the ordinate shows the numbers of cells that could be activated at the indicated stimulus strengths or at higher strengths. (Modified from Foreman et al., 1978a.)

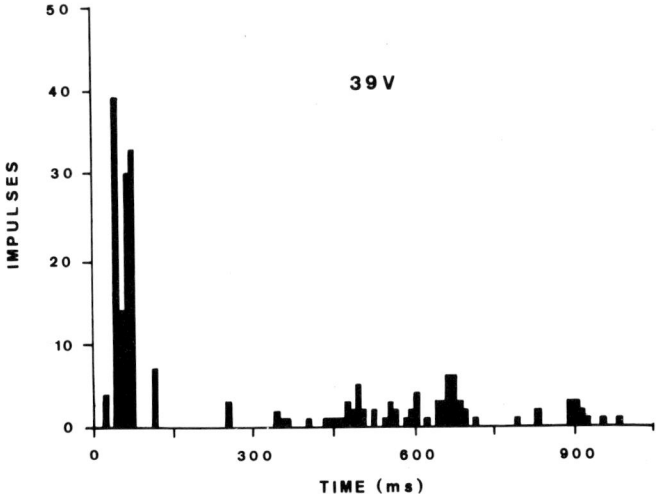

Fig. 3. Responses of a spinothalamic tract cell to an afferent volley in groups II, III and IV. The myelinated muscle afferents in the gastrocnemius-soleus nerve evoked the initial bimodal discharge, whereas the unmyelinated afferents were responsible for much of the late activity. (Modified from Foreman et al., 1978a.)

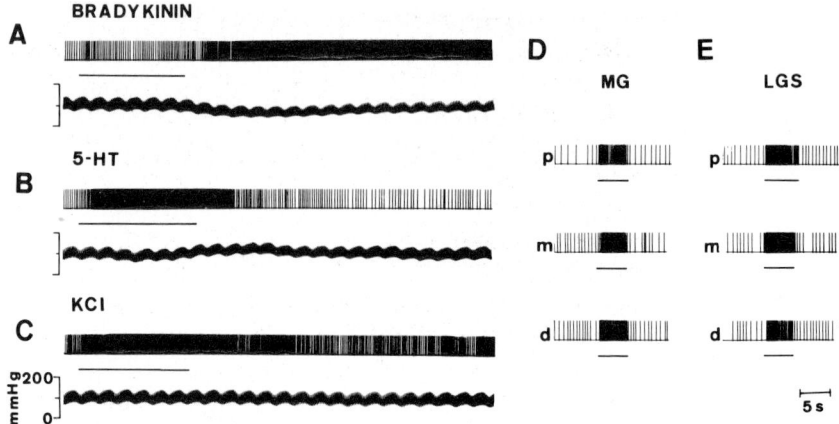

Fig. 4. Responses of a spinothalamic tract neuron to the intra-arterial injection of algesic chemicals. The upper traces in A–C show window discriminator pulses triggered by the discharges of the spinothalamic tract cell. The lower traces show the systemic arterial blood pressure. At the times indicated by the bars, algesic chemicals were injected into the arterial circulation of the gastrocnemius-soleus muscles. The doses were: bradykinin, 26 μg; serotonin (5-HT), 135 μg; KCl, 3.8 mg. In D and E are the responses to mechanical stimulation of the medial gastrocnemius (MG) and the lateral gastrocnemius-soleus (LGS) muscles with a glass probe. The proximal, middle and distal parts of the muscles were stimulated separately (p, m and d). (From Foreman et al., 1978b.)

and Mense, 1976; Mense, 1977). Preliminary evidence in our laboratory indicates that muscle afferents in the monkey respond in much the same way (Foreman et al., 1978b).

When algesic chemicals were introduced into the arterial circulation of the triceps surae muscles of the monkey, it was found that most spinothalamic tract cells (41 of 50) were excited by one or more of these agents. For instance, in Fig. 4A–C, bradykinin, serotonin (5-HT) and KCl all produced a greatly accelerated discharge of the spinothalamic tract neuron under study. The same cell could also be excited when the bellies of the medial gastrocnemius (MG) and lateral gastrocnemius-soleus (LGS) muscles were probed with a glass rod (Fig. 4D–E). The time course of the excitatory action of bradykinin was slower than that of the other agents, as seen both in the records from an individual neuron (Fig. 4A–C) and in the averaged poststimulus time histograms in Fig. 5.

Another technique for activating high threshold afferents is by injection of hypertonic saline into a muscle or its tendon (Lewis, 1942). In the case illustrated in Fig. 6, the injection was into the Achilles tendon. However, similar results were obtained with intramuscular injections. The discharges of many spinothalamic tract neurons were accelerated for 10–15 min following an initial peak of activity just after the injection.

Succinylcholine injections into the arterial circulation of the triceps surae muscles were made while recording from 39 spinothalamic tract cells. There was never a strong effect, although occasionally there was a weak excitation. Generally, this could be mimicked by injecting Tyrode's solution and was perhaps due to the injection pressure.

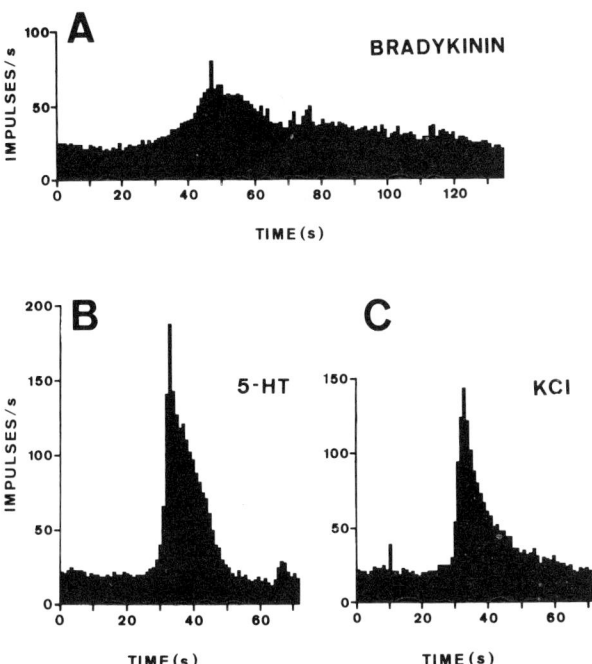

Fig. 5. Time course of excitatory effects of algesic chemicals on spinothalamic tract cells. The histograms are averaged responses from 10 different spinothalamic tract neurons. The first 30 sec of each represents baseline activity. (From Foreman et al., 1978b.)

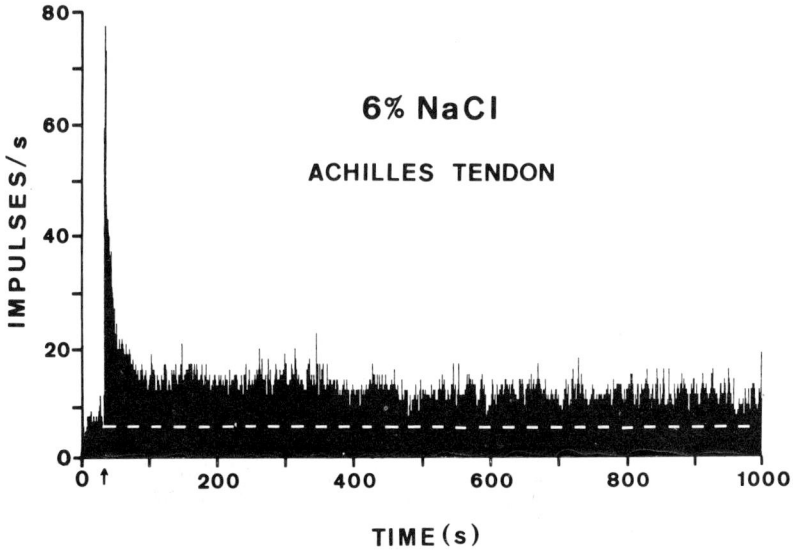

Fig. 6. Effect of hypertonic saline injection into the Achilles tendon on the discharges of a spinothalamic tract neuron. A dose of 0.1 ml of 6% NaCl was injected at the time indicated by the arrow. (From Foreman et al., 1978b.)

DISCUSSION

This study has shown that muscle afferents, especially those belonging to groups II, III and IV, can produce a powerful excitation of many neurons belonging to the spinothalamic tract. A similar conclusion has recently been reached regarding the spinocervical tract (Kniffke et al., 1977). Since the spinothalamic tract cells were antidromically activated by stimulation within the ventral posterior lateral nucleus of the thalamus (Foreman et al., 1978b), it is likely that information originating from the appropriate muscle receptors reaches the somatic sensory areas of the cerebral cortex.

It is not clear what role the group I and group II afferents which excite spinothalamic tract cells might have in proprioception. Only a few of the cells were activated by group I fibers and these not powerfully. Group II fibers commonly had strong excitatory effects, but it must be questioned whether these group II fibers innervated muscle spindles. The spinothalamic neurons tested were not very responsive to intra-arterial injections of succinylcholine, and it is known that some group II fibers (and even group I fibers) in the cat originate from "pressure-pain" endings (Paintal, 1960). Perhaps the actions of group II afferents (and even group I afferents) on spinothalamic tract cells in the monkey reflect an influence of such pressure-pain endings, rather than of muscle spindles.

The excitation of spinothalamic cells by group III and IV afferents, using either electrical or algesic chemical stimuli, is in keeping with the known role of this tract in nociception (Yoss, 1953; White and Sweet, 1955). The ability of comparable stimuli to excite spinocervical tract cells in the cat (Kniffke et al., 1977) raises the question of the involvement of that pathway also in nociception (cf., Kennard, 1954; Cervero et al., 1977).

SUMMARY

The effects of high threshold muscle afferents upon spinothalamic tract cells in the primate were investigated by the use of electrical stimulation of muscle nerves and by chemical stimulation of the sensory endings. Although some spinothalamic tract cells were excited by group I afferents, most of the cells receiving a muscle input required for excitation the inclusion of group II, III or IV afferents in the afferent volleys. Some spinothalamic tract cells appeared not to receive a muscle input, but this could have been a sampling error. Algesic chemicals injected intra-arterially into the circulation of the triceps surae muscles produced a powerful excitation of many spinothalamic tract cells. The agents used included bradykinin, serotonin and potassium ions. A prolonged discharge could also be evoked by the injection of hypertonic saline into the triceps surae muscles or the Achilles tendon. Since intra-arterial succinylcholine injections did not produce any significant effect on spinothalamic tract cells, it is possible that the group II (and I) afferents which excite spinothalamic cells originate from receptors other than muscle spindles. It is speculated that spinothalamic tract cells help mediate muscular pain.

ACKNOWLEDGEMENTS

The author wishes to thank Gail Silver and Kathe Whitten for their expert technical assistance. The work was done in collaboration with Drs. R.D. Foreman, D.R. Kenshalo Jr. and R.F. Schmidt and was supported by NIH research grants NS 09743 and NS 18728, NIH postdoctoral fellowship NS 05698, and by the Deutsche Forschungsgemeinschaft.

REFERENCES

Cervero, F., Iggo, A. and Molony, G. (1977) Responses of spinocervical tract neurones to noxious stimulation of the skin. *J. Physiol. (Lond.)*, 267: 537–558.

Eccles, R.M. and Lundberg, A. (1959) Synaptic actions in motoneurones by afferents which may evoke the flexion reflex. *Arch. ital. Biol.*, 97: 199–221.

Fock, S. and Mense, S. (1976) Excitatory effects of 5-hydroxytryptamine, histamine and potassium ions on muscular group IV afferent units: a comparison with bradykinin. *Brain Res.*, 105: 459–469.

Foreman, R.D., Schmidt, R.F. and Willis, W.D. (1977) Convergence of muscle and cutaneous input onto primate spinothalamic tract beurons. *Brain Res.*, 124: 555–560.

Foreman, R.D., Kenshalo Jr., D.R., Schmidt, R.F. and Willis, W.D. (1979a) Field potentials and excitation of primate spinothalamic neurones in response to volleys in muscle afferents. *J. Physiol. (Lond.)*, 286: 197–213.

Foreman, R.D., Schmidt, R.F. and Willis, W.D. (1979b) Effects of mechanical and chemical stimulation of fine muscle afferents upon primate spinothalamic tract cells. *J. Physiol. (Lond.)*, 286: 215–231.

Franz, M. and Mense, S. (1975) Muscle receptors with group IV afferent fibres responding to application of bradykinin. *Brain Res.*, 92: 369–383.

Kennard, M.A. (1954) The course of ascending fibers in the spinal cord of the cat essential to the recognition of painful stimuli. *J. comp. Neurol.*, 100: 511–524.

Kniffke, D.K., Mense, S. and Schmidt, R.F. (1977) The spinocervical tract as a possible pathway for muscular nociception. *J. Physiol. (Paris)*, 73: 359–366.

Lewis, T. (1942) *Pain*, Macmillan, New York.

Lloyd, D.P.C. and Chang, H.T. (1948) Afferent fibers in muscle nerves. *J. Neurophysiol.*, 11: 199–208.

Lundberg, A. and Oscarsson, O. (1961) Three ascending spinal pathways in the dorsal part of the lateral funiculus. *Acta physiol. scand.*, 51: 1–16.

Lundberg, A. and Oscarsson, O. (1962) Two ascending spinal pathways in the ventral part of the cord. *Acta physiol. scand.*, 54: 270–286.

Mense, S. (1977) Nervous outflow from skeletal muscle following chemical noxious stimulation. *J. Physiol. (Lond.)*, 267: 75–88.

Mense, S. and Schmidt, R.F. (1974) Activation of group IV afferent units from muscle by algesic agents. *Brain Res.*, 72: 305–310.

Oscarsson, O. (1973) Functional organization of spinocerebellar paths. In *Handbook of Sensory Physiology, Vol. II. Somatosensory System.* H. Iggo (Ed.). Springer, New York, pp. 339–380.

Paintal, A.S. (1960) Functional analysis of group III afferent fibres of mammalian muscles. *J. Physiol. (Lond.)*, 152: 250–270.

Trevino, D.L., Coulter, J.D. and Willis, W.D. (1973) Location of cells of origin of spinothalamic tract in lumbar enlargement of the monkey. *J. Neurophysiol.*, 36: 750–761.

White, J.C. and Sweet, W.H. (1955) *Pain, Its Mechanisms and Neurosurgical Control.* Thomas, Springfield, IL.

Yoss, R.E. (1953) Studies of the spinal cord. Part 3. Pathways for deep pain within the spinal cord and brain. *Neurology*, 3: 163–175.

Proprioceptive Influences on Somatosensory and Motor Cortex

P. ZARZECKI and H. ASANUMA

Department of Physiology, Queen's University, Kingston, Ont. K7L 3N6 (Canada); and The Rockefeller University, New York, NY 10021 (U.S.A.)

Neurons of the motor cortex respond to activation of peripheral somatosensory and proprioceptive afferents (Malis et al., 1953; Buser and Imbert, 1961; Brooks et al., 1961a, b). Inputs to particular regions of the motor cortex are related to the motor effects elicited from the same regions (Asanuma et al., 1968; Rosén and Asanuma, 1972), and these inputs are known to modify motor behavior (Conrad et al., 1974; cf. also Murphy et al., 1974, 1975; Lucier et al., 1975). However, despite the probable importance of feedback from the periphery in the control of the output of the motor cortex (for reviews see Brooks and Stoney, 1971; Asanuma, 1975), a determination has not been made of the central pathways relaying precise, somatotopically organized input to the motor cortex. Among suspected routes are thalamo-corticocortical, involving sensory cortex, and direct pathways to the motor cortex from the thalamus. Some studies involving ablation or cooling of the sensory cortex have led to the conclusion that corticocortical connections are not necessary for the activation of the motor cortex by peripheral stimuli (Malis et al., 1953; Murphy et al., 1975), while others have concluded that a sensory cortex relay is involved (Silfvenius, 1970). The peripheral stimuli adequate for activation of thalamic neurons have been examined (e.g., Poggio and Mountcastle, 1963). However, in the one previous experiment where the response properties of identified thalamocortical neurons were examined (Asanuma et al., 1974), neurons of the nucleus ventralis lateralis (VL) did not account for the circumscribed and well-defined receptive fields displayed by many neurons of the motor cortex (Brooks et al., 1961a, b; Asanuma and Rosén, 1971). Therefore, an uncertainty remains as to the central pathways followed by inputs to the motor cortex. In particular, it is not agreed which inputs are relayed through the sensory cortex nor which are transmitted to the motor cortex directly from the thalamus.

This report will describe recently observed features of the peripheral activation of identified corticocortical and thalamocortical neurons which project to the motor cortex.

INPUT RELAYED TO THE MOTOR CORTEX VIA AREA 3A

Connections from the sensory cortex to the motor cortex have been detected in anatomical (Jones and Powell, 1968) and electrophysiological experiments

(Thompson et al., 1970). Furthermore, it has been demonstrated (Grant et al., 1975) that a projection to the motor cortex originates from that part of cortical area 3a which is the receiving area for group I muscle afferents of forelimb nerves (Amassian and Berlin, 1958; Oscarsson and Rosén, 1963).

To investigate the peripheral input which might be relayed to the motor cortex by the latter corticocortical connection, recordings were made from neurons of area 3a while stimulating the motor cortex and a peripheral muscle nerve of the cat. The experimental arrangement is illustrated in Fig. 1A (Zarzecki et al., 1978). The rostral lid of the closed chamber supported an array of stimulating microelectrodes inserted into the focus for group I evoked potentials of the motor cortex (Silfvenius, 1968, 1970). Another microelectrode was inserted into area 3a to record the antidromic responses following microstimulation of the motor cortex. These responses were identified in extracellular recordings in the usual manner, as illustrated in Fig. 1B, C. A group I excitatory input to a corticocortical neuron was detected when precedent peripheral nerve stimulation caused a failure of the antidromic response. Additionally, EPSPs evoked from the peripheral nerve in corticocortical neurons (Fig. 2B) often appeared at a latency compatible with a monosynaptic thalamocortical input (cf. also Oscarsson et al., 1966). It was thus demonstrated that information from group I muscle afferents is relayed to the motor cortex through a thalamo-corticocortical path by way of area 3a.

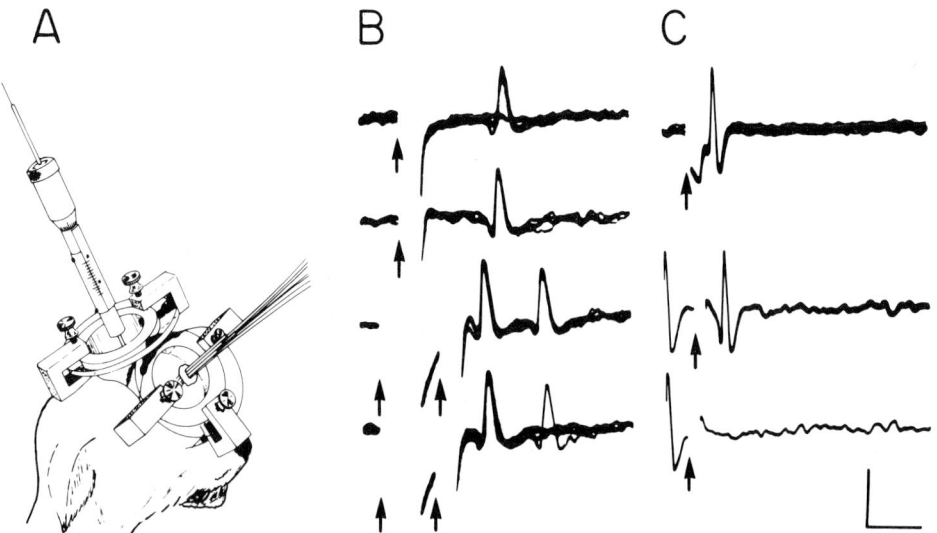

Fig. 1. A) Sketch of the closed chamber system used for electrophysiological investigations of corticocortical and thalamocortical connections in cats. An array of stimulating microelectrodes (right) was inserted into the motor cortex to activate the axons of neurons projecting there from cortical area 3a (Zarzecki et al., 1978a) or the thalamus (Asanuma et al., 1979b) and the antidromic response of these neurons was recorded with the electrode on the left. The chamber was slightly modified for thalamic recording. B) The identification of the antidromic response of a corticocortical neuron by its short, fixed latency at threshold motor cortex stimulation (12 μa, 1st trace), little change in latency when the stimulus was increased to $1.5 \times T$ (2nd trace) and by its short refractory period (1.0 msec, 4th trace). Four sweeps superimposed. Stimulus onset indicated by arrows. C) Failure of the antidromic response of another neuron when the stimulus ($1.5 \times T = 15 \mu A$) follows its spontaneously occurring spikes at short interval (compare 2nd and 3rd traces, single sweeps). Voltage calibration = 200 μV; time calibration = 1.0 msec (B), 2.0 msec (C).

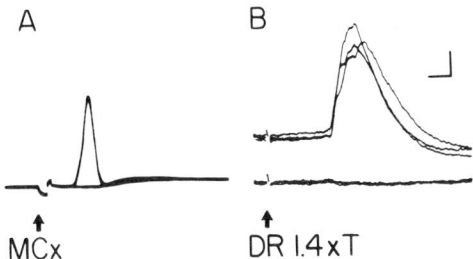

Fig. 2. Intracellular records from a corticocortical neuron of area 3a. Four sweeps superimposed. A) The antidromic response evoked by threshold (20 μA) stimulation of the motor cortex. B) EPSP evoked in the same neuron by group I strength stimulation of the contralateral deep radial nerve. The second trace in B was recorded after withdrawal of the microelectrode from the neuron. Voltage calibration = 10 mV (A) and 2 mV (B). Time calibration = 0.5 msec (A) and 2.0 msec (B). (From Zarzecki et al., 1978a.)

Some unidentified neurons of area 3a receive convergent inputs from several forelimb nerves (Oscarsson et al., 1966), as well as hindlimb and vestibular afferents (Ödkvist et al., 1975). It was therefore of interest to test for convergence of somatosensory and proprioceptive inputs upon identified corticocortical neurons. In a recent series of experiments (Zarzecki and Wiggin, 1979), stimuli were delivered to the deep radial nerve (muscle afferents), the superficial radial nerve (cutaneous afferents), the median nerve (mixed) and the ulnar nerve (mixed), the latter being in most cases divided into its dorsal cutaneous and palmar branches. Collision tests (as illustrated in Fig. 3D–L) were performed to confirm that the spikes evoked from the periphery occurred in the same neuron which was antidromically activated from the motor cortex. As illustrated in Fig. 3, we observed corticocortical neurons which received convergent inputs from nerves innervating dorsal and ventral aspects of the forelimb. Convergence was also observed from group I muscle afferents and low threshold (≤ 1.5 times threshold) cutaneous afferents of nerves having a similar distribution in the forelimb (deep and superficial radial nerves). Other corticocortical neurons could be activated from only one nerve or from none

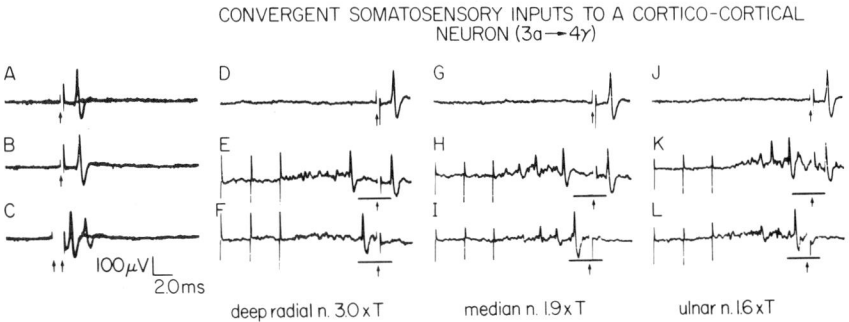

Fig. 3. Convergent sensory inputs to a corticocortical neuron of area 3a. A–C) Identification of the antidromic response in extracellular recordings. Motor cortex stimulation (arrows) = 10 μA in A and 15 μA in B–C. Four sweeps superimposed. E–F, H–I and K–L) Motor cortex stimulation (arrows) preceded by stimulation of indicated peripheral nerve (single sweeps). Intensity of peripheral stimulation in multiple of nerve threshold. Only when the spike evoked from the periphery occurs within the interval equal to two times the latent period plus one refractory period before the expected antidromic response does the latter fail to occur (F, I, L). The critical interval is identified by horizontal line below sweeps. D, G, and J are unconditioned responses for comparison.

(although it should be noted that subliminal excitatory effects are difficult to detect with this method).

The analysis of convergent inputs upon individual corticocortical neurons is continuing with the aim of investigating whether there exist indentifiable subpopulations of these neurons. It is known that corticocortical projections originate from cortical layers II, III and V (Jones et al., 1975, 1978), and we are presently comparing the features of the peripheral activation of neurons in the different layers.

Corticocortical projections to the motor cortex also originate from sensory cortex regions receiving input primarily from cutaneous receptors (Thompson et al., 1970) and connections from more distant cortical regions are known (Jones and Powell, 1968, 1969; Strick and Kim, 1978; Jones et al., 1978). Very little is yet known of the peripheral input activating corticocortical neurons of these regions. However, in the conscious monkey at least some neurons projecting to the motor cortex from parietal association areas respond to joint movement (Zarzecki et al., 1978b). Further investigation is necessary into the effects of these several corticocortical pathways upon the activity of the neurons of the motor cortex.

INPUT RELAYED TO THE MOTOR CORTEX VIA THE THALAMUS

Stimulation of group II afferents of a muscle nerve evokes field potentials at the same latency in both motor and sensory cortices (Asanuma et al., 1979a), as contrasted to the group I evoked potentials which occur with a longer latency in the motor cortex (Silfvenius, 1968, 1970; Zarzecki et al., 1978a). The identical latency inputs to motor and sensory cortices and the finding that potentials evoked in the motor cortex by group II afferents, and also by cutaneous afferents (superficial radial nerve), are not abolished or delayed by extensive removal of the sensory cortex (Asanuma et al., 1979a) suggest that some inputs reach the motor cortex without a relay through the sensory cortex. Among other projections to the motor cortex which might be responsible for the mediation of these responses are those from the nucleus ventralis lateralis of the thalamus (VL) (Amassian and Weiner, 1966; Strick, 1970, 1973; Asanuma et al., 1974). However, the response properties of neurons in VL to peripheral stimuli (Asanuma et al., 1974; Asanuma and Hunsperger, 1975) make them unlikely candidates for a source of the well-circumscribed inputs which reach many motor cortex neurons. Therefore, further investigations were undertaken to identify thalamocortical neurons outside of VL which might carry such information to the motor cortex.

The investigation consisted of two parts, the first of which was an anatomical study of thalamocortical projections in the cat (Larsen and Asanuma, 1979). Injections of horseradish peroxidase into the forelimb region of the motor cortex retrogradely labeled neurons in a "shell" around the rostral portion of the nucleus ventralis posterolateralis (VPL) at its border with VL. These neurons projecting to the motor cortex were distinctly separated from those labeled after injections into the primary somatosensory cortex, but their locations partially overlapped with those labeled by injections into area 3a. This area of the thalamus, which has not previously been identified as projecting to the motor cortex, might be involved in the relay of well-organized somesthetic information to the motor cortex.

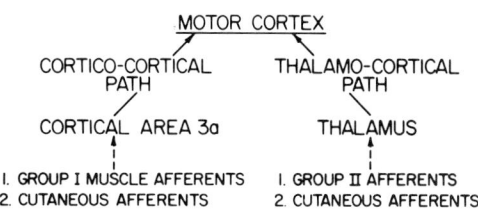

Fig. 4. Summary of investigated pathways to the motor cortex and types of input which have been demonstrated.

In subsequent experiments (Asanuma et al., 1979b), neurons located in the border area between VL and VPL were antidromically activated by microstimulation of the motor cortex of the cat. These identified thalamocortical neurons were found to receive topographically organized somesthetic input arising from skin or deep receptors of the contralateral forelimb. In a few cases comparisons were made with the receptive fields of neurons in the motor cortex where an identified thalamocortical neuron projected. The simultaneously recorded motor cortex and thalamocortical neurons usually received inputs of the same modality from the same body part. Included were pairs of neurons with well-circumscribed, overlapping receptive fields for light touch or pressure. Therefore, these thalamocortical neurons are likely to be contributing to the responses of neurons in the motor cortex to peripheral stimuli.

The input pathways to the motor cortex which have so far been investigated and what is known of the peripheral input which they carry is summarized in Fig. 4. The corticocortical path from area 3a relays group I muscle and cutaneous afferent input to the motor cortex. Individual corticocortical neurons may carry information from both muscle and cutaneous receptors and may receive inputs from nerves innervating dissimilar regions of the forelimb. Inputs from cutaneous afferents and group II afferents of muscle nerves may also reach the motor cortex without a relay through the sensory cortex. Identified thalamocortical neurons in a VL/VPL border region receive inputs from receptors in skin or deep structures, demonstrating that the motor cortex receives some somesthetic inputs directly from the thalamus.

Whether these two input pathways influence the same population of motor cortex neurons or rather function as independent feedback systems remains to be determined. Further information is required regarding the effects of these, and other, input pathways upon identified motor cortex neurons before it can be suggested how they operate in the control of the output of the motor cortex.

SUMMARY

Two input pathways to the motor cortex were investigated in cats, one corticocortical and the other thalamocortical.

1. Group I muscle afferents and cutaneous afferents of forelimb nerves project to cortical area 3a. Identified corticocortical neurons projecting to the motor cortex were among the neurons which these inputs activated. Some individual corticocortical neurons received convergent inputs from low threshold muscle and cutaneous afferents or inputs from dissimilar parts of the forelimb.

2. Group II afferents of forelimb muscle nerves and cutaneous afferents also project to the motor cortex after extensive removal of the sensory cortex. Using horseradish peroxidase a projection to the motor cortex was identified from a "shell" around the rostral portion of the nucleus ventralis posterolateralis (VPL) of the thalamus, at its border with the nucleus ventralis lateralis (VL). Identified thalamocortical neurons of this border region received topographically organized inputs from skin or deep receptors of the forelimb. These inputs were usually of the same modality and from the same body part as those activating neurons of the motor cortex located near the projection of the thalamocortical neuron.

The evidence suggests that these two pathways may be involved in the control of the output of the motor cortex.

ACKNOWLEDGEMENTS

Supported by U.S. NIH grant NS-10705 (H.A.) and by Canadian MRC grants MA-6502 and DG-187 (P.Z.).

REFERENCES

Amassian, V.E. and Berlin, L. (1958) Early cortical projection of group I afferents in the forelimb muscle nerves of the cat. *J. Physiol. (Lond.)*, 143: 61P.

Amassian, V.E. and Weiner, H. (1966) Monosynaptic and polysynaptic activation of pyramidal tract neurons by thalamic stimulation. In *The Thalamus*, D.P. Purpura and M.D. Yahr (Eds.), Columbia University Press, New York, pp. 255–286.

Asanuma, H. (1975) Recent developments in the study of the columnar arrangement of neurons within the motor cortex. *Physiol. Rev.*, 55: 143–156.

Asanuma, H., Fernandez, J., Scheibel, M.E. and Scheibel, A.B. (1974) Characteristics of projections from the nucleus ventralis lateralis to the motor cortex in the cat: an anatomical and physiological study. *Exp. Brain Res.*, 20: 315–330.

Asanuma, H. and Hunsperger, R.W. (1975) Functional significance of projection from the cerebellar nuclei to the motor cortex in the cat. *Brain Res.*, 98: 73–92.

Asanuma, H. and Rosén, I. (1972) Functional role of afferent inputs to the monkey motor cortex. *Brain Res.*, 40: 3–5.

Asanuma, H., Stoney Jr., S.D. and Abzug, C. (1968) Relationship between afferent input and motor outflow in cat motosensory cortex. *J. Neurophysiol.*, 31: 670–681.

Asanuma, H., Larsen, K. and Zarzecki, P. (1979a) Peripheral input pathways projecting to the motor cortex in the cat. *Brain Res.*, in press.

Asanuma, H., Larsen, K. and Yumiya, H. (1979b) Properties of thalamic neurons projecting to the motor cortex in the cat. *Brain Res.*, in press.

Brooks, V.B. and Stoney Jr., S.D. (1971) Motor mechanisms: the role of the pyramidal system in motor control. *Ann. Rev. Physiol.*, 33: 337–392.

Brooks, V.B., Rudomin, P. and Slayman, G.L. (1961a) Sensory activation of neurons in the cat's cerebral cortex. *J. Neurophysiol.*, 24: 286–301.

Brooks, V.B., Rudomin, P. and Slayman, G.L. (1961b) Peripheral receptive fields of neurons in cat's cerebral cortex. *J. Neurophysiol.*, 24: 302–325.

Buser, P. and Imbert, M. (1961) Sensory projections to the motor cortex in cats. In *Sensory Communication*, W.A. Rosenblith (Ed.), Wiley, New York, pp. 607–626.

Conrad, B., Matsunami, K., Meyer-Lohman, J., Wiesendanger, M. and Brooks, V.B. (1974) Cortical load compensation during voluntary elbow movements. *Brain Res.*, 71: 507–514.

Grant, G., Landgren, S. and Silfvenius H. (1975) Columnar distribution of U-fibers from the postcruciate cerebral projection area of the cat's group I muscle afferents. *Exp. Brain Res.*, 24: 57–74.

Jones, E.G. and Powell, T.P.S. (1968) The ipsilateral cortical connexions of the somatic sensory areas in the cat. *Brain Res.*, 9: 71–94.

Jones, E.G. and Powell, T.P.S. (1969) Connexions of the somatic sensory cortex of the rhesus monkey. I. Ipsilateral cortical connections. *Brain.*, 92: 477–502.

Jones, E.G., Burton, H. and Porter, R. (1975) Commissural and cortico-cortical "columns" in the somatic sensory cortex of primates. *Science,* 190: 572–574.

Jones, E.G., Coulter, J.D. and Hendry, S.H.C. (1978) Intracortical connectivity of architectonic fields in the somatic sensory, motor and parietal cortex of monkeys. *J. comp. Neurol.,* 181: 291–348.

Larsen, K. and Asanuma, H. (1979) Thalamic projections to the feline motor cortex studied with horseradish peroxidase. *Brain Res.,* in press.

Lucier, G.E., Rüegg, D.G. and Wiesendanger, M. (1975) Responses of neurons in motor cortex and in area 3a to controlled stretches of forelimb muscles in Cebus monkeys. *J. Physiol. (Lond.),* 251: 833–853.

Malis, L.I., Pribram, K.H. and Kruger, L. (1953) Action potentials in motor cortex evoked by peripheral nerve stimulation. *J. Neurophysiol.,* 16: 161–167.

Murphy, J.T., Wong, W.C. and Kwan, H.C. (1974) Distributed feedback systems for muscle control. *Brain Res.,* 71: 495–505.

Murphy, J.T., Wong, Y.C. and Kwan, H.C. (1975) Afferent-efferent linkages in motor cortex for single forelimb muscles. *J. Neurophysiol.,* 38: 990–1014.

Ödkvist, L.M., Liedgren, S.R.C., Larsby, B. and Jerlvall, L. (1975) Vestibular and somatosensory inflow to the vestibular projection area in the post cruciate dimple region of the cat cerebral cortex. *Exp. Brain Res.,* 22: 185–196.

Oscarsson, O. and Rosén, I. (1963) Projection to cerebral cortex of large muscle spindle afferents in the forelimb nerves of the cat. *J. Physiol. (Lond.),* 169: 924–945.

Oscarsson, O., Rosén, I. and Sulg, I. (1966) Organization of neurones in the cat cerebral cortex that are influenced from group I muscle afferents. *J. Physiol. (Lond.),* 183: 189–210.

Poggio, G.F. and Mountcastle, V.B. (1963) The functional properties of ventrobasal thalamic neurons studied in unanesthetized monkeys. *J. Neurophysiol.,* 26: 775–806.

Rosén, I. and Asanuma, H. (1972) Peripheral afferent inputs to the forelimb area of the monkey motor cortex: input-output relations. *Exp. Brain Res.,* 14: 257–273.

Silfvenius, H. (1968) Cortical projections of large muscle afferents from the cat's forelimb. *Acta physiol. scand.,* 74: 25–26A.

Silfvenius, H. (1970) Projections to the cerebral cortex from afferents of the interosseous nerves of the cat. *Acta physiol. scand.,* 80: 196–214.

Strick, P.L. (1970) Cortical projections of the feline thalamic nucleus ventralis lateralis. *Brain Res.,* 20: 130–134.

Strick, P.L. (1973) Light microscope analysis of the cortical projection of the thalamic ventrolateral nucleus in the cat. *Brain Res.,* 55: 1–24.

Strick, P.L. and Kim, C.C. (1978) Input to primate motor cortex from posterior parietal cortex (area 5). I. Demonstration by retrograde transport. *Brain Res.,* 157: 325–330.

Thompson, W.D., Stoney, S.D. and Asanuma, H. (1970) Characteristics of projections from primary sensory cortex to motor-sensory cortex in cats. *Brain Res.,* 22: 15–27.

Zarzecki, P., Shinoda, Y. and Asanuma, H. (1978a) Projection from area 3a to the motor cortex by neurons activated from group I muscle afferents. *Exp. Brain Res.,* 33: 269–282.

Zarzecki, P., Strick, P.L. and Asanuma, H. (1978b) Input to primate motor cortex from posterior parietal cortex (area 5). II. Identification by antidromic activation. *Brain Res.,* 157: 331–335.

Zarzecki, P. and Wiggin, D. (1979) Convergence of somatosensory and proprioceptive inputs upon cortico-cortical and pyramidal tract neurons. *Neurosci. Abstr.,* in press.

Vestibulospinal, Reticulospinal and Interstitiospinal Pathways in the Cat*

K. FUKUSHIMA, B.W. PETERSON and V.J. WILSON

The Rockefeller University, New York, NY 10021 (U.S.A.)

INTRODUCTION

This review will deal with properties and motor actions of three descending systems: the vestibulospinal tracts, the reticulospinal tracts and the interstitiospinal tract. These systems have three properties in common: i) they terminate most heavily in the ventromedial parts of the ventral horn which are closely associated with motoneuron nuclei that innervate the axial muscles (Kuypers et al., 1962; Nyberg-Hansen, 1966a; Petras, 1967; Sterling and Kuypers, 1967); ii) their interruption produces severe impairment in righting and in the control of postural and antigravity muscles, with little impairment of movement of the distal extremities (Lawrence and Kuypers, 1968) iii) they are linked more closely (monosynaptically) to axial motoneurons, especially to neck motoneurons, than to most limb motoneurons (Wilson and Yoshida, 1969a, b; Wilson et al., 1970; Pitts et al., 1977; Peterson et al., 1978a; Fukushima et al., 1978).

THE VESTIBULOSPINAL TRACTS

The vestibular nuclei project to the spinal cord by three separate pathways: the lateral vestibulospinal tract (LVST), the medial vestibulospinal tract (MVST), and the caudal vestibulospinal tract (CVST; Peterson and Coulter, 1977; Peterson et al., 1978b). Cells in the nuclei receive important input from different endorgans of the vestibular apparatus and the function of vestibulospinal reflexes produced by activation of these receptors is to stabilize the position of the head in space by the vestibulo-neck, -back and -limb reflexes.

Afferent fibers from the semicircular canals and otolith organs directly terminate in the vestibular nuclei, the fastigial nucleus and the vestibulocerebellum (flocculus, nodulus, uvula). Within the vestibular nuclei, canal afferents are found in the superior nucleus, the rostral and perhaps caudal regions of the medial and descending nuclei, and to a much lesser extent in the medial part of Deiters' nucleus. Utricular afferents distribute to Deiters' nucleus, the rostral part of the descending nucleus, and to a

---

*Supported in part by Grants NSF BMS 75–00487 and N.I.H. NS 02619 and EY 02249.

lesser extent to the medial nucleus (Stein and Carpenter, 1967; Gacek, 1969; Sans et al., 1972). Saccular afferents distribute mainly to Deiters' and descending nuclei (Stein and Carpenter, 1967; Gacek, 1969; also Wilson et al., 1978).

Properties of vestibulospinal tract neurons

Lateral vestibulospinal tract (LVST) neurons. The LVST originates in Deiters' nucleus and descends in the ipsilateral ventral funiculus of the spinal cord as far as the sacral segments (Brodal, 1974). LVST fibers are of small, medium-sized to large caliber; the majority are rather thick (Pompeiano and Brodal, 1957; Petras, 1967) and have conduction velocities between 20 and 140 m/sec with a mode of 90–100 m/sec (Ito et al., 1964; Wilson, et al., 1966; Wilson et al., 1967). Within Deiters' nucleus there is a tendency for neurons projecting to the lumbosacral segments (L cells) to be located dorsally and for those projecting to the cervical and thoracic segments (C cells) to be located ventrally (Pompeiano and Brodal, 1957; Ito et al., 1964; Wilson et al., 1967), although there is considerable overlap in the distribution of C and L cells (Wilson et al., 1967). Neurons projecting to lumbar segments may also innervate more rostral ones: 50% of the LVST cells that send terminal branches to the cervical enlargement also have axon branches extending to lumbar levels (Abzug et al., 1974) and therefore are able to act on at least two spinal regions simultaneously. There is also branching of vestibular neurons projecting to the neck segments. The axons of 62% of ipsilateral Deiters' neurons (many of which must have axons in the LVST) not only give off a collateral to C3, but also extend as far as the cervical enlargement; some of these neurons project as far as the upper thoracic cord, but almost none to the lumbar cord (Rapoport et al., 1977a).

Labyrinthine input: LVST cells located in ventral Deiters' nucleus are the main target of monosynaptic excitation from primary vestibular afferents, and monosynaptic input is more prevalent among C cells than L cells (Wilson et al., 1967; Ito et al., 1969; Peterson, 1970; Akaike et al., 1973a). Polysynaptic labyrinthine activation is more widespread (Wilson et al., 1967; Akaike et al., 1973a). Many LVST L and C cells, dorsal and ventral, are affected by lateral tilt (Peterson, 1970; Shimazu and Smith, 1971; Schor, 1974), as expected from anatomical studies. Such tilt-sensitive neurons have been subdivided according to their response patterns (Duensing and Schaeffer, 1959): α and β neurons are excited by tilting the ipsilateral or contralateral side down, respectively; γ and δ neurons are respectively excited or inhibited by tilt in either direction. Second-order vestibular neurons exhibit only α and β responses, and only higher order neurons exhibit γ and δ responses (Fujita et al., 1968; Peterson, 1970). L cells on the average show a γ response, whereas C cells on the average show an α response (Peterson, 1970)

Canal inputs also reach Deiters' neurons (Sans et al., 1972). Recent work from our laboratory has demonstrated that LVST neurons monosynaptically driven by stimulation of the horizontal or anterior canal nerve are found in ventral Deiters' nucleus (Wilson, Peterson, Schor, Hirai and Fukushima, unpublished observations). In response to horizontal rotation some C cells show type II (ipsi —, contra +) and type III (ipsi +, contra +) but not type I (ipsi +, contra —) responses (Precht et al., 1967).

Most LVST neurons receiving otolith input are excited by stimulation of the contralateral vestibular nerve and there is no commissural inhibition (Shimazu and Smith, 1971; Wilson et al., 1978; see also Chan et al., 1977). Very likely LVST neurons receiving canal input are inhibited by stimulation of the contralateral vestibular nerve, since commissural inhibition is prominent among canal-activated neurons (Shimazu and Precht, 1966; Markham, 1968).

Somatosensory input: The activity of Deiters' neurons can be modified by inputs from the periphery (cutaneous, joint, muscle receptors) reaching the nucleus by various ascending pathways, including transcerebellar ones. Input from muscle nerves comes from group II and III fibers, except that there is group I input from quadriceps (Wilson et al., 1966; 1967; Wylie and Felpel, 1971; Allen et al., 1972a, b).

The widespread activity reaching Deiters' neurons through the brain stem or fastigial nucleus is excitatory (Wilson et al., 1966; Ito et al., 1970a) and provides a background that is modulated by inhibitory action relayed through the cerebellar cortex, principally that of the anterior lobe (Ito and Yoshida, 1966; Shimazu and Smith, 1971; Akaike et al., 1973b). When peripheral input reaching Deiters' neurons via climbing fiber afferents was examined, a somatotopic pattern was revealed: C cells and L cells were most effectively inhibited by stimulation of forelimb nerves and hindlimb nerves, respectively (Allen et al., 1972a).

Medial vestibulospinal tract (MVST) neurons. The MVST originates in the medial, descending and Deiters' nuclei and descends in the ventral funiculi of the spinal cord bilaterally, diminishing in numbers below the neck segments (Brodal, 1974; Nyberg-Hansen, 1966a). The fibers from the medial nucleus are of medium to fine caliber (Nyberg-Hansen, 1964) and conduct at 13–76 m/sec, with a mode of 36 m/sec (Wilson et al., 1968; Wilson and Yoshida, 1969b). However, the tract also contains rapidly conducting fibers (Akaike et al., 1973a). Many MVST neurons branch as described above for LVST neurons, giving off collaterals to neck segments and extending as far as the cervical enlargement (Rapoport et al., 1977a).

Labyrinthine input: Stimulation of the labyrinth excites many MVST cells monosynaptically (Wilson et al., 1968; Akaike et al., 1973a). Polysynaptic activation is also widespread (Wilson et al., 1968). MVST neurons receive mainly canal inputs but they are also influenced by otolith inputs (Wilson and Peterson, 1978; Wilson et al., 1978). When MVST neurons were studied by horizontal rotation, type I, type II and type III neurons were found (Precht et al., 1967; Shimazu and Smith, 1971).

Somatic inputs: Information about somatic inputs to MVST neurons is scanty, and will not be considered here (see Wilson, 1972).

Caudal vestibulospinal tract (CVST) neurons. The CVST originates in the caudal poles of the medial and descending nucleus and in cell group f and descends bilaterally at least as far as the lumbar enlargement (Peterson and Coulter, 1977; Peterson et al., 1978b). CVST neurons are mainly small to medium-sized (Peterson and Coulter, 1977) and have a median conduction velocity of 12 m/sec (Peterson et al., 1978b). Unlike LVST and MVST axons, which are located in the ventral funiculi, CVST axons can be found in both the ventral and dorsolateral funiculi on both sides of the spinal cord. Not much information is available either about input to CVST neurons or about synaptic actions of the CVST upon spinal motoneurons.

Action of vestibulospinal tract neurons on spinal motoneurons

A. *LVST.* Stimulation of Deiters' nucleus facilitates activity of extensor muscles (Brodal et al., 1962). Intracellular recordings from limb, axial neck and back motoneurons (Lund and Pompeiano, 1968; Shapovalov, 1966; Wilson and Yoshida, 1969a; Wilson et al., 1970; see also Akaike et al., 1973c) have shown that the LVST is excitatory. The tract makes monosynaptic connections with many neck motoneurons (Wilson and Yoshida, 1969a); with some back extensor motoneurons (Wilson et al., 1970) and with some hindlimb motoneurons, particularly those of quadriceps and gastrocnemius soleus (Lund and Pompeiano, 1968; Wilson and Yoshida, 1969a; Grillner et al., 1970); there is disynaptic reciprocal inhibition of some hindlimb flexor motoneurons (Grillner et al., 1970). Forelimb motoneurons and other hindlimb motoneurons receive polysynaptic excitation (Wilson and Yoshida, 1969a; Grillner et al., 1970). The LVST has also polysynaptic excitatory and inhibitory actions on contralateral limb extensor and flexor motoneurons (Hongo et al., 1971; Maeda et al., 1975). In view of the extensive projection of utricular and saccular afferents to Deiters' nucleus it is likely that connections of the LVST with motoneurons provide a route via which activity of these afferents can influence motor activity (e.g. Wilson et al., 1977, 1978). The LVST also relays canal activity. By transection of this tract it was shown that it transmits anterior canal excitation to ipsilateral dorsal neck motoneurons (Wilson and Maeda, 1974).

MVST. The MVST contains inhibitory as well as excitatory fibers. The existence of long descending inhibitory fibers was demonstrated by Wilson and Yoshida (1969b) who showed that stimulation of the medial nucleus produced monosynaptic inhibition of neck extensor motoneurons. Synaptic action of individual inhibitory MVST neurons has been studied by means of spike-triggered signal averaging in neck extensor motoneurons (Rapoport et al., 1977b), and a typical example is shown in Fig. 1. Mean orthodromic conduction time from the foot of the extracellular spike that triggered the averager, recorded in the vestibular nuclei, was 0.72 msec. Mean synaptic delay was 0.4 msec. IPSPs had a mean time to peak of 0.81 msec and were readily reversed by injection of hyperpolarizing current, as illustrated in Fig. 1D. All vestibular neurons making synapses with neck motoneurons were monosynaptically driven by stimulation of the ipsilateral vestibular nerve (e.g. Fig. 1 A2). Four of seven tested were inhibited by stimulation of the contralateral vestibular nerve (e.g. Fig. 1B). Back motoneurons also receive monosynaptic inhibitory connections from the MVST (Wilson et al., 1970) but limb motoneurons do not (Wilson and Yoshida, 1969a). Inhibitory MVST neurons have medium to slow conduction velocity (Wilson et al., 1970; Akaike et al., 1973c), whereas excitatory MVST neurons conduct rapidly (Akaike et al., 1973c).

Synaptic connections from individual vestibular end-organs to neck motoneurons have been studied extensively (Wilson and Maeda, 1974; Wilson et al., 1977, 1978) and direct pathways conveying semicircular canal inputs have been identified as illustrated in Fig. 2A and B. All semicircular canal ampullae, with the exception of that of the ipsilateral anterior canal, have disynaptic connections with neck extensor motoneurons via the MVST (Fig. 2A). Stimulation of anterior and posterior canal nerves on both sides gives disynaptic excitation and inhibition, respectively; stimulation of the ipsilateral horizontal canal nerve evokes disynaptic inhibition, stimulation of the contralateral horizontal canal nerve evokes disynaptic excitation (Wilson and Maeda, 1974). Neck flexor (sterno-cleidomastoid) motoneurons receive all their disynaptic semicircular canal inputs via the MVST (Fig. 2B). Stimulation of all three

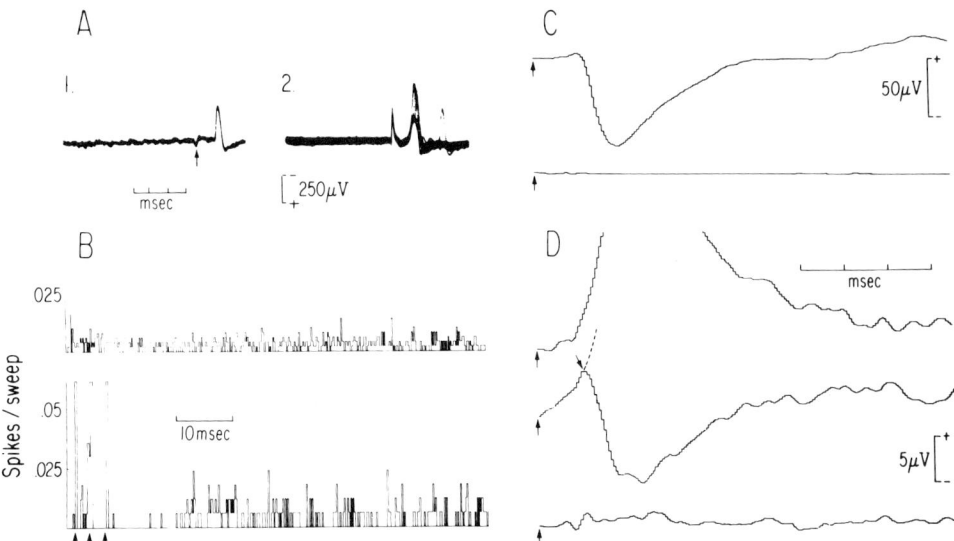

Fig. 1. Properties of an inhibitory neuron in the vestibular nuclei and of unitary IPSPs evoked by activity of this neuron in two motoneurons. A1) Antidromic spike of neuron evoked by an 8 μA stimulus (at upward arrow) to a microelectrode in C3 Biventer-cervicis-complexus motoneuron pool. A2) Monosynaptic response of the neuron to stimulation of the ipsilateral vestibular nerve at 1.4 times N_1 threshold. B) Upper trace shows spontaneous activity of the neuron as a PST histogram (440 sweeps). Lower trace shows inhibitory effect of a 150 μA triple shock to the contralateral vestibular nerve (150 sweeps). C) Unitary IPSP evoked in one motoneuron by activity of the inhibitory neuron; lower trace, extracellular record. D) Unitary IPSP (middle trace) evoked in another motoneuron. IPSP was reversed by injection of 10 nA hyperpolarizing current and part of this reversed IPSP is shown in the upper trace. Downward arrow in middle trace shows divergence between IPSPs recorded with and without current injection. Lower trace, extracellular record. All records in C and D are averages of 600–1000 sweeps; upward arrows indicate time of discriminator output pulses. (From Rapoport et al., 1977b.)

ipsilateral canal nerves gives disynaptic inhibition, of all three contralateral canal nerves, disynaptic excitation (Fukushima, Hirai and Rapoport, unpublished observations).

Stimulation of the utricular nerve evokes IPSPs ipsilaterally and EPSPs contralaterally in neck extensor motoneurons at latencies as short as 2.0 msec (Wilson et al., 1977). Since LVST axons are excitatory, it seems probable that IPSPs evoked by utricular nerve stimulation are produced disynaptically via the MVST. Similarly, since transection of the MLF abolishes all contralateral disynaptic EPSPs evoked in neck motoneurons by stimulation of the whole vestibular nerve (Akaike et al., 1973b) the contralateral utricular-evoked EPSPs are almost surely relayed by the MVST.

Vestibulospinal reflexes

Synaptic connections between the labyrinth and spinal motoneurons have been extensively studied, and most direct pathways have been identified. The MVST appears to be the predominant direct pathway to axial motoneurons and the LVST the only direct pathway to limb motoneurons for all receptors, although much remains to be studied about connections relaying the otolith input to the spinal cord. The pattern of disynaptic excitation and inhibition between individual semicircular canals and neck motoneurons is consistent with the reflex movements expected in response to

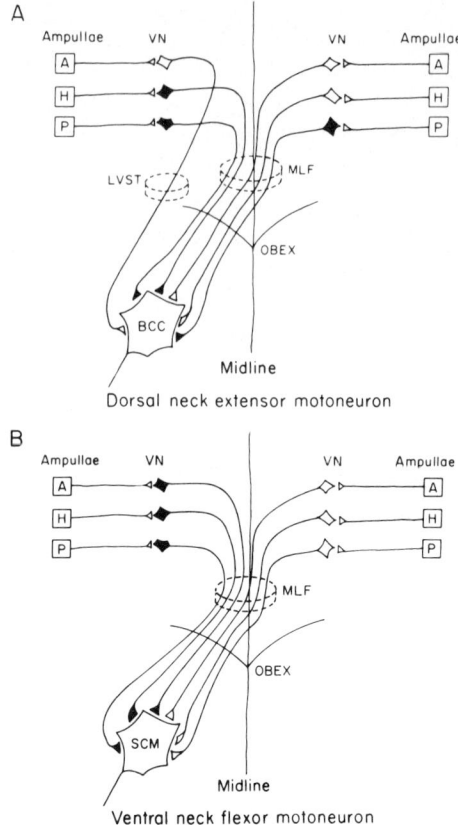

Fig. 2. Diagrams of connections between ipsilateral and contralateral semi-circular canal ampullae and a dorsal neck extensor, biventer cervicis complexus (BCC), motoneuron (A) and a ventral neck flexor, sterno-cleidomastoid (SCM), motoneuron (B). A, H, P are anterior, horizontal and posterior ampullae; VN, vestibular nuclei; MLF, medial longitudinal fasciculus which contains MVST. Inhibitory neurons and their terminals are shown in black, excitatory in white. (A, from Wilson and Maeda, 1974; B, from Fukushima, Hirai and Rapoport, unpublished observations.)

natural stimulation of the semicircular canals (Suzuki and Cohen, 1964; Wilson and Maeda, 1974). The question arises: can functionally meaningful vestibulospinal reflexes be explained by these direct pathways? Dynamic studies of vestibular-neck and vestibular-forelimb reflexes reveal the presence of extensive central processing in these reflexes (Anderson et al., 1977a, b; Ezure and Sasaki, 1978; Ezure et al., 1978; Wilson and Peterson, 1978; Wilson, 1979). This strongly suggests that under many conditions the role of the direct pathways is a minor one, although their precise contribution remains to be determined.

Influence of the cerebellum on vestibulospinal reflexes

A high percentage of LVST and MVST neurons receive Purkinje cell inhibition from the vermis of the anterior lobe, but not from the vestibulocerebellum (Ito and Yoshida, 1966; Shimazu and Smith, 1971; Akaike et al., 1973b), indicating that the anterior vermis can modify activity of vestibulospinal tract neurons and therefore of vestibulospinal reflexes. Since the anterior vermis receives not only somatic input (e.g. Oscarsson, 1973; Berthoz and Llinás, 1974) but also vestibular information

(Anderson and Gernandt, 1954; Berthoz and Llinás, 1974; Pompeiano, 1974, 1975; Precht et al., 1977; Kotchabhakdi and Walberg, 1978a, b), both of these inputs are likely to exert an inhibitory influence on vestibulospinal reflexes via transcerebellar circuits. Such an inhibitory action by somatic inputs has been demonstrated (e.g., Ito, Obata and Ochi, 1966; Allen et al., 1972a, b) and inhibitory modulation of activity of vestibulospinal tract neurons by vestibular inputs has also been shown. Orlovsky and Pavlova (1972) investigated responses of LVST neurons to lateral tilting in cats with and without the cerebellum: with intact cerebellum only dynamic responses were obtained, while without the cerebellum dynamic responses were reduced and static responses appeared in many neurons. Two recent reports provide further evidence that vestibular inputs to the cerebellum, can act to modify vestibulospinal reflexes. In the decerebrate cat with intact cerebellum, rotation of the head (with the neck denervated) produces shortening of the triceps on the side toward which the vertex of the skull is rotated and lengthening in the contralateral medial triceps. In the acutely cerebellectomized cat, rotation of the head in either direction results in simultaneous shortening of the triceps on both sides (Lindsay and Rosenberg, 1977). Similarly, Anderson et al. (1977a) showed that the cerebellum is necessary for the reciprocal behavior seen between the two triceps muscles when utricular afferents are stimulated by horizontal linear acceleration.

THE RETICULOSPINAL TRACTS (RST)

Some properties of reticulospinal tract neurons

The medial ponto-medullary reticular formation gives rise to three groups of descending fibers: one descending in the ventromedial funiculus (RST_m), one in the ipsilateral ventrolateral funiculus (RST_i), and one in the contralateral ventrolateral funiculus (RST_c) (Nyberg-Hansen, 1966a; Petras, 1967; Ito et al., 1970b; Peterson et al., 1975b).

RST_m: The RST_m originates primarily from neurons in the pons and in rostrodorsal nucleus reticularis (n.r.) gigantocellularis although a few RST_m neurons are found more caudally (Ito et al., 1970b; Peterson et al., 1975b). RST_m fibers run in or close to the MLF, and descend almost exclusively ipsilaterally as far as the lumbar segments. RST_m neurons have conduction velocities between 10 and 150 m/sec with a median of 101 m/sec (Peterson et al., 1975b).

RST_i *and* RST_c: RST_i and RST_c originate from neurons in the medullary reticular formation, and descend as far as the lumbar cord, the number of RST_i neurons far exceeding the number of RST_c neurons (Peterson et al., 1975b). There is somatotopic organization in the distribution of RST_i neurons within the medullary reticular formation. RST_i neurons projecting only to the neck (N cells) are found both in the dorsorostral region just behind the abducens nucleus and in the ventrocaudal portion of n.r. gigantocellularis, whereas RST_i neurons that project further tend to cluster in the latter region (Peterson et al., 1975b). Such somatotopic organization has been confirmed using the technique of retrograde labeling with horseradish peroxidase (see Peterson, 1977). RST_i and RST_c neurons have median conduction velocities of 68 and 72 m/sec respectively (Peterson et al., 1975b).

Like vestibulospinal tract neurons, 86% of all three types of reticulospinal tract neurons that send terminal branches to the cervical enlargement also have axon branches extending to lower spinal levels, and therefore are able to act on two or more levels simultaneously. Some RST neurons also project to both sides of the spinal cord (Peterson et al., 1975b) or to dorsal spinal regions (Peacock and Wolstencroft, 1976).

Action of reticulospinal tract neurons on spinal motoneurons

RST_m: RST_m neurons have a direct, monosynaptic excitatory action on spinal motoneurons. Grillner and Lund (1968) first found that stimulation of n.r. pontis caudalis, the dorsal part of n.r. gigantocellularis and the MLF produced direct, monosynaptic excitation of hindlimb motoneurons. Stimuli applied in the MLF within this region gave rise to monosynaptic excitation of motoneurons supplying muscles of the neck, back, forelimbs and hindlimbs (Wilson and Yoshida, 1969a; Wilson et al., 1970). Recently there has been systematic mapping of brain stem regions that produce monosynaptic excitation of motoneurons in different parts of the body (Pitts et al., 1977; Peterson et al., 1978a). Stimuli applied within zone 1 in Fig. 3 produced monosynaptic EPSPs in 67% of neck motoneurons, 50% of back motoneurons, 28% of forelimb motoneurons and 24% of hindlimb motoneurons. Because zone 1 corresponds to the region of origin of RST_m fibers, these fibers are likely to play a major role in producing the widespread excitation of motoneurons evoked by stimulation of zone 1. RST_m fibers also originate from neurons in n.r. pontis oralis (zone 5 in Fig. 3) but stimulation of this region produced virtually no direct excitation of motoneurons (Pitts et al., 1977). RST_m fibers originating in zone 5 therefore appear to be functionally distinct from RST_m fibers originating in zone 1.

RST_i and RST_c: Lateral reticulospinal neurons projecting via RST_i or RST_c establish direct connections preferentially with axial motoneurons that supply muscles of the neck and back (Pitts et al., 1977; Peterson et al., 1978a). As deduced by Ito et al. (1970b) from their observations of synaptic potentials produced in neighboring

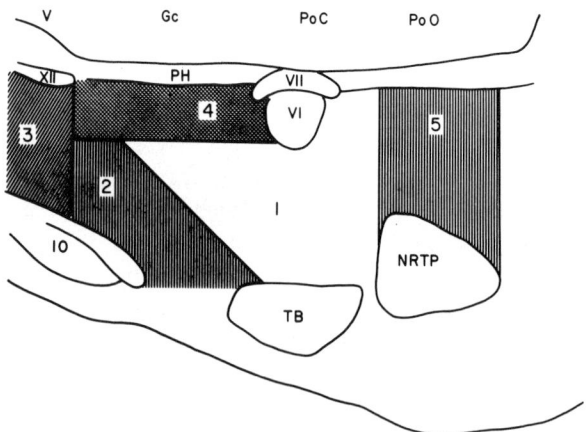

Fig. 3. Schematic parasagittal section showing division of medial pontomedullary reticular formation into five zones. Zone 1 includes n.r. pontis caudalis and the dorsorostral part of n.r. gigantocellularis beginning 2 mm from the dorsal surface of the brainstem. Zone 2 corresponds to the ventrocaudal part of n.r. gigantocellularis, zone 3 to n.r. ventralis, zone 4 to the dorsal part of n.r. gigantocellularis, zone 5 to n.r. pontis oralis. V, Gc, PoC and PoO indicate the approximate levels of reticularis ventralis, gigantocellularis, pontis caudalis and pontis oralis. Other abbreviations: XII hypoglossal nucleus; IO, inferior olivary nucleus; PH, prepositus hypoglossi nucleus; VII, genu of facial nerve; VI, abducens nucleus; TB, trapezoid body; NRTP, nucleus reticularis tegmenti pontis. (From Peterson, 1977.)

reticular neurons by recurrent collaterals of RST_i neurons, the lateral reticulospinal tracts contain both excitatory and inhibitory fibers; the former end on neck and back motoneurons, the latter only on neck motoneurons.

Inhibition: As described by Peterson et al. (1978a), stimuli applied to zone 3 and 4 in Fig. 3 (n.r. ventralis and the dorsal part of n.r. gigantocellularis, respectively) evoke monosynaptic inhibition in more than 90% of neck motoneurons. Since this inhibition survives destruction of the MLF and medial brain stem which interrupts RST_m fibers as well as inhibitory MVST fibers, it appears to be mediated by RST_i or RST_c fibers. Stimulation of this region or other brain stem regions may also produce inhibition of back and limb motoneurons but the properties of this inhibition suggest that it is produced by di- or polysynaptic pathways (Pitts et al., 1977).

Excitation: Stimuli applied to zones 2, 3 and 4 evoke monosynaptic excitation in about 90% of neck motoneurons. About 40% of back motoneurons receive direct excitation from zones 2 and 3, but none receive direct excitation from zone 4. Forelimb and hindlimb motoneurons receive almost no direct excitation from zones 2, 3 and 4. Since the primary reticulospinal projection from zones 2 and 3 is via the RST_i, it is likely that the EPSPs evoked from these regions are produced by RST_i fibers. The bulk of the direct reticulomotor connections are ipsilateral, although stimulation of zones 1–4 occasionally evokes weak, direct EPSPs or IPSPs in contralateral motoneurons. These contralateral direct PSPs have the same topographic organizations as the more prevalent ipsilateral PSPs (Pitts et al., 1977; Peterson et al., 1978a).

To summarize, all three reticulospinal systems (RST_m, RST_i and RST_c) include neurons that establish direct connections with spinal motoneurons. Excitatory RST_m neurons scattered throughout n.r. pontis caudalis and the dorsal part of n.r. gigantocellularis establish direct connections with motoneurons supplying a wide variety of muscles throughout the body. There is no apparent somatotopic organization of RST_m neurons projecting to different motoneuron groups. RST_i and RST_c neurons located in n.r. gigantocellularis and ventralis establish direct inhibitory and excitatory connections with neck and direct excitatory connections with back motoneurons, but do not establish direct connections with limb motoneurons. The RST_i and RST_c systems, therefore, appear to be involved in direct control of the axial musculature with specialized subsets of neurons acting specifically on neck motoneurons. In addition to their direct motor connections, all parts of the medial reticular formation have indirect excitatory and inhibitory actions on a wide variety of motoneurons.

Functional considerations

The function of reticulospinal systems is still not well understood, but the strong motor effects elicited by reticular stimulation (Magoun and Rhines, 1946; Rhines and Magoun, 1946; Sprague and Chambers, 1954) and the presence of direct, topographically organized synaptic connections between pontomedullary reticulospinal neurons and somatic motoneurons (cf. Peterson, 1977, 1979a) indicate that these pathways must be involved in some aspect of motor control.

Reticulospinal systems may be involved in the control of many different types of motor behavior, because they receive major direct inputs from the sensorimotor cortex (Magni and Willis, 1964; Peterson et al., 1974), superior colliculi (Udo and Mano, 1970; Peterson et al., 1974; Peterson et al., 1976), vestibular nuclei (Udo and Mano, 1970; Peterson and Abzug, 1975; Peterson et al., 1975a, 1976), spinal cord (Wolstencroft, 1964; Peterson et al., 1974, 1976; Eccles et al., 1975; Fox and Wolstencroft,

1976) and deep cerebellar nuclei (Ito et al., 1970a; Eccles et al., 1975; Bantli and Bloedel, 1975). Considering these inputs in turn, the strong cortical excitation of reticulospinal neurons suggests that cortico-reticulospinal pathways may make an important contribution to motor behaviour controlled by the cerebral cortex. The visually guided arm movements mediated by uncrossed cortico-motoneuronal pathways in split brained, chiasm-sectioned monkeys (Brinkman and Kuypers, 1973) provide an example of a case where such pathways may play a critical role. The observation by Anderson et al. (1974) that responses obtained in neck motoneurons following stimulation of the superior colliculus were not significantly affected by interruption of tectospinal fibers in the caudal medulla likewise suggests that reticulospinal systems play an important role in transmitting activity originating in the superior colliculus to spinal motoneurons. The point has already been made that direct vestibulospinal systems cannot, by themselves, produce functionally meaningful vestibulospinal reflexes, and it is natural to speculate that reticulospinal systems might therefore play a role in producing such reflexes (cf. Wilson and Peterson, 1978). Indeed, it has been reported that activity of reticulospinal tract neurons is modulated by lateral tilt (Orlovsky and Pavlova, 1972; Spyer et al., 1974) and by stimulation of afferents from the horizontal semicircular canal (Peterson, 1979b). As to the input from the spinal cord, some reticulospinal tract neurons are shown to be involved in the spino-bulbospinal (SBS) reflex (Shimamura and Livingston, 1963), although the role of this reflex is still unknown. Orlovsky (1970) reported that in mesencephalic cats with intact cerebellum activity of reticulospinal tract neurons was modulated in phase with the induced locomotor cycle, and that such modulation was abolished if movement of the legs was artificially restrained, indicating that the modulation depended upon the input from the moving limb.

To summarize, there is considerable evidence that the reticulospinal systems play an important role in the control of somatic musculature, and specific questions of how and in what type of motor behavior the reticulospinal systems are involved remain to be answered.

THE INTERSTITIOSPINAL TRACT (IST)

Some properties of interstitiospinal tract neurons

The IST originates from medium to small neurons in the interstitial nucleus of Cajal (INC) and descends in the dorsal part of the ventral funiculus, primarily ipsilaterally, as far as the sacral segments (Nyberg-Hansen, 1966b). IST fibers are of varying caliber (Nyberg-Hansen, 1966b) and have conduction velocities between 15 and 123 m/sec (Fukushima et al., 1978). Some IST neurons may send branches to more than one level of the spinal cord (Fukushima et al., 1978).

Action of interstitiospinal tract neurons on spinal motoneurons

Action of the IST on spinal motoneurons has been systematically investigated on neck, back, forelimb and hindlimb motoneurons in the cat (Fukushima et al., 1978; Fukushima, Pitts and Peterson, unpublished observations; Fukushima, Hirai and Rapoport, unpublished observations). Stimulation of the ipsilateral INC produces monosynaptic EPSPs consistently in dorsal neck extensor (Biventer-cervicis-complexus, BCC) and ventral flexor (Sterno-cleido-mastoid, SCM) motoneurons, while such EPSPs are observed in about two-thirds of the lateral flexor (splenius, SP) motoneurons and half of the trapezius (TR) motoneurons (shoulder muscle). Stimulation of the contralateral INC produces weak monosynaptic EPSPs in about half the BCC and SCM motoneurons and in a few SP and TR motoneurons.

Effects of INC stimulation on limb and back motoneurons are different from those on neck motoneurons in that none of these motoneurons receives monosynaptic excitation from the INC. No monosynaptic IPSPs were obtained in any of the motoneurons tested. In addition to monosynaptic EPSPs, all types of motoneurons also receive longer latency, polysynaptic PSPs.

Functional considerations

Not much information is available about properties of IST neurons. However, the INC has been considered to be of particular importance for vertical and rotatory eye movements since it sends fibers to all extraocular motor nuclei with the exception of the abducens nucleus and the medial rectus part of the oculomotor nucleus (Szentágothai, 1943; Carpenter et al., 1970). The observations of Schwindt et al. (1974) are consistent with the anatomical findings in showing that stimulation of the INC evoked monosynaptic EPSPs and IPSPs in trochlear motoneurons but failed to evoke a response in abducens motoneurons.

Electrical stimulation of the INC in awake animals results in rotatory shifts of gaze in the frontal plane produced by both eye and head movements (Szentágothai, 1943; Hassler and Hess, 1954; Hyde and Toczek, 1962; Toczek and Hyde, 1963). The head movement component of these gaze shifts could be due to coactivation of neck extensors (BCC) and flexors (SCM), mediated by the direct excitatory connections between the INC and these motoneuron pools. Such coactivation may not always occur, however. It is possible that activation of head flexors and extensors is produced by different populations of IST neurons. If this is so, selective activation of one population might result in upward movements while activation of the other might result in downward movement. The pattern of direct connections by interstitiospinal fibers onto neck motoneurons would then be consistent with the view that the INC plays a role in vertical and rotatory gaze shifts.

SUMMARY

Experiments on the properties and motor actions of three descending systems, the vestibulospinal, reticulospinal and interstitiospinal tracts, have been reviewed. The vestibulospinal tracts are the most direct pathways between the labyrinth and spinal motoneurons. The MVST is the predominant direct pathway to axial motoneurons, the LVST the only direct pathway to limb motoneurons; not much is known about the recently discovered caudal vestibulospinal tract. The role of these direct pathways in functionally meaningful vestibulospinal reflexes remains to be determined.

The reticulospinal tracts consist of three groups of descending fibers: one descending in the ventromedial funiculus (RST_m), one in the ipsilateral ventrolateral funiculus (RST_i), and one in the contralateral ventrolateral funiculus (RST_c). Excitatory RST_m neurons scattered throughout nucleus reticularis (n.r.) pontis candalis and the dorsal part of n.r. gigantocellularis establish direct synaptic connections with motoneurons supplying a wide variety of muscles throughout the body. RST_i and RST_c neurons located in n.r. gigantocellularis and ventralis establish direct inhibitory and excitatory connections with neck and direct excitatory connections with back motoneurons, but do not establish direct connections with limb motoneurons. The reticulospinal systems receive major direct inputs from many different regions including vestibular nuclei, suggesting that they participate in vestibulospinal reflexes. The interstitiospinal tract, which has not been studied extensively, includes neurons that establish direct excitatory connections with neck motoneurons, but do not establish direct connections with limb and back motoneurons.

REFERENCES

Abzug, C., Maeda, M., Peterson, B.W. and Wilson, V.J. (1974) Cervical branching of lumbar vestibulospinal axons. *J. Physiol. (Lond.)*, 243: 499–522.

Akaike, T., Fanardjian, V.V., Ito, M., Kumada, M. and Nakajima, H. (1973a) Electrophysiological analysis of the vestibulospinal reflex pathway of rabbit. I. Classification of tract cells. *Exp. Brain Res.*, 17: 477–496.

Akaike, T., Fanardjian, V.V., Ito, M. and Nakajima, H. (1973b) Cerebellar control of vestibulospinal tract cells in rabbit. *Exp. Brain Res.*, 18: 446–463.

Akaike, T., Fanardjian, V.V., Ito, M. and Ohno, T. (1973c) Electrophysiological analysis of the vestibulospinal reflex pathway of rabbit. II. Synaptic actions upon spinal neurons. *Exp. Brain Res.*, 17: 497–515.

Allen, G.I., Sabah, N.H. and Toyama, K. (1972a) Synaptic actions of peripheral nerve impulses upon Deiters' neurones via the climbing fibre afferents. *J. Physiol. (Lond.)*, 266: 311–333.

Allen, G.I., Sabah, N.H. and Toyama, K. (1972b) Synaptic actions of peripheral nerve impulses upon Deiters' neurones via the mossy fibre afferents. *J. Physiol. (Lond.)*, 226: 335–351.

Anderson, J.H., Soechting, J.F. and Terzuolo, C.A. (1977a) Dynamic relations between natural vestibular inputs and activity of forelimb extensor muscles in the decerebrate cat. I. Motor output during sinusoidal linear acceleration. *Brain Res.*, 120: 1–16.

Anderson, J.H., Soechting, J.F. and Terzuolo, C.A. (1977b) Dynamic relations between natural vestibular inputs and activity of forelimb extensor muscles in the decerebrate cat. II. Motor output during rotations in the horizontal plane. *Brain Res.*, 120: 17–34.

Anderson, M.E., Yoshida, M. and Wilson, V.J. (1974) Influence of superior colliculus on cat neck motoneurons. *J. Neurophysiol.*, 34: 898–907.

Andersson, S. and Gernandt, B.E. (1954) Cortical projection of vestibular nerve in cat. *Acta oto-laryngol. (Stockh.)*, Suppl. 116: 10–18.

Bantli, H. and Bloedel, J.R. (1975) Monosynaptic activation of a direct reticulospinal pathway by the dentate nucleus. *Pflügers Arch.*, 357: 237–242.

Berthoz, A. and Llinás, R. (1974) Afferent neck projection to the cat cerebellar cortex. *Exp. Brain Res.*, 20: 385–401.

Brinkman, J. and Kuypers, H.G.J.M. (1973) Cerebral control of contralateral and ipsilateral arm, hand and finger movements in the split-brain rhesus monkey. *Brain*, 96: 653–674.

Brodal, A. (1974) Anatomy of the vestibular nuclei and their connections. In *Handbook of Sensory Physiology. Vol. 6. Vestibular system, Part 1. Basic Mechanisms.* H.H. Kornhuber (Ed.). Springer, Berlin pp. 239–352.

Brodal, A., Pompeiano, O. and Walberg, F. (1962) *The Vestibular Nuclei and their Connections.* Oliver and Boyd, Edinburgh.

Chan, Y.S., Hwang, J.C. and Cheung, Y.M. (1977) Crossed sacculo-ocular pathway via the Deiters' nucleus in cats. *Brain Res. Bull.*, 2: 1–6.

Carpenter, M.B., Harbison, J.W. and Peter, P. (1970) Accessory oculomotor nuclei in the monkey: Projections and effects of discrete lesions. *J. comp. Neurol.*, 140: 131–154.

Duensing, F. and Schaefer, K.P. (1959) Über die Konvergenz verschiedener labyrinthärer Afferenzen auf einzelner Neurone des Vestibulariskerngebiets. *Arch. Psychiat. Nervenkr.*, 199: 345–391.

Eccles, J.C., Nicoll, R.A., Schwarz, D.W.F., Tabořiková, H. and Willey, T.J. (1975) Reticulospinal neurons with and without monosynaptic inputs from cerebellar nuclei. *J. Neurophysiol.*, 38: 513–530.

Ezure, K. and Sasaki, S. (1978) Frequency-response analysis of vestibular-induced neck reflex in cat. I. Characteristics of neural transmission from horizontal semicircular canal to neck motoneurons. *J. Neurophysiol.*, 41: 445–458.
Ezure, K., Sasaki, S., Uchino, Y. and Wilson, V.J. (1978) Frequency-response analysis of vestibular-induced neck reflex in cat. II. Functional significance of cervical afferents and polysynaptic descending pathways. *J. Neurophysiol.*, 41: 459–471.
Fox, J.E. and Wolstencroft, J.H. (1976) The reduced responsiveness of neurones in nucleus reticularis gigantocellularis following their excitation by peripheral nerve stimulation. *J. Physiol. (Lond.)*, 258: 687–704.
Fujita, Y., Rosenberg, Y. and Segundo, J.P. (1968) Activity of cells in the lateral vestibular nucleus as a function of head position. *J. Physiol. (Lond.)*, 196: 1–18.
Fukushima, K., Pitts, N.G. and Peterson, B.W. (1978) Direct excitation of neck motoneurons by interstitiospinal fibers. *Exp. Brain Res.*, 33: 565–581.
Gacek, R. (1969) The course and central termination of first order neurons supplying vestibular end organs in the cat. *Acta oto-laryngol. (Stockh.)*, Suppl. 254: 1–66.
Grillner, S. and Lund, S. (1968) The origin of a descending pathway with monosynaptic action on flexor motoneurones. *Acta physiol. scand.*, 74: 274–284.
Grillner, S., Hongo, T. and Lund, S. (1970) The vestibulospinal tract. Effects on alpha-motoneurones in the lumbosacral spinal cord in the cat. *Exp. Brain Res.*, 10: 94–120.
Hassler, R. and Hess, W.R. (1954) Experimentelle und anatomische Befunde über die Drehbewegungen und ihre nervösen Apparte. *Arch. Psychiat. Nervenkr.*, 192: 488–526.
Hongo, T., Kudo, N. and Tanaka, R. (1971) Effects from the vestibulospinal tract on the contralateral hindlimb motoneurons in the cat. *Brain Res.*, 31: 220–223.
Hyde, J.E. and Toczek, S. (1962) Functional relation of interstitial nucleus to rotatory movements evoked from zona incerta stimulation. *J. Neurophysiol.*, 25: 455–466.
Ito, M. and Yoshida, M. (1966) The origin of cerebellar induced inhibition of Deiters' neurones. I. Monosynaptic initiation of the synaptic potentials. *Exp. Brain Res.*, 2: 330–349.
Ito, M., Hongo, T. and Okada, Y. (1969) Vestibular-evoked postsynaptic potentials in Deiters' neurones. *Exp. Brain Res.*, 7: 214–230.
Ito, M., Obata, K. and Ochi, R. (1966) The origin of cerebellar-induced inhibition of Deiters' neurones. II. Temporal correlation between the transsynaptic activation of Purkinje cell and the inhibition of Deiters' neurones. *Exp. Brain Res.*, 2: 350–364.
Ito, M., Udo, M., Mano, N. and Kawai, N. (1970a) Synaptic actions of the fastigiobulbar impulses upon neurones in the medullary reticular formation and vestibular nuclei. *Exp. Brain Res.*, 11: 29–47.
Ito, M., Udo, M. and Mano. N. (1970b) Long inhibitory and excitatory pathways converging onto cat reticular and Deiters' neurons and their relevance to reticulofugal axons. *J. Neurophysiol.*, 33: 210–226.
Ito, M., Hongo, T., Yoshida, M., Okada, Y. and Obata, K. (1964) Antidromic and transsynaptic activation of Deiters' neurones induced from the spinal cord. *Jap. J. Physiol.*, 14: 638–658.
Kotchabhakdi, N. and Walberg, F. (1978a) Cerebellar afferent projections from the vestibular nuclei in the cat. An experimental study with the method of retrograde axonal transport of horseradish peroxidase. *Exp. Brain Res.*, 31: 591–604.
Kotchabhakdi, N. and Walberg, F. (1978b) Primary vestibular afferent projections to the cerebellum as demonstrated by retrograde axonal transport of horseradish peroxidase. *Brain Res.*, 142: 142–146.
Kuypers, H.G.J.M., Fleming, W.R. and Farinholt, J.W. (1962) Subcorticospinal projections in the rhesus monkey. *J. comp. Neurol.*, 118: 107–137.
Laurence, D.G. and Kuypers, H.G.J.M. (1968) The functional organization of the motor system in the monkey. II. The effects of lesions of the descending brain-stem pathways. *Brain* 91: 15–36.
Lindsay, K.W. and Rosenberg, J.R. (1977) The effect of cerebellectomy on tonic labyrinth reflexes in the forelimb of the decerebrate cat. *J. Physiol. (Lond.)*, 273: 76p.
Lund, S. and Pompeiano, O. (1968) Monosynaptic excitation of alpha motoneurones from supraspinal structures in the cat. *Acta physiol. scand.*, 73: 1–21.
Maeda, M., Maunz, R.A. and Wilson, V.J. (1975) Labyrinthine influence on cat forelimb motoneurons. *Exp. Brain Res.*, 22: 69–86.
Magni, F. and Willis, W.D. (1964) Cortical control of brain stem reticular neurons. *Arch. ital. Biol.*, 102: 418–433.
Magoun, H.W. and Rhines, R. (1946) An inhibitory mechanism in the bulbar reticular formation. *J. Neurophysiol.*, 9: 165–171.

Markham, C.H. (1968) Midbrain and contralateral labyrinth influences on brain stem vestibular neurons in the cat. *Brain Res.*, 9: 312–333.

Nyberg-Hansen, R. (1964) Origin and termination of fibres from the vestibular nuclei descending in the medial longitudinal fasciculus. An experimental study with silver impregnation methods in the cat. *J. comp. Neurol.*, 122: 355–367.

Nyberg-Hansen, R. (1966a) Functional organization of descending supraspinal fibre systems to the spinal cord. Anatomical observations and physiological correlations. *Ergebn. Anat. Entwickl.-Gesch.*, 39, Heft 2; 1–48.

Nyberg-Hansen, R. (1966b) Sites of termination of interstitiospinal fibers in the cat. An experimental study with silver impregnation methods. *Arch. ital. Biol.*, 104: 98–111.

Orlovsky, G.N. (1970) Work of the reticulospinal during locomotion. *Biofizika*, 15: 728–737 (in translation).

Orlovsky, G.N. and Pavlova, G.A. (1972) Vestibular responses of neurons of different descending pathways in cats with intact cerebellum and in decerebellated ones. *Neurofiziologiya*, 4: 303–310.

Oscarsson, O. (1973) Functional organization of spinocerebellar paths. In *Handbook of Sensory Physiology, Vol. II, Somatosensory System.* A. Iggo (Ed.) Springer, Berlin, pp. 339–380.

Peacock, M.J. and Wolstencroft, J.H. (1976) Projections of reticular neurons to dorsal projections of the spinal cord in the cat. *Neurosci. Lett.*, 2: 7–11.

Peterson, B.W. (1970) Distribution of neural responses to tilting within vestibular nuclei of the cat. *J. Neurophysiol.*, 33: 750–767.

Peterson, B.W. (1977) Identification of reticulospinal projections that may participate in gaze control. In *Control of Gaze by Brain Stem Neurons.* R. Baker and A. Berthoz (Eds.). Elsevier/North-Holland, Biomed Press, Amsterdam, pp. 143–152.

Peterson, B.W. (1979a) Reticulospinal projections to spinal motor nuclei. *Ann. Rev. Physiol.*, in press.

Peterson, B.W. (1979b) Reticulo-motor pathways: their connections and possible roles in motor behavior. In *Integration in the Nervous System.* H. Asanuma and V.J. Wilson (Eds.). Igaku Shoin, Ltd., Tokyo, in press.

Peterson, B.W. and Abzug, C. (1975) Properties of projections from vestibular nuclei to medial reticular formation in the cat. *J. Neurophysiol.*, 38: 1421–1435.

Peterson, B.W. and Coulter, J.D. (1977) A new long spinal projection from the vestibular nuclei in the cat. *Brain Res.*, 122: 351–356.

Peterson, B.W., Anderson, M.E. and Filion, M. (1974) Responses of pontomedullary reticular neurons to cortical, tectal and cutaneous stimuli. *Exp. Brain Res.*, 21: 19–44.

Peterson, B.W., Maunz, R.A. and Fukushima, K. (1978b) Properties of a new vestibulospinal projection, the caudal vestibulospinal tract. *Exp. Brain Res.*, in press.

Peterson, B.W., Filion, M., Felpel, L.P. and Abzug, C. (1975a) Responses of medial reticular neurons to stimulation of the vestibular nerve. *Exp. Brain Res.*, 22: 335–350.

Peterson, B.W., Maunz, R.A., Pitts, N.G. and Mackel, R.G. (1975b) Patterns of projection and branching of reticulospinal neurons. *Exp. Brain Res.*, 23: 333–351.

Peterson, B.W., Frank, J.I., Pitts, N.G. and Daunton, N.G. (1976) Changes in responses of medial pontomedullary reticular neurons during repetitive cutaneous, vestibular, cortical and tectal stimulation. *J. Neurophysiol.*, 39: 564–581.

Peterson, B.W., Pitts, N.G., Fukushima, K. and Mackel, R. (1978a) Reticulospinal excitation and inhibition of neck motoneurons. *Exp. Brain Res.*, 32: 471–489.

Petras, J.M. (1967) Cortical, tectal and tegmental fiber connections in the spinal cord of the cat. *Brain Res.*, 6: 275–324.

Pitts, N.G., Fukushima, K. and Peterson, B.W. (1977) Reticulospinal action on cervical, thoracic and lumbar motoneurons. *Neurosci. Abstr., III:* 276, no. 881.

Pompeiano, O. (1974) Cerebello-vestibular interrelations. In *Handbook of Sensory Physiology, Vol. 6, Vestibular System, Part 1, Basic Mechanisms.* H.H. Kornhuber (Ed.). Springer, Berlin, pp. 417–476.

Pompeiano, O. (1975) Macular input to neurons of the spinoreticulo-cerebellar pathway. *Brain Res.*, 95: 357–368.

Pompeiano, O. and Brodal, A. (1957) The origin of vestibulospinal fibers in the cat. An experimental-anatomical study with comments on the descending medial longitudinal fasciculus. *Arch. ital. Biol.*, 95: 166–195.

Precht, W., Grippo, J. and Wagner, A. (1967) Contribution of different types of central vestibular neurons to the vestibulospinal system. *Brain Res.*, 4: 119–123.

Precht, W., Volkind, R. and Blanks, R.H.I. (1977) Functional organization of the vestibular input to the anterior and posterior cerebellar vermis of the cat. *Exp. Brain Res.,* 27: 143–160

Rapoport, S., Susswein, A., Uchino, Y. and Wilson, V.J. (1977a) Properties of vestibular neurones projecting to neck segments of the cat spinal cord. *J. Physiol. (Lond.),* 268: 493–510.

Rapoport, S., Susswein, A., Uchino, Y. and Wilson, V.J. (1977b) Synaptic actions of individual vestibular neurons on cat neck motoneurons. *J. Physiol. (Lond.),* 272: 367–382.

Rhines, R. and Magoun, H.W. (1946) Brain stem facilitation of cortical motor response. *J. Neurophysiol.,* 9: 219–229.

Sans, A., Raymond, J. and Marty, R. (1972) Projections des cretes ampullaires et de l'utricle dans les noyaux vestibulaires primaires. Étude microphysiologique et corrélations anatomiofunctionelles. *Brain Res.,* 44: 337–356.

Schor, R.H. (1974) Responses of cat vestibular neurons to sinusoidal roll tilt. *Exp. Brain Res.,* 20: 347–362.

Schwindt, P.C., Precht, W. and Richter, A. (1974) Monosynaptic excitatory and inhibitory pathway from medial midbrain nuclei to trochlear motoneurons. *Exp. Brain Res.,* 20: 223–238.

Shapovalov, A.I. (1966) Excitation and inhibition of spinal neurones during supraspinal stimulation. In *Nobel Symposium I. Muscular Afferents and Motor Control.* R. Granit (Ed.). Almqvist and Wiksell, Stockholm, pp. 331–348.

Shimamura, M. and Livingston, R.B. (1963) Longitudinal conduction systems serving spinal and brain stem coordination. *J. Neurophysiol.,* 26: 258–272.

Shimazu, H. and Precht, W. (1966) Inhibition of central vestibular neurons from the contralateral labyrinth and its mediating pathway. *J. Neurophysiol.,* 29: 467–492.

Shimazu, H. and Smith, C. (1971) Cerebellar and labyrinthine influences on single vestibular neurons identified by natural stimuli. *J. Neurophysiol.,* 34: 493–508.

Sprague, J.M. and Chambers, W.W. (1954) Control of posture by reticular formation and cerebellum in the intact, anesthetized and unanesthetized and in the decerebrated cat. *Amer. J. Physiol.,* 176: 52–64.

Spyer, K.M., Ghelarducci, B. and Pompeiano, O. (1974) Gravity responses of the neurons in the main reticular formation. *J. Neurophysiol.,* 37: 705–721.

Stein, B.M. and Carpenter, M.B. (1967) Central projections of portions of the vestibular ganglia innervating specific parts of the labyrinth in the rhesus monkey. *Amer. J. Anat.,* 120: 281–318.

Sterling, P. and Kuypers, H.G.J.M. (1967) Anatomical organization of the brachial spinal cord of the cat II. The motoneuron plexus. *Brain Res.,* 4: 16–32.

Suzuki, J.-I. and Cohen, B. (1964) Head, eye, body and limb movements from semicircular canal nerves. *Exp. Neurol.,* 10: 393–405.

Szentàgothai, J. (1943) Die zentrale Innervations der Augenbewegungen. *Arch. Psychiat. Nervenkr.,* 116: 721–760.

Toczek, S. and Hyde, J.E. (1963) Effect of vestibular nerve section on torsion and on evoked rotatory movements. *Exp. Neurol.,* 8: 143–154.

Udo, M. and Mano, N. (1970) Discrimination of different spinal monosynaptic pathways converging onto reticular neurons. *J. Neurophysiol.,* 33: 227–238.

Wilson, V.J. (1972) Physiological pathways through the vestibular nuclei. *Intern. Rev. Neurobiol.,* 15: 27–81.

Wilson, V.J. (1979) Electrophysiological and dynamic studies of vestibulospinal reflexes. In *Integration in the Nervous System.* H. Asanuma and V.J. Wilson (Eds.). Igaku Shoin, Ltd., Tokyo, in press.

Wilson, V.J. and Maeda, M. (1974) Connections between semicircular canals and neck motoneurons in the cat. *J. Neurophysiol.,* 37: 346–357.

Wilson, V.J. and Peterson, B.W. (1978) Peripheral and central substrates of vestibulospinal reflexes. *Physiol. Rev.,* 58: 80–105.

Wilson, V.J. and Yoshida, M. (1969a) Comparison of effects of stimulation of Deiters' nucleus and medial longitudinal fasciculus on neck, forelimb, and hindlimb motoneurons. *J. Neurophysiol.,* 32: 743–758.

Wilson, V.J. and Yoshida, M. (1969b) Monosynaptic inhibition of neck motoneurons by the medial vestibular nucleus. *Exp. Brain Res.,* 9: 365–380.

Wilson, V.J., Wylie, R.M. and Marco, L.A. (1968) Synaptic inputs to cells in the medial vestibular nucleus. *J. Neurophsyiol.,* 31: 176–185.

Wilson, V.J., Yoshida, M. and Schor, R.H. (1970) Supraspinal monosynaptic excitation and inhibition of thoracic back motoneurons. *Exp. Brain Res.,* 11: 282–295.

Wilson, V.J., Kato, M., Thomas, R.C. and Peterson, B.W. (1966) Excitation of lateral vestibular neurons by peripheral afferent fibers. *J. Neurophysiol.,* 29: 508–529.

Wilson, V.J., Kato, M., Peterson, B.W. and Wylie, R.M. (1967) A single-unit analysis of the organization of Deiters' nucleus. *J. Neurophysiol.*, 30: 603–619.

Wilson, V.J., Gacek, R.R., Maeda, M. and Uchino, Y. (1977) Saccular and utricular input to cat neck motoneurons. *J. Neurophysiol.*, 40: 63–73.

Wilson, V.J., Gacek, R.R., Uchino, Y. and Susswein, A.J. (1978) Properties of central vestibular neurons fired by stimulation of the saccular nerve. *Brain Res.*, 143: 251–261.

Wolstencroft, J.H. (1964) Reticulospinal neurones. *J. Physiol. (Lond.)*, 174: 91–108.

Wylie, R.M. and Felpel, L.P. (1971) The influence of the cerebellum and peripheral somatic nerves on the activity of Deiters' cells in the cat. *Exp. Brain Res.*, 12: 528–546.

Muscle Fields and Response Properties of Primate Corticomotoneuronal Cells

E.E. FETZ and P.D. CHENEY

Department of Physiology and Biophysics, and Regional Primate Research Center, University of Washington, Seattle, WA 98195 (U.S.A.)

INTRODUCTION

The role of motor cortex cells in control of movement and posture depends not only on their activity during relevant behavioral responses but also on the target cells that are affected by this activity. Analysis of precentral cell responses during active and passive movement has revealed two major sources of input, central and peripheral; however, the functional significance of this activity remains uncertain as long as its output destinations are unknown. We have, therefore, investigated those precentral cortex cells whose output effects on the activity of motor units could be confirmed by cross-correlation techniques. Spike-triggered averages (STA's) of rectified EMG activity have revealed that action potentials of certain precentral neurons are followed by a transient postspike facilitation (PSF) of average motor unit activity (Fetz et al., 1976). Such cells would therefore contribute rather directly to control of muscle activity. Since the latency and time course of the PSF suggest that they are mediated by direct corticomotoneuronal (CM) connections, we have referred to these as CM cells. The present review summarizes the evidence concerning the distribution of PSF in different forelimb muscles and the response properties of these cells during controlled ramp-and-hold wrist movements.

METHODS

To provide prolonged periods of coactivation of wrist and finger muscles with cortical cell activity, monkeys were trained to alternately flex and extend the wrist against elastic loads. Thus, displacement of the wrist from a neutral center position required proportional active torques. The hand was held with fingers extended between padded plates which could rotate about a shaft aligned with the wrist. The movement trajectory consisted of a phasic ramp followed by a static hold in an electronically detected hold zone for at least 1 sec (cf. figures). EMG activity of six flexor and six extensor muscles was monitored with pairs of stainless steel electrodes implanted in the belly of each muscle. The muscles were identified both by their anatomical location and by characteristic movements evoked by stimulating through the electrode pairs. As indicated below, recordings were confirmed to be of independent motor units by cross-correlating the EMG activity. The specific muscles that were sampled and illustrated include extensor digitorum communis (EDC); extensors carpi radialis brevis and longus (ECR-B and ECR-L); extensor carpi ulnaris (ECU); flexors digitorum profundus and sublimis (FDP and FDS); flexors carpi radialis and ulnaris (FCR and FCU) and palmaris longus (PL).

Action potentials of precentral cortex cells that fired during flexion or extension were used to compile STA's of the full-wave rectified EMG activity of six covarying agonist muscles. The rectification was intended to eliminate cancellation of positive and negative components of any motor unit potentials that might appear at variable latencies after the cortical spike. The STA's included a minimum of 2,000 events, although more were often included to reduce the noise. The analysis period was typically 30 msec, 5 msec before to 25 msec after the spike.

RESULTS

Distribution of PSF in different forelimb muscles

Examples of cells producing postspike facilitation in coactivated wrist and finger muscles are illustrated in the first three figures. The cell in Fig. 1 fired consistently during extension of the wrist. STA's of two of the six covarying extensor muscles showed clear PSF (EDC and ED 4,5); in addition, two other muscles also showed evidence of a weaker augmentation of EMG activity after the spike. The remaining two muscles showed no significant spike-related effect in the number of events averaged (4,600).

The cell in Fig. 2 covaried with wrist flexion and produced PSF in three of the wrist flexor muscles recorded with implanted electrodes, as well as in EMG recorded with

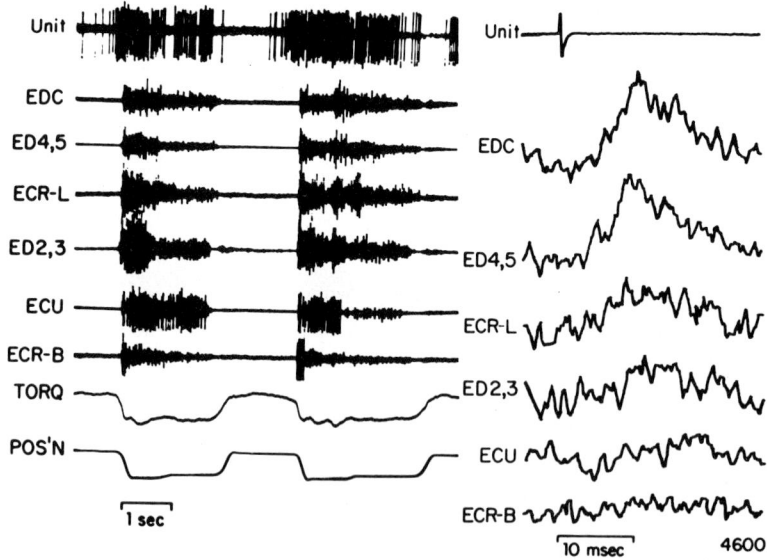

Fig. 1. Responses of an extension-related CM cell during successive wrist movements against a spring-like load (left). Muscles are identified in text; bottom traces illustrate active torque and wrist position. At right are averages of the rectified EMG activity, triggered from 4,600 action potentials of the cell. As shown by position of the unit spike at top, the analysis interval included 5 msec preceding the spike and 25 msec after the spike. In this and subsequent figures the number of events averaged is given beneath the average. (From Fetz and Cheney, 1978).

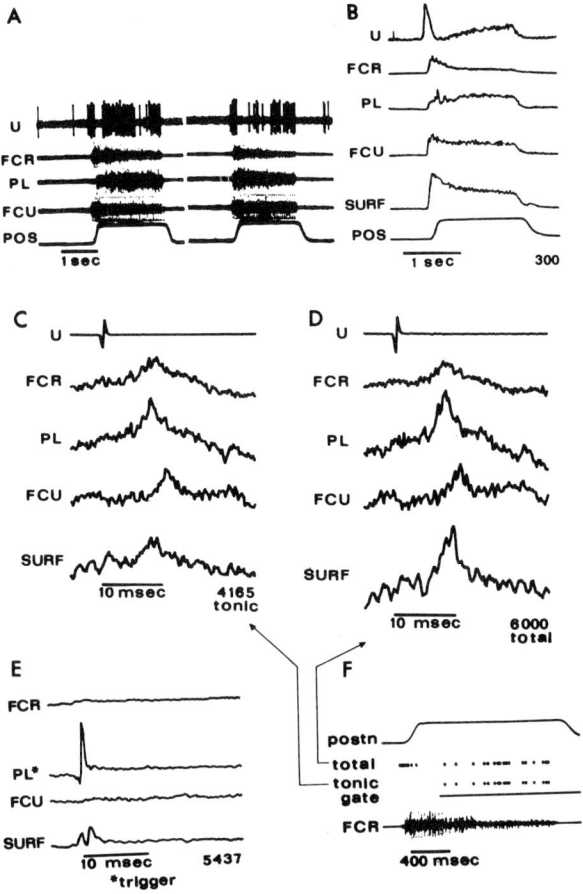

Fig. 2. CM cell related to flexion of the wrist. A) Responses of cortical unit (U), implanted flexor muscles, and wrist position (POS). B) Average of 300 flexion responses, aligned at onset of flexion movement. Unit activity shows pattern characteristic of "phasic-ramp" cells, i.e., a phasic burst of activity at movement onset, followed by gradually increasing discharge during static hold period. Averages of full-wave rectified EMG activity recorded from implanted muscles and by surface electrodes (SURF). C) Spike-triggered averages of rectified EMG activity compiled for action potentials occurring during the static hold period. D) STA's compiled for action potentials occurring during both phasic movement and static hold periods. E) Averages of rectified EMG activity triggered from motor units recorded in PL muscle; as in STA's, analysis interval was 30 msec. F) Single flexion response illustrating gating of unit pulses used to compile STA in C. Dot rasters below position trace show pulses for all action potentials (total) and those occurring during static hold period (tonic); gate signal began 200 msec after end of phasic movement. (From Fetz and Cheney, 1978.)

surface electrodes. Response averages aligned on the phasic flexion movement (Fig. 2B) indicate that this unit exhibited an unusually strong peak of activity at movement onset, when muscle activity was initiated. This unit provides a particularly convincing example of the fact that PSF is not simply an artifactual consequence of averaging non-stationary EMG activity. Excluding from the triggering events all action potentials occurring during the initial peak and compiling STA's only from those action potentials occurring during the static hold period (using the gating procedure illustrated in Fig. 2F) resulted in the clear PSF's shown in Fig. 2C. In comparison, STA's that were triggered from *all* action potentials, including those at movement onset, showed similar effects (Fig. 2D); the larger PSF in the records from PL and SURF in Fig. 2D may be due to the contribution of larger motor units at onset or to enhanced effectiveness of brief interspike intervals. Similar controls with other units confirmed that PSF was not a consequence of changing levels of EMG activity.

In a total of 2,242 STA's, we observed 483 instances of PSF (21%). The average onset latency after the cortical spike was 6.7 ± 2.9 msec (mean $\pm$ SD). This is consistent with known conduction times in pyramidal tract neurons (PTN's) and motoneurons. In the macaque monkey, the latency of antidromic responses evoked from stimulating the pyramidal tract (minimum 0.7 msec) (Evarts, 1965) and from cervical spinal cord (1.3 msec) (Humphrey and Rietz, 1978) indicates that the conduction from cortex to cervical levels could be as fast as 1.1 msec (assuming a 0.2 msec utilization time); minimal conduction for motoneurons to muscles could be 3.5 msec. Most PSF latencies were longer than the sum of these minimal conduction times from cortex to muscle. As shown in Fig. 2C and D, the same cell could produce PSF in different muscles with different latencies, probably attributable to different conduction times of the relevant motoneurons. The rise time of the PSF peak (3.2 ± 1.8 msec) and decay time back to base line (7.0 ± 3.0 msec) would depend on several factors, including the number of motor units affected, their conduction times, the shape of their muscle fiber potentials and the time course of the CM-EPSP's. The time after the cortical spike when facilitation was greatest, i.e., the latency of the PSF peak, was on average 10.2 ± 3.0 msec. The amplitude of this peak relative to base line level was measured for representative samples of PSF. The peaks of the PSF ranged from 12.4% of base line for strong PSF to an average 5.7% for weak PSF. Of course, these values are somewhat arbitrary since they depend on the number of motor units recorded and the amplitude of their recorded potentials as well as on the strength of any underlying cross-correlation.

Most of the cells exhibiting PSF proved to be PTN's when tested. In fact, there was a correlation between the antidromic PT latency and PSF latency: slow PTN's produced PSF at relatively longer latencies after the spike. Fast PTN's produced primarily short-latency PSF but some also produced PSF of longer latency. Of all the PTN's correlated with five or six muscles, 55% exhibited signs of postspike facilitation in at least one of the muscles.

When STA's were compiled simultaneously for multiple muscles, cells that produced PSF generally facilitated more than one muscle. Of 370 precentral neurons that covaried strongly with flexion or extension and were used to compile STA's of five or six covarying forelimb muscles, 160 (43%) produced PSF in at least one muscle and on this basis were defined as CM cells. Of these, 112 (70%) produced PSF in more than one muscle (29% in two muscles, 23% in three, 11% in four, 4% in five and 3% in six).

The set of facilitated muscles represents those muscles whose activity is statistically affected by action potentials of the cell, and may be referred to as the cell's "muscle field" (Fetz and Cheney, 1978). If these PSF's are mediated by monosynaptic CM connections, then the cells that produce them are cortico-motoneuronal cells also in the anatomical sense, and their muscle field would represent their target muscles, i.e., the muscles whose motoneurons are contacted by terminals of the cell. Existence of divergent terminals of PTN's has been demonstrated electrophysiologically in primates by antidromic activation from different motor nuclei (Asanuma et al., 1979). Whether all monosynaptic CM connections are sufficiently potent to produce detectable cross-correlations, and conversely, whether some of the observed correlations may be mediated by indirect connections via other cells, remains an important unresolved issue. At this point, it seems technically simpler to determine the set of muscles whose activity is facilitated by the cell rather than the set of motoneurons that are anatomically contacted by the cell. Indeed, the set of facilitated muscles would seem to be a functionally more significant measure of the output effects of the cortical cell than the distribution of the underlying synaptic connections, which may or may not produce detectable effects

The method used to sample EMG activity will of course influence the extent of the observed muscle field. STA's can only detect effects in muscles that are coactivated with the unit. The extent of facilitation in different muscles could be mistakenly exaggerated if the same motor units were recorded redundantly through different EMG electrode pairs. Therefore, this possibility was routinely tested in all experiments by cross-correlating muscle activity. Compiling averages of rectified EMG activity triggered from motor units of each muscle revealed the extent to which the same units might have been picked up by adjacent leads. For example, Fig. 2E shows that when EMG averages were triggered from motor units in PL, only the surface-recorded activity revealed a peak simultaneously with the triggering peak in PL, confirming that the surface electrodes (over PL) had picked up some of the same units. However, there was no sign of a comparable peak in averages of either of the two adjacent flexor muscles, FCR or FCU. This, plus similar averages triggered from each of the other muscles, confirmed that the implanted electrodes had, indeed, recorded independent motor units. Such "cross-correlation" of muscle recordings was done routinely when multiple PSF's were observed. In a few cases, EMG-triggered averages revealed evidence that some units had been recorded in common; in those cases, one of the redundant EMG records was eliminated from the data base. Thus, the distribution of PSF in different forelimb muscles was not artifactually exaggerated by redundant EMG recording.

On the other hand, since the sample of motor units was limited, the extent of the cells' muscle fields could easily have been underestimated. Implanted leads recorded perhaps 10% of the total motor units of the muscle; if the CM cells facilitated specific motor units within the pool, as suggested by recent evidence (Jankowska et al., 1975), recordings that showed no evidence of PSF might have missed detecting possible facilitation of other motor units in the same muscle. Moreover, additional muscles that were not sampled could also have contained facilitated motor units. Thus, the extent of the cell's muscle field may have been underestimated by the limited sampling of motor unit potentials.

Response patterns of CM cells

Since the PSF represents the average effect of the action potentials of CM cells on muscle activity, the net effect of a given firing pattern of such a cell may be estimated by convolving the PSF with its firing rate. Thus, the response patterns of a CM cell during wrist movements would produce a directly proportional augmentation of the firing probability of its facilitated motor units. To quantify the firing pattern of CM cells during controlled ramp-and-hold wrist movements, monkeys were trained to displace the wrist against elastic loads of different stiffness. On the basis of their phasic response during the ramp movement and their tonic firing during the static hold period, all CM cells could be classified into four basic types. Every CM cell exhibited some degree of activity during the static hold period, firing either at a constant rate ("tonic" types; cf. Figs. 3 and 4) or with gradually increasing rates ("ramp" types; cf. Fig. 2B). In contrast, many other precentral cells fired only phasically at onset of movement, but exhibited no maintained discharge during the hold period; none of these cells ever produced any postspike facilitation in the recorded muscles. The CM cells were further subdivided according to the presence of a phasic peak of activity at the onset of movement (Figs. 2B and 3) or the absence of such enhanced activity at the beginning of movement (Fig. 4). Thus, on the basis of their dynamic and static responses during the ramp-and-hold wrist movements all CM cells could be classified into one of four basic types: phasic-tonic (59%), tonic (28%), phasic-ramp (8%) and ramp (5%). These patterns were related to the torque response, since they occurred whether or not the response involved proportional wrist displacement, i.e., whether the response was isotonic or isometric (Cheney and Fetz, 1978). Clearly those cells with an enhanced phasic peak of activity at onset of movement would provide particularly effective input to activate motoneurons.

Relation of CM cells to active torque

The relation of activity of motor cortex cells to active force and joint displacement was first investigated by Evarts in monkeys trained to displace a handle against different loads. He found that the firing rate of many PTN's was related to the active force or to changes in force (Evarts, 1968). Similar results have been obtained by others

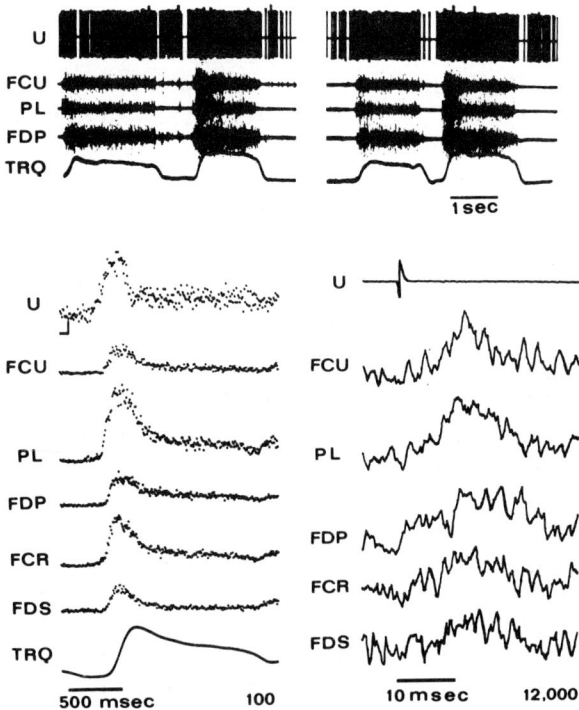

Fig. 3. Flexion-related CM cell facilitating activity of 5 muscles. Top: representative responses during flexion movements, illustrating higher firing rate at greater levels of active torque. Bottom left: response average of unit and rectified EMG activity of implanted flexor muscles, and torque. The firing pattern of this unit is characteristic of "phasic-tonic" CM cells. Bottom right: STA of rectified EMG activity, showing some PSF in each muscle.

(Humphrey et al., 1970; Hepp-Reymond et al., 1978; Smith et al., 1975; Schmidt et al., 1975), although the exact proportions and interpretations differ. Since some of the variation in results may be due to differences in cell types recorded and in their projections, it seemed worth re-investigating this issue with CM cells, whose output effects on specific agonist muscles could be independently confirmed by STA's. Accordingly, monkeys were trained to make the ramp-and-hold wrist movements into the same hold zone but against elastic loads of different stiffness. Separate response averages were compiled for movements involving the same displacement but different degrees of active torque. The firing rate of CM cells was consistently higher when the monkey exerted a greater amount of active torque as illustrated by the single trials in Fig. 3 and the response averages in Fig. 4. Since the onset of movement involved overcoming inertial loads of unknown magnitude, it was difficult to quantitatively interpret the phasic discharge at movement onset. However, the tonic firing rate during the static hold period occurs when activity in central and peripheral circuits has reached a relatively steady-state condition. We therefore measured the tonic firing rate as a function of static torque for CM cells from response averages; representative results are plotted in Fig. 5 for cells related to flexion and extension. All CM cells had a range over which tonic firing rate increased linearly with active static torque. The slope of this linear portion (i.e., the increment in firing rate per increment in static torque) was consistently greater for cells related to extension than for cells related to

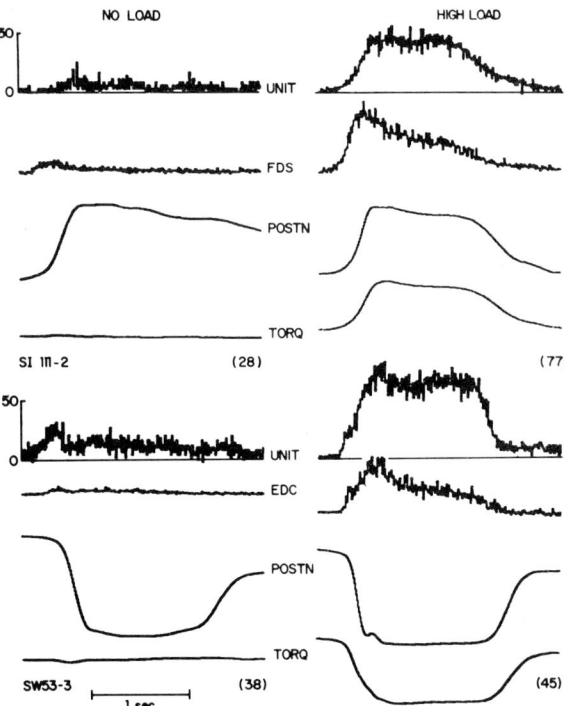

Fig. 4. Relation of CM cell activity to active torque. Response averages of a flexion-related cell (top) and extension-related cell (bottom) during comparable wrist displacements against no external load (left) and against stiff elastic load (right). Also shown is average EMG activity of one of the covarying agonist muscles which exhibited PSF. Response pattern of flexion-related cell (top right) is characteristic of "tonic" CM cells.

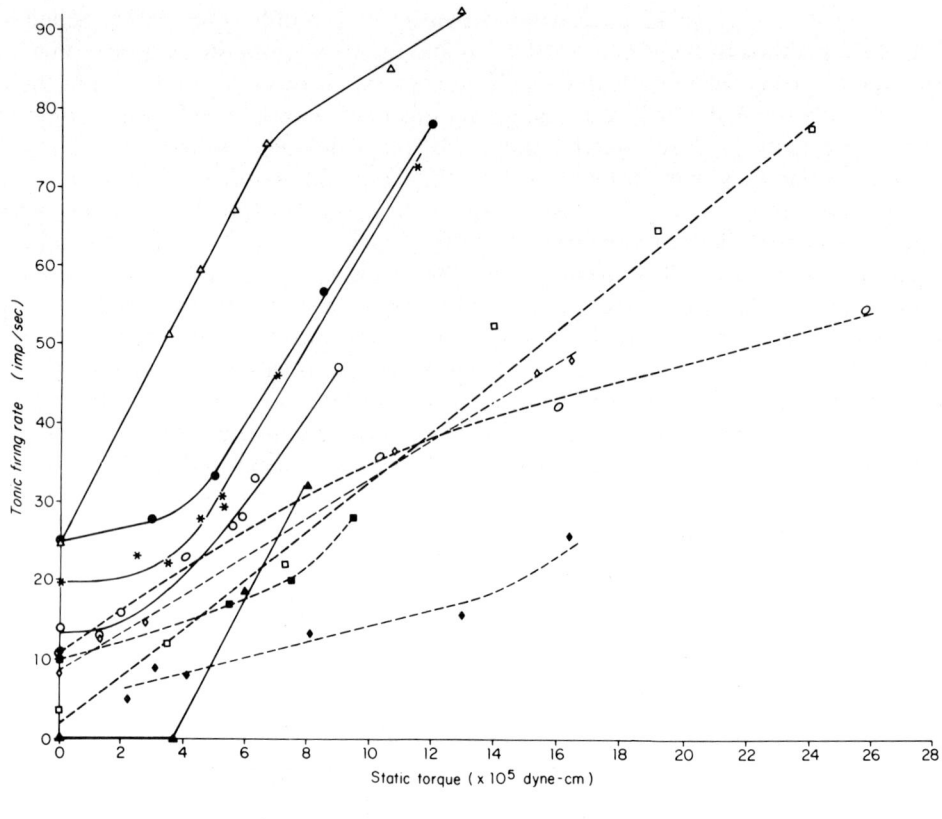

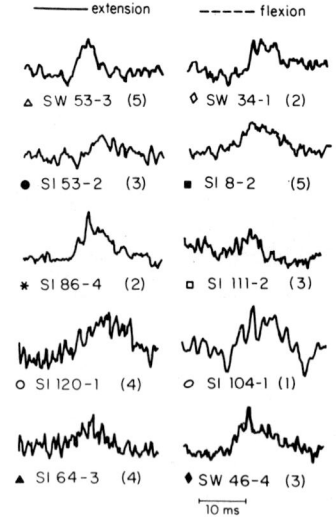

Fig. 5. Tonic firing rate of CM cells during static hold period plotted as function of active torque; representative examples of CM cells related to extension (solid lines) and flexion (dashed lines). Each point represents a measurement from a response average. Below: examples of PSF in one of the facilitated muscles for each cell plotted at left. Number of facilitated muscles is given in parentheses.

flexion (means: 4.8 vs 2.5 imp/sec/10^5 dyne-cm). These differences in load sensitivity between flexion-and extension-related cells may be due partly to differences in mechanical factors such as the radius over which the torque is exerted and to differences in internal loads between flexion and extension. Alternatively, there may also be inherent differences in the effects of CM cells related to flexor and extensor musculature (cf. Clough et al., 1968). Similar differences in load sensitivity were observed under isometric conditions.

Since CM cells have firing rates which increase with active torque and since their activity demonstrably facilitates agonist motoneurons, they clearly contribute causally to producing active force. In the range studied, their contribution to increases in force seems to be more through increments in firing rate than to recruitment of additional CM cells. One clear example of a CM cell recruited into activity at higher load levels is illustrated in Fig. 5; the rest were active to some degree even at the lowest levels of external load. Of course, some of this activity may be related to overcoming internal load, such as stretching the antagonist muscles.

In addition to their role in generating static torque, many CM cells may also respond phasically to perturbations of the wrist, and may participate in a transcortical reflex postulated to subserve load compensation. They responded to passive joint movements which stretched their facilitated muscles and during active movements they responded to load torque pulses at latencies consistent with a contribution to the M2 response (Cheney and Fetz, 1978). All CM cells exhibited at least those responses consistent with the load compensation hypothesis, i.e., they responded to perturbations which lengthened the facilitated muscle. However, half of the CM cells also showed additional responses to load perturbations in which their output would not be appropriate for load compensation. This would suggest that other systems in addition to the cortico-motoneuronal cells could be significantly involved in load compensation.

SUMMARY

In monkeys alternately flexing and extending the wrist against elastic loads, cortico-motoneuronal (CM) cells were identified by characteristic post-spike facilitation (PSF) of averaged forelimb muscle activity. Spike-triggered averages of rectified EMG activity of five to six covarying agonist muscles revealed that action potentials of 43% of strongly covarying motor cortex cells (n = 370) were followed by PSF of at least one muscle. The mean latency of such PSF was 6.7 ± 2.9 msec. The "muscle field" of such CM cells, defined as the set of muscles exhibiting PSF, could include only one muscle (30% of cells) or two (29%) or more (41%).

On the basis of their dynamic and static responses during the ramp-and-hold wrist movements, CM cells could be classified into four basic types: phasic-tonic (59%), tonic (28%), phasic-ramp (8%) and ramp (5%). During the static hold period the "tonic" cells fired at constant rates, while activity of "ramp" cells increased gradually. In addition, some cells of each group exhibited larger phasic responses at onset of movement.

When monkeys exerted different degrees of static torque, the tonic firing rate of CM cells increased linearly as a function of torque over much of the range. In the linear range, the increment in firing rate per increment in static torque was about twice as great for extension-related cells as for flexion-related cells. For the responses studied, the contribution of CM cells to active torque was more through increases in firing rate than recruitment of additional CM cells. Most CM cells also responded to passive movements and load perturbations which stretched their facilitated muscles.

ACKNOWLEDGEMENTS

We gratefully acknowledge the technical assistance of Mr. Jerrold D. Maddocks in many phases of this work and the collaboration of Dr. Dwight German in early stages of this research. This research was supported by Grants RR 00166, NS 5082, NS 12542 and US0966 from the National Institutes of Health, U.S. Public Health Service.

REFERENCES

Asanuma, H., Zarzecki, P., Jankowska, E., Hongo, T. and Marcus, S. (1979) Projections of individual pyramidal tract neurons to lumbar motoneuron pools of the monkey. *Exp. Brain Res.*, 34: 73–89.

Cheney, P.D. and Fetz, E.E. (1977) Comparison of spike-triggered averages and stimulus-triggered averages of forearm muscle activity from identical motor cortex sites in behaving animals. *Neurosci. Abstr.*, 3: 269, no. 852.

Cheney, P.D. and Fetz, E.E. (1978) Functional properties of primate corticomotoneuronal cells. *Neurosci. Abstr.*, 4: 293, no. 918

Clough, J.F.M., Kernell, D. and Phillips, C.G. (1968) The distribution of monosynaptic excitation from the pyramidal tract and from primary spindle afferents to motoneurons of the baboon's hand and forearm. *J. Physiol. (Lond.)*, 198: 145–166.

Evarts, E.V. (1965) Relation of discharge frequency to conduction velocity in pyramidal tract neurons. *J. Neurophysiol.*, 28: 216–228.

Evarts, E.V. (1968) Relation of pyramidal tract activity to force exerted during voluntary movement. *J. Neurophysiol.*, 31: 14–27.

Fetz, E.E. and Cheney, P.D. (1978) Muscle fields of primate corticomotoneuronal cells. *J. Physiol. (Paris)*, 74: 239–245.

Fetz, E.E., Cheney, P.D. and German D.C. (1976) Corticomotoneuronal connections of precentral cells detected by post-spike averages of EMG activity in behaving monkeys. *Brain Res.*, 114: 505–510.

Hepp-Reymond, M.C., Wyss, U.R. and Anner, R. (1978) Neuronal coding of static force in the primate motor cortex. *J. Physiol. (Paris)*, 74: 287–291.

Humphrey, D.R. and Corrie W.S. (1978) Properties of pyramidal tract neuron system within a functionally defined subregion of primate motor cortex. *J. Neurophysiol.*, 41: 216–243.

Humphrey, D.R., Schmidt, E.M. and Thompson, W.D. (1970) Predicting measure of motor performance from multiple cortical spike trains. *Science*, 170: 758–762.

Jankowska, E., Padel, Y. and Tanaka, R. (1975) Projections of pyramidal tract cells to α-motoneurones innervating hindlimb muscles in the monkey. *J. Physiol. (Lond.)*, 249: 637–667.

Schmidt, E.M., Jost, R.G. and Davis, K.K. (1975) Reexamination of the force relationship of cortical cell discharge patterns with conditioned wrist movements. *Brain Res.*, 83: 213–223.

Smith, A.M., Hepp-Reymond, M.C. and Wyss, U.R. (1975) Relation of activity in precentral cortical neurons to force and rate of force change during isometric contractions of finger muscles. *Exp. Brain Res.*, 23: 315–332.

Interpretation of Supraspinal Effects on the Gamma System

R. GRANIT

Nobel Institute for Neurophysiology, Karolinska Institutet, S-104 01 Stockholm (Sweden)

The microelectrode, be it applied for stimulation or recording, has taught us, if we did not know it before, that motor activity is the outcome of highly complex processes hierarchically organized. Best known at the moment are the spinal cord and the motor cortex, when our criterion is understanding of specific definable functions, but we do not doubt that the cerebellum, the basal ganglia and the reticular formation play roles in what Sherrington once called "motricity" that are within reach of comparable understanding though not yet on a par with the two first mentioned.

My task has been restricted to say something about the gamma-spindle system which is a major instrument for handling unexpected encounters with mechanical forces outside the body. The addition of "supraspinal" to the title emphasizes that the muscle spindle is at the disposal of hierarchic governors at different sites. It is not so necessary here to proceed to enumerate these sites. The gamma neurons are focused on the spindles. No other role for them is known. They sensitize spindles and so both the "how" and "why" of the gamma system concern this act of sensitization.

If I begin by being very general, let me first state that motor organizations are built for timing entry and exit of motoneuronal responses. The most perfect timing circuit known in detail today is the spinal centre organizing locomotion by reciprocal innervation. Apparently evolution has put a premium on creating an organization for timing as low down in the neuraxis as possible in order to make it as swift as possible. The Gothenburg group with Lundberg. Hultborn, Jankowska in the lead (see e.g. Lundberg, 1975) has elucidated this organization so well that I think I can assume it familiar to all those present at this Symposium. I have discussed it elsewhere (Granit, 1975).

If it be essential for the spindles to be sensitized in locomotion, the gamma motoneurons would have to be excited already at the spinal level because it is well known that the spinal cord harbours a locomotion centre of its own that can be activated also in the spinal animal. Grillner and his coworkers have devoted a number of relevant experiments to this question. Grillner and Zanger (1974, 1975) and Sjöström and Zanger (1975, 1976) have shown that in the acutely spinalized cat stepping movements can be generated in the deafferented hindlimb. These are a detailed replica of those in non-spinalized animals. The cat is given an injection of 200–500 µg/kg of Clonidine and 100 mg/kg of DOPA or, alternatively, 50 mg/kg Niamid plus 50 mg/kg DOPA. Stepping movements can then be elicited by tonic electrical stimulation at 5 Hz of the two cut dorsal roots. It was found that single gamma efferents to a flexor and an extensor displayed perfect rhythmic activation linked to the supraliminal alpha stepping.

Since DOPA activates static gamma efferents, these at least would have been linked in coactivation with the alpha partners of the stepping generator. The authors hold participation of both static and dynamic gamma motoneurons more likely. The gamma activity generally preceded the alpha activity. Staying for the moment by the timing aspect alone, we conclude that spindle sensitization plays an essential role in this spinal precision instrument for timing entry and exit of different muscles in locomotion.

The existence of a spinal stepping generator in combination with the detailed knowledge of the supraspinal control of the Ia inhibitory interneuron given us by the Gothenburg workers (see e.g., a summary by Lundberg, 1975 and one by Wetzel and Stuart, 1976) means that the mechanism employing reciprocal innervation must be activated in a well-organized manner when started supraspinally, for instance, by mesencephalic stimulation as introduced by Shik et al. (1966). This perfect spinal timing instrument can of course be wrongly approached by a supraspinal governor and this is likely to have happened in the experiments of Brooks (1975) in which local cooling of globus pallidus led to a breakdown of alternating movements. I hope that we shall see much experimentation making use of reciprocal innervation of the limbs as a test of the nature of its supraspinal approaches. And, in particular, one would like to be informed on how the gamma system behaves when disintegration of some kind is noted in the timing machinery of the stepping generator. It is a truism that precise knowledge of the nature of a test favours analytical progress.

I have taken up these experiments on alpha-gamma coactivation in the spinal stepping mechanism also for the light they throw on the nature of alpha-gamma linkage as an anatomical proposition. It means that coactivation, though possible anywhere on a supraspinal route, in many cases could be localized to its final site of action in the spinal cord. Is it hopeless to expect those working with the horseradish peroxidase method to provide precise information on the design of the spinal alpha-gamma link?

That the gamma neurons take precedence over the alpha neurons in coactivation is something we have seen from the beginning of gamma-spindle physiology, even when recording their effect in terms of spindle firing from afferent roots. It is nevertheless worth emphasizing that gamma-spindle precedence now has been confirmed once more with a natural movement as essential as stepping. Grigg and Preston (1971) and Fidone and Preston (1969) in their pyramidal baboon found that cortical stimulation also led to gamma precedence. For the development of such problems it would now be necessary to have recourse to intracellular recording.

In this context it is interesting to recall Hagbarth's dictum, based on his experience with direct recording from spindle afferents in human nerves: "When the skeletomotor system is silent, so is the fusimotor system" (Burke et al., 1976). Silence is then defined in terms of spike generation. His condition for silence, which is complete relaxation, may well involve an act of suppression by inhibition. What happens intracellularly is, and will be of course, unknown when the subject is man. I mention such reservations because for those of us who have had their experiences from spindle studies in the cat, absence of tonic firing in man has come as a surprise. Numerous experiments have shown spindle firing in the absence of alpha activity. Long spindle after-discharges are also quite common in the cat. Need we assume that the cat makes more use of its gamma-spindle system than man? But then the baboon also shows gamma precedence, as stated above. Spindle precedence has been recorded by Burg et al. (1974) from tibial spindle afferents during the Jendrassik manoeuvre, alpha precedence is the general experience of Vallbo (1971) in his studies of human spindle primaries in voluntary contraction of the finger flexors. Are there in man two modes of mobilizing the gamma neurons, a more generalized activation and an exactly timed coactivation in a precise voluntary finger contraction whose spindle firing occurs on its rising phase? Our experience long ago with the cat (Granit and Holmgren, 1955) was that the spindles as well as the gamma motoneurons either could be activated diffusely through a pathway that withstood numerous criss-cross sections in the spinal cord above the segment studied or, alternatively, by a very fast route that was easily interrupted by lateral cuts in the cord. The explanation of the discordant experiences reported above may well lie in the existence of two such possibilities, alternatively, in high conduction velocity of the voluntary monosynaptic component (see Phillips and Porter, 1977).

However, both possibilities imply alpha-gamma coactivation. From the work of Vallbo (1974) and Hagbarth and his colleagues (Burke et al., 1978a, b) we have also learned to look with considerable caution upon reports maintaining lack of alpha-gamma linkage. They have shown how easily an experimenter can be deceived because the selected spindle, that failed to be linked to imitate alpha behaviour, was one that was shunted by activity in a neighbouring muscle. Their warning is especially relevant when experimenters have dealt with freely moving animals or with muscles that have not been properly isolated from one another.

I do not maintain that all reports on absence of alpha-gamma linkage are faulty. Long ago (Granit et al., 1955) we described loss of linked behavior in cooling the cerebellar vermis. Hagbarth tells me (personal communication) that in man he has seen two types of clonus, one with the usual coactivation, another in which the spindles fire on the falling phase and thus behave in the passive fashion of stretch receptors.

What I have said now completes my discourse on a ubiquity of alpha-gamma linkage that elevates it to a principle in motricity. How then should one explain the common gamma precedence? There is the size hypothesis that may be applied but, considering the complex structure of the central nervous system, one should not be forced to swallow the rule of size like a bad pill that has to go down without chewing. Alternatives to be considered are, e.g. a well-developed internuncial apparatus, high synaptic densities and a higher normal level of depolarization than that of the alpha motoneurons which necessarily have to be silent unless used for some specific purpose.

Speaking of gamma precedence in locomotion Sjöström and Zanger (1976) state: "The advantage is an optimal adaptation of the muscle spindle system already at the beginning of the contraction". Their statement made me look up my old views, as formulated in 1954 for the Silliman Lectures (1955): "The interpretation of all these experiments is that the gamma system is there not only to improve the performance of the sense organs but also as an 'ignition mechanism' to initiate movement as well as to maintain tonus... A consequence of this interpretation is that the muscles possessing spindles actually are provided with two motor systems..." "With this arrangement the sense organs in the muscle are immediately ready to 'measure' during the ensuing contraction" (p. 268). Today, later experiments have added substance to these early generalizations.

Significant for the understanding of an important role of coactivity of alpha and gamma motoneurons seem to me particularly two experiments both analyzing an encounter with the unexpected. The earlier one is by Corda et al. (1965) and demonstrates unequivocally the importance of senitized spindles in the response to a sudden obstruction in the even rhythmic flow of respiration. The physiological load compensation initiated by the spindle burst was absent after deafferentation. The other one is by Marsden et al. (1972) and shows load compensation to a halting obstruction in the voluntary contraction of the long thumb flexor. The effect undoubtedly was started by its muscle spindles, even though some background facilitation from skin or joint receptors proved necessary. In both cases we are entitled to assume coactivation. There was direct evidence of it in the former experiment, and, as to the latter, it has been proved beyond doubt that voluntary finger movements are coactivated (e.g. Vallbo, 1971).

In this context I would like to emphasize that it has been so easy to study coactivation in relation to the most sensitive alpha motoneurons that its role in the action of the large and fast alphas has been neglected. They have merely been regarded as the end point in the progress of recruitment, as laid down by the size hypothesis. However, recruitment thresholds have been found variable by several authors, most recently again by Kanda et al. (1977) in the decerebrate cat and by Garnett and Stephens (1978) in man. Thus there are possibilities for studying the link between fast alpha motoneurons and gamma neurons in appropriately designed experiments. Vallbo's experience with alpha preceding may then turn out to be characteristic of fast motoneurons.

As I pointed out in my introduction to this Symposium vibratory stimulation in man has drawn attention to a slow mobilization of a polysynaptically determined rise of excitability ultimately emerging in a contraction, the vibratory reflex (TVR). This process, though held to be spinal (Gillies et al., 1971; Burke et al., 1976), is under strong supraspinal control. It is easily suppressed by an act of will and the ensuing relaxation of the muscle augments the firing rate of the spindles (Marsden et al., 1969; Burke et al., 1976). Hagbarth's group (Burke et al., 1978b) postulates a mechanism of gain regulation, centrally controlled. If gain be centrally put down, the spindles cannot alone re-establish it. The normal operation of the postulated gain control would – I suppose – follow automatically by alpha-gamma linked commands to the muscles, the gamma system contributing its share through the mediation of the loop across the spindles. Setting of the gain would be an alpha affair. Marsden et al. (1976), who reported that gain in their thumb flexor experiment is boosted in proportion to the requirements of opposing force, also refer this effect to the alpha motor system. However, somehow the organism must be informed about the increase in external force and also in this sense the sensitized spindles are bound to add their contribution to the total performance.

In this, as in most other experiments dealing with the gamma-spindle system, one encounters its two major roles, i) the sensory one of information, as delivered by the sensitized spindles, mostly in cooperation with other sensory instruments such as tendon organs, skin and joint receptors, ii) the motor one across the gamma loop on the alpha motoneurons.

The motor role of the gamma loop is more accessible to interpretation than its sensory role as informant. One important reason for this is the known role of the spindles in connexion with reciprocal innervation; they produce monosynaptic excitation and disynaptic inhibition on the antagonist. In voluntary isometric contractions the spindles fire in rough proportion to the effort and therefore, as this increases, become relatively more significant both for monosynaptic "ignition", and maintained load compensation. The gamma loop is likely to be used also for maximizing motor efforts. The range of firing frequencies is not known for spindles in man because of the difficulties in maintaining the flexible tungsten microelectrode in a fixed position. In the cat the range by supraspinal stimulation at a well chosen site was as high as about 160 impulses/sec compared with the basic value of the deefferented spindle at the same length (soleus between 8 and 12 mm muscle extension, according to Eldred et al., 1953).

However also in man the coactivated spindle, despite experimental constraints, has been shown by Hagbarth et al. (1970) to add a substantial amount to the voluntary contraction studied in a finger flexor. After removing the gamma contribution by a Xylocaine block the sustained isometric response was "very much reduced". Full power could then be restored by adding a vibratory stimulus. In the cat Severin (1970) applied a procaine gamma block in experiments on controlled locomotion. The amplitude of the extensor EMG was then found reduced by approximately one half while the rhythm of stepping was essentially unchanged.

The numerous experiments from Hammond (1956) onwards on unexpected loading and the effect of instruction on the stretch reflex have all been concerned with alpha motoneurons. As a rule, parallel records showing how the spindles have behaved in motor volleys succeeding the early monosynaptic "ignition" of the system have not been published. We know that this later activity, the functional stretch reflex of Melvill, Jones and Watt (1971a, b), is the one responsible for muscular force and, on present knowledge, can make a fair guess as to how sensitized spindles have behaved, case for case. The papers by Hagbarth and his coworkers (Burke et al., 1978) provide decisive information.

When in these experiments the subject was ordered to hold a given position of the foot, a spindle burst was elicited by a displacement produced by loading. This led to an alpha load compensating response after a latency of 175–300 msec that with background activity of a stretch reflex was shortened to 40–80 msec. In such experiments dislocation of the microelectrode prevents use of heavy loads but the authors assume that by its short latency the early spindle contribution plays an important role in preparing the ground for an increased voluntary alpha-gamma linked compensation. It is hardly possible fully to evaluate the role of the monosynaptic "ignition" without intracellular studies of its effect on the level of depolarization. Clearly every early effect must be significant for the speed of action of the muscles because this is highly sensitive to the fast firing rates characteristic of the early response of the motoneurons (Buller and Lewis, 1965).

Interesting light is thrown on coactivation by alpha-gamma linkage in isotonic contractions (Vallbo, 1973; Burke et al., 1978b). If the contraction is slow, the spindles may start immediately to discharge in proportion to the anticipated effort instead of, so to speak, waiting for the completed alpha recruitment process or of following it slavishly by a proportionate increase in firing. The least one can say about this experiment is that it really presents a case of gamma precedence in man, unless one is willing to apply psychological concepts such as "effort" and "anticipation" and assume these to be channelled more easily to gamma neurons than to alpha neurons. At any rate it seems permissible to assume that slow isotonic muscular contractions picture the gamma-spindle contribution better than fast ones in which the intrafusal shortening is counteracted by the unloading effect of surrounding extrafusal musculature. If this be accepted, the spindles in slow isotonic contractions measure effort leaving measuring of force to the tendon organs. Velocity and length could be measured by a differential of the frequency responses of the two end organs.

In strong voluntary contractions velocity loses in significance. In this case spindles and tendon organs both respond similarly. Noting that the "mean firing rate" of the spindles follows increasing force rather than decreasing length. Burke et al. (1978b) consider that its motor role then has replaced its sensory functions, or perhaps they mean that the motor role then is wholly dominant. While unwilling to follow them to that length, I am in full agreement about the greater contribution of the spindles to motoneuron depolarization with strong efforts (Granit, 1975). This seems to be an inevitable consequence of its monosynaptic projections. To what extent this effect is counteracted by Golgi tendon organs is dependent upon the state of the interneurons by which their effect is controlled.

Segmental vs. supraspinal is too large an issue for my attempt to interpret some supraspinal effects on the gamma-spindle system. Owing to the labours of the Gothenburg group we now have detailed knowledge not only of some segmental circuits and their organization but also of their supraspinal controlling pathways. We realize that the properties of sensitized spindles may be used in so many different combinations that it hardly makes sense to submit a list of potentialities. Elsewhere I have made some comments on this question (Granit, 1975).

While the experiments on man with direct recording from spindle afferents lately have seemed to me the most interesting ones from the viewpoint of supraspinal control, they also have their limitations that one should be aware of. One has been mentioned: that the intracellular approach is excluded. Another is that the difference between the actions of static and dynamic gamma motoneurons does not stand out well. As to primaries and secondaries, the experience of Vallbo (1974) is that they respond in the same general way. Their different roles have to be elucidated in animal experiments. A further restriction is that the range of possible experiments on man is narrow in comparison with the total resources of his motricity. There is a great deal more to do, I think, for instance, of Nashner's (1976) experiments on adaptability in the stretch reflex. A forced stretch reflex that runs counter to its normal purposiveness adapts to the new situation by gradually disappearing. Would then the gamma neurons follow suit or would they prove even more adaptable than their alpha partners? In view of Buchwald and Eldred's (1962) experiment on learning in the gamma system, the latter alternative seems at least as likely as the former. Finally, I must not suppress a secret suspicion that I have nourished for some time, namely that the voluntary mode of activating gamma motoneurons makes preferential use of one particular supraspinal pathway. There may be others with slightly different properties.

In finishing this brief discourse on the interpretation of supraspinal effects on the gamma system I have left out much that merely would have been a repetition of what I have said over the years, some of my summaries being as late as from 1975 and 1977. And I have already apologized for not going through supraspinal sites from which the gamma neurons can be activated. I have of course abstained from presenting much high-quality work on spindle innervation and internal design that suggests possibilities for differential use of this sensorimotor instrument in the muscles. Our methods have not yet permitted more than a few basic interpretations. Also it seems to me that present knowledge of spinal circuitry and descending pathways of control carries greater promises for interpretation than do the details of spindle design that extend beyond the differentiation of static from dynamic gamma control.

SUMMARY

The paper is written as a review evaluating contributions from the last ten years. It discusses likely interpretations of the functional role of the gamma system in the light of experimental work. Among major themes might be mentioned: spinal alpha-gamma linkage in reciprocal innervation, in relation to slow and fast muscles, the alternatives of alpha respectively gamma precedence in their coactivation, the compensatory response to unexpected loading, gain control in relation to vibratory reflexes, mono- and polysynaptic spindle control of motor activity, the role of the gamma system in maximizing motor effects etc.

REFERENCES

Brooks, V.B. (1975) Roles of cerebellum and basal ganglia in initiation and control of movement. *J. Can. Sci. Neurol.*, 265–277.

Buchwald, J.S. and Eldred, E. (1962) Activity in muscle-spindle circuits during learning. In *Symposium of Muscle Receptors*. D. Baker (Ed.). University Press, Hong Kong, pp. 175–183.

Buller, A.J. and Lewis, D.M. (1965) The rate of tension development in isometric tetanic contractions of mammalian fast and slow skeletal muscle. *J. Physiol. (Lond.)*, 176: 337–354.

Burg, D., Szumski, A.J., Struppler, A. and Velho, F. (1974) Assessment of fusimotor contribution to reflex reinforcement in humans. *J. Neurol.*, 37: 1012–1021.

Burke, D., Hagbarth, K.-E. and Löfstedt, L. (1978a) Muscle spindle responses in man to changes in load during accurate position maintenance. *J. Physiol. (Lond.)*, 276: 159–164.

Burke, D., Hagbarth, K.-E. and Löfstedt, L. (1978b) Muscle spindle activity in man during shortening and lengthening contractions. *J. Physiol. (Lond.)*, 277: 131–142.

Burke, D., Hagbarth, K.-E., Löfstedt, L. and Wallin, B.G. (1976) The responses of human muscle spindle endings to vibration of non-contracting muscles. *J. Physiol. (Lond.)*, 261: 673–693.

Corda, M., Eklund, G. and Von Euler, C. (1965) External intercostal and phrenic α motor responses to changes in respiratory load. *Acta physiol. scand.*, 63: 391–400.

Eldred, E., Granit, R. and Merton, P.A. (1953) Supraspinal control of the muscle spindles and its significance. *J. Physiol. (Lond.)*, 122: 498–523.

Fidone, S.J. and Preston, J.B. (1969) Patterns of motor control of flexor and extensor cat fusimotor neurons. *J. Neurophysiol.*, 32: 103–115.

Garnett, R. and Stephens, J.A. (1978) Changes in the recruitment threshold of motor units in human first dorsal interosseus muscle produced by skin stimulation. *J. Physiol. (Lond.)*, 17P.

Gillies, J.D., Burke, D.J. and Lance, J.W. (1971) Tonic vibration reflex in the cat. *J. Neurophysiol.*, 34: 252–262.

Granit, R. (1955) *Receptors and Sensory Perception. A Discussion of Aims, Means and Results of Electrophysiological Research into the Process of Reception*. Yale University Press.

Granit, R. (1970) *The Basis of Motor Control*. Academic Press, London.

Granit, R. (1975) The functional role of the muscle spindles – facts and hypotheses. *Brain*, 98: 531–556.

Granit, R. (1977) *The Purposive Brain*. The MIT Press, Cambridge, MA.

Granit, R. and Holmgren, B. (1955) Two pathways from brain stem to gamma ventral horncells. *Acta physiol. scand.*, 35: 93–108.

Granit, R., Holmgren, B. and Merton, P.A. (1955) The two routes for excitation of muscle and their subservience to the cerebellum. *J. Physiol. (Lond.)*, 130: 213–224.

Grigg, P. and Preston, J.P. (1971) Baboon flexor and extensor fusimotor neurons and their modulation by motor cortex. *J. Neurophysiol.*, 34: 428–436.

Grillner, S. and Zanger, P. (1974) Locomotor movements generated by the deafferented spinal cord. *Acta physiol. scand.*, 91: 38A–39A.

Grillner, S. and Zanger, P. (1975) How detailed is the central pattern generation for locomotion? *Brain Res.*, 88: 367–371.

Hagbarth, K.-E., Hongell, A. and Wallin, B.G. (1970) The effect of gamma fibre block on afferent muscle nerve activity during voluntary contractions. *Acta physiol. scand.*, 79: 27A–28A.

Hammond, P.H. (1956) The influence of prior instruction to the subject on an apparently involuntary neuromuscular response. *J. Physiol. (Lond.)*, 132: 17–18P.

Kanda, K., Burke, R.E. and Walmsley, B. (1977) Differential control of fast and slow twitch motor units in the decerebrate cat. *Exp. Brain Res.*, 29: 57–74.

Lundberg, A. (1975) Control of spinal mechanisms from the brain. *The Nervous System, Vol. I, The Basic Neurosciences.* D.B. Tower (Ed.). Raven Press, New York.

Marsden, C.D., Meadows, J.C. and Hodgkin, H.J.F. (1969) Observations on the reflex response to muscle vibration in man and its voluntary control. *Brain*, 92: 829–846.

Marsden, C.D., Merton, P.A. and Morton, H.B. (1972) Servo action in human voluntary movement. *Nature (Lond.)*, 238: 140–143.

Marsden, C.D., Merton, P.A. and Morton, H.B. (1976) Servo action in the human thumb. *J. Physiol. (Lond.)*, 257: 1–44.

Melvill Jones, G. and Watt, D.G.D. (1971a) Observations on the control of stepping and hopping movements in man. *J. Physiol. (Lond.)*, 219: 709–727.

Melvill Jones, G. and Watt, D.G.D. (1971b) Muscular control of landing from unexpected falls in man. *J. Physiol. (Lond.)*, 219: 729–737.

Nashner, L.M. (1976) Adapting reflexes controlling the human posture. *Exp. Brain Res.*, 26: 59–72.

Phillips, C.G. and Porter, R. (1977) *Corticospinal Neurones. Their Role in Movement.* Academic Press, London.

Severin, F.V. (1970) The role of the γ-motor systems in the activation of the extensor α-motor neurones during controlled locomotion. *Biofizika*, 15: 1138–1145.

Shik, M.L., Orlovskii, G.N. and Severin, F.V. (1966) Organization of locomotor synergism. *Biofizika*, 11: 879–886.

Sjöström, A. and Zanger, P. (1975) α-γ-linkage in the spinal generator for locomotion in the cat. *Acta physiol. scand.*, 94: 130–132.

Sjöström, A. and Zanger, P. (1976) Muscle spindle control during locomotor movements generated by the deafferented spinal cord. *Acta physiol. scand.*, 97: 281–291.

Vallbo, Å.B. (1971) Muscle spindle response at the onset of isometric voluntary contractions in man. Time difference between fusimotor and skeletomotor effects. *J. Physiol. (Lond.)*, 218: 405–431.

Vallbo, Å.B. (1973) Muscle spindle afferent discharge from resting and contracting muscles in normal human subjects. In *New Developments in Electromyography and Clinical Neurophysiology*, J.E. Desmedt (Ed.). Karger, Basel.

Vallbo, Å.B. (1974) Afferent discharge from human muscle spindles in non-contracting muscles. Steady state impulse frequency as a function of joint angle. *Acta physiol. scand.*, 90: 303–318.

Wetzel, M.C. and Stuart, D.G. (1976) Ensemble characteristics of cat locomotion and its neural control. *Progr. Neurobiol.*, 7: 1–98.

Discharge Rates of Muscle Afferents during Voluntary Movements of Different Speeds

M.McD. LEWIS, A. PROCHAZKA, K.-H. SONTAG and P. WAND

Sherrington School of Physiology, St. Thomas's Hospital Medical School, London SE1 7EH (U.K.); and Max-Planck-Institut für Experimentelle Medizin, Abteilung für Biochemische Pharmakologie, 3400 Göttingen (F.R.G.)

INTRODUCTION

The firing rates of mammalian muscle spindles are modulated mainly by variations in muscle length and variations in fusimotor action. In the absence of fusimotor action, both primary and secondary spindle endings increase and decrease their firing rate with increasing and decreasing muscle length. Indeed, primary endings fall silent even when the rate of extrafusal shortening is very low (1% of the resting length/sec) (Lennerstrand, 1968).

Fusimotor action has been shown, in acute experiments, to be capable of maintaining and even increasing spindle firing rates during muscle shortening (Lennerstrand and Thoden, 1968). However, does this happen in normal voluntary movements?

The evidence from the human neurography experiments of Vallbo (1974) and Hagbarth et al. (1978) strongly supported the idea that fusimotor action, closely coupled with skeletomotor activity, accelerated the firing rates of spindles during muscle contraction. The contractions studied were usually isometric or involved extremely slow movements (e.g., ankle flexion at 0.25–0.5 degrees/sec (Burke et al., 1978).

With the development of techniques for recording from single afferents in conscious cats and monkeys (Taylor and Cody, 1974; Goodwin and Luschei, 1975; Prochazka, 1975), evidence has mounted, that for most movements, variations in muscle length are apparently more powerful than fusimotor action in modulating spindle firing.

In this paper, we shall argue that these views are not contradictory if it is assumed that for muscle velocities in excess of 0.2 resting lengths (RL)/sec, spindle firing is mainly modulated by the length variations, whereas for muscle velocities less than 0.2 RL/sec, fusimotor action may often predominate.

METHODS

Detailed descriptions of the afferent recording technique have appeared elsewhere (Prochazka et al., 1977) and so only a summary is presented here.

Surgery

During one aseptic operation under halothane anaesthesia, pairs of 17 μm wires, insulated except for their tips, were introduced into L6, L7 or S1 spinal roots through small slits in the dura mater. The wires were glued to the dura, and connecting cables were passed subcutaneously to a dental acrylic headpiece, along with a catheter from the jugular vein. Miniature eyelets were attached subcutaneously to the ischium, the head of the tibia and the calcaneum. Flexible wires looped through these eyelets emerged through the skin to provide fixation points for an external length gauge.

Recording sessions

Starting 1 day postoperatively, a small capsule containing 2 FM transmitters was clipped to the animal's head, and miniature plugs were mated with their appropriate sockets. If the implanted dorsal root electrodes happened to be favourably located, the discharge trains of single afferent fibres could now be recorded.

A given afferent was identified by mechanical, electrical and pharmacological tests (Prochazka et al., 1977) during a brief period (5–10 min) of anaesthesia (Epontol, Bayer). If the afferent was found to innervate a knee flexor or an ankle muscle, a mercury-in-rubber length gauge was tied to the appropriate fixation wires so as to be in parallel with the muscle. In the case of tail muscles, the length gauge was attached at one end by adhesive tape to the tail, some 20 mm caudal to the ischium, and at the other end, to the skin overlying the head of the femur. A pair of fine electromyogram (EMG) wires (250 μm diam.) or a concentric EMG needle was inserted into the appropriate muscle percutaneously.

Subsequent recordings in the awake animal generally lasted about 1 hr. An FM cassette recorder stored 3 signals: length, EMG and neurogram, as well as a voice commentary. The data presented in this paper is drawn from some 10 h of taped recordings from 30 identified afferents in 7 cats.

RESULTS

Spindle primary endings

Fig. 1A shows the activity of a spindle primary ending located in a knee flexor. At the beginning of this segment of record, the cat's knee was held in an extended position by the experimenter. The animal moderately resisted this extension, and when the leg was released, active, rapid flexion occurred, as evidenced by the sudden muscle shortening. Prior to the release, the muscle length varied through a range of about 4 mm, at velocities not exceeding 25 mm/sec. The RL of the knee flexors in this cat was about 110 mm between the ischial and tibial fixation points, so the maximum velocity prior to release was about 0.2 RL/sec. During this time, the spindle primary showed modulations of firing which were not strongly related to the length variations. However, during the rapid active flexion movement, which reached a peak muscle velocity of about 160 mm/sec i.e. ca. 1.5 RL/sec, the firing rate dropped in close correspondence with the reduction in length. The cat subsequently held its leg in a fully flexed position, with minimal changes in muscle length, and the spindle resumed firing, reaching a steady level after about 1 sec.

Fig. 1B shows the results of one of the identification tests, namely rapid knee extension during deep Epontol anaesthesia after 200 μg/kg i.v. succinylcholine. The dynamic index of the ending is seen to have exceeded 150 impulses/sec.

Fig. 2 shows the modulations of firing rate of the same knee flexor spindle during stepping at about 1 m/sec. During the swing phase, the muscle lengthened at about

157

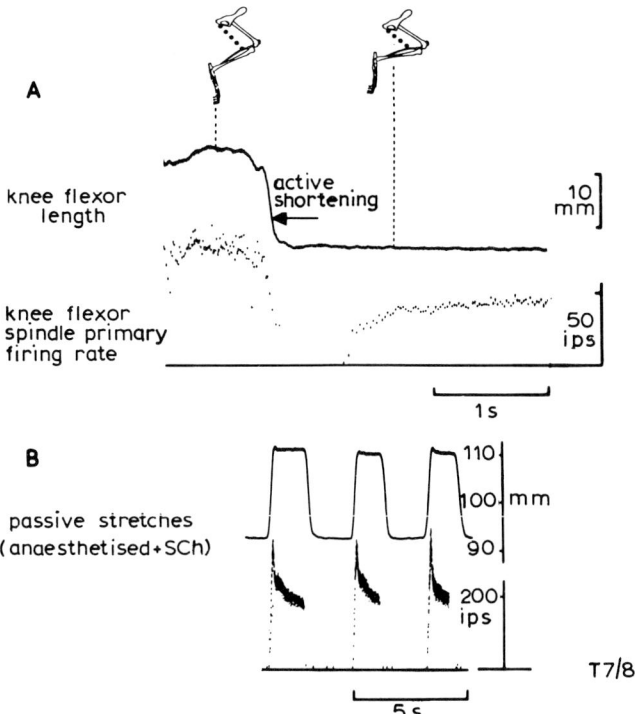

Fig. 1. Knee flexor spindle primary. A) Voluntary flexion movement. Top trace: length from ischium to head of tibia. Bottom trace: instantaneous firing rate of spindle afferent. Cat's knee initially held in extended position by experimenter, then suddenly released. B) Responses to rapid, maintained knee extensions during anaesthesia, after i.v. succinylcholine.

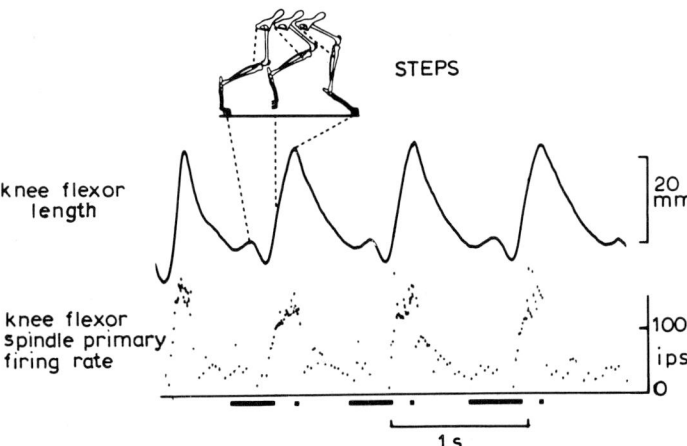

Fig. 2. Knee flexor spindle primary. 4 steps on treadmill showing length variations and instantaneous firing rate of spindle afferent. Bars indicate timing of bursts of knee flexor EMG (From Engberg and Lundberg, 1969.)

150 mm/sec, i.e., ca. 1.4 RL/sec. During the stance phases, the muscle shortened at an average rate of about 60 mm/sec, i.e., ca. 0.5 RL/sec. The firing rate of the spindle was modulated in close correspondence with the length changes throughout.

EMG recordings were not made in this case, but from the data of Endberg and Lundberg (1969), and our own unpublished observations, knee flexors such as semitendinosus and biceps femoris show bursts of electrical activity at the times indicated by the bars at the bottom of the figure. As previously reported by Prochazka et al, (1976, 1977), there is little evidence for significant increases in the firing rate of the knee flexor spindle during these times, other than those due to increases in length. The recordings of Figs. 1 and 2 are, however, consistent with the idea that in movements where the muscle velocities exceed 0.4 RL/sec, modulations in spindle firing frequency are largely due to the length variations.

Spindle secondaries

Fig. 3 shows the firing of a spindle secondary located in an ankle extensor. Muscle palpation and the responses to stimulating through the EMG electrodes (Fig. 3C) indicated that the ending was situated in medial gastrocnemius. The two placing reactions of Fig. 3A were elicited by supporting the animal over a table and moving it horizontally so that the dorsum of the foot touched the side of the table-top. The rates of lengthening of the muscle prior to foot touch-down were about 30 mm/sec, or 0.3 RL/sec, assuming an RL of 100 mm. The rate of shortening just prior to touch-

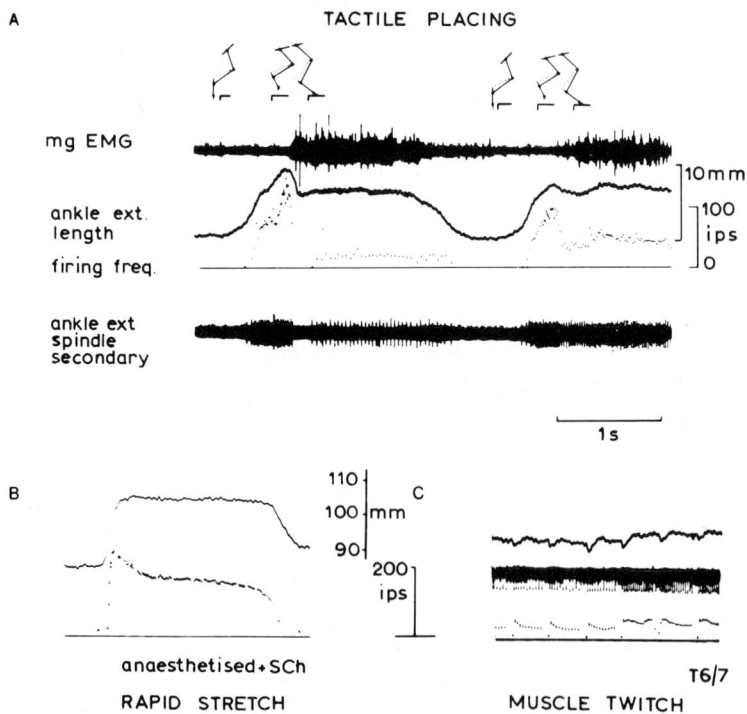

Fig. 3. Ankle extensor spindle secondary. A) 2 tactile placing reactions, elicited by moving the animal horizontally until the paw touched side of table. B) Low dynamic responsiveness to rapid, maintained ankle dorsiflexion during deep anaesthesia, after i.v. succinylcholine. C) Responses to 8 V, 0.1 msec impulses through EMG electrodes.

down was in excess of 0.6 RL/sec in the first placing reaction, but less than 0.1 RL/sec in the second. The maximum speed of shortening in the middle of the record was 0.2 RL/sec.

The firing rate of the secondary ending was closely related to the length changes when these exceeded rates of 0.2 RL/sec; however, below this rate, for example in the middle of the record, the correspondence was not as good. There is little evidence in Fig. 3A for strong static fusimotor coactivation with the skeletomotor activity. Fig. 3B shows that the afferent had a dynamic index of less than 100 impulses/sec after i.v. succinylcholine.

The afferent in Fig. 4 was identified as a spindle secondary located in abductor caudalis externis, a tail abductor. Fig. 4A and B show the modulations in firing rate during slow and fast tail-wags respectively. There is a clear correspondence between the length and firing rate for all but the slowest movements. Assuming that the muscle lengthened by about 20% from a tail angle of 45° left to 45° right, the rate of muscle shortening during the last part of Fig. 4B, where spindle firing was unrelated to the length change, was about 0.06 RL/sec. Again, there was little evidence of strong skeletomotor-fusimotor coactivation, but the fact that in the active movements, the

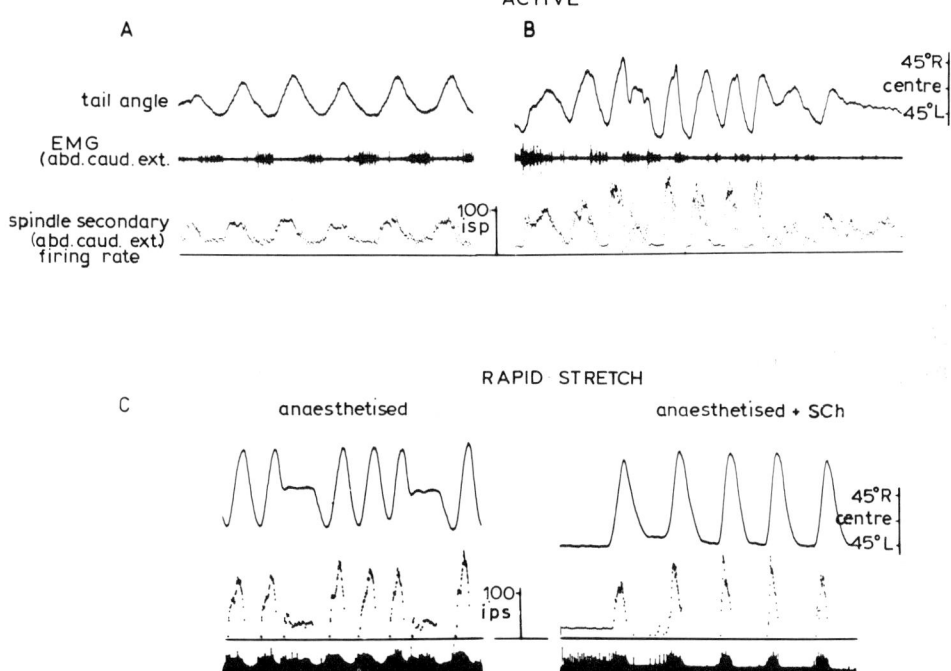

Fig. 4. Tail abductor spindle secondary. A and B) Voluntary tail wags at different rates. Note good correspondence between length and spindle firing rate except at end of B, where muscle velocity is low. C) Rapid tail movements imposed during deep anaesthesia. D) Same as C, after i.v. succinylcholine. Dynamic responsiveness of receptor not appreciably changed.

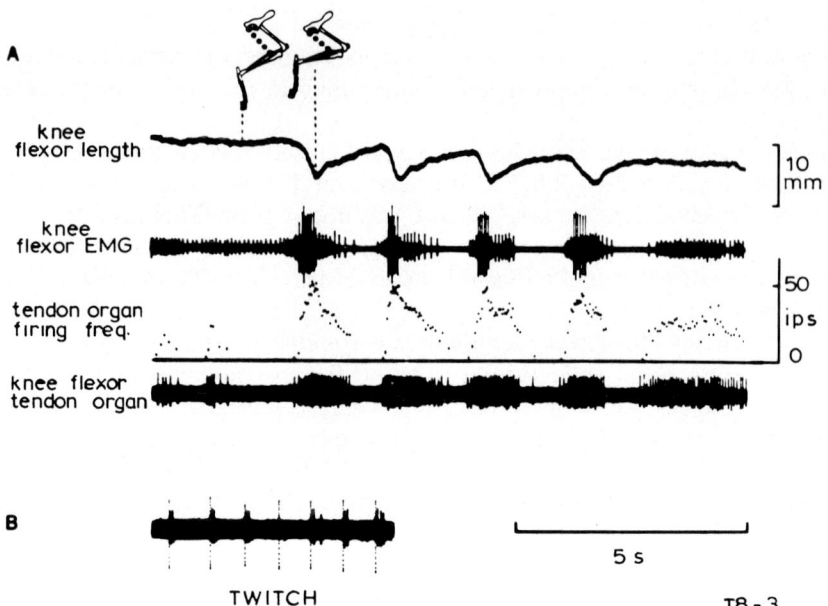

Fig. 5. Knee flexor tendon organ. A) 4 small, voluntary knee flexions elicited by touching paw. EMG trace derived from concentric needle electrode in close proximity to receptor. Note close coupling of EMG and tendon organ activity, despite muscle shortening. B) Responses of tendon organ to 10 V, 0.1 msec impulses through EMG electrode.

firing rate rarely dropped below 10 impulses/sec, suggests that there may have been steady, tonic, fusimotor action.

Tendon organs

By contrast to spindle afferents, the firing of tendon organ afferents was closely related to muscle activity. The record of Fig.5A shows the discharge activity of a knee flexor tendon organ during 4 voluntary, active knee flexions. These were elicited by light touching of the cat's paw. Despite the muscle shortening, the firing rate of the afferent was closely related to the firing of single motor units recorded with a concentric EMG needle. The needle had been carefully positioned by finding the location for which a minimal electrical impulse would accelerate the tendon organ firing (Fig. 5B).

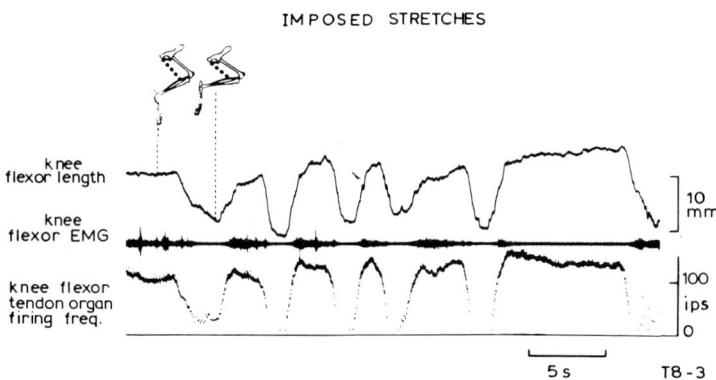

Fig. 6. Knee flexor tendon organ. Knee flexion and extension imposed by experimenter. Small stretch reflexes evidenced by EMG activity.

Fig. 6 shows the firing rate of the same tendon organ afferent during knee movements imposed by the experimenter and moderately resisted by the cat. The EMG electrode was less favourably located than in Fig. 5. It is clear, however, that the receptor was highly sensitive to muscle stretches in the awake animal.

DISCUSSION

The above results indicate that in normal isotonic movements involving muscle velocities above 0.2 RL/sec, the firing rates of both primary and secondary spindle afferents were modulated largely by the length changes. Is this consistent with the respiratory studies of Critchlow and Euler (1963), and previous recordings from spindle afferents in awake cats (Taylor and Cody, 1974), monkeys (Goodwin and Luschei, 1975) and man (Vallbo, 1974; Burke et al., 1978)? Detailed comparisons between these studies were recently published from our laboratory (Prochazka et al., 1979). To summarise these, of all the published human afferent recordings, the fastest muscle velocities were about 0.3 RL/sec (Burke et al., 1978, Fig. 2, assuming tibialis anterior RL ca. 300 mm, 60° ankle flexion = 80 mm length change). At these velocities a "dynamic" spindle ending was modulated largely by the length changes. Most of the human neurography data, however, involved muscle velocities well below 0.1 RL/sec, and so the modulations of spindle firing attributed to fusimotor action are in accordance with our above contention.

In the intercostal studies, the maximum velocities were about 0.2 RL/sec. Under these circumstances, fusimotor action sometimes accelerated spindle discharge despite muscle shortening and this too is consistent with the above.

In the jaw muscle studies of Taylor and Cody (1974) and Goodwin and Luschei (1975), maximum muscle velocities were between 3 and 5 RL/sec. At these rates, spindle firing was strongly modulated by the length variations.

We therefore suggest that the apparent differences in the results of the above groups of experiments can be largely resolved by taking into account the proportional muscle velocities involved. The resultant re-emphasis of the spindle's function as a stretch receptor puts a new slant on the possible roles played by spindle afferents in motor control.

SUMMARY

The discharge trains of single, identified muscle receptors were recorded during voluntary movements of different speeds in cats. The firing rates of both primary and secondary muscle spindles were modulated strongly by the length variations when these involved muscle velocities in excess of 0.2 resting lengths/sec. For movements involving muscle velocities lower than 0.2 resting lengths/sec, there was evidence that fusimotor action could be the predominant modulatory influence. It is argued that spindle afferent recordings in previous studies in cat, monkey and man have shown a similar relationship, and so the above may represent a useful generalisation for future motor control studies.

ACKNOWLEDGEMENTS

The authors thank Dr. J.A. Stephens and Professor A. Taylor for help, criticism and encouragement. This work was supported by St. Thomas's Hospital Endowments Fund and the Deutsche Forschungsgemeinschaft, SFB 33. Dr. M.McD. Lewis was on study leave from Lincoln Institute of Health Sciences, Victoria, Australia.

REFERENCES

Burke, D., Hagbarth, K.-E. and Löfstedt, L. (1978) Muscle spindle activity in man during shortening and lengthening contractions. *J. Physiol. (Lond.)*, 277: 131–142.

Critchlow, V. and Euler, C.V. (1963) Intercostal muscle spindle activity and its γ motor control. *J. Physiol. (Lond.)*, 168: 820–847.

Engberg, I. and Lundberg, A. (1969) An electromyographic analysis of muscular activity in the hindlimb of the cat during unrestrained locomotion. *Acta physiol. scand.*, 75: 614–630.

Goodwin, G.M. and Luschei, E.S. (1975) Discharge of spindle afferents from jaw-closing muscles during chewing in alert monkeys. *J. Neurophysiol.*, 38: 560–571.

Granit, R., Holmgren, B. and Merton, P.A. (1955) The two routes for excitation of muscle and their subservience to the cerebellum. *J. Physiol. (Lond.)*, 130: 213–224.

Hagbarth, K.-E., Wallin, G. and Löfstedt, L. (1975) Muscle spindle activity in man during voluntary fast alternating movements. *J. Neurol. Neurosurg. Psychiat.*, 38: 625–635.

Lennerstrand, G. (1968) Position and velocity sensitivity of muscle spindles in the cat. 1. Primary and secondary endings deprived of fusimotor activation. *Acta physiol. scand.*, 73: 281–299.

Lennerstrand, G. and Thoden, U. (1968) Muscle spindle responses to concomitant variations in length and in fusimotor activation. *Acta physiol. scand.*, 74: 153–165.

Prochazka, A. (1975) A pause in spindle-afferent discharge during rapid active muscle shortening in the freely-moving cat. *Proc. Aust. Physiol. Pharmacol. Soc.*, 6: 183–184.

Prochazka, A., Westerman, R.A. and Ziccone, S.P. (1976) Discharges of single hindlimb afferents in the freely moving cat. *J. Neurophysiol.*, 39: 1090–1104.

Prochazka, A., Westerman, R.A. and Ziccone, S.P. (1977) Ia afferent activity during a variety of voluntary movements in the cat. *J. Physiol. (Lond.)*, 268: 423–448.

Prochazka, A., Stephens, J.A. and Wand, P. (1979) Muscle spindle discharge in normal and obstructed movements. *J. Physiol. (Lond.)*, 287: 57–66.

Taylor, A. and Cody, F.W.J. (1974) Jaw muscle spindle activity in the cat during normal movements of eating and drinking. *Brain Res.*, 71: 523–530.

Vallbo, Å.B. (1974) Human muscle spindle discharge during isometric voluntary contractions. Amplitude relations between spindle frequency and torque. *Acta physiol. scand.*, 90: 319–336.

Supraspinal Control of Ascending Pathways

W.D. WILLIS

Marine Biomedical Institute, University of Texas Medical Branch, Galveston, TX 77550 (U.S.A.)

INTRODUCTION

In addition to modulating, directly or indirectly, the discharges of spinal motoneurons, pathways descending from the brain exert an important influence on the activity of tract cells in the spinal cord which project to the brain. This is true both for pathways which signal information to brain structures concerned with motor control, such as the spinocerebellar tracts, and for pathways engaged in sensory processing, such as the spinothalamic and spinocervical tracts and the dorsal column-medial lemniscus pathway. The significance of the descending pathways in regulating the flow of sensory data to the brain was underscored in the studies of Hagbarth and Kerr (1954) and of Hernández-Peón and his associates (Hernández-Peón et al., 1956). Since that time the attention of a number of workers has been directed to this area of investigation.

Emphasis will be placed here on the descending control of the activity of neurons belonging to the spinothalamic tract in the monkey. However, where possible, comparable observations on the effects of the descending control systems on other ascending pathways will also be described. In the studies in our laboratory, extracellular recordings are made from spinothalamic neurons in the lumbosacral enlargement of the monkey *(Macaca mulatta or M. fascicularis)*. The animals are anesthetized with α-chloralose and a small continuous infusion of sodium pentobarbital. The spinothalamic cells are identified by antidromic activation from the contralateral thalamus (Trevino et al., 1973). Most spinothalamic tract cells have receptive field properties which are consistent with a role in the transmission of nociceptive information (Willis et al., 1974; Foreman et al., 1977).

CEREBRAL CORTEX

A major descending system which influences sensory transmission as well as motor output in the spinal cord originates in the somatosensory cerebral cortex (Kuypers, 1960; Nyberg-Hansen and Brodal, 1963; Liu and Chambers, 1964; Petras, 1967). There appear to be different specific regions of termination of corticospinal projections originating from different cytoarchitectonic regions of the cortex (Coulter and Jones, 1977). Areas 3b, 1 and 2, for example, project largely to the dorsal horn, whereas areas 3a and 4 project to the base of the dorsal horn, the intermediate gray and the lateral ventral horn.

Stimulation of the sensorimotor cortex in the primate can be shown to affect the activity of certain spinothalamic tract neurons (Coulter et al., 1974). For instance, in Fig. 1A, a rapidly adapting discharge of a spinothalamic tract neuron was produced by a step indentation of the skin. When the discharge was preceded by a stimulus to the sensorimotor cortex (Fig. 1B), the discharge was reduced. However, a comparable cortical stimulus failed to affect the enhanced discharge of the same neuron produced by application of a noxious mechanical stimulus (arterial clip) to the skin in the receptive field of the cell (Fig. 1C). The response of another spinothalamic tract neuron to skin indentation is shown in Fig. 1D. Cortical stimulation resulted in a discharge of this cell, followed by inhibition of the response to cutaneous stimulation (Fig. 1E). Again, cortical stimulation did not affect the discharges evoked by noxious

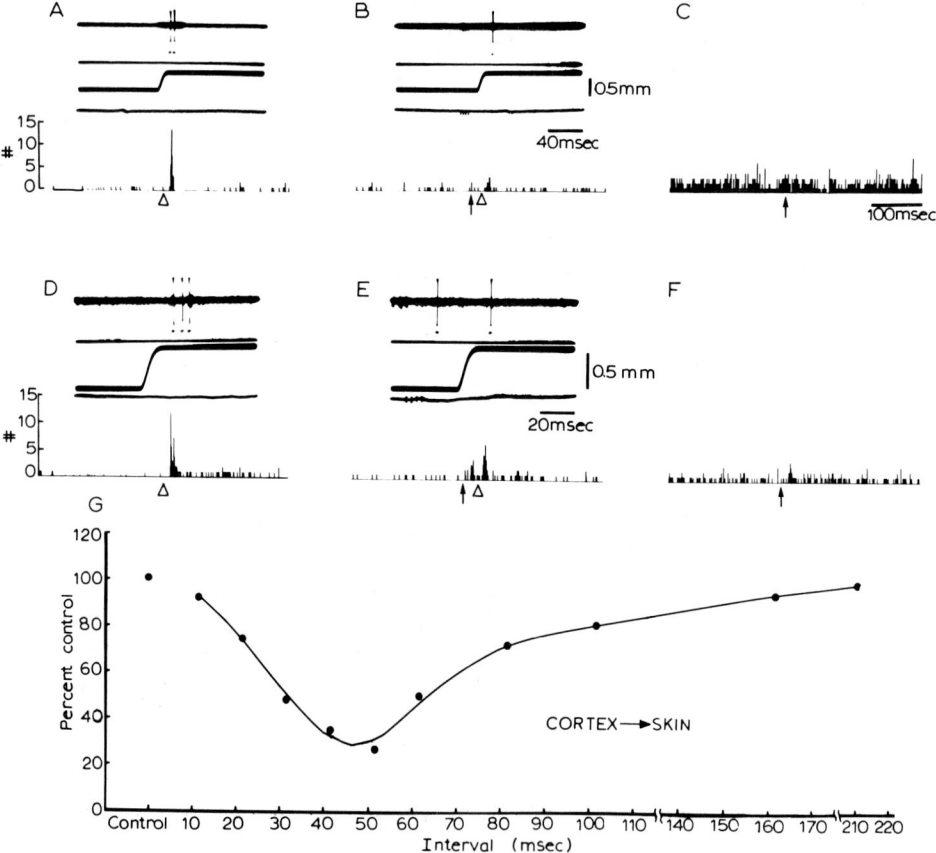

Fig. 1. Effect of stimulation of sensorimotor cortex on spinothalamic neurons. The discharge of a spinothalamic cell shown in A resulted from a step indentation of the skin. The discharge was inhibited (B) following a brief train of conditioning stimuli applied to the sensorimotor cortex. A comparable stimulus failed to inhibit the enhanced background activity (C) produced by application of a noxious stimulus to the skin. Similar results are shown in D–F for another spinothalamic cell, but with the difference that the corticofugal volley produced an initial excitation of the cell. The time course of inhibition of another cell by cortical stimulation is shown in G. (Modified from Coulter et al., 1974.)

stimulation (Fig. 1F). Spinothalamic tract cells activated just by intense stimuli were unaffected by cortical stimulation. Apparently, the corticofugal volleys selectively modulate the input lines to spinothalamic tract cells from rapidly adapting mechanoreceptors. The time course of the inhibition was prolonged (Fig. 1G), suggesting the possibility that presynaptic inhibition is involved. This hypothesis is consistent with reports that stimulation of the sensorimotor cortex in the cat results in primary afferent depolarization in the lumbosacral cord (Carpenter et al., 1963; Andersen et al., 1964a).

Stimulation of the sensorimotor cortex or the pyramid has also been found to affect the activity of other sensory pathways, including the dorsal column-medial lemniscus path (Towe and Jabbur, 1961; Jabbur and Towe, 1961; Andersen et al., 1964b, c; Gordon and Jukes, 1964; Levitt et al., 1964; Winter, 1965) and the spinocervical tract (Lundberg et al., 1963; Fetz, 1963; Brown et al., 1977). Spinocervical tract cells are chiefly inhibited, whereas both excitation and inhibition have been observed in the dorsal column nuclei. It has been proposed that the excitatory actions of corticofugal volleys are restricted to interneurons of the dorsal column nuclei, whereas the inhibitory effects are chiefly on relay cells (Andersen et al., 1964c). However, mixed excitatory and inhibitory actions have been seen in recordings from individual cells (Gordon and Jukes, 1964; Winter, 1965) and at least a few relay cells can be excited (Gordon and Jukes, 1964).

In addition to actions on sensory pathways, the cerebral cortex can modulate the transmission of information through a number of other ascending pathways, including the dorsal spinocerebellar tract (Lundberg et al., 1963; Hongo and Okada, 1967; Hongo et al., 1967), the ventral spinocerebellar tract (Magni and Oscarsson, 1961; Fu et al., 1977) and the spinoreticular tract (Lundberg et al., 1963). In general the corticofugal effects parallel the actions of the flexion reflex afferents on these tract cells.

MEDULLARY RAPHE NUCLEI

It has recently become clear that there is a major projection to the lumbosacral enlargement from the raphe nuclei of the medulla. One portion of the raphe-spinal system descends in the dorsolateral fasciculus. The cells of origin of this pathway can be shown by experimental anatomical methods to be chiefly in the raphe magnus nucleus (Basbaum et al., 1973; Martin et al., 1978). For instance, horseradish peroxidase (HRP) was injected into the spinal cord of a cat at L6 in the experiment illustrated in Fig. 2A. The brain stem was later fixed, sectioned sagittally and reacted histochemically to demonstrate the presence of HRP which had been transported retrogradely from the lumbar cord along the axons of neurons to the cell bodies in the medulla. The dots indicate the locations of labelled cells at or near the midsagittal plane. It can be seen that labelled cells are found in the raphe magnus, pallidus and obscurus nuclei following a lumbar injection of HRP in an animal with an intact spinal cord. However, when the ventral half of the cord was cut at the thoracolumbar junction, restricting retrograde transport to axons in the dorsal part of the cord, as in Fig. 2B, most of the labelled cells were found in the raphe magnus nucleus. The complementary experiment was done by Basbaum et al. (1978). Tritiated leucine was injected into the raphe magnus nucleus. The cells incorporated the amino acid and transported it anterogradely. A pathway could be followed from the raphe magnus

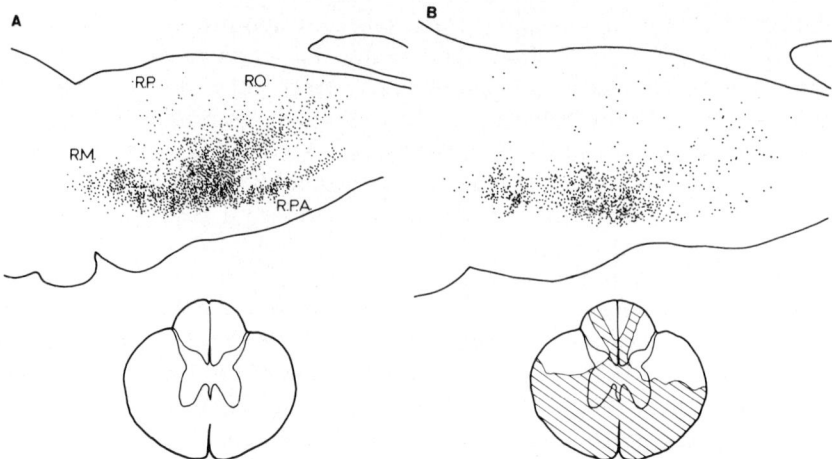

Fig. 2. Distribution of cells in the medullary raphe nuclei labelled with horseradish peroxidase injected into the L6 segment of the spinal cord. In A, the spinal cord was left intact at the thoracolumbar junction. However, in B, the ventral half of the cord was sectioned, so the labelled cells were just those whose axons descended in the dorsolateral fasciculi. The brain stem sections are in the midsagittal plane. R.M., raphe magnus nucleus; R.O., raphe obscurus nucleus; R. PA., raphe pallidus nucleus; R.P., raphe pontis nucleus. (Modified from Martin et al., 1978.)

nucleus to the spinal cord through the dorsolateral fasciculus. Terminals were found in the dorsal horn. These were especially dense in laminae I and II, but there were also endings in laminae V–VII.

It should be pointed out that there are also descending projections from the reticular formation in the dorsolateral fasciculus. These originate from the magnocellular reticular nucleus, which is just lateral to the raphe magnus nucleus (Basbaum et al., 1978; Martin et al., 1978).

Stimulation within or near the raphe magnus nucleus of the monkey results in a powerful inhibition of spinothalamic tract neurons (Willis et al., 1977). For instance, Fig. 3 shows the inhibition of the discharges of a spinothalamic tract cell located in lamina I (Fig. 3D) which resulted from stimulation at the point indicated in Fig. 3F. The discharges were produced by electrical stimulation of the skin within the receptive field of the neuron (Fig. 3E). The threshold stimulus strength producing inhibition was less than $25\mu A$ (Fig. 3B). Inhibition was enhanced by temporal summation (Fig. 3C), and the time course was prolonged (Fig. 3G). The latency of inhibition was short enough that at least part of the inhibitory volley had to be in myelinated axons. Stimulation in the vicinity of the raphe magnus nucleus can inhibit the background discharges of spinothalamic tract neurons (Fig. 4B), as well as the discharges evoked by tactile (Fig. 4A) or noxious stimuli (Fig. 4C). These inhibitory effects are eliminated by sectioning the dorsolateral fasciculi.

There are no studies as yet which show that stimulation in the raphe magnus nucleus inhibits neurons in the dorsal column-medial lemniscus pathway or the spinocervical tract. However, cells of the spinocervical tract are inhibited by axons descending in the dorsolateral fasciculi (Brown et al., 1973). It seems likely that the raphe-spinal projection forms at least part of the tonic descending inhibitory system which has been investigated in detail by Lundberg and his associates (Eccles and Lundberg, 1959; Holmqvist et al., 1960a; Holmqvist and Lundberg, 1961; Carpenter et al., 1965;

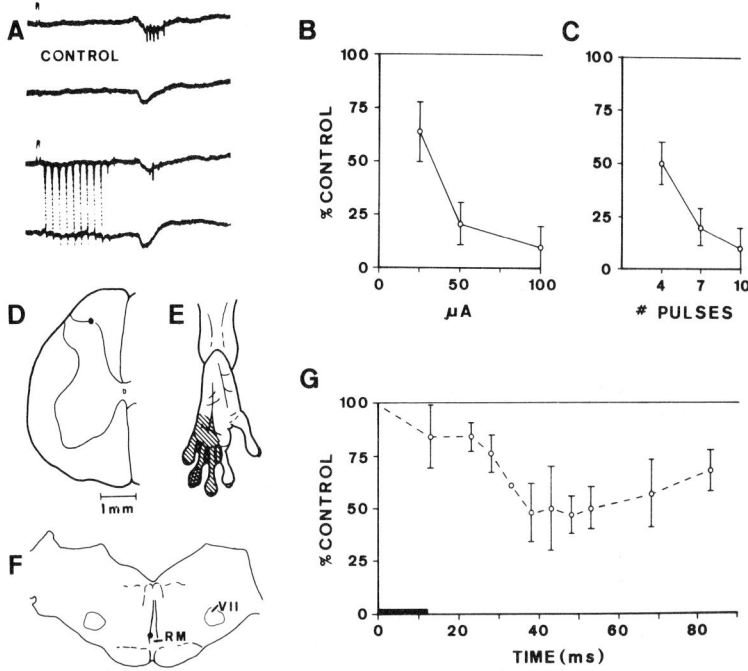

Fig. 3. Inhibition of a spinothalamic tract cell by stimulation in the raphe magnus nucleus. The response of the cell to electrical stimulation of the skin is shown in the control record in A, and inhibition of the response by raphe magnus stimulation in the lower record of A. The relationship between conditioning stimulation strength and amount of inhibition is seen in B and between the number of stimulus pulses and the amount of inhibition in C. The location of the cell is in D, the receptive field in E and the stimulus site in the brain stem in F. The time course of inhibition is in G. (From Willis et al., 1977.)

Engberg et al., 1968). The cells of the spinocervical tract show an increased responsiveness to peripheral stimulation when the tonic descending inhibitory system is interrupted (Brown, 1971; Cervero et al., 1977), and so it can be predicted that these cells would be inhibited by stimulation or the raphe magnus nucleus. However, Lundberg's group suggested that the tonic descending inhibitory system also involves a projection originating in the reticular formation. A likely candidate for the nucleus of origin would be the magnocellular nucleus mentioned above. The ventral spinocerebellar tract is also subject to the tonic descending inhibitory system (Holmqvist et al., 1960a)

PONTOMEDULLARY RETICULAR FORMATION

It is well known that there are abundant projections to the spinal cord from the reticular formation of the medulla and pons which descend through the ventral half of the cord white matter (Nyberg-Hansen, 1965; Petras, 1967; Basbaum et al., 1977). The medullary reticulospinal tract originates in part from the nucleus gigantocellularis. Although not appreciated until recently, there are also direct reticulospinal projections from the midbrain (Castiglioni et al., 1978). These will be discussed in the next section. The terminals of the reticulospinal axons arising from cells in the nucleus

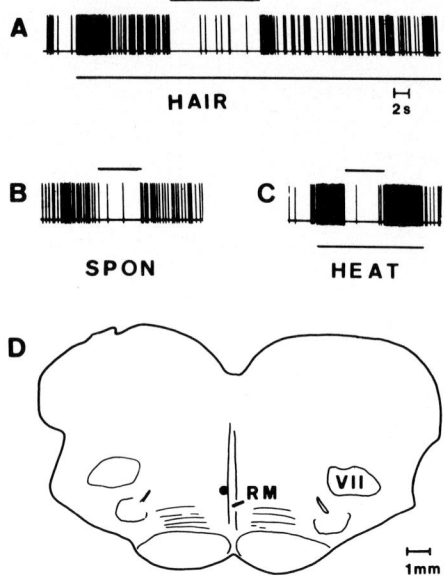

Fig. 4. The inhibition of the activity of a spinothalamic tract cell during stimulation in the region of the raphe magnus nucleus. In A, the background activity of a spinothalamic tract cell was enhanced by directing an air stream across the receptive field during the time indicated by the bar below the record. During the period shown by the bar above the record, a train of stimuli was delivered in the vicinity of the raphe magnus nucleus at the point shown in D. The background activity (SPON) and the activity evoked by a noxious heat stimulus (HEAT), B and C, could also be inhibited by stimulation next to the raphe magnus nucleus. (From Willis et al., 1977.)

gigantocellularis are concentrated in laminae VII and VIII, although there are some endings in the dorsal horn.

Stimulation in the nucleus gigantocellularis may have either excitatory or inhibitory actions on primate spinothalamic tract neurons (Haber et al., 1978). Fig. 5C shows the inhibition of the background discharge of a spinothalamic tract neuron during stimulation in the nucleus gigantocellularis of either side of the brain stem (Fig. 5D). In Fig. 5A are the discharges evoked in the cell by stimulation of the sural nerve. A long stimulus train applied in the ipsilateral reticular formation resulted in a reduction in the number of discharges evoked by the peripheral nerve stimulus (Fig. 5B). Spinothalamic tract cells may instead be excited by comparable stimuli in the nucleus gigantocellularis. No systematic relationship has yet been noted between the sign of the reticular effects and either the location of the stimulating electrode in the nucleus gigantocellularis or the type of spinothalamic tract neuron. The reticular actions are not altered by interruption of the dorsolateral fasciculi.

Reticular formation stimulation is also known to excite or inhibit neurons in the dorsal column nuclei (Cesa-Bianchi and Sotgiu, 1969). It is hypothesized that the excitatory effects are directed at interneurons, which in turn inhibit relay neurons. However, some relay neurons are excited. The nucleus gigantocellularis has been shown to project to the cuneate nucleus (Sotgiu and Margnelli, 1976; Sotgiu and Marini, 1977). It is possible that the inhibition of transmission through the dorsal column nuclei by stimulation of a number of different brain structures is mediated by way of the reticular formation (Sotgiu and Cesa-Bianchi, 1972; Jabbur et al., 1977),

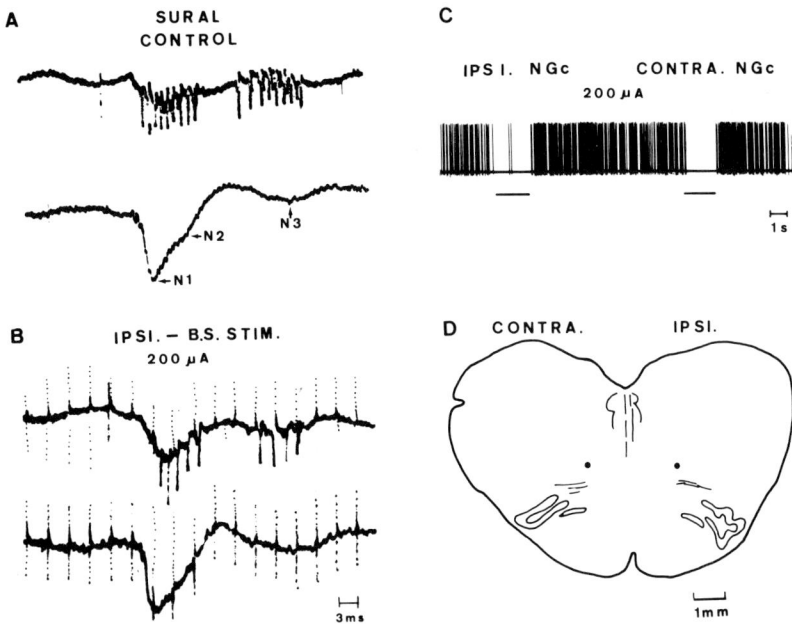

Fig. 5. Inhibition of spinothalamic tract cell by stimulation in the nucleus gigantocellularis. The discharges of a spinothalamic tract cell in response to stimulation of the A$\alpha\beta$ and Aδ fibers of the sural nerve are shown in A. The N1, N2 and N3 waves are indicated in the recording taken from the cord dorsum (lower trace). In B, the discharges of the cell are shown to be inhibited during repetitive stimulation in the nucleus gigantocellularis on the side ipsilateral to the spinothalamic tract cell. In C, it can be seen that the discharges of the cell can be inhibited by reticular formation stimulation on either side. The stimulus points are indicated by the dots in D. (From Haber et al., 1978).

including at least part of the cortical inhibition (Jabbur and Towe, 1961; Levitt et al., 1964).

The reticular formation appears also to regulate activity in the spinocervical tract. Inhibition of spinocervical neurons was observed by Taub (1964) as a result of stimulation in the brain stem tegmentum or the cerebellum. Brown et al. (1973) found that fibers descending in the ventral funiculi produce an inhibition of spinocervical tract cells.

Spinoreticular tract cells, on the other hand, are excited by ventral spinal pathways (Holmqvist et al., 1960b). Part of the excitatory action can be attributed to reticulospinal tracts and part to the lateral vestibulospinal tract (Grillner et al., 1968). The vestibulospinal projection appears to modulate the activity of spinoreticular neurons during tilt (Coulter et al., 1976).

PERIAQUEDUCTAL GRAY AND MIDBRAIN RETICULAR FORMATION

Although there are direct spinal projections from the periaqueductal gray and surrounding reticular formation (Kuypers and Maisky, 1975; Castiglioni et al., 1976), it is not yet clear to what extent these account for the very powerful effects of stimulation in this region upon spinal cord activity. There are also connections between the periaqueductal gray and cells at more caudal levels of the brain stem which in turn project to the cord, including the raphe magnus nucleus (Ruda, 1975; Taber-Pierce et al., 1976).

Stimulation in the periaqueductal gray (and nearby periventricular gray) results in a number of reflex and behavioral alterations which collectively have been called "stimulus produced analgesia" (SPA). Animals undergoing SPA do not evidence the behavioral manifestations generally associated with noxious stimuli (Reynolds, 1969). Flexion reflexes, such as the tail-flick response to noxious heat in the rat and withdrawal from pinch and the jaw opening reflex to tooth pulp stimulation in the cat, are reduced or abolished by such stimulation (Mayer and Liebeskind, 1974; Oliveras et al., 1974). Humans with chronic pain states treated by stimulation in the periaqueductal or periventricular gray report pain relief (Richardson and Akil, 1977a, b; Hosobuchi et al., 1977). The descending pathways involved in SPA appear to include a serotonergic component (Akil and Liebeskind, 1975), and they traverse the dorsolateral fasciculi (Basbaum et al., 1977). At least a part of the SPA system appears to be responsive to the administration of the opiate drugs, either systematically or by microinjection into the region of the periaqueductal gray (Mayer and Hayes, 1975; Proudfit and Anderson, 1975; Akil et al., 1976; Yaksh et al., 1976; Basbaum et al., 1977; Oliveras et al., 1977).

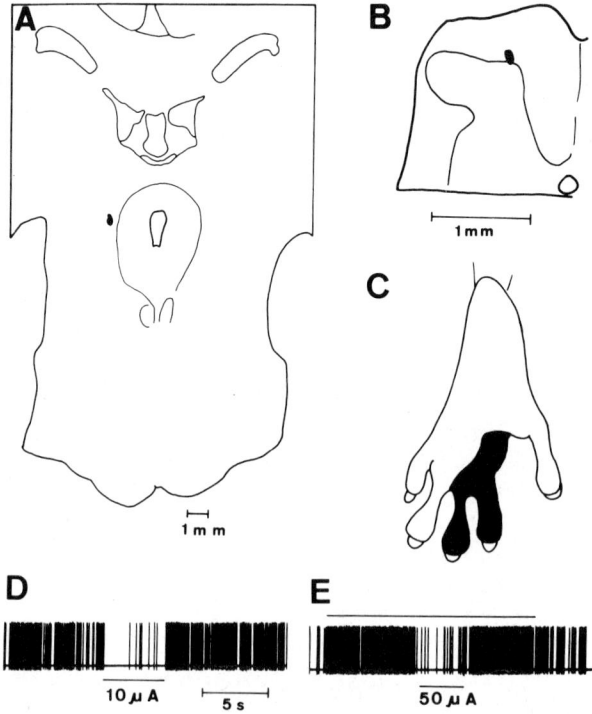

Fig. 6. Inhibition of a spinothalamic tract neuron by stimulation in the midbrain reticular formation. The midbrain stimulus site is shown in A. The spinothalamic tract cell was located in lamina I, as seen in B. The receptive field is drawn in C. The background discharge of the cell is in D, while the enhanced discharge produced by noxious heating of the skin is in E (heat applied during the time indicated by the bar at the top of the record). Stimulation in the midbrain at the strengths indicated and for the periods shown by the bars under the records inhibited the discharges.

Preliminary evidence in our laboratory indicates that primate spinothalamic tract neurons can be inhibited by stimulation in the reticular formation next to the periaqueductal gray. Fig. 6 shows the inhibition of the background discharges (D) and also the enhanced activity of a spinothalamic neuron in lamina I (B) during noxious heat stimulation of the skin (E) by stimulation at the site indicated in A in the midbrain.

It is not yet known what effects, if any, stimulation in the periaqueductal gray would have on various other ascending pathways.

OTHER DESCENDING SYSTEMS

There are other brain structures which have been or will be shown to exert direct or indirect effects upon spinal cord tract cells. These include the rubrospinal and the vestibulospinal tracts. Several descending pathways have only recently been demonstrated by the use of the new anatomical tracing techniques, and there has not yet been time for physiological investigations of their possible contribution to descending control.

DISCUSSION

A number of possible functions have been proposed for the descending control systems which affect transmission in ascending pathways. For instance, Wall and Dubner (1972) point out that the descending systems can modify the gain of the ascending pathways, change the degree of selectivity for different inputs, alter the receptive field size or the inhibitory surround, or switch channels for information flow. Another suggestion is that descending commands may change the signal to noise ratio (Levitt et al., 1965). Such descending activity may relate to the concept of corollary discharges which may inhibit the input predicted to result from a voluntary movement (Von Holst, 1954; Teuber, 1960). Alternatively, the descending inhibition of certain pathways may prevent interference with preprogrammed movements (Dyre-Poulsen, 1975). Descending controls are certainly important factors in the changes which accompany alterations in behavioral states, such as the transition from wakefulness to sleep (Carli et al., 1967; Ghelarducci et al., 1970). Whatever the purpose of a particular descending command upon a particular ascending pathway, it is apparent that the operation of ascending and descending pathways are so interrelated that it is necessary to consider both together.

SUMMARY

The descending control of transmission of information in ascending pathways is important both for motor control and for sensory processing. Experiments on descending pathways which modulate the activity of spinothalamic tract neurons in the primate were discussed as an illustration of some of the major known control systems. Stimulation of the sensorimotor cerebral cortex results in an inhibition of the responses of spinothalamic tract neurons to tactile, but not to noxious stimuli. The raphe nuclei of the medulla send projections to the spinal cord, and the raphe magnus nucleus in particular projects through the dorsolateral fasciculus and terminates in the dorsal horn. Stimulation in the raphe magnus nucleus produces a powerful inhibition of spinothalamic tract neurons, including inhibition of the responses of these cells to noxious mechanical or thermal stimuli. Stimulation within the nucleus reticularis gigantocellularis may either inhibit or excite spinothalamic tract neurons. Preliminary evidence indicates that stimulation in the midbrain reticular formation just lateral to the periaqueductal gray produces a strong inhibition of spinothalamic tract cells. It is likely that some of these inhibitory effects are related to the phenomenon of "stimulation produced analgesia". It is concluded that the actions of descending pathways in modifying afferent transmission could be interpreted in a number of ways (change of gain, switching, changing signal to noise ratio, corollary discharge, etc.) and that the ascending and descending pathways operate in concert.

ACKNOWLEDGEMENTS

The author wishes to thank Gail Silver for her expert technical assistance. The work in the author's laboratory was done in collaboration with Drs. Coulter, Haber, Kenshalo, Leonard, Martin and Maunz. Support was provided by NIH research grant NS 09743.

REFERENCES

Akil, H. and Liebeskind, J.C. (1975) Monoaminergic mechanisms of stimulation-produced analgesia. *Brain Res.*, 94: 279–296.

Akil, H., Mayer, D.J. and Liebeskind, J.C. (1976) Antagonism of stimulation-produced analgesia by naloxone, a narcotic antagonist. *Science*, 191: 961–962.

Andersen, P., Eccles, J.C. and Sears, T.A. (1964a) Cortically evoked depolarization of primary afferent fibers in the spinal cord. *J. Neurophysiol.*, 27: 63–77.

Andersen, P., Eccles, J.C., Schmidt, R.F. and Yokota, T. (1964b) Slow potential waves produced in the cuneate nucleus by cutaneous volleys and by cortical stimulation. *J. Neurophysiol.*, 27: 78–91.

Andersen, P., Eccles, J.C., Schmidt, R.F. and Yokota, T. (1964c) Identification of relay cells and interneurons in the cuneate nucleus. *J. Neurophysiol.*, 27: 1080–1095.

Basbaum, A.I., Clanton, C.H. and Fields, H.L. (1978) Three bulbospinal pathways from the rostral medulla of the cat: an autoradiographic study of pain modulating systems. *J. comp. Neurol.*, 178: 209–224.

Basbaum, A.I., Marley, N.J.E., O'Keefe, J. and Clanton, C.H. (1977) Reversal of morphine and stimulus-produced analgesia by subtotal spinal cord lesions. *Pain*, 3: 43–56.

Brown, A.G. (1971) Effects of descending impulses on transmission through the spinocervical tract. *J. Physiol. (Lond.)*, 219: 103–125.

Brown, A.G., Kirk, E.J. and Martin, H.F. (1973) Descending and segmental inhibition of transmission through the spinocervical tract. *J. Physiol. (Lond.)*, 230: 689–705.

Brown, A.G., Coulter, J.D., Rose, P.K., Short, A.D. and Snow, P.J. (1977) Inhibition of spinocervical tract discharges from localized areas of the sensorimotor cortex in the cat. *J. Physiol. (Lond.)*, 264: 1–16.

Carli, G., Diete-Spiff, K. and Pompeiano, O. (1967) Transmission of sensory information through the lemniscal pathway during sleep. *Arch. ital. Biol.*, 105: 31–51.

Carpenter, D., Lundberg, A. and Norrsell, U. (1963) Primary afferent depolarization evoked from the sensorimotor cortex. *Acta physiol. scand.*, 59: 126–142.

Carpenter, D., Engberg, I. and Lundberg, A. (1965) Differential supraspinal control of inhibitory and excitatory actions from the FRA to ascending spinal pathways. *Acta physiol. scand.*, 63: 103–110.

Castiglioni, A.J., Gallaway, M.C. and Coulter, J.D. (1978) Spinal projections from the midbrain in monkey. *J. comp. Neurol.*, 178: 329–346.

Cervero, F., Iggo, A. and Molony, V. (1977) Responses of spinocervical tract neurones to noxious stimulation of the skin. *J. Physiol. (Lond.)*, 267: 537–558.

Cesa-Bianchi, M.G. and Sotgiu, M.L. (1969) Control by brain stem reticular formation of sensory transmission in Burdach nucleus. Analysis of single units. *Brain Res.*, 13: 129–139.

Coulter, J.D. and Jones, E.G. (1977) Differential distribution of corticospinal projections from individual cytoarchitectonic fields in the monkey. *Brain Res.*, 129: 335–340.

Coulter, J.D., Maunz, R.A. and Willis, W.D. (1974) Effects of stimulation of sensorimotor cortex on primate spinothalamic neurons. *Brain Res.*, 65: 351–356.

Coulter, J.D., Mergner, T. and Pompeiano, O. (1976) Effects of static tilt on cervical spinoreticular tract neurons. *J. Neurophysiol.*, 39: 45–62.

Dyhre-Poulson, P. (1975) Increased vibration threshold before movements in human subjects. *Exp. Neurol.*, 47: 516–522.

Eccles, R.M. and Lundberg, A. (1959) Supraspinal control of interneurones mediating spinal reflexes. *J. Physiol. (Lond.)*, 147: 565–584.

Engberg, I., Lundberg, A. and Ryall, R.W. (1968) Is the tonic decerebrate inhibition of reflex paths mediated by monoaminergic pathways? *Acta physiol. scand.*, 72: 123–133.

Fetz, E.E. (1968) Pyramidal tract effects on interneurons in the cat lumbar dorsal horn. *J. Neurophysiol.*, 31: 69–80.

Foreman, R.D., Schmidt, R.F. and Willis, W.D. (1977) Convergence of muscle and cutaneous input onto primate spinothalamic tract neurons. *Brain Res.*, 124: 555–560.

Fu, T.C., Jankowska, E. and Tanaka, R. (1977) Effects of volleys in cortico-spinal tract fibres on ventral spino-cerebellar tract cells in the cat. *Acta physiol. scand.*, 100: 1–13.

Ghelarducci, B., Pisa, M. and Pompeiano, O. (1970) Transformation of somatic afferent volleys across the prethalamic and thalamic components of the lemniscal system during the rapid eye movements of sleep. *Electroenceph. clin. Neurophysiol.*, 29: 348–357.

Gordon, G. and Jukes, M.G.M. (1964) Descending influences on the exteroceptive organizations of the cat's gracile nucleus. *J. Physiol. (Lond.)*, 173: 291–319.

Grillner, S., Hongo, T. and Lund, S. (1968) The origin of descending fibres monosynaptically activating spinoreticular neurones. *Brain Res.*, 10: 259–262.

Haber, L.H., Martin, R.F., Chatt, A.B. and Willis, W.D. (1978) Effects of stimulation in nucleus reticularis gigantocellularis on the activity of spinothalamic tract neurons in the monkey. *Brain Res.*, in press.

Hagbarth, K.E. and Kerr, D.I.B. (1954) Central influences on spinal afferent conduction. *J. Neurophysiol.*, 17: 295–307.

Hernández-Peón, R., Scherrer, H. and Velasco, M. (1956) Central influences on afferent conduction in the somatic and visual pathways. *Acta neurol. latinoamer.*, 2: 8–22.

Holmqvist, B. and Lundberg, A. (1961) Differential supraspinal control of synaptic actions evoked by volleys in the flexion reflex afferents in alpha motoneurones. *Acta physiol. scand.*, 54, Suppl. 186: 1–51.

Holmqvist, B., Lundberg, A. and Oscarsson, O. (1960a) Supraspinal inhibitory control of transmission to three ascending spinal pathways influenced by the flexion reflex afferents. *Arch. ital. Biol.*, 98: 60–80.

Holmqvist, B., Lundberg, A. and Oscarsson, O. (1960b) A supraspinal control system monosynaptically connected with an ascending spinal pathway. *Arch. ital. Biol.*, 98: 402–422.

Hongo, T. and Okada, Y. (1967) Cortically evoked pre- and postsynaptic inhibition of impulse transmission to the dorsal spinocerebellar tract. *Exp. Brain Res.*, 3: 163–177.

Hongo, T., Okada, Y. and Sato, M. (1967) Corticofugal influences on transmission to the dorsal spinocerebellar tract from hindlimb primary afferents. *Exp. Brain Res.*, 3: 135–149.

Hosobuchi, Y., Adams, J.E. and Linchitz, R. (1977) Pain relief by electrical stimulation of the central gray matter in humans and its reversal by naloxone. *Science*, 197: 183–186.

Jabbur, S.J. and Towe, A.L. (1961) Cortical excitation of neurons in dorsal column nuclei of cat, including an analysis of pathways. *J. Neurophysiol.*, 24: 499–509.

Jabbur, S.J., Harik, S.I. and Hush, J.A. (1977) Caudate influence on transmission in the cuneate nucleus. *Brain Res.*, 120: 559–563.

Kuypers, H.G.J.M. (1960) Central cortical projections to motor and somato-sensory cell groups. *Brain*, 83: 161–184.

Kuypers, H.G.J.M. and Maisky, V.A. (1975) Retrograde axonal transport of horseradish peroxidase from spinal cord to brain stem cell groups in the cat. *Neurosci. Lett.*, 1: 9–14.

Levitt, M., Carreras, M., Liu, C.N. and Chambers, W.W. (1964) Pyramidal and extrapyramidal modulation of somatosensory activity in gracile and cuneate nuclei. *Arch. ital. Biol.*, 102: 197–229.

Liu, C.N. and Chambers. W.W. (1964) An experimental study of the cortico-spinal system in the monkey (Macaca mulatta). *J. Comp. Neurol.*, 123: 257–284.

Lundberg, A., Norrsell, U. and Voorhoeve, P. (1963) Effects from the sensorimotor cortex on ascending spinal pathways. *Acta physiol. scand.*, 59: 462–473.

Magni, F. and Oscarsson, O. (1961) Cerebral control of transmission to the ventral spino-cerebellar tract. *Arch. ital. Biol.*, 99: 369–396.

Martin, R.F., Jordan, L.M. and Willis, W.D. (1978) Differential projections of cat medullary raphe neurons demonstrated by retrograde labelling following spinal cord lesions. *J. comp. Neurol.*, 182: 77–88.

Mayer, D.J. and Liebeskind, J.C. (1974) Pain reduction by focal electrical stimulation of the brain: an anatomical and behavioral analysis. *Brain Res.*, 68: 73–93.

Mayer, D.J. and Hayes, R.L. (1975) Stimulation-produced analgesia: development of tolerance and cross-tolerance to morphine. *Science*, 188: 941–943.

Nyberg-Hansen, R. (1965) Sites and mode of termination of reticulo-spinal fibers in the cat. *J. comp. Neurol.*, 124: 71–100.

Nyberg-Hansen, R. and Brodal, A. (1963) Sites of termination of corticospinal fibers in the cat. An experimental study with silver impregnation methods. *J. comp. Neurol.*, 120: 369–391.

Oliveras, J.L., Besson, J.M., Guilbaud, G. and Liebeskind, J.C. (1974) Behavioral and electrophysiological evidence of pain inhibition from midbrain stimulation in the cat. *Exp. Brain Res.*, 20: 32–44.

Oliveras, J.L., Hosobuchi, Y., Redjemi, F., Guilbaud, G. and Besson, J.M. (1977) Opiate antagonist, naloxone, strongly reduces analgesia induced by stimulation of a raphe nucleus (centralis inferior). *Brain Res.*, 120: 221–229.

Petras, J.M. (1967) Cortical, tectal and tegmental fiber connections in the spinal cord of the cat. *Brain Res.*, 6: 275–324.

Proudfit, H.K. and Anderson, E.G. (1975) Morphine analgesia: blockade by raphe magnus lesions. *Brain Res.*, 98: 612–618.

Reynolds, D.V. (1969) Surgery in the rat during electrical analgesia induced by focal brain stimulation. *Science*, 164: 444–445.

Richardson, D.E. and Akil, H. (1977a) Pain reduction by electrical brain stimulation in man. Part 1. Acute administration in periaqueductal and periventricular sites. *J. Neurosurg.*, 47: 178–183.

Richardson, D.E. and Akil, H. (1977b) Pain reduction by electrical brain stimulation in man. Part 2. Chronic self-administration in the periventricular gray matter. *J. Neurosurg.*, 47: 184–194.

Ruda, M. (1975) *Autoradiographic Study of the Efferent Projections of the Midbrain Central Gray of the Cat*. Ph.D. dissertation, Univ. Pennsylvania, Philadelphia. PA.

Sotgiu, M.L. and Cesa-Bianchi, M.G. (1972) Thalamic and cerebellar influence on single units of the cat cuneate nucleus. *Exp. Neurol.*, 34: 394–408.

Sotgiu, M.L. and Margnelli, M. (1976) Electrophysiological identification of pontomedullary reticular neurons directly projecting into dorsal column nuclei. *Brain Res.*, 103: 443–453.

Sotgiu, M.L. and Marini, G. (1977) Reticulo-cuneate projections as revealed by horseradish peroxidase axonal transport. *Brain Res.*, 128: 341–345.

Taber-Pierce, E., Foote, W.E. and Hobson, J.A. (1976) The efferent connection of the nucleus raphe dorsalis. *Brain Res.*, 107: 137–144.

Taub, A. (1964) Local, segmental and supraspinal interaction with a dorsolateral spinal cutaneous afferent system. *Exp. Neurol.*, 10: 357–374.

Teuber, H.L. (1960) Perception. In *Handbook of Physiology, Section 1, Neurophysiology, Vol. 3*. J. Field (Ed.), Amer. Physiol. Soc., Washington, DC, pp. 1595–1668.

Towe, A.L. and Jabbur, S.J. (1961) Cortical inhibition of neurons in dorsal column nuclei of cat. *J. Neurophysiol.*, 24: 488–498.

Trevino, D.L., Coulter, J.D. and Willis, W.D. (1973) Location of cells of origin of spinothalamic tract in lumbar enlargement of the monkey. *J. Neurophysiol.*, 36: 750–761.

Von Holst, E. (1954) Relations between the central nervous system and the peripheral organs. *Brit. J. Anim. Behav.*, 2: 89–94.

Wall, P.D. and Dubner, R. (1972) Somatosensory pathways. *Ann. Rev. Physiol.*, 34: 315–336.

Willis, W.D., Haber, L.H. and Martin, R.F. (1977) Inhibition of spinothalamic tract cells and interneurons by brain stem stimulation in the monkey. *J. Neurophysiol.*, 40: 968–981.

Willis, W.D., Trevino, D.L., Coulter, J.D. and Maunz, R.A. (1974) Responses of primate spinothalamic tract neurons to natural stimulation of hindlimb. *J. Neurophysiol.*, 37: 358–372.

Winter, D.L. (1965) N. gracilis of cat. Functional organization and corticofugal effects. *J. Neurophysiol.*, 28: 48–70.

Yaksh, T.L., Yeung, J.C. and Rudy, T.A. (1976) Systematic examination in the rat of brain sites sensitive to the direct application of morphine: observation of differential effects within the periaqueductal gray. *Brain Res.*, 114: 83–103.

C

Reflex Control of Posture and Movement

Organization and Programming of Motor Activity during Posture Control

L.M. NASHNER

Neurological Sciences Institute, Good Samaritan Hospital and Medical Center, Portland, OR 97209 (U.S.A.)

INTRODUCTION

A currently accepted concept is that the sensorimotor system expresses purposeful movements, adapts these movements to the external conditions and loads, and maintains the posture and balance of the body utilizing a hierarchy of specialized subsystems (e.g., Gelfand et al., 1971). According to this concept the complexity of the control task is greatly simplified by breaking it down into a combination of much simpler, stereotyped behaviors, each of which is executed independently by a specialized "movement generator". Although some investigators have presented cogent physiological arguments against the predominance of autonomous, specialized subsystems within the sensorimotor system (e.g. Davis, 1976), the concept has nevertheless proved a quite useful one for interpreting a variety of experimental observations, especially those derived from the various spinalized mesencephalic, and decorticate cat preparations locomoting upon treadmills (e.g., Grillner, 1975).

The objective of this brief paper is to synthesize a hierarchical model of posture control from experimental observations of stance posture control of normal human subjects. Apart from the obvious interest in understanding more about ourselves and about the disease processes which afflict our fellow man, a conceptual synthesis of human experimental observations enhances our knowledge about sensorimotor controls derived from the study of animal preparations in several important ways. Studies of chronic and acute cat preparations have revealed much detail about the neural mechanisms within the spinal cord which generate the basic locomotor movement patterns. However, the interruption of many CNS functions in these preparations has thus far prevented any study of the adaptive and postural balance controls, which under normal conditions are integrated into the ongoing movement behaviors. While the limitation to experimentally non-invasive techniques prevents a detailed study of underlying neural mechanisms, standing human subjects can be easily incorporated into highly variable experimental paradigms which focus attention upon what happens to the system during conditions when it must modify a strategy of control and must maintain upright stance under changing external conditions.

The human postural control experiments described here have quantitatively characterized the properties of the earliest functionally useful postural adjustments which occur beginning 100 msec after the onset of a perturbation. (Myotatic reflexes at 40–45 msec latencies were very seldom seen during these experiments and were of no apparent functional consequence when they did.) Because the latencies of earliest functionally useful adjustments are quite long compared to the myotatic reflex, their activity is probably mediated by intersegmental and central structures as well as by the local spinal circuits (e.g., Evarts et al., 1974; Phillips, 1969), although a recent observation suggests that the isolated spinal cord (of the cat) itself produces these so-called "long-loop" or "functional stretch reflexes" (Ghez and Shinoda, 1978). Results presented here, however, suggest that the earliest adjustments are far too complexly interwoven into an organizational structure of leg muscles to be characterized simply as the response of individual muscles to stretch. For lack of a better term these earliest adjustments will be referred to as perturbationally "triggered responses" (TR's). Speculation about their afferent organization will be discussed.

THE EXPERIMENTAL OBSERVATIONS

When a subject stood upon a 6 degree-of-freedom movable platform and unexpectedly experienced a movement perturbation (one out of a number of possible different ones) the first functionally significant changes in surface EMG activity occurred among a group of leg muscles at 100 msec latency. Proprioceptive and joint afferents were necessary inputs to these earliest adjustments, since the latency of activity was always delayed until 180 msec or longer whenever the platform was rotated to stabilize the rotational position of ankle joints during perturbations (Nashner, 1971, 1972). The following paragraphs describe the properties of TR adjustments which occur within the time interval 100–150 msec after the onset of a perturbation. This time interval encompasses "prevoluntary" postural adjustments which require local proprioceptive and somatosensory inputs and probably involve the local "movement generator" circuits of the spinal cord, cerebellum, and brain stem. All of the following results have already been reported in detail in Nashner (1976, 1977) and Nashner et al. (1979).

The strength of the contractile activity of each of four leg muscles (gastrocnemius, G; tibialis anterior, T; hamstrings, H; and quadriceps, Q) was estimated during the interval 100–150 msec after the onset of postural perturbations using the following techniques. Each raw EMG signal (10–5000 Hz bandpass) was full-wave rectified, filtered (0–40 Hz bandpass), and then numerically integrated over a 50 msec interval (100–150 msec) to give a single number in approximate proportion to the contractile strength of the muscle. The gain of each EMG signal was calibrated only once at the beginning of each session so that the statistical significance of relative changes in contractile strength among the four muscles could be addressed throughout each 1-h session.

Fig. 1 shows schematically the four basic modes of movement perturbations used. Horizontal translation of both platforms (A) induced anteroposterior (AP) sway which occurred principally about the ankle joints. Direct rotation of the platforms (B) rotated the ankle joints independent of AP sway motions of the body. Vertical displacement of both platforms (C) flexed or extended both legs together and changed the vertical height of the body. Reciprocal, vertical displacement of the two platforms (D) flexed one leg, extended the other, and caused the body to sway laterally to the side of the lowering leg. In addition to those illustrated, combinations of direct rotational (B), and vertical (C and D) or AP sway (A) perturbations could be given.

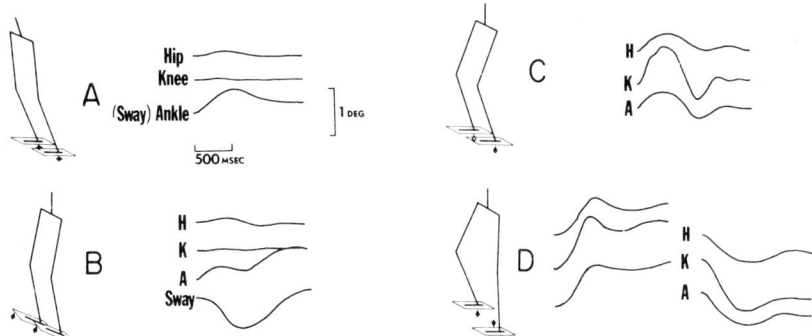

Fig. 1. Four kinds of movements produced by the 6 degree-of-freedom platform.

These combinations changed the relations between ankle joint motions relative to those of the knee joints and the body in space.

TR adjustments were organized into fixed movement specific patterns of activity among the leg muscles. Results reviewed in this section have shown that one of three possible patterns of EMG activity is elicited by a variety of different platform motions. Each pattern of activity is characterized by a fixed ratio of activity among the muscles, is movement specific, and is functionally related to the task of coordinating one kind of postural adjustment.

The organization of activity among the leg muscles was characterized during each TR response by computing the ratios of contractile strength among the four leg muscles. Fig. 2 shows the average ratios computed (sum of 6 patients) for each kind of movement perturbation and schematically illustrates the pattern of muscular activity produced by each.

During AP sway perturbations the stretching ankle muslces (G during forward and T during backward sway) and the proximal leg muscles on the same dorsal or ventral aspect of each leg contracted in fixed preparation. That this pattern of EMG activity is stereotypically fixed and not the result of independent muscle stretch inputs from ankles, knees, and hip rotations during sway was shown by eliciting the exact same pattern of activity with unexpected, direct ankle rotations (a stimulus movement which produced a very different pattern of relative knee and hip joint motions). Functionally, the fixed coupling of contractile activity between G–H and T–Q muscle pairs during sway helps *resist* the ankle joint rotations (thereby stabilizing AP sway under normal conditions). Since the antiphase coupling of ankles and hip joint motions (assuming locked knees) is a characteristic feature of the body during AP sway (e.g. Hemani et al., 1978), the fixed coupling of ankle and hip extensor (G–H) and flexor activities (T–Q) during sway adjustments is a control strategy which reduces these two independent and antiphasically intercoupled degrees-of-freedom of the body into a single coordinated body motion. For this reason, this fixed coupling of leg muscles might appropriately be termed the "sway" synergy.

During the simultaneous vertical perturbations of both feet the stretching ankle muscles (G during upward and T during downward displacements) and the proximal muscle on the *opposite* dorsal or ventral aspect of the leg contracted in fixed proportion. Functionally, this rather different coupling pattern helps *resist* coordinated flex-

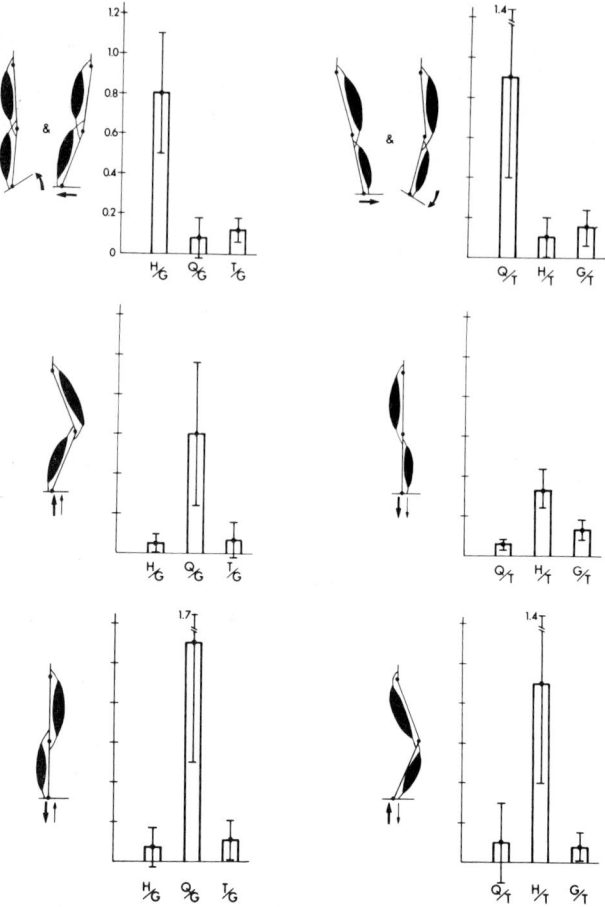

Fig. 2. Coupling ratios (average ±1 SD) between the responding lower leg muscle, the two upper leg muscles, and the antagonist lower leg muscle. Schematic figures at left illustrate the direction of stimulus movement (heavy arrows for the leg shown and light arrows for the other leg). The responding muscles in each case are outlined in heavy black.

ion or extension of the ankle and knee joints. Under normal conditions this pattern of activity helps control the vertical height of the body. For this reason this fixed coupling arrangement might be termed the "suspensory" synergy.

During reciprocal, vertical perturbations of the two feet the coupling between the proximal and distal muscle pairs was the same as that seen during simultaneous, vertical displacements. However, now the shortening rather then the lengthening muscles in each leg were contracted. Functionally, the response of shortening ankle and knee joint muscles in each leg and the reciprocal activation of the two legs helps each leg *follow* the perturbational motion produced by the platform, thereby minimizing any lateral sway of the body. The reciprocal extension and flexion of the two legs resembles in many ways the pattern of activity seen during normal locomotion (Herman et al., 1976). In fact, the imposition of brief, subthreshold cutaneous stimuli to the dorsum of a subject's foot at different phases of reciprocal, extension and flexion movements produced phase-dependent enhancement of response activity very similar to that seen by Forssberg et al. (1977) in the locomoting cat preparation.

In order to show that the "suspensory" coupling of distal and proximal muscles of the legs is stereotypically fixed during vertical perturbations the platforms were rotated during vertical perturbations, to either exaggerate the ankle rotations or to reverse their direction relative to that of the knees. Surprisingly, although ankle rotational inputs alone were sufficient to evoke activity organized into the "sway" synergy, neither of these dramatic changes in ankle rotational inputs had any appreciable effect upon the "suspensory" activity or upon the tendency for the shortening muscles to respond during reciprocal movements (even when the rate of stretch of the stretching ankle muscle was doubled by the platform rotation).

Reorganization of muscular activity from one movement pattern to another occurred within the latent period of a single response. Reorganization of EMG activity among the three different patterns described above always occurred during the *first* trials, even in those instances when a sequence of identical movement perturbations was followed, unexpectedly by a different one.

The strength of sway and suspensory adjustments adapted slowly over several trials following changes in sensory inputs relative to the external environment. Unexpectedly rotating the ankle joints of a standing subject produced activity organized into the "sway" synergy which was essentially identical to that elicited by sway rotations of the body about the ankles. However, during direct ankle rotations such responses were functionally inappropriate and they enhanced rather than helped stabilize the sway motions of the body. Normal subjects adapted to this condition by progressively attenuating the amplitude of activities which resisted ankle joint rotations (Fig. 3). Once they had adapted to expect direct rather than AP sway rotations, subjects did not respond to subsequent, unexpected AP sway perturbations and sway excursions were significantly larger. However, the sway synergic activity was adaptively restored and stability improved following a number of AP sway trials.

ORGANIZATION OF RAPID POSTURAL ADJUSTMENTS

Each rapid TR adjustment of a standing subject is organized into a synergy of leg muscles which is appropriate to the specific movement even when a sequence of perturbations changes unexpectedly from trial to trial. Each synergy modulates a specific subset of proprioceptive inputs (motion inputs from particular leg joints) and also phasically modulates non-specific electrical cutaneous inputs. The rapidity with which an appropriate synergy is established and certain similarities in the mediation of proprioceptive and cutaneous inputs during TR adjustments with those seen during locomotor activities of the spinal cat suggest that postural synergies are organized by spinal "movement generators" very similar to those identified in the spinal cord of the locomoting cat. If this hypothesis is correct we can begin to use observations on standing subjects to functionally distinguish between the processes which generate the basic patterns of movement activity and those which adapt movements to the external conditions and which maintain the postural balance of the body.

So long as the repertoire of "movement generators" is small compared to the

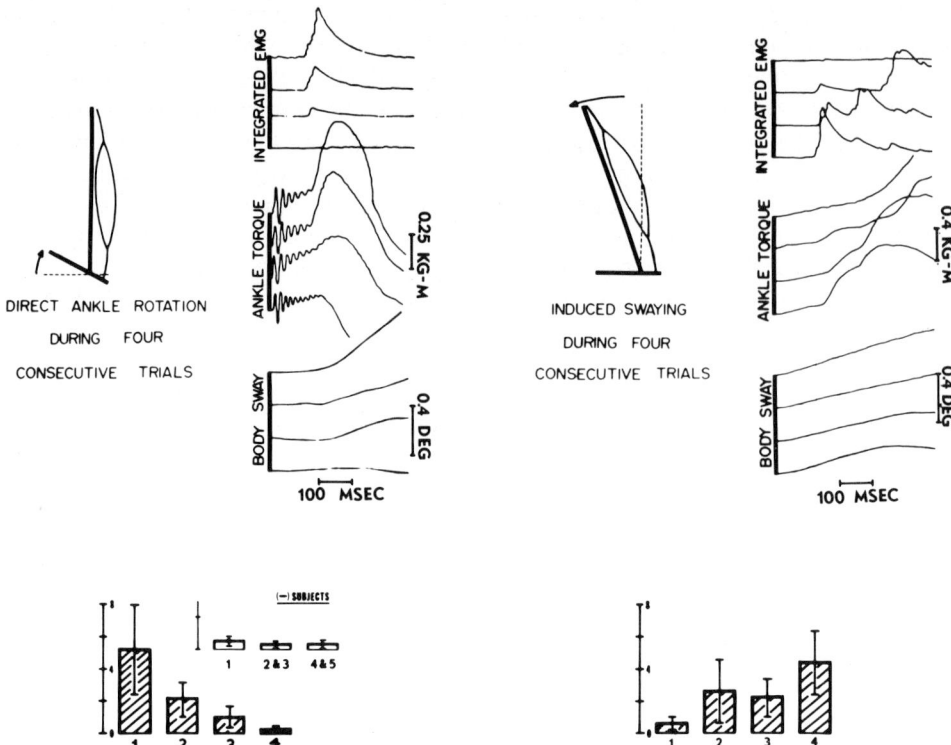

Fig. 3. Adaptive changes in TR strengths elicited by direct ankle rotations and by AP sway rotations. Left: a sequence of 4 direct rotations following unexpectedly, a sequence of AP sway trials. Right: a sequence of 4 AP sway trials following unexpectedly, a sequence of direct rotations. Bar graphs show averaged data (± 1 SD) for three subjects.

number of independent degrees-of-freedom of the body, a hierarchical system has the advantage of greatly simplifying the control task (Bernstein, 1967). However, the hierarchically organized system also has some drawbacks, the most significant (with regard to posture and balance activities) of which is the inability of the local "movement generators" to regulate the more complex "global" variables of the system (e.g. Greene, 1972). For example, postural "movement generators" cannot correctly distinguish between a sway movement of the body (AP sway rotations about the ankle joints) and an equal but opposite rotational movement of the supporting surface relative to the body (direct rotations of the ankle joints). These kinds of distinctions require complex correlations among the local proprioceptive inputs and central vestibular and visual inputs. A hierarchically organized group of localized systems favors the rapid organization of local movement patterns at the expense of executional errors in the regulation of the global variables of posture and balance. Central control subsystems favor the control of global variables at the expense of executional complexity and speed. Clearly, the motor system must carefully delegate its control functions among the different levels of the system.

Fig. 4 graphically outlines an attempt to describe in as simple a way as possible the characteristic features of TR postural adjustments. The model is divided into four subsystems, the motor machinery of each leg, the local spinal organization machinery

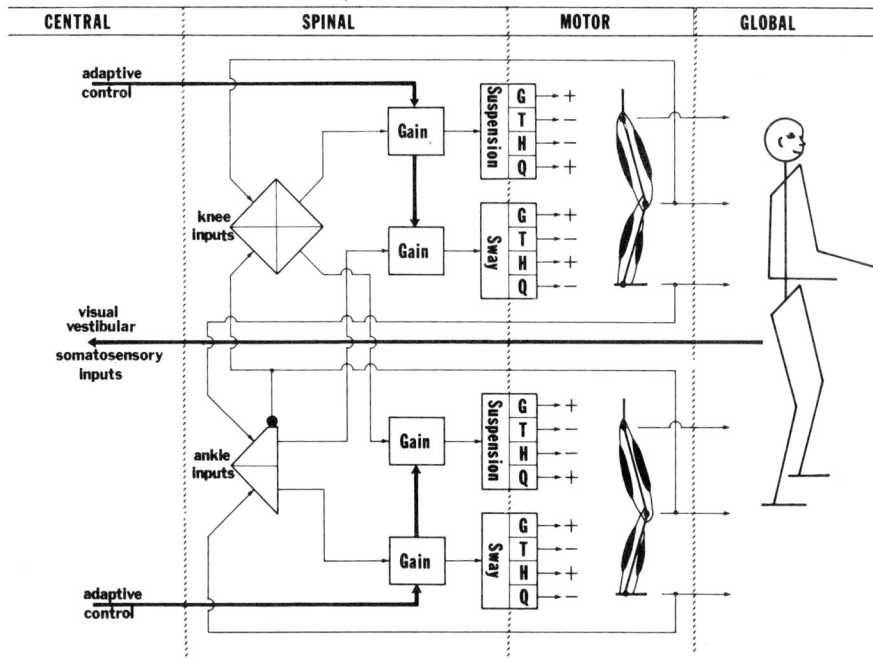

Fig. 4. A conceptual model for the organization of TR postural adjustments of standing subjects. "Global" parameters refer to the overall performance of the system (dynamic balance, orientation, etc.). "Motor" parameters refer to the force generating mechanisms of each leg. "Spinal" parameters refer to the local movement generators of the spinal cord. "Sway" and "Suspension" are the two synergic arrangements of muscle activities. The three and four sided figures represent processes modulating ankle and knee joint inputs respectively. (These functions are explained in the text.) "Central" parameters refer to those which assess the overall performance of the system and adaptively modify the local movement generators.

of the movement generators, and the descending supraspinal influences. Separate generators produce the fixed proportions of muscular activity characteristic of the "sway" and "suspensory" synergies. A separate input to each generator phasically regulates its level of activity. Spinal organizing machinery selects the appropriate synergic generators and provides the phasic activating inputs. The four-sided diamond represents a modulator in which each output is in phase with the *adjacent* input when the two inputs are in-phase. When the two inputs are antiphasically related, each output is phased to the *diagonally opposite* input. The three-sided diamond is a modulator which produces two phase outputs in direct proportion to the common mode (in-phase) portion of the two inputs. Filled circles represent inhibitory action of suspensory inputs upon the sway inputs. Heavy lines represent a descending (adaptive) modulation of sway and suspensory input amplitudes.

Although its simplicity renders it far from complete, this conceptual model can be useful to interpret some of the complex experimental observations. It illustrates the importance of local (relative body) inputs in generating and coordinating basic movements and of multimodal sensory inputs and more complex central processes in the adaptive and balance functions. Also, the conceptual model is providing a format with which to interpret observations of patients with various disorders of posture and movement control.

SUMMARY

The principal objective of this report is to synthesize a conceptual, hierarchical model of postural control from a number of experimental observations of stance postural adjustments performed by normal human subjects. The postural adjustments described are the first functionally useful responses (100 msec latency) elicited by the unexpected movements of a platform upon which a subject stood. Results presented here show that these adjustments are too complexly interwoven into an organizational structure involving many leg muscles to be characterized simply as the response of each individual muscle to its own stretch input. In place of the concept of stretch responses, the hierarchical model suggests that the response activity of each muscle is stereotypically organized into a structure controlled by the pattern of movement inputs from the entire leg.

ACKNOWLEDGEMENTS

The author wishes to acknowledge the assistance of M. Woollacott and G. Tuma, who have been active collaborators in some of the more recent experiments.

Experiments were performed and this manuscript prepared while the author has been supported by Grant NS 00148, and his laboratory by Grant NS 12661, both from the NINCDS of the National Institutes of Health.

REFERENCES

Bernstein, N. (1967) *Coordination and Regulation of Movements*. Pergamon Press, New York.

Davis, W.J. (1976) Organizational concepts in the central motor networks of invertebrates. In *Neural Control of Locomotion, Vol. 18*, R. Herman, S. Grillner, P. Stein and D. Stuart (Eds.). Plenum Press, New York.

Evarts, E.V. and Tanji, J. (1974) Gating of motor cortex reflexes by prior instruction. *Brain Res.*, 71: 479–494.

Forssberg, H., Grillner, S. and Rossignol, S. (1977) Phasic gain control of reflexes from the dorsum of the paw during spinal locomotion. *Brain Res.*, 132: 121–139.

Gelfand, I.M., Gurfinkel, V.S., Fomin, S.V. and Tsetlin, M.L. (1971) *Models of Structural-Functional Organization of Certain Biological Systems*. MIT Press, Cambridge.

Ghez, C. and Shinoda, Y. (1978) Spinal mechanisms of the functional stretch reflex. *Brain Res.*, 32: 55–68.

Greene, P.H. (1972) Problems of organization of motor systems. In *Progress in Theoretical Biology, Vol. 2*. Academic Press, New York, pp. 303–339.

Grillner, S. (1975) Locomotion in vertebrates: central mechanisms and reflex interaction. *Physiol. Rev.*, 55: 247–304.

Herman, R., Wirta, R., Bampton, S. and Finley, F.R. (1976) Human solutions to locomotion. I. Single limb analysis. In *Neural Control of Locomotion, Vol. 18*, R. Herman, S. Grillner, P. Stein and D. Stuart (Eds.). Plenum Press, New York, pp. 13–49.

Hermani, H., Weimer, F.C., Robinson, C.S., Stockwell, C.W. and Cvetkovic, V.S. (1978) Biped stability considerations with vestibular models. *IEEE Trans. Autom. Control*, 23.

Nashner, L.M. (1971) A model describing vestibular detection of body sway motion. *Acta Oto-laryngol. (Stockh.)*, 72: 429–436.

Nashner, L.M. (1972) Vestibular posture control model. *Kybernetik*, 10: 106–110.

Nashner, L.M. (1976) Adapting reflexes controlling the human posture. *Exp. Brain Res.*, 26: 59–72.

Nashner, L.M. (1977) Fixed patterns of rapid postural responses among leg muscles during stance. *Exp. Brain Res.*, 30: 13–24.

Nashner, L.M., Woollacott, M., and Tuma, G. (1979) Organization of rapid responses to postural and locomotor-like perturbations of standing man. *Exp. Brain Res.*, in press.

Phillips, C.G. (1969) Motor apparatus of the baboons hand. *Proc. roy. Soc. B.*, 173: 141–174.

Coupled Stretch Reflexes in Ankle Muscles: An Evaluation of the Contributions of Active Muscle Mechanisms to Human Posture Stability

J.H.J. ALLUM and H.J. BÜDINGEN

Institut für Hirnforschung, Universität Zürich, 8029 Zürich (Switzerland); and Department of Neurology, St. Elisabethan Hospital, Ravensburg (F.R.G.)

INTRODUCTION

Most of our knowledge concerning active muscle mechanisms which contribute to the stretch reflex comes from studies of a single muscle or group of muscles attached to a common tendon. For example, the triceps surae muscles of the cat hindlimb have been extensively studied because of their role in posture and locomotion (Forssberg et al., 1977; Grillner, 1972; Nichols and Houk, 1976; Prochazka et al., 1977). Naturally occurring perturbations to posture, however, produce stretch reflexes in both ankle extensors and flexors (Prochazka et al., 1978). These stretch reflexes are not independent responses for flexor and extensors. Apart from their mechanical interaction across the ankle joint, stretch reflexes are influenced by monosynaptic and mutually interconnected polysynaptic convergence of muscle afferents onto flexor and extensor motoneuron pools (see Jankowska, this volume, chapter IA3; Lundberg, chapter IA2). In addition to these direct interactions, flexor reflex afferents and supraspinal signals indirectly affect stretch reflexes by their influence on the transmission properties of polysynaptic pathways (Burke, 1972; Lundberg, 1966). Our understanding of the active muscle mechanisms contributing to the stretch reflex is therefore enhanced when the responses to postural disturbances are simultaneously studied in ankle flexors and extensors.

The extent to which muscle afferent signals are transmitted to alpha motoneurons and can contribute to the stretch reflex during a postural disturbance remains unclear. Some investigators have suggested that reflex muscle activity in man is effective in correcting for external disturbances (Marsden et al., 1972). Nichols and Houk (1976) proposed that afferent signals improve and linearize the non-linear visco-elastic response of active muscle to stretch and release. Their studies on the decerebrate cat hindlimb showed that muscle afferent signals cause the total stretch reflex to act in a manner similar to a stretched rubber band. In a subsequent study it was proposed that a qualitatively similar action occurs in man (Crago et al., 1976). Other investigators who compared the components of the stretch reflex in man and primates found the visco-elastic response of active muscle to be more important in resisting muscle stretch than the force produced by reflex muscle activity (Allum, 1975; Bizzi et al., 1978). Since this comparison has not been calculated for human ankle muscles, and since this information appeared essential for an understanding of human postural stability, the experiments described in this paper were performed.

The present experiments were designed to examine the involvement of active muscle mechanisms in the stretch reflexes of ankle extensors and flexors during disturbances to quiet standing. The results provided accurate estimates of the timing of reflex responses in ankle flexors and extensors from simultaneous recordings of surface EMG and motor unit potentials in these muscles. Once the order and timing of the short and medium latency parts of the reflexes was known, it was possible to compare the recorded total force with model responses of each of the three force components which contribute to the total reflex force. These components are: i) inertial reaction forces which are proportional to the disturbance acceleration; ii) visco-elastic responses of activated muscle to stretch and release; iii) actively generated reflex forces produced by short- and medium-latency muscle activity.

The analysis of the force which can be provided by each of these components suggests i) that in man short-latency (SL) segmental reflexes have a very minor role in correcting postural disturbances; ii) that any improvements in the stiffness of the stretch reflex are predominantly provided by medium-latency (ML) responses; and iii) because ML responses in ankle flexors and extensors are organized as a time locked coupled response, their net force effect is not inherently destabilizing to the standing posture, nor more important for resisting stretch than the muscles' visco-elastic responses. Our analysis indicates that alternative roles for ML reflexes should be investigated. One alternative is proposed in the discussion.

METHODS

Results were obtained from four healthy male subjects in the ages of 25–37, who gave their informed consent for the experiments. The subjects were asked to stand with locked knees and eyes closed on a platform which could be rotated about an axis colinear with the ankle joints. Applied disturbances to this posture were servo-controlled platform rotations with a terminated ramp profile of 70–90 msec duration and magnitude in the range 1–4 degrees. Thus a stretch velocity of 30–50 degrees/sec was imposed on the ankle muscles. Platform rotations were arranged in pairs. The first disturbance of the pair was of random direction, magnitude, and occurrence time, being either a toe-up or toe-down rotation; designated o-a and o-b respectively. The second disturbance of a pair, designated a-o or b-o, always returned the platform to its original level position 5 sec later.

Forces exerted by the subject on the platform were computed from the output of four strain gauge bridges, one mounted at each corner of the platform. Surface EMG and single unit potentials were recorded from tibialis anterior and triceps surae muscles in each subject's dominant leg with differentially amplified surface electrodes (Beckmann) and concentric needle electrodes (Disa). To improve visualization of reflex responses the EMG signals were filtered between 100 Hz and 2.5 KHz, full wave rectified, and smoothed.

RESULTS

Disturbances to quiet standing which rapidly alter the length of ankle extensor and flexor muscles produced a characteristic activity pattern in triceps surae (TS) and tibialis anterior (TA) muscles. Fig. 1 shows a typical example of muscle and force responses when the platform rotation flexed the ankles, stretching TS muscles and releasing the TA muscle, and then 5 sec later, extended the ankles by returning the platform to the level position. Fig. 2 shows a typical example of responses resulting from a TA stretch – TS release followed again after 5 sec by a TA release – TS stretch. These figures illustrate that changes in activity of the TS muscles, soleus and gastrocnemius, occurred at practically identical latencies. The response to TS stretch of the recorded motor units in gastrocnemius and soleus muscles are not similar at 40 msec after the onset of platform rotation in Figs. 1 and 2. The surface EMG latencies are similar. The dissimilarity in unit activity was probably due to differences between recruitment levels of the recorded units.

Stretch of the TS muscles produced three distinguishable phases of activity in TS and TA muscles. As expected, a short latency response (marked SL in Figs. 1 and 2) occurred in both TS muscles 40 msec after the onset of muscle stretch. The onset of platform rotation was taken as the first inflection on the force record. An arrow on each force record in Figs. 1 and 2 marks the inflection point. Short-latency responses which occur 40 msec following initial muscle stretch are compatible with the fast conducting muscle afferents exciting TS alpha motoneurons. The second phase of muscle activity was a medium-latency (ML) response at approximately 90 msec in the stretched TS muscle. After a further delay of 40 msec a second ML response, an activation of the released TA muscle, was observed. These latter phases of activity are marked by arrows labelled ML in Figs. 1 and 2.

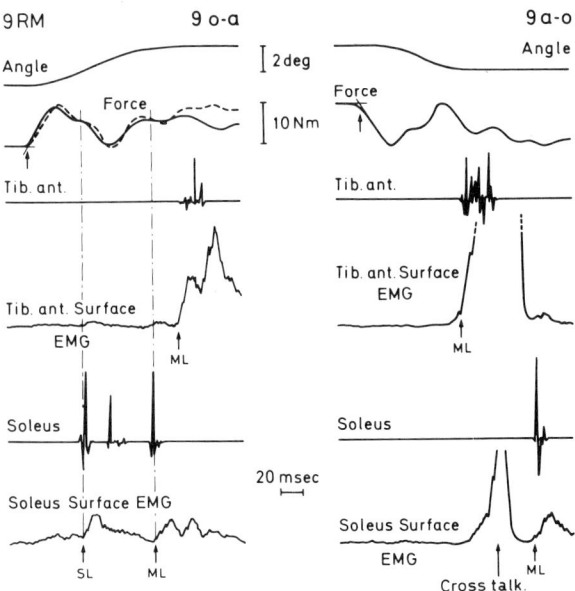

Fig. 1. Single unit and surface EMG responses to toe-up (9 o-a) and toe-down (9 a-o) 2.5 degree platform rotations. The responses of tibialis anterior and soleus muscles to two successive platform rotations are shown. The arrow on each force record marks the point taken as the onset of platform rotation and from which all latencies were measured. Arrows on the integrated surface EMG records mark the onset of short (SL) and medium (ML) latency phases of muscle activity. Note their coincidence with single unit responses. The dashed curve on the left force record is the superimposed curve from the right force trace with its polarity reversed.

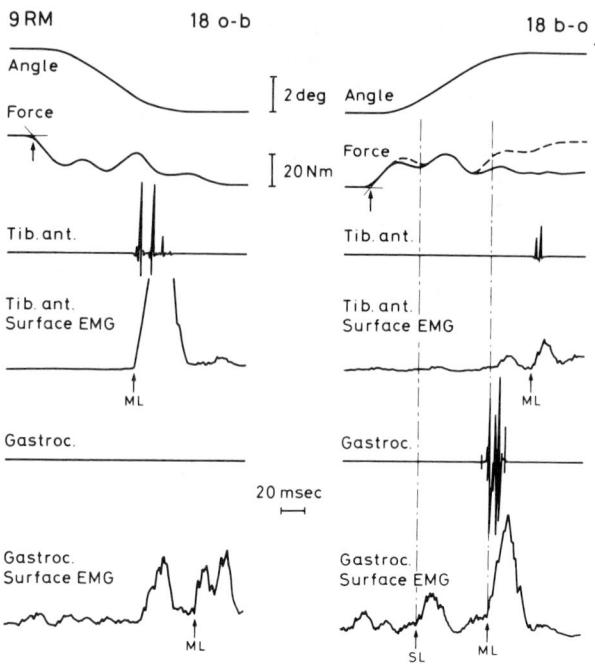

Fig. 2. Single unit and surface EMG responses to toe-down (18 o-b) and toe-up (18 b-o) 3.5 degree platform rotations. The response of tibialis anterior and gastrocnemius muscles to two successive platform rotations are shown. Arrows labelled SL and ML mark the phases of activity associated with simultaneous activation of single units. The first change in the gastrocnemius surface EMG record on toe-down platform rotation is cross talk from tibialis anterior and other ankle flexor muscles. A similar cross talk is observed in Fig. 1. The dashed force curve is the superimposed force curve for the toe-down displacement on the left with its polarity reversed.

Stretch of the TA muscle unexpectedly failed to produce an SL response in the TA muscles. Otherwise, the phases of muscle activity followed the general pattern observed for TS stretches. The first ML response occurred in the stretched TA muscles at 90 msec, and preceded by 40 msec the ML response in the released TS muscles.

By tapping on the TA tendon while subjects stood, it was demonstrated that the absence of an SL response in the TA muscle was not the result of insufficient stretch velocity. Neither single unit nor surface EMG responses were obtained from tendon taps. Prior lengthening of the TA muscle by rotating the platform over 6 degrees toe-down changed the TA activity pattern. In this situation a sudden TA stretch elicited an SL response in the TA muscle. Since Iles (1977) observed SL responses in TA when the ankle angle was 90 degrees but subjects were prone, it is possible that tonic descending vestibular signals modified the responsiveness of TA motoneurons when subjects were standing. The depressing effects of these inputs were presumably overcome during the ML responses or during SL responses when prior muscle stretch was applied.

The characteristic phases of activity in TS and TA muscles to sudden small rotations of the platform were confirmed in all four subjects in a total of 13 experiments. Results from three experiments with three subjects are illustrated in Fig. 3. Fig. 3 shows the latencies of the SL and ML responses plotted as a bar along the time axis. The bar's position represents the average latency for each response and its width is equal to 2 SD. The bar's height is equal to the frequency of occurrence of each type of

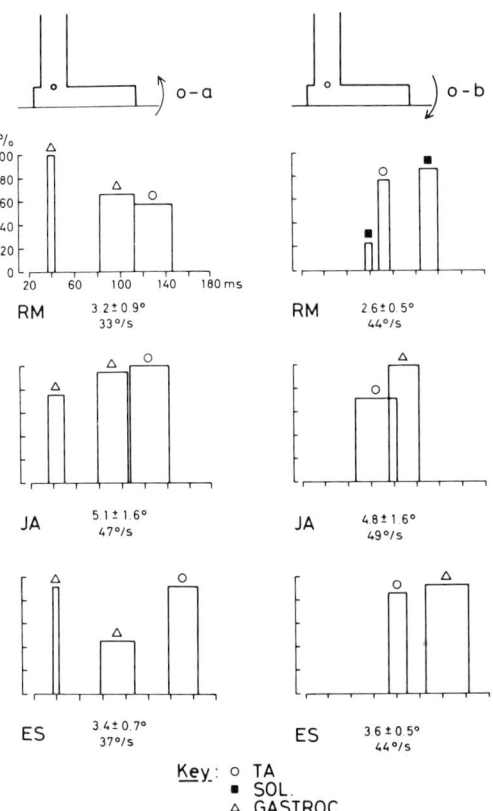

Fig. 3. Surface EMG response latency vs. frequency of occurrence plots. Data for 3 subjects are shown. The height of the columns in each plot is equal to the frequency of occurrence of either SL or ML responses expressed as a percentage of the total number of platform rotations. The position of the column along the time axis is determined by the mean value of the response latency and the column width is equal to two standard deviations of the latencies. A symbol above each column indicates the muscle involved, see key. Below each plot are listed mean amplitude and velocity of the platform rotations.

response expressed as a percentage of the approximately 40 platform rotations applied during an experiment. The frequency of occurrence vs. latency plots of Fig. 3 indicate that the SL and ML phases of muscle activity were not observed for each platform rotation. Generally the ML response in the stretched muscle appeared less frequently as the angle of platform rotation was reduced.

The delay between the ML response in the stretched muscle and that in the released muscle appeared to depend on the subject rather than the velocity or amplitude of platform rotation. For example, individual means and SD for this delay on stretching TA are 36 ± 8, 59 ± 10, 36 ± 14 and 25 ± 9 msec. Because the delay between the ML responses tended to be fixed for individual subjects, we described these responses as "time coupled" responses.

An opportunity to compare the force generated by SL muscle activity with the forces which result from inertial reactions and muscle visco-elasticity is provided by the presence of SL activity when TS is stretched and its absence when TA is stretched. This comparison is shown in Figs. 1 and 2 by plotting the force response to TA stretch with its polarity reversed as a dashed line alongside the force response to TS stretch.

The differences in the dashed and dotted lines should then reflect the force generated by the SL activity in the TS muscles; at least until the onset of ML activity some 50 msec later. Noticeably the comparison shows little, if any, difference between the force curves. Certainly no large force change is associated with the SL phase of TS muscle activity in either Fig. 1 or Fig. 2. Minor variations between the dashed and full curves can be attributed to differences between the acceleration and velocity profiles of each platform rotation and their ensuing effect on the inertial reaction force and muscle visco-elastic responses. In comparing the full force curve for a toe-up platform rotation with the dashed force curve for the succeeding (Fig. 1) or preceding (Fig. 2) toe-down rotation it was assumed that the visco-elastic properties of TS and TA muscles were constant during each pair of rotations. The platform rotations shown in Figs. 1 and 2 are less than 4 degrees which corresponds to a 2% change in TS muscle length (Gurfinkel et al., 1976). Initial length changes of this order do not change the visco-elastic properties of active cat soleus muscle (Joyce et al., 1969; Nichols and Houk, 1976).

The force produced by the coupled ML phases of muscle activity is affected by differences in the amplitudes of each ML phase of activity. This is shown by divergence of the full and dashed force traces in Figs. 1 and 2 once ML activity commenced. The divergence reflects differences in the amplitude of surface EMG activity in each muscle for toe-up and toe-down platform rotations. For example, the amplitude of the TA surface EMG is larger for toe-down displacements in Figs. 1 and 2. Correspondingly a greater number of single units were recruited in TA at medium latency for TA stretches than for TS stretches.

Both sets of force curves in Figs. 1 and 2 are not strongly modulated by the ML phases of muscle activity. The modulations that occurred after the onset of ML activity are less than force modulations prior to the onset of ML activity. As described above these prior forces result almost completely from inertial reaction and muscle visco-elastic forces. The lack of a significantly large force modulation following ML activity is due to the coupled nature of ML responses, i.e., the ML response in the agonist muscle is apposed 40 msec later by an activation of the antagonist muscle. Thus the net action produced by the coupled ML responses is weaker than the action of either ML response acting alone and no destabilizing postural response is observed in the force traces of Figs. 1 or 2.

DISCUSSION

According to the servomechanism theory of load compensation (cf. Marsden et al., 1972), and the Nichols and Houk (1976) proposal for compensation of muscle visco-elasticity, a sudden extension of a muscle should lead to an increase in muscle contraction consequent upon spindle lengthening. The contractile force which develops should resist the disturbance and compensate for the visco-elastic response of active muscle.

A sudden extension of the TS muscles regularly caused SL muscle activity at 40 msecs consistent with increased excitation of motoneurons. However, an extension of TA regularly resulted in no SL activity. It was demonstrated that the forces produced by ankle flexion and extension in the period between 40 and 90 msec after the onset of platform rotation were almost identical. Therefore, SL activity in TS must have produced only a small increase in force, i.e., the gain of monosynaptic and other polysynaptic connections involved in transmitting SL activity is presumably insufficient to either correct for the platform disturbance or improve the visco-elastic muscle response. Other investigators have suggested that SL reflex gains are small (Bizzi, 1978; Vallbo, 1973) or even transiently decreased in the event of a load disturbance (Newsom Davis and Sears, 1970).

If TS and TA muscles were to act as strict load-resisting devices during standing, when either was stretched their rapid contraction would cause subjects to topple over. This does not happen because negligible force is produced by SL activity and because stretch to either muscle produces ML activity in both ankle flexors and extensors which is "time-coupled" together. Although the first part of the coupled ML response in the stretched muscle could destabilize the posture by adding to the sway induced by sudden platform rotation, the second part of the coupled ML response in the released muscle would have the opposite effect and stabilize the posture. A smaller net change in force must result with coupled ML responses than would be produced if only one ML response in the stretched muscle were present.

If, as assumed, the force produced by SL activity is quite small, and that produced by the coupled ML response is not large, what is the function of these responses? A related question concerns the magnitude of the visco-elastic responses of active muscle compared to the active forces produced by muscle contractions. Are the visco-elastic forces alone sufficient to stabilize the posture for sudden ankle rotations of 4 degrees or less? If this is the case, how much linearization of muscle visco-elasticity is required to perform this function?

To facilitate a discussion of these questions, it is essential to consider the passive and active forces which contribute to the total reaction force on the platform. These are the inertial reaction force, the visco-elastic force of activated muscle, and the active forces due to reflex muscle activity. Each of these force components are modelled in Fig. 4. Excluded from this discussion are all forces occurring after 200 msec from the onset of platform rotation because these forces could be assisted by voluntary contractions (Melvill-Jones and Watt, 1971) and vestibular reflexes (Allum and Büdingen, 1979; Nashner, 1972).

The purpose of the model responses shown in Fig. 4 is to illustrate how each force component contributes to the total force response of the stretch reflex shown in Fig. 1. The inertial reaction forces are proportional to the angular acceleration of the platform which is equal to the double time derivative of the angle displacement. The force which the inertial reaction force produces is shown by the biphasic curve under the heading "total force" in Fig. 4.

Visco-elastic responses of active TS muscle to stretch are known for the cat (Joyce et al., 1969; Nichols and Houk, 1976). The data from Nichols and Houk for a low stimulation frequency of the cut ends of ventral root filaments are reproduced in Fig. 4 as model responses for human ankle flexors and extensors. Although the profiles for visco-elastic responses to stretch of human TS muscles are not known, the mechanical properties of active soleus muscle appear to be similar in cat and man (Bawa and Stein, 1976; Mannard and Stein, 1973). In Fig. 4 it has been assumed that visco-elastic responses for human ankle flexors and extensors are similar to those for cat soleus muscle with constant stimulation of 8/sec. In addition to assuming that the mechanical properties of human muscles are similar to these of cat soleus muscle it has also been assumed in Fig. 4 that prior to platform rotation ankle flexors and extensors

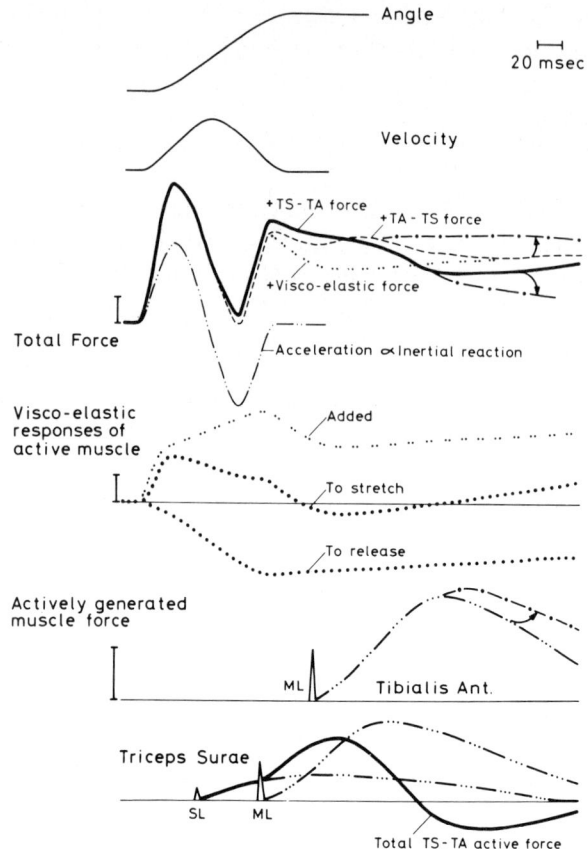

Fig. 4. Contributions of inertial reaction forces, visco-elasticity of activated muscle, and reflex muscle activity to the stretch reflex. The schema indicates how the three components of the stretch reflex can be added together to produce total force responses shown in Fig. 1. Further details are given in the discussion.

were equally active and that reflex muscle activity does not significantly alter the visco-elastic response to stretch or release. Based on these assumptions the resisting force of the ankle extensors (TS) and the compressive force of ankle flexors (TA) for a toe-up platform rotation can be modelled by the responses to stretch and release given in Fig. 4. The total visco-elastic force to a sudden platform rotation is the sum of the magnitudes of the stretch and release response since both stretched and released muscles resist the rotation. The total visco-elastic force is shown by the "added" curve in Fig. 4.

The force produced by inertial reactions and muscle visco-elasticity is labelled "+ visco-elastic force" in Fig. 4. The visco-elastic force was scaled so that this combined curve matched over the first 90 msec the force response in Fig. 1. With respect to the effect of visco-elasticity acting across a joint in an agonist-antagonist muscle pair, two comments can be made. Firstly, the added visco-elastic response is considerably less non-linear than either the agonist or the antagonist responses. Secondly, the effect of inertia is to reduce damping. To fully compensate for the reduced damping the SL response would have to produce a large increase in force just as the inertial reaction force reverses direction. It has been demonstrated that this does not occur.

Consequently, compensation for visco-elastic responses during stretch is functionally useful if the stretch acceleration is small, for example, during sway oscillations of quiet standing (Gurfinkel et al., 1976).

The active force produced by the SL activity in TS muscles and the coupled ML responses was modelled using the twitch contraction curves for human soleus muscles (Bawa and Stein, 1976). It was assumed that the short phases of SL and ML activity (see Figs. 1 and 2) produced contractions with force profiles similar to those of twitch contractions. The amplitudes of the SL and ML force profiles were adjusted so that the total force response (i.e. with inertial reaction and the visco-elastic response) approximated the force responses of Fig. 1. As shown in the lower part of Fig. 4 a small amplitude was used for the SL response in TS, because a minor effect was demonstrated for this response. The ML active forces were modelled with larger amplitudes to produce force curve of significant amplitude. The resultant active force curve is shown in Fig. 4 by the full curve among the group of curves labelled "actively generated muscle force". When this force is added to the inertial reaction and visco-elastic response the full curve labelled "+ TS-TA" in Fig. 4 is obtained. Following a similar procedure for stretch of TA the dashed curve labelled "+ TA-TS" is obtained. Variations in the TA twitch contraction profile and amplitude shown by the arrow in Fig. 4 produced corresponding alterations in the total force curves as indicated by the arrows on the "+ TA-TS" and "+ TS-TA" force curves.

Three aspects of the modelled active forces generated by the ML phases of activity may be noted. Firstly, the coupled ML response produces sufficient force to correct for the most significant non-linearity in the "added" visco-elastic response, i.e., at the termination of the platform rotation. Secondly, the pulse of force produced by the coupled ML response is smaller in amplitude but has a faster rise and fall time than individual twitch contractions for either TS or TA. Thirdly, the net action of coupled ML response is not destabilizing. In fact, small variations in amplitude and rise time of the TA force response can produce a total force response which is almost constant once platform rotation ceased (upward arrow on total force trace in Fig. 4) or slightly corrective (downward arrow in Fig. 4).

This component analysis has shown that when the contributions of active muscle mechanisms to the stretch reflex are considered for an agonist-antagonist pair acting across a joint the contributions are significantly different from those observed for a single muscle. Conclusions about the role of reflex mechanisms in load compensation should also reflect these differences. Visco-elasticity is observed to be far less non-linear with two opposing muscle groups than for a single muscle. ML responses are organized as an excitatory response in the stretched muscle followed 40 msec later by an excitatory response in the released muscle. Presumably the time-locked coupled response is the outcome of interacting reflex arcs in ankle flexors and extensors which are differentially affected by stretch in the agonist muscles and release of the antagonists.

The amount of force provided by the reflex muscle activity matches the functional requirements of the perturbed foot and ankle musculature. The SL response is weak, but in this period the visco-elastic response is spring-like and inertial reaction forces predominate. The coupled medium response provides a moderate amount of force which helps to linearize the visco-elastic response after the completion of platform rotation. When the magnitudes of the visco-elastic and active reflex force components of the stretch reflex were compared in Fig. 4, the visco-elastic force was required to be considerably greater in order to account for the force responses of Fig. 1. In order to account for the postural stability of subjects following 4 degrees or less platform rotations it was not necessary to invoke a major compensating role for reflex muscle activity. However, before concluding that reflex muscle activity has only a minor role in load compensation or the linearization of muscle visco-elasticity in man it is necessary to compare these contributions to the stretch reflex over a wider range of stretch velocities. The stretch velocities used in this study were equal in resting lengths/sec to those normally imposed on isolated triceps surae muscle in decerebrate cats (Matthews, 1972; Nichols and Houk, 1976).

If the visco-elastic forces predominate in the stretch reflex, what is the function of the short small pulse of force provided by the coupled ML response? The pulse of force could be used by the central nervous system as a test pulse. The function of such a test pulse would be to provide the CNS with information about the size, velocity and load of the disturbance, about muscle spindle static and dynamic sensitivity, and the current gating of segmental reflexes. The information which results from the test pulse could be compared by supraspinal structures with the assumed status and gain of the peripheral nervous system before the CNS plans long-latency muscle activity which would return the body into an upright position.

SUMMARY

The aim of this study was to examine the stabilizing role of stretch reflexes acting on the ankle flexors and extensors. For this purpose, the reaction forces and the EMG activity of the ankle musculature were recorded while a platform on which the subjects stood was quickly rotated about the ankles. The force responses correcting the postural disturbances were divided into three components:

1. Forces generated by short (40 msec) and medium (90 or 130 msec) latency muscle activity.
2. Visco-elastic forces of the activated muscles.
3. Inertial reaction forces.

The analysis of the time course and amplitude of each of these components suggests that short-latency segmental reflexes have a minor role in correcting postural disturbances. Second, any improvements in the stiffness of the stretch reflex are predominantly provided by medium-latency responses. Third, because medium-latency responses in ankle flexors and extensors are organized as a time-locked coupled response, their net action stabilizes the standing posture. The force provided by medium latency reflex muscle activity approximately equalled that provided by the inherent visco-elasticity of activated muscles.

ACKNOWLEDGEMENTS

The experimental part of this research was supported by the Deutsche Forschungsgemeinschaft SFB 70; data analysis was supported by the Swiss National Science Foundation grant 3.079-0.76. We thank B. Frei, C. Wüest, H. Kapp and A. Müller for their technical assistance, H. Hauser for typing the manuscript, R. Emch for her graphic assistance and D. Savini for his photographic work.

REFERENCES

Allum, J.H.J. (1975) Responses to load disturbances in human shoulder muscles: the hypothesis that one component is a pulse test information signal. *Exp. Brain Res.*, 22: 307–325.
Allum, J.H.J. and Büdingen, H.J. (1979) Coupled stretch reflexes in ankle flexors and extensors during quiet standing. In preparation.
Bawa, P. and Stein, R.B. (1976) Frequency response of human soleus muscle. *J. Neurophysiol.*, 39: 788–793.
Bizzi, E., Dev, P., Morasso, P. and Polit, A. (1978) The effect of load disturbances during centrally initiated movements. *J. Neurophysiol.*, 41: 542–556.
Burke, R.E. (1972) Control systems operating on spinal reflex mechanisms. In *Neurosciences Research Symposium Summaries, Vol. 9*, E.V. Evarts, E. Bizzi, R.E. Burke, M. DeLong and W.T. Thach (Eds.). MIT Press, Cambridge, pp. 60–85.
Crago, P.E., Houk, J.C. and Hasan, Z. (1976) Regulatory actions of human stretch reflex. *J. Neurophysiol.*, 39: 925–935.
Forssberg, H., Grillner, S. and Rossignol, S. (1977) Phasic gain control of reflexes from the dorsum of the paw during locomotion. *Brain Res.*, 132: 121–139.
Grillner, S. (1972) The role of muscle stiffness in meeting the changing postural and locomotor requirements for force development by the ankle extensors. *Acta physiol. scand.*, 86: 92–108.
Gurfinkel, V.S., Lipshits, M.I., Mori, S. and Popov, K.E. (1976) The state of stretch reflex during quiet standing in man. In *Understanding the Stretch Reflex, Progress in Brain Research, Vol. 44*. S. Homma (Ed.). Elsevier. Amsterdam, Oxford, New York, pp. 473–486.
Iles, J.F. (1977) Responses in human pretibial muscles to sudden stretch and to nerve stimulation. *Exp. Brain Res.*, 30: 451–470.
Joyce, G.C., Rack, P.M.H. and Westbury, D.R. (1969) The mechanical properties of cat soleus muscle during controlled lengthening and shortening movements. *J. Physiol. (Lond.)*, 204: 461–474.
Lundberg, A. (1966) Integration in the reflex pathway. In *Muscular Afferents and Motor Control, Nobel Symposium I*. R. Granit (Ed.). Almqvist and Wiksell, Stockholm, pp. 275–305.
Mannard, A. and Stein, R.B. (1973) Determination of the frequency response of isometric soleus muscle in the cat using random nerve stimulation. *J. Physiol. (Lond.)*, 229: 275–296.
Marsden, C.D., Merton, D.A. and Morton, H.B. (1972) Servo action in human voluntary movement. *Nature*, 238: 140–143.
Matthews, P.B.C. (1972) *Mammalian Muscle Receptors and their Central Actions*. Arnold, London.
Melvill-Jones, G. and Watt, D.G.D. (1971) Observations on the control of stopping and hopping movements in man. *J. Physiol. (Lond.)*, 219: 709–727.
Nashner, L.M. (1972) Vestibular postural control model. *Kybernetik*, 10: 106–110.
Newsom Davis, J. and Sears, T.A. (1970) The proprioceptive reflex control of the intercostal muscles during their voluntary activation. *J. Physiol. (Lond.)*, 209: 711–738.
Nichols, T.R. and Houk, J.C. (1976) Improvement in linearity and regulation of stiffness that results from actions of stretch reflex. *J. Neurophysiol.*, 39: 119–142.
Prochazka, A., Westerman, R.A. and Ziccone, S.P. (1977) Ia afferent activity during a variety of voluntary movements in the cat. *J. Physiol. (Lond.)*, 268, 423–448.
Prochazka, A., Sontag, K.-H. and Wand, P. (1978) Motor reactions to perturbations of gait: proprioceptive and somesthetic involvement. *Neurosci. Lett.*, 7: 35–39.
Vallbo, A.B. (1973) The significance of intramuscular receptors in load compensation during voluntary contractions in man. In *Control of Posture and Locomotion*. R.B. Stein, K.G. Pearson, R.S. Smith and J.B. Redford (Eds.). Plenum Press, New York, pp. 211–226.

The Role of Vision in the Control of Posture During Linear Motion

A. BERTHOZ[1], M. LACOUR[2], J.F. SOECHTING[1] and P.P. VIDAL[1]

[1]*Laboratoire de Physiologie du Travail du CNRS, Département de Physiologie Neurosensorielle, 75005 Paris; and* [2]*Laboratoire de Psychophysiologie, Faculté des Sciences, 13397 Marseille (France)*

The purpose of this paper is to review some findings concerning the role of vision in the control of posture which have been obtained in recent years. We shall first present a short summary of the main ideas which were proposed by other authors and then describe some results obtained in our laboratories concerning more specifically the effect of vision on postural control during linear horizontal or vertical motion of either body or visual surround.

FREQUENCY DOMAIN OF THE INFLUENCE OF VISION IN POSTURAL CONTROL

The role of vision in the control of stance, locomotion, and compensation of various sensory defects such as labyrinthectomy has been known since the pioneer works of early researchers such as Flourens, Purkinje, De Cyon, Magnus etc. It was also described by neurologists such as Thomas (1940) at the beginning of this century. These early works have been reviewed recently (Dichgans and Brandt, 1978; Courion and Jeannerod, this volume, chapter VC4). The specific role of vision in the control of upright posture was demonstrated by many authors (Travis, 1945; Edwards, 1946; Rademaker and Ter Braak, 1948; Wapner and Witkin, 1950; De Haan, 1959) and it was suggested to contribute to locomotory pattern (Davis and Ayers, 1972). In the last ten years methods of frequency analysis of biomechanical parameters of balance, together with recordings of muscular activity, have helped to obtain information concerning the precise dynamics of visual-postural interaction (Baron and Litvinenkova, 1968; Gurfinkel and Elner, 1971; Cernacek and Jagr, 1972; De Wit, 1972; Gantchev et al., 1973; Litvinenkova and Hlavacka, 1973; Walsh, 1974; Bles and De Wit, 1975).

We owe to Gibson (1952, 1958, 1966) a most stimulating theory integrating vision in the overall function of proprioception. In line with this theory the stabilizing role of whole field complex visual scenes was elegantly demonstrated by Lee and his collaborators (Lee and Aronson, 1974; Lee and Lishman, 1974) who showed: a) that small movements of a swinging room surrounding a standing human induce postural oscillations unnoticed by the subject which are particularly great in infants (a finding later confirmed by Brandt et al., 1976, for roll motion, with the additional information that the effect peaks around 3–4 years of age, decreasing up to 16 years); and b) a clear difference between active and passive relative motion between subject and visual surround.

The importance of visual cues was also evaluated by Amblard and Cremieux (1976). In standing humans, destabilization could be induced by stroboscopic illumination at low frequency (3 Hz). This result was interpreted according to the hypothesis that when visual cues are available, they dominate other vestibular or proprioceptive cues. Velocity information is required, however, to efficiently stabilize posture. Stroboscopic light at low frequency only provides position cues, hence, inducing a poor spatial reference.

A major contribution to these problems was offered by an extensive series of experiments performed by Dichgans with several coworkers, who studied thoroughly the stabilizing or destabilizing role of vision for circular movements. These authors placed this problem in the frame of visual-vestibular interaction proposing that "vision improves the speedometer function of the labyrinth in the low frequency range" (Dichgans et al., 1973).

A convenient method to describe the dynamic contribution of vision is to induce body oscillations by moving sinusoidally visual scenes in various directions. This frequency analysis was performed by several groups. Fig. 1 shows a summary of data obtained by authors who compared the subjective sensation of self motion (vection) induced by such scenes with the objective measure of body oscillations induced by

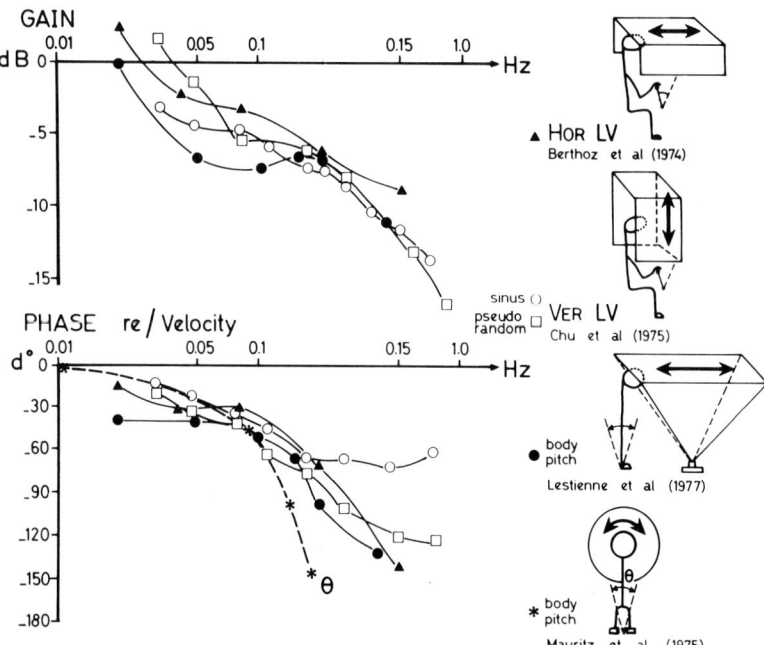

Fig. 1. Comparison between the frequency response curves for linear vection and for postural oscillations induced by moving visual scenes. Gain and phase of linear vection (LV) induced by horizontal and vertical sinusoidal (or pseudo-random) motion of the visual surround. Gain is the ratio between subjective magnitude estimation of LV and visual surround velocity. (From Berthoz et al., 1974; and Chu and Young, 1975.) These subjective evaluations of LV are compared with body pitch induced by visual scenes moving in the horizontal plane (Lestienne et al. 1977) or around the line of sight (Mauritz et al. 1975). In this case the gain is the ratio between body angle θ and visual surround velocity.

visual surround motion. The magnitude estimation curves for horizontal linear vection induced by horizontally moving visual scenes obtained in a seated subject (Berthoz et al., 1975), are plotted and compared with those obtained by Chu and Young (1975) with both sinusoidal and pseudo-random visual surround velocity. Gains have been plotted on an arbitrary scale in order to show the general shape of the slope. In both cases gain and phase indicate vection magnitude estimates vs. image velocity. The two sets of curves are strikingly close although they do not account for a number of important features such as forward-backward or up-down asymmetries which have also been reported by these authors.

It is interesting to note that the curves which represent body pitch induced by linearly moving visual surround (Lestienne et al., 1977) fall very close to the vection curves. This coincidence does not however preclude a dissociation between the two effects (pitch and vection) and subjects may also report vection without postural change, or change their posture without reporting vection. The phase curve obtained by Mauritz et al. (1977) for body pitch induced by roll vection has been plotted for comparison. It also falls within the same range supporting the *general statement of a role of vision in the low frequency range (0–0.2 Hz)* which has been suggested both by linear and circular vection experiments.

Similar conclusions have been reached by Talbott (1974) and Talbott and Brookhart (1976, 1978) by studying the postural response of the dog to horizontal fore and aft motion of visual surround and of the platform on which the animals were standing. They also conclude that the visual component of sensory information may play a dominant role in balance. Inducing a dynamic conflict between information derived by visual pathways and that derived via other sensory pathways by submitting the dog to conflicting visual and body motion, they found a "variable visual gain." For instance, if visual field motion is at 0.3 Hz and platform motion is at 1 Hz, the 0.3 Hz component of the dog's destabilization is more than 10 times that which would be expected on the basis of the response to a 0.3 Hz motion of visual surround given in isolation. In a recent series of papers which are still in press, these authors have demonstrated that peripheral vision is essential in these effects as indicated by both studies using lesions of visual cortex (Mirka et al., 1978) and selective occlusion of parts of the visual field.

In our laboratory the difference between visual field motion given in isolation and/or in combination with linear acceleration has been investigated in standing humans and we shall now describe briefly the essential conclusions of this work.

DYNAMIC EFFECT OF VISION ON LOW FREQUENCY PERTURBATION OF POSTURE

In a previous series of experiments (Lestienne et al., 1977) we have studied the effect of moving visual scenes on the control of posture. These experiments have demonstrated that linear motion of a visual scene induced, in standing humans, postural readjustments observable as a body pitch in the direction of image velocity. The amplitude of the change in body "vertical" was related to the velocity of the visual scene, but also more surprisingly, to its spatial frequency. We suppose that these visual effects may be of even greater magnitude during combined body and visual motion.

The experimental set-up which was designed to verify this assumption is shown in

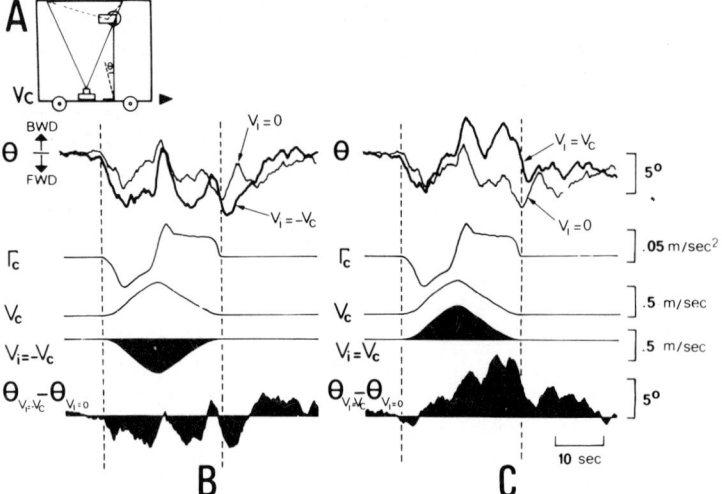

Fig. 2. Influence of linear motion of visual surrounds in response to postural perturbation. A) Experimental set-up: The subject stands erect on a moving platform. The angle of head and body pitch is measured by a potentiometer fixed to the head by means of a rod and crank attachment. An image moving at velocity Vi is projected on the screen which is placed horizontally above the subject as described in Lestienne et al. (1977). Lightweight cardboard blinders obliterate the lateral and inferior aspects of the visual field and permit the subject only a view of the screen S. This ensemble is placed on a cart which can move at velocity Vc. Triangular analog signals are used to drive the cart. The traces in B and C show the average pitch θ of 8 subjects in three experimental conditions. In all three conditions the cart moved in the backward (BWD) direction with a peak velocity Vc = 0.05 m/sec and peak acceleration Γc = 0.05 m/sec². The light traces in B and C were obtained when the image was stationary with respect to the cart (Vi = 0). The dark trace in B was obtained when velocity was in a direction opposite to cart velocity (Vi = −Vc) and thus stationary relative to the ground. The dark trace in C, when image velocity was in the same direction as cart velocity (Vi = Vc). The difference of pitch θ in the two conditions for B and the two conditions for C are shown (lower records). (From Soechting and Berthoz, 1979.)

Fig. 2A. The subjects stood erect on a platform attached to a mobile cart which was driven by servo-controlled torque motors. It was enclosed on all sides. The subject wore lightweight cardboard blinders which blocked the lateral and inferior aspects of their visual field and permitted only restricted views of the cart's ceiling. Details concerning this method are given by Soechting and Berthoz (1979). The visual scene consisting of a black-and-white checkerboard pattern was projected on the ceiling using a film projector and mirror. Accurate control of both image and cart velocity allowed a very precise definition of these stimuli. The subject's pitch angle forward or backward was calculated by potentiometric methods assuming that for these small angles the body was equivalent to an inverted pendulum. This hypothesis was discussed in Lestienne et al. (1977) and has been shown to be a rough but sufficient approximation in the perspective of this work.

Fig. 2B shows records of body pitch θ when the cart accelerated suddenly backward (see acceleration trace Γc) which was induced by a triangular waveform of velocity Vc. Three conditions of visual surround motion were used. In the first case the velocity of visual images was zero (Vi = 0). The visual surround was moving together with the cart, which is equivalent to riding inside a closed illuminated vehicle. This condition was used as a control by comparison with the case when the visual scene motion was a) opposite to cart velocity (Vi = −Vc) as in "natural" motion (Fig. 2B), and b) equal to cart velocity (Vi = Vc) which is a "conflict" situation (Fig. 2C). The records of θ in Fig. 2B, which are averaged curves, show that a clear difference (statistically significant), exists between the two visual conditions Vi = 0 (thin lines) and Vi = −Vc (dark lines), although in both cases θ seems to follow the general waveform of acceleration Γc.

The difference between the two traces has been plotted in the lowest curve ($\theta_{Vi=-Vc} - \theta_{Vi=0}$). It is in the direction of visual image motion (compare dark triangle of image velocity Vi with darkened surface of curve). A similar comparison was made when a conflict was induced between image and cart motion. Fig. 2C shows records obtained when Vi = 0 and Vi = Vc. The influence of visual surround motion is very striking and once more in the direction of image motion. Statistical calculations and precise quantitative measures of peak body pitch have revealed that the effect of visual surround is about twice as great when visual surround and body motion are combined, a finding very much in accordance with the results of Talbott and Brookhart (1976, 1978).

The frequency spectrum of cart and image velocity stimuli given in this series of experiments did not exceed about 0.2 Hz. Consequently, the conclusions drawn from these results are in keeping with the general idea that vision plays a stabilizing, or destabilizing role in the low frequency range. We shall now however, describe some results which have challenged this conclusion and suggest that during rapid postural movements vision may be an essential component of the sensory systems contributing to adequate motor activity.

CONTRIBUTION OF VISION TO RAPID MOTOR REACTIONS DURING POSTURAL PERTURBATIONS

An experiment has been designed to demonstrate conclusively that manipulation of the visual surround, at the time of a rapid postural perturbation, could induce changes in early motor responses which were not predicted by most authors (Nashner and Berthoz, 1978).

The main idea of the experimental set-up, which is shown schematically in Fig. 3, was to suppress visual cues about body motion selectively and only at the very precise moment of a transient backward acceleration of a platform (cart) on which human subjects were standing. For this purpose a sliding box providing a visual surround was fixed to the frame of the cart. The velocity and position of the box could be controlled so that it could either a) move together with the cart, in this case subjects had normal (N) visual cues concerning motion of their body inside the cart (this is equivalent to situation Vi = 0 in Fig. 2), or b) be stabilized (S) with respect to the head through a servo-controlled motor so that at the time of onset of the acceleration, minimal relative motion would occur between head and box, thus suppressing visual motion cues.

Surprisingly visual stabilization induced a striking increase in body pitch and a strong decrease in the earliest motor reaction measured by electromyogram in the triceps surae. This result was at odds with all previous results obtained during free fall in man (Matthews and Whiteside, 1960; Greenwood and Hopkins, 1976, 1977; Melvill-Jones and Watt, 1971), in the monkey (Lacour et al., 1978), or in the cat

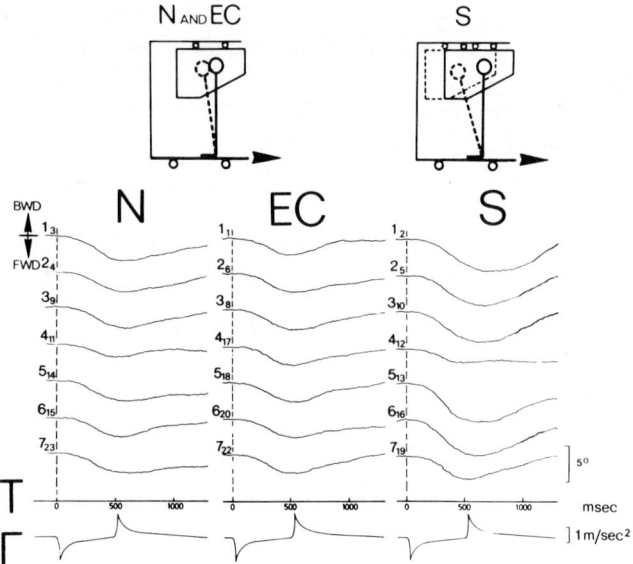

Fig. 3. Influence of stabilized visual surround on body pitch following a perturbation of posture. The experimental set-up has been described by Nashner and Berthoz (1978). The subject is standing on a moving platform which accelerates backward with an acceleration profile Γ. The visual surround is the inside of a sliding box with a high spatial frequency pattern, which is also controlled by a motor and can be either moving together with the cart (N) or translating together with the head (stabilization of visual surround with respect to head) (S). A potentiometer records horizontal fore and aft head motion and controls the movement of the box. A computer and a drive and hold device synchronizes platform (cart) and box motion so that in the case of stabilized condition S, the visual information about body motion is only suppressed at the onset of the perturbation. Body pitch, (backward: BWD; forward: FWD) measured by the potentiometer, is shown in the normal (N), eyes closed (EC) and stabilized (S) conditions. Numbers on each record give the order of the randomized presentation in the experimental sequence. Note the small increase in body pitch in EC condition and the large destabilizing effect of visual stabilization. (From Vidal et al., 1979b.)

(Watt, 1976). All of these authors had compared motor responses occurring previous to, or at the time of landing, and had concluded that there was an absence of vision contribution because repeating the experiments with both eyes closed or in darkness induced similar responses.

In order to study more precisely this question we have recently repeated (Vidal et al., 1978, 1979a, b) our experiments and compared the three conditions as defined above: normal (N), eye closure (EC) and stabilized vision (S). We shall now summarize briefly the main findings of these experiments.

Fig. 3 shows recordings of head displacement (forward and backward) used as a rough estimate of body pitch during a sudden backward acceleration of the cart on which the subject is standing. When the subject has his eyes closed, a small increase of destabilization is evidenced. During visual surround stabilization (S) a striking increase of body pitch occurs.

Quantitative evaluation in which difference in maximum body pitch in the three conditions, N, EC, and S, gives about 5% mean difference (SD, 13%) between N and EC and about 40% difference between S and N (SD, 2%). The underlying muscular response was measured in the triceps surae by standard surface electromyography (EMG). Some example of records are shown on Fig. 4 in which detected and inte-

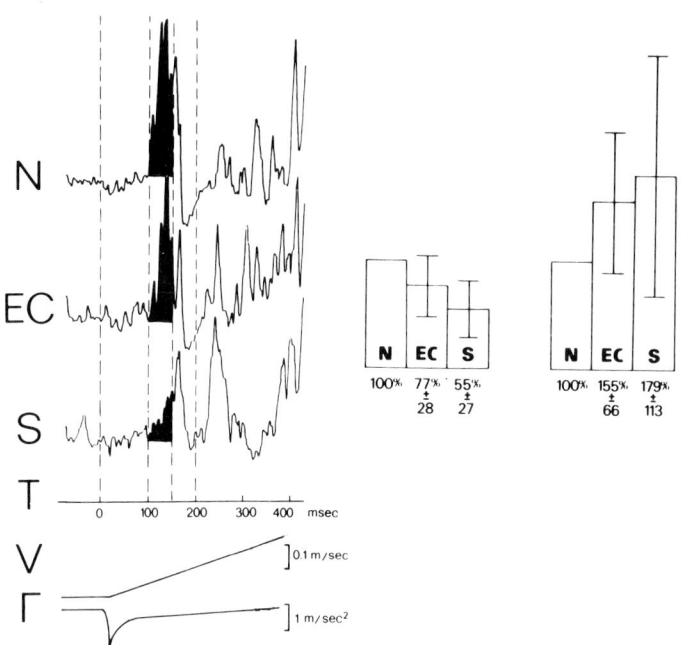

Fig. 4. Influence of stabilization of visual surround on muscle response to a sudden perturbation. Standing subjects are placed in the experimental situation described in Fig. 3. The records show the integrated electromyographic (EMG) activity in the extensor of the ankle (triceps surae), before and immediately after a sudden backward motion of the cart on which the subject is standing (mean value for 3 trials of one subject). The parameters of motion are given by the time scale (T), the velocity of the cart (Vc) and its acceleration (Γc). N, EC and S refer to "normal", "eyes closed" and "stabilized" conditions as defined in Fig. 3. The mean and SD of the surface EMG between 100 and 150 msec (histogram) after the onset of the linear acceleration are shown on the right for 7 subjects. Values are given for each trial and each subject by comparison with the value in the N case (From Vidal et al., 1979b).

grated EMG was obtained in the three conditions, N, EC and S, in the same experimental conditions as those of Fig. 3. Eye closure does not change significantly the first motor response which occurs at a latency of about 100 msec. This large phasic response which is necessary for an efficient maintenance of posture following the perturbation is indeed similar in both cases and most authors have taken argument of this result to negate any role of vision at this early stage. However, as shown by Nashner and Berthoz (1978), and exemplified again in the bottom record of this figure, when visual information about head motion is selectively removed, this early motor reaction is decreased and in this case nearly suppressed.

The statistical measurements shown in the right part of the figure show the average amplitude of the decrease in early EMG component (100–150 msec) which is measured by calculating the surface of integrated EMG. Notice that parallel with this decrease, an increase in amplitude of the subsequent EMG appears in the S condition. This finding is to be compared with the results of Amblard and Crémieux (1976) who, in steady state low frequency stroboscopic illumination, obtained a strong destabilization. The essential point of our result is that visual cues about body motion are only removed at the onset of the perturbation and that other changes in visual parameters are modified.

Having thus demonstrated that visual motion cues have a not yet fully understood, but definitely powerful role in the release of an adequate motor pattern in these conditions of dynamic equilibrium, we were tempted to see whether a similar modification could be induced during free fall. As the basic characteristics of motor activity during free fall had been established in the baboon by Lacour et al. (1978) we designed an experiment to verify whether visual stabilization could also modify early motor responses to which had been attributed a pure vestibular (otolithic) origin.

MODIFICATION OF EARLY MOTOR RESPONSES BY VISUAL STABILIZATION DURING FREE FALL IN THE MONKEY

The experimental set-up for the study of free fall in the baboon (Papio-Papio) is shown in Fig. 5. Total height of the fall was 90 cm. The details of the experiments are described in Lacour et al. (1978). In addition the head of the monkey could be placed in a box which could either provide total darkness, or be illuminated by a small light bulb. A checkerboard pattern of black and white squares whose properties had been studied in Lestienne et al. (1977) was placed inside the box. Fig. 5, modified from Lacour et al. (1978), shows the basic pattern of muscular activity encountered in the extensors of the leg during the fall with normal visual cues (monkey falling in the laboratory without box). It must be emphasized that in this case no landing occurs and mainly vestibular, visual and probably also pressure tactile cues are signalling perti-

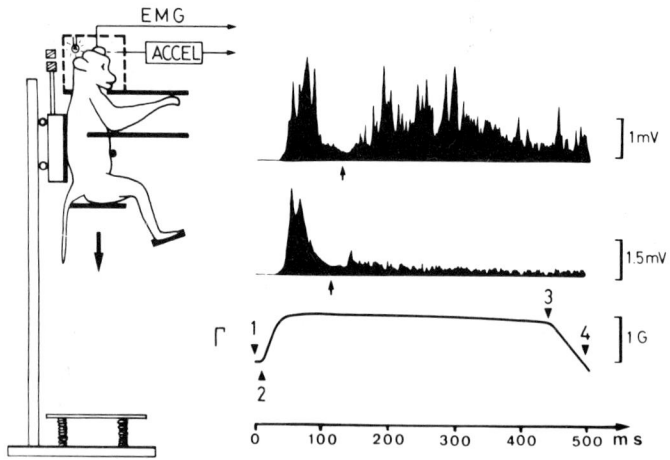

Fig. 5. Experimental set-up and basic responses for the study of motor reactions in the baboon during free fall. A) Experimental set-up: the monkey (Papio-papio) is seated on a chair which can slide down on rails. The chair is maintained in an initial position by electro-magnets. The head is restrained by four hard plates preventing major lateral or vertical head movements and the body is in a plaster cast. The fall occurs either when the monkey is: a) viewing the laboratory (normal condition (N)), b) when a box is surrounding the head (complete darkness (D)), c) when a light illuminates the inside of the box which is covered with a high spatial frequency checkerboard pattern. (Visual surround is stabilized with respect to the head (S).) Landing is smoothed by elastic springs. B) Typical detected and integrated electromyogram from the soleus muscle during free fall (average of ... trials for ... monkeys). The records show the responses obtained during the first (top record) and the 10th (bottom record) fall of an experimental session. Γ is the vertical acceleration record. Arrows indicate initial trigger (1) actual release of the chair (2), and deceleration (3,4).

(Modified from Lacour et al., 1978.)

nent information concerning the fall (practically no neck flexion or extension was allowed by restraint). The top record shows two components of EMG. The first one was shown by Lacour et al. (1978) to be composed of two subcomponents. The long-lasting second component disappears when repeating the fall (lower record is taken after repetition of the fall). It is probably related to the preparation of landing and is suppressed by the monkey. The first rapid component was shown to disappear with bilateral labyrinthectomy by Lacour et al. (1978) and consequently thought to be essentially of otolithic origin. Similar conclusions were drawn by Watt (1976) concerning the motor discharge in gastrocnemius of the cat which occurs about 70 msec after onset of free fall, by Melvill-Jones and Watt (1971) for the 75 msec latency discharge in human gastrocnemius after sudden release of subjects hanging to the ceiling, and Greenwood and Hopkins (1976, 1977) for the responses occurring at about 72 msec in quadriceps, 82 msec in soleus and 81 msec in tibialis of their human subjects during free fall. These last authors attributed the earliest response to a startle-like response as it also disappears when the subject makes a voluntary release. Influence of blindfolding after landing in the cat is studied by Lewis et al. (1979).

The results of comparison between free fall in normal (N) visual conditions, in total

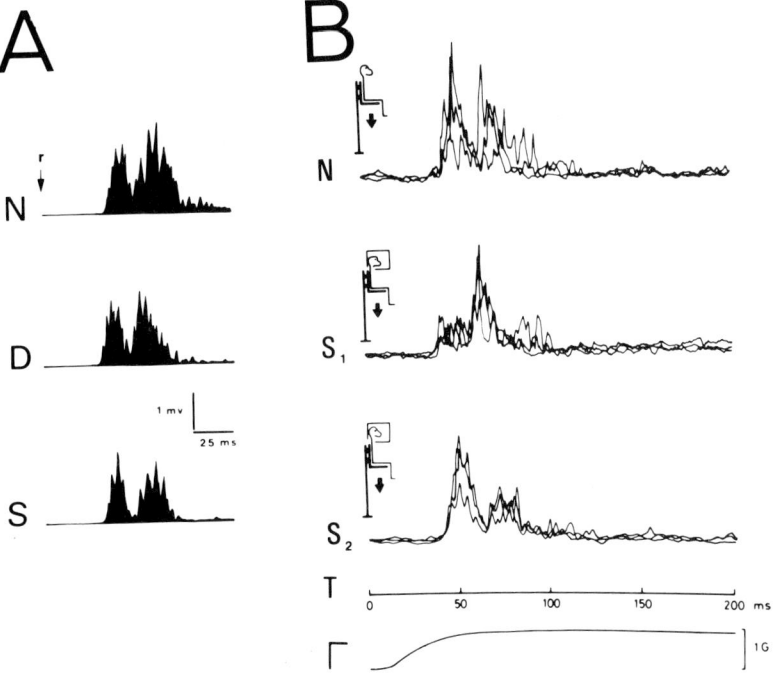

Fig. 6. Decrease of initial motor response in the soleus muscle of the monkey with visual stabilization. Detected and integrated electromyograms of the soleus muscle in the falling baboon (see Fig. 5 for methods) in three conditions: N: normal visual condition (falling in the laboratory); D: darkness; S: stabilized vision. A) Average records from one monkey (90 trials) showing the moderate decrease in the early response in darkness and the strong decrease in S condition. Arrow indicates instant of release which is about 4–8 msec before actual onset of acceleration. B) Typical records of soleus EMG in another monkey for three conditions: N: normal visual condition (falling in the laboratory); S1: three superimposed records in which the first component earliest peak of activity was nearly suppressed; S2: three superimposed records of trials in which both components of earliest peak were decreased but the second was more clearly suppressed. Bottom trace indicates vertical acceleration record Γ. (From Vidal et al., 1979.)

darkness (D) and with visual stabilization are reported in Fig. 6. Recordings were made in various groups of muscles (splenius capitis, soleus, tibialis anterior, quadriceps femoris). In all of them a statistically significant decrease of amplitude (as much as 30%) of the earliest component occurs (Vidal et al., 1979a). An example concerning the soleus is shown in Fig. 6A. In complete darkness (D) the overall amplitude of the EMG response is also decreased although, in some instances, the response is not significantly different from the N condition.

Fig. 6B shows a detailed description of typical records in which the decrease of EMG during stabilization affected either the first or the second subcomponent of the response. This suppression is clearly as powerful as the one described above and in other publications (Nashner and Berthoz, 1978; Vidal et al., 1978b) for the standing man during postural perturbation.

This result demonstrates that, even if there is an important otolithic contribution to the genesis of these early motor responses, the visual surround motion when eyes are open is an important factor in at least determining the gain of the released motor activity. It is not possible at this stage to determine more precisely the nature of this influence.

DISCUSSION

The above results suggest that during linear motion vision is essential not only to regulate the general tone of muscles in the low frequency range, but also to release adequate motor reactions during sudden changes in posture. They also show that the dynamic role of vision has to be studied in conditions when it is a relevant cue as discussed by Granit (1978) on a more general basis.

In spite of the extensive amount of interest raised by the study of postural control recently, the pathways and underlying mechanisms by which vision influences the general motor system are still mostly unknown although a number of possible pathways are suggested by animal investigations (see other contributions in this volume). Most of the available neuronal data concerning visual-vestibular interaction concerns angular rotation and little information has been obtained for linear motion except from the work of Daunton and Thomsen (1976).

The idea that changes in visual surround may modify efferent motor reactions within short latencies is confirmed in the present results, and is compatible with the known latencies after which visual stimulation can influence discharge of neurons in brain stem structures (Baker et al., 1976; Azzena et al., 1977; Keller and Precht, 1979).

It could be argued that the "stabilized vision" condition is artificial and places the subject or the animal in a discordant set of sensory inputs which leads to an unspecific "hold reaction" and a generalized suppression of motor activity. Although this is a possibility, we would rather prefer another interpretation and propose that at the time of the rapid selection of an adapted motor pattern, the release of the adequate motor synergies is dependent upon an expected pattern of congruent sensory inputs (Nashner and Berthoz, 1978). Thomas (1940) had proposed a theory by which once a motor pattern would be learned using a number of sensory cues, any of these would be sufficient to trigger this pattern. This theory would also predict that absence of an essential cue could prevent the onset of the reaction.

We know in vestibulo-ocular physiology, that the sudden appearance of a visual surround, stable with respect to the head of man, monkey or cat, can induce a very rapid suppression of a previously existing vestibular nystagmus (Takemai and Cohen, 1974a, b), which is mediated by visual pathways which through the cerebellum exert a powerful inhibition of the vestibulo-ocular reflex. Our results on posture suggest the existence of a visuo-motor equivalent to the pursuit system which, in the case of eye movements, is responsible for this suppression.

Posture, in this frame, cannot be differentiated from movement and any posture would have to be studied as preparatory to a movement. The question of why the motor reactions have been found to be similar with eyes closed and eyes open cannot yet be precisely answered. It could be proposed that closing the eyes or blindfolding induces a change in the gains of the various segmental sensory motor loops, allowing the maintenance of the same motor performance, but with very different "settings" (see also Gurfinkel and Shik, 1973). However Greenwood and Hopkins (1976, 1977) have tested segmental reflex gains and found no significant difference between eyes open and blindfolding conditions. This point deserves further clarification but we would favor an interpretation which would hypothesize such reorganizations.

Granit has made, in the Sofia Symposium on motor control, a statement which seems particularly adequate to this point: "If a measured response is uninfluenced by the loss of one source of information, it is hardly possible to conclude that this particular source, under any circumstances, could not have contributed to it. Remaining pathways to the central interpretor suffice to throw sentient circuits into action." The logical error, so elegantly denounced in this text, was probably the cause of an underestimation of the role of vision in the rapid control of posture.

SUMMARY

A review of the main theories concerning the influence of visual surround motion on the control of posture is presented. Gains and phases obtained by measuring either linear vection or body pitch induced by sinusoidal motion of visual scenes are compared. These curves seem to confirm the general statement that vision contributes to postural stabilization in the low frequency range of body movements (0 to 0.1–0.2 Hz). However, new experimental findings are described which show successively:

1. Direction-specific influence of visual surround motion when visual and body motion are combined, demonstrating also the enhanced effect of vision when active postural tasks are used.

2. The action of vision on the early motor responses to postural perturbation (within 100–150 msec of onset of perturbation). In this study eye closure is compared with visual stabilization which induces a 40% increase in body pitch.

3. A strong decrease of early motor responses during free fall in the monkey, which is shown to occur with visual stabilization within 50–100 msec from the onset of the fall. These results are interpreted as implying a role of vision in the general setting of the parameters of motor responses during posture and movement.

REFERENCES

Amblard, B. and Cremieux, J. (1976) Role of visual motion information in the maintenance of postural equilibrium in man. *Agressologie* 17C: 25–36.

Azzena, G.B., Tolu, E. and Mameli, O. (1978) Responses of vestibular units to visual input. *Arch. ital. Biol.*, 116: 120–129.

Baker, J., Gibson, A., Glickstein, M. and Stein, J. (1976) Visual cells in the pontine nuclei of the cat. *J. Physiol. (Lond.)*, 225: 415–433.

Baron, J.B., Litvinenokova, V. (1968) Activité tonique posturale et appareil visuel. *C.R. Soc. Biol. (Paris)*, 162: 2098–2103.

Berthoz, A., Pavard, B. and Young, L.R. (1975) Perception of linear horizontal self motion induced by peripheral vision (linear-vection). Basic characteristics and visual-vestibular interactions. *Exp. Brain Res.*, 23: 471–489.

Bles, W. and De Wit, G. (1975) Study of the effects of optic stimuli on standing. *Agressologie*, 17C: 1–5.

Brandt, Th., Wenzel, D. and Dichgans, J. (1976) Die Entwicklung der Visuellen Stabilisation des aufrechten Standes beim Kind: Ein Reifezeichen in der Kinderneurologie. *Arch. Psychiat. Nervenkr.*, 223: 1–13.

Cernacek, J. and Jagr, J. (1972) Posture, vision et dominance motrice. *Agressologie*, 13C: 101–105.

Daunton, N.G. and Thomsen, D.D. (1976) Otolith – visual interaction in single units of cat vestibular nuclei. *Neurosci. Abstr.*, II, 2: 1057, no. 1526.

Davis, W.J. and Ayers, J.L. (1972) Locomotion: Control by positive feedback optokinetic responses. *Science*, 177: 183–185.

De Haan, P. (1959) The significance of optic stimuli in maintaining the equilibrium. *Acta Oto-laryngol. (Stockh.)*, 50: 109–115.

De Wit, G. (1972) Optic versus vestibular and proprioceptive impulses measured by posturometry. *Agressologie*, 13B: 75–79.

Dichgans, J. and Brandt, T. (1978) Visual-vestibular interaction: effects on self-motion perception and postural control. In *Handbook of Sensory Physiology*. H.L. Teuber et al. (Eds.) in press.

Dichgans, J., Schmidt, C.L. and Graf, W. (1973) Visual input improves the speedometer function of the vestibular nuclei in the goldfish. *Exp. Brain Res.*, 18: 319–322.

Edwards, A.S. (1946) Body sway and vision. *J. exp. Psychol.*, 36: 526–535.

Gantchev, S., Dunev, S. and Draganova, N. (1973) On the spontaneous and induced body oscillations. In *Motor Control*. A.A. Gydikov, N.T. Taukov and D.S. Kósazov (Eds). Plenum Press, New York, pp. 179–194.

Gibson, J.J. (1952) The relation between visual and postural determinants of the phenomenal vertical. *Psychol. Rev.*, 59: 370–375.

Gibson, J.J. (1958) Visually controlled locomotion and visual orientation in animals. *Brit. J. Psychol.*, 493: 182–194.

Gibson, J.J. (1966) *The Senses as Perceptual Systems*. Houghton Mifflin, Boston, MA.

Granit, R. (1978) The case for relevance in sensorimotor physiology. *TINS*, Elsevier, 17–18 July 1978.

Greenwood, R. and Hopkins, A. (1976) Muscle responses during sudden falls in man. *J. Physiol. (Lond.)*, 254: 507–518.

Greenwood, R. and Hopkins, A. (1977) Monosynaptic reflexes in falling man. *J. Neurol.*, 40: 448–454.

Gurfinkel, V.S. and Elner, A.M. (1971) Visual information processing and control of motor activity. *Int. Symp., Bulg. Acad. Sciences, Sofia*, 331–335.

Gurfinkel, V.S. and Shik, M.L. (1973) The control of posture and locomotion. In *Motor Control*, A.A. Gydikov et al. (Eds.). Plenum Press, New York, pp. 217–234.

Lacour, M., Xerri, C. and Hugon, M. (1978) Muscle responses and monosynaptic reflexes in the falling monkey: the role of the vestibular system. *J. Physiol. (Paris)*, 74: 427–438.

Lee, D.N. and Aronson, E. (1974a) Visual proprioceptive control of standing in human infants. *Percept. Psychophys.*, 15: 529–532.

Lee, D.N. and Lishman, J.R. (1974b) Visual proprioceptive control of stance. *J. Human Movement Stud.*, 1: 87–95.

Lestienne, F., Soechting, J.F. and Berthoz, A. (1977) Postural readjustments induced by linear motion of visual scenes. *Exp. Brain Res.*, 28: 363–384.

Litvinenkova, V and Hlavacka, F. (1973) The visual feed-back gain influence upon the regulation of the upright posture in man. *Agressologie*, 14C: 95–99.

Maeda, M., Magherini, P.C. and Precht, W. (1977) Functional organization of vestibular and visual inputs to the neck and forelimb motoneurons in the frog. *J. Neurophysiol.*, 40: 225–243.

Matthews, B. and Whiteside, T.C.D. (1960) Tendon reflexes in free fall. *Proc. roy. Soc. B.*, 153: 195–204.

Mauritz, K.H., Dichgans, J. and Hufschmidt, A. (1977). The angle of visual roll motion determines displacement of subjective visual vertical. *Percept. Psychophys.* 22: 557–562.

Melvill-Jones, G. and Watt, D.G.D. (1971) Muscular control of landing from unexpected falls in man. *J. Physiol. (Lond.)*, 219: 729–737.

Mirka, A., Talbott, R.E. and Brookhart, J.M. (1978) Role of the visual cortex in visual feedback for postural control. *Neurosci. Abstr.*, 4: 638, no. 2041.

Nashner, L. and Berthoz, A. (1978) Visual contribution to rapid motor responses during postural control. *Brain Res.*, 150: 403–407.

Paillard, J. (1955) *Réflexes et Régulations d'Origine Proprioceptive chez l'Homme. Etude Neurophysiologique et Psychophysiologique.* Arnette, Paris.

Rademaker, G.G.J. and Ter Braak, J.W.G. (1978) On the central mechanism of some optic reactions. *Brain*, 71: 48–76.

Soechting, J.F. and Berthoz, A. (1979) Role of vision in the dynamic control of posture. Direction-specific effect of moving visual surrounds on the reaction to postural disturbance. *Exp. Brain Res.*, 20, 4, 1–11.

Takemori, S. and Cohen, B. (1974a) Visual suppression of vestibular nystagmus in Rhesus monkeys. *Brain Res.*, 72: 203–212.

Takemori, S. and Cohen, B. (1974b) Loss of visual suppression of vestibular nystagmus after flocculus lesions. *Brain Res.*, 72: 213–224.

Talbott, R.E. (1974) Modification of postural response of the normal dog by blind folding. *J. Physiol. (Lond.)*, 243: 309–320.

Talbott, R.E. and Brookhart, J.M. (1976) Variable postural response of the dog to visual stimuli depending upon the environmental context. *Neurosci. Abstr.*, II, 2: 534, no. 762.

Talbott, R.E. and Brookhart, J.M. (1978) Predicted characteristics of the intact dog's postural control system based on a perturbation analysis of response during blind folding and optokinetic stimulation. *J. Neurophysiol.*, in press.

Thoden, U., Dichgans, J. and Savidis, Th. (1977). Direction-specific optokinetic modulation of monosynaptic hind limb reflexes in cats. *Exp. Brain Res.*, 30: 155–160.

Thomas, A. (1940) *Equilibre et Equilibration.* Masson, Paris.

Travis, R.C. (1945) An experimental analysis of dynamic and static equilibrium. *J. exp. Psychol.*, 35: 216–234.

Vidal, P.P., Gouny, M. and Berthoz, A. (1978) Role de la vision dans le déclenchement de réactions posturales rapides. *Arch. ital. Biol.*, 116: 281–291.

Vidal, P.P., Lacour, M. and Berthoz, A. (1979a) Contribution of vision to rapid muscle responses in the monkey during free fall. *Exp. Brain Res.*, in press.

Vidal, P.P., Millanvoye, M. and Berthoz, A. (1979b) Difference between eye closure and visual stabilization in the control of posture in man. In preparation.

Walsh, E.G. (1974) Standing man, slow rhythmic tilt, importance of vision. *Agressologie*, 14C: 79–85.

Wapner, S. and Witkin, H.A. (1950) The role of visual factors in the maintenance of body balance. *Amer. J. Psychol.*, 63: 385–408.

Watt, D.G.D. (1976) Responses of cats to sudden falls: an otolith originating reflex assisting landing. *J. Neurophysiol.*, 39: 257–265.

Direction-Specific Vestibular and Visual Modulation of Fore- and Hindlimb Reflexes in Cats*

U. THODEN, J. DICHGANS, M. DOERR and Th. SAVIDES

Neurologische Klinik, Abteilung für Neurophysiologie, Universität Freiburg i. Br., 7800 Freiburg i. Br. (F.R.G.)

INTRODUCTION

Modulation of spinal reflexes by neck and vestibular inputs is known since the original work by Magnus (1924). Large field visual motion information, mainly from the periphery of the visual field also exerts influence on posture (Dichgans et al., 1972, 1976). More specifically electrical stimulation of the frog's optic nerve bilaterally yields short latency responses in neck, forelimb and hindlimb motoneurons (Maeda et al., 1977). With quasi physiological motion stimulation it was shown in a previous paper that a large visual stimulus rotating about the animals line of sight tonically modulates the excitability of hindlimb extensor and flexor motoneurons (Thoden et al., 1977). This reflex modulation in terms of directional specificity and amplitude was similar to that induced by body tilt.

Since in the cat vestibular second order neurons are influenced by optokinetic stimuli (Bauer, 1976; Keller and Precht, 1978) a common pathway for both inputs from the brain stem to the spinal cord was assumed (Thoden et al., 1977). The data of Maeda et al. (1977), however, suggest a separate tectospinal and vestibulospinal route in the frog.

A decreasing vestibulospinal influence was found by Gernandt and Gilman (1959) and Thoden et al. (1978) for the forelimb as compared to the hindlimb. Similar results were obtained for the neck-to-spinal input (Wenzel et al., 1978).

In this paper the craniocaudal organization of visuospinal influences was investigated and compared to vestibulospinal effects during body tilt.

METHODS

The experiments were performed in 27 adult cats weighing between 2 and 3.5 kg, anesthetized with a mixture of oxygen, nitrogen and Fluothane during surgery. After surgery animals were immobilized for recording by repeated injections of gallamine triethiodide (Flaxedil®) and artificially respirated after tracheal cannulation. During the recording the animals were under slight halothane anesthesia.

---

*Supported by Sonderforschungsbereich Hirnforschung (SFB 70) der Deutschen Forschungsgemeinschaft (DFG).

In order to study monosynaptic reflexes afferents were electrically stimulated and mass potentials recorded from the efferent motoneuron. For anatomical reasons two different techniques were used for the fore- and hindlimb.

1. For *forelimb reflexes* the cervical spinal cord was exposed by laminectomy from C4 to Th1 in 15 cats. The dura was opened and the dorsal roots C6–C8 were cut. The proximal stump was then put on bipolar stimulation electrodes. Ventral roots were left intact. For recording the following nerves were prepared: the *right* deep radial nerve (DR) to the lateral head of the triceps, paradigmatic for an extensor muscle, and the right mixed ulnar nerve (ULN) to the flexor carpi ulnaris and the ulnar head of the flexor profundus digitorum as flexor muscles. These nerves were then mounted on bipolar silver-silverchloride recording electrodes (interpolar distance 3 mm).

2. For *hindlimb reflexes* the branch of the tibial nerve to the gastrocnemius-soleus muscle (GS) and the deep peroneal nerve (DP) to the tibialis anterior and the extensor digitorum longus muscles were prepared in the *right* popliteal fossa in another 12 cats. This time the peripheral nerve served for electrical stimulation by bipolar silver-silverchloride electrodes and not for recording. Recordings were obtained from the ventral root after the spinal segments L4–S2 were exposed by laminectomy. The dura was opened and ventral roots of the segments L6–S2 were filamented at their exit from the dura, then severed and the proximal stump attached to a silver-silverchloride recording electrode.

The cord and the nerves were covered by a pool of warmed liquid paraffin at constant body temperature. Animals were placed in a modified Horsley-Clark headholder on a tilt table (Fa. Tönnies) with the body fixed at spinal processes.

For steady state *positional stimulation of the vestibular otoliths* the animal was tilted 45° either to the left or to the right around its longitudinal body axis.

For *optokinetic stimulation*, a large visual display rotated about the animals line of sight, 25 cm in front of the eyes at a speed ranging between 2–14 degrees/sec. The surface of the disk was painted with randomly distributed large colored dots that covered 32% of its total surface. The disk subtended 130° of visual angle viewed binocularly.

Both stimuli were combined in physiological and non-physiological directional combinations. Control experiments were performed in an upright position without display motion before each test trial.

For evaluation, 32 reflex responses induced either by stimulation of the hindlimb nerves or the dorsal roots C4–Th1 were averaged. For stimulation, single rectangular 0.2 msec impulses were applied every 3–5 sec. Stimulus intensities were tested at the reflex threshold and 1.3, 1.6, 2, 3 and 5 times the threshold value (T). Stimulation always started 30 sec after the animal was placed in the test position or 30 sec after onset of optokinetic stimulation. The modulation of the average response amplitudes produced by the different head and body tilts and optokinetic stimuli was expressed as a percentage of the average response amplitude in the upright position.

RESULTS

Tonic reflex modulation during animal tilt

Ipsilateral tilt produces an enhancement of monosynaptic extensor reflexes, whereas contralateral tilt produces an enhancement of monosynaptic flexor reflexes (Figs. 1 and 2, on the right).

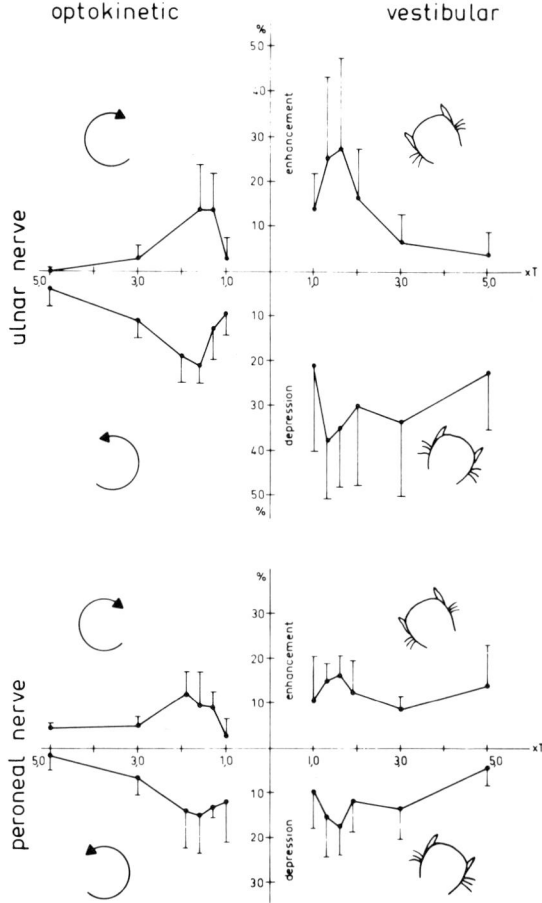

Fig. 1. Reflex modulation of flexor muscles in relation to relative intensity of electrical stimulation (threshold = 1) for pure optokinetic (left) and vestibular (right) stimulation. Average values of 5 cats with SD.

The reverse procedure causes inhibition. The maximal amount of modulation varies between 15 and 50%. It is usually reached at roughly 1.6 T. The amount of reflex modulation induced by vestibular stimuli falls off at higher stimulus intensities, although some modulation can still be detected at up to 5 T. The strongest reflex modulation is found during ipsilateral tilt in monosynaptic extensor reflexes.

The vestibular modulation of monosynaptic flexor and extensor reflexes is generally stronger in the forelimb as compared to the hindlimb. This difference is more pronounced during ipsilateral tilt.

Optokinetic reflex modulation

As described elsewhere, optokinetic stimuli in the erect animal also produce a direction specific reflex modulation (Thoden et al., 1977). In agreement with its behavioral significance an enhancement is caused by contralateral display rotation for extensor reflexes and ipsilateral rotation for monosynaptic flexor reflexes. Standard deviations are similar to the ones observed with stimulation by exclusive body tilt stimulation (Figs. 1 and 2, on the left).

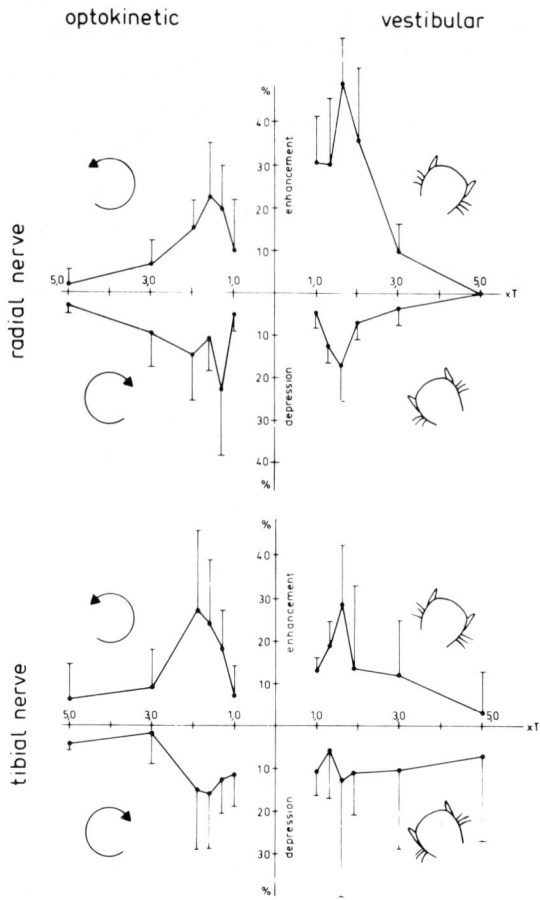

Fig. 2. Reflex modulation of extensor muscles in relation to relative intensity of electrical stimulation (threshold = 1) for roll motion (left) and animal tilt (right). Average values of 5 cats with SD.

Similar amounts of optokinetic reflex modulation are found in the fore- and hindlimb. This contrasts to the craniocaudally decreasing vestibulospinal influence. The directional anisotropy of vestibular reflex modulation through body tilt is not observed with the exclusive application of large field motion of the visual scene.

Reflex modulation by different body tilt angles and different velocities of display rotation

Vestibular modulation of monosynaptic reflexes at 1.6 T starts at tilt angles of about 10° to both sides and increases with larger tilt angles. For the hindlimb the slope of the curve relating body tilt to reflex modulation is steeper for extensor than for monosynaptic flexor reflexes (Fig. 3B,D). Tilt angles exceeding 45° could not be applied for technical reasons.

The *optokinetic modulation* of monosynaptic reflexes depends on the angular velocity of roll motion. It reaches its maximum between 3.5–7 degrees/sec and falls off at velocities exceeding 10 degrees/sec. It seems that in both extensor and flexor monosynaptic reflexes the optokinetic reflex enhancement is larger than the inhibition (Fig. 3A,C).

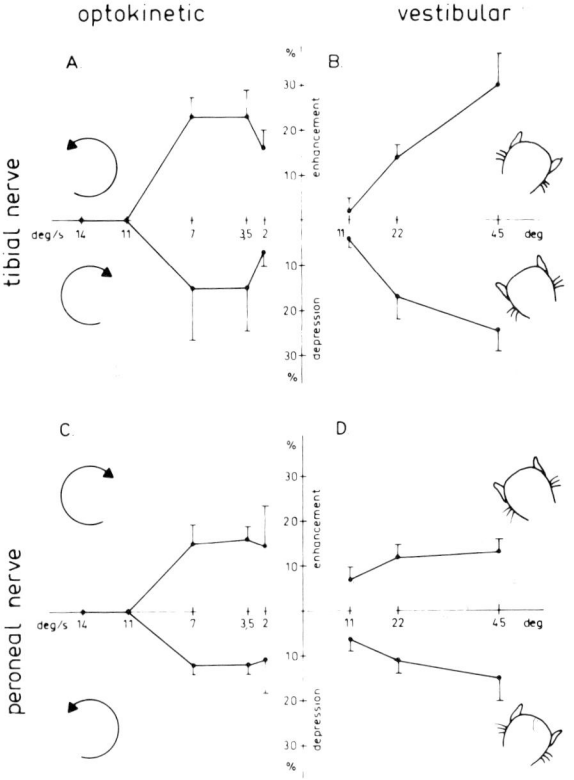

Fig. 3. Reflex modulation by different body tilt angles (right) and different velocities of display rotation (left) for reflexes of 1.6 × the threshold intensity.

Simultaneous optokinetic and vestibular stimulation

For the hindlimb, the physiological combination of both stimuli, e.g. body tilt to the right combined with counterclockwise display rotation results in modulation curves similar to the one obtained with exclusive body tilt (Fig. 4). With the stimulus parameters used summation of optokinetic and vestibular effects, if at all present, could only be detected near the threshold intensity of electrical stimulation as defined by control experiments. But this difference was not significant with the sample taken.

With a *non-physiological combination* of stimulus directions again no interaction could be demonstrated. It seemed as if the reflex modulation whenever using combined optokinetic and vestibular stimulation was exclusively determined by the vestibular signal.

DISCUSSION

The demonstration of an optokinetically induced spinal reflex modulation fits behavioral experiments in man (Dichgans et al., 1972, 1976; Lee and Aronson, 1974; Lestienne et al., 1977). These demonstrate that the body's center of gravity in upright stance may be displaced towards the direction of motion of an ample visual stimulus. By comparing postural sway with eyes open vs. eyes closed it can be shown that physiologically vision stabilizes posture. A frequency analysis of the visual effect

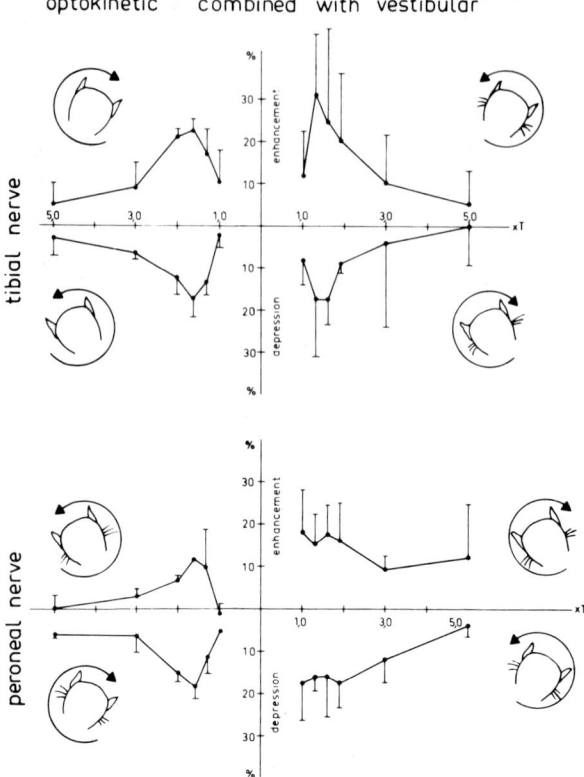

Fig. 4. Simultaneous optokinetic and vestibular stimulation. On the right physiological combination, e.g., body tilt to one side combined with counterclockwise display rotation, compared with the non-physiological combination on the left.

suggests that the domain of visual influences upon body posture is below 1 Hz, whereas that of the vestibulospinal effects covers higher frequencies (Dichgans et al., 1976; Mauritz et al., 1975; Lestienne et al., 1977). These behavioral data fit the results presented in Fig. 3, where the optokinetic modulation reaches saturation at very low velocities.

The underlying mechanism is behaviorally adaptive. An animal inadvertently swaying or falling towards one side is not only protected by vestibulospinal reflexes but also by external visuospinal feedback.

The modulation of spinal reflexes by tilt reflects vestibulospinal connections as previously described (Anderson and Gernandt, 1956; Lund and Pompeiano, 1968; Roberts, 1970; Wilson and Yoshida, 1968). Gernandt investigated the cranio-caudal organization of the decending connections using single shock vestibular nerve stimulation. This results in an early large bilateral response spike, which is immediately followed by a smaller double wave in the forelimb. At the lumbar level only the late response is seen. This indicates more direct, stronger and denser connections influencing the cervical than lumbar segments (Gernandt and Gilman, 1959; Erulkar et al., 1966). The quantification of response modulations by vestibulospinal mechanisms presented above substantiates this finding and allows for a comparison with effects of other postural control mechanisms. The neck input to spinal reflexes is also known to decrease in craniocaudal direction (Wenzel et al., 1978). In so far the craniocaudal distribution of both systems seems similar, but the neck afferents counteract the described vestibulospinal effects (Erhard and Wagner, 1970; Rosenberg and Lindsay, 1973).

Finally, a prevalence of descending connections to the forelimb is also known for the tectospinal tract (Nyberg-Hansen, 1964). The fact that the difference in modulation curves for fore- and hindlimbs cannot be demonstrated for visuospinal mechanisms at first glance suggests a different pathway, as proposed by Maeda et al. (1977). A bias of vestibulospinal information flow by optokinetic stimuli, however, would be expected, since it was shown for semicircular canal dependent neurons in the vestibular nuclei of goldfish (Dichgans et al., 1973), cat (Bauer, 1976; Keller and Precht, 1978) and monkey (Henn et al., 1974) and also for otolith dependent neurons of the cat (Daunton and Thomsen, 1976). Similar conclusions may be drawn from our experiments with combined stimulation of both input channels. They show neither summation nor mutual inhibition, but rather a predominance of vestibular effects. Similar results have been obtained in recording from the vestibular nuclei for canal stimulation in the monkey (Waespe and Henn, 1977). Our data are not sufficient to definitely answer this question. The visuospinal pathway remains unknown until visual effects are directly recorded from identified neurons.

SUMMARY

The influence of large visual stimuli rotating about the animals' line of sight on electrically elicited monosynaptic fore- and hindlimb reflexes was compared with the effect of a 45° static body tilt. Both extensor and flexor reflexes were tested in the forelimb (branch to lateral head of m. triceps and ulnar nerve) and the hindlimb (tibial and peroneal nerve).

Static body tilt enhances ipsilateral monosynaptic extensor reflexes up to 30% and inhibits flexor reflexes to a lesser degree. Contralateral reflexes show reversed effects. Vestibular reflex modulation is stronger in the fore- than in the hindlimb.

Optokinetic roll motion stimuli also produce direction-specific reflex modulations, but with the reverse directional characteristic: equal effects of body tilt, e.g., to the right and pattern motion to the left. The amount of optokinetic modulation is very similar in fore- and hindlimb reflexes. Whereas it is largely identical to vestibular modulation in hindlimb reflexes, the latter is stronger in forelimb reflexes. If combined, the effects of both stimuli do not summate.

The results are interpreted as being due to a vestibulospinal biasing of alpha motoneurons. It is not clear whether optokinetic effects exclusively use these pathways.

REFERENCES

Anderson, S. and Gernandt, B.E. (1956) Ventral root discharge in response to vestibular and proprioceptive stimulation. *J. Neurophysiol.*, 19: 524–543.

Bauer, R. (1976) *Optisch-vestibuläre Interaktion an Einzel-neuronen der Vestibulariskerne und des Vestibulo-cerebellum bei der Katze*. Biol. Diss., Freiburg.

Berthoz, A. and Anderson, J.A. (1971) Frequency analysis of vestibular influence on extensor motoneurons. II. Relationship between neck and forelimb extensors. *Brain Res.*, 34: 376–380.

Daunton, N.G. and Thomsen, D.D. (1976) Otolith-visual interactions in single units of cat vestibular nuclei. *Neurosci. Abstr.*, II: 1057, no. 1526.

Dichgans, J., Schmidt, C.L. and Graf, W. (1973) Visual input improves the speedometer function of the vestibular nuclei in the goldfish. *Exp. Brain Res.*, 18: 319–322.

Dichgans, J., Held, R., Young, L. and Brandt, Th. (1972) Moving visual scenes influence the apparent direction of gravity. *Science*, 178: 1217–1219.

Dichgans, J., Mauritz, K.H., Allum, J.H.H. and Brandt, Th. (1976) Postural sway in normals and atactic patients: analysis of the stabilizing and destabilizing effects of vision. *Agressologie*, 17C: 15–24.

Ehrhardt, K.J. and Wagner, A. (1970) Labyrinthine and neck reflexes recorded from spinal single motoneurons in the cat. *Brain Res.*, 19: 87–104.

Erulkar, S.D., Sprague, J.M., Whitsel, B.L., Dogan, S. and Janetta, P.J. (1966) Organization of the vestibular projection to the spinal cord of the cat. *J. Neurophysiol.*, 29: 626–664.

Gernandt, B.E. and Gilman, S. (1959) Descending vestibular activity and its modulation by proprioceptive, cerebellar and reticular influences. *Exp. Neurol.*, 1: 274–304.

Henn, V., Young, R.L. and Finley, C. (1974) Vestibular nucleus units in alert monkeys are also influenced by moving visual fields. *Brain Res.*, 71: 144–149.

Keller, E.L. and Precht, W. (1978) Persistence of visual response in vestibular nucleus neurons in cerebellectomized cat. *Exp. Brain Res.*, 32: 591–594.

Lee, D.L. and Aronson, E. (1974) Visual proprioceptive control of standing in human infants. *Percept. Psychophys.*, 15: 529–532.

Lestienne, F., Soechting, J. and Berthoz, A. (1977) Postural readjustments induced by linear motion of visual scenes. *Exp. Brain Res.*, 28: 363–384.

Lund, S. and Pompeiano, O. (1968) Monosynaptic excitation of alpha motoneurons from supraspinal structures in the cat. *Acta physiol. scand.*, 73: 1–21.

Maeda, M., Magherini, P.C. and Precht, W. (1977) Functional organization of vestibular and visual inputs to neck and forelimb motoneurons in the frog. *J. Neurophysiol.*, 40: 225–243.

Magnus, R. (1924) *Körperstellung*. Julius Springer, Berlin.

Mauritz, K.H., Dichgans, J., Allum, J.H.J. and Brandt, Th. (1975) Frequency characteristics of postural sway in response to self-induced and conflicting visual stimulation. *Pflügers Arch.*, 355: R 95, no. 189.

Nyberg-Hansen, R. (1964) The location and termination of tectospinal fibres in the cat. *Exp. Neurol.*, 9: 212–227.

Roberts, T.D.M. (1970) Changes in stretch reflexes in limb extensor muscles during position reflexes from the labyrinth in the cat. *J. Physiol (Lond.)*, 211: 5P–6P.

Rosenberg, J.R. and Lindsay, K.W. (1973) Asymmetric tonic labyrinthine reflexes. *Brain Res.*, 63: 347–350.

Thoden, U., Dichgans, J. and Savidis, Th. (1977) Direction specific optokinetic modulation of monosynaptic hindlimb reflexes in cats. *Exp. Brain Res.*, 30: 155–160.

Thoden, U., Dichgans, J., Doerr, M. and Savidis, Th. (1978) Direction specific vestibular and visual modulation of monosynaptic fore- and hindlimb reflexes in cats. *Pflügers Arch.*, Suppl. 373: R 86.

Waespe, W. and Henn, V. (1977) Neuronal activity in the vestibular nuclei of the alert monkey during vestibular and optokinetic stimulation. *Exp. Brain Res.*, 27: 523–538.

Wenzel, D., Thoden, U. and Frank, A. (1978) Forelimb reflexes modulated by tonic neck positions in cats. *Pflügers Arch.*, 374: 107–113.

Wilson, V.J. and Yoshida, M. (1968) Vestibulospinal and reticulospinal effects on hindlimb, forelimb and neck alpha motoneurons of the cat. *Proc. Nat. Acad. Sci. (Wash.)*, 60: 836–840.

Diagonal Stance in Quadrupeds: A Postural Support for Movement

J. MASSION and Y. GAHERY

C.N.R.S., Institut de Neurophysiologie et Psychophysiologie, Département de Neurophysiologie Générale, 13274 Marseille (France)

INTRODUCTION

Living organisms are perpetually submitted to the force of gravity. The whole motor system is organized towards struggling against the effects of that force, both when the animal is immobile and when it is moving.

Postural tone is the main means by which the neuromuscular system compensates for the action of gravity. Under the conditions of a stable posture, as for example standing, the distribution of postural tone is such that it keeps the center of gravity within narrow limits. Several categories of sensory signals are involved in the detection of displacements of the center of gravity. They elicit an adjustment of posture which returns the center of gravity in its initial position. An internal image of normal posture exists which automatically controls the distribution of the postural tone and consequently the center of gravity (see Roberts, 1967; Paillard, 1971).

The sensory messages which are concerned with the regulation of posture as a whole have been extensively investigated in the recent past, and their reflex influence described. Starting with the investigations of Brookhart et al. (1965), Brookhart and Talbot (1974), several groups of investigators have illustrated the contribution of visual, labyrinthine and proprioceptive inputs (Dichgans et al., 1972; Nashner, 1976; Lestienne et al., 1977; Nashner and Berthoz, 1978).

Adjustments of posture are not exclusively the consequence of sensory inputs which signal a disequilibrium. They are also observed when a movement is performed. Investigations on postural changes associated with movement have been performed mainly by Russian scientists in quadrupeds and in humans. Ioffe and Andreyev (1969) and Anokhin (1974) noticed that a diagonal bipedal stance was seen in quadrupeds when a limb movement took place. This postural adjustment was also briefly described by Brookhart et al. (1965). It is still observable after pyramidal tract section or after red nucleus lesion (Ioffe, 1975). Postural changes associated with movement have also been reported in humans (Martin, 1967; Belenkyi et al., 1967; Alexeiev and Naidel, 1973; Gurfinkel and Ebner, 1973). They can be observed during limb movements, or even during the respiratory movements of the thoracic wall (Gurfinkel and Ebner, 1973).

The postural adjustment associated with movement serves two purposes. First, it makes possible the displacement of a limb that was previously supporting a part of the body weight. The flexion of a leg, while standing, needs as a first step that the part of the body weight supported by that leg to be shifted towards the other legs. A second raison d'être of postural adjustment associated with limb movements arises from the fact that each displacement of a body segment is per se a source of disequilibrium, and that an adjustment of posture is needed in order to maintain the center of gravity within limits compatible with equilibrium. When the arm is raised forwards, for example, a backwards displacement of the body takes place which compensated for the disequilibrium which the movement would have produced.

Several features are reported as characteristics of the postural adjustments associated with movements. First, the adjustment of posture starts at the same time or even before the onset of the movement, and thus before that any displacement of the center of gravity has taken place. The adjustment is anticipatory, in the sense that it precedes the changes of position of the gravity center that the movement would have produced (Belenkyi et al., 1967; Gurfinkel and Ebner, 1973; Alexeiev and Naidel, 1973). Second, the intensity of the postural adjustment seems regulated automatically on the basis of sensory cues. For example, the increased myographic activity of triceps muscle of the contralateral leg which takes place in the standing subject when the arm is raised (Belenkyi et al., 1967), is not noticed when the same movement is performed when the subject is lying down. The "gain" of the neural circuit responsible for the postural adjustment is not only regulated by sensory information about the weight supported by the legs but also can be preset on the basis of cognitive cues. According to Gurfinkel (personal communication), the postural reaction is proportional to the estimated value of the weight that the arm will have to raise. Thus the gain of the postural program is adjusted on the basis of information given to the subject before the onset of movements.

If the general features concerning the postural adjustment associated with movement have been analysed and described, the central mechanisms responsible for it have not been well investigated (see nevertheless Ioffe, 1975) and there is a need for an approach to the problem on experimental animals, where various manipulations on the central nervous system such as stimulation, recording or lesion can be performed. For this purpose, the standing quadruped is a particularly good preparation because the animal's weight is approximately distributed on four thrust points, the limbs, and every limb movement is necessarily accompanied by a redistribution of weight on three other limbs by way of a postural adjustment. By measuring the weight of each limb, and its changes under various experimental conditions, one can infer what postural changes are taking place.

DIAGONAL POSTURAL SUPPORT IN QUADRUPEDS

In their experiments performed with dogs, Ioffe and Andreyev (1969) have shown that flexion of one limb is accompanied by a postural adjustment by which the body weight was distributed on the three other limbs. However, it is not a tripodal stance which occurs, but mainly a bipodal stance characterized by a support on two diagonally opposite limbs, the third one acting as a stabilizer.

The same observation was made in the standing cat (Massion et al., 1975; Coulmance et al., 1979). A placing movement of either forelimb was produced by means of moving trays coming into contact with the limbs. Preceding the lift-off of the

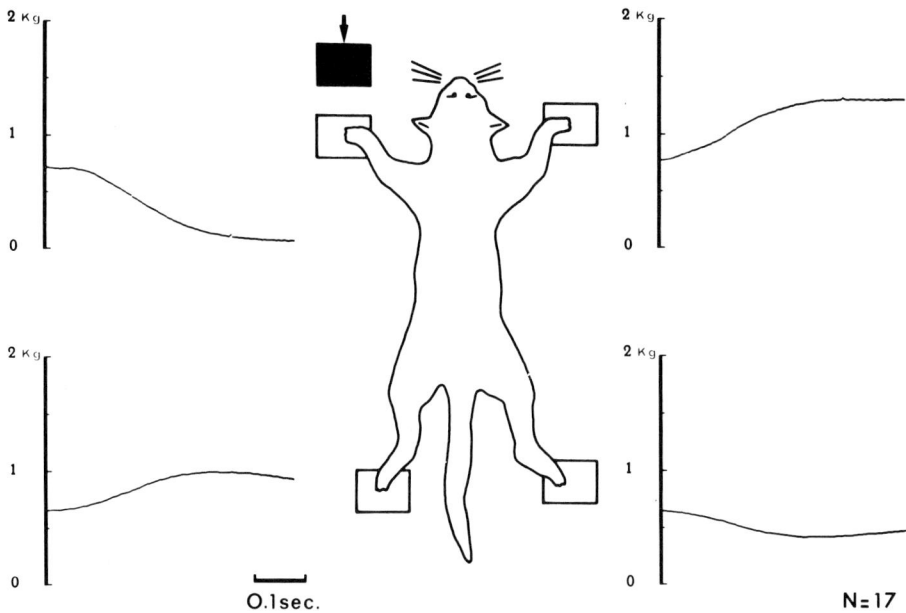

Fig. 1. Forces generated by each animal's limbs before and during the placing reaction of the left forepaw. 17 trials are summated. The change in force for each limb is displayed in respect to time starting with a signal indicating a contact force greater than 20 g of the mobile plate with the ordinate, scaled in kg. The lift-off time of the stimulated paw takes place when the fine trace reaches zero. Notice that nearly 90% of the animal's weight is supported by the two diagonally opposed pair of limbs (contralateral forelimb, ipsilateral hindlimb) at the time of lift-off.

stimulated limb, a postural adjustment could be observed during which not far from 90% of the animal's weight became supported by the contralateral forelimb and the ipsilateral hindlimb (Fig. 1). The diagonal stance was observed with all the cats. In contrast to the marked change in weight distribution which is seen during the postural adjustment, there is only a weak displacement of the projection of the center of gravity which remains close to a central position and shifts slightly sidewards with respect to a diagonal line joining the two supporting limbs. This type of postural adjustment is thus favorable to the maintenance of equilibrium during the performance of movement.

During the prelift-off phase of the placing reaction, the weight supported by the contralateral forelimb increases, whereas the weight supported by the stimulated limb decreases. For most animals, the increase in weight of one forelimb starts before the decrease in weight of the stimulated forelimb. If one admits that the onset of movement coincides with the onset of the decrease in weight of the stimulated forelimb, this result suggests that the postural adjustment is initiated together and even a little earlier than movement.

The postural adjustment associated with a placing movement in the cat is thus initiated before any marked change of the projection of the center of gravity has taken place, and it is anticipatory with respect to the movement in a way comparable to that which has been described in humans.

CENTRAL ORGANIZATION OF THE BIPEDAL STANCE

One of the most commonly accepted schema concerning the central programming of movement is that of Allen and Tsukahara (1974). For "voluntary" movements, a program is supposed to be organized between association cortex and motor cortex using three main pathways. One is represented by direct cortico-cortical connexions, the two others are indirect, relaying within neocerebellum or basal ganglia.

The motor command exerted from motor cortex reaches peripheral musculature through pyramidal and extrapyramidal pathways. The descending messages are also involved in the setting-up of bulbospinal subprograms which contribute to the performance of movement. Feedback information from the periphery or from the motor centers reaches the cerebellum and contributes to the execution of motor performance.

A particular aspect of motor programming hitherto neglected is the site where the postural adjustment associated with movement is organized. Concerning the bipedal stance accompanying movement in the quadruped, the question may be raised as to the contribution of motor cortex which Bard (1933) claimed to be crucial for adjustments of contralateral limbs, such as placing or hopping reactions, and the extent of involvement of structures downstream to motor cortex such as cerebellum and the bulbospinal nuclei.

A first experimental approach used to investigate the site of storage of the postural program associated with limb flexion was to examine the influence of ablation of motor cortex in the standing cat during the placing reaction. It first appeared that unilateral motor cortical ablation abolished only temporarily the placing movement on the other side. After a few days, the placing reappeared but remained hard to elicit. The postural adjustment associated with the contralateral placing was also abnormal. The time required for reaching the diagonal stance was increased, and the stance was reached in several steps. However, the diagonal supporting patterns were maintained. Nevertheless, the postural adjustment associated with ipsilateral placing as well as the placing movement remained unmodified. In particular, during ipsilateral placing the adjustment of the limbs contralateral to the lesion was performed as well as that of ipsilateral limbs. These results suggest that during the diagonal stance motor cortex has no specific influence on the postural adjustment of the contralateral limbs, but that it acts on the diagonal pattern as a whole. Moreover, this type of experiment suggests that movement and associated postural adjustment are centrally linked, but it cannot indicate whether the link is at the level of motor cortex itself or downstream to motor cortex.

A second experimental approach of the problem consisted in directly stimulating the motor cortex of the standing cat. Stimulation of motor cortex induces a contralateral movement, the localization of which depends on the stimulated site within the motor area (Woolsey, 1958; Nieoullon and Rispal-Padel, 1976). When cortical stimulation is applied on the forelimb or on the hindlimb area, while the cat is standing, a contralateral flexion movement is performed without any apparent disequilibrium. Thus, a postural adjustment takes place. Actually, along with the decrease in weight of the limb undergoing flexion, there is an increase in weight of the contralateral limb (Regis et al., 1976a,b; Gahery and Nieoullon, 1978). As shown by Gahery and Nieoullon (1978), the postural adjustment which takes place is characterized by a diagonal support on one forelimb and the opposite hindlimb. This postural support is similar to those occurring during spontaneous movements and movements provoked by natural stimulation. The postural adjustment is not reflexly induced by the afferent messages associated with limb flexion but is centrally commanded in a feedforward manner. This was shown by the fact that for hindlimb induced movements the changes in weight of the forelimb preceded the changes in weight of the hindlimb undergoing flexion (Gahery and Nieoullon, 1978).

Interestingly, stimulation of motor cortex after pyramidal tract section is still effective in producing the diagonal postural adjustment (Nieoullon and Gahery, 1978). Moreover, stimulation within the red nucleus also produces the same diagonal supporting pattern in association with limb flexion (Regis et al., 1976a,b; Massion, 1978). This suggests that the postural program utilized in association with limb flexion is stored somewhere about the bulbospinal level, and that the descending impulses running through the rubrospinal tract or the pyramidal tract are equally effective in the command of postural program.

That the basic network for the diagonal postural pattern would be located at the bulbospinal level is not surprising (Fig. 2). This type of adjustment is used not only during limb movement but also during locomotion and especially trotting (Grillner, 1975). However, these centers do not function in isolation. As shown by Orlovsky (1972) for locomotion, there is an important contribution of the cerebellar loop to the appearance of the diagonal stance associated with movement. It was observed in two chronically decerebellate cats that the postural adjustment associated with a placing movement was deeply depressed, much more than after cortical ablation (Regis et al., 1976a,b). The fact that red nucleus neurons receiving cerebellar input are activated or inhibited during the bipedal stance also indicated the dynamic contribution of the cerebellorubral loop to the postural adjustment (Padel and Steinberg, 1978).

Finally, the fact that motor cortical ablation makes abnormal the placing movement and the whole postural pattern associated with it raises the question as to the specific

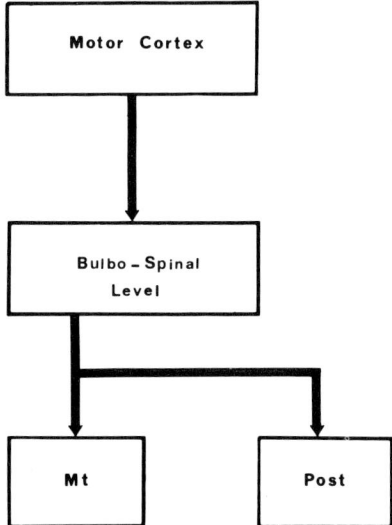

Fig. 2. Diagram illustrating the feedforward command of the associated postural changes together with forelimb movement at the time of lift-off of the stimulated limb.

function of motor cortex in the postural adjustment. A first possibility would be that a gating influence would be exerted from motor cortex on subcortical centers responsible for the movement and its associated adjustment. That such tonic influence exists was shown for the tactile placing reaction by Amassian et al. (1972) and Amassian and Ross (1978a,b). However, there is also a phasic contribution of the cerebellocortical pathway to the associated adjustment, as shown by the changes in unitary activity which are observed at the VL level when the contralateral limbs are involved in the postural adjustment (Smith et al., 1978). The phasic contribution of the cerebellocortical loop might help in adjusting the postural program to peripheral conditions.

FUNCTIONAL SIGNIFICANCE

The diagonal stance in quadrupeds is one type of postural adjustment which is associated with movements along the major body axis. It is probably derived from locomotor patterns which serve to displace the body along the axis. It is used in single limb movements in the immediate surround, performed while standing. The usual types of limb movements, distal and proximal, can probably be supported by this type of postural adjustment.

It is evident that the diagonal supporting pattern is not adapted for numerous types of movements. For example, movements of both forelimbs or both hindlimbs require another type of adjustment, which might be found in the locomotor repertoire. Movement of a limb towards a target which is situated in the surrounding space, laterally with respect to the body axis would require a complementary adjustment which would imply axial musculature in order to permit the orientation of the body towards a target. While in quadrupeds movements made by the limbs in the surrounding space are less common than those which are in the immediate vicinity of the body, the reverse is true in primates and anthropoids, where forelimbs progressively lose their role in the support of the body, and gain freedom for the performance of movements in space. Therefore, the diagonal stance is inadequate for most movements in primates, and a new postural system is needed, which would be adapted to the performance of the new types of movement. The considerable development of the neocerebellum in primates has been related to the need of a new type of postural support for movement (see Massion, 1973; Schultz et al., 1976). There is also evidence that basal ganglia might be implied in the organization of postural changes associated with movement (Martin, 1967). However, new data will be needed before the respective contribution of these two central structures can be defined.

SUMMARY

Under the conditions of a stable posture, as for example during standing, the distribution of the postural tone is such that it keeps the center of gravity within narrow limits. Several types of sensory signals, such as labyrinthine, visual and proprioceptive, are involved in the proper maintenance of the center of gravity.

When a movement is performed, an adjustment of posture takes place, which permits the maintenance of equilibrium during the performance of the movement.

A support of the body weight on two diagonally supporting limbs is the postural adjustment most commonly observed in quadrupeds during limb movement. Several types of experiments such as motor cortical ablation, stimulation of motor cortex or red nucleus, recording of red nucleus and ventrolateral thalamic nucleus and cerebellar ablation have been performed in order to determine where this postural "program" is stored. There is evidence that the basic program is located at the bulbospinal level, that cerebellum and red nucleus contribute to the performance of this adjustment and that the cerebellocortical loop also plays a role in the performance of this type of adjustment. The functional significance of the diagonal postural pattern is discussed with respect to the phylogenetic evolution of motor performance.

REFERENCES

Alexeiev, M.A. and Naidel, A.V. (1973) Rapports entre les éléments volontaires et posturaux d'un acte moteur chez l'homme. *Agressologie*, 14 B: 9–16.

Allen, G.I. and Tsukahara, N. (1974) Cerebro-cerebellar communication systems. *Physiol. Rev.*, 54: 957–1006.

Amassian, V.E. and Ross, R.J. (1978a) Developing role of sensorimotor cortex and pyramidal tract neurons in contact placing in kittens. *J. Physiol. (Paris)*, 74: 165–184.

Amassian, V.E. and Ross, R.J. (1978b) Electrophysiological correlates of the developing higher sensorimotor control system. *J. Physiol. (Paris)*, 74: 185–202.

Amassian, V.E., Ross, R., Wertenbaker, C. and Weiner, H. (1972) Cerebellothalamocortical interrelations in contact placing and other movements in cats. In *Corticothalamic Projections and Sensorimotor Activities*. T. Frigyesi, E. Rinvik and M.D. Yahr (Eds.), Raven Press, New York, pp. 395–444.

Anokhin, P.K. (1974) *Biology and Neurophysiology of the Conditioned Reflex and its Role in Adaptative Behavior*. Pergamon Press, Oxford.

Bard, P. (1933) Studies on the cerebral cortex. I. Localized control of placing and hopping reactions in the cat and their normal management by small cortical remnants. *Arch. Neurol. Psychiat. (Chic.)*, 30: 40–74.

Belenkyi, V.E., Gurfinkel, V.S. and Paltsev, E.I. (1967) On elements of control of voluntary movements. *Biofizika*, 12: 135–141.

Brookhart, J.M. and Talbott, R.E. (1974) The postural response of normal dogs to sinusoidal displacement. *J. Physiol. (Lond.)*, 243: 287–307.

Brookhart, J.M., Parmeggiani, W.A., Petersen, W.A. and Stone, S.A. (1965) Postural stability in the dog. *Amer. J. Physiol.*, 208: 1047–1057.

Coulmance, M., Gahery, Y., Massion, J. and Swett, J.E. (1979) The placing reaction in standing cat: a model for the study of the coordination of posture and movement. In preparation.

Dichgans, J., Held, R., Young, L. and Brandt, T. (1972) Moving visual scenes influence the apparent direction of gravity. *Science*, 178: 1217–1219.

Gahery, Y. and Nieoullon, A. (1978) Postural and kinetic coordination following cortical stimuli which induce flexion movements in the cat's limbs. *Brain Res.*, 155: 25–37.

Grillner, S. (1975) Locomotion in vertebrates: central mechanisms and reflex interaction. *Physiol. Rev.*, 55: 247–304.

Gurfinkel, V.S. and Ebner, A.M. (1973) On two types of static disturbances in patients with local lesions of the brain. *Agressologie*, 14D: 65–72.

Ioffe, M.E. (1975) *Cortico-Spinal Mechanisms of Instrumental Motor Reactions* (Rus.). Nauka, Moscou.

Ioffe, M.E. and Andreyev, A.E. (1969) Interextremities coordination in local motor conditioned reactions of dogs (Rus.). *Zh. Vyssh. Nervn. Deyat. Pavlova*, 19: 557–565.

Lestienne, F., Soechting, J. and Berthoz, A. (1977) Postural readjustments induced by linear motion of visual scenes. *Exp. Brain Res.*, 28: 363–384.

Martin, J.P. (1967) *The Basal Ganglia and Posture*. Pitman, London.

Massion, J. (1973) Intervention des voies cérébello-corticales et cortico-cérébelleuses dans l'organisation et la régulation du mouvement. *J. Physiol. (Paris)*, 67: 117A–170A.

Massion, J. (1979) Postural adjustment associated with movement. *Neuroscience*, in press.

Massion, J., Swett, J.E., Coulmance, M. and Gahery, Y. (1975) Postural adjustment associated with a placing movement. *Exp. Brain Res.*, Suppl., 23: 137. no. 270.

Nashner, L.M. (1976) Adapting reflexes controlling the human posture. *Exp. Brain Res.*, 26: 59–72.

Nashner, L.M. and Berthoz, A. (1978) Visual contribution to rapid motor responses during postural control. *Brain Res.*, 150: 403–407.

Nieoullon, A. and Rispal-Padel, L. (1976) Somatotopic localization in cat motor cortex. *Brain Res.*, 105: 405–422.

Nieoullon, A. and Gahery, Y. (1978) Influence of pyramidotomy on limb flexion movements induced by cortical stimulation and on associated postural adjustment in the cat. *Brain Res.*, 155: 39–52.

Orlovsky, G.N. (1972) Activity of rubrospinal neurons during locomotion. *Brain Res.*, 46: 99–112.

Padel, Y. and Steinberg, R. (1978) Red nucleus cell activity in awake cats during a placing reaction. *J. Physiol. (Paris)*, 74: 265–282.

Paillard, J. (1971) Les déterminants moteurs de l'organisation de l'espace. *Cah. Psychol.*, 14: 261–316.

Regis, H., Trouche, E. and Massion, J. (1967a) Movement and associated postural adjustment. In *The Motor System: Neurophysiology and Muscle Mechanisms*. M. Shahani (Ed.). Elsevier, Amsterdam, pp. 349–361.

Regis, H., Trouche, E. and Massion, J. (1976b) Effet de l'ablation du cortex moteur ou du cervelet sur la coordination posturo-cinétique chez le chat. *Electroenceph. Clin. Neurophysiol.*, 41: 348–356.

Roberts, T.D.M. (1967) *Neurophysiology of Postural Mechanisms*. Butterworths, London.

Schultz, W., Montgomery, E.B. and Marino, R. (1976) Stereotyped flexion of forelimb and hindlimb to microstimulation of dentate nucleus in cebus monkeys. *Brain Res.*, 107: 151–155.

Smith, A.M., Massion, J., Gahery, Y. and Roumieu, J. (1978) Unitary activity of ventrolateral nucleus during a placing movement and on associated postural adjustment. *Brain Res.*, 151: 329–346.

Woolsey, C.N. (1958) Organization of somatic sensory and motor areas of the cerebral cortex. In *Biological and Biochemical Basis of Behaviour*. H.F. Harlow, and C.N. Woolsey (Eds.), University of Wisconsin Press, Madison. pp. 63–81.

Interaction between Central and Peripheral Mechanisms in the Control of Locomotion

S. GRILLNER

Fysiologiska Institutionen III, Karolinska Institutet, S-114 33 Stockholm (Sweden)

INTRODUCTION

To enable an animal to get to a desired place, it is equipped with a neuromuscular propulsive machinery. The animal must be able to: i) "decide" where and if it is desirable to move, ii) initiate, maintain and stop the locomotor movements, iii) meet imposed equilibrium disturbances and iv) adapt them to the environment (i.e., to move around corners, avoid obstacles or predators, etc.) in order to achieve the aim of the mission. Locomotion thus requires "higher nervous functions", but the stereotype pattern of limb or body movements may nevertheless be driven by comparatively simple neuronal networks allowing movements within a certain frequency range. This can be achieved by changing the duration of the different phases and the degree of activation of certain muscles to vary the force output.

At which level of the nervous system and how the stereotypic locomotor movements are ultimately controlled were two questions posed already in the beginning of the last century. Interneuronal networks located in the spinal cord have now been found to control the different limbs. Each network may be called a central pattern generator (CPG). *This term is taken to include all neurons that are primarily responsible for creating an appropriate synaptic drive that results in a desirable output of α- and γ-motoneurones during locomotion.* The word "primarily" gives an intended vagueness in the definition since it is desirable to exclude synaptic inputs that only under certain special conditions may influence the rhythmic motor output and also other secondary effects. There is much information to demonstrate the existence of CPGs located in the spinal cord (cf. Grillner 1975, 1979). In spinal fish the CPGs may be spontaneously active; in mammals like the spinal cat they may be released by an i.v. injection of DOPA or noradrenergic receptor stimulators (Grillner, 1969; Forssberg and Grillner, 1973; Edgerton et al., 1976). Even when the animals are curarized, the motor output resulting from the CPG activity can be recorded in peripheral nerve filaments. This allows traditional neurophysiological techniques to be applied to an animal whose nervous system produces the neuronal correlate of the locomotor behavior, i.e., "fictive locomotion".

Traditional reflex arcs traverse via simple or more complex courses to the motoneurones. Inevitably, the degree of motoneurone excitability must also be influenced from such pathways during locomotion. The monosynaptic connections from the spindle will for example have an effect which depends on the strength of the α-γ

linkage (Sjöström and Zangger, 1976). The polysynaptic pathways will also, if they are open, influence the output. With few exceptions their effects are unknown, since these pathways have not been studied under "fictive" or actual locomotion.

Such direct pathways from periphery to the motoneurones would, however, not be able to influence, e.g., the frequency of stepping or the duration of the different locomotor phases. Such control would require pathways which affect directly the CPGs. Such pathways must, however, exist since spinal and mesencephalic walking animals can adapt to treadmill speed (cf. Grillner, 1975).

In this paper we will deal mainly with two conceptually different types of reflex control during walking: i) the peripheral control of the CPGs and ii) the control of certain reflex pathways by the CPGs.

PERIPHERAL CONTROL OF THE CPG

CPG dependence on position and movement of the hip

If the limb movement is stopped during the support phase the extensor activity will continue, despite continuing locomotor activity in the other limbs (Grillner and Rossignol, 1978a). A short arrest of the hip flexion will in an analogous way prolong to a corresponding degree the actual flexion phase of the entire limb (EMG)(Orlovsky, 1972).

When the limb is subsequently brought backwards during the support phase to near the position in which flexion normally occurs, this will result in a prompt termination of the support and onset of the ensuing flexion. The last effect can be regarded as a position dependent negative feedback (Grillner and Rossignol, 1978a; Andersson et al., 1978b; cf. Sherrington, 1910). Similarly, during fictive locomotion flexed hip positions have been found to promote extensor activity (see Fig. 1, Andersson et al., 1978b; unpublished observations).

A fictive locomotory preparation may display rhythmic flexor and extensor activity even when the limb is denervated except for the hip joint and the small muscles around the hip (prime movers of the hip were denervated). If alternating hip movements are superimposed on the rhythm they will entrain the rhythmic CPG activity over a large frequency range including both lower and higher than resting rates. Flexor activity will occur during imposed flexion and extensor activity during imposed

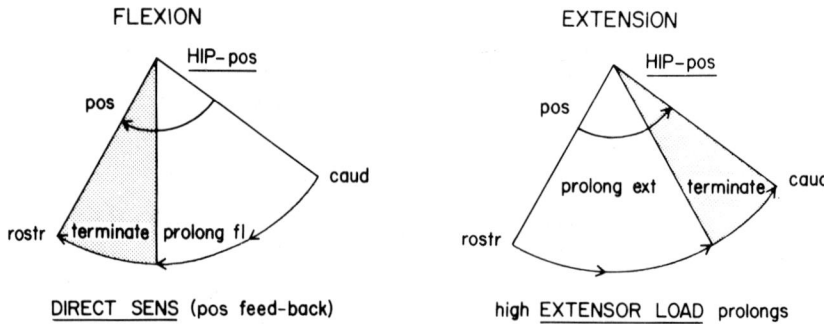

Fig. 1. Schematic representation of the reflex effects exerted from the hip (joint and/or muscle receptors).

extension. With large amplitude movements this could have been explained by the "switching effects" caused by the position dependent feedback in full extension and full flexion, respectively. However, small amplitude movements as low as 2–3° around any mean position from maximal extension to maximal flexion can drive the bursting neural activity. This is explained by a high, directional sensitivity, for example imposed flexion reinforces flexor activity (i.e., positive feedback) and if the direction is reversed the flexor burst is interrupted (Andersson et al., 1978b).

The afferents responsible are not defined. Not only hip joint but afferents from the smaller hip muscles may also be responsible (Andersson et al., 1978b; unpublished observations). The most likely explanation may, however, be that all suitable afferent signals are used for this purpose. These effects are, however, caused by some afferents influenced by the hip position, which carry information about the overall position of the limb in the stepcycle. Whether signals related to other joints of the hindlimb have any effect on the CGP is unknown. It is clear, however, that the "hip effect" dominated over possible effects from other joints (Grillner and Rossignol, 1978a). In addition the knee and ankle joint have a much more complex cycle due to the flexion-extension movement during each support phase.

Analogous position dependent effects are exerted on the late DOPA discharges (Jankowska et al., 1967), as a contralateral strong nerve shock will elicit flexor activity in the contralateral limb if its hip is extended, but extensor activity if it is flexed (Grillner and Rossignol, 1978b). However, the latter effect may also be influenced by stretch of muscles controlling the lower leg (Rossignol and Gauthier, 1977).

Load sensitive feedback in the extensor phase

Pearson and Duysens (1976) have described another input which affects the CPG controlling input. Excessive load exerted on extensors will prevent the normal rhythmic switching from extensor to flexor activity. In the support phase the normal muscle load is maximal at midstance but it decreases as end of the support nears. If, however, the load is forced to remain high, the limb still needs support and the swing is not initiated. The responsible muscle receptors are not yet defined.

General comments

These two different types of peripheral control signals will act to adapt the stepcycle to the external environment. If the limb extends slower than normally, it would be mechanically ineffective to terminate the support phase early. This delay can be caused by both mechanisms described above (hip position and load). Correspondingly, if the limb extends faster than "expected" the support phase would be shortened. These reflex mechanisms will thus assure that flexion is induced at an appropriate time and may be particularly important when the animal changes speed (acceleration or deceleration). For example, a descending control signal for increased speed could just elicit an enhanced extensor activity to result in a larger propulsive force. This propulsive force will depend on a number of different external factors, and cause a variable degree of acceleration. These reflexes could then assure that the duration of the support phase will always be adequate and that it will be terminated when the limb has been extended to an appropriate position (and unloaded). Also if the animal walks in circles such a reflex control could explain the characteristic adaptation of the stepcycle on the "slow" and the "fast side" (Forssberg et al., 1976).

Reflex control in fish

When a fish swims, the forward speed is mainly controlled by the frequency of alternation between the two body halves. The segments along the body are orderly activated with a lag from rostral to caudal. This lag is not a constant time lag but is directly proportional to the duration of the swimming cycle and is thus a *constant phase lag*. After a transection of the spinal cord the caudal part of the body may produce spontaneous swimming in e.g., a dogfish or an eel. These swimming movements have all the general characteristics of the intact fish in regard to the stereotypic movement cycle, frequency range, phase coupling, etc. (Grillner, 1974).

Experiments, in which all movements have been suppressed by means of curarisation, show that the spinal cord is capable of activating the different segmental motoneurons in a coordinated way and tonic stimuli alone can drive the activity into different frequency ranges. The spinal cord may be split into smaller parts, each producing a bursting output with its appropriate intersegmental coordination even after transposing the isolated cord to a Petri dish (Cohen and Wallén, 1978; Poon, unpubl. obs.). Thus, since the motionless animal still produces the output pattern, this provides conclusive evidence that CPGs located in the spinal cord take part in the control of locomotion. A reflex theory would in all cases require a movement (Grillner et al., 1976).

The fact that CPGs have been demonstrated does not of course provide any information on the possible role of afferent input during swimming. To test for this possible role the spontaneous burst activity of a curarised spinal dogfish was recorded both during rest conditions and during the superposition of experimentally induced movements (Grillner and Wallén, 1977). The peripheral feedback acted very powerfully on the CPGs and an imposed movement could entrain the central rhythm to frequencies both higher and lower than the resting activity (see Fig. 2). To be effective the movements may be of much smaller amplitude (2–4 degrees) than those occurring under natural swimming. One factor that causes the driving is a directional sensitivity that induces right side muscle activity during a movement to the right, but blocks activity on the right side when a movement to the left starts. At such a time, activity on the left side also starts. Moreover, the normal extreme position to the left, and to the right can in itself induce the neuronal correlate of a reverse movement, i.e., a position dependent negative feedback also exists, which is analogous to limb effects resulting from maximal flexion and extension positions (Grillner and Wallén, 1977; unpublished observations).

CENTRAL CONTROL OF REFLEX TRANSMISSION IN DIFFERENT PHASES OF THE STEPCYCLE

Cat

If the forward movement of the foot (swing phase) is obstructed by a small or a large object such as a twig, the limb is further elevated. This enables the limb to overcome the object. Cutaneous receptors on the dorsum of the paw even when activated by a localised airjet, will elicit this response (Forssberg, 1979). This stumbling corrective response is due to a spinal reflex arc originating from the dorsum of the paw and which via a few interneurones affects the flexor motoneurones. The response threshold is low during the flexion phase but the response is virtually absent during the support phase. This is partially due to a gain control in reflex arcs since modulation of

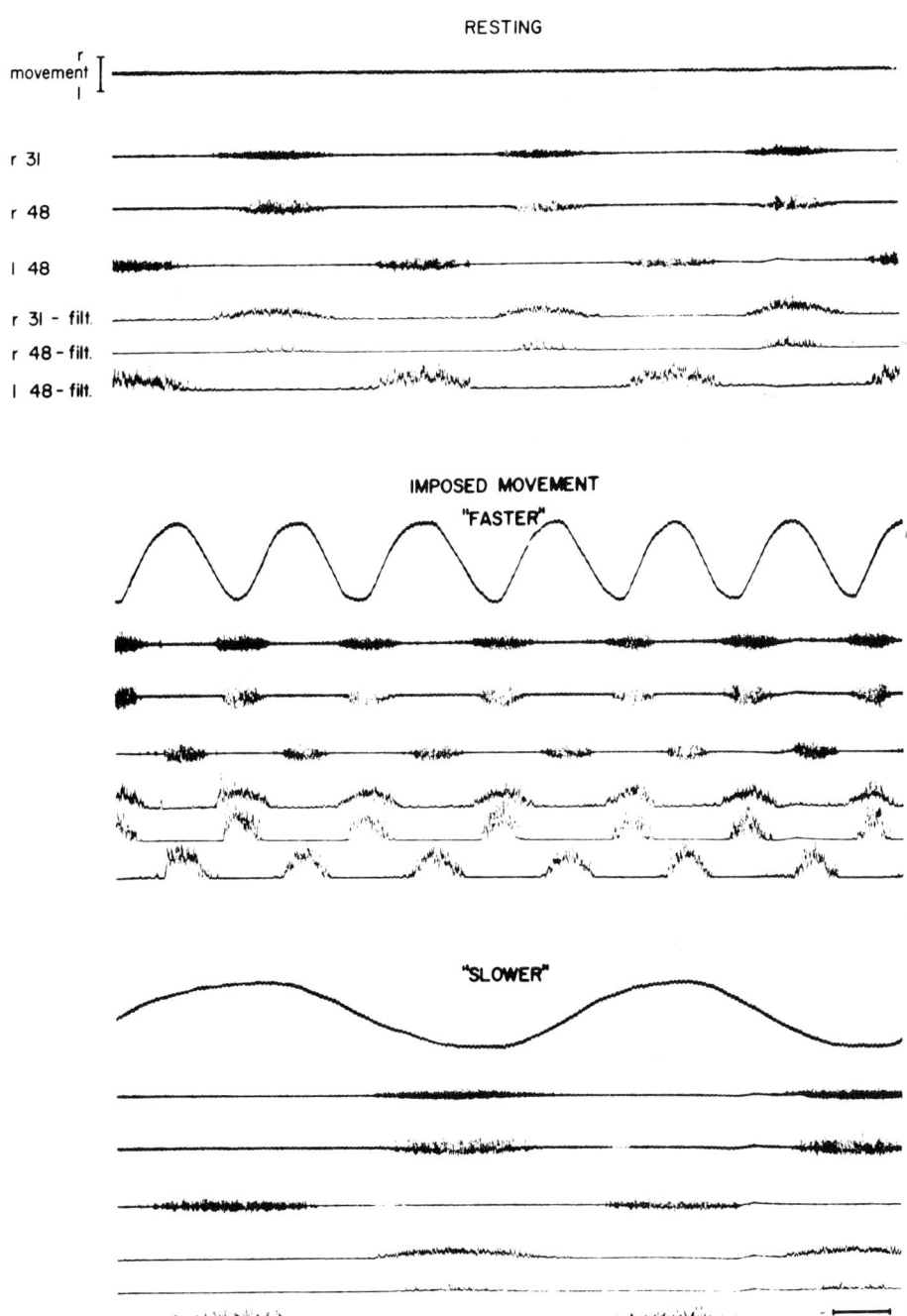

Fig. 2. Recordings from three ventral roots (r VR no. 31, r VR no. 48, l no. 48) in a curarised spinal dogfish. Upper series of records show "fictive locomotion" and the lower two series the effect of superimposed movements at a higher and lower frequency than that of the resting activity. Movement to the left is downwards, the three lowermost traces in each section are the rectified and filtered versions of the direct recordings of three ventral roots. Movement calibration (upper left) equals a 10 cm displacement of the tailfin (the length of the dogfish is approx. 105 cm). (From Grillner and Wallén, 1977.)

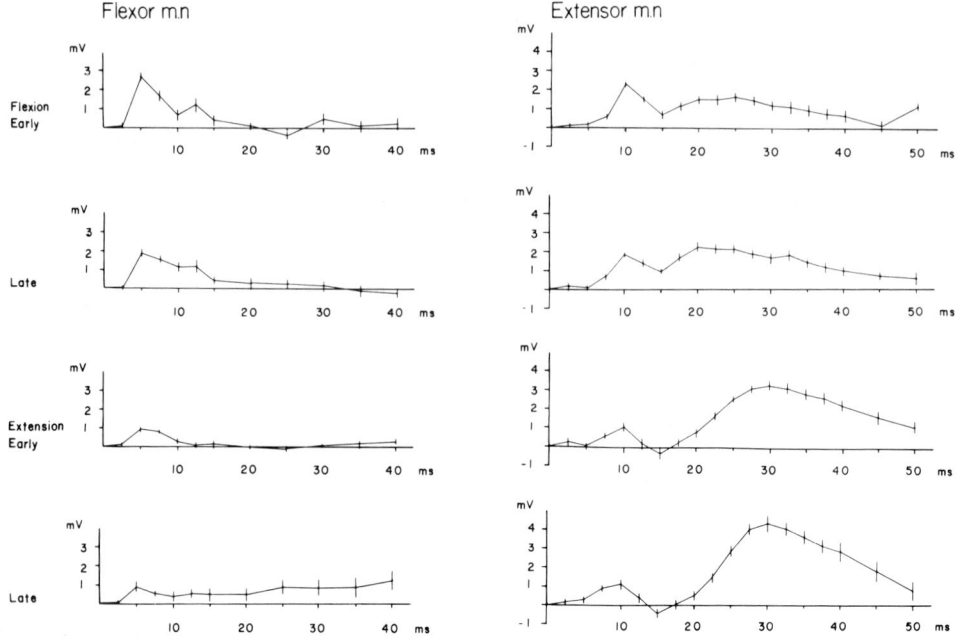

Fig. 3. Averaged responses in a chronic spinal cat. Intracellular recordings from a flexor motoneurone (n = 129) and an extensor motoneurone (n = 78) in a chronic spinal cat during "fictive" locomotion after paw stimulation (5 mA, 5 msec). The responses are divided into 4 groups according to the locomotor activity (see Figs. 1 and 2). In each group the mean and SE of the responses are calculated. The responses in early flexion and late extension are significantly different in both motoneurons ($P < 0.01$). (From Andersson et al., 1978a.)

reflexly evoked EPSPs in flexor motoneurones during "fictive locomotion" has been shown (Fig. 3). Stimulation of the dorsum of the paw thus evokes small EPSPs during extensor bursts but large ones during flexor activity (Andersson et al., 1977, 1978a). Another factor that must influence the efficacy (EMG response) of the cutaneous stimulation is the cyclic variation of the motoneurone excitability in each cycle which will, everything else being equal, discharge more cells during periods of depolarisation as compared to periods of hyperpolarisation (Forssberg et al., 1975, 1977). During the support phase there is thus an absence of flexor response to this stimulus. This is important since it would be disastrous if a light stimulus as an airjet could evoke strong flexion during the support phase, which would of course cause the animal to fall over. During the support phase the same stimulus instead can evoke an EMG response in extensors (Forssberg et al., 1975, 1977), which can be subdivided into early and somewhat later spinal reflex components (Forssberg, 1979) and which can also be modulated phasically (Andersson et al., 1977, 1978a).

The response may be evoked from low threshold receptors in the skin since i) an airjet is sufficient to elicit it during locomotion and ii) the response disappears after local anaesthesia of the skin of the dorsum of the paw.

Similar phase dependence has been demonstrated in reflexes from fore- to hindlimb and vice versa (Miller and Van der Meché, 1977; Schomburg, 1977).

We are thus dealing with a conceptually different type of reflex response, the pathway of which does not include the CPG. Instead, the CPG itself is capable of modulating the transmission in the reflex arcs. The overall effect is that a certain reflex can only be evoked in one part of the stepcycle, i.e., the part in which this response is behaviourally important. Two alternative neuronal mechanisms exist to account for this: i) the reflex pathway is selectively controlled to assure that it is only open in a certain phase; and ii) the CPG and the reflex arc share the last order interneurones. This latter possibility would, everything else being equal, result in a more efficient reflex transmission to e.g. flexors when the CPG is driving flexors, since the last order interneurones would at this time in the stepcycle be under maximal drive.

Fish

Analogous phase dependent reflex effects have been obtained on fish (Grillner et al., 1977; Wallén, 1977; and unpublished observations). A light tactile or electrical stimulation of the tailfin will depending on the phase of the cycle, always activate the side in which there is locomotor activity. The reflex response increases the amplitude and the speed of the ongoing movement, so that the animal moves in a direction away from the stimulus. This phase dependent effect remains after curarisation and is thus not dependent on gating from other movement induced reflexes but rather is under control from the CPGs (compare discussion for cat above).

CONCLUDING REMARKS

Despite the fact that CPGs have been demonstrated conclusively in the last decades, the data discussed above clearly show that these central networks are subject to a variety of control signals including peripheral ones that assures an optimal adaptation of the animal to its environment (see Fig. 4, for components in the cat system). A different type of mechanism is one in which by the CPG there is central phase gating of reflexes, that are needed in only one phase of the movement but suppressed in another.

Despite the obvious differences in the mechanism of the propulsion between fish and cat, it is striking how similar the control structure is.

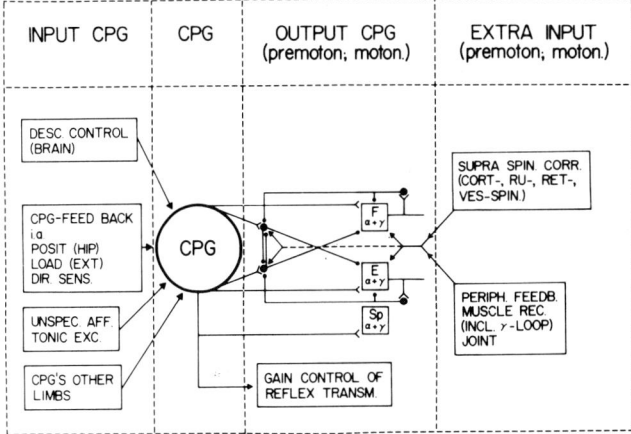

Fig. 4. Schematic representation of the different components related to the control of an individual limb CPG. The different types of input to the CPG are shown, the CPG and the different outputs i.e., to flexor, extensor motor nuclei and "special" motor nuclei (both α- and γ-motoneurones) and also the gain control of the reflexes discussed above. The CPG excites one motor nucleus and does inhibit the antagonists via reciprocal inhibition (see Edgerton et al., 1976). One part will bypass the CPG and may affect the motor nuclei directly as the γ-loop and the descending effects from phasically active descending pathways.

SUMMARY

A short review of peripheral signals fundamental to the control of locomotion is given. In particular, it is shown that afferent signals induced by dynamic hip movements and static hip position in cat and by analogous effects in fish can powerfully drive the central spinal locomotor networks in both cat and fish. The locomotor generator in both cat and fish can in turn exert phasic gain control of short latency reflexes, which act directly on the α-motoneurones.

ACKNOWLEDGEMENTS

The results discussed in this paper have been obtained through support from the Swedish Medical Research Council, project no. 3026 and M. Bergvalls stiftelse. The valuable comments of Professor F. Zajac are gratefully acknowledged as well as the collaboration with Drs. O. Andersson, H. Forssberg and P. Wallén, who have contributed fundamentally to the different studies summarized here.

REFERENCES

Andersson, O., Forssberg, H. and Lindquist, M. (1977) The neural mechanism of the phase dependent reflex reversal. In *Neurophysiological Mechanisms of Locomotion*, Paris.

Andersson, O., Forssberg, H., Grillner, S. and Lindquist, M. (1978a) Phasic gain control of the transmission in cutaneous reflex pathways to motoneurones during "fictive" locomotion. *Brain Res.*, 149: 503–507.

Andersson, O., Grillner, S., Lindquist, M. and Zomlefer, M. (1978b) Peripheral control of the spinal pattern generators for locomotion in cat. *Brain Res.*, 150: 625–630.

Budakova, N.N. (1973) Stepping movements in the spinal cat due to DOPA administration. *Fiziol. Zh. Kiev*, 59: 1190–1198.

Cohen, A.H. and Wallén, P. (1978) Rhythmic locomotor activity induced in an *in vitro* preparation of the lamprey spinal cord. *Neurosci. Lett.*, Suppl. 1, S92.

Edgerton, V.R., Grillner, S., Sjöström, A. and Zangger, P. (1976) Central generation of locomotion in vertebrates. In *Neural Control of Locomotion, Vol. 18*. R. Herman, S. Grillner, P. Stein and D. Stuart (Eds.), Plenum Press, New York, pp. 439–464.

Forssberg, H. (1979) The "Stumbling Corrective Reaction" – a phase dependent compensatory reaction during locomotion. *J. Neurophysiol.*, in press.

Forssberg, H. and Grillner, S. (1973) The locomotion of the acute spinal cat injected with Clonidine i.v. *Brain Res.*, 50: 184–186.

Forssberg, H., Grillner, S. and Rossignol, S. (1975) Phase dependent reflex reversal during walking in chronic spinal cats. *Brain Res.*, 85: 103–107.

Forssberg, H., Grillner, S. and Rossignol, S. (1976) Interaction between the two hindlimbs of a spinal cat walking on a split belt. *Abst. Third Int. Conference on Motor Control*, Albena, Bulgaria, no. 19.

Forssberg, H., Grillner, S. and Rossignol, S. (1977) Phasic gain control of reflexes from the dorsum of the paw during spinal locomotion. *Brain Res.*, 132: 121–139.

Grillner, S. (1969) Supraspinal and segmental control of static and dynamic γ-motoneurones in the cat. *Acta physiol. scand.*, suppl. 327: 1–34.

Grillner, S. (1973) Muscle stiffness and motor control – forces in the ankle during locomotion and standing. In *Motor Control*. A.A. Gydikov et al. (Eds). Plenum Press, New York, pp. 195–215.

Grillner, S. (1974) On the generation of locomotion in the spinal dogfish. *Exp. Brain Res.*, 20: 459–470.

Grillner, S. (1975) Locomotion in vertebrates: central mechanisms and reflex interaction: *Physiol. Rev.*, 55: 247–304.

Grillner, S. (1979) Control of locomotion in bipedes, tetrapodes and fish. In *APS Handbook on Motor Control*. V. Brooks (Ed.) in press.

Grillner, S. and Wallén, P. (1977) Is there a peripheral control of the central pattern generators for swimming in dogfish? *Brain Res.*, 127: 291–295.

Grillner, S. and Rossignol, S. (1978a) On the initiation of the swing phase of locomotion in chronic spinal cats. *Brain Res.*, 146: 269–277.

Grillner, S. and Rossignol, S. (1978b) Contralateral reflex reversal controlled by limb position in the acute spinal cat injected with Clonidine i.v. *Brain Res.*, 144: 411–414.

Grillner, S. and Zangger, P. (1979) On the central generation of locomotion in the low spinal cat. *Exp. Brain Res.*, 34: 241–261.

Grillner, S., Perret, C. and Zangger, P. (1976) Central generation of locomotion in the spinal dogfish. *Brain Res.*, 109: 255–269.

Grillner, S. Rossignol, S. and Wallén, P. (1977) The adaptation of a reflex response to the ongoing phase of locomotion in fish. *Exp. Brain Res.*, 30: 1–11.

Jankowska, E., Jukes, M.G.M., Lund, S. and Lundberg, A. (1967a) The effect of DOPA on the spinal cord. 5. Reciprocal organization of pathways transmitting excitatory action to alpha motoneurones of flexors and extensors. *Acta physiol scand.*, 70: 369–388.

Miller, S., Ruit, J.B. and Van der Meché, F.G.A. (1977) Reversal of sign of long spinal reflexes dependent on the phase of the step cycle in the high decerebrate cat. *Brain Res.*, 128: 447–459.

Orlovsky, G.N. (1972) Activity of rubrospinal neurons during locomotion. *Brain Res.*, 46: 99–112.

Orlovsky, G.N. and Shik, M.L. (1965) Standard elements of cyclic movement. *Biofizika*, 10: 847–854. (Engl. transl. 935–944).

Pearson, K.G. and Duysens. J. (1976) Function of segmental reflexes in the control of stepping in cockroaches and cats. In *Neural Control of Locomotion, Vol. 18*. R. Herman, S. Grillner, P.S.G. Stein, and D. Stuart (Eds.). Plenum Press, New York, pp. 519–538.

Rossignol, S. (1977) The control of crossed extensor and crossed flexor responses. *Neurosci. Abstr.*, III: 277, no. 885.

Sherrington, C.S. (1910) Flexion-reflex of the limb, crossed extension reflex, and reflex stepping and standing. *J. Physiol. (Lond.)*, 40: 28–121.

Sjöström, A. and Zangger, P. (1976) Muscle spindle control during locomotor movements generated by the deafferented spinal cord. *Acta physiol. scand.*, 97: 281–291.

Two Functions of Reflexes in Human Movement: Interaction with Preprograms and Gain of Force

R. JUNG

Neurologische Klinik, Abteilung für Neurophysiologie, Universität Freiburg i.Br., 7800 Freiburg i.Br. (F.R.G.)

INTRODUCTION

For the coordination of supraspinal and spinal mechanisms in man I should like to discuss the functional significance of reflex control in two conditions that were somewhat neglected during this symposium. The first is a reflectory addition to preprogramming extensor innervation compensating side differences, due to hemispheric dominance in falling. The second is a reflectory gain of force after ground contact during locomotion.

Both reflex mechanisms are *additive to voluntary innervation* as demonstrated by recent experiments in our department: telemetric recordings of Dietz and Noth, mentioned in preliminary communications (1977, 1978a,b), or unpublished observations, have shown these compensatory functions in human electromyograms. Of course, voluntary innervation must be measured in *man* since we cannot ask cats or monkeys to apply maximal voluntary forces. Our results demonstrate also a third point: purely physical spring action of muscles and its visco-elastic forces are not sufficient to compensate the impact of ground contact and must be supplemented by reflex activation of motoneurons after external resistance.

REFLEX INTERACTION WITH PREPROGRAMMING

Reflectory compensation of asymmetrical innervation during falling on extended arms. Rectified and averaged electromyograms and mechanical recordings of human arm muscles showed a marked EMG pre-activity in the extensor muscles of the arms that prepare the landing posture of the arms. After the impact of landing on the ground this pre-innervation was increased by stretch responses in the triceps brachii that compensated more or less asymmetrical muscle contraction. The combined reflectory and preprogrammed activation was often stronger than maximal voluntary innervation. This reflex activation after the impact of landing depended upon the strength of pre-innervation and showed in most untrained subjects side differences according to arm dominance. Right-handers usually had a stronger pre-innervation in the right arm extensors with less reflex innervation 20–120 msec after the impact than in the left arm where the strongest reflex activation occurred somewhat later. Conversely, left-handers showed more pre-innervation in their dominant left arm

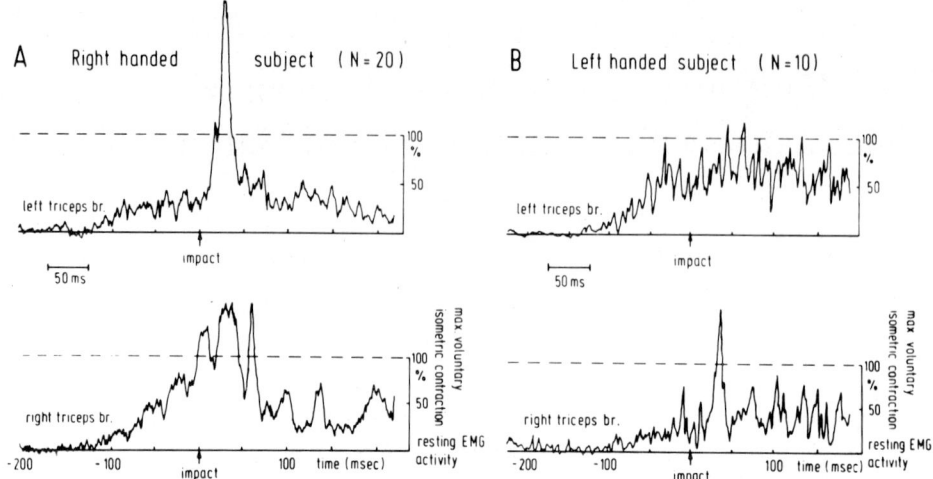

Fig. 1. Reflectory compensation of asymmetrical arm extensor pre-innervation during falling on extended arms. Rectified averaged EMG recordings from both triceps brachii muscles in a right-hander (A) and left-hander (B). The peak of reflex activation is larger when the pre-innervation is smaller in the non-dominant arm. (A) The right-hander shows a stronger EMG pre-innervation of the dominant right extensors. The preprogrammed activity in the right triceps rises continuously before ground contact and exceeds maximal voluntary innervation shortly after the impact. The smaller pre-innervation of the left extensors is compensated by larger early reflectory activation (begin 15 msec and peak 30 msec after impact). EMG average of 20 falls. (B) In the left-hander pre-innervation is more marked on the left dominant side and reflex activity is small. In the right triceps that was less innervated prior to the fall on the hand a higher peak of reflex activation appears 40 msec after impact. EMG average of 10 falls. The activation after ground contact exceeds more or less the maximal isometric voluntary EMG innervation (100% marked by a dotted line). (From experiments of Dietz et al., 1977.)

extensors and more reflex activity in the opposite right arm. The side asymmetries were less marked in sportively trained subjects. Fig. 1 shows the essential findings in two selected untrained subjects to demonstrate that reflex activity compensates insufficient and asymmetrical preprogrammed activity. The early peaks with short latencies (15–35 msec) are more or less synchronized spinal stretch reflexes; later peaks could also be supraspinal reflexes. We may conclude that side differences of preprogrammed cortical motor activation due to hemispheric dominance in man can be compensated by reflexes.

REFLECTORY INCREASE OF MAXIMAL VOLUNTARY INNERVATION

Gain of force during locomotion. Reflex activity enables human motor performance to achieve much higher muscular force than can be executed by maximal voluntary contraction. This increase of innervation, depicted in Fig. 1 for falling, is still higher in human locomotion. In the running man, Dietz found about double up to triple EMG activity of voluntary maximal contraction in extensor muscles during the stance phase that precedes the next step in a running cycle.

Dietz has made detailed studies of locomotion in sprinters and other human subjects to elucidate the interaction of spontaneous and reflex innervation: preprogrammed activation, maximal isometric voluntary contraction in EMG averaging, and

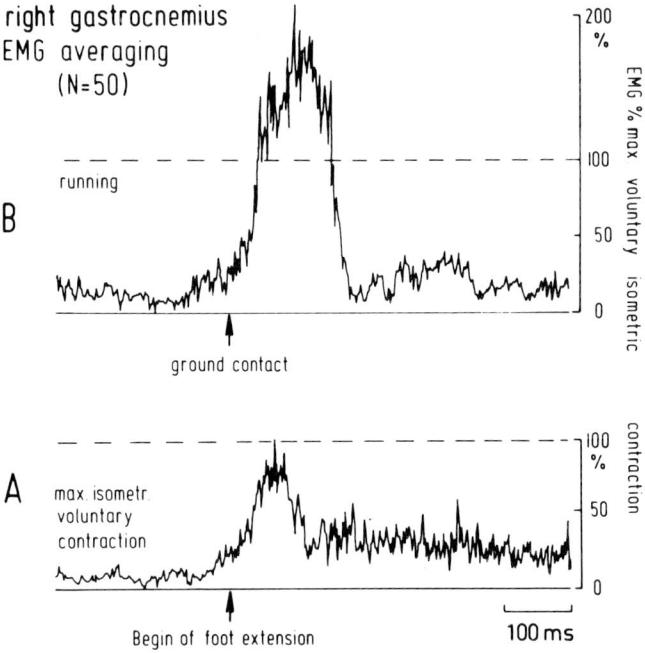

Fig. 2. Reflectory increase of maximal voluntary leg extensor innervation (A) during running (B) demonstrated by EMG integration of right *M. gastrocnemius* in man. (A) Maximal voluntary innervation during isometric contraction of foot extensors against a fixed resistance in sitting posture. (B) Records obtained during running (same subject and electrode placement as in (A) show double EMG-amplitudes after ground contact, compared to maximal voluntary activation (100%). 50 rectified average EMG-recordings from a healthy man, 23 years old. (From Dietz, unpublished observations.)

mechanical force were compared with electrical stimulation effects. Fig. 2 shows an example of voluntary and locomotor innervation of foot extensors. The mechanical force was found to increase from 70 kpond during maximal voluntary contraction to about 160 kpond during the stance phase of running. The preprogrammed EMG-activity alone that always precedes ground contact may be just strong enough to compensate the impact of body weight on one leg but not to push it off to the next step. Hence, it is superimposed by reflexes and this results in a maximal mechanical effect at the end of the stance phase for pushing off the leg. This reflex gain occurs also when the stance phase becomes very short during rapid running. The duration of the gained force after ground contact depends on the speed of running. The timing of the gain of innervation added to the preprogrammed extensor pre-innervation proves that it is achieved by reflexes: it begins 30–40 msec after ground contact and lasts 60–140 msec, varying with locomotion velocity. The gain of force is much smaller when many afferent impulses of muscle and skin receptors are blocked by ischemia (Noth et al., 1978).

These and other experiments demonstrate that the preprogrammed muscle activation of locomotion, be it voluntary or learned, needs additional sources of motoneuron activation to meet ground resistance and to push off the body weight that lasts on one leg during running. Hence, reflexes are necessary for optimal achievements of human locomotion.

CENTRAL COORDINATION, REFLEXES AND MUSCLE MECHANICS

Supraspinal and spinal factors. The rather complex innervation patterns of human locomotion having cortical, subcortical, cerebellar and labyrinthine components make it difficult to analyse the various cerebral and spinal mechanisms and their interaction with reflexes. However, the timing and amplitudes of EMG records allow a few conclusions about spinal and supraspinal mechanisms. According to Hoffmann's (1922) old concept, we had expected that short monosynaptic reflexes were the main mechanisms of extensor activation. However, the typical short-latency synchronized biphasic EMG impulses of the "Eigenreflexe" were less apparent than other spinal and probably also supraspinal stretch responses that appeared with longer latencies beyond 30 msec after touching the ground. These later stretch reflexes seem to contribute mostly to the doubling of maximal voluntary innervation (Fig. 2). The increase of mechanical force appears still later and is prolonged due to the delay of electromechanical coupling and the longer duration of muscle contraction.

Physical and neuronal mechanisms. At this symposium the papers of Allum and Büdingen (chapter IC2) and Roberts and Stenhouse (chapter IVB1) emphasize the physicomechanical nature of "visco-elastic forces" and the "spring action" of muscles. But also if these purely physical mechanisms of the muscle contribute to the first compensation of impact during falling and running a reflectory motoneuron activation after stretch is the essential factor of compensation. The averaged and rectified EMGs of Figs. 1 and 2, clearly prove the *neuronal origin* of the increase of force. The massive rise of EMG activity must be a true activation and synchronization of the motor neuron pool. The first action prior to ground contact is a preprogrammed supraspinal activation of the spinal neuronal circuits to which reflex activation and synchronization is added.

We conclude that an optimal performance of human movements needs several mechanisms in close coordination: voluntary induction, preprogramming innervation, reflex enhancement of muscle contraction and physico-mechanical spring effects of impact following external resistance. The visco-elastic forces of the muscle increase with muscle tone i.e., with innervation. In other words: also the physical spring effects of the muscle depend upon motoneuronal activation. The reflex mechanisms can compensate insufficient or asymmetrical pre-innervation, adapt them to external resistance and thus increase the force of human motor actions.

SUMMARY

Recordings of rectified averaged EMGs from human extensor muscles during falling on extended arms and during running demonstrate two reflex functions which are additive to preprogrammed innervation. These stretch responses after ground contact contribute to active physiological and passive physical mechanisms of muscle resistance.

1. Stretch reflexes compensate asymmetrical arm extensor innervation which is preprogrammed during falling (Fig. 1).
2. Stretch reflexes increase maximal voluntary innervation of leg extensors up to double or triple values during the stance phase of locomotion (Fig. 2).

The reflex enhancement of pre-innervation after ground contact in running shows that the visco-elastic spring action of muscle alone is not a sufficient counter action of impact and needs additional force by peripheral feedback to push off the leg and body weight. This gain of force is of reflectory neuronal origin.

REFERENCES

Dietz, V. and Noth, J. (1978a) Pre-innervation and stretch responses of triceps brachii in man falling with and without visual control. *Brain Res.*, 142: 576–579.

Dietz, V. and Noth, J. (1978b) Spinal stretch reflexes of triceps surae in active and passive movements. *J. Physiol. (Lond.)*, 283: 208 P.

Dietz, V., Noth, J. and Jung, R. (1977) Arm-dominance, motor learning and stretch responses in falling with extended arms. *Pflügers Arch*, Suppl. 368: R 38.

Hoffmann, P. (1922) *Untersuchungen über die Eigenreflexe (Sehnenreflexe) menschlicher Muskeln.* J. Springer, Berlin.

Noth, J., Ledig, T., Schmidtbleicher, D. and Dietz, V. (1978) Ischemic experiments on phasic stretch reflexes in human arm muslces. *Pflügers Arch.*, Suppl. 373: R 71.

SECTION II

PROPRIOCEPTIVE INFLUENCES FROM NECK RECEPTORS

What Are the Proprioceptors of the Neck?

F.J.R. RICHMOND and V.C. ABRAHAMS

Department of Physiology, Queen's University, Kingston, Ontario K7L 3N6, (Canada)

Since the experiments of Magnus and his colleagues (1926), which showed that tonic neck reflexes arose from receptors supplied by upper cervical segments, the neck has been regarded as an important proprioceptive organ for postural processes. Disturbances of gait may be produced in experimental animals by interference with upper cervical sensory supply, either by damaging (Longet, 1845) or anaesthetizing neck muscles (Abrahams and Falchetto, 1969; De Jong et al., 1977) or cutting upper cervical dorsal roots (Cohen, 1961; Richmond et al., 1976). Similarly in man, dizziness and ataxia are seen to follow damage or anaesthesia of neck muscles (Weeks and Travell, 1955; Cope and Ryan, 1959; Jongkees, 1969; De Jong et al., 1977), whiplash injuries (Finneson, 1969), altered cervical vascularity (Sandstrom, 1962) or disturbances of cervical sympathetic tone (Hinoki and Niki, 1975).

The receptors which play an important role in neck reflexes and other postural processes have never been identified. Certain attractive candidates are the large numbers of muscle spindles first described by Voss (1958) and Cooper and Daniel (1963) in dorsal neck muscles of man and the cat. Deprivation of this afferent input is known to provoke postural deficits (Abrahams and Falchetto, 1969). However, its role in the production of tonic reflexes has been discounted by McCouch et al., (1951) who showed that tonic reflexes were not lost following section or denervation of the main neck muscle mass, but could be abolished by cutting nerves which innervated tissue close to intervertebral joints. Hikosaka and Maeda (1973) also showed that modulation of the vestibulo-ocular reflex could be produced by input from neck joints but not from large neck muscles. However, it has been difficult to histologically preserve receptors near bone so that their identity has not been determined.

Two groups of receptors therefore demanded systematic investigation. The first were the receptors of dorsal neck muscles whose high density had no obvious explanation and no clearly understood organization. The second was the unknown receptor population of the upper cervical vertebral column, which although assigned a function, could not be assigned a form.

RECEPTORS OF DORSAL NECK MUSCLES

Muscle spindles

Detailed examination of receptors in dorsal neck muscles (Richmond and Abrahams, 1975b) have extended early observations (Voss, 1958; Cooper and

TABLE I

COMPARISONS OF SPINDLE CONTENTS IN SEVERAL CAT MUSCLES

Muscle	Mean weight (g)	Spindle content	Spindle density (/g)	Content of tandem spindles (% total)	Content of one-bag spindles (% total)	Source
M. gastrocnemius	7.34	46–80	9	11	16	a–c, h
Rectus femoris	8.36	77–132	12	16–20	23	d, k
Soleus	2.49	40–70	23	10	5–13	a–c, h, i
Flexor carpi radialis	1.27	50–57	43–65	12*	3*	f
Extensor digitorum brevis	0.76*	56	74*	21	16	a, c, e, h
Vth interosseus (hind paw)	0.33	22–33	88	–	3	j, i
Splenius	2.92	148–189	47–66	32	27–34*	g
Biventer cervicis	2.11	173–190	74–96	30	10–28*	g
Complexus	2.52	190–254	71–107	33	32–34*	g
Rectus capitis major	0.67	30–57	48–84	5–24	6–31*	g
Occipitoscapularis	0.66	6–15	13–19	0	0*	g

Source: a, Chin et al., 1962; b, Swett and Eldred, 1960; c, Eldred et al., 1962; d, Barker and Chin, 1960; e, Bridgman et al., 1962; f, Gonyea and Ericson, 1977; g, Richmond and Abrahams, 1976b; h, Eldred et al., 1974; i, Boyd, 1962; j, Ip, 1962; k, Barker and Gidumal, 1961; *, Unpublished observations.

Daniel, 1963) that neck muscles are spindle-rich. Densities of spindles in 4 dorsal neck muscles, biventer cervicis (BC), complexus (CM), splenius (SP) and rectus capitis major (RCM) generally exceed 50 spindles/g and often approach or exceed 100 spindles/g (Table I). Similarly high spindle densities are shared by certain small muscles in the cat such as those controlling movements of the paw (Barker and Chin, 1960; Ip, 1961). However, high spindle densities in neck muscles are a particular curiosity when it is recognized that three of the neck muscles are much larger than paw muscles and have large absolute numbers of spindles. A single BC, CM or SP muscle usually contains no fewer than 150 spindles and contents of more than 250 spindles have been observed. Upper cervical nerves which bilaterally serve BC, CM and SP therefore contain afferent fibres from approximately 1000 spindles.

Not all neck muscles, however, have high spindle densities. Occipitoscapularis (OC), which inserts like other dorsal neck muscles on the lamboidal crest but originates from the scapula, contains only a few spindles with densities of only 13–18 spindles/g (Table I) (Richmond and Abrahams, 1975b). Despite its close association with neck muscles subserving head movement, OC is thought to rotate the scapula (Elliott, 1961). Studies are presently underway to determine whether other muscles concerned with shoulder movement show a paucity of spindles compared to muscles subserving head movement despite the fact that they are closely apposed anatomically.

Spindles of neck muscles are also of interest because of their unusually complex organization. In neck muscles with high spindle densities, most spindles form chain-like complexes of 2–6 spindles linked in tandem, paired and parallel groupings (Fig. 1) (Richmond and Abrahams, 1975b). About one-third of spindles in large dorsal neck muscles form tandem linkages in which intrafusal fibres from a "parent" spindle run into other spindles to be contacted by additional primary endings (Table I). The

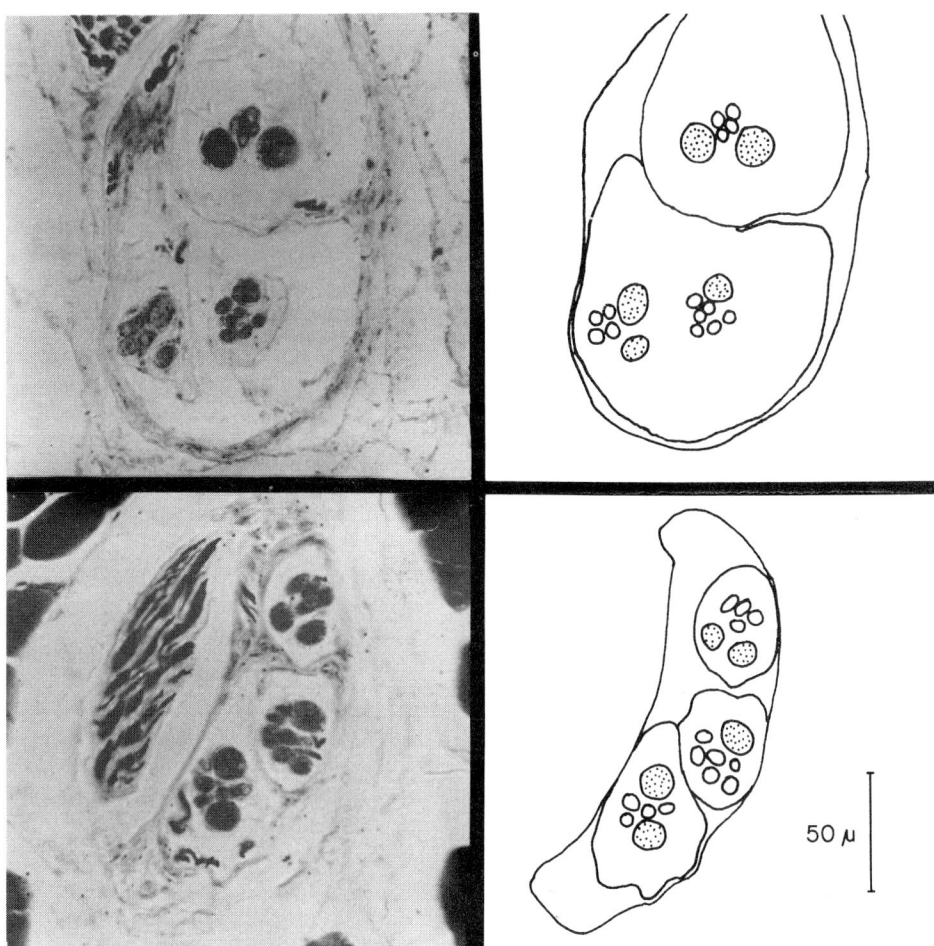

Fig. 1. Photomicrographs of a neck muscle spindle complex at equatorial (top) and polar (bottom) levels. Line drawings illustrate intrafusal fibre and capsular arrangements. At polar regions each spindle is surrounded by a connective tissue capsule. At equatorial regions, the capsular division between 2 spindles is lost, but the third spindle remains separately encapsulated.

functional significance of the tandem complex is not yet understood. It has been suggested that these complexes permit information to be collected over a longer length of muscle (Swett and Eldred, 1960), or that the sharing of intrafusal fibres may lead to a sharing of fusimotor control (Richmond and Abrahams, 1975b).

Many spindles in tandem conjunctions contain only a single nuclear bag fibre plus one to three nuclear chain fibres (Fig. 2). Spindles with only one nuclear bag fibre are also seen as independent receptors with the result that these smaller spindles account for 1 out of every 3 spindles in most BC, SP and CM muscles. Spindles with only one bag fibre are seen less frequently in limb muscles (Table I). Consequently, they have not been subjected to the kind of detailed physiological examinations which have been conducted on the more common spindles of the hindlimb containing two morphologically and enzymically different nuclear bag fibres which both contribute to the primary response (Boyd, 1976; Boyd et al., 1977; Barker et al., 1976, 1977, 1978). An interesting question is whether spindles lacking one of these nuclear bag fibres

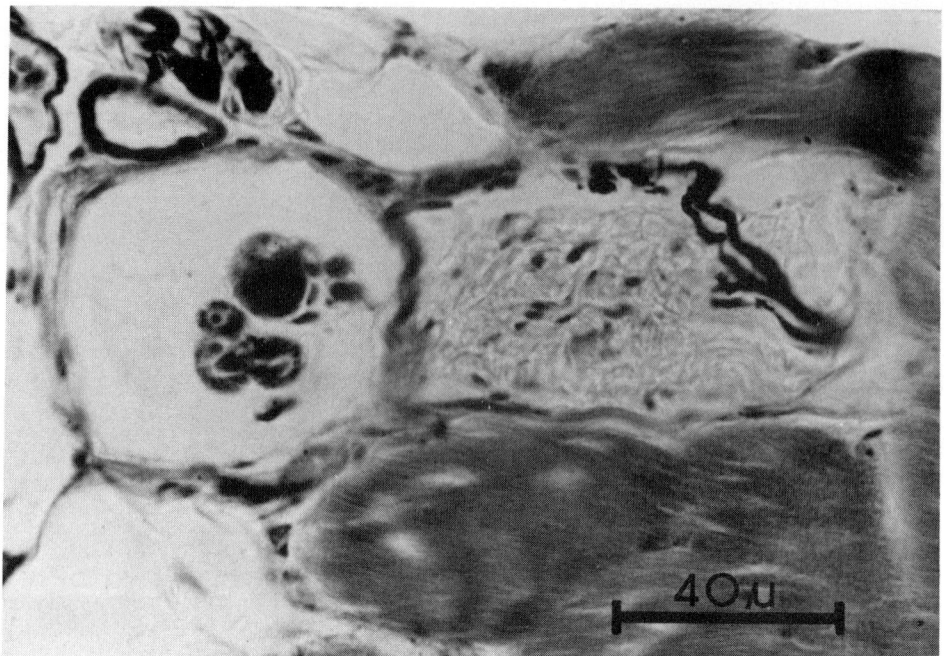

Fig. 2. Morphology of a typical dyad complex. The muscle spindle (left) contains 4 intrafusal fibres, one nuclear bag fibre of large diameter and 3 smaller chain fibres. The Golgi tendon organ (right) is distinguished by its content of collagen fibres interwoven between small nerve fibres.

may produce a different sensory signal than spindles which have two types of bag fibre.

Responses of muscle spindles in BC, CM and SP have recently been examined using single fibre recording techniques (Richmond and Abrahams, 1979). Most neck spindle afferents responded like primary or secondary endings of spindles in hindlimb muscles. Afferents classified as primary endings had marked variability of resting firing frequency with individual values of instantaneous impulse frequency varying from main firing frequency by more than 4%. Primary responses to ramp stretch were characterized by an abrupt acceleration of firing frequency at the onset of ramp stretch and a large dynamic index, usually exceeding 30 imp/sec. In contrast, "secondary" afferents displayed more regular firing patterns in which instantaneous frequency varied from mean firing frequency by less than 4%. These endings showed only a small progressive increase in their firing rate during the dynamic phase of ramp stretch and had small dynamic indices below 20 imp/sec. Some spindle afferents (22/107) had responses which did not match the criteria established to differentiate primary and secondary endings, but showed a mixture of primary and secondary characteristics. Most commonly such endings displayed a large variability of firing frequency like the primary ending, but had a small dynamic response with dynamic indices below 25 imp/sec.

Afferent fibres in C3 neck muscle nerves seldom exceed a diameter of 16 μm (Richmond et al., 1976). Conduction velocities of neck spindle afferents in these nerves were correspondingly low (13–90 m/sec) and most fibres (92%) conducted at less than 70 m/sec (Richmond and Abrahams, 1978). Although primary afferents had significantly higher conduction velocities than secondary afferents in all 3 muscles, only afferents from SP could be divided into subgroups on the basis of conduction velocity. SP primary afferents all conducted faster than 45 m/sec, with average velocities of 65 m/sec, while secondary afferents had values under 40 m/sec. Primary afferents from BC and CM tended to conduct more slowly than those from SP (avg. = 50 m/sec) and the range of primary conduction velocities overlapped with the range of secondary values.

The most obvious feature of neck spindle behaviour is its similarity to that of spindles elsewhere in the cat. Muscle spindles of the neck do not appear to suffer functional deficits or changes which might explain their increased numbers or arrangements in complex forms. The question still remains why so many receptors are needed in the neck muscle system.

One possibility is suggested by the observation that most afferents showed a pause in firing only when extrafusal fibres in the vicinity of the receptor were stimulated to contract. Contractions in muscle regions distant to the receptor usually failed to influence the firing of the spindle. These results suggested that spindles may respond in a sensitive manner to mechanical changes of extrafusal fibres in their immediate vicinity but may have less sensitivity to the behaviour of the main muscle mass unless its action involves extrafusal fibres close to the spindle.

The concept of "compartmentalized" receptor function (see Botterman et al., 1978 for review), may be of particular significance for dorsal neck muscles which have an unusual fibre architecture (Richmond and Abrahams, 1975a). Each large dorsal neck muscle is composed of extrafusal fibres of many lengths, organized in parallel and series arrangements by a matrix of collagen bands (tendinous inscriptions). Each muscle is innervated by a series of separate nerves from different cervical segments (Fig. 3) which enter the muscle at different levels and appear to provide the motor and sensory innervation to a circumscribed muscle region. For example (Fig. 3), muscle spindles with afferent fibres in C3 dorsal roots are confined to a central region of each muscle. The motor units supplied by this nerve have been recently shown using glycogen depletion techniques (Richmond and Scott, unpublished observations) to occupy the same muscle region. This central muscle region is linked in series to rostral and caudal muscle regions supplied by more anterior or posterior nerve branches. It might be suggested that the large numbers of muscle spindles which are spread throughout each muscle may be necessary to provide fine-grain information about the different muscle subsections whose activity must be integrated to provide balanced and effective muscle contractions.

Golgi tendon organs

Golgi tendon organs (GTOs) are frequently observed in dorsal neck muscles, but their numbers have not been systematically examined. In one incomplete BC muscle, 51 GTOs have been counted (Richmond and Abrahams, 1975b). Tendon organs are located not only at tendons of origin and insertion, but also within tendinous inscriptions. They resemble GTOs in other muscles both in structure (Richmond and Abrahams, 1975b) and in response characteristics (Richmond and Abrahams, 1978). Afferents from GTOs rarely fire spontaneously when neck muscles are passively stretched within their optimum length range, but begin to fire regularly and with progressively increasing frequency as the muscle is passively stretched to very long muscle lengths. Burst discharge is readily evoked by twitch contractions of extrafusal fibres near (and probably inserting into) the receptor. However, contractions of

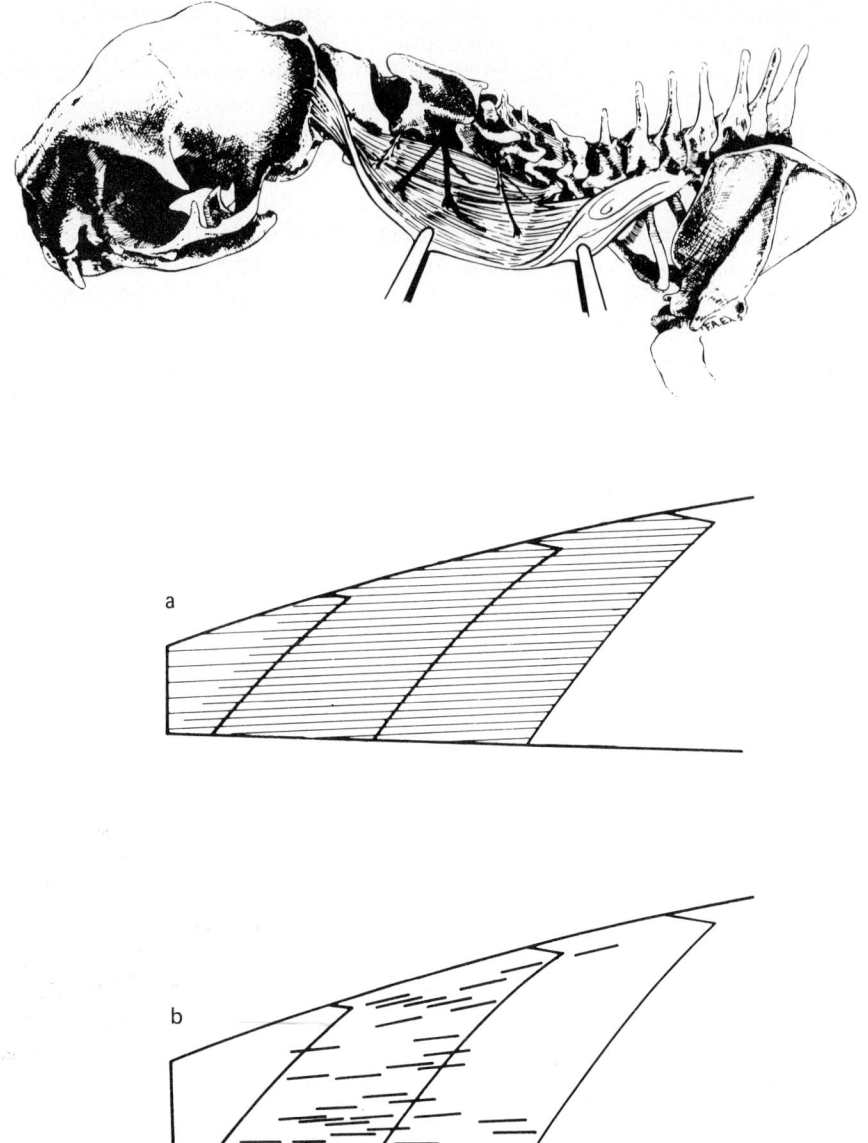

Fig. 3. Distribution of motor units (a) and muscle spindles (b) innervated by C3 muscle nerves. Innervation of BC is shown on the anatomical drawing; C3 nerves are shown in black, C2 (rostral) and C4 (caudal) nerves are white. Motor units supplied by C3 nerves are concentrated between the first and third tendinous inscriptions which are shown as black lines traversing the muscle. Some fibres run rostrally through first inscription to insert at the lamboidal crest. Muscle spindles supplied by C3 nerves are also distributed in the same muscle regions, with the majority of spindles between the first and second inscription.

extrafusal fibres distant to the GTO do not usually cause it to fire. Like primary afferents of spindles in the neck, GTO afferents conduct at relatively low conduction velocities (50–70 m/sec) (Richmond and Abrahams, 1979).

Golgi tendon organs are regularly observed in dyad with muscle spindles (Fig. 2). The functional significance of this complex is not understood.

RECEPTORS AROUND VERTEBRAE

Upper cervical vertebrae are linked both by short muscles and by connective tissue at regions of articulation. This tissue has been examined in serial cross-sections of vertebrae which have been decalcified and sectioned with muscles and connective tissue intact (Fig. 4).

The most prominent receptors of perivertebral tissue are those in muscles around vertebrae. Muscle spindles are particularly numerous; a single intertransversarius muscle contained 200 spindles despite its small estimated weight of 0.4 g. Many spindles are seen in circumscribed muscle regions usually near intramuscular tendons (Fig. 5) where they may be clustered in elaborate complexes of 3–10 spindles (e.g., Fig. 4). In contrast, other regions of the same muscles do not contain spindles. The orientation of spindles in chains between certain intramuscular tendons would suggest that most spindles are in the direct "line of pull" when muscles are stretched. Physiological studies have shown that small flexions of upper cervical joints cause major changes in the firing rate of spindle afferents from perivertebral muscles (Richmond and Abrahams, 1979).

GTOs are also observed at tendinous junctions in regions of high spindle density. Some GTOs occur as isolated receptors, but many are in dyad with spindles, and some

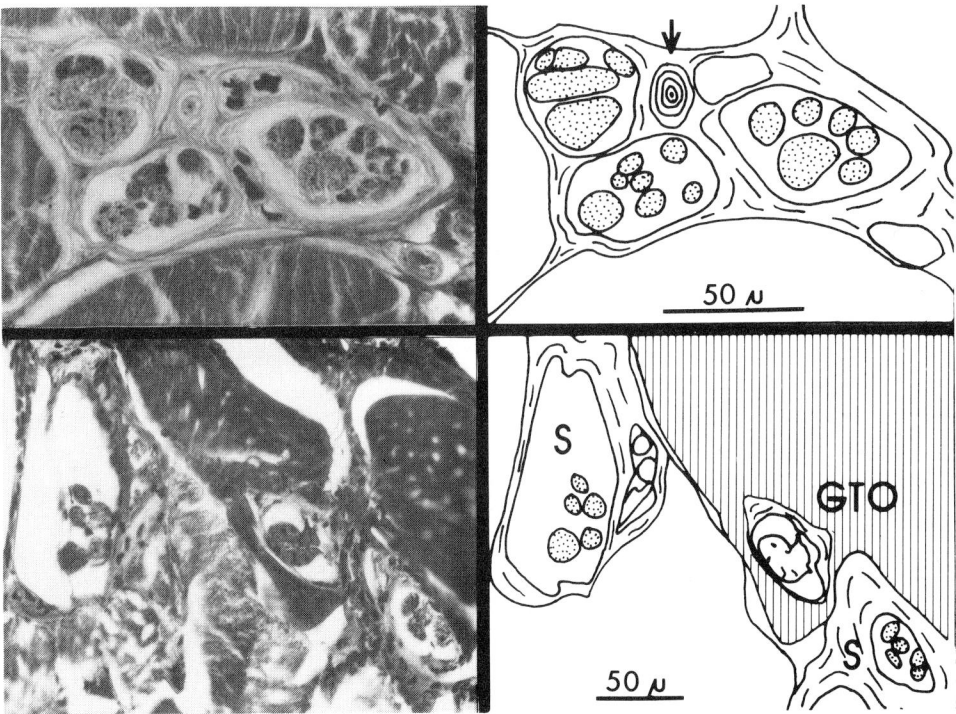

Fig. 4. Photomicrographs and matched line drawings of receptor complexes in perivertebral muscles. In top photomicrograph 3 spindles are grouped with a structure which resembles a small paciniform corpuscle (marked by an arrow). Bottom photomicrograph shows 2 spindles close to their insertion at tendon (shown as hatched region in line drawings). A Golgi tendon organ is located at the neuromuscular junction. Like many GTOs in perivertebral muscles, this receptor appears smaller and less complex than other larger and more typical GTOs such as that in Fig. 2.

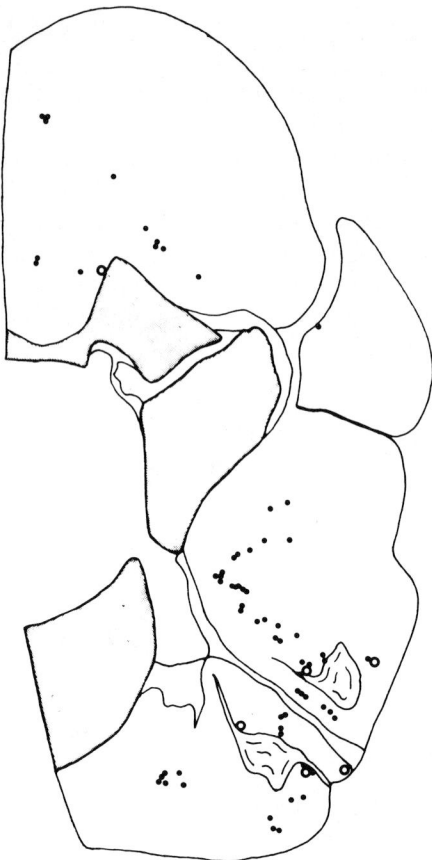

Fig. 5. Projection drawing of a serial section through one half of a decalcified vertebral joint (dorsal surface is top). Bone is shaded grey, muscle and connective tissue are not shaded. Muscle spindles and GTOs at this level of section are shown as black dots and open circles respectively. Spindles are most often clustered in bands and groups in particular regions of muscle often near tendons. GTOs are usually located at musculotendinous junctions in these same regions.

GTOs are grouped in clusters. Most GTOs are morphologically similar to those in other muscles, but some GTO-like structures are smaller (Fig. 4) and appear to serve as insertion points only for intrafusal fibres of a single spindle.

A few paciniform corpuscles were observed close to GTOs or spindle complexes but not outside muscle tissue. Other kinds of muscle and joint receptors were not observed, in part perhaps because receptors without a distinct structure are not easily delineated in transverse sections. Small nerve bundles coursed through connective tissue to ramify near blood vessels. While some nerves were undoubtedly sympathetic, a few nerves appeared to branch into restricted arbours in the connective tissue itself. Whether these nerves supplied free endings or Ruffini-type sprays could not be determined.

These results do not resolve the nature and position of "joint" receptors in cervical segments. They do, however, suggest that any consideration of the "joint receptor" population must include the very large numbers of muscle spindles and GTOs of perivertebral muscles which are strategically located to sense joint position as a function of muscle length or tension.

SUMMARY

Muscles which subserve head movement in the cat contained unusually large numbers of receptors which are linked in a variety of conjunctive forms. Muscle spindles are particularly numerous both in large dorsal neck muscles which usually contain 50–100 spindles /g muscle tissue and in small perivertebral muscles which may contain as many as 500 spindles /g tissue. Neck muscle spindles resemble those described in the hindlimb, with the exception that many neck muscle spindles have a reduced intrafusal fibre complement of only a single nuclear bag fibre and a few nuclear chain fibres. In most neck muscles, the majority of spindles do not occur as single receptors but are arranged in complexes of up to 10 spindles in tandem, paired and parallel linkages.

Most muscle spindle endings from large dorsal neck muscles had physiological properties similar to primary or secondary endings of hindlimb muscle spindles. About 20% of neck spindle endings had a mixture of response properties so that they could not be defined as primary or secondary endings by physiological criteria. Afferents from neck muscles conducted at relatively low velocities ranging from a low of 13 m/sec to a high of 90 m/sec. These conducting velocities are consistent with our anatomical observations that neck muscle afferent fibres from neck muscles seldom exceed a diameter of 16 μm.

Golgi tendon organs are also observed in neck muscles where they frequently occur in dyad with muscle spindles. Paciniform corpuscles are occasionally present near Golgi tendon organs and muscle spindles. Neither Golgi tendon organs nor paciniform corpuscles could be found in the connective tissue around vertebral joints. Although nerve fibres were observed in joint connective tissue, no receptors with a clearly defined morphology could be found in this tissue.

REFERENCES

Abrahams, V.C. and Falchetto, S. (1969) Hind leg ataxia of cervical origin and cervico-lumbar interactions with a supratentorial pathway. *J. Physiol. (Lond.),* 203: 435–447.

Barker, D. and Chin, N.K. (1960) The number and distribution of muscle-spindles in certain muscles of the cat. *J. Anat. (Lond.),* 94: 473–486.

Barker, D. and Gidumal, J.L. (1961) The morphology of intrafusal muscle fibres in the cat. *J. Physiol. (Lond.),* 157: 513–528.

Barker, D., Emonet-Denand, F., Harker, D., Jami, L. and Laporte, Y. (1976) Distribution of fusimotor axons to intrafusal muscle fibres in cat tenuissimus spindles as determined by the glycogen-depletion method. *J. Physiol. (Lond.),* 261: 49–69.

Barker, D., Emonet-Denand, F., Harker, D.W., Jami, L. and Laporte, Y. (1977) Types of intra- and extrafusal muscle fibre innervated by dynamic skeleto-fusimotor axons in cat peroneus brevis and tenuissimus muscles, as determined by the glycogen-depletion method. *J. Physiol. (Lond.),* 266: 713–726.

Barker, D., Bessou, P., Jankowska, E., Pages, B. and Stacey, M.J. (1978) Identification of intrafusal muscle fibres activated by single fusimotor axons and injected with fluorescent dye in cat tenuissimus spindles. *J. Physiol. (Lond.),* 275: 149–165.

Botterman, B.R., Binder, M.D. and Stuart, D.G. (1978) Functional anatomy of the association between motor units and muscle receptors. *Amer. Zool.,* 18: 135–152.

Boyd, I.A. (1962) The structure and innervation of the nuclear bag muscle fibre system and the nuclear chain muscle fibre system in mammalian muscle spindles. *Phil. Trans. R. Soc. B,* 245: 81–136.

Boyd, I.A. (1976) The response of fast and slow nuclear bag fibres and nuclear chain fibres in isolated cat muscle spindles to fusimotor stimulation and the effect of intrafusal contraction on the sensory endings. *Quart. J. exp. Physiol.,* 61: 203–254.

Boyd, I.A., Gladden, M.H., McWilliam, P.N. and Ward, J. (1977) Control of dynamic and static nuclear bag fibres and nuclear chain fibres by gamma and beta axons in isolated cat muscle spindles. *J. Physiol. (Lond.)*, 265: 133–162.

Bridgman, C.F., Eldred, E. and Eldred, B. (1962) Distribution and structure of muscle spindles in the extensor digitorum brevis of the cat. *Anat. Rec.*, 143: 219–227.

Chin, N.K., Cope, M. and Pang, M. (1962) Number and distribution of spindle capsules in seven hindlimb muscles of the cat. In *Symposium on Muscle Receptors*, D. Barker (Ed.), University Press, Hong Kong, pp 241–248.

Cohen, L.A. (1961) Role of eye and neck proprioceptive mechanisms in body orientation and motor coordination. *J. Neurophysiol.*, 24: 1–11.

Cooper, S. and Daniel, P.M. (1963) Muscle spindles in man, their morphology in the lumbricals and the deep muscles of the neck. *Brain*, 86: 563–594.

Cope, S. and Ryan, G.M.S. (1959) Cervical and otolith vertigo. *J. Laryng.* 73: 113–120.

DeJong, P.T.V.M., De Jong, J.M.B.V., Cohen, B. and Jongkees, L.B.W. (1977) Ataxia and nystagmus induced by injection of local anaesthetics in the neck. *Ann. Neurol.*, 1: 240–246.

Eldred, E., Maier, A. and Bridgman, C.F. (1974) Differences in intrafusal fiber content of spindles in several muscles of the cat. *Exp. Neurol.*, 45: 8–18.

Eldred, E., Bridgman, C.F., Swett, J.E. and Eldred, B. (1962) Quantitative comparisons of muscle receptors of the cat's medial gastrocnemius, soleus and extensor digitorum brevis muscles. In *Symposium on Muscle Receptors*, D. Barker, (Ed.), University Press, Hong Kong, pp. 207–213.

Elliott, R. (1961) *Anatomy of the Cat*, 3rd ed., J. Reighard and H.S. Jennings (Eds.), Holt, Rinehart and Winston, New York.

Finneson, B.E. (1969) *Diagnosis and Management of Pain Syndromes*. W.B. Saunders Co., Toronto.

Gonyea, W.J. and Ericson, G.C. (1977) Morphological and histochemical organization of the flexor carpi radialis muscle in the cat. *Amer. J. Anat.*, 148: 329–344.

Hikosaka, O. and Maeda, M. (1973) Cervical effects on abducens motoneurons and their interaction with vestibulo-ocular reflex. *Exp. Brain Res.*, 18: 512–530.

Hinoki, M. and Niki, H. (1975) Neurotological studies on the role of the sympathetic nervous system in the formation of traumatic vertigo of cervical origin. *Acta oto-laryng. (Stockh.)*, Suppl. 330: 185–196.

Ip, M.C. (1961) *The Number and Variety of Proprioceptors in Certain Muscles of the Cat*. M.Sc. thesis, University of Hong Kong.

Jongkees, L.B. (1969) Cervical vertigo. *Laryngoscope (St. Louis.)*, 79: 1473–1484.

Longet, F.A. (1845) Sur les troubles qui surviennent dans l'equilibration, la station et la locomotion des animaux apres la section des parties molles de la nuque. *Gaz. Méd. France*, 13: 565–567.

Magnus, R. (1926) Some results of studies in the physiology of posture (Cameron Prize Lectures). *Lancet*, 211: 531–536.

McCouch, G.P., Deering, I.D. and Ling, T.H. (1951) Location of receptors for tonic neck reflexes. *J. Neurophysiol.*, 14: 191–195.

Richmond, F.J.R. and Abrahams, V.C. (1975a) Morphology and enzyme histochemistry of dorsal muscles of the cat neck. *J. Neurophysiol.*, 38: 1312–1321.

Richmond, F.J.R. and Abrahams, V.C. (1975b) Morphology and distribution of muscle spindles in dorsal muscles of the cat neck. *J. Neurophysiol.*, 38: 1322–1339.

Richmond, F.J.R. and Abrahams, V.C. (1979) Physiological properties of muscle spindles in dorsal neck muscles of the cat. *J. Neurophysiol.*, 42: 604–617.

Richmond, F.J.R., Anstee, G.C.B., Sherwin, E.A. and Abrahams, V.C. (1976) Motor and sensory fibres of neck muscle nerves in the cat. *Canad. J. Physiol. Pharmacol.*, 54: 294–304.

Sandstrom, J. (1962) Cervical syndrome with vestibular symptoms. *Acta oto-laryng. (Stockh.)*, 54: 207–226.

Swett, J.E. and Eldred, E. (1960) Distribution and numbers of stretch receptors in medial gastrocnemius and soleus muscles of the cat. *Anat. Rec.*, 137: 453–460.

Voss, H. (1958) Zahl und Anordnung der Muskelspindeln in den unteren Zungenbeinmuskeln dem M. sternocleidomastoideus und den Bauch- und tiefen Nackenmuskeln. *Anat. Anz.*, 105: 265–275.

Weeks, V.C. and Travell, J. (1955) Postural vertigo due to trigger areas in the sternocleidomastoid muscle. *J. Pediat.*, 47: 315–322.

Proprioceptive and Somatosensory Influences on Neck Muscle Motoneurons

P.K. ROSE and N. SPROTT

Department of Physiology, Queen's University, Kingston, Ontario
K7L 3N6 (Canada)

Much of our understanding of the control of movement is based on studies of two motor systems: one involved in the control of eye movement, the other responsible for movements of the hindlimb. It is well recognized that the functional and anatomical organization of these two systems differ in many respects consistent with their different roles. It is not generally accepted, however, that the organization of the lumbosacral spinal cord may represent a unique arrangement suited to tasks related to hindlimb motor control and that other regions of the spinal cord may be specialized according to their function (Abrahams, 1977). Recent studies of the descending systems which project to neck muscle motoneurons indicate that neck muscle motoneurons are subject to several influences unique to the upper cervical spinal cord (Wilson and Peterson, 1978; Anderson, et al., 1971, 1972). Moreover, the structure of neck muscles displays an organization and complexity not seen in hindlimb muscles (Richmond and Abrahams, 1975a). These studies suggest that some of the mechanisms underlying the control of head movement may differ from those involved in the control of hindlimb movement. The aim of this paper is to examine this suggestion further by describing recent studies on 1) the intrinsic organization of the upper cervical spinal cord, 2) muscle afferent projections to neck muscle motoneurons, and 3) somatosensory projections to neck muscle motoneurons.

ANATOMICAL ORGANIZATION OF THE UPPER CERVICAL SPINAL CORD

Few studies exist which describe the intrinsic anatomical organization of the upper cervical spinal cord. Indeed, the locations of the motoneurons of the dorsal neck muscles were only recently described. Using the retrograde transport of horseradish peroxidase (HRP), Richmond et al. (1978) described the location of motoneurons innervating neck extensor muscles, complexus, biventer cervicis and splenius. Motoneurons of all three muscles are found in the ventromedial portion of the ventral horn. The motoneuron pools, however, are not confined to this region and many motoneurons, particularly those innervating the splenius muscle, are located more laterally and dorsally.

The trapezius muscle which lies lateral and superficial to the other major dorsal neck muscles is commonly described as a shoulder muscle (Elliot, 1963). Recently, however, it was suggested that this muscle, particularly the cranial portion, clavo-trapezius, may also play a role in head movement involving flexion or twisting of the head (J. Keane, personal communication). Trapezius motoneurons were located in the dorsal lateral portion of the ventral horn using electrophysiological techniques (Gura and Limanskii, 1977) confirming older anatomical studies (Pearson, 1938).

Comparatively few synapses are found on the somata of lumbosacral motoneurons, the majority being located on the dendritic tree (Conradi, 1969). The structure of identified neck muscle motoneurons was therefore examined using intracellular injections of HRP (Snow et al., 1976). Fig. 1 shows a photomicrograph of a splenius motoneuron injected with HRP and sectioned horizontally. In this plane, dendrites can be seen which project in all directions from the soma. A striking feature of neck muscle motoneurons is the projection of dendrites into the white matter surrounding the ventral horn. As illustrated in Fig. 1, these dendrites travel for distances of up to 1 mm medial and lateral to the soma. Only rarely however, were these "white matter" dendrites traced to the contralateral spinal cord, a characteristic of motoneurons in the sacral cord (Light and Metz, 1978). Recent studies on the ultrastructure of the ventromedial and ventrolateral funiculi demonstrated the existence of longitudinally arranged sheets of dendrites which receive synaptic input. Some of these dendrites may originate from neck muscle motoneurons, which suggests that synaptic input to neck muscle motoneurons is not confined to those systems which terminate in the grey matter.

Most descriptions of motoneuron morphology are currently based on the distribution of dendrites seen in single 100–200 μm thick sections. For large neurons, like motoneurons, this may result in a gross underestimation of the size and complexity of the dendritic tree. Consequently, complete reconstructions of neck muscle motoneurons were drawn by aligning the processes seen on adjacent serial sections Fig. 2 illustrates the partial reconstruction of a trapezius motoneuron sectioned in the sagittal plane. The rostrally and caudally directed dendrites form a remarkably complex dendritic tree which extends 1,700 μm rostrally and 1,350 μm caudally. Similar measurements of other neck muscle motoneurons indicated that many dendrites exceed 1,000 μm in length. The shortest rostral-caudal dimension observed was 2,150 μm and the largest was 3,500 μm. Although the principal orientation of neck muscle motoneuron dendritic trees is directed rostrally and caudally, aligned parallel to the rostral-caudal axis of the spinal cord, dorsally directed dendrites are also found. These frequently project into lamina VII.

These observations indicate that the structure of motoneuron dendritic trees may be significantly more complicated than was previously assumed. Estimates of the surface area of neck muscle motoneurons combined with observations of the area involved in synaptic contacts suggest that each neck muscle motoneuron may receive in the order of 70,000 synapses. This figure is substantially larger than for lumbosacral motoneurons (Gelfan et al., 1970; Barrett, 1975), but whether this difference is related to a population bias in this sample or to differences in cell staining techniques remains to be determined.

Recently, intracellular injections of HRP have been used to describe the distribution of axon collaterals of hindlimb motoneurons (Cullheim and Kellerth, 1978). Similar studies on neck muscle motoneurons indicate that many neck muscle motoneurons, particularly those innervating the biventer cervicis muscle, do not form local axon collaterals. The axon collaterals found for other motoneurons are all distri-

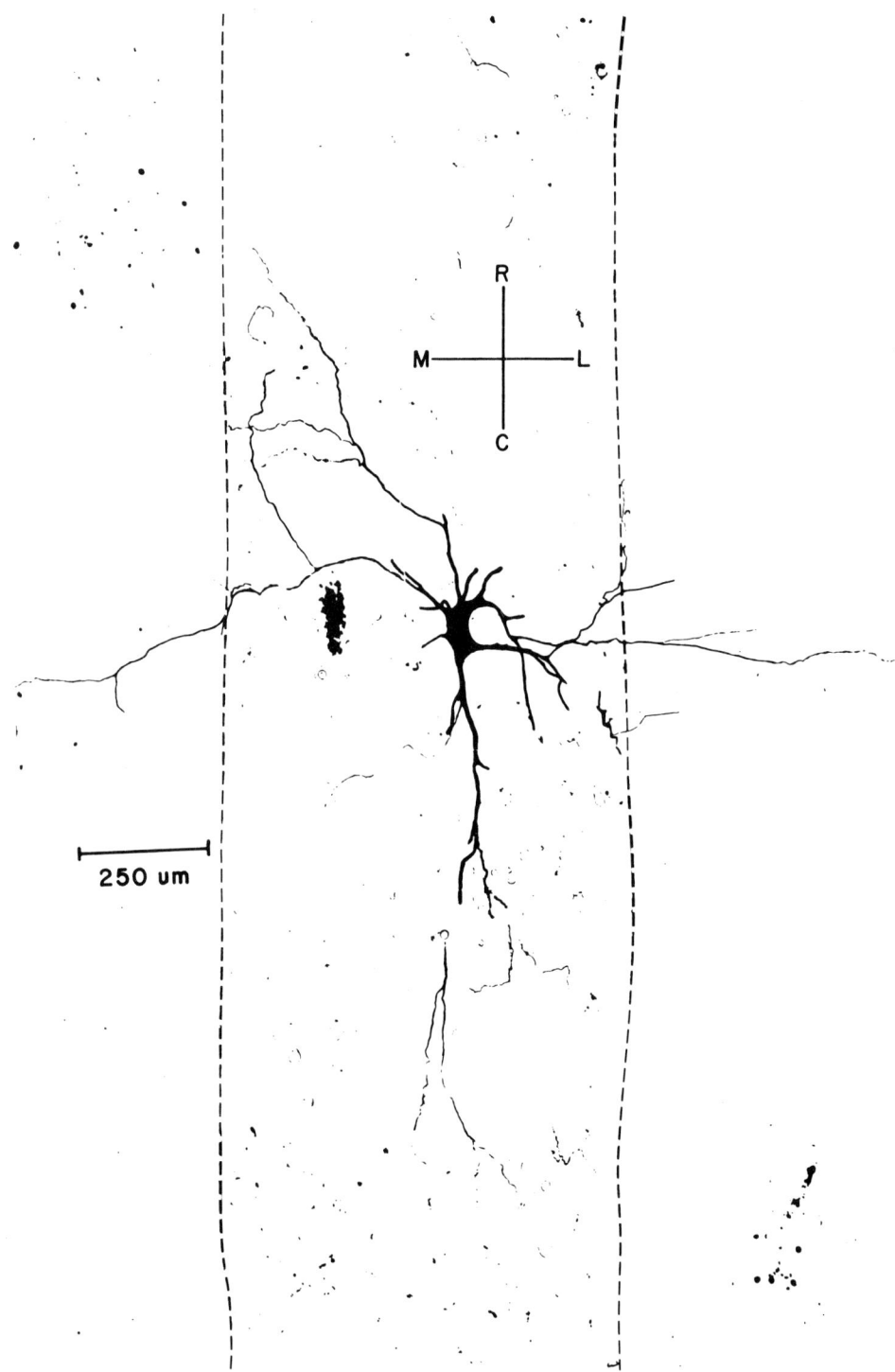

Fig. 1. A photomicrograph montage of a splenius motoneuron intracellularly injected with horseradish peroxidase and sectioned in the horizontal plane. Section thickness 100 μm. Dashed lines indicate the border between grey matter and ventromedial funiculus on the left and grey matter and ventrolateral funiculus on the right. M, medial; L, lateral; R, rostral; C, caudal.

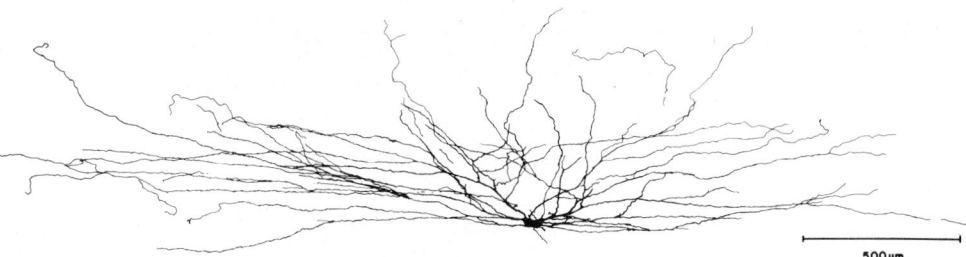

Fig. 2. Reconstruction of part of a trapezius motoneuron. Drawing based on dendritic profiles seen in two adjacent 100 μm thick sections. Remaining part of the dendritic tree spread 900 μm medially and 600 μm laterally.

buted within the motoneuron pool, and no separate projection was found to a "Renshaw cell" area as seen in the lumbosacral cord. This might suggest that either neck muscle motoneurons can interconnect directly (Cullheim et al., 1977) or that, in the upper cervical spinal cord, Renshaw-type cells are located within the motoneuron pool.

MUSCLE AFFERENT PROJECTIONS TO NECK MUSCLE MOTONEURONS

The muscles which move the head are endowed with one of the highest densities of muscle spindles found in any movement system (Richmond and Abrahams, 1975b). Early anatomical studies by Szentagothai (1948) described monosynaptic connections between the upper cervical dorsal roots and cells deep in the ventral horn. By analogy to similar projections seen in the lumbosacral spinal cord, Szentagothai (1948) suggested that these connections represented the anatomical basis for stretch reflexes arising from muscle spindles. However, it was not until recently that monosynaptic connections between group I neck muscle afferents and neck muscle motoneurons were established. Wilson and Maeda (1974) found reciprocal monosynaptic connections between neck extensors, biventer cervicis and complexus. Anderson (1977), in a further study of monosynaptic connections between neck muscles, demonstrated that such connections only exist for homonymous muscle afferents from muscles in the same plane.

Extracellular recordings from neck muscle motoneurons revealed a virtual absence of a monosynaptic discharge (Abrahams et al., 1975). Similar results have recently been reported by Ezure et al. (1978), who found that monosynaptic reflexes were rarely seen and only occurred if the level of spontaneous activity was high. These results suggest that monosynaptic muscle afferent projections to neck muscle motoneurons are substantially weaker than those found in the lumbosacral spinal cord. The finding of long-latency EPSP's and reflexes (Abrahams et al., 1975; Ezure et al., 1978) would suggest that segmental influences are mediated by polysynaptic connections which may involve tectospinal and reticulospinal projections (Anderson et al., 1971, 1972; Abrahams and Rose, 1974; Rose and Abrahams, 1978).

At the present time little information is available on the projections from muscle spindle primary afferents vs. secondary afferents and from Golgi tendon organs. Such studies may be complicated by the difficulty of separating muscle spindle primary and secondary effects due to their overlapping fibre spectra (Richmond et al., 1976). Recently, Rapoport (1977) showed that low intensity stimulation of afferents from sternomastoid and cleidomastoid, antagonists to biventer cervicis, failed to evoke disynaptic inhibition. This result provides further evidence that the segmental organization of muscle afferent projections to neck muscle motoneurons differs from that observed in the lumbosacral spinal cord.

SOMATOSENSORY PROJECTIONS TO NECK MUSCLE MOTONEURONS

Several studies have indicated a close functional relationship between the trigeminal system and upper cervical spinal cord. In 1961 Kerr and Olafsson reported that stimulation of trigeminal afferents excited cells located in the ventral horn of the upper cervical spinal cord. The existence of a "trigemino-neck reflex" was established by Manni et al. (1975) and Sumino and Nozaki (1977), who showed that neck muscle motoneurons were excited by electrical stimulation of several trigeminal afferents. Studies at the single unit level revealed that splenius, biventer cervicis and complexus motoneurons were excited at latencies as short as 7 msec by stimulation of the infraorbital nerve (Abrahams and Richmond, 1977).

Recent experiments in my laboratory have examined the intracellular responses of biventer cervicis-complexus (BC-C), splenius (S) and trapezius (T) motoneurons to stimulation of the infraorbital nerve (IO) which innervate the nose and the supraorbital nerve (SO) which innervates the forehead. Previous intracellular studies were confined to responses of T motoneurons following stimulation of the IO nerves (Gura and Limaskii, 1977).

Fig. 3 shows the response of three neck muscle motoneurons to stimulation of the

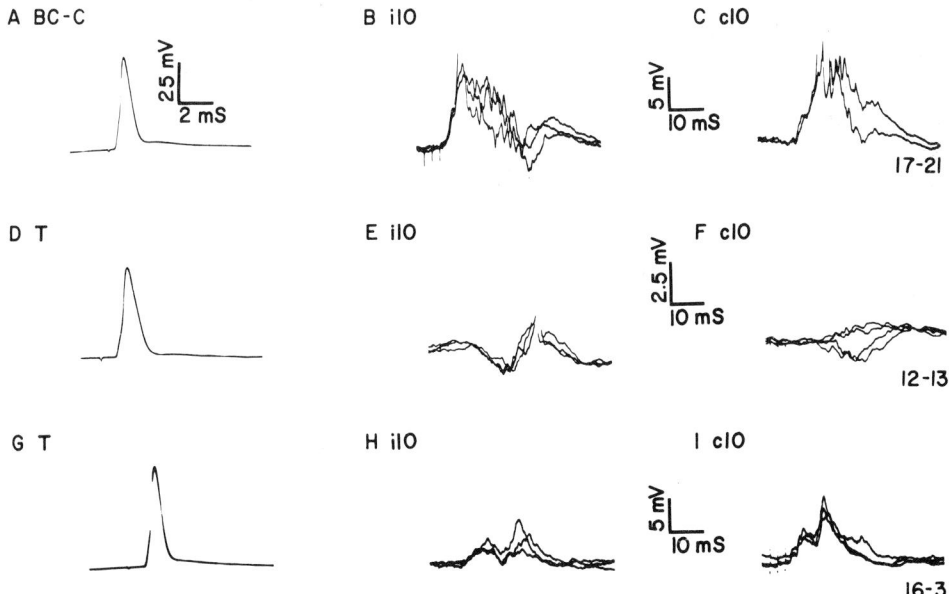

Fig. 3. PSP's evoked by stimulation of the ipsilateral and contralateral IO nerve. A, D and G show the antidromic responses of three neck muscle motoneurons and B, C and E, F and H, I, respectively, illustrate their responses to IO nerve stimulation. iIO, ipsilateral infraorbital; cIO contralateral infraorbital. In F the hyperpolarizing response (lower 2 traces) was recorded with no current passing through the electrode. The depolarizing response (upper 2 traces) was recorded with a hyperpolarizing current of 10 nA applied to the electrode. This reversal indicates that the hyperpolarization is the result of IPSP mechanisms.

ipsilateral and contralateral IO nerve. As illustrated in A, B and C, BC-C motoneurons receive a powerful excitatory input from both IO nerves. In response to high intensity (10 T or greater) single shock stimuli, latencies as short as 2.5 msec were recorded. The responses of S motoneurons (not illustrated) to IO nerve stimulation were similar to those observed for BC-C motoneurons. T motoneurons, as illustrated in D–I of Fig. 3, were either initially excited (H and I) or initially inhibited (E and F) by IO input and consequently no consistent pattern of response emerged.

The pattern of responses following stimulation of the SO nerve differed markedly from that seen following IO nerve stimulation. BC-C and S motoneurons were excited and T motoneurons were inhibited by stimulation of the ipsilateral SO nerve. Stimulation of the contralateral SO nerve elicited inhibitory responses in BC-C motoneurons, both excitatory and inhibitory responses in S motoneurons, and excitatory responses in T motoneurons.

The response of neck muscle motoneurons to IO nerve stimulation is consistent with a movement which would raise the animal's head to avoid a stimulus received on the nose. On the other hand, SO nerve stimulation would result in a complex movement involving twisting of the head away from the stimulus and raising of the head towards the stimulus. Such a movement is similar to both suckling and orientation reflexes and indeed Wall and Taub (1961) have suggested that the spinal nucleus of V may be involved in such behaviour. However, the exact pathways which convey trigeminal input to neck muscle motoneurons are unknown. That the spinal tract of V is at least partly responsible, has been demonstrated by Sumino and Nozaki (1977), who showed that sectioning the spinal tract of V at the level of the obex abolished the response of neck muscle motoneurons to stimulation of the ipsilateral IO nerve.

Unlike trigeminal connections to neck muscle motoneurons, segmental cutaneous connections have only recently been demonstrated. Anderson (1977) reported that electrical stimulation of the ipsilateral great auricular nerve (GA), which innervates the back of the ear, led to predominant excitation in S motoneurons and predominant inhibition in BC and C motoneurons. Responses to contralateral GA nerve stimulation resulted in inhibition of all three motoneuron pools. These results were recently confirmed in our laboratory and extended to include the responses of T motoneurons. Unlike BC-C and S motoneurons, all T motoneurons examined were excited by stimulation of both GA nerves. This finding suggests that responses to stimulation of the ears would result in a lowering of the head, similar to a flexor reflex.

SUMMARY

There is now a substantial body of evidence to suggest that the anatomical and physiological organization of the upper cervical spinal cord differs from that found in the lumbosacral cord. This evidence arises from a consideration of the intrinsic structure of the spinal cord, in particular, the morphology of neck muscle motoneurons whose dendritic trees are both larger and more complex than those of motoneurons innervating hindlimb muscles. Electrophysiological experiments have demonstrated that direct muscle afferent projections to neck muscle motoneurons are surprisingly weak. In contrast, trigeminal afferents are involved in powerful connections to neck muscle motoneurons. This latter projection may be involved in motor behaviour unique to the head movement system, for example, suckling or orientation.

ACKNOWLEDGEMENTS

This work was supported by the Canadian Medical Research Council. The authors gratefully acknowledge the excellent technical assistance of M. Adams. N. Sprott was the recipient of an Ontario Graduate Scholarship.

REFERENCES

Abrahams, V.C. (1977) The physiology of neck muscles; their role in head movement and maintenance of posture. *Canad. J. Physiol.*, 55: 332–338.

Abrahams, V.C. and Rose, P.K. (1975) Projections of extraocular, neck muscle, and retinal afferents to superior colliculus in the cat: Their connections to cells of origin of tectospinal tract. *J. Neurophysiol.*, 38: 10–18.

Abrahams, V.C. and Richmond, F.J.R. (1977) Motor role of the spinal projections of the trigeminal system. In *Pain in the Trigeminal System*, D.J. Anderson and B. Matthews (Eds.). Elsevier/North-Holland Biomedical Press, Amsterdam, pp. 405–411.

Abrahams, V.C., Richmond, F. and Rose, P.K. (1975) Absence of monosynaptic reflex in dorsal neck muscles of the cat. *Brain Res.*, 92: 130–131.

Anderson, M.E. (1977) Segmental reflex imputs to motoneurons innervating dorsal neck musculature in the cat. *Exp. Brain Res.*, 28: 175–187.

Anderson, M.E., Yoshida, M. and Wilson, V.J. (1971) Influence of superior colliculus on cat neck motoneurons. *J. Neurophysiol.*, 34: 898–907.

Anderson, M.E., Yoshida, M. and Wilson, V.J. (1972) Tectal and tegmental influences on cat forelimb and hindlimb motoneurons. *J. Neurophysiol.*, 35: 462–470.

Barrett, J.N. (1975) Motoneuron dendrites: role in synaptic integration. *Fed. Proc.*, 34: 1398–1407.

Conradi, S. (1969) Ultrastructure and distribution of neuronal and glial elements on the motoneuron surface in the lumbosacral spinal cord of the adult cat. *Acta physiol. scand.*, Suppl. 332: 5–48.

Cullheim, S. and Kellerth, J.O. (1978) A morphological study of the axons and recurrent axon collaterals of cat sciatic α-motoneurons after intracellular staining with horseradish peroxidase. *J. comp. Neurol.*, 178: 537–558.

Cullheim, S., Kellerth, J.-O. and Conradi, S. (1977) Evidence for direct synaptic interconnections between cat spinal α-motoneurons via the recurrent axon collaterals: a morphological study using intracellular injection of horseradish peroxidase. *Brain Res.*, 132: 1–10.

Elliott, R. (1961) In *Anatomy of the Cat*, 3rd ed., J. Reighard and H.S. Jennings (Eds.). Holt, Rinehart and Winston, New York.

Ezure, K., Sasaki, S., Uchino, Y. and Wilson, V.J. (1978) Frequency-response analysis of vestibular-induced neck reflex in cat. II. Functional significance of cervical afferents and polysynaptic descending pathways. *J. Neurophysiol.*, 41: 459–471.

Gelfan, S., Kas, G. and Ruchkin, D.S. (1970) The dendritic tree of spinal neurons. *J. comp. Neurol.*, 139: 385–412.

Gura, E.V. and Limanskii, Y.P. (1977) Antidromic and synaptic potentials of motoneurons of the cat accessory nerve nucleus. *Neurophysiology*, 8: 246–248.

Kerr, F.W.L. and Olafsson, R.A. (1961) Trigeminal and cervical volleys: convergence on single units in the spinal grey at C1 and C2. *Arch. Neurol.*, 5: 171–178.

Light, A.R. and Metz, C.B. (1978) The morphology of the spinal cord efferent and afferent neurons contributing to the ventral roots of the cat. *J. comp. Neurol.*, 179: 501–516.

Manni, E., Palmieri, G., Marini, R. and Petterossi, V.E. (1975) Trigeminal influences on extensor muscles of the neck. *Exp. Neurol.*, 47: 330–342.

Pearson, A.A. (1938). The spinal accessory nerve in human embryos. *J. comp. Neurol.*, 68: 243–266

Rapoport, S. (1977) Reflex interactions of muscles controlling head position. *Soc. Neurosci. Abstr.* III: 506, no. 1624.

Richmond, F.J.R. and Abrahams, V.C. (1975a) Morphology and enzyme histochemistry of dorsal muscles of the cat neck. *J. Neurophysiol.*, 38: 1312–1321.

Richmond, F.J.R. and Abrahams, V.C. (1975b) Morphology and distribution of muscle spindles in dorsal muscles of the cat neck. *J. Neurophysiol.*, 38: 1322–1339.

Richmond, F.J.R., Scott, D.A. and Abrahams, V.C. (1978) Distribution of motoneurons to the neck muscles, biventer cervicis, splenius and complexus in the cat. *J. comp. Neurol.*, 181: 451–464.

Richmond, F.J.R., Anstee, G.C.B., Sherwin, E.A. and Abrahams, V.C. (1976) Motor and sensory fibres of neck muscle nerves in the cat. *Canad. J. Physiol. Pharmacol.*, 54: 294–304.

Rose, P.K. and Abrahams, V.C. (1978) Tectospinal and tectoreticular cells. Their distribution and afferent connections. *Canad. J. Physiol. Pharmacol.*, 56: 650–658.

Snow, P.J., Brown, A.G. and Rose, P.K. (1976) Tracing axons and axon collaterals of spinal neurons using intracellular injection of horseradish peroxidase. *Science*, 191: 312–312.

Sumino, R. and Nozaki, S. (1977) Trigemino-neck reflex: its peripheral and central organization. In *Pain in the Trigeminal Region*, D.J. Anderson and B. Matthews (Eds.). Elsevier/North-Holland Biomedical Press, Amsterdam, pp. 365–374.

Szentagothai, J. (1948) Anatomical considerations of monosynaptic reflex arcs. *J. Neurophysiol.*, 11: 445–454.

Wall, P.D. and Taub, A. (1962) Four aspects of trigeminal nucleus and a paradox. *J. Neurophysiol.*, 25: 110–126.

Wilson, V.J. and Maeda, M. (1974) Connections between semicircular canals and neck motoneurons in the cat. *J. Neurophysiol.*, 37: 346–357.

Wilson, V.J. and Peterson, B.W. (1978) Peripheral and central substrates of vestibulospinal reflexes. *Physiol. Rev.*, 58: 80–105.

Cortical, Tectal and Medullary Descending Pathways to the Cervical Spinal Cord

J.D. COULTER, R.M. BOWKER, S.P. WISE, E.A. MURRAY, A.J. CASTIGLIONI and K.N. WESTLUND

Marine Biomedical Institute, Departments of Psychiatry and Behavioral Sciences, and Physiology and Biophysics, University of Texas Medical Branch, Galveston, TX 77550 (U.S.A.)

The importance of descending spinal pathways in the control of head and neck movements has long been recognized particularly with regard to labyrinthine influences on postural reflexes and the coordination of head and eye movements during orienting and visual tracking. However, with the exception of the vestibulospinal system and descending projections from the superior colliculus (Nyberg-Hansen, 1966; Wilson and Peterson, 1978), relatively little is known concerning the precise origins and anatomical organization of descending spinal pathways involved in control of the neck musculature. This is particularly the case for the monkey which is being used increasingly in combined electrophysiological and behavioral studies of the motor and sensory systems.

With the development of sensitive neuroanatomical tracing techniques employing anterograde and retrograde axoplasmic transport of labeled proteins (Cowan and Cuenod, 1975), the organization of descending pathways to the spinal cord has begun to be reinvestigated in recent years (Kuypers and Maisky, 1975; Catsman-Berrovoets and Kuypers, 1976; Saper et al., 1976; Burton and Loewy, 1977; Coulter and Jones, 1977; Basbaum et al., 1978; Castiglioni et al., 1978; Groos et al., 1978; Loewy and Burton, 1978; Kneisley et al., 1978). In addition to the demonstration of a number of previously unknown descending projections, additional features of the organization of spinal descending systems have been revealed.

With the retrograde horseradish peroxidase (HRP) technique (LaVail and LaVail, 1974; Hardy and Heimer, 1977), provided that care is taken to avoid damage to descending fibers of passage, small injections of HRP, confined to part of a single segment or to a few adjacent spinal segments, can be used to determine fairly precisely, the somatotopic (or segmentotopic) organization of the cells of origin of descending spinal pathways.

We have employed this approach, supplemented by anterograde autoradiographic tracing techniques and electrophysiological studies to determine the origins and terminations of descending pathways to the upper cervical spinal cord, their somatotopic pattern, and other features of their organization. While there are extensive regions of the brain stem, cerebral cortex and cerebellum which give rise to descending projections to the upper cervical cord, by comparing the numbers and locations of HRP labeled cells following injections at different spinal levels, those cortical regions and subcortical nuclei which seem to project most heavily to the cervical spinal segments have been identified.

The descending systems to be described in detail here include a corticospinal projection from a heretofore poorly understood region of the medial, parietal association cortex termed the supplementary sensory region (Murray and Coulter, 1977). Among the subcortical systems to be described, is the crossed descending spinal projection from the superior colliculus and the topographic organization of the cells of origin of this pathway in relation to the visual and somesthetic maps of the tectum (Murray et al., 1978). Finally, the somatotopic organization and spinal terminations of a previously unrecognized reticulospinal system originating in the caudal medulla will be described (Gallaway et al., 1977; Bowker and Coulter, 1978).

CORTICAL PROJECTIONS TO THE UPPER CERVICAL SPINAL CORD: THE SUPPLEMENTARY SENSORY REGION

Following an injection of HRP into the upper cervical spinal segments of C2 to C4 in the monkey, labeled neurons are found in five different functionally defined regions of the contralateral cerebral cortical hemisphere (cf. Woolsey, 1958). These are: the primary precentral motor cortex (MI) corresponding to Brodmann's (1909) cytoarchitectonic area 4; the postcentral primary somatic sensory cortex (SI) composed of areas 3a, 3b, 1 and 2; the second somatic sensory region (SII) located in the lateral fissure; the supplementary motor cortex (SM) corresponding to the medial part of area 6 extending into the cingulate sulcus; and a zone lying in the medial posterior parietal cortex corresponding to part of area 5 (and possibly including part of the immediately adjacent area 7) which, following Penfield and Jasper (1959), is termed the "supplementary sensory" (SS) region.

In Fig. 1, representative sagittal sections are shown through each of these regions. All the labeled neurons are confined to cortical layer V and include both large and small pyramidal neurons of this layer. In D, taken just lateral in the hemisphere to the forelimb representation of MI and SI (Woolsey, 1958), large numbers of labeled neurons are seen in the precentral gyrus in area 4, with a few labeled cells scattered anteriorly in area 6. More posteriorly, many labeled neurons are located within each of the subfields, areas 3a, 3b, 1 and 2, of the SI cortex, plus the adjacent parts of area 5. Sections taken more medially through the trunk (Fig. 1C) and hindlimb (Fig. 1B and A) representations in MI and SI do not contain neurons projecting to the upper cervical spinal cord. However, in the medial, buried parts of the hemisphere, anteriorly in area 6, bordering the cingulate sulcus, and posteriorly at the end of the cingulate sulcus and extending into the depths of the medial intraparietal sulcus, in area 5, two distinct populations of labeled neurons are seen. The cells located anteriorly in area 6 correspond in position to part of the supplementary motor region, 6sm (Fig. 1A and B). The neurons located posteriorly in area 5 (5ss, Fig. 1B) comprise a second population of corticospinal neurons projecting to the cervical cord from the medial hemisphere which is separate from the supplementary motor region, from the MI and SI hindlimb regions, and from the lateral parts of area 5. The location of these neurons and their topographic organization with respect to projections to different levels of the spinal cord are consistent with the postulated existence, based upon stimulation and evoked potential studies (Penfield and Jasper, 1954; Hughes and Mazurowski, 1962; Blomquist and Lorenzini, 1965), of a supplementary sensory region representing the somatic sensory periphery in the posterior medial hemisphere of the primate.

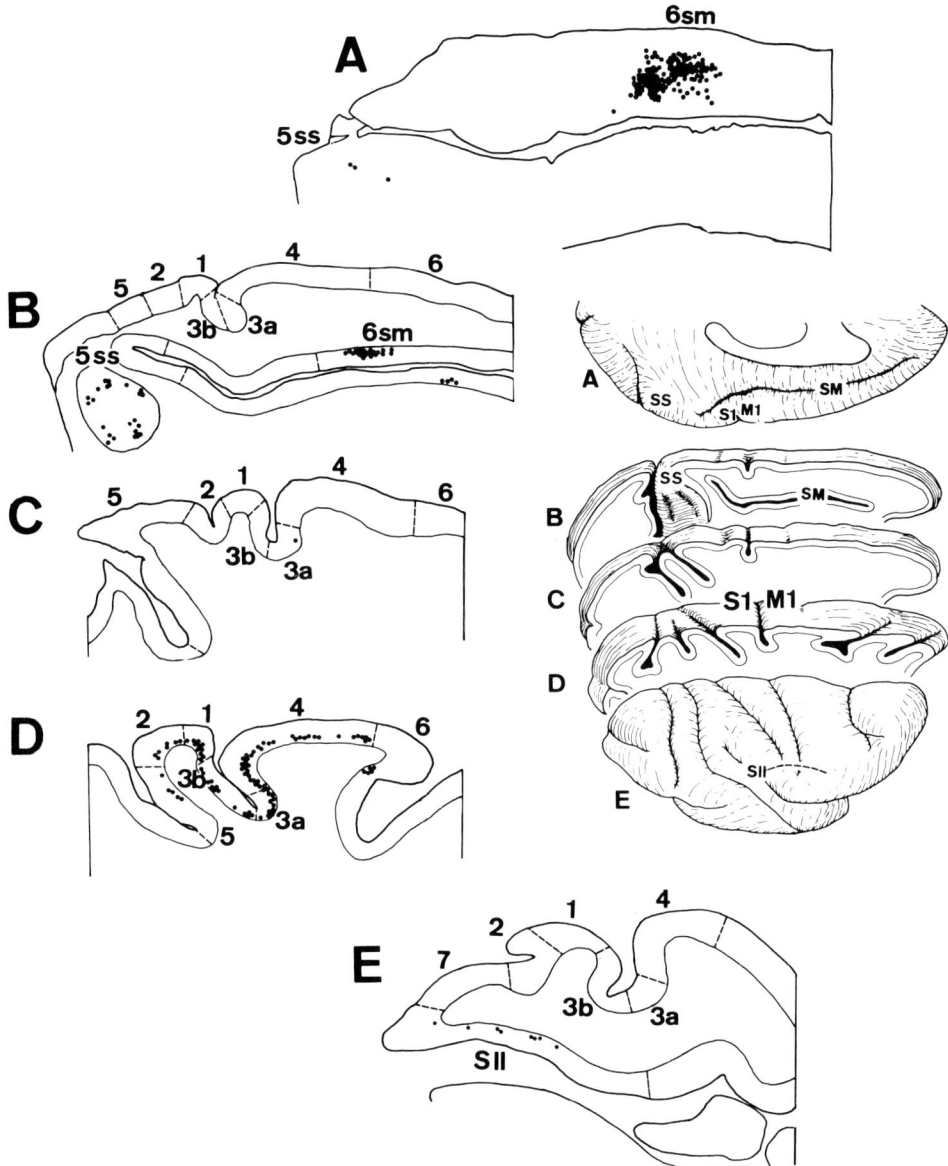

Fig. 1. Corticospinal neurons projecting to the upper cervical cord in monkey. The locations of cortical neurons (dots) in the right hemisphere retrogradely labeled following injection of HRP into the contralateral spinal segments C2–C4. Sagittal sections (50 μm) A–E through the hemisphere correspond approximately to the thick sections shown on the drawing to the right. Note that section A through the medial hemisphere corresponds to A in the drawing which is inverted to show the medial surface of the right hemisphere. In E, the section in the drawing is indicated by the dotted line. For reference, the central sulcus divides the MI (area 4) from SI (areas 3a, 3b, 1 and 2) on the sections and the drawing. Note the position of the supplementary sensory region (SS) on the medial hemisphere and continuing into the anterior bank of the intraparietal sulcus. Cytoarchitectural areas and their approximate borders are indicated by numbers. 5ss: the part of area 5 defined as the supplementary sensory region. 6sm: the part of area 6 defined as the supplementary motor region.

The somatotopic organization of the corticospinal projections from the supplementary sensory region (SS), as well as from the supplementary motor (SM) and hindlimb primary sensory (SI) and motor (MI) regions of the medial hemisphere are illustrated in Fig. 2. The cortical zone in the medial posterior parietal area 5 (5ss) which is devoted to projections to the upper cervical spinal cord (Fig. 2A, D) can be seen to be much more extensive than are the adjacent zones related to either the cervical or lumbosacral spinal levels. Neurons projecting to the upper cervical cord are located more posteriorly in the supplementary sensory region extending around into the depths of the intraparietal sulcus, whereas the neurons projecting to the cervical and lumbosacral enlargements are found progressively more anteriorly in the medial hemisphere around the posterior end of the cingulate sulcus. Thus, the body representation in the supplementary sensory region is essentially reversed in orientation to that seen in the supplementary motor region where the cervical spinal segments are represented anterior to the cortical zones projecting to the lumbosacral cord (Woolsey, 1958).

The corticospinal projections from the supplementary sensory region and their somatotopic organization have been confirmed in electrophysiological studies where pyramidal tract stimulation has been used to identify, by antidromic activation, the descending cortical neurons from this region. The majority of recorded neurons have peripheral somatic receptive fields on the upper extremities, back, neck and face. Compared to neurons in SI, the neurons in the supplementary sensory region have larger receptive fields, are more concerned with proximal, axial and bilateral body areas, and appear to respond to a wider range of somatic sensory stimulus submodalities, including nociceptive inputs.

The existence of corticofugal neurons forming a topographically organized supplementary sensory region which receives a complex type of somatic input from the upper body parts and head is of interest in view of the recently postulated functions of the posterior parietal cortex. It has been suggested that the parietal association cortex, in general, is involved in the formation of spatial maps of the body and immediately surrounding environment and may have an important role in certain aspects of volitional motor control (Mountcastle et al., 1975). Whether this also characterizes the functions of that part of the medial parietal cortex constituting the supplementary sensory region remains to be determined.

ORIGINS OF BRAIN STEM PROJECTIONS TO THE UPPER CERVICAL SPINAL CORD

Fig. 3 shows the locations of retrogradely labeled neurons in the medulla, pons and midbrain of the monkey following an injection of HRP into the upper cervical spinal segments C1–C5. The existence of descending projections from the medullary reticular formation (nucleus gigantocellularis), the nucleus of the solitary tract, the vestibular nuclear complex, the pontine reticular formation, the red nucleus, and the interstitial nucleus of Cajal are well known (see Nyberg-Hansen, 1966; Petras, 1967). In addition, the origins of other descending spinal pathways, which have been described more recently, include certain cells of the dorsal column nuclei (Burton and Loewy, 1977), the nucleus retroambiguus, the caudal medial and inferior vestibular nuclei (Peterson and Coulter, 1970), the raphe magnus and pallidus (Basbaum et al., 1978), the

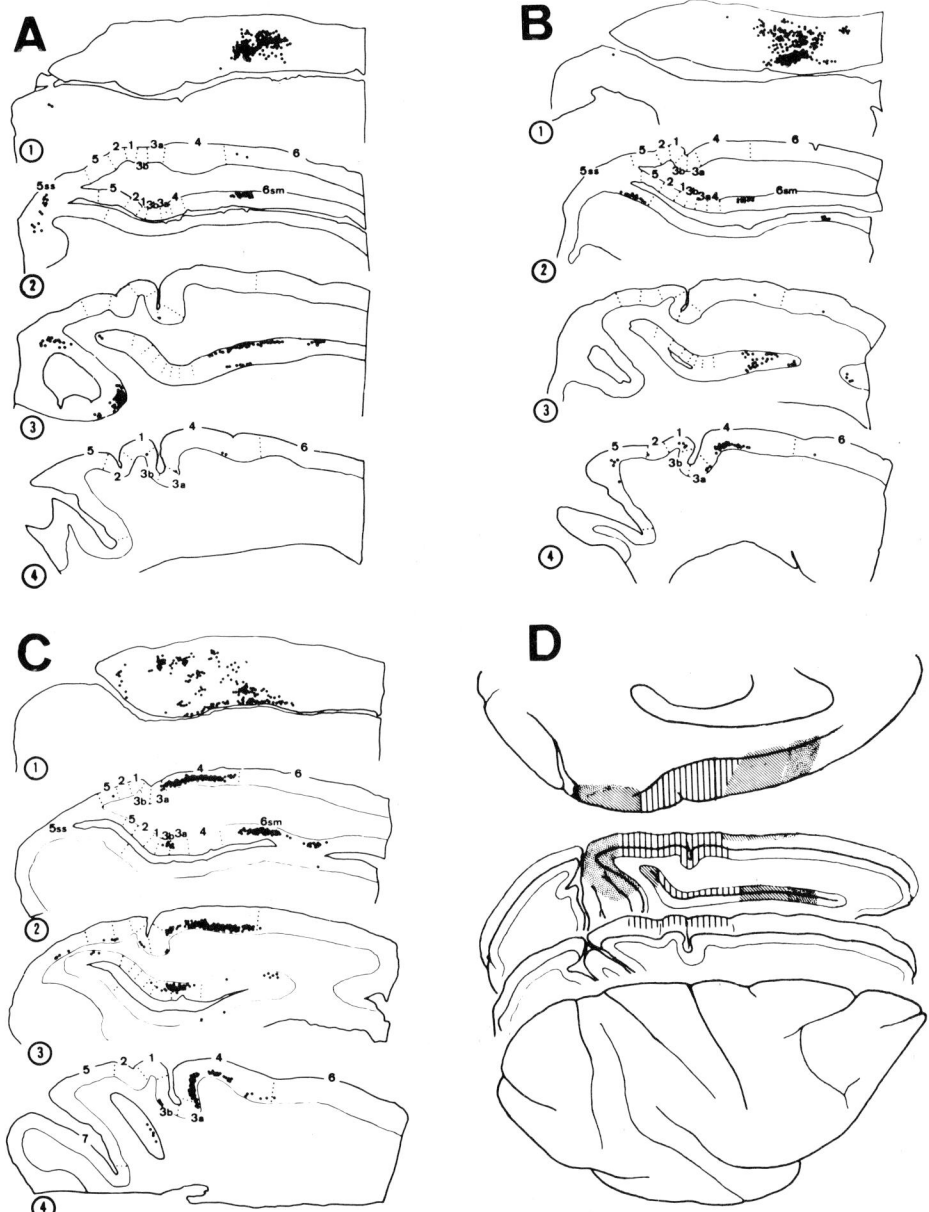

Fig. 2. Somatotopic organization of the corticospinal neurons in the supplementary sensory and supplementary motor regions in the monkey. A) Projection to upper cervical cord. Sagittal sections (50 μm) through the medial hemisphere show the distribution of labeled cells (dots) posterior in 5ss (supplementary sensory region) and anteriorly along the cingulate sulcus in area 6sm (supplementary motor region) following injection of HRP in spinal segments C2–C4 (same animal as Fig. 1). B) Projections to the cervical enlargement. C) Projections to the lumbosacral enlargement. Sections are approximately 1 mm apart beginning (1) at the medial face of the hemisphere moving laterally (4) D) Schematic representation of the topography of the corticospinal projections from the medial hemisphere. Vertical lines: representation of the lumbosacral corticospinal projecting neurons in the SI, MI and adjoining SS and SM regions. Oblique line: representation of the cortical zones projecting to the cervical enlargement in the SS and SM regions. Fine stipple: representation of the cortical zones projecting to the upper cervical spinal cord segments. Note that the "body" representations in the SS and SM regions are essentially reversed in anterior-posterior orientation on the medial hemisphere.

Fig. 3. Locations of brain stem neurons projecting to the upper cervical spinal cord in the monkey. Labeled neurons (dots), following injection of HRP into the right cervical segments C1–C5 are shown on representative sections (50 μm). Labeled neurons from 2 sections approx 0.5 mm apart, are plotted on each drawing. Modified and redrawn in part. (From Castiglioni et al., 1978.) *Abbreviations:* AMB, nucleus ambiguus; CG, central grey; CN, cuneate nucleus; CND, nucleus medulla oblongata centralis, subnucleus dorsalis; CNV, nucleus medulla oblongata centralis, subnucleus ventralis; CUN, nucleus cuneiformis; DK, nucleus of Darkschewitsch; EC, external cuneate nucleus; EW, Edinger-Westphal nucleus; GC, nucleus gigantocel-

locus coeruleus, subcoeruleus and parabrachial nuclei (Kuypers and Maisky, 1975) and various midbrain nuclei, such as the central grey, Edinger–Westphal nucleus and nucleus of Darkschewitsch (Castiglioni et al., 1978a), plus the zona incerta (Castiglioni et al., 1978b) and other cell groups extending into the caudal diencephalon and hypothalamus (Kuypers and Maiskey, 1975; Saper et al., 1976; Kneisley et al., 1978).

While the number of brain stem neurons and discrete nuclei which give rise to descending spinal projections is impressive, and much more extensive than previously recognized, of particular interest here are those descending systems which may be particularly involved in control of the upper cervical spinal cord. Based upon the patterns of HRP labeled neurons following injections of lower spinal levels, certain brain stem nuclei can be seen to have a much more extensive projection to the cervical cord. For certain of the nuclei, there exists little, if any, projection below the level of the cervical enlargement. On these grounds the brain stem nuclei, which appear to be most heavily involved in descending projections to upper cervical spinal levels include: 1) a bilateral projection from neurons located in the extreme caudal medulla, in the nucleus supraspinalis (SSP), (Fig. 3, 1); 2) a contralateral pathway from the medial (fastigial) and adjoining intermediate deep cerebellar nuclei (Fig. 3, 3) (Batton et al., 1977); 3) a contralateral projection from cells located in the far lateral (parvocellular) reticular formation dorsolateral to the facial (VII) nucleus (Fig. 3, 3); 4) a bilateral, but mainly ipsilateral, system arising from the more anterior parts of the medial (Vm) and descending (Vi) vestibular nuclei and adjoining lateral (Deiters') nucleus (Vl) (Fig. 3, 3); 5) an ipsilateral projection from the nucleus cuneiformis, adjacent central grey and midbrain tegmentum (Fig. 3, 6) (Castiglioni et al., 1978a); and 6) a crossed tectospinal pathway originating from the posterolateral part of the superior colliculus (Fig. 3, 7). Descending spinal projections from several of these brain stem nuclei have not been described previously, so there is presently little information regarding their functional relations to the spinal cord. Other pathways, such as the medial vestibulospinal and tectospinal tracts have been directly implicated in motor control of the head, neck and trunk (Anderson et al., 1971; Wilson, 1972; Abrahams and Rose, 1975).

It should be recognized that, while all the descending systems indicated above, provide direct inputs to the cervical spinal cord, they may additionally influence lower cord levels via extensive descending propriospinal pathways which originate at cervical spinal levels (Jankowska et al., 1975; Molenaar and Kuypers, 1978). Also, the existence of a direct projection to the spinal cord from a given brain stem region should not obscure the fact that powerful influences to all levels of the cord, originating particularly from the midbrain and more rostral structures (Lundberg, 1963; Anderson et al., 1972; Peterson et al., 1974), may be mediated indirectly to the spinal cord

---

lularis; GN, gracile nucleus; IC, inferior colliculus; INS, interstitial nucleus of Cajal; Lcoe, locus coeruleus; LRN, lateral reticular nucleus; MLF medial longitudinal fasciculus; NRA, nucleus retroambiguus; PB, parabrachial nuclei; PC, nucleus parvocellularis; PCO, nucleus pontis centràlis oralis; POC, nucleus pontis centralis caudalis; PGD, nucleus paragigantocellularis dorsalis; PRT, pretectal region; R, raphe nuclei; RM, raphe magnus; RP, raphe pallidus; RN, red nucleus; RNp, red nucleus, parvocellular division; SC, superior colliculus; Scoe, Subcoeruleus nuclei; SGI, stratum griseum intermediale, superior colliculus; SGP, stratum griseum profundum, superior colliculus; SOL, solitary nucleus; SSP, nucleus supraspinalis; TEG, central tegmental field; Vi, inferior (descending) vestibular nucleus; Vl, lateral (Deiters') vestibular nucleus; Vm, medial vestibular nucleus; Vs, superior vestibular nucleus; III, oculomotor nucleus; IV, trochlear nucleus; V, trigeminal nucleus; SPV, spinal trigeminal complex; VI, abducens nucleus; VII, facial nucleus; XII, hypoglossal nucleus.

via the numerous reticulospinal paths and other descending systems originating more caudally in the brain stem. Finally, while descending projections to the upper cervical spinal levels tend to suggest functions related to control of the neck musculature and hence movements of the head, the cervical cord also contains other important cell groups which include the phrenic motor neurons involved in respiratory functions, and the lateral cervical nucleus, an important sensory relay in the spinocervical thalamic system.

TOPOGRAPHIC ORGANIZATION OF TECTOSPINAL NEURONS

Following an injection of HRP into the upper cervical cord, large numbers of retrogradely labeled neurons are found in the contralateral superior colliculus. In the monkey (Fig. 3, 7) and cat (Fig. 4B, D), large cells in both the intermediate (stratum grisium intermediale) and deep layers (stratum grisium profundum) are labeled, mainly in the posterolateral portion of the tectum. In the monkey, a few cells of the anteromedial superior colliculus may give rise to an ipsilateral spinal projection. The number of labeled cells projecting to the contralateral cord, however, is much larger, and the distribution of labeled cells in the intermediate and deep tectal layers are approximately equal in numbers and their spatial distribution in the two layers is essentially the same (Fig. 4A, B). In this regard, the labeled cells in the tectum often appear in small groups of 3–6 cells either within a single layer or grouped vertically, perpendicular to the layers. This distribution is thus suggestive of a columnar-like organization. Fewer labeled neurons are found in the superior colliculus following injections in the cervical enlargement (Fig. 4A, C) and they are confined to the extreme posterolateral part. In experiments in rat, cat and monkey no labeled cells have been seen in the tectum following HRP injections below the level of the cervical spinal enlargement.

It is of interest that the origin of the tectospinal projection in the posterolateral part of the colliculus corresponds to the region where neurons of the deep tectal layers are known to be responsive to tactile stimulation of the upper extremities, neck, and parts of the head (Drager and Hubel, 1976; Stein et al., 1976) and where spinotectal fibers (Mehler et al., 1960), as well as the corticotectal projections from the somatosensory cortex terminate (Kuypers and Lawrence, 1967; Wise and Jones, 1977). With regard to the topography of the visual inputs to the overlying superficial tectal layers (Drager and Hubel, 1976; Stein et al., 1976), the origins of the deeper situated tectospinal projections are concentrated in that part of the superior colliculus corresponding to the lower, temporal visual field. This suggests that the topographic distribution pattern of the tectospinal neurons projecting to different cord levels could have a systematic relationship to the retinotopic, somatotopic and, possibly, tonotopic inputs to the superior colliculus.

Evidence for such an organization has been obtained in the rat and cat (Murray et al., 1978) by correlating the previously described retinotopic and somatotopic maps of the tectum with the pattern of cell labeling in the superior colliculus following injections of HRP, in different animals, confined to various levels from the upper cervical segments to the cervical enlargement. As shown in Fig. 4, tectospinal neurons terminating in the cervical enlargement (Fig. 4A, C) are confined to that part of the superior colliculus which receives somatic inputs from the contralateral forelimb and

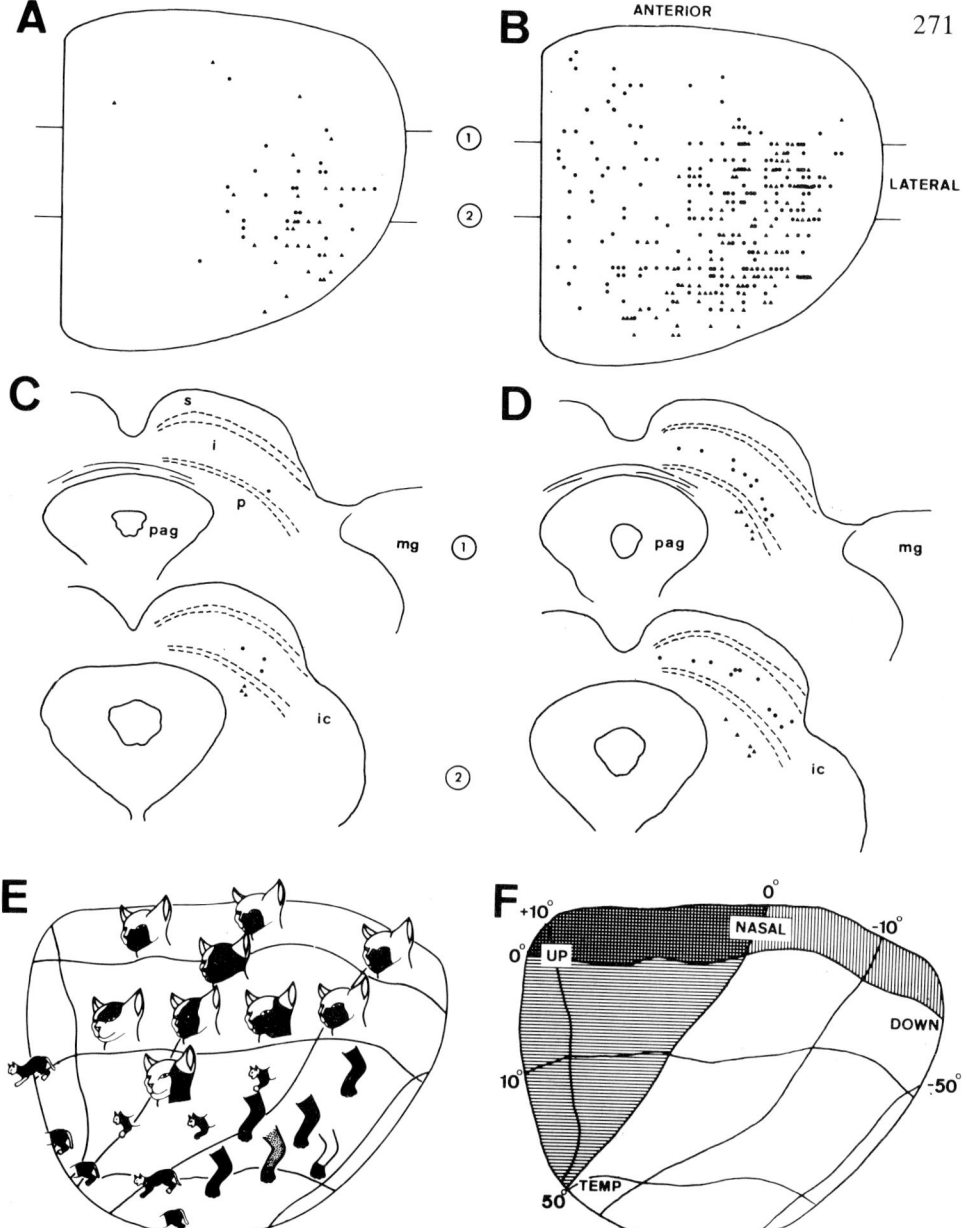

Fig. 4. Topographic organization of tectospinal neurons in the cat superior colliculus. A) Locations of retrogradely labeled neurons (dots and triangles) plotted on a "flattened" representation of the right superior colliculus as viewed from above. HRP was injected into the contralateral left cervical spinal enlargement. Labeled cells are plotted from alternate serial sections (50 μm). B) Locations of HRP labeled neurons following injection of HRP into the contralateral upper cervical spinal segments, C2–C4. C, D) Representative coronal sections through the right superior colliculus from the 2 animals shown in A and B, resp., with the level of section indicated by the numerals in A and B. Dots: labeled cells in the stratum griseum intermediale (i). Triangles: labeled cells in the stratum griseum profundum (p). E) The topographic representation of the somatic sensory periphery in the superior colliculus. Darkened areas indicate receptive fields of responsive neurons. F) The visuotopic representation in the superior colliculus, with horizontal and vertical meridians (0°) indicated for the contralateral monocular visual field. Vertical shading represents the nasal field. Horizontal shading represents the upper (superior) visual field. (E and F are modified and redrawn from Stein et al., 1976.)

trunk (Fig. 4E) and from the contralateral inferior and far temporal part of the visual field (Fig. 4F).

Comparing this distribution pattern to the pattern seen for those tectal neurons projecting to the more rostral cervical spinal levels (Fig. 4B, D) shows the latter to extend into these parts of the colliculus which receive inputs from the contralateral head and trigeminal somesthetic representation (Fig. 4E), in addition to the limb and trunk regions. The cells projecting to the upper cervical cord are also in those more anterior and medial parts of the tectum which receive visual inputs from the upper (superior) and more central parts of the visual field (Fig. 4F). The distribution pattern of labeled neurons following HRP injections into the middle cervical segments is approximately intermediate between that seen for neurons projecting to the more rostral cervical spinal cord and to the cervical enlargement. Thus, tectal neurons projecting more caudally in the spinal cord are located progressively more posterior and lateral towards that part of the tectum which represents the far temporal and lower visual field and the corresponding parts of the body remote from the face and head.

In functional terms, this would suggest that a somatic or visual stimulus delivered to the face and/or the more central part of visual field would activate the corresponding tectospinal neurons in the anteromedial part of the contralateral superior colliculus which project exclusively to the most rostral cervical segments controlling the neck musculature on the side ipsilateral to the stimulus. Eye and head movements would presumably be sufficient to orient towards the stimulus. On the other hand, a stimulus delivered to the upper extremities, trunk or lower body, or falling in the corresponding far inferior and temporal visual field would activate more posterolateral parts of the superior colliculus which in turn give rise to major projections, not only to the upper cervical segments, but more importantly, to the cervical enlargement controlling the forelimb musculature. This would provide a means of turning the whole body, in addition to movements of the eyes and hand, which would be required to orient to this stimulus.

Older data concerning the patterns of motor response evoked by stimulation of the superior colliculus (Hess et al., 1946) seem to suggest such an interpretation. This hypothesis is also consistent with the idea that head and body movements which orient the animal to stimuli in the environment, may be mediated by the superior colliculus and that by virtue of their topographic organization, tectospinal neurons may play a role in the precise guidance of such movements.

TOPOGRAPHIC ORGANIZATION AND SPINAL TERMINATIONS OF A RETICULOSPINAL PROJECTION FROM THE CAUDAL MEDULLA

Neurons of the nucleus gigantocellularis of the medullary reticular formation are generally considered to be the major origin of descending reticulospinal projections from this brain stem region (Bodian, 1946; Torvik and Brodal, 1957; Nyberg-Hansen, 1966). However, in addition to the reticulospinal system originating at this level, there exists a major descending spinal pathway to the upper cervical cord and spinal enlargements which arises bilaterally from cells located in the extreme caudal medulla in the nucleus supraspinalis and adjacent reticular formation just lateral to the decussating fibers of the pyramidal tract (Figs. 3, 1; 5A; 6). The origins of this descending system in the caudal medulla show a clear somatotopic organization unlike the spinal projections which originate from more rostral nuclei of the reticular formation.

In both monkey (Figs. 3 and 6A) and cat (Fig. 5A) the neurons which give rise to ipsilateral projections to the upper cervical cord tend to lie in the more medial and ventral part of the nucleus supraspinalis whereas the neurons projecting contralaterally are more dorsally located in the nucleus. Neurons giving rise to projections to the cervical enlargement (Fig. 5B) or the few neurons extending to the lumbrosacral levels (Fig. 5C) originate ipsilaterally, almost exclusively, and are found in the adjacent subnucleus ventralis (CNV) and dorsalis (CND) of the nucleus medulla oblongata centralis of Olszewski and Baxter (1954).

As shown in Fig. 5, there is a fairly sharp topographic segregation of the neurons projecting to different levels of the spinal cord, from this caudal medullary zone. The neurons projecting to the upper cervical cord are located medially and ventrally in the nucleus supraspinalis (SSP) at caudal medullary levels (Fig. 5A, 5) and extend dorsally in the medial part of the subnucleus ventralis (CNV) more rostrally in the brain stem (Fig. 5A, 1-3). The projections to the cervical enlargement arise lateral to the cells projecting to upper cervical levels and are found in the lateralmost part of the subnucleus ventralis, caudally (Fig. 5B, 4-5), extending dorsally into the lateral and ventral parts of the subnucleus dorsalis (CND) at more rostral levels (Fig. 5B, 1-3). Relatively more neurons projecting to the cervical enlargement are found anteriorly in this medullary region compared to the projections to the upper cervical spinal segments. Only a few neurons of the lower medulla project to the lumbosacral cord (Fig. 5C) and most are located anteriorly, laterally and ventrally to the neurons giving rise to projections to the cervical enlargement. Thus, those neurons extending more caudally in the spinal cord generally originate progressively more lateral and anterior in the caudal medulla.

The course and spinal terminations of the pathway from the region of the nucleus supraspinalis to the upper cervical cord in the monkey have been traced using the anterograde autoradiographic technique. Following a localized injection of a mixture of ^{3}H-leucine and ^{3}H-proline into the nucleus supraspinalis (Fig. 6B), labeled fibers can be traced both ipsilaterally and contralaterally, over and through the fibers of the pyramidal decussation, to descend bilaterally in the ventral funiculus and in the ventral part of the lateral funiculus of the cervical cord. The termination zone includes most of the ventral horn bilaterally from C1–C4 where silver grains overlie many large cells, presumably motoneurons, at these levels. More caudally, at segment C5, the pattern of terminal labeling is much lighter and concentrated more medially in the ventral horn region corresponding to lamina 8 (Rexed, 1954) in the cat. Only a few scattered fibers can be traced in the ventral funiculus to the level of cervical enlargement.

Some additional data have been obtained concerning inputs to the nucleus supraspinalis and adjacent reticular formation from further anatomical observations and from electrophysiological studies where the spinally projecting neurons of this region have been identified by antidromic activation from the cord. The results indicate that, similar to the reticulospinal system originating from the nucleus gigantocellularis of the medulla, the nucleus supraspinalis region receives an ascending spinal input and extensive descending inputs from more rostral structures, including the contralateral tectum and the head and face region of the sensorimotor cerebral cortex.

The actions on spinal neurons originating from the nucleus supraspinalis and adjacent reticular formation have not been investigated as yet. However, in view of

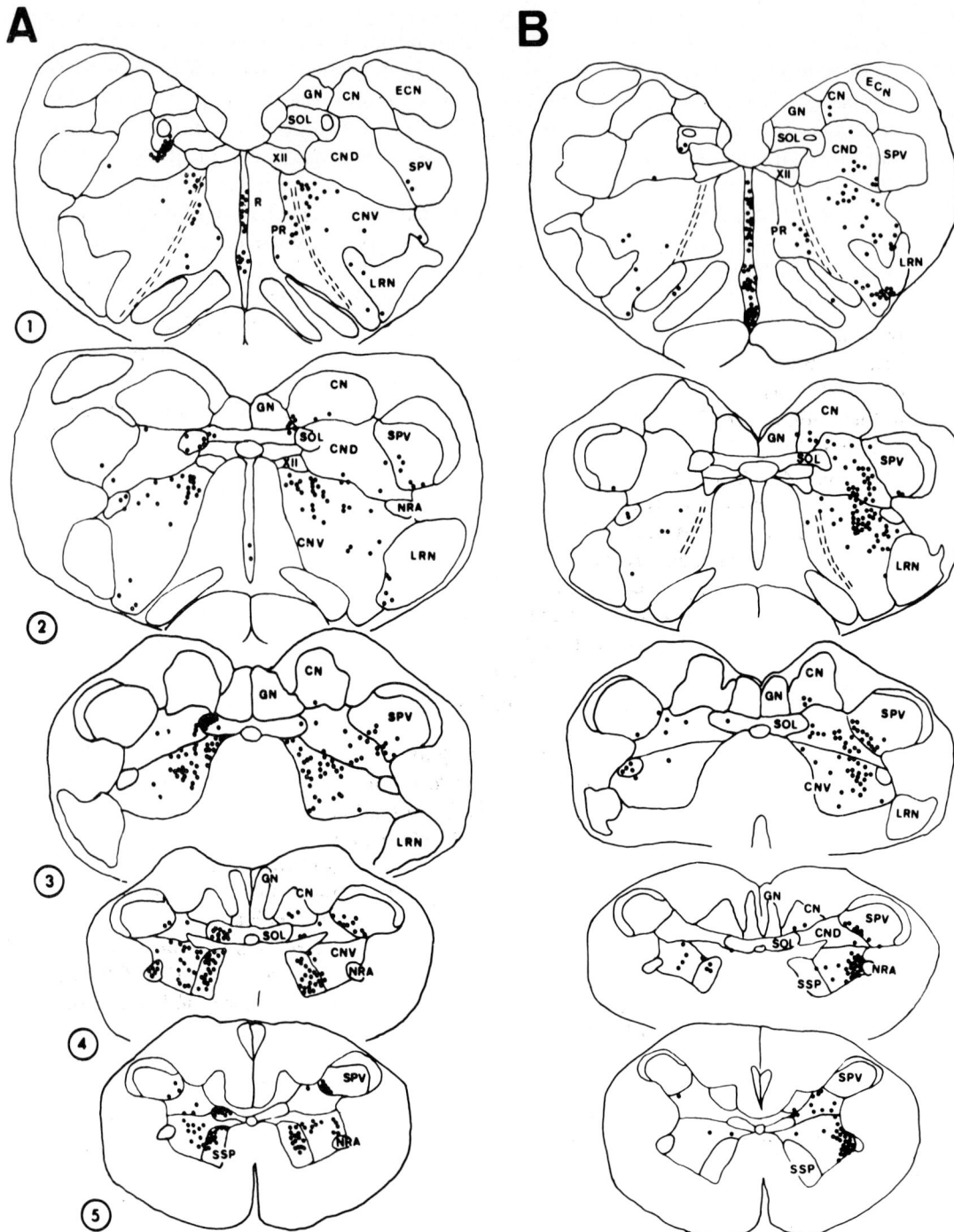

Fig. 5. Somatotopic organization of the reticulospinal projections from the caudal medulla in cat. Sections (50 μm) through the caudal medulla show the locations of retrogradely labeled neurons (dots) in the nucleus supraspinalis (SSP) (lower part of the figure) and in the adjacent reticular formation, anteriorly,

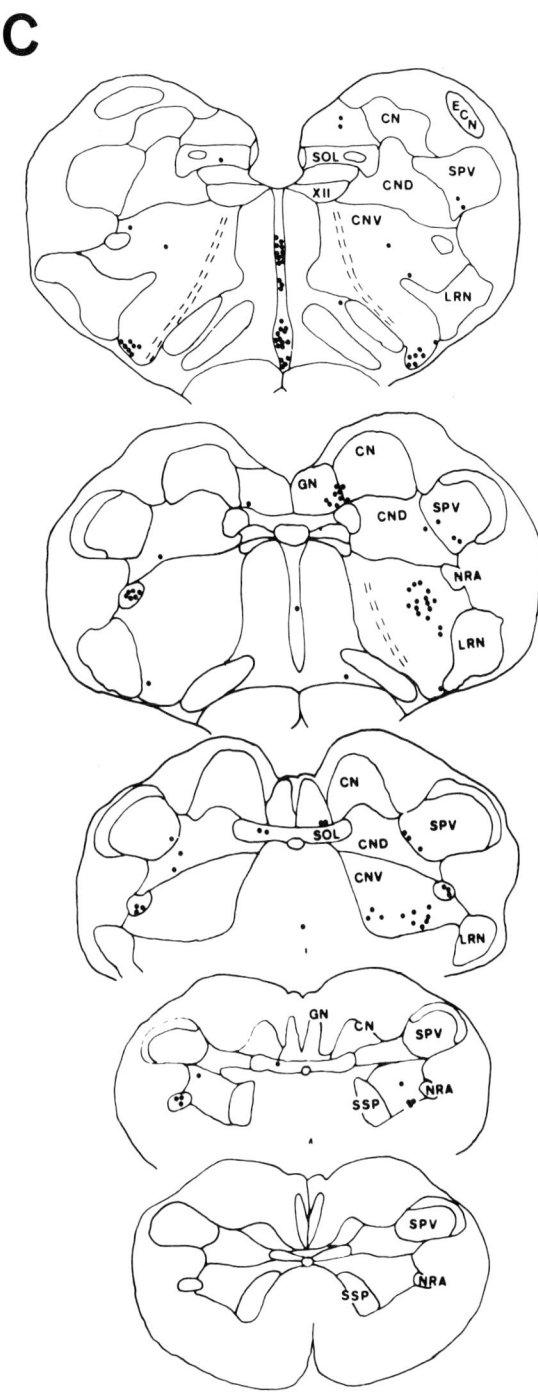

following injection of HRP into the right spinal cord at different levels in 3 animals. A) Neurons labeled from HRP injection at spinal segments C1–C3. B) Neurons labeled from C6–C8. C) Neurons labeled from L6–S1. Abbreviations as in Fig. 3.

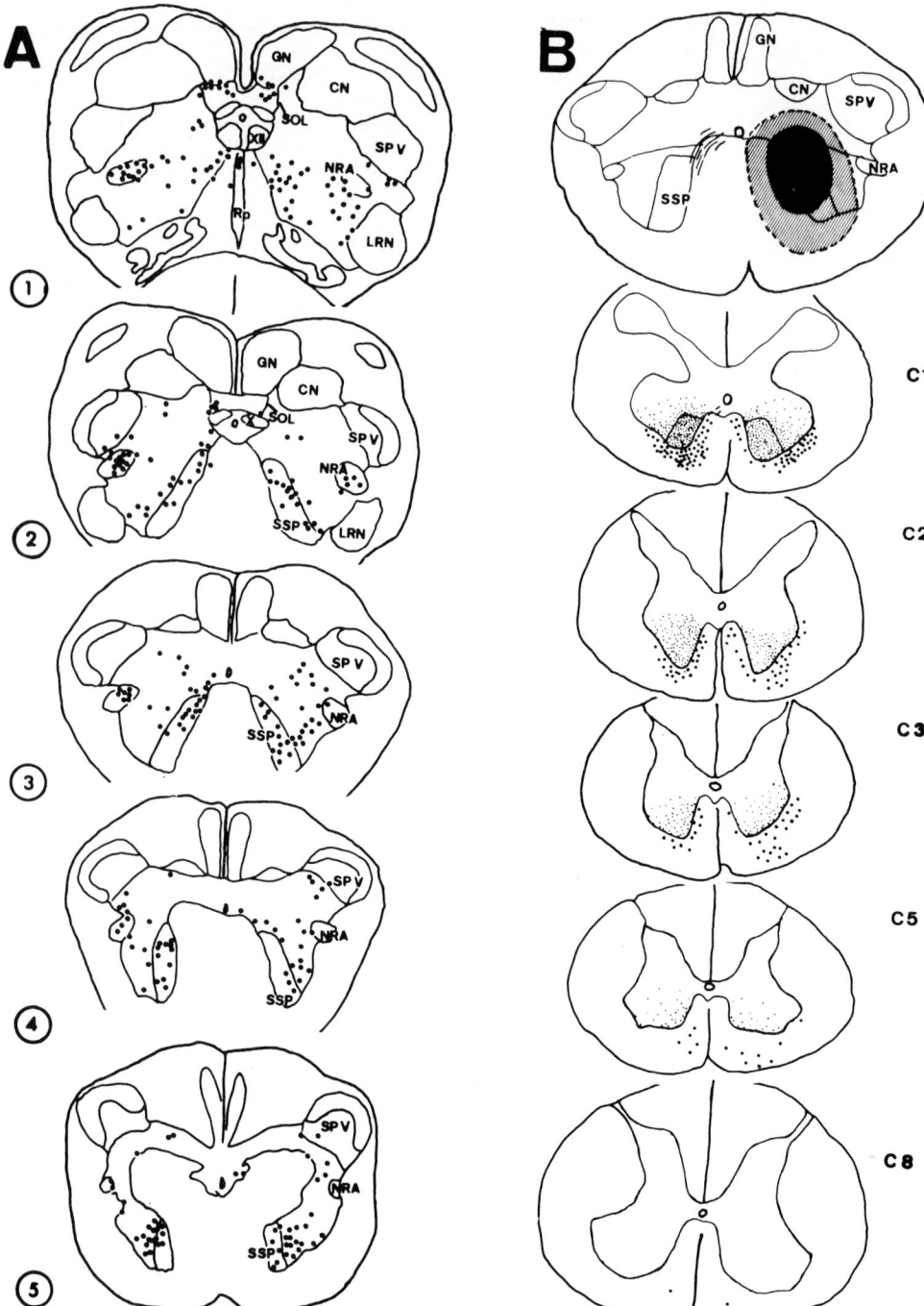

Fig. 6. Locations and spinal terminations of neurons of the caudal medulla projecting to the upper cervical spinal cord in monkey. A) Locations of retrogradely labeled neurons (dots) following injection of HRP into the right spinal segments C1–C5. B) Distribution of anterograde axonal (large dots) and terminal (fine stipple) labeling, demonstrated autoradiographically, following injection of tritiated amino acids into the right medial part of the caudal medulla. The injection site, centered in the nucleus supraspinalis (SSP), is indicated by the heavily shaded areas. Labeling in the spinal segments C1–C8 is shown on representative sections. Abbreviations as in Fig. 3.

the terminations of this pathway in the ventral spinal grey matter, it seems likely to have an important function in mediating descending influences from higher structures involved in movements of the head and neck and upper extremities. In this regard, the descending projections from this caudal medullary region, by virtue of their somatotopic organization would preserve the topographic relations of the cortical and tectal influences to the spinal cord mediated by this pathway. This is generally not considered to be the case for the descending influence relayed from higher structure to the spinal cord via the reticulospinal projection systems originating more rostrally in the brain stem (see, however, Peterson et al., 1974, 1975).

CONCLUSION

Relatively little attention has been given previously to the anatomical organization of descending spinal pathways to the upper cervical spinal levels controlling the neck musculature. The findings presented here have been obtained using recently developed neuroanatomical tracing techniques, supplemented by electrophysiological studies, to identify the origins, spinal terminations, somatotopic pattern and other features of organization of certain of the cortical and brain stem descending pathways projecting to the cervical spinal cord. The findings should provide a basis for future anatomical and physiological studies to discover more precisely the role of these descending spinal pathways in the maintenance of muscle tone, in postural reflexes and balance, in visual tracking and orienting behavior, and in complex volitional motor control.

SUMMARY

Extensive descending projections exist from the cerebral cortex, brain stem and deep cerebellar nuclei to the cervical spinal segments controlling the neck musculature. Among the cortical regions giving rise to corticospinal projections to the cervical cord is a heretofore poorly understood part of the medial posterior parietal cortex termed the supplementary sensory region. The somatotopic organization and other features of this cortical region are described, based upon anatomical experiments employing the retrograde horseradish peroxidase (HRP) technique supplemented by electrophysiological studies. The retrograde HRP method has also been used to identify a number of brain stem nuclei which give rise to descending pathways to the cervical cord. Among them, the topographic organization of the cells of origin of the tectospinal tract and their relation to the retinotopic and somatotopic maps of the tectum are described. Another major descending system to the cervical cord from the ventral caudal medulla is also described. The spinal terminations and somatotopic organization of this descending medullary reticulospinal system have been determined, with the HRP and anterograde autoradiographic techniques.

REFERENCES

Abrahams, V.C. and Rose, P.K. (1975) Projections of extraocular, neck muscle, and retinal afferents to superior colliculus in the cat: their connections to cells of origin of tectospinal tract. *J. Neurophysiol.*, 38: 10–18.

Anderson, M.E., Yoshida, M. and Wilson, V.J. (1971) Influence of superior colliculus on cat neck motoneurons. *J. Neurophysiol.*, 34: 898–907.

Anderson, M.E., Yoshida, M. and Wilson, V.J. (1972) Tectal and tegmental influences on cat forelimb and hindlimb motoneurons. *J. Neurophysiol.*, 35: 462–470.

Basbaum, A.L., Clanton, C.H. and Fields, H.L. (1978) Three bulbospinal pathways from the rostral medulla of the cat. An autoradiographic study of pain modulating systems. *J. comp. Neurol.*, 178: 209–224.

Batton, R.B., Jayaraman, A., Ruggiero, D. and Carpenter, M.B. (1977) Fastigial efferent projections in the monkey: an autoradiographic study. *J. comp. Neurol.*, 174: 271–306.

Blomquist, A.J. and Lorenzini, C.A. (1965) Projection of dorsal roots and sensory nerves to cortical sensory-motor regions of squirrel monkey. *J. Neurophysiol.*, 28: 1195–1205.

Bodian, D. (1946) Spinal projections of brainstem in rhesus monkey, deduced from retrograde chromatolysis. *Anat. Rec.*, 94: 512–513.

Bowker, R.M. and Coulter, J.D. (1978) Studies on descending projections from the caudal medulla in the cat. *Neurosci. Abstr*, 4.

Brodmann, K. (1909) *Vergleichende Lokalisationslehre der Grosshirnrinde in ihren Prinzipien dargestellt auf Grund des Zellenbaues.* Barth, Leipzig.

Burton, H. and Loewy, A.D. (1977) Projections to the spinal cord from medullary somatosensory relay nuclei. *J. comp. Neurol.*, 173: 773–792.

Castiglioni, A.J., Gallaway, M.C. and Coulter, J.D. (1978a) Spinal projections from the midbrain in monkey. *J. comp. Neurol.*, 178: 329–346.

Castiglioni, A.J., Wise, S.P., Murray, E.A. and Coulter, J.D. (1978b) Spinal projections from the midbrain in the rat. *Neurosci. Abstr.*, 4.

Catsman-Berrovoets, C.E. and Kuypers, H.G.J.M. (1976) Cells of origin of cortical projections to dorsal column nuclei, spinal cord and bulbar medial reticular formation in the rhesus monkey. *Neurosci. Lett.*, 3: 245–252.

Cowan, W.M. and Cuénod, M. (1975) *The Use of Axonal Transport for Studies of Neuronal Connectivity.* Elsevier, Amsterdam.

Coulter, J.D. and Jones, E.G. (1977) Corticospinal projections from individual cytoarchitectonic cortical fields in the monkey. *Brain Res.*, 129: 335–340.

Dräger, U. and Hubel, D. (1976) Topography of visual and somatosensory projections to mouse superior colliculus. *J. Neurophysiol.*, 39: 91–101.

Fukushima, K., Peterson, B.W., Uchino, Y., Coulter, J.D. and Wilson, J.J. (1977) Direct fastigiospinal fibers in the cat. *Brain Res.*, 126: 538–542.

Gallaway, M.C., Castiglioni, A.J., Foreman, R.D. and Coulter, J.D. (1977) Origins of spinal projections from the caudal medulla in monkey. *Neurosci. Abstr.*, 3: 271, no. 861.

Groos, W.D., Ewing, L.K., Carter, C.M. and Coulter, J.D. (1978) Organization of corticospinal neurons in the cat. *Brain Res.*, 143: 393–419.

Hardy, H. and Heimer, L. (1977) A safer and more sensitive substitute for diaminobenzidine in the light microscopic demonstration of retrograde and anterograde axonal transport of HRP. *Neurosci. Lett.*, 5: 235–240.

Hess, W.R., Bürgi, S. and Bucher, V. (1946) Motorische Funktion des Tectal und Tegmentalgebietes. *Mschr. Psychiat. Neurol.*, 112: 1–52.

Hughes, J.R. and Mazurowski, J.A. (1962) Studies on the supracallosal mesial cortex of unanesthetized, concious mammals. II. Monkey. A, movements elicited by electrical stimulation. *Electroenceph. clin. Neurophysiol.*, 14: 477–485.

Jankowska, E., Lundberg, A., Roberts, W.J. and Stuart, D. (1974) A long propriospinal system with direct effect on motoneurons and on interneurons in the cat lumbosacral cord. *Exp. Brain Res.*, 21: 169–194.

Kneisley, L.W., Biber, M.P. and LaVail, J.H. (1978) A study of the origin of brainstem projections to the monkey spinal cord using the retrograde transport method. *Exp. Neurol.*, 60: 116–139.

Kuypers, H.G.J.M. and Lawrence, D.G. (1967) Cortical projections to the red nucleus and the brainstem in the rhesus monkey. *Brain Res.*, 4: 151–188.

Kuypers, H.G.J.M. and Maisky, V.S. (1975) Retrograde axonal transport of HRP from spinal cord to brainstem cell groups. *Neurosci. Lett.*, 1: 9–14.

LaVail, J.H. and LaVail, M.M. (1974) The retrograde intraaxonal transport of horseradish peroxidase in the chick visual system: a light and electron microscope study. *J. comp. Neurol.*, 157: 303–358.

Loewy, A.D. and Burton, H. (1978) Nuclei of the solitary tract: efferent projections to the lower brain stem and spinal cord of the cat. *J. comp. Neurol.,* 181: 421–450.

Lundberg, A. (1964) Supraspinal control of transmission in reflex paths to motoneurons and primary afferents. *Progr. Brain Res.,* 12: 197–219.

Mehler, W.R., Feferman, M.E. and Nauta, W.J.H. (1960) Ascending axon degeneration following anterolateral cordotomy. An experimental study in the monkey. *Brain,* 83: 718–752.

Molenaar, I. and Kuypers, H.G.J.M. (1978) Cells of origin of propriospinal fibers and of fibers ascending to supraspinal levels. A HRP study in cat and rhesus monkey. *Brain Res.,* 152: 429–450.

Mountcastle, V.B., Lynch, J.C., Georgopoulos, A., Sakata, H. and Acuna, C. (1975) Posterior parietal association cortex of the monkey: command functions for operations within extrapersonal space. *J. Neurophysiol.,* 38: 871–908.

Murray, E.A. and Coulter, J.D. (1977) Corticospinal projections from the medial cortical hemisphere in monkey. *Neurosci. Abstr.,* 3: 275, no. 878.

Murray, E.A., Westlund, K.N., Watson, D.A. and Castiglioni, A.J. (1978) Organization at the cells of origin of the tectospinal tract. *Anat. Rec.,* 190: 488.

Nyberg-Hansen, R. (1966) Functional organization of descending supraspinal fiber systems to the spinal cord. Anatomical observations and physiological correlations. *Ergebn. Anat. Entwickl.- Gesch.,* 39: Heft 2, 1–48.

Olszewski, J. and Baxter, D. (1954) *Cytoarchitecture of the Human Brain Stem.* Karger, Basel.

Penfield, W. and Jasper, H. (1954) *Epilepsy and the Functional Anatomy of the Human Brain.* Little, Brown and Co., Boston.

Peterson, B.W. and Coulter, J.D. (1977) A new long spinal projection from the vestibular nuclei in the cat. *Brain Res.,* 122: 351–356.

Peterson, B.W., Anderson, M.E. and Filion, M. (1974) Responses of ponto-medullary reticular neurons to cortical, tectal and cutaneous stimuli. *Exp. Brain Res.,* 21: 19–44.

Peterson, B.W., Maunz, R.A., Pitts, N.G. and Mackel, R.G. (1975) Patterns of projection and branching of reticulospinal neurons. *Exp. Brain Res.,* 23: 333–351.

Petras, J.N. (1967) Cortical, tectal and tegmental fiber connections in the spinal cord of the cat. *Brain Res.,* 6: 275–324.

Rexed, B. (1954) A cytoarchitectonic atlas of the spinal cord in the cat. *J. comp. Neurol.,* 100: 297–379.

Saper, C.B., Loewy, A.L., Swanson, L.W. and Cowan, W.M. (1976) Direct hypothalamo-autonomic connections. *Brain Res.,* 117: 305–312.

Stein, B.E., Magalhaes-Castro, B. and Kruger, L. (1976) Relationship between visual and tactile representations in cat superior colliculus. *J. Neurophysiol.,* 39: 401–419.

Torvik, A. and Brodal, A. (1957) The origin of reticulospinal fibers in the cat. An experimental study. *Anat. Rec.,* 128: 113–137.

Wilson, V.J. (1972) Physiological pathways through the vestibular nuclei. *Int. Rev. Neurobiol.,* 15: 27–81.

Wilson, V.J. and Peterson, B.W. (1978) Peripheral and central substrates of vestibulospinal reflexes. *Physiol. Rev.,* 58: 80–105.

Woolsey, C.N. (1958) Organization of somatic sensory and motor areas of the cerebral cortex. In *Biological and Biochemical Bases of Behavior,* H.F. Harlow and C.N. Woolsey (Eds.), University of Wisconsin Press, Madison, pp. 63–81.

Wise, S.P. and Jones, E.G. (1977) Somatotopic and columnar organization of corticotectal projections of somatic sensory cortex. *Brain Res.,* in press.

Tonic Cervical Influences on Forelimb and Hindlimb Monosynaptic Reflexes*

U. THODEN and D. WENZEL

Neurologische Klinik, Abteilung für Neurophysiologie, Universität Freiburg i. Br., 7800 Freiburg i.Br. (F.R.G.)

INTRODUCTION

In bilaterally labyrinthectomized animals the tonic influence of neck afferents on muscle tone can be classified as symmetric and asymmetric tonic neck reflexes (Magnus, 1924). Following lateral flexion or rotation of the head the asymmetric tonic neck reflex shows an extension in the ipsilateral fore- and hindlimb and a flexion of the opposite limbs.

In the symmetric tonic neck reflex an extension of both forelimbs and a flexion of both hindlimbs follows dorsiflexion. Opposite effects are seen after ventriflexion. These symmetric effects seem to be more pronounced in forelimbs than in hindlimbs, and vary in different species (Magnus, 1924).

Both joint (McCouch et al., 1951) and muscle receptors (Richmond and Abrahams, 1975a, b) are discussed as receptors of the neck reflexes. Ascending neck afferents to the formatio reticularis and the vestibular nuclei (Coulter et al., 1975; Fredrickson et al., 1965; Rubin et al., 1975; Thoden et al., 1975) the abducens nucleus (Hikosaka and Maeda 1973; Thoden and Schmidt, 1978) and the cerebellum (Berthoz and Llinas, 1974, Wilson et al., 1975) are demonstrated by electrical stimulation of the neck muscle nerves with low intensities.

As second order vestibular neurons in the cat are modulated by passive neck movements (Fredrickson et al., 1965; Rubin et al., 1975; Thoden et al., 1975; Thoden and Wirbitzky, 1976) a bias of vestibulospinal pathways by tonic neck reflexes seems possible, in contrast to the findings of Magnus (1924). Since vestibulospinal pathways show stronger connections to the cervical level than to the lumbar region (Gernandt and Gilman, 1959) the tonic vestibular influence on spinal reflexes is stronger in forelimbs than in hindlimbs, as recently shown by Thoden et al., (1978).

This paper was aimed to quantify the tonic neck-to-spinal reflexes in order to allow a comparison with other reflex loops stabilizing posture as the vestibulo- and visuo-spinal mechanisms (Thoden et al., 1977).

METHODS

Experiments were performed in cats prepared in 3 ways: 1) cortically intact under slight halothane anesthesia, 2) ischemically decorticated by ligation of both carotid

---

*Supported by Sonderforschungsbereich Hirnforschung (SFB 70) der Deutschen Forschungsgemeinschaft (DFG).

arteries, and 3) precollicularly decerebrated. All animals were curarized and respirated artificially.

For tonic stimulation of neck receptors the animals were mounted in an animal holder with the head inclined downwards at 45° fixed in a stereotaxic frame to exclude variable labyrinthine influences. The body, fixed with slight extension in a caudal direction at the spinal processes of L2 and L7, could be mechanically moved in three planes up to 25° to both sides. The measurement of the tonic influences of neck receptors on monosynaptic reflexes started 30 sec after the final position was adjusted.

The following *forelimb* nerves were mounted for recording on bipolar silver-silverchloride electrodes: the deep radial nerve (DR) to the lateral head of the triceps as an extensor muscle, as well as the mixed ulnar nerve (ULN) to the flexor carpi ulnaris and the ulnar head of the flexor profundus digitorum as flexor muscles. The cervical spinal cord was exposed by laminectomy from C4 to Th1. For stimulation the dorsal roots C7/C8 were cut, the ventral roots were left intact.

At the *hindlimb* the following nerves in the popliteal fossa were prepared for stimulation: the nerve to the lateral gastrocnemius-soleus and to the medial gastrocnemius (GS) as extensor muscles, as well as the deep peroneal nerve (DP) to the tibialis anterior and the extensor digitorum longus as flexor muscles. The reflex responses were recorded from the corresponding ventral roots after appropriate laminectomy.

For stimulation rectangular pulses of 0.2 msec duration with a frequency of 0.2–0.3 Hz were applied. Stimulation intensities were expressed in multiples of the motor threshold (T). The monosynaptic reflexes induced by stimulation of the above mentioned nerves were then amplified with conventional circuits. Thirty-two samples of monosynaptic reflexes were averaged (Nicolet Computer) and stored for further documentation. The modulation of averaged monosynaptic reflex amplitudes by different tonic neck positions was expressed as a percentage of averaged amplitude in normal position.

RESULTS

The asymmetric tonic neck reflex

The test reflexes of the fore- and hindlimb are modulated best in the range up to 3 times the threshold intensity. The reflex modulation starts at about 5° neck flexion and gradually increases with increasing angle of body displacement.

30 msec after a lateral flexion of 20° the monosynaptic DR-extensor reflexes of the forelimb are ipsilaterally increased by 20–25% and contralaterally inhibited by a similar amount in the cortically intact and decorticated cat. The decerebrated animal shows a smaller amount of excitation in the extensor reflex after ipsilateral flexion, and the contralateral inhibition seems slightly more pronounced. Similar effects are found in the GS hindlimb reflexes, but here the amount of modulation never exceeds 10–15% (Fig. 1).

The same procedure of a 20° lateral neck flexion shows reversed effects on the flexor muscle of fore- and hindlimbs (Fig. 2). By ipsilateral body flexion the monosynaptic component of the ULN-reflex is inhibited by between 30–45%. The facilitation of this reflex after contralateral body flexion shows a smaller degree of up to 25% in all the preparations. It thus seems that in flexor muscles the tonic inhibition exceeds the

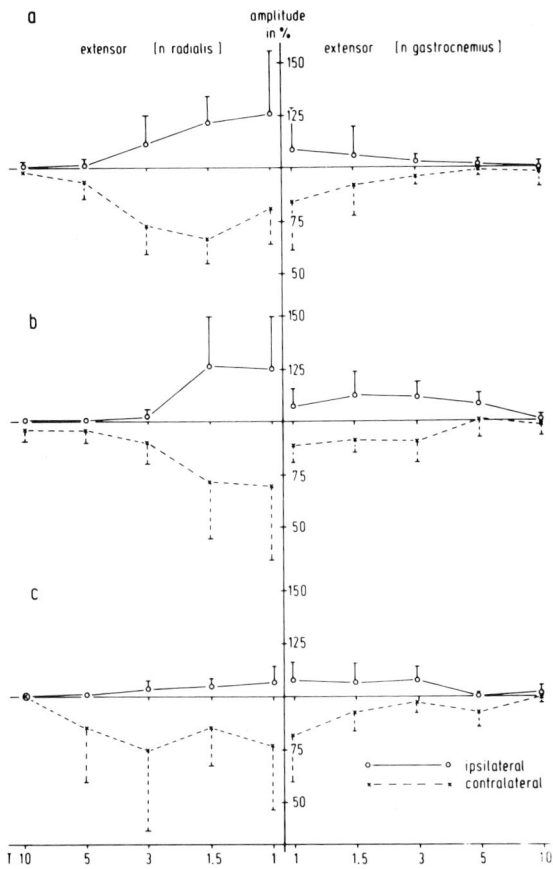

Fig. 1. Modulation of monosynaptic extensor reflexes in fore- and hindlimb following lateral flexion of the body. Modulated amplitudes are expressed as percentage of the reflex amplitude in normal position. The figures refer to the average of 4 experiments, where 32 individual responses were averaged. SD's are indicated by vertical lines. a) cortically intact cats with slight halothane anesthesia; b) ischemically decorticated; c) precollicularly decerebrated cats.

facilitation exerted by neck afferents. The modulation of the reflex of the deep peroneal nerve (DP) does not exceed 10% at its maximum, shown in the ischemically decorticated animal (Fig. 2).

During rotation of the body to both sides up to 25° the amplitudes of the hindlimb reflexes show only a small modulation analogue to the lateral flexion (Wenzel and Thoden, 1977). Forelimb reflexes were not tested by rotation for technical reasons.

The symmetric tonic neck reflex

After a body movement of 20° in the vertical plane up (dorsiflexion) extensor reflexes are slightly excited in the forelimb (DR) and inhibited in the hindlimb (GS) in a range of about 10–15%. The amount of modulation is similar in all three types of preparation (Fig. 3). The reflexes of the flexor muscles are strongly inhibited in the forelimb (up to 40–60%) whereas hindlimb flexors show only a slight excitation of about 10% (Fig. 4)

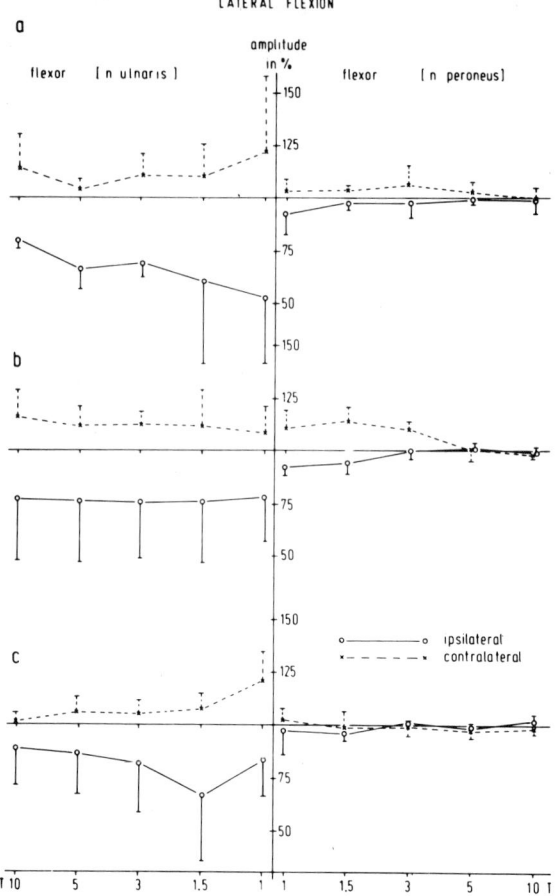

Fig. 2. Modulation of monosynaptic reflex components of flexors in fore- and hindlimb following lateral flexion. Explanation as in Fig. 1.

After ventriflexion of 20° both the hindlimb reflexes (GS and DP) are depressed strongly by 50–70% in all types of preparation and stimulation intensities of up to 10 times the threshold. Only in the decerebrated cat the GS-extensor reflex shows a less pronounced inhibition of 30%. In the forelimb ventriflexion yields opposing effects with inhibition of about 30% of the extensor reflex (DR) and excitation of 25–40% of the ULN-reflex (Figs. 3 and 4).

After acute cerebellectomy the modulation of reflex patterns in reflexes of extensor and flexor muscles is essentially unchanged, but the absolute reflex amplitudes are significantly smaller.

DISCUSSION

The demonstrated tonic neck reflexes on fore- and hindlimbs in the cat are in accordance with the modulation of the muscle tone in the symmetric and asymmetric tonic neck reflexes first described by Magnus (1924). New is the quantification of the

285

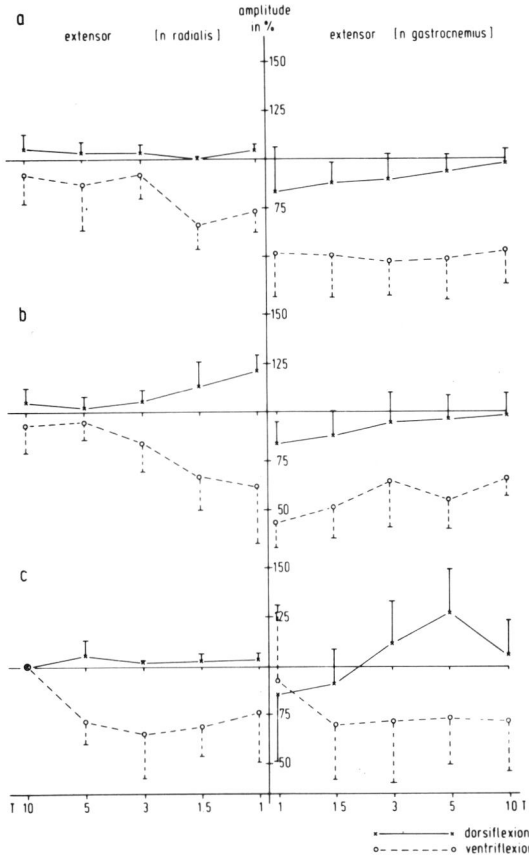

Fig. 3. Modulation of monosynaptic extensor reflexes in fore- and hindlimb following dorsi-ventriflexion of 20°. Explanation as in Fig. 1.

amount of modulation in fore- and hindlimbs allowing a comparison with other stabilizing reflex loops as the visuospinal and vestibulospinal mechanisms. Moreover, Magnus did not mention the asymmetric influence on antagonists in fore- and hindlimbs. The observation of an unchanged modulation after cerebellectomy confirms earlier findings that loss of the cerebellum weakens but does not abolish the effects of neck ventriflexion (Gernandt and Gilman, 1959).

Comparing the reflex modulation in forelimbs with those in hindlimbs similar effects are seen, but the amount of modulation is generally much higher in the forelimb than in the hindlimb following lateral flexion and dorsiflexion. This decreasing descending cervical influence may be compared with vestibulospinal pathways, as tonic neck and labyrinthine reflexes on the spinal cord work together in physiological head movements (Berthoz and Anderson, 1971; Erhardt and Wagner, 1970; Lindsay et al., 1976; Rosenberg and Lindsay, 1973). By means of the vestibulospinal tracts and vestibulo-reticulospinal pathways the vestibular nuclei are enabled to exert influence onto spinal motoneurons and interneurons. The connections between the vestibulospinal tracts and spinal motoneurons decrease from the cervical to the lumbar

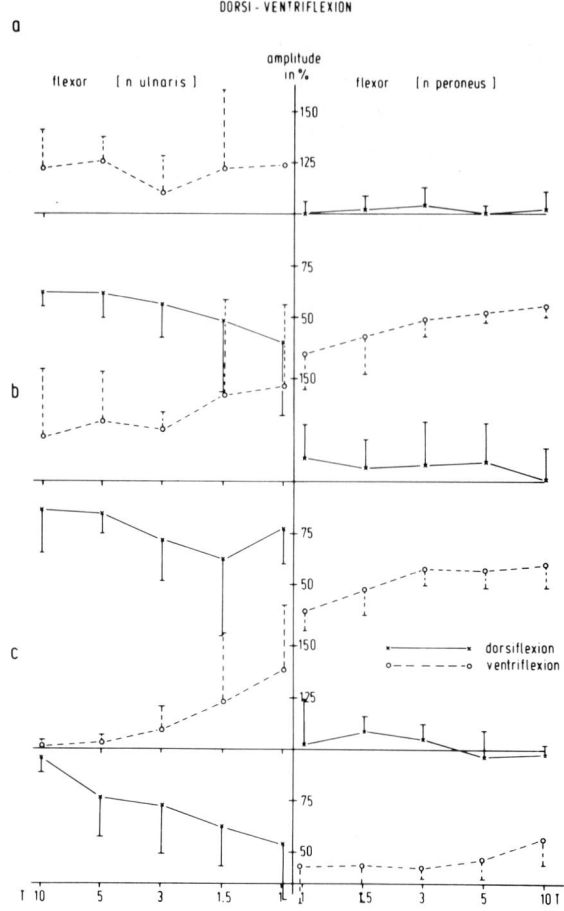

Fig. 4. Modulation of monosynaptic reflexes of flexor muscles in fore– and hindlimbs. For further explanation, see Fig. 1.

level (Precht, 1974; Wilson and Yoshida, 1968). According to Gernandt and Gilman (1959) the discharge of forelimb nerves to single shock electrical nerve stimulation shows an early bilateral spike like response, immediately followed by a late double wave. At lumbar level only this late response is seen, indicating stronger and denser connections influencing the cervical level. Moreover, the differential projection of Deiters' neurons could give further evidence of different postural responses between cervicobrachial and lumbar segments. Deiters' neurons from the rostroventral part projecting to the cervicothoracic region are frequently activated by ipsilateral tilt and inhibited by tilt in the opposite site, whereas most neurons from the dorsocaudal part projecting to the lumbosacral level are activated in response to tilt in either direction (Peterson, 1970; Precht, 1974).

Abrahams et al. (1970) described an involvement of cortical long-loop reflexes after neck muscle nerve stimulation for cervicolumbar reflex interactions. As relays for other long-loop pathways of the tonic neck reflexes the vestibular nuclear complex and the surrounding reticular formation may be discussed. The activity of neurons in all vestibular nuclei can be modulated by slow body movements and tonic neck positions (Fredrickson et al., 1965; Rubin et al., 1975; Thoden and Wirbitzky, 1976). This interaction of neck and labyrinthine activity gives evidence that the vestibular nuclei may be important for integration of descending postural activity. Also in the nearby reticular formation, similar modulated cells are found. It therefore may be argued that the above described attenuation of neck reflexes in fore- and hindlimbs results at least in part from their supraspinal interaction with descending vestibulo- and reticulospinal neurons.

The decreasing cervicospinal influence along the spinal cord proves true for lateral flexion and dorsiflexion of the body. Following ventriflexion at least partly different neuronal mechanisms are to be discussed. In the forelimb the pattern of reflex modulation is in accordance with the symmetric tonic neck reflexes, but the modulation of hindlimb reflexes shows a monotone strong inhibition; described as the vertebra-prominens reflex (Magnus, 1924; Wenzel and Thoden, 1977) A strong inhibitory effect of ventriflexion on monosynaptic responses from the radial nerve recorded before and after C1 spinalization was demonstrated by Gernandt and Gilman (1959). Special intraspinal pathways mediating inhibitory effects after ventriflexion therefore have been assumed.

SUMMARY

Fore- and hindlimb reflexes are modulated after *ipsilateral flexion* of the body with similar pattern. Excitatory (DR, GS) and inhibitory (ULN, DP) reflex modulations are more pronounced in forelimbs than in hindlimbs. The same is true for the contralateral position where reciprocal effects are demonstrated. In flexor muscles the tonic inhibition exceeds the facilitation exerted by neck afferents.

After *dorsiflexion* the DR-reflex is facilitated and the ULN-reflex inhibited. Reverse effects are found in the extensors (GS) and flexors (DP) of the hindlimbs. As in lateral flexion this modulation is more pronounced in the forelimbs.

After *ventriflexion* reciprocal effects are only measured in the forelimbs with facilitation of the ULN-reflex and inhibition of the DR-reflex. No reverse effects are measured on extensors (GS) and flexors (DP) of the hindlimb, where a strong homogeneous depression of about 50% is found for all applied stimulation intensities.

REFERENCES

Abrahams, V.C. (1970) Cervico-lumbar reflex interactions involving a proprioceptive receiving area of the cerebral cortex. *J. Physiol., (Lond.)*, 209: 45–56.

Berthoz, A. and Anderson, J.H. (1971) Frequency analysis of vestibular influence on extensor motoneurones. II. Relationship between neck and forelimb extensor. *Brain Res.*, 34: 376–380.

Berthoz, A. and Llinas, R. (1974) Afferent neck projection to the cat cerebellar cortex. *Exp. Brain Res.*, 20: 385–401.

Coulter, J.D., Mergner, T. and Pompeiano, O. (1977) Integration of afferent inputs from neck muscles and macular labyrinthine receptors within the lateral reticular nucleus. *Arch. ital. Biol.*, 115: 332–354.

Duensing, F. and Schaefer, K.P. (1960) Die Aktivität einzelner Neurone der Formatio reticularis des nicht gefesselten Kaninchens bei Kopfwendungen und vestibulären Reizen. *Arch. Psychiat. Nervenkr.*, 201; 97–122.

Ehrhardt, K.J. and Wagner, A. (1970) Labyrinthine and neck reflexes recorded from spinal motoneurones in the cat. *Brain Res.*, 19: 87–104.

Fredrickson, J.M., Schwarz, D. and Kornhuber, H. (1965) Convergence and interaction of vestibular and deep somatic afferents upon neurones in the vestibular nuclei of the cat. *Acta otolaryng. (Stockh.)*, 61: 168–188.

Gernandt, B.E. and Gilman, S. (1959) Descending vestibular activity and its modulation by proprioceptive, cerebellar and reticular influences. *Exp. Neurol.*, 1: 274–304.

Hikosaka, O. and Maeda, M. (1973) Cervical effects on abducens motoneurones and their interaction with vestibulo-ocular reflex. *Exp. Brain Res.*, 18: 512–530.

Lindsay, K.W., Roberts, T.D.M. and Rosenberg, J.R. (1976) Asymmetric tonic labyrinth reflexes and their interaction with neck reflexes in the decerebrated cat. *J. Physiol. (Lond.)*, 261: 583–601.

Magnus, R. (1924) *Körperstellung*. J. Springer, Berlin.

McCouch, G.P., Deering, I.D. and Ling, T.H. (1951) Location of receptors for tonic neck reflexes. *J. Neurophysiol.*, 14: 191–195.

Peterson, B.W. (1970) Distribution of neuronal responses to tilting within vestibular nuclei of the cat. *J. Neurophysiol.*, 33: 750–767.

Precht, W. (1974) Physiology of the vestibular nuclei. In *Handbook of Sensory Physiology, Vol. VI, Vestibular System, Part 1, Basic Mechanisms*, H.H. Kornhuber (Ed.), Springer Verlag, Berlin-Heidelberg-New York.

Richmond, F.J.R. and Abrahams, V.C. (1975) Morphology and distribution of muscle spindles in dorsal muscles of the cat neck. *J. Neurophysiol.*, 38: 1322–1339.

Rosenburg, J.R. and Lindsay, K.W. (1973) Asymmetric tonic labyrinthine reflexes. *Brain Res.*, 63; 347–350.

Rubin, A.M., Young, J.H., Milne, A.C., Schwarz, D.W.F. and Fredrickson, J.M. (1975) Vestibular neck integration in the vestibular nuclei. *Brain Res.*, 96: 99–102.

Thoden, U. and Wirbitzky, J. (1976) Influence of passive neck movements on eye position and brain stem neurones. *Pflügers Arch.*, 362: R 37, no. 147.

Thoden, U. and Schmidt, P. (1978) Vestibular neck interaction in abducens neurons. *Pflügers Arch.*

Thoden, U., Golsong, R. and Wirbitzky, J. (1975) Cervical influence on single units of vestibular and reticular nuclei in cats. *Pflügers Arch.*, Suppl. 355: R 101, no. 201.

Thoden, U., Dichgans, J. and Savidis, Th. (1977) Direction specific optokinetic modulation of monosynaptic hindlimb reflexes in cats. *Exp. Brain Res.*, 30: 155–160.

Thoden, U., Dichgans, J., Doerr, M. and Savidis, Th. (1978) Direction specific vestibular and visual modulation of monosynaptic fore- and hindlimb reflexes in cats. *Pflügers Arch.*, Suppl. 373: R 86, no. 316.

Thomas, R.C. and Wilson, V.J. (1967) Recurrent interactions between motoneurons of known location in the cervical cord for the cat. *J. Neurophysiol.*, 30: 661–674.

Wenzel, D. and Thoden, U. (1977) Modulation of hindlimb reflexes by tonic neck positions in cats. *Pflügers Arch.*, 370: 277–282.

Wilson, V.J. and Yoshida, M. (1968) Vestibulospinal and reticulospinal effects on hindlimb, forelimb, and neck alpha motoneurons of the cat. *Proc. nat. Acad. Sci. (Wash.)*, 60: 836–840.

Wilson, V.J., Maeda, M. and Franck, J. (1975) Input from neck afferents to the cat flocculus. *Brain Res.*, 89: 133–138.

SECTION III

PROPRIOCEPTIVE INFLUENCES FROM EYE MUSCLE RECEPTORS

Peripheral and Central Organization of the Extraocular Muscle Proprioception in the Ungulata

E. MANNI and R. BORTOLAMI

Istituto di Fisiologia Umana, Università Cattolica del S. Cuore, Roma; and Istituto di Anatomia degli Animali Domestici, Università di Bologna (Italy)

The presence of sensory nerve endings in the extraocular muscles of man and of some other mammals is well known. Muscle spindles in particular have been demonstrated in the eye muscles of man, goat, lamb, pig and wild boar, but not in those of the cat, dog, guinea pig and rabbit, although some types of stretch receptors have been shown in these last animals (literature in Bach-y-Rita, 1975). However, the mode of entry of the ocular proprioceptive fibres to the brain stem and the position of their cell bodies remain a matter of controversy. In fact, it has been claimed that the IIIrd, IVth and VIth cranial nerves conduct such proprioceptive impulses; their perikarya could be located in the brain stem nuclei of the oculomotor, trochlear and abducens nerves intermingled with the motoneurons or could be represented by the few ganglion cells scattered along the trunk of the eye muscle nerves (literature in Manni et al., 1970c). On the other hand, a role of the trigeminal nerve has been postulated since connections between the eye muscle nerves and the Vth nerve have been described by several investigators (literature in Bach-y-Rita, 1975). The somata of such neurons could be contained in the mesencephalic trigeminal nucleus as is the case for the masticatory proprioception, or in the semilunar ganglion.

The present report is based upon the results obtained in investigations carried out in our laboratories and devoted to analysing the peripheral and central course of the proprioceptive afferents from the eye muscles in the ungulata.

The semilunar ganglion contains in the lamb, pig and calf the perikarya of the first-order neurons of the eye muscle proprioception (Manni et al., 1966, 1968, 1970 a–c, 1971a, b, 1972a, b, 1974, 1975, 1976). Units fired by stretching extraocular muscles were found in the medial dorsolateral part of the ganglion. They were unaffected by jaw movements or by stimulation of other trigeminal receptors located on the tongue, on the face or on the teeth (Fig. 1). The firing of the units during stretching was inhibited by contraction of the muscles, a fact showing that activity from the eye muscle spindles was recorded (Fig. 1). Although the responses to excitation of spindles of the extraocular muscles were mainly studied, influences of the Golgi organs were recorded in the mini-pig (Bach-y-Rita, 1975).

Recent investigations carried out by electrophysiological and histological techniques have shown that the first-order neurons of the extraocular muscles proprioception are somatotopically arranged in the medial dorsolateral portion of the semilunar ganglion in the lamb (Manni and Pettorossi, 1976) (Figs, 1, 2, 4,). In fact, responses to stretching of the superior rectus and superior oblique muscles were recorded from the

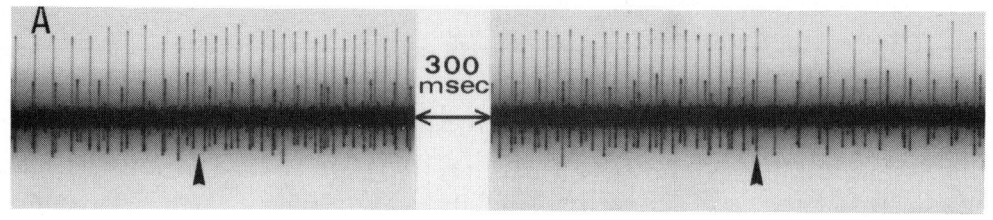

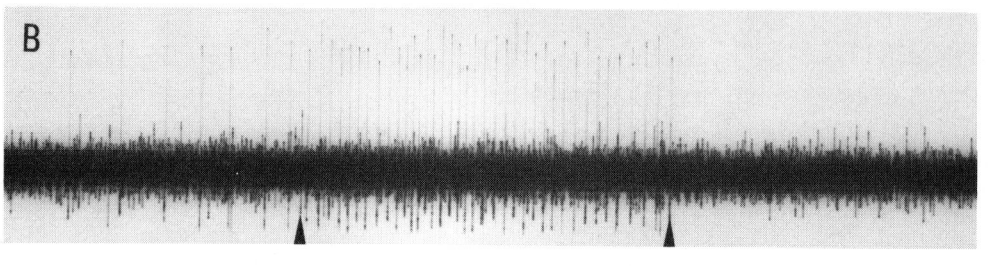

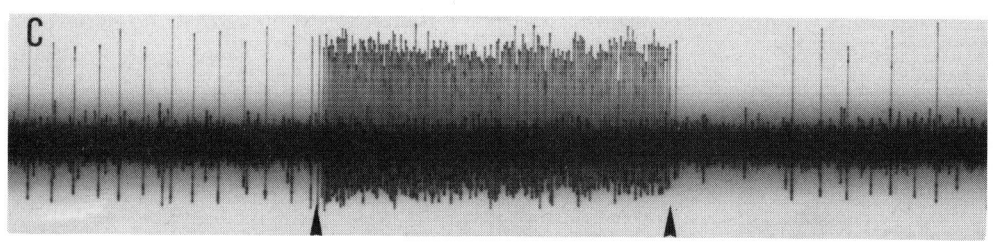

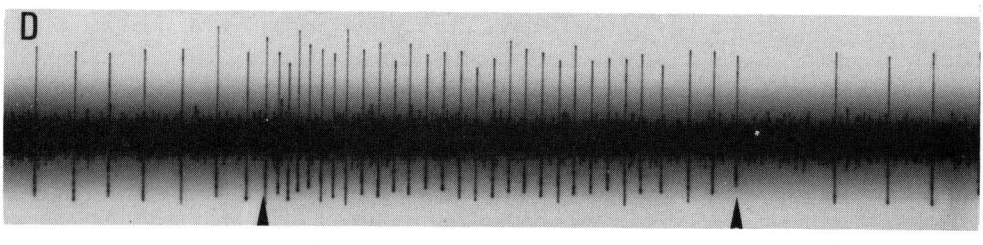

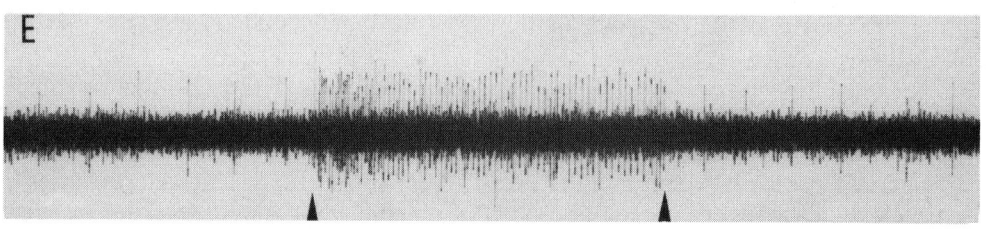

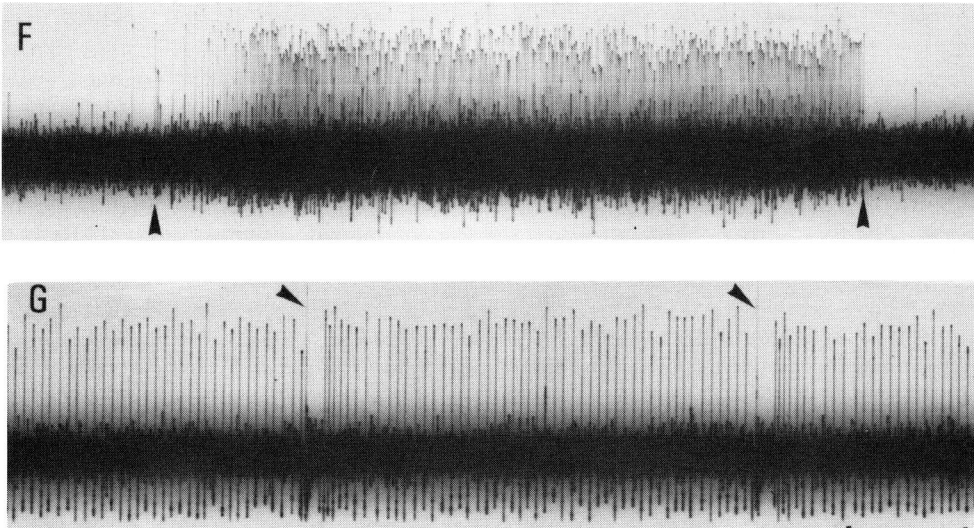

Fig. 1. Responses of gasserian units to stretch of individual eye muscles recorded at different depths and positions from the semilunar ganglion of the lamb. The arrows indicate the beginning and the end of the stretch in A–F. A) lateral rectus (depth 1400 μm); B) superior oblique (1000 μm); C) superior rectus (1200 μm); D) medial rectus (1500 μm); E) inferior oblique (2400 μm); F) inferior rectus (2200 μm). A, B, C, E and F were recorded along a single intermediate dorsoventral penetration on the lateral half of the cellular pool, while D was taken from a penetration in the medial portion. G shows inhibition of the response of a gasserian unit to stretch of the ipsilateral superior rectus elicited by contraction of that muscle. The arrows indicate the stimulus artifact produced by single-shock applied to the corresponding ocular motor nerve. Calibration: time 0.5 sec. (From Manni and Pettorossi, 1976.)

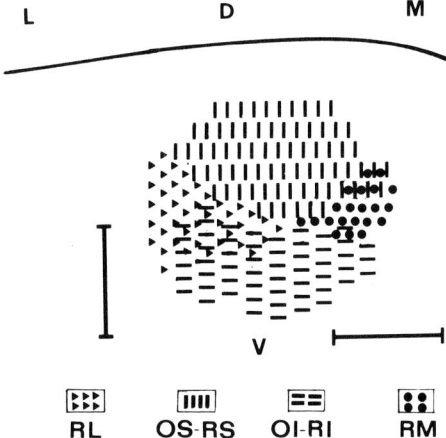

Fig. 2. Schematic diagram of transverse section of the semilunar ganglion of the lamb showing the somatotopic organization of the eye muscle afferents within the proprioceptive cellular pool. L, lateral; D, dorsal; M, medial; V, ventral; RL, lateral rectus; OS, superior oblique; RS, superior rectus; OI, inferior oblique; RI, inferior rectus; RM, medial rectus. Calibration: 1 mm. (From Manni and Pettorossi, 1976.)

dorsal layer of the cellular pool of the semilunar ganglion concerned with the eye muscle proprioception; the inferior rectus and the inferior oblique seemed to be represented in the ventral layer of the same cellular pool, while the medial rectus projected on the medial part and the lateral rectus in the lateral part, wedging between the dorsal and the ventral layers. A similar somatotopic organization could be demonstrated by employing the horseradish peroxidase method (Bortolami et al., 1979). The responses to stretching single ocular muscles were not abolished by acute section of the ipsilateral IIIrd nerve at the basis of the skull between the caverbous sinus and the fossa interpeduncularis while they were suppressed by acute or chronic section of the trigeminal ophthalmic branch which provoked also degeneration of the sensory innervation of the eye muscle spindles. On the other hand, the cutting of the trigeminal root did not abolish the semilunar responses of the eye muscle proprioception and did not provoke degenerations of the ophthalmic branch and of the sensory innervation of the eye muscle spindles, excluding therefore the provenience of such nerve fibres from the mesencephalic trigeminal nucleus. Simultaneous chronic section either of the trigeminal root and of the trochlear nerve or of the trigeminal root and of the abducens nerve did not abolish the semilunar responses to stretching respectively of the superior oblique and the lateral rectus muscle.

Such results are in agreement with those obtained on the IIIrd nerve and show that also the first-order neurons of the abducens and trochlear proprioception are contained in the semilunar ganglion.

In another group of experiments (Manni et al., 1971b, 1972a, 1975b), the cellular pool of the semilunar ganglion, which innervates the eye muscle spindles, was then destroyed by means of a discrete electrolysis. Degenerations occurred not only in the ipsilateral ophthalmic branch and in the sensory innervation of the eye muscle spindles, but also in the medial portion of the trigeminal root and in some fibres directed to the main sensory nucleus and to the oral portion of the spinal trigeminal nucleus.

All such findings support the view that the semilunar ganglion contains nerve cells which send their peripheral processes to the eye muscle spindles through branches of the ophthalmic nerve; in the ungulata, the sensory processes penetrate the muscle from one side, while the motor component reaches it from the opposite side, according to Winckler's description (Winckler, 1956). The central processes enter the brain stem through the trigeminal root and end in the oral portion of the spinal trigeminal nucleus, mainly, and in the main sensory nucleus.

In fact, responses to stretching of individual eye muscles or to single-shock electrical stimulation of the proprioceptive cellular pool of the semilunar ganglion were picked up also from the medial portion of the ipsilateral trigeminal root, from the oral portion of the spinal trigeminal tract and nucleus and from the main sensory nucleus. Some of these responses were positively recorded from the terminals of the central processes of the eye muscle proprioception first-order neurons, since their latencies were in the range of 0.30–0.50 msec.

On the other hand, in some experiments it has been possible to reverse the experimental arrangement to induce antidromic evoked potentials in cells of the medial dorsolateral portion of the semilunar ganglion by electrical stimulation of the spinal trigeminal tract or nucleus and of the main sensory nucleus. However, in other experiments, single-shock electrical stimulation of the eye muscle proprioception nerve cells in the semilunar ganglion evoked responses in the above mentioned trigeminal nuclei, whose latency was 0.90–0.92 msec. One could suppose that one synapse was intercalated. Thus, we were recording unitary discharge from the second-order neurons located in the brain stem.

Recent investigations have shown that a somatotopic arrangement does exist also at the level of the pontine representation of the eye muscle proprioception (Marini and Bortolami, 1979). By stretching single extraocular muscles and by recording the corresponding responses at the level of the oral portion of the spinal trigeminal nucleus and within the main sensory nucleus a somatotopic representation of the eye muscles has been found in the dorsoventral direction. Here also the superior oblique and the superior rectus muscles project on the most dorsal layer of the nuclei; the inferior rectus and the inferior oblique are represented below the two above mentioned muscles while the medial and lateral recti occupy the most ventral position (Figs. 3 and 4). Such anatomical and electrophysiological data on the ending of the central processes of the first-order neurons do not exclude a larger extension of this projection, i.e., as far as the upper cervical cord along the spinal trigeminal tract. In fact, some units responding to stretching extraocular muscles were found at the level of the bulb and of the upper cervical cord, i.e., in the pars interpolaris and caudalis of the spinal trigeminal complex. On the other hand, it must be pointed out that the neck muscles responded to the stretching of either one or all the six extraocular muscles (Easton,

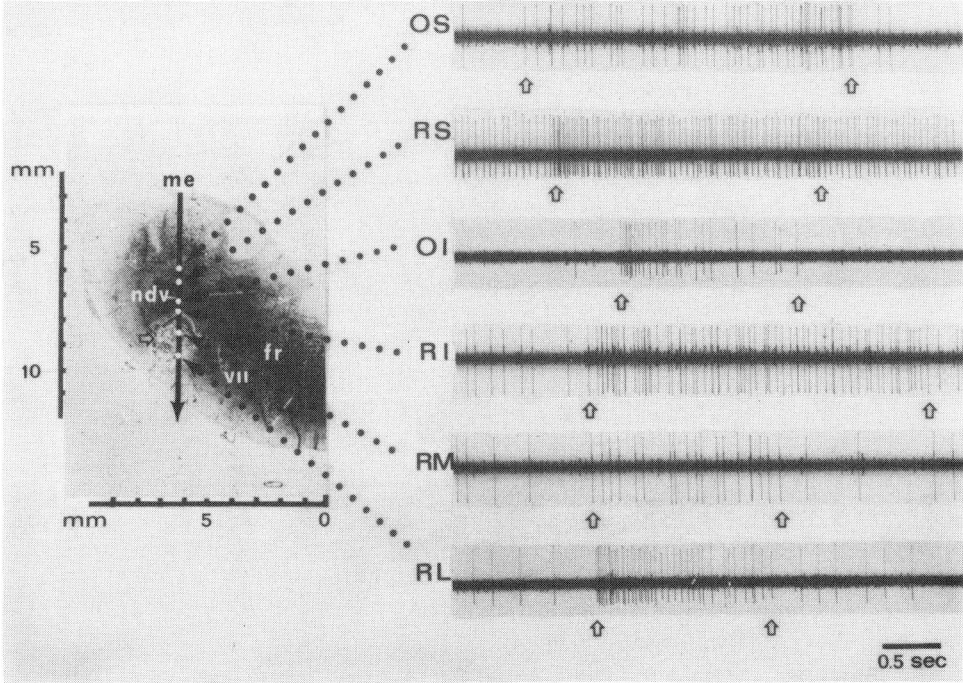

Fig. 3. Depth profile of second-order neurons responding to the stretching of single extraocular muscles in the lamb. On the left: transverse section of the brain stem showing the electrode track from which unitary discharge to stretch of single extraocular muscles was recorded. The responses were localized in the pars oralis of the spinal trigeminal nucleus (white arrow). On the right: records of units from the pars oralis of the spinal trigeminal nucleus during moderate manual stretches of single extrinsic eye muscles. The arrows indicate the duration of the stretch. All records were taken along a single dorso-ventral penetration; me, microelectrode track; ndv, spinal trigeminal nucleus; fr, reticular formation; VII, facial nerve; the other abbreviations are the same as those of Fig. 2. Calibration: time 0.5 sec. (From Marini and Bortolami, 1979.)

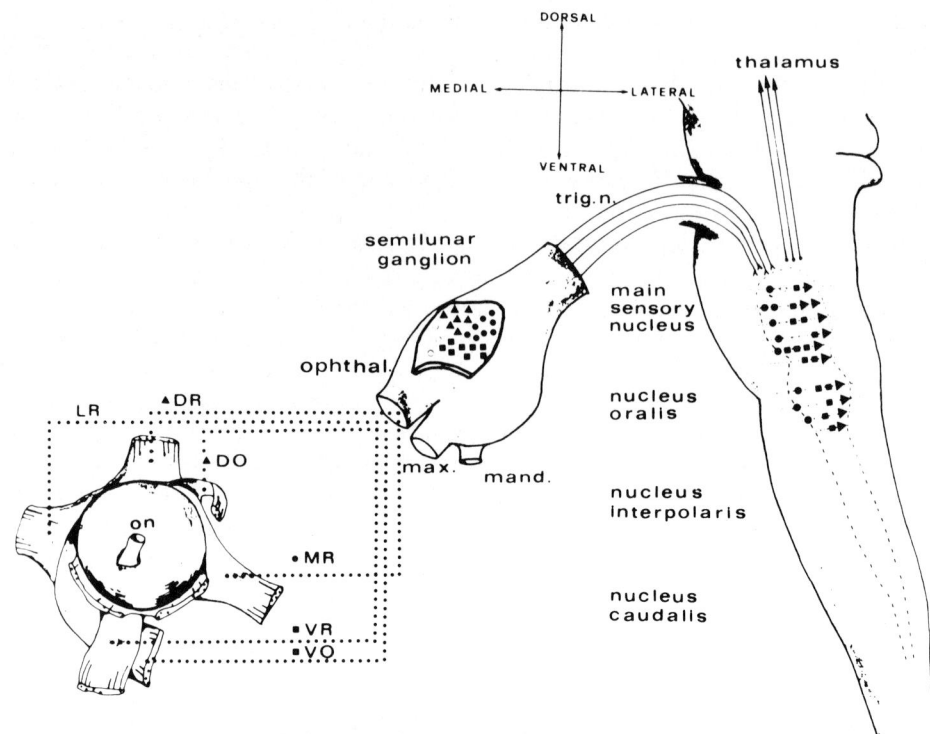

Fig. 4. Schematic diagram showing the somatotopic organization of the eye muscle proprioception within the semilunar ganglion in the main sensory nucleus and in the oral portion of the spinal trigeminal nucleus. LR, lateral rectus; MR, medial rectus; DR, superior rectus; DO, superior oblique; VR, inferior rectus; VO, inferior oblique; mand., mandibular branch of the trigeminal nerve; max., maxillary branch of the trigeminal nerve; ophthal., ophthalmic branch of the trigeminal nerve; trig.n., trigeminal nerve; on, optic nerve. (From Marini and Bortolami, unpublished observations.)

1971). However, the responses of the extensor muscles of the neck to stretching the eye muscles are not a peculiar effect of the extraocular muscle proprioceptors, but they can also be elicited by stimulating other trigeminal afferents. In any case, such a projection may have an important role in the avoiding and defence reactions of the head and of the neck (Manni et al., 1975a).

The second-order neurons of the eye muscle proprioception send their axons to other more rostral nerve formations like the mesencephalon: they terminate in the ventrobasal complex of the thalamus (Manni et al., 1972b, 1974). Other axons end in the cerebellum (Azzena et al., 1970).

The mesodiencephalic projections are supported by the following data: 1) responses to stretching single eye muscles were recorded from the ipsilateral mesencephalic tectum and tegmentum, along the medial lemniscus and from the postero-ventromedial and postero-ventrolateral nuclei of the thalamus; 2) single-shock electrical stimulation of the eye muscle representation in the semilunar ganglion, in the oral portion of the spinal trigeminal nucleus or in the main sensory nucleus elicited evoked potentials in the same ipsilateral mesodiencephalic areas from which responses to stretching single extraocular muscles were recorded. The latencies were shorter when the pons was stimulated (0.33 msec), showing that no synapse was intercalated, and longer when the stimulating electrode tip was in the semilunar ganglion (1–5 msec); in this case at least a synapse was intercalated along the neuronal pathway: 3) chronic destruction of the projection of the eye muscle proprioception in the oral portion of the spinal trigeminal nucleus or in the main sensory nucleus provoked degenerations of the nerve fibres which could be followed along the ipsilateral medial lemniscus and the dorsal trigeminal tract; the degenerations terminated in the postero-ventromedial and postero-ventrolateral nuclei of the ipsilateral thalamus.

Thus it is remarkable that the eye muscle proprioception input attains the same nuclear stations which receive the afferents from the other body receptors.

Like in the semilunar ganglion and in the pons, the mesencephalic representation of the eye muscle proprioception is somatotopically arranged (Marini and Bortolami, 1978). However, here the inferior oblique and the inferior rectus muscles are in dorsal position as regards to the superior rectus and the superior oblique. The two horizontal recti project in the most ventral position (Fig. 5).

However, the eye muscle proprioceptors send afferents also to the cerebellum; units influenced by stretching single extraocular muscles were found in the range of the most lateral part of the lobulus simplex, lobulus ansiformis and in the medius medianus lobe of the lamb (Azzena et al., 1970). Such results are in agreement with other experimental findings according to which such areas are concerned with visual input and with ocular saccadic movements (Precht, 1975).

Finally, in other experiments we have controlled the hypothesis according to which the ganglion cells scattered along the three eye muscle nerves could represent the first-order neurons of the eye muscle proprioception (Manni et al., 1970c). In order to check this possibility we have carried out experiments on the calf since in such species numerous ganglion cells may be consistently found in the intracranial portion of the oculomotor nerve. Thus we have chronically cut the left IIIrd nerve in 9 calves just where it enters the cavernous sinus. Responses to stretching individual eye muscles were recorded from the ipsilateral trigeminal ganglion. They were of the type induced by muscle spindle excitation. On the contrary, no responses to stretching eye muscles were obtained from the intracranial course of the right acutely severed oculomotor nerve. Within the muscles innervated by the left oculomotor nerve, which had been chronically cut, the spindles were normal in several animals whose central oculomotor

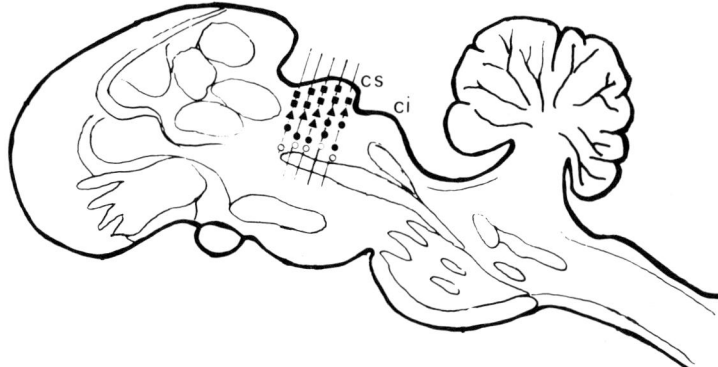

Fig. 5. Parasagittal section of the brain showing the somatotopic organization of the eye muscle proprioception in the superior colliculus of the lamb. Symbols indicating eye muscles as in Fig. 4; cs, superior colliculus; ci, inferior colliculus. (From Marini and Bortolami, 1978.)

stump contained only a few ganglion cells; a few spindles, on the contrary, showed degenerating sensory fibres in the extraocular muscles of the calves in which the intracranial stump of the nerve exhibited more than 100 ganglion cells. Thus, we conclude that the semilunar ganglion of the calf contains the somata of the afferents from the eye muscle spindles as is the case for the lamb and the pig; the ganglion cells along the oculomotor nerve may play only an accessory and negligible role in innervating the eye muscle spindles.

SUMMARY

The present report deals with the anatomical and functional organization of the peripheral and central pathways of the eye muscle proprioception in the ungulata.

On the basis of histological and electrophysiological investigations the first-order neurons of the extraocular muscle proprioception have been localized in the medial dorsolateral portion of the semilunar ganglion. While the peripheral process attains the eye muscle proprioceptors, the central one enters the brain stem through the sensory trigeminal root and terminates ipsilaterally in the oral portion of the spinal trigeminal nucleus and in the main sensory trigeminal nucleus. Such nuclei contain the second-order neurons of the eye muscle proprioception which project on the cerebellum and on the mesodiencephalic areas. Axons reach the ipsilateral ventrobasal nuclear complex of the thalamus through the ipsilateral medial lemniscus and the dorsal trigeminothalamic tract. Collaterals are abandoned in the tectum and in the tegmentum of the mesencephalon.

Finally, a somatotopic arrangement of the eye muscle proprioception has been found in the semilunar ganglion, in the pontine trigeminal nuclei and in the mesencephalon.

ACKNOWLEDGEMENT

This work was supported by the C.N.R.

REFERENCES

Azzena, G.B., Desole, C. and Palmieri, G. (1970) Cerebellar projection of the masticatory and extraocular muscle proprioception. *Exp. Neurol.*, 27: 151–161.

Bach-y-Rita, P. (1975) Structural-functional correlations in eye muscle fibres. Eye muscle proprioception. In *Basic Mechanisms of Ocular Motility and their Clinical Implications*, G. Lennerstrand and P. Bach-y-Rita (Eds.). Pergamon Press, Oxford, pp. 91–111.

Bortolami, R., Manni, E., Lucchi, M.L., Callegari, E., De Pasquale, V. and Lalatta Costerbosa, G. (1979) Labelled trigeminal ganglion cells after injection of horseradish peroxidase in the extraocular muscles and IIIrd nerve of the lamb. *Boll. Soc. it. Biol. sper.*, in press.

Easton, T.A. (1971) Inhibition from cat eye muscle stretch. *Brain Res.*, 25: 633–637.

Manni, E. and Pettorossi, V.E. (1976) Somatotopic localization of the eye muscle afferents in the semilunar ganglion. *Arch. ital. Biol.*, 114: 178–187.

Manni, E., Bortolami, R. and Desole, C. (1966) Eye muscle proprioception and the semilunar ganglion. *Exp. Neurol.*, 16: 226–236.

Manni, E., Bortolami, R. and Desole, C. (1968) Peripheral pathway of the eye muscle proprioception. *Exp. Neurol.*, 22: 1–12.

Manni, E., Bortolami, R. and Deriu, P.L. (1970a) Presence of cell bodies of the afferents from the eye muscles in the semilunar ganglion. *Arch. ital. Biol.*, 108: 106–120.

Manni, E., Bortolami, R. and Deriu, P.L. (1970b) Superior oblique muscle proprioception and the trochlear nerve. *Exp. Neurol.*, 26: 543–550.

Manni, E., Desole, C. and Palmieri, G. (1970c) On whether eye muscle spindles are innervated by ganglion cells located along the oculomotor nerves. *Exp. Neurol.*, 28: 333–343.

Manni, E., Palmieri, G. and Marini, R. (1971a) Peripheral pathway of the proprioceptive afferents from the lateral rectus muscle of the eye. *Exp. Neurol.*, 30: 46–53.

Manni, E., Palmieri, G. and Marini, R. (1971b) Extraocular muscle proprioception and the descending trigeminal nucleus. *Exp. Neurol.*, 33: 195–204.

Manni, E., Palmieri, G. and Marini, R. (1972a) Pontine trigeminal termination of proprioceptive afferents from the eye muscles. *Exp. Neurol.*, 36: 310–318.

Manni, E., Palmieri, G. and Marini, R. (1972b) Mesodiencephalic representation of the eye muscle proprioception. *Exp. Neurol.*, 37: 412–421.

Manni, E., Palmieri, G. and Marini, R. (1974) Central pathway of the extraocular muscle proprioception. *Exp. Neurol.*, 42: 181–190.

Manni, E., Palmieri, G., Marini, R. and Pettorossi, V.E. (1975a) Trigeminal influences on extensor muscles of the neck. *Exp. Neurol.*, 47: 330–342.

Manni, E., Palmieri, G., Marini, R. and Pettorossi, V.E. (1975b) New observations on the representation of the eye muscle proprioception in the descending trigeminal nucleus. *Experientia*, 31: 944–945.

Marini, R. and Bortolami, R. (1979) Somatotopic organization of second-order neurons of the eye muscle proprioception. *Arch. ital. Biol.*, 117: 45–57.

Marini, R. and Bortolami, R. (1978) Localizzazione somatotopica delle afferenze propriocettive oculari nel mesencefalo di agnello. *Boll. Soc. it. Biol. sper.*, 54: Suppl. fasc. 18 bis, no. 132.

Precht, W. (1975) Cerebellar influences on eye movements. In *Basic Mechanisms of Ocular Motility and their Clinical Implications*, G. Lennerstrand and P. Bach-y-Rita (Eds.), Pergamon Press, Oxford, pp. 261–280.

Winckler, G. (1956) L'innervation proprioceptive des muscles extrinséques du globe oculaire chez l'homme, *C.R. Ass. Anat.*, 43: 848–857.

Properties of the Receptors of the Extraocular Muscles

C. BATINI

Laboratoire de Psychophysiologie Sensorielle, Université Pierre et Marie Curie, 75230 Paris (France)

MORPHOLOGICAL PROPERTIES OF THE RECEPTORS

Muscle receptors

Sensory terminals in the extrinsic eye muscles (EOM) are morphologically different from those found in other skeletal muscles. Muscle spindles, however, have been found in the EOM of some but not all mammals. This difference has led to a grouping of the animals having muscle spindles, like man (Cooper and Daniel, 1949, Merrillees et al., 1950; Winkler, 1956), horse (Bonavolontà, 1956), albino mouse (Maharan and Sakle, 1965) and arctyodactyls in general (Cilimbaris, 1910; Cooper and Daniel, 1949; Scalzi and Price, 1970; Harker 1972b; see Barker, 1974), and those lacking muscle spindles like the rabbit, dog, cat, etc. (Cilimbaris, 1910; Fukuda, 1958; see also Hosokawa, 1961, and Barker, 1974). A detailed description of EOM muscle spindles will be found in the work of Harker (1972b) and the review of Barker (1974). On the other hand, electrophysiological recordings made from the muscle nerves of the EOM during passive stretches show responses in both groups of animals: in the goat, sheep and pig (Cooper and Daniel, 1954; 1957; Browne, 1974, 1975; Lennerstrand and Bach-y-Rita, 1974b), as well as in the cat and dog (Cardin and Rigotti, 1947; Cooper and Fillenz, 1955; Bach-y-Rita and Murata, 1964; Bach-y-Rita and Ito, 1966b). To explain the responses elicited in the cat, the free spiral endings found by Cooper and Fillenz (1955) have been claimed to fill the functional role of the spindles. The free spiral endings have been found also in the EOM of men (Daniel, 1946; Sas and Appeltauer, 1963), monkey (Cooper and Fillenz, 1955) and albino mouse (Maharan and Sakla, 1965) also having muscle spindles. Therefore they appear not to be a proprioceptive organ specifically replacing the muscle spindles.

In a recent work Alvarado-Mallart and Pinçon Raymond (1979) have described in the cat the free spiral endings illustrated in Fig. 1. They resemble very strongly the complex terminals drawn by Dogiel (1906) from the EOM of the horse. These structures were found to be located deep in the belly of the muscle, surrounding the central part of a thin muscle fiber which did not receive motor innervation, and did not have cholinesterase activity (Alvarado-Mallart, 1978). Whether the muscle fiber was of the fast or of the slow type (see below) could not be determined. The authors also reported that the spiral endings are present in the four recti and in the two obliqui muscles analysed (the retractor bulbi was not studied) but in each case, to be very few in number. However, the silver impregnation method used may not allow a quantitative

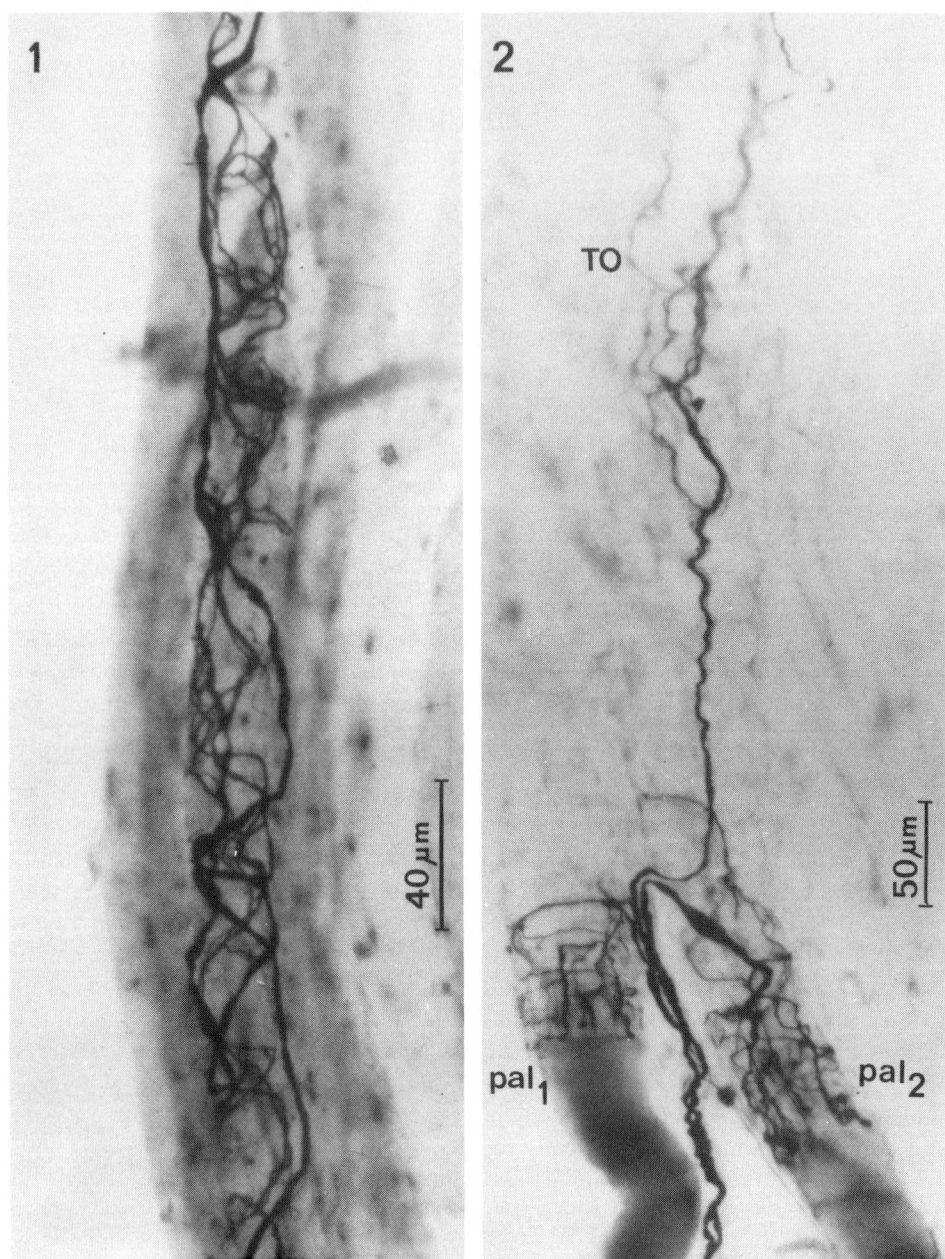

Fig. 1. Free spiral ending in an obliquus inferior muscle of the cat. Silver impregnation. (From Alvarado-Mallart, 1978.)

Fig. 2. Two nearby palisade endings from two separated axons in an obliquus inferior muscle of the cat. Note the collateral fiber ending in the tendon (TO). Silver impregnation. (From Alvarado-Mallart, 1978.)

study since some of the nerve fiber terminals might escape impregnation and detection.

There would be a fundamental difference between muscle spindles and free spiral endings: the free spinal ending lacks a specific motor innervation of the sensory organ. However, one could raise the basic question: is it possible that a muscle fiber lacks motor innervation?

Tendon receptors

The Golgi tendon organs since Golgi (1893) are known to be absent from EOM. However, Huber (1900) first described in the cat and later Crevatin (1902) in the dromedary, a particular type of nerve ending at the musculotendinous junction of the EOM. Soon after Dogiel (1906) showed the same structure in man and other mammals and has called it "palisade" because of the particular shape of the terminals attached to the end of a muscle fiber. Although Tozer and Sherrington (1910) stressed their "obviously receptive (sensorial) type", none of the later works, to the best of our knowledge, has tried to correlate muscle nerve afferent responses to the palisades, perhaps because they resemble the motor basket endings of the lower vertebrates (see Barker, 1974). In fact, their sensory character has been questioned by Sas and Schab (1952) and Cheng (1963). These authors, however, could not demonstrate a motor function for these structures.

Recently, Alvardo-Mallart and Pinçon-Raymond (1979) have also studied the palisade endings by a joint optical and electronmicroscopic study. In addition to the classical description of Dogiel, the authors have observed morphological properties characteristic of a sensory organ, and have identified the muscle fiber to which the palisade is attached.

i) The palisade endings arise from a single axon forming a multiple terminal which attaches around one end of the muscle fiber. Sometimes they give collaterals to the nearby tendon. These collaterals entering the tendon form a structure resembling a Golgi tendon organ. An example of such a palisade is shown in Fig. 2 and its similarity to the drawing of Dogiel is striking.

ii) The morphological properties of the palisade endings with strong presumption of sensory function are illustrated in Figs. 2 and 3. In addition to the fact that a certain number of palisades give collaterals terminating inside the tendon, most of the branches apparently ending around the muscle fibers are enclosed in collagen fascicles. Those of the terminals which make neuromuscular contact do not have interposed basal lamina as found in other sensory organs (Merrillees, 1960; Corvaja et al., 1969; Zelena, 1976). Finally, a thin capsule surrounds all the terminals and the muscle fibers to which they attach. Homologous membranes characteristically encapsulate most of the proprioceptors (see Barker, 1974).

iii) EOM have been shown by Cooper and Eccles (1930) to be the fastest contracting muscles found in mammals. In spite of this character, typical of twitch fibers, they are also extremely sensitive to acetylcholine induced contraction (Duke-Elder and Duke-Elder, 1930) indicating the additional presence of multi-innervated, slow muscle fibers described in other vertebrates (Brown and Harvey, 1941; Kuffler and Vaughan Williams, 1953a, b; Ginsborg, 1960). In fact, multiple or "en grappe" (Retzius, 1892) motor innervation was then described by several authors in EOM of different mammals (Hess, 1961; Hess and Pilar, 1963; Dietest, 1965; Cheng and Breinin, 1966; Cheng et al., 1968; Harker, 1972b; Kaczmarski, 1974; Alvarado and Van

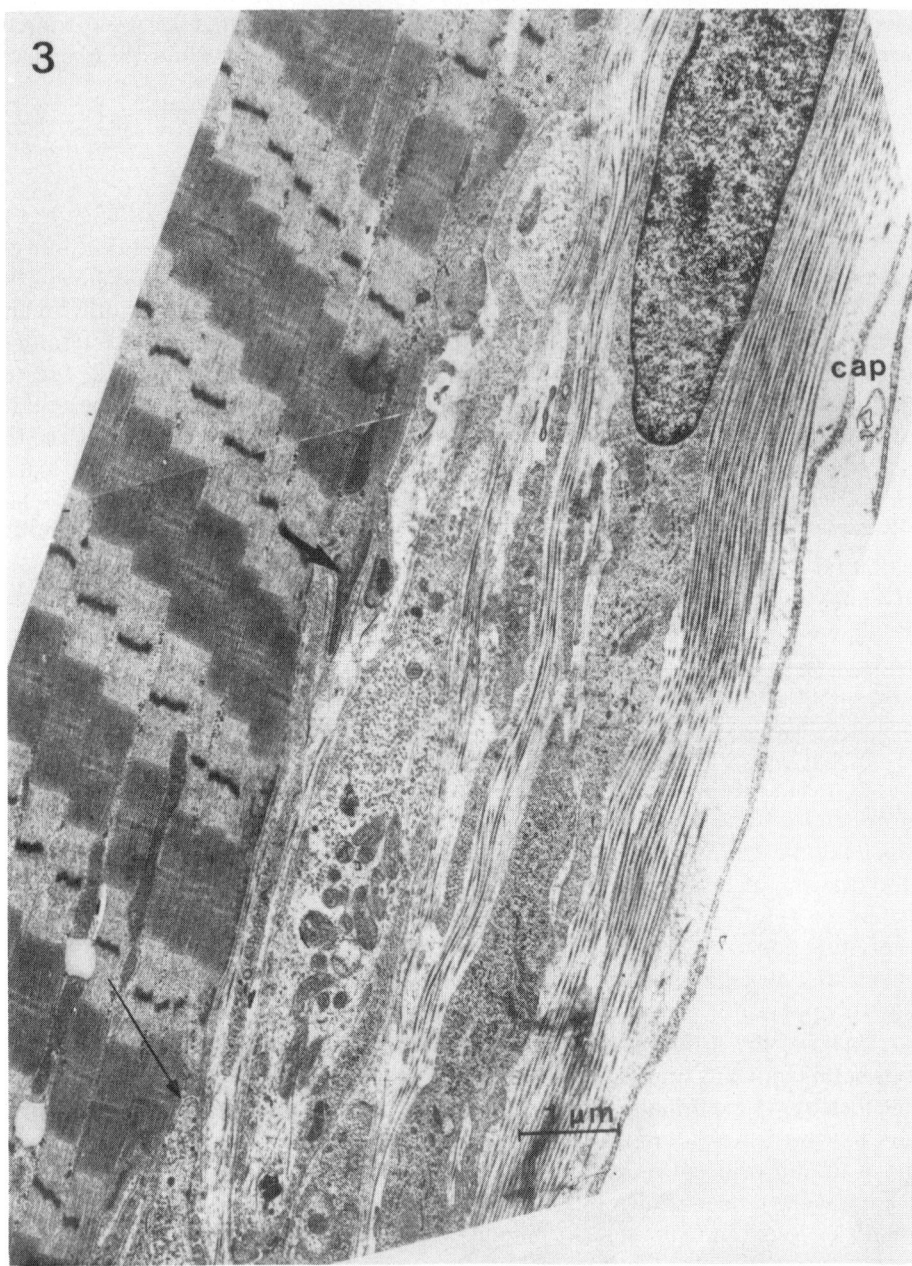

Fig. 3. Electron micrograph of a longitudinal section of a palisade ending. Note the ultrastructural character of the muscle fiber on the left; a protuberance indicates the musculotendinous junction (fat arrow); two terminals of the palisade run along the muscle fiber; one neuromuscular contact (long arrow) is seen without interposed basal lamina; two flat cells belong to the capsule (cap). (From Alvarado-Mallart, 1978.)

Horn, 1975). Functionally, the multi innervated EOM muscle fibers belong to the slow type (Matyuskin, 1961; 1964; Pilar and Hess, 1966; Bach-y-Rita and Ito, 1966a; Pilar, 1967; Lennerstrand and Bach-y-Rita, 1974a; Goldberg and Lennerstrand, 1974; Lennerstrand, 1975). Therefore the EOM have fast twitch fibers and tonic slow fibers. The only exception to this is the retractor bulbi which lacks slow multi-innervated muscle fibers (Bach-y-Rita and Ito, 1965; Alvarado et al., 1967; Bach-y-Rita et al., 1967) and is composed of rather homogeneous motor units of the fast type (Lennerstrand, 1974a). Alvarado-Mallart and Pinçon-Raymond showed that the muscle fibers to which palisades are attached have additional multiple innervation and therefore are of the slow type. More precisely they are the type 4 described in the cat by Alvarado and Van Horn (1975). These authors have equated (see also Chiarandini and Davidowitz, 1978) their type 4 to the type "large G" of sheep, which lacks palisades but has muscle spindles (Harker, 1972a; 1972b), to the type "clear" of the rat (Mayr, 1971) and to the type "slow" of the monkey (Miller, 1973) which have few muscle spindles (Green and Jampel, 1966) but are rich in palisades (Tozer and Sherrington, 1910).

It is interesting to note the following:

i) Palisades are absent from the retractor bulbi (Hosokawa, 1961) which lack multi-innervated muscle fibers.

ii) The type of muscle fiber to which the palisades of cat are connected, have been found, in the animals so far analysed, to be located in the internal or global layer of the EOM, whether the animal possesses palisades, like the cat and the monkey, or not as in sheep (see Barker, 1974). On the contrary, muscle spindles are mostly found, at least in sheep, in the external, or orbital layer (Harker, 1972b) and their intrafusal fibers correspond to the other types of multi-innervated muscle fibers (Barker and Harker, 1971).

iii) In fact the EOM of cats have two types of multi-innervated muscle fibers (Peachey, 1971; Alvarado and Van Horn, 1975). Both are poor in sarcoplasmic reticulum, which indicates slow contraction of the fibers, but they differ in the quantity of mitochondria on which depends the resistance to fatigue. The type 4 has few mitochondria. Lennerstrand (1974b; see also Chiarandini and Davidowitz, 1978) also claim two functional types of slow multi-innervated muscle fibers in the cat, those conducting and those not conducting action potentials, but we do not know to which the palisade is attached.

iv) Contrary to the free spiral endings, relatively more palisades are found in a single muscle: up to 100 have been counted at either end of a rectus inferior muscle (Alvarado-Mallart and Pinçon-Raymond, 1978); their function might, therefore, not be one of minor importance.

v) The position and the morphology of the palisade, with respect to the orientation of the muscle fibers, is clearly of the type "in series", just like the Golgi tendon organ. But the palisade attaches to a single muscle fiber so that any tension changes on this fiber should be sensed by the receptor, whether from stretches or from contractions. Therefore, on the basis of the model of Houk and Henneman (1967), one may hypothesize that the palisade could have an "absolute threshold" lower than that of the tendon organs in the soleus muscle of the cat. From the figures of Goldberg et al. (1976) the intracellular stimulation of a motoneuron corresponding to a slow contracting motor unit, evokes some 15 mg of tension. Even if the EOM have small motor units (innervation ratio 1 : 10 for Torre, 1953), the palisade would sample only part of the tension produced by that motor unit.

PROPERTIES OF THE EOM AFFERENT FIBERS

More emphasis has been given recently to the study of the pathways followed by the proprioceptive fibers of the EOM (see Manni and Bortolami, this volume, chapter III 1) than to their diameter and conduction velocity. In fact, to investigate the properties of nerve fibers one needs to know first where they are.

Nevertheless, we have at our disposal a few data. Bach-y-Rita and Ito (1966b) measured conduction velocities of afferent fibers in the nerve to the obliquus inferior in the cat and found that they range between 6.5 and 52 m/sec. The peak of the distribution falling between 10–15 m/sec (Fig. 4A). Average conduction velocities of some 30–45 m/sec have been measured for the fibers originating in the EOM and passing into the ophthalmic nerve (Batini et al., 1975) in the cat. The diameter of these selected trigeminal fibers have been measured and reported to range between 5 and 20 μm with a peak population between 10 and 15 μm (Buisseret-Delmas, 1976).

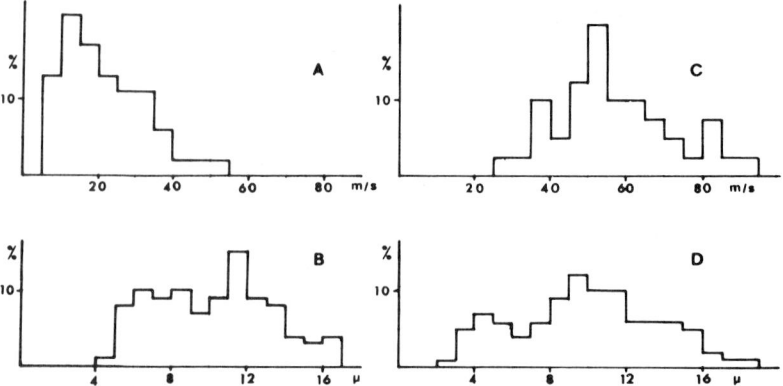

Fig. 4. A) Distribution of conduction velocities for afferent fibers of the nerve to the obliquus inferior muscle of cat (modified from Bach-y-Rita and Ito, 1966b). B) Histogram of the afferent fiber diameters in the nerve from the obliquus inferior passing to the trigeminal nerve in cat (from Buisseret-Delmas, 1976). C and D) Histograms of the conduction velocities (C) and fiber diameters (D) of afferents from muscle spindles of the sheep obliquus superior muscle (modified from Browne, 1975).

These results illustrated in Fig. 4B, indicate that the majority of these afferent fibers would belong with the group II class of fibers and the remaining fibers being group III and group I. If one would apply an average conversion factor of 5 (Boyd and Davey, 1968), the conduction velocities of these fibers would be higher than those measured in the muscle nerve but they would fit values measured in the Vth nerve. The discrepancy may be explained if part of the fibers from the muscle proprioceptors are not passing through the trigeminal nerve. There are fewer fibers from a single EOM passing into the Vth nerve (Buisseret-Delmas, 1976) than there are palisades in the same EOM (Alvarado-Mallart and Pinçon-Raymond, 1978). Also there is lack of knowledge of the relation of fiber diameter to the type of receptor. For the moment, the only available information is the approximate diameters of the preterminal segments of 4–6 μm for the free spiral ending and 2–4 μm for the palisades. If this relation is maintained for the length of the fibers, the free spiral endings would have larger axons that the palisades. It is generally believed that in the sheep, all afferent fibers from the EOM take the trigeminal pathway (see Hosokawa, 1961). Browne (1975) studied the properties of fibers entering the trigeminal nerve and identified those which were afferents from muscle spindles in the sheep. It was found that the conduction velocities ranged between 30 and 110 m/sec (see Fig. 4C). The measured diameters of the corresponding fibers agree very well with the conduction velocities (Fig. 4D).

Sheep EOM spindles have been shown to have type I and II endings (Harker, 1972b). The diameter spectra and conduction velocities obtained by Browne (1974) include both types of endings but neither the diameter spectra nor the conduction velocities indicate the presence of two different populations of afferents. Although the diameter spectra for the EOM afferents are very similar for both the sheep and the cat trigeminal contingent (Fig. 4C, D) their conduction velocities are not. In the sheep the velocities are greater than 20 m/sec (peak 60 m/sec) while in cat the majority of the fibers have velocities less than 20 m/sec. The simplest explanation for this discrepancy is that in the cat the non-trigeminal contingent of fibers must have smaller diameters and thus a slower conduction velocity. If the preterminal diameter relation for the free spiral endings and palisades is maintained it will be possible to conclude that the trigeminal fibers belong to the free spiral endings.

PHYSIOLOGICAL PROPERTIES OF THE RECEPTORS

The functional properties of the EOM receptors have been studied in different mammals mostly by single afferent fiber records in the muscle branch to the EOM (Cooper and Fillenz, 1955; Cooper et al., 1953, 1955; Cooper and Daniel, 1957; Bach-y-Rita and Murata, 1964; Bach-y-Rita and Ito, 1966b; Cardin and Rigotti, 1947). Studies have also been made recording at the level of the ganglionic cell bodies (Cooper and Daniel, 1954; Fillenz, 1955; Manni et al., 1966, 1970; Buisseret et al., 1972; Alvarado-Mallart et al., 1975; Brown, 1974; 1975). This subject has gained some interest in the past decade and has been reviewed by Bach-y-Rita (1971, 1975). We would like to analyse here only the differences between the properties of the EOM muscle spindles and those of the EOM free spiral endings and investigate the possibility that some properties of the cat EOM receptors might be attributed to the palisades.

Muscle receptors
Adequate criteria for the stretch parameters have not yet emerged, so that various EOM receptors might be characterized as to their individual dynamic properties. There are indications that the various receptors are indeed functioning in different modes during a stretch and in fact may be working in relays to maintain an adequate sensitivity over a large dynamic range. In general, the EOM are stretched in such a way as to give an equivalent eye displacement of from 6° to 18° at velocities rarely exceeding 2°–200°/sec. These latter values would correspond to slow pursuit as well as to saccadic eye movements (Crommelink and Roucoux, 1976; see Carpenter, 1977).

In examining the available information, we will take dynamic responses to mean the change in the firing rate of a fiber or nuclear cell to a change in the muscle length. Static responses will mean the firing rate obtained when there are no changes in muscle length, whether in a stretched position or not.

In the first instance, the unit activity found in the mesencephalic nucleus of the Vth nerve to stretch of the rectus lateralis in the cat has both dynamic and static responses

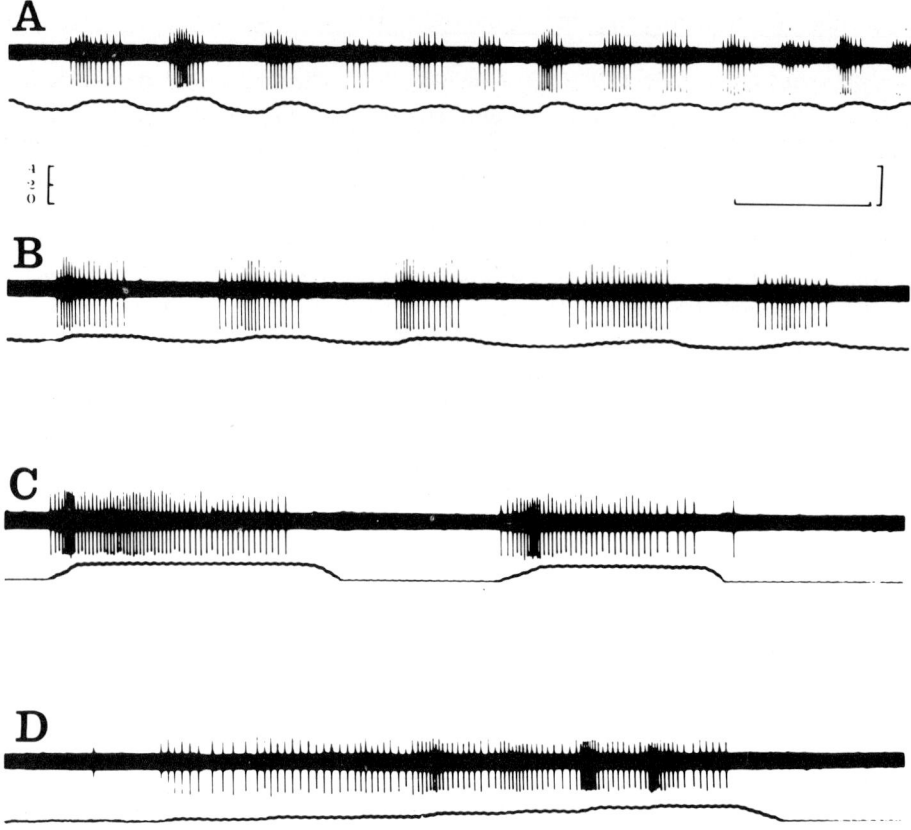

Fig. 5. Responses of a unit in the mesencephalic nucleus of the Vth nerve to various stretches of the rectus lateralis muscle. Upper trace: unit recordings. Lower traces: stretch signal, with an upward deflection corresponding to changes of length calibrated in mm (top left). A and B) Responses to periodic changes of length at different amplitudes. C) More prolonged stretches showing the dynamic phase followed by the static phase. D) Response to a "step-by-step" stretching of the muscle. Note the increased frequency at each change of length. Calibration: 500 μV. Time: 500 msec. (From Alvarado-Mallart et al., 1975.)

similar to those found in the sheep, goat and pig. Fig. 5 is an example of such responses to changes in muscle length showing an increased firing with increased length (see also Fig. 6). On releasing the muscle there is a pause of variable length. The magnitude of the dynamic response has been shown to depend on the initial length of the muscle and on the velocity of the stretch (Bach-y-Rita and Ito, 1966b). Thus, the dynamic responses obtained are very similar whether the animal has muscle spindles (sheep, goat and pig) or few spiral endings and no muscle spindles (cat).

In sheep EOM nerve, Browne (1975) has shown the presence of two extreme responses to a ramp stretch which have been considered homologous to the group Ia and II fibers from the cat soleus muscle spindles. But many intermediate responses were also obtained. In a similar kind of experiment in the cat, Bach-y-Rita and Ito (1966b) have found a great variability in threshold, dynamic and static responses as well as in adaptation of the receptors; all were sensitive to stretch and all fell into the same group. Based on anatomical findings and these results, dynamic and static characteristics of the sheep and pig EOM are attributed to muscle spindles and those in the cat to the free spiral endings. There are, however, several important disparities:

i) There is a large difference in the conduction velocity of the afferent fibers which has been already discussed (see above).

ii) Perhaps the more evident difference in the properties of the receptor is given by the dynamic sensibility. A sudden increase of the discharge frequency at the beginning of the ramp stretch and a sudden decrease at the completion of the stretch is present in the responses of a certain number of receptors in sheep (Browne, 1975). This dynamic property is typical of the Ia endings but has not been found in the cat EOM, even if the dynamic response of the receptor was increased by increasing the velocity of the stretch. In this respect the receptor responses of the cat resemble more the type II-like responses of sheep. In fact there is not a sudden increase in the frequency of discharge at the beginning of a ramp stretch and at the completion of the ramp the firing rate decreases to its static level almost exponentially.

iii) Muscle spindles of sheep (Harker, 1972a), goat (Whitteridge, 1958) and pig (Lennerstrand and Bach-y-Rita, 1974b) have fusimotor innervation. Also the γ innervation could be demonstrated to increase the sensitivity of the EOM receptors when activated by stimulation of the fiber bundles of the muscle nerve (Whitteridge, 1959; Lennerstrand and Bach-y-Rita, 1974b; Browne, 1975). In the cat, on the contrary,

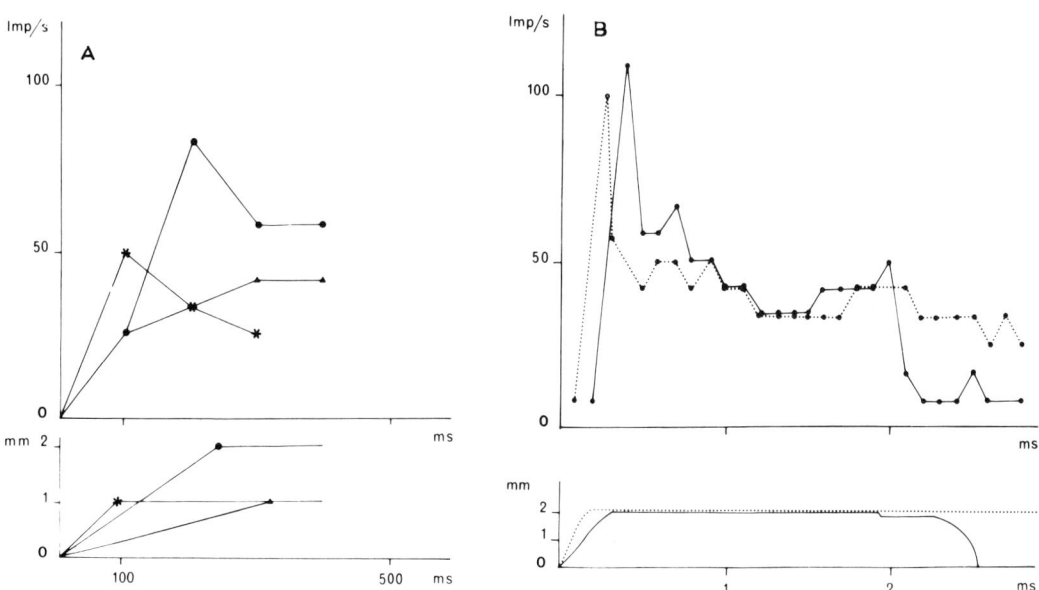

Fig. 6. Frequency analysis of discharges in two units of the mesencephalic nucleus of the Vth nerve. A) Initial velocity sensitive part of the response of one unit to three different stretches of the rectus lateralis muscle. Abscissae: time in msec. Ordinates: upper graph, impulses/sec; lower graph, stretch amplitudes in mm. B) Frequency analysis of a unit's discharges during two sustained stretches. Abscissae: time in msec. Ordinates, as in A.

even with maximal tetanic stimulation of the muscle nerve, no fusimotor innervation was found and in fact the firing rate decreased (Bach-y-Rita and Ito, 1966b).

Other important differences are due to the methodology used. In sheep (Browne, 1975) and pigs (Lennerstrand and Bach-y-Rita, 1974b) only the receptors with properties similar to those described for type Ia and II endings in the soleus muscle have been studied. Therefore, first these receptors were identified and then tested for selective properties. Thus it has been shown that some nerve responses of the Ia type follow a higher frequency of vibration (although not as high as in the cat soleus muscle, Brown et al., 1967) than those of the II type. They also have a higher sensitivity to the succinylcholine-induced contraction of intrafusal fibers. In the cat this distinction between type Ia and II endings is not available perhaps because the different authors were more concerned with demonstrating the presence of responses to stretches of the EOM than in identifying the type of receptor involved. Nevertheless, from the above differences we can hypothesize that the EOM of the cat lacks receptors with type Ia ending properties.

Tendon receptors

Nothing is known about the physiological properties of the palisades as distinct from the Golgi tendon organs and little is known about the tendon receptors in general in the EOM. It is possible that when lacking other tendon organs their functional role is served by the palisades in the EOM.

Cooper and Fillenz (1955) described in the cat and the monkey a second type of stretch-sensitive responses in the afferent nerve fibers, which they ascribed to non-identified tendon receptors. They report them as occurring at long latency, as being quickly adapted and as responding to sharp stretch. It is quite possible that these responses correspond to some of the quickly adapting, high-threshold stretch receptors described by Bach-y-Rita and Ito (1966b). Yet these authors, in their experiment on the inferior oblique muscle, had applied tension, a parameter to which the tendon organs are sensitive (see Hunt, 1974).

Tendon organ responses are distinguished from those of muscle spindles by their increased nerve discharge during the rising phase of twitch contraction of the muscle (Matthews, 1933). The same criteria applied to the EOM of the pig had allowed the identification of only a few tendon receptors (Lennerstrand and Bach-y-Rita, 1974b). Perhaps in the pig, as in the cat, the funtional tendon organs are the palisades and the identification of the palisades would presumably need different criteria since they are only connected with non-twitch fibers.

Bach-y-Rita and Ito (1966b) in the cat classified most of the receptors as being "in parallel", and therefore corresponding to free spiral endings, because they decreased the response to stretches during tetanic contraction of the muscle; only 2 out of 49 receptors increased their response and were therefore classified as "in series", or tendon-like receptors. However, in order to increase the response of an "in series" palisade, the stimulation must specifically reach the motor unit corresponding to the muscle fiber in contact with the palisade. Thus, even if maximum contraction is obtained but the motor unit not activated, it will be impossible to identify the tendon receptor. Furthermore, as demonstrated by Houk and Henneman (1967) in the soleus muscle of the cat, the tendon organ is unloaded and ceases to discharge when other motor units, not connected "in series" with it are stimulated and thus simulate an "in parallel" receptor. Perhaps this is also the correct interpretation of the decrease in firing observed in the response of some of the EOM afferents during contraction of the muscle (Bach-y-Rita and Murata, 1964; Bach-y-Rita and Ito, 1966b). An indication is given by the fact that the two "in series" receptors described in the cat had properties which they shared with other receptors classified as "in parallel": they were rapidly adapting and had threshold of 20 g and 30 g, which for eye muscles (Bach-y-

Rita and Ito, 1966a) is quite high. Therefore we suggest that the second type of response to stretches obtained by Cooper and Fillenz, as well as at least part of the high threshold, rapidly adapting receptors classified as "in parallel" free spiral endings by Bach-y-Rita and Ito, are instead "in series" palisades.

It is possible that for the palisades which are located between the muscle fibers and its tendon, the threshold for passive forces could be lower and the dynamic sensibility greater than for the Golgi tendon organ. Thus, distinguishing between responses from the palisades and the free spiral endings in the EOM would be much more difficult than distinguishing the responses from the stretch receptors in skeletal muscles. We hope that future work will focus attention on the study of the properties of the palisade endings.

SUMMARY

The morphological properties of the EOM have been reviewed with particular reference to the cat. The problem of the existence of the free spiral ending as a receptor replacing the muscle spindles has been raised. The properties of the palisade endings, behaving like the role tendon endings described in the EOM, have been reviewed with emphasis on a possible sensory function. Attention has been focused on the identification of the single, multi-innervated muscle fiber to which the palisade attaches. These are assumed to be a more sensitive tendon receptor than the Golgi tendon organ.

The properties of nerve fibers from different EOM receptors are discussed in terms of the available data on fiber diameter and conduction velocities.

The review of the physiological properties of the EOM receptors is based on available studies in a number of mammals. The differences between the responses of EOM muscle spindles and those of the EOM free spiral endings have been analysed. Concerning the tendon receptors, in spite of the fact that nothing is known about their physiological properties, the possibility has been investigated that some of the responses considered as belonging to the cat free spiral ending could be attributed instead to the palisades.

ACKNOWLEDGEMENT

This work has been supported by CNRS, RCP contract No. 433 and by INSERM, CRL contract No. 76.1.138.6.

REFERENCES

Alvarado, J.A. and Van Horn, C. (1975) Muscle cell types of the cat inferior oblique. In *Basic Mechanisms of Ocular Motility and their Clinical Implications*, G. Lennerstrand and P. Bach-y-Rita (Eds.), Pergamon Press, Oxford, p. 15–43.

Alvarado-Mallart, R.M. (1978) Contribution à l'Etude de l'Innervation de la Musculature Extrinsèque de l'Oeil chez le Chat. Thèse, Paris.

Alvarado-Mallart, R.M. and Pinçon-Raymond, M. (1979) The palisade endings of cat extraocular muscles: a light and electron microscope study. *Tissue and Cell*, in press.

Alvarado, J., Steinacker, A. and Bach-y-Rita, P. (1967) The ultrastructure of the retractor bulbi muscle of the cat. *Invest. Ophthal.*, 6: 548.

Alvarado-Mallart, R.M., Batini, C., Buisseret-Delmas, C., Gueritaud, J.P. and Horcholle-Bossavit, G. (1975) Mesencephalic projections of the rectus lateralis muscle afferents in the cat. *Arch. ital. Biol.*, 113: 1–20.

Bach-y-Rita, P. (1971) Neurophysiology of eye movements. In *The Control of Eye Movements*, P. Bach-y-Rita, C.C. Collins and J.E. Hyda (Eds.), Academic Press, New York, pp. 7–45.

Bach-y-Rita, P. (1975) Structural functional correlation in eye muscle fibers. Eye muscle proprioception. In *Basic Mechanisms of Ocular Motility and their Clinical Implications*, G. Lennerstrand and P. Bach-y-Rita (Eds.), Pergamon Press, Oxford, pp. 91–108.

Bach-y-Rita, P. and Murata, K. (1964) Extraocular proprioceptive responses in the VI nerve of the cat. *Quart. J. exp. Physiol.*, 49: 408–416.

Bach-y-Rita, P. and Ito, F. (1966a) In vivo studies on fast and slow muscle fibres in cat extraocular muscles. *J. gen. Physiol.*, 49: 1177–1198.

Bach-y-Rita, P. and Ito, F. (1966b) Properties of stretch receptors in cat extraocular muscles. *J. Physiol. (Lond.)*, 186: 663–688.

Bach-y-Rita, P., Levy, J.V. and Steinacker, A. (1967) The effect of succinylcholine on the isolated retractor bulbi muscle of the cat. *J. Pharm. Pharmacol. (Lond.)*, 19: 180–181.

Barker, D. (1974) The morphology of muscle receptors. *In Handbook of Sensory Physiology, Vol. III/2 Muscle Receptors*, C.C. Hunt (Ed.), Springer Verlag, Berlin-Heidleberg-New York, pp. 2–190.

Barker, D. and Harker, D.W. (1971) Two types of multiply innervated muscle fibre in the superior rectus muscle of the sheep. *J. Physiol. (Lond.)*, 222: 74–75P.

Batini, C., Buisseret, P. and Buisseret-Delmas, C. (1975) Trigeminal pathway of the extrinsic eye muscle afferents in cat. *Brain Res.*, 85: 74–78.

Bonavolontà, A. (1956) Ricerche comparative sulle espansioni nervose sensitive nei muscoli estrinseci dell'occhio dell'umo e di altri mammiferi. I fusi neuromuscolari. *Quad. Anat. prat.*, 11: 48–79.

Boyd, I.A. and Davey, R.D. (1968) *The Composition of Peripheral Nerves*. E.S. Livingstone, Edinburgh.

Brown, G.L. and Harvey, A.M. (1941) Neuro-muscular transmission in the extrinsic muscles of the eye. *J. Physiol. (Lond.)*, 99: 379–399.

Brown, M.C., Engberg, I. and Matthews, P.B.C. (1967) The relative sensitivity to vibration of muscle receptors of the cat. *J. Physiol. (Lond.)*, 192: 773–800.

Browne, J.S. (1974) The response of de-efferented muscle spindle in sheep extraocular muscles to stretch and vibration. *J. Physiol. (Lond.)*, 242: 60–62P.

Browne, J.S. (1975) The responses of muscle spindles in sheep extraocular muscles. *J. Physiol. (Lond.)*, 251: 483–496.

Buisseret, P., Gueritaud, G.P., Horcholle-Bossavit, G. and Tyc-Dumont, S. (1972) Projections mésencephaliques des afférences proprioceptives de la musculature extrinsèque des yeux. *J. Physiol. (Paris)*, 65: 369A.

Buisseret-Delmas, C. (1976) Parcours trigeminal des fibres sensorielles provenant des muscles extrinsèques de l'oeil chez le chat. *Arch. ital. Biol.*, 114; 341–356.

Cardin, A. and Rigotti, S. (1947) Impulsi afferenti da tensocettori nel III, IV e VI paio di nervi cranici. *Boll. Soc. ital. Biol. sper.*, 23: 56–58.

Carpenter, R.H.S. (1977) Movements of the Eyes. Pion. London.

Cheng, K. (1963) Cholinesterase activity in human extraocular muscles. *Jap. J. Ophthal.*, 7: 174–183.

Cheng, K. and Breinin, G.M. (1966) A comparison of the fine structure of extraocular and interosseous muscles in the monkey. *Invest. Ophthal.*, 5: 535–549.

Cheng-Minoda, K., Davidowitz, J., Liebowitz, A. and Breinin, G. (1968) Fine structure of extraocular muscle in rabbit. *J. Cell Biol.*, 39: 193–197.

Chiarandini, D.J. and Davidowitz, J. (1978) Structure and function of extraocular muscle fibers. In *Current Topics in Eye Research, Vol. 1*, Academic Press, New York, pp. 91–141.

Cilimbaris, P.A. (1910) Histologische Untersuchungen über die Muskelspindeln der Angenmuskeln. *Arch. mikr. Anat.*, 75: 692–747.

Cooper, S. and Eccles, J.C. (1930) The isometric responses of mammalian muscles. *J. Physiol. (Lond.)*, 69: 377–385.

Cooper, S. and Daniel. P.M. (1949) Muscle spindles in human extrinsic eye muscles. *Brain*, 72: 1–24.

Cooper, S. and Daniel, P.M. (1954) Afferent impulses from the muscle spindles of the extrinsic eye muscles and their course within the brainstem. *Trans. ophthal. Soc. U.K.*, 120: 435–440.

Cooper, S. and Filleng, M. (1955) Afferent discharges in response to stretch from the extraocular muscles of the cat and monkey and the innervation of these muscles. *J. Physiol. (Lond.)*, 127: 400–413.

Cooper, S. and Daniel, P.M. (1957) Response from the stretch receptors of the goat's extrinsic eye muscles with an intact motor innervation. *Quart. J. exp. Physiol.*, 42: 222–231.

Cooper, S., Daniel, P.M. and Whitteridge, D. (1953) Nerve impulses in the brainstem of the goat. Short latency responses obtained by stretching the extrinsic eye muscles and the jaw muscles. *J. Physiol. (Lond.)*, 120: 471–490.

Cooper, S., Daniel, P.M. and Whitteridge, D. (1955) Muscle spindles and other sensory endings in the extrinsic eye muscles; the physiology and anatomy of these receptors and of their connexions with the brain-stem. *Brain*, 78: 564–583.

Corvaja, N., Marinozzi, V. and Pompeiano, O. (1969) Muscle spindles in the lumbrical muscle of the cat. *Arch. ital. Biol.*, 107: 365–543.

Crevatin, F. (1902) Su di alcune forme di terminazioni nervose nei muscoli dell'occhio del dromedario. *R.C. Accad. Sci. ist. Bologna*, 6: 57–61.

Crommelinck, M. and Roucoux, A. (1976) Characteristics of cat's eye saccades in different states of alertness. *Brain Res.*, 103: 574–578.

Daniel, P.M. (1946) Spinal nerve endings in the extrinsic eye muscles of man. *J. Anat. (Lond.)*, 80: 189–192.

Dietest, S.E. (1965) The demonstration of different types of muscle fibres in human extraocular muscles by electron microscopy and cholinesterase staining. *Invest. Ophthal.*, 4: 51–63.

Dogiel, A.S. (1906) Die Endigungen der sensiblen Nerven in den Augenmuskeln und deren Schenen beim Menschen und den Säugetieren. *Arch. mikr. Anat.*, 68: 501–526.

Duke-Elder, W.S. and Duke-Elder, P.M. (1930) The contraction of the extrinsic muscles of the eye by choline and nicotine. *Proc. Roy. Soc. B*, 107: 332–343.

Fillenz, M. (1955) Responses in the brainstem of the cat to stretch of extrinsic ocular muscles. *J. Physiol. (Lond.)*, 128: 182–189.

Fukuda, M. (1958) Studies on the nerve endings in the extrinsic eye muscles of the rabbit. *Jap. J. Ophthal.*, 2: 93–102.

Ginsborg, B.L. (1960) Some properties of avian skeletal muscle fibers with multiple neuromuscular junctions. *J. Physiol. (Lond.)*, 154: 581–598.

Goldberg, S.J., Lennerstrand, G. and Hull, C.D. (1976) Motor units responses in the lateral rectus muscle of the cat: intracellular current injection of abducens nucleus neurons. *Acta physiol. scand.*, 96: 58–63.

Greene, T. and Jampel. R. (1966) Muscle spindles in the extraocular muscles of the macaque. *J. comp. Neurol.*, 126: 547–549.

Harker, D.W. (1972a) The structure and innervation of sheep superior rectus and levator palpebrae extraocular muscles. I. Extrafusal muscle fibres. *Invest. Ophthal.*, 11: 956–969.

Harker, D.W. (1972b) The structure and innervation of sheep superior rectus and levator palpebral extraocular muscles. II. Muscle spindles. *Invest. Ophthal.*, 17: 970–979.

Hess, A. (1961) The structure of slow and fast extrafusal muscle fibers in the extraocular muscles and their nerve endings in guinea pigs. *J. cell. comp. Physiol.*, 58: 63–80.

Hess, A. and Pilar, G. (1963) Slow fibers in the extraocular muscles of the cat. *J. Physiol. (Lond.)*, 169: 780–797.

Hosokawa, H. (1961) Proprioceptive innervation of striated muscles in the territory of cranial nerves. *Texas Rep. Biol. Med.*, 19: 405–464.

Houk, J. and Henneman, E. (1967) Responses of Golgi tendon organs to active contractions of the soleus muscle of the cat. *J. Neurophysiol.*, 30: 466–481.

Huber, G.C. (1900) Sensory nerve terminations in the tendons of the extrinsic eye-muscles of the cat. *J. comp. Neurol.*, 10: 152–158.

Hunt, C.C. (1974) The physiology of muscle receptors. In *Handbook of Sensory Physiology, Vol.III/2, Muscle Receptors*, C.C. Hunt (Ed.), Springer-Verlag, Berlin-Heidelberg-New York, pp. 191-234.

Kaczmarski, F. (1974) Motor end-plates in the extraocular muscles of small mammals. *Acta anat.*, 89: 372–386.

Kuffler, S.W. and Vaughan Williams, E.M. (1953b) Properties of the slow skeletal muscle fibres of the frog. *J. Physiol. (Lond.)*, 121: 318–340.
innervate. *J. Physiol. (Lond.)*, 121: 289–317.

Kuffler, S.W. and Vaughan Williams, E.M. (1953b) Properties of the slow skeletal fibres of the frog. *J. Physiol. (Lond.)*, 121: 318–340.

Lennerstrand, G. (1974a) Mechanical studies on the retractor bulbi muscles and its motor units in the cat. *J. Physiol. (Lond.)*, 236: 43–55.

Lennerstrand, G. (1974b) Electrical activity and isometric tension in motor units of the cat's inferior oblique muscles. *Acta physiol. scand.*, 91: 458–474.

Lennerstrand, G. (1975) Motor units in the eye muscles. In *Basic Mechanisms of Ocular Motility and their Clinical Implications.* G. Lennerstrand and P. Bach–y–Rita (Eds.) Pergamon Press, Oxford, pp. 119–143.

Lennerstrand, G. and Bach-y-Rita, P. (1974a) Activation of slow motor units by threshold stimulation of cat eye muscle nerves. *Invest. Ophthal.*, 13: 879–882.

Lennerstrand, G. and Bach-y-Rita, P. (1974b) Spindle responses in pig eye muscles. *Acta physiol. scand.*, 90: 795–797.

Mahran, Z.Y. and Sakla, F.B. (1965) The pattern of innervation of the extrinsic ocular muscles and the intra-orbital ganglia of the albino mouse. *Anat. Rec.*, 152: 173–183.

Manni, E., Bortolami, R. and Desole, C. (1966) Eye muscle proprioception and the semilunar ganglion. *Exp. Neurol.*, 16: 226–236.

Manni, E., Bortolami, R.and Deriu, P.L. (1970) Superior oblique muscle proprioception and the trochlear nerve. *Exp. Neurol.*, 26: 543–550.

Matthews, B.H.C. (1933) Nerve endings in mammalian muscle. *J. Physiol. (Lond.)*, 78: 1–53.

Matyuskin, D.P. (1961) Phasic and tonic neuromotor units in the oculomotor apparatus of the rabbit. *Fiziol. zh. SSSR*, 47: 878–883.

Matyuskin, D.P. (1964) Varieties of tonic muscle fibers in the oculomotor apparatus of the rabbit. *Bull. exp. Biol. Med.*, 55: 235–238.

Mayr, R. (1971) Structure and distribution of fibre types in the external eye muscles of the rat. *Tissue and Cell*, 3: 433–462.

Merrillees, N.C.R. (1960) The fine structure of muscle spindles in the lumbrical muscles of the rat. *J. biophys. biochem. Cytol.*, 7: 725–742.

Merrillees, N.C.R., Sutherland, S. and Hayhow, W. (1950) Neuromuscular spindles in the extraocular muscles in man. *Anat. Rec.*, 108: 23–30.

Miller, J.E. (1973) Recent histologic and electron microscopic findings in extraocular muscles. *Trans. Amer. Acad. Ophthal. Ot.*, 75: 1175–1185.

Peachey, L. (1971) The structure of the extraocular muscle fibers of mammals. In *The Control of Eye Movements,* P. Bach–y–Rita, C.C. Collins and J.E. Hyde (Eds.) Academic Press, New York, pp. 47–76.

Pilar, G. (1967) Further study of the electrical and mechanical responses of slow fibers in cat extraocular muscles. *J. gen. Physiol.*, 50: 2289–2300.

Pilar, G. and Hess, A. (1966) Differences in internal structure and nerve terminals of the slow and twitch muscle fibers in the cat superior oblique. *Anat. Rec.*, 154: 243–252.

Retzius, G. (1892) Zur Kenntniss der motorischen Nervenendigungen. *Biol. Untersuch.*, 3: 41–52.

Sas, J. and Schab, R. (1952) Die Sogenannten Palisaden-Endigungen der Angenmuskeln. *Acta morph. Acad. Sci. hung.*, 2: 259–266.

Sas, J. and Appeltauer, C. (1963) Atypical muscle spindles in the extrinsic eye muscles of man. *Acta anat.*, 55: 311–322.

Scalzi, A. and Price, M. (1970) Ultrastructure of muscle spindles from the extraocular muscles of sheep. *J. Cell. Biol.*, 47: 180a–181a.

Torre, M. (1953) Nombre et dimensions des unités motrices dans les muscles extrinsiques de l'oeil et en general, dans le muscles squelettiques reliés à des organes de sens. *Schweiz. Arch. Neurol. Psychiat.*, 72: 362–376.

Tozer, F.M. and Sherrington, C.S. (1910) Receptors and afferents of the third, fourth and sixth cranial nerves. *Proc. Roy. Soc. B*, 82: 450–457.

Whitteridge, D. (1958) The motor nerve supply to extraocular muscle spindles. *Electroenceph. clin. Neurophysiol.*, 10: 353–353.

Whitteridge, D. (1959) The effect of stimulation of intrafusal muscle fibres on sensitivity to stretch of extraocular muscle spindles. *Quart. J. exp. Physiol.*, 44: 385–393.

Winckler, G. (1956) L'innervation proprioceptive des muscles extrinsèques du globe oculaire chez l'homme. *C.R. Ass. Anat.*, 43: 848–857.

Zelena, J. (1976) Sensory terminals on extrafusal muscle fibres in myotendinous regions of developing rat muscles. *J. Neurocytol.*, 5: 447–463.

Extraocular Muscle Input to the Cerebellar Cortex

C. BATINI

Laboratoire de Psychophysiologie Sensorielle, Université Pierre et Marie Curie, 75230 Paris (France)

The involvement of the cerebellum in eye movement was demonstrated over 100 years ago with the stimulation experiments of Hitzig (1874) and has continued to be studied. Early works on the relations between the cerebellum and eye movement, as well as more recent works, have been well summarized in the reviews of Dow and Moruzzi (1958) and Dow and Manni (1964). These experiments have explored several parts of the cerebellar surface and intracerebellar nuclei but more attention has been given to the vestibulocerebellum because of its direct involvement in the control of the vestibulo-oculomotor system (Ito, 1972). We will consider here only that part of the vermis of the cerebellar cortex where afferents from the extraocular muscles (EOM) have been demonstrated and which are involved in the saccadic oculomotor system.

EYE POSITION DETECTION

Stimulation experiments of the cerebellar vermis have shown that conjugated eye movements can be elicited in several animals (see Dow and Manni, 1964) including cat (Cohen et al., 1965) and monkey (Ron and Robinson, 1973). In spite of the high variability inherent in using electric stimulation in the cerebellum due to non-physiological factors like current intensity, indiscriminate activation of different cell types, parasitic excitation of passing fibers etc. (see Precht, 1975), there seems to be agreement that the vermis of the lobules V, VI, and VII would produce saccades when stimulated. The onset of the saccades produced in monkeys in this way are found to have a 15–30 msec latency. A spatial coding has also been claimed since the direction of the saccades depends on the lobule stimulated: thus lobule VI would elicit ipsilateral horizontal eye movements whereas lobule V would produce upward vertical movements and lobule VII downward vertical movements (Ron and Robinson, 1973). It is not yet clear whether the evoked eye movements are really "Goal directed" as is claimed for the monkey since there is not a reversal of the direction of the eye movement. More interesting for our viewpoint is that the amplitude and direction of the saccades is found to depend on the initial position of the eye (Ron and Robinson, 1973).

The involvement of the posterior vermis in the regulation of saccadic activity is also supported by the experiments that show a correlation between the saccades and

Purkinje cell firing in the cat (Pellet, 1973, Wolfe, 1971; Llinas and Wolfe, 1972; Llinas, 1974). Some discharges arising from the mossy fiber granule cells pathway would fire 25 msec prior to saccade initiation. Moreover a certain number of Purkinje cells have spike activity related to eye movement only in the ipsilateral direction, revealing therefore a positional dependence (see Llinas, 1974).

Kornhuber (1971) hypothesized that the cerebellar vermis could exert its control on the saccadic system by transforming space (distance to be moved) into time (duration of the neuronal discharge which determines the amplitude of the saccade). Space would be measured by the retinal error or the angular retinal distance between the fovea, which is the fixation point, and that part of the peripheral retina where the image of the object falls. The error would be the signal announcing that a saccade should be made of just the amplitude and the direction necessary to perceive the object with the fovea. The posterior cerebellar vermis is in fact, classically, a visual receiving area (Snider and Stowell, 1942, 1944) where the activity of Purkinje cells has been shown to have a certain degree of sensitivity to moving stimuli including direction and speed of the movement (Buchtel et al., 1973).

This hypothesis, however, is somewhat restrictive; it would not apply, for instance, to eye movements directing the gaze toward an object outside the visual field. Neither would it account for saccades of the "orienting type" occurring during acoustic stimulation. Obviously, an analogous "acoustic error" signal could be hypothesized, which would announce the direction and the amplitude of a saccade just right to move the eyes toward the acoustic source. And in fact the posterior vermis is also an auditory receiving area (Stowell and Snider, 1942, Snider and Stowell, 1944). However, the ablation experiments have not substantiated the participation of a retinal error signal in the cerebellar oculomotor program. While saccades can be precipitated by electrical stimulation of the posterior vermis, they were not suppressed by lesions of the same areas (Aschoff and Cohen, 1971; Westheimer and Blair, 1973, 1974; Ritchie, 1976) just as the presumably homologous quick phase of nystagmus is not affected by cerebellectomy (Collewijn, 1970; Robinson, 1974). Changes were observed in the amplitude and direction of the saccadic movements after unilateral lesions (Aschoff and Cohen, 1971; Westheimer and Blair, 1973, 1974). However, such changes mostly indicated an attempted correction of the deficit occurring in the ability to hold eye position or gaze. The saccade is a movement that displaces the eye rapidly to a new position and then leaves it there until the next movement. Any drift away from some holding position will result in a new correction saccade. The clever experiment of Ritchie (1976) demonstrated more clearly that the retinal error signal is not responsible for the dysmetria of the saccades in the monkey after lesion of the cerebellar vermis. He could train animals to make voluntary calibrated saccades either using retinal error signal, or in the dark without any retinal point of reference. In both situations the saccades were identical to those obtained in animals with an intact cerebellum and which had a similar deficit after cerebellar ablations. Therefore, the cerebellar dysmetria observed in the ablation experiments could be attributed to the lack of position information or, as the author pointed out, the impossibility to use this information. Unfortunately, all the lesion experiments have been done only with the monkey so that a comparative analysis is not possible. The areas lesioned usually include portions of the cerebellar cortex and intracerebellar nuclei wider than the receiving area for the EOM afferents (see below). Thus, more experiments are needed before correlations between the cerebellar localization of eye position sensitivity and the EOM proprioceptors can be clearly established.

EOM AFFERENTS TO THE CEREBELLAR CORTEX

While there are indications that the cerebellar vermis might need the eye position sense to exert its oculomotor function, it was reasonable to explore this same area for proprioceptive input. In addition it was demonstrated, in several animals, that the cerebellar vermis is a region also receiving projections from the trigeminal sensory area (Dow and Anderson, 1942; Adrian, 1943; Snider, 1943; Snider and Stowell, 1944; Azzena et al., 1970, 1971; Batini et al., 1974) and we know that at least some part of the EOM afferents makes its way through the Vth nerve (see Manni and Bortolami, this volume, chapter III 1).

Afferents from the EOM to the cerebellar cortex have been first demonstrated in the cat by Fuchs and Kornhuber (1969) who observed evoked potentials in the posterior vermis to a stretch applied to different eye muscles. But their results were contested by Rhan and Zuber (1971) who observed that with stretch of the EOM the evoked response persisted after the cerebellar peduncles had been severed. They explained their results as well as those obtained by Fuchs and Kornhuber, by a selective volume conduction from brain stem structure to the cerebellum. The question of whether or not extraocular proprioception projects to the cerebellum has been solved soon after, by using microelectrodes to record Purkinje cells: stretches of the muscles obliquus inferior and rectus lateralis elicited discharges of unit located in the same vermian region explored by the previous authors (Batini and Buisseret, 1972). These findings have been extended in the cat, for the six pairs of EOM (four recti muscles and two obliqui with the exception of the retractor bulbi) either by electrically stimulating each of the muscle nerves (Baker et al., 1972; Batini and Buisseret, 1972, 1974) or by stretching each muscle itself (Batini et al., 1974; Schwarz and Tomlinson, 1977).

Little information, however, is available for other animals. In rabbit, passive movement of the eye produces cerebellar unit responses (Potter and Smith, 1975). In the lamb, cerebellar unit responses have been elicited by stretching the EOM (Azzena et al., 1970, 1971). Evoked potentials from EOM stretch in the goat and the cat (Cooper et al., 1953; Fillenz, 1955) have been described in the superior cerebellar peduncle, but we do not know if they arise in the incoming or outgoing cerebellar fibers. Therefore the possibility that they may be arising in the dentato-oculomotor pathway (Carpenter and Strominger, 1964; Highstein and Ito, 1971; Highstein et al., 1971) will not be discussed here. Most of the information at hand to be discussed below comes from the cat.

Surface localization of EOM afferents

Fig. 1 illustrates the surface localization of the Purkinje cell responses obtained from the activation of EOM afferents as described by different authors. The figure includes responses to stretches of the rectus lateralis and obliquus inferior muscles and to electrical stimulation of the six muscle nerves (Fig. 1A); to stretches of the four recti muscles (Fig. 1B); to stimulation of the nerve to the obliquus superior only (Fig. 1C) and finally to stretch of some of the other EOM (Fig. 1D). It is interesting to note: i) that there is complete agreement on the lobule VI being the vermian region receiving projections from extraocular afferents; but the extent of involvement of the surfaces of lobules V and VII has varied depending on the author. This might be due in part to the extent of the exploration made by the different authors and also by the

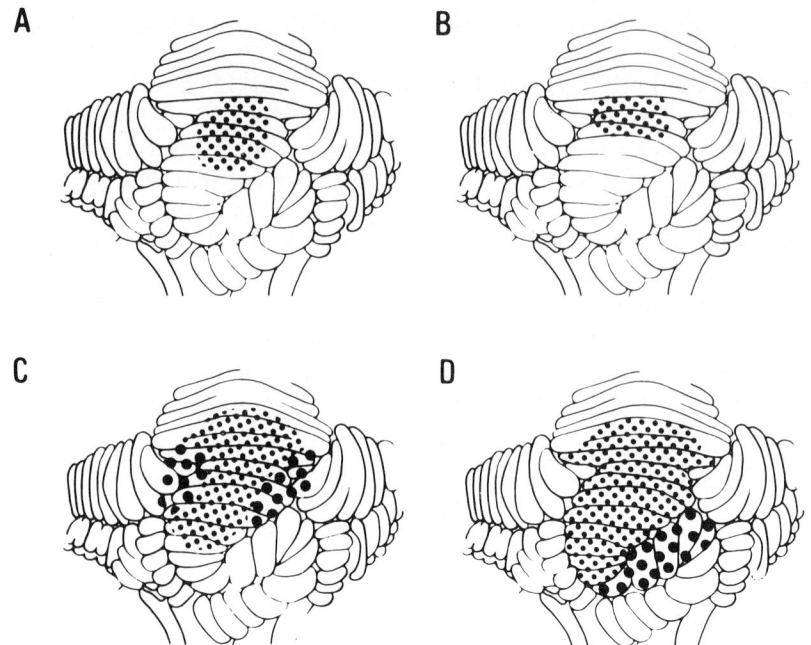

Fig. 1. Localization of EOM afferents in the cerebellar cortex of the cat. A) Unpublished figure from Batini et al. (1972). B) Modified figure from Schwarz and Tomlinson (1977). C) Modified figure from Baker et al. (1972). Small dots indicate area with double input to the Purkinje cells. Large dots indicate area with climbing fiber input only. D) Modified figure from Fuchs and Kornhuber (1969). Small dots indicate area with strong projection, large dots with weaker projection.

use of anesthetized or non-anesthetized preparations. We have no knowledge of other cerebellar regions having been explored. Responses to stretches were found only in the lamb to be localized in two areas, the vermian lobule VIIa and the hemispheral lobule HVI (Azzena et al., 1970, 1971); ii) that, perhaps due to the lack of sufficient information, no somatotopic representations of the different muscles on the surface of the cerebellar cortex have been attempted. Obviously this could not be the case of Fig. 1C since it concerns only one muscle, but it is even more significant that a single muscle is represented in a wider area than all the others together.

Dual afferent pathway to the Purkinje cells

A common result obtained from EOM proprioceptive afferents activated by mechanical or electrical stimuli is that a double input arrives at the cerebellar cortex. In fact, Purkinje cells which are the final path out of the cerebellar cortex, show responses arriving through the mossy fiber-granule cell system and through the climbing fiber system. The two inputs are independent from each other as demonstrated by the fact that a single stimulus can evoke a mossy fiber discharge or a climbing fiber discharge or both, but the climbing fiber discharge always comes second in the sequence (Batini et al., 1974). It has already been shown for other afferents, that impulses reach the cerebellar cortex through both pathways whereas few observations have been reported of afferents arriving through only one of the two pathways (see for references Batini et al., 1974; Strata, 1975). Pure climbing fiber input from EOM is also claimed by Baker et al. (1972) to occur in specific regions situated laterally to the area receiving both inputs (Fig. 1C). However, from our own experience, we are more inclined to believe that single input to Purkinje cells is due to a particular local (such as traumatic or circulatory) or experimental (such as the use of anesthetics) condition. Whether the two channels convey the same or different information, or elicit the same or different afferent function is still an open question.

The response latency

Simply pulling the EOM does not necessarily result in an activation of the stretch receptors until a certain length has been reached. Thus, accurate measurement of the latency between the activation of the EOM receptors and the cerebellar response of the Purkinje cells is obtained by electrical stimulation of the muscle nerves. In addition, nerve fibers corresponding to non-stretch sensitive muscle receptors can also be activated by electrical stimulation. The mean latency in anesthetized and non-anesthetized animals varied slightly with different muscles, probably because of variable nerve length. As an example, mean latencies for the activation of the nerve to the rectus lateralis was 18.4 msec for mossy fibers discharges and 35.8 msec for climbing fiber discharges (shorter latencies have been reported for evoked potentials by Fuchs and Kornhuber (1969)) but large variabilities were also observed (Batini et al., 1974) as might be expected for polysynaptic pathways. We can extrapolate from the stimulation experiments in the cat (see above) that the delay for the cerebellar cortex to move the eye can be as short as 10 msec. Therefore, the total average time of the mossy fiber afferent feedback loop would be less than 30 msec, short enough to arrive before even the shortest saccade (Crommelink and Roucoux, 1976) is over. Thus the proprioceptive afferents could influence, through the cerebellar loop, the amplitude of the saccade, a function well suited for an inhibitory population of neurons like the Purkinje cells (Ito et al., 1964).

The mechanical activation

In this paragraph we will consider only the experiments concerning single unit recording for which we have more information. To activate mechanically the EOM receptors, stretches have been used, either by pulling the tendon of the dissected free muscle (Batini et al., 1974) or by pulling the tendon still attached to the eye ball in a tangential direction (Schwarz and Tomlinson, 1977). The eventual contamination from visual stimuli was eliminated in the first case by enucleation and in the second case by cutting the optic nerve. The differences between the two methods of mechanical stimulation concern the type of receptors involved. In the free, dissected muscle there is a high probability of activating only those receptors inside that muscle whereas pulling on the intact eye produces a complex movement which might involve additional non-identified trigeminal receptors in the orbital cavity. These authors have, however, inactivated, by local anesthesia, the periorbital skin and conjunctive receptors.

The parameters of stimulation were nearly equivalent, the stretches starting from an approximate primary position of the eye implicating an initial tension of 1.5–2.5 g (Schwarz and Tomlinson, 1977), the amplitude, when translated in angular distance, was about 3°–24°, the velocity of the stretch not exceeding 500 degrees/sec but more frequently being of 100–300 degrees/sec. These parameters were chosen to mimic saccades where velocities range in cat from 30–500 degrees/sec, having more frequently an amplitude of 20°–25° (Crommelink and Roucoux, 1976).

Examples of Purkinje cell responses to stretch are illustrated in Figs. 2 and 3 for

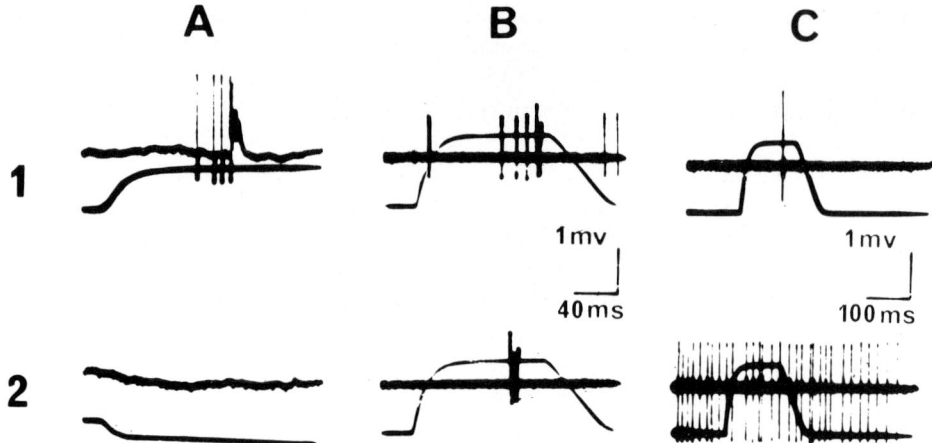

Fig. 2. Double input to the Purkinje cells from extraocular muscle stretch. Upper traces: Purkinje cell activity. Lower traces: changes in muscle length (upward deflection shows stretches and downward deflection shows release of the stretched muscle as a control). A) Stretch of the obliquus inferior (1) with mossy fiber and climbing fiber types of responses in an anesthetized animal. Characteristic absence of response (2) releasing the muscle. B) Stretches of the rectus lateralis in non-anesthetized animal. Mossy and climbing fiber responses (1) and climbing fiber response only (2). C) Responses to stretch of the rectus lateralis in a non-anesthetized animal during low (1) and high (2) firing rate of the Purkinje cell. (From Batini et al., 1974.)

anesthetized and non-anesthetized animals. There is agreement in different experiments on several points.

i) Excitatory responses are elicited through both afferent systems, the mossy fibers-granule cells and the climbing fibers in the usual sequence. Inhibitory responses were also elicited either in the anesthetized or in the non-anesthetized cat, the duration of the suppressed or lowered spontaneous activity being variable. In addition, the sign of the response has been reported to be dependent on the firing rate of the

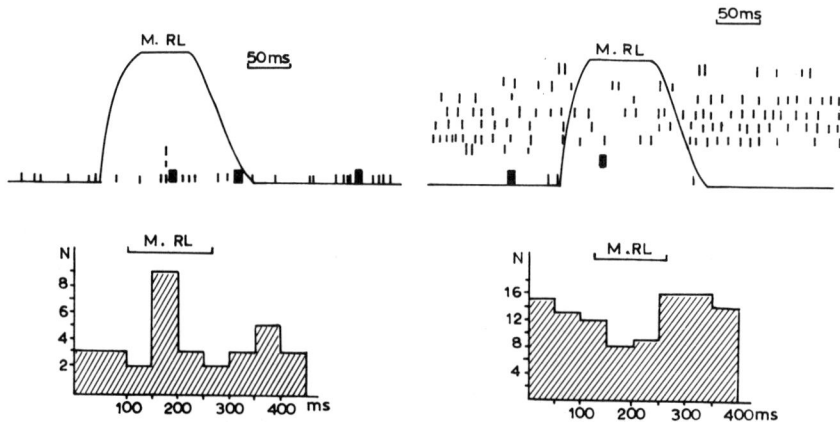

Fig. 3. Frequency dependent responses of two Purkinje cells in a non-anesthetized cat. Left: time course of stretch and scatter diagram of responses to stretch of the rectus lateralis and responses of the same cell shown as a histogram. Narrow bars indicate mossy fiber discharges, broad bars climbing fiber discharges. Muscle stretches are indicated by the line above the histogram. Right: same for a Purkinje cell having a high firing rate showing an inhibition during the stretch. (From Batini et al., 1974.)

neuron, excitatory responses appearing in cells with low frequency of spontaneous discharges and the inhibitory with high frequency of discharge (Fig. 3). A few privileged cells displayed both frequency dependent responses during the recording sessions. The simplest explanation would be that excitation and inhibition both take place at the same Purkinje cell due to the organization of the intracerebellar connections (see Eccles et al., 1967), but in extracellular recording they are only observed when certain conditions are met. Thus, for instance, only during sustained firing would the effect of inhibition be visible.

ii) Whatever the sign of the response, the evoked firing of the Purkinje cells does not depend on the duration of the stretches and thus it is invariably of the "phasic" type. This finding contrasts with the static properties of the receptors observed in muscle nerve fibers (Batini, this volume, chapter III3).

iii) The effective mechanical parameter to obtain responses was the amplitude of the stretch, while the velocity of the stretch did not show a correlation (Schwarz and Tomlinson, 1977). Here again, the velocity information is lost somewhere between the primary fibers and the cerebellum. This is not surprising since the recording sites are separated from the sensory organ by an unknown number of synaptic relays where the transfer process might modify the information content.

iv) Additional interesting information is given by the experiments of Schwarz and Tomlinson (1977). They tested the Purkinje cells for stretch of the four pairs of recti muscles, moving the eyes in the horizontal and the vertical planes. They classified the cells in three categories: "direction-specific" which responded only to those muscle stretches acting in a specific direction of the eye; "plane-specific" which responded to stretches of the muscles acting in a particular plane of eye movement; and finally "complex" which do not have a clear pattern. It seems to us that the best evidence for directionality information is the fact that reciprocal inhibition seems to appear when pulling the antagonist muscles. However, as the authors have pointed out, pulling on a muscle of the intact eye would result in a complex passive movement involving more muscles than only those where the pull is applied. Therefore, this complex situation probably mobilizes more than the EOM proprioceptors. With this method of mechanical eye mobilization, more than the EOM proprioceptors may be activated but it has the advantage of being a more natural eye movement. In fact, the "direction-specific" cells are reminiscent of the Purkinje cell firing observed prior to natural saccades occurring in a specific direction (Llinas, 1974). Essentially, the information collected by single Purkinje cells from the EOM proprioceptive input, is phasic signalling threshold movements. In other words, each Purkinje cell responds to a different amplitude of stretch of an EOM, thus keeping the cerebellum constantly informed of the position of the eye during saccadic movements. That this should be temporally possible is indicated by the fact that the entire feedback loop time can be shorter than the shortest saccade. Such a mechanism would imply that the cerebellum is also informed of the plane and the direction of the movement, although this last information may be due to combined signals coming from the muscles and the intraorbital receptors activated by the eye movement. Interestingly, the proprioceptive systems of the EOM do not respond to the velocity of the movement of the EOM under the present experiment conditions and as viewed from the cerebellar responses. A possibility would be that no velocity control is necessary in the EOM saccadic system. The muscles are position programmed only and go from one position to the next as rapidly as possible. In such a system, eye movements would be obtained by programming many successive small steps. Having to make many steps would lengthen the time needed to move the eye from one position to the next. Also in such a system a highly refined position sensing apparatus will be necessary and its prime requisite would be a high sensitivity or gain in amount of response per unit stretch.

From the stimulation and lesion experiments analyzed above, the cerebellar vermis was expected to receive information about the position of the eye prior to the movement. However, so far, in the mechanical experiments, only passive movements having saccadic velocity have been tested. We lack information about cerebellar responses to slow passive movements. Yet, as already pointed out, saccades consist principally of phasic muscle contraction and secondarily of tonic contraction during gaze holding. Position information could be the result of discrete continuous changes of unit firing in the complex structure of the cerebellar cortex due to the continuous tonic activity from EOM. Such changes could not be detected by the experimental design used so far to activate EOM proprioceptors.

SUMMARY

The possibility that the cerebellar cortex is capable of detecting and using eye position information has been first analyzed in terms of the available literature from stimulation and ablation experiments.

The EOM afferents to the cerebellar cortex have then been analyzed following the data mostly collected in cat and concerning the following findings: a) the surface localization of the projection extend to the lobule VI and, at a lesser extent, to the lobules V and VII; b) the input involves both the mossy and climbing fiber pathways; c) the latency of the projection is compatible with a feedback loop having a delay short enough to influence the late part of the saccade; d) saccadic velocity of the stretch is the mechanical parameter for the EOM receptor excitation producing a phasic excitatory or inhibitory response of the Purkinje cells. No velocity sensitivity has been observed, but there are indications that the cerebellum could be informed of the plane and direction of the eye movement.

The possibility that no velocity control is necessary for the EOM saccadic system is discussed.

ACKNOWLEDGEMENTS

This work has been supported by C.N.R.S., R.C.P. contract No. 433, and by I.N.S.E.R.M., C.R.L. contract No. 76.1.138.6

REFERENCES

Adrian, E.D. (1943) Afferent areas in the cerebellum connected with the limbs. *Brain*, 66: 289–315.
Aschoff, J.C. and Cohen, B. (1971) Changes in saccadic eye movements produced by cerebellar cortical lesions. *Exp. Neurol.*, 32: 123–133.
Azzena, G.B., Desole, C. and Palmieri, G. (1970) Cerebellar projections of the masticatory and extraocular muscle proprioception. *Exp. Neurol.*, 27: 151–171.
Azzena, G.B., Desole, C. and Palmieri, G. (1971) Cerebellar representation of extraocular and masticatory proprioception. *Arch. int. Physiol.*, 79: 407–409.

Baker, R., Precht, W. and Llinas, R. (1972) Mossy and climbing fiber projections of extraocular muscle afferents to the cerebellum. *Brain Res.*, 38: 440–445.

Batini, C. and Buisseret, P. (1972) Projection cérébelleuse et trajet périphérique de la proprioception extraoculaire. *C.R. Acad. Sci. (Paris)*, 275: 2711–2713.

Batini, C. and Buisseret, P. (1974) Sensory peripheral pathway from extrinsic eye muscles. *Arch. ital. Biol.*, 112: 18–32.

Batini, C., Buisseret, P. and Kado, R.T. (1974) Extraocular proprioceptive and trigeminal projection to the Purkinje cells of the cerebellar cortex. *Arch. ital. Biol.*, 112: 1–17.

Buchtel, H. A., Rubia, F.J. and Strata, P. (1973) Cerebellar unitary responses to moving visual stimuli. *Brain Res.*, 50: 463–466.

Carpenter, M.B. and Strominger, N.L. (1964) Cerebello-oculomotor fibers in the Rhesus monkey. *J. comp. Neurol.*, 123: 211–230.

Cohen, B., Goto, K., Shanzer, S. and Weiss, A.H. (1965) Eye movements induced by electrical stimulation of the cerebellum in the alert cat. *Exp. Neurol.*, 13: 145–162.

Collewijn, H. (1970) Dysmetria of fast phase of optokinetic nystagmus in cerebellectomized rabbits. *Exp. Neurol.*, 28: 144–154.

Cooper, S., Daniel, P.M. and Whitteridge, D. (1953) Nerve impulses in the brainstem of the goat: responses with long latencies obtained by stretching the extrinsic eye muscle. *J. Physiol. (Lond.)*, 120: 491–513.

Crommelink, M. and Roucoux, A. (1976) Characteristics of cat's eye saccades in different states of alertness. *Brain Res.*, 103: 574–578.

Dow, R.S. and Anderson, R. (1942) Cerebellar action potentials in response to stimulation of proprioceptors in the rat. *J. Neurophysiol.*, 5: 363–371.

Dow, R.S. and Moruzzi, G. (1958) *The Physiology and Pathology of the Cerebellum*. The University of Minnesota Press, Minneapolis.

Dow, R.S. and Manni, E. (1964) The relationship of the cerebellum to extraocular movements. In *The Oculomotor System*, M.B. Bender (Ed.) Harper and Row, New York, pp. 280–302.

Eccles, J.C, Ito, M. and Szentagothai, J. (1967) *The Cerebellum as a Neuronal Machine*, Springer, New York.

Fillenz, M. (1955) Responses in the brainstem of the cat to stretch of extrinsic ocular muscles. *J. Physiol. (Lond.)*, 128: 182–199.

Fuchs, A.F. and Kornhuber, H.H. (1969) Extraocular afferents to the cerebellum of the cat. *J. Physiol. (Lond.)*, 200: 713–722.

Highstein, S.M. and Ito, M. (1971) Differential localization within the vestibular nuclear complex of the inhibitory and excitatory cells innervating III nucleus oculomotor neurons in rabbit. *Brain Res.*, 29: 358–362.

Highstein, S.M., Ito, M. and Tsuchiya, T. (1971) Synaptic linkage in the vestibulo-ocular reflex pathway of rabbit. *Exp. Brain. Res.*, 13: 306–326.

Hitzig, E. (1874) *Untersuchungen über das Gehirn*. Hirschwald, Berlin.

Ito, M. (1972) Neural design of the cerebellar motor system. *Brain Res.*, 40: 81–84.

Ito, M., Yoshida, M. and Obata, K. (1964) Monosynaptic inhibition of the intracerebellar nuclei induced from the cerebellar cortex. *Experientia*, 20: 575–576.

Kornhuber, H.H. (1971) Motor functions of the cerebellum and basal ganglia: the cerebellocortical saccadic (ballistic) clock, the cerebellonuclear hold regulator and the basal ganglia ramp (voluntary speed smooth movement) generator. *Kibernetic*, 8: 157–162.

Llinas, R. (1974) Motor aspects of cerebellar control. *Physiologist*, 17: 19–46.

Llinas, R. and Wolfe, J.W. (1972) Single cell responses from the cerebellum of rhesus preceding voluntary vestibular and optokinetic saccadic eye movements. *Soc. Neurosci. Abst.*, 2: 201–201.

Pellet, J. (1973) Contribution à l'étude de l'Electrogenèse Spontanée du Cortex Cérébelleux Vermien au cours des Etats de Veille et de Sommeil, Thèse, Aix-Marseille, 1973.

Potter, B.L. and Smith, D.C. (1975) Cerebellar responses to passive eye movement in the rabbit. *J. Physiol. (Lond.)*, 244: 46P–47P.

Precht, W. (1975) Cerebellar influence on eye movements. In *Basic Mechanisms of Ocular Motility and their Clinical Implications*, G. Lennerstrand and P. Bach-y-Rita (Eds.), Pergamon Press, Oxford, pp. 261–280.

Rahn, A.C. and Zuber, B.L. (1971) Cerebellar evoked potentials resulting from extraocular muscle stretch: evidence against a cerebellar origin. *Exp. Neurol.*, 31: 230–238.

Ritchie, L. (1976) Effects of cerebellar lesions on saccadic eye movements. *J. Neurophysiol.*, 39: 1246–1256.

Robinson, D.A. (1974) The effect of cerebellectomy on the cat's vestibulo-ocular integrator. *Brain Res.*, 71: 195–207.

Ron, S. and Robinson, D.A. (1973) Eye movements evoked by cerebellar stimulation in the alert monkey. *J. Neurophysiol.*, 36: 1004–1022.

Schwarz, D.W.F. and Tomlinson, R.D. (1977) Neuronal responses to eye muscle stretch in cerebellar lobule VI of the cat. *Exp. Brain Res.*, 27: 101–111.

Snider, R.S. (1943) A fifth cranial nerve projection to the cerebellum. *Fed. Proc.*, 2: 46–46.

Snider, R.S. and Stowell, A. (1942) Evidence of a projection of the optic system to the cerebellum. *Anat. Rec.*, 82: 448–449.

Snider, R.S. and Stowell, A. (1944) Receiving areas of the tactile auditory and visual system in the cerebellum. *J. Neurophysiol.*, 7: 331–357.

Stowell, A. and Snider, R.S. (1942) Evidence of a representation of auditory sensibility in the cerebellum of the cat. *Fed. Proc.*, 1: 84–84.

Strata, P. (1975) The dual input to the cerebellar cortex. In *Golgi Centennial Symposium*, M. Santini (Ed.), Raven Press, New York, pp. 273–280.

Westheimer, G. and Blair, S.M. (1973) Oculomotor defects in cerebellectomized monkeys. *Invest. Ophthal.*, 12: 618–621.

Westheimer, G. and Blair, S.M. (1974) Functional organization of primate oculomotor system revealed by cerebellectomy. *Exp. Brain Res.*, 21: 463–472.

Wolfe, J.W. (1971) Relationship of cerebellar potentials saccadic eye movements. *Brain Res.*, 30: 204–206.

Proprioceptive Influences from Eye Muscle Receptors on Cells of the Superior Colliculus*

V.C. ABRAHAMS

Department of Physiology, Queen's University, Kingston, Ontario K7L 3N6 (Canada)

It seems barely possible that only 17 years ago, in the proceedings of a symposium on the visual system by Jung and Kornhuber (1961) only 2% of the material was devoted to the superior colliculus. In the 17 years since that symposium there has been an explosive growth in interest in the superior colliculus, but almost all of this recent work has been concerned with the role of visual connections to the superior colliculus as recent reviews attest (cf. Gordon, 1975). Nonetheless many experiments have demonstrated the existence of non-visual connections to the superior colliculus (Ades, 1944; Starzl et al., 1951; Cooper et al., 1953a,b; Fillenz, 1955; Shimazu, 1959; Jassick-Gerschenfeld and Ascher, 1963; Abrahams and Falchetto, 1966; Stein and Arigbede, 1972; Gordon, 1973; Abrahams and Rose, 1975a,b; Drager and Hubel, 1975; Rose and Abrahams, 1975, 1978; Stein et al., 1976). One non-visual connection to the superior colliculus is from receptors in extraocular muscles.

The first evidence that such a connection might exist came from experiments in which unit responses to passive stretch of extraocular muscles were examined in the brain stem of the goat (Cooper et al., 1953a,b). These authors found an occasional response in the superior colliculus. Fillenz (1955) in similar experiments on the cat also showed responses to be present in the deeper layers of the superior colliculus. The extent of the projection from extraocular muscle to the superior colliculus was not recognized for many years. In experiments on cats anaesthetized with chloralose (Abrahams and Rose, 1975a) it was found that electrical stimulation of extraocular muscle nerves could excite more units in the superior colliculus than a visual stimulus although a great deal of convergence between extraocular afferents and visual projections was found. In a sample of 93 units tested (Table I), 61 were found by Abrahams and Rose (1975a) to be activated by extraocular and neck muscle afferents as well as by visual stimuli. One other startling observation which even now has drawn little comment is the fact that the superior colliculus receives a substantial proprioceptive input. It was also possible to show (Abrahams and Rose, 1975b) that proprioceptive afferents from fore- and hindlimb as well as from the extraocular system project to the superior colliculus, and most proprioceptive afferents are able to excite more units than a visual stimulus (Table II), none being more effective than extraocular muscle afferents.

---

*Supported by M.R.C. of Canada.

TABLE I

CONVERGENCE ON UNITS OF THE SUPERIOR COLLICULUS

No. of stimulus types		No. of units
3	Extraocular	61
	Neck	
	Visual	
2	Extraocular and visual	3
	Neck and extraocular	23
	Visual and neck	0
1	Extraocular	4
	Neck	1
	Visual	1
	Total number of units	93

TABLE II

Muscle afferent input	Nc	Nv	Nm	Nv/Nc	Nm/Nv
Extraocular	93	68	91	0.73	1.34
Neck	145	107	124	0.74	1.16
Forelimb	106	77	93	0.73	1.21
Hindlimb	72	54	51	0.75	0.95

Nc = total number of units tested; Nv = number visually excited; Nm = number muscle afferent excited.

We believe that basic to the understanding of the role of proprioceptive input is an understanding of the organization and physiology of the muscles containing the receptors. Extraocular muscles are good examples of what Botterman et al. (1978) call "compartmentalized muscles". That is, muscles where one fibre type will predominate in a given region. In extraocular muscle there is a core of large fast fibres and a periphery of slow fibres (Alvarado and Van Horn, 1975). The peripheral slow fibres are utilized in the maintenance of gaze fixation and slow following movements and both fast and slow fibres are active in a saccade (Collins, 1975). In compartmentalized muscle there is a tendency for receptors to lie among the slow fibres (Yellin, 1969; Richmond and Abrahams, 1975; Maier et al., 1976; Gonyea and Ericson, 1977) and this is the case in extraocular muscle (Cooper et al., 1955). In the cat these receptors are not conventional spindles but simple spiral structure predominantly found in small fibre regions. These receptors differ from spindles in that most are silent at rest. The receptors respond phasically to stretch with an intensity of discharge proportional to the resting tension prior to stretch (Bach-y-Rita and Ito, 1966).

If we are to understand the role of the projection from extraocular receptors to the superior colliculus we must ask what information might be conveyed in the pathway. In initial experiments (Rose and Abrahams, 1975a) we sought the answer by applying passive stretch to the muscles at rates similar to these employed by Bach-y-Rita and Ito (1966) while recording in the superior colliculus. The extraocular muscles were stretched by pulling on a ligature attached to one lateral rectus muscle. This was done in a controlled fashion by attaching the ligature to a vibration exciter. The advantage of the technique, which was borrowed from Fuchs and Kornhuber (1969), was simplicity. The disadvantages were that movement was restricted to the nasal direction and the system was not sufficiently well controlled to permit the study of responses to small movements. Further, although the cornea was occluded and appropriate controls eliminated the retina as a source of input, the possibility remained that the cornea or periorbital structures could contribute to responses.

Our first series of experiments were concerned with responses in the superior colliculus to large rapid nasal displacements. Passive movement of an occluded eye at saccadic velocities proved to be a very effective way of exciting unit discharge within the superior colliculus. Characteristically, the pattern of discharge that was seen in virtually all units examined was a brief burst discharge (Fig. 1). It was found that these bursts signalled the information that the eye was moving at saccadic velocity past a fixed point in the orbit. Fig. 2 shows experimental data that led to that conclusion. It can be seen that increasing the velocity of eye movement reduces the latency of response in a regular fashion. This is to be expected if there is a fixed position which the eye must pass in order to excite the system. This property is reinforced by the experiment illustrated in Fig. 3. The eye in this experiment was moved at saccadic velocity starting from a series of different initial positions. Each additional position was 3° further from the primary positions than the last. Each increment in initial position of 3° was found to decrease the movement required to elicit discharge by 3°. When the initial position displacement is added to the displacements necessary to produce movement it is clear once again that what elicits unit discharge is movement of the eye past a fixed orbital point.

The early experiments were satisfying because they demonstrated a relationship between the nature of a passive eye movement and the response it elicited in the superior colliculus. The experiments were unsatisfactory because as previously mentioned the response to movement in one direction only could be examined; we were unable to look at responses to very small movements, and the controls were not as comprehensive as we would have liked.

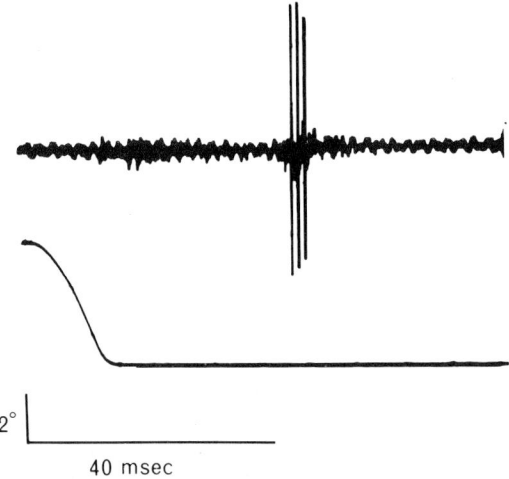

Fig. 1. Typical burst discharge of unit in superior colliculus following passive eye rotation. Lower trace, movement transducer output.

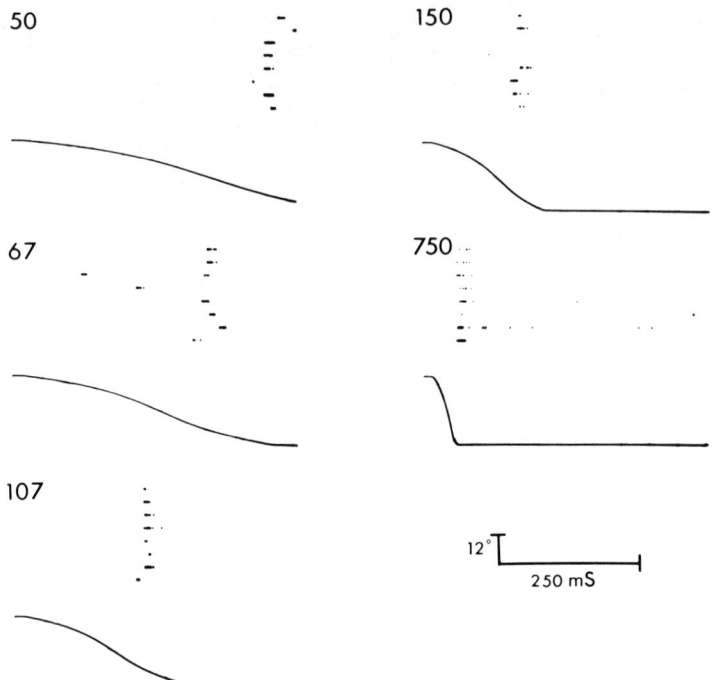

Fig. 2. Raster diagram of unit response recorded in superior colliculus following passive rotation of an occluded eye at velocities (in degrees/sec) indicated.

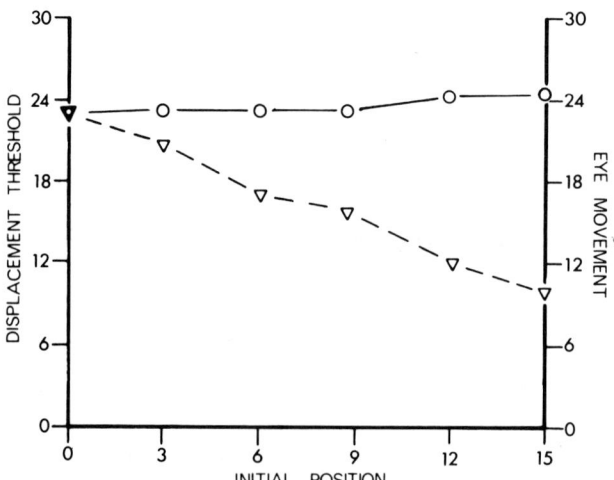

Fig. 3. This diagram illustrates the amplitude of eye movement (scale on right) necessary to elicit discharge from a unit when movement is initiated from a series of positions, each 3° further from initial position (▽--▽). When the initial displacement is added to the movement necessary to elicit discharge (O—O) it can be seen that displacement threshold is constant.

A more sophisticated technique of eye movement was therefore developed to examine the unit responses to small movement which we knew to be present (Rose and Abrahams, 1975) but had not been able to examine in detail. The eye was occluded as before with a lens moulded from aluminum foil. This lens was cemented in place over the cornea with a cyanoacrylic resin and the occlusive lens in turn became the attachment point for one of two mechanical linkages. One linkage enabled the eye to be moved in the vertical plane, the other enabled the eye to be moved in the horizontal plane. With these devices the eye could be moved from any initial position to any second position in that plane and the eye could be moved in either direction. With this apparatus, response of units in the superior colliculus to a wide range of eye movements have been examined (Abrahams and Anstee, 1977, 1979). Controls were also introduced into the experiment to ensure that the observed responses originate with extraocular muscle receptors and not from the retina or cornea nor from periorbital tissues.

We were first able to demonstrate that units of the type that we had previously described (Rose and Abrahams, 1975) with large fixed displacement thresholds responding at saccadic velocities can be activated not only by nasal movement but can be activated by temporal movement and by vertical movement both in an upward and downward direction. Thus, in the execution of a large saccadic movement in any direction the transit of the eye through a series of fixed points in the orbit is continuously signalled to a series of different units within the superior colliculus. The superior colliculus is thus informed with some precision, although at relatively long latencies (10–110 msec) of the progress of rapid eye movements. In the large population of units which we have now examined, about 10% were found to be of this type. Use of the more flexible eye movement system has shown, however, that the majority of units respond to small movements of the eye in a similar manner regardless of the initial position of the eye. Like those units with fixed positional thresholds the response of units activated by small movement is usually a brief burst of impulses. While the movement required to set up activity is small (Fig. 4) it must be relatively rapid, for, as Fig. 5 shows, velocity thresholds are normally at saccadic levels. These units, as well as being insensitive to initial position, are usually insensitive to the direction of movement. Thirty-eight of 43 units examined responded equally well to movement in opposite directions. This should not be interpreted as meaning that there is no directional sensitivity in the system for no systematic analysis was made of firing patterns elicited by different movements nor of the directional sensitivities of the 5 units that did not respond to movement in opposite directions.

Thus, two distinct populations of unit responses can be found in the superior colliculus following passive eye movements. Both require movement at saccadic velocity to be activated, but one is position sensitive and the second is position insensitive. The second type of response appears to contain only one piece of information – that a rapid movement has been initiated. The latency with which this information is conveyed is somewhat shorter (5–65 msec) than information relating to the progress of a saccade. Both types of unit behaviour are consistent with the properties of extraocular muscle receptors described by Bach-y-Rita and Ito (1966). The responsiveness of collicular units to small movements in any direction seems to directly reflect the generally phasic properties of the receptors. The fixed threshold response, as we have previously discussed (Rose and Abrahams, 1975), could be due to the direct rela-

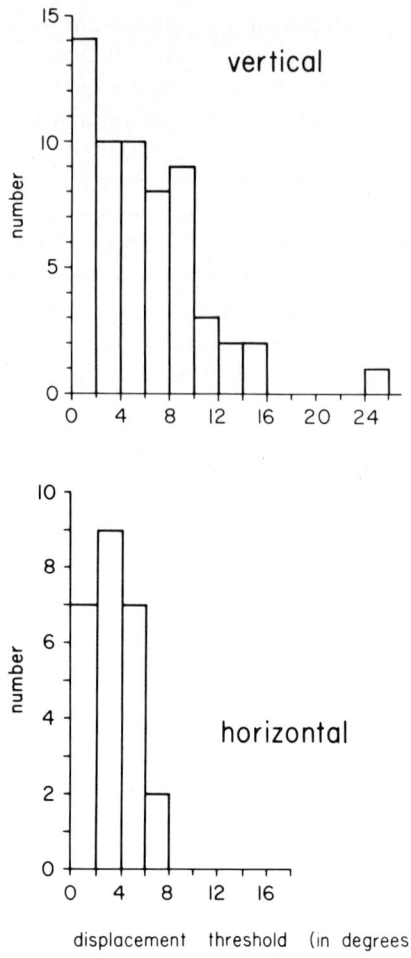

Fig. 4. Displacement thresholds of units responding to passive eye movement in a similar manner regardless of initial eye position.

tionships shown by Bach-y-Rita and Ito (1966) between the time of onset of receptor discharge and duration of pull.

It is also virtually certain now that the responses described here take origin in extraocular muscle receptors. In control experiments the optic nerve was transected to eliminate any possibility of a retinal contribution to the response and to enable corneal stimulation to be tested and eliminated as a possible source of afferents. Lastly, experiments were performed in which extraocular muscles were stretched after enucleation of the eye to eliminate the possibility of activity arising from the periorbital structure. The response pattern of units in the superior colliculus to such movements of the muscle alone was essentially identical to that found to eye movement.

Can these responses in the cat anaesthetized with chloralose be of any significance in the normal physiology of the animal? Experiments recently performed by Straschill and Schick (1977) on non-anesthetized cats describe two populations of unit response in the superior colliculus whose properties fit well with those described here and which from the controls performed appear to originate in extraocular muscle receptors. It

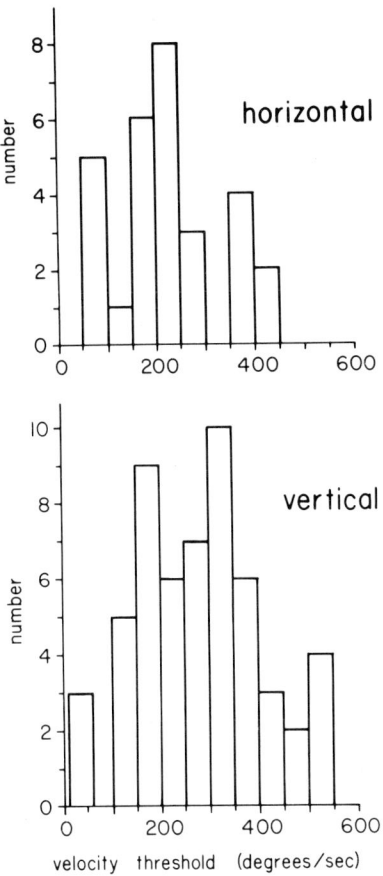

Fig. 5. Velocity thresholds of units responding to small passive eye movements regardless of initial position.

would appear that what has been observed in these experiments relates well to the properties of the superior colliculus in the conscious cat.

What use is made of the information that flows to the superior colliculus from extraocular muscle receptors? Both Batini and Horcholle-Bossavit (1977) and Donaldson and Long (1977) have shown that prior stimulation of extraocular muscle afferents will suppress visual responses in superficial and deep collicular layers. Donaldson and Long (1977) infer that this could be part of a mechanism initiating saccadic suppression. Such a mechanism would explain the kind of signal that is received in the superior colliculus at the initiation of movement at saccadic velocity. It is also likely that input from extraocular muscles is concerned with neck muscle motor control.

We have found that about 45% of the cells of origin of the tectospinal tract within the superior colliculus may be excited or inhibited by extraocular muscle nerve stimulation (Rose and Abrahams, 1978) and units responding to passive movement of the eye in the horizontal plane include cells of origin of the tectospinal tract (Abrahams and Anstee, 1979). It is possible that the pattern of reflex activation that Easton (1972) found in neck muscle following extraocular muscle stretch could be mediated through the superior colliculus.

Is there anything in these experiments to explain the compartmentalized distribution of extraocular muscle receptors? On the surface there seems to be no reason why a saccadic suppressive reflex or a reflex activation of neck muscle should require a compartmentalized receptor distribution. The answer might be in what Wetzel and Stuart (1976) and Binder and Stuart (1978) have called the "ensemble signal". This concept recognizes that although each muscle receptor is capable of generating a unique signal that is determined by the particular events taking place at its location, this information can be combined to provide information about muscle action as a whole. The ensemble information is used in particular situations. Wetzel and Stuart (1976) believe that stereotyped repetitive movements are modulated by the ensemble signal. It seems likely that the superior colliculus utilizes the signal from extraocular muscle receptors for purposes which include saccadic suppression and the initiation of head movement, although within the ensemble signal is embedded information about the progress of a saccade ensemble which could be utilized separately.

The signal from individual receptors in extraocular muscle should be concerned with the fine control of the slow muscle fibres that control gaze fixation and slow following of the eye. Indeed there is good evidence that this happens and Fiorentini and Maffei (1977) have recently shown that input from extraocular receptors is necessary for maintenance of direction of gaze in the dark. The present experiments tend to rule out a role for the superior colliculus in this type of oculomotor control.

It is too early to be definitive about the role of extraocular muscle projections to the superior colliculus, that must await yet further analysis of function within the superior colliculus and may depend on the resolution of the overall role played by proprioceptive projections to that structure.

SUMMARY

Experiments are described in which the behaviour of units in the superior colliculus has been examined during passive vertical and horizontal eye movements in the chloralose anaesthetized cat. In addition to the large, fixed displacement threshold units previously described a population of units is now described which respond to small movements at saccadic velocities. Characteristically these units respond similarly regardless of the initial position of the eye and, in the majority of cases, respond similarly regardless of the direction in which the eye is moved.

Units responding in this way are predominantly located in the intermediate and deep layers of superior colliculus and underlying tegmentum. A significant number of units responding to vertical movement are cells of origin of the tectospinal tract.

Controls showed that such responses are unlikely to originate either from corneal or retinal stimulation or from movement of periorbital tissue. The activity initiated by passive eye movement is therefore believed to arise from receptors in extraocular muscle.

REFERENCES

Abrahams, V.C. and Falchetto, S. (1966) Patterns of unit response in the superior colliculi elicited by non-visual stimuli. *Physiologist*, 9: 128.
Abrahams, V.C. and Rose, P.K. (1975a) Projections of extraocular, neck muscle and retinal afferents to

superior colliculus in the cat: their connections to cells of origin of tectospinal tract. *J. Neurophysiol.*, 38: 10–18.

Abrahams, V.C. and Rose, P.K. (1975b) The spinal course and distribution of fore and hind limb muscle afferent projections to the superior colliculus of the cat. *J. Physiol. (Lond.)*, 247: 117–130.

Abrahams, V.C. and Anstee, G. (1977) Units in the superior colliculus and underlying tegmental structures responding to passive eye movement. *Neurosci. Abstr.*, 3: 153.

Abrahams, V.C. and Anstee, G. (1979) Unit activity in the superior colliculus of the cat following passive eye movements. *Canad. J. Physiol.*, 57: 359–365.

Ades, H.W. (1944) Mid-brain auditory mechanisms in cat. *J. Neurophysiol.*, 7: 415–424.

Alvarado, J.A. and Van Horn, C. (1975) Muscle cell types of the cat inferior oblique. In *Basic Mechanisms of Ocular Motility and their Clinical Implications*, G. Lennerstrand and P. Bach-y-Rita (Eds.) Pergamon Press, Oxford, pp. 15–43.

Bach-y-Rita, P. and Ito, F. (1966) Properties of stretch receptors in cat extraocular muscles. *J. Physiol. (Lond.)*, 186: 663–688.

Batini, C. and Horcholle-Bossavit, G. (1977) Interaction entre activation visuelle et activation proprioceptive au niveau des neurones du colliculus superieur. *C.R. Acad. Sci. (Paris)*, D285: 1491–1493.

Binder, M.D. and Stuart, D.G. (1978) Motor unit-muscle receptor interactions: design features of the neuromuscular control system. In *Motor Control in Man: Suprasegmental Segmental Mechanisms*, J.E. Desmedt (Ed.) Karger, Basel.

Botterman, B.R., Binder, M.C. and Stuart, D.G. (1978) Functional anatomy of the association between motor units and muscle receptors. *Amer. Zool.*, 18: 135–152.

Collins, C.C. (1975) The human oculomotor control system. In *Basic Mechanisms of Ocular Motility and their Clinical Implications*, G. Lennerstrand and P. Bach-y-Rita (Eds.), Pergamon Press, Oxford, pp. 145–180.

Cooper, S., Daniel, P.M. and Whitteridge, D. (1953a) Nerve impulses in the brain stem of the goat. Short latency responses obtained by stretching the extrinsic eye muscles and the jaw muscles. *J. Physiol. (Lond.)*, 120: 471–490.

Cooper, S., Daniel, P.M. and Whitteridge, D. (1953b) Nerve impulses in the brain stem of the goat. Responses with long latencies obtained by stretching the extrinsic eye muscles. *J. Physiol. (Lond.)*, 120: 514–527.

Cooper, S., Daniel, P.M. and Whitteridge, D. (1955) Muscle spindles and other sensory endings in the extrinsic eye muscles; the physiology and anatomy of these receptors and of their connexions with the brain-stem. *Brain*, 78: 564–583.

Donaldson, I.M.L. and Long, A.C. (1977) Suppression of visual responses in the cat superior colliculus following stretch of extraocular muscles. *J. Physiol. (Lond.)*, 272: 94–95P.

Dräger, V.C. and Hubel, D.H. (1975) Responses to visual stimulation and relationship between visual, auditory and somatosensory inputs in mouse superior colliculus. *J. Neurophysiol.*, 38: 690–713.

Easton, T.A. (1972) Patterned inhibition from single eye muscle stretch in the cat. *Exp. Neurol.*, 34: 497–510.

Fillenz, M. (1955) Responses in the brain stem of the cat to stretch of extrinsic ocular muscles. *J. Physiol. (Lond.)*, 128: 182–189.

Fiorentini, A. and Maffei, L. (1977) Instability of the eye in the dark and proprioception. *Nature*, 269: 330–331.

Fuchs, A.F. and Kornhuber, H.H. (1969) Extraocular muscle afferents to the cerebellum. *J. Physiol. (Lond.)*, 200: 713–722.

Gonyea, W.J. and Ericson, G.C. (1977) Morphological and histochemical organisation of the flexor carpi radialis muscle in the cat. *Amer. J. Anat.*, 148: 329–344.

Gordon, B. (1973) Receptive fields in deep layers of cat superior colliculus. *J. Neurophysiol.*, 36: 157–178.

Gordon, B. Superior colliculus: structure, physiology and possible functions. In *MTP International Review of Science*, C.C. Hunt (Ed.), Butterworth, London, pp. 187–230.

Jassik-Gerschenfeld, D. and Ascher, P. (1963) Some responses of the superior colliculus of the cat and their control by the visual cortex. *Experientia*, 19: 655–658.

Jung, R. and Kornhuber, H. (1961) *The Visual System: Neurophysiology and Psychophysics*. Springer-Verlag, Berlin.

Maier, A., Simpson, D.R. and Edgerton, V.R. (1976) Histological and histochemical comparisons of muscle spindles in three hind limb muscles of the guinea pig. *J. Morphol.*, 148: 185–192.

Richmond, F.J.R. and Abrahams, V.C. (1975) Morphology and distribution of muscle spindles in dorsal muscles of the cat neck. *J. Neurophysiol.*, 38: 1322–1339.

Rose, P.K. and Abrahams, V.C. (1975) The effect of passive eye movement on unit discharge in the superior colliculus of the cat. *Brain Res.*, 97: 95–106.

Rose, P.K. and Abrahams, V.C. (1978) Tectospinal and tectoreticular cells: their distribution and afferent connections. *Canad. J. Physiol.*, 56: 650–658.

Shimazu, K. (1959) Superior colliculus, its functional significance relative to the optic and auditory systems. *Med. J. Osaka Univ.*, 10: 39–62.

Starzl, T.E., Taylor, C.W. and Magoun, H.W. (1951) Ascending conduction in reticular activating system with special reference to the diencephalon. *J. Neurophysiol.*, 14: 461–477.

Stein, B.E. and Arigbede, M.O. (1972) Unimodal and multimodal response properties of neurons in the cat's superior colliculus. *Exp. Neurol.*, 36: 179–196.

Stein, B.E., Magalhaes-Castro, B. and Kruger, L. (1976) Relationship between visual and tactile representations in cat superior colliculus. *J. Neurophysiol.*, 39: 401–420.

Straschill, M. and Schick, F. (1977) Discharges of superior colliculus neurons during head and eye movements of the alert cat. *Exp. Brain Res.*, 27: 131–142.

Wetzel, M.C. and Stuart, D.G. (1976) Ensemble characteristics of cat locomotion and its neural control. *Progr. Neurobiol.*, 7: 1–98.

Yellin, H. (1969) A histochemical study of muscle spindles and their relationship to extrafusal fibre types in the rat. *Amer. J. Anat.*, 125: 31–46.

Extraocular Muscle Afferents and Visual Input Interactions in the Superior Colliculus of the Cat

C. BATINI and G. HORCHOLLE-BOSSAVIT

Laboratoire de Psychophysiologie Sensorielle, Université Pierre et Marie Curie, 75230 Paris; and Laboratoire de Physiologie Nerveuse du C.N.R.S., 91190 Gif-sur-Yvette (France)

INTRODUCTION

Since the experiments of Apter (1946), who obtained eye movements by local application of strychnine on the superior colliculus, this structure is considered an integration center where visual information is processed and transformed into motor responses. Retinotopically organized visual information enters the superficial layers (Wilson and Toyne, 1970; Feldon et al., 1970; Kawamura et al., 1974; Graybiel, 1975; Hubel et al., 1975; Harting and Guillery, 1976); the neurons are considered to be movement detectors because they are optimally activated by moving visual stimuli and many are even direction selective (Marchiafava and Pepeu, 1966; McIlwain and Buser, 1968; Sprague et al., 1968; Sterling and Wickelgren, 1969; Rosenquist and Palmer, 1971; Schiller and Koerner, 1971; Gordon, 1973). The motor command exits by the deep collicular layers (Casagrande et al., 1972; Harting et al., 1973; Harting, 1977; Edwards and Henkel, 1978) which upon localized stimulation gives organized conjugated saccadic eye movements (Sika and Radil-Weiss, 1971; Robinson, 1972; Schiller and Stryker, 1972; Straschill and Rieger, 1973; Stryker and Schiller, 1975; Grantyn and Grantyn, 1976), and the discharges of the neurons are temporally related to saccadic eye movements, some occurring prior to saccades (Wurtz and Goldberg, 1971, 1972; Straschill and Hoffman, 1970; Robinson and Jarvis, 1974; Arduini et al., 1974; Sparks, 1975). Other non-visual modalities also enter the superior colliculus (Moore and Goldberg, 1963; Jassik-Gerschenfeld, 1966; Dräger and Hubel, 1975; Antonetti and Webster, 1975; Stein et al., 1976) and a particularly dense projection arises from the extraocular muscle (EOM) afferents (Abrahams and Rose, 1975; Rose and Abrahams, 1975; see Abrahams, this volume, chapter III 4).

The aim of the present work is to study, in the single unit activity of the superior colliculus, the relationship between a specific visual input and EOM proprioceptive input.

In our experimental design, one eye was used for the proprioceptive and the other for the visual stimulation; collicular units were recorded on the side of EOM nerve stimulation. This was possible because i) collicular neurons are activated by ipsilateral EOM afferents (Abrahams and Rose, 1975), and ii) the majority of the collicular neurons are binocularly driven (McIlwain and Buser, 1968; Rosenquist and Palmer, 1971; Sterling and Wickelgren, 1969; Gordon, 1973) so that only a few ipsilaterally driven neurons are lost in our preparation. Our attention has been focused on the proprioceptive input in view of its functional localization in different layers of the superior colliculus and in view of its reciprocal interaction with the particular visual properties of the neurons such as movement, velocity of the movement and direction sensitivity.

METHODS

Experiments were performed in 20 adult cats. Surgery was made under fluothane or ketamine anesthesia. The spinal cord was transected to the C1–C2 level, the animal was immobilized with flaxedil and artificially respired. Rectal temperature was maintained at 37–38°C. The head was fixed in a stereotaxic frame modified to provide the animal with an unobstructed visual field. A craniotomy was performed to stereotaxically introduce vertically oriented recording electrodes. Single units were recorded extracellularly with insulated stainless steel microelectrodes. Conventional amplifying and displaying techniques were used. Marking of the recording sites was made by Fe^{2+} deposit at the tip of the microelectrode and subsequent prussian blue staining. Serial frontal sections were then counterstained with safranine red. The position of the recording sites was identified and the relative location of the successive units was noted in each track.

The ipsilateral eye was prepared for the proprioceptive stimulation: the intramuscular portion of the nerves to the lateral rectus muscle and to the medial rectus muscle was dissected free, mounted on a pair of silver wire electrodes for stimulation and covered with warm paraffin oil. These two nerves correspond to antagonistic muscles acting in the horizontal plane and contain the afferent fibers (Batini and Buisseret, 1972, 1974). Single and burst (200–300/sec, 4 impulses) stimuli were used as required to obtain responses to proprioceptive fiber activation.

The contralateral intact eye was prepared for visual stimulation: the pupil was dilated by local application of 1% atropine sulphate and the cornea was protected with a contact lens. To measure and localize the receptive field, the optic disk of the retina was projected on a tangent screen at 58 cm from the intact eye with a positioned ophthalmoscope (Vakkur et al., 1963). At this distance 1 cm on the screen corresponds to 1°. The position, size and properties of the visual receptive field were first manually searched using the ophthalmoscope. Then, for testing the interaction between visual and proprioceptive inputs, moving controlled visual stimuli were performed using a galvanometer-driven mirror reflecting a slit light on the tangent screen.

RESULTS

The following results are based on data recorded from 94 neurons in different layers of the superior colliculus. The parameters of the two stimuli have been adjusted for each unit to obtain an optimal response for that unit. Neurons not giving clear responses within a particular range (see below) of stimulus parameters were abandoned. The units are then grouped according to the stimulus parameters across experiments to obtain a population of neurons having the same responses to a given set of stimulus parameters.

Proprioceptive responses

Electrical stimulation of the EOM nerves elicited two types of responses in the collicular units. The first was an excitatory response (Fig. 1) which occurred with latencies ranging from about 15 to 100 msec. In spite of this wide variability in latencies, it was not possible to distinguish, in our results, two populations of "early" and "late" responses as reported by Abrahams and Rose (1975). Habituation quickly reduced the response, therefore the interstimulus intervals were kept longer than 10 sec. The second type of response was an inhibition of the background firing (Fig. 1) of variable duration, often followed by a postinhibitory rebound. Complex responses were frequently observed with both excitatory and inhibitory components occurring in succession. Proprioceptive responses were tested from two antagonistic muscles but no clear reciprocal effect has been so far detected. Stimulation of the nerve to the lateral rectus or to the medial rectus muscles usually resulted in a similar effect (Fig. 1C, E, F).

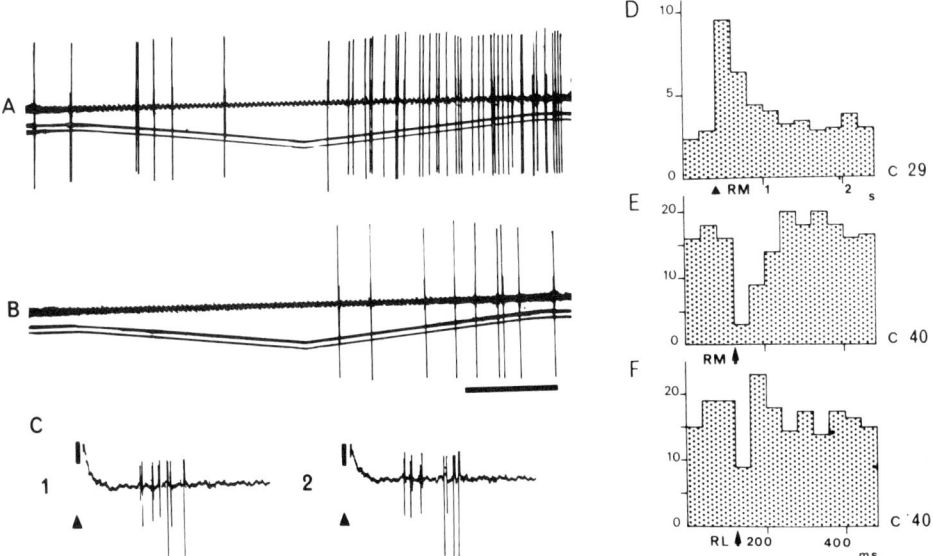

Fig. 1. Responses of collicular units to visual and proprioceptive stimulations. A) Discharge of a direction selective unit. Lower trace is the signal of a horizontal movement from left to right (downward deflection), then from right to left (upward deflection). B) After 5 presentations at 10 sec intervals, discharge shows habituation. C) 1–2: Burst discharges of two simultaneously recorded units evoked by electrical stimulation (black triangle) of the lateral rectus (RL) nerve. Time calibration: 500 msec in A and B, 150 msec in C. On the right, poststimulus histogram showing the sum of 12 responses evoked (every 15 sec) by stimulation of extraocular muscle afferents. D) Excitatory response of a collicular unit to stimulation of the medial rectus (RM) nerve. E) Inhibitory response of another collicular unit to stimulation of the medial rectus nerve and F) to stimulation of the lateral rectus nerve (C).

Visual properties

The criterion to classify a collicular unit as specific with respect to visual input was to respond to at least one of the following parameters: movement of the visual stimulus in the plane of the tangential screen, movement in a particular direction, movement of the stimulus with a particular velocity. Most units recorded in the superior colliculus, responded optimally to moving visual stimuli, with often a clear preference for movement in a particular direction (Fig. 1A, B) and with a particular velocity as already extensively described in anesthetized and non-anesthetized preparations (McIlwain and Buser, 1968; Sprague et al., 1968; Sterling and Wickelgren, 1969; Strashill and Hoffman, 1969; Rosenquist and Palmer, 1971) and chronic preparations (Gordon, 1973). As reported in midpontine and *encephale isolé* preparations, the absence of anesthetics in our experiments, could account for the difficulties we have encountered in defining the parameters of the visual stimulation. Both instability of spontaneous firing and rapid habituation of the responses (Fig. 1A, B) resulted in variability of the size and shape of the receptive field (3° to 30°).

Visuo-proprioceptive convergence

Among the units activated by specific visual stimuli, the majority also received proprioceptive input which was either of the excitatory or of the inhibitory type. Other units not activated by specific visual stimuli, could or could not receive proprioceptive afferents. Of these however, convergence could have escaped detection since i) some units might be those normally activated by ipsilateral monocular visual stimuli not used here, and ii) some others could receive proprioceptive afferents from other EOM not tested in our experiments. Therefore we can assume that convergence of specific visual and proprioceptive input is a general phenomenon taking place at single collicular neurons

Localization

Fig. 2 summarizes the distribution of the various types of units at the surface and at the depth of the superior colliculus. Most of the units responding to specific visual stimuli were found in the stratum griseum superficialis and in the stratum opticum as shown in the previous studies (McIlwain and Buser, 1968; Sterling and Wickelgren,

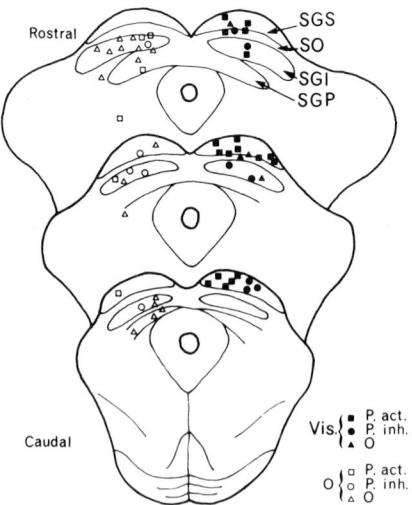

Fig. 2. Schematic representation of the distribution of collicular units histologically localized and electrophysiologically characterized. SGS, stratum griseum superficialis; SO, stratum opticum; SGI, stratum griseum intermedialis; SGP, stratum griseum profundum. Black symbols: visual responding units, tested to proprioceptive stimulation; P. act., excitatory response; P. inh. inhibitory response; O, not responding.
White symbols: units not responding to visual stimulation; P. act. P. inh., and O as in black symbols.

1968; Rosenquist and Palmer, 1971) and few were found in the underlying layers as reported by other authors (Strashill and Hoffman, 1969; Gordon, 1973). On the contrary, most of the units not responding to a specific visual stimuli were found in the stratum griseum intermedium and profundum and only few in the overlying layers. Both groups of units, however, have shown responses (both excitatory and inhibitory) to EOM nerve stimulation.

Proprioceptive influence on the visual response

The interaction experiments were performed by applying short EOM nerve stimulation at different times during specific visual activation. The most prominent result was an inhibition of the response to the specific visual stimulus. Inhibition was observed for units sensitive to moving stimuli, as well as for those sensitive to direction and velocity. This effect was obtained when EOM nerve stimulation elicited an inhibition of the spontaneous activity or when no detectable inhibition could be observed in the extracellular records as illustrated in Fig. 3. Effective inhibition of the visual response was also observed in those units where EOM nerve stimulation alone had elicited an excitatory reponse. In this latter case the inhibition of the visual response usually followed, but eventually preceded the proprioceptive firing as shown in Fig. 4. When direction selective units were involved, interaction occurred for horizontal as well as for vertical directions, although the proprioceptive input was only from the horizontal plane. Excitatory interactions were also sometimes observed, but they usually were simple summations of the two responses appearing simultaneously (Fig. 4D). These results indicate that the proprioceptive influence on the visual response is primarily a

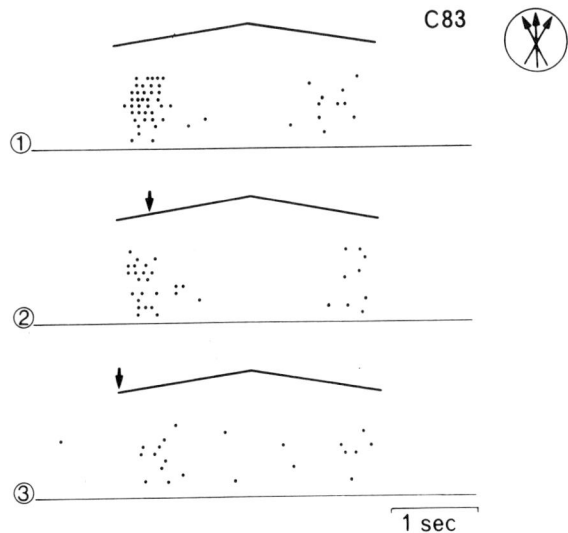

Fig. 3. Behavior of a collicular unit during both visual and proprioceptive stimulations. Each dot corresponds to a spike; successive rows of dots correspond to successive presentation of the visual stimulus signaled by the oblique lines: upward deflection, upward movement of the stimulus; downward deflection, downward movement of the stimulus. 1) Visual discharge evoked by the moving stimulus. 2) Attenuation of the visual discharge by stimulation of the lateral rectus nerve (arrow) 400 msec after the onset of visual stimulation. 3) Suppression of the visual discharge when the proprioceptive stimulus is given at the onset of the visual stimulation.

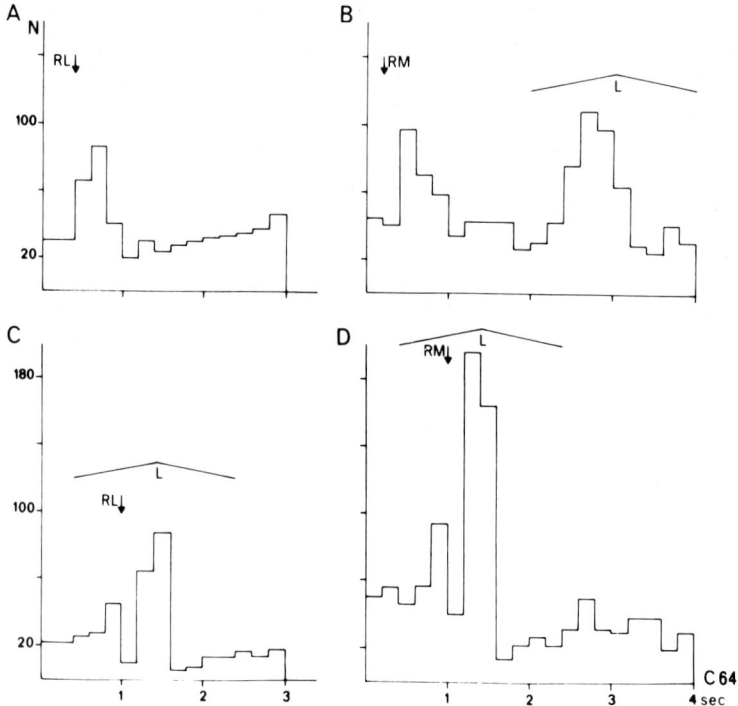

Fig. 4. Poststimulus histograms showing the sum of successive responses evoked by visual and proprioceptive stimulations and their interactions. A) Excitatory response evoked by stimulation of the lateral rectus (RL) nerve at the arrow. B) Excitatory response evoked by stimulation of the medial rectus (RM) nerve (arrow) and visual discharge evoked by a horizontally moving stimulus (upward line, from right to left, downward line, from left to right). C) Inhibition of the visual discharge by lateral rectus nerve stimulation. D) Early inhibition of the visual discharge by medial rectus nerve stimulation is followed by an increase of the visual response.

general inhibition. Furthermore, the inhibitory effects of the EOM afferents appear to be independent of the sign of the EOM response and of the type of the visual response.

Effect of visual stimulation upon proprioceptive response

The interaction experiments have also been performed on units not responding to specific visual stimuli. In these cases the visual stimulus was always a long slit light moving in the horizontal plane which usually did not elicit a response. This aspecific visual stimulus, however, was effective in producing either inhibition or facilitation of the complex firing evoked by the proprioceptive stimulation (Fig. 5). The facilitatory interaction was even more evident in some units in which the proprioceptive stimulation had no effect when applied in total darkness, but produced a clear excitatory response during a moving visual stimulus (Batini and Horcholle-Bossavit, 1977).

DISCUSSION

The present results show that, in *encephale isolé* cats, the electrical stimulation of EOM nerve evokes discharges in collicular units similar in pattern and wide range of

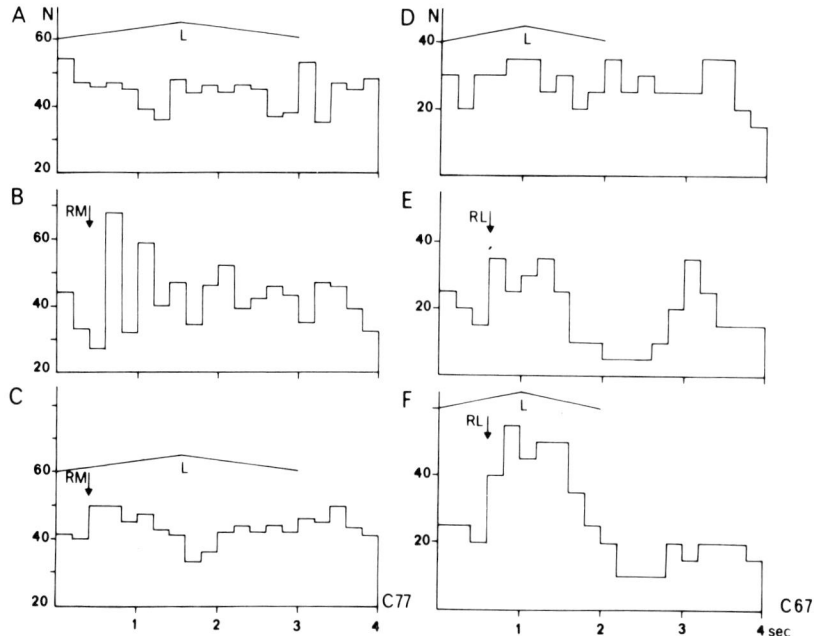

Fig. 5. Poststimulus histograms showing the interactions between proprioceptive and visual stimulations. A, D) Absence of clear response to horizontally moving visual stimulus (oblique lines) in two collicular units as in Fig. 4. B,E) Proprioceptive responses of the two collicular units to medial rectus (RM) and lateral rectus (RL) nerve stimulation. C) Attenuation of the proprioceptive response in the presence of the moving visual stimulation. F) Facilitation of the proprioceptive response in the presence of the moving visual stimulation.

latencies to those described by Fillenz (1955), Rose and Abrahams (1975) and Donaldson and Long (1977) for stretching the EOM. Therefore, we can safely assume that the effect was due to the stimulation of the afferent fibers contained in the muscle nerve and not to stimulation of possible axon collaterals of the motoneurons. The same interpretation holds true for the additional inhibitory response obtained by stretching the EOM (Donaldson and Long, 1977) and found to be the most prominent response in our results.

The inhibitory effect of the proprioceptive input acts upon the spontaneous firing as well as upon the visual response of the neurons but its importance has been brought about particularly by the interaction experiments. Probably in extracellular recordings, when only proprioceptive stimulation is used, the inhibition is underestimated in units with low rates of spontaneous activity. Moreover, the inhibitory effects have been observed to precede, in some cases, the excitatory one. Thus, the long-latency excitatory response observed in our experiments and in those of Abrahams and Rose (1975) might be interpreted as a rebound following the inhibition.

In trying to summarize the present complex results we may assume that: i) input from the EOM reaches the superficial layer of the superior colliculus where they influence, mostly inhibit, the neurons sensitive to the specific visual stimulation. The absence of proprioceptive specificity (for example a vertical direction selective unit is inhibited by afferents coming from a muscle acting in the horizontal plane) may be interpreted as due to the fact that all the EOM cooperate in executing simple eye movements (see Carpenter, 1977); ii) input from the EOM also reaches the deep layers of the superior colliculus where it acts upon neurons not activated by specific visual stimulation but presumably sending the oculomotor efferents. There, the visual input influences the proprioceptive response.

The idea that some collicular neurons may distinguish the retinal image of a moving object from that due to eye movement is based on the observation that their discharges are related to eye movements even in the absence of a visual input (Strashill and Hoffman, 1970), therefore the collicular neuron discharges must be using extraretinal signals (Robinson and Wurtz, 1976; See McKay, 1973, and Carpenter, 1977). However, the question is still open as to whether the extraretinal input is the "corollary discharge" (Sperry, 1950) "efferent copy" (Von Holtz, 1954), or whether it is the EOM proprioception. Our results showing the proprioceptive suppression of the specific visual response strongly support the second hypothesis. As pointed out by Donaldson and Long (1977) the inhibitory effect would reach the superior colliculus before the retinal image. Therefore, the wave of inhibition which suppresses the visual information just before and during the movement will be effective even for very small movements. Such small movements might be present in curarized animals, as mentioned by Richmond and Wurtz (1978) who have shown the participation of the corollary discharge in paralysed monkeys.

The EOM afferents reaching the deep layers of the superior colliculus presumably become involved in complex mechanisms coding the eye position (see Abrahams, this volume, chapter III 4). While they act upon the efferent neurons collecting connections from the upper layers of the colliculus and other afferent systems, it appears that they are also influenced by the visual system.

An open question concerning the visuoproprioceptive interaction is: what is the participation of the visual cortex from which the direction selectivity of the collicular units is largely derived? In fact, recent data have shown that asymmetry in the proprioceptive innervation of the eye modifies the normal proportion of binocularly driven neurons in the visual cortex (Maffei and Fiorentini, 1976) and that afferents from the EOM arrive at the visual cortex in cat (Buisseret and Maffei, 1977). It would appear, therefore, that the EOM proprioceptive afferents play a major role in the processes which are put into effect when the eye is moved.

SUMMARY

Units have been recorded extracellularly and have been histologically localized in the superior colliculus of the cat.

The specificity of the visual responses to moving stimuli was detected by the presentation of the stimuli to the contralateral eye.

The electrical stimulation of the nerves to the ipsilateral rectus medialis and the rectus lateralis muscles containing afferent fibers, was effective in producing excitatory and inhibitory responses.

Proprioceptive responses were observed in units of the superficial layers responding to specific visual stimuli and in units of the deep layers not responding to specific visual stimuli.

Specific visual responses were influenced, mostly inhibited, by a properly timed proprioceptive stimulation. The proprioceptive response was influenced, facilitated or inhibited, by moving lights in some units not responding to specific visual stimuli.

ACKNOWLEDGEMENTS

This work has been supported by C.N.R.S., R.C.P. contract No. 433, A.T.P. No. 651/3621 and by I.N.S.E.R.M., A.T.P. contract No. 29.76.61.

REFERENCES

Abrahams, V.C. and Rose, P.K. (1975) Projections of extraocular neck muscle, and retinal afferents to superior colliculus in the cat: their connections to cells of origin of tectospinal tract. *J. Neurophysiol.*, 38: 10–18.
Antonetti, C.M. and Webster, K.E. (1975) The organization of the spinotectal projection. An experimental study in the rat. *J. comp. Neurol.*, 163: 449–466.
Apter, J.T. (1946) Eye movements following strychninization of the superior colliculus of cats. *J. Neurophysiol.*, 9: 73–86.
Arduini, A., Corazza, R. and Marzollo, P. (1974) Relationship of the neural activity of the superior colliculus to eye movements in the cat. *Brain Res.*, 73: 473–481.
Batini, C. and Buisseret, P. (1974) Sensory peripheral pathway from extrinsic eye muscles. *Arch. ital. Biol.*, 112: 18–23.
Batini, C. and Horcholle-Bossavit, G. (1977) Interaction entre activation visuelle et activation proprioceptive au niveau des neurones du colliculus superieur. *C.R. Acad. Sci. (Paris)*, 285: 1493–1494.
Buisseret, P. and Maffei, L. (1977) Extraocular proprioceptive projections to the visual cortex. *Exp. Brain Res.*, 28: 421–425.
Carpenter, R.H.S. (1977) *Movements of the Eyes*. Plon, London.
Casagrande, V.A., Harting, J.K., Hall, W.C., Diamond, I.T. and Martin, G.F. (1972) Superior colliculus of the tree shrew *(Tupaia glis)*: evidence for a structural and functional subdivision into superficial and deep layers. *Science*, 177: 444–447.
Donaldson, J.M.L. and Long, A.C. (1977) Suppression of visual responses in the cat superior colliculus following stretch of extraocular muscles. *J. Physiol. (Lond.)*, 272: 100P–101P.
Dräger, V.C. and Hubel, D.H. (1975) Responses to visual stimulation and relationship between visual, auditory, and somatosensory input in mouse superior colliculus. *J. Neurophysiol.*, 3: 690–713.
Edwards, S.B. and Henkel, C.K. (1978) Superior colliculus connections with the extraocular motor nuclei in the cat. *J. comp. Neurol.*, 179: 451–468.
Feldon, S., Feldon, P. and Kruger, L. (1970) Topography of the retinal projection upon the superior colliculus of the cat. *Vision Res.*, 10: 135–143.
Fillenz, M. (1955) Responses in the brainstem of the cat to stretches of extrinsic ocular muscles. *J. Physiol. (Lond.)*, 128: 182–199.
Gordon, B. (1973) Receptive fields in deep layers of cat superior colliculus. *J. Neurophysiol.*, 36: 157–178.
Grantyn, A.A. and Grantyn, R. (1976) Synaptic actions of tectofugal pathways on abducens motoneurons in the cat. *Brain Res.*, 105: 269–285.
Graybiel, A.M. (1975) Anatomical organization of retinotectal afferents in the cat: An autoradiographic study. *Brain Res.*, 96: 1–23.
Harting, J.K. (1977) Descending pathways from the superior colliculus: an autoradiographic analysis in the rhesus monkey *(Macaca mulatta)*. *J. comp. Neurol.*, 173: 583–612.
Harting, J.K. and Guillery, R.W. (1976) Organization of retinocollicular pathways in the cat. *J. comp. Neurol.*, 166: 133–144.
Hubel, D.H., Levay, S. and Wiesel, T. (1975) Mode of termination of retinotectal fibers in Macaque monkey: an autoradiographic study. *Brain Res.*, 96: 25–40.
Kawamura, S., Sprague, J.M. and Nümi, K. (1974) Corticofugal projections from the visual cortices to the thalamus, pretectum and superior colliculus in the cat. *J. comp. Neurol.*, 158: 339–362.
Marchiafava, P.L. and Pepeu, G. (1966) The responses of units in the superior colliculus of the cat to a moving visual stimulus. *Experientia*, 22: 51–59.

Maffei, L. and Fiorentini, A. (1976) Asymmetry of motility of the eyes and change of binocular properties of cortical cells in adult cats. *Brain Res.*, 105: 73–78.

MacKay, D.N. (1973) Visual stability and voluntary eye movements. In *Handbook of Sensory Physiology, Vol. VII/3*, R. Jung (Ed.), Springer Verlag, Berlin, pp. 307–331.

McIlwain, J.T. and Buser, P. (1968) Receptive fields of single cells in the cat's superior colliculus. *Exp. Brain Res.*, 5: 314–325.

Moore, R.Y. and Goldberg, J.M. (1963) Ascending projections of the inferior colliculus in the cat. *J. comp. Neurol.*, 121: 109–135.

Richmond, B.J. and Wurtz, R.H. (1978) Visual responses during saccadic eye movement: a corollary discharge to superior colliculus. *Soc. Neurosci. Abstr.*, III: 574, No. 1834.

Robinson, D.A. (1972) Eye movements evoked by collicular stimulation in the alert monkey. *Vision Res.*, 12: 1795–1808.

Robinson, D.L. and Jarvis, C.D. (1974) Superior colliculus neurons studied during head and eye movements of the behaving monkey. *J. Neurophysiol.*, 37: 533–540.

Robinson, D.L. and Wurtz, R.H. (1976) Use of an extraretinal signal by monkey superior colliculus neurons to distinguish real from self-induced stimulus movement. *J. Neurophysiol.*, 39: 852–870.

Rose, P.K. and Abrahams, V.C. (1975) The effect of passive eye movement on unit discharge in the superior colliculus of the cat. *Brain Res.*, 97: 95–106.

Rosenquist, A.C. and Palmer, L.A. (1971) Visual receptive field properties of cells of the superior colliculus after cortical lesions in the cat. *Exp. Neurol.*, 33: 629–652.

Schiller, P.H. and Koerner, F. (1971) Discharge characteristics of single units in superior colliculus of the alert rhesus monkey. *J. Neurophysiol.*, 34: 920–936.

Schiller, P.H. and Stryker, M. (1972) Single unit recording and stimulation in superior colliculus of the alert monkey. *J. Neurophysiol.*, 35: 915–924.

Sika, J. and Radil-Weiss, T. (1971) Electrical stimulation of the tectum in freely moving cats. *Brain Res.*, 28: 567–572.

Sparks, D.L. (1975) Response properties of eye movement related neurons in the monkey superior colliculus. *Brain Res.*, 90: 147–152.

Sperry, R.W. (1950) Neural basis of the spontaneous optokinetic response produced by visual inversion. *J. comp. Physiol. Psychol.*, 43: 482–489.

Sprague, J.M., Marchiafava, P.L. and Rizzolatti, G. (1968) Units responses to visual stimuli in the superior colliculus of the unanesthetized, midpontine cat. *Arch. ital. Biol.*, 106: 169–193.

Stein, B.E., Magalhaes-Castro, B. and Kruger, L. (1976) Relationship between visual and tactile representations in cat superior colliculus. *J. Neurophysiol.*, 39: 401–419.

Sterling, P. and Wickelgren, B.G. (1969) Visual receptive fields in the superior colliculus of the cat. *J. Neurophysiol.*, 32: 1–15.

Straschill, M. and Hoffman, K.P. (1970) Activity of movement sensitive neurons of the cat's tectum opticum during spontaneous eye movements. *Exp. Brain Res.*, 11: 318–326.

Straschill, M. and Rieger, P. (1973) Eye movements evoked by focal stimulation of the cat's superior colliculus. *Brain Res.*, 59: 211–227.

Stryker, M.P. and Schiller, P.H. (1975) Eye and head movements evoked by electrical stimulation of monkey superior colliculus. *Exp. Brain Res.*, 23: 103–112.

Vakkur, G.I., Bishop, P.O. and Kozak, W. (1963) Visual optics in the cat, including posterior nodal distance and retinal landmarks. *Vision Res.*, 3: 289–314.

Von Hol, E. (1954) Relations between the central nervous system and the peripheral organs. *Brit. J. Anim. Behav.*, 2: 89–94.

Wilson, M.E. and Toyne, M.J. (1970) Retino-tectal and cortico-tectal projections in *Macaca mulatta*. *Brain Res.*, 24: 395–406.

Wurtz, R.H. and Goldberg, M.E. (1971) Superior colliculus cell responses related to eye movements in awake monkeys. *Science*, 171: 82–84.

Does Extraocular Proprioception Influence the Development of Visual Processes and the Oculomotor System?

P. BUISSERET

Laboratoire de Neurophysiologie, Collège de France, 75231 Paris (France)

We usually are able to become consciously aware of the position of the different parts of our body relative to each other whether they are moving or not. This awareness has been called "kinaesthesia" during motion and "position sense" during rest. But these two terms have been used interchangeably for a long time, because it was considered that both result from the activation of the same proprioceptors. For the first half of this century, it was thought that receptors of both joints and muscles contribute to kinaesthesia or position sense. Concerning the eye position sense, it was considered according to Helmholtz's view (1866, see also 1925), that awareness of the direction of gaze does not depend upon proprioceptive discharges from extraocular muscles, but is "simply the result of the effort of will involved in trying to alter the adjustment of the eye".

In the 1950s, the inability to detect an evoked potential on the cortical surface when stimulating group 1 skeletal muscle afferents and the development of the hypothesis of the servocontrol of movement through the fusimotor pathway, led to the assumption that information from muscle receptors does not reach consciousness. The attempt to equate the problem of the role of the skeletal muscle receptors with that of the extrinsic eye muscle receptor emphasized this view, particularly when Brindley and Merton (1960) repeated and refined Helmholtz's observations and so established that the eye lacks a position sense. Moreover, experimental evidence progressively supported this view about skeletal muscles (see Goodwin et al., 1972). But it became apparent that extraocular receptors differ morphologically from skeletal muscle spindles and that they are unable to elicit a stretch reflex. However, some workers did not accept the conventional view. Thus, Paillard and Brouchon (1968), Goodwin et al., (1972), Eklund (1972) and Paillard (1973) showed that proprioceptive afferents do contribute to a position sense and to kinaesthesia. In addition, they tried to set apart the role of each type of proprioceptor and to separate position sense from kinaesthesia and conscious from subconscious awareness.

I would first like to review the possible functional roles of extraocular proprioceptors in eye motor control and secondly propose a new role of extraocular proprioceptors based on experiments which demonstrated the need of eye movement for the development of some specific characteristics of the visual cortical neurons. It should also be mentioned that the participation of extraocular proprioceptors has been considered in fixation micronystagmus (Christman and Kupfer, 1963), in correcting overshoot of eye movements and oscillation (Gernandt, 1968), and as negative feedback for small deviations from fixation (Fender and Nye, 1961). More recently, Reinecke and Simons (1965) proposed that the discrepancy between alignment signals from visual-retinal input and from extraocular muscles could be the cause of phoria.

EXTRAOCULAR PROPRIOCEPTORS AND EYE MOTOR CONTROL AND POSITION SENSE

Extraocular muscles do not display stretch reflexes (McCouch and Adler, 1932). Neither muscle contraction consecutive to stretch, nor changes in active muscle tension elicited by opening the feedback loop that could be related to proprioceptive afferents, have been demonstrated. Electrophysiological records do not show excitatory actions exerted by proprioceptive afferents on oculomotoneurons (McIntyre, 1939; Keller and Robinson, 1971; Baker and Precht, 1972).

On the other hand, several reflex actions of extraocular muscle stretch receptors have been reported: Sherrington (1893) observed reflex movements of heterologous eye muscles by stretching the inferior oblique. More recently, Easton (1971, 1972) elicited excitatory and inhibitory actions on neck and forelimb muscles by pulling extraocular muscles. Sasaki (1963) found a hyperpolarization in oculomotoneurons when stimulating a heterologous eye muscle nerve. Baker et al. (1969) and Goldberg (1974) observed orthodromic activation of oculomotoneurons when stimulating the corresponding motor nerve, which they, however, attributed to spreading currents. Also Goldberg et al. (1974) described interneurons with orthodromic activation by afferent fibres, and they also showed (1976) that intracellular stimulation of these interneurons did not produce any muscle activity. However, inhibitory actions cannot be excluded in this case. Furthermore, inhibitory actions have also been reported (Sasaki, 1963; Bach-y-Rita, 1972). In total it appears that despite the fact that a monosynaptic stretch reflex has never been found some reflex actions of the extraocular muscle stretch receptors do exist.

Cerebellar and collicular projections

It is known that afferent fibres from extraocular muscles project to the cerebellum (Fuchs and Kornhuber, 1969; Baker et al., 1972; Batini and Buisseret, 1972; Batini et al., 1974). Different roles have been proposed for these projections such as: providing information about the magnitude of the end point of saccades (Fuchs and Kornhuber, 1969), or about eye position for the cerebellar vermis function in saccadic eye movement (Precht, 1975); for a more detailed review see Batini, chapter III 2.

On the other hand, collicular afferents from extraocular stretch receptors have been clearly demonstrated (Abrahams and Rose, 1975; Rose and Abrahams, 1975; Batini and Horcholle-Bossavit, 1977; Donaldson and Long, 1977; Abrahams and Anstee, 1978). The relationship between colliculus and eye movements is established, but the specific roles of the different types of afferents are not clearly defined (see chapters III 4 and III 5).

Resting eye position

Some new experiments by Fiorentini and Maffei (1977) have shown that section of extraocular afferent fibres induced instability and pendular oscillations of the eye, in the absence of visual and vestibular stimuli. They ascribe this effect to "the opening of the loop of the servo-system device responsible for the maintenance of eye position".

Eye position signal, eye position sense

In 1970, Skavenski and Steinman, using psychophysical methods, showed first that "there is an extraretinal (position) signal of considerable fidelity which can be used to maintain the eye near the primary and other orientations for very long periods of time in the dark". Then Skavenski (1972) showed that the position signal was provided by extraocular proprioceptors and that this "inflow" information could provide a conscious position sense. Although convincing, these experiments are in contradiction with previous results of Irvine and Ludwigh (1936) and Brindley and Merton (1960). Skavenski explains this discrepancy by pointing out the fact that he used more sensitive psychophysical methods. It is also important to consider that when he performed his experiments, it was widely accepted that skeletal muscle receptors do not provide a conscious sense of body position. But almost simultaneously, Goodwin et al. (1972) showed that skeletal muscle receptors do contribute to conscious position sense and kinaesthesia. Further independent observations proved that corollary motor discharges are not, by themselves, sufficient to cause perceived sensations of movement either of skeletal muscles (McClosley and Torda, 1975; Gautier and Hofferer, 1976) or of eye muscles (Brindley et al., 1976; Stevens et al., 1976).

VISUAL CORTEX AND EXTRAOCULAR PROPRIOCEPTION

In 1976 Maffei and Bisti, working on kittens, and Maffei and Fiorentini, working on adult cats, suggested that extraocular proprioceptors could play a role in maintaining the binocularity of single neurons in the visual cortex. It was thus postulated that extraocular proprioceptors exert a direct influence on visual cortical neurons. In our laboratory, experiments have shown that eye movements as well as visual exposure are necessary for the development of the properties of visual neurons specific for orientated visual stimuli. We know that in a 6-week-old dark reared kitten, there is no orientation specificity of the visual cortical neurons (Imbert and Buisseret, 1975; Buisseret and Imbert, 1976). However, normal visual experience during only 6 h is sufficient to produce an almost complete restoration of the orientation specificity (Imbert and Buisseret, 1975; Buisseret et al., 1978). This restoration is not observed when vision only, without any movements, is allowed but it is almost complete when vision plus eye movements only are allowed (Buisseret et al., 1978). Such observations suggest that extraocular proprioceptor activation through eye movement might be involved in the development and restoration of specificity in the visual cortical neurons. Moreover, Gordon and Gummow (1975) found that following the horizontal recti-muscle section in young kittens, the receptive fields of the superior colliculus cells were abnormal. They considered the possibility that impaired proprioceptive feedback from the operated eye might explain this effect.

Donaldson and Long (1977) pointed out an unexpected but attractive new role for some collicular cells, whose activity is closely dependent upon the activation of eye muscle proprioceptors. They proposed a possible involvement of these cells in visual suppression during saccades (see also Batini and Horcholle-Bossavit, this volume, chapter III 5).

But in our experiments and in those of Maffei and Fiorentini and Maffei and Bisti, where defects appear at the visual cortical level, it is necessary to suppose a projection of extraocular proprioceptors to the visual cortex in order to account for the observed defects. Recent experiments have shown that such a projection exists (Buisseret and

Maffei, 1977). Thus, a precisely graded electrical stimulation of each extraocular muscle nerve branch, applied in its intraorbital portion where they are known to contain afferent fibres (Batini and Buisseret, 1974), induced responses in visual cortical neurons. The responses disappeared following the proximal section of the nerve branch. However, in these experiments the possibility of an antidromic stimulation of some motor axon collateral (i.e., the collateral which is supposedly involved in corollary discharge) could not be ruled out. Therefore, stretch experiments were performed which gave the same results. In this latter case, the responses recorded from visual cortical neurons could be temporarily abolished by an intramuscular injection of Xylocaine. Convergence of afferents from different muscles onto the same visual neuron was also found and possible specific arrangements have so far not been demonstrated. The stimulation of only one or the other of a pair of muscle nerves, prepared in a given experiment, activated 25% of the visual neurons; therefore, if we take into account the fact that there are six extraocular muscle nerves, the total proportion of visual cortical neurons that could be activated in these conditions might be more important.

Recent experiments bring even more direct evidence for the role of extraocular proprioceptors in the development of cortical neuron specificity for orientation (Buisseret and Gary-Bobo, 1979). In cats, the extraorbital course of proprioceptive fibres is mainly in the ophthalmic branch of the Vth nerve (Batini and Buisseret, 1972, 1974; Batini et al., 1975); thus, a dissociation of eye movements from the proprioception of eye movement or, at least, from afferents including proprioception is possible. In 6-week-old dark reared kittens, a chronic bilateral section of the ophthalmic branches of the Vth nerve was performed under Nembutal anaesthesia. From 5 to 13 days after the operation, the kittens were allowed 6 h of visual experience. Two series of kittens were exposed under two different sets of conditions: only eye movements could be performed by the kittens of the first series by head and body restraint while free movement and locomotion were allowed to those of the second series. At the end of the exposure time, the kittens were returned to the dark room for 12 h. Then the electrophysiological experiment was performed investigating the properties of the cortical visual cell's receptive field (for details, see Buisseret and Imbert, 1976). In both series of operated kittens the repartition of the tested cells into the three types of cortical visual cells (i.e., cells without orientation specificity = non-specific; cells with some orientational properties = immature; cells meeting all the most severe criteria for specificity = specific) was strikingly different from the repartition found in the non-operated kittens. Of the 155 visual cells tested 93.5% failed to show orientation properties despite the allowance of free eye movements during the 6 h exposure time; only 5.2% of the examined cells were immature and 1.3% specific.

The repartition of 125 visual cells tested in the second series of operated kittens, was again 89% non-specific cells after visual exposure, 9.5% immature and 1.5% specific. By contrast, in a sham-operated kitten, most of the cells (25 out of 37) showed orientation properties the day after visual exposure (Buisseret and Gary-Bobo, 1979).

These experiments show that after the suppression of proprioceptive afferents from extraocular muscles, visual exposure with free eye movements, and even with free motion of the whole body, fails to induce the restoration of orientation specificity in visual cortical cells of dark reared kittens, very much in contrast to what happens when proprioceptive afferents are intact. Thus, it appears that extraocular proprioceptive afferents play a role as a "cofactor" of visual information in the developmental processes of visual cortical neurons. More unexpected are the results obtained in the nerve-operated kittens which were free to move during exposure and in which restoration also failed to appear. One might think that in this situation, head movements and neck proprioception could act as substitutes for at least a partial restoration. This did not occur in our experiments, but the duration of the exposure (6 h), may have been too short to allow an effect from the non-extraocular proprioceptive afferents.

The same types of experiments have been performed by Maffei and Fiorentini, who found that binocular interaction in the visual cortex is significantly reduced after unilateral section of the proprioceptive afferents from the extraocular muscles (Maffei and Fiorentini, 1977, and personal communication).

DISCUSSION

It appears that the postulated roles of the extraocular proprioception have to be divided into two main groups: first, conscious and unconscious motor control and sense of eye position, and second, visual processes. From the first group, nothing is definitely established so far but some experiments point out the necessity of extraocular afferents in the maintenance of eye position for instance (Fiorentini and Maffei, 1977). Also, Skavenski (1972) suggests that a quite sensitive, although unconscious, oculomotor response occurs when loads are applied to the eye. In other instances, the simplest explanation of a phenomenon would be to imply extraocular proprioception. Finally, projections of extraocular proprioceptors to structures implied in oculomotor control (motoneurons, cerebellum, superior colliculus) indicate that they exert some influences at these levels. It appears that with sensitive psychophysical methods it is possible to test the awareness of eye position (Skavenski, 1972) which results from "inflow" information. Moreover, corollary motor discharges are not, by themselves, sufficient to cause perceived sensations of eye movement. In some respects, the function of proprioception in eye muscles might appear not too different from the function of proprioception in skeletal muscles. In the skeletal system, however, information from receptors other than muscle proprioceptors participate in general body position sense and kinaesthesia. Why should it be different in the eye system? Dichgans' group and more recently Berthoz have shown that when particular motions of the visual field (circular-vection or linear-vection) are used, peripheral vision could create illusions and induce compensatory movements (see Berthoz, 1978; Berthoz et al., chapter I C3). He suggests that peripheral vision contributes to the general proprioception, which in other words would imply a "proprioceptive function" of peripheral retinal vision. We would then tentatively propose that eye system proprioception might depend on information from both eye muscle receptors and peripheral retinal vision.

In the second group of postulated roles for extraocular proprioception (those related to vision) some are definitely demonstrated. Thus, Maffei and Fiorentini have shown that information from proprioceptors plays a role in maintaining the binocularity of single neurons in the visual cortex. We have also shown the proprioceptive origin of the role of eye movements as "cofactor" of vision during restoration of orientation specificity in visual cortical neurons.

SUMMARY

The functions of extraocular proprioceptors are reviewed, first for eye motor control and position sense, secondly for visual processes. In the first part, it is pointed out that despite the fact that a monosynaptic stretch reflex has never been found, some reflex actions of extraocular muscles stretch receptors exist. The extraocular proprioceptors are involved in oculomotor control through their projections to structures such as cerebellum and superior colliculus. In particular conditions, these extraocular proprioceptors provide a conscious position sense. In the second part, some observations are reviewed which have led to the assumption that extraocular proprioceptors also have a function in visual processes. Recent experiments are reported which provide evidence for the role of extraocular proprioceptors in the development of orientation specificity in visual cortical neurons.

ACKNOWLEDGEMENTS

This work was supported by grants RCP 348 from C.N.R.S. and CL 76-4-045-6 from I.N.S.E.R.M. I thank Prof. Y. Laporte and Dr. L. Jami for valuable suggestions and observations, Dr. R.T. Kado for the final English form, Mrs. F. Bordeau for the care of the animals, and Mrs. J. Pinot for typing the manuscript.

REFERENCES

Abrahams, V.C. and Rose, R.V. (1975) Projections of extraocular, neck muscle, and retinal afferents to superior colliculus in the cat; their connections to cells of origin of tectospinal tract. *J. Neurophysiol.*, 38: 10–18.

Abrahams, V.C. and Anstee, G. (1978) Units in the superior colliculus and underlying tegmental structures responding to passive eye movement. *J. Physiol. (Lond.)*, 272: 91–92P.

Bach-y-Rita, P. (1972) Extraocular muscle inhibitory stretch reflex during active contraction. *Arch. ital. Biol.*, 110: 1–15.

Baker, R.G. and Precht, W. (1972) Electrophysiological properties of trochlear motoneurons as revealed by IV nerve stimulation. *Exp. Brain Res.*, 14: 124–157.

Baker, R.G., Mano, N. and Shimazu, H. (1969) Post synaptic potentials in abducens motoneurons induced by vestibular stimulation. *Brain Res.*, 15: 577–580.

Baker, R.G., Precht, W. and Llinas, R. (1972) Mossy and climbing fibers projections of extraocular muscle afferents to the cerebellum. *Brain Res.*, 38: 440–445.

Batini, C. and Buisseret, P. (1972) Projections cérébelleuses et trajet périphérique de la proprioception extraoculaire. *C.R. Acad. Sci. (Paris)*, 275: 2711–2713.

Batini, C. and Buisseret, P. (1974) Sensory peripheral pathway from extrinsic eye muscles. *Arch. ital. Biol.*, 112: 18–38.

Batini, C. and Horcholle-Bossavit, G. (1977) Interaction entre activation visuelle et activation proprioceptive au niveau des neurones du colliculus supérieur. *C.R. Acad. Sci. (Paris)*, 285: 1491–1493.

Batini, C., Buisseret, P. and Kado, R.T. (1974) Extraocular proprioceptive and trigeminal projections to the Purkinje cells of the cerebellar cortex. *Arch. ital. Biol.*, 112: 1–17.

Batini, C., Buisseret, P. and Buisseret-Delmas, C. (1975) Trigeminal pathway of the extrinsic eye afferents in cat. *Brain Res.*, 86: 74–78.

Berthoz, A. (1978) Rôle de la proprioception dans le contrôle de la posture et du geste. In *Du Contrôle Moteur à l'Organisation du Geste*, H. Hecaen and M. Jeannerod (Eds.), Masson, Paris, pp. 187–224.

Brindley, G.S. and Merton, P.A. (1960) The absence of position sense in the human eye. *J. Physiol. (Lond.)*, 153: 127–130.

Brindley, G.S., Goodwin, G.M., Kulikowsky, J.J. and Leighton, D. (1976) Stability of vision with a paralysed eye. *J. Physiol. (Lond.)*, 258: 65–66P.

Buisseret, P. and Imbert, M. (1976) Visual cortical cells: their development properties in normal and dark reared kittens. *J. Physiol. (Lond.)*, 255: 511–525.

Buisseret, P. and Maffei, M. (1977) Extraocular proprioceptive projections to the visual cortex. *Exp. Brain Res.*, 28: 421–425.

Buisseret, P. and Gary-Bobo, E. (1979) Development of visual cortical orientation specificity after dark-rearing: role of extraocular proprioception. In preparation.

Buisseret, P., Gary-Bobo, E. and Imbert, M. (1978) Ocular motility and recovery of orientational properties of visual cortical neurons in dark-reared kittens. *Nature (Lond.)*, 272: 816–817.

Chrisman, E. and Kupfer, C. (1963) Proprioceptors in extraocular muscle. *Arch. Ophthal.*, 69: 824–829.

Donaldson, I.M.L. and Long, A.C. (1977) Suppression of visual responses in the cat superior colliculus following stretch of extraocular muscles. *J. Physiol. (Lond.)*, 272: 94–95P.

Easton, T.A (1971) Patterned inhibition from horizontal eye movement in the cat. *Exp. Neurol.*, 31: 419–430.

Easton, T.A. (1972) Patterned inhibition from single eye muscle stretch in the cat. *Exp. Neurol.*, 34: 497–510.

Eklund, G. (1969) Influence of muscle vibration on balance in man. *Acta. Soc. Med. upsalien.*, 74: 113–117.

Fender, D.H. and Nye, P.W. (1961) An investigation of the mechanisms of eye movement control. *Kybernetik*, I: 81–88.

Fiorentini, A. and Maffei, L. (1977) Instability of the eye in the dark and proprioception. *Nature (Lond.)*, 269: 330–331.

Fuchs, A.F. and Kornhuber, H.H. (1969) Extraocular muscle afferents to the cerebellum of the cat. *J. Physiol. (Lond.)*, 200: 713–722.

Gauthier, G.M. and Hofferer, J.M. (1976) Eye tracking of self-moved targets in the absence of vision. *Exp. Brain Res.*, 26: 121–139.

Gernandt, B.E. (1968) Interactions between extraocular myotatic and ascending vestibular activities. *Exp. Neurol.*, 20: 120–134.

Goldberg, S.J., Hull, C.D. and Buchwald, N.A. (1974) Afferent projections in the abducens nerve: an intracellular study. *Brain Res.*, 68: 205–214.

Goldberg, S.J., Lennerstrand, G. and Hull, C.D. (1976) Motor unit responses in the lateral rectus muscle of the cat: intracellular current injection of abducens nucleus neurons. *Acta physiol. scand.*, 96: 58–63.

Goodwin, G.M., McClosey, D.I. and Matthews, P.B.C. (1972) The contribution of muscle afferents to kinaesthesia shown by vibration induced illusions of movements and by effects of paralysing joint afferents. *Brain*, 95: 705–748.

Gordon, B. and Gummow, L. (1975) Effects of extraocular muscle section on receptive fields in cat superior colliculus. *Vision Res.*, 15: 1011–1019.

Imbert, M. and Buisseret, P. (1975) Receptive field characteristics and plastic properties of visual cortical cells in kittens reared with or without visual experience. *Exp. Brain Res.*, 22: 25–36.

Irvine, S.R. and Ludwigh, E.J. (1936) Is ocular proprioceptive sense concerned in vision? *Arch. Ophthal.*, 15: 1037–1049.

Keller, E.L. and Robinson, D.A. (1971) Absence of a stretch reflex in extraocular muscle of the monkey. *J. Neurophysiol.*, 34: 908–919.

McClosey, D.I. and Torda, T.A.G. (1975) Corollary motor discharges and kinaesthesia. *Brain Res.*, 100: 467–470.

McCouch, G.P. and Adler, F.H. (1932) Extraocular reflexes. *Amer. J. Physiol.*, 100: 78–88.

McIntyre, A.K. (1939) The quick component of nystagmus. *J. Physiol. (Lond.)*, 97: 8–16.

Maffei, L. and Bisti, S. (1976) Binocular interaction in strabismic kittens deprived of vision. *Science*, 191: 579–580.

Maffei, L. and Fiorentini, A. (1976) Asymmetry of motility of the eyes and change of binocular properties of cortical cells in adult cats. *Brain Res.*, 105: 73–78.

Maffei, L. and Fiorentini, A. (1977) Oculomotor proprioception in the cat. In *Control of Gaze and Brain Stem Neurons*, R. Baker and A. Berthoz (Eds.), Elsevier/North-Holland Biomed. Press, Amsterdam, pp. 477–481.

Paillard, J. (1973) Proprioception musculaire et le sens de la position. *Arch. ital. Biol.*, 111: 451–461.

Paillard, J. and Brouchon, M. (1968) Active and passive movements in the calibration of position sense. In *The Neuropsychology of Spatially Oriented Behaviour*, S.J. Freedman (Ed.), Dorsey Press, Homewood, pp. 37–55.

Precht, W. (1975) Cerebellar influences on eye movements. In *Basic Mechanisms of Ocular Motility and their Clinical Implications*, G. Lennerstrand and P. Bach-y-Rita (Eds.), Pergamon Press, Oxford, pp. 261–280.

Reinecke, R.D. and Simons, K. (1975) Phoria and EOM afferent; preliminary support for a new theory. In *Basic Mechanisms of Ocular Motility and their Clinical Implications*, G. Lennerstrand and P. Bach-y-Rita (Eds.), Pergamon Press, Oxford, pp. 113–117.

Rose, P.K. and Abrahams, V.C. (1975) The effect of passive eye movement on unit discharge in the superior colliculus of the cat. *Brain Res.*, 97: 95–106.

Sasaki, K. (1963) Electrophysiological studies on oculomotoneurons of the cat. *Jap. J. Physiol.*, 13: 287–302.

Sherrington, C.S. (1893) Further experimental note on the correlation of action of antagonistic muscles. *Proc. roy. Soc. B*, 53: 407–420.

Skavenski, A.A. (1972) Inflow as a source of extraretinal eye position information. *Vision Res.*, 12: 221–229.

Skavenski, A.A. and Steinman, R.M. (1970) Control of eye position in the dark. *Vision Res.*, 10: 193–203.

Stevens, J.K., Emerson, R.C., Gerstein, R.L., Kallos, J., Neufeld, G.R., Nichols, L.W., Rosenquist, A.C. (1976) Paralysis of the awake human: visual perceptions. *Vision Res.*, 16: 93–98.

Von Helmholtz, H. (1866) *Handbuch der Physiologischen Optik*, Woss, Leipzig. (1925) English translation: *Helmholtz's Treatise on Physiological Optics, 3rd ed., Vol. 3*, J.P.C. Southall (Ed.), Optical Society of America, Menasha, Wisconsin.

SECTION IV

LABYRINTHINE INFLUENCES ON THE MOTOR SYSTEM

A

Labyrinthine Receptors and their Influences on the Vestibular Nuclei

Vestibular Receptors in Mammals: Afferent Discharge Characteristics and Efferent Control

J. M. GOLDBERG

University of Chicago, Chicago, IL 60637 (U.S.A.)

In recent years there have been several studies of the discharge characteristics of mammalian vestibular afferents (Goldberg and Fernández, 1971a, b; Fernández and Goldberg, 1971, 1976a–c; Loe et al., 1973; Blanks et al., 1975; Estes et al., 1975; Schneider and Anderson, 1976). On this basis, a reasonably comprehensive picture is available concerning the kinds of information provided to central pathways by the labyrinthine receptors. We shall review these studies and, where possible, draw deductions as to the role of the vestibular end organs in the control of movement and posture. The physiology of the mammalian efferent vestibular system, which is only now being investigated, will also be considered.

GENERAL FEATURES OF THE VESTIBULAR LABYRINTH

There are five sensory regions in the vestibular labyrinth: the cristae ampullares associated with each of the three semicircular canals and the two otolith organs (the utricular and saccular maculae). Before dealing with the separate end organs, it is well to consider some common features of their morphological and functional organization.

Structure of the sensory epithelium and morphological polarization of the hair cells

Each sensory region contains a specialized epithelium, composed of a matrix of sensory and supporting cells (Fig. 1). Two kinds of sensory hair cells are found in mammals and birds (Wersäll and Bagger-Sjöbäck, 1974). The flask-shaped type I hair cell is enveloped by a single calyceal afferent terminal. The type II hair cell is cylindrically shaped and receives bud-shaped terminals from several afferent nerve fibers. The afferent synapses have, for the most part, the morphology of chemically transmitting junctions. Efferent endings, derived from neurons in the brain stem, can be recognized by their highly vesiculated appearance. They terminate presynaptically on type II hair cells and postsynaptically on the calyceal terminals of type I hair cells.

A sensory-hair bundle emerges from the apical or cuticular surface of each hair cell and is made up of 50–100 stereocilia and a single, eccentrically located kinocilium. The sensory hairs are embedded in a gelatinous accessory structure, which is deformed during natural stimulation. The deformation bends the sensory hairs. The eccentric position of the kinocilium defines a morphological polarization which, in turn, determines the directional properties of the hair cell and of the associated

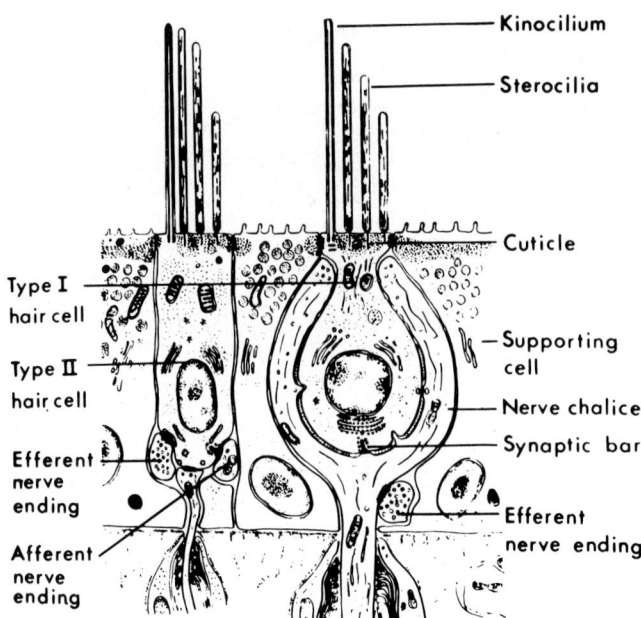

Fig. 1. Ultrastructural organization of sensory epithelium in mammals showing the afferent and efferent innervations of type I and type II hair cells. (From Wersäll and Bagger-Sjöbäck, 1974.)

afferents (Lowenstein and Wersäll, 1959). Movements of the hairs toward the kinocilium increase afferent discharge, oppositely directed movements decrease it. Though the latter response is termed "inhibitory", it is probably the result of a decreased release of excitatory transmitter and, properly speaking, should be considered a disfacilitation.

Resting discharge

Most vestibular afferents are characterized by a resting discharge (Lowenstein, 1956). The background activity in the monkey averages some 90 spikes/sec for semicircular-canal afferents (Goldberg and Fernández, 1971a; Keller, 1976; Louie and Kim, 1976) and 60 spikes/sec for otolith afferents (Fernández and Goldberg, 1976a). Lower values are reported for other mammals (Estes et al., 1975; Schneider and Anderson, 1976). The resting discharge permits the afferents to respond bidirectionally (Lowenstein and Sand, 1940a, b); it reduces or eliminates the sensory threshold (Goldberg and Fernández, 1971a; Fernández and Goldberg, 1976a), and it provides the basis for the tonic influence exerted by the labyrinth on central pathways (Precht et al., 1966).

Fiber caliber and afferent response

Vestibular afferents range in diameter from 1–10 μm (Gacek and Rasmussen, 1961). Thick, medium-sized and thin afferents have different innervation patterns (Lorente de Nó, 1926; Wersäll, 1956; Lindeman, 1969). Thick fibers supply calyceal endings to few neighboring type I hair cells. Medium-sized fibers innervate several hair cells of both varieties. Thin fibers are thought to make bud-shaped contacts with many, widely seprated type II hair cells. Thick fibers supply central regions of the sensory epithelium. Medium-sized and thin fibers form the predominant innervation of peripheral parts of the epithelium and may also contribute to the central zone.

Conduction-time studies demonstrate that thick and thin afferents have different discharge patterns (Goldberg and Fernández, 1977a; Yagi et al., 1978). In thin fibers, there is a regular spacing of action potentials; in thick fibers, the spacing is irregular. Units first characterized as regularly or irregularly discharging differ in other respects as well (Goldberg and Fernández, 1971b; Blanks et al., 1975; Fernández and Goldberg, 1976a, c; Schneider and Anderson, 1976). Regularly discharging (thin) afferents may be considered tonic receptors, their response being closely related to the expected displacement of the accessory structure. Irregularly discharging (thick) afferents are phasic-tonic in their properties; they adapt to prolonged stimulation and they may be sensitive to both the velocity and the displacement of the accessory structure. A related difference concerns the sensitivity to natural stimulation. Thick afferents tend to be somewhat more sensitive than thin afferents to sinusoidal stimuli near 0.1 Hz. The difference in sensitivities grows as frequency is increased, so that at 5 Hz the thick afferents may have gains 5–10 times those of thin afferents. Most voluntary head movements, it should be noted, take place in a frequency range centered about 5 Hz.

SEMICIRCULAR CANAL AFFERENTS

Canal neurons are sensitive to angular accelerations of the head. The afferents can also be affected by constant linear forces (Lowenstein, 1972; Estes et al., 1975), though it may be doubted whether this would occur under entirely physiological circumstances (Goldberg and Fernández, 1975). The response to rotational stimulation is bidirectional (Fig. 2). The directions of excitatory and inhibitory accelerations are the same for all afferents innervating a single canal (Lowenstein and Sand, 1940a; Goldberg and Fernández, 1971a) and are consistent with the morphological polarization of the hair cells (Lowenstein and Wersäll, 1959). A particular canal is only affected by accelerations acting within its plane (Estes et al., 1975). Any acceleration which excites a canal on one side will inhibit the parallel canal of the opposite side. Since the three sets of parallel canals are almost orthogonal to one another, precise information is provided about the magnitude and orientation of any arbitrarily directed angular acceleration.

The mechanics of the semicircular canals may be likened to a heavily damped torsion pendulum (Steinhausen, 1931, 1933). The response to constant angular accelerations should then be largely determined by a single first-order time constant. Some afferents behave in this way (Fig. 2A, B). The time constant of response is about 5 sec. For other units, though, there are major discrepancies from the predictions of the torsion-pendulum model. One discrepancy involves a form of sensory adaptation, reflected by the presence of perstimulus response declines during long accelerations (Fig. 2D) and poststimulus secondary responses following even short accelerations (Fig. 2C). These effects, which are most conspicuous in irregularly discharging afferents (Goldberg and Fernández, 1971b; Blanks et al., 1975), bear a formal resemblance to the response declines and secondary responses seen in human studies of vestibular-induced sensation and nystagmus (Guedry, 1974). Adaptation, it should be noted, would not come into play during naturally occurring head movements.

The afferents' response dynamics deviate in a second way from those of the torsion-pendulum model. The discrepancy is seen during sinusoidal stimulation. Fig. 3

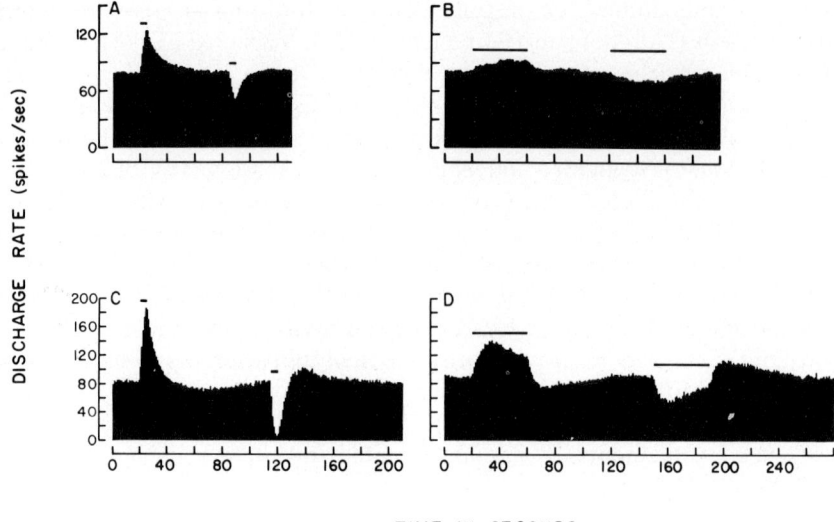

Fig. 2. Responses of semicircular-canal afferents in the squirrel monkey to oppositely directed steps of constant angular acceleration separated by periods of no acceleration. Each stimulus is a velocity trapezoid consisting of the following 5 periods: 1) head stationary; 2) constant angular acceleration (left-hand bar); 3) constant angular velocity; 4) constant angular deceleration (right-hand bar); 5) head stationary. A, B) Responses of a regularly discharging afferent, respectively, to 5-sec, 60 deg/sec^2 and to 40-sec, 7.5 deg/sec^2 steps of acceleration and deceleration. C, D) Responses of an irregularly discharging afferent to the same stimulus conditions. (From Goldberg and Fernández, 1971a.)

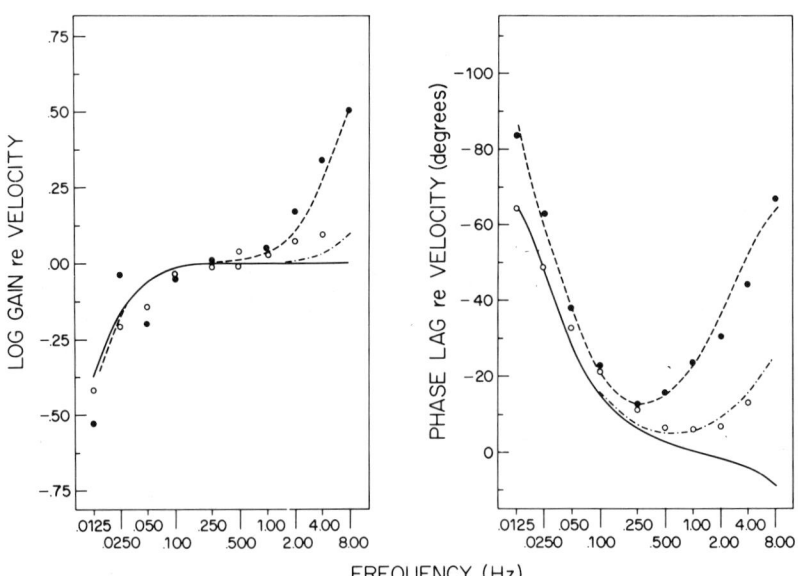

Fig. 3. Responses of semicircular-canal afferents in the squirrel monkey to sinusoidal head rotations. Data are for a regularly discharging (○) and an irregularly discharging (●) unit. Log gains re velocity normalized to gain at 0.25 Hz. Phase leads are shown as negative phase lags. Solid lines are gains and phases predicted from the torsion-pendulum model; the other two curves are based on empirical transfer functions. (From Fernández and Goldberg, 1971a.)

shows the expected response of the model (solid curves), as well as data from two afferents. Based solely on its predicted mechanical behavior, the canal should function as an angular-velocity transducer. The predicted gain re velocity is constant above 0.1 Hz and the expected phase is close to zero. In contrast, the observed gains, rather than remaining constant, begin to increase for frequencies greater than 1 Hz. Corresponding to the gain enhancement, the phase lead reaches a minimum and then starts increasing again. Both the gain enhancement and the phase lead are more conspicuous in irregular units (Goldberg and Fernández, 1971b; Keller, 1976; Louie and Kim, 1976; Schneider and Anderson, 1976). In the squirrel monkey, the two effects can be simulated by assuming that the afferents' response is linearly related to the velocity, as well as to the displacement of the accessory structure (Fernández and Goldberg, 1971). It has been suggested that the velocity sensitivity could compensate for the phase lags and gain attenuations characterizing the high-frequency response of various reflex pathways (Fernández and Goldberg, 1971; Skavenski and Robinson, 1973).

OTOLITH AFFERENTS

The saccular and utricular maculae are both sensors of linear forces acting on the head. Otolith afferents do not respond to even intense angular accelerations (Goldberg and Fernández, 1975).

Functional polarization and response to static tilts

The two otolith organs are disposed at almost right angles to one another. The utricular macula lies in a horizontal plane, the saccular macula in a parasagittal plane (Lindeman, 1969). Each macula is divided in two by a morphologically distinctive zone, the striola. The hair cells to either side of the striola have opposing morphological polarizations (Fig. 4, lower right). The disposition of the two maculae and the arrangement of the hair cells within them have implications for the directional properties of the afferents.

Most mammalian otolith neurons respond in a maintained manner to static tilts (Fernández et al., 1972; Loe et al., 1973; Fernández and Goldberg, 1976a). Fig. 4 shows how the discharge of otolith neurons can be modulated when the head is held in various positions. The modulation takes place around a resting (or zero-force) discharge, which for the unit in Fig. 4B amounts to 80 spikes/sec. The response, taken as the difference between the discharge (d) and the resting discharge (d_0), is an approximately trigonometric function of tilt angle. This is consistent with a linear model. Suppose that each afferent is characterized by a functional polarization vector of unit length. Then were linearity assumed, the response would be

$$d - d_0 = s \cos \theta = sF.$$

θ is the angle between the polarization vector and the gravity vector, s is a sensitivity factor, and $F = \cos \theta$ is the effective force. From this equation, polarization vectors can be calculated from static-tilt data (for details, see Fernández et al., 1972).

The eight units of Fig. 4A–H innervate the utricular macula. In each case, the discharge is close to its resting value when the animal is in a horizontal position and approaches a maximum or minimum when the animal is tilted 90° from the horizontal.

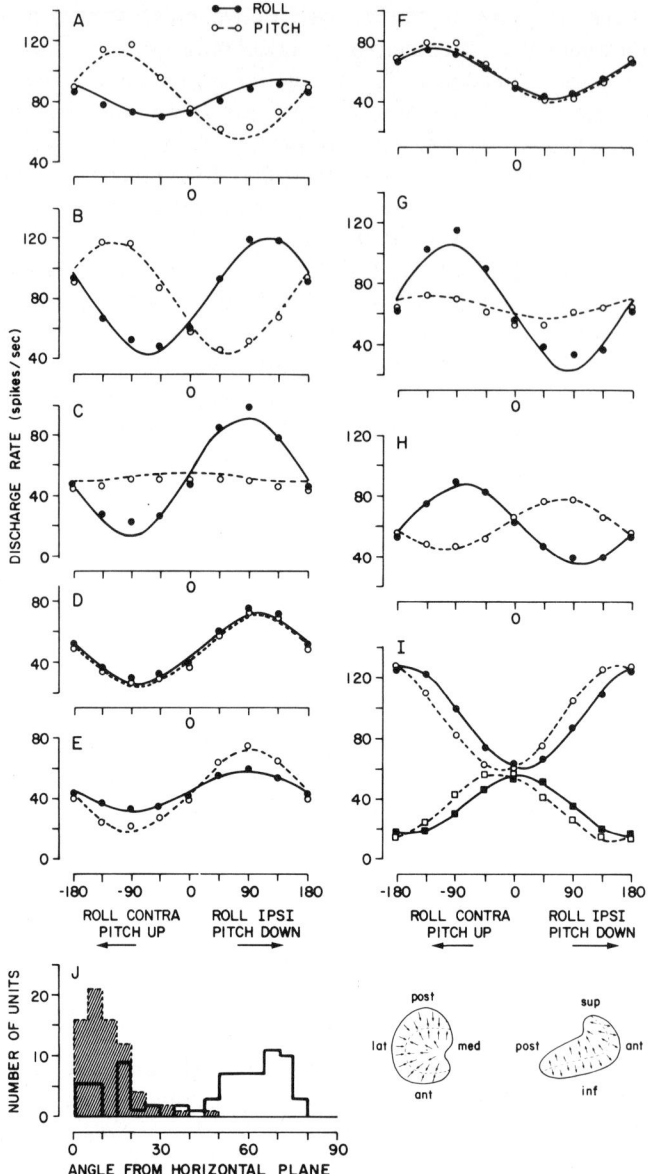

Fig. 4. Responses of otolith afferents in the squirrel monkey to static tilts. A-H) separate utricular afferents; I) 2 saccular afferents. Each point based on a 10-sec sample of steady-state activity. Curves are best-fitting trigonometric functions. Abscissa shows tilt in degrees. J) Distribution of functional polarization vectors for utricular afferents (shaded bars) and saccular afferents (open bars). (Modified from Fernández and Goldberg, 1976a.) Lower right: morphological polarization maps for utricular and saccular maculae of the squirrel monkey are seen, resp., to the left and right. (From Lindeman, 1969.)

These observations imply that gravity is effective only when it acts within the plane of the macula. In short, the utricular end organ is only sensitive to shearing forces. The conclusion is confirmed when variously directed centrifugal forces are used (Fernández and Goldberg, 1976b). Utricular units vary in their directional properties. Some are excited by ipsilateral rolls, others by contralateral rolls, some by downward pitches, others by upward pitches. This diversity is entirely consistent with the morphological polarization map (Fig. 4, lower right). The two units of Fig. 4I are related to the saccular macula. For these, the horizontal positions are points of maximum or minimum discharge, whereas 90° tilts bring the discharge near d_0. Hence, the parasagittally oriented saccular macula also responds only to shearing forces. The two saccular units have static-tilt curves which are mirror images of one another. This reflects the parallel orientation of the hair cells to either side of the striola (Fig. 4, lower right).

The differences in the directional properties of saccular and utricular afferents can be summarized in terms of their functional polarization vectors (Fig. 4J). Most utricular vectors lie within 15° of the horizontal plane, whereas most saccular vectors are disposed more than 50° from this plane (or equivalently, within 40° of the vertical). As a result, utricular units are sensitive to small tilts from the normal head position and to linear forces arising during locomotion in the horizontal plane. Saccular units, in contrast, are sensitive to dorsoventral accelerations, such as occur in jumps and falls. The utricular macula, given the broad distribution of its vectors, can signal information about any arbitrarily directed force acting in the horizontal plane. To the extent that the utricular vectors are confined to this plane, the sensory representation is two dimensional. The saccular macula provides the third dimension.

Response dynamics

Regularly discharging otolith afferents are tonic receptors (Vidal et al., 1971; Goldberg and Fernández, 1976a, c). Their response parallels the applied force, during prolonged stimulation (Fig. 5B) and also during abrupt force transitions (Fig. 6C, D). Irregularly discharging afferents behave in a more phasic manner. They adapt to maintained stimulation (Fig. 5A) and their response to rapid force transitions provides evidence of a velocity sensitivity (Fig. 6A, B).

It may be concluded that the otolith organs subserve both tonic and phasic functions. The two modes of otolith activation may be related to the dual role of these end organs, viz., to monitor static head position and also the dynamic accelerations arising during movement.

EFFERENT VESTIBULAR SYSTEM (EVS)

Peripheral action

In all other hair-cell systems studied, including the mammalian cochlea (Wiederhold and Kiang, 1970) and lateral lines (Russell, 1974), electrical stimulation of efferent pathways results in an inhibition of afferent activity. The recent localization of EVS neurons (Gacek and Lyon, 1974; Warr, 1975) has permitted us to do comparable experiments in the vestibular system (Goldberg and Fernández, 1977b). Remarkably, the predominant action of the EVS is excitatory.

The response of an irregularly discharging afferent to tetanic stimulation of the EVS is seen in Fig. 7A. Abrupt changes in activity occur at the onset and termination of the shock train. During stimulation, there is a gradual build-up of activity. Following the train, the discharge does not immediately return to background levels. Rather there is a persistent excitatory response lasting some 30 sec. The rapid changes in discharge reflect a *fast* response component. The gradual perstimulus build-up and poststimulus decline indicate the presence of a *slow* response component. Except for

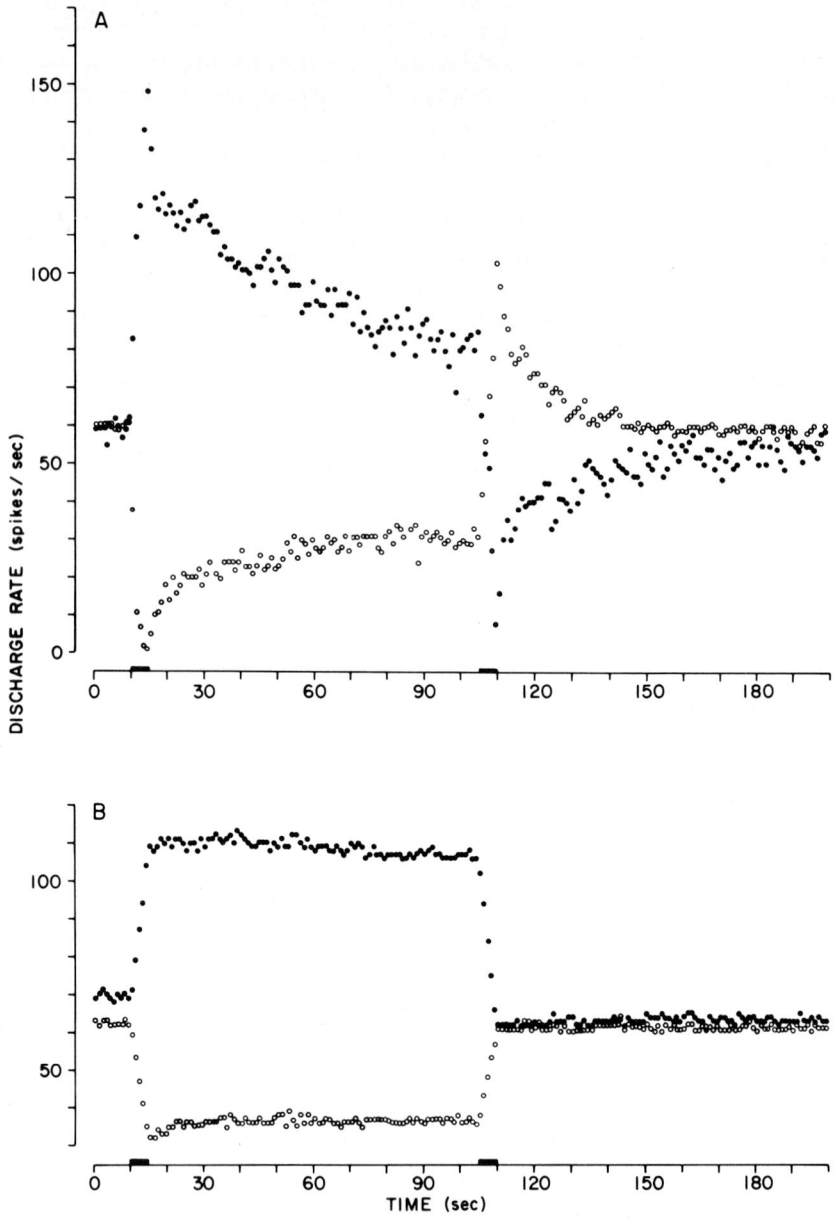

Fig. 5 Responses of otolith afferents in the squirrel monkey to centrifugal-force trapezoids of long duration. A) Irregularly discharging afferent. B) Regularly discharging afferent. Each stimulus consists of the following 5 periods: 1) background force (0.077 g); 2) linear force transition (left-hand bar); 3) constant force (1.23 g); 4) linear force transition (right-hand bar); 5) background force. Responses to excitatory (●) and inhibitory (○) forces directed, respectively, parallel and antiparallel to each unit's functional polarization vector. (Based on Fernández and Goldberg, 1976a.)

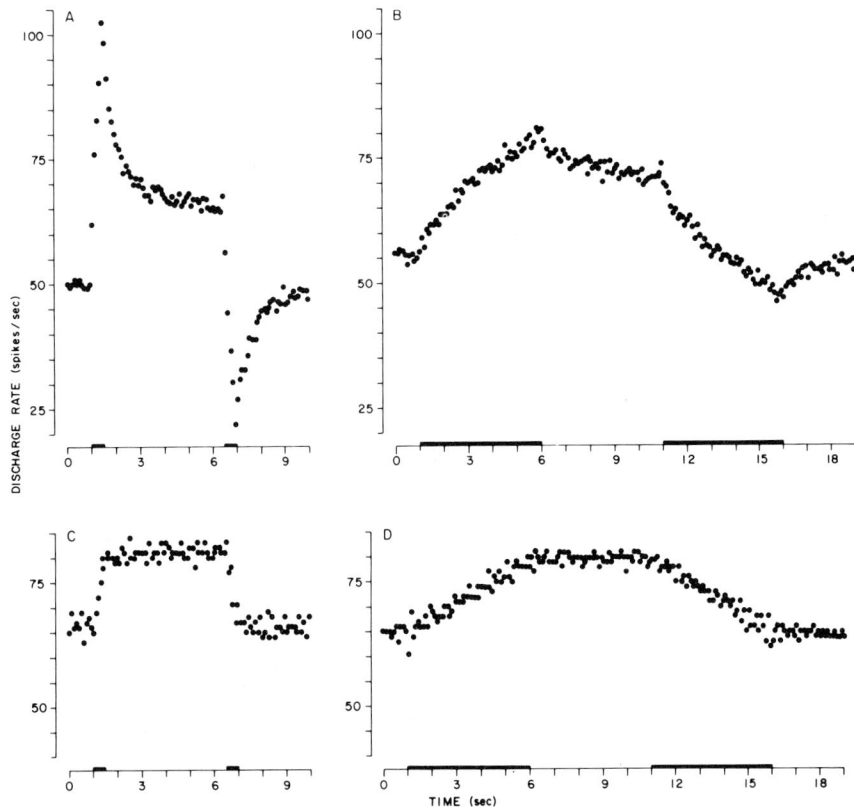

Fig. 6. Response to short-duration centrifugal-force trapezoids, same units as in Fig. 5. A, B) The irregularly discharging afferent shows a phasic response, which is larger during 0.5-sec force transitions (A) than during 5-sec transitions (B). The background and peak forces are each identical for the two trapezoids. C, D) The response of the regularly discharging afferent is determined by the instantaneous force, not by the rate of force application. (Based on Fernández and Goldberg, 1976c.)

the fact that it is excitatory, the *fast* component resembles the efferent responses seen in other hair-cell systems. The *slow* response has a number of unusual features and these would suggest that it is not mediated by a conventional synaptic process.

There is a relation between the magnitude of the EVS responses and the afferents' regularity of discharge. The response of a regularly discharging unit is shown in Fig. 7B. As is typical, the response is 10–20 times smaller than that usually observed in irregular units (Fig. 7A) and further, it is almost entirely composed of a *slow* component. The results may reflect differences in the actions of the postsynaptic efferent innervation of afferent chalices and the presynaptic efferent innervation of type II hair cells. Presumably, the irregular units receive a postsynaptic innervation, the regular units a presynaptic innervation.

Functional considerations

Any discussion of the function of the mammalian EVS must be speculative. One reason is that, given the unphysiological nature of electrical stimulation, only limited deductions can be made from our own studies. Moreover, virtually nothing is known concerning the discharge characteristics of mammalian EVS neurons or about possible efferent modifications of afferent discharge under physiological conditions.

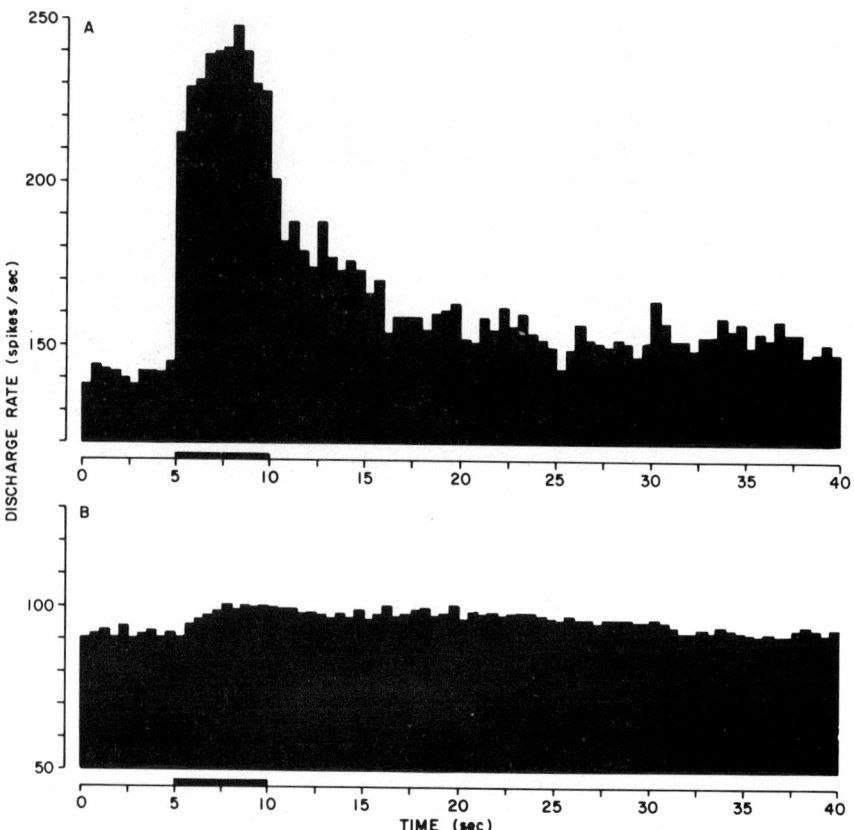

Fig. 7. Response of two vestibular afferents to electrical stimulation of brain-stem efferent pathways in the squirrel monkey. Stimulus (bar) consisted of 5-sec trains of 0.1-msec shocks, 40 µA, delivered at 333/sec. A) Irregular afferent. B) Regular afferent. Both units obtained in same preparation.

In this circumstance, it is helpful to consider the lateral lines of lower vertebrates, since this is the one hair-cell system in which there is some understanding of efferent control (Russell, 1974). Lateral lines are exteroceptors, which are potentially excited by the animal's own movement. The lateral-line efferents fire during movement and they function to prevent self-excitation by inhibiting afferent discharge.

It is unlikely that the EVS works in precisely the same way since the vestibular end organs are proprioceptors, whose function is to monitor the animal's own head movements. Klinke and Schmidt (1970) did propose that the EVS suppresses afferent discharge during movement. That this is not the case would seem proved by the observation that the vestibulo-ocular reflex and, by inference, the vestibular end organs have normal gains during voluntary head movements (Dichgans et al., 1973). As an alternative, it may be suggested that during movements the EVS activates the labyrinths on the two sides. Such a bilateral activation, while it would not lead to vestibular reflex movements or sensation, could improve the performance of the receptors.

The vestibular afferents have a potential dynamic range of 0–400 spikes/sec (Goldberg and Fernández, 1971a). Though the afferents have a high resting discharge, they usually work in the lower part of the range (Goldberg and Fernández,

1971a; Fernández and Goldberg, 1976a, b). The result is that it is easier to silence units than it is to drive them into excitatory saturation. A silencing or inhibitory saturation is of no concern during the small, passive head accelerations that occur when an animal is standing still and the vestibular system is functioning to maintain postural stabilization. But the large accelerations accompanying intended movements could easily lead to afferent silencing and, hence, render the operation of the vestibular system non-linear. A bilateral activation of the EVS, presumably triggered from premotor centers, would reduce or eliminate the non-linearity. At the same time, there would be a symmetric increase in the excitatory drive on vestibular reflex pathways and this should insure that they respond appropriately.

Stated somewhat more generally, it is envisaged that the EVS functions to modulate the dynamic range of the afferents to match expected accelerations. The alternative solution of raising the afferents' resting discharge may simply be too costly in terms of the metabolic requirements of the labyrinth. This notion, though speculative, has two attractions. First, it provides a rationale for the excitatory action of the EVS. And second, it offers a functional explanation as to why the EVS exerts most of its influence on irregularly discharging units. These afferents, being more sensitive than regular units, are more readily driven into inhibitory saturation.

SUMMARY

Vestibular afferents have a resting discharge. This permits the afferents to respond bidirectionally, it reduces or eliminates the sensory threshold, and it provides a tonic input to central pathways. Thick and thin vestibular axons have different discharge characteristics. Thin afferents, which innervate type II hair cells, are characterized by a regular spacing of action potentials and they are tonic receptors. Thick afferents supply type I hair cells; the spacing of action potentials is irregular and the receptors are more phasic in their behavior. Thick (irregular) units are also more sensitive to natural stimuli, particularly to those whose spectrum falls within the bandwidth of naturally occurring head movements. *Semicircular-canal afferents* respond to angular accelerations acting within the plane of the corresponding canal. The three sets of parallel canals provide precise information about the magnitude and direction of any arbitrarily directed angular acceleration. The response dynamics of the afferents deviate from the expected mechanical response of the canals. The afferents show a sensory adaptation and a velocity sensitivity. *Otolith afferents* respond to linear forces. The utricular macula is potentially sensitive to any linear force acting within a horizontal plane, whereas the saccular macula is sensitive to dorsoventrally oriented forces. The otolith organs subserve both tonic and dynamic functions, which may be related to their dual role of monitoring static head position and also the dynamic accelerations arising during movement. Electrical stimulation of the *efferent vestibular system* (EVS) results in an excitation, which is particularly prominent in irregularly discharging (thick) afferents. It is proposed that the EVS functions to modulate the dynamic range of the afferents to match the excepted accelerations occurring during movement.

ACKNOWLEDGEMENTS

The author's research is supported by Grant NS 01330 from the National Institutes of Health and by Grant NGR-14-001-225 from the National Aeronautics and Space Administration.

REFERENCES

Blanks, R.H.I., Estes, M.S. and Markham, C.H. (1975) Physiological characteristics of vestibular first-order canal neurons in the cat. II. Response to constant angular acceleration. *J. Neurophysiol.*, 38: 1250–1268.
Dichgans, J., Bizzi, E., Morasso, P. and Tagliasco, V. (1973) Mechanisms underlying recovery of eye-head coordination following bilateral labyrinthectomy in monkeys. *Exp. Brain Res.*, 18: 548–562.
Estes, M.S., Blanks, R.H.I. and Markham, C.H. (1975) Physiological characteristics of vestibular first-order canal neurons in the cat. I. Response plane determination and resting discharge characteristics. *J. Neurophysiol.*, 38: 1232–1249.
Fernández, C. and Goldberg, J.M. (1971) Physiology of peripheral neurons innervating semicircular canals of the squirrel monkey. II. Response to sinusoidal stimulation and dynamics of peripheral vestibular system. *J. Neurophysiol.*, 34: 661–675.
Fernández, C. and Goldberg, J.M. (1976a) Physiology of peripheral neurons innervating otolith organs of the squirrel monkey. I. Response to static tilts and to long-duration centrifugal force. *J. Neurophysiol.*, 39: 970–984.
Fernández, C. and Goldberg, J.M. (1976b) Physiology of peripheral neurons innervating otolith organs of the squirrel monkey. II. Directional selectivity and force-response relations. *J. Neurophysiol.*, 39: 985–995.
Fernández, C. and Goldberg, J.M. (1976c) Physiology of peripheral neurons innervating otolith organs of the squirrel monkey. III. Response dynamics. *J. Neurophysiol.*, 39: 996–1008.
Fernández, C., Goldberg, J.M. and Abend, W.K. (1972) Response to static tilts of peripheral neurons innervating otolith organs of the squirrel monkey. *J. Neurophysiol.*, 35: 978–997.
Gacek, R.R. and Rasmussen, G.L. (1961) Fiber analysis of statoacoustic nerve of guinea pig, cat, and monkey. *Anat. Rec.*, 139: 455–463.
Gacek, R.R. and Lyon, M. (1974) The localization of vestibular efferent neurons in the kitten with horseradish peroxidase. *Acta oto-laryng. (Stockh.)*, 77: 92–101.
Goldberg, J.M. and Fernández, C. (1971a) Physiology of peripheral neurons innervating semicircular canals of the squirrel monkey. I. Resting discharge and response to constant angular accelerations. *J. Neurophysiol.*, 34: 635–660.
Goldberg, J.M. and Fernández, C. (1971b) Physiology of peripheral neurons innervating semicircular canals of the squirrel monkey. III. Variation among units in their discharge properties. *J. Neurophysiol.*, 34: 676–684.
Goldberg, J.M. and Fernández, C. (1975) Responses of peripheral vestibular neurons to angular and linear accelerations in the squirrel monkey. *Acta oto-laryng., (Stockh.)*, 80: 101–110.
Goldberg, J.M. and Fernández, C. (1977a) Conduction times and background discharge of vestibular afferents. *Brain Res.*, 122: 545–550.
Goldberg, J.M. and Fernández, C. (1977b) Efferent vestibular system in the squirrel monkey. *Neurosci. Abstr.*, 3: 543.
Guedry, F.E., Jr., (1974) Psychophysics of vestibular sensation. In *Handbook of Sensory Physiology, VI, Vestibular System, Part 2, Psychophysics, Applied Aspects and General Interpretations*, H.H. Kornhuber (Ed.), Springer-Verlag, Berlin, pp. 3–154.
Keller, E.L. (1976) Behavior of horizontal semicircular canal afferents in alert monkey during vestibular and optokinetic stimulation. *Exp. Brain Res.*, 24: 459–471.
Klinke, R. and Schmidt, C.L. (1970) Efferent influence on the vestibular organ during active movements of the body. *Pflügers Arch.*, 318: 325–332.
Lindeman, H.H. (1969) Studies on the morphology of the sensory regions of the vestibular apparatus. *Ergebn. Anat. Entwickl.-Gesch.*, 42: 1–113.
Loe, P.R., Tomko, D.L. and Werner, G. (1973) The neural signal of angular head position in primary afferent vestibular nerve axons. *J. Physiol. (Lond.)*, 230: 29–50.
Lorente de Nó, R. (1926) Études sur l'anatomie et la physiologie du labyrinthe de l'oreille et du VIIIe nerf. Deuxième partie. *Trav. Lab. Invest. biol. Univ. Madr.*, 24: 53–153.
Louie, A.W. and Kim, J. (1976) The response of 8th nerve fibers to horizontal sinusoidal oscillations in the alert monkey. *Exp. Brain Res.*, 24: 447–457.
Lowenstein, O. (1956) Peripheral mechanisms of equilibrium. *Brit. med. Bull.*, 12: 114–118.
Lowenstein, O. (1972) Physiology of vestibular receptors. In *Progress in Brain Research, Vol. 37, Basic Aspects of Central Vestibular Mechanisms*, A. Brodal and O. Pompeiano (Eds.), Elsevier, Amsterdam, pp. 19–30.

Lowenstein, O. and Sand, A. (1940a) The individual and integrated activity of the semicircular canals of the elasmobranch labyrinth. *J. Physiol. (Lond.)*, 99: 89–101.

Lowenstein, O. and Sand, A. (1940b) The mechanism of the semicircular canal. A study of the responses of single-fibre preparations to angular accelerations and to rotation of constant speed. *Proc. roy. Soc., B*, 129: 256–275.

Lowenstein, O. and Wersäll, J. (1959) A functional interpretation of the electron-microscopic structure of the sensory hairs in the cristae of the elasmobranch *Raja clavata* in terms of directional sensitivity. *Nature (Lond.)*, 184: 1807–1808.

Precht, W., Shimazu, H. and Markham, C.H. (1966) A mechanism of central compensation of vestibular function following hemilabyrinthectomy. *J. Neurophysiol.*, 29: 996–1010.

Russell, I.J. (1976) Amphibian lateral line receptors. In *Frog Neurobiology. A Handbook*, R. Llinás and W. Precht (Eds.), Springer-Verlag, Berlin, pp. 513–550.

Schneider, L.W. and Anderson, D.J. (1976) Transfer characteristics of first and second order lateral canal vestibular neurons in gerbil. *Brain Res.*, 112: 61–76.

Skavenski, A.A. and Robinson, D.A. (1973) Role of abducens neurons in vestibulo-ocular reflex. *J. Neurophysiol.*, 36: 724–738.

Steinhausen, W. (1931) Über den Nachweis der Bewegung der Cupula in der intakten Bogengansampulle des Labyrinthes bei der natüralichen rotatorischen und calorischen Reizung. *Pflügers Arch.*, 228: 322–328.

Steinhausen, W. (1933) Über die Beobachtung der Cupula in den Bogengansampullen des Labyrinthes des levenden Hechts. *Pflügers Arch.*, 232: 500–512.

Vidal, J., Jeannerod, M., Lifschitz, W., Levitan, H., Rosenberg, J. and Segundo, J.P. (1971) Static and dynamic properties of gravity-sensitive receptors in the cat vestibular system. *Kybernetik*, 9: 205–215.

Warr, W.B. (1975) Olivocochlear and vestibular efferents neurons of the feline brain stem: their location, morphology and number determined by retrograde axonal transport and acetylcholinesterase histochemistry. *J. comp. Neurol.*, 161: 159–182.

Wersäll, J. (1956) Studies on the structure and innervation of the sensory epithelium of the cristae ampullaris in the guinea pig. A light and electron microscopic investigation. *Acta oto-laryng. (Stockh.)*, Suppl. 126: 1–85.

Wersäll, J. and Bagger-Sjöbäck, D. (1974) Morphology of the vestibular sense organ. In *Handbook of Sensory Physiology, VI, Vestibular System, Part 1, Basic Mechanisms*, H.H. Kornhuber (Ed.) Springer-Verlag, Berlin, pp. 123–170.

Wiederhold, M.L. and Kiang, N.Y.S. (1970) Effects of electric stimulation of the crossed olivocochlear bundle on single auditory-nerve fibers in the cat. *J. acoust. Soc. Amer.*, 48: 950–965.

Yagi, T., Simpson, N.E. and Markham, C.H. (1977) The relationship of conduction velocity to other physiological properties of the cat's horizontal canal neurons. *Exp. Brain Res.*, 30: 587–600.

Labyrinthine Influences on the Vestibular Nuclei

W. PRECHT

Max-Planck-Institut für Hirnforschung, Neurobiologische Abteilung, D-6000 Frankfurt/M.-Niederrad (F.R.G.)

The vestibular nuclei form a voluminous nuclear complex in the brain stem in which information arising from the vestibular receptors and other sensory systems such as the somatosensory and visual systems as well as that arising from central motor systems such as the cerebellum and various other supranuclear motor centers is integrated. Through their efferents the vestibular nuclei communicate with many other central systems related to the regulation of posture, locomotion and oculomotion as well as the perception of space. These properties render the vestibular nuclei one of the most interesting structures for the study of sensory motor integration. In this paper, the vestibular canal and otolith inputs to the vestibular nuclei will be reviewed. Other aspects of the system outlined above will be treated by others in this volume. Since there exist several recent review articles that summarize the vestibular input data accumulated in the last decades (Precht, 1974, 1975, 1976, 1978, 1979; Goldberg and Fernández, 1975; Wilson and Peterson, 1978) this account will give only a brief summary of the previous data and rather expand somewhat more on most recent developments.

Quantitatively speaking, the labyrinthine afferents to the vestibular nuclei form the largest input group and their action upon secondary vestibular neurons is certainly of paramount interest for the understanding of the higher order vestibular system since, except for the direct projection to the cerebellum (cf. Precht, 1978), all afferents synapse in the vestibular nuclei.

In the following, I shall start with a brief recapitulation of the main data obtained with electrical stimulation of the VIIIth nerve while recording extra- and intracellularly from vestibular neurons of various species (cat, monkey, frog). This account will be followed by the description of the responses of vestibular neurons to natural stimulation of the semicircular canals as well as otolith organs.

FUNCTIONAL SYNAPTOLOGY AND TOPOGRAPHY OF VESTIBULAR INPUT

Ipsilateral input

Stimulation of the whole VIIIth nerve with brief electrical pulses evokes in the vestibular nuclei of the cat (Shimazu and Precht, 1965), frog (Precht et al., 1974), monkey (Keller and Kamath, 1975) and rabbit (Akaike et al., 1973) typical field potentials that serve as convenient guides for electrode location within the vestibular nuclei during physiological experiments. When single units are recorded, correlation of their response latencies with the various components of the field potential may be used for identification of the units as pre- or postsynaptic (mono- and polysynaptic) in nature (Precht and Shimazu, 1965; Precht et al., 1974; Keller and Kamath, 1975).

As regards the termination sites of the various primary afferents in the vestibular nuclei the results obtained with combined electrical and natural stimulation of *cat* semicircular canals and otolith organs (Shimazu and Precht, 1965; Precht and Shimazu, 1965; Peterson, 1970; Markham, 1968; Hwang and Poon, 1975; Chan et al., 1977) agree well with those obtained from anatomical work (Gacek, 1969): fibers innervating the cristae of the semicircular canals terminate mainly in the superior and rostral part of the medial and descending vestibular nuclei as well as the interstitial nucleus of the VIIIth nerve, whereas utricular fibers end, for the most part, in the rostral descending and medial nuclei and rostroventral parts of the lateral nuclei; saccular afferents terminate mainly in cell group "y" and also in lateral and descending nuclei. The topography of vestibular projections in the *rabbit* is very similar to that in the cat although the anatomical work is not as complete as in the other species discussed (Duensing and Schaefer, 1958, 1959; Ito et al., 1977). As for the termination of canal afferents in the *monkey* both anatomical (Stein and Carpenter, 1967) and physiological (Fuchs and Kimm, 1975; Keller and Kamath, 1975) studies have revealed distributions similar to those described in the cat and rabbit. The physiology of otolith projection is not known in the monkey. Finally, in the *frog* the various subdivisions of the vestibular nuclei described in mammals are too indistinct to allow a precise anatomicofunctional correlation (Gregory, 1972; Precht et al., 1974).

Anatomical studies (Walberg et al., 1958, Gacek, 1969) indicated that many neurons in the vestibular nuclei did not receive primary vestibular afferents. The functional correlate to this finding is the exclusively polysynaptic activation of many tonically active vestibular neurons after VIIIth nerve stimulation (Precht and Shimazu, 1965). Besides the pure polysynaptic and pure monosynaptic units other neurons were found in the horizontal canal system: type I neurons which fired with polysynaptic latencies after weak stimuli and had additional monosynaptic discharge with strong stimulation, and neurons which fired monosynaptically and polysynaptically after weak and strong stimulation, respectively. Finally, some neurons located in the vestibular nuclei were not driven at all. Similar connectivities may exist in vertical canal (Markham, 1968) and otolith projections (Peterson, 1970). Polysynaptic responses are, at least in part, mediated by intranuclear interneurons and/or axon collaterals of directly driven cells. Some important functional implications of the relative importance of mono- and polysynaptic vestibular activation will be discussed in the next section.

Intracellular studies of cat vestibular neurons show that the vestibular input may generate mono- or polysynaptic EPSPs or both (Ito et al., 1969; Kawai et al., 1969). The double firing seen in many vestibular neurons after VIIIth nerve stimulation (Precht and Shimazu, 1965) is thus caused by a sequence of mono- and polysynaptic EPSPs. The synaptic and action current generated by these EPSPs produce the N1 and N2 field potential typically recorded in the vestibular nuclei (Shimazu and Precht, 1965). It is interesting to note that the unitary components of the vestibular EPSPs recorded in different vestibular neurons vary. Thus, non-Deiters' neurons have significantly larger unitary EPSPs as compared to Deiters' neurons. This would suggest that primary vestibular impulses activate some vestibular neurons more effectively than others. In general, however, the transmission from primary afferents is very powerful: many neurons can follow double stimuli with intervals of less than 1 msec with full action potentials. Also of interest is the fact that the earliest latencies of vestibular, evoked EPSPs and action potentials measure 0.6 and 0.7 msec, respectively. Occasionally, polysynaptic IPSPs have also been evoked in vestibular neurons. They may be generated by collaterals of inhibitory secondary neurons.

Contrary to the frog (Precht et al., 1974), fish (Korn et al., 1977), pigeon (Wilson and Wylie, 1970) and rat (Korn et al., 1973) there is no convincing demonstration that cat vestibular neurons are electrotonically coupled or electrically activated by primary afferents. Likewise, dendritic spikes have not been observed in this species. Partial action potentials or dendritic spikes are frequently encountered in frog vestibular neurons and they may be evoked by ipsi- or contralateral VIIIth nerve stimulation (Precht et al., 1974; Ozawa et al., 1974; Dieringer and Precht, 1977). The functional implications of these species differences in the properties of vestibular neurons are presently not well understood.

Contralateral input

The crossed connections between the bilateral, vestibular nuclei have been studied in the cat (Shimazu and Precht, 1966; Precht et al., 1967; Ladpli and Brodal, 1968; Kasahara et al., 1968; Mano et al., 1968; Kasahara and Uchino, 1971; Shimazu and Smith, 1971; Pompeiano et al., 1978), frog (Ozawa et al., 1974; Dieringer and Precht, this volume, chapter IVE4), reptiles (Richter et al., 1975) and monkey (Abend, 1978). The results of these studies will be briefly summarized. In the cat the commissural neurons may be excitatory as well as inhibitory in nature (Fig. 1). Crossed excitation is found in neurons (type II), that mediate inhibition of central canal neurons. Thus, in the canal system the inhibitory type II neurons are excited and, in turn, inhibit the main sensory type I neurons as shown in detail in the circuit diagram of Fig. 1. The latter, however, may also be inhibited directly by commissural fibers (Fig. 1). It was found that this crossed inhibition exists only between functionally synergistic pairs of canals, i.e., bilateral horizontal canals, left anterior and right posterior and right anterior and left posterior canals. Recent studies have

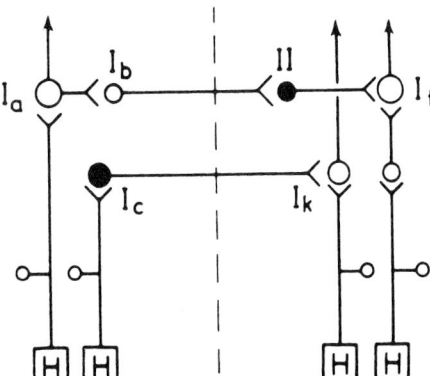

Fig. 1. Diagram showing neuronal circuitry of commissural vestibular pathways. Abbreviations: I and II, type I and II vestibular neuron, H, horizontal canal. (●), Inhibitory neurons, (○), excitatory neurons.

shown that commissural inhibition may be mediated by glycine as well as GABA (Precht et al., 1973).

It appears that crossed inhibition is specific to canal neurons; central otolith neurons are always excited by contralateral stimulation (Precht et al., 1967; Shimazu and Smith, 1971). Comparative physiological studies showed that apparently frogs have a very poorly developed commissural inhibition and, instead, show mainly crossed excitation (see Dieringer and Precht, chapter IVE4). In reptiles, however, crossed inhibition was frequently found. It has been hypothesized that absence of commissural inhibition of canal neurons in the frog may be related to differences in the organization of eye movements (Ozawa et al., 1974). Some of the functional aspects of crossed pathways will be discussed in the following sections.

RESPONSES OF VESTIBULAR NEURONS TO NATURAL LABYRINTHINE STIMULATION

Since the pioneering work of Adrian (1943) who was the first to record from central vestibular neurons during natural vestibular stimulation a great number of studies have confirmed and substantially expanded his work. I have reviewed these studies in various monographs and shall, therefore, only briefly summarize the major points (Precht, 1974, 1975, 1976, 1978, 1979).

Semicircular canal input

Qualitative responses. When a functionally synergistic pair of canals is brought into the plane of rotation of a turntable two major unitary responses may be recorded from vestibular neurons of various species. In the horizontal canal system they have been designated type I and type II by Duensing and Schaefer (1958). Similarly, in the vertical canal system two major groups of neuronal responses were found (Duensing and Schaefer, 1959; Markham, 1968). In addition, two other types of canal responses (types III and IV) were found but they occurred less frequently than type I and II units (Duensing and Schaefer, 1958, 1959; Shimazu and Precht, 1965). It became clear that horizontal type I neurons (or their vertical canal counterpart) were the true secondary neurons, i.e., had the same response polarities as the primary afferents of the corresponding canals in any given head position in space. On the other hand, type II neurons (and their vertical canal counterpart) are the inhibitory neurons projecting onto type I neurons (Fig. 1). It should be noted, however, that pseudo-type II responses may easily be recorded in the vestibular nuclei or nerve when the vertical canals have have an in-plane vector in the horizontal plane of rotation. These type II responses are actually analogous to the type I in the horizontal system and represent the main sensory neurons of the vertical canals in a particular head position. Application of null-point technique allows critical differentiation between real and pseudo-type II responses (Estes et al., 1975; Blanks and Precht, 1976). Essentially similar qualitative results as in the cat and the rabbit were obtained in recordings from frog vestibular neurons (Precht, 1976) except that few type II neurons were encountered in this species (10% only vs. ca. 40–50% in cat and monkey). They were, for the most part, derived from costimulation of the vertical canals. Since, as noted in the preceding section, frogs have a very poorly developed commissural inhibition the dearth of inhibitory type II neurons seems reasonable.

Quantitative responses. In the time domain studies which have been performed in the horizontal canal system of lightly anesthetized rabbit (Duensing and Schaefer, 1958), decerebrate cat (Shimazu and Precht, 1965; Shinoda and Yoshida, 1974), unanesthetized frog (Precht, 1976) and anesthetized monkey (Abend, 1978) vestibular neurons the following major results have emerged: 1) the mean resting rates of central vestibular neurons vary among species being low (< 10 imp/sec) in the frog and relatively high in the alert monkey (Fuchs and Kimm, 1975) and the cat (Keller and Precht, 1979) (ca. 40 and 70 imp/sec in cat and monkey, respectively). Also there are fewer silent neurons in the monkey than in the cat. When compared with the resting rates found in primary afferents (cf. Precht, 1978), these values appear to be slightly smaller in central neurons; 2) thresholds for frequency increase in all species are below 0.5 degree/sec^2; 3) there exist units with linear or non-linear acceleration-frequency increase/decrease relationships. The longer the acceleration lasts the more non-linear the responses become; 4) the distribution of time constants of the responses to constant angular acceleration in the cat is bimodal showing short (ca. 3 sec) and long (ca. 7 sec) time constant groups; they have been called tonic and kinetic units, respectively, and were shown to have predominantly mono- and polysynaptic primary afferent connections, respectively (Precht and Shimazu, 1965). Frog vestibular neurons have mean time constants of ca. 3 sec.; 5) adapting as well as non-adapting units exist in all species studied; 6) the mean sensitivity factor measured in the cat ca. 6 spikes/sec per degree/sec^2 which is significantly higher than the values found in primary afferents (Blanks et al., 1975). Slightly higher values were found in the frog.

In comparing the responses of primary and secondary canal neurons two findings deserve particular emphasis: i) secondary neurons have long time constant units not encountered in the nerve thus indicating that central mechanisms, possibly involving the neural integrator postulated by Robinson (1970), modify the vestibular signal; ii) the fact that the acceleration sensitivity of secondary neurons is much higher than that of the afferents has been shown to be due to commissural inhibition which increases the sensitivity of central neurons (Shimazu and Precht, 1966; Markham et al., 1977; Abend, 1978). As will be shown below, in animals who have no commissural inhibition the sensitivity of primary afferents is much higher than in animals who have it. Besides increasing the sensitivity of vestibular neurons commissural inhibition introduces, by way of saturation of disinhibitory effects, non-linearities in the input-output relationship (Abend, 1978).

In the frequency domain central vestibular neurons of the horizontal canals of the frog (Precht, 1978), cat (anesthetized, Melvill Jones and Milsum, 1970; decerebrate, Shinoda and Yoshida, 1974; alert, Keller and Precht, 1979), anesthetized rat (Kubo et al., 1975) and alert monkey (Fuchs and Kimm, 1975; Keller and Kamath, 1975; Buettner et al., 1979) have been measured. An overview of the mean *phase* and *gain* values of type I neurons obtained in these species is presented in Fig. 2. As for the *phase* secondary neurons of all species, these have mean phases that are slightly larger than those of primary afferents (further details see below); frog's phase values are slightly smaller than those of the cat and monkey and the shape of the phase curves are similar in each species except that the values in the monkey diverge from the others at lower frequencies. The *gain* values increase from primary to secondary neurons in cat and monkey and remain the same or slightly decrease in the frog. Primary afferents in the frog have a much higher gain than cat and monkey afferents; this high peripheral gain

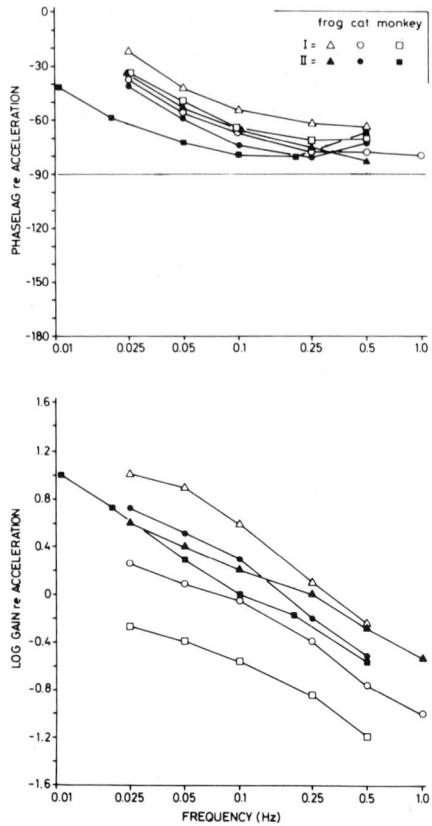

Fig. 2. Phase and gain of primary and secondary type I vestibular neurons of different species. Frog data are from Blanks and Precht (1976) and unpublished work. (Cat central and peripheral data were plotted from the work of Shinoda and Yoshida, 1974; and Anderson et al., 1978, resp.; monkey peripheral and central data are from Fernández and Goldberg, 1971; and Buettner et al., 1979.)

may be required to compensate for the lack of commissural inhibition in this species which – as mentioned above – increases the sensitivity of central neurons.

The data obtained from the alert cat (Keller and Precht, 1979) and monkey (Fuchs and Kimm, 1975; Keller and Kamath, 1975) deserve a closer inspection. In the cat (Figs. 3 and 4) as well as in the monkey, it was noted that central vestibular units may be divided into two groups: i) short time constant (mean of 3 sec in cat), small phase lag (ca. 50°–60° re. acceleration) units, and ii) long time constant (11 sec), large phase lag (ca. 105° in cat and 120°–170° in monkey) neurons. It has also been shown that the second group receives disynaptic vestibular input whereas the short time constant group is monosynaptically driven. Correlation of the neural firing with eye movement revealed that the large phase lag units in the monkey had a burst-tonic character very much like eye motoneurons and were, therefore, related to eye position. No such correlation was found in the small phase lag group; units, however, often paused during saccades (see Kimm and Winfield, this volume, chapter IVA3). In the alert cat, the large phase lag group was not burst-tonic in nature but rather showed a weak eye position relationship also observed in some neurons in the monkey. It must be concluded that the large phase lag units receive signals other than those present in

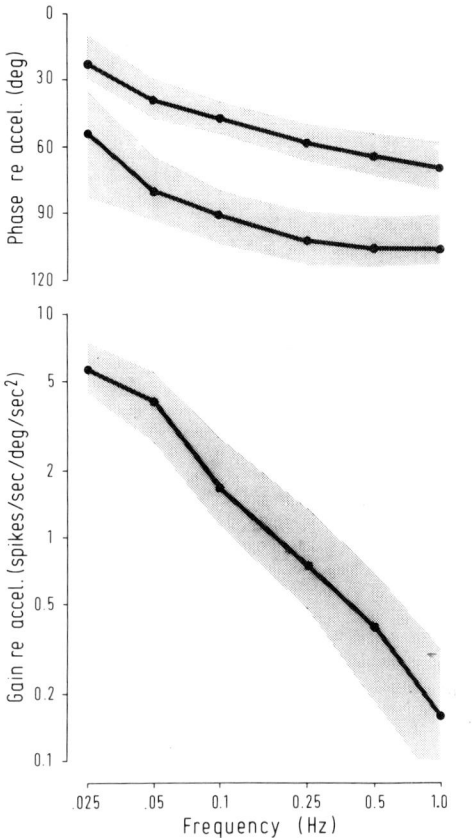

Fig. 3. Phase and gain values of central type I vestibular neurons in the alert cat. (From Keller and Precht, 1979.)

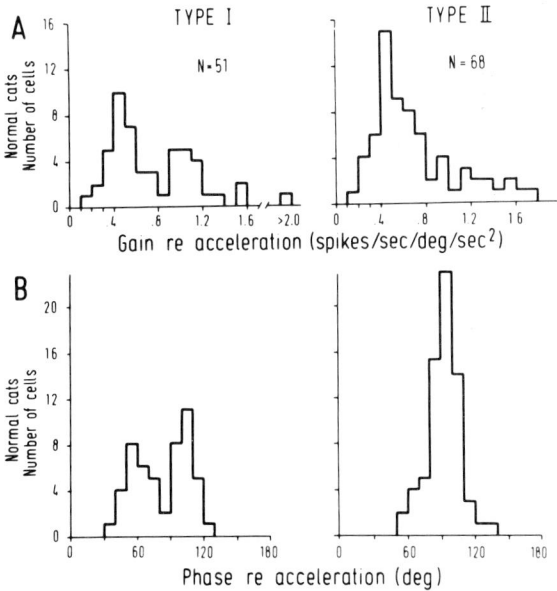

Fig. 4. Frequency distribution of phase and gain of type I and type II vestibular neurons of the alert cat at 0.25 Hz. (From Keller and Precht, 1979.)

primary canal afferents. They are presumably re-afferented from the reticular integrator.

In Fig. 4, a comparison of type I and II phases and gains of the alert cat is given at one frequency (0.25 Hz). It can be seen that contrary to type I the phases of type II are unimodal and closely around 95°, i.e., phases are similar to the large phase lag type I neurons.

Otolith input

Responses of central vestibular neurons to natural stimulation of the otoliths have been studied in rabbit (Duensing and Schaefer, 1959), cat (Hiebert and Fernández, 1965; Fujita et al., 1968; Melvill Jones and Milsum, 1969; Peterson, 1970; Matsuoka et al., 1971; Shimazu and Smith, 1971; Schor, 1974; Blanks et al., 1978) and frog (Precht, 1976).

The major results obtained in these studies may be briefly summarized as follows:

1. Neurons responding to static lateral tilt (mainly utricular stimulation) have been divided into α and β neurons that are excited by tilting the ipsilateral or contralateral side down, respectively; their relative occurrence in the nuclei corresponds to that in the nerve (2/3 α and 1/3 β responses) and reflects the polarization pattern of utricular hair cells. In addition, higher order γ and δ responses were noted which were excited or inhibited, respectively, by tilt in either direction. In general, α and β responses were driven monosynaptically by stimulation of the VIIIth nerve (true secondary neurons) whereas γ and δ responses (higher order neurons) were polysynaptically excited or not driven at all. The location of mono- and polysynaptic responses in the vestibular nuclei is in good agreement with the topography of otolith projections (see first section). Also, monosynaptically driven cells were more sensitive than the polysynaptic units. However, in unanesthetized animals polysynaptic units may also be very sensitive. It has been shown that α and γ units project preferentially to the cervical and lumbar spinal cord, respectively.

2. According to the time course of the responses neurons may be divided into tonic, phasic-tonic and phasic units. While in mammals tonic and phasic-tonic responses predominate, mainly phasic and phasic-tonic units are found in amphibians. Anesthesia and cerebellectomy appear to reduce the frequency of occurrence of phasic responses (Orlovsky and Pavlova, 1972). In many units a given static head position was associated with different firing rates (multi-valuedness) when tested at various times during a complete tilt protocol.

3. That the central otolith units have true dynamic responses has been shown in cats with all semicircular canals rendered non-functional. In this situation, many cells showed greater gains as the frequency of sinusoidal tilt was increased. Frequency analysis of vestibular units with sinusoidally varying linear motion has shown that central responses differ from those recorded in primary afferents in that a frequency-dependent phase lag is introduced centrally. As in the case of canal neurons these phase changes may result from the operation of a leaky integrator (see Precht et al., this volume, chapter IVC4). In Fig. 5 are shown the responses of vestibular neurons projecting to the oculomotor system to sinusoidal rotation in yaw (canal stimulation only) and roll (canal and otolith stimulation). If one subtracts vectorially the yaw from the roll responses the pure otolith response may be obtained (for details, see Precht et al., chapter IVC4): the otolith response phase lags head displacement, the lag being the larger the higher the frequency of stimulation.

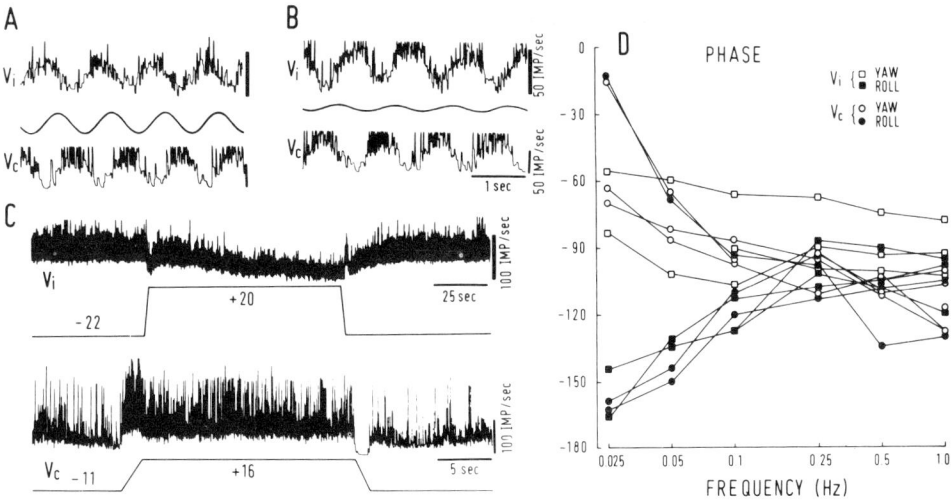

Fig. 5. Responses of vestibular axons presumably terminating in the trochlear nuclei. Responses of a single ipsilateral (Vi) and contralateral (Vc) secondary vestibular axon were recorded during sinusoidal yaw (A ± 5° at 1 Hz) and roll (B ± 2° at 1 Hz) and during ramp changes in roll position (C). For the latter, positive values correspond to ipsilateral side up. In D, the phases and gains are shown for 3 Vi (□, ■) and 3 Vc (○, ●) axons during yaw and roll, resp. Note that the input reference was ipsilateral and contralateral angular acceleration for the Vi and Vc axons, resp. The ordinate scales in B (from 0 to 50 imp/sec) also hold for A. (From Blanks et al., 1978.)

Receptor convergence on vestibular neurons

Canal-canal convergence. Electrophysiological studies employing electrical stimulation of individual canal nerves in the cat (Kasahara and Uchino, 1971; Markham and Curthoys, 1972) and pigeon (Wilson and Felpel, 1972) revealed very little short-latency convergence of ipsilateral (orthogonal) canals on central vestibular neurons. This finding indicates that there is selective projection of canal fibers on to vestibular neurons and if there is significant orthogonal convergence under the condition of natural stimulation, it can only be due to activation of polysynaptic paths which are not activated by single shock stimulation and/or depressed by anesthesia. It should be remembered that coplanar (ipsi-contra) canal pairs converge via the vestibular commissure (see Fig. 1 and section 1). When natural stimulation was employed in the search for convergence in cat (Curthoys and Markham, 1971), rabbit (Duensing and Schaefer, 1959), frog (Precht, 1976) and monkey (Abend, 1978) not only was coplanar canal convergence confirmed but also was ipsi- and contralateral orthogonal canal convergence often observed in addition, of course, to pure single canal responses (ca. one-third). In fact, type III and IV responses may result from such interaction. Since there exists no significant canal-canal convergence at the receptor-primary afferent level the observed convergence must occur centrally. Whereas the function of in-plane canal-canal convergence is to increase the sensitivity of central neurons (see previous section) orthogonal canal convergence may be related to the fact that canal responses are a function of the cosine of the angle between the canal plane and plane of rotation. Such a relationship yields only small differences in response magnitude over a wide range of head positions and may require information from orthogonal canals to enable precise determination of the plane of rotation.

Canal-otolith convergence. Of the horizontal type I and II vestibular neurons about 50% responded also to otolith stimuli which consisted of constant velocity rotations or static head displacements in pitch and roll (Curthoys and Markham, 1971). Frequent occurrence of canal-otolith convergence was also reported in rabbits (Duensing and Schaefer, 1959), frogs (Precht, 1976) and rats (Kubo et al., 1977). In a more recent study, we recorded from a rather selected group of vestibular neurons, namely those projecting to the trochlear nuclei during natural canal or otolith stimulation (Blanks et al., 1978). It was found that ca. 50% of these vestibular axons were carrying both vertical canal and otolith signals. Fig. 5A–C gives sample records of these responses and Fig. 5D shows the frequency response of the units during yaw (□, ○) and roll (■, ●). The position of the head was such that in yaw the posterior/anterior synergistic canal inputs to secondary neurons – Vi from the ipsilateral anterior, and Vc from the contralateral posterior canals – were activated. In yaw the units responded with a small and large phase re. acceleration to low and high frequency of rotation, respectively. In about one half of the units similar responses were observed when tested in roll (Fig. 5D). The other half, however, showed large phase lags at low frequencies and progressively smaller phase lags as the frequency was increased (Fig. 5D). Obviously, in case of stimulation in roll which activates canals and otoliths, the rotating gravity vector is able to change the frequency response of the units in such a way as to compensate for the poor canal performance in the low frequency range (see also Precht et al., chapter IVC4).

SUMMARY

1. The functional synaptology and topography of central vestibular neurons of several species as revealed by electrical and natural stimulation of the bilateral labyrinths is described and correlated with the anatomical findings.

2. The qualitative and quantitative aspects of semicircular canal responses of central neurons in the time and frequency domain of several species (frog, cat, monkey) are described and compared with responses found in the primary neurons.

3. The comparative aspects of central otolith responses are characterized both qualitatively and quantitatively and compared with primary otolith responses.

4. The frequency of occurrence and functional importance of parallel and orthogonal canal convergence is critically evaluated. Canal-otolith convergence in central vestibular neurons is described.

REFERENCES

Abend, W.K. (1978) Response to constant angular accelerations of neurons in the monkey superior vestibular nucleus. *Exp. Brain Res.*, 31: 459–473.

Adrian, E.D. (1943) Discharges from vestibular receptors in the cat. *J. Physiol. (Lond.)*, 101: 389–407.

Akaike, T., Fanardjian, V.V., Ito, M., Kumada, M. and Nakajima, H. (1973) Electrophysiological analysis of the vestibulospinal reflex pathway of rabbit. I. Classification of tract cells. *Exp. Brain Res.*, 17: 477–496.

Anderson, J.H., Blanks, R.H.I. and Precht, W. (1978) Response characteristics of semicircular canal and otolith systems in cat. I. Dynamic responses of primary vestibular fibers. *Exp. Brain Res.*, 32: 491–507.

Blanks, R.H.I. and Precht, W. (1976) Functional characterization of primary vestibular afferents in the frog. *Exp. Brain Res.*, 25: 369–390.

Blanks, R.H.I., Estes, M.S. and Markham, C.H. (1975) Physiological characteristics of vestibular first-order canal neurons in the cat. II. Response to constant angular acceleration. *J. Neurophysiol.*, 38: 1250–1268.

Blanks, R.H.I., Anderson, J.H. and Precht, W. (1978) Response characteristics of semicircular canal and otolith systems in cat. II. Responses of trochlear motoneurons. *Exp. Brian Res..*, 32: 509–528.

Buettner, U.W., Büttner, U. and Henn, V. (1979) Transfer characteristics of neurons in the vestibular nuclei of alert monkey. *J. Neurophysiol.*, in press.

Chan, Y.S., Hwang, J.C. and Chueng, Y.M. (1977) Crossed sacculo-ocular pathway via the Deiters' nucleus in cats. *Brain Res. Bull.*, 2: 1–6.

Curthoys, I.S. and Markham, C.H. (1971) Convergence of labyrinthine influences on units in the vestibular nuclei of the cat. I. Natural stimulation. *Brain Res.*, 35: 469–490.

Dieringer, N. and Precht, W. (1977) Modification of synaptic input following unilateral labyrinthectomy. *Nature (Lond.)*, 269: 431–433.

Duensing, F. and Schaefer, K.P. (1958) Die Aktivität einzelner Neurone im Bereich der Vestibulariskerne bei Horizontalbeschleunigungen unter besonderer Berücksichtigung des vestibulären Nystagmus. *Arch. Psychiat. Nervenkr.*, 198: 225–252.

Duensing, F. and Schaefer, K.P. (1959) Über die Konvergenz verschiedener labyrinthärer Afferenzen auf einzelne Neurone des Vestibulariskerngebietes. *Arch. Psychiatr. Nervenkr.*, 199: 345–371.

Estes, M.S., Blanks, R.H.I. and Markham, C.H. (1975) Physiologic characteristics of vestibular first-order canal neurons in the cat. I. Reponse plane determination and resting discharge characteristics. *J. Neurophysiol.*, 38: 1232–1249.

Fernández, C. and Goldberg, J.M. (1971) Physiology of peripheral neurons innervating semicircular canals of the squirrel monkey. II. Response to sinusoidal stimulation and dynamics of peripheral vestibular system. *J. Neurophysiol.*, 34: 661–675.

Fuchs, A.F. and Kimm, J. (1975) Unit activity in vestibular nucleus of the alert monkey during horizontal angular acceleration and eye movement. *J. Neurophysiol.*, 38: 1140–1161.

Fujita, Y., Rosenberg, J. and Segundo, J.P. (1968) Activity of cells in the lateral vestibular nucleus as a function of head position. *J. Physiol. (Lond.)*, 196: 1–18.

Goldberg, J.M. and Fernández, C. (1975) Vestibular mechanisms. *Ann. Rev. Physiol.*, 37: 129–162.

Gacek, R.R. (1969) The course and central termination of first order neurons supplying vestibular endorgans in the cat. *Acta oto-laryngol. (Stockh.)*, 254: 1–66.

Gregory, K.M. (1972) Central projections of the eighth nerve in frogs. *Brain, Behav. Evol.*, 5: 70–88.

Hiebert, T.G. and Fernández, C. (1965) Deitersian response to tilt. *Acta oto-laryngol. (Stockh.)*, 60: 180–190.

Hwang, J.C. and Poon, W.F. (1975) An electrophysiological study of the sacculo-ocular pathways in cats. *Jap. J. Physiol.*, 25: 241–251.

Ito, M. Hongo, T. and Okada, Y. (1969) Vestibular-evoked postsynaptic potentials in Deiters' neurones. *Exp. Brain Res.*, 7: 214–230.

Ito, M., Nisimaru, N. and Yamamoto, M. (1977) Specific patterns of neuronal connexions involved in the control of the rabbit's vestibuloocular reflexes by the cerebellar flocculus. *J. Physiol. (Lond.)*, 265: 833–854.

Kasahara, M. and Uchino, Y. (1971) Selective mode of commissural inhibition induced by semicircular canal afferents on secondary vestibular neurons in the cat. *Brain Res.*, 34: 366–369.

Kasahara, M., Mano, M., Oshima, T. Ozawa, S. and Shimazu, H. (1968) Contralateral short latency inhibition of central vestibular neurones in the horizontal canal system. *Brain Res.*, 8: 376–378.

Kawai, N., Ito, M. and Nozue, M. (1969) Postsynaptic influences on the vestibular non-deiters nuclei from primary vestibular nerve. *Exp. Brain Res.*, 8: 190–200.

Keller, E.L. and Kamath, B.Y. (1975) Characteristics of head rotation and eye movement-related neurons in alert monkey vestibular nucleus. *Brain Res.*, 100: 182–187.

Keller, E.L. and Precht, W. (1979) Adaptive modification of central vestibular neurons in response to visual stimulation through reversing prisms. *J. Neurophysiol.*, in press.

Korn, H., Sotelo, C. and Crepel, F. (1973) Electrotonic coupling between neurons in the rat lateral vestibular nucleus. *Exp. Brain Res.*, 16: 225–275.

Korn, H., Sotelo, C. and Bennett, M.V.L. (1977) The lateral vestibular nucleus of the toadfish *Opsanus tau*: Ultrastructural and electrophysiological observations with special reference to electrotonic transmission. *Neuroscience*, 2: 851–884.

Kubo, T., Matsunaga, T. and Matano, S. (1975) Effects of sinusoidal rotational stimulation on the vestibular neurons of rats. *Brain Res.*, 88: 543–548.

Kubo, T., Matsunaga, T. and Matano, S. (1977) Convergence of ampullar and macular inputs on vestibular nuclei unit of the rat. *Acta oto-laryngol. (Stockh.)* 84: 166–177.

Ladpli, R. and Brodal, A. (1968) Experimental studies of commissural and reticular formation projections from the vestibular nuclei in the cat. *Brain Res.*, 8: 65–96.

Mano, M., Oshima, T. and Shimazu, H. (1968) Inhibitory commissural fibres interconnecting the bilateral vestibular nuclei. *Brain Res.*, 8: 378–383.

Markham, C.H. (1968) Midbrain and contralateral labyrinth influences on brainstem vestibular neurons in the cat. *Brain Res.*, 9: 312–333.

Markham, C.H., Curthloys, I.S. (1972) Convergence of labyrinthine influences on units in the vestibular nuclei of the cat. II. Electrical stimulation. *Brain Res.*, 43: 383–396.

Markham, C.H. Yagi, T. and Curthoys, I.S. (1977) The contribution of the contralateral to second order vestibular neuronal activity in the cat. *Brain Res.*, 138: 99–109.

Matsuoka, I., Fukuda, N., Takaori, S. and Morimoto, M. (1971) Responses of single neurons of the vestibular nuclei to lateral tilt and caloric stimulation in the intact and hemilabyrinthectomized cats. *Acta oto-laryngol. (Stockh.)*, 72: 182–190.

Melvill Jones, G.E. and Milsum, J.H. (1969) Neural response of the vestibular system to translational acceleration. In *Conference on Systems Analysis Approach to Neurophysiological Problems*. Brainerd, Minnesota.

Melvill Jones, G.E. and Milsum, J.H. (1970) Characteristics of neural transmission from the semicircular canal to the vestibular nuclei of cats. *J. Physiol. (Lond.)*, 209: 295–316.

Orlovsky, G.N. and Pavlova, G.A. (1972) Response of Deiters' neurons to tilt during locomotion. *Brain Res.*, 42: 212–214.

Ozawa, S., Precht, W. and Shimazu, H. (1974) Crossed effects on central vestibular neurons in the horizontal canal system of the frog. *Exp. Brain Res.*, 19: 394–405.

Peterson, B.W. (1970) Distribution of neural responses to tilting within the vestibular nuclei of the cat. *J. Neurophysiol.*, 33: 750–767.

Precht, W. (1974) The physiology of the vestibular nuclei. In *Handbook of Sensory Physiology, Vol. VI, Vestibular System, Part 1, Basic Mechanisms,* H.H. Kornhuber (Ed.) Springer-Verlag, Berlin.

Precht, W. (1975) Vestibular system. In *MTP International Review of Science. Neurophysiology, Phys. Ser. 1, Vol. 3,* A.C. Guyton and C.C. Hunt (Eds.), Butterworths, London and University Park Press, Baltimore, MD.

Precht, W. (1976) Physiology of the peripheral and central vestibular systems. In *Frog Neurobiology,* R. Llinás and W. Precht (Eds.), Springer-Verlag, Berlin-Heidelberg-New York, pp. 481–512.

Precht, W. (1978) Neuronal operations in the vestibular system. In *Studies of Brain Function, Vol. 2,* V. Braitenberg (Ed.), Springer-Verlag, Berlin-Heidelberg-New York, pp. 1–226.

Precht, W. (1979) Vestibular mechanisms. *Ann. Rev. Neurosci.*, 2: 265–289.

Precht, W. and Shimazu, H. (1965) Functional connections of tonic and kinetic vestibular neurons with primary vestibular afferents. *J. Neurophysiol.*, 28: 1014–1028.

Precht, W., Grippo, J. and Wagner, A. (1967) Contribution of different types of central vestibular neurons to the vestibulo-spinal system. *Brain Res.*, 4: 119–123.

Precht, W., Schwindt, P.C. and Baker, R. (1973) Removal of vestibular commissural inhibition by antagonists of GABA and glycine. *Brain Res.*, 62: 222–226.

Precht, W., Richter, A., Ozawa, S. and Shimazu, H. (1974) Intracellular study of frog's vestibular neurons in relation to the labyrinth and spinal cord. *Exp. Brain Res.*, 19: 377–393.

Richter, A., Precht, W. and Ozawa, S. (1975) Responses of neurons of lizard's, Lacerta viridis, vestibular nuclei to electrical stimulation of the ipsi- and contralateral VIIIth nerves. *Pflügers Arch.*, 355: 85–94.

Robinson, D.A. (1970) Oculomotor unit behavior in the monkey. *J. Neurophysiol.*, 33: 393–404.

Schor, R.H. (1974) Responses of cat vestibular neurons to sinusoidal roll tilt. *Exp. Brain Res.*, 20: 347–362.

Shimazu, H. and Precht, W. (1965) Tonic and kinetic responses of cat's vestibular neurons to horizontal angular acceleration. *J. Neurophysiol.*, 28: 991–1013.

Shimazu, H. and Precht, W. (1966) Inhibition of central vestibular neurons from the contralateral labyrinth and its mediating pathway. *J. Neurophysiol.*, 29: 467–492.

Shimazu, H. and Smith, C.M. (1971) Cerebellar and labyrinthine influences on single vestibular neurons identified by natural stimuli. *J. Neurophysiol.*, 34: 493–508.

Shinoda, Y. and Yoshida, K. (1974) Dynamic characteristics of responses to horizontal head angular acceleration in the vestibuloocular pathway in the cat *J. Neurophysiol.*, 37: 653–673.

Stein, B.M. and Carpenter, M.B. (1967) Central projections of portions of the vestibular ganglia innervating specific parts of the labyrinth in the rhesus monkey. *Amer. J. Anat.*, 120: 281–318.

Walberg, F., Bowsher, D. and Brodal, A. (1958) The termination of primary vestibular fibres in the vestibular nuclei in the cat. An experimental study with silver methods. *J. comp. Neurol.*, 110: 391–419.

Wilson, V.J. and Wylie, R.M. (1970) A short-latency labyrinthine input to the vestibular nuclei in the pigeon. *Science*, 168: 124–127.

Wilson, V.J. and Felpel, L.P. (1972) Specificity of semicircular canal input to neurons in the pigeon vestibular nuclei. *J. Neurophysiol.*, 35: 253–264.

Wilson, V.J. and Peterson, B.W. (1978) Peripheral and central substrates of vestibulospinal reflexes. *Physiol. Rev.*, 58: 80–105.

Response of Vestibular and Cerebellar Neurons to Rotational Stimulation

J. KIMM and J.A. WINFIELD*

Departments of Otolaryngology and Physiology and Biophysics, University of Washington, Seattle, WA 98195 (U.S.A.)

INTRODUCTION

The vestibular end organ is responsive to two general classes of stimuli: i) linear accelerations, or gravitational forces, and ii) angular accelerations. This paper will be concerned only with those structures in the brain stem and cerebellum that purportedly are responsive to angular accelerations. The known anatomical connections and physiological responses of neurons in the vestibular nuclear complex and cerebellum will be discussed briefly, and then data will be presented addressing the question of ascending somatic convergence with vestibular processing in the awake pigeon preparation.

CANAL AFFERENTS TO THE CNS

The anatomical distribution of primary vestibular afferents to the central nervous system (CNS) has been studied by several laboratories with a variety of techniques and experimental animal species. The principal targets for VIIIth nerve terminations are the vestibular nuclear complex located in the pontomedullary brain stem, the vestibulocerebellum, and the deep cerebellar nuclei.

Brain stem projections

Walberg et al. (1958) used the Glee and Nauta methods to demonstrate that whereas all of the principal nuclei of the vestibular complex receive primary afferent input, the sites of termination are restricted to specified anatomical loci within each vestibular nucleus. Moreover, the vestibular afferent projection is exclusively ipsilateral for brain stem connections. More recently, the problem of specific canal terminations has been addressed by Stein and Carpenter (1967) in the monkey and by Gacek (1969) in the cat. Following small lesions in the vestibular ganglion, which destroyed ganglion cells subserving specific canals, these studies showed that the canal projection is primarily to the rostral portions of the medial vestibular nucleus and the central region of the superior vestibular nucleus. Some degeneration products were also found in the rostral portion of the lateral vestibular nucleus and the medial region

---

*Visiting student on leave from Albert Einstein College of Medicine, New York, NY.

in the inferior vestibular nucleus, indicating that these areas also receive a direct but sparse canal input.

Primary vestibulo-cerebellar fibers

Succinctly stated, primary vestibular afferents in mammals terminate as mossy fibers in the nodulus, uvula, and the flocculus (Brodal and Høivik, 1964). In addition, in some species the ventral paraflocculus may also receive primary afferents. Terminations of primary vestibular afferents have also been found in the dentate and fastigial nuclei, although some (Brodal, 1972) remain skeptical about primary vestibular afferents in the fastigial nucleus. Most, if not all, of these primary vestibular afferents, which terminate in the vestibulocerebellum, originate from the canal portion of the vestibular end organ.

Responses of vestibular neurons to rotation

Rotational stimulation activates neurons located in the vestibular nuclei. Four types of neuronal responses (Duensing and Schaefer, 1958; and others) have been reported. Type I neurons are excited by ipsilateral rotational stimuli and are also monosynaptically activated by electrical stimulation of the VIIIth nerve (Shimazu and Precht, 1965). Type II neurons are facilitated by contralateral rotation and are polysynaptically activated by electrical stimulation of either the ipsilateral or contralateral VIIIth nerve. Type III and type IV neurons are less numerous than either the type I or type II cells and are excited and inhibited, respectively, by accelerations in either direction.

In the awake monkey, about 50% of the vestibular neurons can be further subdivided with respect to response to eye movements (Fuchs and Kimm, 1975; Keller and Daniels, 1975; and others). The details of these responses have been adequately described elsewhere (Precht and Llinas, 1972; Wilson and Peterson, 1977; Baker and Berthoz, 1977; Precht, this volume, chapter IVA2) and will not be further discussed here.

Response of cerebellar neurons to rotation

Whereas the question of direct primary vestibular afferent input to the fastigial nucleus remains in question, it has been anatomically demonstrated that second-order neurons in the vestibular nuclei do project to the fastigial nucleus (Dow, 1936; Brodal and Torvik, 1957). Single unit recordings that show vestibular sensitivity in the awake monkey (Kimm et al., 1976, Gardner and Fuchs, 1975) and in the chinchilla (Winfield and Kimm, 1979) support these anatomical findings.

Both type I and type II responses in fastigial nucleus have been described and these cell types have been found distributed predominantly caudal to rostral, respectively (Kimm et al., 1975; Furuya et al., 1975). Modulation of the firing rate of fastigial neurons with poorer entrainment (i.e., greater distortion) has been demonstrated at the lower frequencies of sinusoidal oscillation, especially when compared with vestibular neurons. It has been further demonstrated in these studies that the response of fastigial neurons is unrelated to eye movements.

In mammals, primary vestibular afferents have been shown both anatomically (Brodal and Høivik, 1964) and physiologically (Precht and Llinas, 1969; Wilson et al., 1974) to end as mossy fibers in the flocculus. In addition to this ipsilateral input, Shinoda et al. (1975) have demonstrated considerable convergence of bilateral labyrinthine mossy fiber inputs onto individual Purkinje cells. Studies by Lisberger and Fuchs (1977) and Noda et al. (1977) in the awake monkey preparation and studies by Ghelarducci et al. (1975) in the rabbit have shown a vast variety of responses by Purkinje cells, the response depending on the visual vestibular paradigm tested. In addition to having visual receptive fields (Maekawa and Simpson, 1973) Purkinje cell spike activity is modulated by vestibular stimulation and smooth pursuit-tracking, as well as exhibiting bursts or pauses to saccadic eye movements. More unit recording in awake animals is still necessary for determination of the interactions involved, particularly in rabbits, cats and chinchillas, as species differences may also exist.

Unit recordings in awake pigeons

To examine the response properties of units in the brain stem and the cerebellum, we have made recordings from various structures in the awake pigeon. We selected this animal for our studies for a number of practical reasons: it is readily available and hardy and has a low acquistion and maintenance cost. More importantly, however, our choice of this animal was predicated on biological concerns. To date, most experiments dealing with vestibular mechanisms have concentrated on animals with very mobile eyes. In these previous awake animal studies the subject's head was firmly secured to a rotating device, and cell activity and eye movements were recorded. These experiments have provided us with a wealth of exciting new data.

We thought, however, that it would be interesting to use an animal in which eye movements are severely restricted. The ideal choice for these studies would be the great horned owl (*Bubo virginianus*) since the retina in this bird is attached to its skull (Walls, 1942). However, this animal is protected in our country and is not available for experimental use. Nevertheless, we felt that data from birds with restricted eye movements might give us valuable insight into the neural organization of the vestibulomotor system. In birds, the nervous system must be organized differently than in mammals, in which the eyes are very mobile.

Another factor that influenced our choice of birds as an experimental subject was the fact that these animals represent one of the few bipeds available for experimental study. The postural adjustments made in this animal to perturbations imposed on it might therefore be more similar to man's than the quadruped's responses, studied more frequently in the laboratory. This consideration could be of importance since there has been a recent resurgence in clinical medicine of the use of postural adjustments in diagnosing diseases of the vestibulomotor system. Furthermore, since the pigeon must "move its head" in order to "move its eyes," a greater degree of convergence between ascending somatic and vestibular inputs could be expected than is seen in most mammals.

The stimulus used in this study was sinusoidal oscillation imposed in either the horizontal or vertical plane. The stimulus was delivered to either the whole pigeon, as in most experiments with natural vestibular stimulation, or, as in our second condition, to the body alone when the head was held stationary with reference to earth.

Fig. 1 illustrates the response properties of single units recorded from the awake pigeon. These recordings were made from neurons in the cerebellar nuclei (Cb Nuc), processus lateralis cerebellovestibularis (PCV), vestibular nuclei (VDL) or reticular formation (Rpc, RP). (This nomenclature was adopted from Karten and Hodos, 1966.)

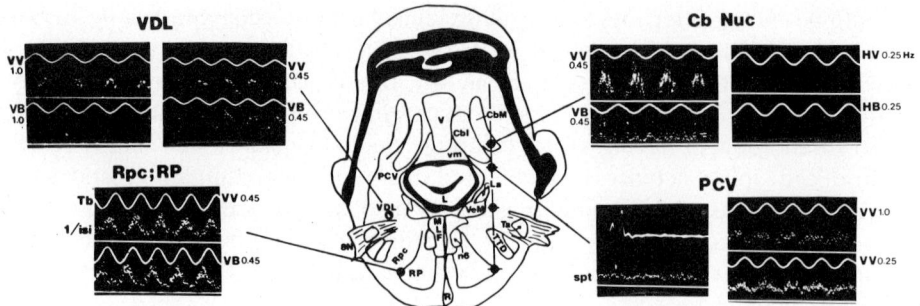

Fig. 1. Response of 6 units to rotational stimulation recorded in a single electrode penetration. The top trace in each frame represents the turntable position (Tb). The second trace is the output from an instantaneous frequency meter (1/isi). Numbers represent rotational frequency in Hz; VV, vertical vestibular; VB, vertical body; HV, horizontal vestibular; HB horizontal body; Cb Nuc, cerebellar nuclei; CbM, nucleus cerebellaris intermedius; Cbl, nucleus cerebellaris internus; PCV, processus lateralis cerebellovestibularis; VDL, nucleus vestibularis dorsolateralis (Sanders); VeM, nucleus vestibularis medialis; Rpc, nucleus reticularis parvocellularis; RP, nucleus reticularis pontis caudalis.

As is shown here, the firing pattern of units in each of these areas is closely related to the appropriate stimulus condition. The turntable position (Tb) is depicted by the upper solid trace and shows that a sinusoidal rotation is being imposed. The dots represent the output of a frequency meter indicating instantaneous spike firing rate. The recordings represented here were made during a single electrode penetration, although for the sake of convenience the vestibular and reticular formation responses in this figure have been drawn as if originating from the opposite side of the brain.

The other symbols represent the various sinusoidal stimulus conditions, as follows: vertical vestibular (VV), horizontal vestibular (HV), vertical body (VB) and horizontal body (HB). The numbers indicate oscillation frequency with the resting rate (spt) also shown in some cases.

As can be seen, all the unit responses recorded from cerebellar structures were correlated with vestibular stimulation only, none with body rotation per se.

The responses of two different Cb Nuc cells are shown here. Some of the Cb Nuc cells responded to vertical vestibular rotation whereas others responded to horizontal. The majority of these Cb Nuc cells responded to rotation in a single plane only but a few showed convergence, responding to both VV and HV oscillations.

Fiber recordings from the PCV tract were also related to vestibular rotation and were not responsive to oscillations of the body alone. The results obtained from a single fiber are also illustrated in Fig. 1.

Unlike the cerebellar unit responses, brain stem neurons often responded to both vestibular and body rotations independently. This effect is depicted in one of the cell responses from the vestibular nuclei (VDL) and again in the reticular formation (Rpc, RP) cell. The modulation of the other unit response shown in Fig. 1 responded to only the vestibular and did not respond to the body only rotations.

The response properties of three cerebellar nuclear cells to vertical oscillations are summarized in the upper half of Fig. 2. The phase response is depicted by the open symbols and the gain response by the filled symbols. Gain as used here is defined as follows: The envelope of the unit's averaged firing rate was obtained for all frequencies of stimulation. The amplitude of the largest averaged envelope in spikes/sec was defined as unity gain for that unit. The averaged envelopes obtained

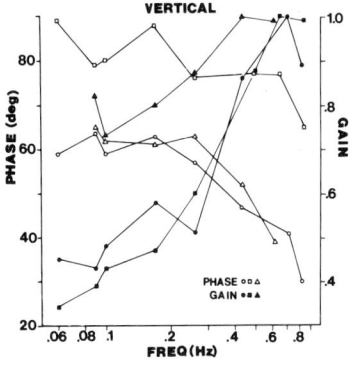
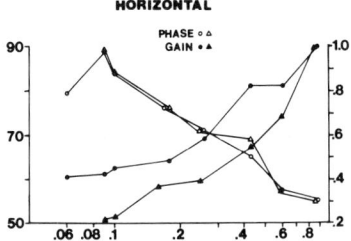

Fig. 2. Phase (○, △, □) and normalized gain (●, ▲, ■) of 3 vertical vestibular related neurons and 2 horizontal vestibular related neurons recorded in the cerebellar nuclei.

from all other frequencies of each unit response were plotted relative to the unity gain as described. This gain is therefore a normalized gain, as distinguished from the gain commonly referred to in Bode analysis.

As can be seen in Fig. 2, the response characteristics of these three cerebellar cells showed a decreasing phase lead as the frequency of oscillation was increased. The normalized-gain values increased with increased stimulation rates. The responses of these units were similar to those reported elsewhere in the monkey (Kimm et al., 1976) and in the chinchilla (Winfield and Kimm, 1979). At the lower rates of oscillation, the unit response was less faithful, so that there was a greater dispersion of the averaged unit response and an increase in distortion as computed from the FFT. The lower portion of this figure shows similar results obtained for two cell responses sensitive to horizontal rotations in the cerebellar nucleus.

In about 25% of the cases the data suggested a convergence of canal input. The responses of these convergence cells showed sensitivity to rotations in both the horizontal and vertical axis. Other cells located as close as 200 μm away in the same electrode tract responded to rotations only in a single plane.

Single fiber recordings obtained from the processus lateralis cerebellovestibularis (PCV) also showed changes correlated with the rotational stimulation. It is our impression that these fibers can be divided into at least two separate groups on the basis solely of maximum firing rate attained to sinusoidal rotation. In one group of fibers a very high rate of firing was observed, approaching 500 spikes/sec. The other group of fibers responded at a much lower maximum rate (about 125/sec). We did not see any fibers with a convergence of canal input since these fibers were sensitive to either the horizontal or the vertical oscillations, never to both. Similarly, none of these PCV fibers responded to oscillation of the body with the head held fixed relative to earth.

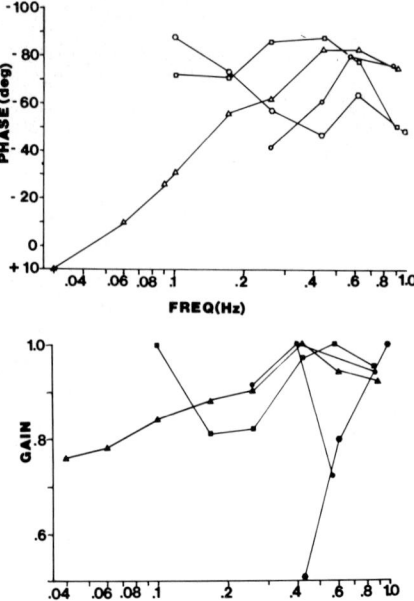

Fig. 3. Phase (○,△,□) and normalized gain (●,▲,■) of 4 units recorded in the PCV fiber track. Phase lead is represented by (+) and phase lag by (−).

Of interest was the change exhibited in the phase response of some units that resulted from a frequency series of oscillation. One fiber's phase response, illustrated in Fig. 3, shows a shift from +10 to −80 (i.e., total of 90°) over a frequency range from 0.03 to about 0.60 Hz. The other fiber responses also shown in the phase plot in this figure were not affected as much. Additionally, the normalized-gain plots demonstrated more variability in these PCV units than the Cb Nuc responses. These results suggest that the difference might represent inputs and outputs of the cerebellum, but further speculation at this time seems unwarranted.

The neurons in the vestibular region also responded to rotational stimulations. Fig. 4 illustrates the responses of 6 units recorded in the vestibular region and shows the phase and normalized gain response of these units. The open and filled triangles are the responses of the same unit to VV and VB, respectively, and the open and filled circles are those from a second unit. It is interesting to note that the phase relationships of the units responsive to VV and VB stimulation are quite similar. About 60% of the cell responses recorded in the vestibular brain stem area were related to both the vestibular and body oscillations. In no cases, however, did we observe any cell that responded to vestibular stimulation in one rotational plane and to body rotation in the other.

We also infrequently encountered units with any eye movement sensitivity. This observation is in stark contrast to the findings in chinchillas (Kimm et al., this volume, chapter VB4) and monkeys (Fuchs and Kimm, 1975; Keller and Daniels, 1975; and others).

The results of our recordings from reticular formation neurons were of considerable interest. Of 25 neurons recorded in Rpc and RP, 25/25 responded to both the vestibular and the body only oscillations. Paired unit response profiles were always observed for the same axis of rotation, i.e., positive for VV and VB or positive for HV

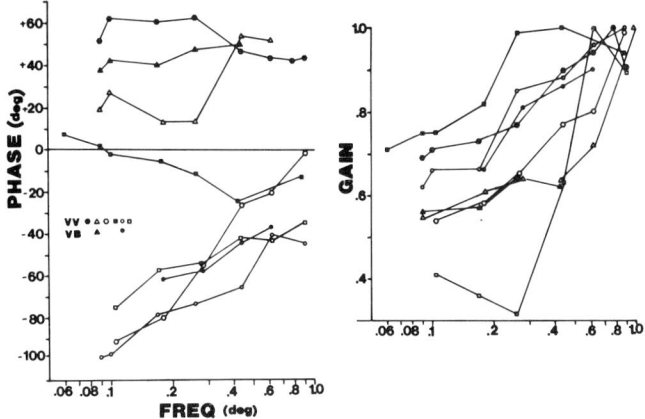

Fig. 4. Phase and normalized gain plots of 6 units recorded in the vestibular nuclear area. △, ▲ and ○, ● represent the response of 2 units, resp. VV, vertical vestibular; VB vertical body; (+) phase lead; (−) phase lag.

and HB. Phase plots showed that the RF units could signal either stimulus position, velocity, or values between these two.

Fig. 5 shows the responses from one of these Rpc neurons. This unit responded to all four conditions tested, VV, VB, HV, and HB. Several interesting points are shown. The depth of modulation for VV and VB was the same whereas the modulation to HV and HB was about one-tenth the response to that for the vertical stimulation but again matched in amplitude. The phase response computed for the four conditions showed a difference in values for vertical and horizontal oscillation but essentially the same values for each pair (VV and VB; HV and HB). Phase matching of vestibular and body oscillation for a given unit was true of all units analyzed.

Finally, since the theme of this symposium is the "Control of Posture and Movement" we feel it is appropriate to present some of the results obtained by us when alcohol was administered to awake pigeons. A dose of 2 g/kg of a 20% solution of alcohol was administered either i.v. or i.p. to the bird. Blood ethanol levels were determined at periodic intervals before and following the alcohol injection as well as unit responses to either rotational or stationary condition of the turntable. Fig. 6 shows a cell's response to sinusoidal stimulation tested at intervals over 4.5 h period following alcohol injection. Turntable position and cell instantaneous firing rate are shown in each example; and the time with respect to alcohol injection in terms of hours and minutes is indicated. Prior to the alcohol injection the cell's firing frequency can be seen to reliably follow the imposed stimulation with maximum firing occurring about 90° phase shifted to table position. Following the i.v. alcohol injection a dramatic change in the cell's firing occurred. As early as 11 min after the alcohol injection, the cell response to the imposed oscillation was altered. A transient increase in the resting rate occurred, accompanied by an increase in the peak-to-peak modulation of the cell's response to the imposed oscillations. This effect was especially apparent at 11, 20, and 29 min following the alcohol injections. Correlated with this increased firing rate to stimulus oscillations was a greater variability in entrainment of the cell's firing rate. This effect is best illustrated in the samples at 45 min and at 1 h and 22 min. Note that the oscilloscope sweep speed has been changed to show this

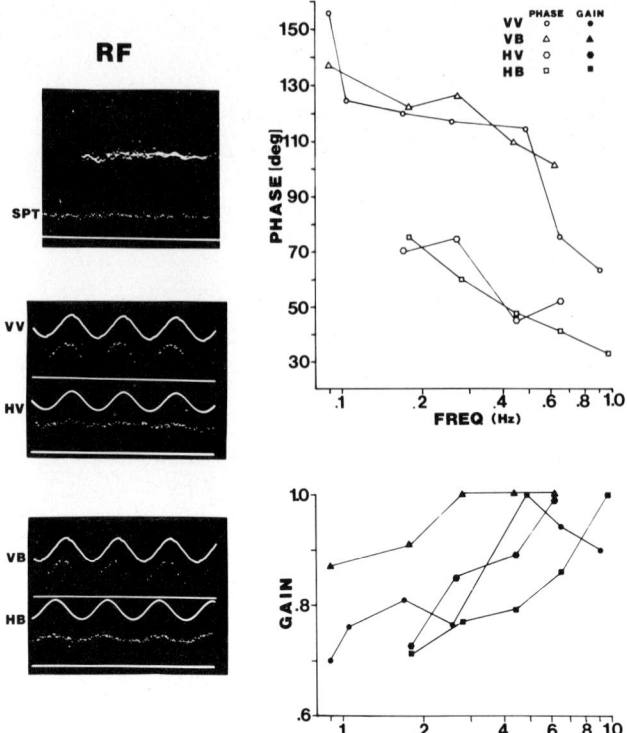

Fig. 5. Response of a single reticular formation neuron (RF) to vertical vestibular (VV), vertical body (VB), horizontal vestibular (HV), and horizontal body (HB) rotation. The 1st and 3rd traces in the two lower left frames represent the turntable position. The 2nd and 4th traces in these two frames represent instantaneous spike frequency. The upper trace above the spontaneous rate trace (SPT) is 3 superimposed sweeps of the single unit illustrated in this figure.

effect more clearly. An increased variability in the unit's response is evident; at times it failed to follow the stimulus input. The examples at the lower righthand corner of this figure show expanded multiple sweeps of the unitary potential as recorded just prior to the alcohol injection and again at 4 h and 38 min later. These results were replicated in six other cell recordings.

In conclusion, we find the bird an interesting animal model for our studies on vestibular function. The occurrence of marked convergence of labyrinthine and ascending somatic informations in the pigeon appears to reflect the extensive head movements required to achieve visual stabilization. Head nystagmus is more pronounced than eye movement nystagmus in birds (Roberts, 1967). Intracellular work by Rabin (1975) suggests the existence of a descending three-neuron arc between the labyrinth and cervical motoneurons. Studies are currently in progress in my laboratory that will include a third rotational paradigm. In addition to rotation stimulation of the whole pigeon and the body while the head is earth-fixed, rotation stimulations will be used while the body is earth-fixed and the head oscillated. This will allow for both the vestibular and the ascending somatic inputs to interact simultaneously on those cells shown to be responsive to the first two paradigms.

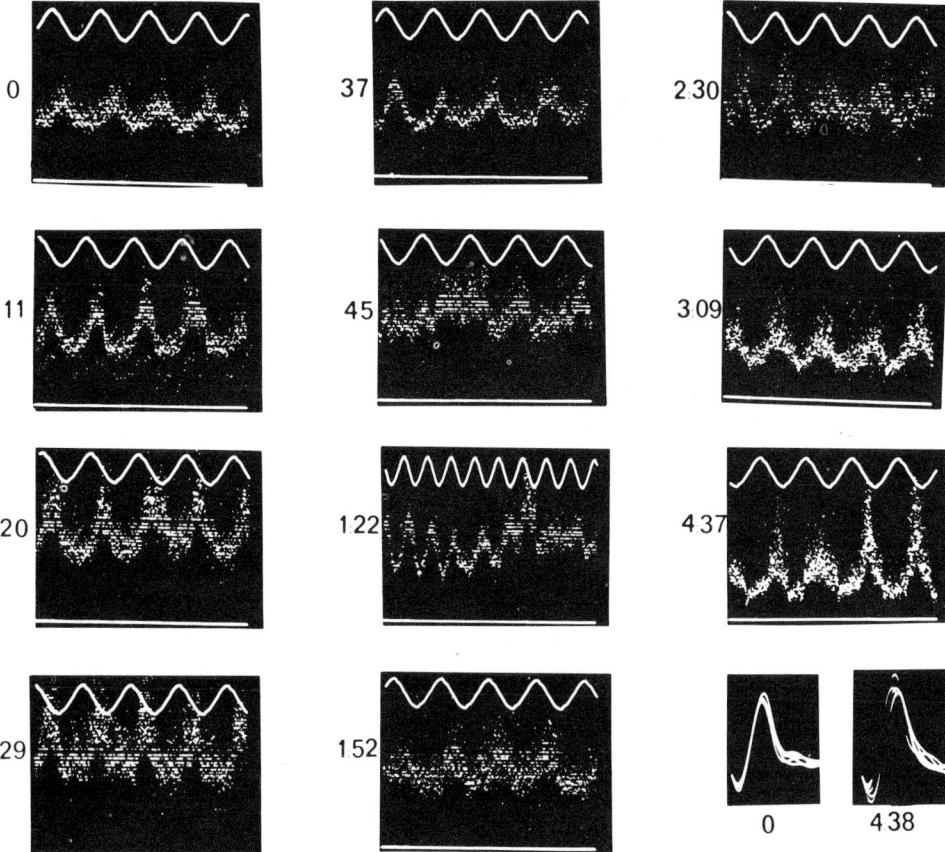

Fig. 6. Dynamic response of a vestibular nucleus neuron to vertical vestibular oscillation at 0.45 Hz. The upper trace in each frame represents turntable position and the lower trace the instantaneous spike frequency. The numbers represent time after an i.v. injection of alcohol.

SUMMARY

The primary vestibular afferent input to the vestibular nuclei and to cerebellar structures was identified and the response properties of cells within these structures were described. In addition, single-unit recordings were made from the cerebellar nuclei, processus lateralis cerebellovestibularis, brain stem vestibular region and reticular formation in awake pigeons under the following experimental conditions: i) either horizontal or vertical rotation, and ii) either rotation of the whole animal in either plane or that of the body alone with the head held fixed relative to earth. In the cerebellar structures single units recorded modulated their firing rate as a function of vestibular stimulation only, never being activated when the body was rotated and the head held fixed relative to earth. In the vestibular nuclei 60% of all cells recorded responded to both vestibular rotational stimulation and rotation of the body with the head held fixed relative to earth. In the reticular formation all cells responded to either rotational stimulation of the whole animal or body rotation alone. We also justify here our use of birds as experimental models for our studies.

REFERENCES

Baker, R. and Berthoz, A. (Eds.) (1977) *Control of Gaze by Brain Stem Neurons*. Elsevier/North-Holland Biomed. Press, Amsterdam.

Brodal, A. (1972) Vestibulocerebellar input in the cat: anatomy. In *Progress in Brain Research, Vol. 37, Basic Aspects of Central Vestibular Mechanisms*, A. Brodal and O. Pompeiano (Eds.), Elsevier, Amsterdam, pp. 315–238.

Brodal, A. and Torvik, A. (1957) Über den Ursprung der sekundaren vestibulocerebellaren Fasern bei der Katze. Eine experimentall-anatomische Studie. *Arch. Psychiat. Nervenkr.*, 195: 550–567.

Brodal, A. and Høivik B. (1964) Site and mode of termination of primary vestibulocerebellar fibers in the cat. *Arch. ital. Biol.*, 102: 1–21.

Dow, R.S. (1936) The fiber connections of the posterior parts of the cerebellum in the cat and rat. *J. comp. Neurol.*, 63: 527–458.

Duensing, F. and Schaefer, K.P. (1958) Die Aktivitat einzelner Neurone un Bereich der Vestibulariskerne bei Horizontalbeschlenigungen unter besonderer Berucksichtigung des vestibularen Nystagmus. *Arch. Psychiat. Nervenkrankh.*, 198: 225–252.

Fuchs, A. and Kimm, J. (1975) Unit activity in the vestibular nucleus of the alert monkey during horizontal angular acceleration and eye movement. *J. Neurophysiol.*, 38: 1140–1161.

Furuya, N., Kawano, K. and Shimazu, H. (1975) Functional organization of vestibulo-fastigial projection in the horizontal semicircular canal system in the cat. *Exp. Brain Res.*, 24: 75–87.

Gacek, R.R. (1969) The course and central termination of first order neurons supplying vestibular endorgans in the cat. *Acta oto-laryngol. (Stockh.)*, Suppl., 254: 1–66.

Gardner, E.P. and Fuchs, A.F. (1975) Single-unit responses to natural vestibular stimuli and eye movements in deep cerebellar nuclei of the alert rhesus monkey. *J. Neurophysiol.*, 38: 627–649.

Ghelarducci, B., Ito. M. and Yagi, N. (1975) Impulse discharges from flocculus Purkinje cells of alert rabbits during visual stimulation combined with horizontal head rotation. *Brain Res.*, 87: 66–72.

Karten, H.J. and Hodos, W. (1966) *A Sterotaxic Atlas of the Brain of the Pigeon (Columba livia)*. The Johns Hopkins Press, Baltimore.

Keller, E.L. and Daniels P.D. (1975) Oculomotor related interaction of vestibular and visual stimulation in vestibular nucleus cells in alert monkey. *Exp. Neurol.*, 46: 187–198.

Kimm, J., Hassul, M. and Cogdell, B. (1976) Fastigial neuronal responses to sinusoidal horizontal rotation. *Exp. Neurol.*, 50: 579–594.

Kimm, J., Samson, H.H. and Winfield, J.A. (1978) The effect of alcohol on vestibular neurons. *Trans. Amer. Acad. Opthal. Otol.*, 86: 72.

Lisberger, S.G. and Fuchs, A.F. (1977) Role of the primate flocculus in smooth pursuit eye movements and rapid behavioral modification of the vestibulo-ocular reflex. In *Control of Gaze by Brain stem Neurons* Baker, R. and Bertnoz, A. (Eds.), Elsevier/North-Holland Biomed Press, Amsterdam-New York, pp. 381–389.

Maekawa, K. and Simpson, J.I. (1973) Climbing fiber responses evoked in the vestibulo-cerebellum of rabbit from visual system. *J. Neurophysiol.*, 36: 649–666.

Noda, H., Asoh, R. and Shibagaki, M. (1977) Floccular unit activity associated with eye movement and fixation. In *Control of Gaze by Brain Stem Neurons*, R. Baker and A. Berthoz (Eds.), Elsevier/North-Holland Biomed. Press, Amsterdam, pp. 371–380.

Precht, W. and Llinas, R. (1969) Functional organization of the vestibular afferents to the cerebellar cortex of frog and cat. *Exp. Brain Res.*, 9: 30–52.

Precht, W. and Llinás, R. (1972) Responses of vestibular nuclear neurons to ampullar input. In *Progress in Brain Research, Vol. 37, Basic Aspects of Central Vestibular Mechanisms*, A. Brodal and O. Pompeiano (Eds.). Elsevier, Amsterdam, pp. 89–108.

Rabin, A. (1975) Labyrinthine and vestibulospinal effects on spinal motoneurons in the pigeon. *Exp. Brain Res.*, 22: 431–448.

Roberts, T.D.M. (1967) *Neurophysiology of Postural Mechanisms*, Plenum Press, New York.

Shimazu, H. and Precht, W. (1965) Tonic and kinetic responses of cat's vestibular neurons to horizontal angular acceleration. *J. Neurophysiol.*, 28: 991–1013.

Shinoda, Y. and Yoshida, K. (1975) Neural pathways from the vestibular labyrinths to the flocculus in the cat. *Exp. Brain. Res.*, 22: 97–111.

Stein, B.M. and Carpenter, M.B. (1967) Central projections of portions of the vestibular ganglia innervating specific parts of the labyrinth in the rhesus monkey. *Amer. J. Anat.*, 120: 281–318.

Walls, G.L. (1972) *The Vertebrate Eye and its Adaptive Radiations*. Cranbrook Inst. of Sci. Bloomfield Hills, MI.
Wilson, V.J., and Peterson, B.W. (1977) Peripheral and central substrates of vestibulospinal reflexes. *Physiol. Rev.*, 58: 80–105.
Winfield, J.A. and Kimm, J. (1979) Dynamic responses of fastigial nucleus neurons in the awake chinchilla. In preparation.
Wilson, V.J., Anderson and Felix, D. (1974) Unit and field potential activity evoked in the pigeon vestibulo-cerebellum by stimulation of individual semicircular canals. *Exp. Brain Res.*, 19: 142–157.

B

Labyrinthine Influences on Spinal Motoneurons

Reactions to Overbalancing

T.D.M. ROBERTS and G. STENHOUSE

Institute of Physiology, University of Glasgow, Glasgow G12 8QQ (Scotland)

It is characteristic of the posture of the terrestrial vertebrates that, except when climbing or flying, the centre of gravity is held above the level of the points of contact with the supporting surface. The condition may be stable or unstable. To distinguish these we need first to define an "area of support" as follows: We take the direction of the resultant of all the contact forces between the body and its supports and then project the positions of the available points of support onto a plane at right angles to this direction. The "area of support" is that polygon, with no re-entrant angles, that just encloses all these projections. Conditions are stable if the projection of the centre of gravity along this same direction of the resultant upthrust falls within the area of support. Overbalancing occurs if the projection of the centre of gravity moves outside the perimeter of the area of support.

By adjusting the activity of the various skeletal muscles an animal can alter the forces with which he pushes against the supports. He can alter the magnitude, direction, and point of application of these forces, and in this way he can impart momentum to the body, either to initiate or to arrest movement relative to the supports.

Before attempting to analyse the mechanisms of motor control involved in balancing we need to identify the nature of the tasks being performed at different times. It is not easy to measure the forces exerted by individual muscles within the body, and changes in the force against the ground are not readily appreciated from simple visual observation of a subject's behaviour. In oneself one can feel changes in the sensations from the feet as weight is transferred from one foot to the other. What we have done is to study balancing and overbalancing behaviour in man using a force platform furnished with strain gauges whose signals are combined by computer and displayed as a moving vector indicating the point of application, magnitude and direction of the resultant contact force between the feet and the ground (for details, see Roberts and Stenhouse, 1977). The moving vector display shows very clearly what is happening (Fig. 1). It reveals that inclined thrusts of two kinds are developed, which can be distinguished according as the line of thrust does or does not pass through the centre of gravity. If the thrust line does not pass through the centre of gravity, torque is needed as well as thrust, if rotation is to be avoided.

When the trunk is swaying to and fro in quiet standing, the point of application of the resultant force against the ground also moves (Roberts and Stenhouse, 1976). The body may be seen to remain momentarily at rest in many different positions each involving a different distribution of weight between the feet, but with the resultant

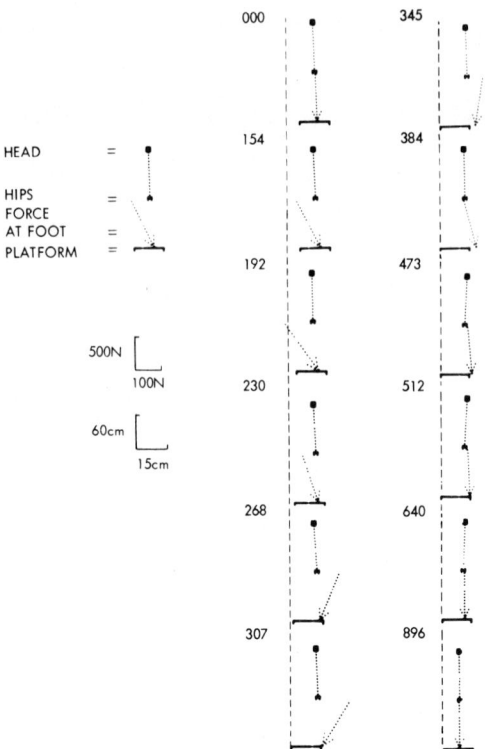

Fig. 1. Successive appearances of the moving vector display, at the times indicated in msec, during a brisk leftward movement of the platform through 6.5 cm. Note that when the platform is exerting horizontal forces first to accelerate and then to decelerate the legs, the point of application of the support thrust maintains its position below the trunk, in spite of the fact that the feet necessarily move with the platform. In the subsequent rescue reaction the thrust point moves out to the right to provide purchase for the leftward acceleration of the trunk. See also Fig. 2.

force vertical in each case. This behaviour was illustrated by a film. Horizontal movements of the trunk are initiated and arrested in apparently sporadic fashion by inclinations of the support vector that involve changes in the point of application of the thrust at the foot (see frame at 384 msec in Fig. 1).

If the subject is standing on a platform which is moved slowly in a horizontal direction, the subject maintains his position over the platform by a somewhat speeded up pattern of swaying movements. These movements, though faster and more frequent than those occurring on a stationary base, do not appear to differ significantly in general character from the swaying movements of quiet standing (illustrated in the film).

If the platform is moved somewhat more briskly, then, even when the total excursion is kept well within the extent of the subject's base, so that there is no real risk of the subject overbalancing, we see a different type of behaviour (Figs. 1 and 2). The support vector shows a rapid and substantial change of inclination (see 154 msec in Fig. 1). The point of application of the resultant thrust moves relative to the foot but remains at first comparatively stationary in space (stabilization phase in Fig. 2, first seven frames in Fig. 1). That is to say, although the foot necessarily moves with the

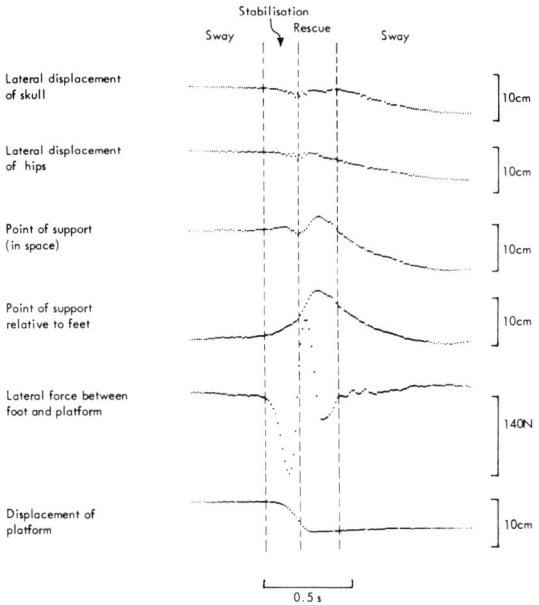

Fig. 2. Time courses of the variables recorded during the manoeuvre of Fig. 1.

platform, the effective point of support is held in its appropriate place below the trunk.

The head and trunk also remain in place in spite of the movement of the legs. For this to happen, there must be a change in the torque exerted at the hips. In the normal upright stance there is a tendency for the legs to slew round under the weight of the trunk and this tendency must be opposed by muscular action to provide a restraining torque. If the restraining system had a finite compliance, relative displacement of the legs would be accompanied by a restoring torque which, in the conditions now being considered, would have the effect of accelerating the trunk in the direction of the platform movement and at the same time of initiating a clockwise rotation of the trunk. But no such acceleration or roll occurs. It follows that the stabilization of the trunk during the brisk movement of the platform implies the development of a torque at the hips, acting clockwise upon the legs.

The relative movement of the support point carries it nearer to the perimeter of the available area of support and this eventually presents a threat of overbalancing. The threat is seen to be countered by a "rescue reaction" akin to that of staggering. The support point is flung out in the direction of the threatened fall (see 384 msec in Fig. 1, rescue phase in Fig. 2) and thus provides a purchase for an inclined thrust which can then accelerate the trunk in the direction of the imposed platform movement so as to keep the trunk in its station over the feet.

A comparison of the lateral force at the foot with the signal from an accelerometer attached to the platform (Fig. 3) reveals that in the early part of the platform movement only about one-third of the mass of the subject's body is being accelerated and that this is being carried passively with the platform. One may suppose this to correspond to the effective mass of the legs pivoting at the hips. The rest of the body remains relatively stationary during this stabilization phase. At about 180 msec or so after the

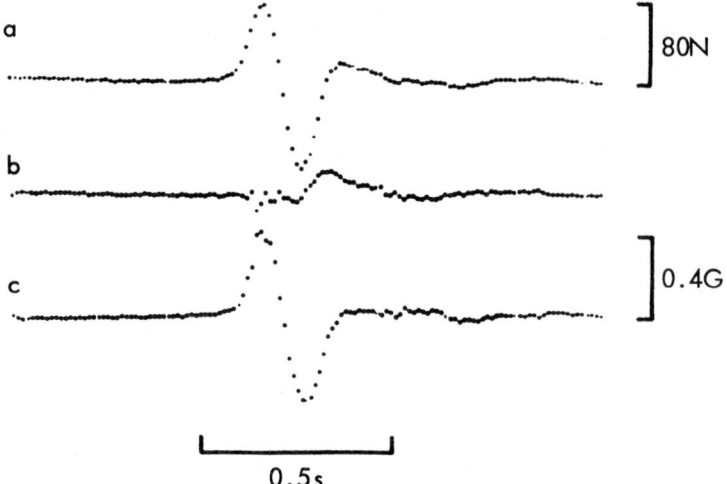

Fig. 3. Comparison between lateral force and platform acceleration during a manoeuvre similar to that of Figs. 1 and 2. a: lateral force at foot; c: acceleration of platform; b: the result of subtracting c from a. Note that this remains substantially unchanged for about 180 msec from the onset of the platform movement.

onset of the platform movement a sudden additional lateral force develops, indicating that a larger mass is now being accelerated. It is at this stage that the support point moves out in the rescue reaction referred to above. The effect of this new lateral force shows itself first at the hips and only later at the head (Fig. 2). After the trunk has begun its movement to "catch up" with the platform, there follows a resumption of the slow sway pattern of quiet standing, with the trunk remaining within the region of stable support over the new position of the feet.

During the early part of the brisk movement of the platform the mass of the trunk appears to be largely uncoupled from the horizontal movement imposed on the feet. The trunk and skull remain in the same place, the legs adopting compensatory attitudes to accommodate the change in relative position of the feet.

Corresponding phases of stabilization, compensation and rescue are also seen in the responses of subjects seated on a tilt table (Purdon Martin, 1967; Roberts, 1978).

The avoidance of toppling on the moving platform involves a movement of the point of application of effective thrust toward the direction of the impending fall. What happens if the support point has already reached the edge of the available area of support? Is a fall inevitable if there is no way of extending the support area by stepping or hopping? We have studied this problem in the following tests (Roberts, 1977, and film illustrating this presentation). The subject stands on a small plank resting on two parallel edges placed about 3.5 cm apart on the force plate. He moves his weight gradually forward until the plank lifts off one of the supports. His task is then to balance on the single remaining edge.

On the single edge there can be no movement of the point of support relative to the feet. Furthermore, so long as the line of action of the supporting force passes through the centre of gravity of the body there is no reason for the nervous system to institute corrective movements as the neural criteria for stability (Roberts, this volume, chapter IVD1) are then met. However, any casual misalignment of the force relative to the position of the centre of gravity will cause the body to start to fall, and because the area of support is restricted to a single line, the usual repertoire of balancing behaviour is ineffective and the subject falls off after only a very few seconds.

Physical considerations might at first suggest that there is no possibility of recovery once the body has started to topple. Suppose that the direction of support thrust for successful balance lies in the direction UP in Fig. 4a, while the centre of gravity of the body is at O, i.e., the subject has begun to topple forward over the edge U. Any thrust applied at U, such as in the direction of X or of Y, will have effects which can be analysed by resolving the applied force into one component along UO and another at right angles to this. The first component can accelerate the centre of gravity of the body only along the line UO. The second produces only a rotation of the body about its centre of gravity, with no movement of the centre of gravity itself. Thus there appears to be no way in which O can be brought nearer to UP by a thrust constrained to pass through U.

The situation of Fig. 4a applies to a rigid body, whereas in man one part of the body may move relative to another. The possibilities illustrated in Fig. 4b thus become available. We suppose the trunk, with centre of gravity at T, to be supported at the hip,

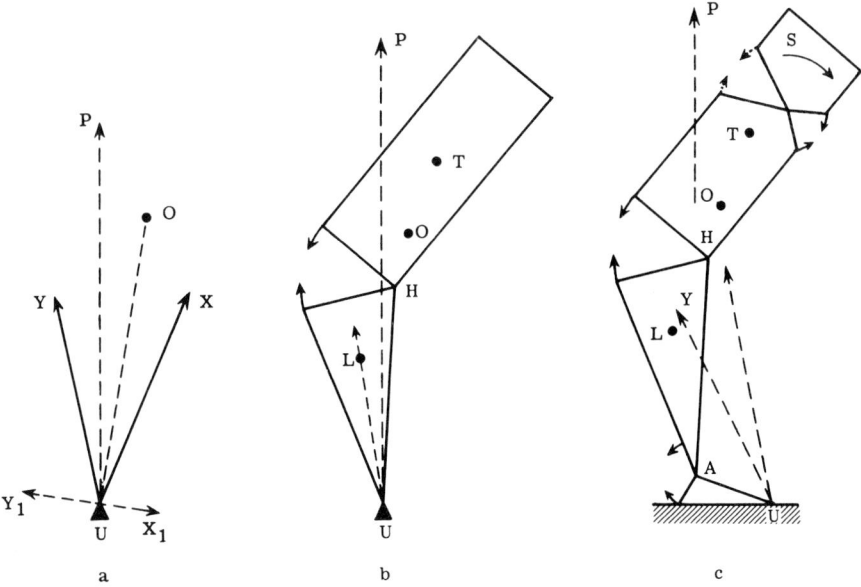

Fig. 4. Illustration of the effects of various forces. a) The centre of gravity, O, of the body lies to the right of the line UP indicating the direction of thrust appropriate to equilibrium. Forces in directions such as UX and UY fail to move O towards UP because the resolved component along UO can move the body only in that direction, while the resolved components UX_1 or UY_1, at right angles to UO, produce rotations about O without displacement. b) Recovery from the brink. O is to the right of UP, indicating unbalance. Relaxation of the torque at the hip allows the trunk, with centre of gravity at T, to fall. The legs, with centre of gravity at L, which is to the left of UP, topple to the left, carrying the trunk also to the left, which is the direction for rescue. c) Effect of torque at the ankle and of sweeping movement of the arms. The point of thrust is moved to U, which is to the right of O. Forward overbalancing is thus corrected even though O is to the right of AP. The movable segment, S, consists of the shoulders, arms, and anything held in the hand. It pivots on the trunk about the sternoclavicular joint. Clockwise torque applied to S is associated with equal anticlockwise torque applied to the trunk. This opposes the forward pitching of the trunk arising when the thrust UY passes to the left of O.

H, by the combined action of thrust at the hip joint together with torque applied by muscles such as the hamstrings and gluteals. The centre of gravity of the legs is at L, and the legs are supported at the single edge at U. The centre of gravity of the whole body is at O, which in our case lies to the right of UP, indicating that the subject has started to overbalance in the forward direction.

A possible strategy for "recovery from the brink" in these conditions is as follows. The hip torque is relaxed. This allows T to fall freely forward. The centre of gravity of the legs alone lies to the left of UP, so that in the absence of hip torque the legs will fall to the left like a felled tree, the hips moving to the left and carrying the trunk with them (illustrated with a model). Thrust against the ground, acting along UL, can move L further to the left so that O also moves to the left, which is the required direction for correcting the overbalancing. Once the desired movement of O has been initiated, the hip torque is turned on again strongly to arrest the fall of the trunk.

The sweeping movements of the arms in certain rescue reactions provide a means of partitioning the angular momentum of the body (see Fig. 4c). When clockwise angular momentum is being developed in the movable segment, S, the anticlockwise reaction torque on the trunk opposes the forward pitching that would otherwise be produced by the thrust UY, whose line of action passes to the left of O. Note that, although O is to the right of AP, the action of the torque at the ankle is to shift the effective point of support to U, which is itself to the right of O, so that the problems associated with a narrow base do not arise in this case. The sweeping manoeuvre is not successful by itself if U is unable to move, as in the fixed edge situation. Habitual arm sweeping does, however, usually accompany attempts to return from the brink. A disadvantage of such sweeping movements when attempting to balance on a fixed narrow base is that the pitching of the trunk reappears at the end of each arm movement when the time comes for the angular momentum of the movable segment to be absorbed. This means that each successive stage in the necessary alternation of corrections, first on one side of the support and then on the other, is interfered with by the consequences of earlier arm movements, adding further difficulty to the task.

Manoeuvres involving movable segments, and using purely internal forces, can have no effect either on the linear momentum of the centre of gravity or on the total angular momentum about that point. They can be of use, however, to adjust the attitude of the body during free fall, as on the trampoline or in an orbiting spacecraft (see Roberts, 1978).

In the normal balance of standing man the legs must be appropriately aimed as well as being stiffened to provide thrust. Stretch reflex activity in the limbs serves to prevent the legs from collapsing under the compressive thrust of weight-bearing. As regards aiming, the direction of the action of the effective strut can be altered by shifting the centre of pressure at the feet. In the fore-and-aft direction this can be done by applying torque at the ankle. In the lateral direction, the weight may be shifted from one leg to the other by changing the torque applied between the pelvis and the trunk. This action is separate from, though often accompanied by, changes in the torque at the hips, such as those used to produce lateral displacement of the trunk, or when the body is supported on one leg only. It is not clear how the sporadic aiming of limb thrusts is organised to produce the normal sway pattern of quiet standing. The rescue reactions of staggering, sweeping, stepping and hopping are invoked only when the success of the normal sway activity in keeping the thrust point within the area of support is threatened by external perturbation. Again it is not clear what initiates the

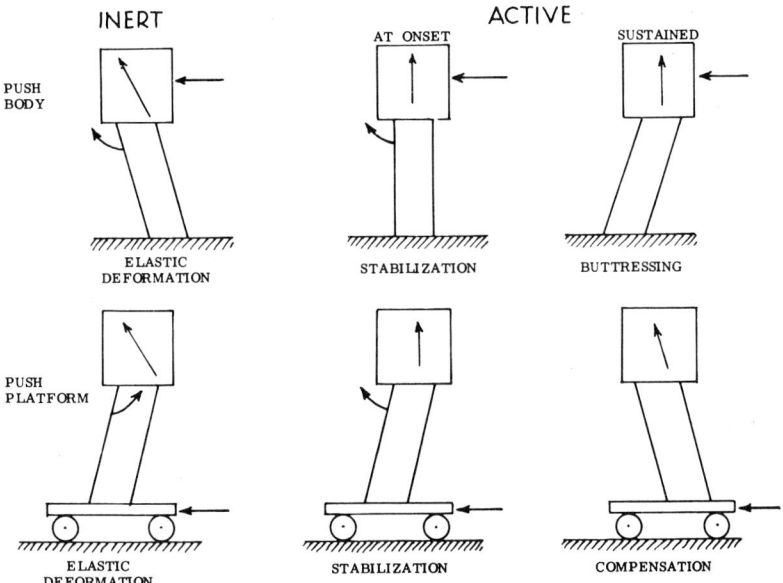

Fig. 5. Comparison of the reactions of inert and active systems subjected to lateral accelerations.

rescue reactions, nor indeed whether they differ in kind as well as in magnitude from the movements involved in normal sway.

Our results show that the body does not behave like an inert system of springs and masses. Fig. 5 illustrates that the application of horizontal forces to an inert system produces opposite torques according as the force is applied to the "trunk" or to the platform. In contrast, a man resists forces applied to the trunk, but initially permits the platform to move under him. In each case, the torque developed is in the same sense, clockwise in the figure, to stabilize the body against threatened leftward acceleration. If the disturbing force to the platform is maintained, the point of thrust moves toward the perimeter of the available area of support and this initiates rescue reactions. These rearrange the relative positions of the feet, by staggering, stepping or hopping, to allow the body take up a new posture aligned with the new direction of the prevailing resultant thrust, and with appropriate compensatory attitudes of the limbs.

If sustained leftward lateral force is applied to the trunk rather than to the platform, the rescue reactions move the point of thrust at the feet to the left so that the limbs can buttress against the external force to produce a vertical resultant.

Not only do the reactions of the body differ from those of an inert structure, but we may see opposite reactions at different times. An initial phase of stabilization allows the platform to move under the body. Then follows a phase of compensation in which the body changes its posture to take account of the new position of the "behavioural vertical" (Roberts, 1975).

SUMMARY

The behaviour of the force against the ground supporting a standing subject is displayed as a moving vector using a computer connected to the strain gauges of a force platform. The normal sway pattern of quiet standing is seen to be speeded up when the platform is slowly moved horizontally. Brisker movements evoke rescue reactions when the point of thrust approaches the perimeter of the available area of support. Stabilizing and compensatory movements follow different time courses and in some conditions produce opposite effects. When the support is restricted to a single edge the usual strategy of balancing reactions is not effective and new skills are called for.

REFERENCES

Purdon Martin, J. (1967) *The Basal Ganglia and Posture.* Pitman, London.
Roberts, T.D.M. (1975) The behavioural vertical. *Fortschr. Zool.*, 23: 192–198.
Roberts, T.D.M. (1977) Balance on a fixed narrow base. *J. Physiol. (Lond.)*, 273: 37–38P.
Roberts, T.D.M. (1978) *Neurophysiology of Postural Mechanisms*, 2nd ed. Butterworths, London.
Roberts, T.D.M. and Stenhouse, G. (1976) The nature of postural sway. *Agressologie*, 17A: 11–14.
Roberts, T.D.M. and Stenhouse, G. (1977) Moving vector display for the study of balance and of the reactions to perturbation. *J. Physiol. (Lond.)*, 273: 8–9P.

Semicircular Canal and Macular Influences on Neck Motoneurons

M. MAEDA

Department of Neurosurgery, School of Medicine, Juntendo University, Bunkyo-ku, Tokyo (Japan)

SEMICIRCULAR CANAL INPUTS TO NECK MOTONEURONS

If animals are free and individual semicircular canal nerves are stimulated, both head and eyes move. When balance is upset, the horizontal position of the head is first restored by vestibular action on the neck muscles (Roberts, 1976). Suzuki and Cohen (1964) have demonstrated in the cat and monkey that semicircular canal stimulation evokes head and eye movements parallel to the plane of the activated canal. Head movements have also been introduced by caloric stimulation in man (Henricksson et al., 1962). Wilson and Yoshida (1969a–c) have studied the neuronal organization of this labyrinthine reflex on the neck musculature and have shown that electrical stimulation of the vestibular nerve evokes disynaptic excitatory and inhibitory postsynaptic potentials in neck motoneurons of cats. It has also been assumed that ipsilateral EPSPs are due to the lateral vestibulospinal tract (LVST), whereas IPSPs are due to the medial vestibulospinal tract (MVST) (Wilson, 1972).

Several questions now arise concerning connections between the labyrinth and neck motoneurons: 1) the pattern of connections between different receptors and neck motoneurons; 2) how the LVST and MVST play in relaying excitation and inhibition to neck motoneurons.

Bipolar stimulation electrodes made of insulated 40 μm stainless steel wire were implanted near the ampullary nerves as described by Suzuki et al. (1969). Adequacy of implantation was tested by observing characteristic eye movements elicited by stimulation of ampullary nerves (Cohen et al., 1964). Intracellular recording from dorsal neck motoneurons (splenius, SP; biventer cervicis, BIV; and complexus, COMP), usually in the C3 segment, was performed in precollicularly decerebrated cats. The muscles that are innervated by BIV and COMP motoneurons are extensors that raise the head and have a remarkably high spindle density (Richmond and Abrahams, 1975). The SP muscle is a lateral flexor, but when acting bilaterally the two splenii act to elevate the head.

Stimulation of the ipsi- and contralateral ampullary nerves evoked EPSPs and IPSPs in all neck motoneurons. As illustrated by the typical potentials in Figs. 1 and 2, EPSPs and IPSPs had very often more than one peak, even with the weaker stimuli. The later components were typically of the same polarity as early potentials. Thresholds of EPSPs and IPSPs were often less than 25 μA. These thresholds were near, or less than 2 times N threshold (Shimazu and Precht, 1965), a strength comparable to that required to activate neurons in the vestibular nuclei monosynaptically

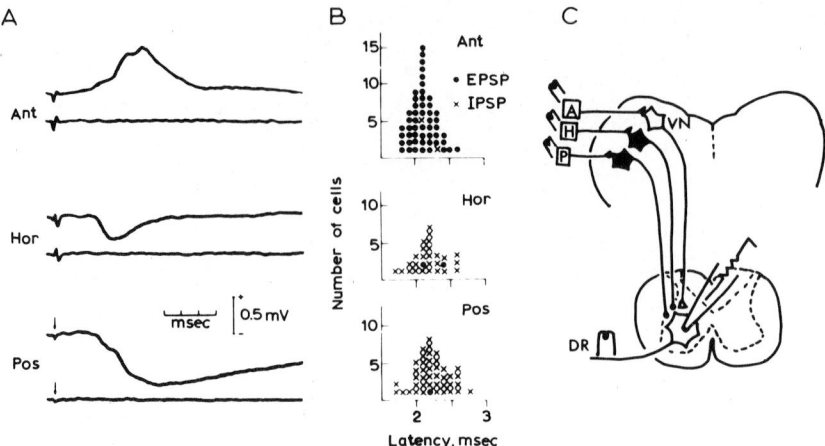

Fig. 1. Typical synaptic potentials evoked in neck motoneuron by stimulation of ipsilateral ampullary nerves (A) and latencies of synaptic potentials (B). In each frame of A the upper trace represents the intracellular response, the lower trace the extracellular field potential. Stimulus strength, 50 μA for anterior canal (Ant), 75 μA for horizontal (Hor), and 20 μA for posterior (Pos). All potentials are averages of 50 sweeps. Arrows indicate time of stimulus. (From Wilson and Maeda, 1974.)

(Wilson, 1972) and to produce disynaptic potentials in neck motoneurons when stimulating the whole vestibular nerve (Wilson and Yoshida, 1969a).

Stimulation of both the ipsilateral and contralateral anterior ampullary nerve evokes EPSPs in BIV and COMP motoneurons (Figs. 1A and 2A). When the different peaks of the synaptic potential were sufficiently separated, the amplitude of the early EPSP was measured. For ipsilateral EPSPs the amplitude was usually between 100 and 500 μV. On the other hand, the amplitude of the peak depolarization ranged from 100 to 2000 μV and was sometimes large enough to cause firing. The effects

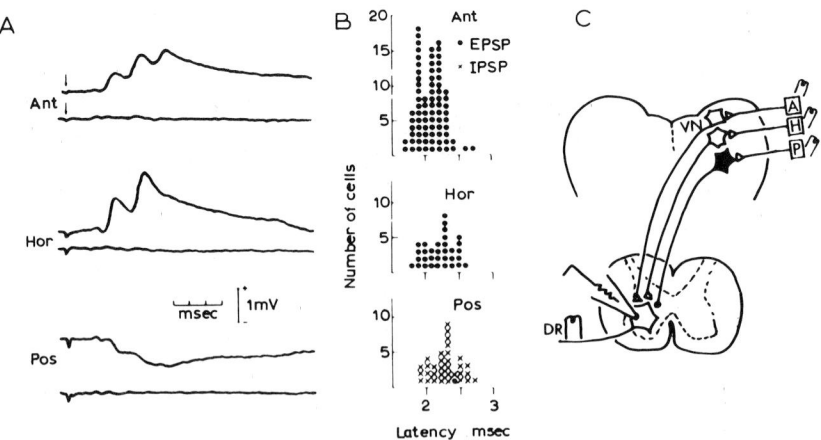

Fig. 2. Response of a BIV motoneuron to stimulation of contralateral ampullary nerves (A) and latency histogram (B). The upper trace shows the intracellular response, the lower trace the extracellular field potential. Stimulus strength, 50 μA for the anterior canal, 80 μA for the horizontal and 75 μA for the posterior. (From Wilson and Maeda, 1974.)

obtained from the two sides were not equivalent and stimulation of the contralateral anterior ampullary nerve produced the bigger disynaptic response and peak depolarization; the ratio of the contralateral to ipsilateral amplitudes usually was 2–4 : 1.

IPSPs were bilaterally produced in neck motoneurons after stimulation of the posterior ampulla (Figs. 1C and 2C). The IPSPs were easily reversed by passing hyperpolarizing current through the recording electrode, suggesting that the inhibitory synapses are near the soma. Stimulation of the horizontal ampulla gives the least consistent effects in BIV and COMP cells and is most effective in motoneurons, in which it leads to contralateral EPSP and ipsilateral IPSP.

The latency of the earliest component of EPSPs and IPSPs ranged from 1.7 to 2.8 msec as shown in Figs. 1B and 2B and these latencies are similar to those obtained with whole-nerve stimulation and shown to be disynaptic (Wilson and Yoshida, 1969a).

The pattern of excitatory and inhibitory connections between ampullae and neck motoneurons is consistent with head movements produced by electrical stimulation of the ampullae (Suzuki and Cohen, 1964). For example, stimulation of the anterior ampullary nerves produces excitation of dorsal neck motoneurons; it evokes EPSP, not only in contralateral but also in ipsilateral neck extensors. This stimulus causes the head to move obliquely upward and backward to the opposite side (Suzuki and Cohen, 1964), since direction of head movement could be determined by the fact that excitation is stronger in contralateral motoneurons.

PATHWAYS LINKING AMPULLARY NERVES TO DORSAL NECK MOTONEURONS

The MVST or LVST was interrupted in order to identify the pathways involved in production of the synaptic potentials. A typical medullary lesion is illustrated in Fig. $3B_1$; the ipsilateral MLF was completely interrupted at the level of the hypoglossal nerve. Fig. $3B_2$ shows the potentials observed in a COMP cell after MLF cut; anterior canal stimulation still produced an EPSP, whereas posterior canal stimulation was ineffective. Following interruption of the LVST in the medulla, sparing the MLF (Fig. $3C_1$), stimulation of the anterior ampullary nerve no longer evoked EPSPs, although horizontal and posterior ampullary nerve stimulation still evoked IPSPs (Fig. $3C_2$). The results of experiments with medullary lesion show that disynaptic EPSPs evoked by ipsilateral anterior ampullary stimulation are mediated by the LVST, and that disynaptic IPSPs elicited from the ipsilateral horizontal or posterior canal are mediated by the MVST as schematically illustrated in Fig. 3A. The small remaining synaptic potentials with longer latencies sometimes observed after medullary lesion, however, could be due to other structures.

Interruption of the MLF ipsilateral to the motoneurons abolished disynaptic EPSPs and IPSPs evoked in neck motoneurons by contralateral ampullary stimulation (Fig. 4). Small late potentials were often seen at latencies of 3.2–5.0 msec.

According to the present results the pathways connecting the ampullae to neck motoneurons can be summarized in Figs. 3A and 4A: all inhibitory fibers are in the MVST, as are contralateral excitatory fibers; ipsilateral excitatory fibers are in the LVST. The results confirm the finding of Akaike et al. (1973) that the MVST

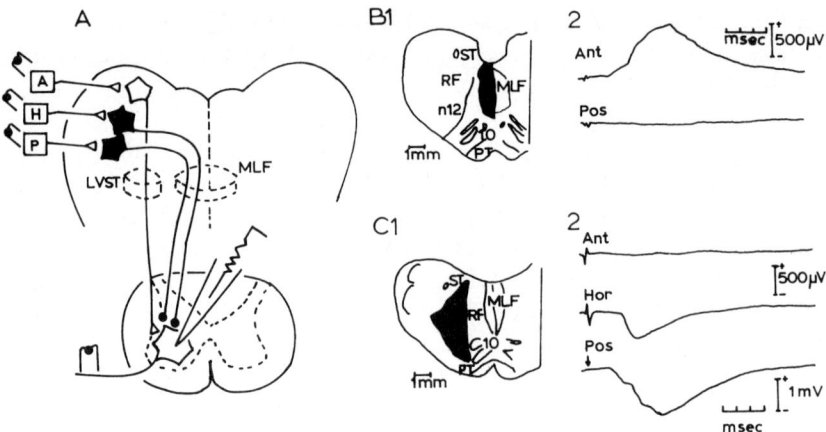

Fig. 3. Effect of MLF and LVST section on synaptic potential by stimulation of the ipsilateral labyrinth. A) Schematic drawing of connections between ipsilateral ampullae and neck motoneurons. Inhibitory neurons and their terminals shown in black, excitatory in white. B) Effect of MLF section. The cut is shown by the hatched area in B_1. Potentials recorded in a COMP cell after the cut is shown in B_2. Potentials obtained by averaging 50 sweeps. C) Effect of LVST section. The medullary lesion is shown by the hatched area in C_1. C_2 shows potentials evoked in a COMP motoneuron after the cut. Abbreviations: A, H, P, anterior, horizontal, and posterior ampullae; VN, vestibular nuclei; ST, solitary tract; RF, reticular formation; n12, hypoglossal nerve; IO, inferior olive; PT, pyramidal tract. (B and C, from Wilson and Maeda, 1974.)

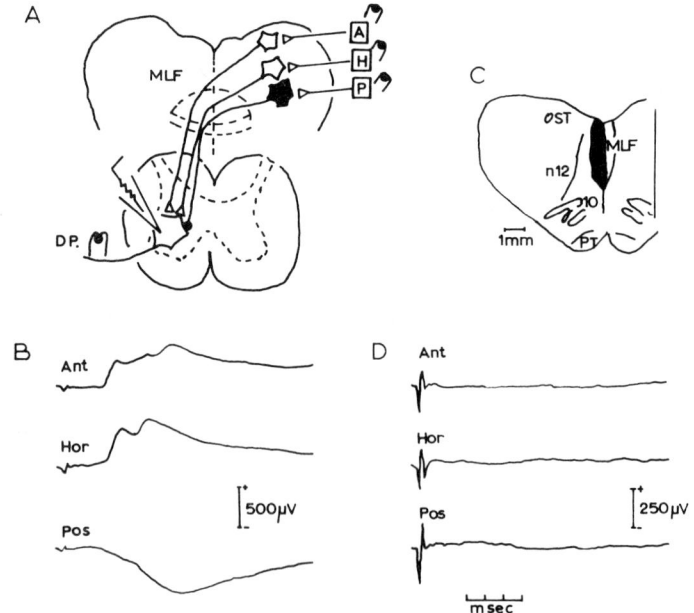

Fig. 4. Effect of MLF cut ipsilateral to motoneurons on response to contralateral ampullary nerves. A) Schematic drawing of connections between contralateral ampullae and neck motoneurons. Inhibitory neurons shown in black, excitatory in white. Abbreviations as in Fig. 3. B) Responses recorded in a COMP motoneuron before the cut. C) Lesion is shown by the dark area. D) Lack of response after the cut. Potentials in B and D are averages of 50 sweeps. (B–D, from Wilson and Maeda, 1974).

contains excitatory fibers and show that these fibers cross in the brain stem and provide the only short-latency excitatory pathway from ampullae to contralateral neck motoneurons. In contrast to the organization of contralateral excitation, excitation of ipsilateral neck motoneurons does not involve the MVST, but it must be relayed by the LVST. Similarly, Akaike et al. (1973), observed disynaptic EPSPs in C_1 neurons ipsilateral to the labyrinth after MLF section.

Although all short-latency effects are relayed by the LVST and MVST, it needs not be assumed that all synaptic potentials produced in neck motoneuron by stimulation of the ampullae are due to these tracts. Some of the late potentials seen after medullary cut, previously also observed by Akaike et al. (1973), may be transmitted by the reticulospinal tract. Recently, Ezure et al. (1978) have made frequency-response analysis of horizontal canal induced neck reflex in cat. With respect to functional significance of descending pathways they have detected no effects on the dynamic characteristics of the frequency response after interruption of the MVST in the MLF, showing the existence of other effective descending pathways which function in response to natural, rotatory stimulation.

SACCULAR AND UTRICULAR INPUTS TO NECK MOTONEURONS

It is generally accepted that the utricle is an important source of vestibulospinal reflexes. Mechanical stimulation of the utricular macula causes contraction of neck muscles (Szentagothai, 1952). Lateral tilt, which activates utricular receptors, excites vestibulospinal neurons (Peterson, 1970) and influences activities of limb muscles. In contrast, there are few electrophysiological analyses of a contribution of the sacculi to vestibulospinal reflexes. Vertical linear acceleration produces activation of neck and limb musculatures in the cat and this response is probably due to activation of the sacculus (Watt, 1976). We have tried to obtain electrophysiological evidence for saccular and utricular input to neck motoneuron (Wilson et al., 1977).

In order to stimulate saccular and utricular afferents, we have transected all branches of vestibular nerve except the one to the sacculus or utriculus, and allowed time for degeneration (Stein and Carpenter, 1967; Gacek, 1969). Thus we have prepared a chronic animal with only one intact nerve branch which could be stimulated in acute experiments (see Fig. 1 of Wilson et al., 1977).

Saccular inputs

In all experiments field potentials evoked by stimulation of the vestibular nerve on the operated and normal sides were first checked in the vestibular nuclei. Field potentials on the normal side consisted of the typical P, N_1, and N_2 waves already described in detail by Shimazu and Precht (1965), as is shown in Fig. 5B. In contrast, potentials evoked by saccular nerve stimulation usually had a different appearance: P-wave was not clearly visible, the negativity was smaller, often somewhat later (foot of negativity of 0.6–1.3 msec) and more slowly rising (Fig. 5A). The negativity resembles that recorded by Hwang and Poon (1975) at a similar location, and will be referred to as the N_1 potential. Stimulation of the saccular nerve evoked in many neck motoneurons synaptic potentials. Both EPSP and IPSPs were observed. Threshold of synaptic potentials ranged from 1–4 times N_1 threshold and was often near $2N_1T$; comparable multiples of N_1T are required to evoke synaptic potentials when

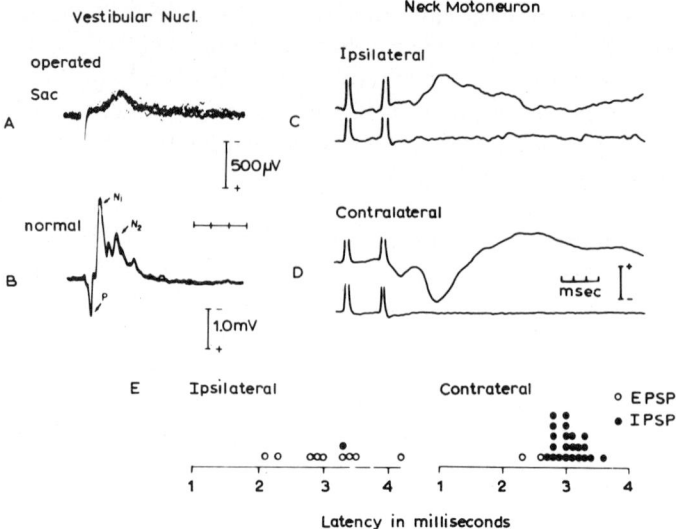

Fig. 5. Saccular effects on neck motoneurons. A, B) Field potentials in the vestibular nuclei by stimulation of the saccular nerve (A) or the whole vestibular nerve (B). C, D) Computer-averaged synaptic potentials evoked in BIV motoneurons by stimulation of the saccular nerve. The upper trace shows the intracellular response; the lower trace, the extracellular field potential. Voltage calibration 100 μV for C, 200 μV for D. E) Latencies of synaptic potentials evoked by stimulation of the saccular nerve. (From Wilson et al., 1977.)

stimulating ampullary nerve. Temporal summation was frequently required to evoke clear PSPs; in contrast, single shock to ampullary nerves is almost always sufficient to evoke a response as described above. Potentials were generally small (IPSPs: 100–150 μV; EPSPs: 50–100 μV), even with two to three shocks. Fig. 5C and D shows the typical sacculus-evoked synaptic potentials; mainly IPSPs were seen contralaterally, and EPSPs ipsilaterally. The latencies of the synaptic potentials were longer than latencies of potentials evoked by stimulation of the normal vestibular nerve, suggesting that some of the potentials are trisynaptic. Since the vestibular fields are small and have a late peak, it is also likely that some potentials are disynaptic. In addition to the di- or trisynaptic potentials described above, later, usually depolarizing, potentials were seen in motoneurons. We have not studied them in this series of experiments.

Utricular inputs

Although it proved difficult to prepare cats with only the utricular nerve intact, there was consistency in the results obtained from these cats. Stimulation of the utricular nerve evoked clear field potentials in the brain stem as in the sacculus experiments; the field potentials were considerably smaller than those of the normal side. Synaptic potentials were evoked in many motoneurons at thresholds that were low multiples of N_1T. The synaptic effects obtained in the utricular series were only IPSPs ipsilaterally, many more EPSPs than IPSPs contralaterally. Whether the mixture of effects in the same neurons or in the same experiment is due to contamination by non-utricular branches or to a heterogeneous population of utricular afferents could not be decided.

The latencies of the typical utriculus-evoked synaptic potentials ranged from 1.8 to 3.5 msec. It seems certain that many of earlier ones are disynaptic, whereas later potentials are very likely trisynaptic (see Figs. 6 and 7 of Wilson et al., 1977).

Since saccular and utricular afferents terminate in the lateral and descending nuclei, in areas known to give rise to lateral and medial vestibulospinal fibers and many of these fibers connect monosynaptically with neck motoneurons (Wilson and Yoshida, 1969a, b), stimulation of macular afferents may be expected to evoke disynaptic potentials in these motoneurons. In fact, many of the utricular- and some of the saccular-evoked potentials are disynaptic. Hwang and Poon (1975) recorded cells in the vestibular nuclei at latencies of 1.0–1.4 msec after stimulation of the saccular macula, in the ketamine-anesthetized cat. In our chloralose-anesthetized cats, latencies of firing of neurons in the vestibular nuclei may often have been later because of the small size and slow rise to peak of the field potential (Fig. 5A). Nevertheless, it is very likely that many of the saccular-evoked potentials were produced via disynaptic connections.

Present results reveal the presence of short-latency connections between the utricular and saccular maculae and dorsal neck motoneurons. Recent experiments with sinusoidal linear and angular acceleration have shown that there is a long phase lag between input acceleration and neck or forelimb muscle responses (Anderson et al., 1977; Ezure et al., 1978). This suggests that pathways more complex than di- or trisynaptic ones sometimes are important in the production of reflex responses to activation of macular or canal receptors. Whatever pathways are important under different conditions, our results confirm previous observations of utricular input to neck muscle (Szentagothai, 1952) and show that it consists in part of short-latency connections and also suggest that the sacculus not only participates in vestibulo-ocular reflexes but may also contribute to vestibulo-neck reflexes (Roberts, 1967; Berthoz and Anderson, 1971; Watt, 1976).

SUMMARY

Synaptic potentials were recorded in cat dorsal neck motoneurons in response to stimulation of individual ampullary nerves or macular nerves.

1. Disynaptic EPSPs and IPSPs were seen in dorsal neck motoneurons by stimulation of individual ampullary nerves. The early disynaptic potentials were usually followed by later components. Dorsal neck motoneurons (biventer cervicis and complexus) are excited by stimulation of the two anterior canals, inhibited from the two posterior ampullae. A majority of these motoneurons were inhibited from the ipsilateral, excited from the contralateral horizontal canal. Splenius motoneurons are influenced most consistently from the horizontal canal. Lesion experiments showed that bilaterally evoked disynaptic inhibition was transmitted by the medial vestibulospinal tract, as was contralateral excitation. Ipsilateral excitation was abolished by interruption of the lateral vestibulospinal tract (Wilson and Maeda, 1974).

2. Stimulation of the saccular nerve usually evoked IPSPs in contralateral, EPSPs in ipsilateral neck motoneurons. Many of the potentials were probably disynaptic, though some were trisynaptic. The effects of utricular nerve were complex, perhaps because of contamination by canal afferents. The predominant effects consisted of

ipsilateral inhibition and contralateral excitation. Many were disynaptic (Wilson et al., 1977).

REFERENCES

Akaike, T., Fanardjian, V.V., Ito, M. and Ohno, T. (1973) Electrophysiological analysis of the vestibulospinal reflex pathway of the rabbit. II. Synaptic actions upon spinal neurones. *Exp. Brain Res.*, 17: 497–515.
Anderson, J.H., Soechting, J.F. and Terzuolo, C.A. (1977) Dynamic relations between natural vestibular inputs and activity of forelimb extensor muscles in the decerebrate cat. I. Motor output during sinusoidal linear accelerations. *Brain Res.*, 120: 1–15.
Berthoz, A. and Anderson, J.H. (1971) Frequency analysis of vestibular influence on extensor motoneurons. II. Relationship between neck and forelimb extensors. *Brain Res.*, 34: 376–380.
Cohen, B., Suzuki, J. and Bender, M.B. (1964) Eye movements from semicircular canal nerve stimulation in the cat. *Ann. Otol. (St. Louis)*, 73: 153–169.
Ezure, K., Sasaki, S., Uchino, Y. and Wilson, V.J. (1978) Frequency-response analysis of vestibular-induced neck reflex in cat. II. Functional significance of cervical afferents and polysynaptic descending pathways. *J. Neurophysiol.*, 41: 459–471.
Gacek, R.R. (1969) The course and central termination of first order neurons supplying vestibular and organs in the cat. *Acta oto-laryngol. (Stockh.)*, Suppl. 254: 1–66.
Henricksson, N.G., Dolowitz, D.A. and Forssman, B. (1962) Studies of cristo-spinal reflexes (laterotorsion) I. A method for objective recording of cristospinal reflexes. *Acta oto-laryngol. (Stockh.)*, 55: 33–40.
Hwang, J.C. and Poon, W.F. (1975) An electrophysiological study of the sacculo-ocular pathways in cats. *Jap. J. Physiol.*, 25: 241–251.
Peterson, B.W. (1970) Distribution of neural responses to tilting within vestibular nuclei of the cat. *J. Neurophysiol.*, 33: 750–767.
Richmond, F.J.R. and Abrahams, V.C. (1975) Morphology and distribution of muscle spindles in dorsal muscles of the cat neck. *J. Neurophysiol.*, 38: 1322–1339.
Roberts, T.D.M. (1967) *Neurophysiology of Postural Mechanisms*. Butterworths, London.
Shimazu, H. and Precht, W. (1965) Tonic and kinetic responses of cat's vestibular neurons to horizontal angular acceleration. *J. Neurophysiol.*, 28: 991–1013.
Stein, B.M. and Carpenter, M.B. (1967) Central projections of portions of the vestibular ganglia innervating specific parts of the labyrinth in the rhesus monkey. *Amer. J. Anat.*, 120: 281–318.
Suzuki, J. and Cohen, B. (1964) Head, eye, body and limb movements from semicircular canal nerves. *Exp. Neurol.*, 10: 393–405.
Suzuki, J., Goto, K., Tokumasu, K. and Cohen, B. (1969) Implantation of electrodes near individual vestibular nerve branches in mammals. *Ann. Otol. (St. Louis)*, 78: 815–826.
Szentagothai, J. (1952) *Die Rolle der Einzelnen Labyrinth-Rezeptoren bei der Orientation von Augen und Kopf in Raume*. Akademiai Kiado, Budapest.
Watt, D.G.D. (1976) Responses of cat to sudden falls: an otolith-originating reflex assisting landing. *J. Neurophysiol.*, 39: 257–265.
Wilson, V.J. (1972) Physiological pathways through the vestibular nuclei. *Int. Rev. Neurobiol.*, 15: 27–81.
Wilson, V.J. and Yoshida, M. (1969a) Comparison of effects of stimulation of Deiters' nucleus and medial longitudinal fasciculus on ncek, forelimb and hindlimb motoneurons. *J. Neurophysiol.*, 32: 743–758.
Wilson, V.J. and Yoshida, M. (1969b) Monosynaptic inhibition of neck motoneurons by the medial vestibular nucleus. *Exp. Brain Res.*, 9: 365–380.
Wilson, V.J. and Yoshida, M. (1969c) Bilateral connections between labyrinths and neck motoneurons. *Brain Res.*, 13 : 603–607.
Wilson, V.J. and Maeda, M. (1974) Connections between semicircular canals and neck motoneurons in the cat. *J. Neurophysiol.*, 37: 346–357.
Wilson, V.J., Gacek, R.R., Maeda, M. and Uchino, Y. (1977) Saccular and utricular input to cat neck motoneurons. *J. Neurophysiol.*, 40: 63–73.

Role of Vestibular Inputs in the Organization of Motor Output to Forelimb Extensors

J.H. ANDERSON, J.F. SOECHTING and C.A. TERZUOLO

Laboratory of Neurophysiology, University of Minnesota Medical School, Minneapolis, MN 55455 (U.S.A.)

INTRODUCTION

The behavioral description of the physiological role of labyrinthine reflex actions upon the antigravity muscles by Magnus and his collaborators (1924) in the early 1900's has remained relatively static in the intervening years, with the exception of the extension and elaboration of the original conclusions provided by Rademaker (1935), Tait and McNally (1934), and more recently by Roberts (1967), among others. This is despite the considerable amount of progress which has been made in characterizing the responses of labyrinthine afferents (e.g., Fernandez and Goldberg, 1971, 1976), the central organization of vestibular functions (cf. Brodal, 1974; Wilson and Peterson, 1978), and the pathways involved (cf. Precht, 1977; Pompeiano, 1972; Wilson, 1972), and even the perceptual correlates (cf. Guedry, 1974). The contrast becomes even more marked when the level of description of vestibular postural reflexes is compared with that available for labyrinthine influences on eye movements (e.g., Chun and Robinson, 1978).

One possible explanation for the relative lack of progress on this problem may lie in the fact that in normal, real-life situations, vestibular effects are subsumed into complex postural responses which reflect also sensory information of other modalities (visual, somatic and proprioceptive), all of which are likely to strongly interact with each other (Nashner, 1970; Young, 1971; Lackner and Graybiel, 1978). Therefore, if one wishes to identify and characterize the labyrinthine contribution to postural mechanisms, it becomes necessary to choose an experimental situation wherein its contribution to the motor response is clearly unambiguous. The preparation utilized by Magnus in his studies, namely the decerebrate cat with the spinal cord cut at the mid-thoracic level, still seems ideal to us for this purpose. It is well known that in this preparation labyrinthine reflexes are enhanced, the high level of activity in the antigravity muscles being due in part to tonic activity arising from the labyrinthine receptors. One further advantage of such tonic activity in the α-motoneurons is that it provides a background level of activity, the modulation of which can be easily detected. Also, the stimulus can be precisely controlled and reasonably well restricted to the labyrinthine receptors.

Once the reflex responses to linear and angular accelerations, mediated by the vestibular receptors, have been characterized in the decerebrate cat, it becomes possible to ask whether or not these responses are appropriate to maintain postural stability. In particular, the problem of whether or not vestibular reflexes function in a similar manner in intact animals, and under what conditions, can then be pursued.

METHOD OF ANALYSIS

In our work, we have addressed ourselves to i) characterizing the kinetic labyrinthine (semi-circular canal) contribution to the motor output and ii) providing a description of the dynamics of the kinetic and so-called static (macular) labyrinthine reflexes to the forelimb extensors by utilizing a systems analysis approach. The advantage of such an approach is that it permits the functional relationships between the two inputs and the motor output to be defined in a quantitative manner. It should be noted that such a line of investigation has been particularly successful in describing the behavior of vestibulo-ocular reflexes (Skavensky and Robinson, 1973; Shinoda and Yoshida, 1974; Baarsma and Collewijn, 1974, 1975; Robinson, 1975) and visual-vestibular interactions (Robinson, 1976, 1977; Melvill Jones et al., 1977; Lau et al., 1978) for compensatory eye movements. In fact, a systems approach appears to be well-suited for studies of vestibular function in general. Firstly, the appropriate inputs, namely linear or angular accelerations, are well-defined and measurable. Secondly, by limiting other sensory inputs, an input-output description relating vestibular inputs to motor output is rather straightforward. Also, multivariate analyses can eventually prove useful in characterizing reflex behavior dependent upon interactions between vestibular and other sensory inputs, such as visual. Finally, since a quantitative description of labyrinthine receptor dynamics is available, the nature of central mechanisms mediating the reflex behavior can be deduced.

Details of our experimental set-up are available elsewhere (Anderson et al., 1977a). Cats were used. Here we wish to emphasize only that relative movements of the head, neck and trunk were minimized by means of a plaster cast and that care was also taken to minimize possible contamination of the motor output by somatic stimuli.

RESULTS

Response of the triceps brachii muscle to either macular or canal inputs

The phase and gain of the motor output (EMG) to continuously varying (sinusoidal) inputs when either the otolith organs or the semicircular canals are selectively stimulated are shown in Fig. 1.

To characterize the behavior due to canal inputs only, the preparations were centered on a turntable such that only the horizontal semicircular canals would be stimulated. The motor response to sinusoidal rotations (yaw) at frequencies from 0.15 to 1.0 Hz shows phase lags of 110°–150°, relative to an acceleration toward the contralateral side, and a gain drop (relative to the magnitude of the acceleration) of about 15–20 dB/decade.

For pure otolith stimulation, the preparations were subjected to linear accelerations along each of the coordinate axes of the cat (Anderson et al., 1977a). The motor output increased when the cats were accelerated in the forward, ipsilateral and downward directions. As shown in Fig. 1, the response is essentially the same for all three directions. In contrast with the response to canal inputs, the phase lags the input acceleration by 0°–60° and the gain decreases about 15–20 dB/decade.

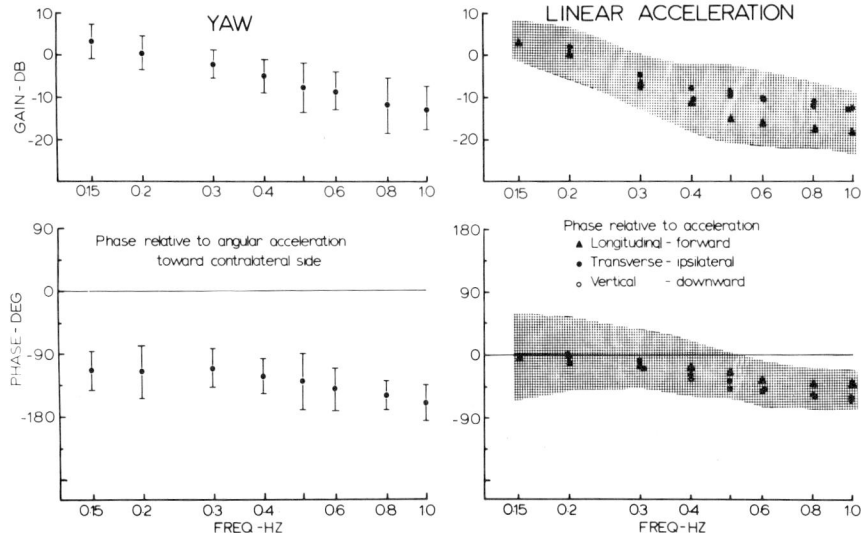

Fig. 1. Response of triceps brachii to pure linear and angular accelerations. The phase and gain of the EMG response to sinusoidal rotations in the horizontal plane (yaw) and linear accelerations along the three body axes are shown. Note that all phase values are relative to acceleration. The gain is given by 20 $\log_{10}(A_0/A_i)$, where A_0 and A_i are the amplitudes of the EMG response and input acceleration, respectively. The zero reference of gain was arbitrarily chosen. The symbols represent the mean values for data from 8 preparations. Error bars and stippled regions encompass ± 1 SD.

From the results it is clear that the motor output differs significantly from that of the vestibular afferents. Indeed, the canal afferents show phase lags (relative to angular acceleration) of less than 90°, while the macular afferents respond either in phase with or lead linear acceleration (or gravity). Therefore, one can conclude that vestibular influences on the motor output to forelimb extensors are subject to extensive central processing (Anderson et al., 1977b) in a manner similar to that for oculomotor outputs (cf. Robinson, 1975).

Responses to rotations in the vertical plane

To gain an understanding into how the reflex responses to combined canal and otolith stimulations, such as they occur in most real-life situations, may be organized, the preparations were subjected to sinusoidal rotations about their longitudinal (roll) and transverse (pitch) axes. Under these experimental conditions, the otolith organs' are stimulated by changes in the orientation of the animal's head relative to gravity.

The upper two diagrams of Fig. 2 illustrate the components of the gravity vector along the animal's coordinate axes for each of the above conditions (A: roll; B: pitch). For small sinusoidal changes in angular position (θ), sine θ also varies approximately sinusoidally at the same frequency as θ, while cos θ varies at twice this frequency. Since the fundamental component is much larger (Soechting et al., 1977), only g_y and g_x – for roll and pitch, respectively – shall be considered here as the effective stimuli for the otoliths.

The lower two diagrams in Fig. 2 illustrate the pattern of activity of the posterior (p) and anterior (a) canal afferents caused by a leftward rotation in roll and a nose downward rotation in pitch, as indicated by the arrows. The head positions shown are those at the time of maximum acceleration.

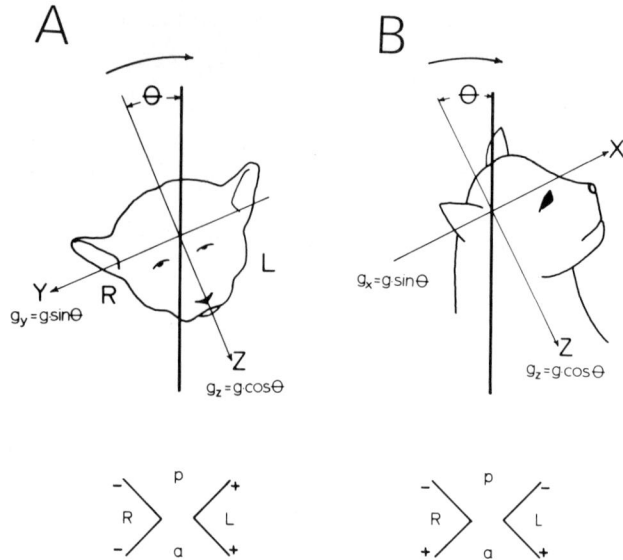

Fig. 2. Vestibular inputs during roll (A) and pitch (B). The figure illustrates the orientation of the cat during roll and pitch rotations, the sign conventions utilized and the forces acting on the otoliths and canals. θ specifies the animal's orientation to the earth vertical. The positions of the cat are those for maximum angular acceleration (indicated by topmost arrows) toward the left in (A) and nose downward in (B). Increase and decrease in the activity of vertical canal afferents (a, anterior and p, posterior) during such accelerations is indicated schematically in the diagrams on the bottom. The relationship of the gravitational stimuli acting on the otoliths to the angular position θ is also indicated.

In Fig. 3 the phases and gains of the motor output obtained during pitch and roll are compared. The phase for roll is relative to an acceleration toward the ipsilateral side. That for pitch is relative to an acceleration tending to push the nose downward. It is clear that the phases and gains are virtually the same under both conditions. The phase ranges from a lead of about 20° at the lower frequency to a lag of 135° at 1.0 Hz; the gain drops nearly 35 dB/decade over the same frequency range. Note that during roll the EMG activity is maximal in the left triceps and minimal in the right triceps at frequencies below 0.2 Hz when the left side is up (Fig. 2A). Also, the activity is maximal in both triceps for a position which is equivalent to a dorsiflexion of the head (Fig. 2B).

The fact that these phases and gains are much different from those shown in Fig. 1 for either pure linear or angular accelerations obviously means that both macular and canal inputs contribute to the motor output during roll and pitch. In particular, at low frequencies the phase tends toward that predicted by the response to linear accelerations, keeping in mind that ipsilateral side up and nose up tilts produce the same response in the macular receptors as linear accelerations toward the ipsilateral side or in the forward direction. At high frequencies the phase lag tends toward that produced by a pure angular acceleration.

In order to estimate the relative contribution by the canal and otolith inputs, a vectorial addition of their responses was made (Anderson et al., 1977b; Soechting et al., 1977), using the results shown in Fig. 1. If the results of such an addition were to predict the experimentally obtained data shown in Fig. 3, then one would conclude

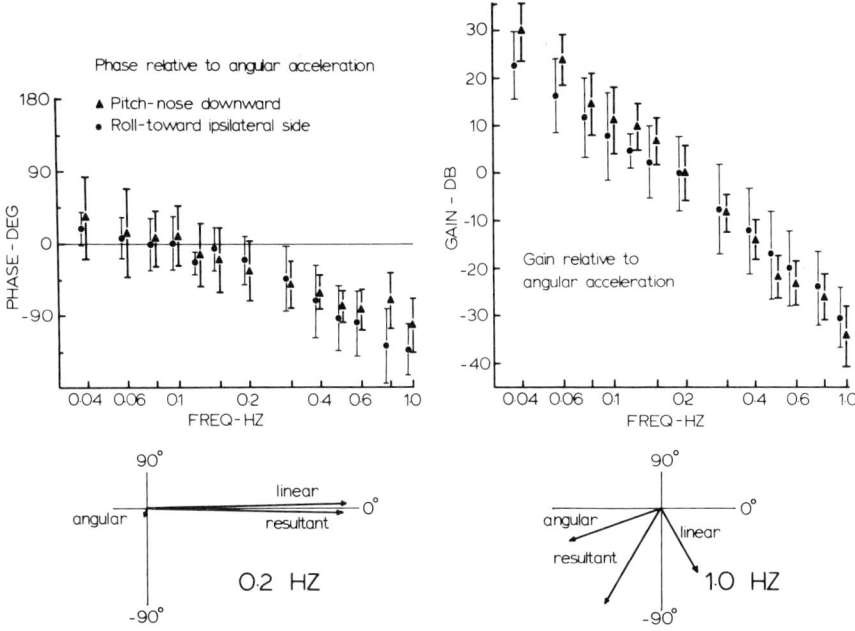

Fig. 3. Response of triceps brachii to rotations in a vertical plane. The phase and gain of the EMG responses to sinusoidal roll (●) and pitch (▲) rotations are shown for frequencies from 0.04 to 1.0 Hz. Note that the phases are relative to angular accelerations causing nose downward (pitch) or ipsilateral side down (roll) rotations. The gains are relative to the angular acceleration magnitude (see Fig. 1, legend). The lower vector diagrams indicate the predicted otolith ("linear") and vertical canal ("angular") contributions to the motor response ("resultant") for rotations at a low (0.2 Hz) and high (1.0 Hz) frequency. Note that the high frequency vectors assume an input amplitude twice that at the lower frequency. The symbols are mean values for data from 3 preparations in pitch and from 7 in roll. Bars encompass ± 1 SD.

that, within the limits of resolution of our data, macular and canal contributions to the motor output add linearly.

The lower diagrams of Fig. 3 illustrate two such vector additions for the lowest and highest frequency for which experimental data are available. The vectors are plotted on a polar diagram. Their length represents the amplitude of the motor response and their direction its phase. The dominance of otolith inputs at 0.2 Hz is clear. The canal input becomes progressively more important as the frequency increases, although even at 1 Hz there is still a large contribution from the macular input.*

The results of such vector additions have previously been shown (Soechting et al., 1977) to predict very adequately the motor output for roll at frequencies above 0.2 Hz. By extending this frequency range toward the static condition and by verifying that the motor output behaves similarly for both pitch and roll, the validity of the conclusion (and its underlying assumptions, cf. Soechting et al., 1977) has now been generalized.

---

*The following organization is suggested by the fact that the canal contribution is the same for both pitch and roll, given their respective references to angular acceleration: the output from each anterior canal would be excitatory to the ipsilateral forelimb extensors while the output from the posterior canal acts as if it were inhibitory to the contralateral forelimb extensor.

DISCUSSION

Before discussing the results along the lines stated in the Introduction, we would like to mention the fact that the data presented here indicate extensive central processing of vestibular inputs. However, since the present data are inadequate to reveal the details of such processing and since this problem has been considered extensively both in the context of labyrinthine influences on spinal and oculomotor systems (Robinson, 1975; Anderson et al., 1977a, b; Ezure et al., 1978; Pola and Robinson, 1978), we shall not discuss it further here.

Instead, we wish to consider the appropriateness of the motor outputs described here toward maintaining postural stability and the role of labyrinthine effects upon antigravity muscles in the context of postural reflexes. On this subject, we would like first to note that the responses to linear accelerations are behaviorally appropriate. Furthermore, qualitatively similar responses, i.e., an increase in limb extensor activity during the downward acceleration (Melvill Jones and Watt, 1971; Greenwood and Hopkins, 1976; Watt, 1976) and a decrease during deceleration (Prochazka et al., 1977) have been observed in man and alert animals during free fall. From the analysis of the dynamics of a fall toward one side provided by Nashner (1970), one can also conclude that the motor output response in roll (Fig. 3) would tend to maintain the body upright following a transient disturbance (Soechting et al., 1977). In brief, during transient accelerations the labyrinthine effects on forelimb antigravity muscles as observed in decerebrate animals with the head fixed, would seem adequate to provide for the postural stability of the intact animal.

By contrast, the dynamic characteristics of the motor output as described here at low frequencies (below 0.2 Hz), are entirely inappropriate from a behavioral viewpoint. For example, during lateral tilting there is a decrease of extensor activity on the side which is tilted downward. (For reasons which are not entirely clear to us, Lindsay et al., 1976, have described diametrically opposite behavior.) In order to account for this discrepancy with respect to behavioral observations, several other well-documented physiological responses should be considered: i) when an animal whose head is unrestrained is tilted toward one side or pitched forwards or backwards, changes in the activity of the antigravity muscles will tend to maintain the body oriented vertically (Tait and McNally, 1934; Roberts, 1967); ii) labyrinthine reflexes acting on the neck muscles will tend to orient the head with respect to the apparent gravitational vertical (Magnus, 1924; Tait and McNally, 1934); iii) these responses are absent in labyrinthectomized animals.

These observations, taken in conjunction with the results we have presented here, indicate that at low frequencies, i.e., as we approach a static condition, neck reflexes dominate the motor output to the limb extensors. This conclusion is in agreement with the scheme which was originally put forth by Magnus (1924).

A few comments are appropriate to place the results presented here in the context of postural control. First, the hypothesis that neck reflexes would be more important at low frequencies and labyrinthine reflexes at higher frequencies appears reasonable since, due to the inertia of the head, relative motion between head and trunk would be attenuated at higher frequencies. This hypothesis can be tested experimentally by studying separately the effect of vestibular inputs upon the neck muscles (head control by vestibular inputs) and the effect of head rotations (in labyrinthectomized animals) upon the forelimb extensors in order to directly evaluate the contribution by the neck and labyrinthine reflexes. In fact, one of us (J.H.A.) is presently applying the systems analysis approach to the study of the organization of the motor output to the neck muscles of alert animals. The results show that the motor responses to rotation (with the head restrained) are those behaviorally appropriate for the control of the head, as seen in non-restrained animals. Moreover, preliminary results obtained by rotating the head implicate neck reflexes as being the basis for the control of limb's posture by head movements.

Secondly, the contributions of reflex effects due to visual and somatic inputs cannot be neglected. Since the information provided by the visual surround can also be quantified (in terms of its velocity and spatial frequencies), it should prove possible to evaluate its influence on motor behavior in certain contexts (Lestienne et al., 1977; Nashner and Berthoz, 1978). Nevertheless, given the multitude of sensory modalities which are involved in maintaining postural stability, the more general problem of defining conditions in which each may be influential seems largely open.

SUMMARY

A quantitative description of the motor output response of forelimb extensors to linear and angular accelerations is provided. Decerebrate cats were used, since labyrinthine reflexes due to such stimuli are enhanced in this preparation. The data showed that: i) the dynamics of the motor output to stimuli exciting primarily either the saccular or utricular receptors are the same, ii) both canal and otolith outputs are subjected to extensive central processing and iii) canal and otolith outputs are combined in a linear manner to generate the motor output. The problem of the relevance of the observed effects to the problem of postural stabilization is addressed. In brief, the data imply that labyrinthine reflex actions on the forelimbs would be adequate to counteract transient perturbations, but that neck reflexes would be more important in mediating tonic labyrinthine reflexes.

ACKNOWLEDGEMENTS

This work was supported by U.S. Public Health Service Grant NS-02567. Computer facilities were made available by the Air Force Office of Scientific Research, AFSC (Grant AFOSR-1221).

REFERENCES

Anderson, J.H., Soechting, J.F. and Terzuolo, C.A. (1977a) Dynamic relations between natural vestibular inputs and activity of forelimb extensors in the decerebrate cat. I. Motor output during sinusoidal linear accelerations. *Brain Res.*, 120: 1–15.

Anderson, J.H., Soechting, J.F. and Terzuolo, C.A. (1977b) Dynamic relations between natural vestibular inputs and activity of forelimb extensors in the decerebrate cat. II. Motor output during rotations in the horizontal plane. *Brain Res.*, 120: 17–33.

Baarsma, E.A. and Collewijn, H. (1974) Vestibulo-ocular and optokinetic reactions to rotation and their interaction in the rabbit. *J. Physiol. (Lond.)*, 238: 603–625.

Baarsma, E.A. and Collewijn, H. (1975) Eye movements due to linear accelerations in the rabbit. *J. Physiol. (Lond.)*, 245: 227–247.

Brodal, A. (1974) Anatomy of the vestibular nuclei and their connections. In *Handbook of Sensory Physiology, Vol. VI/1, Vestibular System, Part 1, Basic Mechanisms*, H. H. Kornhuber (Ed.), Springer-Verlag, Berlin, pp. 239–352.

Chun, K.-S. and Robinson, D.A. (1978) A model of quick phase generation in the vestibulo-ocular reflex. *Biol. Cybernet.*, 28: 209–221.

Ezure, K., Sasaki, S., Uchino, Y. and Wilson, V.J. (1978) Frequency-response analysis of vestibular-induced neck reflex in cat. II. Functional significance of cervical afferents and polysynaptic descending pathways. *J. Neurophysiol.*, 41: 459–471.

Fernandez, C. and Goldberg, J. (1971) Physiology of peripheral neurons innervating semicircular canals of the squirrel monkey. II. Response to sinusoidal stimulation and dynamics of peripheral vestibular system. *J. Neurophysiol.*, 34: 661–675.

Fernandez, C. and Goldberg, J. (1976) Physiology of peripheral neurons innervating otolith organs of the squirrel monkey. III. Response dynamics. *J. Neurophysiol.*, 39: 996–1008.

Greenwood, R.J. and Hopkins, A.P. (1976) Muscle responses during sudden falls in man. *J. Physiol. (Lond.)*, 254: 507–518.

Guedry, F.E. (1974) Psychophysics of vestibular sensation. In *Handbook of Sensory Physiology, Vol. VI/2, Vestibular System. Psychophysics, Applied Aspects and General Interpretations*, H.H. Kornhuber (Ed.), Springer-Verlag, Berlin, pp. 3–154.

Lackner, J.R. and Graybiel, A. (1978) Some influences of touch and pressure cues on human spatial orientation. *Aviat. Space Environ. Med.*, 49: 798–804.

Lau, C.G.Y., Honrubia, V., Jenkins, H.A., Baloh, R.W. and Yee, R.D. (1978) Linear model for visual-vestibular interaction. *Aviat. Space Environ. Med.*, 49: 880–885.

Lestienne, F., Soechting, J.F. and Berthoz, A. (1977) Postural readjustments induced by linear motion of visual scenes. *Exp. Brain Res.*, 28: 363–384.

Lindsay, K.W., Roberts, T.D.M. and Rosenberg, J.R. (1976) Asymmetric tonic labyrinthine reflexes and their interaction with neck reflexes in the decerebrate cat. *J. Physiol. (Lond.)*, 261: 583–601.

Magnus, R. (1924) *Körperstellung*. Springer, Berlin.

Melvill Jones, G. and Watt, D.G.D. (1971) Muscular control of landing from unexpected falls in man. *J. Physiol. (Lond.)*, 219: 729–737.

Melvill Jones, G., Davies, P. and Gonshor, A. (1977) Long-term effects of maintained vision reversal: is vestibulo-ocular adaptation either necessary or sufficient? In *Control of Gaze by Brain Stem Neurons*, R. Baker and A. Berthoz (Eds.), Elsevier, Amsterdam, pp. 59–68.

Nashner, L. (1970) *Sensory Feedback in Human Posture Control*. Sc.D. Thesis, Mass. Inst. Technology, Cambridge.

Nashner, L. and Berthoz, A. (1978) Visual contribution to rapid motor response during postural control. *Brain Res.*, 150: 403–407.

Pola, J. and Robinson, D.A. (1978) Oculomotor signals in the medial longitudinal fasciculus of monkey. *J. Neurophysiol.*, 41: 245–259.

Pompeiano, O. (1972) Vestibulospinal relations: vestibular influences on gamma motoneurons and primary afferents. In *Progress in Brain Research, Vol. 37, Basic Aspects of Central Vestibular Mechanisms*. A. Brodal and O. Pompeiano (Eds.), Elsevier, Amsterdam, pp. 197–232.

Precht, W. (1977) The functional synaptology of brainstem oculomotor pathways. In *Control of Gaze by Brain Stem Neurons*, R. Baker and A. Berthoz (Eds.), Elsevier, Amsterdam, pp. 131–141.

Prochazka, A., Schofield, P., Westerman, R.A. and Ziccione, S.P. (1977) Reflexes in cat ankle muscles after landing from falls. *J. Physiol. (Lond.)*, 272: 705–719.

Rademaker, G.G.J. and Ter Braak, J.W. (1935) Das Umdrehen der fallenden Katze in der Luft. *Acta otolaryngol. (Stockh.)*, 23: 313–343.

Roberts, T.D.M. (1967) *Neurophysiology of Postural Mechanisms*. Butterworth, London.

Robinson, D.A. (1975) Oculomotor control signals. In *Basic Mechanisms of Ocular Motility and Their Clinical Implications*, G. Lennerstrand and P. Bach-y-Rita (Eds.), Pergamon, Oxford, pp. 337–378.

Robinson, D.A. (1976) Adaptive gain control of vestibulo-ocular reflex by the cerebellum. *J. Neurophysiol.*, 39: 954–969.

Robinson, D.A. (1977) Linear addition of optokinetic and vestibular signals in the vestibular nucleus. *Exp. Brain Res.*, 30: 447–450.

Shinoda, Y. and Yoshida, K. (1974) Dynamic characteristics of responses to horizontal head angular acceleration in vestibulo-ocular pathology in the cat. *J. Neurophysiol.*, 37: 653–673.

Skavenski, A. and Robinson, D.A. (1973) Role of abducens neurons in vestibulo-ocular reflex. *J. Neurophysiol.*, 36: 724–738.

Soechting, J.F., Anderson, J.H. and Berthoz, A. (1977) Dynamic relations between natural vestibular inputs and activity of forelimb extensors in the decerebrate cat. III. Motor output during rotations in the vertical plane. *Brain Res.*, 120: 35–47.

Tait, J. and McNally, W.J. (1934) Some features of the action of the utricular maculae (and of the associated action of the semicircular canals) of the frog. *Phil. Trans. B*, 224: 241–286.

Watt, D.G.D. (1976) Responses of cats to sudden falls: an otolith-originating reflex assisting landing. *J. Neurophysiol.*, 39: 257–265.

Wilson, V. (1972) Physiological pathways through the vestibular nuclei. *Int. Rev. Neurobiol.*, 15: 27–81.

Wilson, V.J. and Peterson, B.W. (1978) Peripheral and central substrates of vestibulospinal reflexes. *Physiol. Rev.*, 58: 80–105.

Young, L.R. (1971) *Developments in Modeling Visual-Vestibular Interaction.* AMRL-TR-71-14, Aerospace Medical Res. Lab., Wright-Patterson AFB, Ohio.

Efferent and Afferent Responses During Falling and Landing in Cats

M.McD. LEWIS, A. PROCHAZKA, K.-H. SONTAG and P. WAND

Sherrington School of Physiology, St. Thomas's Hospital Medical School, London SE1 7EH (U.K.); and Max-Planck-Institut für Experimentelle Medizin, Abteilung für Biochemische Pharmakologie, 3400 Göttingen (F.R.G.)

INTRODUCTION

Previous studies have shown that the coordinated muscular contractions associated with a controlled landing from a fall commence well before impact (Matthews and Whiteside, 1960; Melvill Jones and Watt, 1971a, b). Greenwood and Hopkins (1976a, b) found that in human subjects suddenly released from a height of about 1 m, the electromyograms (EMG's) of ankle extensors showed two periods of activation prior to foot contact. The first burst was attributed to a generalised "startle" reaction, because it did not occur if the subjects initiated the fall themselves, and because it could also be recorded in muscles not taking part in the subsequent landing. The second period of muscle activity generally commenced at a fixed interval before impact, and depended on a knowledge of the height of the fall. This second burst, it was suggested, was responsible for the smooth, voluntary control of landing. Similar conclusions were drawn by Watt (1976) in studies on cats, although there were some points of difference regarding the apparent contribution of the otoliths to the two bursts of EMG activity. Greenwood and Hopkins (1976b) presented evidence indicating that after foot contact, the EMG patterns in hindlimb muscles were strongly dependent on afferent input. This was at variance with Melvill Jones and Watt (1971b), who argued against the existence of significant reflexes after landing.

In the cat, Prochazka et al. (1977b) recorded large and apparently functionally effective responses consistent with stretch reflexes in the ankle extensors after landing from falls. Anaesthesia of the foot-pads did not change the timing of the responses, which indicated a proprioceptive mediation. Similar data has recently been published by Laursen et al. (1978), who claimed, however, that the responses after landing were not reflex in origin, but rather were "programmed".

In this paper, we shall present further evidence in support of the contention that these responses are indeed reflex in character.

We shall also show recordings from muscle spindle afferents during free-fall and landing which supplement previous data (Prochazka et al., 1977a) and lend support to the view that fusimotor drive to muscle spindles of the ankle extensors does not strongly accelerate the firing of spindle afferents prior to foot contact.

METHODS

The recording techniques have been largely summarised in another paper in this volume (Lewis et al., this volume, chapter IB8).

EMG records were full-wave rectified and averaged using a Medelec fibre optic recorder with signal delay and averager facilities. A contact plate was constructed, whereby two thin sheets of metal foil, normally separated by about 0.5 mm, were pressed together at the lightest touch. The resulting short circuit was used to trigger the averager.

RESULTS

Efferent responses

In order to test whether visual cues were necessary for the presence of the EMG responses after landing, we compared the averaged, rectified EMG's of ankle extensors and knee flexors in a cat before and after blindfolding.

Fig. 1 (top) shows the averaged EMG of medial gastrocnemius for 36 drops from an average height of 25 cm. As described by Prochazka et al. (1977b), the length of the ankle extensors does not start to increase until some 8–9 msec after landing. This delay was shown by high-speed cinematography, to be due to initial toe dorsiflexion. The EMG trace shows an initial reduction in activity commencing some 8 msec after foot contact, followed by a peak of activity commencing about 9 msec after the onset of muscle stretch.

Fig. 1 (bottom) shows similar records for 36 drops after the cat had been blindfolded. The height of successive drops was varied between 15 and 35 cm. Often, the cat would land awkwardly, indicating that its prediction of height, on the basis of non-visual cues, was poor. Nevertheless, the length of EMG responses in the first 40 msec after landing were similar to those when vision was normal.

Fig. 2 shows the averaged EMG of the knee flexor posterior biceps-semitendinosus (PBSt) under similar conditions. Again, the patterns of response after landing were not significantly altered after blindfolding.

These results provide strong evidence that the EMG activity after landing is not

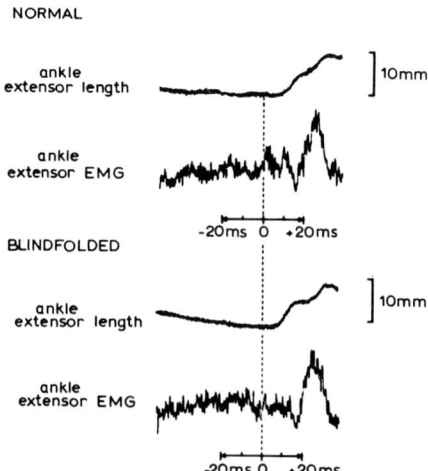

Fig. 1. Top: averaged, rectified EMG of medial gastrocnemius for 36 drops, average height 25 cm. Length monitored between calcaneum and head of tibia: average of 4 drops. Bottom: same as above, after cat was blindfolded.

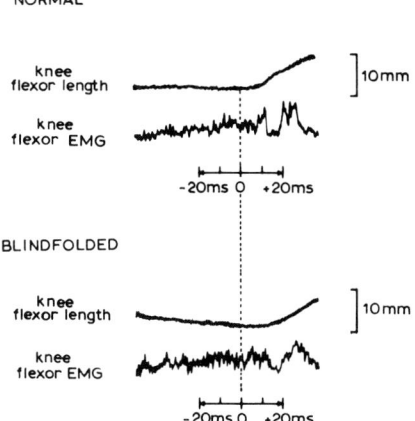

Fig. 2. Top: averaged, rectified EMG of PBSt for 36 drops, average height 25 cm. Length monitored between head of tibia and ischium: average of 4 drops. Bottom: same as above, after cat was blindfolded.

predetermined centrally on the basis of a prediction of the height of the fall. Presumably, the shift in the timing of the responses seen by Laursen et al. (1978), when using a dummy landing surface above a firm platform, was due to afferent activity resulting from contact with the dummy surface.

The responses in PBSt in Fig. 2 are remarkably similar to those in medial gastrocnemius (Fig. 1). The first period of reduced activity in PBSt is unlikely to have resulted from homonymous tendon organ afferent inhibition, a possibility which could not be excluded in the case of the ankle extensor (Prochazka et al., 1977b). This is because PBSt clearly does not undergo stretching soon enough for tendon organ excitation to have a reflex effect. Thus, as previously argued for the ankle extensors, reflex action from afferents in the toe plantarflexors, excited by the initial toe dorsiflexion, may be responsible for the inhibition. The subsequent period of increased activity in PBSt is, as for medial gastrocnemius, presumably a proprioceptive stretch reflex, the functional usefulness of which would be to resist hip flexion. A co-contraction of the knee extensors may occur in order to limit the rate of knee flexion, but this has not been investigated.

Afferent responses

Fig. 3 shows the discharge activity of two muscle spindle afferents for single falls in two separate cats. The upper record shows the activity of a spindle primary ending located in plantaris. The length gauge was attached to the calcaneum at one end, and the head of the tibia at the other. It therefore did not register changes in length due to toe movements. Prior to landing, however, the metatarso-phalangeal joints remain at a fairly constant angle (Prochazka et al., 1977b). The plantaris spindle decreased its firing just prior to landing, presumably as a result of the small degree of ankle plantarflexion evidenced by a reduction in the monitored length, and despite the gradually increasing plantaris EMG. After landing, the afferent responded to muscle stretch with an extremely rapid burst of firing.

The spindle secondary, shown by palpation to be located either in medial gastrocnemius or soleus, showed a small acceleration in firing prior to landing. In this case,

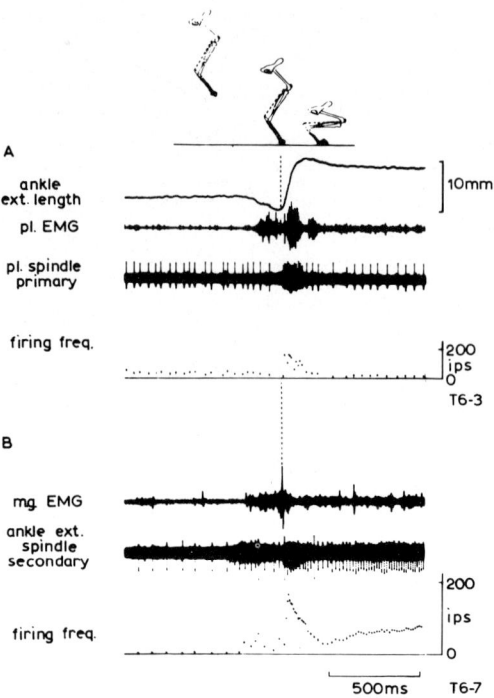

Fig. 3. A) Plantaris spindle primary afferent during free-fall and landing. Length monitored between calcaneum and head of tibia. B) Ankle extensor, probably medial gastrocnemius spindle secondary during free-fall and landing. Length was not monitored, but moment of foot contact was ascertained from contact artifact.

the length changes were not monitored. However, in our experience, it is rare for the ankle extensors to lengthen prior to ground contact, and so the acceleration which was seen consistently in 15 consecutive drops, was presumably due to fusimotor activation of the spindle. This afferent also showed large increases in discharge rate after landing, when the ankle extensors are invariably stretched.

DISCUSSION

Efferent responses

The lack of significant changes in the postlanding responses after blindfolding is further evidence that controlled landing from a fall in cats involves rapid reflex responses in hindlimb muscles. Greenwood and Hopkins (1976b) found consistent peaks in soleus EMG's in man at about 60 msec after landing from a downward step. These peaks were absent after the moment of expected landing during steps which were unexpectedly lengthened, and were thus interpreted as being of reflex origin. Similarly, in a very recent study, Dietz and Noth (1978) reported large increases in triceps surae EMG in man, at latencies after ground contact, consistent with spinally-mediated stretch reflexes. However, the data of Melvill Jones and Watt (1971b) show little evidence for stretch reflexes after landing from unexpected falls.

These results are not necessarily contradictory. In the cat, the heel rarely touches the ground after landing from a fall. Muscles acting about the ankle must therefore provide the appropriate torque to prevent maximal flexion at the ankle, knee and hip. In man, on the other hand, much depends on the initial posture and the strategy of movement control adopted. Subjects may, for example, land on the ball of the foot, with or without subsequent heel contact. Controlled landings with initial heel contact are also possible, and the amount of yielding at the hip, knee and ankle are largely a function of the subject's prior intention.

Afferent responses

As reported previously, there is little evidence here for a very powerful acceleration of ankle extensor spindle afferents prior to landing.

Stimulation of the vestibular nerve has previously been shown to lead to an acceleration of gamma afferents recorded in ventral root filaments (Andersson and Gernandt, 1956). Thus, one might have expected a significant acceleration of spindle firing during the falls.

However, the latency between initial head acceleration and the second component of the prelanding EMG was described by Watt (1976) as being both "long and variable". This author concluded that the response was therefore mediated by a complex polysynaptic pathway rather than by a direct vestibulospinal projection. If the vestibular input merely serves to "trigger" a diffuse startle reaction, descending bulbospinal action could equally well result in a parallel inhibition of static fusimotor neurones and excitation of skeletomotor neurones (Appelberg, 1962).

The afferent of Fig. 3B is the first of 5 ankle extensor spindles observed by us to show a small increase in firing during free-fall. A small, nett fusimotor excitation to some spindles cannot therefore be ruled out, although it should be borne in mind that the effect of sudden weightlessness on spindles within a muscle, might be to stretch some spindles and slacken others (Matthews and Whiteside, 1960; Greenwood and Hopkins, 1977).

SUMMARY

Electrical activity was recorded by telemetry from hindlimb muscles during free-fall and landing in cats. The characteristic responses of ankle extensors and knee flexors were not significantly altered by depriving the cat of visual cues, although its prediction of height was clearly impaired. This provides further evidence that controlled landing from a fall in cats involves rapid reflex responses in hindlimb muscles. The activity of two muscle spindle afferents during free-fall supported the view that there is apparently little increase in fusimotor excitation of spindles prior to landing.

ACKNOWLEDGEMENTS

The authors thank Dr. J.A. Stephens and Professor A. Taylor for help, criticism, and encouragement. This work was supported by St. Thomas's Hospital Endowments Fund and the Deutsche Forschungsgemeinschaft, SFB 33. Dr. M.McD. Lewis was on study leave from Lincoln Institute of Health Sciences, Victoria, Australia.

REFERENCES

Andersson, S.A. and Gernandt, B.E. (1956) Ventral root discharge in response to vestibular and proprioceptive stimulation. *J. Neurophysiol.*, 19: 524–543.

Appelberg, B. (1962) The effect of electrical stimulation in Nucleus Ruber on the response to stretch in primary and secondary muscle spindle afferents. *Acta physiol. scand.*, 56: 140–151.

Dietz, V. and Noth, J. (1978) Spinal stretch reflexes of triceps surae in active and passive movements. *J. Physiol. (Lond.)*, in press.

Greenwood, R. and Hopkins, A. (1976a) Muscle responses during sudden falls in man. *J. Physiol. (Lond.)*, 254: 507–518.

Greenwood, R. and Hopkins, A. (1976b) Landing from an unexpected fall and a voluntary step. *Brain*, 99: 375–386.

Greenwood, R.A. and Hopkins, A. (1977) Monosynaptic reflexes in falling man. *J. Neurol.*, 40: 448–454.

Laursen, A.M., Dyhre-Poulsen, P., Djørup, A. and Jahnsen, H. (1978) Programmed pattern of muscular activity in monkeys landing from a leap. *Acta physiol. scand.*, 102: 492–494.

Matthews, Sir Bryan and Whiteside, T.C.D. (1960) Tendon reflexes in free fall. *Proc. roy. Soc. B*, 153: 195–204.

Melvill Jones, G. and Watt, D.G.D. (1971a) Observations on the control of stepping and hopping movements in man. *J. Physiol. (Lond.)*, 219: 709–727.

Melvill Jones, G. and Watt, D.G.D. (1971b) Muscular control of landing from unexpected falls in man. *J. Physiol. (Lond.)*, 219: 729–737.

Prochazka, A., Westerman, R.A. and Ziccone, S.P. (1977) Ia afferent activity during a variety of voluntary movements in the cat. *J. Physiol. (Lond.)*, 268: 423–448.

Prochazka, A., Schofield, P., Westerman, R.A. and Ziccone, S.P. (1977) Reflexes in cat ankle muscles after landing from falls. *J. Physiol. (Lond.)*, 272: 705–719.

Watt, D.G.D. (1976) Responses of cats to sudden falls: an otolith-originating reflex assisting landing. *J. Neurophysiol.*, 39: 257–265.

C

Labyrinthine Influences on Oculomotor Neurons

Synaptic and Functional Organization of Vestibulo-Ocular Reflex Pathways

S.M. HIGHSTEIN and H. REISINE

Department of Neuroscience, Rose F. Kennedy Center, Albert Einstein College of Medicine, Bronx, NY 10461 (U.S.A.)

The neuronal substrate of the vestibulo-ocular reflex (VOR) has long been a subject of interest. The presumed three neuron arc between vestibular receptors and extraocular muscles is an ideal model system for understanding how a definable set of central nervous system neurons produces compensatory eye movements. Head motion is transduced in the vestibular end organs which then excite central vestibular neurons by impulses carried over the VIIIth nerve into the brain stem. Signals are subsequently relayed as either excitation or inhibition to the appropriate subgroups of extraocular motoneurons. There are three main pathways relaying information from the vestibular nuclei to the six subgroups of extraocular motoneurons controlling the extraocular muscles. They are the median longitudinal fasciculus (MLF), the brachium conjunctivum (BC), and the ascending tract of Deiters' (ATD) (Gacek 1971a, b, 1977; Graybiel and Hartwieg, 1974; Highstein et al., 1971; Ito et al., 1973a, b, c, 1976a, b). Knowledge of the location of the cell bodies of origin and the connectivity of these three pathways is important to the analysis of the physiological behavior of the vestibulo-oculomotor system. Experiments designed to relate the discharge characteristics of various classes of identified brain stem neurons with eye movements in alert animals are by necessity dependent upon the accuracy of pathway studies such as those reported below.

THE VERTICAL VESTIBULO-OCULAR REFLEX

Anterior canal

The vertical vestibulo-ocular reflex originates in the anterior and posterior semicircular canals. Activation of the anterior canal excites the ipsilateral superior rectus and the contralateral inferior oblique extraocular eye muscles in the rabbit, cat and monkey. In the rabbit impulses arising from the anterior canal are relayed as excitation through the anterior dorsal group of superior vestibular nucleus neurons just beneath the BC. Fig. 1 demonstrates direct morphophysiological evidence of this vestibulo-oculomotor projection through the BC (cf. Yamamoto et al., 1978). Horseradish peroxidase (HRP) was iontophoresed into the third nucleus of the rabbit (Fig. $1B_1$, C_1) and retrograde transport through the MLF and ATD was interdicted by a lesion of these pathways in the pontine tegmentum (Fig. $1B_2$). Labeled superior nucleus cells were localized to the dorsal anterior portions of the superior nucleus

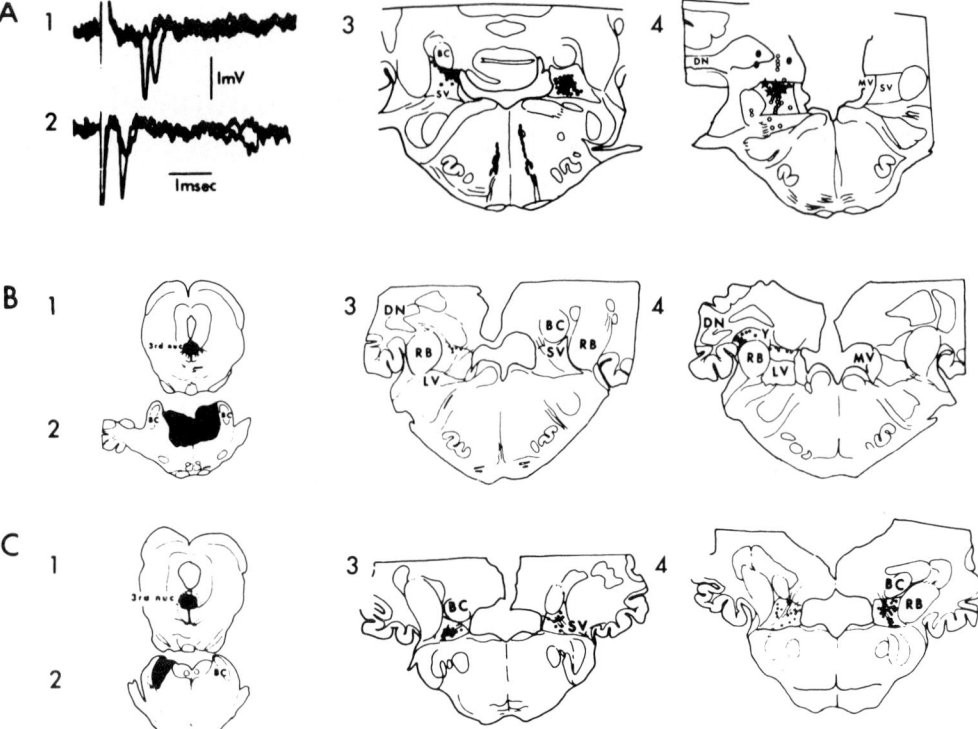

Fig. 1. Morphophysiological evidence on the localization of neurons relaying excitation from the anterior canal to the oculomotor nucleus. A_1) extracellular record of orthodromic activation of a neuron by selective stimulation of the anterior semicircular canal and A_2) antidromic activation of the same neuron by oculomotor nucleus stimulation. $A_{3,4}$, B_{1-4} illustrate coronal sections of the rabbit's brain stem: A_3 plots the location of cells labeled with HRP reaction product following an injection of HRP in the right oculomotor nucleus. A_4 plots the distribution of cells activated ortho- and antidromically as in $A_{1,2}$ (★), orthodromically only (○) or antidromically only (●). B_1, C_1) HRP injection site in the oculomotor nucleus. B_2) Extent of lesion in the pontine tegmentum and C_2) BC. $B_{3,4}$ and $C_{3,4}$ plot the location of labeled cells subsequent to injections and lesions illustrated in $B_{1,2}$ and $C_{1,2}$, resp. BC, brachium conjunctivum; DN, dentate nucleus; MV, medial vestibular nucleus; SV, superior vestibular nucleus; RB, restiform body; LV, lateral vestibular nucleus; y, group y of vestibular nuclei. (Reproduced from Yamamoto et al., 1978.)

(Fig. $1B_{3,4}$). In a complementary experiment, the MLF was left intact and the BC lesioned (Fig. $1C_{1,2}$). As Fig. $1C_{3,4}$ demonstrates the central portions of superior nucleus contain labeled neurons while the dorsal rostral portions are free of labeled cells. Acute electrophysiological experiments employing selective anterior canal and oculomotor nucleus stimulation (for orthodromic and antidromic activation respectively, Fig. $1A_{1,2}$) in the presence of an acutely lesioned MLF and ATD have confirmed this localization of the anterior canal relay neurons (see Fig. $1A_4$, stars).

Previously it was thought that the entire output of the superior vestibular nucleus was inhibitory; however, it is now clear in the rabbit that the excitatory limb of the anterior canal vestibulo-ocular reflex is relayed through the dorsal anterior subgroup of the superior nucleus via the BC to the contralateral inferior oblique and superior rectus motoneurons. Dorsal superior nucleus neurons have been demonstrated to send their axons to the third nucleus via the BC in the cat (Fig. 2A, B) and monkey (J.

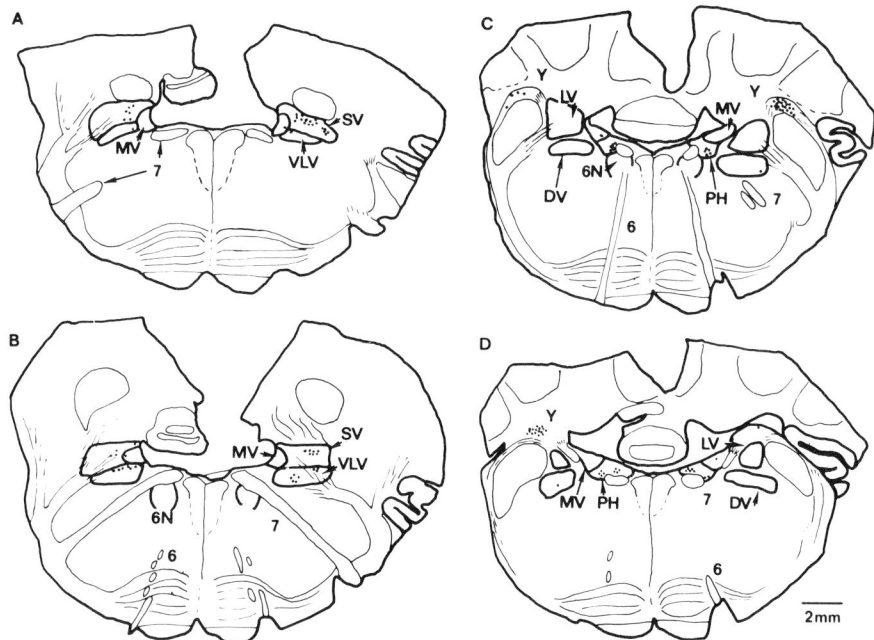

Fig. 2. Cells projecting to the oculomotor nucleus via pathways other than the MLF. Sections A–D are coronal sections of the cat's brain stem. Dots indicate neurons labeled with HRP reaction product subsequent to a large iontophoretic injection of HRP into the oculomotor nucleus. The MLF was acutely transected. Abbreviations as in Fig. 1. pH, prepositus hypoglossi; 6N, sixth nucleus; 6, sixth nerve; 7, seventh nerve.

Büttner-Ennever, personal communication). Abend's (1977) analysis of the superior nucleus in the monkey has demonstrated a dorsal group of neurons which receive anterior canal input. Thus, it is probable that these same superior nucleus neurons relay the excitatory limb of the anterior canal VOR in the cat and monkey as well as the rabbit. By contrast, the superior nucleus cells relaying inhibition via the ipsilateral MLF lie within the central portions of the nucleus.

Posterior canal

Stimulation of the posterior canal activates the contralateral inferior rectus and the ipsilateral superior oblique extraocular muscles. Excitation from the posterior canal is relayed by medial vestibular nucleus neurons whose axons exit the nucleus medially to ascend in the contralateral MLF. The inhibitory signals to antagonist muscles are relayed by the superior nucleus via the ipsilateral MLF.

Thus, the disynaptic vertical vestibulo-ocular reflex utilizes the MLF and BC pathways. All inhibition travels up the MLF while excitation travels up the MLF and BC. In the monkey, lesions of the MLF completely interrupt the vertical vestibulo-ocular reflex as tested behaviorally (Evinger et al., 1977). Vertical eccentric positions of fixation cannot be maintained and vertical smooth pursuit is also impaired; by contrast vertical saccades are relatively normal. One month after MLF lesions, however, there is slight recovery of vertical compensatory eye movements. It was suggested that this slight residual vertical vestibulo-ocular reflex may indicate transmission through the unlesioned BC pathway (Evinger et al., 1977). As Baker and

Berthoz (1974) have pointed out from intracellular recordings of inferior oblique motor neurons during nystagmus, the rhythmic nature of the discharges of these neurons is not completely abolished by MLF lesions. King et al. (1976) and Pola and Robinson (1978) have recorded from vertical VOR fibers in the MLF of alert monkeys. The fibers predominantly encode head velocity but eye position sensitivity was also manifest. Vertical MLF fibers increase their rate of discharge with eye movement into the on direction and pause for all saccades or quick phases of nystagmus, the pause outlasting the rapid eye movement. Baker and Berthoz (1974), recording from identified secondary vestibular axons in the cat's trochlear nucleus during vertical nystagmus have confirmed that these axons carry a head velocity and eye position signal.

The apparent discrepancy between the behavioral effects of an MLF lesion and the electrophysiological data demonstrating rhythmicity in the presence of this lesion is explicable as follows. Lesion of the MLF interrupts three out of four possible disynaptic pathways involved in the vertical vestibulo-ocular reflex; i.e. 1 and 2) the inhibitory pathways from the anterior and posterior canals, and 3) the excitatory pathways from the posterior canals. The only remaining pathways are 4) the excitatory pathways through the BC from the anterior canals. The BC pathways alone are apparently not adequate to generate the organized rhythmic activity necessary to sustain the vertical vestibulo-ocular reflex although they can account for the remaining rhythmicity at the motoneuron somas of the vertical extraocular motoneurons. *Behavioral results notwithstanding the important limb of the vestibulo-ocular reflex travelling through the BC should not be ignored simply because the vertical vestibulo-ocular reflex is abolished by MLF lesions.*

SACCULO-OCULAR PATHWAY

Compensatory eye movements are induced by the utricle and saccule as well as by the semicircular canals. The central organization of otolithic pathways will not be considered here except for the sacculo-ocular reflex which is partially relayed through the y-group of the vestibular nuclei. The morphophysiology of the y-group has been reexamined in the light of recent advances in neuroanatomical methods. This nucleus was described by Fuse (1912) and by Brodal and Pompeiano (1957). On cytoarchitectural grounds the nucleus can be divided into a ventral group of tightly packed fusiform neurons which are in close apposition to the restiform body and a more posterior dorsal group of loosely packed neurons lying between the ventral group and the dentate nucleus of the cerebellum. This latter group of neurons are larger than the former and have a variety of soma shapes from stellate to fusiform (see Fig. 3) but predominantly resemble the neurons in the dentate nucleus (Gacek, 1978). The separation of the two groups of neurons, however, is not complete as spindle shaped cells are often located dorsally and the larger stellate cells can be found ventrally (Figs. 3A–C). Suggestive evidence concerning the inputs and outputs of these two subgroups of y-group neurons has recently been obtained (Highstein, 1971, 1973; Graybiel and Hartwieg, 1974; Gacek, 1971a,b, 1977).

Lorente de Nó (1933) reported that VIIIth nerve fibers of saccular origin terminate centrally in the nucleus cerebellovestibularis, a brain stem area in the guinea pig lying beneath the cerebellar dentate nucleus which we interpret as probably being

435

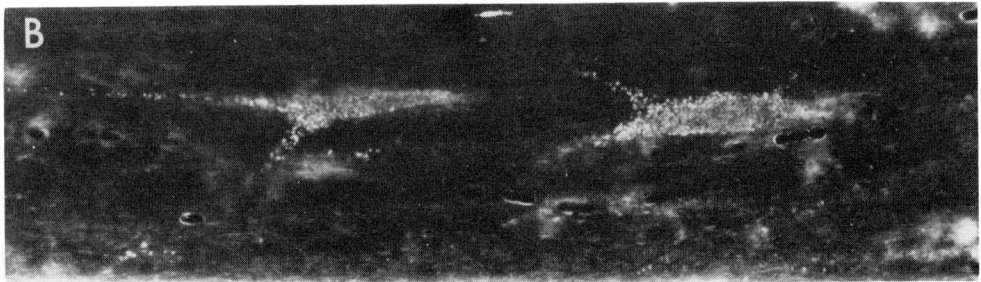

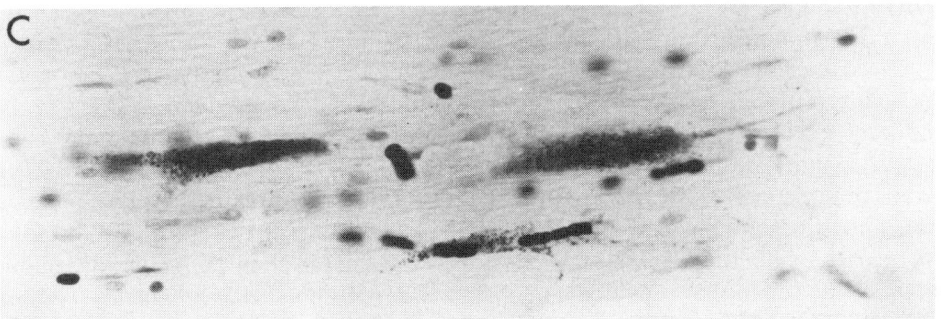

Fig. 3. Soma-dendritic shape and location of y-group neurons projecting to the oculomotor nucleus (same experiment as illustrated in Fig. 2, 3A). Soma-dendritic profiles traced with the aid of a drawing tube under 0 × oil immersion. 3B) Darkfield and 3C) brightfield photomicrographs of fusiform neurons projecting to the oculomotor nucleus. These neurons were located close to the restiform body. The long dimension of the somas of the two neurons in B and C is 50 μm.

equivalent to the y-group in the rabbit, cat, and monkey. According to Gacek (1977, 1978) the dorsal subdivision of the y-group may not receive direct saccular input although the dendrites of these dorsal-posterior neurons are within the range of vestibular primary afferent fibers. The ventral subgroup does, however, receive direct saccular input. Based on the retrograde transport of HRP from either the oculomotor nuclear complex or the contralateral vestibular nuclear complex (Gacek, 1977, 1978; Graybiel and Hartwieg, 1974; Pompeiano et al., 1978; see also Figs. 1–3), it appears that the dorsal subgroup is the projection nucleus reaching oculomotor neurons via the BC, while the ventral subgroup forms a part of the vestibular commissural system. Although this dorsal-ventral separation predominates it is not exclusive as *some* spindle shaped, ventrally located neurons do project to the third nucleus (see Figs. 1–3). Further, Hwang and Poon (1975) using extracellular recording have suggested that the y-group cells which project to the third nucleus are polysynaptically (not monosynaptically) activated by saccular stimulation. To completely clarify the issue of primary vestibular input to the y-group, and y-group neuronal termination site an intracellular microelectrode study is imperative.

THE HORIZONTAL VESTIBULO-OCULAR REFLEX

The horizontal vestibulo-ocular reflex is initiated by a head movement in the horizontal plane which activates the horizontal semicircular canals. Subsequently, the information is transmitted to the motoneurons controlling the horizontal extraocular muscles, the medial and lateral recti. From the ipsilateral horizontal canal signals are relayed in the ipsilateral medial vestibular nucleus to produce disynaptic IPSPs in abducens motoneurons and abducens internuclear neurons (Baker et al., 1969; Maeda et al., 1972; Baker and Highstein 1975; Maciewicz et al., 1977). From the contralateral horizontal canal, signals are relayed through the contralateral medial vestibular nucleus to produce disynaptic EPSPs in abducens motoneurons and internuclear neurons (Baker et al., 1969; Maeda et al., 1972; Baker and Highstein, 1975; Maciewicz et al., 1977). In the cat, stimulation of the horizontal canal ipsilateral to the medial rectus motoneurons produces disynaptic EPSPs in these motoneurons. The PSPs are relayed by neurons in the lateral vestibular nucleus, via the ATD (Baker and Highstein, 1978). With one further synaptic delay an EPSP evoked by VIIIth nerve disynaptic activation of the internuclear neurons of the abducens nucleus is produced in medial rectus cells (Highstein and Baker, 1978). (Internuclear neurons of the abducens nucleus are a unique population of cells lying within the confines of the abducens nucleus; these cells receive the exact same profile of synaptic input from vestibular stimulation as do the abducens motoneurons (Baker and Highstein, 1975). Axons of these internuclear neurons ascend in the contralateral MLF to terminate on medial rectus motoneurons (Highstein, 1977; Highstein and Baker, 1978).

The cat's medial rectus motoneurons are exceptional in that there is no disynaptic canal evoked inhibition evident (Baker and Highstein, 1978). (In the rabbit, stimulation of either VIIIth nerve produces an EPSP-IPSP sequence (Highstein, 1971, 1973).) In the alert cat and monkey records taken from MLF fibers during horizontal eye movements and/or head movement have led to the consensus that the predominant MLF fiber type related to horizontal eye movement is a burst tonic fiber (King et al., 1976; Pola and Robinson, 1978). The horizontal burst tonic fiber recorded in the MLF is undoubtedly the ascending axon of the internuclear neurons of the abducens nucleus (Highstein, 1977). Ascending tract of Deiters' neurons have not yet been studied in the alert animal.

Recently, techniques to correlate patterns of neuronal discharge with eye movements have been combined with techniques using electrical pulse stimulation to antidromically identify neurons. This combination allows the study of *electrophysiologically identified* neurons during eye movement in the alert animal. Internuclear neurons of the abducens nucleus and abducens motoneurons are examples of the use of this combination of techniques (Delgado Garcia et al., 1977; Nakao and Sasaki, 1978). Internuclear neurons discharge strictly in relation to eye movement of the contralateral eye increasing their frequency in response to eye movement in the on-direction of the contralateral medial rectus muscle, bursting with on-direction saccades and pausing for off-direction saccades (Delgado Garcia et al., 1977). Qualitatively, these internuclear neuronal discharges are similar to those of medial rectus motoneurons and it has been assumed (Pola and Robinson, 1978) that these burst-tonic neurons provide *all* of the necessary input for medial rectus motoneuron action. This is clearly not the case for vestibular compensatory eye movements (cf. Baker and Highstein, 1978; and below) and perhaps for gaze as well.

The other ascending pathway to the medial rectus motoneurons (the ATD) must play a prominent role at least in vestibular evoked eye movements. Gacek (1971) described a population of neurons in the ventrolateral vestibular nucleus whose axons do not ascend in the MLF or BC but take an independent middle course through the pontine tegmentum forming a loosely dispersed track which Muskens (1914) originally called the ascending tract of Deiters'. Stimulation of the vestibular nerve ipsilateral to medial rectus motoneurons activates this pathway and produces typical disynaptic vestibular canal evoked EPSPs in identified medial rectus motoneurons (Baker and Highstein, 1978). The burst-tonic internuclear neuron of the abducens nucleus contains an integrated vestibular signal, i.e., eye position (as well as velocity) and an implicit phase relationship to head velocity similar to the abducens motoneurons (Delgado Garcia et al., 1977; King et al., 1976). Because these internuclear neurons project monosynaptically to the contralateral medial rectus motoneurons, they must be important in the organization of horizontal conjugate gaze. In fact, the syndrome of the MLF has now been directly attributed to the interruption of the axons of these internuclear neurons in the MLF (Highstein, 1977; Baker and Highstein, 1978; Highstein and Baker, 1978). The function of the ATD pathway is, however, open to question. It was suggested that the ATD conveys macular signals to the medial rectus motoneuron as it is known that primary afferents of macular origin terminate in the ventrolateral vestibular nucleus (Gacek, 1969). However, primary afferents of canal origin terminate in overlapping areas (Gacek, 1969: Sans et al., 1972). This question of the role of the ATD neurons has been investigated directly in our laboratory.

In decerebrate cats both MLF's and the BC ipsilateral to the recording site were severed at the level of the base of the inferior colliculus. The only remaining ascending vestibular monosynaptic pathway to the oculomotor nucleus should have been the ATD. Ventral lateral vestibular neurons were activated orthodromically (within the monosynaptic range) by stimulation of the ipsilateral horizontal canal (Fig. 4A,C) and antidromically by stimulation of the oculomotor nucleus (Fig. 4A). Commissural inhibition was also present (Fig. 4D). Fig. 2 demonstrates the location of these ventral lateral

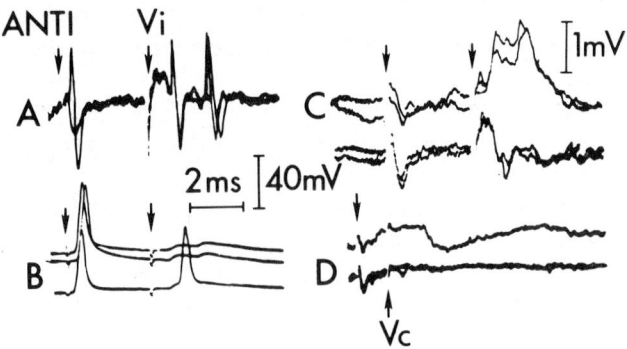

Fig. 4. Intra- and extracellular records from ATD neurons. A) Extracellular records of anti- and orthodromic activation of an ATD neuron following oculomotor nucleus and ipsilateral vestibular nerve stimulation, resp. B) Same neuron as A, but after penetration with microelectrode; records are low gain DC coupled. C) High gain records after impulse initiation was exhausted. First downward arrow indicates oculomotor and second downward arrow ipsilateral vestibular nerve stimulation. Lower records C and D are just extracellular traces. D) Downward arrow as in C, upward arrow indicates contralateral vestibular nerve stimulation.

vestibular neurons. When responses to rotation were examined, a prominent response type was kinetic as described by Shimazu and Precht (1965); i.e., neurons which were monosynaptically activated from the labyrinth, silent at rest and had head velocity sensitivity. However, we also encountered some neurons which lay intermediate in response type between tonic and kinetic. These neurons were monosynaptically activated from the labyrinth, but were tonically active at rest, and had head velocity sensitivity. An example of this type is illustrated in Fig. 5. This neuron, active at rest, was sensitive to ipsilateral rotation in the horizontal plane. The activation of the neuron with ipsilateral horizontal table (and head) movement can be seen. In addition, commissural inhibition was found to be present in the majority of neurons tested. Intracellular records obtained from these ATD neurons are illustrated in Fig. 4.

Several neurons were penetrated with a peroxidase loaded electrode and were stained with peroxidase. These neurons were all found to lie within the ventral lateral vestibular nucleus. Finally, to determine the trajectory of the axons of these ATD neurons a microelectrode was placed in the ventral lateral vestibular nucleus to record

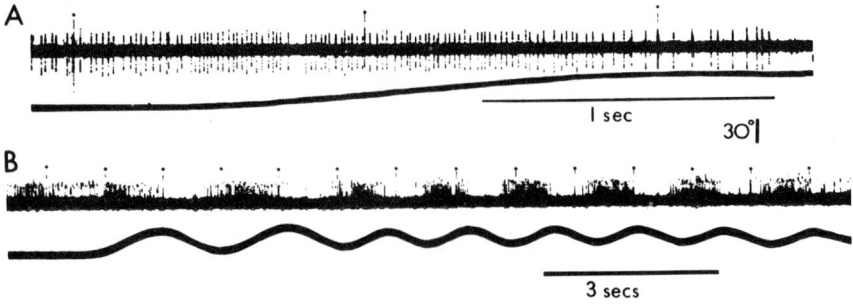

Fig. 5. Head velocity sensitivity of an antidromically identified ascending tract of Deiters' neuron. A, B) Same neuron on two different time bases. Upper records are filtered extracellular unit activity and lower records table position. Dots above upper traces indicate antidromic activation of the neuron at 1/sec. Upward deflection of position trace indicates ipsilateral rotation.

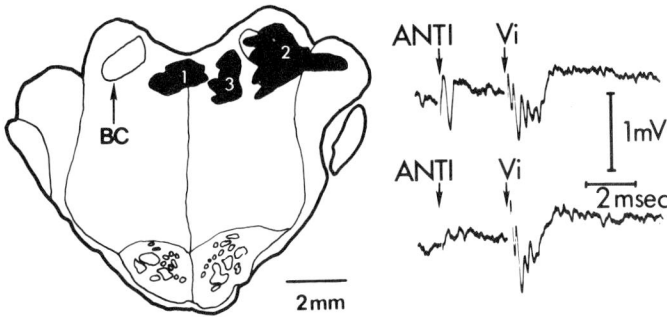

Fig. 6. Location of the ascending tract of Deiters' in the pontine tegmentum. The microelectrode was placed into the ventral lateral vestibular nucleus to record the field potentials following oculomotor and ipsilateral vestibular nerve stimulation. Upper records taken after lesions 1, 2 (indicated by black marks with white numbers on the cross section of the brain stem) and lower records after lesion 3. Note that the antidromic field potential is completely abolished while the vestibular nerve evoked field is unchanged.

the antidromic field evoked by IIIrd nucleus stimulation (Fig. 6, upper traces). When the ATD was cut between the abducens and trochlear nuclei (lesion 3, Fig. 6) the antidromic field potential was completely eliminated (Fig. 6, lower traces).

In conclusion, the ATD neurons encode a head velocity signal proportional to horizontal angular velocity as was previously surmised by Baker and Highstein (1978). We infer that this signal must be utilized to (at least partially) control medial rectus neurons during head rotation. The sensitivity and frequency response of these ATD neurons as well as the presence or absence of gaze related signals are currently under investigation in alert cats.

Considering the vertical oculomotor system, the vertical canal vestibular signals traveling in the MLF and BC carry head velocity and eye position signals and pause for saccades. It has been inferred that burst-tonic signals necessary for the generation of saccades or quick phases descend from above the IIIrd nucleus to the vertical motoneurons. In summary, vertical oculomotor neurons receive input reflecting head velocity, eye velocity and eye position. Neurons in the horizontal system receive similar classes of input. This has been demonstrated for abducens motoneurons. The monosynaptically activated Vc axon carries head velocity information while the monosynaptically activated Vi inhibitory axon carries head velocity and eye position information (Maeda et al., 1972; Shimazu, 1977; Hikosaka et al., 1977). (There is also a burst inhibitory axon present in the abducens nucleus which has not yet been demonstrated for the vertical system (Hikosaka and Kawakami, 1976, 1977).) The pontine reticular neurons which are assumed to project monosynaptically to the abducens nucleus convey saccadic information in the form of a burst with frequency proportional to saccade velocity, and the number of spikes proportional to saccade amplitude. Thus, the abducens motoneuron and internuclear neuron pool receive raw velocity, eye position, and eye velocity signals. The sum of this information is transmitted up the MLF by the internuclear neurons to the medial rectus motoneurons (Highstein, 1977). However, the internuclear neuronal discharge is a prepackaged signal containing no raw head velocity information. This head velocity information is supplied via the ATD neurons. This raw velocity information to medial rectus motoneurons is apparently necessary in order to activate the sluggish oculomotor plant to maintain the appropriate phase relations between eye and head movement (cf. Baker and Highstein, 1978). Whether the ATD neurons are activated for all frequencies of head movement and/or eye movement beginning from all medial-lateral eye positions is still open to investigation.

SUMMARY

The central connections of the individual semicircular canals subserving the vestibulo-ocular reflex (VOR) are elaborated. The three pathways relaying the reflex are the median longitudinal fasciculus (MLF), the brachium conjunctivum (BC), and the ascending tract of Deiters' (ATD). The vertical vestibulo-ocular reflex originates in the anterior and posterior semicircular canals. The MLF contains the axons of the relay neurons for the inhibitory pathways from the anterior and posterior canals and the excitatory pathways from the posterior canals to the ocular motoneurons. The BC contains the axons of the relay neurons for the excitatory pathway from the anterior canal to the IIIrd nucleus. The important limb of the vestibulo-ocular reflex traveling through the BC should not be ignored simply because the vertical vestibulo-ocular reflex is abolished by MLF lesions.

The horizontal VOR originates in the bilateral horizontal semicircular canals. The disynaptic excitatory pathway, from the ipsilateral horizontal canal to the medial rectus motoneurons is relayed via the ATD. The disynaptic excitatory and inhibitory input to abducens motoneurons and internuclear neurons is reviewed.

Recording from internuclear neurons in the alert animal demonstrates that these neurons discharge strictly in relation to eye movement of the contralateral eye, increasing their frequency in response to movement in the on-direction of the contralateral medial rectus muscle, bursting with on-direction saccades and pausing for off-direction saccades. Qualitatively, these internuclear neurons are similar to medial rectus motoneurons in their discharge properties, however, it was pointed out that the other pathway to the medial rectus via the ATD carries a head velocity signal and therefore, must play a prominent role in vestibular evoked eye movements.

Finally, the organization of the saccular ocular pathway through the y-group of the vestibular nuclei was discussed. The y-group has been subdivided into two divisions with the advent of modern morphophysiological techniques.

REFERENCES

Abend, W.K. (1977) Functional organization of the superior vestibular nucleus of the squirrel monkey. *Brain Res.*, 132: 65–84.
Baker, R. and Berthoz, A. (1974) Organization of vestibular nystagmus in oblique oculomotor system. *J. Neurophysiol.*, 37: 195–217.
Baker, R. and Highstein, S.M. (1975) Physiological identification of interneurons and motoneurons in the abducens nucleus. *Brain Res.*, 91: 292–298.
Baker, R. and Highstein, S.M. (1978) Vestibular projections to medial rectus subdivision of oculomotor nucleus. *J. Neurophysiol.*, 41: 1629–46.
Baker, R.G., Mano, N. and Shimazu, H. (1969) Postsynaptic potentials in abducens motoneurons induced by vestibular stimulation. *Brain Res.*, 15: 577–580.
Brodal, A. and Pompeiano, O. (1957) The vestibular nuclei in the cat. *J. Anat. (Lond.)*, 91: 438–454.

Delgado-Garcia, J., Baker, R. and Highstein, S.M. (1977) The activity of internuclear neurons identified within the abducens nucleus of the alert cat. In *Control of Gaze by Brain Stem Neurons*, R. Baker and A. Berthoz (Eds.), Elsevier/North-Holland Biomed. Press, Amsterdam, pp. 291–301.

Evinger, L.C., Fuchs, A.F. and Baker, R. (1977) Bilateral lesions of the medial longitudinal fasciculus in monkeys: effects on the horizontal and vertical components of voluntary and vestibular induced eye movements. *Exp. Brain Res.*, 28: 1–20.

Fuse, G. (1912) *Die innere Abteilung des Kleinhirnstieles (Meynert I, A.K.) und der Deitersche Kern*. Arb. Hirnanat. Inst. (Zürich), 6: 29–267.

Gacek, R.R. (1969) The course and central termination of first order neurons supplying vestibular endorgans in the cat. *Acta oto-laryngol. (Stockh.)*, 254: 1–66.

Gacek, R.R. (1971a) Anatomical demonstration of the vestibulo-ocular projections in the cat. *Acta oto-laryngol. (Stockh.)*, 293: 1–63.

Gacek, R.R. (1971b) Anatomical demonstration of the vestibulo-ocular projections in the cat. *Laryngoscope*, 81: 1559–1595.

Gacek, R.R. (1977) Location of brain stem neurons projecting to the oculomotor nucleus in the cat. *Exp. Neurol.*, 57: 725–749.

Gacek, R.R. (1978) Location of commissural neurons in the vestibular nuclei of the cat. *Exp. Neurol.*, 59: 479–491.

Graybiel, A.M. and Hartwieg, E.A. (1974) Some afferent connections of the oculomotor complex in the cat: an experimental study with tracer techniques. *Brain Res.*, 81: 543–551.

Highstein, S.M. (1971) Organization of the inhibitory and excitatory vestibulo-ocular reflex pathways to the third and fourth nuclei in rabbit. *Brain Res.*, 32: 218–224.

Highstein, S.M. (1973) The organization of the vestibulo-oculomotor and trochlear reflex pathways in the rabbit. *Exp. Brain Res.*, 17: 285–300.

Highstein, S.M. (1977) Abducens to medial rectus pathway in the MLF: a possible cellular basis for the syndrome of internuclear ophthalmoplegia. In *Eye Movements*, B.A. Brooks and F.J. Bajandas (Eds.), Plenum Press, New York.

Highstein, S.M. and Baker, R. (1978) Excitatory termination of abducens internuclear neurons on medial rectus motoneurons; relationship to syndrome of internuclear ophthalmoplegia. *J. Neurophysiol.*, 41: 1647–1661.

Highstein, S.M., Ito, M. and Tsuchiya, T. (1971) Synaptic linkage in the vestibulo-ocular reflex pathways of rabbit. *Exp. Brain Res.*, 13: 306–326.

Hikosaka, O. and Kawakami, T. (1976) Inhibitory interneurons in the reticular formation and their relation to vestibular nystagmus. *Brain Res.*, 117: 513–518.

Hikosaka, O. and Kawakami, T. (1977) Inhibitory reticular neurons related to the quick phase of vestibular nystagmus—their location and projection. *Exp. Brain Res.*, 27: 377–396.

Hikosaka, O., Maeda, M., Nakao, S., Shimazu, H. and Shinoda, Y. (1977) Presynaptic impulses in the abducens nucleus and their relation to postsynaptic potentials in motoneurons during vestibular nystagmus. *Exp. Brain Res.*, 27: 355–376.

Hwang, J.C. and Poon, W.F. (1975) An electrophysiological study of the sacculo-ocular pathways in cats. *Jap. J. Physiol.*, 25: 241–251.

Ito, M., Nisimaru, N. and Yamamoto, M. (1973a) The neural pathways mediating reflex contraction of extraocular muscles during semicircular canal stimulation in rabbits. *Brain Res.*, 55: 183–188.

Ito, M., Nisimaru, N. and Yamamoto, M. (1973b) The neural pathways relaying reflex inhibition from semicircular canals to extraocular muscles of rabbits. *Brain Res.*, 55: 189–193.

Ito, M., Nisimaru, N. and Yamamoto, M. (1973c) Specific neural connections for the cerebellar control of vestibulo-ocular reflexes. *Brain Res.*, 60: 238–242.

Ito, M., Nisimaru, N. and Yamamoto, M. (1976a) Pathways for the vestibulo-ocular reflex excitation arising from semicircular canals of rabbits. *Exp. Brain Res.*, 24: 257–271.

Ito, M., Nisimura, N. and Yamamoto, M. (1976b) Postsynaptic inhibition of oculomotor neurons involved in vestibulo-ocular reflexes arising from semicircular canals of rabbits. *Exp. Brain Res.*, 24: 273–283.

King, W.M., Lisberger, S.G. and Fuchs, A.F. (1976) Responses of fibers in medial longitudinal fasciculus (MLF) of alert monkeys during horizontal and vertical conjugate eye movements evoked by vestibular or visual stimuli. *J. Neurophysiol.*, 39: 1135–1149.

Lorente de Nó, R. (1933) Anatomy of the eighth nerve. *Laryngoscope*, 43: 1–38.

Maciewicz, R.J., Eagen, K., Kaneko, C.R.S. and Highstein, S.M. (1977) Vestibular and medullary brain stem afferents to the abducens nucleus in the cat. *Brain Res.*, 123: 229–240.

Maeda, M., Shimazu, H. and Shinoda, Y. (1972) Nature of synaptic events in cat abducens motoneurons at slow and quick phase of vestibular nystagmus. *J. Neurophysiol.*, 35: 279–296.

Muskens, L.J.J. (1913/14) An anatomico-physiological study of the posterior longitudinal bundle in its relation to forced movements. *Brain*, 36: 352–426.

Nakao, S. and Sasaki, S. (1978) Firing pattern of interneurons in the abducens nucleus related to vestibular nystagmus in the cat. *Brain Res.*, 144: 389–394.

Pola, J. and Robinson, D.A. (1978) Oculomotor signals in medial longitudinal fasciculus of the monkey. *J. Neurophysiol.*, 41: 245–259.

Pompeiano, O., Mergner, T. and Corvaja, N. (1978) Commissural, perihypoglossal and reticular afferent projections to the vestibular nuclei in the cat. An experimental anatomical study with the method of the retrograde transport of horseradish peroxidase. *Arch. ital. Biol.*, 116: 130–172.

Sans, A., Raymond, J. and Marty, R. (1972) Projections des cretes ampullaires et de l'utricule dans les noyaux vestibulaires primaires. Etude microphysiologique et corrélations anatomo-fonctionnelles. *Brain Res.*, 44: 337–355.

Shimazu, H. (1977) Brainstem interneurons modulating motor activity during vestibular nystagmus. In *Control of Gaze by Brain Stem Neurons*, R. Baker and A. Berthoz (Eds.), Elsevier/North-Holland Biomed. Press, Amsterdam, pp. 225–235.

Shimazu, H. and Precht, W. (1965) Tonic and kinetic responses of cat's vestibular neurons to horizontal angular acceleration. *J. Neurophysiol.*, 28: 991–1013.

Yamamoto, M., Shimoyama, I. and Highstein, S.M. (1978) Vestibular nucleus neurons relaying excitation from the anterior canal to the oculomotor nucleus. *Brain Res.*, 148: 31–42.

Labyrinthine Influences on Motoneurons Responsible for Vertical Eye Movements in the Rabbit

B. GHELARDUCCI[1], M. FAVILLA[1] and A. STARITA[2]

[1]*Istituto di Fisiologia Umana, Cattedra II, Università di Pisa; and* [2]*Istituto di Elaborazione dell'Informazione, C.N.R., 56100 Pisa (Italy)*

The role of the vestibulo-ocular reflex (VOR) is to stabilize the image on the retina when the head is moved in space. For head displacements in the horizontal plane the neural signals required for compensatory movements of the eye originate from the ampullar receptors in the horizontal semicircular canals. When the compensation is required for head displacements in other planes, the vertical (anterior and posterior) semicircular canals provide the appropriate velocity signal to drive the oculomotor neurons (OMN). There are, however, other vestibular receptors located in the maculae of the utriculus and sacculus whose specific function is to signal head displacements and their discharge may directly drive the OMN (Lowenstein and Roberts, 1949; Jongkees, 1950; Trincker, 1962). The output from these static receptors must be considered when vertical or rotatory eye movements of vestibular origin are investigated (Magnus, 1924; Lorente de Nó, 1932; Suzuki et al., 1969; Fluur and Mellström, 1971).

There have been many studies on static vestibulo-ocular reflexes and counterrolling (see Cohen, 1974), as well as on the dynamic aspects of vertical eye movements of vestibular origin (Baarsma and Collewijn, 1975; Blanks et al., 1978). However, due to the choice both of the animal and the experimental paradigm, a comparison of the results is sometimes difficult. In fact, in species with front-facing eyes the static component of the reflex is usually rather weak and dominated both by dynamic vestibular reflexes and by vision (Miller, 1967; Krejcova et al., 1973; Uemura and Cohen, 1973); only in species with laterally placed eyes does the otolithic component of the reflex play a significant, compensatory role.

In the rabbit, for instances, when the head is tilted about the longitudinal axis, the eyes rotate so to keep their former orientation in space (Lorente de Nó, 1932; for a review, see Carpenter, 1977). Thus, in this animal, it is possible to elicit almost pure vertical compensatory eye movements which may be attributed, by varying independently the amplitude and the rate of tilt, to the activation of one or both types of vestibular receptors. The vertical vestibulo-oculomotor system of the rabbit may therefore offer a suitable model to study the relative contribution of static and dynamic vestibular proprioceptors in controlling the same motor function.

The following are the results of an analysis of such interactions studied at the level of the IIIrd nerve nucleus of unanaesthetized, encéphale isolé rabbits submitted to sinusoidal lateral tilts of varying amplitudes and rates.

EXPERIMENTAL PLAN

The analysis of the dynamic properties of the discharge of the OMN involved in the vertical VOR was performed by submitting the rabbit, deprived of visual input, to a series of sinusoidal lateral tilts of fixed amplitude ($\pm 15°$) at frequencies varying from 0.013 Hz to 0.2 Hz. The peak values of acceleration corresponding to the frequencies used, varied from 0.1 to 23.7 degrees/sec^2.

There is evidence that the peak acceleration at the highest frequency of tilt used in the present study is well above the threshold for both tonic and kinetic vestibular neurons (Shimazu and Precht, 1965), so that the entire information about head velocity originating from the vertical semicircular canals is forwarded to the OMN. On the other hand, since 0.2 degrees/sec^2 has been reported as the lowest value of threshold acceleration for second order vestibular neurons related to ampullar receptors (Shimazu and Precht, 1965; see also Precht, 1974, for a review), one can assume that at the lowest frequency used in this study any modulation of the OMN discharge observed during sinusoidal tilt should be attributed to the activation of macular receptors, whose adequate stimulus is the variation of head position. The responses of OMN obtained when they are driven only by static vestibular receptors have been correlated with varying amplitudes of tilt ($5°–10°–15°$).

The discharge frequency of each motoneuron was recorded throughout the entire stimulation paradigm for several cycles of tilt. Action potentials and analogue signals of table position and synchronizing pulses were stored on analogue tape. The data were then converted and processed by means of a Hewlett-Packard computer system composed by a 2100A and 5451B units which performed the Fourier and correlation analysis as well as the transfer function of the system.

IDENTIFICATION OF IIIRD NERVE NUCLEUS MOTONEURONS CONTROLLING VERTICAL EYE MUSCLES

Motoneurons recorded within the IIIrd nerve nucleus were attributed to the superior rectus (SR), inferior rectus (IR) and inferior oblique (IO) subnuclei both by their antidromic invasion and by the distribution of the antidromic field potentials evoked by electrical stimulation of each muscle nerve. Their position within the oculomotor nucleus was also determined histologically by marking the recording sites with iontophoretic injection of Pontamine Sky blue (Hellon, 1971). The topography of the three subnuclei within the IIIrd nerve nucleus obtained in our study coincides with that proposed by previous anatomical (Van Biervliet, 1899; Akagi, 1978) and physiological (Highstein, 1973) investigations on the rabbit oculomotor nuclei. A complete survey of the IIIrd nerve nucleus covering all the subnuclei involved in the control of vertical eye movements was justified by the possibility of a different behaviour to vestibular stimulation of motoneurons innervating different, although synergistic, eye muscles such as the IR and the IO. However, since the results obtained in this study have not shown any significant difference between the three groups of oculomotor neurons in the production of compensatory eye movements during lateral tilt, we shall refer to the three subnuclei as one ideal group of motoneurons representing, in this species, the final common path for vertical eye movements of vestibular origin.

The spontaneous discharge frequency of the three groups of ocular motoneurons recorded with the animal in the horizontal position exhibited a bimodal distribution with a peak at 12 imp/sec and a second peak at 35 imp/sec. A similar bimodal distribution of the resting discharge of OMN, also innervating different eye muscles, has been reported by Henn and Cohen (1972) in the monkey and by Yagi (1974) in the rabbit.

RESPONSES OF THE OCULOMOTOR NEURONS TO SINUSOIDAL LATERAL TILT

Response pattern

All the motoneurons clearly identified as belonging to the three subnuclei exhibited a significant modulation of their discharge frequency during sinusoidal lateral tilt. The modulation was in the proper direction to keep the position of the eye and, in this species, of the visual axis constant in space.

Fig. 1A shows the response of an IR motoneuron to 15° tilts at three different frequencies (0.013, 0.055 and 0.2 Hz). The peak of the modulation (arrows) exhibits

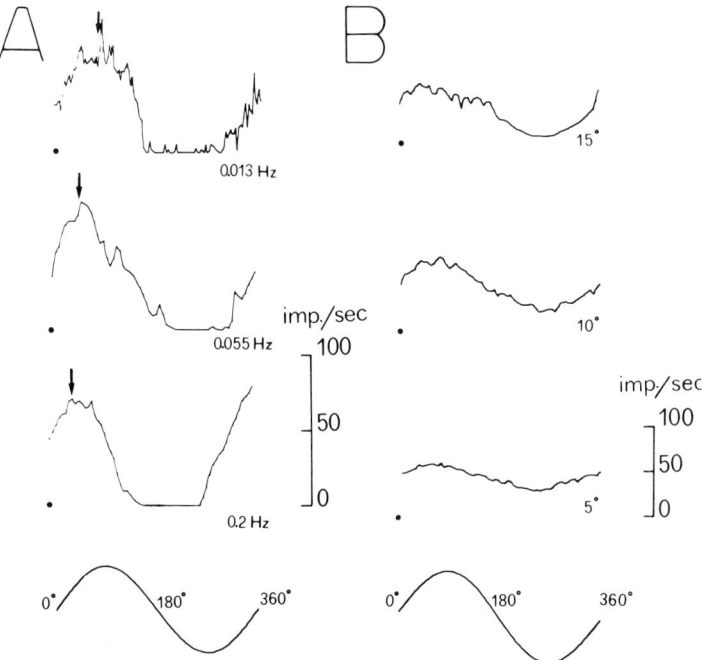

Fig. 1. Response patterns of IIIrd nerve nuclear OMN to sinusoidal lateral tilt. A) The discharge of an IR motoneuron is shown as a sequential pulse density histogram averaged over several cycles of tilt of fixed amplitude (±15°) but of different frequencies corresponding to 0.1, 1.8 and 23.7 degrees/sec^2 peak acceleration. The peak of the discharge indicated by the arrows, is related to table position shown in the bottom line. Upward deflection of the position signal indicates ipsilateral side up of the animal and corresponds to peak ipsilateral angular acceleration. B) Responses of a different IR motoneuron to sinusoidal tilts of varying amplitudes and frequencies yielding the same value of peak acceleration (0.1 degree/sec^2), considered subthreshold for canal-dependent vestibular neurons. The pulse density histograms were averaged over several cycles of stimulation. For all the responses illustrated in A and B the time base has been normalized to the same duration.

a progressive phase shift related to position when the peak acceleration increases. In Fig. 1B the response of another IR motoneuron to varying tilt amplitudes (15°, 10° and 5°) at constant peak acceleration of 0.1 degree/sec² is shown. It appears that the peak to peak amplitude of modulation of the discharge rate, which is almost in phase with position, is related with the magnitude of tilt. This is consistent with the fact that there is a direct relationship between the discharge rate of oculomotor neurons and the tension necessary for a given eye muscle to move the eye ball when compensation for head displacements becomes necessary (Henn and Cohen, 1974; Robinson, 1975).

Amplitude of modulation

In Fig. 2A the average half-peak amplitude of modulation of the discharge of the whole population of recorded OMN is plotted against the frequencies of stimulation used for 15° tilts. The graph shows that the variation of discharge frequency induced by vestibular stimulation on the OMN significantly increases when the frequency of stimulation is raised above 0.025 Hz, corresponding to a peak acceleration of 0.36 degrees/sec², which is in the range of the estimated threshold for canal driven vestibular nuclear neurons (Shimazu and Precht, 1965). This indicates that, with the participation of ampullar vestibular receptors in the vertical VOR, the modulation of the OMN discharge is higher than that displayed when they are driven exclusively by the otoliths. However, when the system works in this range, the amplitude of modulation is related to the magnitude of tilt, although in a non-linear way, as can be seen in Fig. 2B, where the half-peak amplitude of modulation obtained at the value of acceleration subthreshold for canal dependent second order vestibular neurons is plotted against the amplitude of tilt.

Phase of the response

The peak of the modulated discharge exhibited a phase relationship with respect to the extreme inclination of the tilt table which varied with the frequency of the

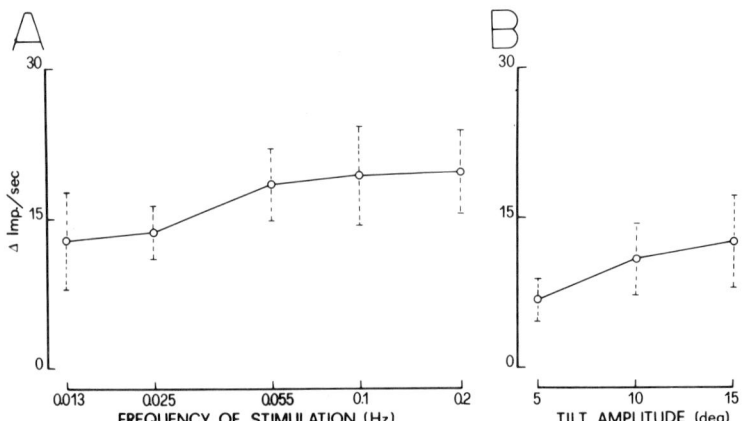

Fig. 2. Amplitude of modulation of discharge of the OMN during sinusoidal lateral tilt as a function of the frequency and amplitude of stimulation. In A, the half-peak amplitude of the response is shown as a function of the frequency of stimulation for a ±15° tilt. In B, the same variable is correlated with increasing amplitudes of tilt at constant peak acceleration (0.1 degree/sec²). Each plotted value represents the mean with SD of the responses of 20 IIIrd nerve nuclear OMN.

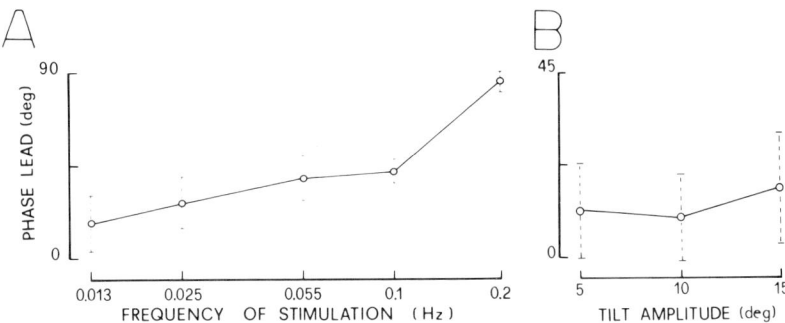

Fig. 3. Phase of the OMN response during sinusoidal lateral tilt. In A, the phase of the peak of the OMN response related to position is plotted as a function of the frequency of stimulation for ±15° tilt. In B, the phase of the response is shown as a function of tilt amplitude at constant peak acceleration (0.1 degree/sec^2).

sinusoidal oscillations. At low frequencies of tilt, when the peak acceleration was below threshold for canal driven second order vestibular neurons, the peak of the response led to extreme position of about 15° as can be seen in Fig. 3A where the phase of the response is shown as a function of tilt frequency for a series of 15° tilts. By increasing the tilt frequency this phase lead increased steadily reaching about 45° at 0.1 Hz. A further increase of the phase lead up to 85° took place at 0.2 Hz. This behaviour of the phase of the response may be explained by the fact that only at frequencies above 0.1 Hz the peak acceleration yielded during a 15° tilt is well above threshold for kinetic second order vestibular neurons (Shimazu and Precht, 1965). During sinusoidal horizontal rotation the kinetic vestibular neurons have been shown to transmit more faithfully than the tonic neurons the phase properties of primary vestibular afferents (Shinoda and Yoshida, 1973; Ezure et al., 1978). The recruitment of this group of vestibular neurons at the highest frequency of stimulation can account for the sharp increase of the phase shift towards peak velocity observed in the OMN response. Fig. 3B shows the phase angle of the response recorded at 3 different amplitudes of tilt at constant peak acceleration of 0.1 degrees/sec^2. It appears from the graph that the phase of the response does not vary systematically with tilt amplitude. This suggests that the otolith-ocular reflex by itself operates without a significant phase shift when driven by an adequate input of variable amplitude. The phase lead of the OMN response related to position is similar to that found for the primary otolith afferents in the cat (Anderson et al., 1978) and in the monkey (Fernandez and Goldberg, 1976). This indicates that macular signals undergo little if any central neural integration before reaching the OMN, at least at very low frequencies of stimulation.

Sensitivity of ocular motoneurons to vestibular stimulation

The sensitivity of the OMN to lateral tilt was computed by means of the transfer function of the system and has been expressed as the ratio between the amplitude of modulation and the magnitude of tilt. This variable was normalized by expressing it as a percentage of the mean discharge of each oculomotor neuron. In Fig. 4A the average sensitivity is plotted on a logarithmic scale against the series of frequencies of stimulation used for 15° tilts. The behaviour of sensitivity in the frequency domain is similar to the one already described for the amplitude of modulation (see Fig. 2A).

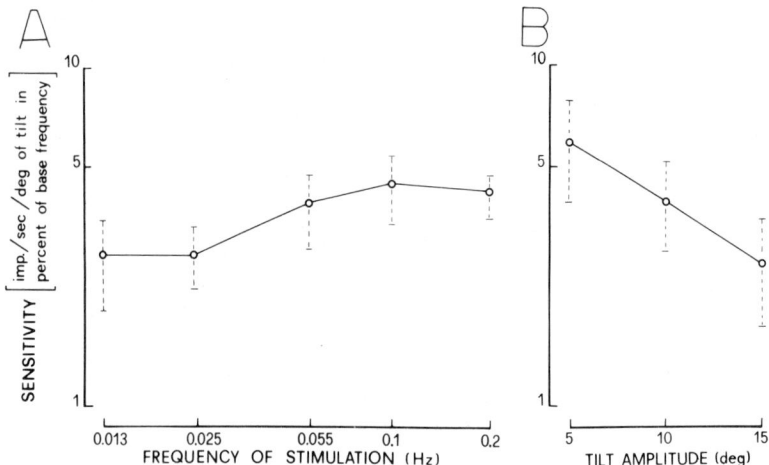

Fig. 4. Sensitivity of the OMN to sinusoidal lateral tilt. In A, the sensitivity is shown on a log-log scale as a function of frequency for fixed tilt amplitude of ±15°. B shows the inverse relationship existing between sensitivity and tilt amplitude at fixed peak acceleration of 0.1 degree/sec^2.

The drop of the sensitivity when only macular receptors are driving the discharge of the OMN is in accordance with the low gain of ocular movements evoked by linear accelerations described in the rabbit by Baarsma and Collewijn (1975).

The synaptic efficacy of the vestibular input is thus greater when both static and dynamic vestibular receptors are active on OMN during lateral tilt. This conforms with the pattern of convergence recently described in trochlear motoneurons of the cat by Blanks et al. (1978) (see also Precht et al., this volume, chapter IVC4).

The graph in Fig. 4B illustrates the relationship that exists between the sensitivity of the OMN and the amplitude of tilt when the vertical VOR is exclusively controlled by the otolith input. The declining slope of the curve indicates that the otolith ocular reflex loses its efficacy with increasing amplitudes of tilt. This is also reflected by the non-linear increase in amplitude of modulation of the OMN discharge shown in Fig. 2B at the same value of peak acceleration.

CONCLUSIONS

The analysis of the dynamic response properties of the OMN innervating three of the four muscles primarily involved in the vertical compensatory movements of the rabbit's eye during sinusoidal lateral tilt has pointed out that both static and dynamic vestibular receptors play a role in the reflex. The different characteristics of vestibular control on OMN can be investigated by choosing the appropriate range of stimulus frequency.

At the lowest frequency of stimulation the peak of the OMN discharge is almost in phase with head position. A progressive phase shift towards head velocity takes place by increasing the frequency of tilt. The amplitude of modulation of the discharge rate as well as the sensitivity of the OMN are also related to the stimulus frequency, exhibiting a significant increase when the frequency is raised above 0.025 Hz, corresponding to a peak acceleration around threshold for canal driven vestibular neurons. Thus, above this frequency, the convergence of the signals originating from the otoliths and from the ampullar receptors of the vertical semicircular canals on OMN affects both the gain and the phase of their response to natural vestibular stimulation. These changes are in the correct direction to improve the efficacy of the vertical VOR.

By lowering the frequency of tilt the analysis of the system was extended into a range of acceleration in which otolithic receptors alone are able to influence the OMN discharge. Since changes in head position represent the adequate stimulus for macular receptors, the pattern of activity of OMN has been correlated with this variable. In the range of the exclusive otolithic drive the phase of the OMN discharge is not influenced systematically by changes in the magnitude of tilt. On the contrary, the amplitude of modulation of the OMN discharge increases with increasing tilt magnitude but not in a linear fashion. Consistent with this finding, the sensitivity of the OMN is inversely related with the magnitude of tilt.

These observations indicate that in the rabbit, during very slow movements of the head about the longitudinal axis, in the absence of vision, a certain degree of compensation in eye position can still be provided by the otoliths. The decrease in gain observed in the OMN response in this range of operation of the vertical VOR implies that the drive exerted by the otoliths alone on the OMN needs to be complemented by other sources of neural activity, possibly visual input, to ensure a perfect compensatory action of the eye muscles.

SUMMARY

The dynamic analysis of the control exerted by macular and ampullar vestibular receptors on oculomotor neurons (OMN) can be investigated by choosing a suitable stimulation paradigm for the vestibular receptors.

The response dynamics of IIIrd nerve nucleus OMN controlling vertical eye movements have been studied in unanaesthetized, encéphale isolé rabbits submitted to sinusoidal lateral tilts of $\pm 15°$ at frequencies varying from 0.013 Hz to 0.2 Hz.

At tilt frequencies yielding a peak angular acceleration below the estimated threshold for second order vestibular neurons related to vertical semicircular canals, the phase of the response slightly led the extreme table position. A progressive phase shift towards head velocity was observed by increasing tilt frequency.

Both the sensitivity and the amplitude of modulation of the OMN significantly increased at frequencies above 0.025 Hz, corresponding to peak accelerations above threshold for canal related vestibular neurons.

In the range of exclusive otolithic drive of the OMN, changing the angle of tilt did not affect the phase of the response. The amplitude of modulation of the OMN discharge was related to the magnitude of tilt, but not linearly. This is reflected by the inverse relationship observed in this frequency range between tilt amplitude and sensitivity of the OMN.

The convergent action of macular and ampullar vestibular receptors in the control of vertical eye movements of the rabbit is discussed in relation to stimulus frequency.

ACKNOWLEDGEMENTS

This investigation was supported by the Public Health Service Research Grant NS 07685–11 from the National Institute of Neurological and Communicative Disorders and Stroke, N.I.H., U.S.A. and by a Research Grant from the Consiglio Nazionale delle Ricerche, Italy.

REFERENCES

Akagi, Y. (1978) The localization of the motor neurons innervating the extraocular muscles in the oculomotor nuclei of the cat and rabbit, using horseradish peroxidase. *J. comp. Neurol.*, 181: 745–762.
Anderson, J.H., Blanks, R.H.I. and Precht, W. (1978) Response characteristics of semicircular canal and otolith systems in cat. I. Dynamic responses of primary vestibular fibers. *Exp. Brain Res.*, 32: 491–507.
Baarsma, E.A. and Collewijn, H. (1975) Eye movements due to linear accelerations in the rabbit. *J. Physiol. (Lond.)*, 245: 227–247.
Blanks, R.H.I., Anderson, J.H. and Precht, W. (1978) Response characteristics of semicircular canal and otolith systems in cat. II. Responses of trochlear motoneurons. *Exp. Brain Res.*, 32: 509–528.
Carpenter, R.H.S. (1977) *Movements of the Eyes*. Pion Ltd., London.
Cohen, B. (1974) The vestibulo-ocular reflex arc. In *Handbook of Sensory Physiology, Vol. VI/1, Vestibular System, Part 1, Basic Mechanisms*, H.H. Kornhuber (Ed.), Springer-Verlag, Berlin, pp. 477–540.
Ezure, K., Schor, R.H. and Yoshida, K. (1978) The response of horizontal semicircular canal afferents to sinusoidal rotation in the cat. *Exp. Brain Res.*, 33: 27–39.
Fernandez, C. and Goldberg, J.M. (1976) Physiology of the peripheral neurons innervating otolith organs of the squirrel monkey. III. Response dynamics. *J. Neurophysiol.*, 39: 996–1008.
Fluur, E. and Mellström, A. (1971) The otolith organs and their influence on oculomotor movements. *Exp. Brain Res.*, 30: 139–147.
Hellon, R.F. (1971) The marking of electrode tip positions in neurons tissue. *J. Physiol. (Lond.)*, 214: 12P.
Henn, V and Cohen, B. (1972) Eye muscle motor neurons with different functional characteristics. *Brain Res.*, 45: 561–568.
Highstein, S.M. (1973) The organization of the vestibulo-oculomotor and trochlear reflex pathways in the rabbit. *Exp. Brain Res.*, 17: 285–300.
Jongkees, L.B.W. (1950) On the function of the saccule. *Acta oto-laryngol. (Stockh.)*, 38: 18–26.
Krejcova, H., Cohen, B. and Highstein, S.M. (1973) Compensatory ocular counter rolling in the monkey. In *The Oculomotor System and Brain Function*, V. Zikmund (Ed.), Butterworth, London, pp. 491–503.
Lorente De Nó, R. (1932) The regulation of eye positions and movements induced by the labyrinth. *Laryngoscope (St. Louis)*, 121: 233–332.
Lowenstein, O. and Roberts, T.D.M. (1949) The equilibrium function of the otoliths organs of the thorn back ray (Raja clavata). *J. Physiol. (Lond.)*, 110: 392–415.
Magnus, R. (1924) *Korperstellung*. Springer, Berlin.
Precht, W. (1974) The physiology of the vestibular nuclei. In *Handbook of Sensory Physiology, Vol. VI/1, Vestibular System, Part 1, Basic Mechanisms*, H.H. Kornhuber (Ed.), Springer-Verlag, Berlin, pp. 353–416.
Robinson, D.A. (1975) Oculomotor control signals. In *Basic Mechanisms of Ocular Motility and their Clinical Implications*, G. Lennerstrand and P. Bach-y-Rita (Eds.), Pergamon Press, New York, pp. 337–374.
Shimazu, H. and Precht, W. (1965) Tonic and kinetic responses of cat's vestibular neurons to horizontal angular acceleration. *J. Neurophysiol.*, 28: 991–1013.
Shinoda, Y. and Yoshida, K. (1974) Dynamic characteristics of responses to horizontal head angular acceleration in vestibuloocular pathway in the cat. *J. Neurophysiol.*, 37: 653–673.
Suzuki, J.I., Tokumasu, K. and Goto, K. (1969) Eye movements from single utricular nerve stimulation in the cat. *Acta oto-laryngol. (Stockh.)*, 68: 350–362.
Trincker, D.E.W. (1962) The transformation of mechanical stimulus into nervous excitation by labyrinthine receptors. *Symp. Soc. exp. Biol.*, 15: 289–315.
Van Biervliet, J. (1899) Noyau d'origine du nerf oculomoteur commune du lapin. *Cellule*, 16: 1–33.
Yagi, T. (1974) Spontaneous and evoked behaviour of single units in the oculomotor nucleus of the rabbit. *Jap. J. Physiol.*, 24: 305–316.

Vestibulo-Ocular Reflex Pathways of Rabbits and their Representation in the Cerebellar Flocculus

M. YAMAMOTO

Department of Physiolgy, Faculty of Medicine, University of Tokyo, 7-3-1 Hongo, Bunkyo-ku, Tokyo
(Japan)

The vestibulo-ocular reflex (VOR) arc contains a number of component pathways connecting labyrinthine end organs to extraocular muscles in a highly specific manner. Relay cells mediating these pathways have either excitatory or inhibitory synaptic action upon oculomotor neurons and are located in various parts of the vestibular nuclear complex and probably also in the lateral nucleus of the cerebellum. They are not entirely independent of each other; an inhibitory interaction occurs in certain combination of them. The cerebellar flocculus is incorporated in the VOR in such a manner that it forms a side path to certain component pathways of the VOR. Evidence has recently been accumulated to indicate that a subgroup of flocculus Purkinje cells bound to a component VOR pathway is localized in a narrow strip of the cortical sheet of the flocculus. Functional localization exists also in the inferior olive corresponding to that in the flocculus. This article introduces recent data concerning construction of the flocculo-vestibulo-ocular system obtained in the albino rabbits.

VOR PATHWAYS IN RABBITS

Table I lists 12 component pathways of the VOR so far identified in the rabbits as arising from semicircular canals.

Excitatory pathways

Signals arising from the anterior canal are relayed by the dorsal part of the superior vestibular nucleus (Yamamoto et al., 1978) and eventually excite motoneurons of the ipsilateral superior rectus and the contralateral inferior oblique muscle. Signals from the horizontal canal are mediated by the medial vestibular nucleus and evoke contraction of the ipsilateral medial and the contralateral lateral rectus muscles (Ito et al., 1976a). Signals from the posterior canal which evoke contraction of the ipsilateral superior oblique and the contralateral inferior rectus muscles are also relayed by the medial vestibular nucleus (Ito et al., 1976a).

Inhibitory pathways

Anterior canal signals relayed by the middle and ventral parts of the superior vestibular nucleus have action to inhibit motoneurons of the ipsilateral inferior rectus

TABLE I

VOR 12-COMPONENT PATHWAYS IN RABBITS

AC, HC and PC, anterior, horizontal and posterior canals on the side of labyrinthine stimulation. AC′, HC′ and PC′, those on the contralateral side. d-, dorsal area of, r-, rostral area of, v-, ventral area of, SV and MV (superior and medial vestibular nuclei); SR, superior rectus; IO, inferior oblique; MR, medial rectus; LR, lateral rectus; SO, superior oblique; IR, inferior rectus; contra and ipsi, mean sides of retinal stimulation relative to the labyrinthine stimulation. (Modified from Ito et al., 1977.)

Reflex pathways	Canal-ocular reflex pathways in the rabbits					
	Receptor canal	Relay cell	Effector muscle	Interreflex inhibition	Flocculus inhibition	Retino-olivary inhibition
Excitatory						
E_1	AC	d-SV	i-SR	PC, PC′	+	−
E_2			c-IO		+	+ contra
E_3	HC	r-MV	i-MR	HC′	+	+ ipsi
E_4			c-LR		−	
E_5	PC	r-MV	i-SO	AC, AC′	−	
E_6			c-IR		−	
Inhibitory						
I_1	AC	v-SV	i-IR	−	+	−
I_2			c-SO	−	+	± ipsi
I_3	HC	r-MV	i-LR	−	+	+ ipsi
I_4		(v-SV)	c-MR	−	−	−
I_5	PC	v-SV	i-IO	−	−	
I_6			c-SR	−	−	

and the contralateral superior oblique muscle. Horizontal canal signals mediated by the medial vestibular nucleus inhibit the ipsilateral lateral rectus muscle. Horizontal canal signals also inhibit the contralateral medial rectus, but this inhibition can be evoked only with a significantly higher threshold of the canal stimulation than that required for the inhibition of the ipsilateral lateral rectus (Ito et al., 1976b), suggesting a certain difference in characteristics of signal transfer through relay cells. Posterior canal signals inhibit the ipsilateral inferior oblique and the contralateral superior rectus muscle through relay cells in the middle and ventral parts of the superior vestibular nucleus (Ito et al., 1976b).

Interreflex interaction

Presence of inhibitory interaction was first shown between relay cells of the VOR arising from horizontal canals of the two sides (Shimazu and Precht, 1966). Such an inhibitory interaction was revealed to occur in the following combinations of conditioning canals and testing reflexes (Ito et al., 1976c), as shown in Table I, i) from the anterior canal to excitatory reflexes evoked from the ipsilateral or contralateral posterior canal; ii) from the posterior canal to excitatory reflexes evoked from the ipsilateral or contralateral anterior canal; iii) from the horizontal canal to excitatory reflexes evoked from the contralateral horizontal canal. No such inhibition was found to occur in the inhibitory reflexes from any canal.

VOR REPRESENTATION IN THE FLOCCULUS

The cerebellar flocculus receives mossy fibre inputs from the labyrinth and in turn projects Purkinje cell axons to vestibular nuclei (Dow, 1938; Angaut and Brodal, 1967), eventually inhibiting relay neurons of the VOR (Ito et al., 1970; Baker et al., 1972; Fukuda et al., 1972). In rabbits, the flocculus inhibits only six of the twelve canal-ocular reflex pathways, i.e., all of the four arising from the anterior canal and two of the four from the horizontal canal (Table I). Relay neurons for the former four are located in the superior vestibular nucleus, and those for the latter two in the medial vestibular nucleus. An analysis of climbing fibre pathways which impinge onto flocculus Purkinje cells suggests that each of the six components of the VOR is under inhibitory control of its own subgroup of Purkinje cells (Ito et al., 1977). For example, relay cells for the excitatory reflex from the anterior canal to the contralateral inferior oblique motoneurons are inhibited by those Purkinje cells which receive climbing fibre inputs from the contralateral retina (Table I). By contrast, relay cells for the excitatory reflex from the anterior canal to the ipsilateral superior rectus motoneurons are inhibited by Purkinje cells whose climbing fibre afferents are not driven from either ipsilateral or contralateral retina (Table I). This provides the basis for discriminating two groups of relay cells in the dorsal area of the superior vestibular nucleus which otherwise behave similarly, being excited by anterior canal signals and inhibited by posterior canal signals.

A question arose as to whether these subgroups of Purkinje cells bound to different components of the VOR are located in the flocculus differentially or diffusely. In order to answer this question, three types of experiments have been performed as described below. In conducting these experiments, it was desirable to normalize the topography of the flocculus which exhibited considerable individual variations. This was done by averaging unfolded maps of the flocculus reconstructed from serial histological sections (Ito, Orlov and Yamamoto, in preparation).

Staining of flocculus Purkinje cells by retrograde transport of horseradish peroxidase (HRP)

When a small amount of HRP was injected electrophoretically into the medial vestibular nucleus, labeled Purkinje cells were found to distribute in a narrow strip covering obliquely the rostral part of the flocculus (Fig. 1A) (Yamamoto and Shimoyama, 1977). Injection into the superior vestibular nucleus labeled Purkinje cells in two narrow strips, the one located rostrally and the other caudally to that stained from the medial vestibular nucleus (Fig. 1B). These three strips occupy the rostral two thirds of the flocculus. The most caudal part of the flocculus was labeled by injection of the HRP to the caudal part of the lateral nucleus of the cerebellum, but not to the medial or superior vestibular nucleus (Fig. 1C) (Yamamoto, 1978). The folium p forming a boundary zone between the flocculus and paraflocculus was labeled by injection to the nucleus praepositus hypoglossi (Fig. 1D). The flocculus of the rabbits can thus be divided at least into five strips according to projection sites of Purkinje cells.

Eye movements evoked by local stimulation of the flocculus

Local stimulation at Purkinje cell layers of the flocculus evoked three major types of eye movements, i.e., abduction or downward movement in the eye ipsilateral to the

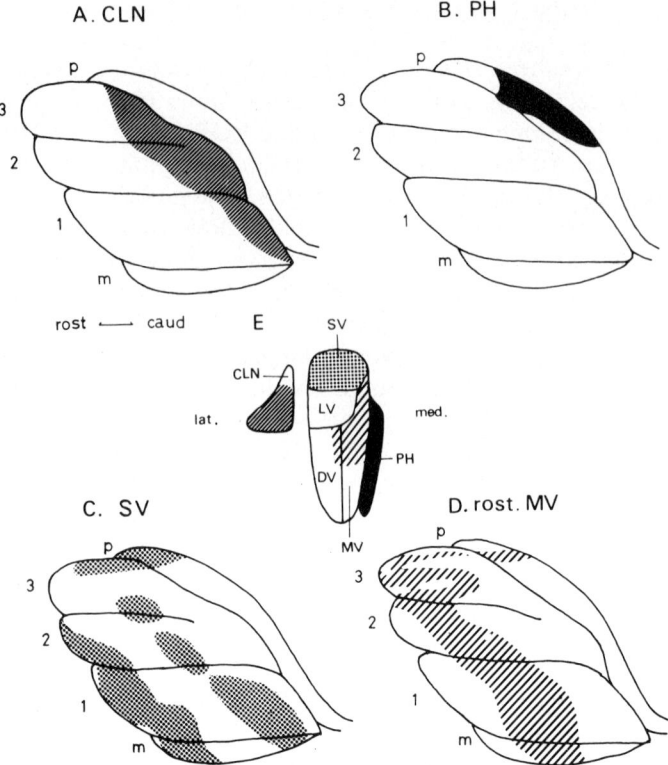

Fig. 1. Efferent projections of the flocculus revealed by means of retrograde axonal transport of HRP. A–D) Surface of the left flocculus. E) Dorsal view of the left vestibular nuclei and the lateral nucleus of the cerebellum. Note that each of the four major injection sites in E and its corresponding projection area on A–D are blackened, shaded or stippled similarly. MV, the medial vestibular nucleus; SV, the superior vestibular nucleus; PH, nucleus praepositus hypoglossi; CLN, lateral nucleus of the cerebellum. (From Yamamoto, 1978.)

stimulation, or rotation in the contralateral eye (Dufossé et al., 1977). These three types of movement are explicable as due to the inhibitory effect of Purkinje cells upon relay cells of the VOR. As would be expected, the area from which abduction of the ipsilateral eye was evoked corresponds well with the distribution of labeled Purkinje cells after injection of the HRP into the medial vestibular nucleus. Downward movement in the ipsilateral eye could be evoked from an area either rostral or caudal relative to that for abduction of the ipsilateral eye, corresponding to the two strips labeled from the superior vestibular nucleus (Fig. 1B). The area for evoking rotatory movements of the contralateral eye was not well surveyed, but in some experiments it was found in the dorsal part of the flocculus. No movement or a small medioventral movement of the ipsilateral eye was evoked from the area which HRP study indicated to project to the lateral nucleus of cerebellum or to the nucleus praepositus hypoglossi.

Testing of individual components of the VOR

Passage of brief current pulses through a glass microelectrode placed in the Purkinje cell layer of the flocculus causes a significant depression in three excitatory

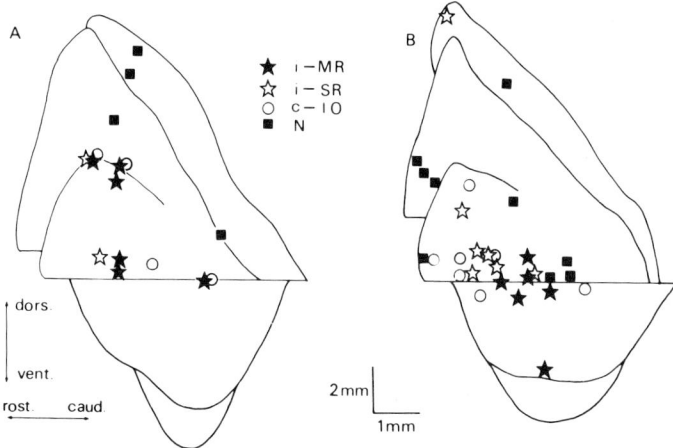

Fig. 2. Mapping of the flocculus for the stimulus effect upon the VOR. Unfolded maps of the flocculus in A and B were obtained from two rabbits. Four different symbols indicate the spots effective in inhibiting the VOR to three muscles as indicated, and those without such an effect (N).

components of the VOR (E_1–E_3 in Table I). When surveyed using pulses of 20 μA intensity, it was possible to locate differentially effective sites for inducing a short-latency inhibition, presumably mediated by Purkinje cell axons, of each of these three component reflexes, as illustrated in Fig. 2 for two experiments. It is interesting that a narrow strip of the cortical sheet of the flocculus can be specified in this way in connection with single extraocular muscles. The strip related to the ipsilateral superior rectus lies rostrally to that related to the ipsilateral medial rectus (Fig. 2A, B). The strip for the contralateral inferior oblique muscle appears to intermingle with these two (B). Determination of such strips for inhibitory components of the VOR (I_1–I_3 in Table I) has been, however, unsuccessful because of technical difficulties.

VOR REPRESENTATION IN THE INFERIOR OLIVE AND PREOLIVARY STRUCTURE

Visual signals transferred through the pretectal area and the inferior olive eventually impinge onto the flocculus as climbing fibre inputs (Maekawa and Simpson, 1973). Signals from the ipsilateral retina are relayed by the optic nucleus of the pretectal area and the caudal half of the dorsal cap of the principal olive, while those from the contralateral retina are mediated sequentially by the pretectal area, the area ventral to the red nucleus, and the rostral half of the dorsal cap and adjacent area of the ventrolateral outgrowth of the principal olive (Maekawa and Takeda, 1977). Multiplicity of the visual climbing fibre projection to the flocculus can be related to the compound structure of the VOR in the following ways.

Reflex testing

Systematic mapping within the inferior olive for the stimulus effect upon three excitatory components of the VOR revealed the differential localization of olivary neurons as related to the VOR (Ito et al., 1978). The caudal half of the dorsal cap contains cells of origin of climbing fibres innervating those Purkinje cells bound to the component reflex to ipsilateral medial rectus muscle. The rostral half of the dorsal cap and the adjacent area of the ventrolateral outgrowth contain those olivary neurons bound via flocculus Purkinje cells to the component reflexes to the ipsilateral superior rectus and contralateral inferior oblique muscles. This is to be expected partly from the fact that signals from the ipsilateral retina inhibit the VOR to the ipsilateral medial rectus muscle, while those from the contralateral retina depress the VOR to the contralateral inferior oblique muscle (Ito et al., 1977). The VOR to the ipsilateral superior rectus muscle could not be inhibited from either the ipsilateral or the contralateral retina. The climbing fibre pathway for this component reflex was traced up to the tectal area, but the connection between the pretectal area and retinae for this component reflex has not yet been revealed.

SUMMARY

Presence of a clear differential localization is demonstrated in both the flocculus and the inferior olive, in connection with certain components of the VOR. Narrow strips thus defined in the flocculus may form functional units of the cerebellar cortex, corresponding to microzones in the vermal cortex (Andersson and Oscarsson, 1978). It is impressive that these strips are related directly to single extraocular muscles. Apparently, the VOR-flocculus system represents a prototype of the functional localization in the cerebellum.

Detailed structural knowledge of the VOR-flocculus system is useful as a basis for the experimental approach to neuronal mechanisms of cerebellar functions. It is necessary to further extend dissection of functional units in the flocculus in connection with not only the excitatory but also inhibitory components of the VOR. It is also desirable to study organization of various inputs to the flocculus, from labyrinths, eyes, neck, etc., in close relationship with the functional localization within the flocculus and the inferior olive.

REFERENCES

Andersson, G. and Oscarsson, O. (1978) Climbing fiber microzones in cerebellar vermis and their projection to different groups of cells in the lateral vestibular nucleus. *Exp. Brain Res.*, 32: 565–579.

Angaut, P. and Brodal, A. (1967) The projection of the vestibulocerebellum onto the vestibular nuclei in the cat. *Arch. ital. Biol.,* 105: 441–479.

Baker, R.G., Precht, W. and Llinás, R. (1972) Cerebellar modulatory action on the vestibulo-trochlear pathways in the cat. *Exp. Brain Res.*, 15: 364–385.

Dow, R.S. (1938) Efferent connections of the flocculo-nodular lobe in Macaca mulatta. *J. comp. Neurol.,* 68: 297–305.

Dufossé, M., Ito, M. and Miyashita, Y. (1977) Functional localization in the rabbit's cerebellar flocculus determined in relationship with eye movements. *Neurosci. Lett.,* 5: 273–277.

Fukuda, J., Highstein, S.M. and Ito, M. (1972) Cerebellar inhibitory control of the vestibulo-ocular reflex investigated in rabbit third nucleus. *Exp. Brain Res.,* 14: 511–526.

Ito, M., Highstein, S.M. and Fukuda, J. (1970) Cerebellar inhibition of the vestibulo-ocular reflex in rabbit and cat and its blockage by picrotoxin. *Brain Res.,* 17: 524–526.

Ito, M., Nisimaru, N. and Yamamoto, M. (1976a) Pathways for the vestibulo-ocular reflex excitation arising from semicircular canals of rabbits. *Exp. Brain Res.,* 24: 257–271.

Ito, M., Nisimaru, N. and Yamamoto, M. (1976b) Postsynaptic inhibition of oculomotor neurons involved in vestibulo-ocular reflexes arising from semicircular canals of rabbits. *Exp. Brain Res.*, 24: 273–283.

Ito, M., Nisimaru, N. and Yamamoto, M. (1976c) Inhibitory interaction between the vestibulo-ocular reflexes arising from semicircular canals of rabbit. *Exp. Brain Res.*, 26: 89–103.

Ito, M., Nisimaru, N. and Yamamoto, M. (1977) Specific patterns of neuronal connections involved in the control of the rabbit's vestibulo-ocular reflexes by the cerebellar flocculus. *J. Physiol. (Lond.)*, 265: 833–854.

Maekawa, K. and Simpson, J.I. (1973) Climbing fibre responses evoked in vestibulocerebellum of rabbit from visual system. *J. Neurophysiol.*, 26: 649–666.

Maekawa, K. and Takeda T. (1977) Afferent pathways from the visual system to the cerebellar flocculus of the rabbit. In *Control of Gaze by Brain Stem Neurons*, R.G. Baker and A. Berthoz (Eds.) Elsevier/North-Holland Biomed. Press, Amsterdam, pp. 187–195.

Shimazu, H. and Precht, W. (1966) Inhibition of central vestibular neurons from the contralateral labyrinth and its mediating pathways. *J. Neurophysiol.*, 29: 467–492.

Yamamoto, M. (1978) Localization of rabbit's flocculus Purkinje cells projecting to the cerebellar lateral nucleus and the nucleus prepositus hypoglossi investigated by means of the horseradish peroxidase retrograde axonal transport. *Neurosci. Lett.*, 7: 197–202.

Yamamoto, M. and Shimoyama I. (1977) Differential localization of rabbit's flocculus Pukinje cells projecting to the medial and superior vestibular nuclei, investigated by means of the horseradish peroxidase retrograde axonal transport. *Neurosci. Lett.*, 5: 279–283.

Yamamoto, M., Shimoyama, I. and Highstein, S.M. (1978) Vestibular nucleus neurons relaying excitation from the anterior canal to the oculomotor nucleus. *Brain Res.*, 148: 31–42.

Canal-Otolith Convergence on Cat Ocular Motoneurons

W. PRECHT, J.H. ANDERSON and R.H.I. BLANKS

Max-Planck-Institut für Hirnforschung, Neurobiologische Abteilung, 6000 Frankfurt M.-Niederrad
(F.R.G.)

In the last decade, considerable progress has been made in our understanding of the response dynamics of vestibulo-ocular reflexes and the central neural systems involved, in particular, for the canal-ocular reflexes (for review, see Schmid and Jeannerod, this volume, chapter IVC6). Relatively little work has been done with the dynamics and neural organization of otolith-ocular reflexes and the importance of canal-otolith convergence for the control of eye movements. In these studies the relationship between acceleratory stimuli excitatory for the otolith receptors, and eye position was studied during ocular-counterrolling under static (Van der Hoeve and De Kleijn, 1917; Fleisch, 1922a) and dynamic (Fleisch, 1922b) lateral tilting of the head. Furthermore, eye movements have been studied during horizontal *linear* acceleration in the rabbit (Fleisch, 1922b; Baarsma and Collewijn, 1975; Kleinschmidt and Collewijn, 1975), cat (Kohut, 1974) and man (Jongkees and Philipszoon, 1964). The contributions made by the different extraocular muscles to these movements (Lorente de Nó, 1932; Suzuki et al., 1968) as well as the receptor groups likely to contribute to activation of the corresponding motoneuron pools (Fluur and Mellström, 1971) have been investigated in a qualitative manner. Of particular interest are the quantitative studies in the rabbit (Baarsma and Collewijn, 1975) which demonstrate a considerable phase lag of eye position relative to linear acceleration at higher frequencies suggesting that the otolith-ocular system performs poorly under dynamic load conditions. Since the responses of otolith afferents in frog, monkey and cat do not show this phase lag (Blanks and Precht, 1976; Fernández and Goldberg, 1976a, b; Anderson et al., 1978), the dynamic characteristics of the otolith-ocular reflexes must be due to either visco-elastic properties of the eyeball-orbit system and/or result from central processing.

One part of this paper will be devoted to the demonstration of central processing of otolithic information, by comparing motoneuron output with primary otolith input. A second part will deal with the importance of canal-otolith convergence for extending the frequency response characteristics of motoneurons involved primarily, but not exclusively, in vertical/rotatory eye movements.

METHODS

Experiments were carried out in cats anesthetized with ketamine. Details of the experimental procedures for natural stimulation and data analysis have been fully described previously and will not be repeated here (Anderson et al., 1978; Blanks et al., 1978; Anderson and Precht, 1979). In short, single identified motoneurons were recorded with glass micropipettes from the abducens and trochlear nuclei during natural stimulation of canal only, otolith only and during combined canal-otolith stimulation. Similar stimuli were applied while recording the EMG activity of the superior and inferior obliques, and medial and lateral recti with flexible wires inserted into the muscles.

RESULTS AND DISCUSSION

As a first step we recorded the responses of trochlear motoneurons to canal and otolith stimulation. These neurons were chosen for the investigation because of their importance for the control of vertical-rotatory eye movements in response to gravitational (static and dynamic) and angular accelerations (Suzuki et al., 1969; Fluur and Mellström, 1971). Secondly, the responses of abducens motoneurons, which are mainly responsible for canal-evoked horizontal eye movements, were studied during gravitational stimuli as well. Finally, to confirm and expand the data obtained from single motoneurons of the vertical-rotatory and horizontal eye muscle systems, we recorded simultaneously multi-unit EMGs from the antagonistic muscle pairs, the superior and inferior obliques and the medial and lateral rectus muscles, during angular and gravitational stimulation. The responses of trochlear and abducens motoneurons to vestibular stimuli will be described first.

Responses of trochlear and abducens motoneurons to canal and otolith stimulation

Determination of inputs. Natural vestibular stimuli were applied to determine which receptors projected to motoneurons. For the *canal projection* this was done by a null-point technique (Blanks et al., 1978). This technique implies systematic changes of the animal's head position until the unit no longer responds to horizontal rotation. With this spatial orientation of the head, the canals providing excitatory inputs are perpendicular to the plane of rotation and yield no response. By comparing the null-points obtained from motoneurons with those determined for the peripheral canal afferents (Estes et al., 1975), it was concluded that *trochlear* motoneurons received their major excitatory input from the contralateral posterior canal and, based on studies in hemilabyrinthectomized cats, an inhibitory input from the ipsilateral anterior canal (Blanks et al., 1978). These data are in full agreement with results obtained from electrical stimulation of individual vestibular nerve branches (Precht and Baker, 1972; Baker et al., 1973). *Abducens* motoneurons received their excitatory and inhibitory inputs from the contra- and ipsilateral horizontal canals, respectively (Richter and Precht, 1968; Baker et al., 1969; Precht et al., 1969; Schwindt et al., 1973; Anderson and Precht, 1979).

It is important to note that the motoneurons showed a very stereotyped input from these coplanar canal pairs and little orthogonal canal-canal convergence when studied with electrical *and* natural stimuli. Only a few units had no null-point, a fact that is indicative of orthogonal convergence (Blanks et al., 1978).

Otolith projections to motoneurons were studied by measuring the firing frequency in various static positions. Fig. 1 exemplifies this procedure in showing the responses

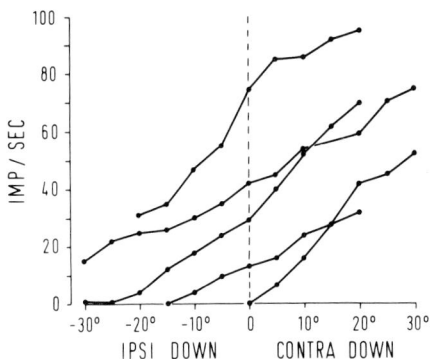

Fig. 1. Responses of trochlear motoneurons (TMns) as a function of static tilt in roll. The discharge for each of 5 different TMns was determined at the different head positions as the average discharge rate over a 30 sec interval, 20 sec after arrival at the new position. Note that all units show a β-otolith response. (From Blanks et al., 1978.)

of five trochlear motoneurons in various roll angles. As the head was tilted laterally towards the recording side the resting rate decreased, whereas the opposite movement caused an increase in firing (β-response). Almost all *trochlear* motoneurons showed this response and only few responded exclusively to canal stimulation. The sensitivity of the response was high (ca. 1 spike/degree/sec), and there was no upper saturation up to 30° of tilt angle. Abducens motoneurons also showed responses to static lateral tilt which consisted of increases and decreases in firing with the recording side down and up, respectively, i.e., an α-otolith response (Anderson and Precht, 1979). Gravity responses were less frequently observed in abducens motoneurons when compared with trochlear motoneurons and, when present, the sensitivity was generally low. The canal and otolith inputs to trochlear and abducens neurons, determined with natural stimulation, are in perfect agreement with electrophysiological studies showing short-latency utricular inputs to ipsilateral abducens (Schwindt et al., 1973) and contralateral trochlear motoneurons (Baker et al., 1973).

It is interesting to note that electrical stimulation of the utricular nerve did not reveal any short-latency inhibitory reflexes to antagonistic motoneurons, as in the case of canal-ocular reflexes (cf. for reference, Precht, 1978). When trochlear motoneurons of hemilabyrinthectomized cats were studied during static tilt, responses were found that could only be explained by assuming an inhibitory otolith-ocular reflex as well (Blanks et al., 1978). Since they had escaped electrophysiological studies it must be assumed that they are polysynaptic in nature and were therefore depressed in anesthetized preparations.

In summary, trochlear and abducens motoneurons receive very specific canal and otolith inputs and show very little orthogonal canal convergence. Contrary to the disynaptic inhibitory and excitatory canal-ocular reflexes, otolith-ocular reflexes appear to be disynaptic only when excitatory.

To corroborate and extend the findings obtained with single motoneurons, simultaneous EMG recordings were obtained from i) the two lateral recti muscles, ii) the medial (MR) and lateral (LR) recti muscles of the same eye, and iii) the superior (SO) and inferior oblique (IO) muscles, during static and/or low frequency sinusoidal roll stimulation. At the low frequency of 0.025 Hz (Fig. 2A–D) the modulation of EMG activity is displacement related and at higher frequencies of

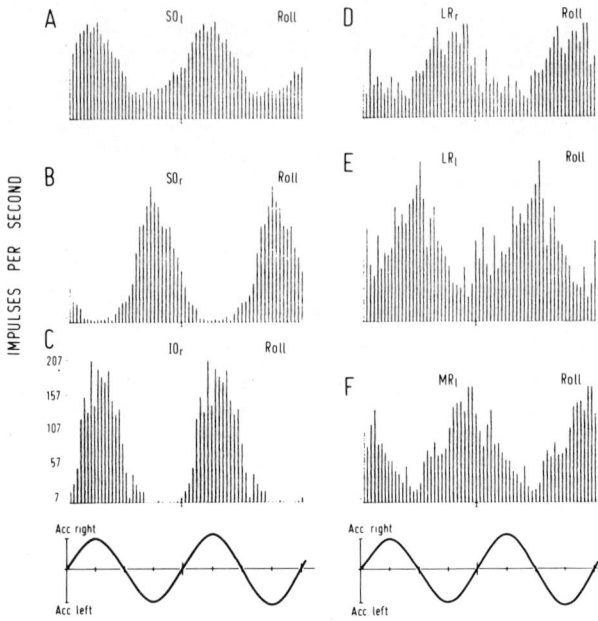

Fig. 2. Reciprocal organization of the oblique and horizontal recti muscles during roll rotations. The response to low frequency rotations shown here demonstrates the pattern of responses to otolith inputs: on side down the EMG of the ipsilateral SO and LR and contralateral IO and MR increase while the contralateral SO and LR decrease, e.g., SO_l(A) and LR_l(E) increase on left side down (i.e., acceleration right). The rotations used were 0.025 Hz, ±20° in A–C and 0.25 Hz, ±20° in D–F. The pulse trains corresponding to the EMG recordings were binned and averaged over 5–10 periods to give the cycle histograms shown. (From Anderson and Precht, 1979.)

rotation (Fig. 2E, F) the horizontal recti show a significant phase lag. Quantitative analysis is presented in Fig. 5. As these records show, both the oblique (Fig. 2A–C) and horizontal (Fig. 2D–F) systems have a reciprocal organization of otolith inputs during roll, i.e., the SO and IO, SO_l and SO_r, LR_l and LR_r are synergistic pairs which can subserve the coordinated eye movements caused by otolith-ocular reflexes. A similar reciprocal relationship was noted in the muscle pairs: MR_l and MR_r, IO_l and IO_r (not shown).

Responses of motoneurons to ramp changes in head position

In the preceding section it has been shown that most trochlear and some abducens motoneurons respond to both canal and static otolithic stimuli. In this and the following sections we shall describe the behavior of motoneurons during stimuli such as ramp changes in head position and sinusoidal rotations (next section) about the longitudinal body axis (roll).

Non-periodic ramp changes in roll-position produce angular acceleration pulses at the beginning and termination of the ramp which can activate the canals. In addition, otolith inputs are present due to the changing orientation of the animal's head with respect to gravity. The latter input predominates with slow changes in head positions whereas canals are recruited as the slopes of the ramps become greater. Typical examples of responses of a *trochlear motoneuron* to ramp changes in head position in roll are shown in Fig. 3. Slow transients of head position (Fig. 3A) produce changes in

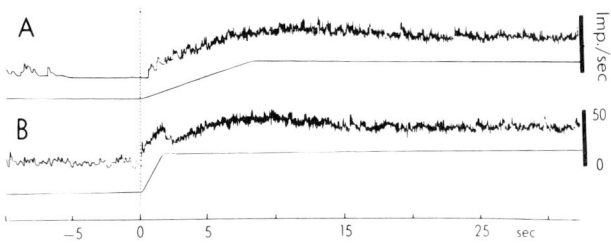

Fig. 3. Response of trochlear motoneurons to ramp changes in roll position. The instantaneous frequency of the discharge of a single motoneuron is shown for changes in angular position from −2° (ipsi side down) to +25° (ipsi side up) with constant velocities of 3.3 (A) and 16.2 degrees/sec (B). (From Anderson et al., 1977.)

instantaneous firing rate which were nearly proportional or slightly lagging head displacement. The maximum discharge level is achieved only *after* arrival of the final displacement and then adapts, as in this neuron, to a new steady state level. With greater slopes of the ramp stimuli (Fig. 3B) and hence greater acceleration pulses at the onset and termination, the threshold for a canal response is exceeded, thereby producing a transient increase in discharge at the onset of the ramp and a decrease of firing following its termination. Following the canal transient, the firing continues to increase due to otolithic stimuli and finally adapts to a new level. Adaptation was observed in about 30% of the neurons. Ramps of the same magnitude but in the opposite direction yielded a slow decrease in firing, along with the fast canal transients. It should be pointed out that the canal induced transients had the same polarity as the otolith-evoked responses at the onset of ipsilateral side-up movements but were of opposite polarity at the termination of the ramp. The opposite pattern of activation was noted with transients from ipsilateral-up to ipsilateral-down.

Abducens motoneurons showed a similar combination of canal and otolithic responses to ramp displacements in roll only when the animal's head was oriented in such a way as to cause costimulation of the horizontal canals. When the latter were brought into a null-plane with respect to roll rotation, only otolith modulation of discharge rate was noted. However, it is important to emphasize that both the sensitivity and the number of otolith responses were smaller than in the trochlear neuron. The importance of this fact will be further discussed in the next section.

In short, the responses of motoneurons to ramp changes have shown that whenever canals are in the stimulating plane and the slope of the ramp is large enough to reach canal threshold, motoneurons are subjected to a combined canal-otolithic input. Gravity responses resulting from ramp changes lag head displacement more than primary otolith afferents (Anderson et al., 1978) indicating central processing of otolith afferent input. Further details of the dynamics of otolith processing will be given in the next section.

Responses of motoneurons to sinusoidal stimulation

The phase-gain response properties of single trochlear and abducens motoneurons as well as the EMGs of the corresponding muscles were recorded during sinusoidal rotation (Anderson et al., 1977; Blanks et al., 1978; Anderson and Precht, 1979). As already shown in the preceding section, roll stimulation activates both canal and otolith inputs to trochlear motoneurons, whereas rotation in yaw leads to canal activation only provided the contralateral posterior and ipsilateral anterior canals have some component in the plane of horizontal rotation. The otolith component of the combined canal-otolith response may then be obtained by vectorially subtracting the yaw (canal only) from the roll (canal and otolith) responses. To test the validity of the subtraction method we have then compared the calculated frequency responses of the otolith input to trochlear motoneurons with that obtained for abducens motoneurons and lateral rectus muscle in roll (an otolith input only, provided the horizontal canals are perpendicular to roll plane).

The averaged responses of trochlear motoneurons during rotation in both yaw and roll are shown in Fig. 4. In yaw, the motoneuron's phase lags increase as the frequency of rotation increases (O, Fig. 4). Also, the phase lags of motoneurons are larger at all frequencies as compared to those of the primary canal afferents (Δ, Fig. 4); the upper-left shaded area in Fig. 4 indicates the central processing (integration) of canal inputs. It should be noted that these experiments were done under light ketamine anesthesia which influences the central integration process resulting in smaller phase values. When the same population of motoneurons were studied in roll (●, Fig. 4), the phase lag decreases as the frequency of rotation increases. Since the phase in both cases was measured with respect to the acceleration excitatory for the contralateral posterior canal, the yaw and roll phase behavior would have been identical if there were only canal input. Therefore, the phase differences between yaw and roll responses can only be due to the otolith input present in roll. It is apparent that at low frequencies, where the canal input is small relative to otolith input, the phase of trochlear motoneurons in roll is largely determined by the otolith input which, from previous sections, have a DC response maximum with ipsilateral side-up position ($-\ddot{\theta}$) in Fig. 4). At higher frequencies, the phase of the roll response is largely determined by the canal input, which increases as the frequency squared. Carefully considering the various problems of the vectorial subtraction procedure (Blanks et al., 1978) the yaw_y response vector ($gain_y$, $phase_y$) was subtracted from the $roll_{(r)}$ vector ($gain_r$, $phase_r$) for each frequency. An example of one such calculation is shown by the insert in Fig. 4. The results of this procedure are shown by the lowest dashed line and ×'s in Fig. 4. It can be seen that the phases of the otolith response show a lag of 10°–90° over the frequency range, 0.025–0.5 Hz. These are relative to angular displacement which is proportional to the gravity input to otoliths. The difference between these calculated values for trochlear neurons and those obtained from primary otolith fibers (▲, Fig. 4) (Anderson et al., 1978) are given by the shaded zone in the lower part of Fig. 4. Obviously, the signal in the otolith-ocular pathway undergoes central transformations similar to those in the canal-ocular path, i.e., phases and gains increase and decrease, respectively, with increasing frequency indicating central integration. The lag dynamics noted in the ramp studies (Fig. 3) are consistent with the results obtained in the frequency domain.

The phase values relative to displacement for the abducens motoneuron response and the EMG activity of lateral and medial recti muscles during sinusoidal stimulation in roll (otolith only) are shown in Fig. 5 (Anderson and Precht, 1979). These values are, indeed, very similar to those calculated for trochlear motoneurons. This similarity indicates that the subtraction method is valid for calculation of otolith input during roll, and that otolith processing is similar for the horizontal and vertical eye movement systems.

The central structures necessary for this processing of the afferent activities are still unknown. However, some data obtained from the vestibular nuclei (Melvill Jones and

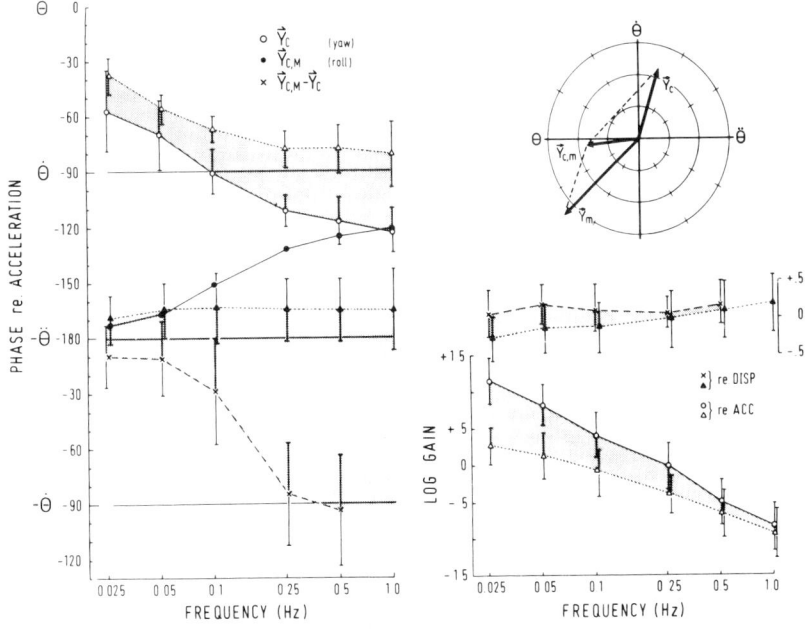

Fig. 4. Canal and otolith contributions to the motor output of trochlear motoneurons (TMns). The responses of 10 TMns during both yaw and roll (mean values shown by ○, ●, resp.) were used to calculate the otolith contribution ($\vec{Y}_m$) to the motor output. This entailed a vectorial subtraction of the yaw ($\vec{Y}_c$, canal only input) from the roll ($\vec{Y}_{cm}$, canal plus macular input) response, as indicated in the insert. The mean ±1 SD of these calculations are shown by the symbol × and the vertical bars. For comparison, the mean phases and gains for the canal (△) and otolith (▲) afferents (Anderson et al., 1978) are shown. The vertical bars include ±1 SD. Note that the stippled areas represent the difference between the afferent inputs and the motoneuron outputs. Thus it is evident that for both the canal and otolith systems there is extensive central processing of the afferent activities. (From Blanks et al., 1978.)

Milsum, 1969) indicates that responses of secondary vestibular neurons already differ significantly from those of vestibular afferents. During the course of our experiments we often recorded from axons of vestibular neurons within the trochlear nucleus which were monosynaptically driven by ipsilateral (Vi) or contralateral (Vc) vestibular nerve stimulation. Since it may be assumed that some of these axons terminate in the trochlear nucleus their response characteristics are of some interest. About half of the Vi and Vc axons showed responses to otolithic in addition to canal stimulation. The polarity of these responses was such that the excitatory Vc and inhibitory Vi axons could modulate the motoneuron discharge as described above. Thus, Vc axons were driven by the contralateral posterior canal and showed a β-otolith response; whereas Vi axons were exclusively activated by the ipsilateral anterior canal, and, when present, showed α-otolith responses. Also, the frequency response of these convergent units during yaw and roll sinusoidal rotations was very similar to those of trochlear motoneurons (Fig. 4) except that their phase lags were somewhat smaller. This finding also supports the notion that secondary vestibular axons projecting to motoneurons carry an otolithic signal that is already different from that of otolith afferents. Interestingly, we did not find any axons that carried a pure otolithic signal without canal components whereas the opposite, i.e., a canal response only, was seen in about half of the cases. Of course, these negative findings do not exclude the possible

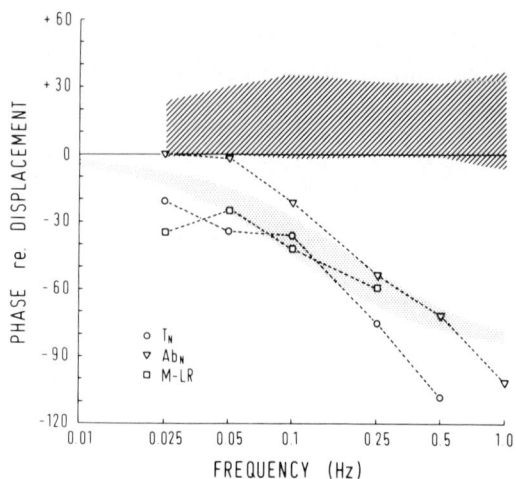

Fig. 5. Phase behavior of otolith afferent and extraocular motor outputs during sinusoidal roll rotations. The calculated otolith-dependent responses of trochlear motoneurons (○, mean of 10 units) are shown together with the measured responses of abducens neurons (▽, mean of 5 units) and the EMGs of the medial and lateral recti (□, mean of 3 animals). The hatched area encompasses the mean ± 1 SD for the responses of the otolith afferents. For all cases, the input reference is angular displacement: ipsilateral side up for the trochlear neurons and medial rectus EMGs and side down for abducens neurons and the lateral rectus EMGs. The stippled region shows the phases of a simple first order system, $1/s + a$, which approximates the motor outputs with time constants, $1/a$ of 0.8–1.6 sec. (From Anderson and Precht, 1979.)

existence of pure otolithic projections. If they were to terminate, for example, as thin fibers on dendritic ramifications of the motoneuron pool they may easily have escaped microelectrode search.

CONCLUSION AND SUMMARY

In the present paper it has been shown that ocular motoneurons of the horizontal and vertical/oblique systems receive both canal and otolithic inputs and that these inputs are reciprocally organized. The functional importance of canal-otolithic convergence in vertical/oblique motoneurons, and probably also in horizontal motoneurons, is to extend the working range over a much wider frequency band than is provided by the simple canal VOR. This is particularly evident in trochlear motoneurons during roll rotation where the canal and otolith responses have the same polarity. At high frequencies, the canal system may, therefore, serve to compensate for the lag present in the otolith system in addition to its compensation for the visco-elastic properties of the eye. When the roll movement is stopped the responses of canal and otolith are of opposite polarities and may cancel one another. In this case, the canal response may serve to prevent the eye from overshooting its final position. On the other hand, the otolith input serves to compensate for the poor canal performance in the low frequency range and, of course, during static head displacement. It is interesting to note that the responses of vertical/rotatory motoneurons in roll, measured in the dark, are comparable to those found in the horizontal system in the presence of vision. Thus, whereas vision improves the insufficiencies of the canal-ocular response in yaw at low frequencies, the otolith contribution in roll may be important for extending the lower frequency range of the vertical/rotatory vestibulo-ocular reflexes. The otolith responses found in the horizontal eye movers may also be important in roll and pitch rotation, because a contraction of the lateral and medial recti muscles, due to their insertions, can facilitate the vertical or rotatory eye movements. However, it should be noted that the otolith signal in the horizontal system was much weaker than that in vertical/rotatory muscles.

When the dynamics of the central otolith signal were compared with those of peripheral afferents, a large difference was apparent suggesting that, as for the canal input, a leaky neural integrator is processing the otolithic afferent input. Pertinent to the present results is the work of Baarsma and Collewijn (1975), who have studied the phase and gain behavior of eye movements in the alert rabbit during horizontal linear accelerations. They found phase lags for the motor output very similar to those shown above for frequencies from 0.023 to 0.5 Hz. They report greater phase lags at the higher frequencies (180° at 1.2 Hz) which suggest that other factors (e.g., eyeball-orbit visco-elastic properties) may be involved.

REFERENCES

Anderson, J.H. and Precht, W. (1979) Otolith responses of extraocular muscles during sinusoidal roll rotations. *Brain Res.*, 160: 150–154.

Anderson, J.H., Precht, W. and Blanks, R.H.I. (1977) Central processing in otolith-ocular reflex pathways. In *Control of Gaze by Brain Stem Neurons*, R. Baker and A. Berthoz (Eds.), Elsevier, Amsterdam, pp. 253–260.

Anderson, J.H., Blanks, R.H.I. and Precht, W. (1978) Response characteristics of semicircular canal and otolith systems in cat. I. Dynamic responses of primary vestibular fibers. *Exp. Brain Res.*, 32: 491–507.

Baarsma, E.A. and Collewijn, H. (1975) Eye movements due to linear accelerations in the rabbit. *J. Physiol. (Lond.)*, 245: 227–247.

Baker, R., Precht, W. and Berthoz, A. (1973) Synaptic connections to trochlear motoneurons determined by individual vestibular nerve branch stimulation in the cat. *Brain Res.*, 64: 402–406.

Blanks, R.H.I. and Precht, W. (1976) Functional characterization of primary vestibular afferents in the frog. *Exp. Brain Res.*, 25: 369–390.

Blanks, R.H.I., Anderson, J.H. and Precht, W. (1978) Response characteristics of semicircular canal and otolith systems in cat. II. Responses of trochlear motoneurons. *Exp. Brain Res.*, 32: 509–528.

Estes, M.S., Blanks, R.H.I. and Markham, C.H. (1975) Physiologic characteristics of vestibular first-order canal neurons in the cat. I. Response plane determination and resting discharge characteristics. *J. Neurophysiol.*, 38: 1232–1249.

Fernández, C. and Goldberg, J.M. (1976a) Physiology of peripheral neurons innervating otolith organs of the squirrel monkey. I. Response to static tilts and to long-duration centrifugal force. *J. Neurophysiol.*, 39: 970–984.

Fernández, C. and Goldberg, J.M. (1976b) Physiology of peripheral neurons innervating otolith organs of the squirrel monkey. III. Response dynamics. *J. Neurophysiol.*, 39: 996–1008.

Fleisch, A. (1922a) Tonische Labyrinthreflexe auf die Augenstellung. *Pflügers Arch.*, 194: 554–573.

Fleisch, A. (1922b) Das Labyrinth als beschleunigungsempfindendes Organ. *Pflügers Arch.*, 195: 499–515.

Fluur, E. and Mellström, A. (1971) The otolith organs and their influence on oculomotor movements. *Exp. Neurol.*, 30: 139–147.

Jongkees, L.B.W. and Philipszoon, A.J. (1964) Electronystagmography. *Acta oto-laryngol. (Stockh.)*, Suppl. 189: 1–111.

Kleinschmidt, H.J. and Collewijn, H. (1975) A search for habituation of vestibulo-ocular reactions to rotatory and linear sinusoidal accelerations in the rabbit. *Exp. Neurol.*, 45: 257–267.

Kohut, R.I. (1974) Vertical linear acceleration (otolithic-ocular responses). *Laryngoscope*, 84: 1627–1662.

Lorente de Nó, R. (1932) The regulation of eye positions and movements induced by the labyrinth. *Laryngoscope*, 42: 233–332.

Melvill Jones, G. and Milsum, J.H. (1969) Neural response of the vestibular system to translational acceleration. In *Conference on Systems Analysis Approach to Neurophysiological Problems*, Brainerd, Minnesota.

Precht, W. (1978) Neuronal operations in the vestibular system. In *Studies of Brain Function, Vol. 2*, V. Braitenberg (Ed.), Springer-Verlag, Berlin-Heidelberg-New York, p. 226.

Precht, W. and Baker, R. (1972) Synaptic organization of the vestibulo-trochlear pathway. *Exp. Brain Res.*, 14: 158–184.

Precht, W., Richter, A. and Grippo, J. (1969) Responses of neurones in cat's abducens nuclei to horizontal angular acceleration. *Pflügers Arch.*, 309: 285–309.

Richter, A. and Precht, W. (1968) Inhibition of abducens motoneurones by vestibular nerve stimulation. *Brain Res.*, 11: 701–705.

Schwindt, P.C., Richter, A. and Precht, W. (1973) Short latency utricular and canal input to ipsilateral abducens motoneurons. *Brain Res.*, 60: 259–262.

Suzuki, J.I., Tokumasu, K. and Goto, K. (1969) Eye movements from single utricular nerve stimulation in the cat. *Acta oto-laryngol. (Stockh.)*, 68: 350–392.

Van der Hoeve, J. and De Kleijn, A. (1917) Tonische Labyrinthreflexe auf die Augen. *Pflügers Arch.*, 169: 241–262.

Vestibular Unit Activity during Nystagmus

H. SHIMAZU

Department of Neurophysiology, Institute of Brain Research, School of Medicine, University of Tokyo, 7-3-1 Hongo, Bunkyo-ku, Tokyo (Japan)

Vestibular nucleus neurons whose activity is closely related to nystagmus have first been found in the rabbit by Duensing and Schaefer (1958). Subsequent studies have confirmed the existence of eye movement-related neurons in the vestibular nuclei of the cat (Horcholle and Tyč-Dumont, 1968) and the monkey (Luschei and Fuchs, 1972; Miles, 1974; Keller and Daniels, 1975; Keller and Kamath, 1975; Fuchs and Kimm, 1975; Waespe et al., 1977). A question arises whether the particular vestibular neurons recorded from are immediate premotor, or their eye movement-related activity is merely corollary or represents a fraction of bulbopontine activity indirectly related to the motor output. As an approach to resolving this problem, unit spikes of presynaptic axons identified as originating from the vestibular nuclei were recorded within the abducens nucleus and their discharge pattern and timing were correlated with abducens nerve activity during nystagmus (Maeda et al., 1971). It was found that identified units exhibited spike activity related to both the slow and quick phases of nystagmus, suggesting that vestibular nucleus neurons are immediate premotor and participate in nystagmic modulation of ocular activity. A similar suggestion was made by observing in the trochlear nucleus presynaptic axon spikes of vestibular nucleus neurons correlated with trochlear motor activity during nystagmus (Baker and Berthoz, 1974). Mergner and Pompeiano (1977) extended this line of study to drug-induced saccadic eye movements (REM) and concluded that the vestibular nuclear complex represents one of the premotor structures responsible for the REM.

The present report will describe a correlative study of presynaptic impulses within the abducens nucleus and postsynaptic potentials in motoneurons during vestibular nystagmus. The results suggest that vestibular nucleus neurons projecting to the abducens nuclei contribute to generation of the rapid change in the postsynaptic potentials in motoneurons at the quick phase. This possibility has been ascertained by recording unit activity of neurons in the medial vestibular nucleus and identifying their direct connection with abducens motoneurons by electrophysiological techniques.

NYSTAGMUS-RELATED DISCHARGE WITHIN THE ABDUCENS NUCLEUS OF PRESYNAPTIC AXONS OF VESTIBULAR NUCLEUS NEURONS

Unit spikes of axons were recorded within the abducens nucleus in the encéphale isolé cat under local anesthesia. Those units which were not activated antidromically from the abducens nerve were selected for study. Units were further selected by the presence of their response to horizontal angular acceleration of the head and by their *monosynaptic* activation following electrical stimulation of the vestibular nerve on the left or right side. These criteria indicate that the units under study are spikes of axons which originate from the secondary vestibular neurons in the horizontal canal system and project to either the ipsilateral or contralateral abducens nucleus.

Nystagmus was induced by high frequency (400 pulses/sec) electrical stimulation of the vestibular nerve. In order to correlate the timing of discharges of presynaptic axons with the postsynaptic potential changes in motoneurons during nystagmus, it was investigated how the extracellular field potentials reflected postsynaptic potentials of the population of motoneurons. The temporal relationship between intra- and extracellular potentials at the quick phase was examined for a large number of nystagmic beats (for detail, see Hikosaka et al., 1977). Fig. 1A exemplifies, from top to bottom, simultaneous recording of a steep depolarization in an abducens motoneuron, the negative deflection of the field potential in the abducens nucleus and abducens nerve discharges on the same side at the quick excitatory phase of the motoneuron. Fig. 1B shows similar recording at the quick inhibitory phase of the motoneuron. The results showed that the onset of the negative or positive field potential at the quick phase was synchronous with the mean onset time of the steep depolarization or hyperpolarization of motoneurons measured in a large number of nystagmic beats. The field potential can therefore be utilized as an indicator to determine the onset time of the steep change in the membrane potential of motoneurons at the quick phase.

Axons monosynaptically activated from the contralateral vestibular nerve (Fig. 2D) are presumed to originate from *excitatory* vestibular type I neurons. They fired invariably in phase with abducens nerve discharges on the same side during nystagmus. Tonic discharges of the axon during the slow phase were abruptly suppressed at the onset of steep positive field potential which reflected intracellular hyperpolarization of motoneurons (Fig. 2A, F). Assuming that the axon terminates on motoneurons, an abrupt decrease in excitatory effects of the axon should cause motoneuronal hyperpolarization due to disfacilitation. In fact, evidence was given in a previous study for the existence of disfacilitation in motoneurons at their quick inhibitory phase (Maeda et al., 1972).

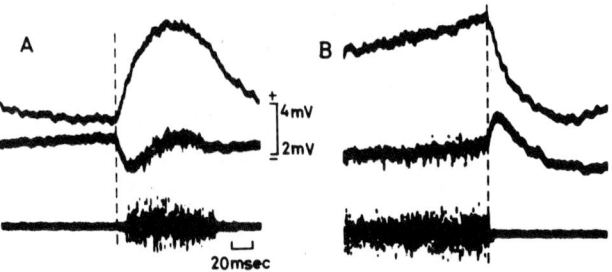

Fig. 1. Temporal relation between intracellular potentials of an abducens motoneuron and extracellular field potentials in the abducens nucleus at the quick phase of nystagmus. A) Simultaneous recording of intracellular (top), extracellular potentials (middle) and abducens nerve discharges (bottom) on the same side at the quick excitatory phase. B) Same arrangement as in A, but for the quick inhibitory phase. (From Hikosaka et al., 1977.)

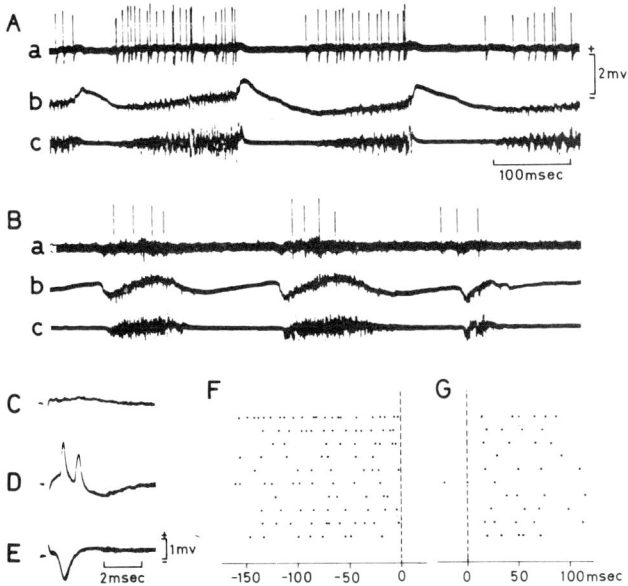

Fig. 2. Nystagmus-related discharge pattern of an axon monosynaptically activated from the contralateral vestibular nerve. A) Simultaneous recording of axonal spikes (a) and the field potential (b) in the abducens nucleus, and abducens nerve discharges with slow excitation followed by quick inhibition (c). All recordings were made on the same side. B) Same as in A, but the direction of nystagmus was reversed. C–E) Responses to single shock stimulation of the ipsilateral vestibular nerve inducing positive field potential alone (C), that of the contralateral vestibular nerve evoking spikes superimposed on negative field potential (D) and that of the abducens nerve inducing antidromic field potential alone (E). F) Dot plots of spike discharges in 10 nystagmic beats during the slow excitatory phase. G) Same as in F, but for the quick excitatory phase. Vertical broken lines in F and G represent the onset time of positive or negative field potential, resp. (From Hikosaka et al., 1977.)

When the direction of nystagmus was reversed (Fig. 2B, G), the same axon was activated during the quick excitatory phase. The firing started usually later than the onset of negative field potential or was often not activated. The synaptic action of these axons presumably contributes to a part of the EPSPs produced in motoneurons at the quick phase described by Maeda et al. (1972).

Axons monosynaptically activated form the ipsilateral vestibular nerve (Fig. 3C) are presumed to originate from *inhibitory* vestibular type I neurons. During nystagmus they fired periodically in phase with silent period of abducens nerve activity on the same side. The spike frequency usually increased gradually during the slow phase and was abruptly suppressed at the onset of steep negative deflection of the field potential which was a counterpart of intracellular depolarization (Fig. 3A, F). Given that these axons terminate on motoneurons, gradually increasing IPSPs in motoneurons during their slow inhibitory phase are attributed to synaptic action of these axons. Abrupt suppression of tonic discharges of these axons should cause disinhibition which exists in motoneurons at the quick excitatory phase (Maeda et al., 1972). A remarkable functional role of disinhibition in bringing about facilitation of motoneurons has been well known since the work of Wilson and Burgess (1962). Abrupt suppression of spike activity of inhibitory vestibular neurons

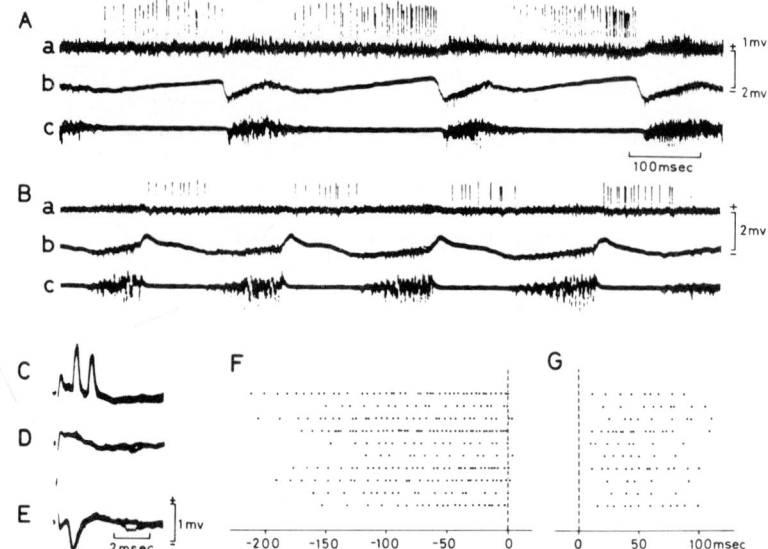

Fig. 3. Nystagmus-related discharge pattern of an axon monosynaptically activated from the ipsilateral vestibular nerve. Same arrangement as in Fig. 2, but spikes are induced during the silent phase of abducens nerve activity. (From Hikosaka et al., 1977.)

therefore fulfils an immediate premotor function for the generation of quick excitation of motoneurons.

The same axon as in Fig. 3A was activated during the quick inhibitory phase of motoneurons as well (Fig. 3B, G). Spike initiation of this axon at the quick phase tended to be slightly later than the onset of positive field potential, but they fired until the end of silent period of abducens nerve activity (more than 100 msec). The IPSPs observed in motoneurons at the quick inhibitory phase (Maeda et al., 1972) are in part attributed to the synaptic action of these vestibular neurons.

NYSTAGMUS-RELATED DISCHARGE OF PREMOTOR NEURONS IN THE MEDIAL VESTIBULAR NUCLEUS

In the study described above there was no evidence for the *termination* of nystagmus-related axons in the abducens nucleus. Moreover, the site of origin of these axons was presumed to be the vestibular nuclei because of their monosynaptic activation from the vestibular nerve, but the precise location of individual neurons in the vestibular nuclear complex could not be determined.

Type I neuron

In a recent study of Schor et al. (1977), extracellular unit spikes were recorded from neurons in the rostral part of the medial vestibular nucleus and were identified as secondary vestibular type I neurons in the horizontal canal system on the basis of their response to rotation and their monosynaptic activation following stimulation of the ipsilateral vestibular nerve. Neurons sending their axons to the contralateral abducens nucleus were selected by their antidromic response to microstimulation (less than 15 μA) in the nucleus. This technique, however, could not distinguish between an axon which terminated within the abducens nucleus and a passing fiber which merely traversed the nucleus. During successive tracks through the contralateral abducens nucleus, the vestibular type I neuron could typically be activated from a variety of scattered sites within the nucleus, with intervening ineffective sites. This suggested that the axon branched extensively and terminated within the nucleus. More direct evidence for the excitatory *connection* of a projecting vestibular neuron to abducens motoneurons was obtained by the use of postspike averaging of abducens nerve discharges triggered from spikes of a single vestibular neuron. The validity of this technique was described elsewhere in detail (Hikosaka et al., 1978). These identified vestibular type I neurons projecting to the contralateral abducens nucleus were found almost invariably to exhibit a nystagmic modulation of their spike activity (Fig. 4A). The discharge pattern was similar to that of presynaptic axons in the abducens nucleus shown in Fig. 2A.

A similar study was performed on secondary type I neurons in the medial vestibular nucleus which projected to the ipsilateral abducens nucleus. By the aid of antidromic microstimulation technique for tracing the course of the axon and postspike averaging of abducens nerve discharges, the direct inhibitory action of these vestibular neurons on ipsilateral abducens motoneurons was confirmed. Most of the neurons thus identified showed a nystagmic modulation of firing. The discharges were in phase with the inhibitory period of ipsilateral abducens motoneurons and exhibited a similar pattern to that of presynaptic axons in the abducens nucleus shown in Fig. 3.

Type II neuron

The nystagmic modulation of activity in the secondary vestibular neurons was caused by a periodic production of IPSPs at the onset of the quick phase (Hikosaka, Nakao and Shimazu, unpublished observation). As a candidate for inhibitory interneurons responsible for this periodic inhibition, vestibular type II neurons were selected for study (Schor et al., 1977), since they have been suggested as inhibitory neurons acting on type I neurons in the same nucleus (Shimazu and Precht, 1966). Type II neurons in the medial vestibular nucleus, characterized by their response to

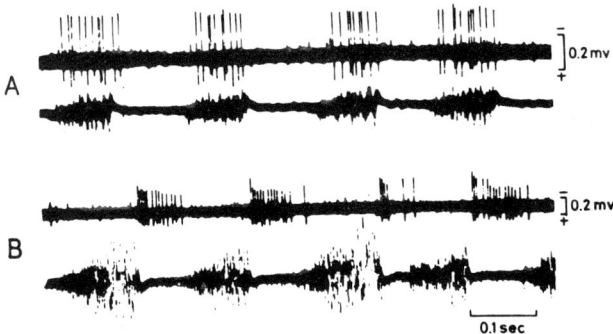

Fig. 4. Nystagmic modulation of spike activity of horizontal type I and type II neurons recorded in the medial vestibular nucleus. A) Spikes of a type I neuron projecting to the contralateral abducens nucleus. B) Spikes of a type II neuron. Bottom record in A and B represents abducens nerve discharges on the contralateral side. (From Nakao, Schor and Shimazu, unpublished observation.)

horizontal rotation (increased firing with contralateral angular acceleration and decreased firing with ipsilateral acceleration), were further selected by their activation at short latencies from the contralateral labyrinth. Most of these neurons had a nystagmic rhythm, showing an abrupt increase in discharge frequency at the onset of the quick inhibitory phase of contralateral abducens motoneurons, when type I neurons were showing a quick suppression of activity (Fig. 4B).

The firing pattern of vestibular type II neurons during nystagmus was consistent with the hypothesis that they are contributing to inhibition of type I activity at the quick phase. This was more directly confirmed by postspike averaging of the membrane potential of a type I neuron triggered from spikes of a single type II neuron, revealing an IPSP with a monosynaptic latency. It is therefore concluded that the vestibular type II neurons mediating commissural inhibition play a role in nystagmic modulation of activity observed in type I neurons which make direct connections with abducens motoneurons and contribute to the generation of nystagmic rhythm of abducens nerve activity.

DISCUSSION

The role of the vestibular nuclei in production of the quick eye movements has long been a subject of dispute. Until very recently, many oculomotor physiologists have been concerned about the pontine reticular formation as the site of origin of quick eye movements (cf. Raphan and Cohen, 1978). The term of "site of origin" of movements may need some explanation. An explicit working definition has been given by Cohen and Henn (1972) who state that by site of origin is meant the location of the *immediate* supranuclear neural mechanism which generates these movements. Given this definition, the site of origin (i.e., the location of "immediate premotor neurons") of quick phases of nystagmus may not necessarily be a single site, but comprise multiple structures which make direct connections with ocular motoneurons to lead to excitation or inhibition related to nystagmic rhythm of motor output.

The present results have shown that immediate premotor neurons participating in generation of not only the slow phase but also the quick phase of nystagmus are located in the vestibular nuclei. Obviously, however, this does not exclude other premotor neurons responsible for the quick phases. In fact, recent studies have provided evidence that interneurons in the abducens nucleus projecting to the contralateral oculomotor nucleus are also premotor neurons causing excitation of medial rectus motoneurons during both slow and quick eye movements (Delgado-Garcia et al., 1977; Nakao and Sasaki, 1978). A group of burst neurons in the dorsomedial reticular formation caudal to the abducens nucleus, are also immediate premotor neurons to give rise to inhibition of contralateral abducens motoneurons at the quick phase (Hikosaka and Kawakami, 1977; Hikosaka et al., 1978). The paramedian pontine reticular formation has also been proposed as a site of origin of quick phases of nystagmus in the horizontal plane (Bender and Shanzer, 1964; Cohen and Henn, 1972). This hypothesis is supported by lesion and stimulation experiments. Unit activity of burst neurons recorded in this area is closely related to saccades and quick phases of nystagmus. Since anatomical studies have shown that this area projects to the ipsilateral abducens nucleus (Büttner-Ennever and Henn, 1976; Graybiel, 1977), burst neurons in this area may make direct connections with abducens motoneurons. It is of importance to prove this possibility by examining specific connections between abducens motoneurons and individual burst neurons.

Finally, the well coordinated timing of impulse frequency change observed in vestibular and reticular neurons at the quick phase of nystagmus may imply the existence of significant vestibuloreticular or reticulovestibular interaction. To understand the mechanism of generation of nystagmus-related pattern in various premotor neurons, it seems therefore essential to know how functionally identified neurons in these two structures are coupled with each other.

SUMMARY

1. Unit spikes of axons were recorded within the abducens nucleus in the encéphale isolé cat under local anesthesia. Units were identified by the absence of antidromic activation from the abducens nerve and the presence of response to horizontal rotation and monosynaptic activation from the vestibular nerve, indicating that the units under study were spikes of axons of vestibular nucleus neurons projecting to the abducens nucleus. Axons monosynaptically activated from the contralateral vestibular nerve fired in phase with abducens nerve activity and those from the ipsilateral vestibular nerve fired during the silent period of abducens nerve activity. Abrupt suppression of their tonic discharges occurred at the quick phase coincidently with disfacilitation or disinhibition in motoneurons.

2. Type I neurons in the medial vestibular nucleus were selected by the response to horizontal rotation and monsynaptic activation from the ipsilateral labyrinth. Neurons sending their axons to the contralateral or ipsilateral abducens nucleus were identified by their antidromic response to microstimulation in the nucleus and their excitatory or inhibitory synaptic connection with motoneurons was confirmed by postspike averaging of abducens nerve discharges triggered from spikes of a single vestibular neuron. Spikes of thus identified vestibular type I neurons showed a nystagmic modulation, similar to that of presynaptic axons in the abducens nucleus.

3. Type II neurons in the medial vestibular nucleus were identified by their response to horizontal rotation and their activation at short latencies from the contralateral labyrinth. Most of these neurons had a nystagmic rhythm, showing a spike burst when type I neuron activity was suppressed.

4. It is concluded that type I neurons in the medial vestibular nucleus are immediate premotor neurons which participate in generation of nystagmic rhythm in abducens motoneurons. Type II neurons which mediate commissural inhibition play a role in nystagmic modulation of type I neuron activity by their periodic spike burst at the quick phase.

REFERENCES

Baker, R. and Berthoz, A. (1974) Organization of vestibular nystagmus in oblique oculomotor system. *J. Neurophysiol.*, 37: 195–217.

Bender, M.B. and Shanzer, S. (1964) Oculomotor pathways defined by electrical stimulation and lesions in the brainstem of monkey. In *The Oculomotor System*, M.B. Bender (Ed.), Harper and Row, New York, pp. 81–140.

Büttner-Ennever, J.A. and Henn, V. (1976) An autoradiographic study of the pathways from the pontine reticular formation involved in horizontal eye movements. *Brain Res.*, 108: 155–164.

Cohen, B. and Henn, V. (1972) The origin of quick phases of nystagmus in the horizontal plane. *Bibl. ophthal., (Basel)*, 82: 36–55.

Delgado-Garcia, J., Baker, R. and Highstein, S.M. (1977) The activity of internuclear neurons identified within the abducens nucleus of the alert cat. In *Control of Gaze by Brain Stem Neurons*, R. Baker and A. Berthoz (Eds.), Elsevier, Amsterdam-New York, pp. 291–300.

Duensing, F. and Schaefer, K.P. (1958) Die Aktivität einzelner Neurone im Bereich der Vestibulariskerne bei Horizontalbeschleunigungen unter besonderer Berücksichtigung des vestibulären Nystagmus. *Arch. Psychiat. Nervenkr.*, 198: 225–252.

Fuchs, A.F. and Kimm, J. (1975) Unit activity in vestibular nucleus of the alert monkey during horizontal angular acceleration and eye movement. *J. Neurophysiol.*, 38: 1140–1161.

Graybiel, A.M. (1977) Direct and indirect preoculomotor pathways of the brainstem: an autoradiographic study of the pontine reticular formation in the cat. *J. comp. Neurol.*, 175: 37–78.

Hikosaka, O. and Kawakami, T. (1977) Inhibitory reticular neurons related to the quick phase of vestibular nystagmus — their location and projection. *Exp. Brain Res.*, 27: 377–396.

Hikosaka, O., Igusa, Y., Nakao, S. and Shimazu, H. (1978) Direct inhibitory synaptic linkage of pontomedullary reticular burst neurons with abducens motoneurons in the cat. *Exp. Brain Res.*, 33: 337–352.

Hikosaka, O., Maeda, M., Nakao, S., Shimazu, H. and Shinoda, Y. (1977) Presynaptic impulses in the abducens nucleus and their relation to postsynaptic potentials in motoneurons during vestibular nystagmus. *Exp. Brain Res.*, 27: 355–376.

Norcholle, G. and Tyč-Dumont, S. (1968) Activités unitaires des neurones vestibulaires et oculomoteurs an cours du nystagmus. *Exp. Brain Res.*, 5: 16–31.

Keller, E.L. and Daniels, P.D. (1975) Oculomotor related interaction of vestibular and visual stimulation in vestibular nucleus cells in alert monkey. *Exp. Neurol.*, 46: 187–198.

Keller, E.L. and Kamath, B.Y. (1975) Characteristics of head rotation and eye movement-related neurons in alert monkey vestibular nucleus. *Brain Res.*, 100: 182–187.

Luschei, E.S. and Fuchs, A.F. (1972) Activity of brain stem neurons during eye movements of alert monkeys. *J. Neurophysiol.*, 35: 445–461.

Maeda, M., Shimazu, H. and Shinoda, Y. (1971) Rhythmic activities of secondary vestibular efferent fibers recorded within the abducens nucleus during vestibular nystagmus. *Brain Res.*, 34: 361–365.

Maeda, M., Shimazu, H. and Shinoda, Y. (1972) Nature of synaptic events in cat abducens motoneurons at slow and quick phase of vestibular nystagmus. *J. Neurophysiol.*, 35: 279–296.

Mergner, T. and Pompeiano, O. (1977) Neurons in the vestibular nuclei related to saccadic eye movements in the decerebrate cat. In *Control of Gaze by Brain Stem Neurons*, R. Baker and A. Berthoz (Eds.), Elsevier, Amsterdam-New York, pp. 243–251.

Miles, F.A. (1974) Single unit firing patterns in the vestibular nuclei related to voluntary eye movements and passive body rotation in conscious monkeys. *Brain Res.*, 71: 215–224.

Nakao, S. and Sasaki, S. (1978) Firing pattern of interneurons in the abducens nucleus related to vestibular nystagmus in the cat. *Brain Res.*, 144: 389–394.

Raphan, T. and Cohen, B. (1978) Brainstem mechanisms for rapid and slow eye movements. *Ann. Rev. Physiol.*, 40: 527–552.

Schor, R.H., Nakao, S. and Shimazu, H. (1977) Responses of medial vestibular nucleus neurons during vestibular nystagmus. *Neurosci. Abstr.*, 3: 545.

Shimazu, H. and Precht, W. (1966) Inhibition of central vestibular neurons from the contralateral labyrinth and its mediating pathway. *J. Neurophysiol.*, 29: 467–492.

Waespe, W., Henn, V. and Miles, T.S. (1977) Activity in the vestibular nuclei of the alert monkey during spontaneous eye movements and vestibular or optokinetic stimulation. In *Control of Gaze by Brain Stem Neurons*, R. Baker, and A. Berthoz (Eds.), Elsevier, Amsterdam-New York, pp. 269–278.

Wilson, V.J. and Burgess, P.R. (1962) Disinhibition in the cat spinal cord. *J. Neurophysiol.*, 25: 392–404.

Organization and Control of the Vestibulo-Ocular Reflex*

R. SCHMID and M. JEANNEROD

Istituto di Elettronica, Università di Pavia, 27100 Pavia (Italy); and Laboratoire de Neuropsychologie Expérimentale, I.N.S.E.R.M. Unité 94, 69500 Bron (France)

The specific function of the vestibulo-ocular reflex (VOR) is that of compensating head rotations by eye rotations in the opposite direction in order to maintain the image of the external world stable on the retina. This function is mainly accomplished through numerous three-neuron arcs connecting the semicircular canals to the extraocular muscles via the medial longitudinal fasciculus (Szentágothai, 1950). In addition to this basic circuit, any semicircular canal can establish functional contact with any one of the extraocular muscles by means of complex polysynaptic pathways through the reticular formation (Lorente de Nó, 1933). Vestibulo-cerebello-ocular pathways have also been proved (Angaut and Brodal, 1967; Highstein et al., 1971; Baker et al., 1972; Ito, 1974). An excellent review of the basic anatomy and physiology of VOR has recently been published by Demanez (1977). In order to establish a functional linkage among the tremendous amount of experimental data which have been made available in vestibular and oculomotor physiology, many attempts to describe the VOR in terms of mathematical models have been made in the last years (Young and Oman, 1968; Malcolm, 1968; Young, 1969; Outerbridge, 1969; Sugie and Melvill Jones, 1971; Robinson, 1971, 1972, 1975, 1977; Schmid, 1970, 1975; Schmid et al., 1971, 1975; Barnes and Benson, 1973; Schmid and Lardini, 1976; Montella et al., 1977; Raphan et al., 1977; Benson and Barnes, 1978; Chun and Robinson, 1978). All the proposed models are, perforce, oversimplified descriptions of the complex circuits organizing vestibularly induced eye movements. Nevertheless, most of them have been proved to be useful for programming new experiments or to suggest meaningful parameters for clinical evaluations of VOR functionality (Schmid et al., 1971; Zee et al., 1974; Schmid, 1975; Mira et al., 1975; Stefanelli et al., 1978).

Two generations of VOR models can be distinguished. The first generation models describe the VOR as an isolated system with uncontrollable characteristics. The second generation models can, in turns, be divided in two classes, models which consider the interaction between the vestibular and other sensory systems, namely the visual (Robinson, 1975, 1977; Raphan et al., 1977; Benson and Barnes, 1978) and the acoustic system (Berthoz et al., 1978), and models which take into account the

---

*Work supported by C.N.R. (Italy), Special Project on Biomedical Engineering, and by I.N.S.E.R.M. (France).

possibility of a parametric control of VOR in order to adapt it to unusual environmental conditions or to repair internal damages (Robinson, 1975; Jeannerod et al., 1976; Schmid et al., 1979).

This paper is divided in three parts. Part I describes a first generation model of VOR. Part II reports some experimental results proving the plasticity of VOR, and part III presents a second generation model of VOR which can interpret its plastic behaviour.

I. A FIRST GENERATION MODEL OF VOR

Most of the first generation models of VOR are limited to a description of the smooth compensatory eye movements induced by vestibular stimulations. Only few of them (Schmid, 1970; Sugie and Melvill Jones, 1971; Barnes and Benson, 1973; Schmid and Lardini, 1976; Chun and Robinson, 1978) try to interpret the generation of vestibular nystagmus for what concerns both its slow and its fast phases. In the latter, two distinct pathways are generally assumed, one controlling eye movement during the slow phase of nystagmus and the other programming the fast phase saccades. The major controversies regard the description of the saccadic mechanism for which further anatomical and physiological knowledge is needed. In the absence of any conclusive experimental evidence it is not possible to determine which of the proposed models gives the correct configuration.

A first generation model of VOR describing the basic organization of vestibularly induced eye movements is shown in Fig. 1.

The dynamics of the semicircular canals (SCC) is described by the second order transfer function derived by Jones and Milsum (1965) from Steinhaüsen's equation (Steinhaüsen, 1931) likening the cupula-endolymph system to an overdamped torsion

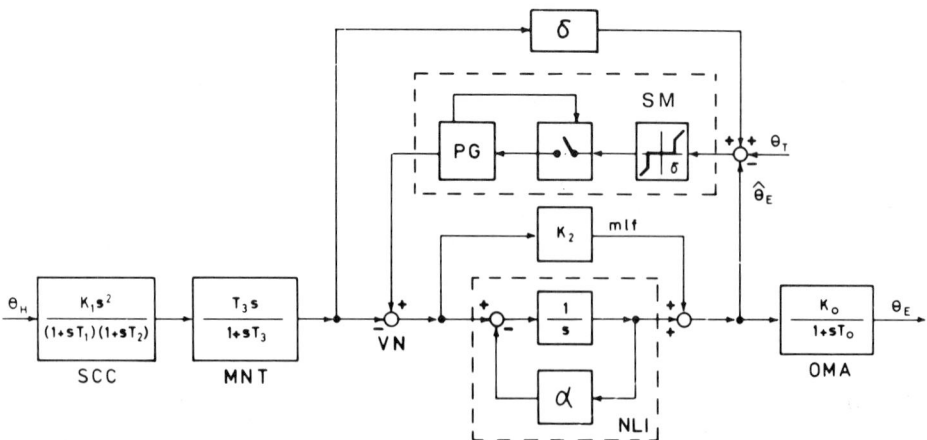

Fig. 1. A first generation model of the vestibulo-ocular reflex. SCC, semicircular canals; MNT, mechanoneural transduction; VN, vestibular nuclei; mlf, medial longitudinal fasciculus; NLI, neural leaky integrator; OMA, oculomotor apparatus; SM, saccadic mechanism; PG, pulse generator; θ_H, head rotation; θ_E, eye rotation in skull; $\hat{\theta}_E$, efference copy of eye position; θ_T, target position; T_1 and T_2, short and long time constants of the semicircular canals; T_3, adaptation time constant of vestibular receptors; T_0, time constant of the oculomotor apparatus; α, leaky factor of the central neural integrator; K_0, K_1, K_2 and σ, grains; δ, threshold.

pendulum. The mechanoneural transduction (MNT) is described in a simplified form according to the results obtained by Fernandez and Goldberg (1971) on the squirrel monkey. Owing to the values of the time constants T_1, T_2 and T_3 ($T_1 \simeq 0.005$ sec, $T_2 = 10$–20 sec and $T_3 \simeq 80$ sec in humans), a frequency domain analysis of the peripheral vestibular system leads to the conclusion that the system behaves in the range of natural head movements like an angular velocity transducer (Jones and Milsum, 1965). The head velocity signal produced by the vestibular receptors seems to be repeated by many second-order cells in the vestibular nuclei (VN) (Melvill Jones and Milsum, 1970, 1971).

Since the neural command to the oculomotor nuclei has to be proportional to head position, an integration of the incoming vestibular signal should take place somewhere between the vestibular and the oculomotor nuclei. By definition, the direct pathway running in the medial longitudinal fasciculus (mlf) cannot do it. Thus, it is reasonable to suspect that integration takes place in the polysynaptic pathways of the pontine reticular formation. In Fig. 1 the central neural integrator has been modelled as a leaky integrator (NLI) obtained through a negative feedback around a pure integrator. Carpenter's finding (Carpenter, 1972) that cerebellectomy causes a phase advance and a marked gain decrease in the frequency response of VOR should probably be better interpreted as evidence of a cerebellar control of NLI characteristics (e.g., of its leaky factor α) rather than proof of a cerebellar localization of it. The experimental evidence for a pontine localization of NLI has been reviewed by Robinson (1975) who also provided strong arguments in support of the hypothesis that it is the same integrator that operates in the generation of the slow and the fast phase of vestibular nystagmus.

The neural pathway in the mlf has been represented in Fig. 1 as a direct connection with a gain K_2 for which the value of 0.2 was suggested (Robinson, 1971). Since this pathway carries on a signal proportional to head angular velocity, its function might be that of fastening the dynamics of VOR, in particular of compensating the dynamics of the oculomotor apparatus (OMA) described in Fig. 1 by the simplified first order transfer function proposed by Robinson in 1972 ($T_0 = 0.15$ sec).

The peripheral and central mechanisms so far illustrated form the basic vestibulo-ocular reflex producing compensatory smooth eye movements. A fairly constant amplitude ratio and an approximately 180° phase shift between head and eye rotation exist over the frequency range (0.1–1.0 Hz) of most natural head movements (Meiry, 1965).

In order to account for the fact that in the absence of visual references, as in darkness, the smooth compensatory eye movements produced by the VOR are rhythmically interrupted by saccades in the anticompensatory direction giving rise to the so-called vestibular nystagmus, the upper part of the block diagram in Fig. 1 has been introduced. This part of the model gives a simplified description of the saccadic mechanism (SM).

The input to SM is the difference between the position of a possible visual target (θ_T) and that of the eyes (let us forget for a moment the pathway with gain σ). According to the "efference copy theory" (Von Holst, 1954) it is assumed that the information about eye position (θ_E) at a given instant is made available in the central nervous system as an efference copy ($\hat{\theta}_E$) of the neural command to the oculomotor nuclei at that instant. If the error $E = \theta_T - \hat{\theta}_E$ exceeds a threshold δ and the sampler is closed, the pulse generator (PG) produces a pulse with a delay of about 200 msec.

The duration of the pulse is proportional to the amplitude of the error E. Entering into the direct pathway with gain K_2 this pulse generates the burst component of the neural command of saccadic eye movements (Fuchs and Luschei, 1970; Robinson, 1970; Schiller, 1970) while the same signal passing through the NLI produces the tonic component.

As for the sampler control it can be assumed (Schmid and Lardini, 1976) that the SM cannot receive a new input (sampler open) until the last saccade has been programmed and an output signal generated. Actually, the SM must perform some operations on each input data in order to convert the contained spatial information (target relative position) into a temporal information (duration of the burst discharge) (Kornhüber, 1971). Therefore, the signal that sets the input line can be produced only after a fixed delay (200 msec) which represents the reaction time of SM. Once the input line has been set, the sampler remains closed until the threshold is exceeded again and a new input is produced. The main difference with previous sampled data models (Young, 1962; Sugie and Melvill Jones, 1971) consists in the fact that only the minimum interval between two saccades becomes fixed, while a continuous range of intersaccadic intervals is allowed beyond that minimum. This is what has been basically observed for the intersaccadic interval distribution in vestibular nystagmus (Schmid and Lardini, 1976).

If the head oscillates with the eyes fixating upon a stationary target, it results $\theta_T = -\theta_H = \hat{\theta}_E$. Therefore, no saccade is produced at least within the range of frequencies and amplitudes for which perfect compensation of head rotation by eye rotation occurs. In darkness θ_T can be assumed equal to zero. Therefore, when the head oscillates in this condition, the eye deviation during the vestibularly induced compensatory eye movement coincides with the error $\theta_T - \hat{\theta}_E$, and it is periodically corrected by saccades in the anticompensatory direction. A nystagmoid pattern of eye movement takes place with a slow and a fast phase. The existence of a pathway from the VN to the SM carrying on a signal proportional to head angular velocity was suggested by Schmid and Lardini (1976) in order to justify the predominance of anticompensatory eye movements in vestibular nystagmus. Owing to the value that can be assigned to the gain σ ($\sigma = 0.3$), the presence of this pathway does not modify significantly the basic behaviour of the model.

Recently, Chun and Robinson (1978) presented a model of fast phase generation in the VOR which contains the same essential mode of operation as the model presented in this section which, in turns, has been derived from a previous model by Schmid and Lardini (1976). In addition, Chun and Robinson introduced two noisy signals which enter into the SM and give a random component to its behaviour. In this way, a fine description of the statistical fluctuations in the amplitude and timing of the fast phases was obtained.

II. PLASTICITY OF VOR

The plasticity of VOR has been investigated by several authors under experimental situations of visual-vestibular conflict. Gonshor and Melvill Jones (1973, 1976) examined VOR modifications in subjects wearing reversing prisms. In this situation the normal vestibulo-ocular response becomes anticompensatory. If the wearing of prisms is prolonged in time, a decrease of the vestibulo-ocular response in the dark is first observed. The response is then abolished in 4–7 days, and finally reversed in about one month.

Similar experiments have been made by Robinson (1976) on the cat with the following results. The gain of the reflex decreases by about 90% in one week and then remains stationary. The return to normal visual conditions is followed by a progressive increase of VOR gain, which resumes its initial value in about one week. When the vestibulocerebellum was removed, these plastic changes were completely abolished.

Miles and Fuller (1974) have adapted telescope lenses on monkeys which resulted in magnifying or reducing visual objects. After a few days the gain of VOR was adapted so as to hold images still on the retina when the head moved. Gauthier and Robinson (1975) found similar results in human subjects and Ito et al. (1974) in rabbits.

Exposure to unusual visual or vestibular conditions is thus one way to force plastic changes of VOR characteristics. Remarkable changes can also be observed during the recovery processes following internal damages.

We have investigated VOR plasticity in the cat by examining two different processes, namely vestibular habituation to repeated rotatory stimulations (Jeannerod et al., 1976) and vestibular compensation after unilateral labyrinthectomy (Courjon et al., 1977). Another aspect of VOR plasticity in human subjects has been studied by examining the compensation of the vestibulo-ocular response in patients suffering from unilateral Meniere's disease (Stefanelli et al., 1978). Only the main results of the first two studies will be briefly reviewed in this section in order to provide the experimental support to the second generation model of VOR proposed in the next section.

Habituation was induced by submitting cats to angular velocity steps of 160 degrees/sec in the dark. Ten steps directed in the CW and CCW direction were alternated during each session. In a first group of cats sessions were repeated every day for 5 days. In a second group sessions were repeated every day for 3 days; they were followed by a 4 day rest, and then repeated again, for 3 weeks. Nystagmic responses were processed to compute the peak amplitude (C_M) of the slow cumulative eye position (SCEP), that (V_M) of the slow phase velocity (SPEV), and the time t_o corresponding to the peak of the SCEP. The parameter t_o roughly represents the duration of the primary phase of the postrotational nystagmus.

The evolution of the parameters C_M, V_M and t_o for one cat of the second group is shown in Fig. 2. The development of habituation can be followed either in terms of acquisition within a session or in terms of retention after a period of rest of 1 or 4 days. A progressive inhibition of the beginning of the response leading to a total suppression of it for periods up to 2–3 sec could also be observed.

It is worth noting that the progressive decrease of t_o indicates that habituation brings about not only modifications of VOR gain, but also remarkable changes of its dynamic characteristics. Thus, any model considering only gain adaptation (Ito, 1972; Robinson, 1975) appears to be inadequate.

The role of vision in the development of habituation was investigated by modifying the stimulation protocol so that every day the cats received the first 4 steps in the dark, the next 8 in the light, and the last 2 again in the dark. It was proved that the presence of a period of vision makes the habituation of the responses produced in the dark more rapid.

Vestibular compensation was examined in two groups of right hemi-

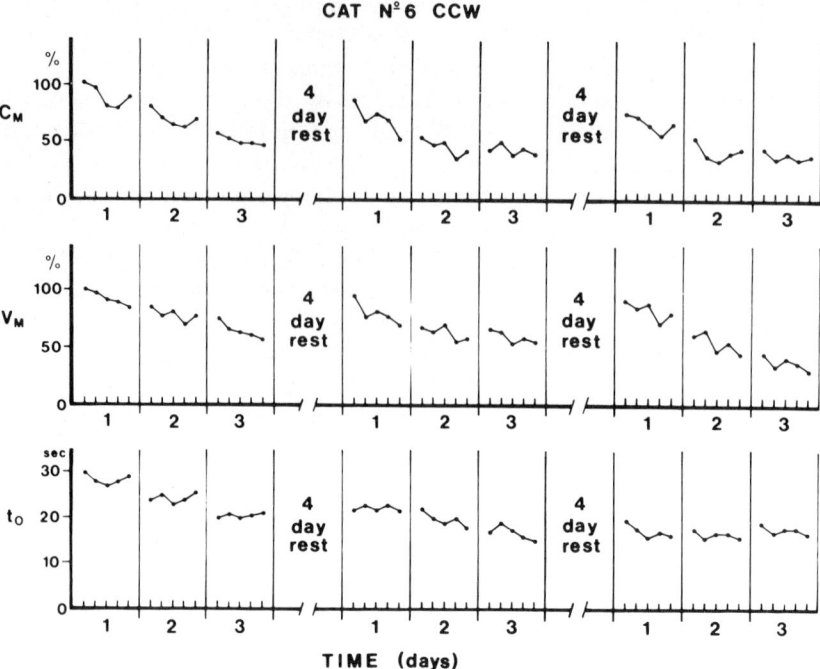

Fig. 2. Evolution of the parameters C_M (peak amplitude of the slow cumulative eye position, SCEP), V_M (peak amplitude of the slow phase eye velocity, SPEV), and t_o (time of SCEP peak amplitude) during vestibular habituation in the cat.

labyrinthectomized cats, one recovering in normal laboratory conditions and the other maintained in the dark for one month after the operation. The cats were periodically submitted to one CW and one CCW angular velocity step of 160 degrees/sec in the dark. The cats of the second group received the first step one day before the re-exposure to light. The same parameters C_M, V_M and t_o as considered in the study of habituation were computed from the nystagmic responses. Obviously, in order to obtain the SCEP from the responses recorded during the acute stage, when a spontaneous nystagmus was present, the effect of spontaneous nystagmus was removed and only the induced responses were considered.

The progressive variations of the parameters C_M, V_M and t_o for two cats of the first group is shown in Fig. 3. Inspection of the diagrams of C_M and V_M reveals the existence of three phases in the compensation process. During the first phase corresponding to the acute postoperative stage (first 2–3 days) the system reacts to the imbalance produced by the operation through a strong inhibition of the intact side and a disinhibition of the injured side. Spontaneous nystagmus disappears during this phase. During the second phase, lasting about 10 days, the system remains stationary. Finally, during the third phase a progressive disinhibition of the intact side takes place and an increase of both CW and CCW responses occurs.

These findings are in good agreement with the results reported by Precht et al. (1966) on the modifications occurring in VN activity during compensation after hemilabyrinthectomy.

The influence of vision on vestibular compensation was investigated in the second

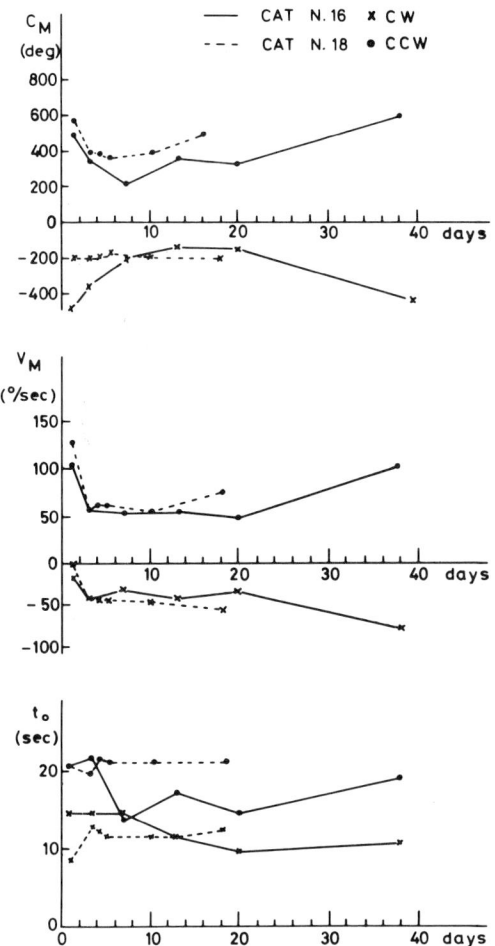

Fig. 3. Evolution of the parameters C_M, V_M and t_o during vestibular compensation after right hemilabyrinthectomy in the cat.

group of cats. A weak spontaneous nystagmus was still present at the end of the period of darkness and disappeared rapidly after re-exposure to light. Moreover, the evolution of the parameters C_M and V_M seems to suggest the conclusion that vision is a necessary condition for the third phase of compensation, that is disinhibition of VN, to take place.

III. A SECOND GENERATION MODEL OF VOR

The results reviewed in the preceding section on VOR plasticity during habituation and compensation processes suggest a multilevel hierarchical organization of the system controlling vestibularly induced eye movements. A three-level model is shown in Fig. 4.

The third and highest level, which receives information from the external environment and from the lower levels, decides the goal of the overall control system.

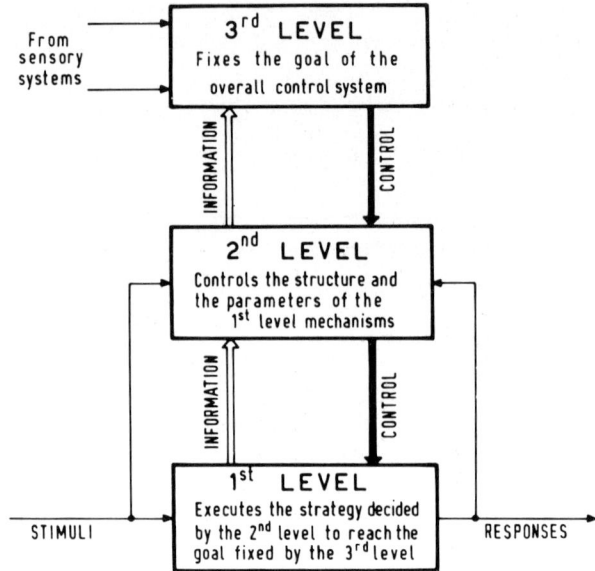

Fig. 4. Hierarchical organization of the system controlling vestibularly induced eye movements.

The second level, which can identify the state of the first level through information concerning stimuli and corresponding responses, sets up the best strategy to reach the goal fixed by the third level. Finally, the first level executes the strategy decided by the second level.

In the case of habituation, the third level would recognize the unphysiological nature of the repeated stimuli and the unusefulness of the corresponding vestibulo-ocular responses. It would thus decide that responses should be reduced or possibly abolished. The strategy set up by the second level would then consist in a progressive and symmetric inhibition of VOR. Retention after a period of rest would represent a state of presetting of the system, which makes habituation to progress faster once it has been recognized that the situation did not change.

In the case of compensation after hemilabyrinthectomy the third level would recognize from the presence of spontaneous nystagmus and from the asymmetry between CW and CCW responses that a static and a dynamic imbalance are present in the system. It would thus decide, as a goal for the overall control system, to rebalance VOR. After hemilabyrinthectomy, rotations towards the intact side are perceived as an increase of activity in the VN of the same side, while rotations towards the operated side are perceived as a decrease of activity in the VN of the intact side as well. Thus, the strategy set up by the second level would consist in reducing the excitation and increasing the inhibition produced on the VN of the intact side by the vestibular stimulations. This strategy would then be adapted to the fact that spontaneous activity in the de-afferented VN progressively regenerates and, therefore, contralateral inhibition on the VN of the intact side increases.

Vision would play a different role in habituation and compensation. In the case of habituation the vision of the environment during the vestibular habituation would make the response further undesirable and suppression will thus be reinforced. In the case of compensation, vision is an essential factor to estimate the state of imbalance of the system. Moreover, vision provides a feedback information on the progress of compensation. It is, therefore, not surprising that cats deprived of vision after hemilabyrinthectomy do not recover completely.

At present, there are several anatomical and physiological evidences supporting the theory of a multilevel hierarchical organization of the system controlling vestibularly induced eye movements. The best candidate to play the role of second level controller is the vestibulocerebellum (Ito, 1972, 1976; Carpenter, 1972; Robinson, 1976) which receives inputs from primary and secondary vestibular neurons (Brodal and Torvik, 1967; Precht and Llinás, 1969) and produces an output to the VN (Angaut and Brodal, 1967).

In the case of habituation, where the goal and the strategy of the system are more easily definable, it has been possible to identify the dynamics of the neural mechanisms of the second level (Jeannerod et al., 1976).

Under the assumption that vestibular habituation is produced through an increasing cerebellar inhibition on the VN and that an algebraic summation of vestibular and cerebellar inputs takes place in these nuclei, the cerebellar action at each stage of habituation can be appreciated by making the difference between the SPEV of the first unhabituated response and the SPEV of the response recorded at that stage. Actually, SPEV gives a measure of the output of the VN (see section II). Fig. 5 shows the results of such an operation. The upper diagrams are the SPEV of the first CCW response of the first day (d_1) and the last CCW response of the third day (d_3) of the same cat. The continuous line (d_2) in the lower part of the figure is the difference between d_1 and d_3. The signal d_2 seems to be decomposable into the two components d_t and d_p indicated by dashed lines. The simplest way to justify the generation of the signals d_t and d_p is to assume the existence in the cerebellum of two distinct neural mechanisms receiving the same input from the primary vestibular neurons and acting

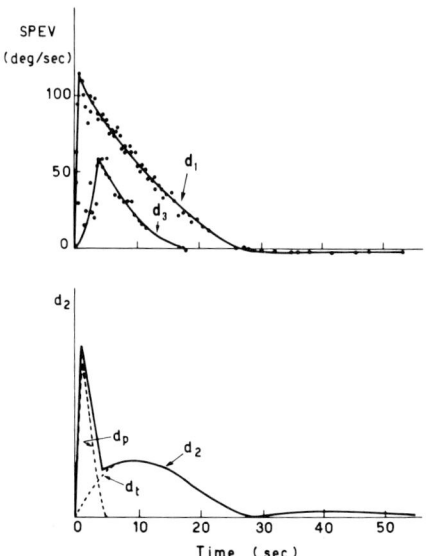

Fig. 5. Computation of the inhibitory signal d_2 produced by the cerebellum on the vestibular nuclei. Upper diagrams are the SPEV of the first CCW response of the 1st day (d_1) and the last of the 3rd day (d_3). The signal d_2 in the lower diagram is the absolute value of the difference d_1–d_3. The signal d_2 is decomposable into the components d_t and d_p shown by dashed lines. (From Jeannerod et al., 1976.)

in parallel on the VN. Since the signal transmitted by the primary vestibular neurons is known, the dynamics of the two cerebellar mechanisms, for which input and output are thus available, can be identified. The following transfer functions

$$\frac{D_t(s)}{D_1(s)} = \frac{K_t}{1 + sT_t} \qquad (1)$$

and

$$\frac{D_p(s)}{D_1(s)} = \frac{K_p s}{1 + sT_p} \qquad (2)$$

have been derived for the mechanisms generating d_t and d_p, respectively (Jeannerod et al., 1976).

Transfer functions (1) and (2) seem to be consistent with the finding by Llinás et al. (1971) of tonic and phasic responses to horizontal constant angular accelerations in Purkinje cells of the vestibulocerebellum of frogs. As claimed by Llinás (1971), these tonic and phasic responses are produced by cells which tend to function as integrators and differentiators of a particular input, respectively. Integration and differentiation are indeed the operations basically performed by mechanisms with transfer functions (1) and (2).

By assuming that the gain K_t and K_p of the two cerebellar mechanisms are progressively increased during habituation, an evolution of the parameters C_M, V_M and t_o of the nystagmic response as that observed experimentally is obtained. Thus the three-level model in Fig. 6 can be proposed to describe vestibular habituation. Further analysis is needed to prove that this model, which can be considered as the second generation, is adequate to explain also the results observed in vestibular compensation after hemilabyrinthectomy. Obviously, a different strategy should be defined for the third level.

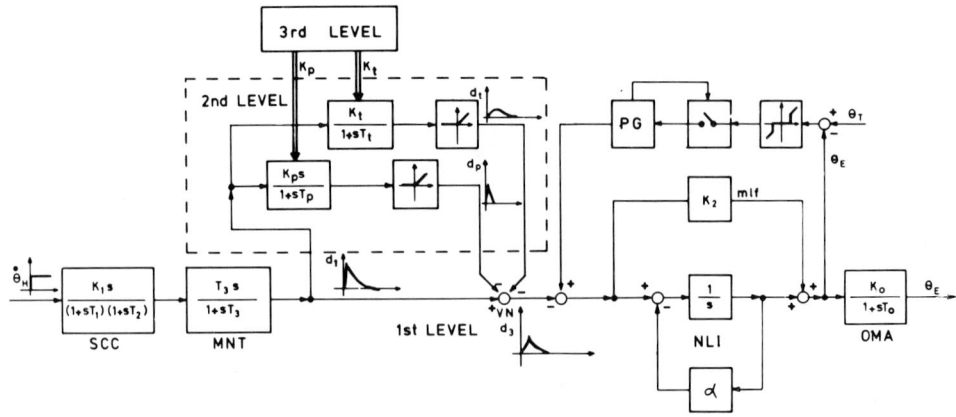

Fig. 6. A second generation model of VOR.

CONCLUSION

A progressive and exciting evolution in VOR modelling is going on parallel to the acquisition of new and more precise knowledge of the basic anatomy and physiology of VOR, its functional relationships with other sensory systems, and its tremendous adaptability to unusual environmental conditions or to structural modifications following internal damages.

In early models the dynamics of VOR was assumed to coincide with that of the semicircular canals. Then, a finer description of VOR was obtained by introducing the dynamics of the mechanoneural transduction and that of the central neural mechanism integrating the output of the vestibular system and producing on the oculomotor nuclei a signal proportional to head rotation. A further improvement in VOR modelling was obtained by the description of the mechanisms involved in the generation of the fast phase of vestibular nystagmus. The time has then come to take into account the experimental evidence that the VOR does not behave as an isolated system but interacts with other sensory systems, and it is continuously maintained under the strict control of higher neural centers. The second generation of VOR models began, and complex notions such as adaptation, sensitivity, identification, optimization, multilevel and hierarchical control were brought about. It is likely that the simple mathematical tool of the Laplace transform which has characterized the models of the first generation will be proved to be inadequate for the second generation models.

Attempts will be made to develop new models or to modify previous ones in order to explain as many characteristics of vestibularly induced eye movements as possible. Any new or revised model will give a challenge to the physiologist for programming oriented experiments which, in turns, will provide additional data for further modelling efforts. The models presented in this paper have to be considered as a contribution to this trial and error process. They do not represent anything definite. On the contrary, they are open to any change suggested by new experimental evidences. The temptation of forcing the interpretation of new experimental data according to a previous model becoming inadequate should be avoided as the most insidious danger in modelling physical systems.

SUMMARY

The mathematical models of the vestibulo-ocular reflex (VOR) so far proposed in the literature can be divided into models which consider the VOR as an isolated system with uncontrollable characteristics (first generation models), and models which take into account the interaction between VOR and other sensory systems and/or the control that can be made on it in order to adapt its characteristics to unusual environmental conditions or to repair internal damages (second generation models). A first generation model illustrating the basic organization of VOR is first presented in this paper. Then, some experimental results proving VOR plasticity are reviewed. Finally, a second generation model of VOR is proposed.

REFERENCES

Angaut, P. and Broadal, A. (1967) The projection of the "vestibulocerebellum" into the vestibular nuclei in the cat. *Arch. ital. Biol.*, 105: 441–479.

Baker, R., Precht, W. and Llinás, R. (1972) Cerebellar modulatory action on the vestibulo-trochlear pathway in the cat. *Exp. Brain Res.*, 15: 364–385.

Barnes, G.R. and Benson, A.J. (1973) A model for the prediction of the nystagmic response to angular and linear acceleration stimuli. In *The Use of Nystagmography in Aviation Medicine, NATO AGARD CCP-128, A23*, pp. 1–13.

Benson, A.J. and Barnes, G.R. (1978) Vision during angular oscillation: the dynamic interaction of visual and vestibular mechanisms. *Aviat. Space environm. Med.*, 340–345.

Berthoz, A., Buizza, A., Schmid, R. and Zambarbieri, D. (1978) Acoustic-vestibular interaction in eye tracking movements. *Neurosci. Lett.*, Suppl. 1: S10g.

Brodal, A. and Torvick, A. (1967) Über den Ursprung der secondaren vestibolo-cerebellaren Fasern bei der Katze. Eine experimentall-anatomische Studie. *Arch. Psychiat. Nerven. r.*, 195: 550–567.

Carpenter, R.H.S. (1972) Cerebellectomy and transfer function of the vestibulo-ocular reflex in the decerebrate cat. *Proc. roy. Soc. B*, 181: 353–374.

Chun, K.S. and Robinson, D.A. (1978) A model of quick phase generation in the vestibuloocular reflex. *Biol. Cybernet.*, 28: 209–221.

Courjon, J.M., Jeannerod, M., Ossuzio, I. and Schmid, R. (1977) The role of vision in compensation of vestibulo-ocular reflex after hemilabyrinthectomy in the cat. *Exp. Brain Res.*, 28: 235–248.

Demanez, J.P. (1977) Anatomie et physiologie du réflexe vestibulo-oculaire. *Acta oto-rhinolaryngol. belg.*, 31: 315–535.

Fernandez, C. and Goldberg, J.M. (1971) Physiology of peripheral neurons innervating semicircular canals of the squirrel monkey. II. Response to sinusoidal stimulation and dynamics of the peripheral vestibular system. *J. Neurophysiol.*, 82: 661–675.

Fuchs, A.F. and Luschei, E.S. (1970) Firing patterns of abducens neurons of alert monkey in relationship to horizontal eye movement. *J. Neurophysiol.*, 33: 382–392.

Gauthier, G.M. and Robinson, D.A. (1975) Adaptation of the human vestibulo-ocular reflex to magnifying lenses. *Brain Res.*, 92: 331–335.

Gonshor, A. and Melvill Jones, G. (1973) Changes of human vestibulo-ocular response induced by vision-reversal during head rotation. *J. Physiol. (Lond.)*, 234: 102–103.

Gonshor, A. and Melvill Jones, G. (1976) Extreme vestibulo-ocular adaptation induced by prolonged optical reversal of vision. *J. Physiol. (Lond.)*, 256: 381–414.

Highstein, S., Ito, M and Tschuchiya, T. (1971) Synaptic linkage in the vestibulo-ocular reflex pathway of rabbit. *Exp. Brain Res.*, 18: 306–326.

Ito, M. (1972) Neural design of the cerebellar motor control system. *Brain Res.*, 40: 81–84.

Ito, M. (1974) The control mechanism of cerebellar motor systems. In *Neurosciences Third Study Program*, F.O. Schmitt and F.G. Worden (Eds.), MIT Press, Cambridge, MA, pp. 293–303.

Ito, M. (1976) Cerebellar learning control of vestibulo-ocular reflex mechanisms. In *Mechanisms in Transmission of Signals for Conscious Behaviour*, T. Desiraju (Ed.), Elsevier, Amsterdam, pp. 1–21.

Ito, M. Shiida, T., Yagi, N. and Yamamoto, M. (1974) The cerebellar modification of rabbit's horizontal vestibulo-ocular reflex induced by sustained head rotation combined with visual stimulation. *Proc. Jap. Acad.*, 50: 85–89.

Jeannerod, M., Magnin, M., Schmid, R. and Stefanelli, M. (1976) Vestibular habituation to angular velocity steps in the cat. *Biol. Cybernet.*, 22: 39–48.

Jones, G.M. and Milsum, J.H. (1965) Spatial and dynamic aspects of visual fixation. *IEEE Trans. biomed. Eng.*, 12: 54–62.

Kornhüber, H.H. (1971) Motor function of cerebellum and basal ganglia: the cerebellocortical saccadic (ballistic) clock, the cerebellonuclear hold regulator, and the basal ganglia rap (voluntary speed smooth movement) generator. *Kybernetik*, 8: 157–162.

Llinás, R. (1971) Frog cerebellum: biological basis for a computer model. *Math. Biosci.*, 11: 137–151.

Llinás, R., Precht, W. and Clarke, M. (1971) Cerebellar Purkinje cell responses to physiological stimulation of the vestibular system in the frog. *Exp. Brain Res.*, 13: 408–431.

Lorente de Nó, R. (1933) Vestibulo-ocular reflex arc. *Arch. Neurol. Psychiat. (Chic.)*, 30: 245–291.

Malcolm, R. (1968) A quantitative study of vestibular adaptation in humans. In *The Role of the Vestibular Organs in Space Exploration*, NASA SP-187, pp. 369–380.

Meiry, J.L. (1965) *The Vestibular System and Human Dynamic Space Orientation*, Sc. D. Thesis, MIT, Cambridge, MA.

Melvill Jones, G. and Milsum, J.H. (1970) Characteristics of neural transmission from the semicircular canal to the vestibular nuclei of cats. *J. Physiol. (Lond.)*, 209: 295–316.

Melvill Jones, G. and Milsum, J.H. (1971) Frequency response analysis of central vestibular unit activity resulting from rotational stimulation of the semicircular canals. *J. Physiol. (Lond.)*, 219: 191–215.

Miles, F.A. and Fuller, J.H. (1974) Adaptive plasticity in the vestibulo-ocular responses of the rhesus monkey. *Brain Res.*, 80: 512–516.
Mira, E., Schmid, R. and Stefanelli, M. (1975) Application clinique d'un modèle mathématique du système vestibulo-oculomotoeur, *Acta oto-rhinolaryngol. belg.*, 29: 24–55.
Montella, R., Ghilardi, P.L., Guglielmino. S. and Giampietro, C. (1977) A mathematical model of the vestibulo-oculomotor reflex in phsyiological and pathological conditions, *Acta oto-laryngol. (Stockh.)*, 346: 1–18.
Outerbridge, J. (1969) *Experimental and Theoretical Study of Reflex Vestibular Control of Head and Eye Movement*, Ph. D. Thesis, McGill University, Montreal.
Precht, W. and Llinás, R. (1969) Functional organization of the vestibulo afferents to the cerebellar cortex of the frog and cat. *Exp. Brain Res.*, 9: 30–52.
Precht, W., Shimazu, H. and Markham, C.H. (1966) A mechanism of central compensation of vestibular function following hemilabyrinthectomy. *J. Neurophysiol.*, 29: 996–1010.
Raphan, T., Cohen, B. and Matsuo, V. (1977) A velocity-storage mechanism responsible for optokinetic nystagmus (OKN), optokinetic afternystagmus (OKAN) and vestibular nystagmus. In *Control of Gaze by Brain Stem Neurons*, R. Baker and A. Berthoz (Eds.), Elsevier/North-Holland Biomed. Press, Amsterdam, pp. 37–48.
Robinson, D.A. (1970) Oculomotor unit behavior in the monkey. *J. Neurophysiol.*, 33: 393–404.
Robinson, D.A. (1971) Models of oculomotor neural organization. In *The Control of Eye Movements* P. Bach–y–Rita, C.C. Collins and I.E. Hyde (Eds.), Academic Press, New York, pp. 519–538.
Robinson, D.A. (1972) Progress in models of eye movement control. *Proc. Int. Conf. on Cybernetics and Society, Washington, D.C.*, pp. 19–24.
Robinson, D.A. (1975) Oculomotor control signals. In *Basic Mechanisms of Ocular Motility and their Clinical Implications*, G. Lennerstrand and P. Bach–y–Rita (Eds.) Pergamon Press, Oxford, pp. 337–374.
Robinson, D.A. (1976) Adaptive gain control of the vestibulo-ocular reflex by the cerebellum. *J. Neurophysiol.*, 39: 954–969.
Robinson, D.A. (1977) Vestibular and optokinetic symbiosis: an example of explaining by modelling. In *Control of Gaze by Brain Stem Neurons*, R. Baker and A. Berthoz (Eds.) Elsevier/North-Holland, Biomed. Press, Amsterdam, pp. 49–58.
Schiller, P. (1970) The discharge characteristics of single units in the oculomotor and abducens nuclei of the unanesthetized monkey. *Exp. Brain Res.*, 10: 347–362.
Schmid, R. (1970) Systems analysis of the vestibulo-ocular system. In *The Role of the Vestibular Organs in Space Exploration*, NASA SP-314, pp. 237–249.
Schmid, R. (1975) Clinical application of a mathematical model of the vestibulo-ocular reflex. *Pol. Acad. Sci. (Warsaw)*, 26: 23–59.
Schmid, R. and Lardini, F. (1976) On the predominance of anti-compensatory eye movements in vestibular nystagmus. *Biol. Cybernet.*, 23: 135–148.
Schmid, R., Stefanelli, M. and Mira, E. (1971) Mathematical modelling: a contribution to clinical vestibular analysis. *Acta oto-laryngol. (Stock.)*, 72: 292–302.
Schmid, R., Mira, E. and Stefanelli, M. (1975) Un modèle du système vestibulo-oculomoteur. *Acta oto-rhinolaryngol. belg.*, 29: 9–28.
Schmid, R., Jeannerod, M. and Mira, E. (1979) Plasticity of the vestibulo-ocular reflex. In preparation.
Stefanelli, M., Mira, E. and Lombardi, R. (1978) Quantification of vestibular compensation in unilateral Ménière's disease. *Acta oto-laryngol., (Stockh.)*, 85: 411–419.
Steinhausen, W. (1931) Uber den Nachweis der Bewegung der Cupula in der intakten Bogengansampulle des Labyrinthes bei der naturlichen rotatorischen und calorischen Reizung. *Pflügers Arch.*, 228: 322–328.
Sugie, N. and Melvill Jones, G. (1971) A model of eye movements induced by head rotation. *IEEE Trans. Syst. Man Cybernet., SMC-1:* 251–260.
Szentágothai, J. (1950) The elementary vestibulo-ocular reflex arc. *J. Neurophysiol.*, 13: 395–407.
Von Holst, E. (1954) Relations between the central nervous system and peripheral organs. *Brit. J. Anim. Behav.*, 2: 89–94.
Young, L.R. (1969) Current status of vestibular system models. *Automatica*, 5: 369–383.
Young, L.R. and Oman, C.M. (1968) A model for vestibular adaptation to horizontal rotation. In *The Role of the Vestibular Organs in Space Exploration*, NASA SP-187, pp. 363–368.
Zee, D.S., Friendlich, A.R. and Robinson, D.A. (1974) The mechanism of downbeat nystagmus. *Arch. Neurol.*, 30: 227–237.

D

Interaction of Labyrinthine and Proprioceptive Neck Inputs

Otoliths and Uprightness

T. D. M. ROBERTS

Institute of Physiology, University of Glasgow, Glasgow G12 8QQ (Scotland)

Traditional medical teaching on this topic contains two erroneous elements: firstly, the otolith apparatus is said to incorporate gravity sensors; secondly, the positional reflexes from the labyrinth upon the limbs are said to produce similar effects in all four limbs (Magnus, 1924, pp. 60–61). Both of these propositions turn out to be incorrect.

The requirements for stable balance demand that disturbances from the standard "upright" posture should be met by some system of necessarily asymmetric restoring forces. A set of reflexes that produce symmetrical effects in the limbs when the head is tilted could play no part in stabilization. A study of the profiles of animals in various attitudes of the head and neck (Roberts, 1968) establishes that asymmetric attitudes of the limbs are not attributable solely to deformations of the neck, as Magnus maintained, but can be seen to be associated with tilting of the head even when the neck is straight. Asymmetric reflex changes have now been demonstrated in the extensor muscles of the limbs of the decerebrate cat in response to tilting of the skull about various horizontal axes, after eliminating neck reflexes by denervation and immobilization. The changes in the limbs occur in a stabilizing pattern, describable by the rubric "downhill limbs extend", with the corollary that uphill limbs flex (Roberts, 1963, 1970; Lindsay, 1975; Lindsay et al., 1976).

The pattern of neck reflexes described by Magnus and De Kleijn (1912) has been confirmed. Neck movements at any or all of the intervertebral joints may be of three kinds. Flexion may be from side to side or in a dorsoventral direction. A small amount of torsion may occur between any two vertebrae as well as the extensive rotations possible at the atlanto-axial joint. In each case, the reflex effect upon the limbs is in such a direction as to move the body to straighten the neck (see Roberts, 1978). Magnus repeatedly emphasized the contrast between the asymmetric limb changes occurring in the neck reflexes and the symmetrical effects attributed to the labyrinth.

In many situations a head movement will evoke positional effects both from the labyrinth and from the neck but with opposite effects upon the limbs (Fig. 1). For example, a head tilt toward left side down produces limb extension on the left side with flexion on the right by labyrinth reflexes but the same movement leads, in the neck reflexes, to flexion on the left side and extension on the right. In consequence, if an animal keeps its body still and moves its head, the effects of tilting the head are offset by the effects of moving the head on the neck, the conflict being resolved centrally. Thus an animal is free to move its head about in any direction without reflex restraint. There is no need to postulate that any of the positional reflexes are switched off during voluntary movement of the head. If the trunk is tilted, or if the animal is standing on uneven ground, the combined effect of the neck and labyrinth reflexes is such as to stabilize the trunk, even though the relevant directional sensor lies in the skull and not in the trunk (Roberts, 1973). The head is still free to move without affecting the compensatory pose of the limbs, because when a head movement leads to a demand for more extension in a particular limb from labyrinth reflexes, the neck reflex calls for less extension in that same limb, and so on.

In the experiment illustrated by Fig. 1, the first two cervical nerve trunks had both been cut on each side. Relative movement between the axis vertebra and the skull consequently produced no effect either during skull movement with the axis vertebra clamped (to elicit labyrinth reflexes) or during movement of the axis vertebra with the skull clamped (to elicit neck reflexes). If the cervical nerves were left intact, head tilting in this experimental arrangement did not result in any change in the limb muscles (Fig. 2). In some experiments C1 was first cut on both sides, leaving C2 intact for a test of the response to head tilting, and then C2 was cut on both sides. In other experiments C2 alone was cut first on both sides, leaving C1 intact during a test for the effect of tilting. The response in the labyrinth reflex appeared only after both C1 and C2 had been cut on both sides. It did not matter which had been cut first (Lindsay et al., 1976). These experiments show that there are several alternative sources for the initiation of the neck reflexes.

The asymmetric character of the positional reflexes from the labyrinth upon the limbs appears to depend upon the integrity of the cerebellum. When the cerebellum was removed acutely, in the experimental arrangement referred to above for testing labyrinth reflexes without neck reflexes, a head tilt to either side produced contraction of the extensor muscles simultaneously on both sides (Lindsay and Rosenberg, 1977). In man it appears that the basal ganglia are also involved in so far as patients with

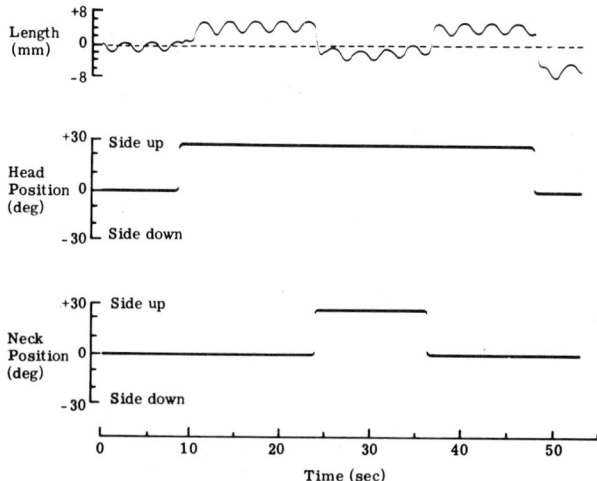

Fig. 1. Interaction between neck reflexes and labyrinth reflexes. An extensor muscle (medial triceps) of the right forelimb of a decerebrate cat is pulled rhythmically by a compliant system (Roberts, 1963). Skull and axis vertebra are supported by independent clamps which are tilted separately. C1 and C2 are cut on both sides. The response to tilting the skull disappears when the axis vertebra is tilted similarly. (From Lindsay, 1975.)

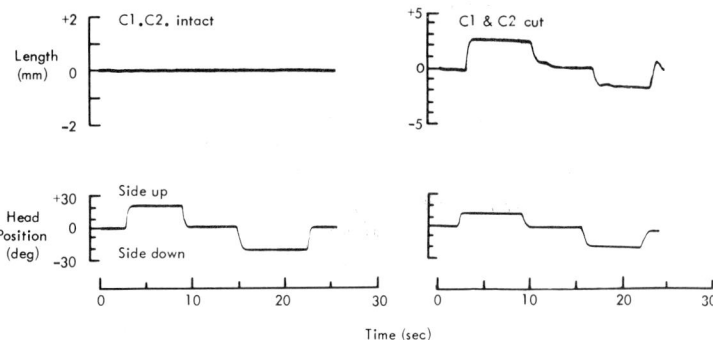

Fig. 2. Isotonic length changes in medial triceps in response to head tilting with the axis vertebra clamped: A, before and B, after cutting C1 and C2 on both sides. Decerebrate cat. The response in the labyrinth reflex appeared only after both C1 and C2 had been cut on each side, no matter which was cut first. (From Lindsay et al., 1976.)

certain types of basal ganglion disease show characteristic defects of the balancing reactions when tested on a tilt table (Purdon Martin, 1967).

Symmetrical reactions in a different pattern appeared when the labyrinth of one side was destroyed acutely. The responses to tilting remained unchanged in the ipsilateral limb but they were reversed from normal in the contralateral limb. A head tilt toward the intact side now produced contraction of the extensors on both sides, while a tilt toward the operated side produced flexion on both sides (Lindsay and Rosenberg, 1978). Cats tested some eight weeks after unilateral labyrinthectomy showed stabilizing limb responses on the tilt table although these animals characteristically hold their heads tilted operated side down (Schaefer and Meyer, 1974).

In many studies of the effects on the labyrinth produced by tilting the head there is an implied assumption that the relevant stimulus is the change in the orientation of the otolith apparatus to the gravitational field. The neuromasts coupled to the otolith membrane are taken to be "gravity sensors" in the same way that instruments such as the plumb line and spirit level are said to indicate the gravitational vertical. However, gravity is not the only force involved. If the support of a plumb line is moved about, the string no longer gives any consistent indication. If the weight is swung in a circle, the string hangs at an angle to the vertical. It is not gravity that keeps the string taut. If we let go, the string becomes slack, while gravity continues to act, accelerating both string and weight toward the centre of the earth.

The force that keeps the string taut is of a different kind from gravitational force. The tension in the string is an aspect of the intermolecular forces that characterize the solid state and which are responsible for the rigid properties of solid objects. The structure of the material composing the string is deformed when we pull on it to prevent the weight from falling. It exhibits tensile stress and associated strain and all parts of the string are similarly affected. If we hold the weight in our hand, the tissues of the hand undergo compressive strain while developing the necessary thrust to support the weight. Gravity, on the other hand, is a force acting at a distance. It does not have to be transmitted by the molecular architecture of a rigid structure. There are no stresses and strains associated with the unopposed action of gravity in free fall conditions.

We are accustomed to thinking of forces in terms of the diagrams of elementary mechanics. Each force is represented by a line indicating its direction, and we designate a geometrical point in the diagram as the "point of application" of such and such a force. However convenient this representation may be, it is for the present purpose important to remember the nature of the situation that is being represented. Contact forces are not restricted to geometrical points of application. A surface of contact between two structures is the site of stresses and strains distributed through the materials involved in ways that depend at each point upon the local pattern of relevant molecular architecture. If we take any arbitrarily chosen plane of section through a structure the parts of the structure to each side of this plane exert equal and opposite stress forces upon one another across this plane.

When we examine the structure of the otolith apparatus we find that it has components that can act as a differential-density accelerometer in the same way as a spirit level responds to the contact forces acting on its case. The relationship to gravity is, in principle, the same as for the plumb line. The direction indicated is that of the prevailing contact force. If the indicator is at rest on the earth's surface, the supporting contact force is opposite in direction and approximately equal in magnitude to the gravitational attraction toward the centre of the earth. This equality is not that between the members of a Newtonian force-pair. The resultant contact force may be altered at any time by the movement of other bodies, without change in the gravitational force.

It is clear from the behaviour of a pendulum swung in a circle, or of an aircraft or cyclist moving in a curved path, that the avoidance of falling calls for balancing about the direction of the supporting contact force, rather than about the direction of the gravitational field. It is accordingly appropriate that the otolith apparatus used in the regulation of balancing behaviour should report the direction of the supporting contact force, as an accelerometer does, rather than that of the gravitational vector itself.

This conclusion has important consequences in the context of the analysis of balancing behaviour. For example, the series of experiments illustrated in Fig. 3 was undertaken to determine what aspect of the stimulating situation acts as the trigger for initiating a hop when the subject, standing on one leg, is slowly but forcibly pulled off balance. Hops occur similarly in conditions b, c, and d, but not in conditions a, e, and f (Roberts, 1977). This result shows that balancing is classified by the central nervous system as successful, that is to say, no hop is initiated so long as the resultant upthrust vector remains within the area of support. The direction of the support vector is presumably indicated by the otolith apparatus, the frame of reference being transferred from the skull to the trunk using information from proprioceptors in the various intervertebral joints of the neck. The line of action of the thrust is assessed by its effects on proprioceptors at a number of places in the body. Its relation to the area of support requires cutaneous information as well as proprioception. The decision whether to hop or not thus clearly depends upon a quite complicated act of recognition, based on information from many different sources. Gravity is relevant only in so far as, on the surface of the earth, punishing collision of the head with the ground occurs if the strategy of continually varying thrusts against the ground is not organized in such a way as to avoid such impacts. If the subject is moving or being transported in a vehicle, the appropriate direction for the aiming of support thrusts depends upon the motion and not upon the gravitational field directly.

In condition e of Fig. 3, the thrust vector is inclined to the gravitational vertical but its line of action passes safely through the support at the feet. As well as being

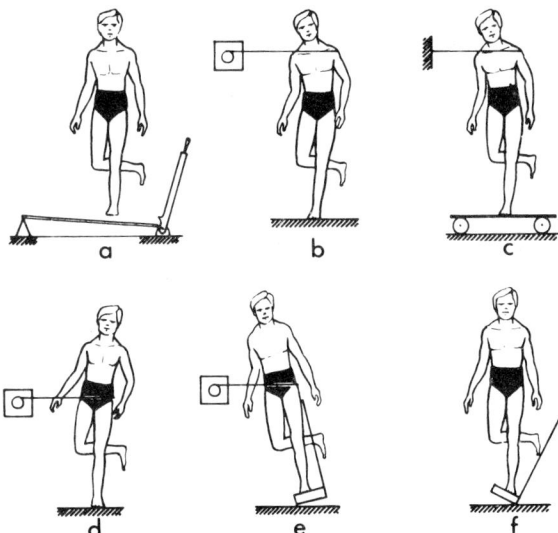

Fig. 3. Tests to localize the site of initiation of the hopping response to overbalancing. The subject stands on one leg in each case. a) Sudden drop, platform released by a catch. b) Lateral pull applied by winching in a rope attached to the shoulder. c) Shoulder tethered to the wall while the platform moves away. d) Lateral pull applied at the hip. e) Foot support tilts with the subject. f) Tilt of foot support alone. Hopping occurs identically in conditions (b), (c) and (d), but not in conditions (a), (e) and (f). (From Roberts, 1977.)

supported against the pull of gravity, the body is being accelerated horizontally. If this horizontal acceleration were to be maintained, by suitable movement of the support platform, the observed behaviour (absence of hopping) would be appropriate. The conditions would correspond to those in an aircraft executing a correctly banked turn.

It should be noted that the organization of muscular activity to maintain the upright posture does not involve the principles used in inertial guidance systems. We do not have a reference direction from which deviations can be detected for correction by servomechanisms. The directional signal available from the labyrinth indicates what is actually happening in the nature of applied contact force. There is no error signal to indicate that the prevailing attitude is in some way inappropriate. Corrective behaviour is initiated on the basis of recognition of the nature of the prevailing conditions and of the way these are changing.

Where the signals from the relevant sense organs are changing slowly with time, there is room for prediction to play a part, with the possibility of very early intervention, whereas reflex mechanisms necessarily involve delays for neural conduction and synaptic transmission. To take a particular example, in the case of hopping induced by forced overbalancing one can distinguish two types of response according as the increase in upthrust needed for take off is, or is not, preceded by a momentary diminution of upthrust. In a hop executed on command, the subject allows his body to drop momentarily before take off. He goes down first. This pattern is also seen in forced overbalancing when the subject is attending to the stimulus and preparing himself. A different pattern is seen in the time-course of the upthrust if the subject is relaxed, distracted, and free of apprehension. No dip in the trace then occurs. The response of the relaxed subject may be taken to be "reflex". The hop with no preliminary dip in the force record cannot be produced on command.

If we take the inclination of the support vector as an indicator of the effective stimulus, we find that the threshold for reflex hops is higher than that for hops of "voluntary" pattern (with a dip in the trace). The voluntary hop is thus an example of an "anticipatory pre-emptive action" (Roberts, 1976). Anticipatory in the sense that its initiation depends upon a recognition that a particular situation is about to arise; pre-emptive in the sense that, once the response has been initiated, the conditions never reach the threshold at which reflex mechanisms would be brought into play.

The normal maintenance of the upright posture appears to depend on a succession of anticipatory pre-emptive actions based upon complicated recognition processes that take account of changing signals from a great number of sources, including both the otolith apparatus and the proprioceptors in the intervertebral joints of the neck. When the animal is moving about, as in locomotion and many other activities, the head is in continual motion. The contact forces exerted on the skull by the muscles of the neck are continually changing both in magnitude and in direction. The discharges from many of the relevant receptors, such as those in the otolith apparatus (Lowenstein and Roberts, 1949) and also those in the intervertebral joints, are influenced by rate-of-change of deformation as well as by deformation itself. Furthermore, the sensitivity of the receptors in the labyrinth is under the continual influence of changing activity in the efferent system (Klinke and Galley, 1974). It follows that no system of dead reckoning (such as is used in inertial guidance) can recover the direction of a "reference vertical" from the available sensory information.

An alternative role can be proposed for the sensory apparatus of the labyrinth, namely that the receptors indicate some aspect of the changes occurring in the momentum of the skull, both linear and angular (Roberts, 1976). Such information would be appropriate to the use of the skull, on the end of the flexible neck, as an "inertia paddle" to assist in the performance of certain of the tasks involved in locomotion. Examples of such skull movements are: the brisk nodding movements made during fast walking, and the plunging head movements made when an animal is negotiating obstacles or getting up after lying on the ground. In these movements the repartition of momentum between the skull and the trunk is used at strategic moments to relieve the necessity for support by a particular limb for just long enough to allow that limb to be lifted and repositioned. The successful execution of such manoeuvres clearly depends on learning and on a great deal of practice.

With this view of otolith function, "uprightness" represents a conclusion about the suitability of parts of the environment for use as supports in the avoidance of impact of the skull with the ground, rather than a directly sensed orientation of the body.

SUMMARY

Positional reflexes from the labyrinth upon the limbs are shown to be asymmetric in their effects. The combination of labyrinth reflexes with neck reflexes serves to provide stabilization of the trunk. The initiation of neck reflexes is not confined to specific nerve roots. The pattern of labyrinth reflexes is altered by acute cerebellectomy and by destroying one labyrinth.

The otolith apparatus responds to the stresses imposed by contact forces rather than to gravitational force directly. Balancing behaviour is organized around the direction of thrust against the supports, not in relation to the gravitational vertical. In changing conditions anticipatory pre-emptive actions provide early correction, avoiding the delays inherent in reflex responses, as illustrated by the initiation of a hop during forced overbalancing.

In place of the traditional view that the otolith apparatus reports orientation with respect to gravity, it is suggested that the role of the labyrinth is to indicate changes in skull momentum, this information being relevant to the use of the skull as an inertia paddle in locomotion and other activities.

REFERENCES

Klinke, R. and Galley, N. (1974) The efferent innervation of the vestibular and auditory receptors. *Physiol. Rev.*, 54: 316–357.

Lindsay, K.W. (1975) *A Study of Labyrinthine and Neck Reflexes in the Decerebrate Cat*, Ph.D. Thesis, University of Glasgow.

Lindsay, K.W. and Rosenberg, J.R. (1977) The effect of cerebellectomy on tonic labyrinth reflexes in the forelimb of the decerebrate cat. *J. Physiol. (Lond)*, 273: 76–77P.

Lindsay, K.W. and Rosenberg, J.R. (1978) Tonic labyrinth reflexes in the forelimb of the acute and chronic hemilabyrinthectomized cat. *J. Physiol. (Lond.)*, 275: 43–44P.

Lindsay, K.W., Roberts, T.D.M. and Rosenberg, J.R. (1976) Asymmetric tonic labyrinth reflexes and their interaction with neck reflexes in the decerebrate cat. *J. Physiol. (Lond.)*, 261: 583–601.

Lowenstein, O. and Roberts, T.D.M. (1949) The equilibrium function of the otolith organs of the Thornback Ray *(Raja clavata)*. *J. Physiol. (Lond.)*, 110: 392–415.

Magnus, R. (1924) *Körperstellung*. Julius Springer, Berlin.

Magnus, R. and De Kleijn, A. (1912) Die Abhängigkeit des Tonus der Extremitatenmuskeln von der Kopfstellung. *Pflügers Arch*, 145: 455–548.

Purdon Martin, J. (1967) *The Basal Ganglia and Posture*. Pitman, London.

Roberts, T.D.M. (1963) Rhythmic excitation of a stretch reflex, revealing (a) hysteresis and (b) a difference between the responses to pulling and to stretching. *Quart. J. exp. Physiol*, 48: 328–345.

Roberts, T. D. M. (1968) Labyrinthine control of the postural muscles. In *Third Symposium on the Role of the Vestibular Organs in Space Exploration*, NASA SP-152, Washington, DC, pp. 149–168.

Roberts, T.D.M. (1970) Changes in stretch reflexes in limb extensor muscles during positional reflexes from the labyrinth. *J. Physiol., (Lond.)*, 211: 5–6P.

Roberts, T.D.M. (1973) Reflex balance. *Nature (Lond.)*, 244: 156–158.

Roberts, T.D.M. (1976) The role of vestibular and neck receptors in locomotion. In *Neural Control of Locomotion, Vol. 18*, R.M. Herman, S. Grillner, P. Stein and D. Stuart (Eds.). Plenum Press, New York, pp. 539–560.

Roberts, T.D.M. (1977) The relationship of gravity to vestibular studies. In *Life Sciences Research in Space*, ESA SP-130, Paris, pp. 139–145.

Roberts, T.D.M. (1978) *Neurophysiology of Postural Mechanism*, 2nd Ed. Butterworths, London.

Schaefer, K.P and Meyer, D.L. (1974) Compensation of vestibular lesions. In *Handbook of Sensory Physiology, Vol. 61, Vestibular System, Part 2, Psychophysics, Applied Aspects and General Interpretations*, H.H. Kornhüber (Ed.), Springer-Verlag, Berlin-Heidelberg-New York, pp. 463–490.

Neck and Macular Labyrinthine Influences on the Cervical Spino-Reticulocerebellar Pathway

O. POMPEIANO

Istituto di Fisiologia Umana, Cattedra II, Università di Pisa, 56100 Pisa (Italy)

INTRODUCTION

Experimental anatomical observations have shown that the precerebellar lateral reticular nucleus (NRL) receives both uncrossed and crossed spinoreticular fibers (Corvaja et al., 1977a). The uncrossed spinoreticular fibers originate from cells located more dorsolaterally within the spinal grey matter than those giving rise to the crossed spinoreticular fibers. The segregation of the crossed and uncrossed spinoreticular neurons may be related to different sensory modalities, peripheral receptive fields, and/or supraspinal control mechanisms.

There is physiological evidence for the subdivision of the spinoreticular pathway in two systems. The corresponding neurons are polysynaptically influenced by cutaneous and high threshold muscle afferents (cf. Oscarsson, 1973). However, most of the presumably uncrossed spinoreticular neurons, particularly located in the cervical cord, have an ipsilateral receptive field which is apparently restricted to parts of the ipsilateral forelimb (Clendenin et al., 1974a), while the presumably crossed spinoreticular neurons have a bilateral receptive field which often include all four limbs (Lundberg and Oscarsson, 1962; Grant et al., 1966; Rosén and Scheid, 1973a–c; Clendenin et al., 1974c). Moreover, the dorsolateral part of the NRL, which receives the presumably uncrossed spinoreticular path from the cervical cord, projects ipsilaterally to the intermediate cortex of the cerebellar anterior lobe and the paramedian lobule. On the contrary, the ventral portion of the NRL, which receives the presumably crossed spinoreticular path, projects bilaterally to the vermis and pars intermedia of the anterior lobe and sparsely to the pyramis and the ipsilateral paramedian lobule (Clendenin et al., 1974b). These findings should be related to the observation that the vermal cortex of the cerebellum influences mainly spinal extensor motoneurons, via the lateral vestibular nucleus of Deiters (LVN), whereas the paravermal cortex affects mainly spinal flexor motoneurons, via the red nucleus (Pompeiano, 1967).

The experiments reviewed in this paper were performed to find out whether the macular labyrinthine input and the neck input may influence the activity of neurons of the spino-reticulocerebellar pathway, and whether this pathway contributes to the vestibular and the cervical control of posture and reflex movements.

MACULAR LABYRINTHINE INPUT ON NEURONS OF THE SPINO-RETICULOCEREBELLAR PATHWAY

A first group of experiments was performed to determine whether changes in head and body position leading to selective stimulation of macular receptors, can modify the activity of cervical spinoreticular tract neurons (Coulter et al., 1974, 1976). Among 106 units recorded from the lateral funiculus of the cervical segments C2–C3 of the cord in decerebrate cats, 42 units attributed to ascending axons, responded to lateral tilt of the whole animal. This response consisted of steady changes in frequency of discharge which persisted with little adaptation as long as the position of the cat was maintained (Fig. 1G–I). In particular 22 units were excited by tilting the ipsilateral side down (α-response), while 17 units were excited by tilting the contralateral side down (β-response); moreover, 3 units were excited (γ-response), but no unit was inhibited (δ-response) by tilt in either direction. In all instances the magnitude of the responses was related to the degree of tilt. It is of interest that 25 out of these 42 axons affected by tilt could be antidromically excited by stimulating the ipsilateral NRL (Fig. 1A–C). Control experiments indicated that the spontaneous discharge of these spinoreticular tract neurons as well as their responses to tilt depended on influences arising from macular labyrinthine receptors.

In contrast with the observations made on the cervical spinoreticular tract neurons, only a small number of lumbar spinoreticular tract neurons responded to natural stimulation of macular labyrinthine receptors (Mergner et al., 1977).

In addition to the findings reported above, experiments were performed to determine whether changes in head and body position could modify the activity of NRL neurons (Ghelarducci et al., 1974a, b; Coulter et al., 1977a). In all, the activity of 221 NRL

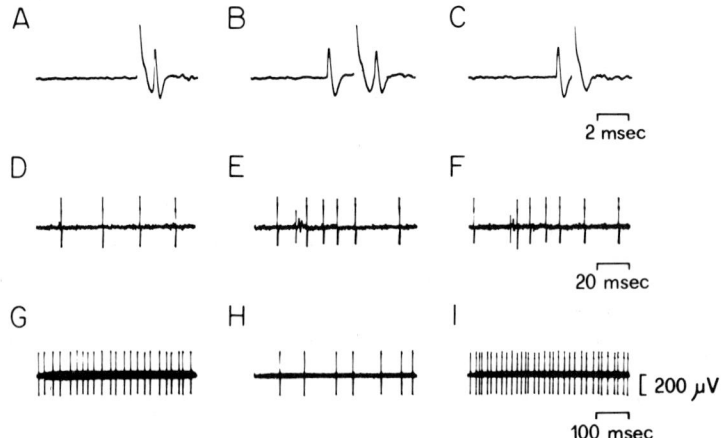

Fig. 1. Criteria for identification of a cervical spinoreticular tract unit. A–C) Antidromic action potential recorded extracellularly from one cervical spinoreticular tract neuron, following single shock stimulation of the ipsilateral NRL with 0.1 msec pulse, 10 V (A,B). The antidromic spike which follows the shock artifact at the latency of 0.79 msec collides with an orthodromic spike occurring 1.77 msec before the expected antidromic response (C). D–F) Spontaneous discharge (D) and response to stimulation of the ipsilateral (E) and contralateral (F) median nerve (0.2 msec pulses, 7.1 and 11.5 times the threshold for the ingoing volley resp. G–I) Samples of discharge recorded from the same unit while the animal was in the horizontal plane (G) and during 15° tilt on the ipsilateral (H) and contralateral side (I). (From Coulter et al., 1974.)

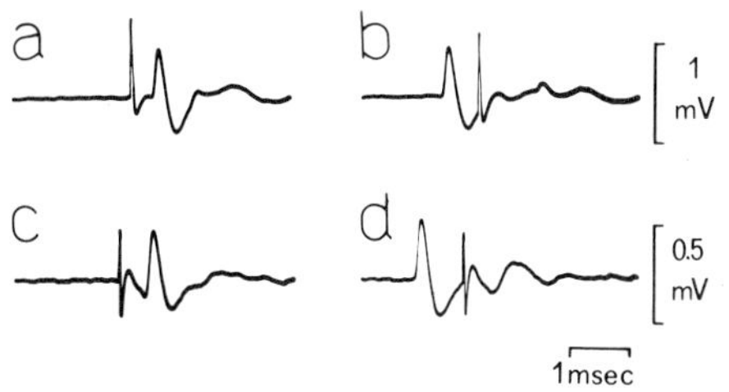

Fig. 2. Identification of reticulocerebellar neurons. Units recorded in precollicular decerebrate cats, following section of the right VIIIth nerve ipsilateral to the recording side, and complete section of the spinal cord at T10–T12. a,b) Unit recorded from ventral part of the right NRL, showing an α-response to tilt. Stimulation of the ipsilateral vermal cortex of the cerebellar anterior lobe with 0.05 msec pulse, 3 V evoked an antidromic spike at the latency of 0.39 msec (a), which collided with an orthodromic spike (b). c,d) Unit recorded from the dorsal part of the right NRL, showing an α-response to tilt. Stimulation of the cerebellar vermis with the same parameters as above produced an antidromic spike at the latency of 0.43 msec (c), which collided with an orthodromic spike (d). (From Pompeiano and Hoshino, 1977b.)

neurons was recorded in decerebrate cats with the cerebellum intact and the effects of lateral tilt on their activity investigated. Several of these neurons projected to the cerebellum, as they could be activated antidromically by electrical stimulation of the vermal cortex of the cerebellar anterior lobe (Fig. 2). In addition 53 neurons were recorded in decerebrate, cerebellectomized animals. On the whole, 94 out of 274 NRL units responded to lateral tilt of the head and body. In particular, 38 units showed an α-response, 49 units a β-response, 4 units a γ-response, and finally 3 units a δ-response. In every case the response was characterized by steady changes in unitary discharges which were maintained with varying degrees of adaptation for the duration of the tilt. Moreover, the magnitude of the response was related to the degree of tilt. Control experiments indicated that the responses depended upon stimulation of macular labyrinthine receptors. In summary, it appears that the labyrinthine input originating from macular receptors may influence the activity of cervical spino-reticulocerebellar neurons projecting to the vermal cortex of the cerebellar anterior lobe.

CROSSED MECHANISM TRANSMITTING THE MACULAR INPUT TO NEURONS OF THE SPINO-RETICULOCEREBELLAR PATHWAY

Lateral tilting acts predominantly on the utricular macula, whose major central projection is on the LVN (cf. Peterson, 1970). There are at least two mechanisms through which this nucleus may affect the NRL neurons. The first is by way of a vestibuloreticular projection, the second by vestibulospinal excitation of spino-reticular neurons terminating within the NRL. The first projection is considered to be uncrossed, since degenerating fibers appear within the NRL following lesion of the ipsilateral Deiters nucleus (Ladpli and Brodal, 1968). On the other hand, the second projection is assumed to be mainly crossed, since the majority of the crossed spino-reticular neurons are anatomically located within Rexed's laminae VII, VIII and IX (Corvaja et al., 1977a), i.e., in the same area where the lateral vestibulospinal tract (LVST) fibers terminate (Nyberg-Hansen and Mascitti, 1964). Moreover, stimulation experiments have shown that the LVST excites monosynaptically neurons of the presumably crossed spinoreticular pathway (Grillner et al., 1968; Rosén and Scheid,

1973b; Clendenin et al., 1974c), whereas the presumably uncrossed spinoreticular neurons escape this monosynaptic control (Clendenin et al., 1974a; cf. however Coulter et al., 1974, 1976).

The possibility that the LVN activates the ipsilateral NRL by way of a direct vestibuloreticular projection (Ladpli and Brodal, 1968) was not supported by the results of recent anatomical (Corvaja et al., 1977a) and physiological experiments (Pompeiano and Hoshino, 1977a, b).

The alternative possibility, i.e., that Deiters' nucleus affects the NRL by a crossed pathway, was explored by studying the response of NRL units to tilt in animals with section of the ipsilateral VIIIth nerve (Pompeiano and Hoshino, 1977a,b). In a first group of experiments, the animals were also submitted to complete section of the spinal cord at T10–T12. In a second group of experiments, in addition to the surgical procedures reported above, the animals were submitted to bilateral ablation of the cerebellar vermis and the fastigial nuclei ("cerebellectomized" preparations). In a third group of experiments, the animals were submitted to cerebellectomy as described above, but the spinal cord was cut between C1 and C2.

Among 101 units recorded from the right NRL, 36 units responded to lateral tilts of the head and body, in spite of the complete section of the VIII nerve ipsilateral to the recording side. In particular, responses to tilt were found in 23 of the 39 NRL neurons recorded in cats with the cerebellum intact (Fig. 3A) and in 13 of 29 neurons recorded after cerebellectomy (Fig. 3B). However, none of the 33 units recorded in cerebellectomized preparations with complete section of the spinal cord at C1–C2 responded to tilt (Fig. 3C). These findings indicate that the macular input can be transmitted from the intact labyrinth to the contralateral NRL by a crossed pathway, which does not necessarily pass through the cerebellum. The observation that crossed macular responses could still be observed in cerebellectomized preparations with transection of the spinal cord at T10–T12, but suppressed after section of the spinal cord C1–C2, indicates that the response of reticular neurons to volleys originating from the contralateral intact labyrinth utilizes a vestibulospinal mechanism, which acts on neurons of the crossed spinoreticular pathway located in the upper segments of the spinal cord.

NECK AFFERENT INPUT ON NEURONS OF THE SPINO-RETICULOCEREBELLAR PATHWAY

Coulter et al. (1974, 1976) had shown that all the 42 units attributed to the cervical spinoreticular tract neurons which were affected by tilt, except one, responded to stimulation of both ipsilateral and contralateral forelimb nerves (Fig. 1D–F). One of the units responsive to tilt was affected by stimulation of only the contralateral forelimb nerve; none of the units affected by tilt were exclusively activated by ipsilateral forelimb nerve stimulation. The same units which were influenced by stimulation of forelimb nerves of both sides, were also affected by stimulation of both ipsilateral and contralateral splenius nerves with stimulus intensities suprathreshold for the group II muscular afferents. The latency as well as the variability of these responses suggested that they depended on polysynaptic excitation of these neurons.

In addition to these findings, experiments were performed to find out whether reticulocerebellar neurons located in the NRL and submitted to tilt received a somatic

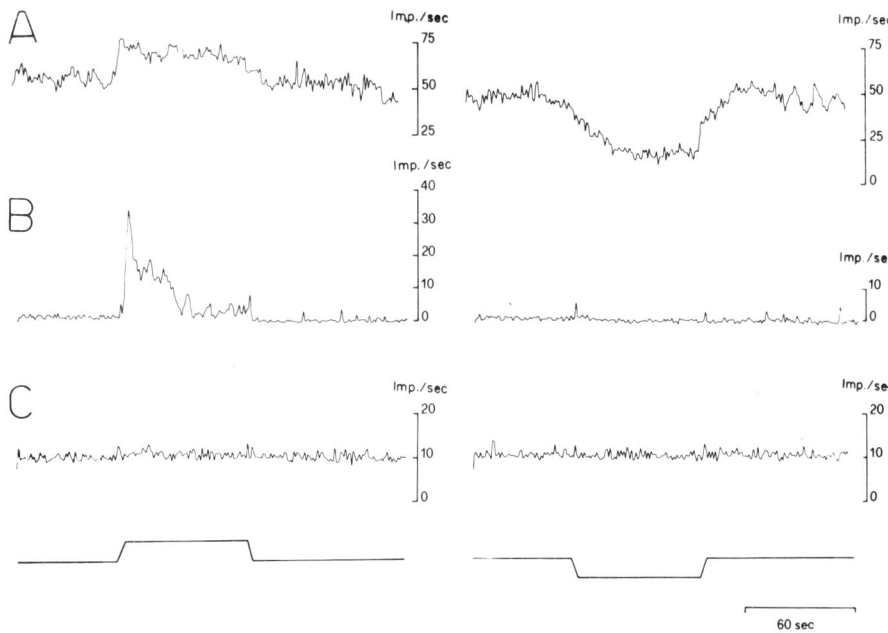

Fig. 3. Typical responses to lateral tilt of NRL units ipsilateral to the labyrinthine deafferentation. All the experiments were performed in decerebrate cats with section of the right VIIIth nerve. In A, the animal was also submitted to complete section of the spinal cord at T10–T12, while the cerebellum was intact (group I experiment). In B, the animal was submitted to section of the spinal cord at T10–T12, but in addition the vermal cortex of the cerebellum and the fastigial nucleus of both sides were completely removed (group II experiment). Finally, in C the animal was submitted to cerebellar ablation as described above and to complete section of the spinal cord between C1 and C2 (group III experiment). All units were recorded from the right NRL. The responses of these units to 20° tilt to the ipsilateral side (left records) and the contralateral side (right records) are shown as average sequential pulse density histograms, using 256 bins with 800 msec sampling intervals. Each record is the result of 3 trials. A) Typical α-response of a single unit located in the ventral part of the NRL. B) Asymmetrical response of a single unit located in the dorsal part of the NRL. C) Absence of response to tilt of a single unit located in the ventral part of the NRL. (From Pompeiano and Hoshino, 1977a.)

input not only from the limbs but also from the neck muscles (Coulter et al., 1977b). Among 110 NRL units affected by stimulation of forelimb and/or hindlimb nerves of the ispilateral and/or the contralateral side, 63 units (57.3%) responded also to stimulation of the splenius nerve of one or both sides. Moreover, most of these units (45/63) received a bilateral convergent input from the nuchal musculature, while the remaining units were affected only by stimulation of the ipsilateral or of the contralateral splenius nerve.

The threshold stimulus intensities which were able to elicit these unit responses were suprathreshold for the group II muscle afferents. After identification of the response of NRL neurons to peripheral nerve stimulation, the units were tested as to their responses to stimulation of macular labyrinthine receptors. In all, 46 out of the 110 units located in NRL (41.8%) responded to lateral tilt of the animal. In addition, 25 of these units were also influenced by stimulation of the nuchal afferents. If the NRL units were subdivided in two groups, the first one being affected by stimulation of the splenius nerves (63 units), the other not affected by this stimulation (47 units), it appeared that the proportion of the units influenced by tilting was not much different in the first group (25/63, 39.7%) with respect to the second (21/47, 44.7%). It is of interest that within the NRL all the units affected by tilt were influenced either by stimulation of the splenius nerves of both sides or by stimulation of the contralateral splenius nerve. In no instance were the units affected only by stimulation of the ipsilateral splenius nerve. In contrast, among the NRL units which were not affected by tilt, there were also units which responded to selective stimulation of the ipsilateral splenius nerve.

The neurons which were influenced by the neck input were located within both the magnocellular and the parvicellular part of the NRL. On the other hand, the area which received convergence of both the macular input and the somatosensory input from the neck musculature was located within the central part of the magnocellular NRL.

EFFECTS OF UNILATERAL LESIONS OF THE NRL OR UNILATERAL DEAFFERENTATION OF THE NECK ON POSTURE AND REFLEX MOVEMENTS

Experiments were performed to study the postural and motor deficits which occur following lesion of the NRL and to find out whether specific features of this syndrome could be attributed to interruption of tonic labyrinthine or cervical inputs impinging upon this nucleus (Corvaja et al., 1977b; Manzoni et al., 1978).

Unilateral lesions of the NRL produced marked motor deficits involving the entire body (Corvaja et al., 1977b). In particular, the cats were unable to stand or walk during the early, postoperative period. They began to right themselves and stand within 3 days, but these efforts frequently ended by falling in the contralateral direction (i.e., contralateral to the side of the lesion). By the fourth day, they attempted to walk in short steps with legs abducted; the cat often sought support of a wall on the contralateral side, but when unsupported, it showed circling movements.

This behavior was due in part to an asymmetric posture involving the whole limb musculature. In particular, there was an increased tonic contraction of the extensor muscles of the ipsilateral limbs (i.e., ipsilateral to the side of the lesion), which were thereby extended and abducted. The contralateral limbs, however, showed a decreased tonic contraction of the extensor muscles and even some tonic contraction of the flexor muscles (Fig. 4A, B). There was a moderate degree of head tilting to the ipsilateral side, being associated with a tonic contraction of the posterior cervical muscles of the ipsilateral side.

The tactile placing reactions were usually abolished in both ipsilateral limbs, but they were still present in the contralateral limbs. The proprioceptive placing reactions were depressed and more easily fatigued in the ipsilateral than in the contralateral limbs. These changes were more prominent in forelimbs than in hindlimbs.

The postural asymmetry described above was greatly reduced within 1–3 weeks after the NRL lesion, but a full compensation did not occur within the maximum period of observation (154 days). In spite of the improvement of posture, deficits in the tactile placing reactions usually persisted on the ipsilateral side throughout the survival period.

Specific aspects of the syndrome produced by selective lesion of the NRL could also be elicited following unilateral section of the dorsal roots C1–C3 (Manzoni et al.,

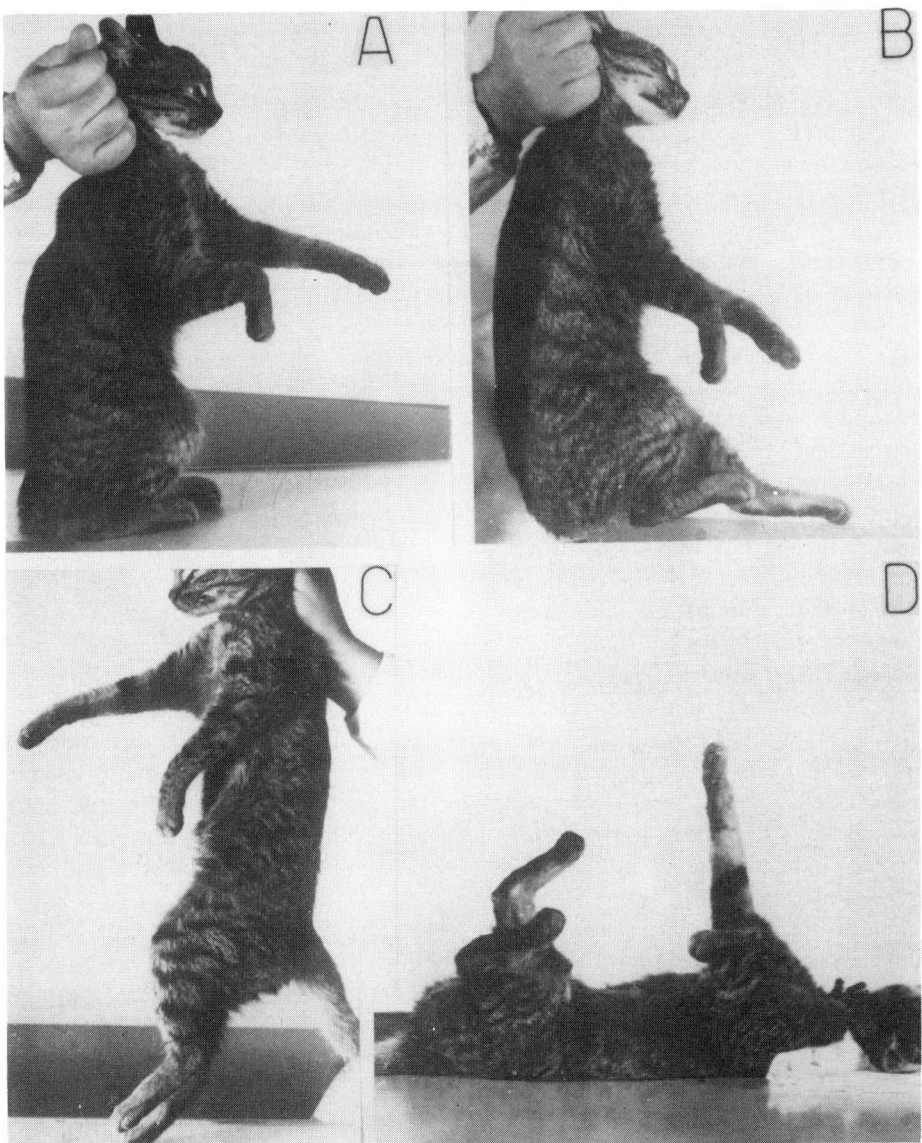

Fig. 4. Effects of unilateral lesion of the lateral reticular nucleus on posture in the unanesthetized as well as in the decerebrate preparation. Cat no. 10, submitted under Nembutal anesthesia to electrolytic lesion of the left NRL. A,B) 2 days after this lesion, the animal showed an increase of the extensor tonus in the ipsilateral limbs and a decrease of the extensor tonus associated with a mild flexor tonus in the contralateral limbs. Soon afterwards, the animal was lightly anesthetized with Nembutal again, and then submitted to unilateral ablation of the right hemivermis of the cerebellar anterior lobe. C) 1 day after the corticocerebellar ablation the animal showed a reversal of the postural asymmetry produced by the NRL lesion. D) The same cat following decerebration performed 2 days after the cerebellar lesion; the extensor rigidity was particularly present in the right limbs ipsilateral to the cerebellar ablation, but it was suppressed in the left limbs. (From Corvaja et al., 1977b.)

1978). In particular one day after the operation the cats were unable to stand and walk. They began to right themselves and stand within 1–2 days; however, the legs contralateral to the side of the deafferentation were unable to support the body weight under extension. When the animals attempted to walk, they tended to fall in the contralateral direction. By the third to the fifth day the animal was able to walk without much impairment, although the direction of the locomotor activity was still oriented towards the intact side.

Even in this group of experiments the behavior of the animal was in part due to an asymmetric posture, which was similar to that elicited by unilateral lesion of the NRL. In particular soon after the unilateral deafferentation of the neck, there was an increased tonic contraction of the expensor muscles of the ipsilateral limbs, and a decreased tonic contraction of the extensor muscles of the contralateral limbs (Fig. 5C–E). However, the head was slightly tilted to the contralateral side, due to tonic contraction of the posterior neck muscles of that side not compensated by the decreased tonic activity of the ipsilateral posterior neck muscles produced by the cervical deafferentation (Fig. 5A,B).

While these changes in head position correspond to those already described in the literature (cf. McNally, 1926), the postural asymmetry which involved the limb musculature in our experiments contrasts with that described by previous authors, who found that unilateral cervical deafferentation produced a decrease in postural activity of the ipsilateral limbs, and an increase in postural activity of the contralateral limbs (Biemond and De Jong, 1969). Our experiments therefore contradict their conclusion, i.e., that the postural asymmetry of the four limbs produced by unilateral cervical deafferentation was similar to that elicited by unilateral labyrinthectomy.

In addition to the postural asymmetry described above, the animal showed a suppression of the tactile placing reaction in both the fore- and the hindlimb ipsilateral to the side of the cervical deafferentation. On the other hand, the tactile placing reaction was still present in both the contralateral limbs. The proprioceptive placing reactions were depressed and more easily fatigued in the ipsilateral than in the contralateral limbs.

The postural changes described above were greatly compensated within 20 days after the cervical deafferentation, while the deficits in the tactile placing reaction were still present after this period. The postural asymmetry which involved the limb musculature following either lesion of the NRL or section of the dorsal roots C1–C3 of one side was reversed following: i) section of the ipsilateral VIII nerve, ii) electrolytic lesion of the ipsilateral LVN, or iii) ablation of the contralateral vermal cortex of the cerebellar anterior lobe (Fig. 4C,D). In contrast to these findings, the deficit in the tactile placing reaction was unchanged on the ipsilateral side throughout all of these subsequent operations. Moreover, the reversal of the postural asymmetry following the lesions reported above was still observed after decerebration.

DISCUSSION

Control of posture

The experiments reported above have shown that in decerebrate cats neurons of the cervical spino-reticulocerebellar pathway, which received a bilateral somatosensory input from the periphery, responded to macular labyrinthine stimulation (Coulter et

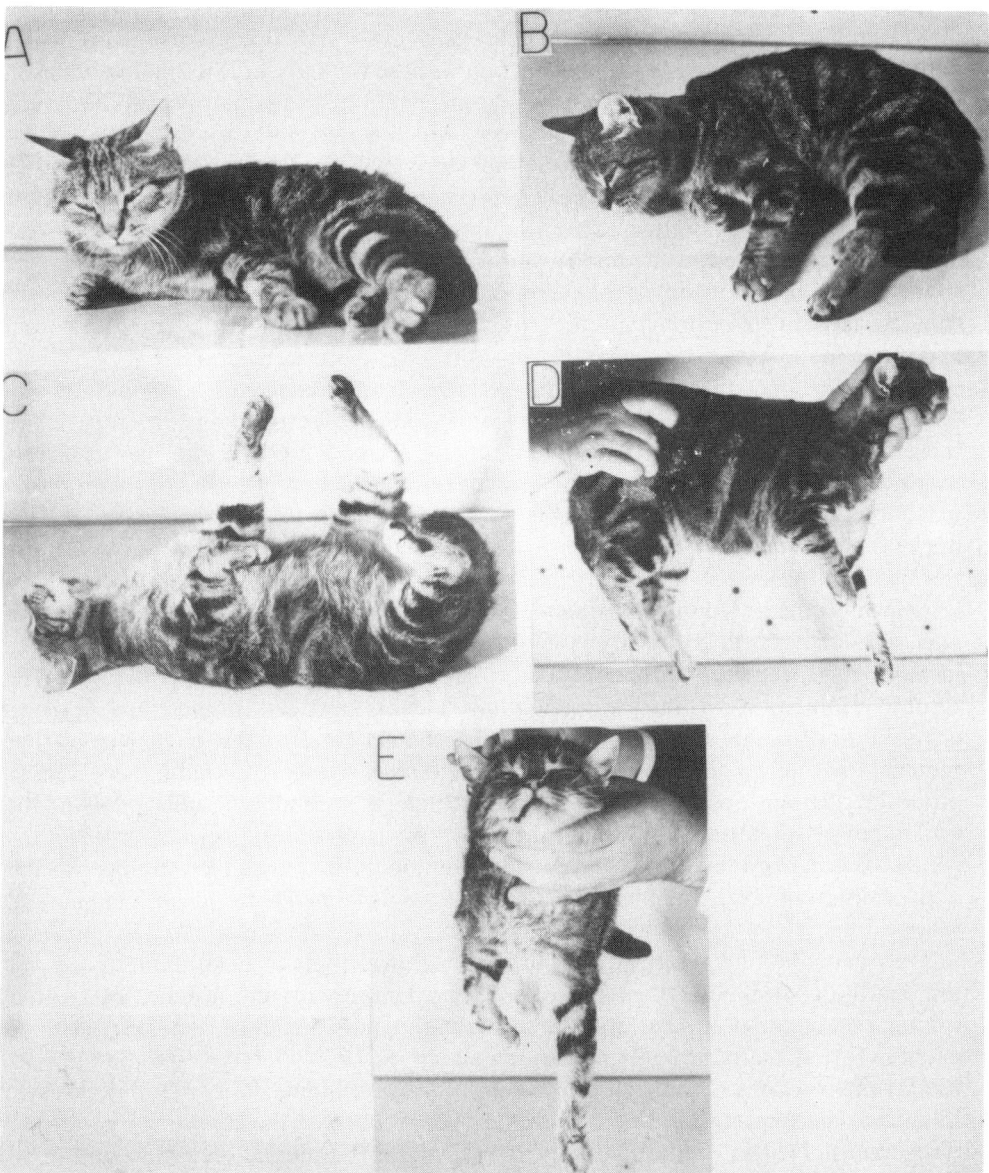

Fig. 5. Effects of unilateral deafferentation of the neck on posture in the unanesthetized preparation. Cat no. 3, submitted under Nembutal anesthesia to section of the dorsal roots C1–C3 of the left side. Photos taken 1 day after the cervical deafferentation. A,B) The animal tended to lie on the right side, with the contralateral right limbs flexed and the ipsilateral left limbs extended. There was also tilting of the head with the right side down. C–E) Both in supine (C) and in prone (D,E) position the animal showed an increase of the extensor tonus in the ipsilateral limbs and a decrease of the extensor tonus associated with a mild flexor tonus in the contralateral limbs. This postural asymmetry affected both fore- and hindlimbs and was observed also when the head of the animal was kept in symmetric position (C–E). (From Manzoni et al., 1978.)

al., 1974, 1976, 1977a; Ghelarducci et al., 1974a,b; cf. Pompeiano, 1975b). In particular, the macular labyrinthine input of one side affected the contralateral NRL by vestibulospinal volleys acting on neurons of the cervical crossed spinoreticular pathway (Pompeiano and Hoshino, 1977a,b). These findings indicate that in addition to the excitatory three-neuronal pathway represented by the primary vestibular afferents-LVN-ipsilateral extensor motoneurons, which is under the control of the macular labyrinthine receptors (Lund and Pompeiano, 1968; cf. Pompeiano, 1975a), there is a parallel three-neuronal pathway represented by the primary vestibular afferents-LVN-crossed spinoreticular tract neurons, through which the macular input of one side is transmitted to the contralateral NRL. This pathway is tonically active, as shown by the fact that unilateral lesion of the NRL produced a postural asymmetry characterized by hypertonia in the extensor muscles of the ipsilateral limbs and hypotonia in the extensor muscles of the contralateral limbs (Corvaja et al., 1977b).

There is evidence that the part of the NRL which receives the crossed spinoreticular path (Clendenin et al., 1974c) projects mainly to the corresponding vermal cortex of the cerebellar anterior lobe (Clendenin et al., 1974b). We postulated that unilateral lesion of the NRL produces a decreased background discharge of the ipsilateral vermal cortex of the anterior lobe, leading to disinhibition of the corresponding Deiters' nucleus. Since this last structure exerts a direct excitatory influence on ipsilateral extensor motoneurons, the activity in the extensor musculature ipsilateral to the NRL lesion will increase. On the other hand, the increased discharge of this Deiters nucleus may also activate neurons of the crossed spino-reticulocerebellar pathway, thus increasing the activity of the contralateral vermal cortex of the anterior lobe. The resulting inhibition of the corresponding Deiters nucleus would then lead to disfacilitation of the extensor motoneurons contralateral to the NRL lesion, thus reducing the activity in the extensor musculature of the corresponding side. These proposed relationships are supported by the results of secondary lesions involving the ipsilateral VIIIth nerve, the ipsilateral LVN or the contralateral vermal cortex of the cerebellar anterior lobe. In these cases interruption of the crossed vestibulocerebellar loop at different levels fully reversed the postural asymmetry produced by unilateral lesion of the NRL.

The existence of a long-loop reflex arc, through which the macular input of one side inhibits the contralateral Deiters nucleus, compensates for the absence of a direct inhibitory commissural connection between the Deiters nuclei of both sides, as indicated in recent anatomical (Pompeiano et al., 1978, 1979; Gacek, 1978) and physiological observations (Schimazu and Smith, 1971; cf. Shimazu, 1972). The crossed inhibitory mechanism described above may contribute to the reciprocal pattern of discharge of Deiters' neurons during lateral tilt of the animal (Hoshino and Pompeiano, 1977), as well as to the reciprocal changes in tonic contraction of the limb extensors which occur either during the asymmetric tonic labyrinthine reflexes (cf. Lindsay et al., 1976) or during the asymmetric posture produced by unilateral section of the VIIIth nerve (cf. Winkler, 1918) or by unilateral lesion of the cerebellar fastigial nucleus (Moruzzi and Pompeiano, 1956, 1957).

The NRL and the corresponding cerebellar vermis of the anterior lobe are under the tonic control not only of macular labyrinthine receptors but also of cervical receptors. Electrophysiological experiments have in fact shown that a large number of neurons located within the magnocellular and the parvicellular parts of the NRL, which project to the forelimb and the hindlimb regions of the cerebellar vermis (cf.

Corvaja et al., 1977a), responded to electrical stimulation of the neck muscle nerves of both sides (Coulter et al., 1977b). Moreover, lesion experiments have shown that a postural asymmetry of the limb musculature similar to that produced by unilateral lesion of the NRL could also be elicited by unilateral section of the dorsal roots C1–C3 (Manzoni et al., 1978). It is of interest that even in this instance the hypertonia in the extensor muscles of the ipsilateral limbs was apparently due to disfacilitation of the ipsilateral vermal cortex of the cerebellar anterior lobe, leading to disinhibition of the corresponding Deiters nucleus, whereas the hypotonia in the extensor muscles of the contralateral limbs was attributed to vestibulospinal activation of the crossed spinoreticulocerebellar pathway, leading to inhibition of the contralateral Deiters nucleus. These findings indicate that the neck input of one side exerts a tonic excitatory influence on the ipsilateral vermal cortex of the cerebellar anterior lobe, while the contralateral vermal cortex is depressed.

In summary, it appears that the macular labyrinthine input of one side exerts a predominant tonic excitatory influence on the contralateral NRL and the corresponding vermal cortex of the cerebellar anterior lobe, thus inhibiting the *contralateral* LVN, whereas the cervical input of one side exerts a predominant tonic excitatory influence on the ipsilateral NRL and the corresponding cerebellar vermis, thus inhibiting the *ipsilateral* LVN. This conclusion may explain why the response of the limbs in the tonic neck reflexes is just in the opposite sense to that in the tonic labyrinthine reflexes for the same direction of head and neck rotation. In fact, a side-down rotation of the neck produces relaxation of ipsilateral and contraction of contralateral limb extensors whereas side-down rotation of the head produces contraction of the ipsilateral and relaxation of the contralateral limb extensors (cf. Lindsay et al., 1976).

Control of reflex movements

In contrast with the neurons of the crossed cervical spino-reticulocerebellar pathway which received a bilateral somatosensory input from the periphery and were influenced by macular labyrinthine stimulation, neurons of the uncrossed cervical spino-reticulocerebellar pathway received only an ipsilateral somatosensory input from the periphery, and were not affected by the macular input (Coulter et al., 1974, 1976, 1977a; Ghelarducci et al., 1974a,b; cf. Pompeiano, 1975b). Interruption of this pathway following lesion of the NRL abolished the tactile placing reaction and greatly depressed the proprioceptive placing reaction ipsilaterally to the side of the lesion. This finding did not depend upon the postural changes produced by the NRL lesion since the reversal of the postural asymmetry produced by i) section of the VIIIth nerve ipsilateral to the NRL lesion, ii) lesion of the ipsilateral LVN, or iii) ablation of the vermal cortex of the contralateral anterior lobe, was not associated with a parallel reversal in the pattern of the tactile placing reactions produced by the reticular lesion.

There is evidence that the part of the NRL which receives the uncrossed spinoreticular path (Clendenin et al., 1974a) projects mainly to the ipsilateral intermediate cortex of the cerebellar anterior lobe (Clendenin et al., 1974b) and the underlying interpositus nucleus. In agreement with this finding is the observation that the tactile placing reaction depends in part upon the anatomical integrity of the ipsilateral interpositus nucleus (cf. Corvaja et al., 1977b, for references), which is known to exert an excitatory influence on the ipsilateral flexor motoneurons via the contralateral red nucleus (Pompeiano, 1967). On the other hand, lesions limited to the intermediate part of the anterior lobe cortex enhance the ipsilateral tactile placing reaction due to suppression of the inhibitory influence that this corticocerebellar region exerts on the interpositus nucleus (cf. Pompeiano, 1967).

It is of interest that a loss of the tactile placing reaction, similar to that elicited by unilateral lesion of the NRL, was also evoked by ipsilateral deafferentation of the neck. This effect can be attributed to disfacilitation of the ipsilateral NRL as well as of the corresponding interpositus nucleus, from which the interposito-rubrospinal pathway originates. This last effect would only partially be compensated by the disfacilitation of the intermediate cortex of the cerebellar anterior lobe, whose decreased activity would lead to some disinhibition of the corresponding interpositus nucleus. Recent experiments have shown that neurons located within the rostral part of the interpositus nucleus as well as Purkinje neurons located in the intermediate cortex of the cerebellar anterior lobe undergo steady changes in firing rate which are related to changes in neck position (Boyle and Pompeiano, 1979). If we assume that the neck input of one side projects on both the fore- and hindlimb regions of the ipsilateral NRL, one may explain why the response of these nuclear regions to the somatic input originating from the ipsilateral fore- and hindlimb are greatly impaired following ipsilateral deafferentation of the neck, thus leading to a suppression of the tactile placing reaction in the corresponding limbs.

SUMMARY

Neurons of the crossed spino-reticulocerebellar pathway originating from the cervical cord and passing through the NRL, responded to stimulation of macular labyrinthine receptors. This effect was attributed to vestibulospinal volleys acting on neurons of the crossed spino-reticulocerebellar pathway. The same neurons also received a bilateral somatosensory input from the periphery, including that from the neck musculature. In contrast with these findings, neurons of the uncrossed cervical spino-reticulocerebellar pathway did not respond to stimulation of macular receptors. These neurons received apparently an ipsilateral somatosensory input from the periphery including that from the ipsilateral neck muscles.

Unilateral lesion of the NRL produced a postural asymmetry, characterized by ipsilateral hypertonia and contralateral hypotonia of the limb extensor muscles. This effect was apparently due to interruption within the NRL of the crossed spinoreticulocerebellar pathway acting on the vermal cortex of the cerebellar anterior lobe. In particular, the ipsilateral hypertonus was attributed to disfacilitation of the ipsilateral vermal cortex of the anterior lobe, leading to disinhibition of the corresponding Deiters nucleus; on the other hand, the contralateral hypotonia was attributed to vestibulospinal activation of the crossed spino-reticulocerebellar pathway, which increased the activity in the contralateral vermal cortex of the anterior lobe thus inhibiting the corresponding Deiters nucleus.

A postural asymmetry similar to that elicited by unilateral lesion of the NRL was also elicited by section of the ipsilateral dorsal roots C1–C3.

It appears therefore that the macular input of one side exerts a predominant excitatory influence on the contralateral NRL and the corresponding vermal cortex of the cerebellar anterior lobe, thus inhibiting the contralateral Deiters nucleus, whereas the neck input of one side exerts a predominant excitatory influence on the ipsilateral NRL and the corresponding cerebellar vermis, thus inhibiting the ipsilateral Deiters nucleus.

In addition to the postural asymmetry described above, unilateral lesion of the NRL produced in the ipsilateral limbs a suppression of the tactile placing reaction and a reduction of the proprioceptive placing reactions. Similar results were also obtained following section of the ipsilateral dorsal roots C1–C3. This effect was attributed to interruption or disfacilitation of the uncrossed spinoreticulocerebellar pathway acting on the intermediate cortex of the cerebellar anterior lobe and the interpositus nucleus.

The NRL appears therefore to be composed of two more or less independent parts: a crossed one, related to major changes in postural tone, is under the combined control of both macular and cervical receptors, while the uncrossed one, restricted to discrete movements of the ipsilateral limbs, is under the selective control of the cervical receptors.

ACKNOWLEDGEMENTS

This investigation was supported by the Public Service Research Grant NS 07685-10 from the National Institute of Neurological and Communicative Disorders and Stroke, N.I.H., U.S.A. and by a research grant from the Consiglio Nazionale delle Ricerche, Italy.

REFERENCES

Biemond, A. and De Jong, J.M.B.V. (1969) On cervical nystagmus and related disorders. *Brain*, 92: 437–458.

Boyle, R. and Pompeiano, O. (1979) Sensitivity of interpositus neurons to neck afferent stimulation. *Brain Res.*, in press.

Clendenin, M., Ekerot, C.-F. and Oscarsson, O. (1974a) The lateral reticular nucleus in the cat. III. Organization of component activated from ipsilateral forelimb tract. *Exp. Brain Res.*, 21: 501–513.

Clendenin, M., Ekerot, C.-F., Oscarsson, O. and Rosén, I. (1974b) The lateral reticular nucleus in the cat. I. Mossy fiber distribution in cerebellar cortex. *Exp. Brain Res.*, 21: 473–486.

Clendenin, M., Ekerot, C.-F., Oscarsson, O. and Rosén, I. (1974c) The lateral reticular nucleus in the cat. II. Organization of component activated from bilateral ventral flexor reflex tract (b VFRT). *Exp. Brain Res.* 21: 487–500.

Corvaja, N., Grofová, I., Pompeiano, O. and Walberg, F. (1977a) The lateral reticular nucleus in the cat. I. An experimental anatomical study of its spinal and supraspinal afferent connections. *Neuroscience*, 2: 537–553.

Corvaja, N., Grofová, I., Pompeiano, O. and Walberg, F. (1977b) The lateral reticular nucleus in the cat. II. Effects of lateral reticular lesions on posture and reflex movements. *Neuroscience*, 2: 929–943.

Coulter, J.D., Mergner, T. and Pompeiano, O. (1974) Macular influences an ascending spinoreticular neurons located in the cervical cord. *Brain Res.*, 82: 322–327.

Coulter, J.D., Mergner, T. and Pompeiano, O. (1976) Effects of static tilt on cervical spinoreticular tract neurons. *J. Neurophysiol.* 39: 45–62.

Coulter, J.D., Mergner, T. and Pompeiano, O. (1977a) Effect of tilting on the responses of lateral reticular nucleus neurons to somatic afferent stimulation. *Arch. ital. Biol.*, 115: 294–331.

Coulter, J.D., Mergner, T. and Pompeiano, O. (1977b) Integration of afferent inputs from neck muscles and macular labyrinthine receptors within the lateral reticular nucleus. *Arch. ital. Biol.*, 115: 332–354.

Gacek, R.R. (1978) Location of commissural neurons in the vestibular nuclei of the cat. *Exp. Neurol.*, 59: 479–491.

Ghelarducci, B., Pompeiano, O. and Spyer, K.M. (1974a) Macular input to precerebellar reticular neurones. *Pflügers Arch.* 346: 223–231.

Ghelarducci, B., Pompeiano, O. and Spyer, K.M. (1974b) Activity of precerebellar reticular neurones as a function of head position. *Arch. ital. Biol.*, 112: 98–125.

Grant, G., Oscarsson, O. and Rosén, I. (1966) Functional organization of the spinoreticulocerebellar path with identification of its spinal component. *Exp. Brain Res.*, 1: 306–319.

Hoshino, K. and Pompeiano, O. (1977) Responses of lateral vestibular neurons to stimulation of contralateral macular labyrinthine receptors. *Arch. ital. Biol.*, 115: 237–261.

Ladpli, R. and Brodal, A. (1968) Experimental studies of commissural and reticular formation projections from the vestibular nuclei in the cat. *Brain Res.*, 8: 65–96.

Lindsay, K.W., Roberts, T.D.M. and Rosenberg, J.R. (1976) Asymmetric tonic labyrinth reflexes and their interaction with neck reflexes in the decerebrate cat. *J. Physiol. (Lond.)*, 261: 583–601.

Lund, S. and Pompeiano, O. (1968) Monosynaptic excitation of alpha motoneurones from supraspinal structures in the cat. *Acta physiol. scand.*, 73: 1–21.

Lundberg, A. and Oscarsson, O. (1962) Two ascending spinal pathways in the ventral part of the cord. *Acta physiol. scand.*, 54: 270–286.

MacNally, W.J. (1926) Welche cervicalen Wurzeln beteiligen sich an dem Zustandekommen der Kopfdrehung nach einseitiger Labyrinthexstirpation? *Pflügers Arch.*, 213: 673–684.

Manzoni, D., Pompeiano, O. and Stampacchia, L. (1979) Tonic cervical influences on posture and reflex movements. *Arch. ital. Biol.*, 117: 81–110.

Mergner, T., Pompeiano, O. and Coulter, J.D. (1977) Response of lumbosacral spinoreticular tract neurons to body tilt. *Proc. XXVII int. Congr. physiol. Sci., Paris*, XIII: 502, no. 1485.

Moruzzi, G. and Pompeiano, O. (1956) Crossed fastigial influence on decerebrate rigidity. *J. comp. Neurol.*, 106: 371–392.

Moruzzi, G. and Pompeiano, O. (1957) Inhibitory mechanisms underlying the collapse of decerebrate rigidity after unilateral fastigial lesion. *J. comp. Neurol*, 107: 1–25.

Nyberg-Hansen, R. and Mascitti, T.A. (1964) Sites and mode of termination of fibers of the vestibulospinal tract in the cat. An experimental study with silver impregnation methods. *J. comp. Neurol*, 122: 369–387.

Oscarsson, O. (1973) Functional organization of spinocerebellar paths. In *Handbook of Sensory Physiology, Vol. II, Somatosensory System*, A. Iggo (Ed.). Springer-Verlag, Berlin, pp. 339–380.

Oscarsson, O. and Rosén, I. (1966) Response characteristics of reticulocerebellar neurones activated from spinal afferents. *Exp. Brain Res.* 1: 320–328.

Peterson, B.W. (1970) Distribution of neural responses to tilting within vestibular nuclei of the cat. *J. Neurophysiol.*, 33: 750–767.

Pompeiano, O. (1967) Functional organization of the cerebellar projections to the spinal cord. In *Progress in Brain Research, Vol. 25, The Cerebellum*, C.A. Fox and R.S. Snider (Eds.), Elsevier, Amsterdam, pp. 282–321.

Pompeiano, O. (1975a) Vestibulo-spinal relationships. In *The Vestibular System*, R.F. Naunton (Ed.) Academic Press, New York, pp. 147–184.

Pompeiano, O. (1975b) Macular input to neurons of the spinoreticulocerebellar pathway. *Brain Res.*, 95: 351–368.

Pompeiano, O. and Hoshino, K. (1977a) Macular labyrinthine input to the contralateral lateral reticular nucleus. *Brain Res.*, 131: 147–151.

Pompeiano, O. and Hoshino, K. (1977b) Responses to static tilt of lateral reticular neurons mediated by contralateral labyrinthine receptors. *Arch. ital. Biol.*, 115: 211–236.

Pompeiano, O., Mergner, T. and Corvaja, N. (1978) Commissural, perihypoglossal and reticular afferent projections to the vestibular nuclei in the cat. An experimental anatomical study with the method of the retrograde transport of horseradish peroxidase. *Arch. ital. Biol.*, 116: 130–172.

Pompeiano, O., Mergner, T. and Corvaja, N. (1979) Commissural connections between the vestibular nuclei studied with the method of retrograde transport of horseradish peroxidase. In *Structure and Function of Cerebral Commissures*, I. Steele Russell, M.W. van Hof and G. Berlucchi (Eds.) Macmillan, New York, pp. 319–332.

Rosén, I. and Scheid, P. (1973a) Patterns of afferent input to the lateral reticular nucleus of the cat. *Exp. Brain Res.*, 18: 242–255.

Rosén, I. and Scheid, P. (1973b) Responses to nerve stimulation in the bilateral ventral flexor reflex tract (bVFRT) of the cat. *Exp. Brain Res.*, 18: 256–267

Rosén, I. and Scheid, P. (1973c) Responses in the spino-reticulo-cerebellar pathway to stimulation of cutaneous mechanoreceptors. *Exp. Brain Res.*, 18: 268–278.

Shimazu, H. (1972) Organization of the commissural connections: physiology. In *Progress in Brain Research, Vol. 37, Basic Aspects of Central Vestibular Mechanisms*, A. Brodal and O. Pompeiano (Eds.), Elsevier, Amsterdam, pp. 177–190.

Shimazu, H. and Smith, C.M. (1971) Cerebellar and labyrinthine influences on single vestibular neurons identified by natural stimuli. *J. Neurophysiol*, 34: 493–508.

Winkler, C. (1918) The central course of the nervus octavus and its influence on motility. In *Opera Omnia, IV*. Bohn, Haarlem, pp. 357–481.

Neck and Macular Labyrinthine Influences on the Purkinje Cells of the Cerebellar Vermis

F. DENOTH[2], P. C. MAGHERINI[1], O. POMPEIANO[1] and M. STANOJEVIĆ[1]

[1]*Istituto di Fisiologia Umana, Cattedra II, Università di Pisa; and* [2]*Istituto di Elaborazione dell' Informazione, C.N.R., 56100 Pisa (Italy)*

INTRODUCTION

There is evidence that the tonic labyrinthine reflexes may act asymmetrically on forelimb extensors, in that side-down rotation of the head produces contraction, whereas side-up rotation of the head results in relaxation of the ipsilateral forelimb extensors (Lindsay et al., 1976). The converse responses can be observed during the tonic neck reflexes; in fact, side-down rotation of the neck results in a relaxation, whereas side-up rotation of the neck produces contraction of the ipsilateral forelimb extensors (Magnus and De Kleijn, 1912; McCouch et al., 1951; Lindsay et al., 1976; cf. also Thoden and Wenzel, this volume, chapter II4. In agreement with these findings are the results of lesion experiments showing that the postural asymmetry produced by unilateral section of the VIIIth nerve is characterized by ipsilateral extensor hypotonia and contralateral hypertonia, while that elicited by unilateral section of the dorsal roots C1–C3 is just opposite in sign (cf. Manzoni et al., 1979).

The observation that the neck and labyrinthine reflexes act in opposition may explain why changes in the position of the head leave the position of the limbs unchanged. The intervention of the opposed action of the labyrinthine and neck reflexes would in fact provide an effective stabilizing action of the limbs on the trunk, while permitting movements of the head with respect to the body (Roberts, 1973).

The experiments summarized in the present report were performed to investigate whether the vermal cortex of the cerebellar anterior lobe is able of integrating both signals which originate from macular receptors as well as those which originate from the neck as a result of the displacement of the head (Magherini et al., 1978; Denoth et al., 1979a, b). Afferent volleys originating from both neck (Berthoz and Llinás, 1974) and macular labyrinthine receptors (cf. Erway et al., 1978) may indeed reach the vermal cortex of the cerebellar anterior lobe; however, no attempt was made so far to find out whether the conflict between the asymmetrical influences originating from the labyrinth and the neck was in part at least resolved at the level of the cerebellar Purkinje(P) cells.

METHODS

The experiments were performed in 25 cats, subjected to precollicular decerebration under ether anesthesia. The dorsal neck muscles of both sides were disconnected from the occipital bone and partially removed in order to visualize the dorsal apophysis of the second cervical vertebra, which was later used for clamping the vertebral axis. The ventral rami of C1–C3 were then crushed in order to denervate the remaining large neck muscles as well as the skin of the neck.

The head was fixed in a stereotaxic frame and the axis vertebra was held at the level of the spinous process of the second cervical vertebra by a clamp rigidly fixed to the tilting table. The trunk was mounted on a spinal cord frame and both fore- and hindlimbs were extended and clamped, thus avoiding effects due to stimulation of receptors other than the labyrinthine and the neck receptors.

Fig. 1 schematically represents the three stimulating procedures used in the experiments. Rotating the neck clamp and table simultaneously, while maintaining the horizontal position of the head, elicited a neck input probably arising from the atlanto-occipital and atlanto-axial joints and/or from small neck muscles, in the absence of a labyrinthine input (Fig. 1, neck input). Rotating the stereotaxic equipment and table together along the longitudinal axis of the animal, elicited a labyrinthine input in the absence of a neck input (Fig. 1, macular input). Finally, if the head was rotated while the vertebral clamp remained fixed on the table in the horizontal position, both labyrinthine and neck inputs were elicited (Fig. 1, neck + macular inputs).

Sinusoidal rotation of the neck and/or the head along the longitudinal axis was made possible through independent hydraulic systems which were driven electronically. Control experiments were taken in cats paralyzed with gallamine triethiodide

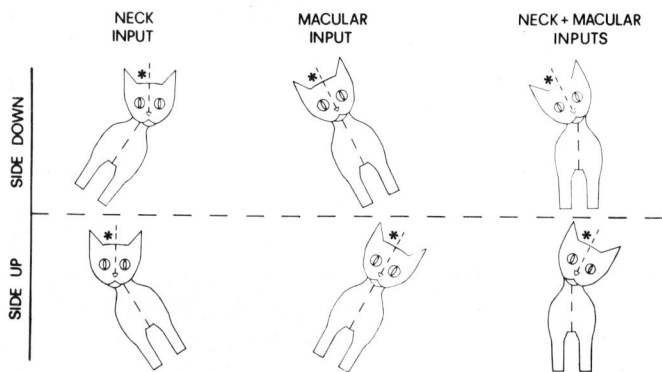

Fig. 1. Schematic representation of different head-body positions of the animal leading to selective stimulation of neck receptors, macular receptors and both neck and macular receptors. Neck input: the head of the animal was fixed in horizontal position with a head-holder, while the spinous process of the second cervical vertebra, held through the tilting table by a vertebral clamp, was moved about the longitudinal axis of the animal, thus leading to selective stimulation of neck receptors. Macular input: both the head-holder and the vertebral clamp were fixed on the tilting table, which was then moved about the longitudinal axis of the animal, thus leading to selective stimulation of macular receptors. Neck + macular inputs: the head of the animal was rotated, while the vertebral clamp, fixed on the tilting table, was maintained in horizontal position. In this case head movement led to combined stimulation of both neck and macular receptors, due to tilting of the head and rotation of cervical axis vertebra. Asterisks indicate the side of the recording, whereas the upper and the lower figures represent the relative position of the head with respect to the body during side-down or side-up rotation of the neck (neck input), of the head (macular input), and of both the neck and the head (neck + macular inputs). (From Denoth et al., 1979b.)

(Flaxedil, 3–4 mg/kg, i.v.) and artificially ventilated. Body temperature was maintained within the range of 37°–38°C throughout the experiment.

A stimulating bipolar electrode was stereotaxically positioned in the ipsilateral lateral vestibular (Deiters') nucleus (LVN), in order to activate antidromically some of the recorded P-cells and its position was verified at the beginning of the experiment by pronounced ipsilateral forelimb extension during stimulation, as well as by histological controls taken at the end of the experiment.

Neural activity from the vermal cortex of the cerebellar anterior lobe was recorded extracellularly with glass microelectrodes (5–10 MΩ impedance). The units were identified as P-cells by the presence of single and complex spikes, due to the activity of the mossy fiber(MF) and the climbing fiber(CF) pathways, respectively. These units were then tested during neck, macular or neck + macular stimulation.

The frequency of stimulation varied from 0.015 Hz up to 0.15 Hz, for amplitudes ranging from 1° to 10° in both directions of the median plane for the neck input and from 2° to 15° for the macular input. The unit activity was selected through a window discriminator and analyzed by a digital signal averager (Correlatron 1024, Laben), which provided sequential pulse density histograms. Synchronizing pulses permitted superimposition and averaging of several identical stimulus cycles, while the neck or the head position was recorded in a separate channel of the Correlatron. The analogue output of this signal averager was plotted on an X–Y plotter (Hewlett-Packard 7035B) while the digital data printed through a punch-tape. Further processing of the data was performed off-line by means of a computer system HP 2100 A equipped with a HP 5451 B Fast Fourier Analyzer. The computer evaluated the base (mean) discharge frequency of the unit during any experimental condition. A spectral analysis of the input (neck or head rotation) and of the output (unit activity) was then performed and the correlation parameters between the two spectra (in particular the sensitivity and the phase shift of the first harmonic of the output with respect to the input) were evaluated by cross-spectral analysis.

The sensitivity of the response was expressed in percentage change of the base frequency per degree of displacement, while the phase of the response was evaluated with respect to the peak of the ipsilateral side-down position of the neck or of the head. We considered as positive only those units whose response to successive cycles showed a coherence corresponding to or higher than 0.8 and a sensitivity of the first harmonic equal to or greater than 0.5.

RESULTS

The electrical activity of P-cells was recorded from the right hemivermis of the cerebellar anterior lobe. Most of the recorded units were located within a longitudinal parasagittal zone, 0.5–1.5 mm in width, from which an antidromic field potential could be elicited following single shock stimulation of the ipsilateral LVN. This zone contributes to the cerebellar corticovestibular projection to LVN(Corvaja and Pompeiano, 1979), as shown by the fact that some of the P-cells recorded from this longitudinal strip could be antidromically activated following single shock stimulation of Deiters' nucleus. The criteria used to determine whether a recorded unit was activated antidromically were: i) a fixed latency at threshold strength, ii) the ability to follow high frequency, iii) the collision of antidromic spikes with orthodromically evoked discharges.

Response of Purkinje cells to the neck input

The responses of 95 P-cells during sinusoidal rotation of the axis vertebra were examined. Among them, 83 units were tested for a MF-response, while 12 units were tested for a CF-response. 31 out of 83 units showed a MF-response, while 4 out of 12 units showed a CF-response to the neck input at the frequency of stimulation of 0.026 Hz and peak displacement amplitude of 5° or 10°. The responses consisted in a periodic modulation of the firing rate of the P-cells in response to the sinusoidal neck input. Fig. 2 illustrates the MF-response of a P-cell to the neck input at the parameters of stimulation of 0.026 Hz, 5° and the corresponding power spectrum, while Fig. 3 (left column) shows the first harmonic of the response as evaluated by the computer.

In all the P-cells affected by the neck input, the threshold amplitude of rotation responsible for the MF-induced responses varied from 1° to 3° at the frequency of 0.026 Hz. Moreover, the amplitude of the modulation increased by increasing the amplitude of the neck displacement, so that the sensitivity of the units did not change for increasing amplitude of stimulation from 1° to 10° (Fig. 4, left column).

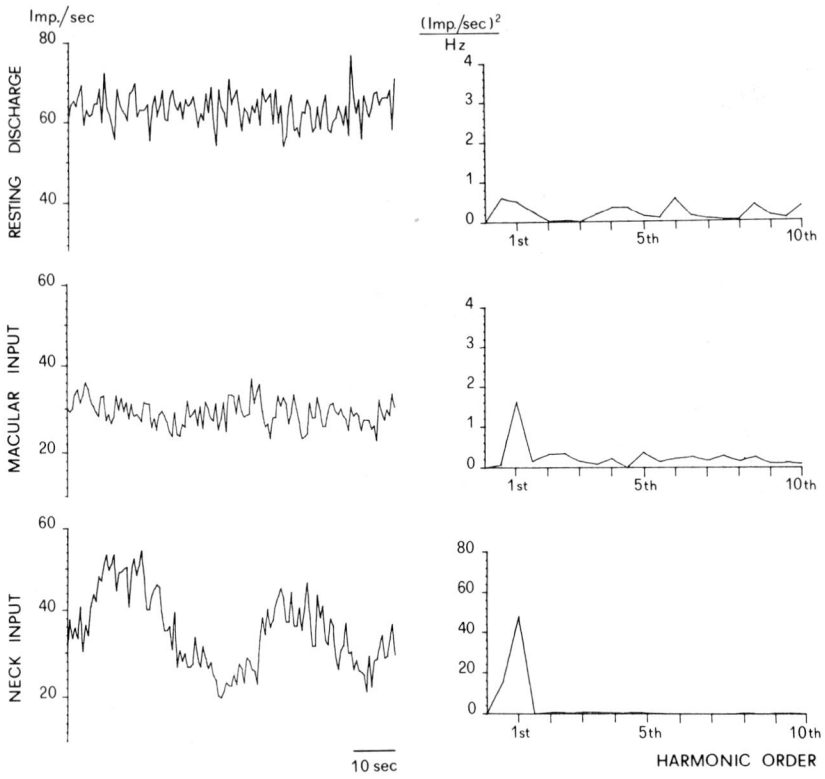

Fig. 2. Power spectrum of the MF-response of a P-cell to sinusoidal stimulation of neck receptors and macular receptors. Left records: sequential pulse density histograms showing the MF-discharge of a P-cell at rest as well as the MF-response of the same unit to sinusoidal rotation of the head (macular input) or of the neck (neck input). In both instances the frequency of stimulation corresponded to 0.026 Hz for peak amplitudes of 10° (macular input) and 5° (neck input) resp. 5 sweeps were accumulated for each of the computer records, using 128 bins with the dwell time of 0.6 sec/bin. Right diagrams: power spectrum up to the tenth harmonic of the previously illustrated records showing the preponderance of the first harmonic on higher order harmonics and the greatest magnitude of the neck response with respect to the macular response.

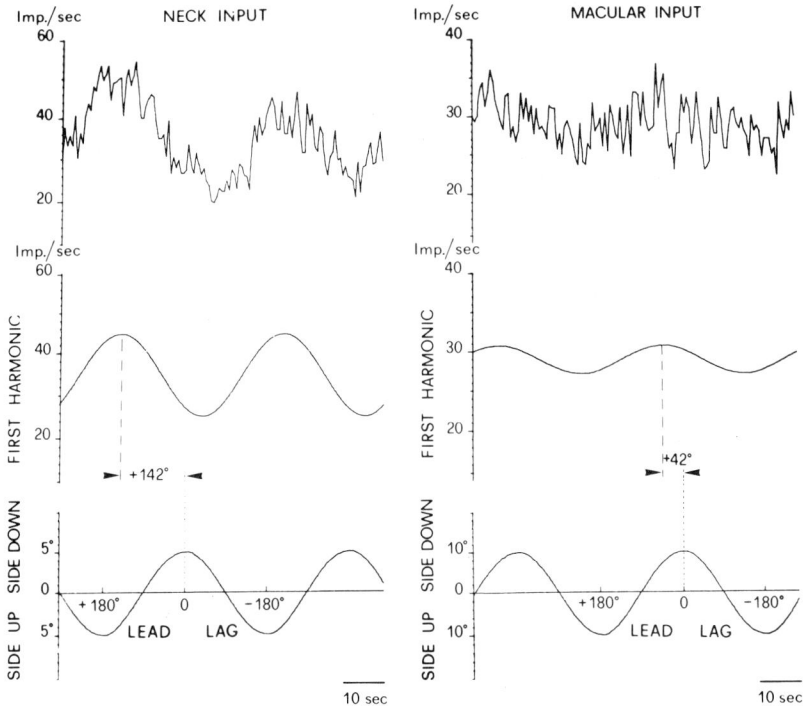

Fig. 3. Evaluation of the pattern and magnitude of the MF-response of an individual P-cell to sinusoidal stimulation of neck receptors and macular receptors. Upper records: MF-response of the same P-cell illustrated in Fig. 2 following sinusoidal stimulation of neck receptors or macular receptors at the frequency of 0.026 Hz, for peak amplitudes of 5° and 10° resp. Middle records: first harmonic of the MF-response of the P-cell to the neck input and to the macular input as indicated above. Lower records: traces indicating the displacement of the neck or of the head. Side-down rotations of the neck (neck input) or of the head (macular input) are indicated by upward deflections in the sine wave. The sensitivity of the MF-response of the P-cell to the neck input corresponded to 5.6, while that for the macular input corresponded to 0.6 (the base frequency being 35 imp/sec during neck stimulation and 29 imp/sec during macular stimulation). The phase (ϕ) of the neck response had a lead of +142° with respect to the side-down rotation of the neck, whereas that of the macular response has a lead of +42° with respect to the side-down rotation of the head.

The sensitivity of the MF-response of the P-cells to neck rotation evaluated at the frequency of stimulation of 0.026 Hz, 5°–10°, varied from 0.8 to 8.0 imp/sec/degree expressed in percent of the base frequency, with a mean value of 2.71 ± 1.67(SD).

In addition to the sensitivity of the unit response to the neck input, the phase of the response relative to the side-down displacement of the neck has also been evaluated (Fig. 3, left column). Both the sensitivity as well as the phase shift characterizing any MF-response to sinusoidal rotation of the neck at the frequency of stimulation of 0.026 Hz, 5°–10°, have been plotted in a polar diagram (Fig. 5). 17 MF-units increased their firing rate during side-down rotation of the neck (the phase of the response ranging from +45° lead to −45° lag) whereas 10 MF-units were excited by a side-up rotation of the neck (the phase of the response ranging from +135° lead to −135° lag). The remaining unit responses were not related to the position but rather to the velocity of neck displacement, since the corresponding phase relative to side-down position of the neck varied from +45° to +135° lead or else from −45° to −135°

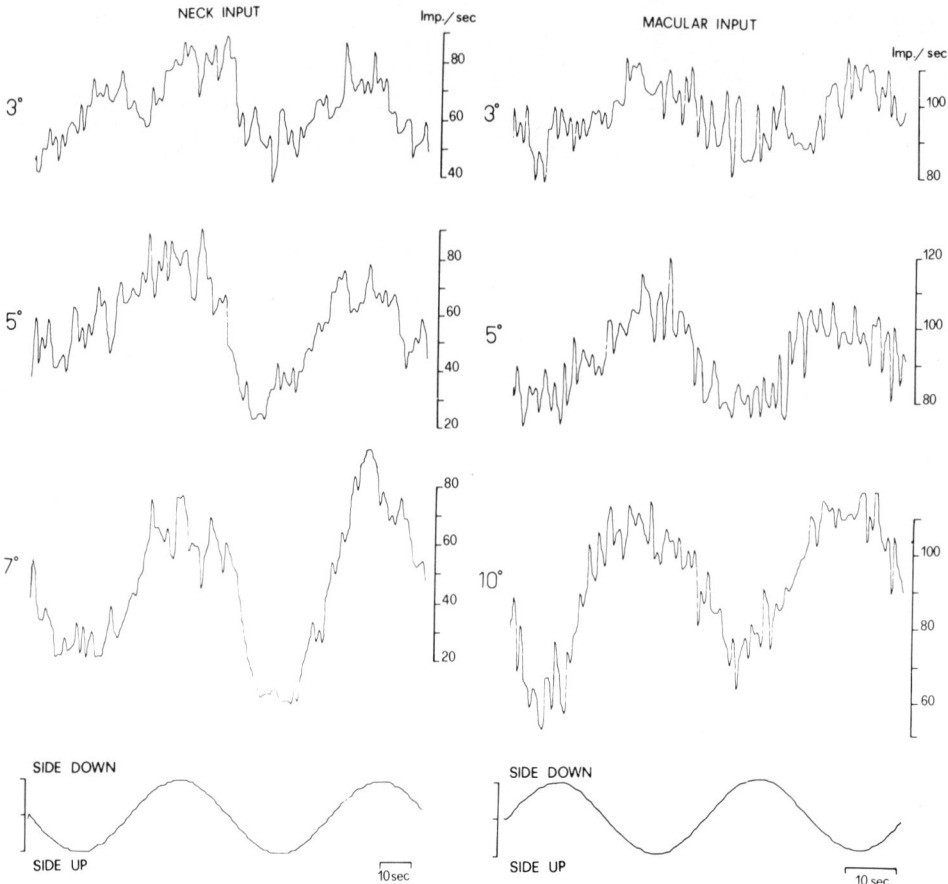

Fig. 4. MF-response of a P-cell to increasing amplitudes of neck and macular stimulation. Left column: sequential pulse density histograms showing the MF-response of a P-cell during sinusoidal rotation of the neck at the frequency of 0.015 Hz and at progressively increasing amplitudes. 6 sweeps were accumulated for each of the computer records, using 128 bins with the dwell time of 1 sec/bin. The unit showed an excitation during side-down rotation of the neck. For 3°, 5° and 7° of peak displacement the sensitivity of the response corresponded on the average to 7.2, while ϕ corresponded to $+26°$, $+32°$ and $+2°$ resp. Right column: sequential pulse density histograms showing the MF-response of the same P-cell during sinusoidal rotation of the head at the frequency of 0.026 Hz and at progressively increasing amplitudes. The unit showed a depression during side-down rotation of the animal. For 3°, 5° and 10° of tilting the sensitivity of the response corresponded on the average to 2.6, while ϕ corresponded to $-167°$, $-147°$ and $-150°$. 10 sweeps were accumulated for each of the computer records, using 128 bins with the dwell time of 0.6 sec/bin.

lag. Finally, among the 4 P-cells showing a CF-response to the neck input, 3 were excited by side-down rotation of the neck, while 1 was excited by side-up rotation.

Changes in amplitude of stimulation from 1° to 10° did not modify the phase of the unit response relative to the side-down position of the neck (Fig. 4, left column).

Response of Purkinje cells to the macular input

Among the 95 P-cells tested for neck stimulation, 70 P-cells were tested during sinusoidal stimulation of labyrinthine receptors. 19 out of 65 units showed a MF-response, while 2 out of 5 units showed a CF-response to the labyrinthine input at the frequency of stimulation at 0.026 Hz, and at the peak amplitude of displacement of

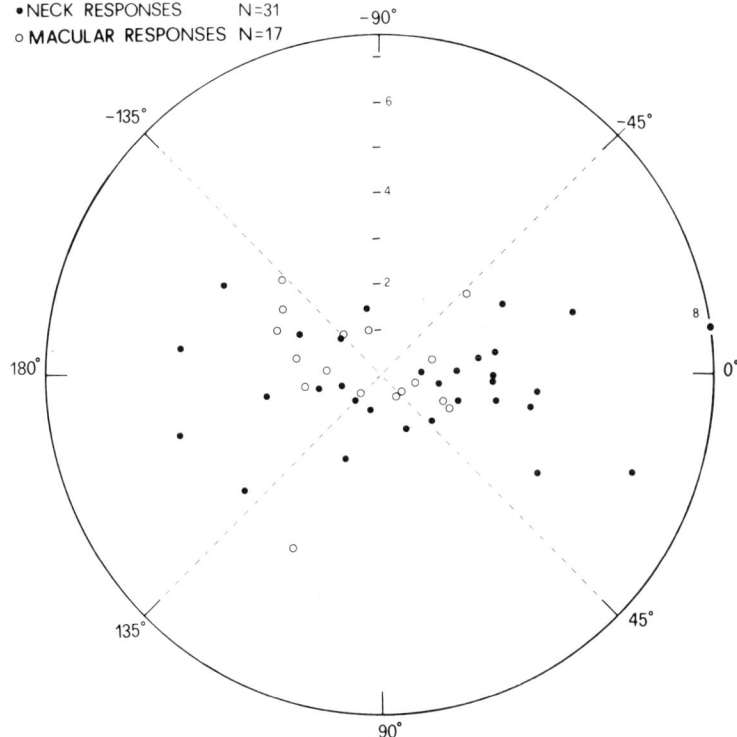

Fig. 5. Polar diagram showing the sensitivity and the phase of the MF-responses of P-cells to neck and macular labyrinthine stimulation. The sensitivity of the MF-response of the P-cells to neck (●) and macular labyrinthine stimulation (○) is indicated by the relative distance of the symbols from the center of the diagram (only one unit had the sensitivity of 8, i.e., higher than that indicated in the scale along the vertical meridian). The relative position of the symbols with respect to the 0° meridian indicates, in degrees, the phase lead (positive values) or the phase lag (negative values) of the response with respect to the peak of the ipsilateral side-down displacement of the neck or the head. On the whole the responses of 31 P-cells to neck stimulation and 17 P-cells to macular stimulation have been plotted. The responses were evaluated at the frequency of stimulation of 0.026 Hz and at 5° or 10° of peak displacement, resp. (From Denoth et al., 1979b.)

10° or 15°. The responses consisted in a periodic modulation of the firing rate of the P-cells related to the sinusoidal tilt. Fig. 2 illustrates the MF-response of a P-cell to the labyrinthine input at the parameters of stimulation of 0.026 Hz, 10° and the corresponding power spectrum, while Fig. 3 (right column) shows the first harmonic of the response.

The threshold amplitude of tilt eliciting MF-induced responses varied from 2° to 5° at the frequency reported above. Moreover, the amplitude of the modulation was closely related to the amplitude of tilt (Fig. 4, right column), so that the sensitivity of the units did not change for increasing amplitudes of stimulation from 2° to 15°.

The sensitivity of the MF-response of the P-cells to the animal's tilt evaluated at the frequency of stimulation of 0.026 Hz, 10°, varied from 0.6 to 4.3 imp/sec/degree expressed in percent of the base frequency, with a mean value of 1.71 ± 1.01(SD). These findings indicate that the sensitivity of the P-cells to the labyrinthine input is lower than that for the neck input, this difference being statistically significant (Student t-test, $P < 0.05$).

The polar diagram shown in Fig. 5 illustrates the sensitivity and the phase of the MF-response of the P-cells relative to the side-down position of the whole animal. This figure shows that only 6 MF-units were excited by side-down tilt of the animal, whereas 8 MF-units were excited by side-up tilt; these responses were attributed to stimulation of macular receptors. On the other hand, 3 units did not respond to position but rather to velocity of animal displacement; 2 units were not plotted in the polar diagram since they were excited during both side-down and side-up tilts. Finally, among the 2 P-cells showing a CF-response to the labyrinthine input, one unit was excited by side-up tilt and the other unit responded to velocity of tilt. Changes in amplitude of tilting did not modify the phase of the unit response relative to the side-down position of the whole animal (Fig. 4, right column).

Conflicting neck-macular vestibular stimulation

Among the 83 MF-units tested during neck stimulation, 64 units were also tested to macular labyrinthine stimulation. 30 out of these units were sensitive to these inputs; in particular, 12 units (40%) responded only to stimulation of the neck receptors, 4 units (13.3%) to stimulation of the macular receptors and 14 units (46.7%) to stimulation of both neck and macular receptors. Even within this last population of units the sensitivity of the MF-response to the neck input (mean ± SD: 2.44 ± 1.56) was on the average higher than that obtained by the same units during macular stimulation (mean ± SD: 1.53 ± 0.75). The difference between the two mean values was statistically significant (Student t-test, $P < 0.05$).

Most of the MF-units which received a convergent input from neck and macular receptors showed a reciprocal behavior in response to the two inputs, the majority of the units being excited during side-down rotation of the neck but inhibited by side-down tilt of the whole animal. However, the responses were never 180° out of phase with respect to each other.

P-cells which showed a MF-response to the neck input and to the macular input were also submitted to rotation of the head alone. In this instance, both labyrinthine and neck receptors were stimulated. An example of the interacting influences of the two inputs on an individual P-cell is shown in Fig. 6. In this figure the MF-responses of a P-cell to independent neck and macular stimulation are illustrated; the unit was excited by side-down rotation of the neck (sensitivity 6.0; phase lead +21°) but it was inhibited by side-down tilt of the whole animal (sensitivity 2.6; phase lag −147°). Side-down rotation of the head alone, leading now to conflicting stimulation of both neck and macular receptors, produced a response whose sensitivity (3.2) closely corresponded to the algebraic summation of the two responses. However, the peak of the response occurred in this case with a lead of +60° with respect to the side-down rotation of the neck, thus becoming more related to the velocity of neck rotation than to the neck position.

DISCUSSION

Stimulation of macular labyrinthine receptors produced by tilt of the animal elicits strong influences on extensor motoneurons innervating the neck, the trunk and the limb musculature, thus contributing to stabilize the head position and maintain the upright posture. It is assumed that the vestibulospinal reflex arc, which is composed of

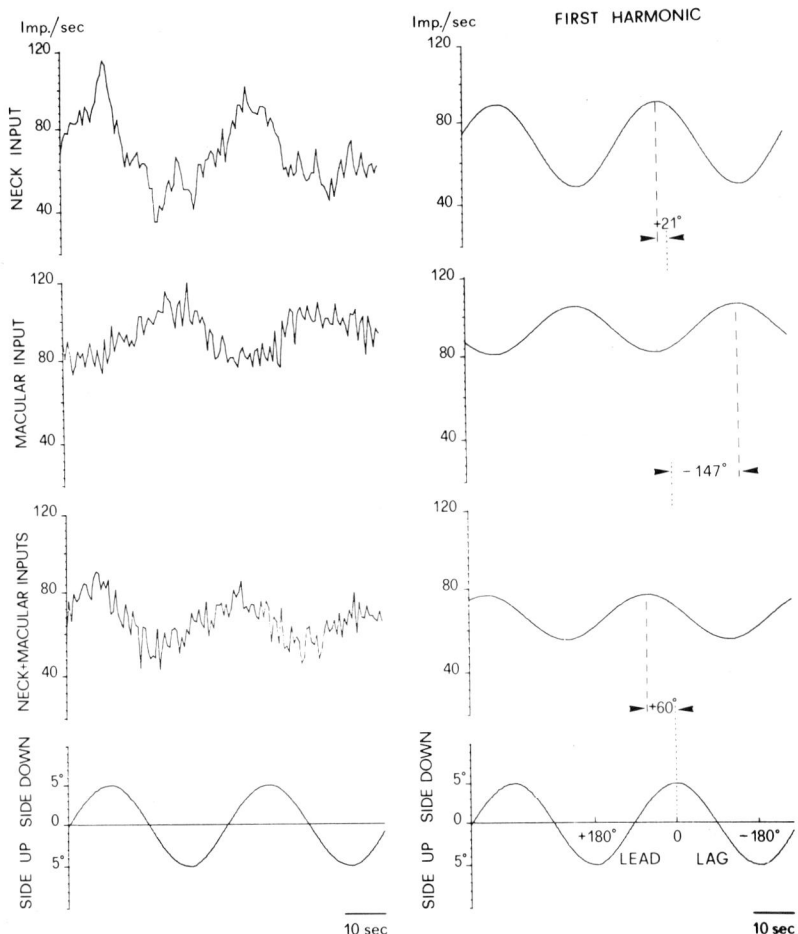

Fig. 6. Interaction of the neck input and the macular input on the MF-discharge of a P-cell. Left records: sequential pulse density histograms showing the MF-response of a P-cell during sinusoidal rotation at 0.026 Hz, 5° of the neck (neck input), of the head (macular input) and or both the neck and the head (neck + macular inputs). 5 sweeps were accumulated for each of the computer records, using 128 bins with the dwell time of 0.6 sec/bin. Upward deflections in the sine wave indicate a side-down rotation of the neck, of the head and of both the neck and the head. Right records: first harmonic of the averaged responses illustrated on the left side. The unit showed a positional sensitivity to the neck input and to the macular input; however, it was excited by side-down rotation of the neck (neck input) and inhibited by side-down rotation of the head (macular input). The peak of the neck response (sensitivity 6.0 imp/sec/degree in % of the base frequency) showed in fact a phase lead of +21° with respect to the peak of the side-down rotation of the neck, whereas the peak of the macular response (sensitivity 2.6 imp/sec/degree in % of the base frequency) showed a phase lag of −147° with respect to the peak of the side-down rotation of the head. Side-down rotation of both the neck and head, leading to combined stimulation of neck and macular receptors, produced a response whose sensitivity (3.2 imp/sec/degree in % of the base frequency) almost corresponded to the algebraic summation of the two responses; moreover, the response showed a phase lead of +60° with respect to the peak of the side-down rotation of the neck, thus becoming more related to the velocity of neck rotation than to the neck position. (From Denoth et al., 1979b.)

primary vestibular afferents, LVN neurons and spinal extensor motoneurons (Lund and Pompeiano, 1968; cf. Pompeiano, 1975b) acts as a closed-loop system in the control of head position. Any change in the head position is in fact detected by the vestibular organ, whose signals will in turn evoke contraction in the neck musculature and restore head and body to their normal attitude.

It was postulated that this closed-loop mechanism does not require a control from the cerebellum (Ito, 1970, 1972). In contrast with this hypothesis the experimental evidence accumulated recently (Pompeiano, 1975a; Erway et al., 1978) led to the conclusion that the macular input may reach the vermal cortex of the cerebellar anterior lobe, thus having the possibility of modifying the activity of LVN neurons. The present experiments have indeed shown that P-cells located in the longitudinal parasagittal zone of the vermal cortex of the anterior lobe, which projects to LVN (Corvaja and Pompeiano, 1979), respond to natural stimulation of macular labyrinthine receptors. In particular, slow sinusoidal rotation along the longitudinal axis of the animal produced a periodic modulation in firing rate of both the MF- and the CF-discharge of the P-neurons, which was related in most instances to the position of the animal for the parameters of stimulation of 0.026 Hz, $\pm 10°$. Moreover, some of these units could be antidromically activated by stimulation of the ipsilateral LVN, thus indicating that the macular input may affect the discharge of cerebellar corticovestibular neurons projecting to Deiters' nucleus. These findings indicate that during the tonic labyrinthine reflexes the discharge of LVN neurons induced by macular stimulation may be modified by the inhibitory action of P-neurons converging upon them.

It is worth noticing that while most of the LVN neurons, similarly to the macular receptors, are excited by side-down rotation of the whole animal (cf. Peterson 1970), most of the MF-units located in the vermal cortex of the cerebellar anterior lobe decrease their firing rate during this stimulus. These findings suggest that the macular input of one side may keep under its excitatory control the contralateral vermis, thus having the possibility of inhibiting the contralateral LVN. The pathway and the mechanism through which the macular input of one side can be transmitted to the contralateral cerebellar vermis have been described elsewhere (Pompeiano, this volume, chapter IV D2).

The present experiments indicate that the increased activity of the extensor muscles in the ipsilateral fore- and hindlimbs which occurs during side-down tilt of the animal (Lindsay et al., 1976) is due not only to labyrinthine activation of the ipsilateral LVN, but also to disfacilitation of the ipsilateral P-neurons leading to disinhibition of Deiters' nucleus. On the other hand, the reduced postural activity in the contralateral fore- and hindlimb extensor muscles (Lindsay et al., 1976) can be attributed not only to labyrinthine disfacilitation of the contralateral LVN, but also to activation of contralateral P-neurons leading to inhibition of the corresponding/Deiters' nucleus. This hypothesis is supported by the results of recent experiments showing that the anatomical integrity of the cerebellum is important in order to produce reciprocal and symmetrical responses of LVN neurons to ipsilateral and contralateral tilts in preparations with section of the VIIIth nerve ipsilateral to the recording side (Hoshino and Pompeiano, 1977). One may also explain why the asymmetric character of the positional reflexes from the labyrinth upon the limbs disappears after removal of the cerebellum (Lindsay and Rosenberg, 1977).

The P-cells located in the vermal cortex of the cerebellar anterior lobe exhibited also a periodic modulation of their MF- and CF-discharge during sinusoidal rotation of the axis vertebra. Most of these MF-units showed an increase in firing rate during side-down rotation of the neck and a decrease in firing rate during side-up rotation of the neck. This finding contrasts with the predominant pattern of response of the MF-units to the macular input, which showed just the opposite behavior. These findings indicate that the neck input of one side exerts a predominant excitatory influence on the ipsilateral cerebellar vermis, in contrast to the macular input which tends to excite mainly the contralateral cerebellar vermis. Indeed, most of the MF-units which received a convergent input from both the neck and the macular receptors were excited by side-down rotation of the neck, while they were inhibited by side-down tilt of the whole animal. However, the peaks of the MF-response of the same P-cells to macular and neck inputs were never 180° out of phase. Moreover, the sensitivity of the MF-response to neck input was always higher than that of the same units to macular input for the same amount of angular displacement.

A final observation concerns the interaction which occurs in these units when the head of the animal was rotated while the vertebral clamp and the animal's trunk remained fixed in horizontal position. In this instance, rotation of the head produced a conflict between neck and macular inputs, which was only partially resolved at the level of the P-neurons. In fact, due to the predominance of the neck over the macular response, the magnitude of the combined response was in most instances similar to the difference between the two individual responses. Moreover, the phase of the resulting response became more related to the velocity of neck rotation than to the neck position. These findings should be related to the observations that in the intact decerebrate animal normal movements of the head, leading to stimulation of both labyrinthine and neck proprioceptors, resulted in neither net reflex change in the posture of the limbs (Ajala and Poppele, 1967) nor changes of the motoneuronal discharge (Erhardt and Wagner, 1970). A final integration of both labyrinth and neck inputs, possibly leading to suppression of the conflicting responses, may occur either at medullary (LVN) or at spinal cord level.

SUMMARY

1. The present experiments have shown that P-cells located in the vermal cortex of the cerebellar anterior lobe, particularly in the longitudinal parasagittal zone which projects to the ipsilateral LVN, can monitor both the signals which originate from the macular receptors, as well as those which originate from the cervical receptors as a result of the displacement of the neck. MF- and/or CF-responses of the same or of different P-cells to the two inputs were observed.

2. The sensitivity of the MF-response of the P-cells to the neck input elicited by sinusoidal rotation of the neck was higher than that of the MF-response of the P-cells to the macular input elicited by sinusoidal tilt along the longitudinal axis of the whole animal.

3. Most of the MF-response of the P-cells to the neck input were characterized by an excitation during side-down rotation of the neck and by an inhibition during side-up rotation, whereas most of the MF-response of the P-cells to the macular input showed just the opposite behaviour.

4. Units which received a convergence from both neck and macular receptors and

showed an antagonistic pattern of response to the two inputs were tested during rotation of the head alone, in order to simultaneously excite the two kinds of receptors. Due to the higher sensitivity of the neck over the macular response, the magnitude of the combined response tended to be similar to the difference between the individual ones. Moreover, the phase of the resulting response became in some instances more related to the velocity of neck rotation than to the neck position.

5. These findings indicate that the conflict between the neck input and the macular input is only partially resolved at corticocerebellar level and that the suppression of the conflicting response may occur either at medullary (LVN) or at spinal cord level.

ACKNOWLEDGEMENTS

This investigation was supported by the Public Health Service Research Grant NS 07685-10 from the National Institute of Neurological and Communicative Disorders and Stroke, N. I. H., U.S.A., by a research grant from the Consiglio Nazionale delle Ricerche, and by a grant in "Biologia e Medicina Spaziale" from the C.N.R., Italy. Dr. Milka Stanojević is a postdoctoral fellow of the Scuola Normale Superiore, Pisa, on leave of absence from the Institute for Medical Research, Belgrade, Yugoslavia.

REFERENCES

Ajala, G.F. and Poppele, R.E. (1967) Some problems in the central actions of vestibular inputs. In *Neurophysiological Basis of Normal and Abnormal Motor Activities*, M.D. Yahr, and D.P. Purpura (Eds.), Raven Press, New York, pp. 141–154.

Berthoz, A. and Llinás, R. (1974) Afferent neck projection to the cerebellar cortex. *Exp. Brain Res.*, 20: 385–401.

Corvaja, N. and Pompeiano, O. (1979) Identification of cerebellar cortico-vestibular neurons retrogradely labeled with horseradish peroxidase. *Neuroscience*, in press.

Denoth, F., Magherini, P.C., Pompeiano, O. and Stanojević, M. (1979) Responses of Purkinje cells of the cerebellar vermis to sinusoidal rotation of the neck. In preparation.

Denoth, F., Magherini, P.C., Pompeiano, O. and Stanojević, M. (1979) Responses of Purkinje neurons of the cerebellar vermis to neck and macular vestibular inputs. *Pflügers Arch.*, in press.

Ehrhardt, K.J. and Wagner, A. (1970) Labyrinthine and neck reflexes recorded from spinal single motoneurons in the cat. *Brain Res.*, 19: 87–104.

Erway, L., Ghelarducci, B., Pompeiano, O. and Stanojević, M. (1978) Response of Purkinje and fastigial nucleus neurons to sinusoidal tilt. *Midwinter Research Meeting. Ass. Res. Otolaryng.*, St. Petersburg Beach, Florida, Jan. 30–Febr. 1.

Hoshino, K. and Pompeiano, O. (1977) Responses of lateral vestibular neurons to stimulation of contralateral macular labyrinthine receptors. *Arch. ital. Biol.*, 115: 237–261.

Ito, M. (1970) Neurophysiological aspects of the cerebellar motor control system. *Int. J. Neurol. (Montevideo)*, 7: 162–176.

Ito, M. (1972) Neural design of the cerebellar motor control system. *Brain Res.*, 40: 81–84.

Lindsay, K.W., Roberts, T.D.M. and Rosenberg, J.R. (1967) Asymmetric tonic labyrinth reflexes and their interaction with neck reflexes in the decerebrate cat. *J. Physiol. (Lond.)*, 261: 583–601.

Lindsay, K.W. and Rosenberg, J.R. (1977) The effect of cerebellectomy on tonic labyrinth reflexes in the forelimb of the decerebrate cat. *J. Physiol. (Lond.)*, 273: 76–77P.

Lund, S. and Pompeiano, O. (1968) Monosynaptic excitation of alpha motoneurones from supraspinal structures in the cat. *Acta physiol. scand.*, 73: 1–21.

Magherini, P.C., Pompeiano, O. and Stanojević, M. (1978) Convergence of macular and neck afferent inputs on Purkinje cells of the cerebellar anterior lobe. *Neuroscience Lett.*, Suppl., S355.

Magnus, R. and De Kleijn, A. (1912) Die Abhängigkeit den Tonus der Extremitätenmuskeln von der Kopfstellung. *Pflügers Arch.*, 145: 455–548.

Manzoni, D., Pompeiano, O. and Stampacchia, G. (1979) Tonic cervical influences on posture and reflex movements. *Arch. ital. Biol.*, 117: 81–110.

McCouch, G.P., Deering, I.D. and Ling, T.H. (1951) Location of receptors for tonic neck reflexes. *J. Neurophysiol.*, 14: 191–195.

Peterson, B.W. (1970) Distribution of neural responses to tilting within vestibular nuclei of the cat. *J. Neurophysiol.*, 33: 750–767.

Pompeiano, O. (1975a) Macular input to neurons of the spinoreticulocerebellar pathway. *Brain Res.*, 95: 351–368.

Pompeiano, O. (1975b) Vestibulo-spinal relationships. In *The Vestibular System*, R.F. Naunton (Ed.), Academic Press, New York, pp. 147–184.

Roberts, T.D.M. (1973) Reflex balance. *Nature (Lond.)*, 244: 156–158.

The Neck and Labyrinthine Influences on Cervical Spinocerebellar Tract Neurones of the Central Cervical Nucleus in the Cat

N. HIRAI*, T. HONGO, S. SASAKI and K. YOSHIDA

Laboratory of Physiology, Institute of Basic Medical Sciences, University of Tsukuba, Niihari-gun, Ibaraki-Ken 300-31 (Japan)

Afferent signals from the neck have long been known to play an important role in the reflex regulation of the balance of the body as well as the eye movements. If these regulatory actions are controlled by the cerebellum, as the labyrinthine reflexes, the cerebellum should be provided with afferent information from the neck. In fact, projection from neck afferents has recently been demonstrated to the vermal lobules V and VI (Berthoz and Llinas, 1974) and to the flocculus (Wilson et al., 1976). Neurones in Brodal and Pompeiano's group x were shown to relay the projection to the flocculus, but little information has so far been available on contribution, if any, of spinocerebellar tracts. However, retrograde labelling technique using horseradish peroxidase (HRP) has revealed that one of spinocerebellar tracts originates from the central cervical nucleus (CCN) (Matsushita and Ikeda, 1975; Wiksten, 1975; Petras, 1977; Cummings and Petras, 1977) where dorsal root fibres of the upper cervical segments terminate (Ranson et al., 1932; Shriver et al., 1968; Imai and Kusama, 1969; Cummings and Petras, 1977). This spinocerebellar tract might thus be involved in conveying afferent information from the neck to the cerebellum. The present study has shown that this is indeed the case. It showed also that the CCN-spinocerebellar tract (CCN-SCT) in the cat transmits sensory signals not only from the neck but also from the labyrinth, mainly to the vermis of the anterior lobe (Hirai et al., 1978).

TRAJECTORY AND CEREBELLAR PROJECTION OF AXONS OF CCN NEURONES

Cerebellar stimulation evoked antidromic field potentials in the region of the central cervical nucleus in C1–C3 as shown in Fig. 1A. The illustrated potentials were of composite nature with single unit activity superimposed on field potentials. In somewhat more caudal or rostral penetrations there were only few or no single spikes at the corresponding depths. This observation is in agreement with the anatomical finding that the CCN exhibits a beady structure (Matsushita and Ikeda, 1975, Commings and Petras, 1977). Location of single cells antidromically activated from the cerebellum coincided approximately with the CCN, although some neurones were encountered

---

*Present address: The Rockefeller University, 1230 York Avenue, New York, NY 10021, U.S.A.

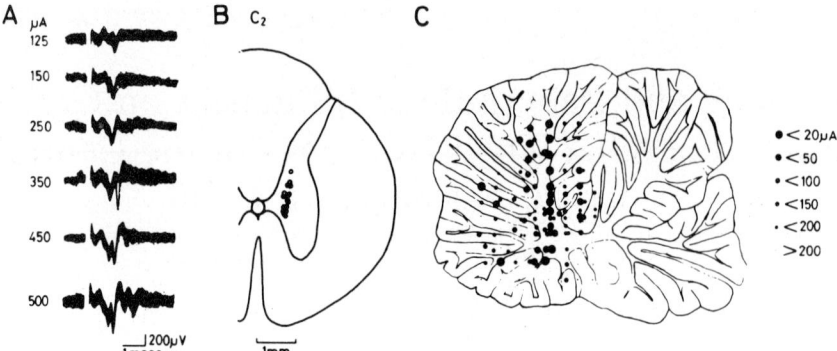

Fig. 1. A) Antidromic field potentials evoked by cerebellar stimulation with different current intensities. B) Location of spinocerebellar tract cells at C2 sampled in one cat. C) Distribution of thresholds for antidromic activation of 10 cells in the midsagittal plane of the cerebellum. See text for further details.

more dorsally or more ventrally as shown in Fig. 1B (cf. Matsushita and Ikeda 1975, Petras and Cummings 1977).

The trajectory of axons of CCN neurones and the region of their cerebellar projection were examined by mapping those sites in the spinal cord, brain stem and the cerebellum from which the individual neurones were antidromically excited with low threshold stimuli. Such a mapping showed that axons of about 90% of CCN neurones crossed the midline and ascended in the contralateral ventrolateral funiculus towards the restiform body to enter the cerebellum. The crossed projections have also been demonstrated histologically (Cummings and Petras, 1977; Matsushita et al., 1979; Wiksten, personal communication). Within the cerebellum the axons were found to project bilaterally predominantly to the medial part of the vermis of nearly the entire anterior lobe (lobule II–V). This is shown in Fig. 1C illustrating the results of threshold measurements for antidromic activation of 10 cells at a midsagittal plane. The mapping experiments further showed that CCN neurones projected also to the posterior lobe including the paramedian lobe; the same neurones could often be activated antidromically from different lobules of the anterior and posterior lobes, indicating multiple terminations of single axons.

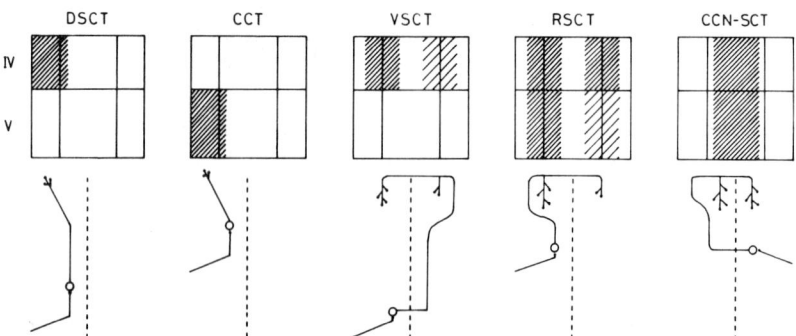

Fig. 2. Course and projection areas of the direct spinocerebellar paths. The dorsal and the ventral spinocerebellar tracts (DSCT and VSCT) convey information from the hindlimb, the cuneocerebellar and the rostral spinocerebellar tracts (CCT and RSCT) from the forelimb, and CCN-SCT from the neck. Projections to Larsell's lobules IV and V of the anterior lobe are shown. (Modified from Oscarsson, 1973.)

The course of axons of the CCN-SCT and their projection areas in the cerebellar anterior lobe can be compared with those of other previously analysed spinocerebellar tracts in Fig. 2. Note that the projection areas of the CCN-SCT are the most medial. Note also that the cervical spinocerebellar tracts hitherto physiologically identified are uncrossed and that the CCN-SCT is the first one found to cross.

INPUT FROM SPINAL NERVES OF THE CERVICAL SEGMENTS

The great majority of CCN-SCT neurones were excited with short latencies when C2 and C3 dorsal roots were stimulated, the effects from C1 not being tested. Only a few neurones were excited from C4–C8 dorsal roots. Correspondingly, stimulation of forelimb nerves (deep radial, superficial radial, median, ulnar, biceps brachii, triceps brachii etc.) did not evoke any short-latency excitation in CCN neurones. Thus, the main spinal input to the CCN would be through the upper cervical dorsal roots.

The excitatory effects from C2 and C3 dorsal roots were potent enough to fire the neurones. Latencies of action potentials evoked in them are shown in the histogram of Fig. 3C (○) together with those of EPSPs (●). Clearly, the shortest path appeared to be monosynaptic. No similar effects could, however, be reproduced from the peripheral nerve branches of the same root or even from the portion of the dorsal ramus just peripheral to the spinal ganglion. Accordingly, large orthodromic field potentials were evoked in the region of CCN from the C2 dorsal root (Fig. 3A, upper trace) but hardly any from a muscle nerve branch of the same root (Fig. 3A, lower trace). These observations suggest that the afferent fibres responsible for the short-latency excitation of CCN neurones join the dorsal roots at the level of the spinal ganglion (cf. McCouch et al., 1951; Hikosaka and Maeda, 1973; Berthoz and Llians, 1974; Wilson et al., 1976).

We have not been successful in identifying those nerve branches which include afferent fibres supplying the CCN neurones with the excitatory input. However, electrical stimulation with bipolar electrodes of the lateral surface of the vertebral bones

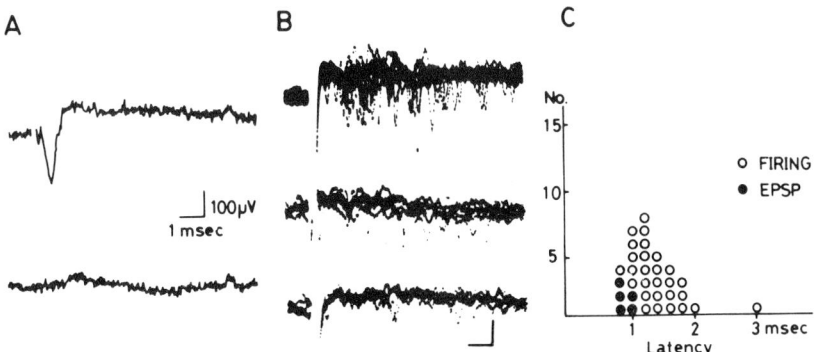

Fig. 3. A) Field potential evoked from the C2 dorsal root (upper trace) and lack of effects to stimulation of its muscle nerve branch (lower trace). B) Responses of CCN cells to stimulation of the surface of the rostral part of C2 vertebra (top trace) and of the main branch of the C2 dorsal ramus (middle trace). The excitatory responses as in the top record were abolished after cutting the C2 dorsal root (bottom trace). Extracellular records. C) Distribution of latencies of spike potentials in extracellular records (○) and of EPSPs (●) evoked in CCN cells by stimulation of the upper cervical dorsal roots. See text for further details.

near the joint region was found to produce excitation of CCN neurones as exemplified in Fig. 3B (top trace), whereas no excitatory effects were evoked from the main nerve branch of the same segment (middle trace). In this case, the joint region was stimulated with 2 V, which was far below the threshold for current spread to the spinal ganglion; the current spread occurred only with 9 V as evidenced by records from the main peripheral nerve trunk of the same dorsal root. Further, the effect of vertebral stimulation disappeared after section of the corresponding dorsal root (Fig. 3B, bottom trace), indicating that the effect was produced by afferent fibres entering the dorsal root and not by current spread to intraspinal elements. On the basis of these observations we suggest that the fibres responsible for exciting CCN cells originate from receptors in or around the region of the vertebral joint.

INPUT FROM THE LABYRINTH

It appeared to be a typical feature of CCN neurones that they receive synaptic inputs not only from the neck afferents but also from the labyrinth. The excitatory convergence from the two sources occurred in nearly all (ca 90%) neurones examined.

Effects from the whole vestibular nerve

The VIIIth nerve was stimulated through two silver ball electrodes placed at the oval and the round windows. Excitatory effects with short latencies were evoked only from the contralateral VIIIth nerve, the excitation being often followed by a phase of depression of firing (Fig. 4A). By contrast, stimulation of the ipsilateral VIIIth nerve usually evoked only a depression of the tonic activity (Fig. 4B, the same cell as in Fig. 4A).

The firing of CCN cells from the contralateral VIIIth nerve occurred with latencies of 2.2–9.5 msec, as shown in Fig. 4D (o). Short-latency responses were also evoked by stimulating the region of the contralateral vestibular nuclei and the medial longi-

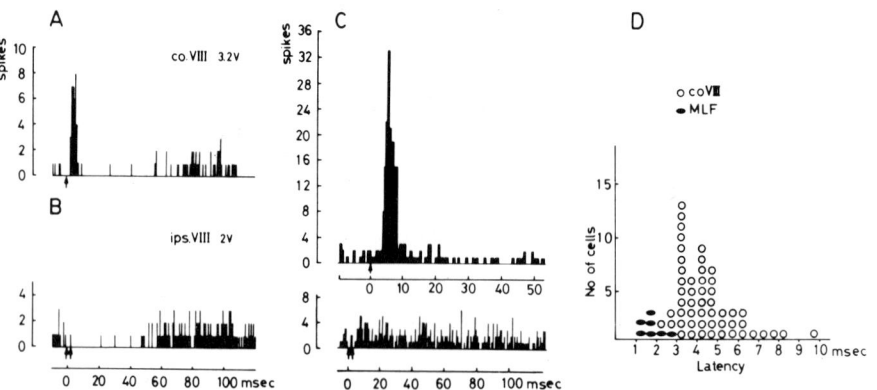

Fig. 4. Effects from the VIIIth nerve and the MLF on CCN cells. A, B) Poststimulus time histograms showing responses to stimulation of the contralateral (A) and ipsilateral VIIIth nerve (B) in 50 consecutive trials. Arrows indicate stimuli C) Similar histograms showing effects of contralateral VIIIth nerve stimulation before (upper histogram) and after (lower histogram) transection of MLF. D) Frequency histogram of latencies of excitation from the contralateral VIIIth nerve (o) and from MLF (●). See text for further details.

tudinal fasciculus (MLF) (spike latency 1.0–2.5 msec, ● in Fig. 4D), and the earliest effects from these regions indicate monosynaptic coupling with at least some CCN cells. The excitation from the contralateral VIIIth nerve was abolished by transection of the MLF at a level of few mm rostral to the obex as illustrated in Fig. 4C with the upper poststimulus histogram before and the lower after MLF lesion. These observations suggest that the excitatory effect from the contralateral VIIIth nerve is mediated primarily by neurones with axons descending in the MLF. That the effect originated from the vestibular component of the VIIIth nerve was indicated also by results of experiments in which the three ampullary nerves were separately stimulated.

Effects from individual ampullary nerves

On the basis of experiments with the whole VIIIth nerve stimulation we cannot conclude as to whether the effects on CCN cells were evoked from the canal or the otolith afferents, or even from the cochlear nerve. In order to define their origin and to find out if there is any specific pattern of responses evoked from different labyrinthine receptors, individual ampullary nerves were stimulated with the technique described by Suzuki et al. (1969).

Effects from the three contralateral ampullary nerves on CCN cells are exemplified in Fig. 5A, B and D. Note that in the cell in A the responses were evoked from the anterior canal nerve and that stimulation of the other two ampullary nerves had no effect. Excitation from only the anterior canal nerve, as in this case, was found in about half of the examined cells. The second major group of cells responded to stimulation of only the posterior canal nerve, as illustrated in D and the third group was excited from both the anterior and the posterior canal nerves. Short-latency excitation from the anterior and inhibition from the posterior canal nerve, or vice versa, were found in an additional small number of cells. Surprisingly, no excitatory

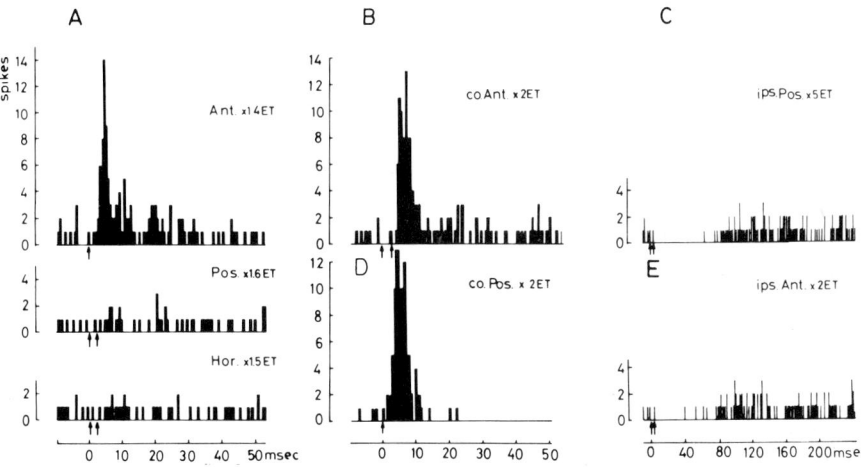

Fig. 5. Poststimulus time histograms showing responses to stimulation of individual ampullary nerves in 50 trials. A) Responses of a CCN cell to stimulation of the contralateral anterior (top), posterior (middle) and horizontal (bottom) canal nerves. Strengths of stimuli are indicated in multiples of threshold for the characteristic eye movements (ET). B, C) Histograms showing excitation from the contralateral anterior canal nerve and inhibition from the ipsilateral posterior canal nerve in another CCN cell. D, E) As in B and C, but with excitation from the contralateral posterior canal and inhibition from the ipsilateral anterior canal in a third cell. See text for further details.

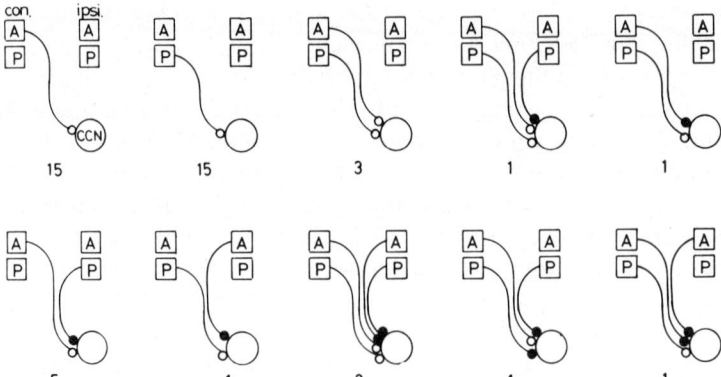

Fig. 6. Patterns of effects from anterior and posterior canal nerves on CCN cells. ○ and ● indicate excitatory and inhibitory input, resp. Figures below the diagrams indicate numbers of CCN cells with a given pattern of labyrinthine inputs. Effects from the horizontal canal nerves was not taken into account since they were absent in all the cells. See text for further details.

input from the horizontal canal nerve was found in any of the 45 cells so far analysed in C1–C3.

The early excitation was often followed by a phase of depression in response to individual nerve stimulation as well as to the whole nerve stimulation. Since descending fibres in the MLF exhibited a similar pattern of response, i.e., excitation followed by depression, the depression of CCN cells after stimulation of the contralateral labyrinth was presumably due, at least in part, to removal of excitatory impingement from the medial vestibulospinal tract. The short-latency inhibition from the contralateral posterior or anterior canal found in some cells was, on the other hand, postsynaptic at the CCN cell level, as revealed by intracellular recording.

Stimulation of the ipsilateral canal nerves resulted in inhibition of CCN cells or else in no effect onto them. The origin of the ipsilateral inhibition appeared to be as specific as the origin of the contralateral excitation. The inhibition was produced from either the anterior or the posterior canal nerve, and again not from the horizontal canal nerve. Furthermore, the inhibition was plane-specific, i.e., inhibition from the posterior canal nerve occurred only in cells with excitatory input from the contralateral anterior nerve (Fig. 5B, C), or vice versa (Fig. 5D, E). When a CCN neurone received excitatory inputs from both the anterior and the posterior canal nerves of the opposite side, one of them usually had a stronger effect. Under such conditions, the ipsilateral inhibition was stronger from a canal nerve that was plane-specific to the origin of the stronger excitatory input. The patterns of convergence of effects from individual canal nerves, and frequency of their occurrence are summarized in Fig. 6.

The plane-specific inhibition from the ipsilateral individual nerves occurred with short latencies (2–3 msec). Intracellular recordings showed that at least the early phase of the inhibition was due to postsynaptic inhibition at the CCN cell level.

DISCUSSION

The CCN-spinocerebellar tract has now been identified as i) receiving synaptic inputs from both the neck and the labyrinths, and ii) projecting to the vermis of the anterior lobe and to the posterior lobe.

Our experiments indicate that the input from the neck to CCN-SCT cells is primarily of joint origin probably from the same afferents which initiate tonic neck reflexes (McCouch et al., 1951), influence extraocular motoneurones (Hikosaka and Maeda, 1973), and project to the flocculus (Wilson et al., 1976). Wilson et al. (1976) have shown that the joint afferent impulses are relayed by neurones of Brodal and Pompeiano's group x; the latter neurones do not receive input from the labyrinth in contrast with the CCN cells, the majority of which are co-excited by impulses from both the neck and vestibular afferents.

In regard to the receptor origin of the vestibular input to CCN cells we have so far analysed only the input from the semicircular canals. The effects of stimulation of individual ampullary nerves occurred at current intensities 1.5–2.0 times the thresholds for characteristic eye movements, and hence current spread to other labyrinthine afferents seems unlikely (Tokumasu et al., 1971; Markham and Curthoys, 1972). Since only one of the three pairs of electrodes used to stimulate the contralateral ampullary nerves was usually effective in individual cells and in addition excitation and inhibition from the two sides were produced in a manner conforming with the plane-specificity of the semicircular canals, we conclude that the semicircular canal system is an important source of input to the CCN-SCT. On the other hand, nothing definite can yet be said about the input from the otolith organs. There are indications that impulses from the otoliths are relayed by neurones in the lateral vestibular nucleus, the origin of the lateral vestibulospinal tract (e.g., Fujita et al., 1968, Peterson, 1970, Shimazu and Smith, 1971). Our observations that effects of stimulation of the whole vestibular nerve were abolished by cutting the medial vestibulospinal tract within the bulbar MLF and that no significant effects could be evoked by stimulation of the lateral vestibulospinal tract (unpublished) do not give any indication for contribution of the otolith organs to the input to CCN cells; such observations are, however, not sufficient to exclude it. It is of interest in this context that Coulter et al. (1976) reported macular influences, via the lateral vestibulospinal tract, on spinoreticular tract cells which in turn relay them to the cerebellum via neurones in the lateral reticular nucleus.

SUMMARY

1. The trajectory, the cerebellar projection areas and the input to the spinocerebellar tract originating from the central cervical nucleus (CCN) were investigated electrophysiologically in the cat.

2. The majority of CCN neurones was found to have axons crossing at the level of cell body and ascending in the contralateral ventrolateral funiculus and the restiform body before entering the cerebellum.

3. The axons projected mainly to the medial part of vermis of nearly the entire anterior lobe, and less extensively to the posterior lobe including the paramedian lobe.

4. Potent excitatory input was supplied by the dorsal roots of the upper cervical segments (C2–C3). The main source of this input appeared to be afferents from joints or ligaments between the upper cervical vertebrae.

5. CCN-SCT neurones received excitatory input from the contralateral and inhibitory input from the ipsilateral VIIIth nerve. Stimulation of individual canal nerves showed that these inputs were from the vertical but not from the horizontal canals. In individual cells the contralateral excitation was usually evoked from either the

anterior or the posterior canal, and the ipsilateral inhibition, when present, from a canal that was plane-specific to the canal of the excitatory input on the contralateral side.

6. Some aspects of neck and labyrinthine inputs to CCN-SCT neurones are discussed.

REFERENCES

Berthoz, A. and Llinás, R. (1974) Afferent neck projection to the cat cerebellar cortex. *Exp. Brain Res.*, 20: 385–401.

Coulter, J.D., Mergner, T. and Pompeiano, O. (1976) Effects of static tilt on cervical spinoreticular tract neurons. *J. Neurophysiol.*, 39: 45–62.

Cummings, J.F. and Petras, J.M. (1977) The origin of spinocerebellar pathways. I. The nucleus cervicalis centralis of the cranial cervical spinal cord. *J. comp. Neurol.*, 173: 655–692.

Fujita, Y., Rosenberg, J. and Segundo, J.P. (1968) Activity of cells in the lateral vestibular nucleus as a function of head position. *J. Physiol. (Lond.)*, 1–18.

Hikosaka, O. and Maeda, M. (1973) Cervical effects on abducens motoneurons and their interaction with vestibulo-ocular reflex. *Exp. Brain Res.*, 18: 512–530.

Hirai, N., Hongo, T. and Sasaki, S. (1978) Cerebellar projection and input organizations of the spinocerebellar tract arising from the central cervical nucleus in the cat. *Brain Res.*, 157: 341–345.

Imai, Y. and Kusama, T. (1969) Distribution of the dorsal root fibers in the cat. An experimental study with the Nauta method. *Brain Res.*, 13: 338–359.

Markham, C.H. and Curthoys, I.S. (1972) Labyrinthine convergence on vestibular neurons using natural and electrical stimulation. In *Progress in Brain Research, Vol. 37, Basic Aspects of Central Vestibular Mechanisms*, A. Brodal and O. Pompeiano (Eds.), Elsevier, Amsterdam, pp. 121–137.

Matsushita, M. and Ikeda, M. (1975) The central cervical nucleus as cell origin of a spinocerebellar tract arising from the cervical cord: a study in the cat using horseradish peroxidase. *Brain Res.*, 100: 412–417.

Matsushita, M., Hosoya, Y. and Ikeda, M. (1979) Anatomical organization of the spinocerebellar system in the cat, as studied by retrograde transport of horseradish peroxidase. *J. comp. Neurol.*, 184: 63–80.

McCouch, G.P., Deering, I.D. and Ling, T.H. (1951) Location of receptors for tonic neck reflexes. *J. Neurophysiol.*, 14, 191–195.

Oscarsson, O. (1973) Functional organization of spinocerebellar paths. In *Handbook of Sensory-Physiology, Vol. II, Somatosensory System*, A. Iggo (Ed.), Springer-Verlag, Berlin, pp. 339–380.

Peterson, B.W. (1970) Distribution of neural responses to tilting within vestibular nuclei of the cat. *J. Neurophysiol.*, 33: 750–767.

Petras, J.M. (1977) Spinocerebellar neurons in the rhesus monkey. *Brain Res.*, 130: 146–151.

Petras, J.M. and Cummings, J.F. (1977) The origin of spinocerebellar pathways. II. The nucleus centrobasalis of the cervical enlargement and the nucleus dorsalis of the thoracolumbar spinal cord. *J. comp. Neurol.* 173; 693–716.

Ranson, S.W., Davenport, H.K. and Doles, E.A. (1932) Intramedullary course of the dorsal root fibers of the first three cervical fibers. *J. comp. Neurol.*, 54; 1–12.

Shimazu, H. and Smith, C.M. (1971) Cerebellar and labyrinthine influences on single vestibular neurons identified by natural stimuli. *J. Neurophysiol.*, 34: 493–508.

Shriver, J.E., Stein, B.M. and Carpenter, M.B. (1968) Central projections of spinal dorsal roots in monkey. I. Cervical and upper thoracic dorsal roots. *Amer. J. Anat.*, 123: 27–74.

Suzuki, J.-I., Goto, K., Tokumasu, K. and Cohen, B. (1969) Implantation of electrodes near individual vestibular nerve branches in mammals. *Ann. oto-rhino-laryngol. (St. Louis)*, 78: 815–826.

Tokumasu, K., Suzuki, J.-I. and Goto, K. (1971) A study of the current spread on electric stimulation of the individual utricular and ampullary nerves. *Acta oto-laryngol. (Stockh.)*, 71: 313–318.

Wiksten, B. (1975) The central cervical nucleus — a source of spinocerebellar fibres demonstrated by retrograde transport of horseradish peroxidase. *Neurosci. Lett.*, 1: 81–84.

Wilson, V.J., Maeda, M., Franck, J.I. and Shimazu, H. (1976) Mossy fiber neck and second-order labyrinthine projections to cat flocculus. *J. Neurophysiol.*, 39: 301–310.

A Role of Neck Afferents on Vestibulocollic Reflex Elicited by Dynamic Labyrinthine Stimulation

K. EZURE*, S. SASAKI, Y. UCHINO** and V. J. WILSON***

Laboratory of Physiology, Institute of Basic Medical Sciences, University of Tsukuba, Niihari-gun, Ibaraki-ken 300—31 (Japan)

INTRODUCTION

Semicircular canal stimulation induces head and eye movements parallel to the activated canal (Suzuki and Cohen, 1964). With respect to the horizontal canal system, stimulation of a horizontal canal nerve produces contralateral head movement in the horizontal plane and causes a circling of the body to the opposite side. The vestibular induced head movement, the vestibulocollic reflex, has been observed in many animals. Connections between vestibular nucleus neurones and spinal motoneurones have been studied anatomically (Nyberg-Hansen, 1964, 1966) and physiologically (Wilson and Yoshida, 1969a, b; Wilson and Maeda, 1974). In a vestibulospinal system stimulation of the labyrinth or vestibular nuclei activates both α and γ motoneurones (Carli et al., 1967; Poppele, 1967; Grillner et al. 1969; Kato and Tanji, 1971) and it has been considered that coactivation of α and γ motoneurones occurs in the reflex. However, the dynamic characteristics of the vestibulospinal reflex were little known except for the studies of Berthoz and Anderson (1971) and Anderson et al. (1977). Since muscle spindles are known to be abundant in neck muscles (Cooper and Daniel, 1963; Tompson, 1970; Richmond and Abrahams, 1975), we have been interested in how the γ-fiber spindle loop takes part in control of the vestibulocollic reflex. This has been done by quantifying the dynamic characteristics of single motor units and compound EMGs of neck extensor muscle in response to sinusoidal stimulation of the horizontal canal.

METHODS

The present experiments have been performed on decerebrate unanesthetized cats. The head of the animal was mounted on a stereotaxic frame and was placed at the center of the turntable. Recordings were made from single motor units of the splenius (SP), biventer cervices (BIV) and complexus (COMP) muscles with acupuncture

---

Present addresses:
*Department of Physiology, Institute of Brain Research, School of Medicine, University of Tokyo, Japan.
**Department of Physiology, Kyorin University, School of Medicine, Mitaka, Tokyo, Japan.
***The Rockefeller University, New York, NY 10021, U.S.A.

electrodes insulated except for the very tips. For recording compound EMGs, similar electrodes without insulation were used bipolarly, the interpolar distance being 3 cm. The horizontal canal was selectively stimulated by sinusoidal oscillation of the turntable in the horizontal plane. The stimulus frequency ranged 0.011–0.5 Hz and the stimulus amplitude of oscillation ranged from 0.24 to 180 degrees/sec².

RESULTS

Response of single motor units

The spontaneous firing rates of motor units ranged from 0 to 40 spikes/sec in the decerebrate condition. When the turntable was rotated in the direction contralateral to the recording side, the firing rate of the units increased, and decreased with ipsilateral rotation as exemplified in Fig. 1A. An example of averaged responses over a range of one cycle is shown in Fig. 1B.

The firing rates at all five frequencies are found to be approximately sinusoidally modulated (all-round type response). However, the response patterns of the motor units were deeply affected by their spontaneous firing rate and their inherent maximal firing rate. When the spontaneous firing rate was relatively low, the response was clipped at zero level (Fig. 1D). This type of response will be called a cut-off type response. When the spontaneous firing rate was relatively high or the stimulus amplitude was large, the response was often saturated at its peak (Fig. 1E). When the spontaneous firing rate was at its inherent maximal firing rate, the firing rate was scarcely modulated even with fairly large stimulation (Fig. 1F).

The saturation level of firing rate, or maximal firing rate, was examined in 65 motor units by applying a large amplitude oscillation. It ranged from 8 to 52 spikes/sec. In the present experiments, motor units were tentatively classified into two classes, the high frequency(HF) units which could fire above 20 spikes/sec and low frequency(LF) units whose maximal firing rate were below 20 spikes/sec.

The gain and phase of the frequency response were expressed relative to angular acceleration of the head in the present experiments. The phase lag was determined by measuring the difference between the peak of angular acceleration and the peak of the firing rate and expressed in degrees. The gain was defined by the ratio of response amplitude to the amplitude of angular acceleration and expressed in decibels. When the response was a cut-off type or saturated at its peak, the response amplitude was estimated by measuring the slope of the response curve at a given point near the base line, assuming that the response at this point is part of a sine wave (Fig. 1 in Ezure and Sasaki, 1978).

The linearity of the system was examined by changing the stimulus amplitude (7°–45°) at a fixed stimulus frequency. Both the phase and gain were fairly constant, and linearity was maintained in most cases. Some non-linearity was observed with high stimulus frequencies, however, and a stimulus amplitude, as small as possible, was employed to minimize their effect.

The response of a low frequency (LF) unit to sinusoidal oscillation of the turntable at five different frequencies is exemplified in Fig. 1B, and the gain and the phase lag (relative to the angular acceleration) of five representative units (3 LF and 2 HF units) are shown in the Bode diagram of Fig. 1C. Note that the gain decreased with a slope of 40 dB/decade at stimulus frequencies from 0.02 to 0.4 Hz. The phase lag

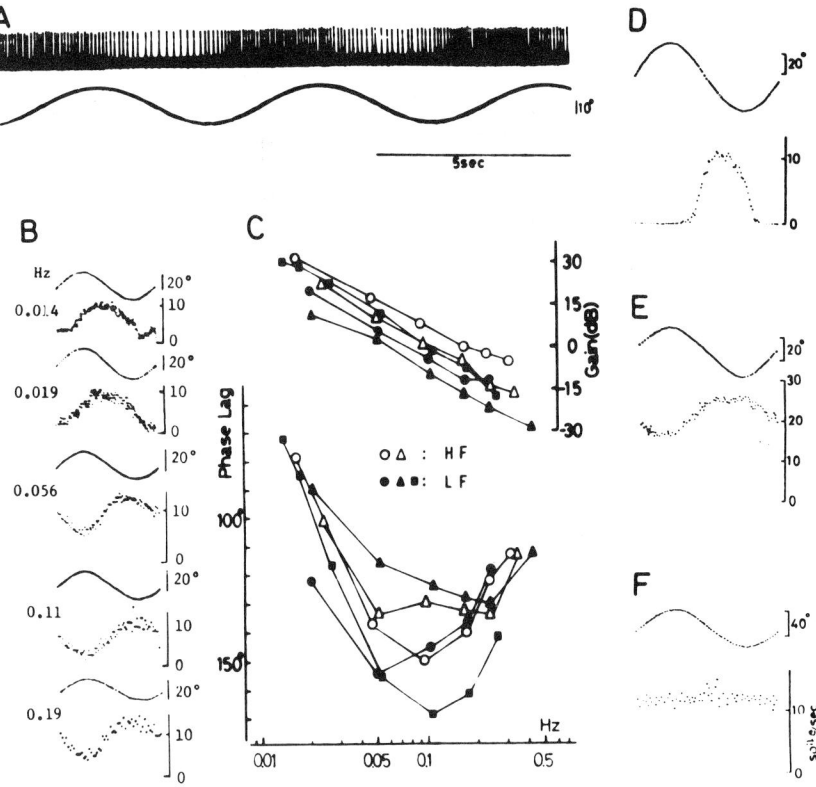

Fig. 1. A) Response of a single motor unit to horizontal oscillation of the turntable. Record was obtained from the right SP muscle. Lower trace indicates the position of turntable upward displacement indicating leftward rotation. B) Example of computer averaged responses of a motor unit to 5 different stimulus frequencies. The same unit as represented with ■ in C. In this and following records, motor units were from right muscles and upper sine curve indicated the position of the turntable, its upward displacement being rightward rotation. C) Bode diagram of 5 motor units; 2 HF units with open symbols and 3 LF units with filled symbols. The gain and the phase lag are plotted as decibels and degrees, resp., against frequencies. D–F) Various pattern of computer averaged responses. D) Cut-off type response of a unit having low spontaneous firing rate. E) Saturated type response of a unit having high spontaneous firing rate. F) No response even to stimulus of large amplitude.

relative to the angular acceleration was approximately 140° in the frequency range from 0.05 to 0.2 Hz and it gradually decreased with increasing stimulus frequencies in both HF and LF units. In Fig. 2A, the relation between the maximal firing rate and the gain at 0.1 Hz is illustrated. There is a highly significant positive correlation ($r = 0.75$, $P < 0.0005$, t-test) indicating that the motor units with higher firing rates exhibited higher gain than those with lower firing rates. The HF and LF units described above are divided by the vertical broken line at 20 spikes/sec in Fig. 2A. On the other hand, there was no significant correlation between the phase lag and maximal firing rate ($P > 0.2$, t-test, Fig. 2A, inset). At fixed frequencies of 0.1 and 0.17 Hz, the relation between the gain and the phase lag was examined in HF and LF units, respectively (Fig. 2B, C). There is a clear, positive correlation between them: The correlation coefficient was r = 0.6 for HF units, r = 0.42 for LF units at 0.1 Hz, r = 0.78, for HF units, and r = 0.79, for LF units at 0.17 Hz. The distribution of HF

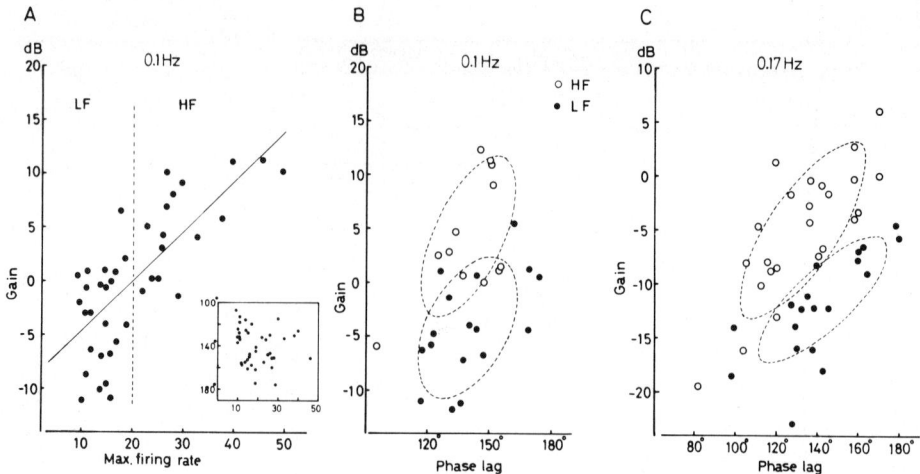

Fig. 2. Classification of motor units and relationships among the maximum firing rate, gain and phase lag. A) Relation between the maximum firing rate and the gain at 0.1 Hz. The solid line is the linear regression. Motor units were classified into HF and LF units at 20 spikes/sec of their maximum firing rates (vertical broken line). Inset: the relationship between the maximum firing rate and the phase lag, showing no correlation. B, C) Relationship between the phase lag and the gain at 0.1 and 0.17 Hz. HF and LF units are represented by ○ and ●, resp. Ellipses: probability density ellipses drawn at the levels of SDs along the principal axes. Slope of linear regression lines: 0.19 and 0.12 for HF and LF units, resp., at 0.1 Hz; and 0.20 and 0.15 for HF and LF units, resp., at 0.17 Hz.

and LF units is shown with probability density ellipses, indicating clear separation of the two groups ($P < 0.001$; t-test). The slope of the linear regression line which approximately corresponded to major axis of the probability density ellipses did not differ significantly between HF and LF units ($P > 0.1$, t-test). This may suggest that the both HF and LF units receive similar effects from the vestibular input and that the differences in the gain between them depend on the properties of motoneurones (see Ezure and Sasaki, 1978).

Response of compound EMGs

The response of compound EMGs which presumably represent whole muscle activity was examined. The inset of Fig. 3A shows the compound EMGs whose amplitude was modulated by oscillation of the turntable. EMGs were increased in amplitude when the turntable was rotated in the direction contralateral to the recording side and decreased with ipsilateral rotation, thus exhibiting the same mode of response as that of motor units. The phase lag was defined by the value from the peak of acceleration to the peak of the response which were rectified and averaged by a computer. The gain was defined by the ratio of amplitude of the responses to that of the angular acceleration. To calculate the gain, the compound EMG response was rectified and integrated through a low pass filter (time constant, 0.3 sec) and averaged with a computer. The activity of the compound EMGs in response to sinusoidal oscillation of turntable usually showed sinusoidal modulation as shown in Fig. 3A. The linearity of the system was always maintained at small stimulus amplitudes.

The response of compound EMGs to various frequencies of oscillation is represented by the Bode diagram in Fig. 3A. It was found that the phase characteristics of the frequency response obtained from 11 cats were similar to those of motor units whose averaged phase lag were plotted by open circles in the diagram for the sake of

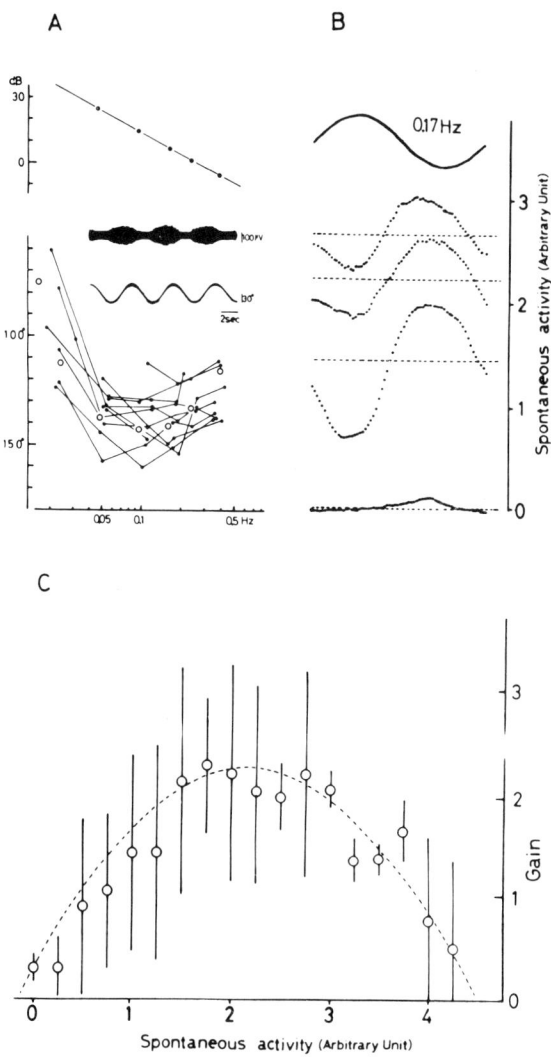

Fig. 3. Response of compound EMGs to sinusoidal oscillation. A) Bode diagram. Phase lags were plotted from 11 cats, ○ being the averaged phase lag of motor units. Gain was obtained from a cat whose spontaneous activity was maintained constant during the whole period of recording (normalized at 0.25 Hz). Inset shows the response of compound EMGs to sinusoidal oscillation of turntable. B, C) Relation between spontaneous activity and the gain of compound EMGs. At a fixed frequency and amplitude (0.17 Hz, 35°), the gain was measured at various levels of spontaneous activity. B) Example of responses from one and the same cat. Top sine curve indicates the position of turntable. C) Averaged gains in 8 cats plotted against the spontaneous activity. Mean and SD are indicated by ○ and bars, resp.

comparison. The gain of the compound EMGs also decreased with a slope of 40 dB/decade in a cat whose spontaneous EMG activity remained relatively constant during the oscillation (see below).

The relation between the gain and the spontaneous activity was systematically examined at a fixed frequency (0.17 Hz) and angular amplitude (35°) of oscillation. The gain of the compound EMG response varied with the level of its spontaneous activity, as exemplified in Fig. 3B which illustrates responses obtained from one and the same cat at four different levels of spontaneous activity (dotted line). In Fig. 3C, averaged gains obtained from eight cats are plotted against the various levels of spontaneous activity. The gain of the compound EMG response was maximal at around the level of 2 (arbitrary units). When the spontaneous activity level decreased below the optimal level, the gain decreased due to the decrease of the number of motor units involved as well as to a decrease in their firing rate, because the gain of individual motor units remained almost constant, independent of their spontaneous activity. As spontaneous activity increased above the optimal level, the gain decreased and finally became near zero at extremely high spontaneous activity levels. This may be attributed to the saturation of the firing rate of individual motor units (cf. Fig. 1F).

Transfer function

The above results that the phase lag of motor units and compound EMGs with respect to angular acceleration was about 140° from 0.05 to 0.2 Hz, and that the gain relative to the angular acceleration decreased with a slope of about 40 dB/decade in frequency range of 0.01–0.4 Hz, indicate that there exists a double integrator in the vestibulocollic reflex arc. From analysis of the responses of the primary afferents in monkey or of vestibular neurones in cats to sinusoidal oscillation of the head, it is known that the first integration was taking place at the cupula and is approximated by a first order lag system (Steinhausen, 1933; Goldberg and Fernandez, 1971; Fernandez and Goldberg, 1971; Melvill Jones and Milsum, 1970, 1971; Shinoda and Yoshida, 1974). Thus, the second integrator should be in the central pathway from vestibular nucleus neurones to neck motoneurones.

The transfer function of the vestibulocollic reflex T(s) may be expressed as follows:

$$T(s) = V(s) \cdot N(s)$$

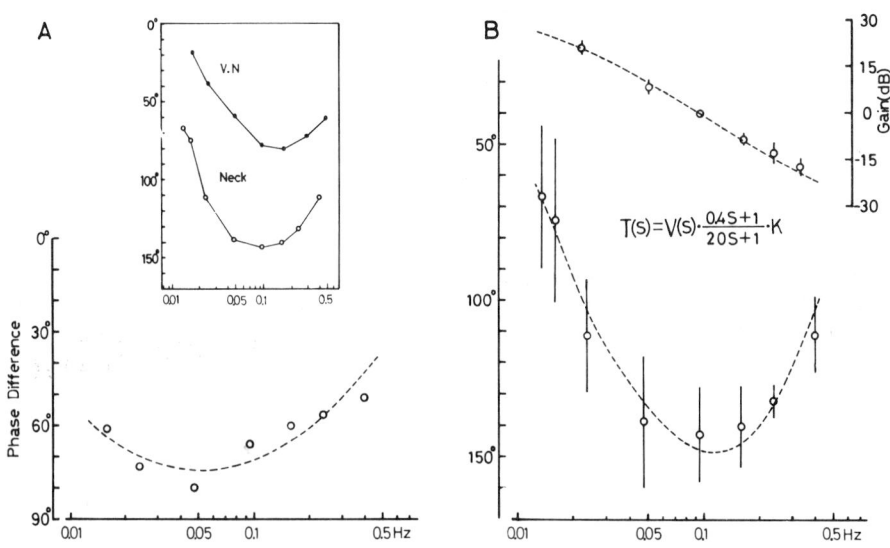

Fig. 4. Determination of the transfer function. A) Averaged phase differences between vestibular nucleus neurones and neck motor units. The best-fit curve is drawn with least-square methods (broken line). Inset: upper curve (from data of Shinoda and Yoshida); lower curve, present data of neck motor unit responses. B) Broken lines indicate reconstruction of the gain and phase lag of the vestibulocollic reflex on the basis of inset transfer function T(s). The gain was normalized at 0.1 Hz. V(s) is the transfer function of vestibular nucleus neurones. K is dimensionless constant.

where V(s) is the transfer function of vestibular nucleus neurones with respect to input angular acceleration (cf. Shinoda and Yoshida, 1974; Ezure and Sasaki, 1978), and N(s) is the transfer function from vestibular nuclei to neck motoneurones.

The inset of Fig. 4A illustrates the average phase diagram of vestibular nucleus neurones (●) (Shinoda and Yoshida, 1974) and that of the neck motor units (○).

The difference between the two phase diagrams is plotted in Fig. 4A. The phase difference was about 80° around 0.05 Hz but it decreased at higher frequencies. Taking this phase lead component into consideration, the transfer function N(s) will be expressed as

$$N(s) = \frac{\tau_2 s + 1}{\tau_1 s + 1} K$$

The broken line in Fig. 4B shows the best-fit curve of N(s) determined by the least square method (τ_1 was 20 sec and τ_2 was 0.4 sec). In the Bode diagram of Fig. 4B, T(s) was represented by the broken line including the characteristics of the gain which was normalized at 0.1 Hz. The agreement between the transfer function and the actual results is fairly good. Constant K (dimensionless) was estimated to be 10 and 4 for HF and LF units, respectively.

Functional significance of cervical afferents in vestibulocollic reflex

Since muscle spindles are known to be abundant in neck muscles (Cooper and Daniel, 1963; Richmond and Abrahams, 1975), and α and γ motoneurones are co-activated by stimulation of labyrinth or vestibular nuclei (Carli et al., 1967; Poppele, 1967; Grillner et al., 1969), it would be expected that the gamma loop takes part in control of vestibulocollic reflex. To examine how the γ-loop contributes to control of the vestibulocollic reflex, the C1–C4 dorsal roots were sectioned ipsilaterally to the recording side. As expected, the rate of spontaneous firing of the motor units was markedly decreased after deafferentation, so that those exhibiting cut-off responses were predominantly encountered. Unexpectedly, however, their gain and phase lag were not changed by deafferentation as shown in Fig. 5A, where the mean and SD of the phase lag and the gain of motor units are plotted, before (○, n = 11) and after (● n = 10) deafferentation.

It has been reported that the amplitude of the maximal Ia monosynaptic EPSPs is larger in the slow (type S) than in the fast (type F) motoneurones (Burke, 1967; Burke et al., 1973). There is a possibility that different types of motor units are differently affected by deafferentation. Hence, we examined the effects of deafferentation on the frequency responses by comparing diagrams relating the phase lag and the gain before and after deafferentation separately for HF and LF motor units (Fig. 5B). The points before (○) and after (●) deafferentation are completely intermingled, and no significant differences were found between their scatters in either HF or LF units ($P > 0.2$, t-test).

Effects of deafferentation on the spontaneous activity, phase lag and gain were examined in compound EMGs as well. Spontaneous activity was consistently decreased after deafferentation. The phase lag was not altered, but the gain was markedly affected, the change being closely related to the level of spontaneous activity before the dorsal root section. In Fig. 6, illustrating the relation between the gain

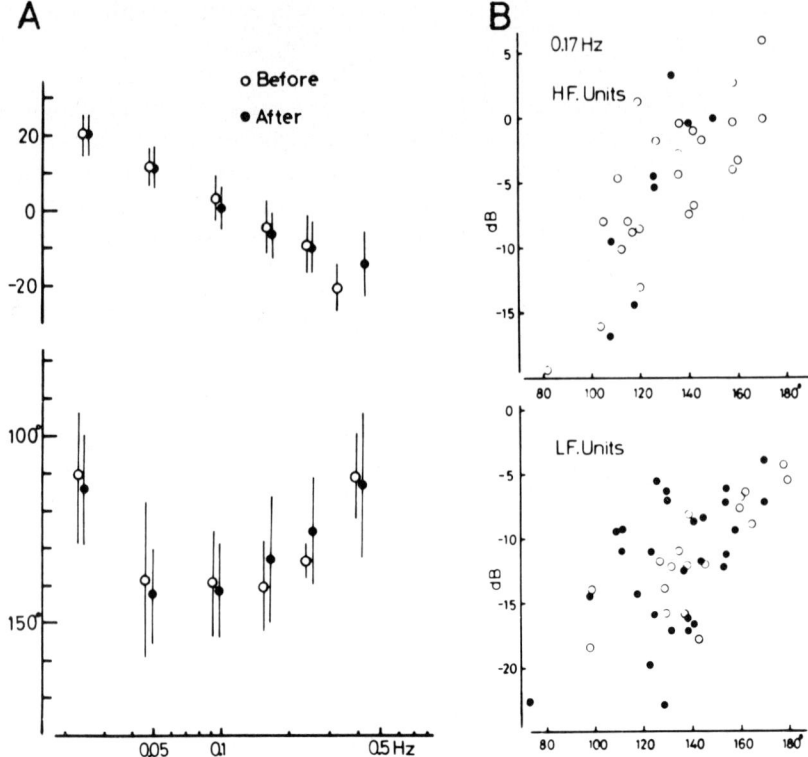

Fig. 5. Effect of dorsal root section on neck extensor motor unit response. A) Mean and SD of the gain and the phase lag at each frequency are plotted on the Bode diagram, before (o) and after (•) deafferentation. B) Effect of deafferentation on HF and LF motor units at 0.17 Hz. In each figure, the abscissa represents phase lag in degrees, the ordinate represents the gain in decibels.

and the spontaneous activity, each pair of open (before) and filled (after deafferentation) circles derived from one animal. Note that when spontaneous activity had been at the intermediate level, the gain of the compound EMG response was decreased. When the spontaneous activity had been high, the gain became larger in one case, but usually decreased with deafferentation. It should be noted that these

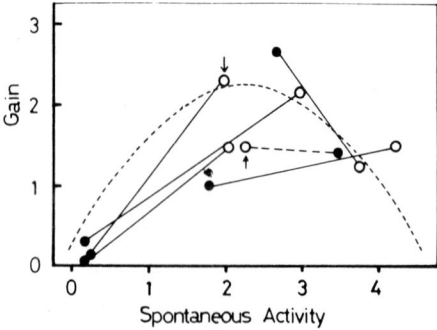

Fig. 6. Effect of deafferentation on the compound EMG response. o and • represent data before and after deafferentation resp. Arrows show data from the same cat. The convex curve,- - - ; relation between the spontaneous activity and the gain, which was obtained in Fig. 3.

alterations took place along the relation between the gain and spontaneous activity before deafferentation (- - - Fig. 6).

From these results it may be concluded that the cervical afferent system, including muscle spindle afferents, regulates the number of motor units participating the vestibulocollic reflex, but it does not directly affect the gain and the phase lag of motor units.

The question may be raised why the gain of the individual motor unit is not changed by deafferentation despite the presence of numerous muscle spindles in neck muscles. The following two points were examined: 1) whether or not the cervical afferents have such a strong monosynaptic connection with motoneurones as to modulate their activity, and 2) whether or not the activity of muscle spindle afferents is modulated during the oscillation of turntable.

Effects of muscle afferents on motoneurones

Since it has been reported that monosynaptic reflex discharge is absent in neck motoneurones (Abrahams et al. 1975), intracellular recording was made from motoneurones innervating dorsal neck muscles. Fig. 7A shows the monosynaptic EPSP and

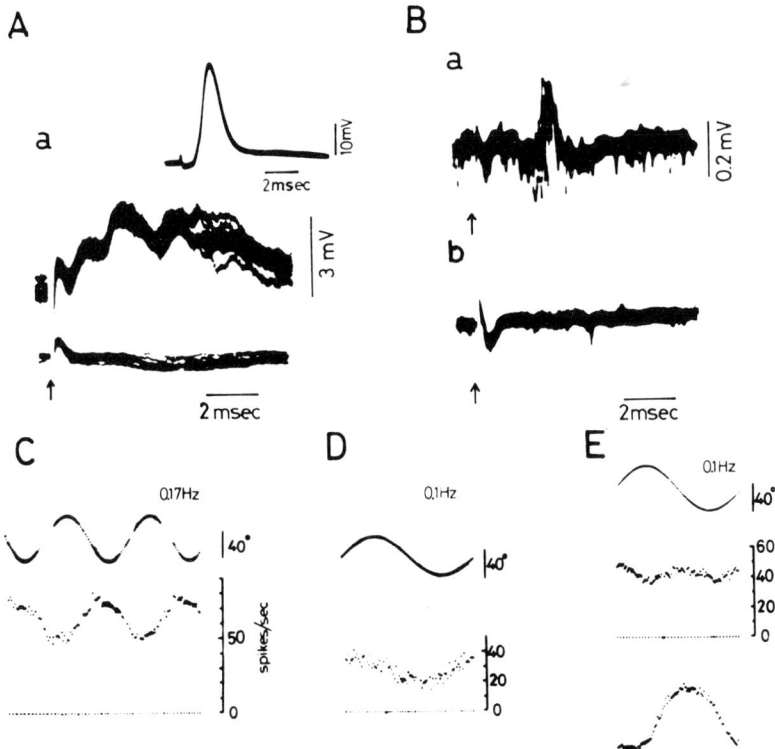

Fig. 7. A) Intracellular recording from C3 motoneurone. Motoneurone was identified by antidromic activation from muscle nerves which innervated SP, BIV and COMP. Monosynaptic EPSP can be seen in response to stimulation of C3 dorsal root. Lower trace: juxtacellular field potential. B) Monosynaptic reflex discharge from C2 nerves in response to stimulation of ipsilateral C3 nerves; nerves were those to SP, BIV and COMP. Note that monosynaptic discharge was observed only when spontaneous activity of nerves was observed (a). C–E) Responses of muscle spindle afferents to sinusoidal oscillation of turntable. In-phase (C). reversed-phase (D) and biphase (E) responses. Lowest trace in E shows the response of compound EMGs.

polysynaptic EPSP with a latency of 10–15 msec when the C3 dorsal root was stimulated at a strength maximal for group II (Wilson and Maeda, 1974; Anderson, 1977). The amplitude of monosynaptic EPSPs was 0.5–3 mV, rather smaller than those observed in triceps motoneurones (Burke, 1973), although in these experiments all Ia afferents from C1–C4 were not stimulated.

Fig. 7B shows the monosynaptic reflex discharges recorded from C2 muscle nerves innervating SP, BIV and COMP muscles, when the C3 dorsal nerve innervating the same muscles was stimulated. Monosynaptic reflex discharges were closely related to the level of the spontaneous activity of motoneurones. When the spontaneous discharge in the nerve was markedly high, they were large (Fig. 7B$_a$), but no monosynaptic reflex discharge could be observed when there was little spontaneous discharge (Fig. 7B$_b$).

Responses of the muscle spindle afferents

Response of muscle spindle afferents to sinusoidal oscillation of the turntable was examined. Recordings were made from single fibers of the C3 dorsal root, the fiber being identified as a spindle afferent from neck extensor muscles (SP, BIV and COMP) with the usual methods (Ezure et al., 1978). Although group I or group II spindle afferents could not be discriminated, 16 afferent fibers were examined at the angular frequencies 0.1 and 0.17 Hz. All of them were spontaneously active (above 10 spikes/sec). Their firing rate was modulated with sinusoidal oscillation of the turntable and three types of responses were observed; in-phase, reversed-phase and biphase responses. Fig. 7C–E shows in-phase (the same mode as EMG response), reversed-phase and biphased responses to oscillation, respectively; the lowest trace in E illustrates the phase characteristics of the compound EMG response recorded simultaneously. In-phase responses were obtained from 10 of 16 spindle afferents examined. Reversed-phase and biphase responses were each seen in 3 of 16. The existence of in-phase responses indicates that the firing rate of spindle afferents is modulated through the γ-system; however, the other types of responses suggest the possibility that intrafusal muscle fibers are passively extended or shortened since the muscles are not maintained fully isometrically under the present experimental condition.

It may be concluded that the γ-loop in the vestibulocollic reflex does not affect the phase lag and the gain of motor units despite the presence of α and γ co-activation and of spindle afferents modulated through the γ-loop.

DISCUSSION

From the frequency response analysis of the vestibulocollic reflex, the existence of a neural integrator was suggested in the pathway between vestibular nucleus neurones and neck motoneurones. The pathways participating in this integration are still unknown. However, in our previous experiments (Ezure et al., 1978) interruption of the direct pathway from vestibular nuclei to neck motoneurones in the MLF (Wilson and Maeda, 1974) caused no detectable effects either on the phase lag or the gain of the motor units, or on the responses of compound EMGs. On the other hand, injection of a small dose of Nembutal (3 mg/kg) caused remarkable phase advancement so that the response was close to the phase of vestibular neurones. This may be attributed to reduction of the activity of polysynaptic pathways which presumably participate in the integration. The neural network involved in this integration has been considered as a reverberating circuit in the brain stem reticular formation from the studies of the vestibulo-ocular reflex (Lorente de Nó, 1933; Robinson, 1971; Skavensky and Robinson, 1973; Shinoda and Yoshida, 1974) the characteristics of which are similar to those of the vestibulocollic reflex (Ezure and Sasaki, 1978). The finding of a positive correlation between the phase lag and the gain of the motor units is expected and compatible with the idea that neural integration was performed by a reverberating circuit in the brain stem. Thus, it appears that such a reverberating circuit may also be functioning in vestibulocollic reflex.

In the present study, motor units in neck extensor muscle were tentatively classified into two groups on the basis of a maximal firing rate of 20 spikes/sec. In this classification the HF units showed, on the average, larger gain than LF units. On the other hand, Burke et al. (1967) have classified motoneurones of triceps surae into FF, FR and S types on the basis of twitch time and fatiguability. S type motoneurones have a lower threshold and are recruited earlier than F type motoneurones. They suggested that maximal instantaneous firing rate of F type motoneurones may be higher than those of S type motoneurones, with a separation of around 25–30 spikes/sec. Since Richmond and Abrahams (1975) in a histochemical study showed that the neck extensor muscles (SP, BIV and COMP) are also composed of three type of muscles (cf. Anderson, 1977), it is hoped that a correspondence between the present classification of the motor units and that of Burke et al. (1967) will be proved in future experiments.

The functional significance of the γ-loop in motor control has been discussed by many investigators since the length follow-up servo hypothesis (Merton, 1953; Granit, 1970). In jaw closing muscles, removal of spindle afferents did not cause any change in chewing movement, suggesting that the contribution of spindle afferents to excitation of motoneurones is not large (Goodwin and Lushei, 1974). Other studies (Vallbo, 1971; Bizzi et al., 1975; Cody and Taylor, 1973; Anderson et al., 1977) also indicate that movements can be accounted for without assuming a large contribution of the γ-route. In the present experiment, neither the gain nor the phase lag of single motor unit responses were affected by deafferentation, suggesting that the γ-loop does not directly control the dynamic characteristics of the vestibulocollic reflex, presumably because the loop has a low gain. This result agrees well above results. On the other hand, firing rate of motor units and spontaneous activity of compound EMGs were usually decreased by deafferentation. The gain of compound EMGs was greatly affected by this decrease. The change of the gain of compound EMG response, however, can only be due to the change in the number of motor units participating in the reflex. The decrease of the DC component of the response induced by deafferentation may not be attributed only to spindle afferents, since in the present experiments the dorsal roots C1–C4 were sectioned as a whole, thereby severing other afferents such as those from joints and skin.

The role of cervical afferents in the vestibulocollic reflex may be summarized as follows for the frequency range employed in this study. In the motoneurone pool, there is a wide variety of motoneurones having different thresholds (Creed et al., 1932; Henneman et al., 1965) and different firing patterns (Granit, 1970; Burke, 1968). The movements of the muscle as a whole are induced by a summed activity of these units, as reflected by compound EMGs. The afferent system may control the gain of whole muscle activity by changing the number of motor units participating the reflex as well as by changing the DC component of individual motor unit firing from all-round to cut off type, but not by changing the gain of individual motor unit response.

SUMMARY

1. Vestibular-induced neck movement (vestibulocollic reflex) was studied with the frequency response method in decerebrate cats. The horizontal semicircular canal was stimulated by sinusoidal oscillation of the turntable. Recordings were made from single motor units or compound EMGs of dorsal neck muscles.

2. Motor units were classified into two groups on the basis of their maximum firing rate: HF (high frequency) and LF (low frequency) units. HF units had a larger gain than LF units.

3. The phase lag of the motor units relative to angular acceleration was about 140° at a frequency range from 0.05 to 0.2 Hz and the gain was decreased at a slope of about 40 dB/decade at a frequency range from 0.01 to 0.4 Hz. A positive correlation was observed between the phase lag and the gain of each motor units. These results suggested the existence of a neural double integrator in the vestibulocollic reflex arc.

4. The gain of compound EMG responses depended on the spontaneous activity. When the spontaneous activity was low or too high, the gain was small. There was an intermediate spontaneous activity level at which the gain became maximal.

5. The dorsal roots, C1–C4, were cut in order to open the feedback loop through γ-fiber spindle afferent system. No effects were observed on the gain and the phase lag of the motor unit response. However, the DC components of the response were consistently decreased by deafferentation both in motor unit and compound EMG responses. Thus, it is suggested that the cervical afferent system controls the gain of the reflex by changing the number of the motor units participating in the vestibulocollic reflex.

REFERENCES

Abrahams, V.C., Richmond, F. and Rose, P.K. (1975) Absence of monosynaptic reflex in dorsal neck muscle of the cat. *Brain Res.*, 92: 130–131.

Anderson, M.E. (1977) Segmental reflex input to motoneurones innervating dorsal neck musculature in the cat. *Exp. Brain Res.*, 28: 175–187.

Anderson, J.H., Soeching, J.F. and Terzuolo, C.A. (1977) Dynamic relations between natural vestibular inputs and activity of forelimb extensor muscles in the decerebrate cat. II. Motor output during rotations in the horizontal plane. *Brain Res.*, 120: 17–33.

Anderson, J.H., Berthoz, A., Soechting, J.F. and Terzuolo, C.A. (1977) Motor output to deafferented forelimb extensors in the decerebrate cat during natural vestibular stimulation. *Brain Res.*, 122: 150–153.

Berthoz, A. and Anderson, J.H. (1971) Frequency analysis of vestibular influence on extensor motoneurons. II. Relationship between neck and forelimb extensors. *Brain Res.*, 34: 376–380.

Bizzi, E., Polit, A. and Morasso, P. (1976) Mechanisms underlying achievement of final head position. *J. Neurophysiol.*, 39: 435–444.

Burke, R.E. (1967) Motor unit types of cat triceps surae muscle. *J. Physiol. (Lond.)*, 193: 141–160.

Burke, R.E., Levine, D.N., Tsairis, P. and Zajac, F.E. (1973) Physiological types and histochemical profiles in motor units of the cat gastrocnemius. *J. Physiol. (Lond.)*, 234: 723–748.

Carli, G., Diete-Spiff, K. and Pompeiano, O. (1967) Responses of the muscle spindles and of the extrafusal fibers in an extensor muscle to stimulation of the lateral vestibular nucleus in the cat. *Arch. ital. Biol.*, 105: 209–242.

Cody, F. W. J. and Taylor, A. (1973) The behaviour of spindles in the jaw-closing muscles during eating and drinking in the cat. *J. Physiol. (Lond.)*, 231: 49P–50P.

Cooper, S. and Daniel, P.M. (1949) Muscle spindles in human extrinsic eye muscles. *Brain*, 72: 1–24.

Ezure, K. and Sasaki, S. (1978) Frequency-response analysis of vestibular-induced neck reflex in cat. I. Characteristics of neural transmission from horizontal semicircular canal to neck motoneurons. *J. Neurophysiol.*, 41: 445–458.

Ezure, K., Sasaki, S., Uchino, Y., and Wilson, V.J. (1978) Frequency-response analysis of vestibular-induced neck reflex in cat. II. Functional significance of cervical afferents and polysynaptic descending pathways. *J. Neurophysiol* 41: 459–471.

Goodwin, G.M. and Luschei, E.S. (1974) Effects of destroying spindle afferents from jaw muscles on mastication in monkeys. *J. Neurophysiol.*, 37: 967–981.

Granit, R. (1970) *The Basis of Motor Control*. Academic Press, New York.

Grillner, S., Hongo, T. and Lund, S. (1969) Descending Monosynaptic and reflex control of γ-motoneurons. *Acta physiol. scand.*, 75: 592–613.

Henneman, E., Somjen, G. and Carpenter, D.O. (1965) Functional significance of cell size in spinal motoneurons. *J. Neurophysiol.*, 28: 560–580.

Kato, M. and Tanji, J. (1971) The effects of electrical stimulation of Deiters' nucleus upon hind limb γ-motoneurons in the cat. *Brain Res.*, 30: 385–395.

Lorente de Nó, R. (1933) Vestibulo-ocular reflex arc. *Arch. Neurol. Psychiat. (Chic.)*, 30: 245–291.

Melvill Jones, G. and Milsum, J.H. (1970) Characteristics of neural transmission from the semicircular canal to the vestibular nuclei of cats. *J. Physiol., (Lond.)*, 209: 295–316.

Melvill Jones, G. and Milsum, J.H. (1971) Frequency-response analysis of central vestibular unit activity resulting from rotational stimulation of the semicircular canals. *J. Physiol. (Lond.)*, 219: 191–215.

Merton, P.A. (1953) Speculation on the servo-control of movement. In *The Spinal Cord*, G.E.W. Wolstenholme (Ed.), Churchill, London, pp. 247–255.

Nyberg-Hansen, R. (1964) Origin and termination of fibers from the vestibular nuclei descending in medial longitudinal fasciculus. An experimental study with silver impregnation methods in the cat. *J. comp. Neurol.*, 122: 355–367.

Nyberg-Hansen, R. (1966) Functional organization of descending supraspinal fiber system to spinal cord. Anatomical observations and physiological correlations. *Ergebn. Anat. Entwickl-Gesch.*, 39: 1–48.

Poppele, R.E. (1967) Response of gamma and alpha motor systems to phasic and tonic vestibular inputs. *Brain Res.*, 6: 535–547.

Richmond, F.J.R. and Abrahams, V.C. (1975) Morphology and enzyme histochemistry of dorsal muscles of the cat neck. *J. Neurophysiol.*, 38: 1312–1321.

Robinson, D.A. (1971) Models of oculomotor neural organization. In *The Control of Eye Movements*, P. Bach-y-Rita and C.C. Collins (Eds.), Academic Press, New York, pp. 519–538.

Shinoda, Y. and Yoshida, K. (1974) Dynamic characteristics of responses to horizontal head angular acceleration in vestibuloocular pathway in the cat. *J. Neurophysiol.*, 37: 653–673.

Skavenski, A.A. and Robinson, D.A. (1973) Role of abducens neurons in vestibuloocular reflex. *J. Neurophysiol.*, 36: 724–738.

Suzuki, J. and Cohen, B. (1964) Head, eye, body and limb movements from semicircular canal nerves. *Exp. Neurol.*, 10: 393–405.

Thompson, J. (1970) Parallel spindle system in the small muscles of the rat tail. *J. Physiol. (Lond.)*, 211: 781–799.

Vallbo, A.B. (1971) Muscle spindle response at the onset of isometric voluntary contraction in man. Time difference between fusimotor and skeletomotor effects. *J. Physiol. (Lond.)*, 318: 405–431.

Wilson, V.J. and Maeda, M. (1974) Connections between semicircular canals and neck motoneurons in the cat. *J. Neurophysiol.*, 37: 346–357.

Wilson, V.J. and Yoshida, M. (1969) Comparison of effects of stimulation of Deiters' nucleus and medial longitudinal fasciculus on neck, forelimb and hindlimb motoneurons. *J. Neurophysiol.*, 32: 743–758.

Wilson, V.J. and Yoshida, M. (1969) Bilateral connections between labyrinths and neck motoneurons. *Brain Res.*, 13: 603–607.

Neck Influences on the Vestibulo-Ocular Reflex Arc and the Vestibulocerebellum

M. MAEDA

Department of Neurosurgery, School of Medicine, Juntendo University, 3-1-3 Hongo, Bunkyo-ku, Tokyo (Japan)

NECK INFLUENCES ON VESTIBULO-OCULAR REFLEX ARC

Since the early work of Magnus and his collaborators (Magnus and De Kleihn, 1913; Magnus, 1924), there have been several investigations on the neck afferent system in postural adjustment (Abrahams and Falchetto, 1969) and its close relation to vestibular function (Fredrickson et al., 1966). The actual origin of the neck afferents involved in this regulation was shown by McCouch et al. (1951) to be the joint receptors for the first three cervical vertebrae. It has also been suggested that the nuchal afferents may play a role not unlike that of the vestibular system (Cohen, 1961).

Besides the body orientation reflex, the neck proprioceptors play an important role in control of eye position (Magnus, 1924). Extraocular muscle tension is influenced by spinal nerve stimulation in the rabbit (Suzuki and Takemori, 1971). Frenzel (1928) studied the influence of neck torsion on nystagmus in man and reported a shift of the beating field of nystagmus. There have been few studies, however, on the neuronal organization between neck afferents and ocular motoneurons. In this respect, the effect of neck afferents upon abducens motoneurons and their interaction with the vestibulo-aducens reflex arc were studied in chloralose anesthetized cats.

Cervical influences on vestibulo-abducens reflex

Action potentials were induced in the left abducens nerve when a single or double shock was applied to the contralateral vestibular nerve. As amplitude of the reflex response (the test reflex) often fluctuated, the test reflex was averaged over 5–10 sweeps (Fig. 1A). Prior to the test shocks the right dorsal root at the level of C2 or C3 segment was stimulated (contralateral conditioning), the amplitude of the test response was thereby markedly reduced (compare Fig. 1A and B). On the other hand, it was remarkably increased by conditioning stimulation of the left (ipsilateral) C2 or C3 dorsal root (Fig. 1C). Conditioning shocks at the level of C2 or C3 segment were most effective for provoking these effects, compared to the effects from more caudal level.

Stimulation was applied to the nerve innervating the ipsi- or contralateral biventer cervicis and complexus muscles. This did not produce any suppression or facilitation of the test reflex. After removal of the dorsal neck muscle, stimulating electrodes were placed on the region of upper neck joints. Contralateral conditioning shocks inhibited the test reflex (Fig. 1E), whereas ipsilateral shocks facilitated it (Fig. 1F). The effects

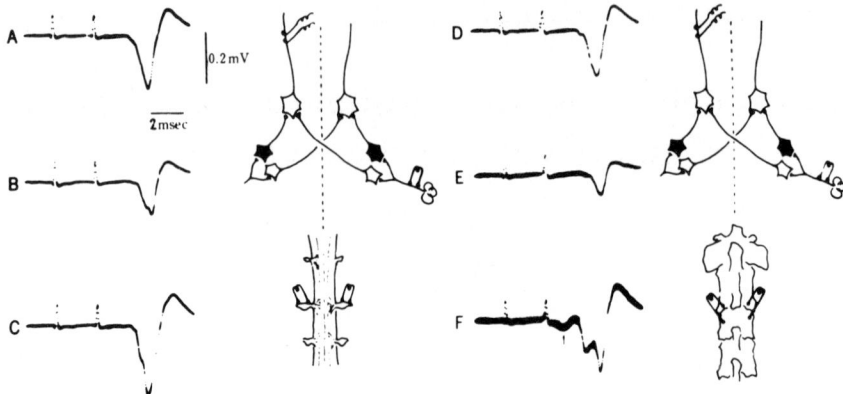

Fig. 1. Effects of cervical afferent volleys on vestibulo-abducens reflex. A, D) Test responses in left abducens nerve to double shocks to right vestibular nerve. Control for B–C and E–F, resp. B, C) Conditioning single shock (not shown on trace) were applied to right (contralateral) (B) and left (ipsilateral) (C) dorsal root at C2 segment. E, F) Conditioning triple shocks were applied to right (E) and left (F) neck joint at C2–C3. All potentials are averages of 5 sweeps. (From Hikosaka and Maeda, 1973.)

elicited by a conditioning shock to the neck joint were similar to those induced from the dorsal root. The inhibitory or the facilitatory effects was blocked by local application of procaine to the stimulated region, indicating that the origin which caused these effects was most likely upper neck joints.

Postsynaptic potentials produced in abducens motoneurons by stimulation of dorsal root and neck joint region at C2–C3

The abducens motoneurons impaled were identified by their antidromic responses to stimulation of the ipsilateral abducens nerve (Fig. 2A). Contralateral dorsal root stimulation produced a hyperpolarizing potential (Fig. 2B). The hyperpolarization was inverted into a depolarizing potential by Cl⁻ ions injection through the recording micro-electrode as shown in Fig. 2C. The reversal of the hyperpolarization to a depolarization was obtained during intracellular passage of hyperpolarizing current as well. Thus the hyperpolarization should represent an IPSP. Similar IPSP were produced by stimulation of the contralateral neck joint.

A depolarizing potential was induced in abducens motoneurons by ipsilateral dorsal root or neck joint stimulation (Fig. 2D). The depolarization was not appreciably changed after Cl⁻ injection into the cell (compare Fig. 2D and E), thus indicating that the depolarizing potential was mainly due to EPSP.

Fig. 2F represents histograms of latencies of the IPSPs and the EPSPs induced by stimulation of the dorsal root and the neck joint. Inspection of the histograms reveals that the latencies of the EPSPs and IPSPs distribute within approximately the same range, suggesting that the pathway mediating the excitation and the inhibition are not very much different in complexity. The histograms also reveal a tendency for the IPSPs and the EPSPs induced from the dorsal root to start slightly earlier than neck joint-induced PSPs. The latency difference, 0.5 msec on the average, was statistically significant.

With respect to the receptive area responsible for neck-induced motor activities, McCouch et al. (1951) limited the receptive field in the region of the upper neck joints, especially at the atlanto-axial and atlanto-occipital joints. In agreement with

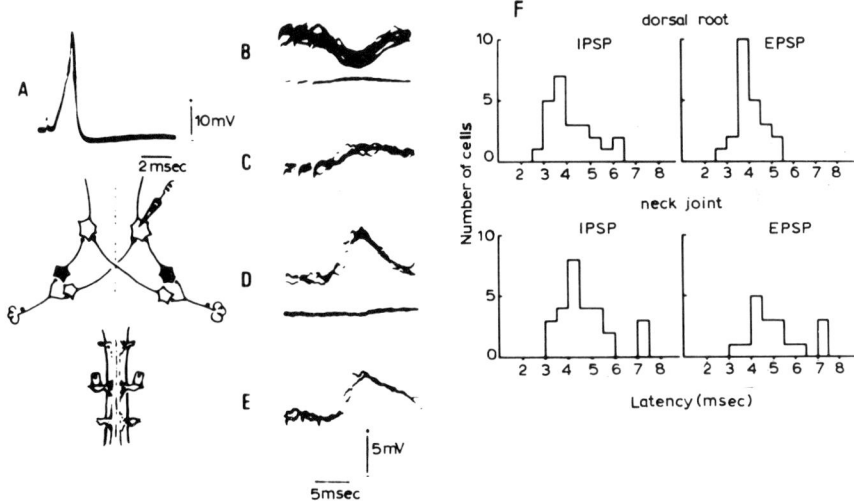

Fig. 2. Synaptic potentials evoked in abducens motoneuron by stimulation of cervical afferents. A) Antidromic response to stimulation of right abducens nerve. B) IPSPs induced by stimulation of left dorsal root (C2), C) Same as in B but after Cl⁻ injection into the cell. D) EPSPs evoked by right dorsal root stimulation. Lower traces in B and D represent extracellular field potentials. E) Same as in D but after Cl⁻ injection. F) Histograms of latencies of PSPs evoked in abducens motoneurons by stimulation of dorsal root and neck joint. (From Hikosaka and Maeda, 1973.)

the previous studies (McCouch et al., 1951; Biemond and De Jong, 1969), the effective volleys in the present experiments may be attributed to those from the upper neck joint, probably intervertebral mechanoreceptors, though some contribution of muscle or fascial afferents cannot completely be excluded.

Interaction of vestibular and cervical effects on abducens motoneurons

Stimulation of the contralateral and ipsilateral vestibular nerve evokes an EPSP and IPSP, respectively, in the abducens motoneurons and these PSPs are disynaptically produced from the nerve, the interneurons being located mainly in the rostral part of the medial vestibular nucleus (Baker et al, 1969). It was investigated whether there is any interaction between vestibular and cervical volleys at the level of the vestibular nuclei. Fig. 3A shows a control disynaptic IPSP (reversed in polarity after Cl⁻ injection) evoked in an abducens motoneurons by stimulation of the ipsilateral vestibular nerve. When conditioned by contralateral cervical stimulation that was so adjusted as to produce only a small IPSP (Fig. 3B), the same test vestibular volley produced a larger disynaptic IPSP (Fig. 3C) than the algebraical sum (Fig. 3C, dotted line). Fig. 3D–F shows a facilitatory convergence in the excitatory pathway to abducens motoneuron from the labyrinth and from the neck. These findings indicate that the sites of labyrinthine and cervical interaction are within the vestibular nuclei. The vestibular nuclei neurons projecting to the abducens nuclei exhibited responses to neck stimulation as expected from the above results (see Fig. 8 of Hikosaka and Maeda, 1973). These results confirmed the previous works suggesting that there is an interaction between labyrinthine and cervical influences on motor performance. In fact, Cohen (1961) noted even more marked labyrinthine deficit-like disturbances in monkeys following section or local anesthesia of the C1–C3 dorsal root.

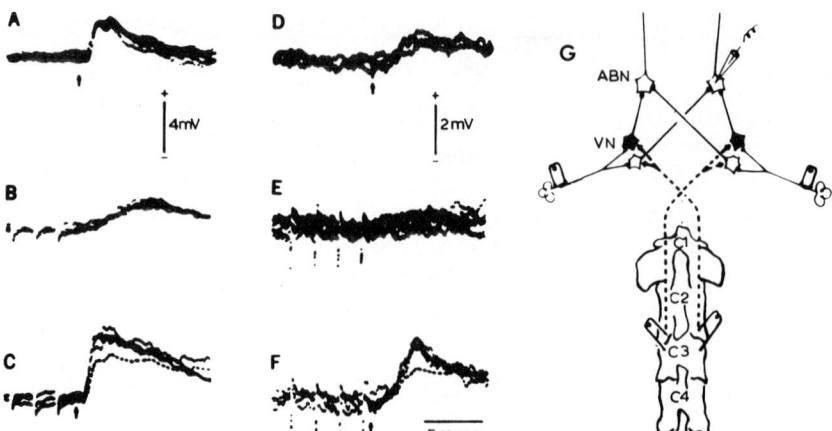

Fig. 3. Interaction of vestibular and cervical effects on abducens motoneurons. A–C) Intracellular records from a right abducens motoneuron. A) Disynaptic IPSPs (reversed in polarity after Cl⁻ injection) evoked by stimulation of right vestibular nerve. B) Response to left neck joint stimulation. C) The same test stimulation as in A was conditioned by neck joint stimulation. Dotted line indicates an algebraical summation of record A and B. D–F) Intracellular record from another right abducens motoneuron. D) Disynaptic EPSPs evoked by left vestibular nerve stimulation. E) Response to right neck joint stimulation. F) The same test stimulation as in D was conditioned by neck joint stimulation. Dotted line indicates an algebraical summation of record D and E. G) Schematic drawing of simplified pathways from neck joints to abducens motoneurons and their interaction with the vestibulo-abducens reflex arc. Broken lines indicate cervical afferent pathways which converge on secondary vestibular neurons and facilitate them. ABN, abducens nucleus; VN, vestibular nucleus. Inhibitory neurons are filled in black and excitatory neurons open. (From Hikosaka and Maeda, 1973.)

Effects of brain stem lesions on cervico-abducens reflex

A longitudinal incision of the dorsal brain stem was made between the left vestibular nuclei and the left abducens nucleus, in order to interrupt the connections between them. After the incision was made, inhibition of the test response in the left abducens nerve by conditioning stimulation of the right neck joint was completely abolished, supporting the view that the inhibitory effects on abducens motoneurons from the neck joint are mediated through the ipsilateral vestibular nuclei. The facilitatory effects were also remarkably decreased after interrupting the vestibulo-abducens excitatory pathway. Transverse hemisection of the spinal cord and longitudinal midline incision in the brain stem indicate that afferent volleys from the neck joint ascend ipsilaterally in the spinal cord, cross to the contralateral side in the brain stem, and eventually project to the vestibular nuclei, thus interacting with the vestibulo-ocular reflex activity (Fig. 3G). Although the precise tract and the location of synapses along the pathway from the spinal cord to the vestibular nuclei remain to be studied, it may not be a multisynaptic, but a fairly direct route, considering that the shortest latency of evoked spikes of vestibular neurons projecting to the abducens nuclei was 2.0 msec after stimulation of the contralateral dorsal root (see Fig. 8 of Hikosaka and Maeda, 1973). The pathway concerned may be different from the spinal ascending route projecting to the Deiters nucleus so far studied. According to Ito et al. (1964), short-latency, presumably monosynaptic, EPSPs were produced in Deiters' neurons after spinal cord (C3) stimulation on the ipsilateral side, instead of the contralateral side, whereas the contralaterally induced PSPs had a fairly long latency such as 8 msec.

It has long been known that compensatory eye movements, a movement which is counter to that of the head and perfectly compensate for it, are critically influenced by vestibular (Szentagothai, 1950) and neck proprioceptive inputs (De Klein, 1918) and play an important role for clear vision and discrimination of visual targets during head movements. Bizzi and his collaborators (Dichgans et al., 1974) investigated the relative contribution of the vestibular and neck afferents to the compensatory eye movement and the mechanisms underlying recovery of compensatory eye movements following bilateral labyrinthectomy in monkeys (Dichgans et al., 1973). They have shown potentiation of the neck-to-eye loop as one of the three main mechanisms underlying recovery after bilateral labyrinthectomy.

The cervico-ocular reflex pathway analyzed here may function conjointly with the vestibulo-ocular reflex to carry the compensatory eye movement and may participate in the neck-to-eye loop underlying recovery after labyrinthectomy.

NECK INFLUENCES ON THE VESTIBULOCEREBELLUM

It has been shown that Purkinje cells in the vestibulocerebellum receive vestibular (Precht and Llinas, 1969; Lisberg and Fuchs, 1974; Ghelarducci et al., 1975) and visual (Maekawa and Simpson, 1973) inputs, and that Purkinje cells in turn inhibit monosynaptically those vestibular neurons which participate in certain vestibuloocular reflexes (Fukuda et al., 1972; Ito et al., 1974). From these results, it is postulated that one of the major function of the flocculus is to control the performance of the vestibulo-ocular reflex. It is, therefore, interesting to investigate whether the vestibulocerebellum is involved in the regulation of cervico-ocular (abducens) reflexes (analyzed here) as well. To begin with, we have searched for an input from neck afferents to the vestibulocerebellum (Wilson et al., 1975; Precht et al., 1976) and have tried to determine the receptors of origin and the brain stem relay of the neck input (Wilson et al., 1976).

Cervical inputs to the flocculus and the nodulus

The ipsilateral vestibular and optic nerves were stimulated. Neck afferents were activated either by stimulation of the upper cervical dorsal rami or C2–C4 dorsal root ganglia. Field potentials were recorded in the flocculus and nodulus, identified by fast green dye marks in frozen sections.

Stimulation of the C3–C4 dorsal rami or dorsal root ganglia usually evoked climbing fiber (CF) field potentials in the flocculus. In almost every instance, the field potentials produced by stimulation of the neck afferents had depth profiles similar to those of optic nerve-evoked potentials, which are the same as previously described for a CF input (Eccles et al., 1968). The latency of the CF potentials produced by ipsilateral or contralateral neck afferent stimulation varied within the range of 6.5–11 msec, but in most cases was 8–10 msec. In some experiments when the C2 ganglion was stimulated, a mossy fiber (MF) input as well as a CF input was observed in the rostral part of the flocculus. The early potentials in the granular and molecular layers were essentially unaffected when the frequency of stimulation was raised from 1/sec to 10/sec, whereas the later, CF, activity was abolished (see Fig. 4 of Wilson et al., 1975). The onset of the N2 potentials was usually at 2.0–3.5 msec and the latency of the N3 was 4–6 msec. Although a MF-potential was sometimes seen on stimulation of the C3 ganglion, it was much more likely to be present when the C2 ganglion was stimulated.

Stimulation of the ipsilateral C2 ganglion evoked field potentials also in the nodulus. These field potentials had depth profiles similar to those evoked by vestibular stimulation (see Fig. 8 of Precht et al., 1976) and may, therefore, be ascribed to the MF pathways. In contrast to the floccusus, which receives MF and CF inputs from upper cervical regions, CF-evoked potentials were not seen in the nodulus after stimulation of C3 ganglion. Berthoz and Llinás (1974) have shown that stimulation of neck afferents causes activity in the anterior lobe by MF and CF pathways. In this literature, stimulation of cervical afferents also causes activity in the flocculus and the nodulus. This cervical inputs to the vestibulocerebellum may participate in control of vestibulo-ocular and cervico-ocular reflexes, both of which are essentially an open loop control system (Ito, 1972).

Locations of relay neurons in the brain stem of MF-cervical projections to the flocculus

Neck afferents reach the flocculus by both MF and CF pathway as described above. As part of an investigation of the role of the flocculus, we concentrated on MF pathways and studied the brain stem relay of the cervical input.

Fig. 4H is a diagram of the experimental arrangement. Stimulation of the ipsilateral C2 ganglion produced a field potential in a discrete area lateral and caudal in the

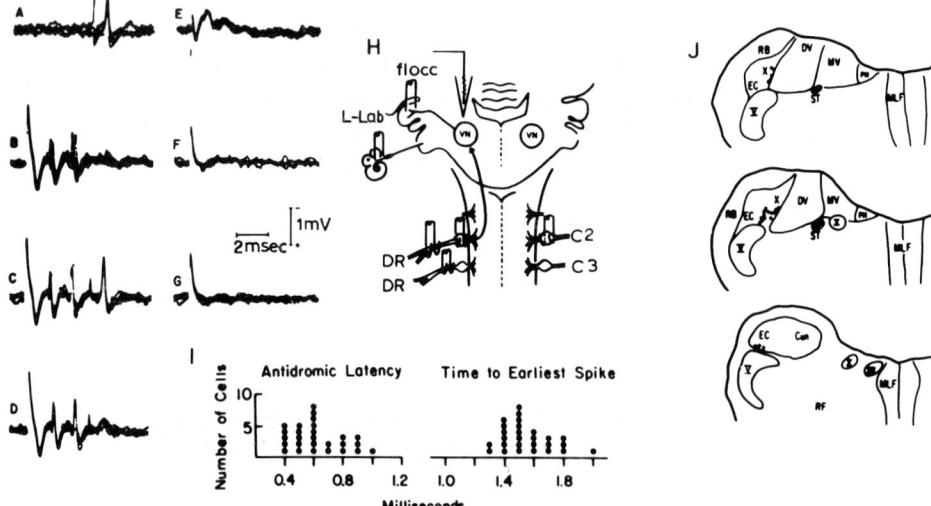

Fig. 4. Response of brain stem neuron to stimulation of the C2 ganglion. A) Antidromic response to flocculus stimulation. B) Stimulation of the left C2 ganglion. C) Stimulus to the ganglion precedes flocculus stimulation, at this interval response to the latter stimulus is not blocked. D) At a close interval the antidromic spike is blocked by collision with the second of the ganglion evoked spikes. E–G) The following three frames show that this neuron does not respond to stimulation of the left vestibular nerve (E), C_2DR (F) or C_3DR (G). H) Experimental arrangement. I) Response latencies of neurons fired orthodromically by stimulation of neck afferents. The left column gives antidromic latencies to stimulation of flocculus. The right column gives the earliest latency of the orthodromic response. J) Location of neurons fired by stimulation of neck afferents. Cun, cuneate nucleus; DV, descending vestibular nucleus; EC, external cuneate nucleus; MLF, medial longitudinal fasciculus; MV, medial vestibular nucleus; PH, nucleus praepositus hypoglossi; RB, restiform body; RF, reticular formation; ST, solitary tract; x, group x; V; spinal tract of the trigeminal. (From Wilson et al., 1976.)

medulla. The positive peak of this potential had a latency of 0.7–1.0 msec. Conduction time between the C2 ganglion and the vestibular nuclei is approximately 0.4–0.5 msec, even for rapidly conducting fibers (Wilson and Yoshida, 1969) and the field potential, therefore, must represent the arrival of a volley in primary afferent fibers. We have studied 27 neurons that responded to stimulation of the C2 ganglion and were fired antidromically by stimulation of the flocculus. The response of a typical cell is shown in Fig. 4. The same unit responded antidromically to floccular stimulation (Fig. 4A) and orthodromically to C2 ganglion stimulation (Fig. 4B). The antidromic response was blocked by collision (Fig. 4D). The earliest orthodromic latency ranged from 1.3 to 2.0 msec (Fig. 4I), 0.5–1.1 msec later than the positive peak of the afferent valley. Most of the neurons therefore responded monosynaptically to C2 ganglion stimulation. Neurons responding to ipsilateral C2 ganglion almost never responded to stimulation of the contralateral C2 ganglion, vestibular nerve, dorsal rami (Fig. 4E–G) or the deep or superficial radial nerve and if a vestibular N1 field potential could be recorded in the vicinity of the neurons it was usually small (Fig. 4E).

The receptor of origin in neck afferents was tested, and by stimulation of the atlanto-axial joint area and by the experiment with incision around the stimulating electrode, it was revealed that neck activity arises, at least in part, in joint receptors (see Fig. 4 of Wilson et al., 1976).

Fig. 4J shows the location of fast green dye marks made after recording from 23 neurons fired by stimulation of the C2 dorsal root ganglion. Most of the neck relay cells are in the region called group x by Brodal and Pompeiano (1957) and in the ventrolateral part of the descending vestibular nucleus. Pompeiano and Brodal (1957) noted abundant degeneration in group x following a lesion at C1 and restriction of degeneration to the dorsal part of group x with lesion at L3. This cell group, thus, is known to receive spinal afferents and it has been suggested by Brodal (1974) that it serves to relay spinal impulses to the vestibulocerebellum.

The pathway from neck to the flocculus via group x and the ventrolateral part of the descending nucleus is quite specific. Input from neck receptors is ipsilateral and there is no obvious interaction, at the brain stem level, between neck and either labyrinthine or forelimb impulses. With regard to cervicolabyrinthine interaction, this pathway

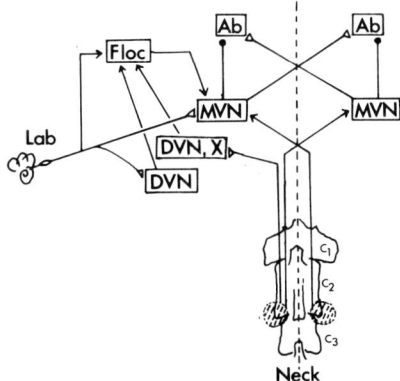

Fig. 5. Schematic diagram summarizing neck influences on the vestibulo-abducens reflex arc and the vestibulocerebellum (flocculus). Ab, abducens motoneuron; DVN, descending vestibular nucleus; Floc, flocculus; Lab, vestibular nerve; MVN, medial vestibular nucleus.

differs from cervico-ocular reflex pathway, in which neck joint and labyrinthine impulses converge at the level of the vestibular nuclei. The neck and labyrinthine MF relays are separate at the brain stem level (Fig. 5), but cause activity in the same areas of the flocculus. More information at the unit level is required for profitable speculation about the role of the MF and the CF neck information reaching the flocculus. Activity in neck afferents causes cervico-ocular reflex and participates in eye-head cordination under some conditions (Dichgan et al., 1973). The flocculus is involved in eye movement control, and neck as well as visual information (Maekawa and Simpson, 1973) may be required for the proper performance of the control.

SUMMARY

Contralateral and ipsilateral neck joint (C2–C3) stimulation induced IPSPs and EPSPs, respectively, in abducens motoneurons of the cat. The labyrinthine induced disynaptic IPSP or EPSP was facilitated by conditioning stimulation of the contralateral and ipsilateral neck joint, respectively. It was thus shown that the cervico-abducens and vestibulo-abducens reflex pathways converge upon common inhibitory or excitatory interneurons in the vestibular nuclei. Lesion experiments in the brain stem indicated that afferent volleys from the neck joint ascend ipsilaterally in the spinal cord, cross to the contralateral side in the brain stem, and project to the vestibular nuclei, thus interacting with the vestibulo-abducens reflex activity (Hikosaka and Maeda, 1973).

Stimulation of the C2–C4 dorsal rami or dorsal root ganglia usually evoked mossy fiber (MF) and climbing fiber (CF) field potentials in the flocculus, which receives inputs from vestibular and visual afferent (Wilson et al., 1975). In the nodulus, stimulation of C2 ganglion evoked MF field potnetials. In contrast to the flocculus, no CF evoked field potentials were seen in the nodulus after C2 ganglion stimulation (Precht et al., 1976).

With respect to MF pathways to the flocculus, we have tried to determine the receptors of origin and the brain stem relay of the neck input. Neck activity arises, at least in part, in joint receptors and the relay is located principally in Brodal and Pompeiano's group x and in the ventrolateral part of the descending nucleus (Wilson et al., 1976).

REFERENCES

Abrahams, V.X. and Falchetto, S. (1969) Hind leg ataxia of cervical origin and cervico-lumbar spinal interactions with a supratentorial pathway. *J. Physiol. (Lond.)*, 203: 435–447.

Baker, R.G., Mano, N. and Shimazu, H., (1969) Postsynaptic potentials in abducens motoneurons induced by vestibular stimulation. *Brain Res.*, 15: 577–580.

Berthoz, A. and Llinás, R. (1974) Afferent neck projection to cat cerebellar cotex. *Exp. Brain Res.*, 20: 385–402.

Biemond, A. and De Jong, J.M.B.V. (1969) On cervical nystagmus and related disorders. *Brain*, 92: 437–458.

Brodal, A. (1974) Anatomy of the vestibular nuclei and their connections. In *Handbook of Sensory Physiology, Vol. VIII, Vestibular System, Part 1, Basic Mechanisms* H.H. Kornhuber (Ed.), Springer-Verlag, Berlin, pp. 239–352.

Brodal, A. and Pompeiano, O. (1957) The vestibular nuclei in the cat. *J. Anat. (Lond.)*, 91: 438–454.

Cohen, L.A. (1961) Role of eye and neck proprioceptive mechanisms in body orientation and motor coordination. *J. Neurophysiol.*, 24: 1–11.
De Klein, A. (1918) Action reflexes du labyrinthe et du cou sur les muscles de l'oeil, *Arch. néerl. Physiol.*, 2: 644–649.
Dichgans, J., Bizzi, E., Morasso, P. and Tagliasco, V. (1973) Mechanisms underlying recovery of eye-head coordination following bilateral labyrinthectomy in monkeys. *Exp. Brain Res.*, 18: 548–562.
Dichgans, J., Bizzi, E., Morasso, P. and Tagliasco, V. (1974), The role of vestibular and neck afferents during eye-head coordination in the monkey. *Brain Res.*, 71: 225–232.
Eccles, J.C., Provini, L., Strata, P. and Taborikova, H., (1968) Analysis of electrical potentials evoked in the cerebellar anterior lobe by stimulation of hindlimb and forelimb neurones. *Exp. Brain Res.*, 6: 171–194.
Fredrickson, J.M., Schwarz, D. and Kornhuber, H.H. (1966) Convergence and interaction of vestibular and deep somatic afferents upon neurons in the vestibular nuclei of the cat. *Acta oto-laryngol. (Stockh.)*, 61: 168–188.
Frenzel, H. (1928) Rucknystagmus als Halsreflex und Schlagfeldverlagerung des labyrinthären Drehnystagmus durch Halsreflexe. *Z. Hals, Nasen. Ohrenheilk.*, 21: 177–187.
Fukuda, J., Highstein, S. and Ito, M. (1972) Cerebellar inhibitory control of the vestibulo-ocular reflex investigated in rabbit IIIrd nucleus. *Exp. Brain Res.*, 14: 511–526.
Ghelarducci, B., Ito, M. and Yagi, N. (1975) Impulse discharges from flocculus Purkinje cells of alert rabbits during visual stimulation combined with horizontal head rotation. *Brain Res.*, 87: 66–72.
Hikosaka, O. and Maeda, M. (1973) Cervical effects on abducens motoneurons and their interaction with vestibulo-ocular reflex. *Exp. Brain Res.*, 18: 512–530.
Ito, M. (1972) Neural design of the cerebellar motor control system. *Brain Res.*, 40: 81–85.
Ito, M., Hongo, T., Yoshida, M., Okada, Y. and Obata, K. (1964) Antidromic and transsynaptic activation of Deiters' neurones induced from the spinal cord. *Jap. J. Physiol.*, 14: 638–658.
Lisberg, S.G. and Fuchs, A.F. (1974) Response of flocculus Purkinje cells to adequate vestibular stimulation in the alert monkey: fixation vs. compensatory eye movements. *Brain Res.*, 69: 347–353.
Maekawa, K. and Simpson, J.I. (1973) Climbing fiber responses evoked in vestibulocerebellum of rabbit from visual system. *J. Neurophysiol.*, 36: 649–666.
Magnus, R. (1924) *Körperstellung.* Julius Springer, Berlin.
Magnus, R. and De Kleijn, A. (1913) Analyse der Folgezustande einseitiger Labyrinth-Extirpation mit besonderer Berucksichtigung der Rolle der Tonischen Halsreflexe. *Pflügers Arch.*, 154: 178–306.
McCouch, G.P., Deering, I.D. and Ling, T.H. (1951) Location of receptors for tonic neck reflexes. *J. Neurophysiol.*, 14: 191–195.
Pompeiano, O. and Brodal, A. (1957) Spinovestibular fibers in the cat. An experimental study. *J. comp. Neurol.*, 108: 353–382.
Precht, W. and Llinás, R. (1969) Functional organization of the vestibular afferents to the cerebellar cortex of frog and cat. *Exp. Brain Res.*, 9: 30–52.
Precht, W., Volkind, R., Maeda, M. and Giretti, M.L. (1976) The effects of stimulating the cerebellar nodulus in the cat on the responses of vestibular neurons. *Neuroscience*, 1: 301–312.
Suzuki, J. and Takemori, S. (1971) Eye movements induced from the spinal nerves. *Equilibrium Res.*, Suppl., 2: 33–40.
Szentagothai, J. (1950) The elementary vestibulo-ocular reflex arc. *J. Neurophysiol.*, 13: 395–407.
Wilson, V.J. and Yoshida, M. (1969) Comparison of effects of stimulating of Deiters' nucleus and medial longitudinal fasciculus on neck, forelimb hindlimb motoneurons. *J. Neurophysiol.*, 32: 743–758.
Wilson, V.J., Maeda, M. and Franck, J.I. (1975) Input from neck afferents to the cat flocculus. *Brain Res.*, 89: 133–138.
Wilson, V.J., Maeda, M., Franck, J.I. and Shimazu, H. (1976) Mossy fiber neck and second order labyrinthine projections to cat flocculus. *J. Neurophysiol.*, 39: 301–310.

Vestibular-Neck Interaction in Abducens Neurons*

U. THODEN and P. SCHMIDT

Neurologische Klinik, Abteilung für Neurophysiologie, Universität Freiburg i. Br., 7800 Freiburg i. Br. (F.R.G.)

INTRODUCTION

During a movement of the head the labyrinthine and neck proprioceptors interact to produce compensatory eye movements in order to maintain stable vision (Magnus, 1924). This effect seems to be achieved by the convergence of activity in the cervico-ocular and vestibulo-ocular pathways upon common inhibitory or excitatory interneurons in the vestibular nuclei (Hikosaka and Maeda, 1977). In the case of neurons in the vestibular nuclei and the surrounding reticular formation an interaction of vestibular and cervical input by natural stimulation is known (Fredrickson et al., 1965; Rubin et al., 1975; Thoden and Wirbitzky, 1976). The mode of interaction of both receptor systems still seems to be a matter for discussion. In the case of spinal motoneurons opponent effects were demonstrated during turning of the head (Erhardt and Wagner, 1970; Rosenberg and Lindsay, 1973), although for compensatory eye movements a summation of both inputs was already assumed by Magnus (1924).

At the level of the vestibular nuclei an additive as well as an opponent type of interaction was found experimentally for both inputs during head movements (Rubin et al. 1975; Thoden and Wirbitzky, 1976). It therefore may be assumed that opposing effects take place in vestibular-spinal neurones, whereas ascending pathways are synchronously activated by both receptor systems.

The present study is intended to elucidate the effects of neck afferents and their interaction with the vestibulo-ocular reflex upon abducens motoneurons, by means of natural stimulation with comparable parameters.

METHODS

The results were obtained with precollicularly decerebrated adult cats. In addition, medial parts of the cerebellum were removed by gentle suction in order to visualize the floor of the 4th ventricle. The recording was commenced about 2 h after cessation of anesthesia. The right eye was enucleated and the abducens nerve dissected free in the orbit to facilitate bipolar electrical stimulation (Precht et al., 1969).

---

*Supported by Sonderforschungsbereich Hirnforschung (SFB 70) der Deutschen Forschungsgemeinschaft (DFG).

All animals were fixed with the head in the prone position in a standard stereotaxic head holder. The body was supported by a frame allowing movements of 25° to each side in the horizontal position. The spinal processes were clamped in order to extend the animal caudally. This frame could be driven sinusoidally with variable amplitudes and frequencies. The whole apparatus was then mounted in the center of a turntable (Tönnies) with the animal's head placed in the center of rotation. The equipment thus permitted a sinusoidal swing of the whole turntable as well as sinusoidal body movements with the fixed head. During analysis the amplitude and frequency of vestibular and neck signals were of comparable size. Extracellular action potentials and field potentials were recorded from the abducens nucleus using glass micropipettes. The electrodes were inserted in the dorso-ventral direction with a micromanipulator into the region of the nucleus abducens. The units were localized by antidromic stimulation according to the method described by Precht et al. (1969).

The action potentials from single units were selected by an amplitude discriminator and their sequence analysed during a period of sinusoidal movements.

RESULTS

Identification of abducens neurons

As previously shown by Precht et al. (1969) antidromic stimulation of the VIth nerve allows a quick identification of the abducens nucleus on the floor of the 4th ventricle. The field potentials (Fig. 1) evoked by single shock stimulation are characterized by a positive wave after 0.6 msec, as long as the tip is above the ventral border of the nucleus. On penetration of the nucleus a large negative wave is recorded after 1 msec peak latency. Tracking through the VIth nucleus it seems that preponderance of positive wave marks the border region of the nucleus. The negative wave is enhanced if the electrode tip reaches the nucleus (Fig. 1). As postulated by Precht et al. (1969), the short latency of this negative wave, the response to short-interval double shocks and the histological correlations make it reasonable to assume that the negativity is due to antidromic activation of the pool of abducens motoneurons. With increased stimulus strength unitary action potentials were often superimposed on the negative fields. The input from the horizontal semicircular canal during sinusoidal animal rotation was compared with the input from neck receptors during horizontal sinusoidal body movements, with the head fixed. Frequencies for both inputs were between 0.2 and 1 Hz and amplitudes ranged up to 40°. Thus, a comparison of the effects of the two receptor systems was possible.

Vestibular influence

68 abducens neurons were found to respond to head rotation. Of these, 62 cells were modulated by input from the horizontal semicircular canal during sinusoidal animal rotation, increasing their discharge during contralateral movement and decreasing it during ipsilateral torsion (type IIa). Six cells responded with an opposite pattern (type Ia). Among the labyrinthine dependent cells 20 neurons were spontaneously silent, but were activated during rotation with a type IIa pattern.

A directional non-specific modulation with bilateral activation or inhibition was not found.

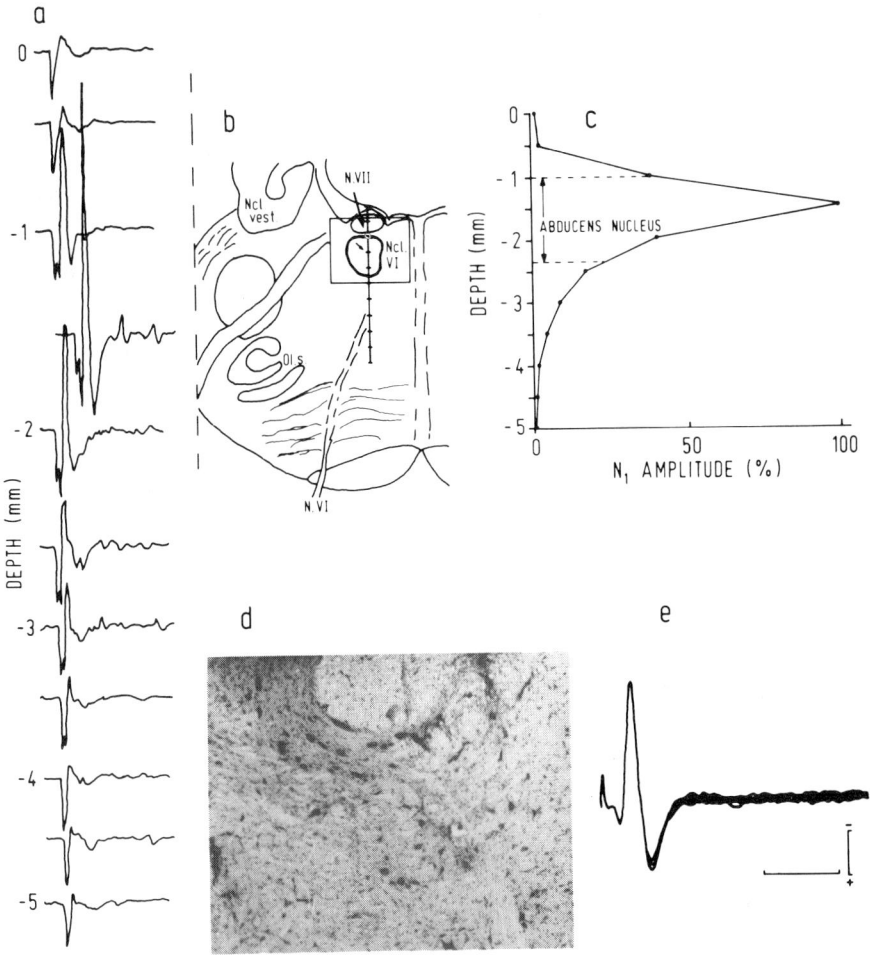

Fig. 1. Field potentials of the abducens nucleus evoked by VIth nerve stimulation. One microelectrode track is represented with its field potentials in steps of 500 μm (a), the line drawing (b) and the histological section (d) of the brain stem; c shows the relation of the negative field potentials to the nucleus abducens; e, a typical field potential. In all records upward deflection indicates negativity of the microelectrode. Time: 2 msec; calibration: 300 μV.

Neck input

Out of the neurons with labyrinthine modulation 42 cells (61.8%) responded to both vestibular and neck inputs, 13 of these during contralateral "head" movement in phase with the velocity signal, and in phase with the vestibular input (Fig. 2a, b; Table I). Another 11 neurons were activated with their maximal discharge rate around the ipsilateral position (Fig. 2c, d). This last response type does not seem to be due to a position sensitivity, since these cells did not respond tonically during extreme neck torsion.

Of the above-mentioned 20 silent motoneurons 5 showed a neck input, increasing their discharge during contralateral head movement.

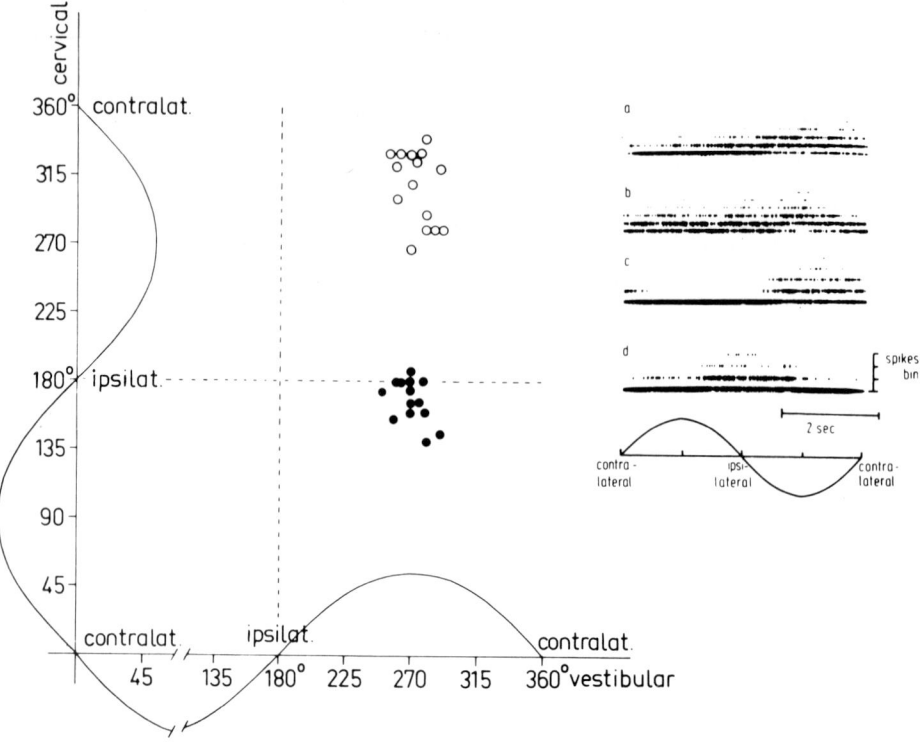

Fig. 2. Maximum discharge rate of abducens motoneurons (type IIa) during isolated neck and labyrinthine input by sinusoidal swing at 0.2 Hz with amplitudes of 40° At the right, different types of neck-vestibular interaction on type IIa abducens motoneurons. a, c) Labyrinthine response of two different motoneurons. b,d) Corresponding response by neck input alone.

TABLE I

LABYRINTHINE MODULATION

		Ipsilateral	Acceleration Contralateral Spontaneously active	Silent motoneurons
Cervical modulation	Ipsilateral	1	11	3
	During contralateral movement	1	13	2
	No modulation	3	13	14
	Lost	1	5	1
	Total	6	42	20

DISCUSSION

The elementary vestibulo-ocular reflex arc is composed of excitatory as well as inhibitory parts, acting on the abducens nucleus disynaptically from the primary vestibular nerve (Baker et al., 1969; Lorente de Nó, 1933, Szenthagotai, 1950; Tarlov, 1970). The second-order vestibular neurons interposed in this reflex arc cross the midline at the level of the medial longitudinal fascicles to reach the abducens motoneuron pool. Above this direct connection multisynaptic reticular pathways can transmit vestibular activity to the ocular motoneurons.

Precht et al. (1969) subdivided cells in the abducens nucleus into two main groups according to their response to horizontal angular acceleration. The discharge rate of the motoneurons was increased with contralateral acceleration and decreased with ipsilateral acceleration (type IIa). These motoneurons are activated antidromically from the abducens nerve and receive their input from the contralateral semicircular canal. Another group of units, whose axons do not project to the eye muscle, showed a reverse pattern during horizontal acceleration (type Ia). They were never excited by antidromic stimulation from the abducens nerve. These type Ia units, excited on stimulation by the ipsilateral vestibular nerve, consist probably of two groups, axons of type I vestibular neurons, and interneurons. Richter and Precht (1969) concluded that type Ia units are inhibitory, thus providing disynaptic as well as polysynaptic inhibitory pathways from the ipsilateral vestibular nerve to the motoneurons of the abducens nucleus. Thus, inhibitory control of ipsilateral motoneurons complements predominance of excitatory innervation of the contralateral neurons in a pattern of reciprocal innervation.

Our results confirm that the main excitatory labyrinthine input to the abducens neurons arises in the contralateral semicircular canal. As all the cervical-dependent abducens units receive a labyrinthine input, the two pathways probably converge at the level of the vestibular nuclei, where an interaction is known (Fredrickson et al, 1965; Rubin et al., 1975; Thoden and Wirbitzky, 1976). Our results thus favor the postulate of Hikosaka and Maeda (1977), that the cervical-abducens and vestibular-abducens reflex pathways converge upon common inhibitory or excitatory interneurons in the vestibular nuclei.

According to our data for cervico-ocular activation, the predominant influence is exerted by contralateral vestibular cells activating the abducens motoneurons. The effect via ipsilateral vestibular cells with type Ia activation seems negligible.

Thus the influence of the cervico-ocular reflex on the abducens nucleus does not have the same antagonistic inhibitory power as that known to be exerted by the vestibulo-ocular reflex. But the cervico-ocular reflex also works with two different mechanisms, one with its maximum activation during contralateral head movement in phase with velocity and the vestibular input, and the second with an activation around the ipsilateral position, possibly in phase with contralateral acceleration. The first type of neck influence seems to be mediated via contralateral vestibular type I neurons. From the study of Hikosaka and Maeda (1977) it seems that excitation during ipsilateral head turn is due to activation of ipsilateral interneurons converging on to the same abducens motoneurons.

SUMMARY

Single cell activity in the abducens nucleus, identified by antidromic nerve stimulation was extracellularly recorded in decerebrate unanesthetized cats.

The input from the horizontal semicircular canals during sinusoidal animal rotation was compared with the input from neck receptors during horizontal sinusoidal body movement while the head was fixed. Frequencies of body oscillation were between 0.2 and 1 Hz and amplitudes ranged up to 40°.

Among the 68 neurons with labyrinthine input, 62 were activated by contralateral and 6 by ipsilateral head motion. With rare exceptions only the contralaterally activated cells were also modulated by neck inputs. Among those, 54% were in phase with velocity and 46% in phase with acceleration. Modulation by neck afferents was never found in cells without labyrinthine input.

REFERENCES

Baker, R.G., Mano, N. and Shimazu, H. (1969) Postsynaptic potentials in abducens motoneurons induced by vestibular stimulation. *Brain Res.*, 15: 577–580.

Ehrhardt, K.J. and Wagner, A. (1970) Labyrinthine and neck reflexes recorded from spinal single motoneurons in the cat. *Brain Res.*, 19: 87–104.

Fredrickson, J.M., Schwarz, D. and Kornhuber, H.H. (1965) Convergence and interaction of vestibular and deep somatic afferents upon neurons in the vestibular nuclei of the cat. *Acta oto-laryngol. (Stockh.)*, 61: 168–188.

Hikosaka, O. and Maeda, M. (1973) Cervical effects on abducens motoneurons and their interaction with vestibulo-ocular reflex. *Exp. Brain Res.*, 18: 512–530.

Lorente de Nó, R. (1933) Vestibulo-ocular reflex arc. *Arch. Neurol. Psychiat. (Chic.)*, 30: 245–291.

Magnus, R. (1924) *Körperstellung*. Julius Springer, Berlin.

Precht, W., Richter, A. and Grippo, J. (1969) Responses of neurones in cat abducens nuclei to horizontal angular acceleration. *Pflügers Arch.*, 309: 285–309.

Rosenberg, J.R. and Lindsay, K.W. (1973) Asymmetric tonic labyrinthine reflexes. *Brain Res.*, 63: 347–350.

Rubin, A.M., Young, J.H., Milne, A.C., Schwarz, D.W.F. and Fredrickson, J.M. (1975) Vestibular-neck integration in the vestibular nuclei. *Brain Res.*, 96: 99–102.

Szentagothai, J. (1950) The elementary vestibulo-ocular reflex arc. *J. Neurophysiol.*, 13: 395–407.

Tarlov, E. (1970) Organization of vestibulo-oculomotor projections in the cat. *Brain Res.*, 20: 159–172.

Thoden, U. and Wirbitzky, J. (1976) Influence of passive neck movements on eye position and brain stem neurons. *Pflügers Arch.*, Suppl. 362: R 37, no. 147.

Vestibular Influences on the Cat's Cerebral Cortex

T. MERGNER

Neurologische Klinik, Universität Ulm, Ulm-Donau (F.R.G.)

The existence of true vestibular sensation in close cooperation with somatic and visual senses had been anticipated by Mach and contemporaries at the turn of the last century from observations of deaf mutes and operated animals (cf. Mach, 1902). Conclusive evidence for a cortical representation receiving vestibular input, however, has been obtained only recently from animal experiments. Walzl and Mountcastle (1949) evoked cortical field potentials in the region of the anterior suprasylvian sulcus (ASSS) of the cat upon electrical stimulation of the vestibular nerve. Subsequently, a cortical vestibular field has been described in the rhesus monkey at the tip of the intraparietal sulcus (Fredrickson et al., 1966). Surveys of current literature about vestibular influences on thalamic structures and on the cerebral cortex have been given a few years ago by Copack et al. (1972), Fredrickson et al. (1974), Kornhuber (1972) and Schwarz and Fredrickson (1974). The present article gives a short account of recent studies on this topic done in the cat including data from our laboratory (Becker et al., 1979; Deecke et al., 1979; Wagner et al., 1978). For data on vestibular cortex in the rhesus monkey see Büttner and Lang (this volume, chapter IV D9).

CORTICAL VESTIBULAR REPRESENTATIONS

Field potential studies, using well-defined electrical stimulation applied to the vestibular nerve or its branches, established the existence of a vestibular representation in the anterior bank of the ASSS (Andersson and Gernandt, 1954; Landgren et al., 1967; Mickle and Ades, 1952). In the lightly anesthetized cat the evoked responses appear with a latency as short as 3.5 msec (Mills and Taylor, 1974) (see also Fig. 1Aa). This finding suggests an oligosynaptic vestibulo-thalamocortical pathway. Single-neuron recordings in the ASSS field revealed specific responses to galvanic and caloric labyrinthine stimulation (Kornhuber and daFonseca, 1964). A second vestibular representation in the cerebral cortex of the cat has been found in the region of the postcruciate dimple (Sans et al., 1970) being located within the somatosensory forelimb region of area 3a (Ödkvist et al., 1975a). Also neurons in the pericruciate cortex (motor and premotor areas) have been found to respond to galvanic and caloric stimulation of the labyrinths (Kornhuber and Aschoff, 1964) and to natural vestibular stimulation (lateral tilt; Boisacq-Schepens and Roucoux-Hanus, 1975). Vestibular evoked activity

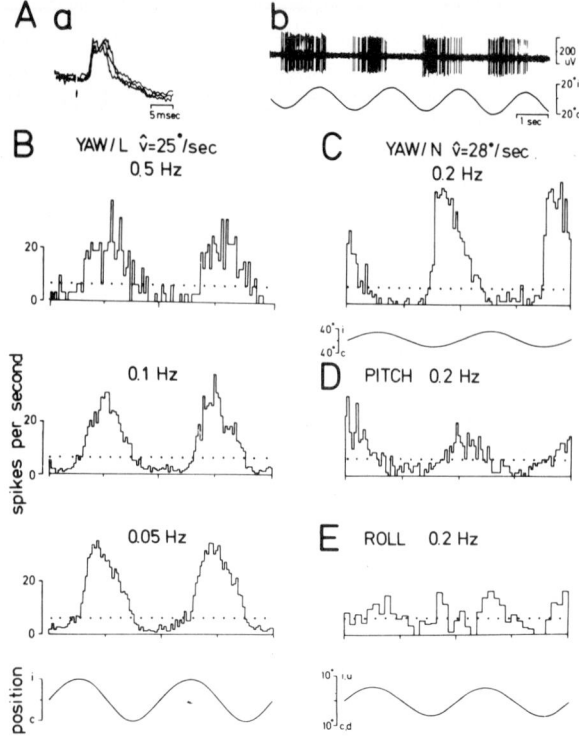

Fig. 1. Aa) Early negative field potential recorded intracortically from the anterior bank of the ASSS following electrical stimulation of the contralateral labyrinth. Ab) Original responses of type I unit recorded extracellularly from the left ASSS region to sinusoidal rotation in the horizontal plane in darkness. B–E) Averaged date from a representative type II neuron. Responses to horizontal rotation of the whole animal (YAW/L) at constant peak angular velocity of 25 degrees/sec and at different stimulus frequencies as indicated (B). Note phase advance at 0.05 Hz. The unit also responded to horizontal rotation of the trunk alone (neck stimulus; YAW/N; cf.C), and to sinusoidal rotation in pitch (D). No response was evoked by rotational stimuli in roll (E). The data are presented as instantaneous frequency histograms which were averaged over 8 double cycles of the stimulus. Response at 0.5 Hz in B appears rippled because of the short bin width used. Dotted horizontal lines indicate discharge frequency at rest. Marks on abscissae give instances of peak angular velocity. Sine waves at bottom show chair position; upward refers to ipsilateral (i) or nose-up (u), and downward to contralateral (c) or nose-down (d) excursions, resp.

in the primary visual cortex (Grüsser and Grüsser-Cornehls, 1972) has longer and variable latencies and is considered to be "non-specific". Up to now detailed information about response characteristics of the cat's cortical vestibular neurons to natural stimulation has been missing.

In the study of Becker et al. (1979) and Deecke et al. (1979) chronically prepared cats were used, which were paralysed during the recording sessions by muscular relaxation. Electrical stimulation studies on the ASSS vestibular field were repeated and extended to natural vestibular stimuli. Intracortical field potentials upon single electrical shocks applied to round window electrodes were recorded in a narrow region within the ASSS anterior bank at latencies of 3.5 msec (Fig. 1Aa). Neurons recorded extracellularly from this region responded to single electrical shocks and galvanic stimuli applied to the labyrinths of both sides, and to natural vestibular stimuli. Single-shock responses consisted of few if any early spikes on top of the first negative field (weak "excitation"), a subsequent pause in spontaneous firing rate ("inhibition") and a massive discharge mostly associated with the second negative potential deflection (strong "rebound"). Galvanic stimuli yielded specific responses corresponding to those of Kornhuber and daFonseca (1964). In agreement with earlier authors the effects observed were stronger upon contralateral stimulation as compared to ipsilateral stimulation.

Consistent responses to natural vestibular stimuli were obtained only in the awake cat after termination of anesthesia (Fluothane/N_2O). Only a minor proportion of the neurons investigated in the cortical field responded to natural vestibular stimulation, which usually consisted of sinusoidal rotation in the horizontal plane (yaw) in darkness. Among the responsive neurons (n = 110), about half showed an increase in firing rate during ipsilateral rotation and a decrease during contralateral rotation (type I response according to Duensing and Schaefer, 1958; cf. also Fig. 1Ab), while the other half revealed the reciprocal response pattern (type II; cf. Fig. 1B). Neurons with either increase or decrease in firing rate during rotation in both directions (type III and type IV responses, respectively) were rarely observed.

No major differences in response characteristics or spontaneous rate (on the average 9 spikes/sec) were observed between type I and type II units. The sinusoidal stimulus evoked a sine-like frequency modulation around resting discharge in only a minor portion of the neuronal population investigated. The majority of neurons showed responses with "cut-off" of the inhibitory half wave. However, neuronal firing was usually not completely silenced, but there remained a certain level of minimum firing rate (cf. Fig. 1B). Furthermore, the neurons showed skewness of the slopes and, occasionally, "DC-shifts". These non-linearities observed for cortical vestibular units appeared to exaggerate those described in vestibular nuclear neurons (cf. Melvill Jones and Milsum, 1970; Shinoda and Yoshida, 1974). Due to the early "cut-off" of the inhibitory half wave, many ASSS neurons essentially coded rotation in the excitatory direction only. Still, both directions of rotation were well represented in the population as a whole because of equal numbers of such type I and type II units. Maximum firing rate above resting discharge increased almost linearly with growing stimulus strength in the low velocity range (0–20 degrees/sec), but usually saturated at higher velocities (up to 60 degrees/sec). At a sinusoidal frequency of 0.2 Hz and peak velocity of 25 degrees/sec, excitatory sensitivity – defined as maximum frequency deviation from resting discharge divided by peak angular velocity – averaged 0.55 spikes $\cdot$ sec^{-1}/degree $\cdot$ sec^{-1}.

On average, the maximum modulation of neuronal firing was found to be roughly in phase with peak angular velocity at 0.2 Hz or higher stimulus frequencies. At lower frequencies it advanced in phase (cf. Figs. 1B and 3) and occasionally reached peak angular acceleration at 0.03 Hz. A theoretical first-order linear system ($\tau = 5.3$ sec) did not fit such phase characteristics.

Most neurons sensitive to rotation in yaw were also observed during rotation in the two other planes of space (roll and pitch; cf. Fig. 1D and E). More than half of them showed an additional sensitivity to one of these stimuli. The plane of optimal sensitivity of such neurons appeared to be a combination of the two planes tested. Neurons sensitive to static tilts were occasionally observed; however, the effect of concomitant somatosensory stimulation was difficult to exclude under these conditions. In summary, the neuronal population investigated in the ASSS region essentially codes *angular velocity of rotation in the different planes of space.*

Animal experiments might help to understand the central processes leading to vestibular sensation in humans. From the above-described results in the cat it seems not surprising that human subjects submitted to horizontal oscillatory displacements have the sensation of velocity changing in a sine-like manner. Differences in response characteristics between cortical vestibular neurons and vestibulo-oculomotor neurons might be analogous to divergences of human sensation cupulogram from nystagmus cupulogram found under appropriate experimental conditions (cf. review of Guedry, 1974). Generally, cortical representation of vestibular input appears to be poor compared to that of other modalities. This observation in animal experiments might have its equivalent in the low "channel capacity" for human perception of angular acceleration (cf. Kongehl and Kornhuber, 1969).

CONVERGENCES WITH OTHER MODALITIES

Convergence of labyrinthine input with other modalities is known to occur already at the level of the vestibular nuclei in cats as has been shown for proprioceptive and, to a minor extent, for cutaneous or hair afferent input by Fredrickson et al. (1965; also Rubin et al., 1977), as well as for visual input by Keller and Precht (1978; optokinetic stimuli). Furthermore, certain populations of vestibular nuclear neurons in cat have been shown to carry horizontal eye movement signals (Mergner and Pompeiano, 1977). Further convergences may be added at higher levels. Vestibular units with bi- or multimodal input have been observed in thalamic structures (Abraham et al., 1975; Wepsic, 1966) and in the cortical vestibular fields (Kornhuber and daFonseca, 1964; Landgren et al., 1967; Mickle and Ades, 1952; Ödkvist et al., 1975b; Roucoux-Hanus and Boisacq-Schepens, 1974).

In the above cited own study on the ASSS vestibular field the majority of vestibular neurons responded in a direction-specific way to neck displacements (Fig. 1C) and to limb movements, similar to those described for vestibular nuclear neurons (cf. Fredrickson et al. 1965). Convergences with hair or cutaneous input were rarely encountered. Modifications of rotational responses were observed when the lights were turned on (additional optokinetic stimulus), but no quantitative analysis was attempted because of the functional interruption of the cat's optokinetic system due to muscular relaxation. In the neuronal population investigated there was no convincing evidence that eye movement signals were superimposed on the labyrinthine responses. The study focused mainly on the interaction of neck and labyrinthine input on ASSS neurons.

In higher vertebrates with the head moving independently of the trunk, neck receptors play an important role for posture and movement control. A major function of neck reflexes seems to be to counteract or to modify vestibularly induced reflexes during rotation of the head alone (cf. Kornhuber, 1966). Neck reflexes arise predominantly from joint receptors of the uppermost cervical vertebrae (McCouch et al., 1951; cf. Richmond and Abrahams, this volume, chapter II 1), and they include prominent kinetic components (Erhardt and Wagner, 1970) in addition to the tonic ones ("tonic neck reflexes"). They appear to be involved in passive as well as in active head movements (Dusser de Barenne, 1914).

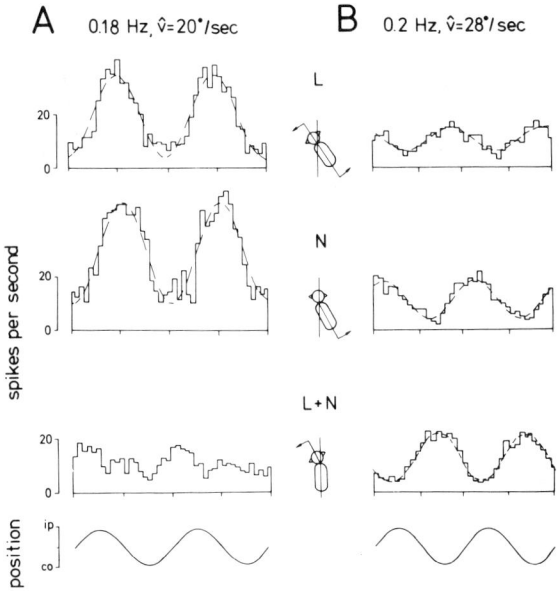

Fig. 2. Two different modes of vestibular and neck interaction observed in ASSS neurons, as obtained by sinusoidal rotation in the horizontal plane. Stimulus conditions are sketched in insets. In this and the following figures, L refers to rotation of the whole animal (labyrinthine stimulus), N to rotation of the trunk alone (neck stimulus), and L + N to rotation of the head alone (combined labyrinthine and neck stimulation; note that neck stimulation is now opposite in sign compared to experimental condition N). Dashed sine waves in histograms represent fundamental waves according to Fourier analysis of the responses. Otherwise as in Fig. 1B–E. A) Labyrinthine and neck stimulation alone give similar responses, which are roughly in phase with angular velocity. The responses are almost completely extinguished during combined stimulation, when the head is rotating alone ("extinction" mode of interaction). B) Neck response is lagging in phase behind labyrinthine response. During rotation of the head alone a prominent response occurs, which is shifted in phase towards head angular displacement ("head position" mode of interaction).

Interaction of labyrinthine and neck receptors were studied with rotational stimuli in the horizontal plane. Stimulation conditions consisted of rotation of the whole animal (labyrinthine stimulus), rotation of the trunk alone with the head stationary (neck stimulus; the limbs were fixed to the trunk) and rotation of the head alone with the trunk stationary (labyrinthine and neck stimulus, the latter being now opposite in sign; cf. insets in Fig. 2). Among neurons with either labyrinthine or neck influences, other units with strong convergence of the two inputs were encountered. Two predominant modes of interaction will be described here. Neurons of the first type showed similar responses which were approximately in phase with peak angular velocity when rotating the whole animal or the trunk alone in the same direction. The stimulation of neck and vestibular afferents occurring during movements of the head alone resulted in extinction of the responses (*"extinction" mode* of interaction: Fig. 2A and model in Fig. 5A). Thus, such neurons appear to code angular velocity of the trunk relative to the external world independently of head movements.

A consistent observation in such neurons was, however, that neck responses were larger in amplitude than labyrinthine responses. Consequently, extinction was never complete; instead there remained a "head alone response", which was much smaller and opposite in sign compared to the neck and labyrinthine responses, respectively (cf. Figs. 2A and 4B). The effect did not depend on the depth of muscular relaxation (obtained by pancuronium bromide or gallamine triethiodide), nor whether the axis of head-trunk deflection was shifted from C1 to C3. A speculative explanation for this phenomenon would be that amplitudes and directions of neuronal responses, obtained under the different stimulus conditions, are related to the amounts and directions of intended active counteractions which the non-paralysed cat would have to perform in order to account for the imposed shifts in its center of gravity.

In the second group of neurons, the neck response was clearly lacking in phase behind the labyrinthine response at 0.2 Hz. Rotation of the head alone led to a clear response, which was shifted further towards the peak of chair angular position (Fig. 2B). However, labyrinthine and neck responses differed in their phase characteristics when tested at stimulus frequencies of 0.05–1 Hz (Fig. 3). Consequently, the responses to the combined stimulation during rotation of the head alone depended on the stimulus frequency used. In the example of Fig. 4B, labyrinthine and neck responses cancel during combined neck and labyrinthine stimulation at 1 Hz, as could have been predicted from the intersection of the two phase characteristics at this frequency in Fig. 3. In fact, the neck response consisted of a kinetic as well as of a static component (cf. Fig. 4C). The static neck component modified the labyrinthine induced responses in a direction-specific way, in that static head deflection in the labyrinthine off-direction enhanced the mean discharge rate and increased the amplitude of the labyrinthine response, whereas head deflection to the opposite side depressed the labyrinthine response (Fig. 4A). Trapezoidal angular displacement of the head alone mainly led to the static neck response, while the kinetic neck and labyrinthine responses appeared to be subtracted from each other ("*head position*" *mode* of interaction; cf. Fig. 4C and model in Fig. 5B). The latter stimulus wave form was considered to resemble natural head movement. As stated above, the fundamental mechanisms underlying the interaction of neck and labyrinthine input appear to be already present at vestibular nuclear level (compare also Schaefer et al., and Maeda, this volume, chapters VI2 and IVD6, respectively).

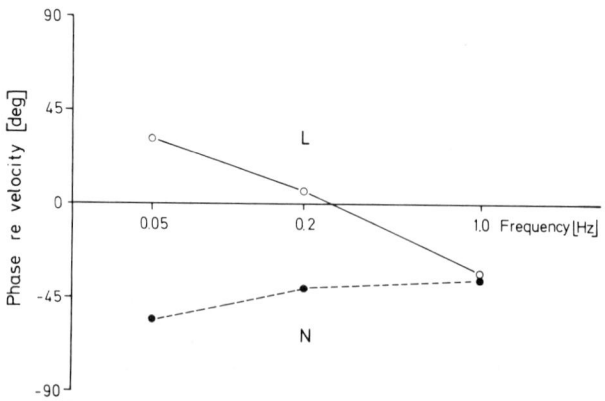

Fig. 3. Bode plot showing different relations between phase of labyrinthine (L) and neck (N) induced responses (ordinate; referred to peak angular velocity) and frequency of sinusoidal rotation (abscissa). At 0.2 Hz stimulus frequency the neuron had shown a "head position" mode of interaction. Differences in the phase characteristics suggest, however, that the effects of combined stimulation during rotation of the head alone depend on the stimulus frequency used. The intersection of both curves at 1 Hz predicts extinction of the two responses at this frequency (cf. Fig. 4B). Peak angular velocity was 10 degrees/sec in all examples.

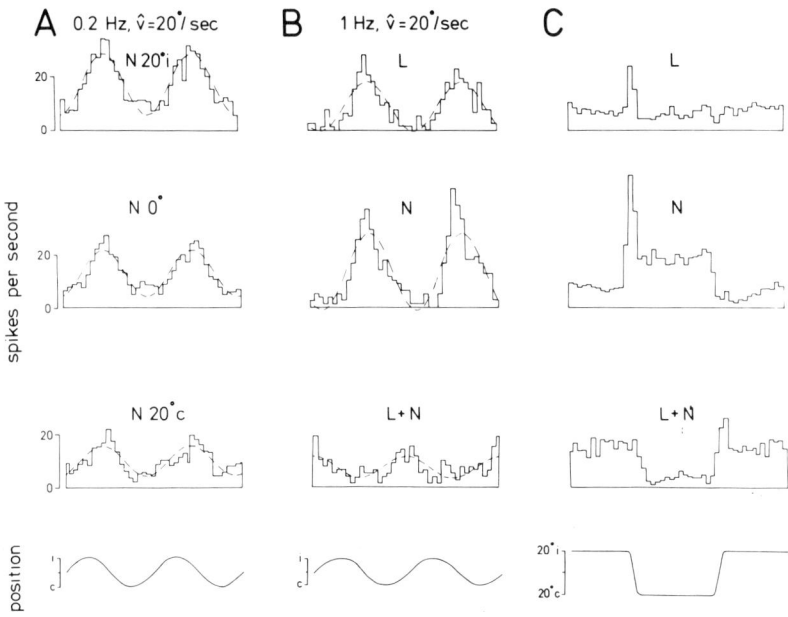

Fig. 4. Effects of interaction between neck and labyrinthine input in a "head position" ASSS neuron to three different stimulus conditions. Presentation of data as in foregoing figures. A) Static head deflexion (N 20° i, 20° ipsilateral deflexion; N 0°, head in straight forward position; N 20° c, 20° contralateral deflexion) modifies labyrinthine induced responses obtained at 0.2 Hz sinusoidal rotation in a direction specific way. B) At 1 Hz stimulus frequency, labyrinthine response is almost in phase with neck response. When rotating the head alone (combination of the 2 stimuli) the responses are almost completely extinguished (cf. also Fig. 3). C) When using position trapezoid as stimulus wave form, the labyrinthine response is purely kinetic, whereas the neck response reveals a kinetic and a static component. When moving the head alone, there remains predominantly the static neck response (head position signal), whereas the kinetic responses appear to be subtracted from each other (trapezoid duration, 20 sec).

So far, speculations can only be made about possible functions of the ASSS vestibular field. Such considerations should take into account the extensive convergence of other modalities on the vestibular neurons. Also neuronal activity in the immediate vicinity of the cortical vestibular neurons was found to be affected by a variety of exteroceptive and proprioceptive stimuli, e.g., by visual stimuli applied to the contralateral eye, by contralateral whisker displacements and by hair and muscle or joint proprioceptive stimulation. Histologically, the region appears to be located at the transition zone between area 2 pr.i (Hassler and Muhs-Clement, 1964) and area 5, whereas physiologically it overlaps with the third somatosensory area in the cat's cerebral cortex (compare location in Tanji et al., 1978).* Discrimination among various stimuli is known to be most conspicuously impaired after lesions of this cortical region. Its role might include the perception of vestibular stimuli and their integration in complex interpretational processes, as well as the perception of vestibular illusions, which may result from appropriate optokinetic, acoustic and somatic stimuli.

---

*The location of vestibular neurons described did not correspond to the projection area of the group I muscle afferents in the second somatosensory field, as indicated by Landgren et al. (1967).

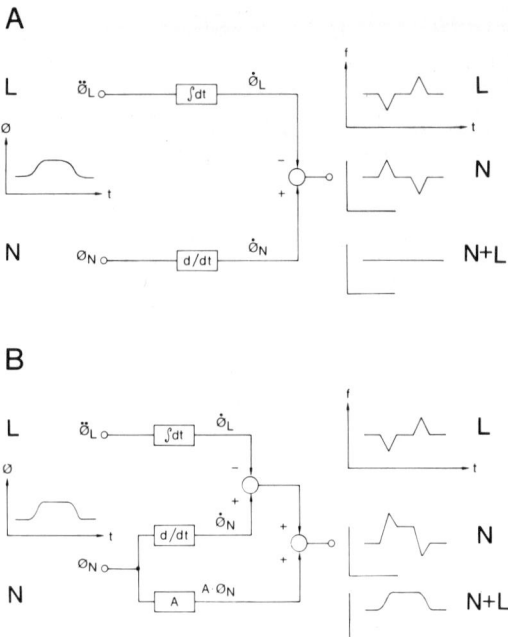

Fig. 5. Models for two different modes of interaction between labyrinthine and neck input in yaw as observed for ASSS neurons. Trapezoidal position (θ) changes are used as input signals, since such a stimulus wave form might resemble active head movements. A) "Extinction" mode of interaction: the integrated labyrinthine signal ($\dot{\theta}_L$) is subtracted from the differentiated neck signal ($\dot{\theta}_N$), so that combined stimulation of the two inputs result in extinction of the individual responses. B) "Head position" mode of interaction: the velocity signals of labyrinthine and neck origin are subtracted from each other as in A, so that during combined stimulation of labyrinthine and neck input only the neck position response ($A \cdot \theta_N$) is obtained as output signal. Other abbreviations as in foregoing figures.

ASCENDING PATHWAYS

Vestibulothalamic projections in the cat appear to run mainly lateral and ventral to the ascending medial longitudinal fasciculus (MLF), as has been shown in anatomical (Hassler, 1948, 1956; Raymond et al., 1976) and physiological studies (Abraham et al., 1977; Mickle and Ades, 1954; Roucoux-Hanus and Boisacq-Schepens, 1977; Sans et al., 1976). The pathways seem to run apart from the lateral lemnisci, but not with them, as has been indicated by earlier studies (Gernandt, 1950). Additional projections via the MLF, which is known to carry the major bundle of ascending fibers from the vestibular nuclei (Brodal and Pompeiano, 1957), are unlikely, since few, if any such fibers have been traced into thalamic structures (Carpenter and Hanna, 1962; Gacek, 1973; cf. also Brodal et al., 1962). Furthermore, MLF lesions did not abolish vestibular field potentials recorded in the ASSS and the postcruciate regions (Spiegel et al., 1965; Abraham et al., 1977). Additional long loop pathways via the cerebellum have been proposed (Ruwaldt and Snider, 1965; Sans et al., 1970, 1976), but they are questioned by other authors (Abraham et al., 1977). Also indirect pathways via the reticular formation have been taken into account.

The magnocellular portion of the medial geniculate body (mcMGB), which is known to receive multimodal afferents, is considered to represent one important thalamic relay station for vestibular afferents, as indicated by field potential studies (Spiegel et al., 1965; Abraham et al., 1977; Roucoux-Hanus and Boisacq-Schepens, 1977; for a location more medial cf. Mickle and Ades, 1954) and single-neuron recordings (Abraham et al., 1975; Wepsic, 1976). Lesioning of this nuclear region was found to diminish vestibular field potentials recorded from the ASSS field (Copack et al., 1972), and ascending pathways from the mcMGB to the ASSS vestibular field have been described (Liedgren et al., 1976a; cf. also Heath and Jones, 1971). In addition, the region of the ventral posterior lateral (VPL) and ventral lateral (VL) nuclei have also been shown to receive vestibular input (Sans et al., 1970; Copack et al., 1972).

Ascending vestibular fiber projections towards this thalamic region have been described (Raymond et al., 1976), and labeled neurons have been found in the VPL after injection of horseradish peroxidase (HRP) into the postcruciate vestibular field (Liedgren et al., 1976a). Possibly the ventro-intermedial nucleus (Vim) of Hassler (1948, 1956), which he found to represent the target of ascending vestibular fibers, corresponds to this thalamic region taken as VL and dorsal border of VPL (cf. Hassler in Copack et al., 1972). Such observations, including findings in monkey and other animal species, led to the concept of a dual vestibulo-thalamocortical projection, one associated with the medial lemniscal system to the somatosensory cortex, the other parallel to the lateral lemniscus towards the "association" cortex (cf. Schwarz and Fredrickson, 1974; for similar evidence in man cf. Hawrylyshyn et al., 1978).

Vestibular input to other thalamic regions and to basal ganglia has been reported (for some of the literature cf. Copack et al., 1972). Single neurons in the reticular thalamic nucleus and the ventral lateral geniculate body were found to respond to natural vestibular stimuli, but appear to carry predominantly eye movement signals (Magnin and Putkonen, 1978).

In recent experiments of our laboratory (Wagner et al., 1978; Deecke et al., 1979) HRP was injected into the ASSS field where the above described vestibular units were encountered. In the thalamus, some labeled neurons were found in the mcMGB. The vast majority, however, was localized in an array of patches along and within the ventral-most region of the lateral posterior nucleus (LP; cf. also Liedgren et al., 1976a; Tanji et al., 1978). Scanning from medial to lateral, the patches reached from the nucleus centralis medialis (CM), the nucleus centralis lateralis (CL), the dorsal-most parts of the VL and of the nucleus suprageniculatus (SG), along the ventral part of the LP immediately dorsal to the ventrobasal complex, towards the border region of the geniculate complex. The broad scattering of labeled neurons included several thalamic nuclei in which, or close to which, vestibular field potentials have been found (CM, SG, border region between VL and VPL, LP as well as mcMGB). It opens the possibility that not only one thalamic nucleus, i.e., the mcMGB, but several functionally distinct neuronal pools from different thalamic areas convey vestibular input to the ASSS field. Such a concept would be in accordance with findings of Liedgren et al. (1976b), who did not find convincing evidence for a specific vestibular thalamic nucleus in squirrel monkey, but rather recorded neuronal responses to vestibular nerve stimulation over large thalamic areas.

In the same study, HRP injections of small diameter were made in the thalamus, aiming at the ventral part of the LP bordering the dorsal VL and VPL (so far, vestibular fibers have only been anterogradely traced towards this thalamic region, cf. Raymond et al., 1976). However, no significant HRP labeling among vestibular nuclear neurons was observed. This negative finding indicates either that the vestibular thalamic region was not included in the injection site, or that the vestibular neurons were scattered over much larger areas. Larger HRP injections, indeed, gave positive results. In these cases, some labeled neurons were encountered on the contralateral side in the caudal parts of the medial and descending vestibular nuclei, the group y, the interstitial nucleus of the VIIIth nerve and the nucleus intercalatus of Staderini (the latter being closely linked, anatomically as well as physiologically, to the vestibular nuclear complex, cf. Mergner et al., 1977). Other neurons were scattered over the four main vestibular nuclei of both sides including transition zones to the somatosensory medullary relay nuclei and to the reticular formation. Heavy labeling was present in the contralateral and, to a lesser extent, in the ipsilateral small cell group z. This small nucleus is considered a medullary relay for group I muscle afferents (Landgren and Silvenius, 1971; Magherini et al., 1975), and is not attributed to vestibular function. On the whole, the number of labeled neurons in the vestibular nuclei after thalamic HRP injections was small and their location scattered, which possibly explains earlier disagreement about the existence of direct vestibulothalamic connections (cf. also Nakatani and Matano, 1978).

The pericruciate cortical areas might possibly receive vestibular input directly via the thalamus (cf. Hassler in Copack et al., 1972) or indirectly via the postcruciate vestibular fields. Finally, it should be mentioned that the cerebral cortex in its turn may influence unit activity in the vestibular nuclei as has been shown by Arslan and Molinari (1965), Gildenberg and Hassler (1971) and Manni and Giretti (1968).

SUMMARY

In the cerebral cortex of the cat, vestibular input mainly reaches the anterior bank of the anterior suprasylvian sulcus (ASSS) and the postcruciate dimple area, but also precruciate regions. Response characteristics of cortical vestibular neurons are so far only known in the ASSS vestibular field. There the neuronal population appears mainly to code velocity of angular displacement of the animal in different planes of space. These vestibular neurons receive convergences from other sensory modalities. One major input stems from neck proprioceptors. By way of interaction between labyrinthine and neck influences, some of the ASSS neurons distinguish between trunk angular rotation and displacement of the head alone.

Ascending vestibulothalamic pathways apparently do not project within the MLF, but lateral and ventral to it. The pathways stem from scattered neurons in the four main vestibular nuclei of both sides, with contralateral projections predominating, and from some small vestibular cell groups of the contralateral side. Possible indirect loops to thalamus and cortex via the cerebellum or the reticular formation have not yet been established. At thalamic level, two main regions are considered to represent relays for cortical vestibular projections, i.e., the dorsal border region of the ventral-lateral (VL) and ventral-posterior-lateral (VPL) nuclei, and the magnocellular portion of the medial geniculate body (mcMGB). The mcMGB is thought to relay vestibular information to the ASSS vestibular field, the dorsal VPL/VL region to the postcruciate area. The current concept of two vestibular pathways to the cortex, one parallel to the lateral lemniscal system, the other along with the medial lemniscal system, might represent too much of a simplification of a more complex vestibular thalamocortical organization.

ACKNOWLEDGEMENT

The author wishes to thank his colleagues W. Becker, L. Deecke, R. Jürgens and H.-J. Wagner, who have shared the effort to obtain the experimental data. This work was supported by Deutsche Forschungsgemeinschaft, SFB 70.

REFERENCES

Abraham, L., Blum, P., and Gilman, S. (1975) Multimodal responses of cells in the medial geniculate body of the cat. *Neuroscience*, 1: 222.
Abraham, L., Copack, P.B. and Gilman, S. (1977) Brain stem pathways for vestibular projections to cerebral cortex in cat. *Exp. Neurol.*, 55: 436–448.
Andersson, S. and Gernandt, B.E. (1954) Cortical projection of vestibular nerve in cat. *Acta oto-laryngol. (Stockh.)*, 116 Suppl.: 10–18.
Arslan, M. and Milinari, G.A. (1965) Modifications of the activity of the vestibular nuclei in the cat following stimulation of the temporal lobe. *Acta oto-laryngol. (Stockh.)*, 59: 338–344.
Becker, W., Deecke, L. and Mergner, T. (1979) Neuronal responses to natural vestibular and neck stimulation in the anterior suprasylvian gyrus of the cat. *Brain Res.*, 165: 139–143.
Boisacq-Schepens, N. and Roucoux-Hanus, M. (1975) Responses of motor cortex cells to tilt in the 'encephal isolé' in the cat. *Brain Res.*, 29: 149–152.
Brodal, A. and Pompeiano, O. (1957) The origin of ascending fibers in the medial longitudinal fasciculus from the vestibular nuclei. An experimental study in the cat. *Acta morphol. neerl.-scand.*, 1: 306–328.
Brodal, A., Pompeiano, O. and Walberg, F. (1962) *The Vestibular Nuclei and their Connections, Ramsay Henderson Trust Lectures*. Oliver and Boyd, Edinburgh.
Carpenter, M.B. and Hanna, G.R. (1962) Lesions of the medial longitudinal fasciculus in cat. *Amer. J. Anat.*, 110: 307–332.
Copack, P., Dafny, N. and Gilman, S. (1972) Neurophysiological evidence of vestibular projections to the thalamus, basal ganglia and cerebral cortex. In *Cortico-Thalamic Projections and Sensorimotor Activities*. T. Friggesi, E. Rinvich and M.D. Yahr (Eds.), Raven Press, New York, pp. 309–339.
Deecke, L., Mergner, T. and Becker, W. (1979) Neuronal responses to natural vestibular stimuli in the cat's anterior suprasylvian gyrus. *Adv. oto-rhino-laryngol. (Basel)*, 25: in press.
Duensing, F. and Schaefer, K.P. (1958) Die Aktivität einzelner Neurone im Bereich der Vestibulariskerne bei Horizontal-Beschleunigungen unter besonderer Berücksichtigung des vestibulären Nystagmus. *Arch. Psychiat. Nervenkr.*, 198: 225–252.
Dusser de Barenne, J.G. (1914) Nachweis, dass die Magnus-de Kleinschen Reflexe bei der erwachsenen Katze mit intaktem Zentralnervensystem bei passiven und aktiven Kopf-resp. Halsbewegungen auftreten, und somit im normalen Leben der Tiere eine Rolle spielen. *Folia neurobiol. (Lpz.)*, 8: 413.
Erhardt, K.J. and Wagner, A. (1970) Labyrinthine and neck reflexes recorded from spinal single motoneurons in the cat. *Brain Res.*, 19: 87–104.
Fredrickson, J.M., Schwarz, D. and Kornhuber, H.H. (1965) Convergence and interaction of vestibular and deep somatic afferents upon neurons in the vestibular nuclei of the cat. *Acta oto-laryngol. (Stockh.)*, 61: 168–188.
Fredrickson, J.M., Kornhuber, H.H. and Schwarz, D.W.F. (1974) Cortical projections of the vestibular nerve. In *Handbook of Sensory Physiology, Vol. VI/1, Vestibular System, Part 1, Basic Mechanisms*, H.H. Kornhuber (Ed.), Springer-Verlag, New York, pp. 565–582.
Fredrickson, J.M., Figge, U., Scheid, P. and Kornhuber, H.H. (1966) Vestibular nerve projection to the cerebral cortex of the rhesus monkey. *Exp. Brain Res.*, 2: 318–327.
Gacek, R.R. (1973) Anatomical studies of the vestibulo ocular pathways in cat *Adv. oto-rhino-laryngol. (Basel)*, 19: 66–75.
Gernandt, B. (1950) Midbrain activity in response to vestibular stimulation. *Acta physiol. scand.*, 21: 73–81.
Gildenberg, P.L. and Hassler, R. (1971) Influence of stimulation of the cerebral cortex on vestibular nuclei units in the cat. *Exp. Brain Res.*, 14: 77–94.

Grüsser, O.J. and Grüsser-Cornehls, U. (1972) Interaction of vestibular and visual inputs and the visual system. In *Progress in Brain Research Vol. 37, Basic Aspects of Central Vestibular Mechanisms*, A. Brodal and O. Pompeiano (Eds.), Elsevier, Amsterdam, pp. 573–583.

Guedry, F.E. (1974) Psychophysics of vestibular sensation. In *Handbook of Sensory Physiology, Vol. VI/1, Vestibular System, Part 2, Psychophysics, applied Aspects and General Interpretations*, H.H. Kornhuber (Ed.), Springer-Verlag, New York, pp. 3–154.

Hassler, R. (1948) Forels Haubenfaszikel als vestibuläre Empfindungsbahn mit Bemerkungen über einige andere sekundäre Bahnen des Vestibularis und Trigeminus. *Arch. Psychiat. Nervenkr.*, 180: 23–53.

Hassler, R. (1956) Die zentralen Apparate der Wendebewegungen. I. und II. *Arch. Psychiat. Nervenkr.*, 194: 456–516.

Hassler, R. and Muhs-Clemens, K. (1964) Architektonischer Aufbau des sensomotorischen und parietalen Cortex der Katze. *J. Hirnforsch.*, 6: 377–420.

Hawrylyshyn, P.A., Rubin, A.M., Sasker, R.R., Organ, L.W. and Fredrickson, J.M. (1978) Vestibulothalamic projections in man – a sixth primary sensory pathway. *J. Neurophysiol.*, 41: 394–401.

Heath, C.J. and Jones, E.G. (1971) An experimental study of ascending connections from the posterior group of thalamic nuclei in the cat. *J. comp. Neurol.*, 141: 397–426.

Keller, E.L. and Precht, W. (1978) Persistence of visual response in vestibular nucleus neurons in cerebellectomized cat. *Exp. Brain Res.*, 32: 591–594.

Kongehl, G. and Kornhuber, H.H. (1964) Die Kanalkapazität des vestibulären Systems für Wahrnehmung von Drehbeschleunigungen. *Pflügers Arch.*, 307: 129–130.

Kornhuber, H.H. (1966) Physiologie und Klinik des zentralvestibulären Systems (Blick- und Stützmotorik). In *Hals-Nasen-Ohren-heilkunde Handbuch, Vol. III, Part 3*, J. Berendes, R. Link and F. Zöllner (Eds.), Thieme, Stuttgart, pp. 2150–2351.

Kornhuber, H.H. (1972) Vestibular influences on the vestibular and the somatosensory cortex. In *Progress in Brain Research, Vol. 37, Basic Aspects of Central Vestibular Mechanisms*, A. Brodal and O. Pompeiano, (Eds.), Elsevier, Amsterdam, pp. 567–572.

Kornhuber, H.H. and Aschoff, J.C. (1964) Somatisch-vestibuläre Integration an Neuronen des motorischen Cortex. *Naturwissenschaften*, 51: 62–63.

Kornhuber, H.H. and daFonseca, J.S. (1964) Optovestibular integration in the cats cortex: a study of sensory convergence on cortical neurons. In *The Oculomotor System*, M.B. Bender, (Ed.). Hoeber, New York, pp. 239–279.

Landgren, S. and Silfvenius, H. (1971) Nucleus Z, the medullary relay in the projection path to the cerebral cortex of group I muscle afferents from the cat's hind limb. *J. Physiol. (Lond.)*, 218: 551–571.

Landgren, S., Silfvenius, H. and Wolsk, D. (1967) Vestibular, cochlear and trigeminal projections to the cortex in the anterior suprasylvian sulcus of the cat. *J. Physiol. (Lond.)*, 191: 561–573.

Liedgren, S.R.C., Kristensson, K., Larsby, B. and Ödkvist, L.M. (1976a) Projection of thalamic neurons to cat primary vestibular cortical fields studied by means of retrograde axonal transport of horseradish peroxidase. *Exp. Brain Res.*, 24: 237–243.

Liedgren, S.R.C. Milne, A.C., Rubin, A.M., Schwarz, D.W.F. and Tomlinson, R.D. (1976b) Representation of vestibular afferents in somatosensory thalamic nuclei of the squirrel monkey (*Saimiri sciureus*). *J. Neurophysiol.*, 39: 601–612.

Mach, E. (1902) *The Analysis of Sensations*. Republication of 5th ed. (1959). Dover, New York.

Magherini, P.C., Pompeiano, O. and Seguin, J.J. (1975) Responses of nucleus Z neurons to vibrations of hindlimb extensor muscles in the decerebrate cat. *Arch. ital. Biol.*, 113: 150–187.

Magnin, M. and Putkonen, P.T.S. (1978) A new vestibular thalamic area: electrophysiological study of the thalamic reticular nucleus and of the ventral lateral geniculate complex of the cat. *Exp. Brain Res.*, 32, 91–104.

Manni, E. and Giretti, M.L. (1968) Vestibular units influenced by labyrinthine and cerebral nystagmogenic impulses. *Exp. Neurol*, 22: 145–157.

McCouch, G.P., Deering, J.D. and Ling, T.H. (1951) Locations of receptors for tonic neck reflexes. *J. Neurophysiol.*, 14: 191–195.

Melvill Jones, G. and Milsum, J.H. (1970) Characteristics of neural transmission from semicircular canal to the vestibular nuclei of cats. *J. Physiol. (Lond.)*, 209: 295: 295–316.

Mergner, T. and Pompeiano, O. (1977) Neurons in the vestibular nuclei related to saccadic eye movements in the decerebrate cat. In *Control of Gaze by Brain Stem Neurons*, R. Baker and A. Berthoz (Eds.), Elsevier/North-Holland Biomed. Press, Amsterdam, pp. 243–251.

Mergner, T., Pompeiano, O. and Corvaja, N. (1977) Vestibular projections to the nucleus intercalatus of Staderini mapped by retrograde transport of horseradish peroxidase. *Neurosci. Lett.*, 5: 309–313.

Mickle, W.A. and Ades, H.W. (1952) A composite sensory projection area in the cerebral cortex of the cat *Amer. J. Physiol.*, 170: 682–689.
Mickle, W.A. and Ades, H.W. (1954) Rostral projection pathway of the vestibular system. *Amer. J. Physiol.*, 176: 243–252.
Mills, K.R. and Taylor, A. (1974) The projection of the vestibular nerve to the cerebral cortex in the cat. *J. Physiol. (Lond.)*, 239: 165–178.
Nakatani, J. and Matano, S. (1978) Vestibulo-thalamo-cortical connections in the cat — an experimental study with horseradish peroxidase. *IBRO News*, 6: 12.
Ödkvist, L.M., Larsby, B. and Fredrickson, J.M. (1975a) Projection of the vestibular nerve to the SI forelimb field in the cerebral cortex of the cat. *Acta oto-laryngol. (Stockh.)*, 79: 88–95.
Ödkvist, L.M., Liedgren, S.R.C., Larsby, B. and Jervall, L. (1975b) Vestibular and somatosensory inflow to the vestibular projection area in the postcruciate dimple region of the cat cerebral cortex. *Exp. Brain Res.*, 22: 185–196.
Raymond, J., Dememes, D. and Marty, R. (1976) Voies et projections vestibulaires ascendantes émanant des noyaux primaires: étude radioautographique. *Brain Res.*, 111: 1–12.
Roucoux-Hanus, M. and Boisacq-Schepens, N. (1974) Projections vestibulaires au niveau des aires corticales suprasylvienne et postcruciée chez le chat anesthesié au chloralose. *Arch. ital. Biol.*, 112: 60–76.
Roucoux-Hanus, M. and Boisacq-Schepens, N. (1977) Ascending vestibular projections: Further results at cortical and thalamic levels in the cat. *Exp. Brain Res.*, 29: 283–292.
Rubin, A.M., Liedgren, S.R.C., Milne, A.C., Young, J.A. and Fredrickson, J.M. (1977) Vestibular and somatonsensory interaction in the cat vestibular nuclei. *Pflügers Arch.*, 371: 155–160.
Ruwaldt, M.M. and Snider, R.S. (1965) Projection of vestibular areas of cerebellum to the cerebrum. *J. comp. Neurol.*, 104: 387–401.
Sans, A., Raymond, J. and Marty, R. (1970) Réponses thalamiques et corticales à la stimulation électrique du nerf vestibulaire chez le chat. *Exp. Brain Res.*, 10: 265–275.
Sans, A., Raymond, J. and Marty, R. (1976) A vestibulo-thalamic pathway: electrophysiological demonstration in the cat by localized cooling. *J. Neurosci. Res.*, 2: 167–174.
Schwarz, D.W.F. and Fredrickson, J.M. (1974) The clinical significance of vestibular projection to the parietal lobe: a review. *Can. J. Oto-Laryngol.*, 3: 381–392.
Shinoda, Y. and Yoshida, K. (1974) Dynamic characteristics of responses to horizontal head angular acceleration in vestibuloocular pathways in the cat. *J. Neurophysiol.*, 37: 653–673.
Spiegel, E.A., Szekely, E.G. and Gildenberg, P.L. (1965) Vestibular responses in midbrain, thalamus and basal ganglia. *Arch. Neurol. Psychiat. (Chic.)*, 12: 258–269.
Tanji, D.G., Wise, S.P., Dykes, R.W. and Jones, E.G. (1978) Cytoarchitecture and thalamic connectivity of third somatosensory area of cat cerebral cortex. *J. Neurophysiol.*, 41: 268–284.
Wagner, H.-J., Mergner, T. and Deecke, L. (1978) Rostral vestibular projection to thalamus and cortex of cat as revealed by retrograde transport of horseradish peroxidase. *Neurosci. Let.*, Suppl. 1: S 358.
Walzl, E.M. and Mountcastle, V.B. (1949) Projection of vestibular nerve to cerebral cortex of the cat. *Amer. J. Physiol.*, 159: 595.
Wepsic, J.G. (1966) Multimodal sensory activation of cells in the magnocellular medial geniculate nucleus. *Exp. Neurol.*, 15: 299–318.

The Vestibulocortical Pathway: Neurophysiological and Anatomical Studies in the Monkey*

U. BÜTTNER and W. LANG

Neurologische Klinik und Institut für Hirnforschung, Universität Zürich, 8091 Zürich (Switzerland)

INTRODUCTION

In humans an activation of the vestibular system leads to a sensation of motion. Given that the cortex participates in the perception of sensations this suggests the existence of a vestibulocortical pathway. The exact course and the functional significance of this pathway is still poorly understood. During recent years two separate vestibulo-thalamocortical pathways have been proposed both in cat and monkey (Fig. 1). In these experiments, the vestibular nerve was electrically stimulated, and latencies and response magnitudes of evoked potentials were measured in thalamic and cortical areas. Latencies for the thalamic and cortical areas indicate, respectively, a two or three neuron pathway. The vestibular thalamic areas so defined are i) the nucleus ventroposterior (VP), mainly the oral lateral division (VPL_o), and to a lesser extent the inferior division (VPI) (Liedgren et al., 1976b; Deecke et al., 1977), and ii) the posterior group (PO) consisting of the magnocellular part of the medial geniculate body, the nucleus posterior, and the nucleus suprageniculatus (Liedgren et al., 1976b). The cortical projection sites in the monkey are the area 2v, at the lower end of the intraparietal sulcus (*Macaca mulatta*: Fredrickson et al., 1966; Schwarz and Fredrickson, 1971), and the area 3a (*Saimiri sciureus:* Oedquist et al., 1974). Short-latency responses have also been obtained in corresponding cortical areas of the cat in the anterior suprasylvian sulcus (ASSS: Mickle and Ades, 1952), equivalent to area 2v in the monkey, and the postcruciate dimple, which is a homologue of area 3a (Sans et al., 1970).

Anatomical support for these electrophysiologically determined pathways is scarce. Tarlov (1969), in a detailed degeneration study, found no evidence for a vestibulothalamic pathway in the monkey. The availability of more sensitive anatomical methods prompted us to reinvestigate the vestibulocortical pathway in the monkey in combination with single neuron recordings.

Alert monkeys were exposed to natural vestibular and optokinetic stimuli, while single neurons were recorded in the thalamus and cortex. At the conclusion of experiments, various anatomical tracer substances were injected in these physiologically defined areas in order to establish their anatomical connections.

---

*Supported by Swiss National Foundation for Scientific Research (3.672.77 and 3.636.75).

ANATOMY

Anterograde (radioactive amino acids: ^{3}H-proline and ^{3}H-leucine) as well as retrograde (horseradish peroxidase, HRP) tracer substances were injected into the vestibular nuclei, the thalamus and the cortex. Prior to the injections single-neuron activity was recorded for the exact localization of the micropipette.

The main results are summarized in Fig. 1. The injection of anterograde tracer substances in the vestibular nuclei leads to labelling in the ipsilateral and contralateral VPL_o. A few endings are also found in the VPI and VL_c (nucleus ventrolateralis pars caudalis). Though definite, the number of projection sites is very small in comparison to the heavy vestibular projection to the oculomotor nuclei. This sparse projection might well be below the resolution of the degeneration technique, and explain the negative results of Tarlov (1969). A distinct feature is also the distribution of the projections within the VPL_o: they are not localized in a small restricted area, but scattered in small patches over wide areas of the nucleus. The ascending vestibulothalamic fibers run mainly outside the medial longitudinal fasciculus (mlf), the main ascending vestibular pathway to the oculomotor system. Ipsilateral fibers originate predominantly but not exclusively in the lateral vestibular nucleus and are contained in the ascending tract of Deiters' nucleus before entering the thalamus. Most fibers ascending contralaterally, arise from the superior vestibular nucleus and cross the midline mainly dorsal to the nucleus reticularis tegmenti pontis of Bechterew. Further rostrally these now contralateral fibers can be followed around the red nucleus and through the fields of Forel. It should be emphasized that the vestibulothalamic fibers

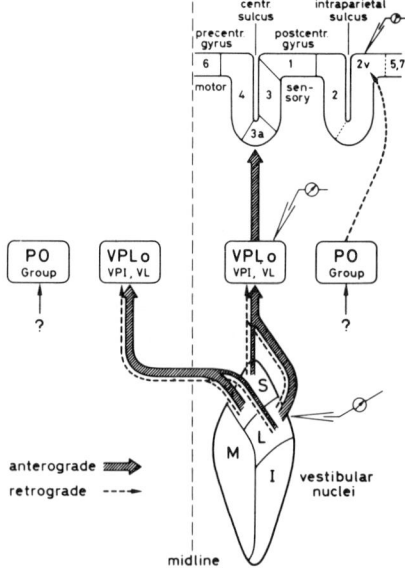

Fig. 1. Diagram of the ascending vestibulo-thalamocortical projections investigated with antero- and retrograde tracer substances in the monkey (*Macaca mulatta*). The vestibular nuclei project mainly to the ipsi- and contralateral nucleus ventroposterior lateralis pars oralis (VPL_o). No projection from the vestibular nuclei to the posterior group (PO) was found with anterograde tracer substances. VPL_o projects to area 3a (anterograde tracer substances), and the PO group to area 2v (retrograde tracer substances). The electrode symbols indicate the nuclei in which vestibular related neuronal activity was recorded in the alert monkey, during natural vestibular stimulation.

do not ascend in one compact bundle, but are widely dispersed. Therefore, the above-mentioned routes have to be considered as the most prominent but not the sole ones. These afferent pathways have also been confirmed by retrograde labelling (HRP injection in the thalamus). Anterograde tracer substances into the vestibular nuclei failed to label the PO group (Fig. 1).

Other experiments were directed towards determining the projection sites of the ascending vestibular thalamocortical fibers. The injection of anterograde tracer substances in VPL_o led to labelling in area 3a, but not in area 2v, confirming the known projection from VPL_o to area 3a (Jones and Powell, 1970); however, anatomical evidence for a projection from VPL_o to 2v had not been found. Injection of HRP into 2v leads to labelling of cells in the PO group (Fig. 2). A similar interconnection of thalamic and cortical vestibular areas has been shown in the cat (HRP method: Liedgren et al., 1976a).

The anatomical studies in the monkey demonstrate the existence of two parallel vestibulocortical pathways: one projects from the vestibular nuclei to the VPL_o and to area 3a, the other projects via the PO group to area 2v. No direct projection from the vestibular nuclei to the PO group could be found. Thus, it is unclear how vestibular information reaches the PO group. Electrical stimulation of the vestibular nerve leads to responses in the PO group within 2–4 msec, which indicates a disynaptic pathway (Liedgren et al., 1976b; Abraham et al., 1977). A possible, alternative route via the cerebellum was ruled out in the cat after cerebellectomy, which does not abolish the responses (Abraham et al., 1977). A small contingent of vestibular fibers projects directly into the lateral reticular formation outside the vestibular nuclei (Hauglie-Hanssen, 1968). Whether this projection is part of the pathway to the PO group has yet to be investigated.

NEUROPHYSIOLOGY

Neurons in the vestibular nuclei, thalamus (VPL_o and VPI), and cortex (2v) were investigated with natural vestibular stimulation (Fig. 1). Alert monkeys were seated

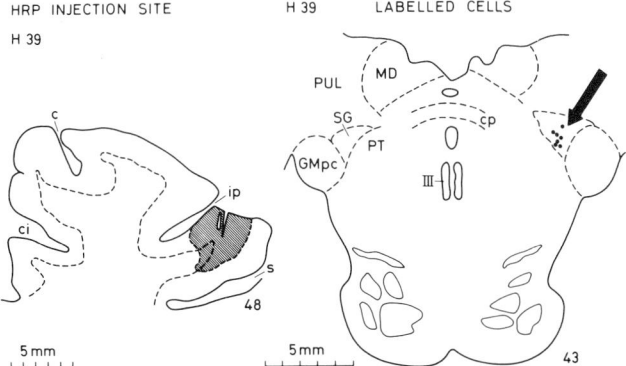

Fig. 2. The location of labelled cells (dots) in the ipsilateral suprageniculate nucleus (SG, part of the posterior group) after injection of retrograde tracer substances (HRP) in area 2v (hatched area shown on the left); the micropipette for injection was placed in this area after recording vestibular neurons. Labelled cells in one 40 μm section are shown. c, sulcus centralis; ci, sulcus cinguli; cp, commissura posterior; GMpc, nucleus geniculatus medialis parvocellularis; ip, sulcus intraparietalis; MD, nucleus medialis dorsalis; PT, area praetectalis; PUL, pulvinar; s, fissura sylvii; SG, nucleus suprageniculatus; III, nucleus oculomotorius.

upright in a primate chair with their heads fixed. For vestibular stimulation they were rotated sinusoidally about a vertical axis in complete darkness at frequencies between 0.002 and 1 Hz. Eye position was recorded through chronically implanted DC silver–silver chloride electrodes, and neurons were recorded with tungsten microelectrodes. For quantitative analysis, the phase and gain of neuronal activity, relative to turntable velocity, was determined using a fast Fourier analysis program (Büttner et al., 1977).

Neurons in the thalamus and cortex were also tested during optokinetic stimulation. A cylinder covered on the inside with black and white stripes was rotated around the stationary monkey. Such stimulation leads to specific activity changes of vestibular nuclei neurons (Waespe and Henn, 1977).

Only a small percentage (less than 10%) of neurons in the VPL_o and VPI respond to rotation about a vertical axis. These neurons are distributed over wide areas and not concentrated within certain regions (Büttner and Henn, 1976; Büttner et al., 1977; Magnin and Fuchs, 1977). A similar pattern of distribution was found in electrophysiological (Liedgren et al., 1976b; Deecke et al., 1977) and in our anatomical studies.

Cortical neurons were recorded at the lower end of the intraparietal sulcus (area 2v), which is located in the transition zone of areas 2, 5 and 7 (Brodman, 1905). Here, as in the thalamus, only a minority of neurons responded to rotation of the monkey about a vertical axis.

Vestibular neurons in the thalamus and cortex can be classified as type I and type II neurons (Büttner et al., 1977; Magnin and Fuchs, 1977; Büttner and Buettner, 1978). According to the nomenclature of Duensing and Schafer (1958), type I neurons are activated during rotation to the ipsilateral side and inhibited during rotation to the contralateral side. Type II neurons show a mirror-like behavior. About 60% of the neurons are type I and about 40% type II neurons. Vestibular neurons in the thalamus and cortex respond regularly to each stimulus. They reliably show a constant phase and gain relationship at all stimulus frequencies (Figs. 3 and 4). Quantitative analysis (Fig. 4) reveals a phase advance of 0°–20° of neuronal activity relative to the turntable velocity at 0.1–1.0 Hz, and an increasing phase advance towards lower frequencies. Vestibular nuclei neurons recorded under identical conditions behave in a similar manner (Buettner et al., 1978).

Vestibular thalamic and cortical neurons could also be influenced by whole field optokinetic stimulation (Büttner and Henn, 1976; Büttner and Buettner, 1978). The optokinetic cylinder and the turntable have to move in opposite directions to elicit nystagmus in the same direction (Fig. 5). These stimulus conditions usually lead to identical changes of neuronal activity. A few neurons, however, are activated when the cylinder and turntable are rotated into the same direction giving rise to nystagmus into opposite directions. This indicates a more complex visual-vestibular interaction in the thalamus and cortex, since such a pattern is not regularly found in the vestibular nuclei (Waespe and Henn, 1977).

None of the vestibular neurons in the thalamus and cortex show an additional modulation with single eye movements, a feature present in about 50% of the neurons in the vestibular nuclei (Fuchs and Kimm, 1975). Electrophysiological studies have shown that vestibulothalamic and vestibulocortical neurons also respond to proprioceptive inputs, like muscle pressure or joint movements (thalamus: Liedgren et al., 1976b; Deecke et al., 1977; cortex: Schwarz and Fredrickson, 1971). This was

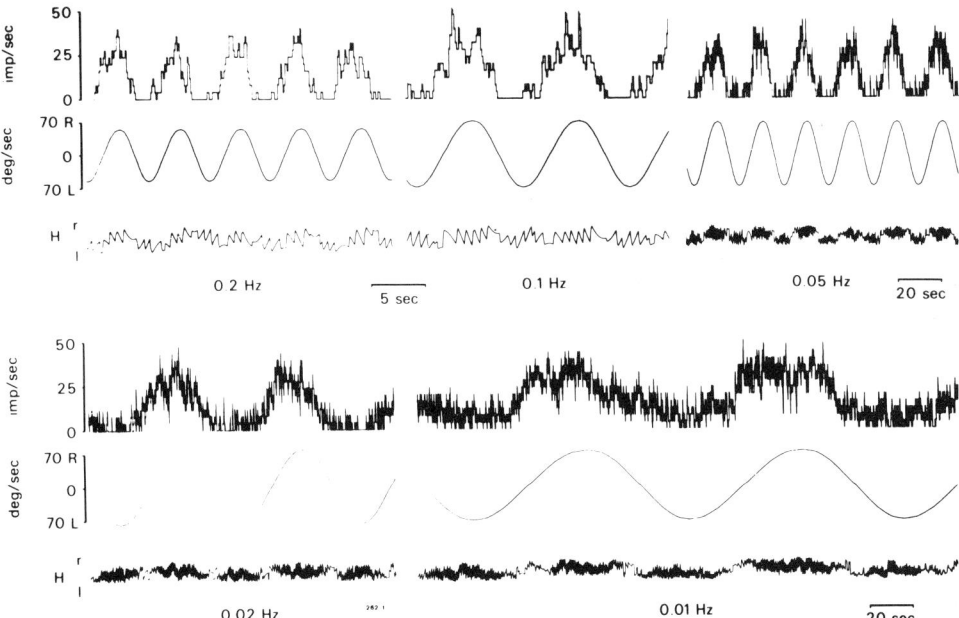

Fig. 3. Type I neuron in the thalamus (VPL$_o$) of the alert monkey during sinusoidal rotation (0.2–0.01 Hz) about a vertical axis in complete darkness. Upper trace: neuronal activity (running average over 250 msec, with a display updated every 100 msec). Middle trace: turntable velocity. Lower trace: horizontal eye position (r, rightward movement). The neuronal activity shows a small phase advance relative to maximal turntable velocity (R, right), which only slightly increases at lower frequencies.

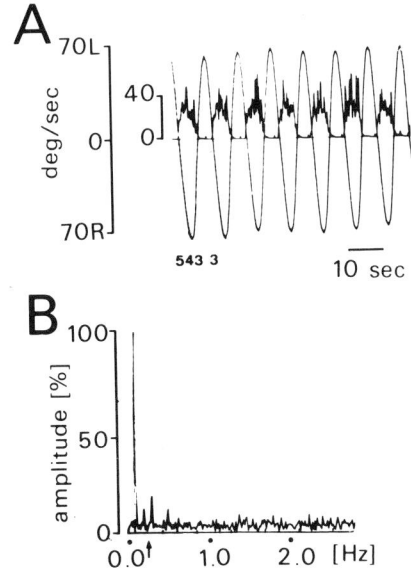

Fig. 4. Type I vestibular neuron recorded in the cortex (area 2v) during sinusoidal rotation of the alert monkey at 0.1 Hz in complete darkness. A shows the neuronal activity and turntable velocity, in digital form, used for the Fourier analysis. The amplitude distribution of the Fourier analysis shown in B is dominated by the first harmonic (fundamental, 0.1 Hz). The 2nd and 3rd harmonics (arrow) can be clearly separated. A phase advance for the neuronal activity (1st harmonic) of 6° relative to turntable velocity was determined.

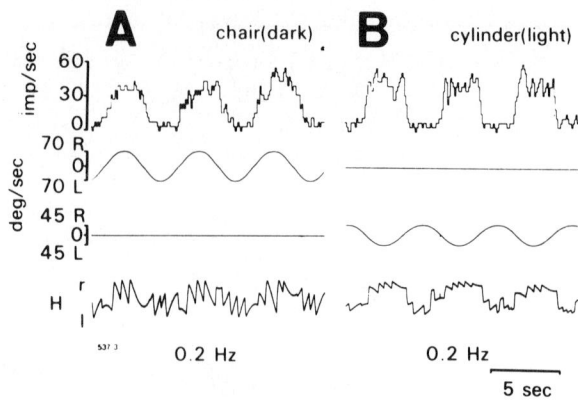

Fig. 5. Type 1 vestibular neuron recorded in area 2v during vestibular (A) and optokinetic (B) stimulation. Upper trace is neuronal activity (running average over 250 msec, with a display every 100 msec), second is turntable velocity, third is cylinder velocity, and fourth is horizontal eye position with movements to the right represented as upward deflections. Sinusoidal rotation in the dark at 0.2 Hz (A) evokes an activation similar to whole field visual rotation at the same frequency with the monkey stationary (B). Activation is obtained when the turntable or cylinder move in opposite directions. In both instances nystagmus to the right is present, vestibular in (A) and optokinetic in (B). The neuronal activity is in phase with turntable and cylinder velocity.

also found in the alert monkey, but no quantitative investigation was carried out. Recently, the interaction of vestibular and neck afferents in cortical (ASSS) neurons has been investigated in the cat also using natural vestibular stimuli (Mergner, this volume, chapter IV D8).

DISCUSSION

All sensory systems have a cortical projection area. The evidence for a vestibular projection was mainly based on electrophysiological studies. With sensitive tracer techniques a direct vestibulothalamic pathway can now be demonstrated anatomically. Neurophysiological studies in the alert monkey show that a precise vestibular signal can be recorded in the thalamus as well as in the cortex. The major functions of the vestibular system are to convey a sense of motion and to participate in the regulation of posture and eye movements. As none of the vestibular neurons in the thalamus or cortex are modulated with single eye movements, these pathways are probably not involved in oculomotor control. However, most ascending vestibular fibers in the VPL_o are in close contact and intermingled with proprioceptive relay cells, which are known to provide a major input to area 3a (Jones and Powell, 1970). Whether area 2v is more involved in posture control or motion sensation can not be concluded from the existing data so far.

Responses to optokinetic and proprioceptive stimuli have been demonstrated in vestibular neurons of the thalamus and cortex in addition to the vestibular responses. This has also been reported in the vestibular nuclei (Rubin et al., 1977; Waespe and Henn, this volume, Chapter VB1). Therefore, information carried in the ascending vestibular pathways is not an isolated signal from the peripheral vestibular organ, but already a very complex signal receiving information from different sensory organs about motion and posture. Still, when keeping all other inputs constant, a reliable signal from the horizontal semicircular canals can be traced via the vestibular nuclei and thalamus to the cortex. Besides being involved in the regulation of posture and movement, we suggest that this pathway provides a cortical projection which might be a necessary basis for subjective sensation.

SUMMARY

The vestibulo-thalamocortical pathway was investigated in the monkey (*Macaca mulatta*) with anatomical and neurophysiological methods. Previous electrophysiological studies by other authors have shown that electrical stimulation of the vestibular nerve leads to short-latency responses at the thalamic level in the ventroposterior nucleus (nucl. ventroposterior lateralis pars oralis, VPL_o; nucl. ventroposterior inferioris, VPI) and the posterior group (PO), and at the cortical level in area 2v and 3a.

With anterograde (radioactive amino acids) and retrograde (horseradish peroxidase) tracer substances a pathway from the vestibular nuclei via VPL_o to area 3a, and from the PO group to area 2v can be demonstrated. No projection was found from the vestibular nuclei to the PO group.

In neurophysiological studies neuronal activity was recorded in the VPL_o and VPI, and area 2v, while the alert monkey was exposed to natural vestibular and optokinetic stimulation. In these structures neurons were found, which responded clearly to vestibular stimulation, with patterns similar to those found in the vestibular nuclei. In most cases, vestibular neurons in the thalamus and cortex could also be activated by optokinetic stimuli.

It is suggested that the ascending vestibulocortical pathways contain information, which is used for motor control and/or motion sensation.

REFERENCES

Abraham, L., Copack, P.B. and Gilman, S. (1977) Brain stem pathways for vestibular projections to cerebral cortex in the cat. *Exp. Neurol.*, 55: 436–448.

Brodman, K. (1905) Beiträge zur histologischen Lokalisation der Grosshirnrinde. Dritte Mitteilung: Die Rindenfelder des niederen Affen. *J. Psychol. Neurol*, 4: 177–226.

Buettner, U.W., Büttner, U. and Henn, V. (1978) Vestibular nuclei activity in the alert monkey during sinusoidal rotation in the dark. *J. Neurophysiol.*, 41: 1614–1628.

Büttner, U. and Henn, V. (1976) Thalamic unit activity in the alert monkey during natural vestibular stimulation. *Brain Res.*, 103: 127–132.

Büttner, U. and Buettner, U.W. (1978) Parietal cortex (2v) neuronal activity in the alert monkey during natural vestibular and optokinetic stimulation. *Brain Res.*, 153: 392–397.

Büttner, U., Henn, V. and Oswald, H.P. (1977) Vestibular-related neuronal activity in the thalamus of the alert monkey during sinusoidal rotation in the dark. *Exp. Brain Res.*, 30: 435–444.

Deecke, L., Schwarz, D.W.F. and Fredrickson, J.M. (1977) Vestibular responses in the rhesus monkey ventroposterior thalamus. II. Vestibulo-proprioceptive convergence at thalamic neurons. *Exp. Brain Res.*, 30: 219–232.

Duensing, F. and Schaefer, K.P. (1958) Die Aktivität einzelner Neurone im Bereich der Vestibulariskerne bei Horizontalbeschleunigungen unter besonderer Berücksichtigung des vestibulären Nystagmus. *Arch. Psychiat. Nervenkr.*, 198: 225–252.

Fredrickson, J.M., Figge, U., Scheid, P. and Kornhuber, H.H. (1966) Vestibular nerve projection to the cerebral cortex of the rhesus monkey. *Exp. Brain Res.*, 2: 318–327.

Fuchs, A.F. and Kimm, J. (1975) Unit activity in vestibular nucleus of the alert monkey during horizontal angular acceleration and eye movement. *J. Neurophysiol.*, 38: 1140–1161.

Hauglie-Hanssen, E. (1968) Intrinsic neuronal organization of the vestibular nucleus complex in the cat. A Golgi study. *Ergebn. Anat. Entwickl.-Gesch.*, 40: 1–105.

Jones, E.G. and Powell, T.P.S. (1970) Connexions of the somatic sensory cortex of the rhesus monkey. III. Thalamic connexions. *Brain*, 93: 37–56.

Liedgren, S.R.C., Kristensson, K., Larsby, B. and Oedkvist, L.M. (1976a) Projection of thalamic neurons to cat primary vestibular cortical fields studied by means of retrograde axonal transport of horseradish peroxidase. *Exp. Brain Res.*, 24: 237–243.

Liedgren, S.R.C., Milne, A.C., Rubin, A.M., Schwarz, D.W.F. and Tomlinson, R.D. (1976b) Representation of vestibular afferents in somatosensory thalamic nuclei of the squirrel monkey (*Saimiri sciureus*). *J. Neurophysiol.*, 39: 601–612.

Magnin, M. and Fuchs, A.F. (1977) Discharge properties of neurons in the monkey thalamus tested with angular acceleration, eye movement and visual stimuli. *Exp. Brain Res.*, 28: 293–299.

Mickle, W.A. and Ades, H.W. (1952) A composite sensory projection area in the cerebral cortex of the cat. *Amer. J. Physiol.*, 170: 682–689.

Ödkvist, L.M., Schwarz, D.W.F., Fredrickson, J.M. and Hassler, R. (1974) Projection of the vestibular nerve to the area 3a arm field in the squirrel monkey (*Saimiri sciureus*). *Exp. Brain Res.*, 21: 97–105.

Rubin, A.M., Liedgren, S.R.C., Milne, A.C., Young, J.A. and Fredrickson, J.M. (1977) Vestibular and somatosensory interaction in the cat vestibular nuclei. *Pflügers Arch.* 371: 155–160.

Sans, A., Raymond, J. and Marty, R. (1970) Réponse thalamique et corticales à la stimulation électrique du nerf vestibulaire chez le chat. *Exp. Brain Res.*, 10: 265–275.

Schwarz, D.W.F. and Fredrickson, J.M. (1971) Rhesus monkey vestibular cortex: a bimodal primary projection field. *Science*, 172: 280–281.

Tarlov, E. (1969) The rostral projections of the primate vest. nuclei: an experimental study in Macaque, Baboon and Chimpanzee. *J. comp. Neurol.*, 135: 27–56.

Waespe, W. and Henn, V. (1977) Neuronal activity in the vestibular nuclei of the alert monkey during vestibular and optokinetic stimulation. *Exp. Brain Res.*, 27: 523–538.

E

Compensation of Labyrinthine Functions

Somatosensory and Cerebellar Influences on Compensation of Labyrinthine Lesions

K.-P. SCHAEFER, D. L. MEYER and G. WILHELMS

Neurobiology Unit, Department of Psychiatry, University of Göttingen, 3400 Göttingen (F.R.G.); and Neurobiology Unit, Department of Neurosciences, University of California, La Jolla, San Diego, CA 92093 (U.S.A.)

INTRODUCTION

As many investigations have revealed (e.g., Von Holst, 1935a,b; Witkin, 1959; Meyer et al., 1976a,b, 1977; Meyer and Bullock, 1977), equilibrium control is a multimodal function and thus guided by a variety of sense organs. One of the sensory systems of special significance for postural control is the somatosensory system. In certain cases it can even be dominant over the vestibular system with respect to the guidance of body posture during normal behavior (Meyer et al., 1976b). Afferents from this system reach vestibular circuits directly from the spinal cord as well as via the cerebellum (Pompeiano and Brodal, 1957; Fredrickson et al., 1965; Ebbesson, 1969; Brodal, 1974; and others).

To investigate what significance these inputs have for vestibular functions we used guinea pigs that were unilaterally "labyrinthectomized" by an injection of chloroform into the middle ear. Our study was concerned with the influences exerted by somatosensory afferents on compensation of labyrinthectomy symptoms. In order to distinguish the effects of somatosensory afferents that are directly fed into vestibular circuits from those relayed in the cerebellum before reaching that system we also studied vestibular compensation in animals with various cerebellar lesions.

RESULTS

Considering the close functional connections between vestibular and motor systems, it is understandable that unilateral labyrinthectomy does not only result in tonus asymmetries in the oculomotor system but also causes postural asymmetries in the trunk region and of the extremities. Furthermore, spinal cord transection and other interferences with somatosensory afferents (e.g., suspending an animal in air and thereby eliminating its ground contact) exert significant influences on vestibular functions, in particular, on mechanisms compensating for labyrinthine lesions. In this context, it is also of relevance that some labyrinthectomy symptoms may reappear

---

*Supported by the "Deutsche Forschungsgemeinschaft" (SFB 33, Schaefer) and NSF, NIH and NASA grants to Dr. T. H. Bullock.

after compensation for the lesion has already been accomplished if the spinal cord is transected (release phenomena; see Azzena, 1969; Schaefer and Meyer, 1973, 1974).

Our interest was focused on the region of the upper cervical cord which mediates neck-proprioceptive inputs through C1-C3. As the significance of this region for vestibular compensation cannot be tested by spinal transection experiments, we studied the influence of various head positions on the time course of compensation for unilateral labyrinthectomy. For this investigation, we used 4 groups of 10 animals each that were unilaterally labyrinthectomized on the right side. The first group served as a control. The heads of the animals in the other three groups were restrained in the following three positions for a period of 5 h after the labyrinthine lesion: i) 60° to the right, ii) mid-position, iii) 60° to the left. The time of restraining the head was chosen from pilot studies which revealed this time span to be optimal for our purpose as, on the one hand, it was sufficiently long to cause measurable effects on vestibular compensation and, on the other hand, the control animals still displayed significant lesion symptoms at this time. The latter rendered it possible to also detect accelerative effects of restraining of the head on the compensation time course.

As compared with the controls, compensation was faster in those animals which had the head restrained 60° towards the side contralateral to the labyrinthine lesion.

After removal of the head restraining device 5 h after the labyrinthectomy, the values measured for the *head turning in the horizontal plane* were significantly lower than in the control group ($2P < 0.01$). As shown in Fig. 1, the head turning towards the lesioned labyrinth present after 5 h was practically identical to that of the control group after 11 h of compensation. Fixating the head in the midposition resulted in the finding of a head turning of 58° when the head restraining was terminated. Also, in this case a significant acceleration for this labyrinthine lesion symptom was thus present. Data for compensation of head turning in the horizontal plane in the group which had the head restrained 60° towards the lesioned labyrinth did not differ from the controls.

Compensation of the *ocular nystagmus* resulting from the labyrinthine lesion was less affected by restraining the head in the three positions. A slight but not statistically significant acceleration of compensation was noted in the group that had the head

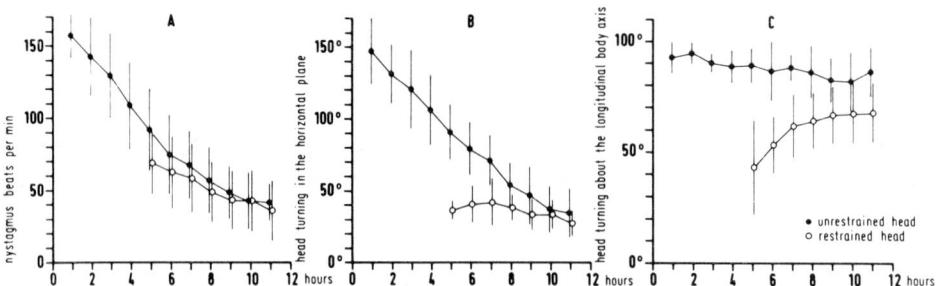

Fig. 1. Compensation of labyrinthectomy symptoms under normal conditions (unrestrained head) and with 5 h of restraining the head at a position 60° towards the side contralateral to the lesion. A) Ocular nystagmus measured after termination of restraining the head is almost equivalent to controls. B) The symptom of head turning in the horizontal plane is significantly decreased in animals that had their head restrained towards the side contralateral to the labyrinthectomy. C) Head- (and body-) turning about the longitudinal body axis measured while the animals were lifted off the ground is not compensated for (see: unrestrained head). After the head restraining procedure this symptom is weaker than in the controls, but increases with time.

restrained towards the side contralateral to the vestibular lesion (Fig. 1). Nevertheless, statistically different values were noted between the group with the head contralateral and the group that had the head fixated towards the side ipsilateral to the lesion ($2P < 0.05$). In the latter group of animals, ocular nystagmus compensation was insignificantly slower as compared with the controls.

Most surprising were the observations of the *head turning about the longitudinal body axis* present when a labyrinthectomized animal is lifted off the ground (for reference, see Schaefer and Meyer, 1974). This vestibular lesion symptom is not compensated for under normal conditions. Fig. 1 demonstrates that this symptom was significantly less pronounced ($2P < 0.05$) in those animals that had their heads fixated 60° towards the side contralateral to the side of the labyrinthectomy for 5 h. The head turning about the longitudinal body axis slowly increased after the head was free again but was still significantly lower than in the controls after 11 h. Within the subsequent 12 h the values became identical to those of the control group.

The fact that practically no differences from the control group were noted in animals that had the head restrained towards the lesioned side, suggests that *dynamic mechanisms* do not facilitate compensation of the symptoms studied here under normal conditions. Nevertheless, it should be stated that other symptoms, such as head nystagmus, rolling about the longitudinal axis, and circular movements were noted when the head restraining was terminated and the animals were free to move again. These remained present for several hours. Compensation for such vestibular lesion symptoms seemed retarded.

In order to determine whether or not the neck-proprioceptive influences on vestibular compensation described above are mediated by the cerebellum, we repeated the same experiments with animals that had been *cerebellectomized* two weeks before the labyrinthine lesion was inflicted. As values for ocular nystagmus and for head turning in the horizontal plane are different in cerebellectomized animals in comparison with controls, conclusive evidence for the role of the cerebellum in compensating for these symptoms could not be obtained. But our data strongly suggest that an influence of the head position was absent in animals without cerebellum with respect to the compensation of nystagmus and head deviation in the horizontal plane.

With regard to the head turning about the longitudinal body axis we could obtain clear evidence for an involvement of the cerebellum in mediating neck-proprioceptive influences on compensation, since this symptom is displayed practically unchanged by cerebellectomized specimens. Such animals did not display different time courses of the strength of that symptom anymore when head fixation in the three positions had been performed. They did not differ from the controls. Experiments with partial cerebellar lesions similar to the ones that will be described below, have not yet been carried out in connections with studies of neck-proprioceptive influences on vestibular compensation.

The vestibular system and the cerebellum are phylogenetically closely related. After ablation of the cerebellum vestibular nystagmus is diminished and body postural reflexes are disturbed. Direct spinoreticular pathways to postural control centers are obviously unable to take over the functions of indirect spinocerebellar-brain stem connections completely. Accordingly, a retardation of compensation for some vestibular lesion symptoms has been found (see Schaefer and Meyer, 1974). Such investigations revealed that the different labyrinthectomy symptoms are affected differently by cerebellar lesions. In guinea pigs the decrease of head turning in the horizontal plane is still slower than normal when the cerebellectomy has been carried out 6 months before a labyrinthectomy is inflicted. Ocular nystagmus compensation is clearly retarded during the first few days after cerebellar ablation only; this retardation may be a consequence of general trauma rather than being due to interferences with specific mechanisms (Schaefer and Meyer, 1973, 1974).

Previous to more recent unpublished investigations of ours, we assumed that a retardation of vestibular compensation in cerebellectomized animals might mainly result from ablations of structures of the archicerebellum. These areas have particularly close connections to vestibular circuits. In contrast to our hypothesis, we found that selective ablation of the flocculus, the paraflocculus, the nodulus (see Fig. 2), and of the caudal part of the uvula had rather little effects on vestibular compensation in our animals that were labyrinthectomized two weeks after the cerebellar lesions had been inflicted. (Each of the above cerebellar lesions was carried out on ten animals.) Ablation of the left cerebellar hemisphere resulted in more pronounced labyrinthectomy symptoms than in the controls when the vestibular lesion was inflicted on the right side. Under such circumstances the head deviation in the horizontal plane amounted to more than 170° whereas the control animals only displayed a head turning in the horizontal plane of 140°–150° 1 h after unilateral labyrinthectomy. The ocular nystagmus measured 1 h after the vestibular lesion was 280 beats/min as compared to 180 beats/min in the controls. Naturally, the stronger symptoms took longer to be abolished by compensation. It is of relevance that, only in the case of head turning in the horizontal plane, a true retardation of compensatory mechanisms was found. Whereas the compensation plot for the ocular nystagmus in the test animals ran parallel to the one of the control group, the compensation curve for the head turning was not only parallel shifted upward, but also decreased with a slightly lower time constant. When labyrinthectomizing an animal on the right side that previously lost its right cerebellar hemisphere, some of the vestibular lesion symptoms were insignificantly less pronounced than in the controls. This is interpreted as being due to

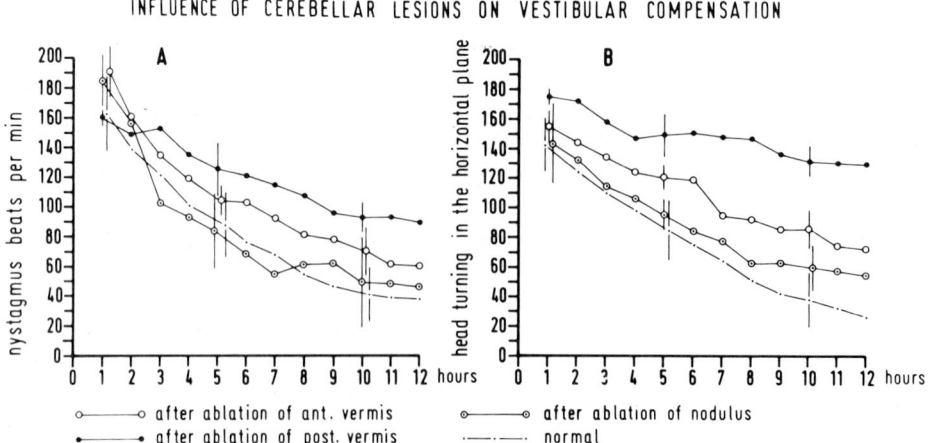

Fig. 2. Influence of various cerebellar lesions on compensation of labyrinthine lesion-induced ocular nystagmus and head turning in the horizontal plane. Whereas compensation of ocular nystagmus A) is only slightly affected by all cerebellar lesions, compensation of head turning in the horizontal plane B) is distinctly retarded after lesioning of the posterior vermis.

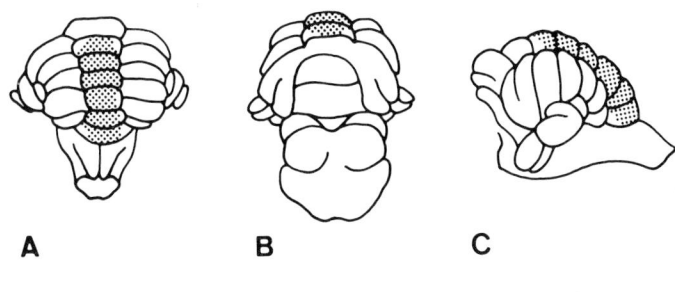

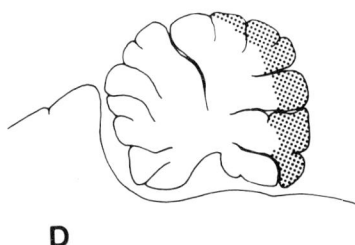

Fig. 3. Histologically verified lesion of the posterior vermis of one of our specimens. The lesion (dotted area) involves neither the nodulus nor cerebellar or vestibular nuclei. A) View from caudal. B) View from anterior. C) View from lateral. D) Sagittal section near cerebellar midline.

two artificially induced tonus asymmetries of opposite sign counteracting each other.

Further experiments revealed that an ablation of the posterior vermis (Fig. 3) caused a distinct retardation of vestibular compensation, in particular of the compensation of head turning in the horizontal plane (Fig. 2). Again unilateral labyrinthectomy was performed two weeks after the cerebellar lesion (10 animals). The interference with labyrinthine lesion compensation was almost equivalent to the one present when a complete cerebellectomy was carried out. This is surprising since body postural reflexes in guinea pigs are only slightly impaired by posterior vermis ablation.

The consequences of posterior vermis ablation cannot simply be interpreted as disinhibition phenomena, since tonic inhibitory influences on vestibular circuits mainly originate from the anterior vermis. Lesioning of that area resulted in a significantly less distinct retardation of vestibular compensation as compared with posterior vermis ablations. From the fact that the posterior vermis plays an important role in the integration of spinal afferents and considering its connections to the vestibular nuclei via intermediate nuclei, one may conclude that an interruption of spinocerebellar-vestibular pathways by posterior vermis lesions deprives the vestibular compensation mechanisms of an important source of afferents. Thus, the retardation of compensation observed seems understandable.

DISCUSSION

Under normal conditions, compensation of head turning in the horizontal plane and of ocular nystagmus resulting from unilateral labyrinthectomy have similar time courses in guinea pigs. This is in contrast with the findings Llinás and Walton (1977) obtained in rats. Possibly species differences are responsible. Also, in guinea pigs the two functions can be dissociated. After spinal cord transection, cerebellar lesions, hemispherectomy, and after application of various drugs, nystagmus and head deviation are compensated at different rates in this animal (see Schaefer and Meyer, 1974). The experimentally induced interferences with vestibular compensation mechanisms always affect the symptom of head turning in the horizontal plane more distinctly than the nystagmus. This leads to the conclusion that compensation of head deviation is highly dependent on afferents from other CNS systems. Somatosensory afferents seem to be of particular relevance. This is emphasized by the observation that the symptom of head deviation can reappear when a CNS lesion is inflicted after vestibular compensation has been completed (release phenomenon).

Compensation of ocular nystagmus can be influenced by CNS lesions and by pharmacological means to a significantly smaller degree (Schaefer and Meyer, 1974). Even ablation of the vestibulo-cerebellum, in particular, ablation of the flocculus has no striking effect on nystagmus compensation after labyrinthine lesions. This is astonishing as the flocculus is considered to be involved in gaze stabilization and plasticity in the oculomotor system and receives visual afferences (Ito, 1977; Maekawa and Takeda, 1977). Only in the case of posterior vermis ablation ocular nystagmus compensation is slightly influenced (Fig. 2). At this point it has to remain uncertain whether or not this is due to an interruption of visual projections to the folium and the tuber vermis. As nystagmus compensation is also hardly influenced by drugs this phenomenon could be thought to be mediated by just vestibular circuits. Buchanan (1940) described that compensation of nystagmus is still present if the nystagmus resulted from a unilateral lesion of the vestibular nuclei. Thus, it appears unlikely that nystagmus compensation after labyrinthine lesions is achieved by vestibular mechanisms alone.

The head (and body) turning about the longitudinal body axis seen when the animal is lifted off the ground is one of the labyrinthectomy symptoms that are not compensated for by the organism. Probably short vestibulospinal connections and roll substrates (Hassler and Hess, 1954) are involved which cannot be influenced by other CNS structures. This symptom also remains after spinal lesions, cerebellectomy, and hemispherectomy. By pharmacological means (Schaefer and Wehner, 1966) and by head fixation towards the side of the intact labyrinth (see above) at least some temporary influences can be exerted. The head turning about the longitudinal body axis instantaneously disappears in animals that have been labyrinthectomized some time before they regain contact with the ground. This effect is not present in the cerebellectomized animals as well as after ablation of just the posterior vermis. From this we conclude that the cerebellum does not play an active role in the compensation process. We thus agree with Llinás et al. (1975) who considered the cerebellum's function in vestibular compensation to be a passive one.

SUMMARY

Fixation of the head in positions towards the left, right, and in the midposition during the initial 5 h after unilateral labyrinthectomy demonstrated the significance of neck-proprioceptive afferents for vestibular compensation. Fixation of the head towards the side of the intact labyrinth caused an acceleration of compensation for head deviation in the horizontal plane, as seen when the head restraining was terminated. Ocular nystagmus compensation was insignificantly faster than in controls. The symptom of head turning about the longitudinal body axis present when lifting a unilaterally labyrinthectomized animal off the ground was less pronounced than in the control animals. It increased in strength with time once the head was free again and matched the values measured in control specimens after about 20 h.

Experiments with discrete cerebellar lesions demonstrated that the strongest effects on vestibular compensation mechanisms are present after ablation of the posterior vermis. These are almost as strong as those resulting from a complete cerebellectomy.

We conclude that compensation of the labyrinthectomy symptom of head turning in the horizontal plane is highly dependent on somatosensory afferents. These influences seem to be mediated through the cerebellum. As far as the symptoms considered here are concerned, the cerebellum is thus thought to play a passive role in vestibular compensation only.

REFERENCES

Azzena, G. B. (1969) Role of the spinal cord in compensating the effects of hemilabyrinthectomy. *Arch. ital. Biol.*, 107; 43–53.

Brodal, A. (1974) Anatomy of vestibular nuclei and their connections. In *Handbook of Sensory Physiology, Vol. VI/1, Vestibular System, Part 1, Basic Mechanisms*, H. H. Kornhuber (Ed.), Springer-Verlag, Berlin-Heidelberg-New York, pp. 239–352.

Buchanan, A. R. (1940) Nystagmus and eye deviations in the guinea pig with lesions in the brainstem. *Laryngoscope*, 50: 1002–1011.

Ebbesson, S. O. E. (1969) Brain stem afferents from the spinal cord in a sample of reptilian and amphibian species. *Ann. N.Y. Acad. Sci.*, 167: 60–101.

Fredrickson, J. M., Schwartz, D. and Kornhuber, H. H. (1965) Convergence and interaction of vestibular and deep somatic afferents upon neurons in the vestibular nuclei of the cat. *Acta oto-laryngol. (Stockh.).* 61: 168–188.

Hassler, R. and Hess, W. R. (1954) Experimentelle und anatomische Befunde über die Drehbewegungen und ihre nervösen Apparate. *Arch. Psychiat. Nervenkr.*, 192: 488–526.

Ito, M. (1977) Neuronal events in the cerebellar flocculus associated with an adaptive modification of the vestibulo-ocular reflex of the rabbit. In *Control of Gaze by Brain Stem Neurons*, R. Baker and A. Berthoz (Eds.), Elsevier, Amsterdam-New York, pp. 391–398.

Llinás, R. and Walton, K. (1977) Significance of the olivo-cerebellar system in compensation of ocular position following unilateral labyrinthectomy. In *Control of Gaze by Brain Stem Neurons*, R. Baker and A. Berthoz, (Eds.), Elsevier, Amsterdam-New York, pp. 399–408.

Llinás, R., Walton, D., Hillman, D. E. and Sotelo, C. (1975). Inferior olive: Its role in motor learning. *Science*, 190: 1230–1231.

Maekawa, K. and Takeda, T. (1977) Afferent pathways from the visual system to the cerebellar flocculus of the rabbit. In *Control of Gaze by Brain Stem Neurons*, R. Baker and A. Berthoz (Eds.), Elsevier, Amsterdam-New York, pp. 187–195.

Meyer, D.L., Heiligenberg, W. and Bullock, T.H. (1976) The ventral substrate response. A new postural control mechanism in fishes. *J. comp. Physiol.*, 109: 59–68.

Meyer, D. L., Platt, C. and Distel, H. J. (1976) Postural control mechanisms in the upside-down catfish (*Synodontis nigriventris*). *J. comp. Physiol.*, 110: 323–331.

Meyer, D.L., Becker, R. and Graf, W. (1977) The ventral substrate response of fishes. *J. comp. Physiol.*, 117: 209–217.

Pompeiano, O. and Brodal, A. (1957) Spino-vestibular fibres in the cat. An experimental study. *J. comp. Neurol.* 108: 353–378.

Schaefer, K.-P. and Wehner H. (1966) Zur pharmakologischen Beeinflussung zentralnervöser Kompensationsvorgänge nach einseitiger Labyrinthausschaltung durch Krampfgifte und andere erregende Substanzen. *Naunyn-Schmiedebergs Arch. Pharmak. exp. Path.*, 254: 1–17.

Schaefer, K.-P., Wilhelms, G. and Meyer, D. L. (1978) Der Einfluß von Alkohol auf die zentralnervösen Ausgleichsvorgänge nach Labyrinthausschaltung. *Z. Rechtsmed.*, 81: 249–260.

Schaefer, K.-P. and Meyer, D. L. (1973) Compensatory mechanisms following labyrinthine lesions in the guinea pig. A simple model of learning. In *Memory and Transfer of Information*, H. P. Zippel (Ed.), Plenum Press, New York-London, pp. 203–232.

Schaefer, K.-P. and Meyer, D. L. (1974) Compensation of vestibular lesions. In *Handbook of Sensory Physiology, Vol. VI/1, Vestibular System, Part 2, Psychophysics, Applied Aspects and General Interpretations*, H. Kornhuber (Ed.), Springer-Verlag, Berlin-Heidelberg-New York, pp. 463–490.

Von Holst, E. (1935a) Über den Lichtrückenreflex bei Fischen. *Publ. Staz. Zool. Napoli*, 15: 143–158.

Von Holst, E. (1935b) Die Gleichgewichtssinne der Fische. *Verh. dtsch. Zool. Ges.*, 108–114.

Witkin, H. A. (1959) The perception of the upright. *Sci. Amer.*, 200: 50–56.

Cerebellar Contribution in Compensating the Vestibular Function

G. B. AZZENA, O. MAMELI and E. TOLU

Istituto di Fisiologia Umana, Università di Sassari, 07100 Sassari (Italy)

INTRODUCTION

Unilateral lesion of the labyrinth produces severe postural asymmetries involving the eye, trunk and limb muscles. However, the central nervous system is able to overcome these deficiencies during a compensation period, which is followed by the complete disappearance of the motor deficits (Spiegel and Démétriades, 1925; Azzena, 1969; Schaefer and Mayer, 1974). This powerful plasticity is shown by the fact that, in a compensated animal, a mid-thoracic transection of the spinal cord will induce the reappearance of almost all the symptoms elicited by the previous hemilabyrinthectomy. Thus, the lesion of the spinal cord disrupts the compensated symmetry, inducing a "decompensation" stage (Azzena, 1969).

The neuronal correlations for the vestibular compensation have been recently analysed. It has been shown that transection of the spinal cord, in guinea pigs with compensation following a previous unilateral labyrinthectomy, modifies the electrical activity of the vestibular nuclei. Recordings of the field potentials, generated in the vestibular nuclear complexes by stimulating the intact ampullar receptors in hemilabyrinthectomized and compensated guinea pigs, indicate that following transection of the spinal cord there is a depression of vestibular nuclei ipsilateral to the vestibular deafferentation and a facilitation of the contralateral vestibular nuclei (Azzena et al., 1976). Similar results have been obtained by recording the unitary discharge of lateral vestibular neurons. The discharge frequency of vestibular cells ipsilateral to the vestibular deafferentation is reduced, while that of the contralateral vestibular neurons is increased (Azzena et al., 1977). It appears, therefore, that ascending spinal volleys are essential for correcting the balance of the vestibular output. Since these results were mainly obtained by recording the unit discharge from the lateral vestibular nuclei, which are involved in the control of posture and are under the direct influence of the cerebellum, experiments were performed to find out whether the ascending spinal volleys which are involved in the compensation of the labyrinthine syndrome act on the vestibular neurons directly or via the cerebellum (Wilkems and Schaefer, 1976). In fact, according to Llinàs et al. (1975) the integrity of the olivocerebellar pathways is essential in producing a compensation of the vestibular symptoms.

The aim of the present research was then to study the compensatory relationships between spinal cord and cerebellum. This purpose was achieved by recording the unitary discharge of lateral vestibular neurons in hemilabyrinthectomized and compensated guinea pigs during the stage of decompensation evoked either by combining the transection of the spinal cord with the lesion of the inferior olive or by cooling of the cerebellum.

METHODS

The experiments were performed on 71 guinea pigs. The animals were submitted, under ether anaesthesia, to lesion of the left labyrinth according to the method of Simonelli (1923); in fact, total degeneration of the acoustic and labyrinthine receptors is provoked by injecting a solution of oil and chloroform in the middle ear (Menzio, 1952). The animals were then allowed to recover and compensate the vestibular symptoms.

After compensation, one group was anaesthetized by i.m. administration of 10 mg/kg of ketamine (Ketalar, Parke Davis) and 10 mg/kg of diazepam (Valium, Roche). Following tracheotomy, each animal of this group was fixed in a stereotaxic apparatus, paralyzed with Flaxedil and artificially respirated. The cerebellum and the spinal cord were exposed by performing a craniotomy and a laminectomy between C3–C4 and T4–T5. Recordings of the unitary discharge from the lateral vestibular nucleus were carried out by means of tungsten microelectrodes (700–900 kΩ) advanced through the cerebellum and connected with conventional preamplifiers. Signals were recorded on magnetic tape for later analysis of the discharge frequency by a Schmitt trigger and a 4672 Ortec instantaneous frequency/timer meter. Only negative or negative-positive spikes were recorded from units, which were identified at random by antidromic stimulation at C3 level. The decompensation was induced by i.p. injection of 80 mg/kg of 3-acetyl-pyridine (3-AP) which provokes a complete destruction of the inferior olive neurons (Desclin, 1974; Desclin and Escubi, 1974). Thus, records of vestibular unitary discharge were taken before and at least 3 h after the administration of the drug (see also Results). Once the 3-AP was effective, as shown by a clear modification of the frequency rate of the cells, the animals were submitted to the interruption of the spinal cord that was performed by transverse transection or by application of dry-ice at the mid-thoracic level (T4–T5).

In a second group of animals, procedure was the same except that the cerebellum was blocked by applying dry-ice over the cerebellar surface instead of injecting 3-AP. Cerebellar temperature was controlled by a thermocouple inserted into the cerebellar white matter. In this group, the recordings of the unitary discharge were performed before and after the cooling of the cerebellum and also after the subsequent transection of the spinal cord. In all the animals, the recorded sites were marked for subsequent histological control.

RESULTS

The present report is based upon the analysis of the responses of 63 deitersian units; 36 of them were recorded from the left lateral vestibular nucleus (i.e., the deafferented side) and 27 from the right nucleus.

This paper reports the pattern of deitersian unit responses related to the discharge frequency modifications provoked by the following procedures. In the first group of animals, recordings of normal electrical activity were taken in the compensated stage, after administration of 3-AP and after transection of the spinal cord. In the second group, recordings were similarly taken during the compensation period, during the cooling of the cerebellum, after the transection of spinal cord with the rewarmed cerebellar surface, and after a second period of cerebellar cooling. However, the effects induced on the vestibular units by the 3-AP reported here are those observed at least 3 h after the injection of the drug, since it has been observed that the short-term effects of 3-AP result prevalently in excitation of the olivary neurons. This fact, that could be attributed to an initial sensibilization of the cells of the inferior olive by the drug, is shown in Fig. 1. Fig. 1 A–D illustrates the field potentials recorded from the molecular layer in the pars intermedia of lobule V, and Fig. 1E–H shows the poststimulus time histograms of the response of a single Purkinje cell following electrical stimulation of a forelimb nerve (Allen et al., 1974; Eccles et al., 1967, 1968). Fig. 1A and E are the normal responses, while B–F, C–G and D–H are the responses recorded 40, 130 and 220 min after administration of 3-AP. It is evident that the field potentials and the peak of the poststimulus time histograms generated by the climbing fibres increase in B and F, but, as recording proceeds, there is a total cut-off of the climbing fibre peak in the poststimulus time histograms (Fig. 1H) since the small field potential observed in D is not sufficient to trigger the response of the Purkinje cell. On the other hand, the effects of the 3-AP on the mossy fibre activity were negligible. Fig. 2 illustrates the pattern of activity of a right deitersian unit following the injection of 3-AP. As shown in Fig. 2A and B, the discharge rate of this cell during the compenstated stage ranged from 10 to 55 imp/sec. After 20 min the discharge frequency became lower (C) and reached its lowest value after 100 min when it was firing at

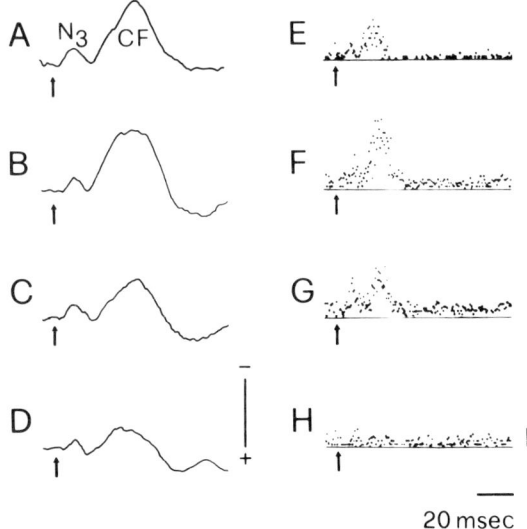

Fig. 1. Influence of 3-AP on field potentials and single Purkinje cell response to electrical stimulation of ipsilateral radial nerve. A–D) Averaged field potentials of 16 sweeps at 1 Hz recorded from the surface of lobule V of the pars intermedia. E–H) Poststimulus time histograms of responses of a single Purkinje cell obtained after 128 sweeps by recording from the same cerebellar lamella of A–D. A, E) Responses in normal conditions. B and F, C and G, D and H) Records taken resp. 40, 130 and 220 min after the i.p. administration of 3-AP. Calibration of 0.5 mV applies to A–D, 20 counts to E–H and 20 msec to all traces.

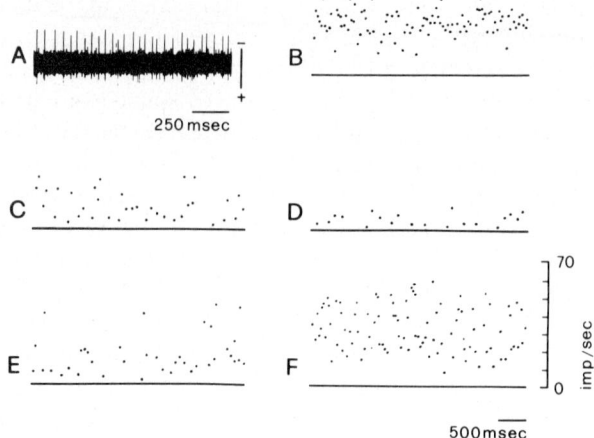

Fig. 2. Electrical activity of a right deitersian unit in a hemilabyrinthectomized and compensated guinea pig. A) Specimen record and B) instantaneous discharge frequency in the compensated stage: C-F) 20, 100, 140 and 220 min after the i.p. administration of 3-AP.

1-10 imp/sec (D). After 140 min the cell reversed its pattern of activity (E), reaching a high degree of facilitation after 220 min (F). Fig. 3 reports the behaviour of a left deitersian unit following i.p. administration of 3-AP and the subsequent transection of the spinal cord. Fig. 3A and B respectively are the specimen record and the instantaneous discharge frequency in normal conditions following compensation. After 190 min from the injection of 3-AP, the cell increases the frequency rate (C) which is further facilitated by the spinal interruption (D).

The effects of the combined removal of the spinal and cerebellar influences on a left deitersian unit are shown in Fig. 4 which illustrates an inhibition by removing the spinal cord and then a facilitation after cooling of the cerebellar cortex. Fig. 5 summarizes the effects induced upon all the vestibular cells of both sides during the course of these experiments. In particular, A represents the percentage of the units responding with excitation (unshaded column) or inhibition (shaded column) to the transection of the spinal cord as reported in a previous paper (Azzena et al., 1977). B represents the late effects of the 3-AP on 34 deitersian cells; 20 out of 34 units were recorded from

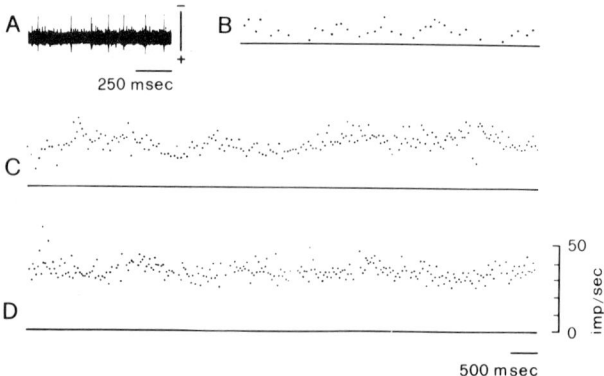

Fig. 3. Electrical activity of a left deitersian unit in a hemilabyrinthectomized guinea pig. A) Specimen record and B) instantaneous discharge frequency in the compensated stage. C) 190 min after i.p. administration of 3-AP; D) 4 min after the transverse section of the spinal cord.

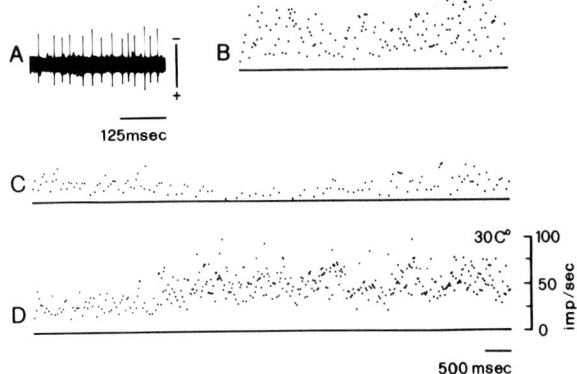

Fig. 4. Electrical activity of a left deitersian unit in a hemilabyrinthectomized and compensated guinea pig. A) Specimen record and B) instantaneous discharge frequency in the compensated stage; C) following the transverse section of the spinal cord; D) during the cooling of the cerebellar cortex at 30° C.

the left and 14 from the right nucleus. C represents the further changes observed on the same cells following spinalization. The effects of cooling of the cerebellar cortex were analysed on 29 deitersian cells (16 from the left and 13 from the right Deiters nuclei). The pattern of activity of these units was similar as reported in Fig. 5B, the only difference being represented by a higher level of excitation.

DISCUSSION

The analysis of the discharge frequency of the vestibular neurons seems to be an important aid in elucidating the mechanisms involved in compensating the symptoms provoked by the unilateral lesion of the labyrinth. As far as the immediate consequence of the hemilabyrinthectomy is concerned, it has been shown that the unitary

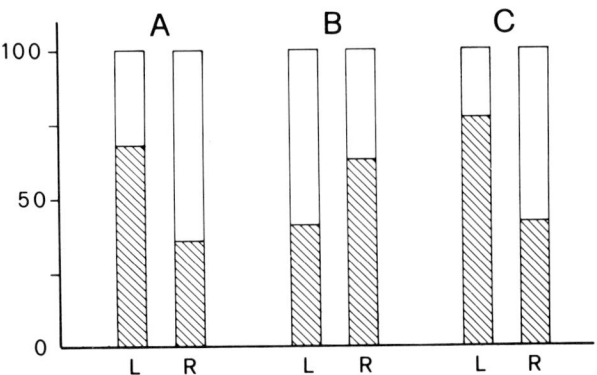

Fig. 5. A) Percentage of units of left (L) and right (R) lateral vestibular nuclei, in hemilabyrinthectomized and compensated animals, responding with excitation (unshaded column) or inhibition (shaded column) to spinalization as reported in Azzena et al. (1977). B and C report the results of present experiments by indicating the percentage of activation or inhibition (same symbols as in A) evoked on cells of left (L) and right (R) lateral vestibular nuclei, in hemilabyrinthectomized and compensated animals, following i.p. administration of 3-AP (B) and the subsequent transection of spinal cord (C).

discharge of cells located in the vestibular nuclei is typically upset. In fact, it has been observed that, following the vestibular lesion, the activity of the deafferented vestibular neurons is dramatically reduced while the activity of the neurons of the contralateral side is exaggerated (Spiegel and Démétriades, 1925; Precht et al., 1966; McCabe and Ryu, 1969; McCabe et al., 1972; Schaefer and Mayer, 1974). This different level of activity has been attributed to the imbalance which takes place at the moment of the vestibular deafferentation, i.e., afferent excitatory impulses travelling through the VIIIth nerve on the side of the lesion are eliminated and, by contrast, those of the vestibular nerve of the intact side are still present. This effect will cause a different flow of inhibitory impulses through the commissural system mediating influences between the vestibular nuclei of both sides. If the vestibular deafferentation has been performed, for example, at the left side, the left vestibular nuclei are under a strong right inhibition while the right nuclei are disinhibited owing to the left inactivation. Compensation after hemilabyrinthectomy must then be directed to recover the balance between the two sides by introducing a strong inhibition upon the "intact" nuclei together with an excitation upon the vestibular cells of the deafferented side. Therefore, the problem regards the mechanism which controls these excitatory and inhibitory regulations.

For this task an important role has been attributed to the spinal cord. In hemilabyrinthectomized and compensated guinea pigs it has been observed that the discharge frequency of the deafferented vestibular neurons drops after the transverse section of spinal cord, while that of the "intact" vestibular cells increases (Azzena et al., 1977). This pattern of activity, therefore, is similar to that recorded in the acute stage of hemilabyrinthectomy. On the other hand, these results corroborate the data of previous experiments which showed reappearance of the vestibular asymmetries in compensated animals after spinalization (Azzena, 1969). In conclusion, these studies demonstrate that the spinal cord provides an excitatory input to the vestibular cells of the deafferented side and inhibitory input to the cells of the "intact" side. The different spinal activity could be attributed to the ascending pathways connected with the high threshold systems for the cutaneous and articular receptors which are asymmetrically stimulated by the distorted posture of the animal, as reported in a previous paper (Azzena et al., 1976).

In addition to these findings, the experiments of the present research show that also the cerebellum functions in re-establishing the normal posture in an animal which underwent the lesion of one labyrinth. In fact, the electrical activity of the deitersian units is altered by the lesion of the olivocerebellar system or by the functional inactivation of the cerebellar cortex. However, it is to be noted that the cerebellar volleys act in an opposite manner upon the vestibular nuclei, as compared to the ascending spinal volleys. The release of the vestibular symptoms following partial severance of the cerebellar afferents (Fig. 5B) or functional removal of the cerebellar cortex, is shown by an excitation of the deafferented vestibular cells and by an equal proportion of excitatory and inhibitory effects upon the cells of the "intact" side. It appears that the cerebellum is mainly concerned in regulating the deafferented vestibular nuclei. They are, in fact, receiving the strong spinal excitation which tends to restore the unitary activity to the previous normal level, which then depends upon the cerebellar modulation. At the same time, the higher activity that rises in the deafferented nuclei could be regarded as the major source for the inhibition to the contralateral nuclei, since they do not show a predominant effect following the removal of the cerebellum. This concept has been proposed also by Sanchez Robles and Anderson (1978) who attribute a primary role to the commissural system in compensating the labyrinthine syndrome. An additional contribution to the cerebellar regulation may come from the visual input. Recent papers have pointed out that vision is necessary for compensating the static and dynamic deficits of hemilabyrinthectomy in the cat (Courjon et al., 1977; Putkonen et al., 1977). The possibility exists that vision may compensate the vestibular deficits either by utilizing the cerebellar loop (Maekawa and Simpson, 1973) or by acting directly on the vestibular nuclei. Impulses travelling through the optic pathways may in fact impinge upon the vestibular nuclei with latencies which exclude the cerebellar circuitry (Azzena et al., 1978).

SUMMARY

The unitary discharges of cells located within the lateral vestibular nuclei of hemilabyrinthectomized and compensated guinea pigs were recorded. The effects of destruction of the inferior olive neurons or functional inactivation of the cerebellar cortex were investigated and compared with those evoked by transverse section of the spinal cord which induces a stage of decompensation. It has been shown that the cerebellum compensates the deficits of unilateral labyrinthectomy by acting on the vestibular cells in an opposite way compared to that exerted by the spinal cord, particularly by modulating the excitatory spinal volleys directed to the deafferented vestibular nuclei.

REFERENCES

Allen, G.I., Azzena, G.B., and Ohno, T. (1974) Cerebellar Purkyně cell responses to input from sensorimotor cortex. *Exp. Brain Res.*, 20: 239–254.

Azzena, G.B., (1969) Role of the spinal cord in compensating the effects of hemilabyrinthectomy. *Arch. ital. Biol.*, 107: 43–53.

Azzena, G.B., Mameli, O. and Tolu, E. (1976) Vestibular nuclei of hemilabyrinthectomized guinea pigs during decompensation. *Arch. ital. Biol.*, 114: 389–398.

Azzena, G.B., Mameli, O. and Tolu, E. (1977) Vestibular units during decompensation. *Experientia*, 33: 234–235.

Azzena, G.B., Tolu, E. and Mameli, O. (1978) Responses of vestibular units to visual input. *Arch. ital. Biol.*, 116: 120–129.

Courjon, J.H., Jeannerod, M., Ossuzio, I. and Schmid, R. (1977) The role of vision in compensation of vestibulo ocular reflex after hemilabyrinthectomy in the cat. *Exp. Brain Res.*, 28: 235–248.

Desclin, J. (1974) Histological evidence supporting the inferior olive as the major source of cerebellar climbing fibers in the rat. *Brain Res.*, 77: 365–384.

Desclin, J. and Escubi, J. (1974) Effects of 3-acetylpyridine on the central nervous system of the rat, as demonstrated by silver methods. *Brain Res.*, 77: 349–364.

Eccles, J.C., Ito, M. and Szentagothai J. (1967) *The Cerebellum as a Neuronal Machine*. Springer-Verlag, Berlin-Heidelberg-New York.

Eccles, J.C., Provini, L., Strata, P. and Taborikova, H. (1968) Analysis of electrical potentials evoked in the cerebellar anterior lobe by stimulation of hind-limb and forelimb nerves. *Exp. Brain Res.*, 6: 171–194.

Llinás, R., Walton, K., Hillman, D.E. and Sotelo, C. (1975) Inferior olive: its role in motor learning. *Science*, 190: 1230–1231.

Maekawa, K. and Simpson, J. (1973) Climbing fiber responses evoked in vestibulo cerebellum of rabbit from visual system. *J. Neurophysiol.*, 36: 649–666.

Mano, N., Oshima, T. and Shimazu, H. (1968) Inhibitory commissural fibers interconnecting the bilateral vestibular nuclei. *Brain Res.*, 8: 378–382.

McCabe, B.F. and Ryu, J.H. (1969) Experiments on vestibular compensation. *Laryngoscope*, 79: 1728–1736.

McCabe, B.F., Ryu, J.H. and Sekitani, T. (1972) Further experiments on vestibular compensation. *Laryngoscope*, 82: 381–396.

Menzio, P. (1952) Reperti istologici dell'orecchio interno in animali emislabirintati con applicazione endocanalicolare di spugna di fibrina e con iniezione di olio e cloroformio. *Boll. soc. ital. Biol. sper.*, 28: 1657–1659.

Precht, W., Shimazu, H. and Markham, C.H. (1966) A mechanism of central compensation of vestibular function following hemilabyrinthectomy. *J. Neurophysiol.*, 29: 996–1010.

Putkonen, P.T.S., Courjon, J.H. and Jeannerod, M. (1977) Compensation of postural effects of hemilabyrinthectomy in the cat. A sensory substitution process? *Exp. Brain Res.*, 28: 249–257.

Sanchez Robles, S. and Anderson, J.H. (1978) Compensation of vestibular deficits in the cat. *Brain Res.* 147: 183–187.

Schaefer, K.P. and Mayer, D.L. (1974) Compensation of vestibular lesions. In *Handbook of Sensory Physiology, Vol. VI/2, Vestibular System, Psychophysics, Applied Aspects and General Interpretations*, H.H. Kornhuber (Ed.), Springer-Verlag, Berlin-Heidelberg-New York, pp. 463–490.

Simonelli, G. (1923) Un metodo di distruzione chimica del labirinto. *Arch. Fisiol.*, 21: 231–233.

Spiegel, E.A. and Demetriades, T.D. (1925) Die zentrale Kompensation des Labyrinthverlustes. *Pflügers Arch.*, 210: 215–222.

Wilkems, G. and Schaefer, K.P. (1976) Cerebellar influences in compensation for vestibular lesions. *Pflügers Arch.*, Suppl. 362: R 49, no. 194.

Synaptic Mechanisms Involved in Compensation of Vestibular Function Following Hemilabyrinthectomy*

N. DIERINGER and W. PRECHT

Max-Planck-Institut für Hirnforschung, Neurobiologische Abteilung, 6000 Frankfurt/M.-Niederrad (F.R.G.)

The behavioral deficits resulting from acute hemilabyrinthectomy are similar in all vertebrates. However, there exist also interesting differences. Thus, in the frog ocular nystagmus cannot be observed, and tonic head and especially eye deviations are small in comparison to mammalian species (Dieringer and Precht, 1979a,b). Furthermore, the time courses of compensation of the tonic head deviation are quite different ranging from a few days in mammals (Schaefer and Meyer, 1974) to some weeks in amphibians (Kolb, 1955). These differences may in part be due to qualitative as well as quantitative differences in the synaptic circuitry of these species. Thus, the vestibular commissure is inhibitory in the cat (Shimazu and Precht, 1966) and excitatory in the frog (Ozawa et al., 1974), and cerebellar responses to natural stimulation of the horizontal canals are strong and mainly type III in the frog (Llinás et al., 1971), but weak and type I or II in mammals (Ghelarducci et al., 1975; Precht et al., 1976). Optokinetic modulation of central vestibular neurons is common in goldfish (Allum et al., 1976) and several mammalian species (Dichgans and Brandt, 1972; Waespe and Henn, 1976; Bauer, 1976) but rare in the frog (Dieringer and Precht, unpubl. obs.).

Since there is no regeneration of the labyrinth compensation of behavioral deficits implies central modifications. The mechanisms underlying these modifications can be assumed to involve changes in the efficacy of existing excitatory and inhibitory synaptic connections. Compensation of a given deficit may occur through similar basic mechanisms in different species although the pathways involved may be different or similar pathways may be subjected to different modifications. Specifically, to account for the functional requirements in static as well as in dynamic situations the newly acquired resting rate of the deafferented side should be under excitatory as well as inhibitory control from the intact side. That this is, indeed, the case will be illustrated below.

In the following, we compare the synaptic efficacy of excitatory and inhibitory terminals ending on partially deafferented vestibular neurons in the acutely and chronically operated frog (Dieringer and Precht, 1977, 1979a,b). This comparison will be divided into four sections: (1) analysis of the excitatory input via commissural fibres; (2) description of the crossed and uncrossed inhibition of vestibular neurons; (3) modifications in the cerebellar output during compensation and (4) the temporal sequence of these modifications.

---

*Supported by the Deutsche Forschungsgemeinschaft (Pr 158/1).

Microelectrode recordings were obtained from animals that were hemilabyrinthectomized either acutely (up to 12 h) or 3 days or chronically (at least 60 days) before the experiment. Recording and stimulation techniques were similar to those described earlier (Precht et al., 1974). More specific procedures are described in Dieringer and Precht (1979a,b).

INCREASED COMMISSURAL EXCITATION

Electrical stimulation of the vestibular nerve on the intact side evokes a negative field potential (N_1) in the ipsilateral vestibular nucleus and via commissural fibers an N_1 potential in the contralateral vestibular nucleus. With stimulus intensities restricted to 5 times the threshold of the N_1 potential on the ipsilateral side, N_1 potentials in the partially deafferented vestibular nucleus are small in acute but significantly larger in chronic animals. Bursts of spikes are typically superimposed upon N_1 potentials in chronic but only rarely in the acute animals. Most single units were silent in the acutely but very often spontaneously active in the chronically deafferented vestibular nucleus.

Excitatory postsynaptic potentials provoke in vestibular neurons two types of active responses: partial, presumably dendritic spikes, and full action potentials (Precht et al., 1974). These responses occur spontaneously (Fig. 1A), after activation of primary afferents (Fig. 1B) or commissural fibers (Fig. 1C). Thresholds of dendritic spikes and full action potentials are different and were, therefore, used as an indicator of excitability of the neuron. The primary afferent input is powerfully excitatory: with the stimulus intensity used typically a burst of spikes is evoked. This burst evokes in most of the vestibular neurons of the contralateral nucleus small EPSPs that have a few partial spikes superimposed, occasionally an action potential and in very rare cases a burst of spikes (Ozawa et al., 1974). In chronic animals a very similar output pattern of ipsilateral vestibular neurons evokes in the contralateral nucleus many more active responses in a much higher percentage of vestibular neurons (Table I).

To compare the shape indices of the underlying EPSPs, stimulation intensity was kept below threshold of active responses and evoked EPSPs were averaged (16 sweeps) and corrected for the extracellular field potential. EPSPs evoked from primary afferent fibers had a significantly shorter time to peak and duration at half width than EPSPs evoked from commissural fibers (Fig. 2). In addition, the amplitude of primary afferent EPSPs was nearly twice as large as those evoked from commissural fibers. These differences

TABLE I
CHANGES IN THE OUTPUT PATTERNS OF VESTIBULAR AND CEREBELLAR NEURONS

	Percentage of neurons in acute animals	*Percentage of neurons in chronic animals*
Bursts of spikes activated in part. deaff. VN	2.6	46.4
Inhibition of part. deaff. VN	2.6	28.2
Disfacilitation of cerebellar PC	8.0	30.0

VN, vestibular neurons; PC, Purkinje cells

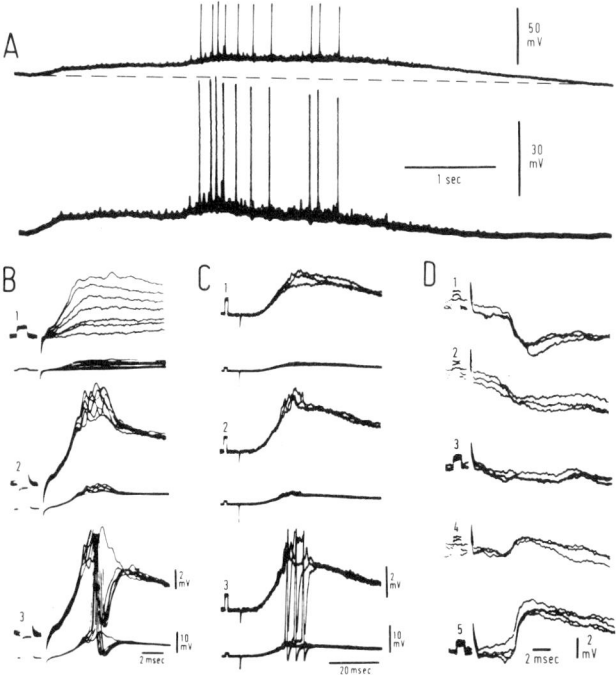

Fig. 1. Response patterns of normal and deafferented vestibular neurons to VIIIth nerve stimulation. Excitatory postsynaptic potentials are superimposed by dendritic spikes of varying amplitudes and by full action potentials (A–C). These responses occur either spontaneously (A), after stimulation of ipsilateral primary afferent vestibular fibers (B) or they are evoked by commissural fibers (C). In all cases dendritic spikes have a lower threshold than full action potentials. Inhibitory postsynaptic potentials (D) are mainly observed in partially deafferented vestibular neurons of chronic animals. They can be reversed easily by injection of hyperpolarizing current. In D, hyperpolarization was gradually increased from 0 to 5 nA. Upper trace in A and lower traces in B and C were recorded with low gain d.c. coupled amplifiers, other traces with high gain a.c. coupled amplifiers (300 msec time constant).

were also found in the same neuron when the VIIIth nerve of either side was stimulated. In chronic frogs EPSPs of similar shape were found on the ipsilateral side. In partially deafferented vestibular neurons, however, commissurally evoked EPSPs had a significantly shorter time to peak and duration at half width (Fig. 2) and a significantly larger amplitude than comparable EPSPs in acute animals. The ratio of duration to rise time was close to 3 for EPSPs evoked by primary afferent and by commissural fibers in acute animals, but 3.5 for EPSPs evoked by commissural fibers in chronic animals. These differences in the shape indices of the EPSPs suggest differences in the electrotonic length between the sites of the synaptic input and the site of recording, presumably the soma (Rall, 1967). Accordingly, (i) primary afferent fibers would terminate on the average closer to the soma than commissural fibers and (ii) after degeneration of primary afferent fiber terminals synapses of commissural fibers would end on the average closer to the soma than in normal or in acute animals. Changes in input resistance or increased chemosensitivity cannot explain all the observed modifications. In fact, the range of input resistances (0.7–4.2 MΩ) was not different in acute and chronic animals and the time to peak of evoked EPSPs did not shift systematically with increasing stimulus intensity.

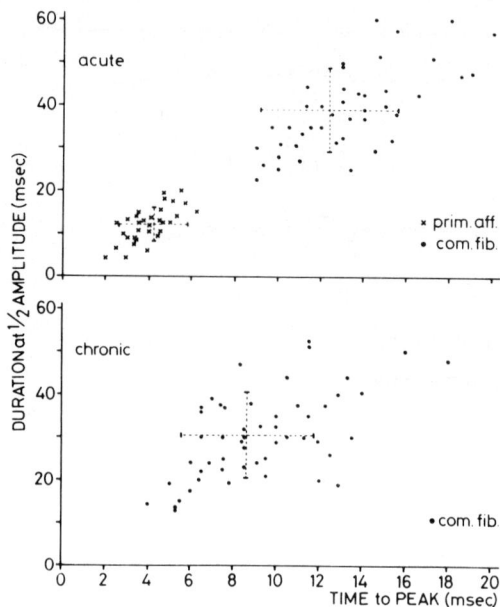

Fig. 2. Shape indices of EPSPs evoked in vestibular neurons. Upper part: In acute animals, EPSPs evoked by primary afferent vestibular fibers (x) have a significantly shorter duration and time to peak than EPSPs evoked via commissural fibers (●). Lower part: In chronic animals, EPSPs evoked by a very similar pattern of commissural fiber activity on the average have a significantly shorter duration and time to peak than comparable EPSPs recorded in acute animals. Dashed lines indicate means and SD. (From Dieringer and Precht, 1979a.)

INCREASED CROSSED AND UNCROSSED INHIBITION

Contrary to the crossed excitatory responses, suppression of resting discharge and IPSPs were only rarely observed in acute animals on either side of the brain stem. Similar results were obtained by Precht et al. (1974) and by Ozawa et al. (1974). In chronic animals, however, the percentage of inhibited vestibular neurons was moderately increased on the intact side and strongly enhanced on the partially deafferented side (Table I). Suppression of spontaneous activity could be preceded by a short burst of evoked spikes. Subsequent intracellular recordings revealed EPSP-IPSP sequences. The IPSPs could be easily reversed by either passing hyperpolarizing current through the recording electrode or injection of chloride (Fig.1D). The distribution of the latencies of these IPSPs was bimodal with ranges between 4–10 and 18–25 msec. A very similar bimodal distribution was also observed for the latencies of IPSPs on the intact side.

Because of the close vestibulocerebellar relationship throughout phylogeny (Larsell, 1967) and the inhibitory cerebellovestibular feedback loop also present in the frog (Magherini et al., 1975), some of the observed IPSPs can be assumed to originate from cerebellar Purkinje cells (see Fig. 3). The latency of Purkinje cell simple spike responses was around 19 msec on the average, except for Purkinje cells located in the ipsilateral auricular lobe (see below). Thus, only IPSPs belonging to the second latency group in the bimodal distribution of vestibular inhibitions can be attributed to activity in Purkinje cells. Removal of the cerebellum significantly reduced the number of observed IPSPs. Latencies of remaining IPSPs were shorter than 10 msec. Interestingly, after systemic

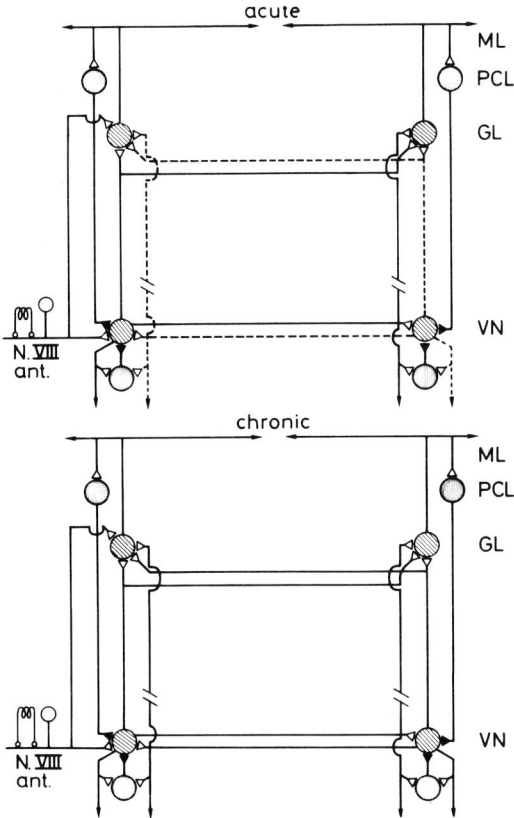

Fig. 3. Schematic diagram of the major pathways of the vestibulo-vestibular and the vestibulo-cerebellovestibular loops activated by stimulation of the vestibular nerve (N.VIII ant.) in acutely (upper part) and chronically (lower part) hemilabyrinthectomized frogs. Granule cells (GL) in the cerebellar cortex are activated by ipsilateral primary afferents and secondary vestibular neurons. The latter reach granule cells of the contralateral cerebellar cortex via a cerebellar and brain stem vestibular commissure. Purkinje cells (PCL) receive their input predominantly from parallel fibers of the ipsilateral cerebellar cortex and inhibit (dark terminals) neurons in the homolateral vestibular nucleus (VN). A second inhibitory feedback loop is represented by interneurons located close to vestibular neurons. Dashed lines in the acute animal indicate a very weak activation in comparison to the full lines. In the chronic animal, the synaptic efficacy of the vestibular commissure in the brain stem is increased on the lesioned side (right). Thereby the output of partially deafferented vestibular neurons is increased. Note the wide-spread effects of this change in synaptic transmission for the cerebellar, vestibular and spinal (downward arrows) inputs in the chronic animals. Interrupted lines indicate the possibility of involvement of interneurons.

injection of picrotoxin (0.3 mg/frog) IPSPs were no longer observed suggesting that both cerebellar as well as brain stem mediated inhibition are gabanergic.

The increased inhibition of vestibular neurons in chronic frogs could result either from an increased input to inhibitory neurons e.g., Purkinje cells and/or brain stem neurons or from a stronger inhibitory efficacy of these neurons. We tried to investigate this problem by recording in the cerebellum the input (N_2, N_3 potentials) and the output (Purkinje cells) in acute and chronic frogs. Such measurements should indicate whether Purkinje cells in chronic frogs are more readily excited than in control animals.

CHANGES IN THE CEREBELLAR OUTPUT

Latencies of Purkinje cell spike responses are very sensitive to stimulus intensities (Llinás et al., 1969). With the stimulus intensities used in this study, Purkinje cells in the ipsilateral auricular lobe responded after 10–20 msec (mean 14.5±4.1 msec). Purkinje cells in the region of the dorsal rim (Llinás et al., 1971) of the bilateral cerebellar corpus as well as in the contralateral auricular lobe had slightly longer latencies (mean 19.6±3.3 msec). The N_3 potentials in the molecular layer started on the ipsilateral side of the cerebellar corpus already after 2.5–3.0 msec and on the contralateral side after 7.0–8.0 msec. The long delay between Purkinje cell input and its spike output is explained by the small and slowly rising EPSPs evoked by parallel fibers. Intracellular records showed that the EPSPs always started at least 5–10 msec earlier than the first evoked spike recorded from the same cell extracellularly (Dieringer and Precht, 1979c).

The N_3 potential in the contralateral cerebellum is preceded by a strong N_2 potential that starts in the granular layer about 3 msec earlier. This contralateral mossy fiber input is mediated via secondary vestibular and polysynaptic pathways that cross the midline in the cerebellum as well as in the brain stem. Parallel fibers activated in the ipsilateral corpus seem to contribute only little to the contralateral N_3 potential as judged from lesion experiments and recordings from the molecular layer in the cerebellar midline. These different vestibulocerebellar connections are shown in the schema of Fig. 3.

Latencies of Purkinje cell responses recorded in various parts of the cerebellum of acute and chronic frogs had a very similar range and did not differ significantly. The firing rate of these cells remained increased for well over 100 msec, and in chronic frogs Purkinje cells on either side of the cerebellum responded with a few spikes more than in acute frogs. Besides these excitatory responses a few Purkinje cells of acute animals stopped firing after stimulation. The percentage of cells responding in this fashion was strongly increased in chronic animals (Table I). Suppression of resting rate was observed to be equally frequent on either side of the cerebellum as well as in the auricular lobes. Intracellular recordings from Purkinje cells responding with suppression never showed an IPSP, even though the membrane potential was partially depolarized. After reduction of firing frequency by artificial hyperpolarization in some cases similar periods of suppression could be observed as before impalement of the neuron. This suggests that the majority of Purkinje cells was disfacilitated and not inhibited. This result is in agreement with the small number of inhibitory interneurons present in the cerebellar cortex of the frog (Llinás, 1976). The onset of disfacilitation occurred between 10 and 30 msec after stimulation and was thus slightly later than the onset of IPSPs in the bilateral vestibular nuclei. Thus, it seems that in the chronic frog cerebellar granule cells are not only excited bilaterally stronger, but that they are also disfacilitated from an increased number of vestibular and other neurons (Fig. 3). Depending on the pattern of convergence of these two different types of inputs then, Purkinje cells would be more or less strongly excited or disfacilitated. These results suggest that the total number of Purkinje cells activated in chronic animals is not increased and that their excitation on the average is only weakly enhanced. To account for the increased number of cerebellarly mediated inhibitions observed in chronic animals, one may therefore assume that in addition to the icreased excitation the synaptic efficacy of Purkinje cell terminals became stronger as well.

ONSET OF CENTRAL MODIFICATIONS

The results obtained in acute and chronic frogs were compared with data from animals operated 3 days before the experiment. At this time, degeneration of boutons of primary afferent fibers can already be observed in the vestibular nucleus (Hillman, 1972). Partially deafferented vestibular neurons exhibited commissurally evoked EPSPa with slightly shorter times to peak and larger amplitudes than in acute animals. These EPSPs produced in more cells partial spikes and action potentials at a slightly lower threshold than in acute animals. However, most of these values were statistically not significant. Thus, in comparison to chronic animals synaptic efficacy was still small even though already increased. Similarly, the number of partially deafferented neurons receiving an IPSP 3 days after the operation was slightly higher than in acute or intact animals. These data suggest that: (1) parallel to degeneration of primary afferent boutons the shape indices of commissurally evoked EPSPs change; (2) these modified EPSPs initiate more dendritic spikes and thereby lower the threshold for action potentials, and (3) the increase in inhibition develops in parallel to the enhancement of the excitatory synaptic efficacy.

Assuming that these synaptic modifications are related to the behaviorally observed compensation of tonic and dynamic, vestibular reflexes, one may compare the onset of compensation with the results reported here. The beginning of behaviorally relevant central modifications can be determined by the earliest occurrence of Bechterew's symptom (1883) which is produced by the removal of the remaining labyrinth. It was found that the shortest time interval between bilateral, successive hemilabyrinthectomies that produced this symptom, was 5 days (unpublished observations). This time period is very similar to that revealing the first signs of changes in excitatory and inhibitory transmission.

DISCUSSION

Central to the interpretation of the data presented are the changes in the shape indices of the evoked EPSPs. These parameters are generally accepted as a reliable indicator of synaptic input location. In the case of compound EPSP, evoked in different dendritic compartments, these parameters can only indicate an averaged synaptic site. In vestibular neurons this average site is closer to the soma for primary afferent than for commissural fiber terminals. Our data suggest that, in the chronic frog, the average site of commissural fiber terminals should be closer to the soma than in intact or acute animals. One mechanism that could produce such a change is reactive synaptogenesis (Cotman and Lynch, 1976) where the formation of new, functional synapses from closeby intact presynaptic elements is provoked by the degeneration of other terminals. In this process, vacant postsynaptic receptor sites will become reoccupied. In the case of excitatory vestibular commissural terminals new boutons would synapse now closer to and at the soma. The shape indices of their evoked EPSPs would then depend on the combination of simultaneously depolarized dendritic compartments. As a result of the spatially more distributed nature of this synapse, the ratio between time to peak and duration of the EPSPs will become larger (Rall, 1967) and more synaptic current will flow across the membrane whereby dendrites are more strongly depolarized, initiate more dendritic spikes which in turn, facilitate generation of action potentials. Even though our data are in good agreement with similar results obtained in neurons where sprouting was morphologically shown to occur (Nakamura et al., 1974; Tsukahara et al., 1975), ultimately only anatomical studies can show whether our hypothesis is correct.

Besides the effects mediated by the vestibular commissural system other structures have to provide excitation to partially deafferented vestibular neurons particularly to reestablish symmetrical resting rate. That the remaining labyrinth alone cannot achieve this is already indicated by the occurrence of the Bechterew symptom. Ultimately, it will be the combination of all of these excitatory and inhibitory influences that will be crucial for the reestablishment of functionally well- or mal-adapted vestibular reflexes. As far as the vestibular commissure is concerned, its excitation will decrease the asymmetry in the activity of the bilateral vestibular nuclei at rest and will provide the proper vestibular signal for some dynamic reflexes such as in linear forward motion or during pitch. For other reflexes such as compensatory movements generated by horizontal acceleration, crossed and uncrossed inhibition may be essential. As indicated in Fig. 3 the increase in the excitatory or inhibitory synaptic efficacy of one pathway already alters the output of a nucleus sufficiently to produce a whole chain of effects in various moto- and interneuronal networks. For further functional interpretation of these modifications, of course, experiments with natural stimulation have to be done.

SUMMARY

Following hemilabyrinthectomy primary afferent terminals ending on central vestibular neurons degenerate. Terminals from other pathways such as the vestibular commissure, cerebellar Purkinje cells and inhibitory brain stem neurons, increase their synaptic efficacy. The characteristic changes in the shape indices of the excitatory post-synaptic potentials recorded in the partially deafferented vestibular nucleus suggest that reactive synaptogenesis may have occurred. These synaptic mechanisms can be assumed to present part of the basis for the behaviorally observed modifications in the tonic and dynamic, vestibular reflexes during compensation.

REFERENCES

Allum, J. H. J., Graf, W., Dichgans, J. and Schmidt, C. L. (1975) Visual-vestibular interactions in the vestibular nuclei of the goldfish. *Exp. Brain Res.*, 26: 463–485.

Bauer, R. (1976) *Optisch-vestibuläre Interaktion an Einzelneuronen der Vestibulariskerne und des Vestibulo-Cerebellum bei der Katze*. Biol. Diss., Freiburg.

Bechterew, W. von (1883) Ergebnisse der Durchschneidung des N. acusticus, nebst Erörterung der Bedeutung der semicirculären Kanäle für das Körpergleichgewicht. *Pflügers Arch.*, 30: 312–347.

Cotman, C.W. and Lynch, G.S. (1976) Reactive synaptogenesis in the adult nervous system in *Neuronal Recognition*, S. H. Barondes (Ed.) Plenum Press, New York, pp. 69–108.

Dieringer, N. and Precht, W. (1977) Modification of synaptic input following unilateral labyrinthectomy. *Nature*, 269: 431–433.

Dieringer, N. and Precht, W. (1979a) Mechanisms of compensation for vestibular deficits in the frog. I. Modification of the excitatory commissural system. *Exp. Brain Res.*, 36: in press.

Dieringer, N. and Precht, W. (1979b) Mechanisms of compensation for vestibular deficits in the frog. II. Modification of the inhibitory pathways. *Exp. Brain Res.*, 36: in press.

Dieringer, N. and Precht, W. (1979c) Timing of bilateral cerebellar output evoked by unilateral vestibular stimulation in the frog. *Pflügers Arch.*, in press.

Dichgans, J. and Brandt, Th. (1972) Visual-vestibular interaction and motion perception. *Bibl. ophthal. (Basel)*, 82: 327–338.

Ghelarducci, B., Ito, M. and Yagi, N. (1975) Impulse discharges from flocculus Purkinje cells of alert rabbits during visual stimulation combined with horizontal head rotation. *Brain Res.*, 87: 66–72.

Hillman, D. E. (1972) Vestibulocerebellar input in the frog: Anatomy. In *Progress in Brain Research, Vol. 37, Basic Aspects of Central Vestibular Mechanisms*, A. Brodal and O. Pompeiano (Eds.), Elsevier, Amsterdam, pp. 329–339.

Kolb, E. (1955) Untersuchungen über zentrale Kompensation und Kompensationsbewegungen einseitig enstateter Frösche. *Z. vergl. Physiol.*, 37: 136–160.

Larsell, O. (1967) *Comparative Anatomy and Histology of the Cerebellum from Myxinoids through Birds*, J. Jansen (Ed.) University of Minnesota Press, Minneapolis.

Llinás, R. (1976) Cerebellar physiology. In *Frog Neurobiology*, R. Llinás and W. Precht (Eds.), Springer-Verlag, Berlin-Heidelberg-New York, pp. 892–923.

Llinás, R., Bloedel, J. R. and Hillman, D. E. (1969) Functional characterization of the neuronal circuitry of the frog cerebellar cortex. *J. Neurophysiol.*, 32: 847–870.

Llinás, R., Precht, W. and Clarke, M. (1971) Physiological responses of frog vestibular fibers to horizontal angular rotation. *Exp. Brain Res.*, 13: 408–431.

Magherini, P. C., Giretti, M. L. and Precht, W. (1975) Cerebellar control of vestibular neurons of the frog. *Pflügers Arch.*, 356: 99–109.

Nakamura, Y., Mizuno, N., Kanishi, A. and Sato, M. (1974) Synaptic reorganization of the red nucleus after chronic deafferentation from cerebellorubral fibers: An electron microscopic study in the cat. *Brain Res.*, 82: 298–301.

Ozawa, S., Precht, W. and Shimazu, H. (1974) Crossed effects on central vestibular neurons in the horizontal canal system of the frog. *Exp. Brain Res.*, 19: 394–405.

Precht, W., Simpson, J. I. and Llinás, R. (1976) Responses of Purkinje cells in rabbit nodulus and uvula to natural vestibular and visual stimuli. *Pflügers Arch.*, 367: 1–6.

Precht, W., Richter, A., Ozawa, S. and Shimazu, H. (1974) Intracellular study of frog's vestibular neurons in relation to the labyrinth and spinal cord. *Exp. Brain Res.*, 19: 377–393.

Rall, W. (1967) Distinguishing theoretical synaptic potentials computed for different soma-dendritic distributions of synaptic input. *J. Neurophysiol.*, 30: 1138–1168.

Schaefer, K. P. and Meyer, D. L. (1974) Compensation of vestibular lesions. In *Handbook of Sensory Physiology, Vol. VI/2*, H.H. Kornhuber (Ed.), Springer-Verlag, Berlin-Heidelberg-NewYork, pp. 463–490.

Shimazu, H. and Precht, W. (1966) Inhibition of central vestibular neurons from the contralateral labyrinth and its mediating pathway. *J. Neurophysiol.*, 29: 467–492.

Tsukahara, N., Hultborn, H., Murakami, F. and Fujito, Y. (1975) Electrophysiological study of the formation of new synapses and collateral sprouting in red nucleus neurons after partial denervation. *J. Neurophysiol.*, 38: 1359–1372.

Waespe, W. and Henn, V. (1977) Neuronal activity in the vestibular nuclei of the alert monkey during vestibular and optokinetic stimulation. *Exp. Brain Res.*, 27: 523–538.

SECTION V

VISUAL INFLUENCES ON THE MOTOR SYSTEM

A

Reticular Control of Eye Movements

Organization of Reticular Projections onto Oculomotor Neurones

J. A. BÜTTNER-ENNEVER

Institut für Hirnforschung, Universität Zürich, 8029 Zürich (Switzerland)

Parts of the pontine, mesencephalic and medullary reticular formation have been found to subserve very specific functions in the control of eye movements. For example, the paramedian part of the pontine reticular formation (PPRF), shown in Fig. 1, is essential for the initiation of fast eye movements in the horizontal plane (for review, see Raphan and Cohen, 1978). Small unilateral lesions there lead to a total paralysis of all fast conjugate eye movements to the ipsilateral side, leaving vertical saccades and blinks intact.

This localized and specific function stands in sharp contrast to the more diffuse roles usually associated with the pontine brain stem reticular formation such as a source of the ascending reticular activating system involved in arousal (Moruzzi and Magoun, 1949), or a trigger zone for paradoxical sleep (Jouvet, 1962) and pontine-geniculo-occipital (PGO) waves (Jouvet, 1965; Laurent et al., 1974). Earlier studies on the anatomical connections of the reticular formation were stimulated by the discovery of these more general functions and in turn revealed the diffuse nature of the reticular connectivity (Nauta and Kuypers, 1958; Scheibel and Scheibel, 1958; Valverde, 1961). The well-known Golgi studies on newborn mammals, in which incidently coordinated eye movements are absent, have very impressive tracings of reticular axons with countless ramifications stretching from the spinal cord to the thalamus, and including branches within the motor nuclei of the extraocular eye muscles. In spite of the extreme complexity of the reticular formation connectivity, a few pathways can be isolated out and assigned a specific function in oculomotor control (Büttner-Ennever

---

Abbreviations used in figures and text: III, oculomotor nucleus; IV, trochlear nucleus; VI, abducens nucleus; cg, central gray; CTT, central tegmental tract; g, group g of vestibular complex; h, habenular complex; HRP, horseradish peroxidase; iC, interstitial nucleus of Cajal; iMLF, see rostral iMLF; io, inferior olive; iv, inferior vestibular nucleus; lv, lateral vestibular nucleus; m, midline; mb, mammillary body; md, nucleus medialis dorsalis; med RF, medullary reticular formation; MLF medial longitudinal fasciculus; mr, medial rectus subgroup of III; MRF, mesencephalic reticular formation; MT, mammillothalamic track; mv, medial vestibular nucleus; N III, rootlets of the oculomotor nerve; N IV, trochlear nerve; N VI, rootlets of the abducens nerve; N VII, facial nerve; nD, nucleus of Darkschewitsch; ni, nucleus intercalatus of Staderini; nrt, nucleus reticularis tegmenti pontis; PC, posterior commissure; ppH, nucleus praepositus hypoglossi; PPRF, paramedian pontine reticular formation; pt, pretectum; PT, pyramidal tract; rostral iMLF, rostral interstitial nucleus of the MLF; rn, red nucleus; RS, reticulospinal tract; sc, superior colliculus; sn, substantia nigra; so, superior olive; sv, superior vestibular nucleus; t, thalamus; TR, tractus retroflexus or habenulo-interpeduncular tract.

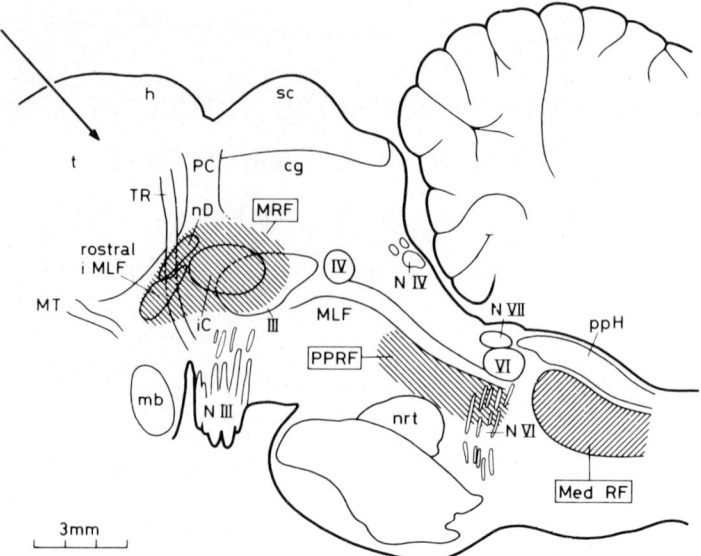

Fig. 1. A sagittal view of the monkey brain stem to show the overall topography of structures discussed in the text. The 3 shaded areas represent the location of the mesencephalic (MRF), medullary (med RF), and paramedian pontine (PPRF) reticular formation. The arrow indicates the Horsley-Clark plane of section for *Macaca fascicularis* used in Fig. 2.

and Henn, 1976; Graybiel, 1977b; Büttner-Ennever and Büttner, 1978). This recent progress has been mainly due to two factors. First, the introduction of the autoradiographic tracing methods which do not label fibres of passage (Lasek et al., 1968; Edwards, 1972), an unavoidable source of confusion in degeneration tracing techniques. Second, the development of single unit recording in alert animals, which has revealed in many areas an oculomotor function that is absent under anaesthesia or even sleep.

Oculomotor regions defined with the above methods do not correlate with the loose cytoarchitectural boundaries of the reticular formation. However, the oculomotor, pontine and medullary zones discussed below do coincide with the areas receiving afferents from the fastigial nucleus, the superior colliculus, and the cerebral cortex (Kawamura et al., 1974; Grantyn et al., 1977; Graybiel, 1977b) as well as those giving rise to reticulospinal pathways (Torvik and Brodal, 1957; Peterson, 1977). No attempt will be made here to explain the interaction between the afferent and efferent systems which converge in the reticular formation. This article will be confined to describing some specific pathways from the reticular formation onto oculomotor neurones, along with the physiological evidence for their role in the control of fast eye movements.

PARAMEDIAN PONTINE RETICULAR FORMATION (PPRF)

The term PPRF was coined by Goebel et al. (1971) to describe the small area which, if lesioned, led to a paralysis of all fast horizontal conjugate eye movements to the ipsilateral side. Single unit studies revealed that the neurones in PPRF encode the parameters of a subsequent eye movement very precisely in the alert monkey (for review, see Raphan and Cohen, 1978). We undertook a study to trace the pathways from PPRF to the extraocular motoneurones using anterograde (^{3}H-proline and ^{3}H-leucine) and retrograde (HRP) tracer techniques (for technical details, see Büttner-Ennever and Büttner, 1978). These results, and those of similar studies (Goebel et al., 1971; Graybiel and Hartwieg, 1974; Graybiel, 1977b) revealed that only abducens motoneurones receive a direct input from PPRF. The direct connection arises mainly from caudal and not rostral PPRF in the monkey (Büttner-Ennever and Henn, 1976; Büttner-Ennever, 1977) and cat (Graybiel, 1977b) and terminates throughout the abducens (VI) nucleus.

There are many different sets of neurones within the confines of the abducens nucleus (Baker and Highstein, 1975; Graybiel, 1977a); the largest group of non-motoneurones are the internuclear neurons, which send axons across the midline at the level of VI, to ascend within MLF and terminate on the contralateral medial rectus subgroup of the oculomotor nucleus (III) (Fig. 5) (Fuse, 1912; Graybiel and Hartwieg, 1974; Büttner-Ennever and Henn, 1976).

The control of medial rectus from PPRF utilizes this internuclear pathway and, as theory would predict, lesions of VI paralyse the lateral rectus as well as the medial rectus, with the exception of its involvement in convergence movements (Carpenter et al., 1963; Bender and Shanzer, 1964). Lesions of the MLF will also paralyse the medial rectus, producing the condition known as internuclear ophthalmoplegia. Thus, it is now possible to define some pathways through which PPRF controls the horizontal eye muscles (Fig. 5).

In an attempt to quantify the number of internuclear and motoneurones in the cat abducens nucleus, and also study the location of the two cell populations and test for the existence of collaterals from abducens motoneurones to III, double retrograde labelling experiments were carried out (Steiger and Büttner-Ennever, 1978). The results (Fig. 3) reveal two overlapping cell pools, and thereby destroy the text book hypothesis that there is a separate area (parabducens nucleus) controlling medial rectus and not lateral rectus (for review, see Baker, 1979). The proportion of motoneurones to interneurones in this study on the cat was about 2 : 1, and no evidence for collaterals from the axons of the motoneurones up to III was found.

There is one clear difference in oculomotor organization between cat and monkey. The subgroups of the oculomotor nucleus which control individual extraocular eye muscles in the monkey are extremely well demarcated (Warwick, 1953). Any afferent system terminating within III, e.g., vestibular, abducens, *rostral* iMLF, will label some compact individual subgroups, leaving others totally devoid of label. This is not so in the cat (Gacek, 1974). Using autoradiographic anterograde tracing techniques in the monkey, not only the medial rectus area in III is labelled from VI but also a second more caudal region lying within the inferior rectus area of Warwick (1953) (Büttner-Ennever, 1977). The function of this "inferior rectus" patch is not clear and is being investigated.

MESENCEPHALIC RETICULAR FORMATION (MRF)

Premotor areas subserving vertical eye movements are known from clinical and experimental studies to lie in the mesencephalic reticular formation. Although the exact location of premotor functions is far from clear, a number of studies suggest that the nuclei of the posterior commissure, the interstitial nucleus of Cajal (iC) and also structures further rostrally at the mesodiencephalic junction are involved (André-Thomas et al., 1933; Szentagothai, 1943; Hess, 1954; Duensing et al., 1963; Bender and Shanzer, 1964; Markham et al., 1966; Pasik et al., 1969; Carpenter et al., 1970; Mabuchi and Kusama, 1970; Jacobs et al., 1973; Christoff, 1974; Cogan, 1974; Schwindt et al., 1974; King, 1976; Büttner et al., 1977; Büttner-Ennever and Büttner, 1978). However PPRF is also very important for the generation of vertical eye movements, as lesion and unit studies show. Unilateral lesions in PPRF reveal that it is essential for the generation of horizontal eye movements, but bilateral lesions in PPRF give rise to a paralysis of both horizontal and vertical eye movements (Bender and Shanzer, 1964; Christoff, 1974). Furthermore, single unit recordings show that the neurones in PPRF encode parameters of both 'vertical' and horizontal fast eye movements (Luschei and Fuchs, 1972; Henn and Cohen, 1976). Until recently, the role of PPRF in vertical eye movements was difficult to explain. However, one possible hypothesis is as follows: it is in PPRF that the multiplicity of inputs determine whether or not a fast eye movement, in any direction, will occur and then from PPRF a coordinating signal to the two widely separated vertical and horizontal premotor centres is sent to produce the activation of the component extraocular muscles.

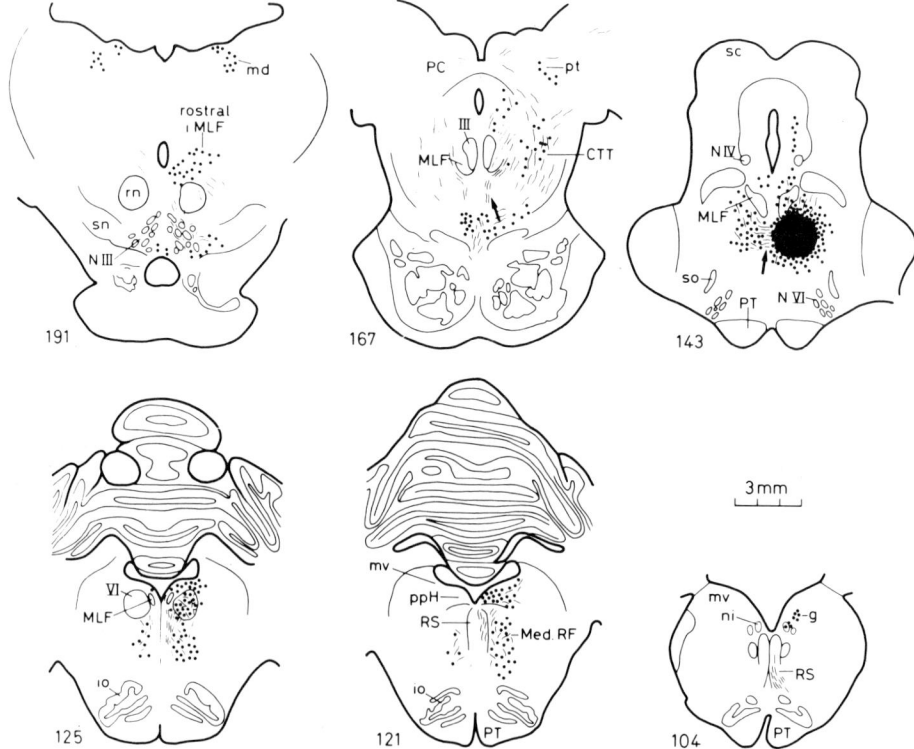

Fig. 2. Transverse sections from *Macaca mulatta* brain stem (experiment 14) to show the distribution of anterograde tracer substances (^{3}H-proline and ^{3}H-leucine – 4 days survival time) after injection into PPRF. Lines indicate labelled axons, and dots the diffuse pattern of silver grains lying over labelled axon terminals. The arrow in section 143 points to the commissural connections of PPRF; in 167 it indicates the compact ascending pathway described in the text. (From Büttner-Ennever and Henn, 1976.)

Injections of radioactive anterograde tracer substances into PPRF revealed a compact pathway, ascending to the mesencephalon (Büttner-Ennever and Henn, 1976). It originates from neurones near the midline in PPRF, some of which are probably omnipause units (King et al., 1978), and ascends ipsilaterally, ventral to MLF (see Fig. 2, section 167). The fibres circumscribe the oculomotor nucleus, iC and nD to terminate in a cell group of MRF rostral to iC, and referred to here as the *rostral* interstitial nucleus of the medial longitudinal fasciculus (*rostral* iMLF). It is a nucleus which has received very little attention in the literature, and has only recently been defined as a cell group involved in vertical eye movements (Büttner-Ennever, 1977; Büttner et al., 1977; Graybiel, 1977a; Büttner-Ennever and Büttner, 1978). However, there is now a large amount of evidence to support this. First, the neurones in rostral iMLF encode very precisely the parameters of the vertical component of an eye

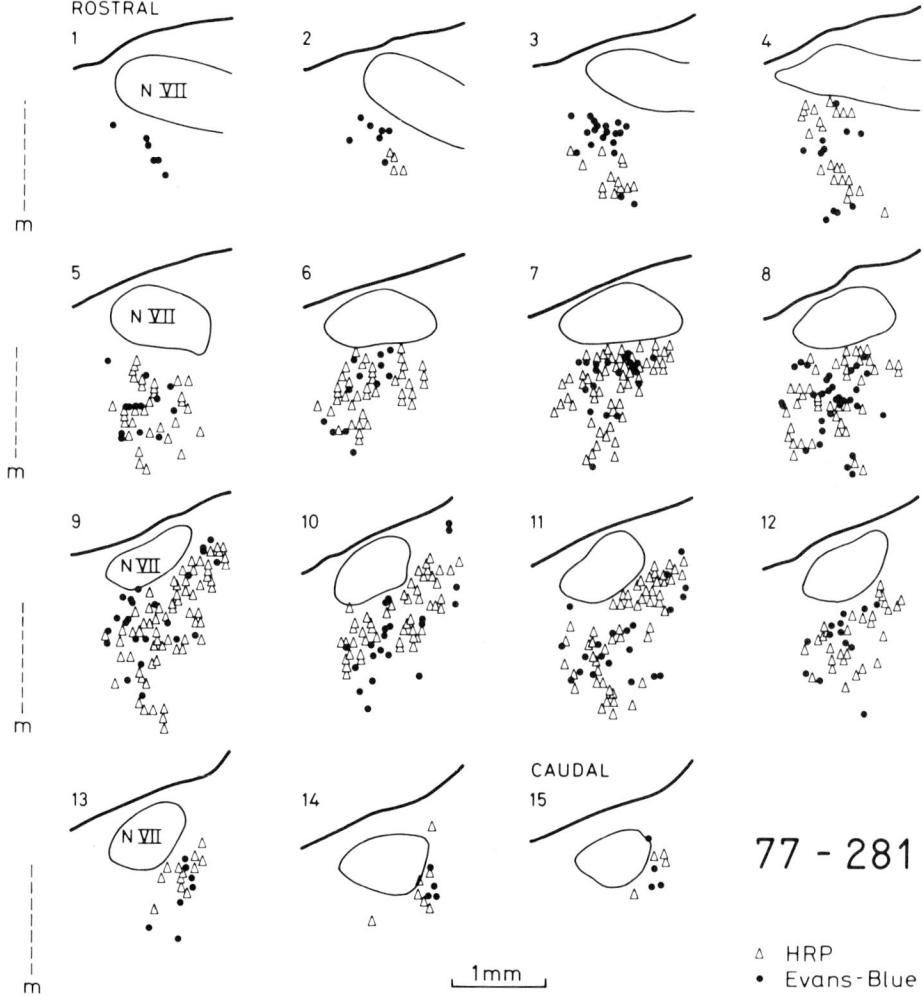

Fig. 3. Drawings of every third (40 μm) section on the right abducens nucleus of a cat. Dots are the internuclear neurones labelled by retrograde transport of Evans blue, injected into the contralateral oculomotor nucleus: triangles are abducens motoneurones labelled by retrograde transport of horseradish peroxidase from the lateral rectus muscle of the same cat (From Steiger and Büttner-Ennever, 1978a.)

movement, and they are active before the oculomotor neurones (King, 1976; Büttner et al., 1977). Bilateral lesions of rostral iMLF cause a permanent impairment of fast vertical eye movements, usually in the downward direction (Kömpf et al., 1977; Kömpf, 1978; and personal communication). Finally, experiments using HRP demonstrated a direct, and predominantly ipsilateral, pathway between rostral iMLF and the oculomotor nucleus (Graybiel, 1977a; Büttner-Ennever and Büttner, 1978); and we have recently been able to show, using anterograde tracing techniques, that rostral iMLF sends afferents into the trochlear nucleus, and the three subgroups of the oculomotor nucleus which supply vertical eye muscles (IO, SR, IR, Fig. 5).

The main input to rostral iMLF that has been found so far is from PPRF. We previously reported that it also receives a direct input from the vestibular nuclei (Büttner-Ennever and Büttner, 1978). The extent and origin of this input has now been examined in detail. The results of eight injections of radioactive anterograde tracer into the vestibular nuclei are summarized in Fig. 4. The vestibular input to rostral iMLF is definite, but weak compared with that to iC; it terminates mainly in the caudal, medial part of the nucleus. This result raises the interesting possibility of functional divisions within rostral iMLF, a suggestion which has also been made by Graybiel (personal communication). The ipsilateral projection arises mainly from the superior nucleus (sv), while the contralateral pathway originates around the lateral border of the medial vestibular nucleus (mv). This dual origin of the vestibular affer-

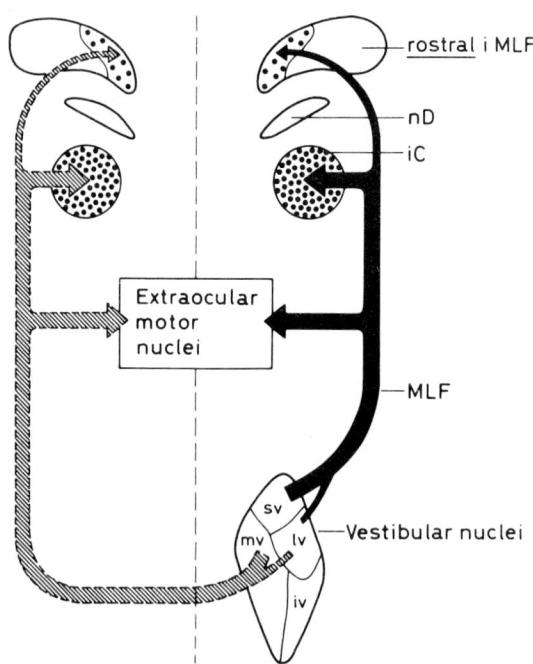

Vestibulo-tegmental Connections

Fig. 4. The origins and extent of the ascending projection from the vestibular nuclei to *rostral* iMLF. It receives an ipsi- and contralateral projection, which are both markedly weaker than those to the interstitial nucleus of Cajal (iC), and only partially fill the nucleus. No evidence for a vestibular projection to nD was found.

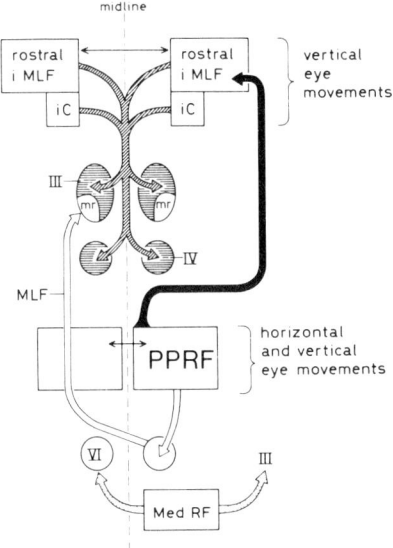

Fig. 5. Interconnectivity between areas involved in the supranuclear control of fast eye movements. Signals from PPRF are sent to the vertical premotor centre in the rostral mesencephalon (black arrow) and also caudally to the ipsilateral VI from which lateral rectus, and, via an ascending crossed pathway, the medial rectus are activated. The vertical premotor areas project onto the motoneurones of 'vertical' eye muscles (IO, IR, SR, SO). A possible role for the dorsal med RF which is also controlled from PPRF is discussed in the text. Commissural connections between *rostral* iMLF and PPRF are represented by double-headed arrows.

ents is very similar to the system described by Tarlov (1972) for the source of crossed and uncrossed vestibulo-oculomotor afferents running within the MLF, and it is possible that the rostral iMLF afferents are part of the same ascending vestibular system.

Vertical gaze is represented bilaterally. For example, unilateral lesions do not produce vertical gaze paralysis, whereas bilateral lesions for example in PPRF or rostral iMLF can (Bender and Shanzer, 1964; Christoff, 1974; Kömpf et al., 1978). This bilateral principle for vertical eye movements was clearly stated by Bender (1960). A partial explanation for this could be provided by the commissural connections between areas involved in vertical eye movements. Commissural fibres interconnect PPRF on each side (Fig. 2): rostral iMLF is similarly interconnected. Our autoradiographic experiments show that these are not diffuse commissures, but compact pathways. Multiple commissural connections, for example in PPRF, rostral iMLF and other stages of vertical eye movement information processing, would ensure vertical coordination of the two eyes, as well as a bilateral representation, which could not be destroyed by a unilateral lesion.

The functions of this small medial rostral mesencephalic region, rostral iMLF, have been discussed by Hess (1954), Hassler (1972) under the name of nucleus praestitialis in the cat, and Graybiel (1977a) under the eponym "NPRF" nucleus of the prerubral field. Only its role in vertical eye movements is considered here, but it is clear that this area is involved in eye-head coordination, as is the closely related nucleus lying just caudal to it, the interstitial nucleus of Cajal.

MEDULLARY RETICULAR FORMATION (med RF)

The involvement of med RF in the control of eye movements has not been as widely considered as that of PPRF or MRF. A general hypothesis for the role of part of med RF in oculomotor control has been put forward by Graybiel (1977b). She suggests that it may deal with information encoded in terms of visual and eye-head coordinates, whereas PPRF may encode more specifically the 'visual map'. This suggestion rests mainly on the fact that tectoreticular afferents supply both regions but vestibular and spinal afferents terminate mainly in the med RF zone.

Several papers have recently produced detailed evidence for monosynaptic connections to the oculomotor and abducens nuclei from the rostrodorsal part of med RF (also referred to as the dorsomedial gigantocellular tegmental field of the medulla (FTG)). This area is contained within the shaded medullary area in Fig. 1. Graybiel (1977a) showed that in the cat and particularly monkey, cells in the dorsal med RF were labelled after the injection of HRP into the oculomotor nucleus. Our HRP experiments on the monkey (Steiger and Büttner-Ennever, 1979), and physiological data in the cat (for review, see Baker, 1977), confirmed these observations. Using anterograde tracer techniques we have further been able to show that in the monkey there is a projection from the region of the dorsal med RF to the ipsilateral subdivisions of the oculomotor nucleus, including the medial rectus group. It is exactly these subgroups which are supplied by the abducens internuclear system. At this stage we still cannot distinguish between the dorsal med RF and the caudal nucleus praepositus as the fibre source, since both were included in the injection site. There is additional evidence from both anterograde and retrograde studies that the rostromedial part of the dorsal med RF sends fibres to the contralateral abducens nucleus (Grantyn et al., 1977, Graybiel, 1977b; Hikosaka and Kawakami, 1977; Maciewicz et al., 1977). The physiological experiments of Hikosaka and Kawakami (1977) in addition show that the projection is inhibitory. It is possible then that the projection from the dorsal med RF region to the oculomotor nucleus observed in the monkey is also inhibitory. This being the case, the dorsal med RF zone contains a system which could inhibit the eye muscles executing horizontal conjugate movements to the contralateral side. PPRF has a similar set of connections which EXCITE the motoneurones responsible for horizontal conjugate gaze to the ipsilateral side. Since the dorsal med RF area receives a direct input from PPRF (Fig. 2) (Büttner-Ennever and Henn, 1976; Graybiel, 1977b), a signal originating in PPRF can activate the motoneurones which move the eyes horizontally to the ipsilateral side, and, via the dorsal med RF and perhaps caudal ppH, simultaneously inhibit the antagonists.

This hypothesis was tentatively put forward by Baker (1977), but since there was no evidence for the specific control of the ipsilateral medial rectus in the cat from the dorsal med RF area, he considered its involvement in the inhibition of the horizontal antagonists to be improbable.

Our studies show that in the monkey the neurones projecting to the contralateral abducens nucleus are mainly further rostral than the cells supplying ipsilateral medial rectus subdivision of the oculomotor nucleus, but we cannot yet say whether the ipsilateral projection arises from the dorsal med RF or ppH, or both, and its extent and function are still uncertain. For these reasons, it is not included in Fig. 5. However, the suggestion that the inhibition of the horizontal antagonists are in some way controlled from dorsal med RF does gain support from the lesion study of Uemura and Cohen (1973).

To summarize, small specific parts of the mesencephalic, pontine and medullary reticular formation have been shown to be interconnected and project directly to the oculomotor complex. An attempt has been made to relate these few pathways to specific oculomotor functions, even though it is abundantly clear that each of these areas subserve many additional and interrelated functions.

One working hypothesis for the involvement of these areas in the premotor control of fast eye movements is the following (see Fig. 5): information required to generate a fast eye movement converges on the PPRF. A compact pathway originating from the caudal and medial parts of PPRF ascends beneath the ipsilateral MLF and carries a coordinating signal to the rostral iMLF, a nucleus lying in the rostral mesencephalic reticular formation. Here, and in the surrounding area, which includes iC, the vertical components of the fast eye movements are elaborated and fed through a system of direct and indirect connections onto the oculomotor subgroups in III subserving vertical eye movements and IV. The horizontal components of the fast eye movements are transmitted through a 'cascade' of short axonal systems from the rostral to the caudal PPRF and thence directly to the abducens nucleus, where the lateral rectus motoneurones are activated. The activating signal for the synergistic medial rectus is relayed through internuclear neurones which also lie within the abducens nucleus. Their axons cross the midline at the level of the VIth nucleus and ascend, within the contralateral MLF, to the medial rectus motoneurones in the oculomotor nucleus. Simultaneously, PPRF drives cells in the dorsal part of the medullary reticular formation (and ppH), which could possibly mediate the inhibition of the horizontal antagonists.

SUMMARY

The functional and anatomical connections from small parts of the pontine, mesencephalic and medullary reticular formation to the oculomotor complex are discussed. There is evidence that the paramedian pontine reticular formation (PPRF) is a major premotor area in which fast eye movements are generated. It controls the horizontal eye muscles through a projection from mainly caudal PPRF to the ipsilateral abducens nucleus (VI). Motoneurones of the lateral rectus, lying within VI, are intermingled with a set of internuclear neurones that project rostrally within MLF to the contralateral medial rectus motoneurones in the oculomotor nucleus (III). PPRF utilizes this internuclear system to activate medial rectus.

The vertical eye movement premotor area lies in the rostral mesencephalon. A direct and compact pathway from PPRF to one nucleus in this region (*rostral* iMLF) could carry the coordinating signal by which PPRF can control the generation of vertical eye movements. Both rostral iMLF and the neighbouring interstitial nucleus of Cajal send direct projections to the motoneurones of inferior and superior recti, and inferior and superior oblique.

A possible role for the dorsal medullary reticular formation in the inhibition of horizontal antagonists is discussed.

ACKNOWLEDGEMENTS

This work was supported by the Swiss National Science Foundation, grant No. 3.636.75 and the Dr. Eric Slack-Gyr Foundation in Zurich.

REFERENCES

André-Thomas, J., Schaeffer, H. and Bertrand, I. (1933) Paralysie de l'abaissement du regard paralysié des inférogyres hypertonie des supérogyres et des releveurs des paupières. *Rev. neurol.*, 40: 535–542.

Baker, R. (1977) Anatomical and physiological organization of brain stem pathways underlying the control of gaze. In *Control of Gaze by Brain Stem Neurons. Vol. 1*, R. Baker and A. Berthoz (Eds.), Elsevier/North-Holland Biomed. Press, Amsterdam, pp. 207–222.

Baker, R. (1979) The parabducens nucleus. In *Integration in the Nervous System*, V. Wilson and H. Asanuma (Eds.), Igakaku Shoin Ltd. In press.

Baker, R. and Highstein, S. (1975) Physiological identification of interneurons and motoneurons in the abducens nucleus. *Brain Res*, 91: 292–298.

Bender. M. B. (1960) Comments on the physiology and pathology of eye movements in the vertical plane. *J. Nerv. Ment. Dis.*, 130: 456–466.

Bender, M. B. and Shanzer, S. (1964) Oculomotor pathways defined by electrical stimulation and lesions in the brainstem of monkey. In *The Oculomotor System*, M. B. Bender (Ed.), Harper and Row, New York, pp. 81–140.

Büttner-Ennever, J. A. (1977) Pathways from the pontine reticular formation to structures controlling horizontal and vertical eye movements in the monkey. In *Control of Gaze by Brain Stem Neurons, Vol. 1*, R. Baker and A. Berthoz (Eds.), Elsevier/North-Holland Biomed. Press, Amsterdam, pp. 89–98.

Büttner-Ennever, J. A. and Henn, V. (1976) An autoradiographic study of the pathways from the pontine reticular formation involved in horizontal eye movements. *Brain Res.*, 108: 155–164.

Büttner-Ennever, J. A. and Büttner, U. (1978) A cell group associated with vertical eye movements in the rostral mesencephalic reticular formation of the monkey. *Brain Res.*, 151: 31–47.

Büttner, U., Büttner-Ennever, J. A. and Henn, V. (1977) Vertical eye movement related unit activity in the rostral mesencephalic reticular formation of the alert monkey. *Brain Res.*, 130: 239–252.

Carpenter, M. B., Harbison, J. W. and Peter, P. (1970) Accessory oculomotor nuclei in the monkey: projections and effects of discrete lesions. *J. Comp. Neurol.*, 140: 131–154.

Carpenter, M. B., McMasters, R. E. and Hanna, G. R. (1963) Disturbances of conjugate horizontal eye movements in the monkey. I. Physiological effects and anatomical degeneration from lesions of the abducens nucleus and nerve. *Arch. Neurol. Psychiat. (Chic.)*, 8: 231–247.

Christoff, N. A. (1974) A clinicopathologic study of vertical eye movements. *Arch. Neurol. Psychiat. (Chic.)*, 31: 1–8.

Cogan, D. C. (1974) Paralysis of down-gaze. *Arch. Ophthal.* 91: 192–199.

Duensing, F., Schaefer, K. P. and Trevison, C. (1963) Die Raddrehung vermittelnde Neurone in der zentralen Funktionsstruktur der Labyrinthstellreflexe auf Kopf und Augen. *Arch. Psychiat. Nervenkr.*, 204: 113–132.

Edwards, S. B. (1972) The ascending and descending projections of the red nucleus in the cat: an experimental study using an autoradiographic tracing method. *Brain Res.*, 48: 45–63.

Fuse, G. (1912) Ueber den Abduzenskern der Säuger. *Arb. Hirnanat. Inst. Univ. Zürich*, VI: 401–447.

Gacek, R. R. (1974) Localization of neurons supplying the extraocular eye muscles in the kitten using horseradish peroxidase. *Exp. Neurol.*, 44: 381–403.

Goebel, H. H., Komatsuzaki, A., Bender, M. B. and Cohen, B. (1971) Lesions of the pontine tegmentum and conjugate gaze paralysis. *Arch. Neurol. Psychiat., (Chic.)*, 24: 431–440.

Grantyn, A., Grantyn, R. and Robiné, K. P. (1977) Neuronal organization of tecto-oculomotor pathways. In *Control of Gaze by Brain Stem Neurons, Vol. 1*, R. Baker and A. Berthoz (Eds.), Elsevier/North-Holland Biomed. Press, Amsterdam, pp. 197–206.

Graybiel, A. M. (1977a) Organization of oculomotor pathways in the cat and rhesus monkey. In *Control of Gaze by Brain Stem Neurons, Vol. 1*, R. Baker and A. Berthoz (Ed.), Elsevier/North-Holland Biomed. Press, Amsterdam, pp. 79–88.

Graybiel, A.M. (1977b) Direct and indirect preoculomotor pathways of the brainstem: an autoradiographic study of the pontine reticular formation in the cat. *J. Comp. Neurol.*, 175: 37–78.

Graybiel, A. M. and Hartwieg, E. A. (1974) Some afferent connections of the oculomotor complex in the cat: an experimental study with tracer techniques. *Brain Res.*, 81: 543–551.

Hassler, R. (1972) Supranuclear structures regulating binocular eye and head movements. *Bibl. Ophthal. (Basel)*, 82: 207–219.

Henn, V. and Cohen, B. (1976) Coding of information about rapid eye movements in the pontine reticular formation of alert monkeys. *Brain Res.*, 108: 307–325.

Hess, W. R. (1954) *Das Zwischenhirn; Syndrome, Localisation, Funktionen.* Schwabe and Co., Basel.

Hikosaka, O. and Kawakami, T. (1977) Inhibitory reticular neurons related to the quick phase of vestibular nystagmus – their location and projection. *Exp. Brain Res.*, 27: 377–396.

Jacobs, L., Anderson, P. J. and Bender, M. B. (1973) The lesions producing paralysis of downward but not upward gaze. *Arch. Neurol. Psychiat. (Chic.)*, 28: 319–323.

Jouvet, M. (1962) Recherches sur les structures nerveuses et les mécanismes responsables des différentes phases du sommeil physiologique. *Arch. ital. Biol.*, 100: 125–206.

Jouvet, M. (1965) Pardoxical sleep – a study of its nature and mechanisms. In *Progress in Brain Research. Vol. 18, Sleep Mechanisms.* K. Akert, C. Bally and J. P. Schadé (Eds.), Elsevier, Amsterdam, pp. 20–62.

Kawamura, K., Brodal, A. and Hoddevik, G. (1974) The projection of the superior colliculus onto the reticular formation of the brain-stem. An experimental anatomical study in the cat. *Exp. Brain Res.* 19: 1–19.

King, W. M. (1976) *Quantitative Analysis of the Activity of Neurons in the Accessory Oculomotor Nuclei and the Mesencephalic Reticular Formation of Alert Monkeys in Relation to Vertical Eye Movements Induced by Visual and Vestibular Stimulation.* Doct. Diss., Univ. of Washington, Seattle.

King, W. M., Precht, W. and Dieringer, N. (1978) Connections of behaviorally identified cat omnipause neurons. *Exp. Brain Res.*, 32: 435–438.

Kömpf, D (1978) Vertikale Augenbewegungen. *Nervenarzt*, 49: 377–384.

Kömpf, D., Pasik, T. and Pasik, P. (1977) Critical structures for downward gaze in monkeys. *Neurosci. Abstr.*, 3: 156, no. 480.

Lasek, R., Joseph, B.S. and Whitlock, D.G. (1968) Evaluation of a radioautographic neuroanatomical tracing method. *Brain Res.*, 8: 319–336.

Laurent, J.-P., Cespuglio, R. and Jouvet, M. (1974) Délimitation des voies ascendantes de l'activité pontogeniculo-occipitale chez le chat. *Brain Res.*, 65: 29–52.

Luschei, E. S. and Fuchs, A. F. (1972) Activity of brain stem neurons during eye movements of alert monkeys. *J. Neurophysiol.*, 35: 445–461.

Mabuchi, M. and Kusama, T. (1970) Mesodiencephalic projections to the inferior olive and the vestibular and perihypoglossal nuclei. *Brain Res.*, 17: 133–136.

Maciewicz, R. J., Eagen, K., Kaneko, C. R. S. and Highstein, S. M. (1977) Vestibular and medullary brainstem afferents to the abducens nucleus in the cat. *Brain Res.*, 123: 225–240.

Markham, C. H., Precht, W. and Shimazu, H. (1966) Effect of stimulation of interstitial nucleus of Cajal on vestibular unit activity in the cat. *J. Neurophysiol.*, 29: 493–507.

Moruzzi, G. and Magoun, H. W. (1949) Brain stem reticular formation and activation of the EEG. *Electroenceph. clin. Neurophysiol.*, 1: 455–473.

Nauta, W. J. H. and Kuypers, H. G. J. M. (1958) Some ascending pathways in the brain stem reticular formation. In *Reticular Formation of the Brain. Henry Ford Hospital Symposium*, H. H. Jasper, L. D. Proctor, R. S. Knighton, W. C. Noshay and R. T. Costello (Eds.), Little Brown and Co., Boston, pp. 3–30.

Pasik, P., Pasik, T. and Bender, M. B. (1969) The pretectal syndrome in monkeys. I. Disturbances of gaze and body posture. *Brain*, 92: 521–534.

Peterson, B.W. (1977) Identification of reticulospinal projections that may participate in gaze control. In *Control of Gaze by Brain Stem Neurons, Vol. 1*, Elsevier/North-Holland Biomed. Press, Amsterdam, pp. 143–152.

Raphan, T. and Cohen, B. (1978) Brainstem mechanisms for rapid and slow eye movements. *Ann. Rev. Physiol.*, 40: 527–552.

Scheibel, M. E. and Scheibel, A. B. (1958) Structural substrates for integrative patterns in the brain stem reticular core. In *Reticular Formation of the Brain. Henry Ford Hospital Symposium*, H.H. Jasper, L.D. Proctor, R.S. Knighton, W.C. Noshay and R.T. Costello (Eds.), Little Brown and Co., Boston, pp. 31–55.

Schwindt, P. C., Precht, W. and Richter, A. (1974) Monosynaptic excitatory and inhibitory pathways from medial midbrain nuclei to trochlear motoneurons. *Exp. Brain Res.*, 20: 223–238.

Shimazu, H. Brain stem interneurons modulating abducens motor activity during vestibular nystagmus. In Baker, R. and Berthoz, A. (Eds.). *Control of Gaze by Brain Stem Neurons. Developments in Neuroscience*. Vol. 1, pp. 225–233. Amsterdam-New York. Elsevier/North Holland, Biomed. Press, 1977.

Steiger, H.-J. and Büttner-Ennever, J.A. (1978) Relationship between motoneurons and internuclear neurons in the abducens nucleus: a double retrograde tracer study in the cat. *Brain Res.*, 148: 181–188.

Steiger, H.J. and Büttner-Ennever, J.A. (1979) Oculomotor nucleus afferents in the monkey demonstrated with horseradish peroxidase. *Brain Res.*, 160: 1–15.

Szentagothai, J. (1943) Die zentrale Innervation der Augenbewegungen. *Arch. Psychiat. Nervenkr.*, 116: 721–760.

Tarlov, E. (1972) Anatomy of the two vestibulo-oculomotor projection systems. In *Progress in Brain Research, Vol. 37, Basic Aspects of Central Vestibular Mechanisms,* A. Brodal and O. Pompeiano (Eds.), Elsevier, Amsterdam, pp. 471–491.

Torvik, A. and Brodal, A. (1957) The origin of reticulospinal fibers in the cat. *Anat. Rec.*, 128: 113–137.

Uemura, T. and Cohen, B. (1973) Effects of vestibular nuclei lesions on vestibulo-ocular reflexes and posture in monkeys. *Acta oto-laryngol. (Stockh.)*, Suppl. 315.

Valverde, F. (1961) Reticular formation of the pons and medulla oblongata. A Golgi study. *J. Comp. Neurol.*, 116: 71–100.

Warwick, R. (1953) Representation of the extra-ocular muscles in the oculomotor nuclei of the monkey. *J. Comp. Neurol.*, 98: 449–504.

Organization of Reticular Projections to the Vestibular Nuclei in the Cat

N. CORVAJA, T. MERGNER and O. POMPEIANO

Istituto di Fisiologia Umana, Cattedra II, Università di Pisa, 56100 Pisa (Italy)

INTRODUCTION

The anatomical organization of the reticulovestibular connections has been investigated by some authors, who used either the Golgi method (Cajal, 1909–1911; Lorente de Nó, 1933; Scheibel and Scheibel, 1958; Leontovich and Zhukova, 1963; Hauglie-Hanssen, 1968) or the method of anterograde degeneration (Mehler, 1968; Hellstrøm Hoddevik et al., 1975) to trace axons or their collaterals from neurons in the reticular formation (RF) to the vestibular nuclei. In particular, Mehler (1968) described degenerating fibers to the lateral vestibular nucleus following reticular lesions in rats. On the other hand, Hellstrøm Hoddevik et al. (1975) found in cats that the reticular nuclei gigantocellularis, pontis caudalis and pontis oralis project bilaterally to the four main vestibular nuclei with an ipsilateral overweight. However, no projection to the vestibular nuclei was observed from the mesencephalic RF (cf. also Pompeiano and Walberg, 1957) and the nucleus reticularis ventralis. There was a tendency for rostrally located RF neurons to project to the rostral-most parts of the vestibular nuclear complex; in particular, the maximal terminal field after lesions of the nucleus reticularis pontis oralis was found within the superior vestibular nucleus (SVN), while the lateral (LVN), medial (MVN) and descending vestibular nucleus (DVN) were preferred sites of termination of fibers from the nuclei reticularis gigantocellularis and pontis caudalis. Moreover, the termination areas within the vestibular complex were rather diffuse, although in the SVN and LVN there was a preponderance ventrally. The exact contribution of the different reticular nuclei to the reticulovestibular projections could hardly be evaluated in this study, since lesions of one of them might have affected fibers emanating from the others as well as ascending and descending afferents originating from sources different from the reticular nuclei.

The conclusion from this study, i.e., that most of the reticulovestibular projections are not specifically organized, contrasts with the results of autoradiographic experiments, which suggest a specificity in the zone of origin of reticulovestibular afferents (Büttner-Ennever and Henn, 1976; Graybiel, 1977a). In particular, in the experiments of Graybiel (1977a), injections of tritium-labeled amino acids were placed in the brain stem RF of cats: (i) at the border between nucleus reticularis pontis oralis and caudalis, (ii) in more rostral and dorsal parts of the pontine tegmentum, (iii) at the pontomesencephalic border, and (iv) at the pontomedullary border. Very sparse labeling of the ipsilateral vestibular complex, mainly of the MVN, was present in the group

I experiments, but not after more rostral injections as in the groups II and III experiments (cf. also Büttner-Ennever and Henn, 1976). On the other hand, the nuclei of the contralateral vestibular complex were labeled in the group IV experiments, particularly after the most dorsal injections. Unfortunately, some of these labeled fibers may belong either to commissural vestibular fibers or to vestibular afferents originating from contralateral perihypoglossal nuclei.

In physiological experiments, reticular neurons could be antidromically activated following stimulation of the vestibular nuclei (cf. Remmel and Skinner, 1976; Remmel et al., 1977a,b). In addition to the reticulovestibular projections, the possibility should be considered, i.e., that the perihypoglossal nuclei project to the vestibular nuclei.

The present report describes recent findings from a detailed study in which the reticular and perihypoglossal neurons projecting to the vestibular nuclei were identified following injection of horseradish peroxidase (HRP) in the vestibular nuclear complex (Pompeiano et al., 1978). This method was also used to localize the reticular and vestibular neurons projecting to the perihypoglossal nuclei (Mergner et al., 1977).

METHODS

The experimental material was prepared according to the method described by Mergner et al. (1977). In brief, HRP of the Sigma type VI dissolved in physiological saline was injected in various places of the vestibular and the perihypoglossal nuclei in cats by using glass micropipettes with a diameter at the tip of 10–20 μm. Small injections were achieved by a combined electrophoretic (Graybiel and Devor, 1974) and mechanical method, as described in our previous study (Mergner et al., 1977). Details of surgical procedures, amount and concentration of HRP fluid injected, survival time, localization and extension of the HRP staining at the injection site and the detailed mapping of labeled neurons are available in the original studies (Mergner et al., 1977; Pompeiano et al., 1978). Since endogenous peroxidase activities were reported in the brain of normal animals (Keefer and Christ, 1976; Wong-Riley, 1976), a normal cat's brain was also treated by the same method and used as control for the presence of endogenous labeling. However, labeling of neurons in the reticular and perihypoglossal nuclei of normal cats was not observed under the condition and the method used in this laboratory.

RESULTS

Detailed information about the localization within the reticular and the perihypoglossal nuclei of neurons projecting to the vestibular nuclear complex of one side was obtained following injections of HRP within different components of the vestibular complex. In one experiment (Cat 16) the HRP was selectively injected within the rostral part of the vestibular nuclear complex, i.e. within the SVN, LVN and the dorsorostral aspect of both the MVN and DVN, including the group y. Labeled neurons of small and medium size were found bilaterally in the paramedian pontine and medullary RF, but there was hardly any labeling present in the mesencephalic RF. In particular the neurons in the pontine and medullary RF were located within the nucleus reticularis gigantocellularis of both sides, the dorsolateral part of the contralateral nucleus reticularis pontis caudalis, just caudal to the abducens (VI) nucleus, and the nucleus reticularis parvicellularis ipsilateral to the side of the injection. Some labeled cells were also found among the fibers of the medial longitudinal bundles. On the other hand, no labeled neuron was observed in the precerebellar reticular nuclei.

In two other experiments (Cats 14 and 17) the injection involved the same vestibular nuclei reported in the previous case, except for the MVN which was spared by the injection. Labeled neurons were observed within the same paramedian regions of the pontine and medullary RF reported in the previous experiment. In particular in cat 17, where the injection involved the whole SVN, LVN, the rostral pole of the DVN and the group y, there was a bilateral distribution of the labeled neurons within the nucleus reticularis gigantocellularis and a prominent distribution of labeled neurons within the contralateral nucleus reticularis pontis caudalis extending into the periabducens zone and also within the ipsilateral nucleus reticularis parvicellularis (Fig. 1, left-side). Finally, in cat 14, where the injection involved the dorsal aspect only of the SVN, LVN, and rostral half of the DVN including the group y (Fig. 1, right side), labeled neurons were observed mainly within the ipsilateral nucleus reticularis parvicellularis and the contralateral nucleus reticularis gigantocellularis. Only few labeled neurons were found in the ipsilateral nucleus reticularis gigantocellularis, and in the nucleus reticularis pontis caudalis of both sides, while no labeled neuron was observed within the periabducens zones. In the experiments reported above (particularly in cats 16 and 17) very few labeled neurons were also observed in the nucleus praepositus hypoglossi (p.h.) of both sides particularly in the peripheral zone of the contralateral side; however, no labeled cells were found in the remaining perihypoglossal nuclei, i.e., the nucleus intercalatus of Staderini (i.c.) and the nucleus of Roller (Ro).

The reticular and the perihypoglossal projections to the MVN were studied in experiments in which the injection site was limited to this nucleus. Fig. 2 illustrates the results of one representative experiment (cat 5) in which the brown reaction product was restricted to the caudal half of the MVN, i.e., neither the surrounding DVN and the p.h., nor the neighbouring RF were involved (see also Fig. 3A). Positive neurons were found asymmetrically on both sides of the medullary and pontine RF, but hardly any in the mesencephalic RF and none in the precerebellar reticular nuclei. In particular a densely packed group of neurons was located *contralaterally* to the side of the injection, in the dorsal part of the paramedian medullary tegmentum caudal to the VIth nucleus (Fig. 2, sections 7, 8). On the *ipsilateral* side, labeled neurons were highly concentrated in a first group of neurons located in the dorsal part of the paramedian region of the pontine RF (PPRF), within the tegmentum of the nucleus reticularis pontis caudalis, just rostral to the ipsilateral VIth nucleus (Fig. 2, sections 1–3). A second group of more scattered neurons was located in the periabducens zone of the RF, between the rostral part of the MVN and the *genu facialis* (Fig. 2, sections 5, 6). Finally, a third group of labeled neurons was found in the nucleus reticularis lateralis, ventromedially to the motor nucleus of the facial (VII) nerve and dorsolaterally to the medullary pyramis (Fig. 2, section 7'). Fig. 3C–E illustrates some of the labeled neurons located in the medullary and pontine RF in this experiment. Most of these positive reticular neurons projecting to the MVN were of small size.

In the same experiment numerous labeled neurons were found within the peripheral zone of the p.h., the i.c. and the Ro nuclei of the contralateral side (see Fig.

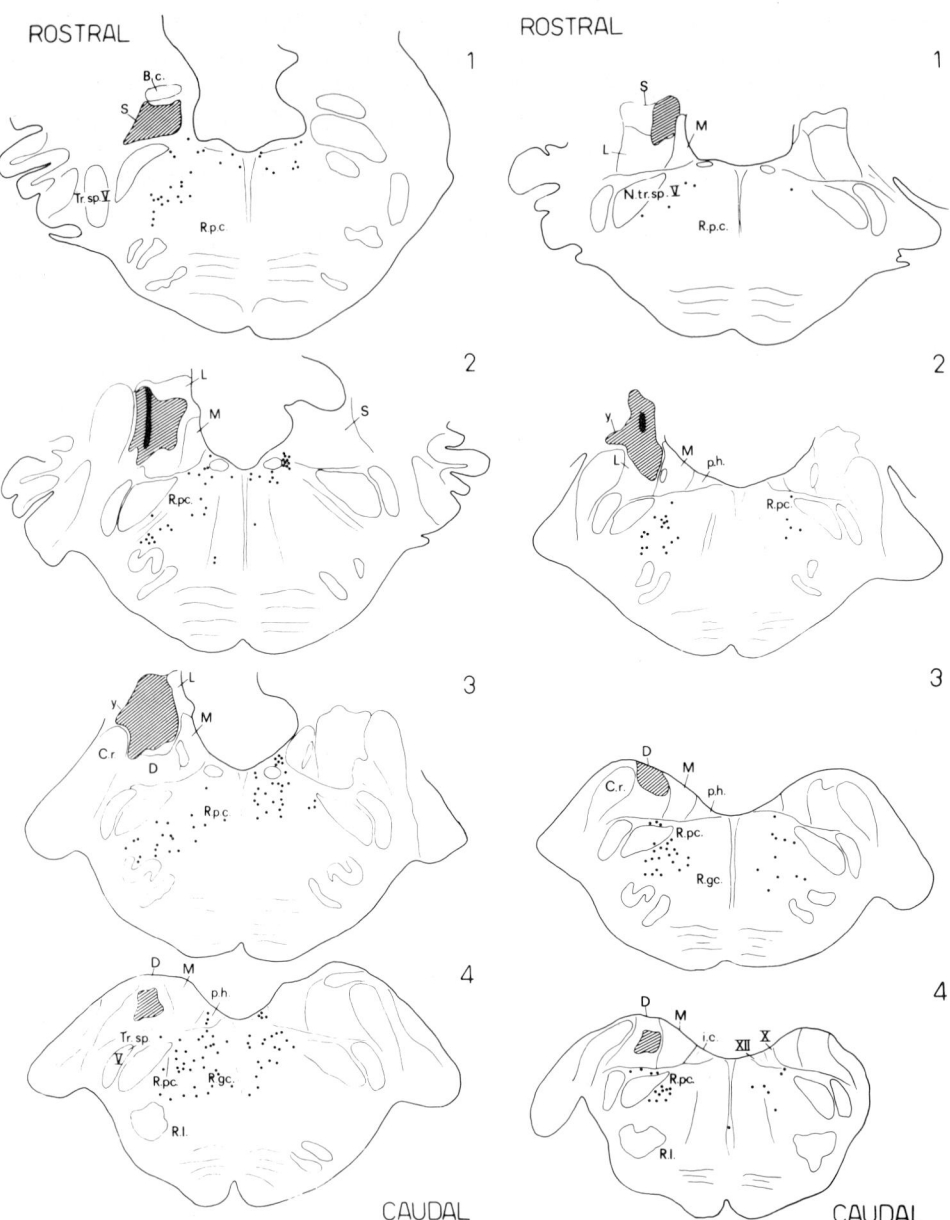

Fig. 1. Distribution of labeled neurons in the reticular and perihypoglossal nuclei of both sides following unilateral injections of HRP within the SVN, LVN, the rostral pole of the DVN and the group y. Left: cat 17. Drawings of transverse sections through the medulla taken at regular intervals of 1 mm from rostral to caudal levels. Each labeled neuron is indicated by a dot. Positive cells on each drawing are taken from 4 out of 20 serial sections, 50 μm thick. Right: cat 14. Drawings of transverse sections through the medulla taken at regular intervals of 1 mm from rostral to caudal levels. Positive cells on each drawing are taken from 2 out of 20 serial sections, 50 μm thick.

Abbreviations for all figures: B.c., Br.c., brachium conjunctivum; C.r., restiform body; D, descending vestibular nucleus; f, small cell group f in the caudal part of the descending vestibular nucleus (Brodal and Pompeiano); i.c., nucleus intercalatus (Staderini); L, lateral vestibular nucleus (Deiters); M, medial vestibu-

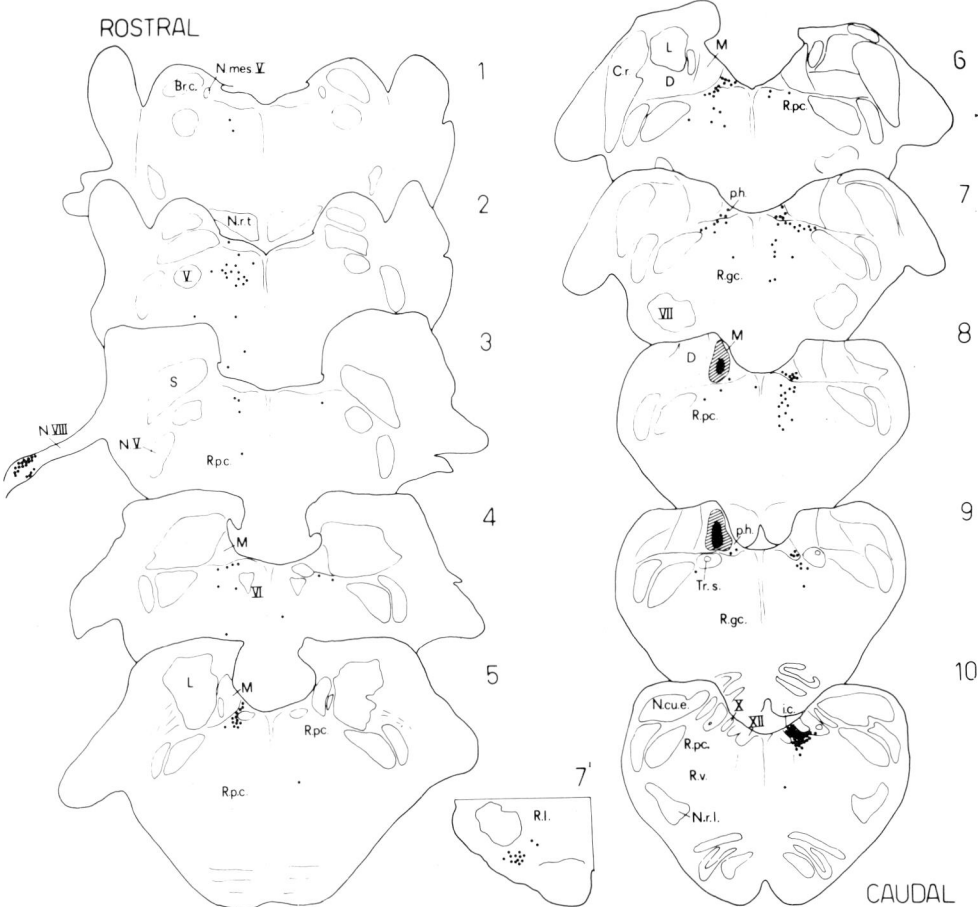

Fig. 2. Distribution of labeled neurons in the reticular and perihypoglossal nuclei of both sides following selective injection of HRP within the MVN of one side. Cat 5. Drawings of transverse sections of the brain stem taken at regular intervals of 1 mm from sections 3 to 10 (750 μm from sections 1 to 3). Labeled neurons are indicated by dots. Positive cells on each of these drawings are taken from 4 out of 20 serial sections, 50 μm thick. Section 7' represents a detail of section 7, to illustrate a small group of labeled reticular neurons located dorsolaterally to the pyramidal tract and ventromedially to the motor nucleus of the VIIth cranial nerve; no labeled cells were found in this region at more rostral or caudal levels. (From Pompeiano et al., 1978).

◁ lar nucleus (Schwalbe); N.cu.e., external (accessory) cuneate nucleus; N.mes.V, mesencephalic nucleus of N.V; N.r.l., lateral reticular nucleus (nucleus of lateral funiculus); N.r.t., nucleus reticularis tegmenti pontis; N.tr.sp.V, nucleus of spinal tract of trigeminal nerve; N.V, N.VIII, cranial nerves V and VIII: Ol.i., inferior olive; p.h., nucleus praepositus hypoglossi; R.gc. nucleus reticularis gigantocellularis; R.l., nucleus reticularis lateralis (Meessen and Olszewski); R.pc., nucleus reticularis parvicellularis; R.p.c., nucleus reticularis pontis caudalis; R.v., nucleus reticularis ventralis; S, superior vestibular nucleus (Bechterew); Tr.s., solitary tract; Tr.sp.V, spinal tract of trigeminal nerve; x, small cell group x lateral to the descending vestibular nucleus (Brodal and Pompeiano); y, small cell group y dorsal to the restiform body (Brodal and Pompeiano); V, VI, VII, X, XII, motor cranial nerve nuclei. (From Pompeiano et al., 1978.)

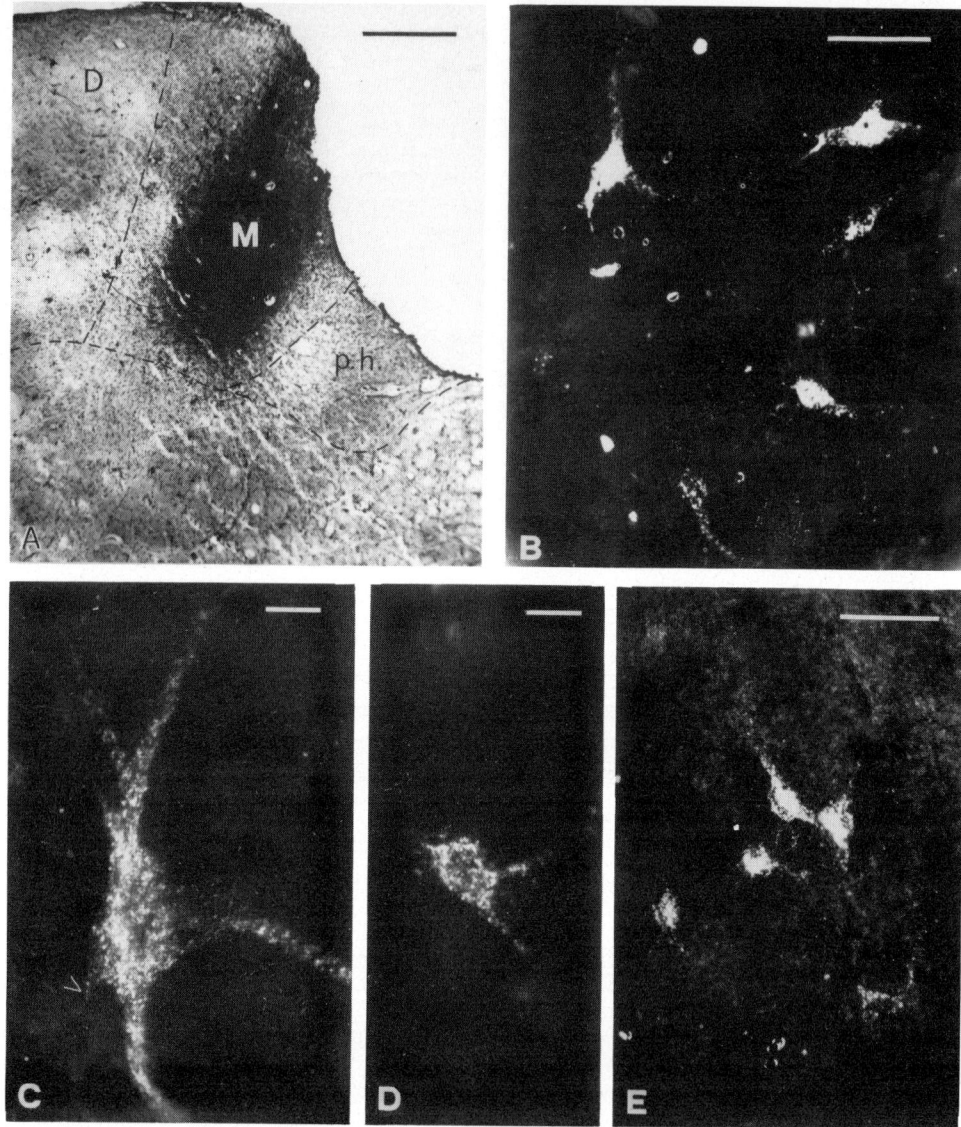

Fig. 3. Photomicrographs illustrating a representative case of HRP injection within the MVN (bright field), with some of the corresponding retrogradely labeled neurons located in representative perihypoglossal and reticular structures of both sides. Cat 5. (A) Microphotograph, taken at low magnification illustrating the injection site. (B) labeled neurons located in the peripheral zone of the contralateral p.h. taken at medium magnification. (C–E) Labeled neurons located in the contralateral paramedian medullary RF (C) and the ipsilateral paramedian pontine RF (D), as well as in the ipsilateral nucleus reticularis lateralis of the medulla, just ventral to the motor nucleus of the VII cranial nerve (E). Arrow-head in C points on an axon. Magnification scales: 0.5 mm in A; 50 µm in B and E; 10 µm in C and D. (From Pompeiano et al., 1978.)

3B). Some labeled neurons were also found within the p.h., but not within the i.c. and Ro nuclei of the ipsilateral side.

The location of the neurons projecting to the DVN was finally investigated. In our experiments only small injections of HRP strictly located within the DVN were performed. No labeled neuron was found within the medullary and pontine RF.

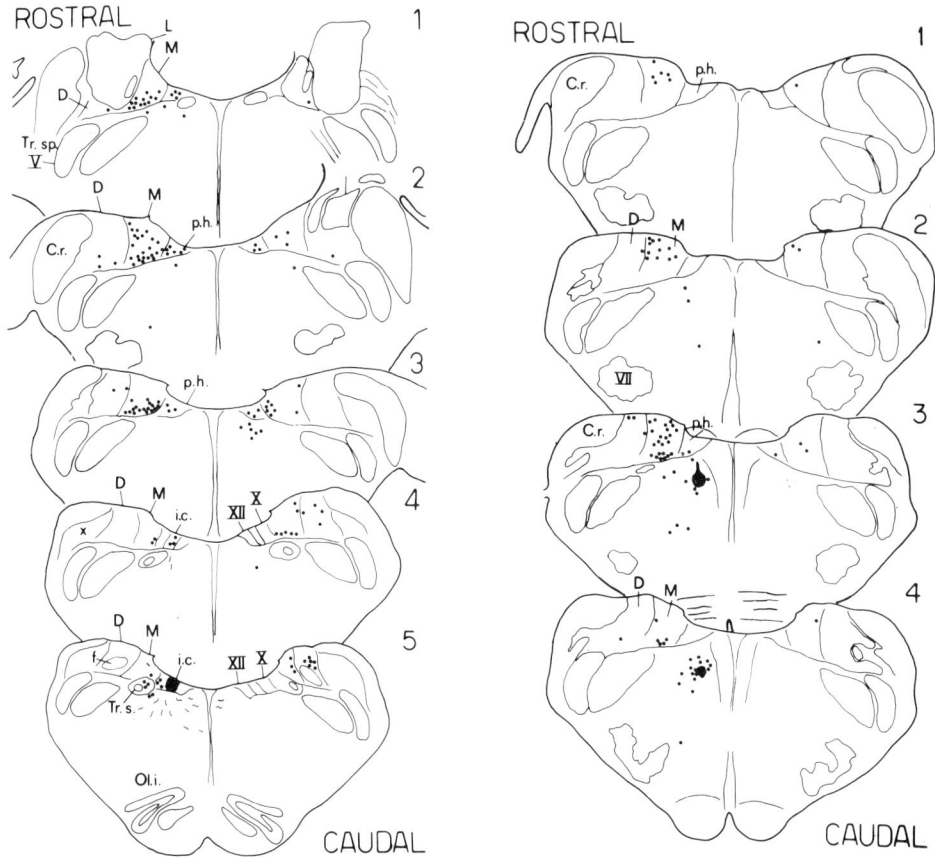

Fig. 4. Distribution of labeled neurons in the medullary structures of both sides following small injections of HRP in the nucleus intercalatus of Staderini (left side) and in the dorsal part of the paramedian medullary tegmentum (right). Left: cat 12. Drawings of transverse sections of the medulla taken at regular intervals of 1000 μm from rostral to caudal levels. Each drawing represents the results obtained in 4 out of 20 sections, 50 μm thick. In this as well as in the following case the black regions correspond to the injection site, while labeled neurons are indicated with dots. Right: cat 10. Drawings of transverse sections of the medulla taken at regular intervals of 500 μm from rostral to caudal levels. Each drawing represents the results obtained in 2 out of 10 sections, 50 μm thick. (From Mergner et al., 1977.)

However, few labeled neurons were observed in both the ipsilateral and the contralateral p.h., but not in the remaining perihypoglossal nuclei. In all the experiments in which the HRP was injected within the vestibular nuclei of one side, labeled neurons were also observed in the ipsilateral ganglion of Scarpa (see Fig. 2, section 3).

Projections from the pontine and medullary RF to the perihypoglossal nuclei appear to be scarce compared to those to the vestibular nuclei. Localized injections of HRP in the i.c. (Fig. 4, left side) gave rise to abundant labeling in the vestibular nuclei of both sides, predominantly in the ventral part of the ipsilateral MVN, whereas labeling of reticular neurons was rather poor. Only one group of positive neurons was observed which was located in the dorsal part of the paramedian medullary tegmentum of the contralateral side. The location corresponded essentially to that of the reticular neurons projecting to the contralateral MVN (cf. p. 633). In additional experiments, HRP was injected also into this dorsal tegmental region, i.e., just ventromedially to the p.h. and caudal to the VIth nucleus, in order to verify the specificity of the afferent projections found for the perihypoglossal nuclei. In the example illustrated in Fig. 4 (right side), a major number of the labeled neurons was localized within the ipsilateral MVN – but now predominantly in its dorsal half – and few neurons in the contralateral MVN, whereas no labeling was found in the neurons of the contralateral paramedian medullary tegmentum. It cannot be entirely excluded in these experiments that vestibular fibers projecting into the MLF were injured during the injection procedure, and that they took up HRP, thus contributing to the labeling in the MVN. However, from physiological experiments it appears that neurons in the paramedian medullary tegmentum responded to electrical stimulation of the ipsilateral VIIIth nerve, whereas responses from the contralateral side were rarely found (cf. Hikosaka and Kawakami, 1977).

DISCUSSION

Projections from the reticular formation to the vestibular nuclei

The present experiments have shown that the reticular neurons projecting to the vestibular nuclei were located in discrete regions of the main medullary and pontine RF. On the other hand, neither the mesencephalic RF (cf. also Pompeiano and Walberg, 1957; Hellstrøm Hoddevik et al., 1975), nor the precerebellar reticular nuclei contributed to the reticulovestibular projections; this last observation is in contrast with previous findings (Hellstrøm Hoddevik et al., 1975).

When the HRP was injected mainly within the SVN and LVN of one side, labeled neurons were found predominantly within the nucleus reticularis gigantocellularis of both sides, the dorsolateral part of the contralateral nucleus reticularis pontis caudalis just caudal to the VIth nucleus and the ipsilateral nucleus reticularis parvicellularis. The presence of crossed and uncrossed reticulovestibular projections to the LVN may explain the bilateral convergence of the spinal input to this nucleus (Wilson et al., 1966). These projections may also transmit the macular labyrinthine input of one side to the SVN and LVN of the contralateral side (Shimazu, 1972; Hoshino and Pompeiano, 1977); this finding is supported by the results of anatomical (Ladpli and Brodal, 1968) and physiological experiments (Gernandt et al., 1959; Duensing and Schaefer 1960; Spyer et al., 1974; Peterson and Abzug, 1975; Peterson et al., 1975), showing that some of the reticular regions projecting to the vestibular nuclei receive in their turn afferents from the vestibular nuclei of both sides.

Very little can be said from our experiments about the existence of reticulovestibular projections to the DVN, since the absence of labeled reticular neurons in the experiments in which HRP was injected within this nucleus could be due to the small size of the injections.

A major point of interest in our study was the localization of the reticular neurons projecting to the MVN, since this vestibular nucleus plays not only a role in the horizontal labyrinthine nystagmus, but also in horizontal rapid eye movements of extralabyrinthine origin (cf. Pompeiano, 1972; Mergner and Pompeiano, 1978). There is evidence that the MVN gives off major projections to the VIth nucleus (McMasters et al., 1966; Tarlov, 1970; Gacek, 1971; Graybiel and Hartwieg, 1974; Graybiel, 1977b; Maciewicz et al., 1977). Moreover, microstimulation of the MVN generates monosynaptic excitation in contralateral abducens motoneurons and monosynaptic inhibition in ipsilateral abducens motoneurons (Baker et al., 1969; Highstein, 1973; cf. Precht, 1975), whereas stimulation of the SVN has no such effect, instead, it inhibits oculomotor and trochlear motoneurons (Highstein, 1971; cf. Precht, 1975).

Four groups of reticular neurons project to the MVN. The first group is located caudal to the contralateral VIth nucleus in the dorsal part of the paramedian medullary tegmentum. This region appears to project also to the contralateral perihypoglossal (i.c.) nucleus and may be in its turn under the control of the ipsilateral MVN. The projection originating from this region might be related to the decussating fiber system which has recently been found to project not only to the contralateral MVN (Graybiel, 1977a), but also to the region of the contralateral VIth nucleus (Graybiel, 1977a,b; Maciewicz et al., 1977). Physiological experiments indicate that this system exerts an inhibitory influence on the contralateral abducens nucleus (Baker, 1977; Shimazu, 1977; Grantyn et al., 1977; Hikosaka and Kawakami, 1977). However, the role that the crossed reticulovestibular and reticuloperihypoglossal neurons play during the horizontal labyrinthine nystagmus as well as during the rapid eye movements of extralabyrinthine origin (cf. Pompeiano, 1972; Mergner and Pompeiano, 1978), remains to be elucidated.

A second group of RF neurons projecting to the MVN is located rostral to the ipsilateral VIth nucleus in the dorsolateral part of the PPRF, within the tegmentum of the nucleus reticularis pontis caudalis. This ipsilateral tegmentovestibular projection probably corresponds to that described by authors who used autoradiographic techniques (Büttner-Ennever and Henn, 1976; Graybiel, 1977a). In the monkey, the PPRF represents an important premotor structure initiating horizontal eye movements to the ipsilateral side (cf. Cohen and Henn, 1972; Henn and Cohen, 1975). Moreover, rhythmic neural activity appears in this region of the cat during the rapid eye movements typical of desynchronized sleep (Hobson et al., 1974a,b; McCarley and Hobson, 1975; Hoshino et al., 1976; Pivik et al., 1977). Physiological and anatomical observations have shown the existence of a direct monosynaptic excitatory pathway from this region to the ipsilateral VIth nucleus, innervating the lateral rectus of the same side (Briggs and Kaelber, 1971; Büttner-Ennever et al., 1975; Graybiel, 1975, 1977a,b; Kaneko et al., 1975; Büttner-Ennever and Henn, 1976; Highstein et al., 1976; Baker, 1977; Büttner-Ennever, 1977); however, there is no clear-cut anatomical evidence for the monosynaptic connection from the PPRF to the contralateral IIIrd nucleus supplying the medial rectus of the other side (Graybiel and Hartwieg, 1974; Büttner-Ennever et al., 1975; Graybiel, 1975, 1977a; Büttner-Ennever and Henn, 1976; Büttner-Ennever, 1977; cf. however Goebel et al., 1971; Highstein et al., 1974; Kaneko et al., 1975). It has been recently shown that the PPRF exerts an excitatory influence not only on motoneurons, but also on internuclear neurons located within the ipsilateral VIth nucleus which project to the medial rectus division of the contralateral IIIrd nucleus, thus being able to exert an excitatory influence on this structure (cf. Pompeiano et al., 1978, for references).

Since the PPRF sends also a direct projection to the dorsal part of the ipsilateral paramedian medullary tegmentum (Büttner-Ennever and Henn, 1976; Graybiel, 1977b), impulses originating in the PPRF can excite the motoneurons responsible for horizontal conjugate eye movements to the ipsilateral side and via the paramedian medullary tegmentum inhibit the motoneurons innervating the antagonists. It would be of interest to know how the ipsilateral tegmentovestibular projection collaborates with the ipsilateral tegmento-abducens projection in the reticular control of the horizontal eye movements. Interaction between the PPRF of both sides may also occur by utilizing crossed reticuloreticular projections (Walberg, 1974; Edwards, 1975; Büttner-Ennever and Henn, 1976; Graybiel, 1977a).

Another group of ipsilaterally projecting reticulovestibular neurons is located in the periabducens zone.

Finally, the last group of ipsilateral reticulovestibular neurons is located in the nucleus reticularis lateralis, between the motor nucleus of the VIIth nerve and the medullary pyramis. The nature and the origin of the afferent volleys transmitted by these reticular regions to the MVN remains to be identified.

Projections from the perihypoglossal to the vestibular nuclei

Afferent projections to the vestibular nuclei of both sides originate also from the perihypoglossal nuclei, i.e., the peripheral zone of the p.h., the i.c. and Ro nuclei. The peripheral zone of the p.h. projects mainly to the contralateral and to a smaller extent also to the ipsilateral MVN and DVN; it is doubtful, however, whether the same region projects to other components of the vestibular complex, such as the SVN and the group y. On the other hand, the i.c. and Ro nuclei send their major projections to the contralateral MVN, particularly at caudal level. So far a projection of p.h. neurons to the MVN of both sides has not been sufficiently studied in electrophysiological experiments (cf. Remmel and Skinner, 1976).

These observations should be related to the finding that specific regions of the MVN and DVN project in their turn to the perihypoglossal nuclei, namely to the i.c. of both sides (Mergner et al., 1977). This is in agreement with recent physiological studies, in which p.h. neurons could be reciprocally influenced by vestibular nerve stimulation (Baker and Berthoz, 1975; Baker et al., 1975; Gresty and Baker, 1976) and by natural stimulation of the horizontal canals (Blanks et al., 1977). It is well known that the vestibular nuclei are reciprocally connected by commissural fibers (Ladpli and Brodal, 1968; Pompeiano et al., 1978, 1979; Gacek, 1978); commissural connections exist also between the perihypoglossal nuclei of both sides, as shown by the fact that the p.h. nucleus projects to the contralateral perihypoglossal nuclei, namely to the i.c. (Mergner et al., 1977).

It is of interest that among the afferent inputs to the perihypoglossal nuclei, an important projection originates from the ipsilateral PPRF (Gacek, 1971; Büttner-Ennever and Henn, 1976; Graybiel, 1975, 1977a,b).

According to the present findings afferent fibers appear to stem also from the contralateral paramedian medullary tegmentum. On the other hand, among the efferent fibers originating from the perihypoglossal nuclei, some project to the motoneurons innervating the extrinsic eye muscles (Graybiel, 1974, 1977b; Graybiel and Hartwieg, 1974; Baker et al., 1977a,b; Maciewicz et al., 1977). The efferent projection of the PPRF to the ipsilateral perihypoglossal nuclei parallels that from the same reticular region to the ipsilateral VIth nucleus, while the projection of the paramedian medullary tegmentum to the contralateral i.c. nucleus parallels that from this reticular region to the contralateral VIth nucleus. Thus it appears that a close functional linkage exists between the perihypoglossal and the VIth nuclei and that the former nuclei, like the vestibular nuclei, may serve as premotor structures for eye movements.

SUMMARY

The reticular and perihypoglossal afferent projections to the vestibular nuclei have been studied using the method of retrograde axonal transport of HRP.

Reticulovestibular fibers originate from different pontine and medullary reticular nuclei of both sides. When the HRP is injected mainly within the SVN and the LVN of one side, labeled neurons are found predominantly within the nucleus reticularis gigantocellularis of both sides, the dorsolateral part of the contralateral nucleus reticularis pontis caudalis just caudal to the VIth nucleus and the ipsilateral nucleus reticularis parvicellularis.

Very little can be said from our experiments about the existence of reticulovestibular projections to the DVN, due to the small size of the injections made within this structure. On the other hand, the MVN receives specific projections from the contralateral paramedian medullary tegmentum caudal to the VIth nucleus. Furthermore, there are three distinct sources of ipsilateral reticular projections to the MVN: the first is located in the paramedian region of the pontine RF within the tegmentum of the nucleus reticularis pontis caudalis, just rostral to the VIth nucleus. The second one overlies like a cap the dorsolateral border of the VIth nucleus. The third one is located in the nucleus reticularis lateralis, ventromedially to the motor nucleus of the VIIth nerve.

The perihypoglossal nuclei which receive vestibular and reticular afferents from both sides in their turn give rise to crossed and uncrossed afferent projections to the vestibular nuclei. The corresponding neurons are mainly located in the peripheral border region of the p.h. and in the adjacent i.c. nucleus. In particular, major targets of the efferent projections from the p.h. are the MVN and DVN of both sides, particularly of the contralateral side; it is doubtful whether the p.h. projects to other components of the vestibular complex, such as the SVN and the group y; on the other hand, the i.c. sends major projections to the contralateral MVN.

The possible role that the reticular and the perihypoglossal projections to the vestibular nuclei play in the control of the oculomotor activity is discussed.

ACKNOWLEDGEMENTS

This investigation was supported by the Public Health Service Research Grant NS 07685–10 from the National Institute of Neurological and Communicative Disorders and Stroke, N.I.H., U.S.A. and by a research grant from the Consiglio Nazionale delle Ricerche, Italy. T. Mergner is a postdoctoral fellow of the Scuola Normale Superiore, Pisa, on leave of absence from the Physiologisches Institut der Universität, Münster, F.R.G.

REFERENCES

Baker, R. (1977) Anatomical and physiological organization of brain stem pathways underlying the control of gaze. In *Control of Gaze by Brain Stem Neurons*, R. Baker and A. Berthoz (Eds.), Elsevier/North-Holland Biomed. Press, Amsterdam-New York, pp. 207–222.

Baker, R. and Berthoz, A. (1975) Is the prepositus hypoglossi nucleus the source of another vestibulo-ocular pathway? *Brain Res.*, 86: 121–127.

Baker, R., Mano, N. and Shimazu, H. (1969) Postsynaptic potentials in abducens motoneurons induced by vestibular stimulation. *Brain Res.*, 15: 577–580.

Baker, R., Berthoz, A. and Gresty, M. (1975). Activity of prepositus neurons during eye movements and vestibular stimulation in the cat. *Exp. Brain Res.,* Suppl. 23: 14, no. 23.

Baker, R., Berthoz, A. and Delgado-Garcia, J. (1977a) Monosynaptic excitation of trochlear motoneurons following electrical stimulation of the prepositus hypoglossi nucleus. *Brain Res.,* 121: 157–161.

Baker, R., Delgado-Garcia, J. and Alley, K. (1977b) Morphological and physiological demonstration that prepositus hypoglossi neurons terminate on medial rectus motor neurons. *Proc. XXVII int. Congr. physiol. Sci., Paris,* XIII: 46, no. 119.

Blanks, R. H. I., Volkind, R., Precht, W. and Baker, R. (1977) Responses of cat prepositus hypoglossi neurons to horizontal angular acceleration. *Neuroscience,* 2: 391–403.

Briggs, T. L. and Kaelber, W. W. (1971) Efferent fibre connections of the dorsal and deep tegmental nuclei of Gudden. An experimental study in the cat. *Brain Res.,* 29: 17–29.

Büttner-Ennever, J. A. (1977) Pathways from the pontine reticular formation to structures controlling horizontal and vertical eye movements in the monkey. In *Control of Gaze by Brain Stem Neurons,* R. Baker and A. Berthoz (Eds.). Elsevier/North-Holland Biomed. Press, Amsterdam-New York, pp. 89–98.

Büttner-Ennever, J. A. and Henn, V. (1976) An autoradiographic study of the pathways from the pontine reticular formation involved in horizontal eye movements. *Brain Res.,* 108: 155–164.

Büttner-Ennever, J. A., Miles, T. S. and Henn, V. (1975) Role of the pontine reticular formation in oculomotor function. *Exp. Brain Res.,* Suppl. 23: 31, no. 58.

Cajal, S. R. y (1909/1911) *Histologie du Système Nerveux de l'Homme et des Vertébrés.* Maloine, Paris.

Cohen, B. and Henn, V. (1972) The origin of quick phases of nystagmus in the horizontal plane. *Bibl. ophthal., (Basel),* 82: 36–55.

Duensing, F. und Schaefer, K.P. (1960) Die Aktivität einzelner Neurone der Formatio reticularis des nicht gefesselten Kaninchens bei Kopfwendungen und vestibulären Reizen. *Arch. Psychiat. Nervenkr.,* 201: 97–122.

Edwards, S. B. (1975) Autoradiographic studies of the projections of the midbrain reticular formation: descending projections of nucleus cuneiformis. *J. comp. Neurol.,* 161: 341–358.

Gacek, R. R. (1971) Anatomical demonstration of the vestibulo-ocular projections in the cat. *Acta otolaryngol. (Stockh.),* Suppl. 293: 1–63.

Gacek, R. R. (1978) Location of commissural neurons in the vestibular nuclei of the cat. *Exp. Neurol.,* 59: 479–491.

Gernandt, B. E., Iranyi, M. and Livingston, R. B. (1959) Vestibular influences in spinal mechanisms. *Exp. Neurol.,* 1: 248–273.

Goebel, H.H., Komatsuzaki, A., Bender, M.B. and Cohen, B. (1971) Lesions of the pontine tegmentum and conjugate gaze paralysis. *Arch. Neurol. (Chic.),* 24: 431–440.

Grantyn, A., Grantyn, R. and Robiné, K.-P. (1977) Neuronal organization of the tecto-oculomotor pathways. In *Control of Gaze by Brain Stem Neurons,* R. Baker and A. Berthoz (Eds.), Elsevier/North-Holland Biomed. Press, Amsterdam-New York, pp. 197–206.

Graybiel, A.M. (1974) Some afferent connections of the oculomotor complex in the cat. *Soc. Neurosci., IV Annual Meeting,* 253.

Graybiel, A. M. (1975) Anatomical pathways in the brain stem oculomotor system. In *Eye Movements and Movement Perception, Ninth Symposium of the Center for Visual Science, Rochester, N.Y.,* pp. 37–38.

Graybiel, A.M. (1977a) Direct and indirect preoculomotor pathways of the brainstem: an autoradiographic study of the pontine reticular formation in the cat. *J. comp. Neurol.,* 175: 37–78.

Graybiel, A. M. (1977b) Organization of oculomotor pathways in the cat and rhesus monkey. In *Control of Gaze By Brain Stem Neurons,* R. Baker and A. Berthoz (Eds.), Elsevier/North-Holland Biomed. Press, Amsterdam-New York, pp. 79–88.

Graybiel, A. M. and Devor, M. (1974) A microelectrophoretic delivery technique for use with horseradish peroxidase. *Brain Res.,* 68: 167–173.

Graybiel, A. M. and Hartwieg, E. A. (1974) Some afferent connections of the oculomotor complex in the cat: an experimental study with tracer techniques. *Brain Res.,* 81: 543–551.

Gresty, M. and Baker, R. (1976) Neurons with visual receptive field, eye movement and neck displacement sensitivity within and around the nucleus prepositus hypoglossi in the alert cat. *Exp. Brain Res.,* 24: 429–433.

Hauglie-Hanssen, E. (1968) Intrinsic neuronal organization of the vestibular nuclear complex in the cat. A Golgy study. *Ergebn. Anat. Entwickl.-Gesch.,* 40: 1–105.

Hellstrøm Hoddevik, G., Brodal, A. and Walberg, F. (1975) The reticulo-vestibular projection in the cat. An experimental study with silver impregnation methods. *Brain Res.,* 94: 383–399.

Henn, V. and Cohen, B. (1975) Activity in eye muscle motoneurons and brainstem units during eye movements. In *Basic Mechanisms of Ocular Motility and their Clinical Implications*, G. Lennerstrand and P. Bach-y-Rita (Eds.), Pergamon Press, Oxford, pp. 303–345.
Highstein, S. M. (1971) Organization of the inhibitory and excitatory vestibulo-ocular reflex pathways to the third and fourth nuclei in rabbit. *Brain Res.*, 32: 218–224.
Highstein, S. M. (1973) Synaptic linkage in the vestibulo-ocular and cerebellovestibular pathways to the VIth nucleus in the rabbit. *Exp. Brain Res.*, 17: 301–314.
Highstein, S. M., Cohen, B., Matsunami, K. (1974) Monosynaptic projections from the pontine reticular formation to the IIIrd nucleus in the cat. *Brain Res.*, 75: 340–344.
Highstein, S. M., Maekawa, K., Steinacker, A. and Cohen, B. (1976) Synaptic input from the pontine reticular nuclei to abducens motoneurons and internuclear neurons in the cat. *Brain Res.*, 112: 162–167.
Hikosaka, O. and Kawakami, T. (1977) Inhibitory reticular neurons related to the quick phase of vestibular nystagmus. Their location and projection. *Exp. Brain Res.*, 27: 377–396.
Hobson, J. A., McCarley, R. W., Freedman, R. and Pivik, R. T. (1974a) Time course of discharge rate changes by cat pontine brain stem neurons during sleep cycle. *J. Neurophysiol.*, 37: 1297–1309.
Hobson, J.A., McCarley, R.W., Pivik, R.T. and Freedman, R. (1974b) Selective firing by cat pontine brain stem neurons in desynchronized sleep. *J. Neurophysiol.*, 37: 497–511.
Hoshino, K. and Pompeiano, O. (1977) Responses of lateral vestibular neurons to stimulation of contralateral macular labyrinthine receptors. *Arch. ital. Biol.*, 115: 237–261.
Hoshino, K., Pompeiano, O., Magherini, P. L. and Mergner, T. (1976) The oscillatory system responsible for the oculomotor activity during the bursts of REM. *Arch. ital. Biol.*, 114: 278–309.
Kaneko, C. R. S., Steinacker, A., Cohen, B., Maciewicz, R. and Highstein, S. M. (1975) Synaptic linkage of the reticulo-ocular pathway in cat. *Neurosci. Abst.*, I: 225, no. 349.
Keefer, D. A. and Christ, J. F. (1976) Distribution of endogenous diamino-benzidine-staining cells in the normal rat brain. *Brain Res.*, 116: 312–316.
Ladpli, R. and Brodal, A. (1968) Experimental studies of commissural and reticular formation projections from the vestibular nuclei in the cat. *Brain Res.*, 8: 65–96.
Leontovich, T. A. and Zhukova, G. P. (1963) The specificity of the neuronal structure and topography of the reticular formation in the brain and spinal cord of carnivora. *J. comp. Neurol.*, 21: 347–379.
Lorente de Nó, R. (1933) Vestibulo-ocular reflex arc. *Arch. Neurol. Psychiat. (Chic.)*, 30: 245–291.
Maciewicz, R. J., Eagen, K., Kaneko, C. R. S. and Highstein, S. M. (1977) Vestibular and medullary brain stem afferents to the abducens nucleus in the cat. *Brain Res.*, 123: 229–240.
McCarley, R. W. and Hobson, J. A. (1975) Discharge patterns of cat pontine brain stem neurons during desynchronized sleep. *J. Neurophysiol.*, 38: 751–766.
McMasters, R. E., Weiss, A. H. and Carpenter, M. B. (1966) Vestibular projections to the nuclei of the extraocular muscles. Degeneration resulting from discrete partial lesion of the vestibular nuclei in the monkey. *Amer. J. Anat.*, 118: 163–194.
Mehler, W. R. (1968) Reticulovestibular connections compared with spino-vestibular connections in the rat. *Anat. Rec.*, 160: 485, abstr.
Mergner, T. and Pompeiano, O. (1978) Single unit firing patterns in the vestibular nuclei related to saccadic eye movements in the decerebrate cat. *Arch. ital. Biol.*, 116: 91–119.
Mergner, T., Pompeiano, O. and Corvaja, N. (1977) Vestibular projections to the nucleus intercalatus of Staderini mapped by retrograde transport of horseradish peroxidase. *Neurosci. Lett.*, 5: 309–313.
Peterson, B. W. and Abzug, C. (1975) Properties of projections from vestibular nuclei to medial reticular formation in the cat. *J. Neurophysiol.*, 38: 1421–1435.
Peterson, B. W., Filion, M., Felpel, L. P. and Abzug, C. (1975) Responses of medial reticular neurons to stimulation of the vestibular nerve. *Exp. Brain Res.*, 22: 335–350.
Pivik, R. T., McCarley, R. W. and Hobson, J. A. (1977) Eye movement-associated discharge in brain stem neurons during desynchronized sleep. *Brain Res.*, 121: 59–76.
Pompeiano, O. (1972) Reticular control of the vestibular nuclei. Physiology and pharmacology. In *Progress in Brain Research, Vol. 37, Basic Aspects of Central Vestibular Mechanisms*, A. Brodal and O. Pompeiano (Eds.). Elsevier, Amsterdam, pp. 601–618.
Pompeiano, O., Mergner, T. and Corvaja, N. (1978) Commissural, perihypoglossal and reticular afferent projections to the vestibular nuclei in the cat. An experimental anatomical study with the method of the retrograde transport of horseradish peroxidase. *Arch. ital. Biol.*, 116: 130–172.

Pompeiano, O., Mergner, T. and Corvaja, N. (1979) Commissural connections between the vestibular nuclei studied with the method of retrograde transport of horseradish peroxidase. In *Structure and Function of the Cerebral Commissures*, I. Steele Russell, M. W. van Hof and G. Berlucchi (Eds.), Macmillan, New York, pp. 319–332.

Pompeiano, O. and Walberg, F. (1957) Descending connections to the vestibular nuclei. An experimental study in the cat. *J. comp. Neurol.*, 108: 465–503.

Precht, W. (1975) Vestibular system. In *MTP International Review of Science, Physiology Series One, Vol. 3, Neurophysiology*, A. C. Guyton and C. C. Hunt (Eds.). Butterworths, London, pp. 81–149.

Remmel, R. S. and Skinner, R. D. (1976) An electrophysiological study of cat reticulovestibular neurons. *Neurosci. Abstr.*, II, 1: 281, no. 403.

Remmel, R. S., Skinner, R. D. and Pola, J. (1977a) Cat pontomedullary reticular neurons projecting to the regions of the ascending MLF and the vestibular nuclei. In *Control of Gaze by Brain Stem Neurons*, R. Baker and A. Berthoz, (Eds.), Elsevier/North-Holland Biomed. Press, Amsterdam-New York, pp. 163–166.

Remmel, R. S., Skinner, R. D., Yelvington, D. B. and Sadikin, L. (1977b) Cat reticular neurons projecting to the region of the medial vestibular nucleus. *Neurosci. Lett.*, 6: 223–229.

Scheibel, M. E. and Scheibel, A. B. (1958) Structural substrates for integrative patterns in the brain stem reticular core. In *Reticular Formation of the Brain. Henry Ford Hosp. Int. Symp.*, H.H. Jasper, L.D. Proctor, R.S. Knighton, W.C. Noshay and R.T. Costello (Eds.). Little, Brown and Co., Boston, MA, pp. 31–55.

Shimazu, H. (1972) Organization of the commissural connections: physiology. In *Progress in Brain Research, Vol. 37, Basic Aspects of Central Vestibular Mechanisms*. A. Brodal and O. Pompeiano (Eds.), Elsevier/North-Holland Biomed. Press, Amsterdam, pp. 177–190.

Shimazu, H. (1977) Brain stem interneurons modulating abducens motor activity during vestibular nystagmus. In *Control of Gaze by Brain Stem Neurons*, R. Baker and A. Berthoz (Eds.), Elsevier/North-Holland Biomed. Press, Amsterdam, pp. 225–233.

Spyer, K. M., Ghelarducci, B. and Pompeiano, O. (1974) Gravity responses of neurons in the main reticular formation. *J. Neurophysiol.*, 37: 705–721.

Tarlov, E. (1970) Organization of the vestibulo-oculomotor projections in the cat. *Brain Res.*, 20: 159–179.

Walberg, F. (1974) Crossed reticuloreticular projections in the medulla, pons and mesencephalon. An autoradiographic study in the cat. *Z. Anat. Entwickl. Gesch.*, 143: 127–134.

Wilson, V. J., Kato, M., Thomas, R. C. and Peterson, B. W. (1966) Excitation of lateral vestibular neurons by peripheral afferent fibers. *J. Neurophysiol.*, 29: 508–529.

Wong-Riley, M. T. T. (1976) Endogenous peroxidatic activity in brain stem neurons as demonstrated by their staining with diaminobenzidine in normal squirrel monkeys. *Brain Res.*, 108: 257–277.

Neuronal Activity Preceding Rapid Eye Movements in the Brain Stem of the Alert Monkey

K. HEPP and V. HENN

Theoretische Physik, Eidgenössische Technische Hochschule, 8093 Zürich; and Neurologische Klinik, Universität, 8091 Zürich (Switzerland)

INTRODUCTION

The motor control system of higher animals appears to have both a hierarchical structure and the characteristics of parallel information processing with sensory and internal feedback. The long-term aim of our investigation is to work out this structure for the oculomotor system of the monkey, where saccadic eye movement neurons can be easily identified at many levels in the central nervous system, and quantitatively related to the few degrees of freedom of the eye. In the past years we have (together with U. Büttner, B. Cohen, J. Jaeger, and T. S. Miles) recorded extracellularly from more than 1000 eye movement related neurons in the brain stem and cerebellum of awake untrained Rhesus monkeys (Henn and Cohen, 1976; Büttner et al., 1977). We have tested many of the neurons under several natural conditions, which include spontaneous eye movements in the light or in darkness, optokinetic or vestibular nystagmus, visual tracking (presentation of novel and edible objects), and different behavioral states (alertness, drowsiness, light sleep). We will give a systematic classification of eye movement related neurons from the brain stem which have been described in the literature with different terminologies (Sparks and Travis, 1971; Luschei and Fuchs, 1972; Keller, 1974; King and Fuchs, 1977). Then we shall discuss a simple flow model for the causal relations between these neuron populations, and we shall speculate about higher order control mechanisms of eye movements.

METHODS

Rhesus monkeys (*Macaca mulatta*) were chronically prepared with implanted electrodes to monitor eye position, bolts to fixate the head, and a cylinder for the microdrive over a trephine hole in the skull. Neurons were recorded extracellularly with varnished tungsten electrodes. Monkeys were placed on a turntable, which was enclosed by an optokinetic cylinder (cf. Waespe and Henn 1977). Relevant data were stored on a 7 channel FM tape machine for off-line analysis (cf. Henn and Cohen 1976).

Recordings have been taken from the motor nuclei, the reticular formation of the pons, mesencephalon, and medulla oblongata. The paramedian pontine reticular formation (PPRF), and the rostral interstitial nucleus of the medial longitudinal fasciculus (rostral iMLF) were thoroughly explored. In addition, recordings were made from the fastigial interpositus and dentate nucleus and overlying cerebellar cortical regions. All recording sites were checked histologically (cf. Cohen and Henn 1972; Büttner et al. 1977).

RESULTS

Motoneurons and the immediate premotor system

The *motoneurons* (MN) as the final common pathway of the oculomotor system give important criteria on the functional relation of central neuronal firing patterns and motor output. Their discharge characteristics have been described extensively in the literature (Fuchs and Luschei 1970; Robinson 1970; Henn and Cohen 1973). For eye movements not too far from the primary position conjugate eye movements are completely characterized by the horizontal and vertical angular coordinates of the axis of gaze. The pulling directions of the motoneurons of the six eye muscles are horizontal and about $\pm 10°$ off-vertical. In the awake monkey, the orthogonal projection of the position of the eye on the pulling direction of the muscle, $p(t)$, is related to the firing frequency, $f(t)$, of the motoneuron approximately by $f(t) = a(p(t) - p_o) + b \dot{p}(t)$ (with the left side of the equation zero, if the right side is negative). The motoneurons range between *phasic* and *tonic* units, depending on the relative strength of the velocity component and on the threshold p_o. Tonic motoneurons should contribute more to the holding of the eye, while the phasic ones are better for the fast and precise execution of saccades. For rapid acceleration and deceleration, phasic motoneurons have some additional, high frequency spikes at the beginning and an early fall-off of the burst at the end of the saccade. This firing pattern provides the rapid breaking which in the somatic motor system is affected by an activation of the antagonist. In motoneurons the inhibition during movement in the off-direction is symmetric to the excitation in the on-direction, both with an on-latency of -6 to -4 sec.

Burst tonic neurons (BT) occur in many families with different firing properties. The on-directions are the same as those of the MN, the firing is generally rougher, excitation and inhibition are not necessarily symmetrical, and late latency bursts can occur for saccades in the off-direction. In phasic BT, the burst frequency can exceed 1000 Hz, and the tonic activity can be a non-linear function of eye position.

Most of the *short lead bursters* (S) are completely inhibited except for a burst, starting at -10 to -4 msec before the saccade and stopping before the end of the saccade. We distinguish S-bursters by their movement field which can be either a direction or a sector in eye displacement space. A *directed* S-burster generally has the same on-direction as a motoneuron. Its firing pattern is often related to the velocity component of the saccade along the on-direction, with some additional spikes at the beginning and some missing spikes at the end. Such bursters usually give very few spikes for movements into the opposite hemifield, although more isotropic patterns exist. Hence the total number of spikes is proportional to the position change, and the mean frequency to the cosine of the angular difference relative to the on-direction. A *transient* S-burster shows a strong direction dependent high-frequency component at the beginning of the saccade (up to 1000 Hz for rarely longer than 40 msec) and a rapid decrease of firing at the end. The high-frequency burst is always present in small saccades, but can be much weaker for large saccades, e.g., during nystagmus. Tran-

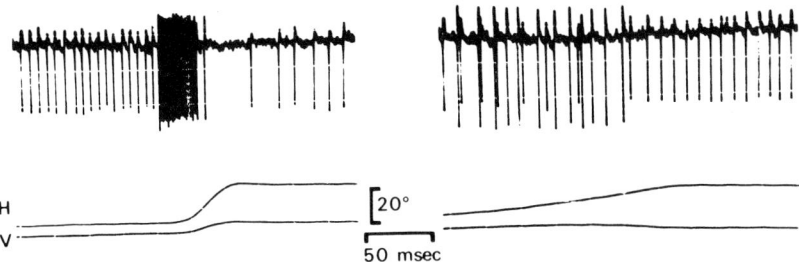

Fig. 1. Pause and burst neuron recorded simultaneously in the PPRF. From above: neuronal activity, horizontal and vertical eye position. On the left side the monkey was fully alert, and the 2 neurons behave in a reciprocal manner. On the right side the monkey is drowsy and makes very slow saccades, and the burst frequency is reduced at the same time as the incomplete inhibition of the pause neuron.

sient S-bursters often code the direction of the saccade better than the position change. *Sectorial* S-bursters have a whole sector of on-directions. If this is a hemifield with an on-latency of −10 to −4 msec, then we shall call it a *trigger burster* (T). In the other important class of the *duration bursters* (D) the firing is isotropic in all directions. The on-latency can be late (∼0 msec) and often the end of large saccades is well coded, even if the eye is slowed down during a blink (see Fig. 5b).

Pause neurons (P) fire tonically with 100–150 Hz except for pauses during all saccades or at least into a hemifield. The firing of the S-bursters and the onset of the pauses are well synchronized in the awake animal, and in the drowsy state the activity of both P and S becomes simultaneously diffuse (see Fig. 1).

Long lead bursters (L) and *medium lead bursters* (M) differ from S-bursters by their irregular firing pattern and by a significant early increase of activity for saccades in the on-direction. A frequent type found in the PPRF is the *directed* L-burster, with horizontal and off-vertical on-directions. The early activity and the maximal frequency grow with the size of the saccade in the on-direction (see Fig. 2, lower part). Then only saccades with amplitudes larger than ∼10° are well coded. In the M-bursters the components of large and small saccades are equally well represented with a direction-coding high frequency burst at −15 to −8 msec and a less significant early activity. Other L-bursters in the PPRF are long lead *trigger* bursters as in Fig. 2 (upper part). L-bursters with a movement field (which does not lie outside of the physiological range of eye movements, as it does for the directed L-bursters) are rare in the PPRF, but they have been found in the MRF lateral to the oculomotor nucleus and, of course, in the superior colliculus. These neurons code, by their frequency, the *eye displacement* vector of the forthcoming saccade at a time when the S-bursters are still completely inhibited. We have never encountered neurons in the brain stem and cerebellum which at this early stage reliably code components of the desired *eye position*, as it would be required in Robinson's model (Van Gisbergen and Robinson, 1977).

Basic model

We propose the following "whisper-bang" model for the causal interdependence of these neuron populations (Fig. 3). The "whispering" early activity of the L-bursters with movement fields is the earliest manifestation of the central command of a saccade. In the reticular formation a recoding from movement to muscle coordinates occurs, where the directed L-bursters in the PPRF can incorporate the duration of

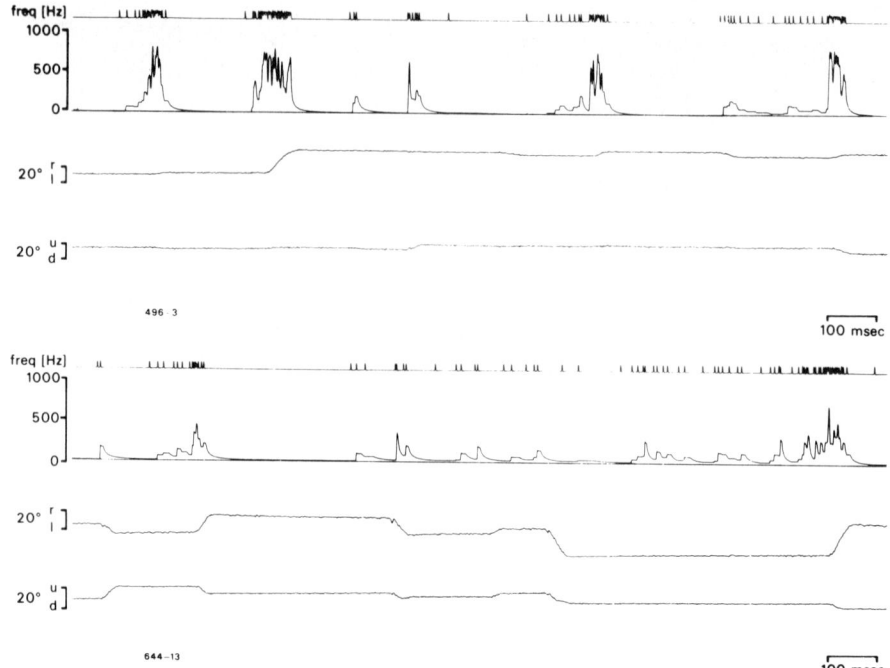

Fig. 2. Two examples of long lead burst neurons from the PPRF. From above: time marks to indicate occurrence of spike, instantaneous neuronal frequency, horizontal and vertical eye position. Upper parts: trigger burst, which reaches a similar frequency for every movement into the right hemifield independent of saccadic size. Lower parts: burst activity increases with saccade size to the right.

larger saccades as an intermediate stage from L-bursters with spatial and without temporal coding to the M-bursters with a burst pattern, which is appropriate to drive the S-bursters. After the integration of the eye displacement burst in a reverberatory network of BT-neurons (the "integrator") the MN receive their finely graded signal via internuclear neurons (IN). Phasic and tonic MN are probably fed by different S-

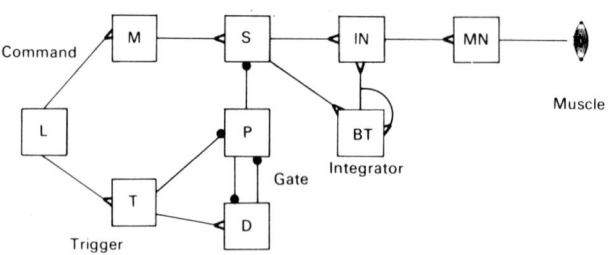

Fig. 3. Flow model of rapid eye movement generation. L, long lead burst neurons: M, medium lead burst neurons; S, short lead burst neurons; BT, burst-tonic neurons; IN, internuclear neurons; MN, motoneurons; D, duration burst neurons; P, pause neurons. As described in the text, the earliest information concerning a saccade is present in L. The immediate premotor organization consists of two parts. One pathway determines the amount of position change along the pulling direction of each eye muscle via L, M, and S neurons. In the late stage, this information requires an integration to obtain a position signal (integrator loop with BT, IN, and S neurons). The second pathway (D, P neurons) determines the exact onset and duration of saccades in all directions by providing a gate for the burst activity.

and BT-populations. The pausers form a "gate", which filters the early activity of the L-bursters. The latter inhibit the pausers by the trigger pathway, and the duration of the pause is controlled by the D-population.

There is good evidence for the qualitative validity of such a model: the simultaneous decay of the P- and S-profile during drowsiness (Fig. 1) is as impressive as the continuing activity of the T-population (which acts by disinhibition) when the drowsy monkey no longer makes any saccades. Keller (1977) has stimulated the P-region and could interrupt saccades in flight, which are continued for stimulations shorter than 40 msec (the approximate duration of high frequency). Another attractive feature of this model is that the smooth pursuit system can easily be incorporated by appropriate S-burster populations which code eye velocity also for tracking movements. Such neurons can be found in the periabducens region for the horizontal direction. The common integrator then provides the new holding position.

Some higher order control systems

The general problem for eye movement control seems to be the temporal coordination of different inputs in relation to motoneuron output. One question is, how the duration of saccades is controlled. The L-bursters with small movement fields and the transient S-bursters exhibit a pronounced activity during small saccades, which seems necessary in order to break the inhibition of the pausers. In order to be acceptable for the phasic MN, the duration of the burst has to be matched with the amplitude of the intended saccade. There are many types of *follow bursters* (F) in the reticular formation, which could inhibit the burst at the end of the saccade. Many F-bursters have some similarity to pausers with rebound excitation.

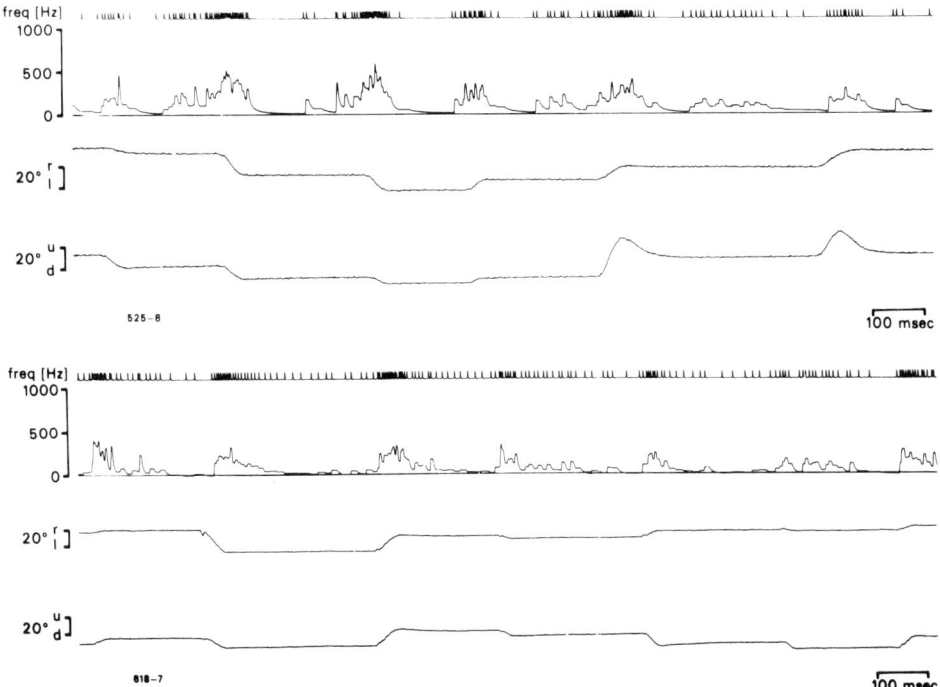

Fig. 4. Medium-lead burst neuron M and follow burst neuron F from the fastigial nucleus.

To test at what levels the cerebellum is involved in oculomotor control, we investigated neurons with eye movement related activity in the cerebellar nuclei (CNB).

Fig. 4 shows two neurons in the fastigial nucleus. The upper one is of medium lead type and the lower one is an F-burster. In our laboratory we have found a great variety of neurons in the CBN, whose firing patterns are consistent with the hypothesis that the cerebellum is active in a "bang-bang" control of visually evoked and rapid shifting saccades, coupled to the transient S-bursters and F-bursters in the brain stem. Consistent with such a modulating cerebellar loop is the observation that neurons, like those in Fig. 4, are sculptured by excitation and inhibition before, during and after the saccade. The cerebellum could receive information about the eye movement at the early planning stage from L-bursters with large and small movement fields, at the late stage through eye muscle receptors and vision, and at the immediate premotor stage through S- and BT- fibers.

A second interesting control structure is the coding of long saccadic duration and the coordination of the horizontal and vertical saccade components. A rough coordination can already be found in the incoming central command, since the directional L-bursters code the length of a saccade by the duration of the high-frequency burst. There is, however, evidence that the duration and direction of large saccades along oblique angles is another important oculomotor output of the cerebellar nuclei. Fig. 5 (lower part) shows a D-burst from the CBN, which has an earlier on-latency than the reticular D-burst in Fig. 5 (upper part) and a similar enhancement during blinks.

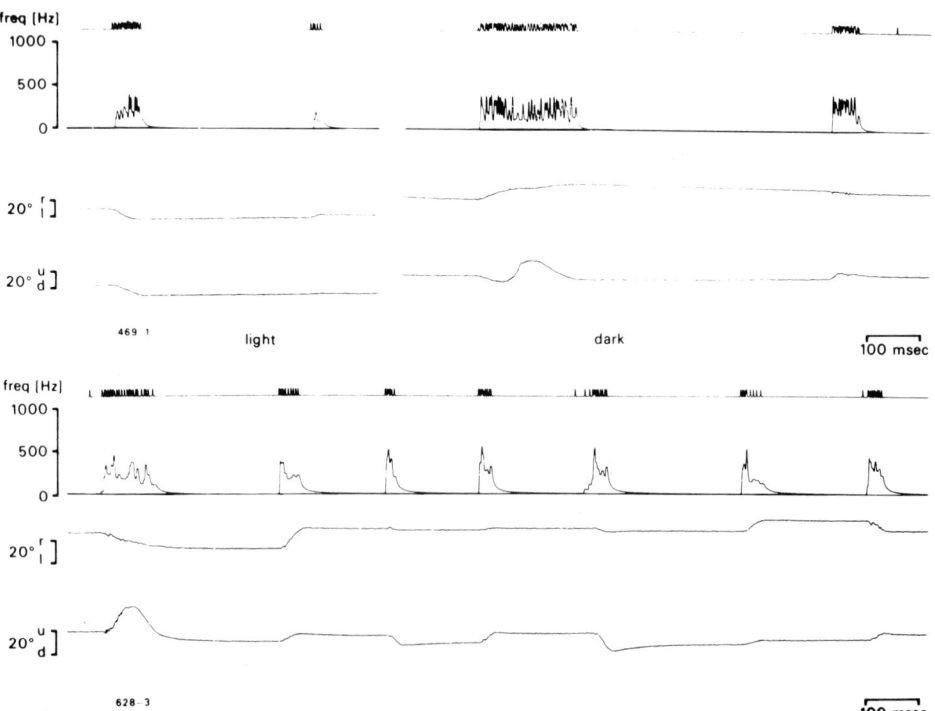

Fig. 5. Duration burst neuron D from rostral iMLF (upper parts) and cerebellum (lower parts). In both instances neurons are active during the whole length of the saccade, even if it is slowed down, when the monkey is in total darkness, or during blinks.

These few examples are shown here to emphasize that information about direction and duration of saccades is present in the cerebellum in the early planning stage of a saccade as well as during the time of its execution.

CONCLUSION

At present the theoretical interpretation of our experimental results is speculative and shows, like an archeological museum, only fragments of the functional organization of the oculomotor system. However, it is a fact that anatomically and physiologically identifiable neuron populations have their characteristic firing properties and can be consistently connected in a spatiotemporal flow diagram. A quantitative mathematical model has not even been found for the basic vestibulo-ocular reflex, and in this sense the oculomotor system in all its complexity is still far removed from a realistic computer simulation. However, the interpretation of the data of the past and the planning of interesting new experiments requires a detailed logical scheme for the transformations of signals in the central nervous system. "Neurologics" is an important intermediate step between "annotated anatomy" and mathematical modeling, and we are optimistic that this type of approach will reveal interesting details of the hierarchical multiloop structure of the central nervous system.

SUMMARY

A survey of neuronal populations involved in the generation and control of rapid eye movements is given. Neurons are grouped according to temporal relations (activity in relation to the onset of movement) and the spatial coding characteristics (position change, direction and size of eye displacement in retinal or eye muscle coordinates). A flow model based on the physiological characteristics of known neuron populations is introduced.

ACKNOWLEDGEMENTS

We would like to thank V. Isoviita for technical, J. Jaeger for experimental and C. Lasner for programming assistance. K. H. is grateful to Prof. G. Baumgartner for his kind hospitality in his laboratory.

REFERENCES

Büttner, U., Büttner-Ennever, J. A. and Henn, V. (1977) Vertical eye movement related unit activity in the rostral mesencephalic reticular formation of the alert monkey. *Brain Res.*, 130: 239–252.

Cohen, B. and Henn, V. (1972) Unit activity in the pontine reticular formation associated with eye movements. *Brain Res.*, 46: 403–410.

Fuchs, A. F. and Luschei, E. S. (1970) Firing patterns of abducens neurons of alert monkeys in relation to horizontal eye movements. *J. Neurophysiol.*, 33: 382–392.

Henn, V. and Cohen, B. (1973) Quantitative analysis of activity in eye muscle motoneurons during saccadic eye movements and positions of fixation. *J. Neurophysiol.*, 36: 115–126.

Henn, V. and Cohen, B. (1976) Coding of information about rapid eye movements in the pontine reticular formation of alert monkeys. *Brain Res.*, 108: 307–325.

Keller, E. L. (1974) Participation of medial pontine reticular formation in eye movement generation in monkey. *J. Neurophysiol.*, 37: 316–332.

Keller, E. L. (1977) Control of saccadic eye movements by midline brain stem neurons. In *Control of Gaze by Brain Stem Neurons*, R. Baker and A. Berthoz (Eds.), Elsevier, Amsterdam, pp. 327–336.

King, W. M. and Fuchs, A. F. (1977) Neuronal activity in the mesencephalon related to vertical eye movements. In *Control of Gaze by Brain Stem Neurons*, R. Baker and A. Berthoz (Eds.), Elsevier, Amsterdam, pp. 319–326.

Luschei, E. S. and Fuchs, A. F. (1972) Activity of brain stem neurons during eye movements of alert monkeys. *J. Neurophysiol.*, 35: 455–461.

Robinson, D. A. (1970) Oculomotor unit behavior in the monkey. *J. Neurophysiol.*, 33: 393–404.

Sparks, D. L. and Travis, R. P. (1971) Firing patterns of reticular formation neurons during horizontal eye movements. *Brain Res.*, 33: 477–481.

Van Gisbergen, J. A. M. and Robinson, D. A. (1977) Generation of micro- and macro-saccades by burst neurons in the monkey. In *Control of Gaze by Brain Stem Neurons*, R. Baker and A. Berthoz (Eds.) Elsevier, Amsterdam, pp. 301–308.

Waespe, W. and Henn, V. (1977) Neuronal activity in the vestibular nuclei of the alert monkey during vestibular and optokinetic stimulation. *Exp. Brain Res.*, 27: 523–538.

Afferent and Efferent Organization of the Prepositus Hypoglossi Nucleus

R. A. McCREA[1], R. BAKER[1] and J. DELGADO-GARCIA[2]

[1]*Department of Physiology and Biophysics, New York University Medical Center, New York, NY 10016 (U.S.A.); and* [2]*Departamento de Fisiologia, Facultad de Medicina, Seville (Spain)*

Since the first morphophysiological demonstration that the prepositus hypoglossi nucleus sends axons rostrally towards the oculomotor complex (Graybiel and Hartwieg, 1975; Baker and Berthoz, 1975), *every* new study in the past few years has supported the hypothesis that this nucleus is concerned with the control of gaze. When the older literature is viewed in retrospect, it is easy to find many studies, such as that by Graux (1878) which suggested, possibly for the first time, that this medullary area was involved with both horizontal and vertical gaze. There were many other corroborative studies (see Baker, 1977, for literature review); however, it was clear that the newer axonal tracing techniques coupled with physiological studies of identified neurons would have to be utilized in order to obtain a better idea of the functional role of the prepositus.

In the past year, we have employed a number of experimental techniques in order to study the anatomy and physiology of the perihypoglossi nuclei. The afferents to the prepositus have been investigated by injecting horseradish peroxidase (HRP) extracellularly into the nucleus and by intracellularly recording the responses of prepositus neurons following stimulation of the vestibular nerve, the cerebellum and various other brain stem sites. The efferent organization of the perihypoglossi nuclei has been studied by injections of tritiated leucine, by antidromic activation of neurons later identified by intracellular injections of HRP, and by injecting HRP extracellularly into other brain stem and cerebellar sites. These data have been supplemented by electrophysiological experiments designed to ascertain the synaptic effect of prepositus neurons on target neurons, especially in the oculomotor nuclei and cerebellum. The technique of intracellular HRP injection has also been utilized to characterize the morphology of the dendrites and initial trajectory (2–10 mm) of the axons of the various types of neurons in the prepositus and then to correlate the morphology of prepositus neurons with the electrophysiological responses evoked by vestibular, oculomotor or cerebellar stimulation. Finally, to further define the possible roles the prepositus may have during eye movement, single units have been recorded in the alert cat with the use of the magnetic search coil to measure eye movement (Lopez-Barneo et al., this volume, chapter VA5). Because of their detail, the results of the above studies will also be presented elsewhere. This paper will concentrate on providing a general outline of our findings regarding the afferent and efferent organization of the prepositus nucleus.

AFFERENTS TO THE PREPOSITUS

Extracellular injections of HRP into the prepositus nucleus (Fig. 1A) have revealed a number of possible sources of afferents to this nucleus. In these experiments, labeled cells have been found primarily in areas where eye or head movement related activity has been recorded. The vestibular nuclei are quantitatively one of the most important sources of input to the prepositus. Labeled cells are found bilaterally in all of the major subdivisions of the vestibular complex; however, the heaviest projections originated from neurons in the medial and descending subdivisions. Obviously, this finding agrees well with all physiological studies, employing electrical stimulation of the vestibular nerve (Baker and Berthoz, 1975, 1976; Fukushima et al., 1977, Hikosaka et al, 1978) and natural stimulation of the horizontal semicircular canals (Blanks et al., 1977; Fukushima et al., 1977, Hikosaka et al., 1978). All physiological studies suggest that the majority of prepositus neurons respond in a type II fashion (i.e., contralateral EPSP and ipsilateral IPSP). The vestibular input appears to be symmetrically distributed, which may indicate that response profiles are a result of synaptic connectivity, not density.

A second important input to the prepositus originates from neurons located in the reticular formation. Labeled neurons are found bilaterally throughout the reticular formation extending from the caudal medulla to diencephalic levels. The heaviest concentrations of labeled cells are found ipsilaterally in the caudal medullary reticular formation, the paramedian pontine reticular formation, the rostral interstitial nucleus of the MLF and the fields of Forel. Interestingly, these projections were quite difficult to demonstrate when diaminobenzidine (DAB) was used as a reaction substrate. Only when tetramethylbenzine was used as the substrate (DeOlmos et al., 1978) were large numbers of neurons in these areas labeled (Fig. 1B). These findings corroborate the autoradiographic studies of Graybiel (1977) and Büttner-Ennever and Henn (1976), which demonstrated projections from pontine reticular areas to the ipsilateral prepositus area. The visual cortex-superior colliculus-reticular-prepositus circuit is obviously well developed and therefore it may underlie part of the visual-motor role the prepositus appears to play (Baker, 1977; Lopez-Barneo et al., chapter VA5).

One of the most intriguing and significant projections to the prepositus emanates from the oculomotor and the accessory oculomotor nuclei. Neurons in and around each of the 12 ocular nuclei are labeled following HRP injections in the prepositus nucleus (McCrea and Baker, 1978). Especially noticeable are neurons in the central brain stem gray overlying the oculomotor complex (Edwards and Henkel, 1978) as well as a considerable number of cells situated rostrally in the interstitial nucleus of Cajal (Fig. 1D) and the rostral interstitial nucleus of the MLF (Büttner-Ennever, 1978). Stimulation of these areas evokes antidromic responses and monosynaptic EPSPs in many prepositus neurons. It is likely that these reciprocal circuits are of considerable importance in coordinating both horizontal and vertical eye movement.

Our retrograde HRP experiments confirm and extend previous observations (Walberg, 1961; Angaut and Brodal, 1967; Alley, 1977) concerning projections from the

---

Fig. 1. Photomicrographs of neurons retrogradely labeled following HRP injection into the prepositus nucleus. A) Injection site in the prepositus nucleus. B) Neurons in the rostral pontis caudalis area of the ipsilateral reticular formation. C) Neurons in the lateral part of the contralateral prepositus nucleus. D) Cells in the interstitial nucleus of Cajal. Calibrations are 1 mm in A and 0.1 mm for B–D.

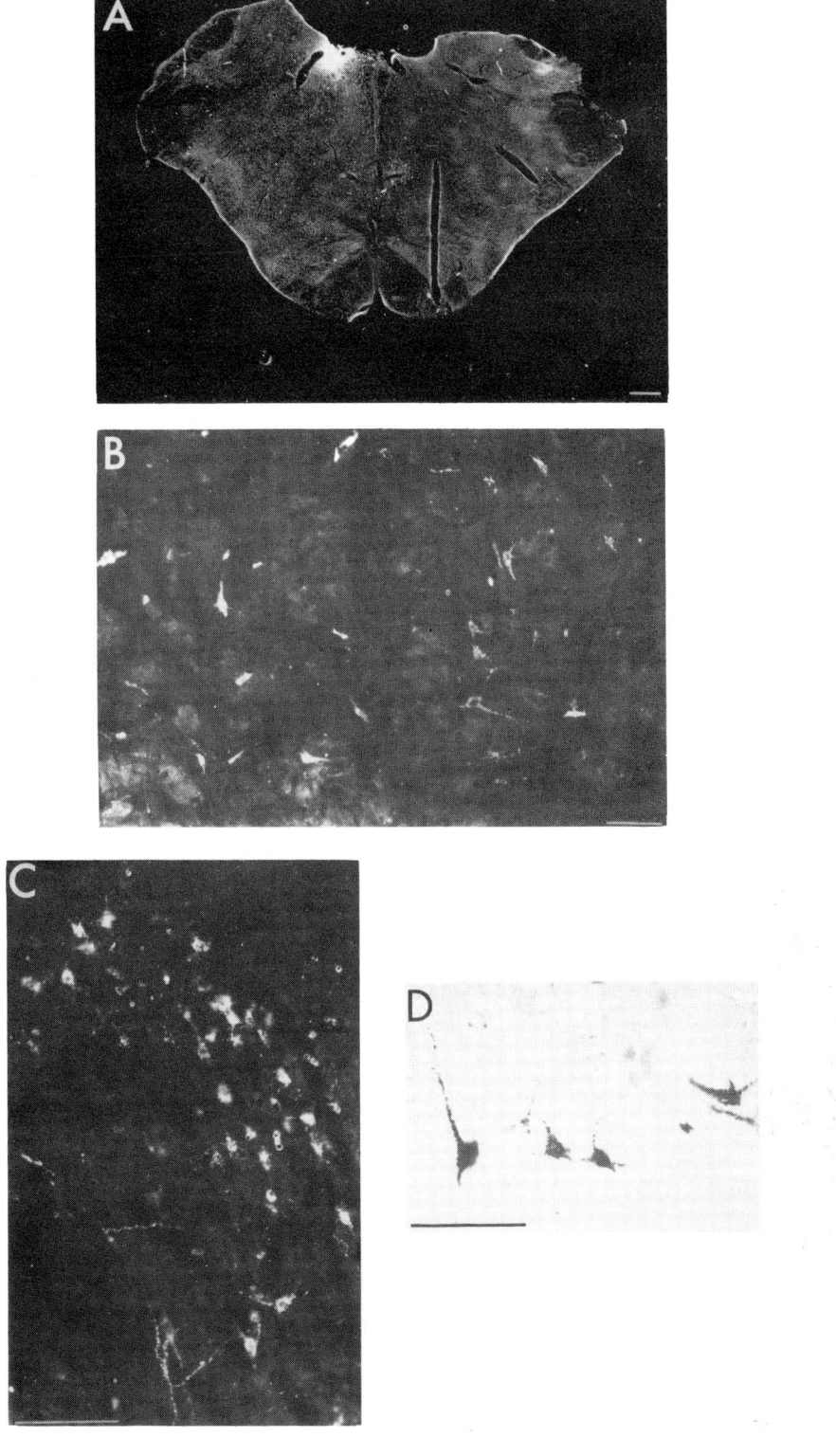

cerebellum to the prepositus. As expected, Purkinje cells in the ipsilateral cerebellar cortex were labeled following prepositus injections. The largest concentration of labeled Purkinje cells was found in the flocculus, but many labeled cells were also found in the dorsal and ventral paraflocculus (Fig. 2A) and in Crus I. This latter finding was most surprising and has been observed in four experiments. The nodulus and uvula were not available for study in recent experiments, but extrapolating from the work of Angaut and Brodal (1967), we would anticipate a strong projection from these areas. The fastigial nuclei were also bilaterally labeled following even very localized prepositus injections, thereby substantiating a sizeable efferent cerebello-prepositus connection (Walberg 1961). Clearly, there is a convergence within the prepositus of two cerebellar circuits (vermis and flocculus) intimately involved with all aspects of eye movement.

It is significant that the densest labeling following HRP injections in the prepositus was in the contralateral prepositus nucleus (Fig. 1C). Many labeled cells were also found bilaterally in the nucleus intercalatus of Staderini and the nucleus of Roller (Pompeiano et al., 1978). Conversely, when ^{3}H-leucine was injected into the prepositus, heavy labeling was observed in the contralateral prepositus, and bilaterally in the intercalatus and Roller nuclei. The magnitude of these intrinsic perihypoglossal connections is reminiscent of the massive commissural connections between the bilateral vestibular nuclei (Shimazu and Precht, 1966; Pompeiano et al., 1978) and

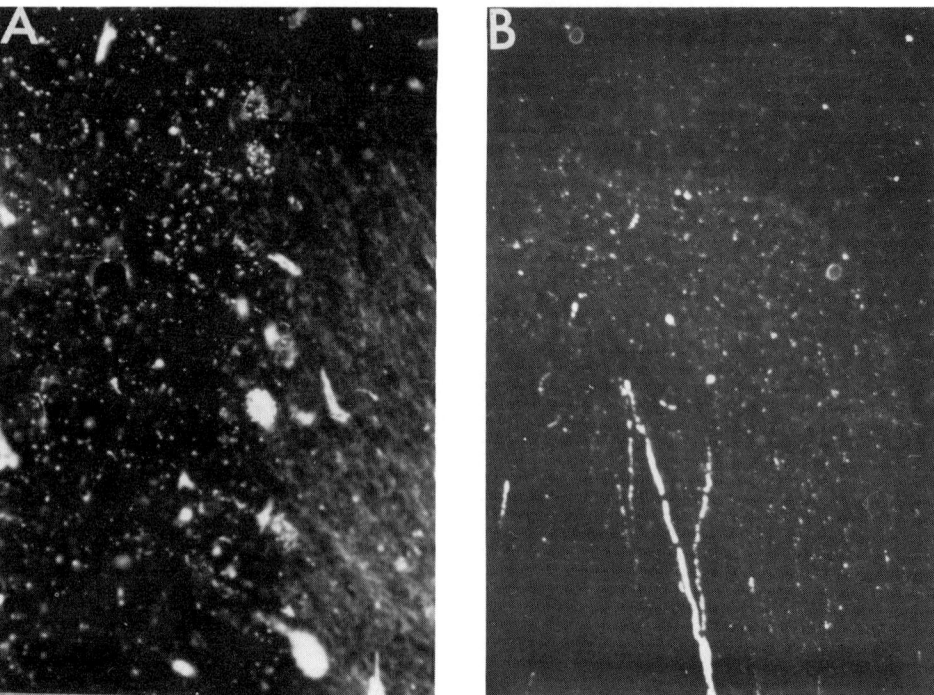

Fig. 2. Afferent and efferent connections of the prepositus nucleus with the cerebellar cortex. A) A row of Purkinje cells in the dorsal paraflocculus following an injection of HRP into the prepositus. B) Axons of prepositus neurons entering a vermal folium to terminate as mossy fibers in the granule layer. The latter was obtained following an injection of ^{3}H-leucine into prepositus nucleus. Calibration bars are 0.2 mm in both cases.

may serve roles similar to those suggested for vestibular commissural connections (Precht, 1974).

Finally, several other areas consistently contained a few labeled cells following prepositus HRP injections. They include the pretectal nuclei, the superior colliculus, the cervical spinal cord and the spinal trigeminal nucleus. The complete results of the extracellular HRP experiments will be amplified upon in the papers to follow on this subject.

EFFERENT ORGANIZATION

Torvik and Brodal (1954) first demonstrated that the prepositus nucleus sent axons towards the cerebellum. In recent years, many new studies utilizing retrograde axonal transport of HRP have commented on various aspects of this projection (Alley et al., 1975, 1976; Eller and Chan-Palay, 1976; Frankfurter et al., 1977; Kotchabhadki, 1977; McCrea et al., 1977; Ruggiero et al., 1977; Kotchabhadki et al., 1978). The most lengthy study is that of Kotchabhadki and Walberg (1978) which in summary suggests that the prepositus projects to every vermal lobule, the intermediate and lateral parts of the anterior lobe, the paraflocculus and the flocculus. The fastigial and interpositus nuclei appear to receive a projection from the prepositus (McCrea et al., 1977; Kotchabhadki and Walberg, 1978). Eller and Chan-Palay (1976) found labeled cells in the prepositus following HRP injections in the lateral nucleus of the rat, but there is no evidence that this projection exists in the cat (Bishop et al., 1976, Kotchabhadki and Walberg, 1978). If any conclusion stands out at present regarding the topography of the cerebellar projections from the prepositus, it would be simply that the prepositus tends to be associated with areas of the cerebellum important in eye movement, i.e., the posterior vermis, the nodulus, uvula and flocculus.

Thus far, we have the results of six experiments in which ^{3}H-leucine was injected into the prepositus. In no case was the injection completely confined to the nucleus, thus the results discussed below are somewhat preliminary in nature, but in most cases they have been corroborated by intracellular and extracellular HRP injections. Axons leave prepositus in several directions. Fig. 3 is a schematic diagram of the initial trajectory of prepositus axons as determined from intracellular HRP and autoradiographic studies. A surprising finding was that a considerable percentage of prepositus neurons have axons which descend on the midline (Figs. 3 and 4), traverse the pyramid, inferior olive or trapezoid body and ascend to the ipsi- or contralateral inferior cerebellar peduncle as external arcuate fibers. Our intracellular HRP data suggest that the neurons contributing to this pathway are the larger multipolar neurons in the ventral part of the prepositus. A second group of axons establish connections with the vestibular nuclei bilaterally before entering the ipsi- or contralateral inferior cerebellar peduncle (Fig. 3, 1 and 3). Another group of axons (Fig. 3, 1) crosses the midline more dorsally, establishes commissural connections with the contralateral prepositus and intercalatus and ascends toward the contralateral abducens nucleus. It appears that the large multipolar neurons in the central and ventral areas of the prepositus project exclusively to the cerebellum while the smaller reticular shaped neurons (Fig. 4) in the lateral and medial most parts of the nucleus appear to be responsible both for commissural connections with the contralateral prepositus and for the rostral projections of this nucleus. Although we have not

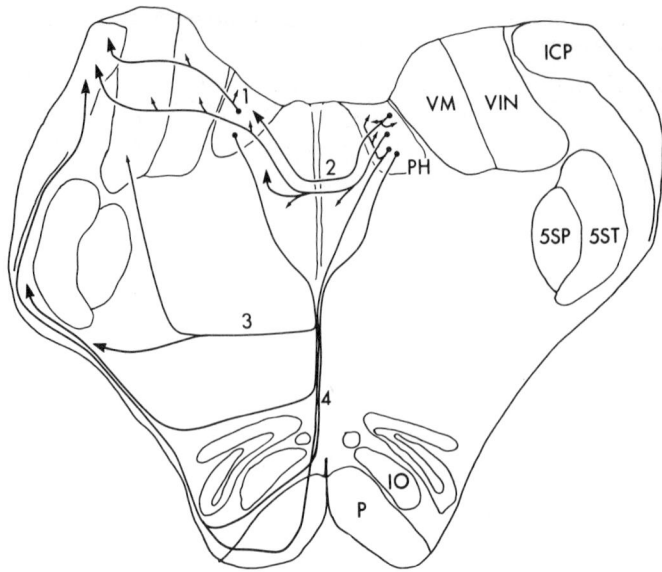

Fig. 3. Schematic illustration of the initial trajectory of axons exiting the prepositus nucleus as ascertained from intracellular HRP and autoradiographic experiments. To assist clarity, most of the pathways are illustrated only on the right side of the diagram. Pathways on the left side would be nearly identical. Up to now, evidence obtained from intracellular HRP experiments indicates that pathways 1–4 are each formed from morphologically identifiable classes of prepositus neurons. The text contains further explanation of the cell types.

analysed the cerebellar projections from the prepositus in detail, the results of extracellular HRP and autoradiographic experiments indicate that the prepositus projects bilaterally to the cortex of the cerebellar vermis, flocculus and paraflocculus as mossy fibers (Fig. 2B), and to the interposed and fastigial nuclei. Based on the magnitude of the latter projection, the prepositus could be considered to be largely a precerebellar structure which is involved both in the circuitry of the vermis-fastigial-reticular loop and the floccular-vestibular-prepositus pathway. A combination of the above areas is not represented as clearly in any other brain stem structure as it is in this nucleus.

One of the largest efferent projections of the prepositus within the brain stem is to the bilateral vestibular nuclei. Our autoradiographic studies confirm the observations of Pompeiano et al. (1978) that the prepositus projects bilaterally to the medial and descending vestibular nuclei. The results of these experiments also indicate that there is a small bilateral projection from neurons in the area of prepositus to parts of the superior and lateral vestibular nuclei. The trajectory taken by prepositus axons toward the medial and descending nuclei is indicated in Figs. 3 and 4. A drawing of a reconstructed prepositus neuron whose axon belongs to the second group of cells indicated in Fig. 3 is shown in Fig. 4. Prepositus neurons whose axons cross the midline dorsally and project toward the abducens nucleus like the one illustrated in Fig. 4, typically have relatively sparsely branched dendritic trees, which often are studded by many long pedunculated spines. The axons of these cells give off many collaterals, some of which terminate in the ipsi- or contralateral dorsomedial reticular formation and

Fig. 4. Reconstruction of a prepositus neuron intracellularly injected with HRP. The upper part of the figure shows the location and distribution of the somadendritic tree in respect to the boundaries of the prepositus nucleus. Note the sparsely branched dendritic tree and numerous spines. The bottom inset shows a reconstruction of part of the initial course of the axon. In this cell, the parent axon gave rise to at least 10 collaterals in the medulla; however, the concentration of HRP in the axon was not sufficient to follow all of the collaterals to termination. At node 1, the axon gave rise to two fine collaterals which terminated in the ipsilateral dorsal medial reticular formation. After crossing the midline, the axon splits into two major branches and two fine collaterals at node 2. The first major branch coursed rostrally toward the abducens nucleus (rostral branch) giving off several branches in its course. Just caudal to the abducens nucleus, the rostral branch produced several fine collaterals before coursing dorsally, laterally and rostrally through the rostral part of the MVN and the LVN at which point it could no longer be followed. The second major collateral originating at node 2 (caudal branch) coursed dorsally and laterally sending a collateral to the dorsal medial reticular formation (4). The caudal branch entered the contralateral PH giving off a fine collateral (5) in the lateral aspect of the nucleus which appeared to terminate in the MVN. From this point, the caudal branch coursed caudally and laterally through the MVN to the DVN at which point it could no longer be followed. The large arrow near the cell soma in the upper part of the diagram indicates the axon hillock. Ventrally the course of the axon is indicated by a small case a. The small arrows indicate axonal branching points.

others which either pass through or terminate in the contralateral prepositus and vestibular nuclei (more specific details are given in the legend of Fig. 4). In our autoradiographic studies, axons with similar trajectories project to the cerebellum; thus, it is likely that both of the main branches of the axon of the neuron illustrated in Fig. 4 continue on to the contralateral cerebellum. It is probable that a single prepositus neuron can simultaneously influence activity in a number of brain stem loci, including several of the vestibular nuclei, the reticular formation and the cerebellum. The synaptic effect(s) of these connections have not yet been studied electrophysiologically, but it is clear that the prepositus nucleus plays an influential role in determining the activity of neurons in the vestibular nuclei, both directly and indirectly via the cerebellum.

From ^{3}H-leucine experiments, it appears that axons of neurons in the vicinity of the prepositus project bilaterally and rostrally in a position ventral and lateral to the MLF. As these axons ascend toward the oculomotor complex, they appear to terminate in the abducens nucleus, the paramedian pontine reticular formation and the trochlear nucleus. The topographical projection from the prepositus to the pontine reticular formation is still being studied, but this pathway is another example of the reciprocal connectivity that the prepositus establishes with the brain stem areas which send afferents to it.

The parabigeminal nucleus appears to be one exception. Just caudal to the oculomotor complex, a group of fibers course ventrally, laterally and then dorsally to terminate in the periphery of the parabigeminal nucleus. Baleydier and Magnin (1978) found that prepositus neurons were labeled following HRP injections in the parabigeminal nucleus. Graybiel (1977) and Sherk (1977) have provided morphophysiological evidence for a reciprocal parabigeminal-collicular pathway which now includes an equally distinct perihypoglossal connection. This above combination suggests yet another interesting circuit in addition to the reticuloprepositus one involving the superior colliculus, namely, the prepositus-parabigeminal-superior colliculus pathway. Physiologically, the above connection directly introduces eye movement activity into the early levels of visual processing. As expected, prepositus neurons appear to terminate in the accessory oculomotor areas, specifically, the rostral interstitial nucleus of the MLF and the interstitial nucleus of Cajal. However, an interesting story seems to be developing regarding the efferent connections of the prepositus nucleus with the oculomotor nuclei. A number of recent morphophysiological studies are at variance concerning this important question and the remainder of this paper addresses this topic.

Following the physiological demonstration that the prepositus neurons exhibited a strong disynaptic vestibular input and the axons of some such cells were directed toward the oculomotor complex (Baker and Berthoz, 1975), a series of acute and chronic experiments were carried out to record intracellularly from vertical oculomotor neurons within the oculomotor complex to determine whether or not monosynaptic connections between these two structures existed (Baker and Berthoz, 1976). After exclusion of most axon collaterals (at least definitely those from the vestibular nuclei), it was concluded that a monosynaptic EPSP could be found in all vertical oculomotor neurons with a synaptic weighting towards the upward eye movers. For instance, large EPSPs were recorded from superior rectus motoneurons even when the interstitial nucleus of Cajal was lesioned (Fig. 5C). The monosynaptic excitatory connection was most convincingly shown in the trochlear motoneurons

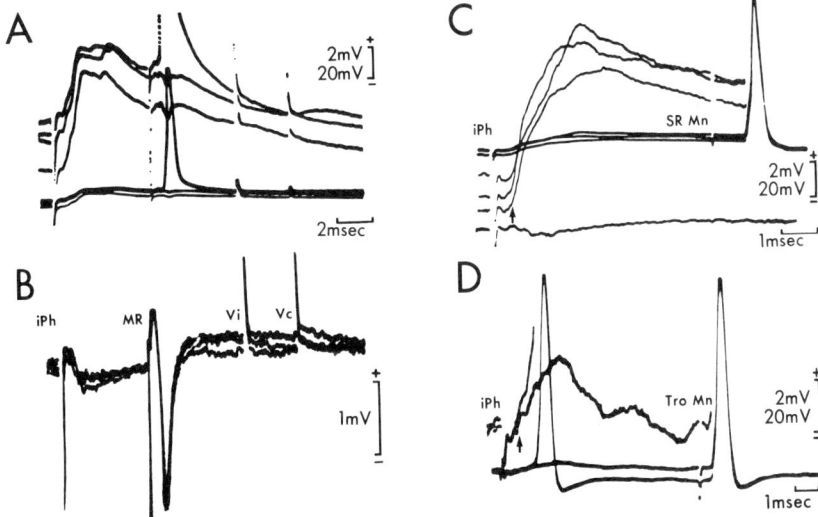

Fig. 5. Intracellular records from oculomotor neurons following electrical stimulation of the prepositus nucleus. All records were obtained 1 week after producing the brain stem lesions shown in Fig. 6. Intracellular recording was made from the side ipsilateral to prepositus stimulation (site indicated by the star in 6D). A) An intracellular record from a medial rectus motoneuron. B) The extracellular field potentials at higher magnification and indicating stimulation points. Note the absence of any vestibular evoked synaptic potential but the presence of a large short-latency EPSP from ipsilateral prepositus stimulation. C, D) intracellular records from a superior rectus and trochlear motoneuron, resp. The small arrows indicate synaptic latency for the depolarization. Calibrations are indicated and both high and low gain AC and DC records are included.

following a variety of chronic brain stem lesions (Fig. 5D) (Baker et al., 1977). However, it was obvious that given extensive reciprocal connectivity between the interstitial nucleus of Cajal, the rostral interstitial nucleus of MLF and the reticular nuclei, additional chronic studies needed to be carried out to exclude possible axon collaterals from the above areas. Even so, other electrophysiological evidence (Baker et al., 1977) indicated monosynaptic excitation in medial rectus motoneurons even when the vestibular pathway and internuclear pathway had been compromised. Fig. 5A and B shows a large EPSP in a medial rectus motoneuron in the absence of any vestibular synaptic potential. As shown in Fig. 6, part of the internuclear and all of vestibular pathways were compromised in this experiment thereby strongly suggesting a strong ipsilateral projection to medial rectus motoneurons. The ipsilateral nature of the projection was similar to that found for the vertical oculomotor neurons. These synaptic potentials could be greatly reduced by lesions in the area of the MLF (3 or 4 mm in diameter), providing further evidence that prepositus axons travel near the MLF.

The above results are consistent with our morphological observations. Recent anterograde ^{3}H-amino acid studies have also shown a bilateral projection from the prepositus to the oculomotor and abducens nuclei. In this respect, even though prepositus neurons projecting to the oculomotor nuclei are distributed throughout the rostral caudal extent of the prepositus nucleus, they are predominantly concentrated in the rostral part of the prepositus in the so-called supragenual area (Brodal, 1952). Our retrograde HRP, anterograde ^{3}H-leucine and electrophysiological experiments indicate that the projection from the prepositus to the abducens nucleus is heavier

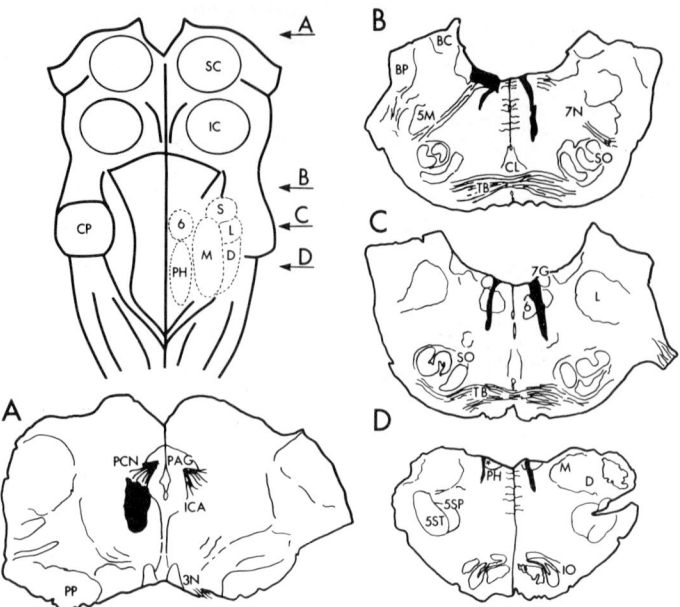

Fig. 6. A diagram showing the extent of lesions for the electrophysiological experiment shown in Fig. 5. A–D are coronal sections at the level indicated in the diagram on the upper left. One week prior to the experiment, the lesion in A was produced by electrolytic means and the ones in B–D via a small sharpened blade following removal of the lobules IX and X. As demonstrated in Fig. 5, the lesions effectively separated the vestibular nuclei from the oculomotor complex and that included the ascending tract of Deiters to medial rectus motoneurons (see 5B). The ipsilateral interstitial nucleus of Cajal and rostral interstitial nucleus of the MLF were damaged to the extent shown in A. As seen in C, the internuclear pathway from the abducens to medial rectus subdivision was compromised only in part. BP, brachium pontis; BC, brachium conjunctivum; 7N, facial nerve; SO, superior olive; CL, inferior central nucleus; TB, trapezoid body; 5M, motor trigeminal nucleus; 7G, genu of the facial nerve; 6, abducens nucleus; L, M, S, D, lateral, medial, superior and descending vestibular nuclei; 5SP, spinal trigeminal nucleus; 5ST, spinal trigeminal tract; IO, inferior olive; PH, prepositus hypoglossi nucleus; SC, superior colliculus; IC, inferior colliculus; CP, cerebellar peduncle; PCN, nucleus of posterior commissure; PAG, periaqueductal gray; ICA, interstitial nucleus of Cajal; 3N, oculomotor nerve; PP, pes peduncle.

contralaterally than ipsilaterally. There is sparse termination in the trochlear and oculomotor complex which is in agreement with the electrophysiological result, but not so much in terms of a weighted ipsilateral connection. The latter, in part, may be explained by the fact that in earlier experiments (Baker and Berthoz, 1976; Baker et al., 1977) the rostral tip of the prepositus nucleus was stimulated and in many of the chronic experiments, lesions would not have compromised the ventral decussation of the contralateral system. The relatively heavy prepositus projection to the contralateral abducens nucleus also could explain the inability of Hikosaka et al., (1978) to correlate neurons in the prepositus nucleus with the ipsilateral abducens nucleus discharge by either signal averaging or antidromic activation by microstimulation of the abducens nucleus. Perhaps the best answer to the question of whether prepositus axons terminate on abducens or oculomotor neurons will be obtained in studies which combine amino acid injections in the prepositus and intracellular HRP injections in abducens motoneurons in order to directly demonstrate terminal boutons on motoneuronal dendritic trees.

The number of circuits the perihypoglossal complex is linked with is truly remarkable. When one considers the diversity of afferents as well as efferents from the prepositus, it is apparent that this nucleus is more than a simple site for preoculomotor or precerebellar activity. We believe it is essential for both vertical and horizontal eye movement and especially significant for fixation (i.e., eye position). It may be of equal importance for both visual and vestibular control of eye movement, with the likelihood that it is at the same time both participating in the ongoing activity in a true premotor sense and acting as an "afferent source" for many other brain stem structures which either are in the stage of passing information onto oculomotor nuclei or are involved in the planning of future activity in the gaze system.

We feel that the apparent complexity of the prepositus nucleus is resolvable, especially with respect to ascertaining the fine topographical and cyto-architectonic organization of the system. It is obvious that due to the distributed nature of most eye movement signals in the posterior brain stem, it will not be easy to come to grips with what the functional role of any given pathway may be, but without detailed information concerning the circuitry, it would not be reasonable to even ask such questions.

SUMMARY

The afferents and efferents of the nucleus prepositus were studied in a series of anatomical and electrophysiological experiments. From retrograde HRP experiments, it appears that, qualitatively, the important inputs to this nucleus come from the vestibular nuclei, perihypoglossal nuclei, reticular formation, extraocular nuclei, accessory oculomotor nuclei and the cerebellum. The efferents of the prepositus hypoglossi were studied with autoradiographic, electrophysiological and both intra- and extracellular HRP techniques. The prepositus appears to establish efferent connections bilaterally with the cerebellar cortex, the interposed and medial cerebellar nuclei, the vestibular nuclei (especially the medial and descending subdivisions), the perihypoglossal nuclei, the medial medullary and pontine reticular formation, the extraocular motor nuclei, the accessory occulomotor nuclei and the region around the parabigeminal nucleus. Intracellular injections of HRP reveal that a single prepositus neuron sends collaterals to several of the above brain stem areas. Evidence is also presented which indicates that there is a direct projection from prepositus to ocular motoneurons and that it is excitatory in effect. Such a wealth of anatomical connections suggests that the prepositus is much more than a simple site for either preoculomotor or precerebellar activity.

REFERENCES

Alley, K. (1977) Anatomical basis for interaction between cerebellar flocculus and brainstem. In *Control of Gaze by Brain Stem Neurons*, R. Baker and A. Berthoz (Eds.), Elsevier North-Holland Biomed. Press, Amsterdam, pp. 109–117.

Alley, K., Baker, R. and Simpson, J.I. (1975) Afferents to the vestibulocerebellum and the origin of the visual climbing fibers in the rabbit. *Brain Res.*, 98: 582–589.

Alley, K., Baker, R. and Simpson, J.I. (1976) Projections from the peri-hypoglossal nucleus to the vestibulo-cerebellum. *Neurosci. Abstr.*, 2: 104, no. 144.

Angaut, P. and Brodal, A. (1967) The projection of the "vestibulocerebellum" onto the vestibular nuclei in the cat. *Arch. ital. Biol*, 105: 441–479.

Baker, R. (1977a) Anatomical and physiological organization of brainstem pathways underlying the control of gaze. In *Control of Gaze by Brain Stem Neurons*, R. Baker and A. Berthoz (Eds.), Elsevier North-Holland Biomed. Press, Amsterdam, pp. 207–224.

Baker, R. (1977b) The nucleus prepositus hypoglossi. In *Eye Movements*, B. Brooks and F.J. Bajandes (Eds.), pp. 145–178.

Baker, R. and Berthoz, A. (1975) Is the prepositus hypoglossi nucleus the source of another vestibular ocular pathway? *Brain Res.*, 86: 121–127.

Baker, R. and Berthoz, A. (1976) Prepositus neurons synapse directly on vertical oculomotoneurons. *Fed. Proc.*, 35: 562.

Baker, R. and Berthoz, A. (Eds.) (1977) *Control of Gaze by Brain Stem Neurons*. Elsevier/North Holland Biomed. Press, Amsterdam.

Baker, R., Gresty, M. and Berthoz, A. (1976) Neuronal activity in the prepositus hypoglossi nucleus correlated with vertical and horizontal eye movement in the cat. *Brain Res.*, 101: 366–371.

Baker, R., Delgado-Garcia, J. and Alley, K. (1977) Morphological and physiological demonstration that prepositus hypoglossi neurons terminate on medial rectus motoneurons. *Proc. XXVII int. Congr. Physiol.*, XIII: 46, no. 119.

Baleydier, C. and Magnin, M. (1979) Afferent and efferent connections of the parabigeminal nucleus in the cat revealed by retrograde axonal transport of HRP. *Brain Res.*, 161: 187–198.

Bishop, G.A., McCrea, R.A. and Kitai, S.T. (1976) Afferent projections to the nucleus interpositus anterior (NIa) and lateral nucleus (LN) of the cat cerebellum. *Anat. Rec.*, 184: 360.

Blanks, R.H.I., Volkind, R., Precht, W. and Baker, R. (1977) Responses of cat prepositus hypoglossi neurons to horizontal angular acceleration. *Neuroscience*, 2: 391–403.

Brodal, A. (1952) Experimental demonstration of cerebellar connections from the perihypoglossal nuclei (nucleus intercalatus, nucleus praepositus hypoglossi and nucleus of Roller) in the cat. *J. Anat. (Lond.)*, 86: 110–120.

Büttner-Ennever, J.A. and Henn, U. (1976) An autoradiographic study of pathways from the pontine reticular formation involved in horizontal eye movements. *Brain Res.*, 108: 155–164.

Büttner-Ennever, J.A. and Buttner, O. (1978) A cell group associated with vertical eye movements in the rostral mesencephalic reticular formation of the monkey. *Brain Res.*, 151: 31–47.

DeOlmos, J., Hardy, H. and Heimer, L. (1978) The afferent connections of the main and the accessory olfactory bulb formations in the rat: an experimental HRP study. *J. comp. Neurol.*, 181: 213–244.

Edwards, S.E. and Henkel, C.K. (1978) Superior colliculus connections with the extraocular motor nuclei in the cat. *J. comp. Neurol.*, in press.

Eller, T. and Chan-Palay, V. (1976) Afferents to the cerebellar lateral nucleus. Evidence from retrograde transport of horseradish peroxidase after pressure injections through micropipettes. *J. comp. Neurol.*, 166: 285–302.

Frankfurter, A., Weber, J.T. and Harting, J.K. (1977) Brain stem projections to lobule VII of the posterior vermis in the squirrel monkey: as demonstrated by the retrograde axonal transport of tritiated horseradish peroxidase. *Brain Res.*, 124: 135–139.

Fuchs, A.F. (1977) Role of the vestibular and reticular nuclei in the control of gaze. B: Reticular, prepositus and other internuclear neuronal activity. In *Control of Gaze by Brain Stem Neurons*, R. Baker and A. Berthoz (Eds.), Elsevier/North-Holland Biomed. Press, Amsterdam, pp. 341–348.

Fukushima, Y., Igusa, Y. and Yoshida, K. (1977) Characteristics of response of medial brain stem neurons to horizontal head angular acceleration and electrical stimulation of the labyrinth in the cat. *Brain Res.*, 120: 564–570.

Gacek, R.R. (1977) Location of brain stem neurons projecting to the oculomotor nucleus in the cat. *Exp. Neurol.*, 57: 725–749.

Graux, G. (1878) *De la Paralysie du Moteur Oculaire Externe avec Deviation Conjugée*. Thesis. A. Parent, Paris.

Graybiel, A.M. (1977) Direct and indirect preoculomotor pathways from the brainstem: An autoradiographic study of the pontine reticular formation in the cat. *J. comp. Neurol.*, 175: 37–73.

Graybiel, A.M. (1978) A satellite system of the superior colliculus: The parabigeminal nucleus and its projections to the superficial collicular layers. *Brain Res.*, 145: 365–374.

Graybiel, A.M. and Hartwieg, E.A. (1974) Some afferent connections of the oculomotor complex in the cat: an experimental study with tracer techniques. *Brain Res.*, 81: 543–551.

Gresty, M. and Baker, R. (1976) Neurons with visual receptive field, eye movement, neck displacement sensitivity within and around the nucleus prepositus hypoglossi in the alert cat. *Exp. Brain Res.*, 24: 429–433.

Hikosaka, O., Igusa, Y. and Imai, H. (1978) Firing pattern of prepositus hypoglossi and adjacent reticular neurons related to vestibular nystagmus in the cat. *Brain Res.*, 144: 395–403.

Kotchabhadki, N. and Walberg, F. (1978) Cerebellar afferent projections from the vestibular nuclei in the cat: an experimental study with the method of retrograde axonal transport of horseradish peroxidase. *Exp. Brain Res.*, 31: 591–604.

McCrea, R.A. and Baker, R. (1978) Neurons in the oculomotor, trochlear and abducens nuclei project caudally in the MLF to the prepositus nucleus. *Neurosci. Abstr.*, 4: 166, no. 507.

McCrea, R.A., Bishop, G.A. and Kitai, S.T. (1977) Electrophysiological and horseradish peroxidase studies of precerebellar afferents to the nucleus interpositus anterior. II. Mossy fiber system. *Brain Res.*, 122: 215–228.

Mergner, T., Pompeiano, O. and Corvaja, N. (1977) Vestibular projections to the nucleus intercalatus of staderini mapped by retrograde transport of horseradish peroxidase. *Neurosci. Lett.* 5: 309–313.

Pompeiano, O., Mergner, T. and Corvaja, N. (1978) Commissural, perihypoglossal and reticular afferent projections to the vestibular nuclei in the cat: An experimental anatomical study with the method of the retrograde transport of horseradish peroxidase. *Arch. ital. Biol.*, 116: 130–172.

Precht, W. (1974) The physiology of the vestibular nuclei. In *Handbook of Sensory Physiology, Vol. VI/1, Vestibular System, Part 1, Basic Mechanisms*, H.H. Kornhuber (Ed.), Springer-Verlag, Berlin-Heidelberg-New York, pp. 353–416.

Ruggiero, D., Batton, III, R.R., Jayaraman, A. and Carpenter, M.B. (1977) Brain stem afferents to the fastigial nucleus in the cat demonstrated by transport of horseradish peroxidase. *J. comp. Neurol.*, 172: 189–210.

Sherk, H. (1978) Visual response properties and visual field topography in the cat's parabigeminal nucleus. *Brain Res.*, 145: 375–379.

Shimazu, H. and Precht, W. (1966) Inhibition of central vestibular neurons from the contralateral labyrinth and its mediating pathway. *J. Neurophysiol.*, 29: 467–492.

Tagaki, J. (1925) Studien zur vergleichenden Anatomie des Nucleus vestibularis triangularis. I. Der Nucleus intercalatus und der Nucleus prepositus hypoglossi. *Arb. Neurol.*, 27: 157–188.

Torvik, A. and Brodal, A. (1954a) The cerebellar projection of the peri-hypoglossi nucleus (nucleus intercalatus, nucleus prepositus hypoglossi and nucleus of Roller) in the cat. *J. Neuropath.*, 13: 515–527.

Uemura, T. and Cohen, B. (1973) Effects of vestibular nucleus lesions on vestibuloocular reflexes and posture in monkeys. *Acta oto-laryngol. (Stockh.)*, Suppl., 315: 1–71.

Walberg, F. (1961) Fastigiofugal fibers to the perihypoglossal nuclei in the cat. *Exp. Neurol.*, 3: 525–541.

Functional Role of the Prepositus Hypoglossi Nucleus in the Control of Gaze

J. LOPEZ-BARNEO, C. DARLOT and A. BERTHOZ

Laboratoire de Physiologie du Travail du CNRS, Département de Physiologie Neuro-sensorielle, 75005 Paris (France)

INTRODUCTION

The first report concerning a possible role of the perihypoglossal area in the control of gaze is due to Hyde and Eliasson (1957) who suggested the existence of a "hind brain center for horizontal gaze" located caudal to abducens nucleus. Further evidence was obtained by lesion experiments in the monkey (Uemura and Cohen, 1973) which induced a deficit in OKN (optokinetic nystagmus) and OKAN (optokinetic after-nystagmus) together with gaze nystagmus. The relation of the nucleus prepositus hypoglossi (Ph) with the oculomotor and vestibular systems was directly evidenced by anatomical (Mabuchi and Kusama, 1970; Graybiel and Hartwieg, 1974), electrophysiological (Baker and Berthoz, 1975), and neurophysiological (Baker et al., 1976) investigations.

Since these first studies, extensive information has been obtained concerning both the afferent and efferent connections of this nucleus (McCrea et al., this volume, chapter VA4). Various classes of neurons have been found in Ph in the alert cat (Baker et al., 1976) and rabbit (Schaefer et al., 1977) which are directly related to eye movements irrespective of their origin (optokinetic, vestibular or spontaneous), and justified further studies concerning the information carried by Ph neurons during pure vestibular stimulation. Two studies made in the ketamine anesthetized (Blanks et al., 1977) and decerebrate (Fukushima et al., 1977) cat addressed this question.

Only a few years after the discovery of the involvement of the Ph nucleus in the control of gaze a great number of hypotheses have already been made concerning its function. Although some have been reviewed recently (Baker, 1977), it may be useful at this point to summarize them briefly. Two main types of function have been suggested according to the postulated direction of the flow of information in Ph nucleus. They depend upon whether the nucleus is considered as a *"premotor center"* carrying motor commands, or as a *"corollary* discharge" type of structure to transfer signals from motor centers to other areas of the brain.

When viewed as a premotor structure, the Ph nucleus can be thought to play a role in the *integration along the vestibulo-ocular pathway*, namely, the mathematical transformation of head velocity input from the labyrinth into an eye position command (see Fig. 5). Taking in consideration the anatomical connections between Ph and sensorimotor cerebral cortex (Sousa-Pinto, 1970) this nucleus could also integrate, in a Sherringtonian sense, vestibular with other sensory inputs. In particular, it was suggested that it could be an important station for the transformation of signals coded in retinal coordinates into head coordinates. Along with these ideas, a *visuomotor* role has been suggested because of the presence of cells with visual receptive fields in or around the nucleus (Gresty and Baker, 1976), and the existence of a powerful tectal projection to the reticular formation underlying Ph (Kawamura et al., 1974).

Finally, the demonstration that horizontal and vertical components of eye movements are generated in different areas of brain stem reticular formation, together with the existence of Ph neurons that code saccade parameters and eye position in these two different main directions, has led to the idea that Ph could contribute to the coordination of these two components during oblique gaze.

The possibility for Ph to be premotor has been challenged however, by the fact that most Ph neurons activated by horizontal rotation could not be antidromically activated by microstimulation of ipsilateral abducens nucleus in the cat (Hikosaka, personal communication). In fact, no clear-cut conclusion has been reached to date concerning the role of the monosynaptic projection from the Ph zone to trochlear (Baker et al., 1977) and medial rectus (Baker and Delgado-Garcia, unpublished observation) motoneurons. The powerful projections of Ph neurons to various central structures, in particular the cerebellum, has consequently stimulated speculation concerning a possible "corollary discharge" type of function.

Considered as a precerebellar nucleus, Ph could provide eye position or eye velocity (Lisberger and Fuchs, 1977) signals to the cerebellar cortex and hence participate in its regulatory function during visual and vestibular control of eye movement. Another possibility is that the eye position signal contained in Ph neurons be the required input to the so-called vestibular plus eye-position neurons in the vestibular nuclei which have now been demonstrated in various species (see review in Pola and Robinson, 1978, for the monkey). This hypothesis would fit with the suggestion made earlier (Berthoz et al., 1974) that the modulation of vestibular nucleus neuron, during vestibular nystagmus, required a feedback from eye position related neurons in reticular structures.

At the present stage of prepositus story, we feel that there is some urgent need for further quantitative studies of the behavior of Ph neurons in the alert animal before any emphasis can be put on any of these ideas. The data obtained in our study from about 150 Ph neurons confirms the general pictures reported by us previously (Baker et al., 1976). In the present paper, we will briefly summarize some new findings obtained with a more precise eye movement measurement concerning those Ph neurons very closely related to eye movements. Bursters with/or visual receptive field and those so-called long-lead burst tonic neurons have not been retained in the present study. We have only considered here these neurons (about 70% of the total number of neurons recorded) which discharged with ipsilateral eye movements and showed a type II discharge pattern during vestibular stimulation in the dark. In a subsequent paper (Berthoz et al., in preparation) the physiological properties of more than 300 Ph neurons, in relation with eye movements, will be extensively described.

METHODS

Recording and stimulation

The experiments have been carried out in four alert cats, with heads restrained, in which eye movements were recorded by the search coil technique (Robinson, 1963).

Search coils were implanted chronically three weeks to one month previous to the recording sessions. The animals were placed in a magnetic field. The overall precision of the eye movement recording was about 10' of arc and the frequency band-width 400 Hz.

Extracellular neuronal activity was recorded by glass micropipettes filled with 2 M sodium acetate saturated with pontamine sky blue. Electrode impedance varied from 1 to 7 mΩ. Low impedance electrodes were generally preferred in order to reduce the probability of recording from axons. The Ph nucleus was reached through the intact cerebellum via a small opening (3–4 mm diameter) made chronically in the bone above the posterior cerebellar vermis. This opening was covered by inert material and plugged with bone wax between recording sessions.

Ph neurons were identified according to: (i) anatomical, (ii) electrophysiological and (iii) functional criteria. (i) Stereotaxic coordinates were used according to the Bergman (1968) atlas and their location and depth were calculated in this frame of reference. For verification by histology, pontamine marks were left at the end of some electrode tracks. (ii) A stimulating electrode was chronically implanted through the cerebral cortex to stimulate abducens nerve at its exits from the brain stem (Delgado-Garcia et al., 1977; Darlot et al., 1977). The depth and extent of the antidromic field potential so obtained in abducens nucleus was studied and the location of Ph nucleus inferred. (iii) Functional identification and selection of Ph neurons was made on the basis of their clear modulation during eye movements. Because of this selection no attempt was made to calculate the proportions of Ph neurons not related with eye movements. These criteria allowed localization of Ph neurons within 200–300 μm; this is not enough to distinguish the fine mediolateral and the superficial depth organization shown by morphology (McCrea et al., chapter VA4). Further experiments will have to refine this identification. Most of the recordings reported here were made in the rostral 2/3 of Ph which extended from abducens nucleus to 2 mm caudal and occasionally as far as 3.5 mm caudal. The caudal third of the nucleus has been extensively investigated by our previous studies (Baker et al., 1976).

Neuronal activity was studied during spontaneous eye movements and when the animal was rotated in the dark. Results of visual-vestibular interaction will be reported elsewhere. Vestibular stimulation consisted of sinusoidal oscillations (0.05–0.5 Hz) by a turntable on which the animal was mounted. In all cases the head of the cat was rotated 21° nose down to stimulate horizontal semi-circular canals (Blanks et al., 1972).

Data processing

Although to date most of the data have been processed, those reported here were calculated during fixations and saccades. During fixations discharge rate-position plots were made for both horizontal and vertical components of eye movement. Correlation coefficients and regression lines, as well as 95 per cent confidence intervals, were computed by a standard statistical analysis.

Because many Ph neurons show a particular direction in which they are activated (see below), an attempt was made to plot this behavior within the oculomotor range as defined by Crommelinck et al., (1977). The oculomotor range in our experimental conditions was obtained from the storage oscilloscope image of about 2000 consecutive spontaneous fixations of each animal.

The preferred direction was defined by the two points obtained by plotting the respective values for the horizontal and vertical positions in which the lowest and highest neuronal discharges during fixation were observed. Details will be given in the Results section.

RESULTS

About 70% of neurons reported in the present work were closely related to eye movements. All these neurons fall into the categories previously called tonic and burst-tonic (Luschei and Fuchs, 1972). Although we are conscious of the fact that this terminology already implies that a variable amount of eye velocity information is coded in the discharge frequency of the neurons, we would like to give a different terminology to describe Ph neurons. As we shall see below they all seem to contain some eye velocity information, and thus we shall place them between two extremes: mainly eye position related neurons (position plus velocity) and mainly eye velocity related neurons (velocity plus position).

Discharge patterns of Ph neurons during spontaneous fixations

During spontaneous eye movements neuronal discharge was correlated to absolute eye position in the orbit during fixation. In about 10% of the neurons, discharge rate was mainly coupled with the vertical, and in 90% with the horizontal components of eye movements, but this percentage will probably turn out to be linked with the areas explored. From our data we can not yet deduce any orderly arrangement of neurons within Ph nucleus, but they clearly appear in clusters of various properties. Our results on this particular point confirm previous results (Delgado-Garcia et al., 1977) concerning the existence of a group of vertical neurons dorsal and caudal to the abducens nucleus.

Some PV neurons never became silent and their discharge rate was precisely modulated by eye position. The behavior of such a position-velocity (PV) neuron located 2.5 mm caudal to abducens nucleus and about 900 μm below the surface of brain stem showed neither pause nor threshold and its rate-horizontal and vertical position plots are shown in Fig. 1. The main features of this neuron are: regular discharge, small eye velocity component and absence of any relationship between neuronal discharge and vertical fixations. This type of neuron had firing rates not higher than 100–120 imp/sec.

Apart from this particular category of neurons, most other PV neurons encountered showed some clear eye velocity component which induced an increase of firing rate for "on" direction and a pause during "off" direction saccades.

Careful examination of the records showed that on many occasions cells which seemed to have good correlation of their discharge rate with horizontal position were also influenced by vertical components. Fig. 2 shows an example of such a neuron recorded in the left Ph 3 mm caudal to abducens nucleus and about 200 μm below the surface of brain stem whose main "on" direction is to the left. The threshold zero frequency was for eccentric eye position in the orbit. Maximum frequency in our sample (133 imp/sec mean value) was reached when the eye was near the primary position. (The actual maximum was not reached in this sample.) However, the records shown in Fig. 2A indicate that during a vertical down saccade the discharge frequency increased and decreased after upward saccades (filled stars in Figs. 2A and

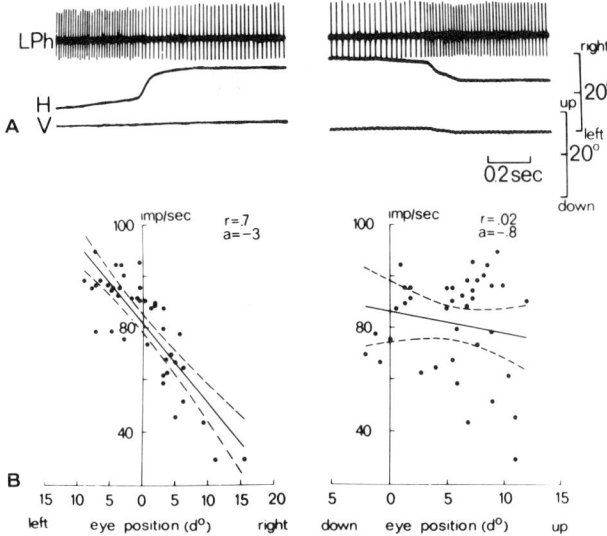

Fig. 1. Prepositus hypoglossi neuron mainly coding horizontal eye position. (A) Behavior of a neuron (recorded in left Ph) during fixation and horizontal saccades in the "on" (left) and "off" directions. Note the regular interspike interval during eye fixations and the absence of pauses during saccades in the "off" directions. Horizontal (H) and vertical (V) components of eye movements are shown below the discharge pattern of the neuron. (B) Rate-position plots for this neuron show the relationship between spike frequency and horizontal or vertical eye position. —, regression lines; - - - -, confidence interval (95%); r, correlation coefficient; a mean rate-position slope in imp/sec/degree.

B). Following these observations, the data points in the rate-position plots for horizontal eye movements were divided into classes of vertical eye position. A "preferred direction" was defined by a vector joining two points within the oculomotor range. Vertical and horizontal axes were divided into segments of 2.5 mm. In each particular neuron, the mean frequency corresponding to each class was calculated. The onset and the end of the vector represented plots of the respective values for the horizontal and vertical classes in which the lowest and highest mean neuronal discharges were observed. Fig. 2C shows an example of how the data points separate if only two classes (1°–7° and 7°–13°) of vertical eye positions are considered. This simple separation leads to two distinct regression lines with an improved correlation coefficient.

Note that the mean rate-position slope is the same in both cases. In conclusion, the coding of eye position for this neuron was much more precise than when predicted by the simple correlation with horizontal component only.

Fig. 2D shows an attempt to describe schematically the relationship of the firing rate of this neuron according to the position of the eye in the oculomotor range (dotted line). When the eye is in the upper-right quadrant the neuron shows a low firing rate. Firing frequency increases when the eye moves downward and to the left from this area. It must be noted that this vector represents only the mean values of the firing rate for a particular sample of eye positions. Using this method we could show that a number of Ph neurons have a preferred oblique direction. The existence of such a variety of specific directions in the discharge rate of Ph neurons precludes any attempt to correlate their activity with vestibular stimulus without first assessing, for each neuron, its "preferred direction".

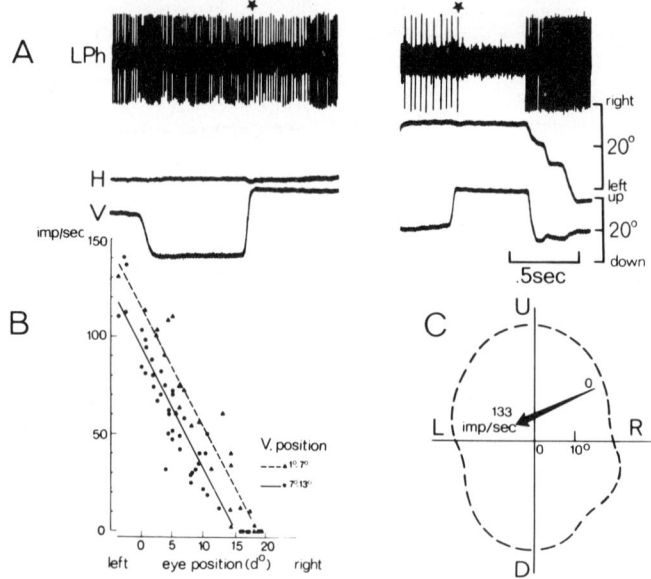

Fig. 2. Behavior of an oblique position-velocity neuron recorded in left prepositus hypoglossi nucleus during spontaneous eye movements. (A) Although the discharge rate of this neuron sems to be related only to horizontal eye positions, in the selected records of this figure changes in eye position occurring only in the vertical direction (black stars) also modified the firing of the neuron. (B) Rate-position curves for the same neuron showing the existence of two separate populations of data points belonging to two different vertical classes of eye position (1°–7° and 7°–13° upwards). The regression lines have been plotted for these two classes. (C) Schematic representation, within the oculomotor range (– – – –), of the preferred direction along which the neuron is activated. 0 imp/sec at the thin end of the arrow indicates area of firing threshold; 133 imp/sec at the thick end indicates the maximum mean firing rate encountered in the most leftward 2.5° class of horizontal position, for this particular sample of eye movements.

Discharge patterns of Ph neurons during spontaneous saccades

During spontaneously occurring saccades in the light, the discharge pattern of Ph neurons was studied in relation to several questions.

Firstly, we tried to assess the latency of onset and cessation of neuronal discharge in order to evaluate the possibility for Ph neurons to have a premotor role. The calculation of latency for those neurons which showed little influence of eye velocity (position plus velocity) was found to be irrelevant because the onset of activity in such neurons was only related to the eye position threshold irrespective of the velocity at which the eye moved. In other words, irrespective of the eye velocity, the onset of discharge as well as cessation was always correlated (within 3°–5°) with the static threshold of firing as evidenced by the rate-position curve. These results will be reported in a subsequent paper. Latency was only calculated for neurons showing a high velocity sensitivity. As previously described (Baker et al., 1976), such neurons fired about 5 msec (mean value) prior to the onset of the "on" direction saccade. The latency for the pause before an "off" direction saccade shows obviously much more variability because, for instance, when the eye was in the "off" direction the corresponding low discharge rate introduced a great error in determination of "off" latency.

Secondly, we computed for velocity plus position neurons the relationship of mean intra-burst frequency versus amplitude of the saccade and initial eye position. The

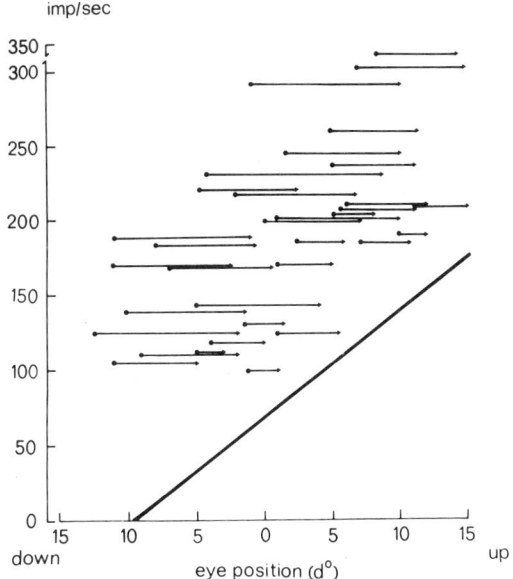

Fig. 3. Behavior of a vertical velocity plus position neuron during spontaneous saccades in the light. The rate-position regression line for fixations is drawn for reference (———). Each arrow represents one of 33 saccades. Origin and end of arrows show vertical component of initial and final eye position. Each arrow is drawn at a level equal to the mean intraburst firing frequency during the corresponding saccade. Note that the general orientation of this cloud of arrows is parallel to the regression line, which suggests an algebraic summation of position and velocity discharge components.

main purpose of this study was to test whether the behavior of Ph neurons was in any way similar to either motoneurons or interneurons in the extraocular muscle motor nuclei. Fig. 3 shows an example of a particularly interesting Ph neuron which was recorded 200 μm above the antidromic field potential of the abducens nucleus. It belongs to the group of "vertical" Ph neurons found above the VIth nucleus which was observed also by Delgado-Garcia et al. (1977). The mean value of burst frequency (intraburst frequency) during 32 saccades in the "on" direction is shown. Intraburst frequency seems to depend upon the intial position of the eye in the orbit, in the sense that the tonic has been introduced by the eye position dependent mean firing rate (see rate position curve, shown as a solid line). This seems to add to the burst characteristics in a manner similar to the interneurons in the VIth nucleus in contrast to the burst reticular neurons studied by Henn and collaborators. Note that intraburst frequency is in general not higher than 350 imp/sec.

Response of Ph neurons to vestibular stimulation

Before presenting some preliminary findings concerning the behavior of Ph neurons during vestibular stimulation, we shall summarize the main findings of this problem. The simplified but convenient tool of calculating the phase between input head acceleration and output neuronal firing rate gives an indication of the type of mechanical parameters coded in neurons during vestibular stimulation. Fig. 4 summarizes the phase characteristics which have been obtained for vestibular nuclei, prepositus hypoglossi; abducens motoneurons and eye movements in the cat, either in the dark or in preparations without visual input.

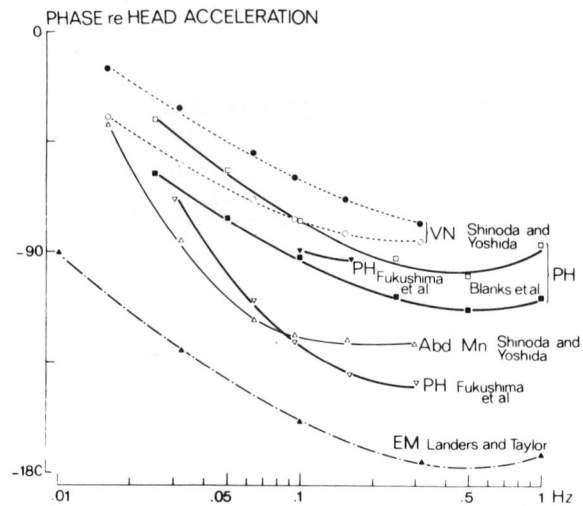

Fig. 4. Summary of previous data on the dynamic characteristics of vestibulo-ocular reflex. The phase of either neuronal discharge rate or eye angular displacement, with respect to head acceleration during vestibular stimulation in the dark is plotted from various authors. Oscillation frequency is indicated in abscissa. From top to bottom: (1) Data from Shinoda and Yoshida (1974) on two different populations (●, short-phase ○, long-phase) of type I neurons recorded in vestibular nuclei (VN). (2) Results of Blanks et al. (1977) for type I (□) and type II (■) Ph neurons, each referred to peak acceleration in its on-direction. (3) Those of Fukushima et al. (1977) which show two populations of type II Ph neurons, in phase with either head velocity (▼) or abducens nerve activity (▽). (4) Data from Shinoda and Yoshida (1974) on abducens motoneurons (Abd Mn) (△). (5) Phase relation to eye movement (EM) (▲) after Landers and Taylor (1975). See text for experimental conditions of each author.

Shinoda and Yoshida (1974) have calculated the phase of short and long phase type I cells in the vestibular nucleus, and of abducens motoneurons, during horizontal sinusoidal rotation of decerebrate cats. For comparison we have drawn the phase of eye angular displacement obtained by Landers and Taylor (1972) in alert cats. In Ph nucleus a surprising phase lag was obtained by Blanks et al. (1977) in Ketamine anesthetized cats and by Fukushima et al. (1977) in decerebrate cats using also sinusoidal oscillation. These authors consequently suggested that the Ph was part of the integrating network in the vestibulo-ocular arc. However Fukushima et al. (1977) concluded that two distinct classes of neurons were present in the Ph, one of them being clearly similar to abducens motoneurons and the other, less numerous, having a phase which fell in the range of the two classes obtained by Blanks et al. (1977). This last group showed only a small phase lag with respect to vestibular nuclei and contained apparently the same head velocity component.

In our experiments, using vestibular stimulation in complete darkness, we have found discharge patterns which confirm the findings of Fukushima et al. (1977). The response of Ph neurons to horizontal angular acceleration was only studied in those neurons which had a clear horizontal direction-specific response according to the analysis described in the first part of this paper. No formal analysis was made at this point of the discharge rate because we used amplitudes of vestibular stimulation which induced saccadic activity, and found that the relation of firing rate with the head acceleration (input) varied according to the state of alertness of the animal. This is exemplified by the recordings of Figs. 5 and 6 for two neurons. The Ph neuron shown

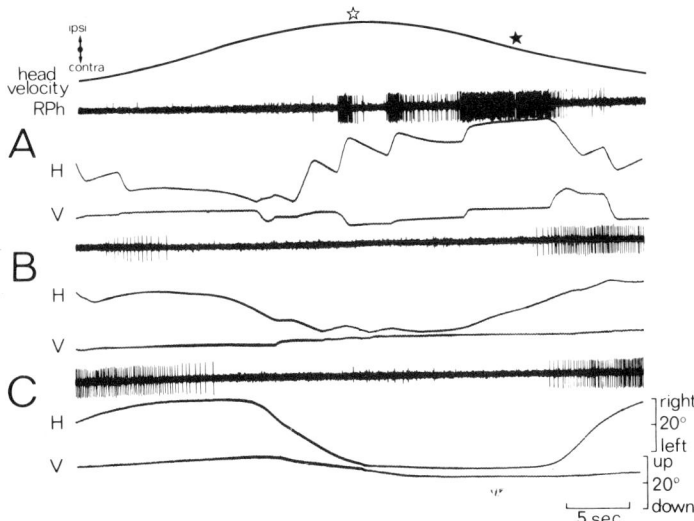

Fig. 5. Behavior of a horizontal position plus velocity prepositus neuron during vestibular nystagmus in the dark. From top to bottom: Head velocity (maximum indicated by white star), to the ipsilateral (ipsi) and contralateral (contra) side with respect to recording site (right prepositus: RPh). Unit activity, horizontal (H) and vertical (V) eye position. Vestibular stimulation was given by sinusoidal horizontal rotation (0.2 Hz and 50° peak to peak amplitude) in three different states of alertness. In the alert state, vestibular nystagmus is strong and the neuronal discharge is closely related to extreme ipsilateral eye position corresponding to contralateral head position (black star). The neuronal discharge is modulated by nystagmus and the appearance of neural firing depends on a rather precise eye position threshold. In B and C the animal becomes drowsy and the corrective effects of quick phase is lost. The absolute position of eyes in the orbit differs from the one obtained in the alert state. Note that neuronal discharge remains closely coupled with eye position without any influences of head velocity position. White star indicates maximum head velocity. See text for detailed description.

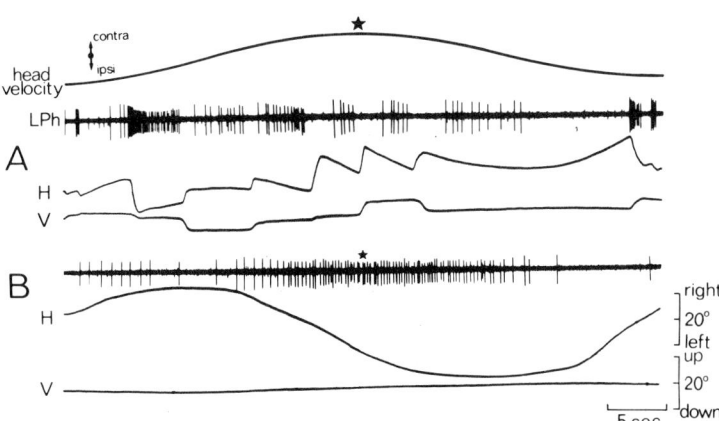

Fig. 6. Behavior of a horizontal velocity plus position prepositus neuron during vestibular nystagmus in the dark. Same recordings as in Fig. 5. Neuronal discharge during vestibular stimulation (0.2 Hz and 50° peak to peak amplitude and 30 degrees/sec peak velocity) in two different states of alertness. In the alert state (A) the neuronal discharge is modulated by both eye velocity and position. Peak discharge rate is during peak ipsilateral head velocity. In the drowsy state (B) the peak discharge rate is still related to maximum eye velocity in the "on" direction (small star) but to contralateral peak head velocity (big star). See text for detailed description.

in Fig. 5 was recorded 1.8 mm caudal to the abducens nucleus and its firing pattern is shown here during three states of alertness. In Fig. 5A the cat was alert, nystagmus brisk. The discharge rate of this neuron during spontaneous eye movements was related to eye position (correlation coefficient with horizontal component 0.85, rate-position). "On" direction is to the right and its threshold was near the primary position. Whatever the state of the cat, the discharge rate, during vestibular nystagmus, is also only correlated with eye position. When saccades disappear due to drowsiness, the phase of the neuron in relation to the vestibular input changes drastically, probably because the amount of head velocity information contained in the discharge rate is minimal compared to the influence of eye position. Quantitative analysis has revealed a small effect of head movement (10% of change of firing rate at primary position for variations of head velocity from 10 to 60 degrees/sec). It could be concluded that this class of neurons is very independent from head acceleration although a clear synaptic input may exist from the labyrinth. This fact could of course be interpreted as either proving that these neurons receive only a weak influence from the labyrinth or that the labyrinth input is *integrated* and does not show, with its initial phase in the *output* of these neurons, which is the measured parameter. Only further experiments will clarify this question.

Fig. 6 shows an example of a very different type of neuron which has a clear eye velocity component in its discharge pattern. This neuron shows also a modulation which is mainly related to *eye velocity* and not to head velocity as demonstrated by the fact that when the animal becomes drowsy the maximum discharge rate follows the phase of maximum eye velocity. Quantitative analysis of the head velocity and acceleration component in this neuron will be reported in a subsequent paper, but the clear difference of its behavior with the neuron shown in Fig. 5 points to the existence of extreme, if not distinct types of neurons with their behavior relative to head velocity during vestibular stimulation.

DISCUSSION

By using precise methods of eye movement measurement and relating the firing rate to the direction of gaze during fixations and saccades, we have established that both the tonic discharge frequency during fixations and the burst characteristics during saccades, of eye movement related neurons, are dependent upon absolute eye position in the orbit coded in head coordinates.

In many aspects, Ph neurons have discharge characteristics close to both motoneurons and internuclear neurons (see review of these properties in Baker and Berthoz, 1977). It is now clear, very much in accordance with the findings of Fukushima et al. (1977), that at least two extreme classes of neurons exist: those coding mainly eye position and exhibiting a very weak eye velocity signal (which during vestibular stimulation show little head velocity influence) and those coding both eye velocity and eye position. But we cannot yet know if the difference between these two classes corresponds to any of the types of neurons found with morphological methods by McCrea et al. (chapter VA4), or if this is a continuously scaled property of one type of neuron.

A second main conclusion which can be drawn from this study is that eye movement related Ph neurons show a change of firing rate which is direction-specific with respect to eye movements. Horizontal, vertical and oblique "preferred directions" have been found and can be measured by a vector analysis, placing the vector in the oculomotor range. The details of this quantitative study will be reported in a subsequent paper but we can already state some of the questions which arise from this finding. It may be important to know whether the preferred direction, when oblique, falls either in the main pulling direction of the extraocular muscles as advocated by Henn and Cohen (1976), or in the plane of the semicircular canals as found in the medial terminal nucleus by Simpson et al. (this volume, chapter VB5), or evenly distributed in space. This last hypothesis would be consistent with a role of Ph as one more of a sensorimotor map involved in visuovestibulo motor coordination.

Obviously, more work is necessary to answer the basic question put forward in the title of this paper, but the exquisite relation of Ph neurons to eye position, and strong projection to the cerebellum may indeed support the idea that even if this structure plays some integrating or coordinating role in premotor function, it is a very good candidate for providing a corollary discharge type of signal either to the cerebellum or to the vestibular nuclei themselves. Pathways for this last possibility have recently been evidenced (Pompeiano et al., 1978).

SUMMARY

The main hypothesis concerning the functional role of the prepositus hypoglossi (Ph) nucleus are reviewed. Experimental results obtained in the alert cat using eye movement recording by the search coil technique and extracellular neuronal recording during spontaneous eye movements or vestibular stimulation are described. Ph neurons code eye position with various degrees of eye velocity information. They have a preferred direction which is either horizontal, vertical or oblique. During vestibular stimulation those neurons with mainly eye position information show little head velocity component. Both an immediate premotor role and a corollary discharge type of function involving feedforward of eye position information to the cerebellum or feed back to the vestibular nuclei are compatible with the results.

ACKNOWLEDGEMENTS

This research was supported by CNRS ATP No. 3626. Dr. Lopez Barneo was supported by a ETP BBR training award. Mrs. A. M. Madariaga developed the technique of marking with pontamine and determined, in a separate study, how long the animals could be kept with the marks for chronic recording. We gratefully acknowledge her participation.

REFERENCES

Baker, R. (1977) The nucleus prepositus hypoglossi. In *Eye Movements*, B.A. Brooks and F.J. Bajandas (Eds.), Plenum Press, New York, pp. 145–178.
Baker, R. and Berthoz, A. (1975) Is the prepositus hypoglossi nucleus the source of another vestibular ocular pathway? *Brain Res.*, 86: 121–127.
Baker, R., Gresty, M. and Berthoz, A. (1976) Neuronal activity in the prepositus hypoglossi nucleus correlated with vertical and horizontal eye movement in the cat. *Brain Res.*, 101: 366–371.

Baker, R., Berthoz, A. and Delgado-Garcia, J. (1977) Monosynaptic excitation of trochlear motoneurons following electrical stimulation of the prepositus hypoglossi nucleus. *Brain Res.*, 121: 157–161.

Berman, A.L. (1968) *The Brain Stem of the Cat: a Cytoarchitectonic Atlas with Stereotaxic Coordinates.* University of Wisconsin Press, Madison.

Berthoz, A., Baker, R. and Goldberg, A. (1974) Neuronal activity underlying vestibular nystagmus in the oculomotor system of the cat. *Brain Res.*, 71: 233–238.

Blanks, R.H.I., Estes, M.S. and Markham, C.H. (1972) Planar relationships of semicircular canals in the cat. *Amer. J. Physiol.*, 223: 55–62.

Blanks, R.H.I., Volkind, R., Precht, W. and Baker, R. (1977) Responses of cat prepositus hypoglossi neurons to horizontal angular acceleration. *Neuroscience*, 2: 391–403.

Büttner, W., Hepp, K. and Henn, V. (1977) Neurons in the rostral mesencephalic and paramedian pontine reticular formation generating fast eye movements. In *Control of Gaze by Brain Stem Neurons*, R. Baker and A. Berthoz (Eds.) Elsevier/North-Holland Biomed. Press, Amsterdam, pp. 309–318.

Crommelinck, M., Guitton, D. and Roucoux, A. (1977) La position primaire de l'oeil en relation avec le champ oculomoteur chez le chat. Association des Physiologistes Lyon.

Darlot, C., Berthoz, A. and Baker, R. (1977) Une nouvelle méthode d'étude des propriétés des muscles extraoculaires chez l'animal éveillé. *J. Physiol. (Paris)*, 100A.

Delgado-Garcia, J., Baker, R. and Highstein, S.M. (1977) The activity of internuclear neurons identified within the abducens nucleus of the alert cat. In *Control of Gaze by Brain Stem Neurons*, R. Baker and A. Berthoz (Eds.) Elsevier/North-Holland Biomed. Press, Amsterdam, pp. 291–300.

Fukushima, Y., Igusa, Y. and Yoshida, K. (1977) Characteristics of responses of medial brain stem neurons to horizontal head angular acceleration and electrical stimulation of the labyrinth in the cat. *Brain Res.*, 120: 564–570.

Graybiel, A.M. and Hartweig, E.A. (1974) Some afferent connections of the oculomotor complex in the cat: an experimental study with tracer techniques. *Brain Res.*, 81: 543–551.

Gresty, M., and Baker, R. (1976) Neurons with visual receptive field, eye movement and neck displacement sensitivity within and around the nucleus prepositus hypoglossi in the alert cat. *Exp. Brain Res.*, 24: 429–433.

Henn, V. and Cohen, B. (1976) Coding of information about rapid eye movements in the pontine reticular formation of alert monkeys. *Brain Res.*, 108: 307–325.

Hyde, J.E. and Eliasson, S.G. (1957) Brain stem induced eye movements in the cat. *J. comp. Neurol.*, 108: 139–172.

Kawamura, K., Brodal, A., and Hoddevick, G. (1974) The projection of the superior colliculus onto the reticular formation of the brain stem. An experimental anatomical study in the cat. *Exp. Brain Res.*, 19: 1–19.

Landers, P.H. and Taylor, A. (1972) Transfer function analysis of the vestibuloocular reflex in the conscious cat. In *Basic Mechanisms of Ocular Motility and their Clinical Implications*. G. Lennerstrand and P. Bach-y-Rita (Eds.), Pergamon Press, Oxford, pp. 505–508.

Lisberger, S.G. and Fuchs, A.P. (1977) Role of the primate flocculus in smooth pursuit eye movements and rapid behavioral modification of the vestibuloocular reflex. In *Control of Gaze by Brain Stem Neurons*, R. Baker and A. Berthoz (Eds.), Elsevier/North-Holland Biomed. Press, Amsterdam, pp. 381–390.

Luschei, E.S. and Fuchs, A.F. (1972) Activity of brain stem neurons during eye movements of alert monkeys. *J. Neurophysiol.*, 35: 445–461.

Mabuchi, M. and Kusama, T. (1970) Mesodiencephalic projections to the inferior olive and the vestibular and perihypoglossal nuclei. *Brain Res.*, 17: 133–136.

Pola, J. and Robinson, D.A. (1978) Oculomotor signals in medial longitudinal fasciculus of the monkey. *J. Neurophysiol.*, 41: 245–259.

Pompeiano, O., Mergner, T. and Corvaja, N. (1978) Commissural, perihypoglossal and reticular afferent projections to the vestibular nuclei in the cat. *Arch. ital. Biol.*, 116: 130–172.

Robinson, D.A. (1963) A method of measuring eye movement using a scleral search coil in a magnetic field. *IEEE Trans. Bio Med. Electron.*, BME-10: 137–145.

Schaefer, K.P., Zierau, H. and Suss, K.J. (1977) Differentiation of neuronal activity in the vestibular nuclei of rabbits. In *Control of Gaze by Brain Stem Neurons*, R. Baker and A. Berthoz (Eds.), Elsevier/North-Holland Biomed. Press, Amsterdam, pp. 257–260.

Shinoda, Y. and Yoshida, K. (1974) Dynamic characteristics of responses to horizontal head angular acceleration in vestibulo-ocular pathway in the cat. *J. Neurophysiol.*, 37: 653–673.

Sousa-Pinto, A. (1970) The cortical projection onto the paramedian reticular and perihypoglossal nuclei (nucleus prepositus hypoglossi, nucleus intercalatus and nucleus of roller) of the medulla oblongata of the cat. An experimental-anatomical study. *Brain Res.*, 18: 77–91.

Uemura, T. and Cohen, B. (1973) Effects of vestibular nuclei lesions on vestibulo-ocular reflexes and posture in monkeys. *Acta oto-laryngol. (Stockh.)*, Suppl. 315: 1–71.

B

Visual Control of Eye Movements: Interaction of Visual and Labyrinthine Inputs

Motion Information in the Vestibular Nuclei of Alert Monkeys: Visual and Vestibular Input vs. Optomotor Output

W. WAESPE and V. HENN

Neurologische Klinik, Universität Zürich, 8091 Zürich (Switzerland)

INTRODUCTION

For many decades research in classical vestibular physiology was concerned with the transformation of the acceleration stimulus in the semicircular canals and the otoliths, how it is transmitted centrally and relayed in the vestibular nuclei. In his monograph on "Outlines of a Theory of Motion Sensation" (1875) Ernst Mach included the visual and other senses as important contributors to any motion sensation. The question whether and where the inputs from these different sensory systems interact, has been taken up by neurophysiologists only recently.

Vestibular influence has been found in classical visual areas like the occipital and parietal lobe (Grüsser et al., 1959; Kornhuber and Da Fonseca, 1964; Horn and Hill, 1969; Denney and Adorjani, 1972), and superior colliculus (Bisti et al., 1974). Visual influences have been found in vestibular areas like the vestibular nuclei (Dichgans and Brandt, 1972; Azzena et al., 1974; Henn et al., 1974; Allum et al., 1976; Daunton and Thomson, 1976; Waespe and Henn, 1977a, b), the flocculus (Miles and Fuller, 1974; Gherladucci et al., 1976; Ansorge and Grüsser-Cornehls, 1977; Lisberger and Fuchs, 1974, 1978a, b), and the ascending vestibulocortical pathway (Büttner and Henn, 1976; Büttner and Buettner, 1978). Also, possible interaction sites in the brain stem have been traced electrophysiologically (Kubo et al., 1978). Having established this close interaction of different sensory inputs at many different levels, the problem has to be redefined: what is the motion signal in cells of the vestibular nuclei and through which sensory system did it come? It has been shown that one input comes from the visual system, and another one from the peripheral vestibular organ. The acoustic and somatosensory systems also contribute. A velocity dependent modulation of vestibular neuron activity was shown, when the animal's body was rotated relative to the stationary head (Fredrickson et al., 1965; Rubin et al., 1977). The vestibular nuclei are therefore part of the central system which detects motion, and which makes use of inputs from many different sensory systems, the peripheral vestibular organ just being one of them.

This introductory characterization of the vestibular nuclei is still a simplified one, because in about 50% of all neurons there is in addition a motor signal present. This motor signal can be related to eye movements; other possible motor signals have not been investigated, because in all experiments animals were restrained and unable to move around.

In the present paper we will investigate the interaction of just two inputs from the visual and the peripheral vestibular system and relate it to the oculomotor output. Other possible inputs and outputs were kept constant.

METHODS

Experiments were done in chronically prepared juvenile Rhesus monkeys (*Macaca mulatta*). During experiments animals were alert sitting in a primate chair with their head bolted to a frame (further details: Waespe and Henn, 1977a). Single neuron activity was recorded extracellularly.

For pure vestibular stimulation the monkey was rotated in complete darkness on a servo-controlled turntable, about a vertical axis, within a light-proof cylinder. For visual stimulation the cylinder (diam. 124 cm, height 86 cm) covered with black and white stripes (stripe width 7.5°) was rotated around the stationary monkey. During normal combined visual-vestibular stimulation the monkey was rotated in the light within the stationary cylinder. The cylinder could also be coupled to the turntable resulting in conflicting visual-vestibular stimulation. The conflict is that during rotation there is no visual displacement. Rotatory stimuli consisted of trapezoid velocity profiles with accelerations between 1.25 and 20 degrees/sec^2 and constant velocities between 25 and 200 degrees/sec.

RESULTS

General characteristics of vestibular nuclei neurons

172 neurons, each receiving its main input from the horizontal semicircular canals, were recorded in 9 monkeys. Neurons were located within the rostral pole of the medial, the lateral and superior vestibular nuclei. 64.5% were type I neurons (Duensing and Schaefer, 1958), i.e. activated by rotation to the ipsilateral side. 32.5% were type II neurons showing a mirror-like behavior. Only 3% were type III neurons and activated with acceleration into both directions. In the absence of motion stimulation the activity of 40.5% of neurons was modulated with fast eye movements and/or eye position (Miles, 1974; Fuchs and Kimm, 1975; Keller and Daniels, 1975; Waespe et al., 1977). 50% of the type I neurons, but only 23% of the type II neurons, were modulated with eye movements. Average resting discharge was 41.9 Hz (n = 171; SD, 23.7 Hz). Eye movement related vestibular neurons had on average a higher resting discharge of 47.0 Hz (SD, 23.0 Hz; Fuchs and Kimm, 1975) as compared to other vestibular neurons (38.4 Hz, SD, 23.7). Eye movement related vestibular neurons showed qualitatively the same behavior as pure vestibular neurons during vestibular or visual stimulation; certain quantitative differences will be discussed below. Not every neuron could be tested under all stimulation conditions. However, neuronal responses were so reliable that results are probably characteristic of vestibular nuclei neurons in general.

Neuronal activity during vestibular stimulation

During pure vestibular stimulation monkeys were rotated about the vertical axis in complete darkness. Following periods of acceleration or deceleration neuronal activity returned to baseline levels with a dominant time constant of 10–40 sec (Buettner et al.,

1978b; Waespe and Henn, 1977a). The time constant was defined as time elapsed between maximum frequency change at the end of acceleration and the point frequency has returned to 37% (1/e) above resting discharge.

Slow-phase velocity of nystagmus showed similar time constants, both varying together. If secondary nystagmus was observed, there was a similar change of neuronal activity into the opposite direction strictly parallel to the nystagmus. During prolonged constant acceleration (up to 40 sec) in the excitatory direction neuronal activity first increased with rising velocity. After about 5–10 sec the increase became successively smaller (Fig. 5). Sensitivity (frequency change in response to acceleration) of type II neurons was generally less compared with type I neurons. Those type I neurons, which were additionally modulated with eye movements, showed the greatest sensitivity.

In conclusion, the adequate vestibular stimulus is an angular acceleration; the neuronal signal is centrally transformed so that neuronal activity decays with a dominant time constant of 15–30 sec.

Neuronal activity during visual (optokinetic) stimulation

All vestibular nuclei neurons receiving an input from horizontal semicircular canals can also be influenced by rotating the visual surround around the stationary monkey. With only rare exceptions, neuronal activation is direction specific. Neurons are activated (or inhibited) if vestibular or optokinetic nystagmus goes into the same direction. During constant velocity rotation, neuronal activity remains constant as long as the stimulus is applied. Therefore, the visual input has the dimension of a velocity. To determine the working range of this velocity input, the visual surround was rotated with different values of accelerations and velocities. Many type I neurons exhibited only little response to vestibular stimulation with acceleration values of less than 2.5 degrees/sec^2, or type II neurons with values of less than 5 degrees/sec^2. However, these neurons show a clear response, if the visual surround is rotated at these or lower accelerations. They respond whenever the velocity exceeds a certain value independent of the applied acceleration (Figs. 5 and 6). The visual velocity threshold is usually around 2–5 degrees/sec, and the saturation level for constant velocity visual stimulation around 60 degrees/sec. Within that range neuronal activity faithfully follows any change of visual velocity, if it is slower than 5 degrees/sec^2. Optokinetic nystagmus saturates at much higher velocities pointing at a dissociation between saturation levels of vestibular neuron activity and nystagmus velocity (Fig. 1).

Fig. 2 shows the behavior of different neurons during slowly increasing velocity of the visual surround. Visual acceleration was 2.5 degrees/sec^2, a value low enough that neuronal activity depends only on actual velocity. Sensitivity was measured as frequency increase relative to velocity increase. To avoid possible saturation effects, only the first 12 sec (velocity range 0–30 degrees/sec) were used for calculation, the results of which are shown in Table I. The most sensitive neurons are type I modulated with eye movements, whereas type II neurons are less sensitive. If similar calculations are performed for vestibular stimulation, values for sensitivity are very similar for all subgroups (with the possible exception of type II neurons).

To estimate thresholds, visual acceleration of 1.25 degrees/sec^2 was used. Neuronal frequency increased linearly with visual velocities over about 16 sec up to 20–30 degrees/sec. The intercept of the best fitting regression line (r between 0.87 and 0.98) with the abscissa was taken as velocity threshold for each unit. On average the intercept was 3.3 degrees/sec (SD 2.9; range 0.3–11.5 degrees/sec; n = 15).

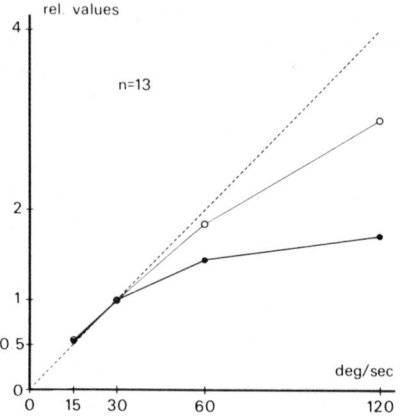

Fig. 1. Neuronal activity (average response from 13 neurons) and nystagmus velocity are plotted against optokinetic stimulus velocity. Abscissa is constant optokinetic stimulus velocity, ordinate normalized neuronal response (———, ●), and nystagmus velocity (———, ○). Note the dissociation of nystagmus velocity and neuronal activity, which occurs at about 60 degrees/sec, and becomes more prominent with higher stimulus velocities.

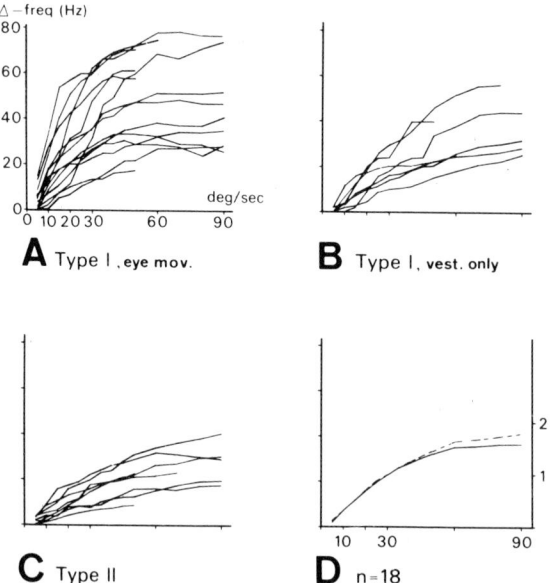

Fig. 2. Vestibular neuron activity in response to increasing optokinetic stimulation. Abscissa is optokinetic stimulus velocity, which is continuously increased with 2.5 degrees/sec^2 from 0 to 90 degrees/sec. Ordinate is frequency increase. A) Type I neurons, which were additionally modulated with eye movements. B) Type I, vestibular-only neurons. C) Type II neurons, 2 of which had a small amount of eye movement modulation. D) Average response from all neurons, which were tested over the whole stimulus range. Initial frequency increase is strongest and saturation prominent for type I eye movement neurons. The averaged curve in D shows that most neurons respond to constant velocities of about 5 degrees/sec and increase their response up to 60 degrees/sec. Continuous line represents absolute values (left ordinate), dashed line normalized values (right ordinate).

TABLE I

SENSITIVITY OF NEURONS

Δ − frequency (Hz)/Δ − velocity (degrees/sec)

	Visual stimulation	Vestibular stimulation
Type I, eye movement related, Fig. 2A, n = 15	1.26 (SD 0.56)	1.32 (SD 0.57)
Type I, vestibular only, Fig. 2B, n = 7	0.62 (SD 0.24)	0.68 (SD 0.35)
Type II, all subgroups combined, Fig. 2C, n = 7	0.51 (SD 0.20)	0.39 (SD 0.17)

In conclusion, the visual input to the vestibular nuclei has the dimension of a velocity and its working range is from 3–60 degrees/sec. The range of optokinetic nystagmus differs, as it extends well over 120 degrees/sec (Cohen et al., 1977).

If lights are switched off during visual stimulation, optokinetic nystagmus (OKN) continues as optokinetic after-nystagmus (OKAN). As OKAN velocity declines, neuronal frequency in parallel returns to baseline activity with time constants similar to these for vestibular stimulation. If secondary nystagmus develops, it is also reflected in neuronal activity (Waespe and Henn, 1977b).

Neuronal activity during conflicting visual-vestibular stimulation

If the optokinetic drum is mechanically coupled to the turntable and both rotated together, the monkey will experience angular acceleration in the abscence of visual

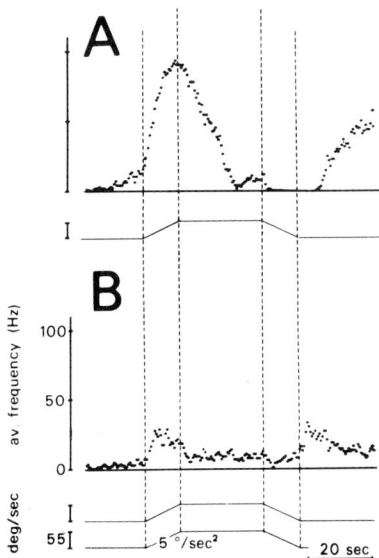

Fig. 3. Type I neuron during vestibular (A, rotation of the monkey in the dark) and during conflicting visual-vestibular stimulation (B, turntable and optokinetic drum rotating together into the same direction). Neuronal frequency averaged over 1 sec, and displayed every 0.5 sec. Below stimulus profile with acceleration of 5 degrees/sec^2 and constant velocity of 55 degrees/sec. During the conflict stimulation in B the response is strongly attenuated.

displacement, which creates a sensory conflict. With accelerations up to 10 degrees/sec² vestibular nystagmus was reduced to less than 10% of chair velocity, or totally suppressed. Neuronal activity was also attenuated compared to pure vestibular stimulation in the dark (Fig. 3); more specifically, the thresholds were raised in comparison to pure vestibular stimulation, sensitivities were lowered, and time constants were reduced to about 5 sec (Waespe and Henn, 1978). Attenuation of neuronal activity was always less than that of nystagmus. This points to another dissociation of neuronal activity and the nystagmus slow-phase velocity.

Neuronal activity during combined visual-vestibular stimulation

The normal visual-vestibular interaction takes place when the monkey is rotated in the light within the stationary optokinetic cylinder. Then vestibular and visual information reinforce each other. In a previous paper (Waespe and Henn, 1977a), results have been reported for a single value of acceleration (8 degrees/sec²): neuronal activity follows the vestibular input during the period of acceleration, and the visual input during constant velocity rotation (Fig. 4A). Stimulus parameters have now been extended. The interaction is non-linear and is mainly determined by the working ranges of the visual and

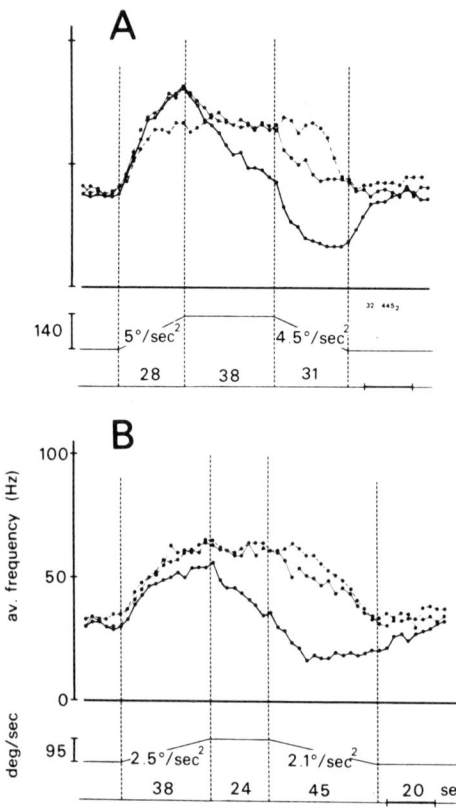

Fig. 4. Activity of a type I neuron during vestibular (———), visual (-----) and combined visual-vestibular (———) stimulation. Neuronal activity is averaged over 1 sec, and displayed every 3 sec; below trapezoid velocity profile indicates acceleration, constant velocity rotation, and deceleration. At the bottom time-scale. In A during combined stimulation the activity is dominated by the vestibular input during the period of acceleration (with 5 degrees/sec²). In B with lower acceleration, the visual input dominates the response.

vestibular inputs. For low accelerations in the light, which are near or below vestibular threshold, the visual input completely dominates the response (Fig. 4B). On the other hand, with higher values of acceleration, the visual input might be saturated, so that during constant velocity rotation, neuronal activity is less than during the period of acceleration (Fig. 4A). The functional significance of this dual input with different, but overlapping working ranges, could be to provide the vestibular system with a true velocity signal independent of the values of acceleration.

Fig. 5 shows the basic features for type I neurons and Fig. 6 for type II neurons (averaged responses, n = 4). Neuronal frequency increase is plotted against actual velocity for different values of acceleration. During vestibular stimulation alone (Figs. 5A and 6A) neuronal activity increase is higher with higher values of acceleration. During visual stimulation alone (Fig. 5B) neurons saturate at a velocity of about 40 degrees/sec (or 60 degrees/sec for the neurons in Fig. 6B), and were also unable to follow visual accelerations of more than 2.5 degrees/sec^2 (5 degrees/sec^2 for neurons in Fig. 6B). If the monkey is rotated in the light, both inputs reinforce each other resulting in a much improved response, which is linear over a wider range than for

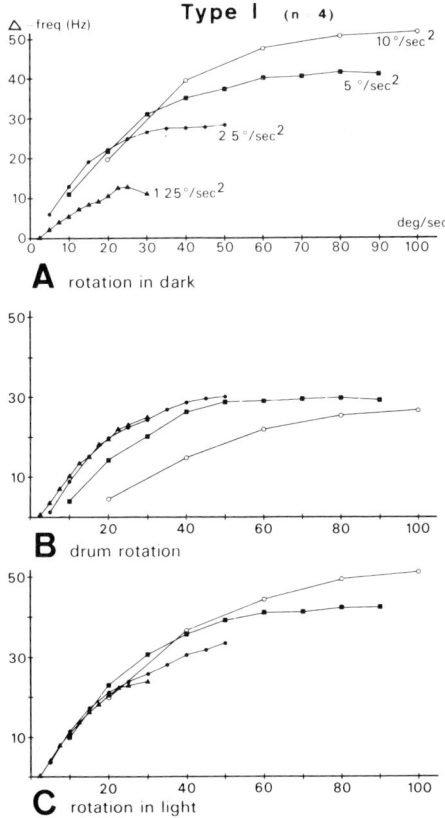

Fig. 5. Activity of type I neurons (n = 4, averaged curves) during vestibular (A), visual (B) and combined stimulation (C) over periods of different stimulus accelerations. Ordinate is frequency increase above resting level, abscissa is instantaneous velocity of turntable (A,C) or cylinder (B). During combined visual-vestibular stimulation (C), all 4 curves are superimposed over the low velocity range. In this range neuronal frequency is a function of velocity, independent of acceleration.

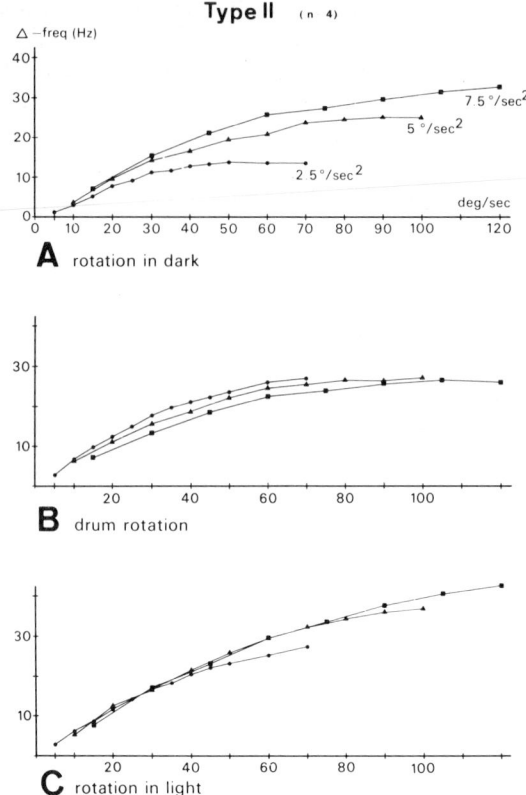

Fig. 6. Activity of type II neurons (n = 4). Same display as in Fig. 5. Again, during combined stimulation (C), curves are superimposed, showing that frequency is a function of stimulus velocity independent of the value of acceleration. For type II neurons the linear range of this velocity function is greatly extended, compared to the type I neurons.

each input alone, independent of the values of acceleration. In particular, for type I neurons (Fig. 5C) curves of different accelerations are superimposed on each other in the low velocity range. Curves have a relatively high slope indicating a great sensitivity to changes in that velocity range. For type II neurons (Fig. 6C), the slope is less, but the linear range of the curve is extended up to higher velocities. These results suggest some specialization: type I neurons are more sensitive and have their linear response in the low velocity range, whereas type II neurons are less sensitive, but have a much extended working range.

DISCUSSION

The interaction between visual and peripheral vestibular inputs onto vestibular nuclei neurons seems to be firmly established. In order to be compatible, these two inputs should code motion using the same dimensions. Although the adequate vestibular signal for semicircular canal excitation is angular acceleration, there is a twofold partial integration of this signal, first by the mechanics of the cupula-endolymph system, secondly by central mechanisms. Therefore, dominant time constants of central vestibular neurons to pulses of acceleration are about 20 sec. During any naturally occurring head movements, which do not last more than several seconds, the central vestibular signal therefore represents head velocity instead of acceleration. It had been shown that the visually mediated motion information also uses a velocity code. The directions of this visually and vestibularly mediated motion information coincide. The velocity range of the visual input has been determined and normally lies between 3 and 60 degrees/sec. The velocity range of the vestibular input is less easily described, because duration of acceleration plays an important role. However, the only "natural and adequate" stimulation for any sensory system to detect motion, is to rotate the animal in the light. In laboratory terms, this is the combined visual-vestibular stimulation, which results in a response that is a function of instantaneous velocity only. As shown in Figs. 5C and 6C, vestibular neurons transmit a signal about actual velocity, which is independent of the value of acceleration. Type I neurons are more sensitive for low velocity ranges and saturate at lower velocities. Type II neurons are less sensitive, but have a linear response range which extends up to higher velocities. The anatomical and physiological mechanisms for this interaction are still unclear, and have been discussed elsewhere (Dichgans and Brandt, 1972; Azzena et al., 1974; Allum et al., 1976; Waespe and Henn, 1977a; Kubo et al., 1978).

Several models have been proposed to simulate or explain visual-vestibular interaction (Allum et al., 1976; Raphan et al., 1977; Robinson, 1977). So far, they have been successful in describing special instances of visual-vestibular interaction, but not the whole dynamic range of responses. Therefore, a non-linear model involving a switch (Young, 1970) or, in a refined version, a soft switch has been proposed (Zacharias and Young, 1979).

In our experiments with untrained animals, the different optokinetic and vestibular stimuli always elicited nystagmus. It had been shown that activity in the vestibular nuclei and nystagmus velocity can always be dissociated either by applying high stimulus velocities or during conflicting visual-vestibular stimulation. As about 50% of all vestibular neurons are additionally modulated with individual eye movements, the question arises, whether their activity already encodes the motor signal. Surprisingly, these neurons saturate to optokinetic stimulation already at low velocities, and show a large degree of dissociation with nystagmus during conflicting visual-vestibular stimulation. To pursue this question further, experiments have been extended to include trained animals, which suppress their nystagmus during vestibular or visual stimulation (Buettner et al., 1978a). Activity in most neurons was attenuated during the suppression of nystagmus. This demonstrates that any concepts, which introduce sharp separations between sensory and motor mechanisms, are artificial. It further stresses that a general description of neuronal activity in the vestibular nuclei cannot yet be given. There are still too many experimental paradigms which have not yet been tested. Therefore, the above descriptions should only be understood as phenomenological descriptions of events observed under certain controlled stimulus conditions.

SUMMARY

Activity of vestibular nuclei neurons has been recorded during angular rotation in darkness, optokinetic stimulation or combined stimulation. When monkeys were rotated in the light, neuronal activity was related to actual velocity independent of the rate of acceleration. Relation to oculomotor output is complex; the vestibular signal alone cannot be responsible for either eliciting or suppressing nystagmus.

ACKNOWLEDGEMENTS

We thank Ms V. Isoviita for assistance during experiments and Mr V. Corti for building and maintaining part of the electronic equipment. This work was supported in part by Swiss National Foundation for Scientific Research 3.672-0.77 and the European Training Program for Brain and Behavior Research.

REFERENCES

Allum, J.H.J., Graf. W., Dichgans, J. and Schmidt, C.L. (1976) Visual-vestibular interactions in the vestibular nuclei of the goldfish. *Exp. Brain Res.*, 26, 463–485.
Ansorge, K. and Grüsser-Cornehls, U. (1977) Visual and visual-vestibular responses of frog cerebellar neurons. *Exp. Brain Res.*, 29: 445–465.
Azzena, G.B., Azzena, M.T. and Marini, R. (1974) Optokinetic nystagmus and the vestibular nuclei. *Exp. Neurol.*, 42: 158–168.
Bisti, S., Maffei, L. and Piccolino, L. (1974) Visuovestibular interactions in the cat superior colliculus. *J. Neurophysiol.*, 37: 146–155.
Buettner, U.W., Büttner, U. and Henn, V. (1978a) Neuronal activity in the vestibular nuclei of monkeys trained to suppress their nystagmus during vestibular and optokinetic stimulation. *Neurosci. Lett.*, Suppl. 1:S351.
Buettner, U.W., Büttner, U. and Henn, V. (1978b) Vestibular nuclei activity in the alert monkey during sinusoidal rotation in the dark. *J. Neurophysiol.*, 41: 1614–1628.
Büttner, U. and Henn, V. (1976) Thalamic unit activity in the alert monkey during natural vestibular stimulation. *Brain Res.*, 103: 127–132.
Büttner, U. and Buettner, U.W. (1978) Parietal cortex (2v) neuronal activity in the alert monkey during natural vestibular and optokinetic stimulation. *Brain Res.*, 153: 392–397.
Cohen, B., Matsuo, V. and Raphan, Th. (1977) Quantitative analysis of the velocity characteristics of optokinetic nystagmus and optokinetic afternystagmus. *J. Physiol (Lond.)*, 270: 321–344.
Daunton, N.G. and Thomson, D.D. (1976) Otolith-visual interactions in single units of cat vestibular nuclei. *Neurosci. Abstr.*, II: 1057, no. 1526.
Denney, D. and Adorjani, C. (1972) Orientation specificity of visual cortex neurons after head tilt. *Exp. Brain Res.*, 14: 312–317.
Dichgans, J. and Brandt, Th. (1972) Visual-vestibular interaction and motion perception. *Bibl. ophthal. (Basel)*, 81: 327–338.
Duensing, F. and Schaefer, K.P. (1958) Die Aktivität einzelner Neurone im Bereiche der Vestibulariskerne bei Horizontalbeschleunigungen unter besonderer Berücksichtigung des vestibulären Nystagmus. *Arch. Psychiatr. Nervenkr.*, 198: 224–252.
Fredrickson, J.M., Schwarz, D. and Kornhuber, H.H. (1965) Convergence and interaction of vestibular and deep somatic afferents upon neurons in the vestibular nuclei of the cat. *Acta oto-laryngol. (Stockh.)*, 61: 168–188.
Fuchs, F.F. and Kimm, J. (1975) Unit activity in vestibular nucleus of the alert monkey during horizontal angular acceleration and eye movement. *J. Neurophysiol*, 38: 1140–1161.
Ghelarducci, B., Ito, M. and Yagi, N. (1975) Impulse discharges from flocculus cells of alert rabbits during visual stimulation combined with horizontal head rotation. *Brain Res.*, 87: 66–72.
Grüsser, O.J., Grüsser-Cornehls, U. and Saur, G. (1959) Reaktionen einzelner Neurone im optischen Cortex der Katze nach elektrischer Polarisation des Labyrinthes. *Pflügers Arch.*, 269: 593–612.
Henn, V., Young, L. and Finley, C. (1974) Vestibular nucleus units in alert monkeys are also influenced by moving visual fields. *Brain Res.*, 71: 144–149.
Horn, G. and Hill, R.M. (1969) Modifications of receptive fields of cells in the visual cortex occurring spontaneously and associated with bodily tilt. *Nature (Lond.)* 221: 186–188.
Keller, E. and Daniels, P. (1975) Oculomotor related interaction of vestibular and visual stimulation in vestibular nucleus cells in alert monkey. *Exp. Neurol.*, 46: 187–198.
Kornhuber, H.H. and Da Fonseca, J.S. (1964) Optovestibular integration in the cat's cortex: a study of sensory convergence on cortical neurons. In *The Oculomotor System*, B. Bender (Ed.), pp. 239–277.
Kubo, T., Matsunaga, T. and Hayashi, Y. (1978) Convergence of visual and vestibular inputs on pontine reticular formation of the rabbit. *Brain Res.*, 147: 177–182.
Lisberger, S.G. and Fuchs, A.F. (1974) Response of flocculus Purkinje cells to adequate vestibular stimulation in the alert monkey: fixation vs. compensatory eye movements. *Brain Res.*, 69: 347–353.

Lisberger, S.G. and Fuchs, A.F. (1978a) Role of primate flocculus during rapid behavioral modification of vestibuloocular reflex. I. Purkinje cell activity during visually guided horizontal smooth-pursuit eye movements and passive head rotation. *J. Neurophysiol.*, 41: 733–763.

Lisberger, S.G. and Fuchs, A.F. (1978b) Role of primate flocculus during rapid behavioral modification of vestibuloocular reflex. II. Mossy fiber firing patterns during horizontal head rotation and eye movement. *J. Neurophysiol.*, 41: 764–777.

Mach E. (1875) Grundlinien der Lehre von den Bewegungsempfindungen. Engelmann, Leipzig (1967, Bonset, Amsterdam).

Miles, F.A. (1974) Single unit firing patterns in the vestibular nuclei related to voluntary eye movements and passive body rotation in conscious monkeys. *Brain Res.*, 71: 215–224.

Miles, F.A. and Fuller, J.H. (1975) Visual tracking and the primate flocculus. *Science*, 189: 1000–1002.

Raphan, Th., Cohen, B. and Matsuo, V. (1977) A velocity-storage mechanism responsible for optokinetic nystagmus (OKN), optokinetic after-nystagmus (OKAN) and vestibular nystagmus. In *Control of Gaze by Brain Stem Neurons*, R. Baker and A. Berthoz (Eds.), Elsevier, Amsterdam, pp. 37–47.

Robinson, D.A. (1977) Linear addition of optokinetic and vestibular signals in the vestibular nuclei. *Exp. Brain Res.*, 30: 447–450.

Rubin, A.M., Liedgren, S.R., Milne, A.C., Young, J.A. and Fredrickson, J.M. (1977) Vestibular and somatosensory interaction in the cat vestibular nuclei. *Pflügers Arch.*, 371: 155–160.

Waespe, W. and Henn, V. (1977a) Neuronal activity in the vestibular nuclei of the alert monkey during vestibular and optokinetic stimulation. *Exp. Brain Res.*, 27: 523–538.

Waespe, W. and Henn, V. (1977b) Vestibular nuclei activity during optokinetic after-nystagmus (OKAN) in the alert monkey. *Exp. Brain Res.*, 30: 323–330.

Waespe, W. and Henn, V. (1978) Conflicting visual-vestibular stimulation and vestibular nucleus activity in alert monkeys. *Exp. Brain Res.*, 33: 203–211.

Waespe, W., Henn, V. and Miles, T.S. (1977) Activity in the vestibular nuclei of the alert monkey during spontaneous eye movements and vestibular or optokinetic stimulation. In *Control of Gaze by Brain Stem Neurons*, R. Baker and A. Berthoz (Eds.), Elsevier, Amsterdam, pp. 269–278.

Young, L.R. (1970) On visual-vestibular interaction. *Fifth Symposium on the Role of the Vestibular Organs in Space Explorations.* NASA SP-314, pp. 205–210.

Zacharias, G.L. and Young, L.R. (1979) Influence of combined visual and vestibular cues on human perception and control of horizontal rotation. *Exp. Brain Res.*, in press.

Interaction of Visual and Canal Inputs on the Oculomotor System via the Cerebellar Flocculus

Y. MIYASHITA

*Department of Physiology, Faculty of Medicine,
University of Tokyo, 7-3-1 Hongo, Bunkyo-ku, Tokyo (Japan)*

In order to stabilize retinal images during movements, an animal uses two cooperative ocular reflex mechanisms; vestibulo-ocular reflex and optokinetic response. Several lines of evidence obtained in various animal species have established that the cerebellar flocculus plays an important role in these two reflex mechanisms. To specify the role of the flocculus, three modes of study, i.e., i) dissection of neuronal circuitry, ii) recording of neuronal activity, and iii) analysis of dynamics of eye movements, should be advanced conjointly, preferably on one and the same animal species (Ito, 1975). This article introduces knowledge obtained along these three lines of investigations on the albino rabbits as to how vestibular and visual signals interact with each other via the cerebellar flocculus.

LINEAR INTERACTION OF EYE MOVEMENTS BETWEEN VESTIBULO-OCULAR REFLEX AND OPTOKINETIC RESPONSE

In the experiment by Batini et al. (1979), an alert albino rabbit was rotated sinusoidally on the motor-driven horizontal turntable by $5°–30°$ (peak-to-peak) at a frequency of 0.03–0.5 Hz. A narrow vertical slit light presented in front of the eye provided optokinetic stimulus. Evoked horizontal eye movements were observed by means of a closed circuit television monitor system. The net horizontal vestibulo-ocular reflex (HVOR) was obtained by rotation of the turntable in darkness and the net optokinetic response (OKR) by movement of the slit light. The transfer function G for the HVOR is defined as the complex ratio of the eye movement vs. turntable rotation, and that for the OKR as the complex ratio of the eye movement vs. slit light movement. It can be expressed in the complex form: $G = g\, e^{i\rho}$, where g is the gain and ρ is the phase shift of the eye movement relative to the stimulus.

Dynamic characteristics of the net HVOR in the albino rabbits closely resemble those described for Dutch belted rabbits (Baarsma and Collewijn, 1974). In both of the two strains of rabbits, the HVOR is considerably smaller than that obtained in monkeys and cats (Carpenter, 1972; Robinson, 1976). The input-output relationship of the HVOR of the albino rabbits is not quite linear, as the HVOR gain at 0.5–0.1 Hz increases slightly but significantly at larger amplitudes of turntable rotation compared with those at 5°. On the other hand, as pointed out by Baarsma and Collewijn (1974), the gain of the OKR can be related uniquely to the maximum velocity of the slit light movement regardless frequencies and amplitudes. Rotation of the turntable combined with presentation of the stationary slit light immediately increased the gain and reduced the phase shift of the HVOR. Movement of the slit light in phase with the turntable by an angular amplitude twice that of the turntable reduced the gain and advanced the phase of the HVOR. These changes occur in the direction of stabilizing retinal images of the slit light. Baarsma and Collewijn (1974) originally postulated that modification of the HVOR by vision is due to linear combination of the HVOR and the OKR. However, in testing this postulate only gain relationships were considered, whereas phase relationships were entirely neglected. This might be one of the possible reasons why the calculation did not agree with measurements. Indeed, when phase relationships were taken into account, a good agreement was obtained between the calculated and measured eye movements, as shown below (Batini et al., 1979).

The transfer function, gain, and phase, of the net HVOR are represented by G_v, g_v and ρ_v, those of the HVOR under the influence of the fixed slit light by G_f, g_f and ρ_f, those of the HVOR under the moving slit light by G_m, g_m and ρ_m, and those of the net OKR by G_o, g_o and ρ_o. In the postulated linear interaction between the HVOR and the OKR, the eye is initially driven by the HVOR, and then the remaining retinal errors induce the OKR. Hence, G_f and G_m are given by the following equations:

$$G_f = G_v + G_o (1 - G_v) \tag{1}$$
$$G_m = G_v - G_o (1 + G_v) \tag{2}$$

These are essentially the same as the equations derived by Baarsma and Collewijn (1974). From these equations, we can derive the following equations by which g_f, ρ_f, g_m and ρ_m are given as function of g_v, ρ_v, g_o and ρ_o.

$$g_f = \sqrt{g_v^2 + g_o^2 + g_v^2 g_o^2 + 2 g_v g_o \cos(\rho_v - \rho_o) - 2 g_o g_v^2 \cos \rho_o - 2 g_o^2 g_v \cos \rho_v}, \tag{3}$$

$$\rho_f = \tan^{-1} \frac{g_v \sin \rho_v + g_o \sin \rho_o - g_o g_v \sin(\rho_o + \rho_v)}{g_v \cos \rho_v + g_o \cos \rho_o - g_o g_v \cos(\rho_o + \rho_v)}, \tag{4}$$

$$g_m = \sqrt{g_v^2 + g_o^2 + g_o^2 g_v^2 - 2 g_o g_v \cos(\rho_v - \rho_o) - 2 g_o g_v^2 \cos \rho_o + 2 g_o^2 g_v \cos \rho_v}, \tag{5}$$

$$\rho_m = \tan^{-1} \frac{g_v \sin \rho_v - g_o \sin \rho_o - g_o g_v \sin (\rho_o + \rho_v)}{g_v \cos \rho_v - g_o \cos \rho_o - g_o g_v \cos(\rho_o + \rho_v)}. \tag{6}$$

Since G_o is amplitude-dependent, values of G_o to be used for computing G_f or G_m should vary depending on the retinal error to be opposed by the OKR, precise value of which was not known at the time of run; for convenience, we used slit light movement by 2.5° (for G_f) and 7.5° (for G_m) for measuring G_o, which corresponded to an assumption, $g_v = 0.5$ and $\rho_v = 0$. In a few cases, where the actual value of g_v seemed to be relatively small (less than 0.2), 5° was adopted for measurement of the G_o to be used for calculating both G_f and G_m. Values of these transfer functions exhibited considerable individual and daily fluctuation. Therefore, they were measured during continuous runs in one and the same preparations. For both G_f and G_m, calculated values by Eqs. (3)–(6) agree very closely with the actual ones, as exemplified in Fig. 1, thus supporting the postulate of the linear interaction between the HVOR and OKR.

Effects of flocculectomy upon dynamic characteristics of the HVOR and OKR

Involvement of the flocculus in the visual-vestibular interaction has been demonstrated by chronic ablation of the flocculus which depresses the rapid modification of the

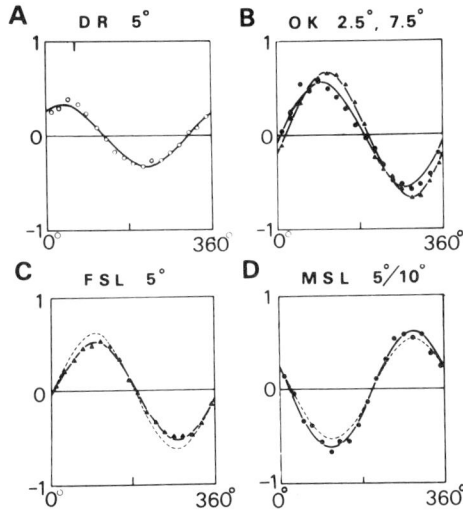

Fig. 1. Calculation of the eye movements induced by combined vestibular and optokinetic stimulation. Averaged curves of eye rotation obtained during continuous trials on one and the same rabbit, with 4 different stimulus conditions. Ordinates: displacement of the left eye. Full compensation is taken as unity. Abscissae: angular displacement of the turntable (A,C,D) or the slit light (B). A) 5° turntable rotation in darkness at 0.03 Hz. B) ▲, 2.5° movement of the slit light at 0.03 Hz around the stationary rabbit; ●, 7.5° slit light movement at the same frequency. C) 5° turntable rotation at 0.03 Hz with the fixed slit light. D) That with 10° in-phase movement of the slit light. Solid lines are best-fit sine curves derived with a Fourier analysis. Broken lines in C and D indicate the time course of eye movements calculated from measurements in A and B. (Modified from Batini et al., 1979).

VOR with vision (Ito et al., 1974; Takemori and Cohen, 1974). In the albino rabbits, this effect is specific to flocculectomy, and is not reproduced by lesions of any other parts of the cerebellum, i.e., nodulus, uvula, lobules VI and VII, or paraflocculus (Ito, Jastreboff and Miyashita, in preparation). In accordance with the above described results indicating that the visual-vestibular interaction is due to linear combination of the VOR with OKR, the OKR gain was found to be reduced significantly by flocculectomy, to less than a half at all tested frequencies from 0.1 to 0.03 Hz. Linearity in combination of the HVOR and OKR was maintained after flocculectomy, as tested at relatively low frequencies of head and slit light rotations.

The area of the flocculus specifically related to the HVOR

In accordance with Anderssen and Oscarsson's (1978) cerebellar microzones in the vermal cortex, a narrow strip of a mm wide and a few mm long has been localized on the cortical sheet of the flocculus as specifically related to the HVOR. This strip was revealed by labeling of Purkinje cells with horseradish peroxidase (HRP) injected into the rostral area of the medial vestibular nucleus which contains relay cells of the HVOR (Yamamoto and Shimoyama, 1977; Yamamoto, 1978). Local stimulation at this strip evokes horizontal abduction of the ipsilateral eye, apparently due to Purkinje cell inhibition of relay cells for the HVOR (Dufossé et al., 1977). It has been confirmed further that local stimulation at this strip inhibits the reflex contraction evoked in the medial rectus muscle by the ipsilateral labyrinthine stimulation (Ito, Orlov and Yamamoto, in preparation). The inhibition occurs with latencies brief enough to indicate that it is mediated by monosynaptic innervation by Purkinje cells of relay cells of the HVOR.

IMPULSE DISCHARGES FROM FLOCCULUS PURKINJE CELLS DURING THE HVOR AND OKR

When an alert rabbit was rotated in darkness on the turntable sinusoidally at 0.1 Hz by 5° or 10° (peak-to-peak), many flocculus Purkinje cells exhibited significant modulation in discharge frequencies of their simple spikes. By a Fourier analysis, the amplitude of modulation was estimated relative to the average discharge frequency and the phase angle relative to the head velocity (Ghelarducci et al., 1975). Modulation with the phase angle around 0° (±45°) was defined as in-phase type; that around 180° (±45°) as out-phase type; and that around 90° (±45°) and 270° (±45°) as intermediate type. Relevance of a sampled Purkinje cell to the HVOR was examined by applying electric pulse trains (frequency, 500/sec; duration, 1 sec; pulse width, 0.2 msec; current intensity, 5–30 µA) through the recording glass microelectrode. The area of the flocculus from which an abduction of the ipsilateral eye was induced was called the H-zone, and that from which a downward movement of the ipsilateral eye was caused the V-zone. If no movement was induced in the ipsilateral eye, the stimulated area was taken as the N-zone. The characteristic features of the H-zone Purkinje cells are shown in polar diagrams of Fig. 2. In most of the H-zone Purkinje cells sampled (84 of 93 cells), head rotation in darkness induced significant modulation (Dufossé et al., 1978). Rather surprising was the fact that a significant modulation was induced also in many Purkinje cells sampled

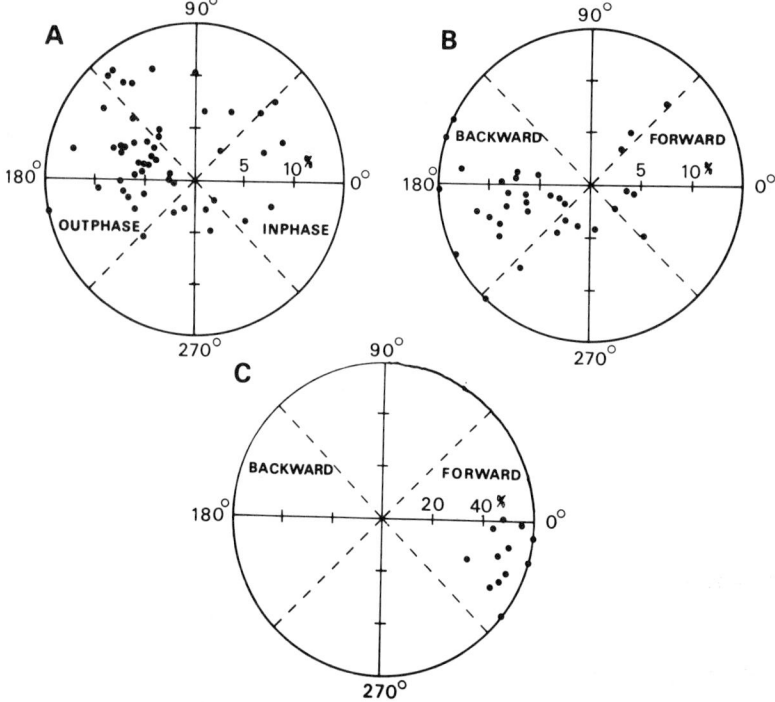

Fig. 2. Modulation of Purkinje cell discharge in H-zone of the flocculus. Broken lines passing the center of the polar diagrams with 45° inclination separate the modulation types as indicated. A) Simple spike modulation induced by head rotation in darkness at 0.1 Hz by 5° (peak-to-peak). B) Simple spike modulation induced by slit light movement at 0.1 Hz by 2.5°. C) Complex spike modulation obtained under the same condition as B. (From Dufossé, Ito, Jastreboff and Miyashita, in preparation.)

from other zones (80 of 135 V-zone cells, 72 of 123 N-zone cells). The H-zone, however, was characterized by the out-phase modulation; more than a half of the H-zone Purkinje cells exhibited out-phase modulation (49 of 82 cells), and more than half of Purkinje cells with out-phase modulation obtained through the three zones belonged to the H-zone (49 of 86 cells). By contrast, in-phase modulation was frequently found in the V- and N-zones, but rather rarely in the H-zone. In relay cells of the HVOR, out-phase modulated impulses of inhibitory Purkinje cells should facilitate the excitatory action of primary vestibular afferents from the horizontal canal which modulate in-phase. Hence, the dominance of out-phase modulation in the H-zone suggests that the flocculus normally facilitates the HVOR. This agrees with the fact that the HVOR gain in the albino rabbits is reduced after flocculectomy (Ito et al., in preparation).

Many flocculus Purkinje cells also responded to sinusoidal movement of the slit light at 0.1 Hz by 2.5° (peak-to-peak). When the phase shift of simple spike modulation is within 180° (±45°) relative to the slit light velocity, this is defined as backward type, as the firing rate increases at backward movement of the slit light. When the phase shift is within 0° (±45°), this is defined as forward type. About half of the Purkinje cells in the H-zone are of the backward type (21 of 41 cells) as shown in Fig. 2B. Since electrical stimulation of H-zone Purkinje cells induces abduction of the ipsilateral eye as shown above, it is likely that the backward type of firing in H-zone Purkinje cells contributes to the abduction of the eye evoked by the OKR. This view is in agreement with the fact that the OKR gain in the rabbit is reduced after flocculectomy.

When the fixed slit light is turned on during the turntable rotation, the contralateral head rotation is accompanied with backward movement of the slit light relative to the head. Consequently, the majority of H-zone Purkinje cells exhibited an increase of the out-phase modulation due to head rotation as soon as the slit light was turned on. This would, at least in part, account for the instantaneous increase of the HVOR gain under the fixed slit light. Similar consideration applies to the reduction of the out-phase modulation of H-zone Purkinje cells under the moving slit light which could be closely related to the reduction of the OKR gain.

Reciprocal to simple spikes, complex spikes of H-zone Purkinje cells exhibited almost exclusively a modulation of forward type (11 of 12 cells) during the OKR, as shown in Fig. 2C. Except for one N-zone cell which showed an in-phase modulation, no cell showed modulation of complex spike discharges during the OKR in either the V-zone (9 cells tested) or the N-zone (7 cells).

RESPONSES OF H-ZONE PURKINJE CELL TO FLASH LIGHT STIMULATION

Impulse discharges from H-zone Purkinje cell are also modulated by flash light stimulation in an alert rabbit (Miyashita, in preparation). Poststimulus histograms (PST-histograms) were constructed during 200–800 sweeps repeated at 2 Hz. Fig. 3C shows depression in simple spike discharges from an H-zone Purkinje cell evoked by flash light stimulation. The latency of onset of this response is 21.5 msec. In the same Purkinje cell complex spikes were evoked very effectively with a slight delay to onset of the depression of simple spikes as shown in Fig. 3D. This Purkinje cell was of backward type in its simple spike modulation by the slit light movement, as shown in Fig. 3B. Another type of response seen in H-zone P-cells was a facilitation of simple

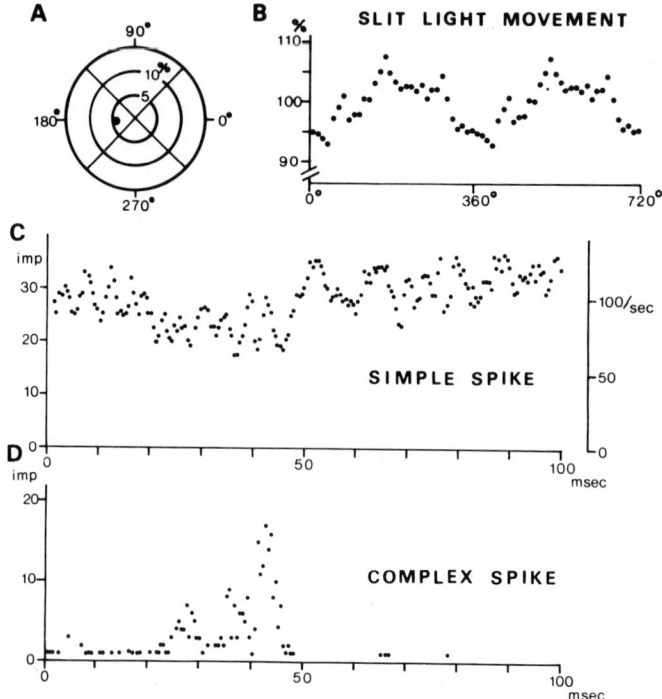

Fig. 3. Responses of an H-zone Purkinje cell to flash light stimulation and slit light movement. A) Polar diagram expression of the simple spike response to slit light movement. B) Spike density histogram of simple spikes averaged over 10 periods of sinusoidal slit movement at 0.1 Hz by 2.5°. Ordinate: number of simple spikes per bin, relative to the mean spike number over the whole period of rotation. Abscissa: angular displacement of the slit light. C) PST-histogram of simple spikes to flash light stimulation during 700 sweeps repeated at 2 Hz. The scale on the left side: no. of spikes/bin. That on the right side: instantaneous frequency. Bin width: 500 μsec. D) In the same cell and under the same stimulus conditions as in C, but for complex spikes in 200 sweeps.

spike discharges which occurred with a latency of 23–24 msec after flash light stimulation, which accompanied no response of complex spikes. As summarized in Table IA, those H-zone Purkinje cells which increased their simple spike discharges in response to backward slit light movement (12) responded to the flash light exclusively with depression of simple spike discharges. Purkinje cells modulated with forward type by slit light movement, responded with facilitation in simple spike discharges or did not respond at all to flash light stimulation. The relationship between simple spike modulation by slit light movement and complex spike response to flash light stimulation is indicated in Table IB. Here, facilitation in simple spike discharges during backward slit light movement is closely correlated with excitation in complex spikes by flash light (12 of 13 cells). When simple spike modulation by slit light was of the forward type, complex spikes did not respond to flash light. It was common that responses in simple and complex spikes to flash light were reciprocal to each other, as exemplified by Fig. 3C and D. The close correlation between simple spike responses to slit light movement and those to flash light indicates an important role played by visual inputs in determining discharge patterns of H-zone Purkinje cells during the OKR. Above results are in part confirmatory of those in an anesthetized rabbit by Simpson and

TABLE I

RESPONSE TO FLASH LIGHT STIMULATION VS. SIMPLE SPIKE RESPONSE TO SLIT MOVEMENT

A

Flash light – simple spike

Slit movement – simple spike		Inhibition	Excitation	No response
	Backward	12	0	0
	Forward	0	2	3
	Others	0	3	2

B

Flash light – complex spike

Slit movement – simple spike		Inhibition	Excitation	No response
	Backward	0	12	1
	Forward	0	0	3
	Others	2	0	2

Hess (1977), although their visual stimuli did not evoke the OKR even in an alert state.

A question arises as to whether the depression of simple spike discharges by flash light is affected through mossy fiber pathway or it is a side effect of initiation of complex spike, as an inhibition in simple spike discharges is known to occur through climbing fiber collaterals innervating Golgi cells and basket cells (Bloedel and Roberts, 1971). The fact that the onset of simple spike depression after flash light stimulation often (5 of 12 cells) precedes that of complex spike excitation suggests that the simple spike depression is mediated, at least partly, by mossy fiber pathway. In anesthetized rabbits, it was confirmed that impulses evoked by electrical stimulation of the ipsilateral retina reached the H-zone of the flocculus via not only climbing but also mossy fiber pathways (Ito, Orlov and Yamamoto, in preparation).

CONCLUSION AND SUMMARY

This article presents three lines of evidence supporting the view that the flocculus is a center of visual-vestibular interaction in eye movements: 1) the HVOR and OKR interact linearly with each other, and the flocculectomy affects this interaction by reducing the gain of the OKR; 2) impulse discharge patterns of H-zone Purkinje cells are modulated during the VOR, OKR and their combinations; 3) H-zone Purkinje cells respond to flash light visual stimulation.

The relationship between eye movements and patterns of concomitant Purkinje cell discharges in the H-zone of the flocculus indicates that these Purkinje cells contribute to initiate eye movements during both the HVOR and OKR. However, possible influences of eye velocity inputs on flocculus Purkinje cells have been pointed out in monkeys (Lisberger and Fuchs, 1978), and this opens a question as to whether a modulation of Purkinje cell discharges is not a cause but a result of eye movements. Nevertheless, the present results obtained with flash light stimulation indicate an important contribution of visual signals to modulation in H-zone Purkinje cells of both simple and complex spike discharges. Dichotomy of the pathways for modulation of the two types of spikes, i.e., mossy fiber pathway for simple spikes and climbing fiber pathway for complex spikes is also suggested by the flash stimulation experiment. Yet, it remains for a future task to evaluate the possible contribution of the climbing fiber collateral pathways to modulation of simple spike discharges, and the relative contribution of visual and eye velocity inputs to simple spike modulation in H-zone Purkinje cells during the OKR.

REFERENCES

Anderssen, G. and Oscarsson, O. (1978) Climbing fibre microzones in cerebellar vermis and their projection to different group of cells in the lateral vestibular nucleus. *Exp. Brain Res.*, in press.

Baarsma, E.A. and Collewijn, H. (1974) Vestibulo-ocular and optokinetic reactions to rotation and their interaction in the rabbit. *J. Physiol. (Lond.)*, 238: 603–625.

Batini, C., Ito, M., Kado, R.T., Jastreboff, P.J. and Miyashita, Y. (1979) Interaction between the horizontal vestibulo-ocular reflex and optokinetic response in rabbits. *Exp. Brain Res.*, in press.

Bloedel, J.R. and Roberts, W.J. (1971) Action of climbing fibres in cerebellar cortex of the cat. *J. Neurophysiol.*, 34: 17–31.

Carpenter, H.R.S. (1972) Cerebellectomy and the transfer function of the vestibulo-ocular reflex in the decerebrate cat. *Proc. roy. Soc. B*, 181: 353–374.

Dufossé, M., Ito, M. and Miyashita, Y. (1977) Functional localization in the rabbit's flocculus determined in relationship with eye movements. *Neurosci. Lett.*, 5: 273–277.

Dufossé, M., Ito, M., Jastreboff, P.J. and Miyashita, Y. (1978) A neuronal correlate in rabbit's cerebellum to adaptive modification of the vestibulo-ocular reflex. *Brain Res.*, 150: 611–616.

Ghelarducci, B., Ito, M. and Yagi, N. (1975) Impulse discharge from flocculus Purkinje cells of alert rabbits during visual stimulation combined with horizontal head rotation. *Brain Res.*, 87: 66–72.

Ito, M. (1975) Learning control mechanisms by the cerebellum investigated in the flocculo-vestibulo-ocular system. In *The Nervous System. Vol./1, The Basic Neurosciences*, D.B. Tower (Ed.). Raven Press, New York, pp. 245–252.

Ito, M., Shiida, T., Yagi, N. and Yamamoto, M. (1974) Visual influence on rabbit horizontal vestibulo-ocular reflex presumably effected via the cerebellar flocculus. *Brain Res.*, 65: 170–174.

Lisberger, S.G. and Fuchs, A.F. (1978) Role of primate flocculus during rapid behavioral modification of vestibulo-ocular reflex. I. Purkinje cell activity during visually guided horizontal smooth-pursuit eye movements and passive head rotation. *J. Neurophysiol.*, 41: 733–763.

Robinson, D.A. (1976) Adaptive gain control of vestibulo-ocular reflex by the cerebellum. *J. Neurophysiol.*, 36: 954–969.

Simpson, J.I. and Hess, R. (1977) Complex and simple visual messages in the flocculus. In *Control of Gaze by Brain Stem Neurons*, R. Baker and A. Berthoz (Eds.) Elsevier, Amsterdam, New York, pp. 351–360.

Takemori, S. and Cohen, B. (1974) Loss of visual suppression of vestibular nystagmus after flocculus lesions. *Brain Res.*, 72: 213–224.

Yamamoto, M. (1978) Localization of rabbit's flocculus Purkinje cells projecting to the cerebellar lateral nucleus and the nucleus prepositus hypoglossi investigated by means of the horseradish peroxidase retrograde axonal transport. *Neurosci. Lett.*, 7: 197–202.

Yamamoto, M. and Shimoyama, I. (1977) Differential localization of rabbit's flocculus Purkinje cells projecting to the medial and superior vestibular nuclei, investigated by means of the horseradish peroxidase retrograde axonal transport. *Neurosci. Lett.*, 5: 279–283.

Visual-Vestibular Interactions and the Role of the Flocculus in the Vestibulo-Ocular Reflex

J. KIMM, J.A. WINFIELD* and A.E. HENDRICKSON

Departments of Otolaryngology and Physiology and Biophysics, University of Washington, Seattle; and Department of Ophthalmology, University of Washington, Seattle, WA 98195 (U.S.A.)

When the head moves in space the visual and vestibular systems interact to maintain a constant retinal input. With head rotation the eyes move in the opposite direction with a magnitude of angular displacement about equal to the head movement. These compensatory eye movements are called the vestibulo-ocular reflex (VOR). Recent experiments have shown that the reflex is affected by inputs to either the visual system and/or the vestibular system and is modified by the cerebellum as well.

In anesthetized or decerebrate animals, stimulation of cerebellar structures has produced inhibitory or facilitatory effects on vestibular neurons depending on the location of the stimulating electrode within the cerebellum. For example, stimulation of the flocculus inhibits (Shimazu and Smith, 1971; Ito et al., 1972; Baker et al., 1972; Fukuda et al., 1972; and others) vestibular neurons, some of which have been identified to project to oculomotor neurons, whereas similar stimulation of the nodulus and uvula has no significant inhibitory effect on the vestibulo-ocular pathway (Fukuda et al., 1972). In contrast to these findings with cerebellar cortical stimulation, Shimazu and Smith (1971) reported that some vestibular neurons are monosynaptically excited by stimulation of the ipsilateral fastigial nucleus, whereas cells in the vestibular nuclei are both facilitated and inhibited by stimulation of the contralateral fastigial nucleus. Kimm et al. (1976) have reported similar effects in the awake monkey.

These are the results of only some of the studies on cerebellar vestibulo-ocular relations at the single unit level. Other experiments have demonstrated participation of the cerebellum in vestibulo-ocular interactions. For instance, Purkinje cells located in the monkey's flocculus have been shown to respond to eye movements and/or vestibular stimulation (Noda et al., 1977; Lisberger et al., 1977; and others). It has also been shown that the firing rate of flocculus Purkinje cells in the rabbit is related to both the vestibular and visual stimuli (Ghelarducci et al., 1975). Other examples of visual-vestibular interactions have been discussed by Ito and others (Ito et al., 1974; Ito, 1975) as part of the long-term changes in the VOR following modifications in the visual input purportedly dependent on an intact cerebellum. Furthermore, in an experiment by Takemori and Cohen (1974) visual suppression of an induced nystagmus was shown to be lost following flocculus lesions. The results of these studies provide an empirical basis for the assumption that the flocculus is involved in the VOR.

---

* Visiting student on leave from Albert Einstein College of Medicine, New York.

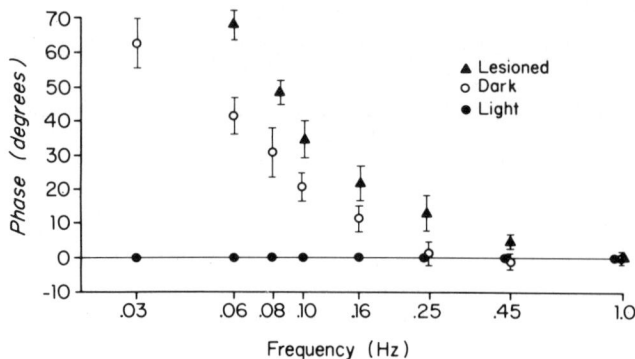

Fig. 1. Bode phase curves (eye position re normalized table position) for normal and bilateral flocculectomized chinchillas. ●, the mean light-on response of 8 animals at the amplitudes of ±10°; bars represent SDs. ○, mean dark responses of the same eight animals at ±10° table amplitude. ▲, mean response of bilateral flocculectomized chinchillas. The responses in the light-on and light-off conditions were essentially identical.

Further support for this idea is provided by the effects of floccular lesions on the normal pattern of visual modification of the VOR. Immediate and striking visual effects on the VOR have been demonstrated in normal rabbits (Baarsma and Collewijn, 1974) and in chinchillas by my laboratory (Hassul et al., 1976; Daniels et al., 1978). In these animals, the dynamics of the VOR have been studied in the dark and in the presence of a fixed visual surround. The difference in the VOR between these two conditions is summarized in Fig. 1. When a fixed visual surround is present, the eye movements are 180° out of phase with head movements at all frequencies of oscillation tested. In the dark and at frequencies below 0.16 Hz, eye movements lead head movements by up to 60°. The results of the experiment by Baarsma and Collewijn (1974) also show that the gain of the reflex decreased as phase lead increased.

When we lesioned the flocculus bilaterally in chinchillas, this normal pattern of visual influence on the VOR immediately changed. When an animal was oscillated at low frequencies in the presence of a fixed visual surround, the normally observed visual effect on the phase and gain of the VOR was eliminated. This resulted in a decrease in the mean time constant of the phase response in the dark, from 3 sec in the normal animal to 2 sec in the animals with lesions. This finding can be interpreted from Fig. 1 by estimation of the time constant from the increase in the phase lead at and below the frequencies of 0.25 Hz in the animals with bilateral flocculus lesions. The results of these studies lead us to believe that the flocculus plays an important role in visual-vestibular interactions.

VISUAL INPUT TO THE FLOCCULUS

There have been a number of physiological and anatomical studies delineating the visual input to the rabbit's flocculus (Maekawa and Simpson, 1972; Simpson and Alley, 1974; Maekawa and Takeda, 1975, 1976; and others). The salient points of these studies have shown that each retina provides input via both mossy fibers (MF) and climbing fibers (CF) to each flocculus. The rostrodorsal portion of the flocculus receives retinal input from an ipsilateral mossy fiber system and a contralateral mossy and climbing fiber pathway. In the cerebellar cortical area adjacent to this region, ipsilateral optic nerve stimulation results in both MF and CF responses whereas contralateral optic nerve stimulation results in only a MF response. In other areas of the flocculus outside of this dorsorostral region, ipsilateral optic nerve stimulation produces CF responses with little or no contribution from the MF. In the cortical area beyond the rostrodorsal area neither CF nor MF responses resulted following contralateral optic nerve stimulation. Fig. 2 summarizes these visual pathways to the flocculus.

Climbing fiber pathway

The anatomical sites subserving the CF pathway have been identified. The ipsilateral projection is from the retina to the contralateral accessory optic tract to relays in the lateral pretectal area, these being the dorsal terminal nucleus and nucleus of the accessory optic tract. Cells from these nuclei project via the central tegmental tract to the medial accessory olive (dorsal cap of Kooy) on the same side. Cells within the dorsal cap of Kooy then project to the contralateral flocculus. The contralateral CF pathway results from projections from the retina that cross at the optic chiasm to the accessory optic tract (Maekawa and Takeda, 1976). These fibers then recross the midline through the posterior commissure to terminate in the dorsal cap of Kooy. As stated before, the cells from this structure project to the contralateral flocculus. The visual CF input to the flocculus therefore carries information from both eyes. Details of the retinal projections to this pathway have yet to be elucidated.

Mossy fiber pathway

The MF pathway to the flocculus has been characterized physiologically, and recently the anatomical route has also been described. A disynaptic anatomical connection from the retina to the cerebellum has been described in the pigeon by Brauth and Karten (1977) and in the chinchilla by my laboratory (Winfield et al., 1978). In these two studies portions of the cerebellum were injected with horseradish peroxidase (HRP). Cells projecting to these areas were then labeled by retrograde transport of endocytosed HRP reaction products. In the pigeon study, a disynaptic projection from the retina to the cerebellum by way of the contralateral nucleus of the basal optic root (nBOR) was reported. The cerebellar sites involved were the uvula and flocculonodular lobes (folia IX_c, IX_d). Following HRP injections into the flocculus of the chinchilla, we (Winfield et al., 1978) reported the occurrence of HRP reaction products in cells within the medial terminal nucleus (MTN) of the accessory optic tract. In mammals, this nucleus is the homologue of the nBOR of pigeon. A direct retinal projection to the chinchillas' MTN was also demonstrated by means of orthograde transport of tritiated protein following intravitreal injections of ^3H precursor.

The retinal projections to the MTN are bilateral with the primary route by way of the contralateral pathway. Silver grain counts in the MTN following the intravitreal injections, showed a distribution ratio of 14 : 1 contra- to ipsilateral, respectively. Cells within the MTN then project bilaterally to the flocculus with the principal projection to the contralateral side. These anatomical results are also summarized in Fig. 2.

The retinal components of the MF pathway were identified in another series of HRP injection experiments. Fig. 3 shows that some of the displaced ganglion cells as well as ganglion cells in the normal ganglion cell layers were labeled after injection of the MTN. All labeled ganglion cells were approximately of the 65–75th percentile in size. They were distributed chiefly in the central retina and were as large or larger than their immediate neighbors. The largest ganglion cells in the chinchillas' retina are found in the

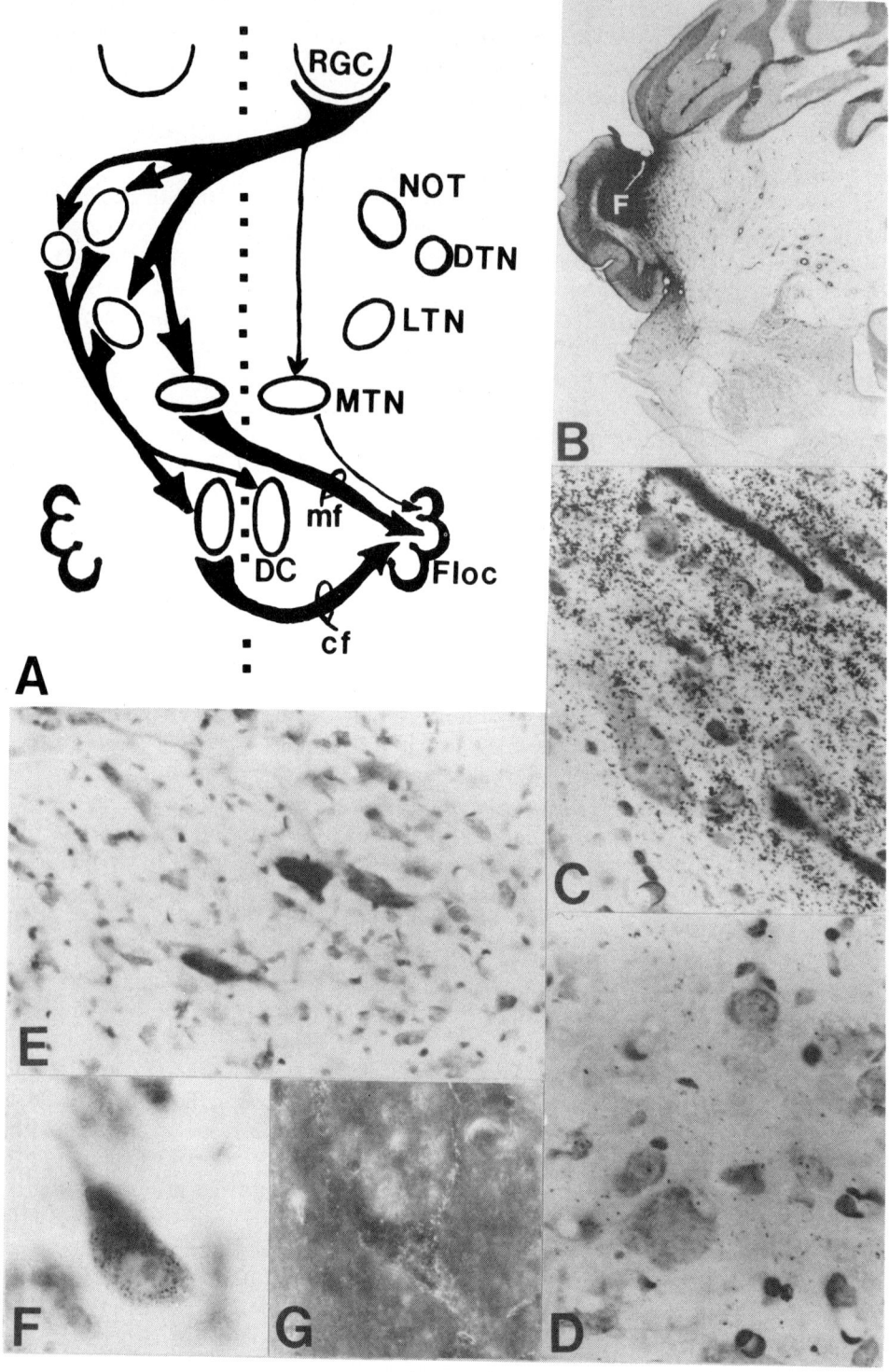

peripheral regions. Furthermore, no ganglion cells in the peripheral retina were labeled following MTN injections.

In contrast to the MTN injections, control injections of HRP in the superior colliculus labeled ganglion cells of many different sizes in both the central and peripheral retina. Only cells in the normal ganglion cell layers were labeled by the control injection; none of the displaced ganglion cells were labeled.

The demonstration of two discreet pathways, MF and CF, for visual input to the flocculus is particularly exciting and suggests that separate functions may be subserved by these two pathways. Their clear anatomical separation facilitates experimental testing of each pathway independent of the other.

Experiments presently being conducted in my laboratory make use of the dramatic and quantifiable change in the VOR produced by visual input as a means of examining the separate functions of the MF and CF pathways. Two of these experiments will be discussed below. The first is an investigation of the alterations in eye movements after the MF pathway has been interrupted owing to lesions of the MTN. The second set of experiments describes changes in single unit activity in the brain stem under conditions of varying visual and vestibular stimuli.

MTN LESION EFFECTS

Our initial results indicated that bilateral MTN lesions do not produce an effect on the VOR. Lesioned animals show the same pattern of phase changes of eye movements during dark rotation as control animals do. In normal chinchillas, visual suppression of the VOR occurs at the low frequencies of oscillations when the visual background is rotated in concert with body rotation. In this experimental paradigm the compensatory eye movements begin to reappear at about 0.16 Hz, and the reflex is readily apparent at frequencies above 0.45 Hz. The eye movement results of normal and bilateral MTN-lesioned animals in this situation are very similar. The lower half of Fig. 4 illustrates the eye movements from one animal with bilateral MTN lesions. The locations of the lesions are also shown.

On the other hand, the eye movements evoked by a moving visual world in the absence of body rotation (i.e., OKN response) are markedly affected by these lesions. The upper portion of Fig. 4 shows the OKN response in the same animal for which the suppression was shown. The typical OKN response in the chinchilla is similar to that reported for the rabbit (Baarsma and Collewijn, 1974). As the velocity of the OKN stimuli is increased, the eye movement velocity and the number of nystagmic beats/sec increase in a monotonic manner. At velocities greater than 30 degrees/sec, the number of nystagmic beats/sec appears to reach its maximum. In animals with MTN lesions, the OKN

Fig. 2. A) Diagrammatic scheme summarizing the retinal input to the flocculus. B) Horseradish peroxidase (HRP) injection site (F) confined to the floccular cortex. 8×. C) Autoradiograph of MTN contralateral to the eye injected with ^{3}H-proline showing very heavy labeling of retinal synapses. 390×. D) Autoradiograph of MTN ipsilateral to the eye injected with ^{3}H-proline showing light but definite labeling of retinal synapses. 390×. E) Three large basophilic neurons in MTN which were labeled with HRP after a floccular injection. The numerous small neurons were not labeled. 300×. F) High power detail of a MTN neuron containing HRP reaction product. 840×. G) Another MTN neuron in high power darkfield showing the HRP reaction product in its cell body and dendrites. 520×.

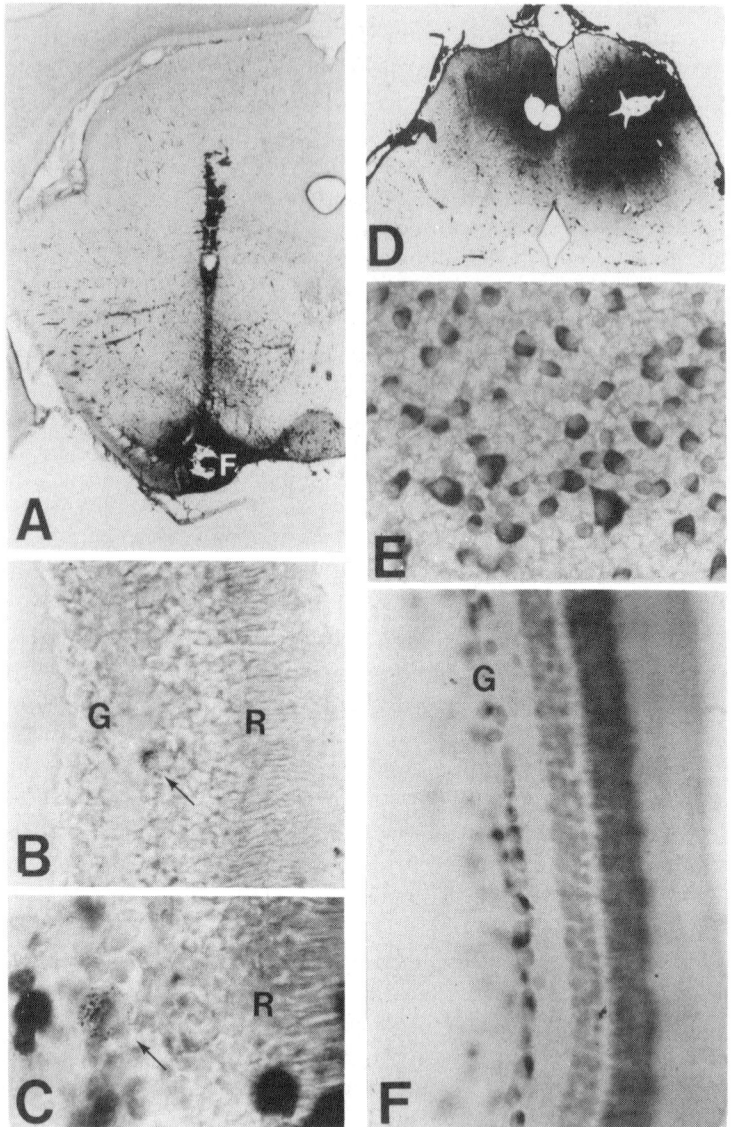

Fig. 3. A) Injection track and focus (F) depositing HRP into MTN. 8×. B) Unstained section of a chinchilla retina showing a HRP-labeled displaced retinal ganglion cell (arrow) at the inner edge of the inner nuclear layer after the HRP injection shown in A. The ganglion cell layer (G) is to the left and the photoreceptors (R) to the right. 320×. C) Higher power of another HRP-labeled displaced retinal ganglion cell (arrow) lying between the amacrines and bipolars of the inner nuclear layer. Photoreceptors (R) are to the right. 840×. D) Bilateral HRP injection into the superior colliculus. 8×. E) Whole mount stained preparation of the chinchilla retina showing at least three sizes of retinal ganglion cells. 340×. F) Section stained with cresyl violet showing that almost every retinal ganglion cell (G) is labeled with HRP after a superior colliculus injection, but there are no labeled displaced ganglion cells. 260×.

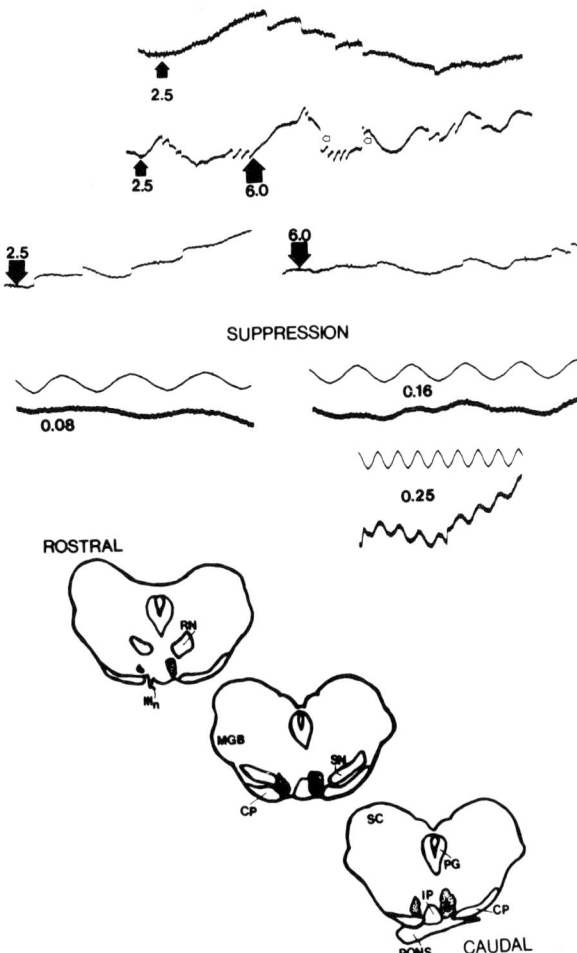

Fig. 4. MTN lesion effects. The upper traces represent eye movements evoked by OKN stimulation. The numbers represent OKN drum velocity with an upward arrow indicating drum rotation to the right and a downward arrow signifying drum rotation to the left. The open horizontal arrows indicate an adjustment made in the EOG trace by the experimenters to offset EOG drift. Suppression indicates the drum is coupled and rotating with the turntable. The brain sections are redrawn from histological sections of the MTN lesions in the chinchilla's brain stem.

response is evident with the number of beats/sec being similar to that in the normal animals at the lower drum velocities. At higher drum rotation velocities the OKN response falls off significantly. In one animal the OKN response failed at rotational velocities above 2.5 degrees/sec.

A curious result observed in these animals with bilateral MTN lesions is that the eyes do not maintain a relatively constant neutral position during OKN but drift first in one direction and then in the other. Furthermore, the shape of the nystagmus tends to round off. That is, rather than the typical sharp sawtooth waveform that is evident in the normal animal's response just prior to the quick phases, the eye velocity seems to slow or roll off. It appears as if the position maintenance system has been affected in the MTN-lesioned animals.

BRAIN STEM RECORDINGS

We made single-unit recordings in the brain stem of alert chinchillas to study the mechanisms underlying the difference in eye movements during rotation in the light and rotation in the dark. We attached a chamber designed to accept a microdrive assembly to the skull to enable us to make electrode penetrations through the vestibular nuclei. We also implanted EOG electrodes to record eye movements.

When we isolated a single unit potential we tested several visual and vestibular paradigms, as follows:

i) Determining the unit's response to right and left OKN stimulations at different velocities.

ii) Determining the unit's response to sinusoidal oscillation ($\pm 10°$ amplitude at frequencies from 0.06 Hz to 0.86 Hz) in the presence or absence of a fixed visual surround. The former condition was met by rotation of the animal inside a lighted, stationary optokinetic drum that provided a fixed visual stimulus (LON). The latter condition was met by simply rotating the animal in the dark (LOFF).

iii) Determining the unit response during sinusoidal oscillation while the drum was coupled to the turntable (SUPP). In this condition, the visual environment moved at the same amplitude and frequency as the turntable.

We recorded from 115 units in the brain stem of the normal awake chinchilla. Only those units found by postmortem anatomical examination to be located within the superior and medial vestibular nuclei are included in this discussion. In addition to the usual classification of vestibular neurons as type I or type II, the unit responses could be further divided into those that were or were not correlated with eye movements. Units related to vestibular stimulation only displayed resting rate activity similar to the tonic cells reported by Shimazu and Precht (1965) and showed no change in firing rate when eye movements were evoked by OKN stimulation. No changes in the gain or phase response of the units's firing rate were detectable between the LON or LOFF paradigms at any of the frequencies tested. Furthermore, the firing patterns of these units were unaltered even when the EOGs indicated complete suppression of eye movements during oscillation at 0.08 Hz in the SUPP paradigm.

The distribution of these units in the brain stem of the chinchilla was in marked contrast to that in the monkey (Fuchs and Kimm, 1976). Approximately 50% of the vestibular unit responses were of this category in the monkey; in the chinchilla less than 10% were of this category, whereas the majority of units were also related to eye movements.

During OKN stimulation the units of the chinchilla related to eye movements either showed an eye position sensitivity (40%) and paused for all saccades, or displayed a burst-tonic relationship (60%). The burst-tonic units responded with a burst of activity to an on-direction saccade and paused to an off-direction saccade. Unlike the units related to vestibular stimulation only, the eye movement-related units exhibited a striking difference in their responses during the LON vs. LOFF vestibular oscillation paradigm. There were two main classes of unit responses to the LON, LOFF conditions. In the first group no alteration in the phase of the unit's response relative to the phase of the table rotation occurred during changes from LON to LOFF or LOFF to LON at any frequency of table rotation tested. However, the normalized gain of these units was consistently lower during the LOFF paradigm, as can be seen in Fig. 5. In addition to the observed changes in normalized gain, some units showed a DC shift in activity that resulted in a clipping of the unit response during LOFF oscillation. These two changes occurred either

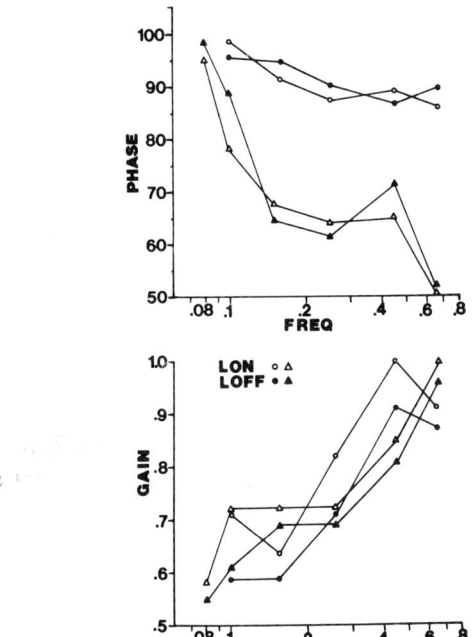

Fig. 5. Bode phase and normalized gain plots summarizing two unit responses recorded from the vestibular nucleus. ○, △ represent the unit response recorded in the light-on; ●, ▲, in the light-off condition.

separately or in combination, depending on the particular unit response being studied. This group of units was comprised of both position and burst-tonic cells.

In the second group of units, marked phase shifts occurred between the LON vs. LOFF paradigm, as can be seen in Fig. 6, in addition to a change in gain similar to that of the first group.

These phase changes typically occurred at frequencies below 0.25 Hz, converging toward a common phase lead at higher frequencies. The differences observed in the unit phase responses during the LON vs. LOFF conditions were not necessarily correlated

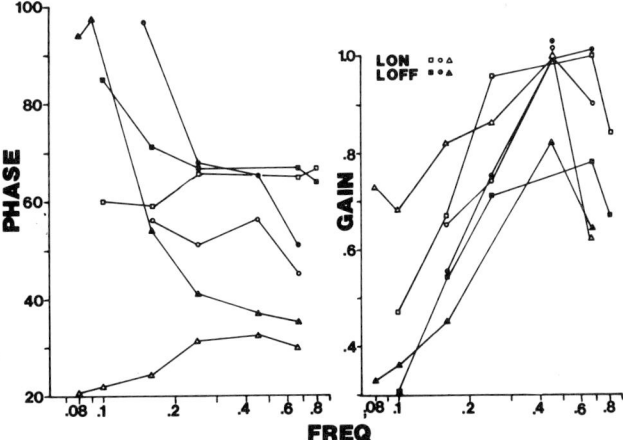

Fig. 6. Bode phase and normalized gain plots as in Fig. 5.

with the changes seen in the EOG responses. In some instances, the change in the unit firing rate was closely coupled to the observed EOG changes whereas in other instances it was not.

During the suppression paradigm, eye movements as measured by EOGs were attenuated, and at low frequencies were suppressed altogether. The responses of all the eye movement-related units that were tested were also reduced. This decrease was strictly correlated with the decline in amplitude of the EOGs. At higher frequencies where the animal was unable to suppress its eye movements, the EOG amplitude increased, and so did the activity of the units.

These results pose several interesting questions. The number of units in the chinchilla vestibular nucleus that are eye movement related is relatively large in comparison to the reported data in alert monkeys. The chinchilla, however, has special problems in achieving visual stabilization at low acceleration rates of head movements. Perhaps the increased integration of visual and vestibular information in the chinchilla's vestibular nucleus is a way of compensating for inferior dynamic responses of its vestibular system.

A second question concerns the identification of the vestibular neurons projecting to the oculomotor system. One likely candidate in the chinchilla would be the burst-tonic neuron; neurons of this type showed phase shifts at the lower frequencies. However, confirmation of this hypothesis would require at least antidromic and collision type experiments indicating that this is indeed the case. Finally, some of the observed light-dark difference in responses in unit activity at the level of the vestibular nuclei may be reflections of floccular activity. Recording from single units in the brain stem of bilateral flocculectomized chinchillas is currently under investigation in my laboratory; the results may allow us to speculate further on the role of the flocculus in the VOR.

SUMMARY

The eye movement response in the chinchilla to an imposed sinusoidal rotation was studied under three experimental conditions: 1) in the presence of a fixed visual surround, 2) in the dark, and 3) following bilateral lesions of the flocculus. In the first condition the eye movements essentially completely phase-compensated for head movements; the relative movements were 180° out of phase. In the dark, the eyes phase-led head movements by up to 60° at the lower oscillation rates. Bilateral flocculectomy eliminated this difference in eye movement under these two visual-vestibular conditions. A visual mossy fiber pathway has been described. Retinal ganglion cells have been identified projecting bilaterally to the medial terminal nucleus of the accessory optic tract, with primary input to the contralateral side. Cells within the MTN then project to the contralateral flocculus. To investigate the possible functional role this pathway subserves, we bilaterally lesioned the medial terminal nucleus and observed the resultant eye movements to various stimulus conditions; these are described. Cell activity in the vestibular nuclei was observed under various visual-vestibular stimulus conditions; the results are also described.

REFERENCES

Baarsma, E.A. and Collewign, H. (1974) Vestibulo-ocular and optokinetic reactions to rotation and their interaction in the rabbit. *J. Physiol. (Lond.)*, 238: 603–625.

Baker, R., Precht, W. and Llinas, R. (1972) Cerebellar modulatory action on the vestibulo-trochlear pathway in the cat. *Exp. Brain Res.*, 15: 364–385.

Brauth, S.E. and Karten, H.J. (1977) Direct accessory optic projections to the vestibulo-cerebellum: A possible channel for oculomotor control systems. *Exp. Brain Res.*, 28: 73–84.

Daniels, P.D., Hassul, M. and Kimm, J. (1978) Dynamic analysis of the vestibulo-ocular reflex in the normal and flocculectomized chinchilla. *Exp. Neurol.*, 58: 32–45.

Fukuda, J., Highstein, S.M. and Ito, M. (1972) Cerebellar inhibitory control of the vestibulo-ocular reflex investigated in the rabbit IIIrd nucleus. *Exp. Brain Res.*, 14: 511–526.

Ghelarducci, B., Ito, M. and Yagi, N. (1975) Impulse discharges from flocculus Purkinje cells of alert rabbits during visual stimulation combined with horizontal head rotation. *Brain Res.*, 87: 66–72.

Hassul, M., Daniels, P.D. and Kimm, J. (1976) Effects of bilateral flocculectomy on the vestibulo-ocular reflex in the chinchilla. *Brain Res.*, 118: 339–343.

Ito, M. (1975) Learning control mechanisms by the cerebellum investigated in the flocculo-vestibulo-ocular system. In *The Nervous System*, B. Tower (Ed.), Raven Press, New York.

Ito, M., Nishimaru, N. and Yamamoto, M. (1973) Specific neural connections for the cerebellar control of vestibulo-ocular reflexes. *Brain Res.*, 60: 238–243.

Ito, M., Shiida, T., Yagi, N. and Yamamota, M. (1974) The cerebellar modification of rabbit's horizontal vestibulo-ocular reflex induced by sustained head rotation combined with visual stimulation. *Proc. Japan Acad.*, 50: 85–89.

Kimm, J., Hassul, M. and Cogdell, B. (1976) Fastigial neuronal responses to sinusoidal horizontal rotation. *Exp. Neurol.*, 50: 579–594.

Lisberger, S.G. and Fuchs, A.F. (1974) Response of flocculus Purkinje cells to adequate vestibular stimulation in the alert monkey: fixation vs. compensatory eye movements. *Brain Res.*, 69: 347–353.

Maekawa, K. and Simpson, J.I. (1973) Climbing fiber responses evoked in the vestibulo-cerebellum of rabbit from visual system. *J. Neurophysiol.*, 36: 649–666.

Maekawa, K. and Kumura, K. (1974) Inhibition of climbing fiber responses of rabbit's flocculus Purkinje cells induced by light stimulation of the retina. *Brain Res.*, 65: 347–350.

Maekawa, K. and Takeda, T. (1975) Mossy fiber responses evoked in the cerebellar flocculus of rabbits by stimulation of the optic pathway. *Brain Res.*, 98: 590–595.

Maekawa, K. and Takeda, T. (1976) Electrophysiological identification of the climbing and mossy fiber pathways from the rabbits retina to the contralateral cerebellar flocculus. *Brain Res.*, 109: 169–174.

Noda, H., Asoh, R. and Shibagaki, M. (1977) Floccular unit activity associated with eye movement and fixation. In *Control of Gaze by Brain Stem Neurons*, R. Baker and A. Berthoz (Eds.), Elsevier/North-Holland Biomed. Press, Amsterdam.

Shimazu, H. and Precht, W. (1965) Tonic and kinetic responses of cat's vestibular neurons to horizontal angular acceleration. *J. Neurophysiol.*, 28: 991–1013.

Shimazu, H. and Precht, W. (1966) Inhibition of central vestibular neurons from the contralateral labyrinth and its mediating pathway. *J. Neurophysiol.*, 29: 467–492.

Shimazu, H. and Smith, C.M. (1971) Cerebellar and labyrinthine influences on single vestibular neurons identified by natural stimuli. *J. Neurophysiol.*, 34: 493–508.

Simpson, J.I. and Alley, K.I. (1974) Visual climbing fiber input to rabbit vestibulo-cerebellum: A source of direction-specific information. *Brain Res.*, 82: 302–308.

Takemori, S. and Cohen, B. (1974) Loss of visual suppression of vestibular nystagmus after flocculus lesions. *Brain Res.*, 72: 213–224.

The Accessory Optic System and its Relation to the Vestibulocerebellum*

J. I. SIMPSON, R. E. SOODAK and R. HESS

Department of Physiology and Biophysics, New York University Medical Center, New York, NY 10016 (U.S.A.); and Max-Planck-Institut für Biophysikalische Chemie, Neurobiologische Abteilung, 3400 Göttingen-Nikolausberg (F.R.G.)

INTRODUCTION

The accessory optic system (AOS), which exists in all vertebrate classes (Marg, 1964, 1973; Ebbesson, 1970), consists of a group of nuclei at the mesodiencephalic border innervated by optic fibers, which constitute the accessory optic tract (AOT). A substantial anatomical literature on the AOS has accumulated since Gudden (1870) presented the first detailed description of the major component of the mammalian AOT. In comparison, physiological investigations are far less numerous. The work of Marg and coworkers in rabbit (Hamasaki and Marg, 1960, 1962; Hill and Marg, 1963; Walley, 1967) plus the one reported here, also in rabbit, constitute the full extent of published single unit electrophysiological investigations of the AOS proper. A variety of functions has been proposed for the AOS (Marg, 1964, 1973), ranging from mediation of light induced changes in endocrine activity (Moore et al., 1967) to participation in various aspects of oculomotor behavior (Walley, 1963; Lázár, 1973; Westheimer and Blair, 1974; Brauth and Karten, 1977). The results of the present study are consistent with a role for the AOS in compensatory oculomotor behavior. In brief, the AOS processes visual activity signaling self-motion. In addition, consideration of the projection of the AOS via the inferior olive to the vestibulocerebellum leads to the proposition that the AOS is a visual system organized in vestibular coordinates.

ANATOMY OF THE ACCESSORY OPTIC SYSTEM

Prior to presenting our experimental results, it seems in order to briefly describe the AOS, since even some visual physiologists are unfamiliar with it. Over a century ago, Gudden (1870) described a superficial tract, termed the transpeduncular tract, arising at the anterior edge of the superior colliculus, running ventrolaterally over the cerebral peduncle and ending near the exit of the third nerve. Later, Gudden (1881) demonstrated that the integrity of the transpeduncular tract depends upon the presence of the retina. Since that time, various terms have been used to describe the components of the mammalian AOS. The terminology introduced by Hayhow (1959,

---

*This study was supported by NINCDS Grant NS–13742 and the Max-Planck Society.

1966) is the one most widely used today. In most mammalian species, the AOS is composed of three paired terminal nuclei called the dorsal, lateral and medial terminal nuclei (Fig. 1). The largest of these, the medial terminal nucleus (MTN), is located at the base of the midbrain, medial and dorsal to the substantia nigra, ventral to the red nucleus and just anterior to the exit of the third nerve. The lateral terminal nucleus (LTN) is, in part, embedded in the transverse peduncular tract and is located ventral to the posterior edge of the medial geniculate at the dorsal edge of the cerebral peduncle. The dorsal terminal nucleus (DTN) lies on the surface and within the transverse peduncular tract along the ventral part of the anterior edge of the superior colliculus. The DTN may, in fact, be a continuation of a portion of the adjacent nucleus of the optic tract located in the lateral pretectum.

The optic fibers which innervate the three terminal nuclei constitute the AOT, which is divided into two principal fasciculi, an inferior and a superior fasciculus. The superior fasciculus is, in turn, further divided into three branches, a posterior, a middle and an anterior branch. The transpeduncular tract of Gudden is identical to the posterior branch of the superior fasciculus. The optic fibers of the inferior fasciculus separate from the optic tract proper just posterior to the chiasm and course more or less directly caudally to reach the MTN. The thin medial and anterior branches of the superior fasciculus leave

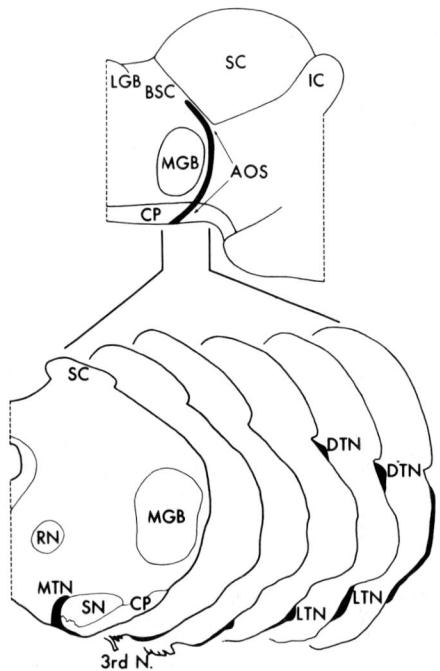

Fig. 1. Schematic of the anatomy of the accessory optic system of the rabbit. Only the principal component of the accessory optic tract is shown, i.e., the transpeduncular tract (Gudden) or posterior bundle of the superior fasciculus (Hayhow). The top panel represents a lateral view of the mesodiencephalic portion of the brain while the bottom panel consists of a series of transverse sections illustrating the locations of the three accessory optic terminal nuclei. Description is in the text. AOS, accessory optic system; BSC, brachium of the superior colliculus; CP, cerebral peduncle; DTN, dorsal terminal nucleus; IC, inferior colliculus; LGB, lateral geniculate body; LTN, lateral terminal nucleus; MGB, medial geniculate body; MTN, medial terminal nucleus; RN, red nucleus; SC, superior colliculus; SN, substantia nigra; 3rd N., third nerve.

the optic tract just prior to and at the level of the lateral geniculate body and terminate in the MTN. The number of accessory optic fibers is relatively small. In the rabbit, an animal in which the AOS is well developed, approximately 2,000 retinal fibers course in the transpeduncular tract to the MTN (Giolli, 1961). The input to the accessory optic terminal nuclei is predominantly from the contralateral eye although a small projection from the ipsilateral eye to the MTN exists (Takahashi et al., 1977; Winfield et al., 1978).

The accessory optic system is present in all mammals, but to differing degrees. For example, in the cat, there appears to be no inferior fasciculus (Hayhow, 1959), while in primates the AOS is compressed in that retinal fibers do not project to the region occupied by the MTN in other species (Giolli, 1963; Campos-Ortega and Glees, 1967; Tigges and Tigges, 1969; Tigges et al., 1977). In primates, the nucleus occupying the place of the MTN of other species is called the nucleus of the transpeduncular tract and it probably receives fibers from the accessory optic nuclei proper.

Information of the projections of the terminal nuclei of the AOS is limited, but the available data suggest that the signals from the AOS ultimately participate in processes of visual-vestibular interaction. Definite evidence that the AOS is intimately involved in visual-vestibular interaction was provided by Maekawa and Simpson (1972, 1973) who showed that the optic fibers relevant to the visual climbing fiber input to the vestibulocerebellum are located in the posterior bundle of the superior fasciculus of the AOT. The DTN and/or part of the lateral pretectum was proposed to be the first intracranial synaptic relay of this visual pathway to the vestibulocerebellum. More recently, Maekawa and Takeda (1977; Takeda and Maekawa, 1976) have shown that the three accessory optic terminal nuclei, along with part of the nucleus of the optic tract, project differentially to the dorsal cap of the inferior olive. This portion of the inferior olive is known to send climbing fibers to the flocculonodular lobe (Alley et al., 1975; Hoddevik and Brodal, 1977). Evidence also exists for a direct projection from the MTN to the cerebellar flocculus in the chinchilla (Winfield et al., 1978), while the homologue of the MTN in the pigeon was previously shown to project directly to the uvula and paraflocculus (Brauth and Karten, 1977).

NEURAL RESPONSES OF THE ACCESSORY OPTIC NUCLEI

Extracellular single unit recordings from cells of each of the three accessory optic nuclei were obtained in pigmented rabbits initially anesthetized with α-chloralose (60 mg/kg) plus Nembutal (10 mg/kg). The animals were paralyzed and artificially respirated; anesthesia was continuously infused during the experiment. Most commonly, the visual stimulus was a large (70° by 70°), highly textured pattern which could be moved behind a masked tangent screen at various speeds and in various directions.

Since the most extensive investigations were conducted on cells of the MTN, they will be described first. MTN cells typically have a high background activity (25–50 spikes/sec) and respond to electrical stimulation of the optic chiasm at a latency of 2–2.5 msec (Fig. 2). All cells influenced by moving patterns (n = 72) showed both direction and speed selectivity (Figs. 2 and 3). Receptive fields are large, averaging about 40° vertical by 60° horizontal. Presentation of large (about 20° by 20°), slowly moving textured patterns is required to produce a strong modulation. Other visual stimuli such as single spots or bars have comparatively little effect on cell activity.

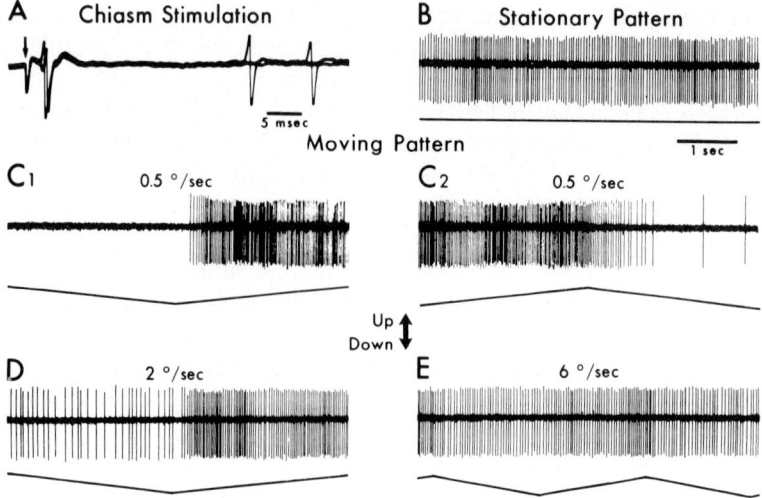

Fig. 2. Example of the response of a cell in the medial terminal nucleus (MTN) to electrical stimulation of the optic chiasm and to movement of a large, textured pattern presented to the contralateral eye. The latency of response to chiasm stimulation is 2–2.5 msec (A) which, after allowance for synaptic delay, translates into a conduction velocity of about 17 m/sec for the optic fibers. MTN neurons typically have a substantial background discharge rate (B) and they exhibit true direction selectivity in that their activity both increases and decreases with appropriate visual stimuli. Comparison of the responses to the pattern moving at different speeds (C, D and E) clearly shows the strong preference of MTN neurons for quite slow speeds.

Best modulation occurs at speeds ranging from 0.1–1 degree/sec. Activity increases in a sustained manner 2–3 times over background for preferred direction movement and can be silenced for null direction movement. The sensitivity to low speeds is quite remarkable; speeds as slow as a few hundredths of a degree per second still effectively modulate cell activity.

Preferred directions are vertical with a posterior component; cells preferring upward movement are twice as numerous as those preferring downward movement. Interestingly, the preferred and null directions of MTN cells are not 180° apart. For example, if

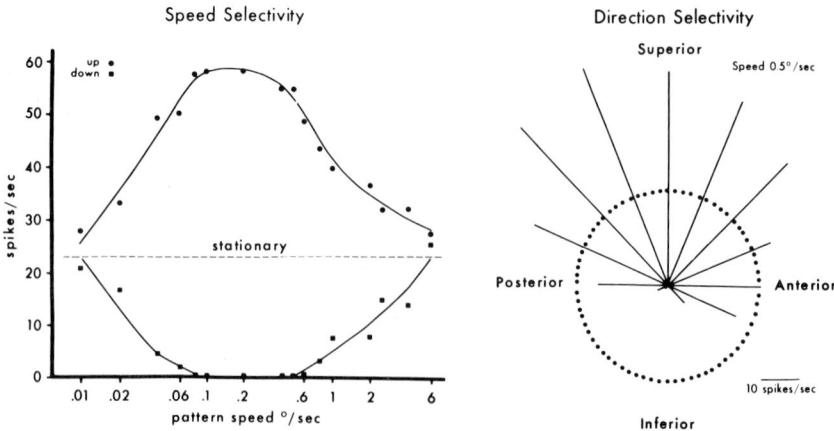

Fig. 3. Details of the speed and direction selectivity of an MTN neuron. All neurons are best modulated by slow movement as indicated on the left. In addition, all neurons are best modulated by vertical movement with a posterior component. The preferred direction is either inferior with a posterior component or superior with a posterior component, as in the case illustrated. The best null direction is not opposite to the best preferred direction. Both null and preferred directions have a posterior component in visual space.

the preferred direction is up with a posterior component, the null direction is down with a posterior component. The fact that the preferred and null axes are not 180° apart is most apparent at non-optimal speeds when inhibition is less complete. In addition to the direction and speed selectivity properties described above, MTN cells respond primarily only at the onset of illuminations of a stationary pattern. Although all units so far encountered by us in the rabbit MTN are directionally selective, a small minority, approximately 5%, also show tonic changes in activity with changes in the ambient level of illumination.

The response properties of neurons of the LTN and DTN are identical to those of the MTN with the exception of the preferred direction orientation in space. Whereas cells of the MTN which prefer movement up and posterior are twice as numerous as those which prefer down and posterior, nearly all LTN cells so far examined (7 of 8) preferred movement down and posterior. The remaining cell preferred movement up and posterior. In contrast to both MTN and LTN cells, DTN cells (5 of 5) are best modulated by horizontal movement and are excited by movement from posterior to anterior in the visual field.

RELATION OF THE ACCESSORY OPTIC SYSTEM TO THE VESTIBULOCEREBELLUM

In rabbit, the AOS projects indirectly to the vestibulocerebellum via a climbing fiber input arising from the dorsal cap of the inferior olive (Maekawa and Simpson, 1972, 1973; Alley et al., 1975; Takeda and Maekawa, 1976; Maekawa and Takeda, 1977). In addition, it is possible, but not yet confirmed that the MTN projects directly to the rabbit vestibulocerebellum as a mossy fiber input (see Brauth and Karten, 1977; Winfield et al., 1978). In our initial investigation (Simpson and Alley, 1974) of the type of information conveyed to the vestibulocerebellum by visual climbing fibers, we found that the predominant class of this input is optimally responsive in a directionally selective manner to large, textured patterns slowly moving at less than 1 degree/sec. Further investigation has revealed that, collectively, the visual climbing fiber input to the rabbit's flocculus describes three preferred directions in visual space: anterior; up with a posterior component; and down with a posterior component. Climbing fibers related to the eye ipsilateral to the flocculus respond best for movement in one of two preferred directions, either anterior or up and posterior. Climbing fibers driven by visual stimuli presented to the contralateral eye respond best for movement down and posterior, as illustrated in Fig. 4. The preferred and null axes are not 180° apart; this arrangement is precisely the same as that for the vertically selective cells of the terminal nuclei of the AOS.

DISCUSSION

Recordings from neurons in the three terminal nuclei of the AOS have revealed that this visual system processes information about the speed and direction of movement of large parts of the visual world. Certain of our findings could have been anticipated, in part, from the work of Walley (1963, 1967) on the rabbit MTN. Using simple spot or bar stimuli, Walley found that over half of the cells in the MTN were direction selective in response to vertically moving stimuli and that the vast majority of these cells preferred

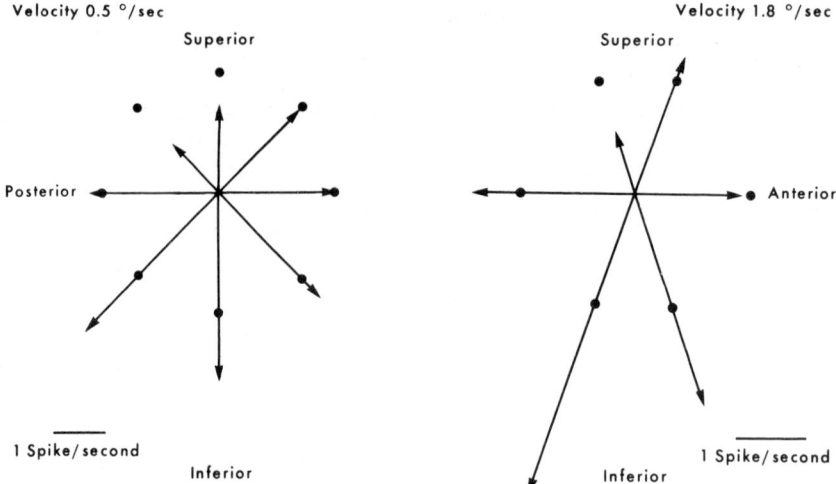

Fig. 4. Two examples of the direction selectivity of the visual climbing fiber responses of floccular Purkinje cells driven through the contralateral eye. The preferred direction is inferior with a posterior component. The null direction is not opposite the preferred direction, but is superior with a posterior component. This type of relation between the preferred and null axes is also characteristic of vertically selective cells of the accessory optic nuclei.

upward movement. Walley's use of the term directionally selective differs from ours since he did not find reversal of the direction of modulation with reversal of the direction of stimulus movement. We suspect this difference arises as a result of differences in composition and speed of the visual stimuli; with large, slowly moving textured targets we found a far deeper modulation of activity than did Walley.

DTN cells respond to visual stimuli in a manner similar to that of a subcategory of cells found in the rabbit nucleus of the optic tract by Collewijn (1975). In the nucleus of the optic tract, the vast majority of cells respond to movement in the horizontal direction; the preferred direction is from posterior to anterior. The similarity of the properties of DTN cells to those of some cells of the nucleus of the optic tract suggests that accessory optic fibers project to the nucleus of the optic tract.

Response properties of accessory optic terminal nuclei cells are in several ways markedly similar to those of a class of directionally selective ganglion cells which has been found in the rabbit retina (Barlow et al., 1964; Oyster and Barlow, 1967; Oyster, 1968; Oyster et al., 1972). Taken collectively, the preferred directions of these retinal ganglion cells define three directions in visual space: anterior, up with a posterior component, and down with a posterior component (Fig. 5). These ganglion cells respond only at the onset of steady illumination and are thus called on-direction selective ganglion cells. The speed selectivity of these ganglion cells is identical to that of cells in the terminal nuclei of the AOS (compare Fig. 5 with Fig. 3). It is clear that the preferred direction and speed selectivity of the AOS results exclusively from an input from on-direction selective ganglion cells. Since both the preferred and null directions of vertically selective cells of the AOS correspond to a preferred direction of the on-direction selective ganglion cell class, it is likely that null inhibition is mediated by an inhibitory interneuron rather than simply due to silencing of the ganglion cells providing the excitatory preferred direction input.

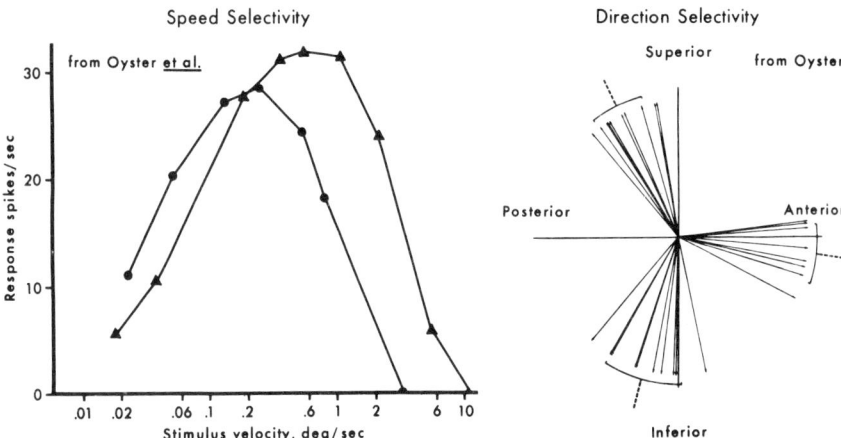

Fig. 5. Direction and speed selectivity of the on-direction selective class of retinal ganglion cells in the rabbit retina. The speed selectivity curves on the left represent the extremes found by Oyster et al., (1972) for movement in the preferred direction. The similarity of the speed selectivity for ganglion cells and MTN cells (Fig. 3) is quite remarkable. The plot on the right illustrates the distribution of the preferred directions for a number of ganglion cells (Oyster, 1968) and shows that collectively they define three directions in visual space. These same three directions are found in the terminal nuclei of the accessory optic system. The horizontal direction is represented in the dorsal terminal nucleus and the two off-vertical directions are represented in the medial and lateral terminal nuclei.

The simple and uniform nature of the response properties of the cells of the AOS allow us to ascribe to it, with reasonable confidence, a definite role in the life of the animal. In an animal's natural environment, movement of large parts of the visual world will only occur as a result of movement of the animal itself. Since a large, textured moving stimulus results in a strong modulation of activity, we conclude that the AOS serves to signal self-motion, a function similar to that of the vestibular system. The sensitivity of AOS cells to speeds as slow as a few hundredths of a degree per second would make the AOS an ideal complement to the vestibular system. This contention is supported by the fact that signals from the AOS converge with vestibular signals in portions of the cerebellum.

Convergence of accessory optic and vestibular inputs raises the question of how signals of self-motion from two different sensory modalities interact. We propose that the three preferred directions described in visual space by on-direction selective ganglion cells, AOS cells and climbing fibers are functionally related to and derived from the three principal axes of the semicircular canals, allowing for dynamic visual-vestibular interactions in one basic coordinate system. The argument for this proposition was originally presented by Simpson and Hess (1977); it is repeated and updated here. Consider the movement of an optic axis with the eyes fixed in the rabbit's head and confine movement to the region of visual space from which the distribution of preferred directions of ganglion cells was drawn (longitudes anterior 20° to posterior 20° and latitudes 0° to 40° superior) (Oyster, 1968). Rotation of the head about a vertical axis such that an optic axis moves in space from posterior to anterior corresponds to the direction of movement which would reduce activity of the ipsilateral primary vestibular afferents. Rotation about the axis of the ipsilateral anterior canal so as to reduce afferent discharge would result in movement of the optic axis in an arc directed upward and somewhat posteriorly;

a similar rotation about the ipsilateral posterior canal axis would result in movement of the optic axis upward and somewhat anteriorly. The optic axis movements associated with these maneuvers about the ipsilateral horizontal and anterior canal axes correspond to two of the three preferred directions of the retinal ganglion cells and AOS cells and to the two preferred directions found for floccular climbing fibers related to the ipsilateral eye. The optic axis movement associated with the posterior canal is opposite to that found for ganglion cells and for AOS neurons. However, since the rabbit's eyes are laterally placed, movement of the optic axis of one eye upward and somewhat anteriorly is associated with movement of the optic axis of the other eye downward and somewhat posteriorly. This latter direction is one of the three preferred directions for the ganglion cells and AOS cells and, more importantly, this direction is precisely the one preferred by the floccular climbing fiber input driven from the contralateral eye. Thus, from the vestibulocerebellum we have evidence indicating that each of the three preferred directions of the AOS in visual space is predictably related to one of the three principal axes of the semicircular canals.

The separation of visual kinesthetic signals into three preferred directions is spatially maintained within the flocculus. Groenewegen and Voogd (1977) have found three termination zones within the flocculus following injection of radioactive amino acid into the dorsal cap of the inferior olive and Ito and collaborators (Dufossé et al., 1977) have found that microstimulation in the flocculus delineates three classes of Purkinje cells on the basis of the direction of the evoked eye movement. The three directions in visual space defined by the preferred directions of the climbing fiber input have an internally consistent relation to the three directions of eye movement determined by flocculus stimulation.

Our studies of the AOS and its projection via the inferior olive to the vestibulocerebellum indicate that visual and vestibular signals which produce compensatory eye movements are organized about a common set of axes derived from the orientation of the semicircular canals. It is important to learn more of this organizational scheme since other multimodal sensorimotor relations may be more clearly understood if their specific common coordinates are known. For the particular case of the accessory optic and vestibular systems, the term coordinates can be readily associated with actual spatial dimensions, but for other cases, the term coordinates will probably have a less literal geometrical meaning. However, the coordinates may, in general, be the functional correlates of the longitudinal strips which characterize the climbing fiber projection to the cerebellum.

SUMMARY

The accessory optic system (AOS) in mammals is, in general, composed of three terminal nuclei (dorsal, DTN; lateral, LTN; and medial, MTN) innervated by primary optic fibers. These three nuclei provide a visual input to the dorsal cap of the inferior olive, which is a source of climbing fibers to the flocculonodular lobe of the cerebellum. Investigation of the visual response properties of neurons in the three terminal nuclei of the rabbit AOS has revealed that this visual system processes information about the speed and direction of movement of large parts of the visual world. Best modulation of terminal nuclei neurons requires use of large (20° by 20°), textured stimuli and occurs at low speeds with peak sensitivity at 0.1–1 degree/sec. For MTN and LTN neurons, the preferred directions in space are vertical, but with a posterior component. For DTN neurons, preference is for horizontal movement from posterior to anterior. Taken collectively, the preferred directions for terminal nuclei neurons define three directions in visual space: anterior, up with a posterior component and down with a posterior component. The speed and direction selectivity of neurons of the accessory optic terminal nuclei are identical to those of the on-direction selective class of rabbit retinal ganglion cells and to those of visually driven climbing fibers in the cerebellar flocculus. It is most likely that the AOS serves to signal self-motion; the sensitivity of AOS neurons to slow speeds makes the AOS an ideal complement to the vestibular system. Indeed, consideration of the projections of the AOS via the inferior olive to the flocculus leads to the proposition that the AOS is a visual system organized in vestibular coordinates.

REFERENCES

Alley, K., Baker, R. and Simpson, J.I. (1975) Afferents to the vestibulo-cerebellum and the origin of the visual climbing fibers in the rabbit. *Brain Res.*, 98: 582–589.

Barlow, H.B., Hill, R.M. and Levik, W.R. (1964) Retinal ganglion cells responding selectively to direction and speed of image motion in the rabbit. *J. Physiol. (Lond.)*, 173: 377–407.

Brauth, S.E. and Karten, H.J. (1977) Direct accessory optic projections to the vestibulo-cerebellum. A possible channel for oculomotor control systems. *Exp. Brain Res.*, 27: 73–84.

Campos-Ortega, J.A. and Glees, P. (1967) The subcortical distribution of optic fibers in *Saimiri sciureus* (squirrel monkey). *J. comp. Neurol.*, 131: 131–142.

Collewijn, H. (1975) Direction-selective units in the rabbit's nucleus of the optic tract. *Brain Res.*, 100: 489–508.

Dufossé, M., Ito, M. and Miyashita, Y. (1977) Functional localization in the rabbit's cerebellar flocculus determined in relationship with eye movements. *Neurosci. Lett.*, 5: 273–277.

Ebbesson, S.O. (1970) On the organization of central visual pathways in vertebrates. *Brain Behav. Evol.*, 3: 178–194.

Giolli, R.A. (1961) An experimental study of the accessory optic tracts (transpeduncular tracts and anterior accessory optic tracts) in the rabbit. *J. Comp. Neurol.* 117: 77–95.

Giolli, R.A. (1963) An experimental study of the accessory optic system in the Cynomolgus monkey. *J. comp. Neurol.*, 121: 89–108.

Gudden, B. (1870) Ueber einen bisher nicht beschriebenen Nervenfasernstrang im Gehirne der Säugethiere und des Menschen. *Arch. Psychiat.*, 2: 364–366.

Gudden B. (1881) Ueber den Tractus peduncularis transversus. *Arch. Psychiat.*, 11: 415–423.

Hamasaki, D. and Marg, E. (1960) Electrophysiological study of the posterior accessory optic tract. *Am. J. Physiol.*, 199: 522–528.

Hamasaki, D. and Marg, E. (1962) Microelectrode study of accessory optic tract in the rabbit. *Am. J. Physiol.*, 202: 480–486.

Hayhow, W.R. (1959) An experimental study of the accessory optic fiber system in the cat. *J. Comp. Neurol.*, 113: 281–313.

Hayhow, W.R. (1966) The accessory optic system in the marsupial phalanger, *Trichosurus vulpecula*. An experimental degeneration study. *J. comp. Neurol.*, 126: 653–672.

Hill, R.M. and Marg, E. (1963) Single-cell responses of the nucleus of the transpeduncular tract in the rabbit to monochromatic light on the retina. *J. Neurophysiol.*, 26: 249–257.

Hoddevik, G.H. and Brodal, A. (1977) The olivocerebellar projection studied with the method of retrograde axonal transport of horseradish peroxidase. V. The projections to the flocculonodular lobe and the paraflocculus in the rabbit. *J. comp. Neurol.*, 176: 269–280.

Lázár, G. (1973) Role of the accessory optic system in the optokinetic nystagmus of the frog. *Brain Behav. Evol.*, 5: 443–460.

Maekawa, K. and Simpson, J.I. (1972) Climbing fiber activation of Purkinje cells in the flocculus by impulses transferred through the visual pathway. *Brain Res.*, 39: 245–251.

Maekawa, K. and Simpson, J.I. (1973) Climbing fiber response evoked in vestibulocerebellum of rabbit from visual system. *J. Neurophysiol.*, 36: 649–666.

Maekawa, K. and Takeda, T. (1977) Afferent pathways from the visual system to the cerebellar flocculus of the rabbit. In *Control of Gaze by Brain Stem Neurons,* R. Baker and A. Berthoz (Eds.) Elsevier/North-Holland Biomed. Press, Amsterdam, pp. 187–196.

Marg, E. (1964) The accessory optic system. *Ann. N.Y. Acad. Sci.*, 117: 35–52.

Marg, E. (1973) Neurophysiology of the accessory optic system. In *Handbook of Sensory Physiology. Vol. VII/3.* Central Processing of Visual Information, R. Jung (Ed.), Springer-Verlag, Berlin, pp. 103–111.

Moore, R.Y., Heller, A. Wurtman, R.J. and Axelrod, J. (1967) Visual pathway mediating pineal response to environmental light. *Science,* 155: 220–223.

Oyster, C.W. (1968) The analysis of image motion by the rabbit retina. *J. Physiol. (Lond.),* 199: 613–635.

Oyster, C.W. and Barlow, H.B. (1967) Direction-selective units in rabbit retina: Distribution of preferred directions. *Science,* 155: 841–842.

Oyster, C.W., Takahashi, E. and Collewijn, H. (1972) Direction selective retinal ganglion cells and control of optokinetic nystagmus in the rabbit. *Vision Res.,* 12: 183–193.

Simpson, J.I. and Alley, K.E. (1974) Visual climbing fiber input to rabbit vestibulocerebellum: A source of direction-specific information. *Brain Res.,* 82: 302–308.

Simpson, J.E. and Hess, R. (1977) Complex and simple visual messages in the flocculus. In *Control of Gaze by Brain Stem Neurons*, R. Baker and A. Berthoz (Eds.), Elsevier/North-Holland Biomed. Press, Amsterdam, pp. 351–360.

Takahashi, E.S., Hickey, T.L. and Oyster, C. (1977) Retinogeniculate projections in the rabbit: An autoradiographic study. *J. comp. Neurol.,* 175: 1–12.

Takeda, T. and Maekawa, K. (1976) The origin of the pretecto-olivary tract. A study using the horseradish peroxidase method. *Brain Res.,* 117: 319–325.

Tigges, J. and Tigges, M. (1969) The accessory optic system in *Erinacues* (Insectivora) and *Galago* (Primates). *J. comp. Neurol.,* 137: 59–70.

Tigges, J., Bos, J. and Tigges, M. (1977) An autoradiographic investigation of the subcortical visual system in chimpanzees. *J. comp. Neurol.,* 172: 367–380.

Walley, R.E. (1963) *Receptive Fields in the Accessory Optic System of the Rabbit*. Ph.D. dissertation, Univ. of California, Berkeley, CA.

Walley, R.E. (1967) Receptive fields in the accessory optic system of the rabbit. *Exp. Neurol.,* 17: 27–43.

Westheimer, G. and Blair, S.M. (1974) Unit activity in accessory optic system in alert monkeys. *Invest. Ophthal.,* 13: 533–534.

Winfield, J.A., Hendrickson, A. and Kimm, J. (1978) Anatomical evidence that the medial terminal nucleus of the accessory optic tract in mammals provides a visual mossy fiber input to the flocculus. *Brain Res.,* 151: 175–182.

Colliculoreticular Organization in the Oculomotor System

E. L. KELLER

Department of Electrical Engineering and Computer Sciences, and Electronics Research Laboratory, University of California, Berkeley, CA 94720 (U.S.A.)

A variety of recent experiments support the notion that the superior colliculus (SC) is involved in the generation of saccadic eye movements for foveal acquisition of peripheral objects of interest. At the same time, the brain stem reticular formation (RF), and in particular its paramedian zone in the pons and rostral medulla, has been shown to be exceedingly important for the immediate supranuclear control of saccadic eye movements. More recently, there have been a series of anatomical and physiological studies of the colliculoreticular pathways. The purpose of the present paper is to summarize briefly the details concerning saccadic eye movement-related discharge in SC neurons and in median pontomedullary RF neurons and review the organization of colliculoreticular pathways, particularly with a view towards clarifying the mechanisms involved in controlling saccadic eye movements in primates. In the course of the review some new data on the functional type of SC input to the RF will be presented.

SACCADIC-RELATED DISCHARGE IN SUPERIOR COLLICULUS

Detailed reports on discharge patterns just before and during saccadic eye movements of neurons located in the deeper layers of the SC have been published by Mohler and Wurtz (1976), Schiller and Koerner (1971), Schiller and Stryker (1972), Sparks et al., (1976, 1977), and Wurtz and Goldberg (1972). The fact that this SC discharge is specifically related to eye movements and not head movements has been shown by Robinson and Jarvis (1974). This material has been recently reviewed by Sparks and Pollack (1977), so that only salient points will be discussed here.

The most important feature of the discharge characteristics of saccadic-related SC neurons is their spatially organized movement fields (Sparks and Pollack, 1977). Thus, neurons discharging for small contralateral saccades are located rostrally while those for large saccades are located caudally. Cells located near the midline discharge for saccades with up components and lateral cells fire for movements with down components. Specifity or tuning for movement amplitude and direction is not sharp, especially for more caudally placed cells with large movement preference. These neurons are normally quiescent or discharge sporadically at low rates during fixation and show a burst of discharge for appropriate saccades. A vigorous discharge precedes movements to the center of the movement field, but reduced responses occur when the movement occurs to an eccentric field position. The intensity of discharge is not related to initial eye position.

The temporal pattern of discharge of these units falls into two classes. The majority showed a gradual build-up of activity beginning about 80–100 msec before an appropriate saccade leading to a maximum discharge rate just before the saccade followed by a gradual decrease in activity during and outlasting the saccade. Another smaller group of cells started a ragged build-up of discharge about 80–100 msec before appropriate saccades, but then discharged a relatively discrete burst of high-frequency (up to 1000 spikes/sec) activity beginning about 20 msec before the saccade. In a sample of cells recorded at various topographic sites across the SC, neither the duration of the burst nor the discharge intensity was correlated with cell location when saccades were made to the center of each cell's movement field. Thus, a rostral SC cell with a movement field center at 3° and a caudal cell with a field center at 20° both were characterized by the same duration burst (about 35–40 msec) for separate saccades of 3° (duration about 20 msec) and 20° (duration about 50 msec), respectively.

The temporal pattern of cell discharge did change for saccades which landed eccentric to a given neuron's movement field center. Earlier, as well as more intense discharge, was associated with movements to the center of the field when compared to the activity generated in the same cell for movements to off-center locations.

In a conditioned situation in which the appearance of a visual stimulus sometimes elicits an appropriate saccade and sometimes fails to evoke any movement, the appearance of the high-frequency burst of activity was found to be exactly linked to the trials on which a saccade did occur. The lower-frequency earlier build-up of discharge still occurred on trials which did not result in a saccade (Sparks and Pollack, 1977).

Thus, before large contralateral saccades, one finds a large population of deeper layer SC neurons located over a considerable extent of caudal collicular area activated. Cells in the center of the active area show an earlier build-up of activity and a certain subset begin a high-frequency burst of activity about 20 msec before the saccade. Before a smaller saccade a more rostrally placed population of SC neurons, with some possible overlap with the previous population, begin a discharge build up. Cells located in the spatial center of this group begin earlier discharge and some also show a high-frequency burst 20 msec before the saccade.

One additional type of SC neuron has been reported by Sparks et al. (1977). These cells, called quasi-visual neurons, appear to have visual receptive fields and show increased discharge rates following appropriate target appearance whether or not a saccade is made to the target. However, based on their preliminary report, it appears that the temporal pattern of discharge is quite different for the two cases. When a saccade is made which moves the target onto the fovea, unit discharge, which began about 70 msec after the appearance of the visual stimulus, is rapidly reduced to the quiescent state during the saccade and remains in this state. When no saccade is made, unit discharge begins as before shortly after stimulus presentation but now continues with a gradual decline over a period of about 200 msec beyond normal saccadic latency (had a saccade occurred in response to the onset of visual stimulation). Double saccade trials suggest that this discharge is coded in head or body spatial coordinates as opposed to a retinotopic frame of reference (see Sparks et al., 1977, for details).

SACCADIC-RELATED DISCHARGE IN BRAIN STEM RETICULAR FORMATION

The patterns of activity of RF units, which show discharge patterns correlated with saccades in alert monkey, have been closely studied (Sparks and Travis, 1971; Cohen and Henn, 1972a; Luschei and Fuchs, 1972; Keller, 1974; Büttner, et al., 1977; Van Gisbergen and Robinson, 1977), and the results of this work and past lesion and anatomical studies have been extensively reviewed (Cohen and Henn, 1972b; Henn and Cohen, 1975; Keller, 1977a, b). Therefore, only the nomenclature used to classify RF eye movement-related cells and some very recent results on the temporal discharge parameters of long-lead burst cells will be included in the present review.

Reticular formation neurons which are characterized by a relatively steady discharge at a rate which is proportional to fixation position in a particular direction, usually the ipsilateral horizontal, are called tonic cells. Burst-tonic units are cells with similar eye position-coded discharge, but in addition show a burst of activity above that associated with the change in eye position during saccades in the preferred direction. The burst typically precedes saccadic onset with a short lead of about 8–10 msec and hence these cells closely resemble oculomotor neurons and abducens interneurons in both pattern of discharge and saccadic lead time. Both types of fixation-related neurons are only rarely encountered in the pontomedullary RF but within this region the highest concentration is found in more caudal areas, especially in the parabducens region. Both horizontal and vertically coded burst-tonic units are found in this region.

Burst units are characterized by brief phasic episodes of discharge for saccadic eye movements but are otherwise silent or only discharge irregularly. Neurons which begin a discrete burst of high-frequency discharge just preceding the burst of motoneurons and burst-tonic cells are called medium-lead burst neurons (MLn). Neurons which begin a period of uneven but gradually increasing discharge as early as 250 msec, but more typically 50–80 msec, before saccades are called long-lead burst neurons (LLn). A few cells burst only after the start of saccadic eye movements and hence are called following burst neurons (Fn).

Finally, a number of neurons are found in this region which are characterized by tonic firing punctuated by discrete pauses in discharge during saccades. The firing rate of some of these units is modulated by eye position or natural vestibular stimulation. Frequently for this type of unit, pauses occur only for saccades with a movement component in a specific direction. A dense accumulation of omnidirectional pausers occurs in a compact pool located on the midline and extending from the rostral pole of the abducens nucleus forward to just caudal of the trochlear nucleus. The discharge of these cells is not modulated by eye position or vestibular stimuli. The midline location of this group of pause neurons places them in or very close to the dorsal raphe nucleus complex (Taber et al., 1960). The universal feature uniting all of these diverse RF cells is the temporal relationships of their various discharge parameters with saccadic eye movements and the lack of any spatial organization encoding size or direction of saccades. An exception appears to be the location of vertical coding MLn in the mesencephalic RF (Büttner et al., 1977; King and Fuchs, 1977). As an example of the temporal coding of RF neurons, during a saccade the burst duration of horizontal MLn is nearly identical to the burst duration of oculomotor neuron discharge, and hence horizontal saccade duration. Maximum intraburst frequency is closely correlated with maximum saccadic velocity (thus saccade size) and instantaneous burst frequency with instantaneous saccadic velocity in MLn.

Only in the LLn is the clear relationship between unit discharge and the temporal parameters of oculomotor neurons not so apparent. Most neurons of this type do show a more discrete pulse of activity beginning about 20 msec before a saccade superimposed on the ragged earlier build-up of discharge and the duration of this pulse of activity is correlated with saccade duration. However, variability in both burst duration and intensity is large, even for a group of identical saccades. Such units are found distributed throughout this region of the RF and intermixed with MLn and tonic cells without any apparent spatial organization.

In a recent series of experiments similar to those reported by Sparks and Pollack (1977), I have been able to dissociate the earlier build-up of activity in LLn from the more discrete burst of activity occurring just before and during the saccade. Monkeys were trained to make saccades from a central fixation point to an eccentric target lamp. When the duration of the eccentric stimulus appearance is made very short, a condition is obtained where a saccade is only made on a certain number of the trial presentations but not others for identical target presentations. As can be seen in Fig. 1, an early discharge occurred following each target presentation, but the unit only made a transition into the high-frequency burst state on trials in which saccades were made (Fig. 1A, C). Thus, this group of neurons displayed very similar properties to discrete burst SC cells during the period of early discharge and in the tight correlation

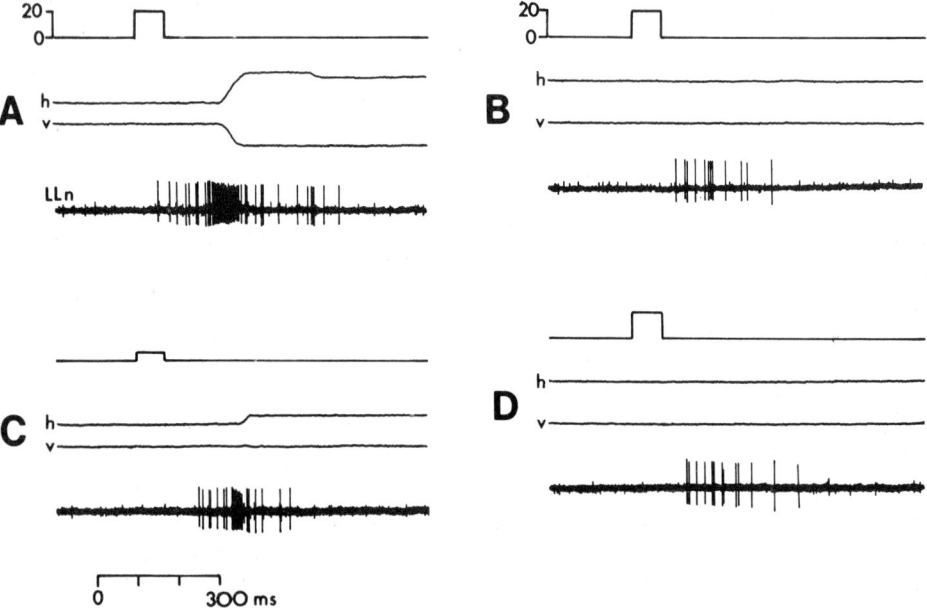

Fig. 1. Discharge patterns of a PRF long-lead burst neuron (lowest trace in each panel) following appearance of an eccentric visual target (upper trace in each panel) for a brief period of time (about 70 msec). A, C) Trials in which a saccade was made after target appearance to foveate the stimulus. B, D) No saccade trials. Calibration: 20° for target position and horizontal (h) and vertical (v) eye position. Time calibration shown in C applies to all records.

of the burst appearance to saccade occurrence. In contrast to these similarities, however, was the clear increase in burst duration with larger saccades (Fig. 1A) found in LLn.

COLLICULORETICULAR CONNECTIONS

Quite recently a number of anatomical and electrophysiological reports on SC pathways to specific RF locations have been published (Kawamura et al., 1974; Peterson et al., 1974; Precht et al., 1974; Grantyn and Grantyn, 1976; Graham, 1977; Grantyn et al., 1977; Harting, 1977). The general consensus reached on the basis of these studies is that the medial pontomedullary RF receives a direct input from the deeper layers of the contralateral colliculus. This projection originates from the entire spatial extent of SC, crosses the midline in the dorsal tegmental decussation and descends in the predorsal bundle. Bilateral terminations appear in caudal portions of the dorsal raphe complex while extensive contralateral terminations appear in nuclei pontis oralis and caudalis. Excitatory postsynaptic effects are elicited in RF neurons particularly in the area just ventral and rostral to the abducens nucleus. Since a major component of reticulospinal neurons is also located in this same region, it was essential to know which, if any, of these RF terminations were directed to supranuclear oculomotor cells. A suggestion that some of these projections were involved in oculomotor control was provided by intracellular stimulation and dye injection studies (Grantyn and Grantyn, 1976).

The question then remained as to which functional type of RF eye movement-related cells (LLn, MLn, Fn, tonic, or pause neurons) received SC input, since all these types are found closely intermingled within these same regions of the RF. We recently have completed a joint RF microelectrode recording and SC stimulating study in alert monkey to examine this question (Raybourn and Keller, 1977).

Among RF eye movement-related cells, the LLn type received the most direct and powerful input from SC. In most LLn the SC input was monosynaptic and very effective (even single-shock SC stimulations elicited spike responses in LLn). Each LLn could be activated from widely separated contralateral SC stimulating sites with no consistent relationship of stimulus threshold to collicular location. Triple-pulse stimulation of the SC evoked a reverberation of LLn discharge which considerably outlasted stimulus duration even at stimulation currents below that required to evoke a saccadic eye movement from the SC site.

Omnidirectional pause units also received a short-latency, excitatory input from the colliculus (in some cases monosynaptic), but the activation of spike discharge was less secure than in LLn. There was also a wide, including bilateral, spatial convergence from SC to each pause neuron. Triple-pulse stimulation of the colliculus led to initial activation of the pauser neurons, but this was quickly followed (at 6–8 msec following the onset of stimulation) by a brief cessation of these units' normal high-frequency ongoing discharge. The duration of this discharge inhibition was about 30 msec when the stimulus was increased to just threshold for evoking saccades. The pause duration increased for larger currents, as did the duration of the evoked eye movement. The similar time cause of LLn after discharge and pauser inhibition following SC stimulation suggested that the former units play a role in control of the pauser inhibition. The initial excitation of pauser units from the SC may represent the stimulation of a

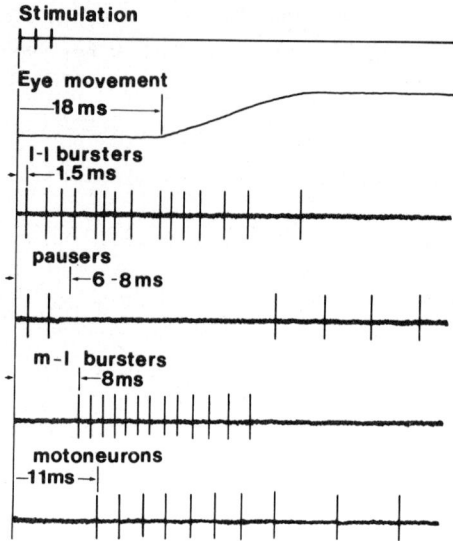

Fig. 2. Schematic representation of the average patterns of neural activities present in PRF neuron population types following focal triple-pulse stimulation (upper trace) of the deeper layers of the SC. A saccadic eye movement (second trace) was induced following the onset of stimulation. Only the initial discharge of long-lead burst neurons (l-l bursters) and omnidirectional pause neurons was time locked closely to the stimulus.

parallel path from the colliculus carrying non-specific tonic excitation to pauser neurons.

In the alert monkey, MLn, burst-tonic, and tonic neurons appear not to be directly driven by SC efferents. Units of these types and abducens motoneurons could all be activated following collicular triple-pulse stimulation, but only at current levels above the threshold for evoking saccades and then the evoked unit discharge was more closely locked temporally to saccade onset than to the stimulus onset. These results are summarized in Fig. 2, which shows in sequence the chain of RF neural events set in motion following SC stimulation.

Quite recently, I have begun another series of experiments to determine which of the eye movement-related types of SC cells form the RF connections just described. Some preliminary results of these experiments are reported here. Chronic stimulating electrodes were implanted in the midline RF pauser area and in the slightly more lateral saccadic bursting area in two monkeys (Fig. 3A). Microelectrode recordings were then made in the deeper collicular layers in the alert animal. When a collicular eye movement-related cell was isolated, its temporal behavior and movement field were roughly determined with the cooperation of the animals who were trained to track a small spot of light for liquid reinforcement. Then each typed SC unit was tested for possible direct projection to the RF areas containing the chronically implanted electrodes by delivering brief, single-pulse electrical stimuli on the implanted electrodes and by looking for antidromic responses in the collicular unit being recorded The possible occurrence of orthodromically activated discharges in collicular cells was also studied.

In confirmation of the results reported by Sparks et al. (1976), we have been able to divide the eye movement-related discharge of SC cells into two classes: 1) neurons

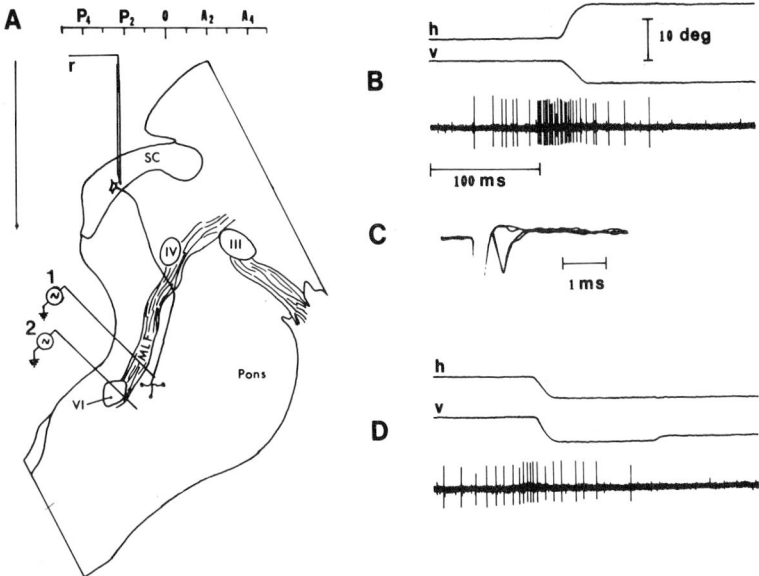

Fig. 3. Antidromic activation of SC neurons from the medial PRF. A) Schematic parasagittal section of monkey brain stem showing the location of the recording microelectrode (r) in the deeper layers of the SC and the stimulating electrodes: 1, located in the omnidirectional pauser area; and 2, located in the saccadic burst area of the RF. Vertical arrow shows the direction of the vertical stereotaxic plane. B) Response pattern of a discrete burst type SC eye movement neuron (lower trace) associated with a saccade (vertical and horizontal eye movement components shown in upper traces) to the center of the unit's movement field. C) Antidromic activation of unit shown in B from stimulating electrode 2. D) Response pattern of a non-bursting type of SC eye movement neuron associated with a saccade to the center of this unit's movement field. Unit was not antidromically activated from either stimulating site (not shown).

which show a period of uneven discharge and then a discrete burst of activity just preceding saccades to the movement field of the neuron (Fig. 3B, 10 cells), and 2) another type which was characterized by a gradual build-up of activity to a maximum just preceding saccades to the movement field but no discrete burst activity (Fig. 3D, 12 cells). Unfortunately, no quasi-visual cells (Sparks et al., 1977) have been located yet.

All 10 burst cells were antidromically activated (Fig. 3C) from both RF stimulating sites at similar current thresholds (30 μA) and with similar conduction latencies (0.35–0.5 msec). Only one of the 12 non-bursting SC cells appeared to be antidromically activated. No short-latency orthodromic responses were observed in either type of SC cell. Based on these results it is proposed that the burst type of SC eye movement field cells forms at least one of the efferent components of SC projection to the RF. No definite conclusions can be reached on the basis of these results as to the precise RF target (pause cells or burst cells) of this SC output, since the RF regions stimulated are only about 2 mm apart and the SC fibers destined for the more lateral burst area may pass through or very close to the midline pauser area. However, based on the similarity of discharge patterns of SC burst cells and LLn, it seems more likely that this type of RF neuron receives this collicular input.

In summary it is hypothesized (Fig. 4) that LLn receives input from SC burst cells located over the entire spatial extent of the colliculus, although only a subset of these

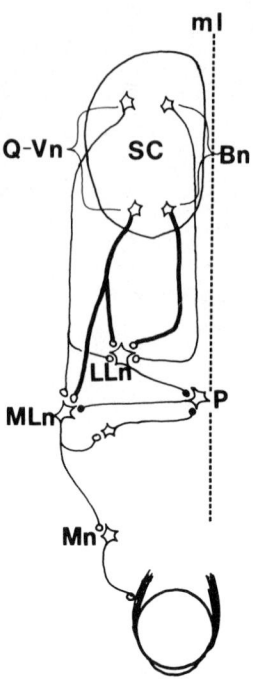

Fig. 4. Hypothesized organization of the colliculoreticular saccadic generator. Output of the SC consists of spatially organized populations of two types of cells: quasi-visual neurons (Q-Vn) and discrete burst neurons (Bn). For a small saccade only a rostal group of each type is active (shown by single uppermost SC neurons). For a large saccade only a caudal group of each type is active (shown by lower SC neurons). Bn from all spatial locations in colliculus project to long-lead burst neurons (LLn). Q-Vn project to both LLn and medium-lead burst neurons (MLn). Larger caudal population of SC neurons active for big saccades shown by heavier projection lines. LLn in turn generate inhibition (filled synaptic endings) in pauser neurons (P) when triggered by sufficient SC input. Pausers inhibit MLn. Pause duration, and hence saccade duration, controlled by P-MLn interaction and Q-Vn input (error signal) to MLn. MLn directly generate saccadic burst in oculomotor neurons (Mn). All connections shown on same side of midline (ml) for simplicity.

SC neurons (those within the movement field) are active for a particular saccade. This class of RF neurons in turn inhibit midline pause cells. When a decision has been reached to make a saccade, the intense burst of a spatially localized group of SC neurons initiates a burst in LLn, and through these latter neurons, a brief synchronized inhibition of pause neurons. Since SC burst neurons are not temporally coded (they fire the same burst duration for all size saccades) this signal could not provide the coding for the duration of pauser inhibition (and hence saccade size). Hence one must hypothesize other collicular input signals to RF neurons, if indeed the colliculus does control and not just initiate saccades.

A possible signal source for this coding role is the SC quasi-visual cell (Sparks et al., 1977). This SC neuron type appears to represent an error signal resulting from the difference of target position (or a centrally created memory of target position) coded in a head frame of reference and current eye position in head. This error signal is reduced to zero and hence unit discharge is nulled as an appropriate saccade to the target occurs. In this hypothesis this type of neuron plays the role of the error detector in the Robinson (1975) model of saccadic control with the difference that it is located in the SC instead of the RF. In addition, different spatial errors in the present hypothesis are encoded by different sets of quasi-visual neurons. The next necessary step in our study is to record from these SC neurons and determine by the use of RF stimulation if they do, in fact, project to this area of the RF.

SUMMARY

The discharge patterns of saccadic eye movement related neurons recorded in alert monkey superior colliculus (SC) is reviewed. A new interpretation is placed on the possible role played in saccadic generation by a class of neurons called quasi-visual cells. Saccade-related discharge of brain stem reticular formation (RF) neurons is discussed and comparisons are made between presaccadic activity in neurons in SC and RF. Studies showing the existence and nature of synaptic input from the SC to specific functional types of RF cells are reviewed and new preliminary results indicating the functionally identified type of SC neuron supplying at least a portion of these connections are shown. The possible role played by the neurons and synaptic connections described in these two anatomic structures is hypothesized in the form of saccadic control model.

ACKNOWLEDGEMENT

The author wishes to thank Professor David Sparks for helpful discussion on SC unit discharge characteristics. Research sponsored by the National Institutes of Health Grant EY00955-07.

REFERENCES

Büttner, U., Hepp, K. and Henn, V. (1977) Neurons in the rostral mesencephalic and paramedian pontine reticular formation generating fast eye movements. In *Control of Gaze by Brain Stem Neurons*, R. Baker and A. Berthoz (Eds.), Elsevier, Amsterdam, pp. 309–318.

Cohen, B. and Henn, V. (1972a) Unit activity in the pontine reticular formation associated with eye movements. *Brain Res.*, 46: 403–410.

Cohen, B. and Henn, V. (1972b) The origin of the quick phases of nystagmus in the horizontal plane. In *Cerebral Control of Eye Movements and Motion Perception*, J. Dichgans and E. Bizzi (Eds.), Karger, Basel, pp. 36–55.

Graham, J. (1977) An autoradiographic study of the efferent connections of the superior colliculus in the cat *J. comp. Neurol.*, 173: 629–654.

Grantyn, A.A. and Grantyn, R. (1976) Synaptic actions of tectofugal pathways on abducens motoneurons in the cat. *Brain Res.*, 105: 269–285.

Grantyn, A., Grantyn, R. and Robine, K.P. (1977) Neural organization of the tecto-oculomotor pathways. In *Control of Gaze by Brain Stem Neurons*, R. Baker and A. Berthoz (Eds.), Elsevier, Amsterdam, pp. 197–206.

Harting, J.K. (1977) Descending pathways from the superior colliculus: an autoradiographic analysis in the rhesus monkey *(Macaca mulatta)*. *J. comp. Neurol*, 173: 583–612.

Henn, V. and Cohen, V. (1975) Activity in eye muscle motoneurons and brainstem units during eye movements. In *Basic Mechanisms of Ocular Motility and their Clinical Implications*, G. Lennerstrand and P. Bach-y-Rita (Eds.), Pergamon Press, New York, pp. 303–324.

Kawamura, K., Brodal, A. and Hoddevik, G. (1974) The projection of the superior colliculus onto the reticular formation of the brain stem. An experimental anatomical study in the cat. *Exp. Brain Res.*, 19: 1–19.

Keller, E.L. (1974) Participation of medial pontine reticular formation in eye movement generation in monkey. *J. Neurophysiol.*, 37: 316–332.

Keller, E.L. (1977a) The role of the brain stem reticular formation in eye movement control. In *Eye Movements*, B.A. Brooks and F.J. Bajandas (Eds.), Plenum Press, New York, pp. 105–126.

Keller, E.L. (1977b) Control of saccadic eye movements by midline brain stem neurons. In *Control of Gaze by Brain Stem Neurons*, R. Baker and A. Berthoz (Eds.), Elsevier, Amsterdam, pp. 327–336.

King, W.M. and Fuchs, A.F. (1977) Neuronal activity in the mesencephalon related to vertical eye movements. In *Control of Gaze by Brain Stem Neurons*, R. Baker and A. Berthoz (Eds.), Elsevier, Amsterdam, pp. 319–326.

Luschei, E.S. and Fuchs, A.F. (1972) Activity of brain stem neurons during eye movements of alert monkey. *J. Neurophysiol.*, 35: 445–461.

Mohler, C.W. and Wurtz, R.H. (1976) Organization of monkey superior colliculus: intermediate layer cells discharging before eye movements. *J. Neurophysiol.*, 39: 722–744.

Peterson, B.W., Anderson, M.E. and Filion, M. (1974) Responses of pontomedullary reticular neurons to cortical, tectal, and cutaneous stimuli. *Exp. Brain Res.*, 21: 19–44.

Precht, W., Schwindt, P.C. and Magherini, P.C. (1974) Tectal influences on cat ocular motoneurons. *Brain Res.*, 82: 27–40.

Raybourn, M.S. and Keller, E.L. (1977) Colliculoreticular organization in primate oculomotor system. *J. Neurophysiol.*, 40: 861–878.

Robinson, D.A. (1975) Oculomotor control signals. In *Basic Mechanisms of Ocular Motility and their Clinical Implications*, G. Lennerstrand and P. Bach-y-Rita (Eds.), Pergamon Press, New York, pp. 337–374.

Robinson, D.L. and Jarvis, C.D. (1974) Superior colliculus neurons studied during head and eye movements of the behaving monkey. *J. Neurophysiol.*, 37: 533–540.

Schiller, P.H. and Koerner, F. (1971) Discharge characteristics of single units in superior colliculus of the alert Rhesus monkey. *J. Neurophysiol.*, 34: 920–936.

Schiller, P.H. and Stryker, M. (1972) Single-unit recording and stimulation in superior colliculus of the alert Rhesus monkey. *J. Neurophysiol.*, 35: 915–924.

Sparks, D.L. and Travis, R.P. (1971) Firing patterns of reticular neurons during horizontal eye movements. *Brain Res.*, 33: 477–481.

Sparks, D.L., Holland, R. and Guthrie, B.L. (1976) Size and distribution of movement fields in the monkey superior colliculus. *Brain Res.*, 113: 21–34.

Sparks, D.L. and Pollack, J.G. (1977) The neural control of saccadic eye movements: Role of the superior colliculus. In *Eye Movements*, Brooks, B.A. and Bajandas, F.J. (Eds.), Plenum Press, New York, pp. 179–219.

Sparks, D.L., Mays, L.E. and Pollack, J.G. (1977) Saccade-related unit activity in the monkey superior colliculus. In *Control of Gaze by Brain Stem Neurons*, R. Baker and A. Berthoz (Eds.), Elsevier, Amsterdam, pp. 437–444.

Taber, E., Brodal, A. and Walberg, F. (1960) The raphe nuclei of the brain stem in the cat. *J. comp. Neurol.*, 114: 161–187.

Van Gisbergen, J.A.M. and Robinson, D.A. (1977) Generation of micro- and macro-saccades by burst neurons in the monkey. In *Control of Gaze by Brain Stem Neurons*, R. Baker and A. Berthoz (Eds.), Elsevier, Amsterdam, pp. 301–308.

Wurtz, R.H. and Goldberg, M.E. (1972) Activity of superior colliculus in behaving monkey. III. Cells discharging before eye movements. *J. Neurophysiol.*, 35: 575–586.

Labyrinthine and Visual Inputs to the Superior Colliculus Neurons

M. MAEDA[1], T. SHIBAZAKI[2] and K. YOSHIDA[3]

[1]*Department of Neurosurgery, Juntendo University School of Medicine, Tokyo;* [2]*Department of Neurophysiology, Institute of Brain Research, Tokyo University School of Medicine, Tokyo; and* [3]*Department of Physiology, Institute of Basic Medical Sciences, University of Tsukuba, Ibaraki-ken 300-31 (Japan)*

The superior colliculus receive substantial input from both the retina and visual cortex as well as from the contralateral superior colliculus in the cat and the rat. The prominence and precision of the retinotectal and corticotectal projections suggest that the superior colliculus is important in the mediation and integration of appropriate visually guided behavior in the cat (Sprague, 1966). It has been also suggested that the superior colliculus plays an important role in the coordination of eye and head movements during gaze. This conclusion is based on several major lines of evidence. Cells within the superior colliculus give high frequency discharges prior to saccadic eye movements (Schiller and Stryker, 1972; Wurtz and Goldberg, 1972; Robinson and Jarvis, 1974; Mohler and Wurtz, 1976). Electrical stimulation of the superior colliculus leads to movement not only of the eye (Robinson, 1972; Schiller and Stryker, 1972; Straschill and Rieger, 1973; Precht et al., 1974; Grantyn and Grantyn, 1976; Roucoux and Crommelinck, 1976) but also the neck and body (Anderson et al., 1971). Ablation of the superior colliculus can produce profound deficits in visually guided behaviors (Sprague and Meikle, 1965). On the other hand, the vestibular system has also been known to regulate eye and head movements (Lorente de Nó, 1933; Wilson and Maeda, 1974; Maeda et al., 1977). In this respect, studies of the neuronal organization between the vestibular and retinotectal system and the interaction between them (Bisti et al., 1974) are required for elucidating the mechanism for gaze control.

LABYRINTHINE AND VISUAL INPUTS TO TECTAL NEURONS

Bipolar electrodes (fine Ag-AgCl wires) were implanted in the ipsi- and contralateral labyrinths to stimulate the vestibular nerve. In some cases the cochlear nerve was cut on one side in the internal auditory meatus to rule out the possibility that the observed responses in tectal neurons to labyrinthine stimulation were due to stimulus spread to the cochlear nerve. Bipolar electrodes were used for stimulation of the optic disk on both sides. After the surface of the superior colliculus was exposed by removal of the occipital lobe, intracellular recording from tectal neurons was performed on chloralose-anesthetized cats. The projection areas of the neurons impaled were identified by their antidromic responses to stimulation of the contralateral abducens nucleus and adjacent area, the contralateral prepositus hypoglossi or the C_2 segment with bipolar tungsten electrodes. We classified cells projecting to the abducens, the prepositus hypoglossi or

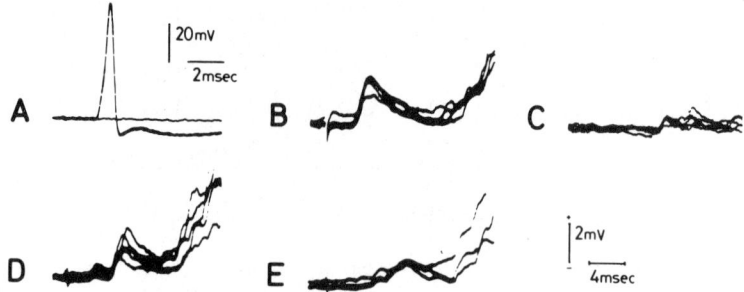

Fig. 1. Intracellular records from tectospinal neurons in response to optic disk and vestibular nerve stimulation. A) Antidromic response to stimulation of contralateral cervical cord (C2). B) EPSPs induced by single shocks to the contralateral optic disk. C) Small EPSPs with long latency induced by ipsilateral optic disk stimulation. D) EPSPs evoked from the contralateral vestibular nerve. Note the early EPSPs are followed by late larger depolarization. E) Slowly rising EPSPs after stimulation of the ipsilateral vestibular nerve. Voltage calibration and time scale for E also apply to B–D.

the vicinity of these nuclei and not to the cervical segment as tectoreticular neurons (TR neurons). Cells projecting to the cervical region were classified as tectospinal neurons (TS neurons). The location of the neurons recorded intracellularly was reconstructed from the reference points (2–3 fast green dye marks made during and at the end of the experiment). A representative antidromic spike potential is shown in Fig. 1A.

Tectospinal neurons

Single-shock stimulation of the contralateral optic disk (ODc) typically caused depolarization in TS neurons. Their amplitudes varied widely with stimulus strength and with the number of stimuli and occasionally generated action potentials. The early part of the response could be followed by a later, larger depolarization (Fig. 1B). In a number of cases the response was increased in amplitude by passing a hyperpolarizing current and decreased by depolarization through the recording electrode, although little or no effect was observed in some cases. The most likely explanation of these results is that the response is predominantly an EPSP. EPSPs were recorded in all TS neurons impaled after contralateral OD stimulation. In contrast to the contralaterally evoked response, stimulation of the ipsilateral optic disk (ODi) produced late small depolarizations (Fig. 1C), and in almost all cases the spike potentials were not induced. By quantitative mapping with the electron microscope, Sterling (1973) concluded that the ipsilateral retinocollicular terminals, rather than representing 20% of the total as gauged by light microscopy, must be considerably less than 1%. Therefore, it is likely that late small PSPs evoked in TS neuron after ODi-stimulation are probably due to ipsilateral sparse retinocollicular connections or through other structures.

Following single shock stimulation of the contralateral vestibular nerve (Vc), EPSPs were recorded (Fig. 1D) in 23 cells out of 25 TS neurons impaled, and action potentials were occasionally generated. In many cases fast-rising, short-latency EPSPs were followed by a later, long-lasting depolarization. Such short-latency potentials varied in amplitude with the strength of stimulation and responses to successive shocks in the train typically increased in size, as well as PSPs elicited by stimulation of the optic disk even with double or triple shocks. In a few cases mixed effects, i.e., depolarization-hyperpolarization sequences, were also obtained. The most common potential changes induced from the ipsilateral vestibular nerve (Vi) were slowly-rising small EPSPs (Fig. 1E). Multiple stimuli were usually necessary to elicit PSPs and the responses to successive shocks were markedly facilitated as well. In summary, tectospinal neurons were influenced most consistently and strongly from the contralateral optic disk and vestibular nerve.

The latencies of the EPSPs evoked in TS neurons showed the following ranges: ODc, 2.2–2.7 msec (median 3.0 msec); ODi, 4.0–12.0 msec (7.0 msec); Vc, 2.4–12.0 msec (3.6 msec); Vi, 4.0–13.0 msec (7.0 msec). PSP latency histogram revealed a tendency for Vc- or ODc-latencies to be shorter than those evoked by Vi or ODi. This tendency was confirmed by comparing latencies of Vc- or ODc- and Vi- or ODi-evoked synaptic potentials recorded in the same cells.

Tectoreticular neurons

Stimulation of the contralateral optic disk with single shock produced EPSPs with short latencies in almost all of TR neurons. In sharp contrast to ODc-evoked response, stimulation of the ipsilateral optic disk gave little effects or later, very slowly rising depolarization.

Following contralateral vestibular nerve stimulation EPSPs were produced in TR neurons. The early response was followed by a later, large depolarization, and action potentials were initiated from later depolarization in many cases. In some cells the response consisted of both depo- and hyperpolarization sequences like those observed in TS neurons. The most common potential change evoked, ipsilaterally, in TR neurons, was very slowly rising depolarization (EPSP) although an IPSP or a response consisting of a mixture of EPSP and IPSP was observed in some cells (7/45). Thus, the characteristics and patterns of synaptic potentials evoked in TR neurons were very similar to those obtained in TS neurons. They were also influenced more consistently and strongly from ODc and Vc than from ODi and Vi.

When the patterns of Vc- and Vi-evoked PSPs in a given cell were compared, reciprocal effects consisting of contralateral excitation and ipsilateral inhibition were observed in 4 TR neurons. On the other hand, such reciprocal effects were not observed in TS neurons.

The latencies of the EPSPs induced in TR neurons showed the following ranges: ODc, 2.1–9.0 msec (median 2.8 msec); ODi, 3.2–11.0 msec (7.0 msec); Vc, 2.2–10.0 msec (3.6 msec); Vi, 2.4–12.0 msec (5.4 msec). The latencies of the IPSP elicited after stimulation of Vi ranged from 4.0 to 6.0 msec.

In the case of neurons in the superficial layer, which have been known to send their axons to the deeper layers (Kanaseki and Sprague, 1974), stimulation of ODc induced monosynaptic EPSPs with latencies of 1.4–1.9 msec and usually generated action potentials shortly (0.2–0.4 msec) after the onset of EPSPs. Addition to these values of about 0.4 msec for synaptic delay in the superior colliculus gives an expected shortest latency range of 2.0–2.5 msec for a disynaptic pathway between the contralateral retina and neurons located in the intermediate or deep layer. Clearly then, the early EPSPs recorded in TS or TR neurons indicate disynaptic transmission. ODi-induced EPSPs had a latency of more than 3.2 msec, suggesting that these EPSPs were evoked via a pathway containing more synapses.

We performed experiments in which synaptic potentials were evoked in the same neurons by stimulation of the vestibular nerve and nucleus. Stimulation of the contralateral vestibular nucleus induced EPSPs in these tectal neurons with latencies ranging from 1.8 to 2.5 msec and clear temporal summation was observed, indicating the existence of

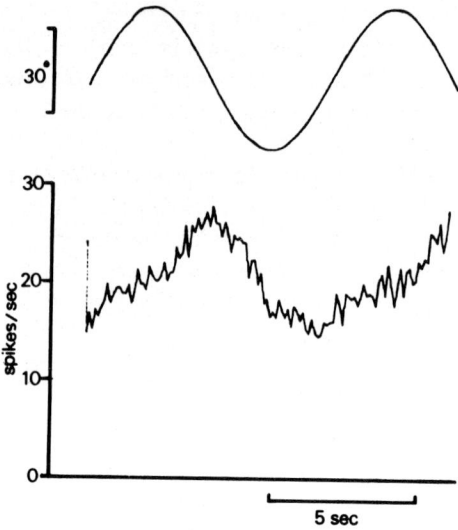

Fig. 2. Responses of the tectal neuron in the right side to sinusoidal oscillation. In upper trace, downward displacement of the curve indicates the leftward movement of the turn table. Unit spike was counted at each interval of 100 msec. The computer-averaged response represents the mean value of 30 successive sweeps.

disynaptic connections between them. The latency differences between the vestibular nerve and nucleus-evoked EPSPs measured 0.5–2.8 msec. These results suggest the existence of trisynaptic pathway linking the contralateral labyrinth and tectal neurons. In this pathway, the first synapse is located in the vestibular nuclei and the second synapses outside the vestibular nuclei. The latency ranges and the properties of IPSPs evoked in TR neurons by stimulation of the ipsilateral vestibular nerve suggest the presence of both trisynaptic and polysynaptic inhibitory connections between them.

RESPONSE TO SINUSOIDAL OSCILLATION

The animal was mounted on a stereotaxic apparatus located on a rotating device. Extracellular spike potentials were recorded with glass micropipettes filled with Ringer solution saturated with fast green FCF. We have recorded 11 tectal neurons (8 horizontal canal-related cells and 3 vertical canal-related cells) located in the intermediate and the deep layers. Out of 8 horizontal canal-related cells, 5 cells increased their firing frequency with contralateral angular acceleration and decreased their frequency with ipsilateral acceleration. These neurons were called type II neurons. We have also recorded type I and type III tectal neurons. These results indicate the presence of, at least, semicircular canal inputs to tectal neurons. Fig. 2 represents responses of a horizontal type II tectal neuron to sinusoidal oscillation.

INTERACTION OF VESTIBULAR AND VISUAL EFFECTS UPON TECTAL NEURONS

Horn and Hill (1969) and subsequently Denney and Adorjani (1972) have found that a few visual cortical cells alter their preferred axis when the animal tilts and the alteration is such as to compensate for the tilt. Recently, Bisti et al. (1974) have shown that a large percentage of the superficial layer cells of the superior colliculus alter their visual responses as a function of the body tilt of the animal. All these data motivated us to look at the interaction of vestibular and visual inputs upon the output cells of the colliculus, i.e., TS or TR neurons located in the two deep gray layers. Fig. 3A shows a control *trisynaptic* EPSP evoked in a TR neuron by ODc-stimulation. When conditioned by Vc stimulation (Fig. 3B), the same test ODc volley induced a larger EPSP (Fig. 3C) than the algebraic summation of the two potentials. This finding indicates that the Vc volley converges on and facilitates interneurons that mediate trisynaptic excitation from the ODc as schematically drawn in Fig. 3D. These interneurons may play an important role in the control of gaze-saccade accompanied by head movement. The precise locations of these interneurons remained to be studied (see also Fig. 4).

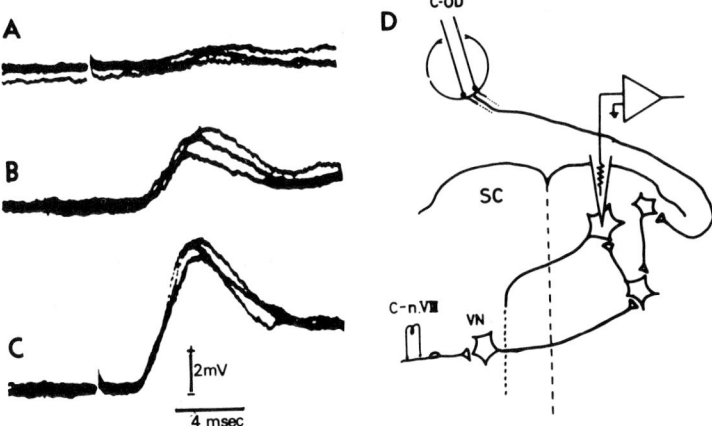

Fig. 3. Facilitation of the optic disk-evoked trisynaptic EPSPs by vestibular stimulation. A) Small trisynaptic EPSPs evoked by stimulation of the contralateral optic disk. B) Response to contralateral vestibular. C) The same test stimulation as in A was conditioned by single shock to the vestibular nerve (as in B). ODc volley induced a larger EPSP than the algebraic summation. D) Schematic drawing of simplified pathways from ODc and Vc to tectal neurons and the interaction between them. Voltage calibration and time scale for C apply to A and B.

LOCATION OF TS AND TR NEURONS

We recorded 27 TS and 61 TR neurons. The location of these neurons were reconstructed from the reference point as described before. 16 neurons out of 27 TS neurons were found at the intermediate layer, and the remaining 11 cells were at the deep layer. 34 cells out of 61 TR neurons were found at the intermediate layer and 27 cells at the deep layer. None of TS or TR neurons were located in the superficial layer. The large cells of laminae IV and VI are known to give rise to axons in the crossed predorsal bundle and in the tectospinal tract. Thus, our results are consistent with the anatomical data (Sprague et al., 1963; Kanaseki and Sprague, 1974).

EFFECTS OF CONTRALATERAL SUPERIOR COLLICULUS STIMULATION

Sprague (1966) has proposed, on the basis of behavioral experiments in the cat, that the functional state of the superior colliculus may be influenced by both ipsilateral cortical facilitation and contralateral tectal inhibition. The phenomenon described by Sprague has been confirmed in behavioral studies (Sherman, 1974) and reciprocal tectal inhibition has been reported, on the basis of field analysis and extracellular unit recording, in the rat (Goodale, 1973) and in the cat (Hoffmann and Straschill, 1971). Synaptic mechanism underlying the tectotectal interaction has not been studied.

Stimulation of the contralateral superior colliculus (the intermediate or deep layer) evoked IPSPs (Fig. 4D) in both TS and TR neurons, which were easily reversed by Cl^- ion injection. In some cases IPSPs were preceded by small EPSPs. The latencies of IPSPs ranged from 0.7 to 1.4 msec. We recorded the commissural neurons projecting to the contralateral superior colliculus (Magalhaes-Castro, 1978) (see also Fig. 4), and the latencies of antidromic spikes of these neurons ranged from 0.4 to 1.2 msec. Addition to these values of about 0.4 msec for synaptic delay gives an expected shortest latency range of 0.8–1.6 msec for a monosynaptic pathway. Therefore, the IPSPs evoked from the contralateral superior colliculus should be monosynaptic. It is very likely that these monosynaptic IPSPs at least partly participate in the production of tectotectal inhibition as was suggested from lesion experiments.

In four commissural neurons recorded intracellularly, stimulation of the contralateral colliculus evoked IPSPs after deterioration of antidromic spike potentials, suggesting that intertectal fibers mediate an effect of mutual suppression between the two colliculi.

TECTOTECTAL PROJECTIONS REVEALED BY RETROGRADE TRANSPORT OF HRP

In three cats under sodium pentobarbital anesthesia (40 mg/kg), horseradish peroxidase (HRP, Sigma Type VI, 50% in 0.9% saline) was injected hydraulically into the unilateral superior colliculus. Retrograde cell-labeling (Kristensen and Olsson, 1971) was visualized after 2 days survival time. Fig. 4 illustrates the injection site (right-superior colliculus) and the HRP-labeled cells in the superior colliculus and other structures between frontal plane A-4 and P-8. With respect to the distribution of all labeled cells along the collicular rostrocaudal extension, it was observed that the majority of the cell were located in the rostral portion of the superior colliculus. It can be noted that HRP-positive cell bodies found in the colliculus contralateral to the injected side were distributed over collicular layers III, IV and VI (Fig. $4B_{A4}$). Most labeled cells were small and medium sized. These results are consistent with the locations of fast green dye marks of commissural neurons electrophysiologically identified and confirmed those obtained by Edwards (1977) and Magalhaes-Castro (1978). It is also worthwhile noting the distribution of the labeled cells in the brain stem. As illustrated in Fig. 4B, the HRP-positive cells were found in the substantia nigra, the parabigeminal nucleus, the mesencephalic reticular formation, nucleus reticularis pontis, and nucleus tractus spinalis n. trigemini (pars oralis) (see also Grofova et al., 1978). Fig. 4B also shows that there are few HRP-labeled cells in the vestibular nuclei and this would be in agreement with the physiological data, indicating no direct connections between the vestibular nuclei and the superior colliculus (see also Fig. 3).

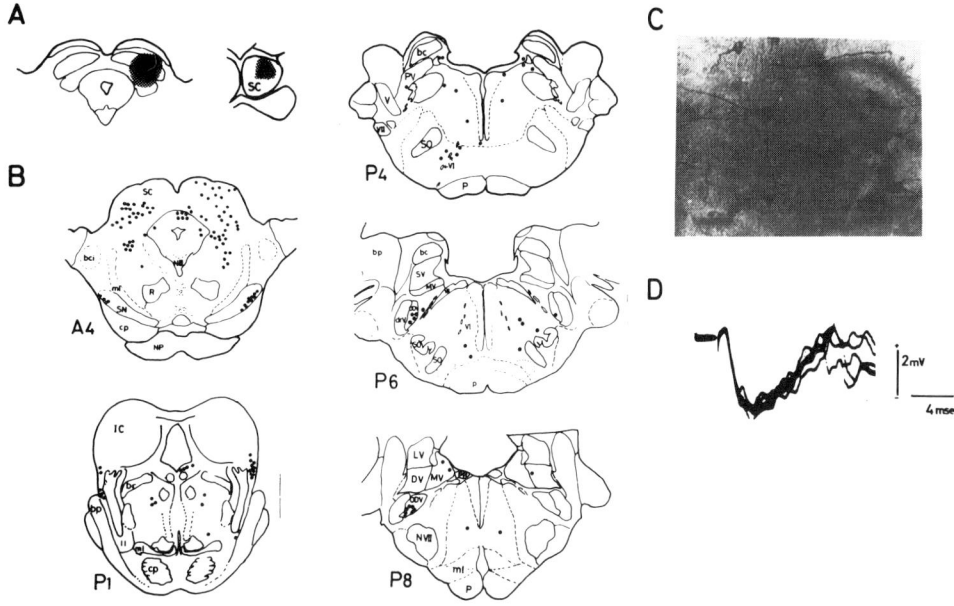

Fig. 4. Drawings of selected sections from one HRP experiment arranged rostral to caudal illustrating the distribution of HRP-labeled cells. A) The core (black area) of the HRP injection site in the right superior colliculus. The shaded area represents the medium brown halo surrounding the core of the injection site. B) Drawing showing the distribution of labeled cells in the mesencephalon and the pons following unilateral injection involving the superior colliculus on the right side as shown in A. Each drawing represents one section, and one dot on cell-labeled neuron. C) Photomicrograph showing labeled cells located in the intermediate layer of the contralateral superior colliculus. D) Monosynaptically evoked IPSPs in a tectospinal neuron by stimulation of the contralateral superior colliculus. bc; brachium conjunctivum; bci; brachium colliculi inferioris; bp; brachium pontis; cp; cerebral peduncle; drV; descending root of the trigeminal nerve; IC; inferior colliculus; LV; lateral vestibular nucleus; ml; medial lemniscus; MV; medial vestibular nucleus; NP; nuclei pontis basales; ODV; nucleus tractus spinalis n. trigemini (pars oralis); P; pyramid; PR; nucleus praepositus hypoglossi; SC; superior colliculus; SN; substantia nigra; SO; superior olive.

In conclusion, the superior colliculus neurons, which project to the vicinity of the abducens nucleus or to the upper cervical cord, presumably participating in control of eye and head movement, receive excitatory inputs predominantly from the contralateral optic disk and vestibular nerve. Volleys from vestibular afferents converge on and facilitate interneurons which mediate retinotectal transmission. Thus, visuovestibular interactions occurring at the levels of the interneurons and the output cells in deep gray layers, may play an important role in production of visually guided eye movement toward contralateral hemifields accompanied by head movement.

TR and TS neurons receive monosynaptic IPSPs from the contralateral superior colliculus, and the commissural neurons also receive monosynaptic IPSPs from the contralateral side. Since the two colliculi operate to orient the animal to opposite hemifields (Edward, 1978), this effect could serve to prevent competing responses in the opposite direction. Such a mutually inhibiting effect would be comparable to the mechanism of reciprocal inhibition in the spinal cord. Whatever function is under this system, our results show that the commissural neurons mediate suppressive effects on the output of the contralateral superior colliculus and provide evidence for the intertectal suppression proposed by Sprague (1966) and Sherman (1974).

SUMMARY

Synaptic potentials were recorded in cat superior colliculus neurons (tectospinal, TS and tectorecticular, TR) in response to stimulation of the optic disk, vestibular nerves and contralateral superior colliculus.

1. Disynaptic and polysynaptic EPSPs were seen in TS and TR neurons by stimulation of the contralateral optic disk. Ipsilateral optic disk stimulation induced only polysynaptic EPSPs. Trisynaptic EPSPs were recorded following stimulation of the contralateral vestibular nerve. Stimulation of the ipsilateral vestibular nerve evoked polysynaptic EPSPs and EPSP-IPSP sequences. TS and TR neurons received such excitatory inputs more consistently and strongly from the contralateral optic disc and vestibular nerve than from the ipsilateral.

2. Vestibular afferent volleys converge on and facilitate interneurons which mediate trisynaptic retinotectal transmission. Thus, visuovestibular interaction at the level of the interneurons may play an important role for visually guided eye movement accompanied by head movement.

3. Stimulation of the contralateral superior colliculus evoked IPSPs monosynaptically in TS and TR neurons. It is very likely that these IPSPs participate in the production of tectotectal inhibition proposed by lesion experiments. The locations of the commissural neurons presumably mediating the intertectal inhibition were electrophysiologically and histologically identified by using microstimulation and HRP injection.

REFERENCES

Anderson, M.E., Yoshida, M. and Wilson, V. (1971) Influence of superior colliculus on cat neck motoneurons. *J. Neurophysiol.*, 34: 898–907.

Berman, A.L. (1968) *The Brain Stem of the Cat. A Cytoarchitectonic Atlas with Stereotaxic Coordinates.* Univ. of Wisconsin Press, Madison.

Bisti, S., Maffei, L. and Piccolino, M. (1974) Visuovestibular interactions in the cat superior colliculus. *J. Neurophysiol.*, 37: 146–155.

Denney, D. and Adorjani, C. (1972) Orientation specificity of visual cortical neurons after head tilt. *Exp. Brain Res.*, 14: 312–317.

Edwards, S.B. (1977) The commissural projection of the superior colliculus in the cat. *J. Comp. Neurol.*, 173: 23–40.

Edwards, S.B. and Henkel, C.D. (1978) Superior colliculus connections with the extraocular motor nuclei in the cat. *J. Comp. Neurol.*, 179: 451–508.

Goodale, M.A. (1973) Cortico-tectal and intertectal modulation of visual responses in the rat's superior colliculus. *Exp. Brain Res.*, 17: 75–86.

Grantyn, A.A. and Grantyn, R. (1976) Synaptic actions of tectofugal pathways on abducens motoneurons in the cat. *Brain Res.*, 105: 269–286.

Grofova, I., Ottersen, O.P. and Rinvik, E. (1978) Mesencephalic and diencephalic afferents to the superior colliculus and pereaqueductal gray substance demonstrated by retrograde axonal transport of horseradish peroxidase in the cat. *Brain Res.*, 146: 205–220.

Hoffmann, K.P. and Straschill, M. (1973) Influences of corticotectal and intertectal connections on visual responses in the cat's superior colliculus. *Exp. Brain Res.*, 17: 75–86.

Horn, G. and Hill, R.M. (1969) Modification of receptive fields in cells in the visual corter occurring spontaneously and associated with body tilt. *Nature*, 221: 186–188.

Kanaseki, T. and Sprague, J.M. (1974) Anatomical organization of pretectal nuclei and tectal laminae in the cat. *J. Comp. Neurol.*, 158: 319–338.

Kristensson, K. and Olsson, Y. (1971) Retrograde axonal transport of protein. *Brain Res.*, 29: 363–365.

Lorente de No, R. (1933) Vestibulo-ocular reflex arc. *Arch. Neurol. Psychiat, (Chic.)* 30: 245–291.

Maeda, M., Magherini, P.C. and Precht, W. (1977) Functional organization of vestibular and optic inputs to neck and forelimb motoneurons in the frog. *J. Neurophysiol.*, 40: 225–243.

Magalhaes-Castro, B. (1978) Horseradish peroxidase labeling of cat tectotectal cells. *Brain Res.*, 148: 1–13.

Melvill Jones, G. and Milsam, J.H. (1971) Frequency-response analysis of central vestibular unit activity from rotational stimulation of the semicircular canals. *J. Physiol. (Lond.)* 219: 191–215.

Mohler, C.W. and Wurtz, R.H. (1976) Organization of monkey superior colliculus: intermediate layer cells discharging before eye movements. *J. Neurophysiol.*, 39: 722–744.

Precht, W., Schwindt, P.C. and Magherini, P.C. (1974) Tectal influences on cat oculomotor neurons. *Brain Res.*, 82: 27–40.

Robinson, D.A. (1972) Eye movement evoked by collicular stimulation in the alert monkey. *Vision Res.*, 12: 1795–1808.

Robinson, D.L. and Jarvis, C.D. (1974) Superior colliculus neurons studied during head and eye movements of the behaving monkey. *J. Neurophysiol.*, 37: 533–540.

Roucoux, A. and Crommelinck, M. (1976) Eye movements evoked by superior colliculus stimulation in the alert cat. *Brain Res.*, 106: 349–363.

Schiller, P.H. and Stryker, M. (1972) Single-unit recording and stimulation in superior colliculus of the alert rhesus monkey. *J. Neurophysiol.*, 35: 915–924.

Sherman, S.M. (1974) Visual field of cat with cortical and tectal lesions. *Science*, 185: 355–357.

Sprague, J.M. (1966) Interaction of cortex and superior colliculus in mediation of visually guided behavior in the cat. *Science*, 153: 1544–1547.

Sprague, J.M. and Meikle Jr. (1965) The role of the superior colliculus in visually guided behavior. *Exp. Neurol.*, 11: 115–146.

Sprague, J.M., Levitt, M., Robson, K., Liu, C.N., Stellar, E. and Chambers, W.W. (1963) A neuroanatomical and behavioral analysis of the syndromes resulting from midbrain lemniscal and reticular lesions in the cat. *Arch. Ital. Biol.*, 101: 225–295.

Sterling, P. (1973) Quantitative mapping with the electron microscope: retinal terminals in the superior collilculus. *Brain Res.*, 54: 347–354.

Straschill, M. and Rieger, P. (1973) Eye movements evoked by focal stimulation of the cat's superior collliculus. *Brain Res.*, 59: 211–227.

Wilson, V.J. and Maeda, M. (1974) Connections between semicircular canals and neck motoneurons in the cat. *J. Neurophysiol.*, 37: 346–357.

Wurtz, R.H. and Goldberg, M.E. (1972) Activity of superior collliculus in behaving monkey. III. Cells discharging before eye movements. *J. Neurophysiol.*, 35: 375–586.

Visual Fixation: a Collicular Reflex?

A. ROUCOUX*, M. CROMMELINCK and M. MEULDERS

Laboratoire de Neurophysiologie, Université Catholique de Louvain, 1200 Bruxelles (Belgium)

Superior colliculus (SC) is undoubtedly concerned with gaze orientation. Quite demonstrative experiments in the monkey (Robinson, 1972; Schiller and Stryker, 1972) have given a strong support to the "foveation hypothesis". According to this model, when a stimulus suddenly appears within the animal's visual field or when a particular detail of the visual environment is chosen by the animal as a target, a restricted zone within the sensory upper collicular layers is activated. To the retinotopy of these upper layers corresponds a similar spatial organization in the deep "motor" or efferent layers, a "motor map". The focal activation of the sensory layers is transmitted to the deep layers which are in register. After an adequate signal processing, a still mysterious spatiotemporal translation (Robinson, 1973), the signal is fed to the ocular motoneurons who manage to bring the image of the stimulus onto the fovea. It is to be noted that, in this model, the collicular signal merely acts as a trigger of a preprogrammed ballistic saccade. Direction and amplitude of this saccade are spatially or, better, topographically encoded in the SC.

The direct transmission of information from upper sensory to lower premotor layers is however questioned by some observations (Mohler and Wurtz, 1976; Wurtz and Mohler, 1976). These authors assert that the temporal sequence of neuronal events they record within the collicular layers is in contradiction with this straightforward foveation hypothesis. However, for the monkeys used in these experiments, unavoidably overtrained, stimuli are hardly sudden and unexpected and their behavioral significance is far different from the naturally occurring visual stimulus which is supposed to trigger a fixation reflex in response to a "what is that?". The contradiction between an attention-shifting and motor command mechanism thus appears rather artificial. It is obvious that the choice of a new target is necessarily correlated with an attention shift. We may admit that the primate SC is involved in reflex orientation of the eyes and operates some sort of sensory-motor point to point translation, be it with the help of a downward, upward flow of information or both.

This attractive foveation hypothesis and the fact that monkey's SC does not seem to directly participate in head orientation (Robinson and Jarvis, 1974; Stryker and Schiller, 1975) maybe has somewhat obscured the implications of old and well established observations in non-primates. The famous "visual grasp reflex" (Hess et al., 1946) is a

---

*Chargé de Recherches, F.N.R.S., Belgium.

synchronous movement of eye and head, both cooperating in the achievement of a goal-directed gaze shift. Let us remember that in all species in which SC or tectum has been electrically stimulated, body, head, eye and even pinnae orienting movements have been observed (see bibliography in Roucoux and Crommelinck, 1976). Eye saccade is only one component of a whole sequence of orienting movements. Let us also note that, in the cat, for example, the range of eye movements is only 20°–23° from the primary position (Crommelinck et al., 1977a), and that the retinotopic projection onto the SC covers 70° and more of visual field (Feldon et al., 1970).

Thus obviously, the simple and attractive foveation model involving the eye alone does not apply as such to subprimates. The divergence of results of SC stimulation in cats and monkeys constitutes another argument and will be illustrated here. In cats, SC seems to be divided into two distinct regions (Roucoux and Crommelinck, 1976; Crommelinck et al., 1977b). In approximately the anterior half, the organization of eye saccades obtained by electrical stimulation of deep layers is roughly similar to that found in the monkey. A short pulse train (50–100 msec) evokes eye saccades whose amplitude and direction are independent of eye orientation in the orbit (see Figs. 1B and 2B). Direction and amplitude are solely determined by the collicular site being stimulated, according to the retinotopic projection. Increasing the duration of the pulse train evokes a steplike series of identical saccades (Fig. 1A). In the posterior half, evoked saccades are clearly goal-directed. Their direction and amplitude depend on the initial orientation of the eye

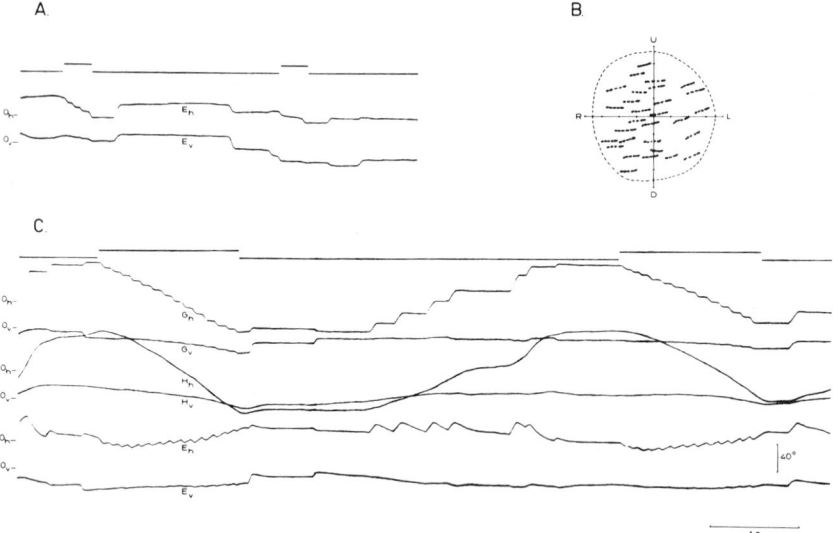

Fig. 1. Eye and head movements evoked by stimulating the anterior part of right SC in alert cat. A) Eye saccades evoked by a 350 msec stimulus train. Head is fixed. Upper trace: stimulation periods. Eh and Ev: horizontal and vertical components of eye movement. Oh and Ov: horizontal and vertical zero position of the eye in the orbit. Upward deflection is right and up. B) Eye saccades evoked by a 50 msec stimulus train. Head is fixed. Saccades were displayed on the screen of a memory oscilloscope in X–Y mode. Z input was modulated with 1 msec pulses at 400 Hz for the duration of saccades. The dotted contour corresponds to the limits of the oculomotor range. Graduations every 5°. U, up; D, down; R, right; L, left for the cat. C) Eye and head movements evoked by 1.6 sec stimulation train. Gh and Gv: horizontal and vertical components of eye position in space. Hh and Hv: horizontal and vertical components of head orientation. Eh and Ev: horizontal and vertical components of eye position in the orbit. Oh and Ov: horizontal and vertical zero position of gaze, head and eye.

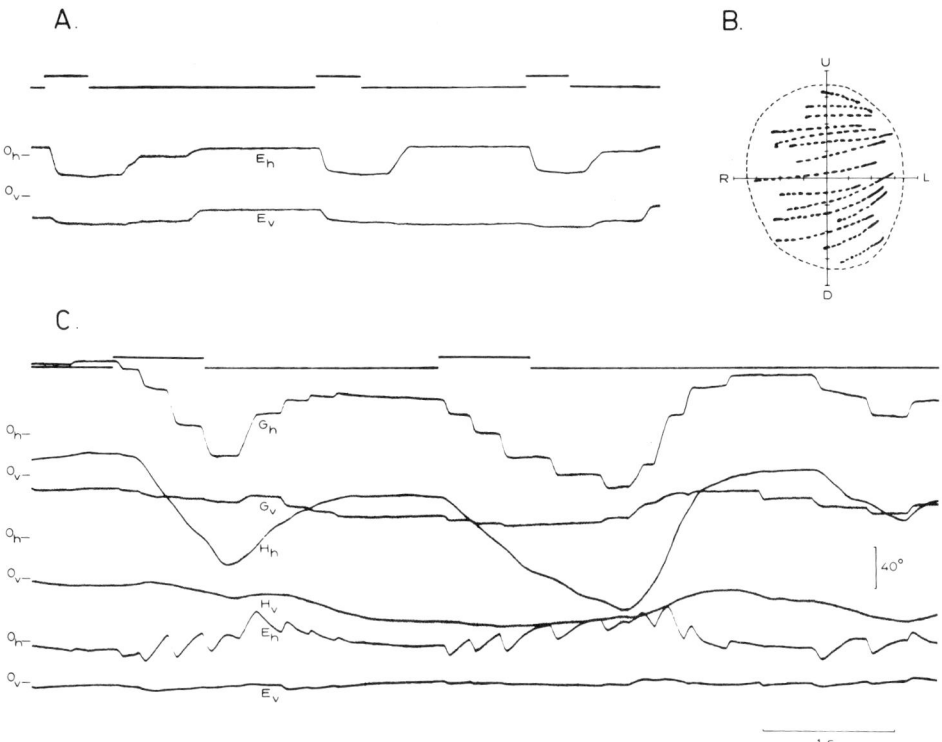

Fig. 2. Eye and head movements evoked by stimulating anterior part of cat's right SC. Stimulation site is more caudal than in Fig. 1. Same indications as in Fig. 1. In B, train duration is 100 msec. Note that, in the head-fixed condition (A), no "staircase" is evoked, due to the large amplitude of the unitary saccade.

in the orbit so as to bring the line of sight in one particular direction specified by the site of stimulation (Fig. 3). The goal, always situated in the visual hemifield contralateral to the stimulated colliculus can be reached by contraversive or ipsiversive saccades. In the latter case, latency is longer (80–100 msec vs. 20–30 msec). For long stimulation trains, the eye, once brought on the goal, is kept there for the time of stimulation. The collicular region in which such saccades can be evoked approximately covers the external 30°–60° of the retinotopically projected visual field. A point worth remembering here is that the oculomotor range of the cat is limited to about 23° from a central position. Goals being situated well within the oculomotor range limits (ipsiversive saccades towards the goal are possible), their position does not correspond to the retinotopic map. Only their direction is specified by the map; their eccentricity does not vary much from rostral to more caudal stimulation sites. This obviously raises a crucial question we shall examine later: what is the significance of these goal-directed saccades?

Following our idea expressed above that the cat's SC participates in orienting gaze shifts with the help of coordinated eye and head movements, we stimulated SC of alert cats free to move their heads. Eye and head movements were recorded using the electromagnetic technique. Gaze direction (eye in space) was obtained by electrically subtracting the signals from the eye and head coils. Electromyographic activity (EMG) in the neck muscles contralateral to the SC being stimulated was also monitored.

Head movements evoked in the anterior half of the SC (corresponding to the central 20°–30° of retinotopic map) are reminiscent of what has been shown in the monkey by

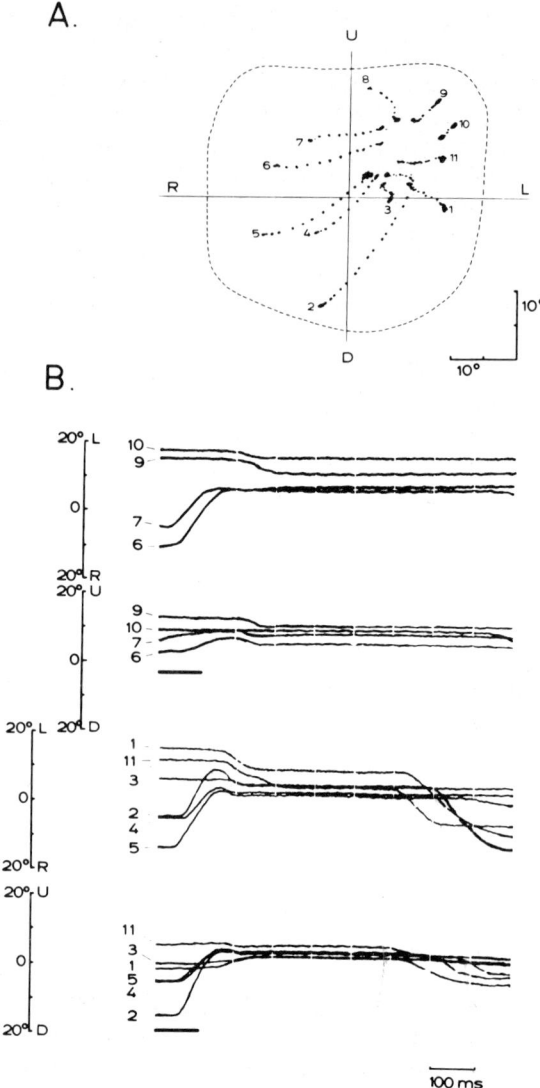

Fig. 3. Example of goal-directed eye saccades evoked in right posterior SC by a 100 msec stimulus train, with head fixed. Saccades are numbered. Stimulation periods are indicated by black bars in B. Note that the ipsiversive saccades (1, 9, 10 and 11) have longer latencies than contraversive ones.

Stryker and Schiller (1975); though in the cat, it is not required that the eye reaches an extreme orientation in the orbit to start a head movement. For a short stimulation train evoking only one eye saccade, the head moves only imperceptibly. If stimulus is long enough to evoke several successive saccades, the head begins to move in the same direction with a rather low velocity. Interestingly, this velocity matches the mean eye velocity in the saccadic step sequence evoked with head fixed (Figs. 1A, C and 2A, C). The head movement appears to start when the eye in the orbit has just crossed its midposition. The latency of the head movement (best illustrated by the latency of the neck EMG discharge) is dependent of the initial eye position in the orbit (Fig. 4). The mean threshold for triggering EMG clearly lies at about 3° contralateral to the midline.

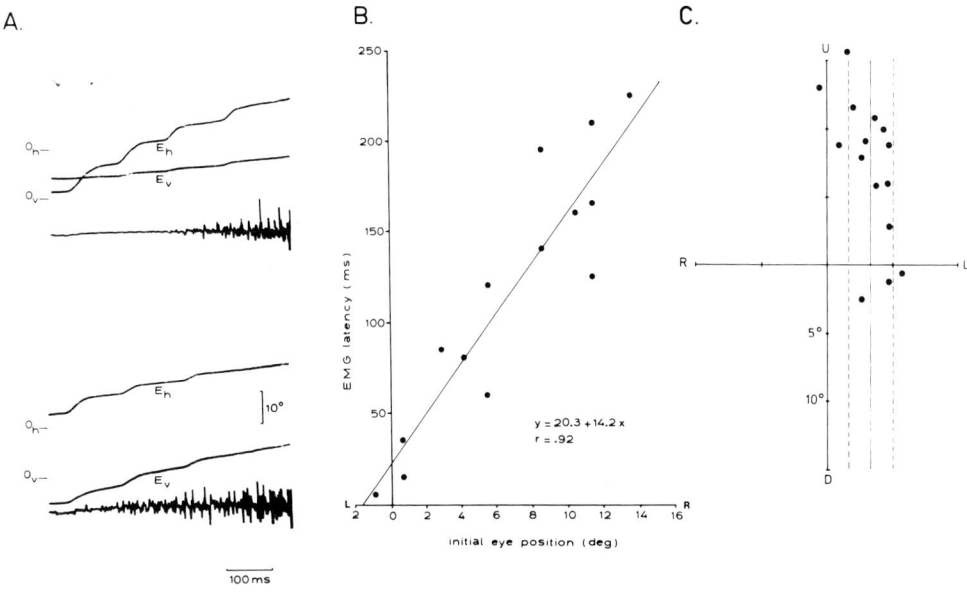

Fig. 4. A) Eye saccades evoked in right SC anterior part, by 600 msec stimulus train. The latency of EMG activity recorded from the left biventer cervicis muscle (lower trace) depends on the initial horizontal eye position. Upper example: the eye starts from the right side of its range; lower example: initial position is slightly to the left of the center of the range. B) EMG latency plotted against different initial eye positions. Note the high correlation coefficient between the two variables. C) Each dot represents the position attained by the eye in the orbit when the EMG begins to burst. The mean position is 3.2° to the left. SD, indicated by the two dotted lines, is 1.8°.

Like in the monkey, where its latency is quite variable and where it only appears for large eye eccentricities, the head movement in this region of the cat's SC seems to be indirectly triggered by the eye movement itself. This will be discussed later on.

Another remarkable fact is that, when the head is free, the evoked successive saccades of gaze have the same amplitude as the staircase of eye saccades with head fixed. This implies that the vestibulo-ocular reflex (VOR) is constantly operating with a gain close to 1 during the whole sequence of evoked movements. Again, this type of eye-head coordination is similar to what has been shown in the monkey (Morasso et al., 1973).

Head movements evoked in the posterior part of cat's SC are another story. Here, stimulation evokes a short latency, rather high velocity head saccade (up to 700 degrees.sec^{-1}) synchronous with an eye saccade in the same direction. Head starts to move slightly after the eye but neck EMG onset just precedes the beginning of the eye movement. Frequently, if stimulation lasts long enough, a second saccade of the head is initiated, accompanied by another eye saccade (Fig. 5). Amplitude and direction of evoked head saccades approximately correspond to the retinotopic map (amplitude of 40°–70°).

Considering the high velocity attained by the head during the evoked movements and observing that the eye simultaneously turns in the orbit at a quasi normal velocity in the same direction, and still reaches the goal, we are led to admit that the VOR is cancelled during the evoked eye saccade. Otherwise, supposing a simple addition of VOR slow phase and eye motor command as it is realized in the monkey or by stimulating the anterior part of cat's SC, the eye would at least remain stationary in the orbit or even move

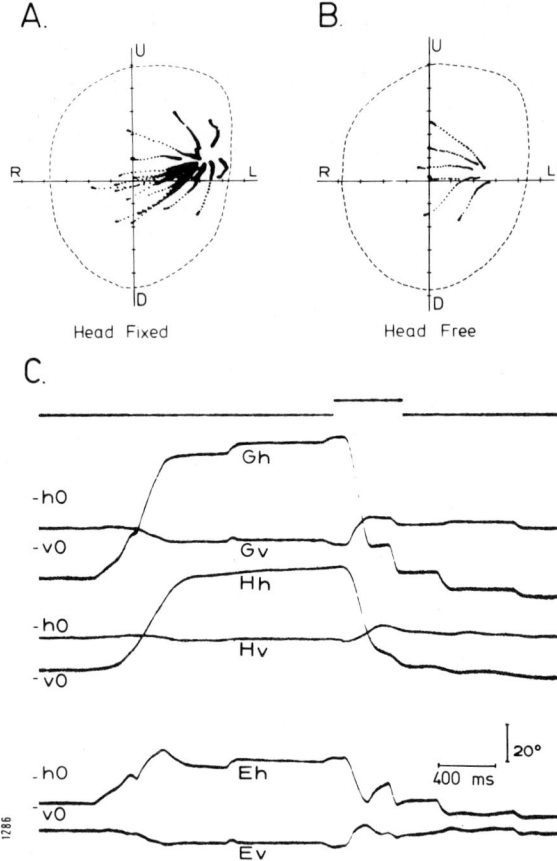

Fig. 5. A, B) Eye movements evoked in right posterior SC by 400 msec stimulus train with head fixed and free. When the head is free to move, saccades remain goal directed. C) Eye and head movements evoked at the same site. Maximum head velocity attains 350 degrees.sec^{-1} and eye in the orbit, 200 degrees.sec^{-1}. Note the beginning of a second evoked head saccade.

in the opposite direction. This type of eye-head coordination has not been described in the monkey.

We shall not give here further details on the results of these stimulations of SC in cats. These elements will enable us to emit some hypotheses on the role of SC in the visual fixation reflex. Several questions may be raised: 1) why would SC organization differ in cat and monkey? 2) is eye-head coordination different in cat and monkey? 3) what is the significance of goal-directed saccades?

To try to answer the first question, let us imagine some primitive animal with two lateral eyes, no fovea, and no neck (head immobile on the trunk), maybe some sort of fish. To get a clear view of things, despite the numerous movements of his body, this animal has devised two stabilizing mechanisms: the VOR and the optokinetic (OK) system. Both of them work in cooperation (Robinson, 1977) and generate compensatory eye movements (slow phases), as well as anticompensatory saccades (quick phases). Slow phases stabilize and quick phases bring the eyes back, keeping it around a central position in the orbit; a most interesting situation because, at this moment, retinal and body coordinate systems coincide. When this animal orients towards a prey, he cannot move his eyes or his head, he must turn his whole body and the first quick phase helps him to acquire the target. This visual grasp reflex is mediated by the tectum (its stimulation evokes body turning). This way of behaving is however quite time consuming and pretty exhausting. So, more clever animals have decided to move their head independently of their body.

Let us consider a rabbit. He still possesses the same basic outfit: VOR and OK system, but he has developed his visual abilities: he has a specialized central region on his retina (though still rather large) and some binocular vision. To take advantage of this, he has learned to pursuit with his head: his two eye stabilizing mechanisms with their quick and slow phases are just good enough to maintain the target's image within the best zone on his retina. He cannot move his eyes without moving his head. His visual grasp reflex usually consists of a head turning and synchronous eye saccade, itself followed by a compensatory slow phase (Collewijn, 1977). The target is acquired earlier than with a head movement alone and gaze locked on it during the rather long deceleration of head. What is the nature of this synchronous eye saccade: quick phase or voluntary command? God knows. Some evidence in favor of a voluntary movement has been shown (Collewijn, 1977), though, in practice, the distinction might reveal meaningless. Stimulation of rabbit's SC evokes conjugate head, eye and pinnae movements (Schaefer, 1970). This fixation reflex appears to be organized according to a map in a head coordinate system though no precise recording of eye and head movements have been done.

We already know how the monkey manages to acquire visual targets. He executes, like the rabbit, a coordinated eye and head movement (the neck muscles beginning to discharge slightly prior to the onset of the saccade) (Bizzi et al., 1972). But, rather surprisingly, stimulation of his SC only evokes eye saccades. To complete the picture, let us add that the monkey has, over the rabbit, the advantage of possessing a highly specialized fovea, binocular vision, a pursuit mechanism for the eye, and can make saccades whenever he wants without moving his head.

From all this, it clearly appears that SC has progressively evolved from a structure governing body orienting reflexes into a head fixation reflex center and, lastly into an eye orienting device, but he has always remained a "controller" of gaze orientation. Our results in the cat, placed in this perspective, might appear a little less puzzling. The cat indeed can be safely placed midway between the rabbit and the monkey as far as his visual and oculomotor aptitudes are concerned. His colliculus is composite: half-rabbit, half-monkey.

Our second question concerned the differences in eye-head coordination we have evoked between cat and monkey. The cat indeed appears to possess two modes of eye-head coordination. The first one would be similar to that of the monkey. In this mode, only the eye receives the adequate motor command in order to annul a retinal error. This error must be within the reach of the eye alone (within the oculomotor range). As soon as the eye leaves some central zone (whose dimensions may be different in cat and monkey according to the extent of their respective oculomotor range), it triggers a rather slow displacement of the head. This oculocephalic reflex appears quite plausible in the cat. Abrahams and Rose (1975) have indeed discovered an eye muscle proprioceptive input onto collicular cells which are at the origin of the tectospinal tract. This proprioceptive input may be activated by displacements of the eye in the orbit as small as 2° (Abrahams and Anstee, 1977). In this mode, head movement need not be precisely programmed: the VOR is sufficient to keep gaze displacement accurate. This type of eye-head coordination is illustrated by the results of stimulation of the anterior part of cat's SC and by the observations of Bizzi's group in the monkey. The other strategy is illustrated by stimulating the posterior part of cat's SC. In this case, the target is too eccentric to be acquired by an eye movement alone. An adequate command is given to the head (in a head coordinate system). The simultaneous eye saccade may have two origins: either it is an anticompensatory eye movement (quick phase) triggered via the labyrinth and then it should at least slightly follow the onset of head movement, or it is a saccade commanded by a copy of the head motor order sent simultaneously to the oculomotor centers. Our data in the cat suggest the second possibility may be plausible. Indeed, the saccade begins slightly before the head movement (excluding a quick phase) and, if the head is prevented from moving, the saccade is still evoked (goal directed saccade with head fixed). As already underlined above, the velocity of the eye saccade remains little affected when the head is allowed to move at high speed. This suggests that the VOR gain is drastically reduced at this moment though coming back to 1 near the end of the eye saccade (Fig. 5).

To answer our third question, goal-directed movements could be interpreted as the copy of the head motor order sent to the eye, in head coordinates. To be of any value, this copy has to be adapted to the eye movement range. Combination of the head saccade, goal-directed eye movement and brief cancellation of the VOR (via another copy of the head command for example) would result in an adequate target acquisition. The role of the goal-directed saccade would be double: i) to decrease the delay of target acquisition, and ii) to permit a precise foveation even if the eye is not initially centered: the goal, indeed, is in the same direction as the head movement. No overshoot of the target is possible if the eye saccade duration is noticeably shorter than the head movement, what is apparently the case.

In the latest accomplishments of evolution, three gaze shifting mechanisms are superimposed. They are designed to cooperate, each in its speciality. This implies that rather complex and multiple coordinations between them exist. No wonder then, that more than one mode of eye-head coordination may exist in the cat as well as in the rabbit (Collewijn, 1977), the monkey (Bizzi et al., 1972) and man (Barnes, 1976). It is not surprising either that SC, which has been a center of reflex gaze orientation for millions of years reveals a multiple organization at some levels of its evolution.

SUMMARY

Electrical stimulation of superior colliculus (SC) in alert cats with their head free reveals a double organization: 1) The anterior part of the structure, which receives a retinotopic input corresponding to the central 20°–30° of the visual field, is essentially involved in the generation of fixation *eye* saccades. The data are quite comparable to those obtained in the monkey. 2) The posterior part of the structure, corresponding to the peripheral retina, is, on the other hand, primarily implicated in the control of *head* fixation movements. This has not been shown in the monkey.

These results show that, in the cat, SC constitutes a major piece in the mechanism of the visual fixation reflex. It commands gaze shifts by means of two types of eye-head coordination strategies. The presence of this dual operating collicular mechanism in the cat is discussed in the light of evolutionary processes. It is suggested that, from fish to monkey, in parallel with the development of visual performances, tectal structures have evolved from a reflex body-turning device into an eye-orienting machine. The basic function, visual grasping, has been preserved, but the tools have become more and more efficient and precise: movement of the whole body, then of one of its segments, the head, and, finally, of the sensory organ itself, the eye.

ACKNOWLEDGEMENTS

The authors are grateful to Dr. D. Guitton who devised the eye and head recording apparatus and participated in all the experiments.

REFERENCES

Abrahams, V.C. and Anstee, G. (1977) Units in the superior colliculus and underlying tegmental structures responding to passive eye movement. *Neurosci. Abstr.*, III: 153, no 467.
Abrahams, V.C. and Rose, P.K. (1975) Projections of extraocular, neck muscle, and retinal afferents to superior colliculus in the cat: their connections to cells of origin of tectospinal tract. *J. Neurophysiol.*, 38: 10–18.
Barnes, G.R. (1976) The role of the vestibulo-ocular reflex in visual target acquisition. *J. Physiol. (Lond.)*, 258: 64–65P.
Bizzi, E., Kalil, R.E. and Morasso, P. (1972) Two modes of active eye-head coordination in monkeys. *Brain Res.*, 40: 45–48.
Collewijn, H. (1977) Eye- and head-movements in freely moving rabbits. *J. Physiol. (Lond.)*, 266: 471–498.
Crommelinck, M., Guitton, D. and Roucoux, A. (1977a) La position primaire de l'oeil en relation avec le champ oculomoteur chez le chat. *J. Physiol. (Paris)*, 73: 71A.
Crommelinck, M., Guitton, D. and Roucoux, A. (1977b) Retinotopic versus spatial coding of saccades: clues obtained by stimulating deep layers of cat's superior colliculus. In *Control of Gaze by Brain Stem Neurons*, R. Baker and A. Berthoz, (Eds.), Elsevier, Amsterdam, pp. 425–435.
Feldon, S., Feldon, P. and Kruger, L. (1970) Topography of the retinal projection upon the superior colliculus of the cat. *Vision Res.*, 10: 135–143.
Hess, W.R., Bürgi, S. und Bucher, V. (1946) Motorische Funktion des Tektal- und Tegmentalgebietes. *Mschr. Psychiat. Neurol.*, 112: 1–52.
Mohler, C.W. and Wurtz, R.H. (1976) Organization of monkey superior colliculus: intermediate layer cells discharging before eye movements. *J. Neurophysiol.*, 39: 722–744.
Morasso, P., Bizzi, E. and Dichgans, J. (1973) Adjustment of saccade characteristics during head movements. *Exp. Brain Res.*, 16: 492–500.
Robinson, D.A. (1972) Eye movements evoked by collicular stimulation in the alert monkey. *Vision Res.*, 12: 1795–1808.
Robinson, D.A. (1973) Models of the saccadic eye movement control system. *Kybernetik*, 14: 71–83.
Robinson, D.A. (1977) Vestibular and optokinetic symbiosis: an example of explaining by modelling. In *Control of Gaze by Brain Stem Neurons*, R. Baker and A. Berthoz (Eds.), Elsevier, Amsterdam, pp. 49–58.
Robinson, D.L. and Jarvis, C.D. (1974) Superior colliculus neurons studied during head and eye movements of the behaving monkey. *J. Neurophysiol.*, 37: 533–540.
Roucoux, A. and Crommelinck, M. (1976) Eye movements evoked by superior colliculus stimulation in the alert cat. *Brain Res.*, 106: 349–363.
Schaefer, K.P. (1970) Unit analysis and electrical stimulation in the optic tectum of rabbits and cats. *Brain Behav. Evol.*, 3: 222–240.
Schiller, P.H. and Stryker, M. (1972) Single-unit recording and stimulation in superior colliculus of the alert rhesus monkey. *J. Neurophysiol.*, 35: 915–924.
Stryker, M.P. and Schiller, P.H. (1975) Eye and head movements evoked by electrical stimulation of monkey superior colliculus. *Exp. Brain Res.*, 23: 103–112.
Wurtz, R.H. and Mohler, C.W. (1976) Organization of monkey superior colliculus: enhanced visual response of superficial layer cells. *J. Neurophysiol.*, 39: 745–765.

C

Compensation and Adaptation of Labyrinthine Functions by Visual Input

Adaptive Modification of the Vestibulo-Ocular Reflex in Rabbits Affected by Visual Inputs and its Possible Neuronal Mechanisms

M. ITO

Department of Physiology, Faculty of Medicine, University of Tokyo, 7-3-1 Hongo, Bunkyo-ku, Tokyo (Japan)

The hypothesis that the cerebellar flocculus is a site of visually-induced adaptive and plastic modification of the vestibulo-ocular reflex has been derived from a cybernetic interpretation of the neuronal diagram of the cerebello-vestibular system (Ito, 1970, 1972). The finding in human subjects of a remarkable adaptiveness of the horizontal vestibulo-ocular reflex (HVOR) under certain conditions of visual-vestibular conflict (Gonshor and Melvill-Jones, 1974), and subsequent demonstration that similar adaptiveness in rabbits (Ito et al., 1974a, b) and cats (Robinson, 1976) is impaired by flocculectomy provides a good support for the hypothesis. Further, it has been shown that Purkinje cells in the rabbit's flocculus modulate frequencies of their simple spike discharges in response to vestibular and optokinetic stimuli in the manner to be expected from the hypothesis (Ghelarducci et al., 1975; Dufossé et al., 1978b). Thus, three lines of studies, i.e., dissection of neuronal diagrams, analyses of eye movements, and recording from Purkinje cells, conjointly support the hypothesis.

However, recording from primate Purkinje cells revealed eye velocity inputs to the flocculus (Miles and Fuller, 1975). It has thus been assumed that visual inputs to the primate flocculus are of minor importance and that the primate flocculus is a pathway for feeding eye velocity signals to the oculomotor system in the manner of positive feedback (Lisberger and Fuchs, 1978), but not to be a site for visual-vestibular interaction such as has been postulated in the rabbits. Before ascribing this seeming contradiction to a difference of animal species, it is necessary to obtain more information about visual and eye velocity inputs to the flocculus. In the rabbits, existence of visual inputs to the flocculus not only via climbing fiber afferents (Maekawa and Simpson, 1973) but also via mossy fiber afferents (Maekawa and Takeda, 1975), has been demonstrated by using electrical stimulation of visual pathways as well as flush light stimulation of the retina. Recently, it has been confirmed that visual signals, via both climbing and mossy fiber pathways, reach the narrow zone of the flocculus which, with its stimulus effect on eye movements, is specifically related to the HVOR (Ito et al., unpublished observations; Miyashita, this volume, chapter VB3). The relay site of the visual mossy fiber pathway in rabbits has been recently located in the nucleus reticularis tegments pontis (Bechterew) by Maekawa and Takeda (1978). By contrast, there is no good evidence indicating eye velocity inputs to the rabbit's flocculus. Since rabbits do not move their eyes in darkness as often as monkeys or cats do, it is difficult to evaluate the possible eye velocity inputs in isolation from visual or vestibular inputs. As for primate experiments, it is necessary to confirm whether eye velocity inputs impinge on those Purkinje cells related to the

HVOR, since two-thirds of flocculus Purkinje cells are unrelated to the HVOR (Dufossé et al., 1978b). It may also be pointed out that, when using natural stimuli such as the movements of a visual target, time resolution of observations is not always good enough to judge which one of the two correlated events, i.e., in the present case, Purkinje cell discharges and eye movements, is the cause for, or the result of the other. This is particularly so when flocculus Purkinje cells are connected with oculomotor neurons only through two synapses (Fukuda et al., 1972). The fact that Purkinje cells discharge in the primate flocculus in parallel with eye velocity, but not with retinal errors (Lisberger and Fuchs, 1978), does not necessarily exclude the possibility that Purkinje cells are driven by retinal errors in a complex, non-linear fashion, thereafter leading to changes of eye velocity. Apparently, further work is needed to determine the reason for the seemingly contradictory flocculus function.

The hypothesis concerning the role of the flocculus in controlling the HVOR also proposes that visual climbing fiber inputs act as error signals which inform the flocculus whether the HVOR is appropriately compensating for head movements and which thereby trigger a plastic modification of intrinsic parameters of the flocculus so as to improve the performance of the HVOR (Ito, 1970, 1972). According to the current cerebellar theories (Marr, 1969; Albus, 1971), such a modification should occur in the transmission efficacy at synapses from granule cells to dendrites of Purkinje cells. The finding that vestibular modulation of simple spike discharges from Purkinje cells is altered in parallel with an adaptive modification of the HVOR (Dufossé et al., 1978), is supporting evidence for this view. However, it is still not possible to test the postulated changes in granule cell-to-Purkinje cell synapses directly. One difficulty lies in identifying the Purkinje cells and their synapses actually involved in the adaptive and plastic modification of the HVOR. It is not improbable that only a small number of synapses on a small number of Purkinje cells exhibit the postulated modification of the transmission efficacy. The demonstration of a functional localization in the flocculus is an important step for overcoming this difficulty (see Yamamoto, this volume, chapter IVC3). Another difficulty arises from our present ignorance of the manner in which such a modification of synaptic transmission efficacy is affected. It could accompany a certain morphological change in synaptic structures, or it could be a purely functional issue describable only in physiological or chemical terms. Thus, the study of the flocculus involves fundamental problems generally applicable to plasticity of the central nervous system.

An experimental approach to understanding of functional roles of the climbing fiber afferents to the cerebellar cortex, which is now a central problem of cerebellar physiology, may be provided by studying functional disorders arising specifically from destruction of the inferior olive (Llinás et al., 1975; Soechting et al., 1976). It is surprising, however, to see that destruction of the inferior olive rapidly causes depression in the inhibitory efferent synaptic action of Purkinje cells (Dufossé et al., 1978a; Ito et al., 1978, 1979), as described below. It follows that olivectomy in its effect is equivalent to cerebellar corticotomy, and therefore that it does not reveal functional roles of climbing fiber afferents in isolation.

In the experiment by Dufossé et al. (1978a), effects of local stimulation within the flocculus were tested in those rabbits having electrolytic lesions in the inferior olive. In normal rabbits, local stimulation in rostral areas of the flocculus evokes abduction or downward shift of the ipsilateral eye, depending on the location of the stimulating electrode (see Yamamoto, chapter IVC3). When a lesion had been placed in the rostral portion of the dorsal cap and adjacent area of the ventrolateral outgrowth of the principal olive, the stimulus effect of evoking downward eye movements diminished or even reversed in polarity, whereas that of evoking abduction was preserved normally. With a lesion placed in the caudal portion of the dorsal cap, the effect of evoking abduction, but not downward shift, was specifically affected. The rostral portion of the dorsal cap and its adjacent area of ventrolateral outgrowth of the principal olive send climbing fiber afferents to those Purkinje cells which inhibit relay cells of the vestibulo-ocular reflex's vertical component. The caudal portion of the dorsal cap sends the afferents to those Purkinje cells which inhibit relay cells of the horizontal component. Therefore, it is indicated that electric stimulation of the flocculus fails to actuate the inhibitory synaptic action of those Purkinje cells of which the climbing fiber afferents are deprived.

In the succeeding experiment by Ito et al. (1978), the inferior olive of albino rats was destroyed by administration of 3-acetylpyridine (Desclin and Escubi, 1974) in combination with harmaline and Niacynamide (Llinás et al., 1975). In the normal rats, stimulation at a relatively deep lamella of the vermis of lobule IV or V induced inhibition in about 70% of Deiters neurons, which was represented in extracellular recording by drastic suppression of spontaneous discharges and in intracellular recording by the appearance of the inhibitory postsynaptic potentials (IPSPs) with brief latencies (Akaike et al., 1972; Ito and Yoshida, 1964). Two to four months after poisoning with 3-acetylpyridine such inhibition could be found only rarely, in 2% of sampled Deiters neurons. Similar results were obtained even in an earlier period, 3–7 days after the intoxication. Sampling at 24–32 h after the intoxication revealed that, even though the IPSPs were still preserved in 50% of Deiters neurons, their amplitudes were significantly smaller than in the normal rats. Therefore, it can be concluded that destruction of the inferior olive not only removes climbing fiber afferents from the cerebellar cortex, but also affects Purkinje cells in such a way that electrical stimulation no longer actuates their inhibitory efferent action upon target neurons.

In the further experiment by Ito et al. (1979), discharges from rabbit Deiters neurons evoked by electrical stimulation of the labyrinth were recorded from the ventrolateral funiculus of the first cervical segment of the spinal cord. Normally, stimulation of a lamella of lobule IV or V caused marked depression, by 40% or so in magnitude, of these discharges, presumably due to inhibitory action of Purkinje cells (Akaike et al., 1972). It was thus revealed that, following electrolytic destruction of the inferior olive, this depression rapidly became ineffective, reaching a low plateau level of 10% or so in 2–5 h. Recording from the nucleus of Deiters' and also from the Purkinje cell layer of lobule IV or V revealed no change in the electrical excitability and conductivity in Purkinje cell axons. Hence, following destruction of the inferior olive, there should be either depression of release of the inhibitory transmitter, presumably GABA (cf. Obata, 1977) from axon terminals of Purkinje cells, or reduction of reactivity of Deiters neurons to this transmitter. Activation of Purkinje cells via climbing fiber afferents stimulated in their intracerebellar course also became gradually ineffective after destruction of the inferior olive in a time course comparable with that of the depression of Purkinje cell inhibition on Deiters neurons. These effects of destruction of the inferior olive could not be reproduced by completely stopping impulse activities of the inferior olive by local application of the tetrodotoxin, indicating that these effects are mediated by a non-impulse process, such as an axonal flow. It is probable that the death of olivary neurons produces a disturbance of the axonal flow in the climbing fiber afferents which, across the climbing fiber synapses on Purkinje cells, influences the axonal flow in Purkinje cell axons. Fast components of axonal flow have a velocity as high as 400 mm/day (Ochs, 1969), and so would cover the distance from the inferior olive to Deiters neurons via Purkinje cells in about 1 h, fast enough to account for the rapid transfer of the effect of destruction of the inferior olive to Deiters neurons.

The above described finding not only urges reinterpretation of the symptoms arising from destruction of the inferior olive, but also provides an interesting example of remote control of synaptic transmission efficacy in neuronal chains probably mediated by axonal flow. Whether or not it is relevant to adaptive and plastic processes in the cerebellum should be determined in a future experiment.

SUMMARY

Vision-guided adaptive modification of the vestibulo-ocular reflex has now been established as a good model representing cerebellar adaptive control functions. Neuronal mechanisms of this reflex adaptation have been investigated by dissection of relevant neuronal connections and recording of neuronal signals from cerebellar Purkinje cells. An interesting recent finding emerged from these investigations is that destruction of the inferior olive leads to rapid depression of the inhibitory efferent action of Purkinje cells on their target neurons. This effect is conveyed by a non-impulse process, presumably an axonal flow, in climbing fiber afferents and Purkinje cell axons.

REFERENCES

Akaike, T., Fanardjian, V. V., Ito, M. and Nakajima, H. (1973) Cerebellar control of the vestibulospinal tract cells in rabbit. *Exp. Brain Res.*, 18: 446–463.
Albus, J. S. (1971) A theory of cerebellar function, *Math. Biosci.*, 10: 25–61.
Desclin, J. C. and Escubi, J. (1974) Effects of 3-Acetylopyridine on the central nervous system of the rat, as demonstrated by silver methods. *Brain Res.*, 77: 349–364.
Dufossé, M., Ito, M. and Miyashita, Y. (1977) Functional localization in the rabbit's flocculus determined in relationship with eye movements. *Neurosci. Lett.*, 5: 273–277.
Dufossé, M., Ito, M. and Miyashita, Y. (1978a) Diminution and reversal of eye movements induced by local stimulation of rabbit cerebellar flocculus after partial destruction of the inferior life. *Exp. Brain Res.*, 33: 139–141.
Dufossé, M., Ito, M., Jastreboff, P. J. and Miyashita, Y. (1978b) A neuronal correlate in rabbit's cerebellum to adaptive modification of the vestibulo-ocular reflex. *Brain Res.*, 150: 611–616.
Fukuda, J., Highstein, S. M. and Ito, M. (1972) Cerebellar inhibitory control of the vestibulo-ocular reflex investigated in rabbit IIIrd nucleus. *Exp. Brain Res.*, 14: 511–526.
Ghelarducci, B., Ito, M. and Yagi, N. (1975) Impulse discharge from flocculus Purkinje cells of alert rabbits during visual stimulation combined with horizontal head rotation. *Brain Res.*, 87: 66–72.
Ito, M. (1970) Neurophysiological aspects of the cerebellar motor control system. *Int. J. Neurol.*, 7. 162–176.
Ito, M. (1972) Neural design of the cerebellar motor control system. *Brain Res.*, 40: 81–84.
Ito, M. and Yoshida, M. (1964) The cerebellar-evoked monosynaptic inhibition of Deiters' neurons. *Experientia*, 20: 515–516.
Ito, M., Orlov, I. and Shimoyama, I. (1978) Reduction of the cerebellar stimulus effect on rat Deiters neurons after chemical destruction of the inferior olive. *Exp. Brain Res.*, 33: 143–145.
Ito, M., Nisimaru, N. and Shibuki, K. (1979) Rapid depression in synaptic action of cerebellar Purkinje cells induced by destruction of the inferior olive. *Nature*, 277: 568–569.
Ito, M., Shiida, T., Yagi, N. and Yamamoto, M. (1974a) Visual influence on rabbit horizontal vestibulo-ocular reflex presumably effected via the cerebellar flocculus. *Brain Res.*, 65: 170–174.
Ito, M., Shiida, T., Yagi, N. and Yamamoto, M. (1974b). The cerebellar modification of rabbit's horizontal vestibulo-ocular reflex induced by sustained head rotation combined with visual stimulation. *Proc. Acad. Jap.*, 50: 85–89.

Llinás, R., Walton, K., Hillman, D.E. and Sotelo, C. (1975) Inferior olive: its role in motor learning. *Science*, 190: 1230–1231.

Lisberger, S. G. and Fuchs, A. F. (1978) Role of primate flocculus during rapid behavioral modification of vestibulo-ocular reflex. I. Purkinje cell activity during visually guided horizontal smooth-pursuit eye movements and passive head rotation. *J. Neurophysiol.*, 41: 733–763.

Maekawa, K. and Simpson, J. I. (1973) Climbing fiber responses evoked in vestibulocerebellum of rabbit from visual system. *J. Neurophysiol.*, 36: 649–666.

Maekawa, K. and Takeda, T. (1975) mossy fiber responses evoked in cerebellar flocculus of rabbits by stimulation of the optic pathway. *Brain Res.*, 98: 590–595.

Maekawa, K. and Takeda, T. (1978) Origin of the mossy fiber projection to the cerebellar flocculus from the optic nerves in rabbits. In *Integrative Control Functions of the Brain, Vol. 1*, M. Ito, et al. (Eds.), Kodansha-Scientific, Tokyo, pp. 110–112.

Marr, D. (1969) A theory of cerebellar cortex. *J. Physiol., (Lond.)*, 202: 437–470.

Miles, F. A. and Fuller, J. H. (1975) Visual tracking and the primate flocculus. *Science*, 189: 1000–1002.

Obata, K. (1977) Biochemistry and physiology of amino acid transmitters. In *Handbook of Physiology Section, 1, The Nervous System*, J. M. Brookhart et al. (Eds.) pp. 625–650.

Ochs, S. (1972) Rate of fast axoplasmic transport in mammalian nerve fibers. *J. Physiol. (Lond.)*, 227: 627–645.

Robinson, D. A. (1976) Adaptive gain control of vestibulo-ocular reflex by the cerebellum. *J. Neurophysiol.*, 36: 954–969.

Soechting, J. F., Ranish, N. A., Palminteri, R. and Terzuolo, C. A. (1976) Changes in a motor pattern following cerebellar and olivary lesions in the squirrel monkey. *Brain Res.*, 105: 21–44.

Modification of Central Vestibular Neuron Response by Conflicting Visual-Vestibular Stimulation

E.L. KELLER and W. PRECHT

Department of Electrical Engineering and Computer Sciences, University of California, Berkeley, CA 94720 (U.S.A.); and Max-Planck-Institut für Hirnforschung, Neurobiologische Abteilung, 6000 Frankfurt/M.- Niederrad (F.R.G.)

The vestibulo-ocular reflex (VOR) serves to stabilize retinal images during self and environmentally induced motions of the head by producing eye movements that compensate for angular motions of the head in space. The reflex functions stably throughout our development and lifetime, and yet it is clear that the basic neural circuit mediating the behavior is an open-loop system, that is, visual image motion on the retina does not modify the response of the vestibular organ which transduces head accelerations. Thus, it has been hypothesized that visual feedback must act at some central neuronal site to aid in the proper maintenance of the basic reflex pathways (Ito, 1970).

This hypothesis has received support from the observations that the VOR can be adaptively modified by artificial alteration of the normal visual stimulation produced during rotation of the head (Gauthier and Robinson, 1975; Gonshor and Melvill Jones, 1971, 1976a,b; Ito et al., 1974; Miles and Fuller, 1974). It has been further hypothesized that the anatomical location of the neural plasticity underlying this behavioral adaptation resides in floccular lobes of the cerebellum (Ito, 1970). Experimental support for this view has been provided by ablation studies which show that visual modification of the VOR can no longer be produced in animals with total ablations of the floccular lobes (Ito et al., 1974; Robinson, 1976). More recently, single-unit recording studies in rabbits made during visual modification of the VOR show changes in Purkinje cell discharge patterns that are in line with changes expected on the basis of the floccular plasticity theory (Ito, 1977).

Since the flocculus does not project directly to oculomotor neurons, which serve as the neural output for the VOR, this plastic modification, if it does occur exclusively in the cerebellum, must make the adaptation manifest through other neural substrates mediating the VOR. It is known that floccular Purkinje cells exert a direct influence on neurons of the vestibular nuclei (Baker et al., 1972; Fukuda et al., 1972) which in turn serve as the most direct neuronal pathway of the VOR. Therefore, we recorded from neurons in this nuclear complex in animals with a visually modified state of VOR to ascertain what changes had occurred in the adapted state in the basic brain stem pathways mediating the VOR.

To study this question we modified the gain of the normal VOR in the alert cat with the use of visual reversing prisms and then recorded the response of a large sample of vestibular nuclei neurons (Vn) to natural vestibular stimulation. In order to provide comparative data from the same animal and the same area of the vestibular nuclei, we also recorded the response of a numerically similar number of Vn with the cat in the normal-gain VOR state.

METHODS

Cats were prepared for chronic microelectrode recordings under sterile, anesthetic surgery conditions (Keller and Precht, 1978). During subsequent conditioning and recording sessions (starting about 4 days after surgery), the animals were placed in a loose, restraining box with their heads bolted to a framework mounted on the box. The restraining box was then mounted on a three-axis Toennies turntable with the animal's head directly over the vertical axis of the table. The head was tilted down by 10° to bring the horizontal semicircular canals approximately into the plane of table rotation.

Microelectrode recordings from single neurons located in the vestibular nuclei were made by driving glass pipettes with DC resistances of 2–5 mΩ through the stereotaxically located chamber and intact cerebellum into the brain stem.

A gridwork of electrode penetrations separated by about 0.5 mm were made into the region of the rostral medial vestibular nucleus and the superior vestibular nucleus with the animal in one of two functional states – with normal vestibulo-ocular reflex (VOR), and with VOR modified by visual experience through reversing prisms.

In each cat, after running a set of penetrations with the animal in one state, another complete set of tracks at the same stereotaxic locations were made with the animal in the other state. The order of states was varied among the animals.

Left-right reversed vision was obtained by placing Dove prisms mounted in a specially machined holder in front of the animal's face. The holder fit closely to the face and prevented non-prism vision. Before the animal had obtained any experience with vision through the prisms the gain of the VOR was measured in total darkness. The animal was oscillated at 0.25 Hz ($\pm 20°$) about a vertical axis and the peak velocity of the compensatory horizontal eye movement was measured from recordings obtained with chronically implanted silver/silver-chloride EOG electrodes. We defined gain of the VOR as the ratio, peak eye velocity to peak head velocity. In agreement with the results of Robinson (1976), all our cats showed VOR gains in total darkness of approximately one at this frequency. Phase of the VOR was consistently in phase or slightly leading by a few degrees the phase of the head oscillation.

After this initial measurement of VOR gain the animal viewed a highly structured, brightly illuminated visual field through the prisms and was then force-rotated en bloc about a vertical axis at a frequency of about 0.1 Hz through a peak amplitude of 10° (peak head velocity = 12.6 degrees/sec). The gain of the VOR was remeasured every hour or second hour in the dark. In agreement with the results reported by others (Melvill Jones and Davies, 1976; Robinson, 1976) we found in all four animals a consistent pattern of decreasing gain of the VOR with increasing duration of rotation with prism vision. On the average, the cats showed a decrease in VOR (the ratio present VOR gain to the control value measured before experience with reversed vision) to 0.38 after an average of 4.5 h of forced rotation. The response of all units isolated in the vestibular nucleus which were selectively modulated by horizontal angular accelerations of the head, was first tested at a frequency of oscillation of 0.25 Hz in total darkness to avoid any visual contribution to Vn modulation. Quantitative analysis on each unit was done by directing the frequency modulated pulse train output from a window detector to an on-line PDP-12 computer. The analyses on the data performed by the computer are similar to those described by Anderson et al. (1978). Briefly, the neural response was averaged over 10 cycles of sinusoidal stimulation. Each period of stimulation was divided into equal time interval bins, usually 32, and the average number of spikes occurring in each bin was calculated. A Fourier analysis of these averaged, binned values was then made so as to obtain the phase and gain of the response relative to head acceleration. The phase is defined as the difference, in degrees, between the maxima of the input acceleration and the maxima of the calculated fundamental component of the neural response. The gain is defined as the ratio, amplitude of the calculated fundamental of neural response to the amplitude of the head acceleration (spikes/sec per degree/sec^2).

RESULTS

With the animal in the normal gain VOR state (gain close to unity at 0.25 Hz) we made 35 electrode penetrations in four cats that passed through the area of the vestibular nuclei and isolated 119 neurons that responded selectively to horizontal accelerations. Thus about 3.4 cells responding selectively to horizontal angular acceleration were isolated per penetration in the normal cat. A horizontal oscillation at a frequency of 0.25 Hz, ±10° was continuously applied as the electrode was advanced as a searching stimulus to insure that non-spontaneously active Vn were missed. Only 3 neurons out of the 119 were not spontaneously active.

From the entire sample of 119 neurons 51 cells were classified as type I and 68 were classified as type II after the nomenclature of Duensing and Schaefer (1958). In this alert preparation the resting discharge rates were relatively high with an overall average of 35.4 spikes/sec (±39 SD) and with no significant difference between type I and II unit resting rates.

The gains and phases of these 119 neurons with respect to angular acceleration were calculated by Fourier analysis as described in the Methods section and the results are shown in Fig. 1 in separate columns for type I and II neurons. The results are similar to those obtained by Shinoda and Yoshida (1974) in the decerebrate cat. For type I, the individual unit values of gain (Fig. 1A) show a distribution skewed toward higher values with a mean of 0.75 spikes/sec per degree/sec^2 (±1.4 SD). Most of the population variance is contributed by the presence of those units with higher gains giving the suggestion of a second peak in the distribution above about 0.9. These cells will arbitrarily be called high-gain units (gain > 0.9 spikes/sec per degree/sec^2). Such type I units in the normal cat constitute 40% of the sample (20 out of 51 cells).

The histogram of phases for type I units in the normal cat (Fig. 1B, upper left) clearly shows a bimodal distribution in agreement with the findings of Shinoda and Yoshida (1974). Approximately equal numbers of cells are found distributed around the peak at 55° of phase lag re ipsilateral acceleration and around another peak at 105° of phase lag. The mean value of phase for the whole population of type I units was 82.5° lagging ipsilateral acceleration.

For type II cells the histogram of unit gains (Fig. 1A, upper right) is unimodally distributed but skewed toward units with higher gain (mean gain = 0.69 spikes/sec per degree/sec^2, ±1.3 SD). High-gain type II units are found, but only constitute 25% of the sample (17 out of 68 cells).

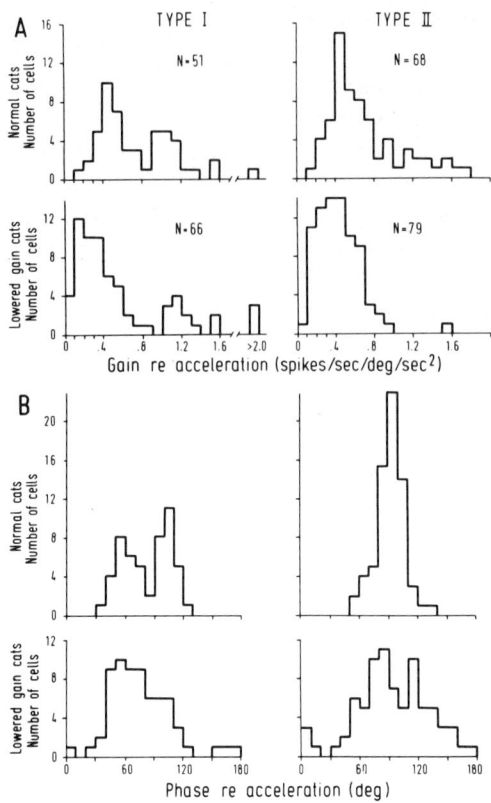

Fig. 1. Histograms of unit responses in vestibular nucleus to horizontal sinusoidal oscillation of the head at 0.25 Hz. A) Unit gains re head acceleration; units in cats with normal VOR (upper plates); units in same animals with lowered-gain VOR (lower plates). B) Unit response phases re acceleration; normal VOR (upper plates); lowered gain VOR (lower plates). All plates in left-hand column – type I responses; all plates in right-hand column – type II responses.

The histogram of phases for type II units in the normal cat (Fig. 1B, upper right) is distributed tightly around a single peak at 95° (mean = 91°) close to the same value of the phase lag noted for the larger phase lag type of unit in the type I population.

In the same four cats, and from nearly identical anatomical locations (as explained in Methods section), having lowered-gain VORs we made 40 penetrations and isolated 145 Vn that responded selectively to horizontal accelerations. Thus we obtained an almost identical ratio (3.6) of horizontal sensitive neurons per electrode penetration as we found in the cats with normal VOR gain. Thus, at the onset it can be said that whatever mechanism lowers VOR gain, it does not cause a total inhibition of even a subset of the entire Vn population of horizontal-acceleration sensitive neurons. Moreover the mean resting discharge rate (38.3 spikes/sec, ±42 SD) of units in the lowered-gain VOR animals was almost identical to that found in the normal animals.

On the other hand, the gains and phases of the horizontal-acceleration sensitive population of Vn were altered clearly and reproducibly by the short-term periods of visual-vestibular conflict caused by the reversing prism vision.

Figure 1A shows that for type I units there is a shift of the lower gain modal peak to lower values (overall population mean = 0.57 spikes/sec per degree/sec^2, ±1.3 SD).

This shift is, we believe, rather significant because the most frequently encountered value of neuronal gain was now between 0.1 and 0.2, a value that hardly ever occurs (only 1 cell out of 51) in normal cat vestibular nucleus. High-gain type I cells were still found in the lowered-gain animals, but less frequently (15 out of 66 cells – 23%) than in normal animals.

The histogram of phases for type I units (Fig. 1B) shows a unimodal distribution with a peak at 55° lagging ipsilateral acceleration (population mean = 76°). These two sets of histograms show that a total loss of high-gain type I units with large phase lags has occurred in the lowered-gain VOR animals.

The histogram of unit gain for type II units in the lowered-gain VOR animals (Fig. 1B) shows a similar shift to smaller values (overall population mean = 0.43 spikes/sec per degree/sec^2, ±1.3 SD). The most striking feature of the distribution in lowered-gain animals is, however, the total or almost total disappearance of high-gain units. Only two such cells with gains above 0.9 were found in all four animals (less than 3% of the population).

The histogram for phases of type II units in the prism vision cats is widely spread over a range of phases from a value in phase with contralateral acceleration to 180° lagging this acceleration (overall mean = 88°). This wide spread is in marked contrast to the narrowly distributed range of phase lags shown by type II units in normal animals.

DISCUSSION

Based on the results of ablation studies, the hypothesis has been advanced that the cerebellar flocculus is an essential part of the neural substrate which undergoes adaptive change during modification of the VOR by visual stimulation (Ito et al., 1974; Robinson, 1976). Single-unit recording studies in the flocculus of rabbits with visually adapted VORs tend to support the idea that the synaptic modification occurs in the cerebellum (Ito, 1977). In this study, it was reported that in the rabbit the major change occurring among floccular Purkinje cells (P-cells) after short-term visual stimulation with conditions, which in effect, are the same as those produced by reversing prisms, is a decrease in the depth of modulation of type II P-cells during sinusoidal vestibular stimulation. These changes occur for the most part in specific regions of the flocculus (H-zone) from which inhibition of the direct vestibular (Vi neurons) pathway to the ipsilateral lateral rectus muscle can be produced by electrical stimulation (Ito et al., 1977). In this latter study, it was demonstrated that the direct excitatory vestibular input (Vc) to the lateral rectus was not under floccular control.

The decrease in flocculus, type II, P-cell modulation would be expected to result in a decreased disinhibition of a selective subset of type I, secondary vestibular neurons, namely the Vi neurons. This would lead in turn to a decreased modulation of inhibition of abducens neurons and a decreased gain of the horizontal VOR. Vestibular neurons carrying excitation to abducens neurons (Vc neurons) would be unaffected.

Our data recorded in the vestibular nuclei of cats with visually lowered-gain VORs partially agrees with these expectations. We found an attenuation of the modulation of some type I vestibular neurons and this effect was relatively selective for those Vn with large gains and phase lags with respect to ipsilateral accelerations of the head. That at least some of this class of Vn are the cell bodies of Vi axons, was shown by recording from a sample of Vi axons in the abducens nucleus in these same cats (Keller and Precht, manuscript in preparation). The other population of type I neurons, also with some large gains but with small phase lags and intermixed anatomically with these former cells, were relatively unaffected by the visual experience with reversing prisms. These cells could be the cell bodies of Vc axons, but unfortunately, no representative sample of Vc axons could be recorded in the alert animal.

On the other hand, a discrepancy does result from this hypothesized arrangement of P-cell synaptic connections. Since under this scheme a reduction in outphase P-cell modulation leads to a reduction in the gain of the VOR, one would predict a large reduction in VOR gain immediately following surgical removal of the flocculus. In fact, at a frequency of 0.5 Hz, in the cat, the gain of the VOR stays about the same or even increases to values greater than unity (Robinson, 1976) following flocculectomy.

The gain reductions observed in type II vestibular neurons in the present study could be a result of the decreased modulation seen in type I vestibular cells, since type II cells receive excitatory input from contralateral type I neurons via the commissural pathways (Shimazu and Precht, 1966). On the other hand, the large phase shifts observed in some type II neurons (up to 90° leading and lagging normal values) cannot be produced by the observed gain changes and only small phase changes recorded in type I neurons which are essentially 180° out of phase with type II neurons in normal animals.

One way out of this difficulty is to propose an inhibitory floccular input to type II vestibular neurons in addition to the floccular input to type I cells. To produce the observed gain reductions in ipsilateral type II neurons, this floccular input would have to be from the outphase type Purkinje cell. This hypothesized floccular input would have to go primarily to a selective subset of type II neurons, namely the high-gain type. Finally, an increased input to type II vestibular neurons from intermediate-phase type Purkinje cells (Ghelarducci et al., 1975) would be required to produce the observed large phase changes recorded in type II neurons in the adapted cats.

SUMMARY

1. The response of central vestibular neurons (Vn) to natural vestibular stimulation was recorded in alert cats which had undergone visual modification of their vestibulo-ocular reflex (VOR) through the use of visual reversing prisms. On the average, forced rotation of the animals with visual input through the reversing prisms produced a reduction in gain of the VOR to about one-third of control value over a period of about 4–5 h. As a comparative control, responses of additional Vn were recorded from the same cats and from similar anatomical locations, but with the animal possessing a normal VOR unmodified by the reversing prisms.

2. In addition, responses were recorded from presumed axon terminations of central, vestibular neurons within or closely adjacent to the abducens nucleus with the animals in both VOR behavioral states.

3. By careful comparison of the population responses recorded from Vn in both VOR states a number of conclusions about the possible neural changes underlying visual modification of the VOR were reached.

4. Resting rates of Vn were not modified during VOR adaptation nor were the modulations of the whole population of Vn reduced to one-third of normal value to

correspond with the change produced in VOR. Instead, a selective modification of a subset of Vn neurons occurred. The modulations of high-gain, large-phase lag type I neurons, as well as vestibular axons projecting to the abducens nucleus were greatly reduced and a tendency for a small phase shift toward values of lower phase lag with respect to ipsilateral acceleration of the head occurred. The modulations of high-gain type II neurons were also significantly reduced. This change was accompanied by a drastic modification in the phases of the type II units from a population almost entirely in phase with contralateral head velocity in the normal animals to a population in which individual units showed phase leads or lags up to 90° displaced from contralateral velocity in the lowered gain VOR animals.

5. The changes observed in type I neurons and Vn axons projecting to the abducens nucleus are in accordance with the idea that the adaptive changes, induced by visual input through the reversing prisms in floccular Purkinje cell discharge, lead to a decrease in the modulation of inhibitory Vn projecting to the abducens nuclei.

6. The changes observed in type II neurons, especially the large modifications of phase, require additional assumptions about adaptive modification of floccular dynamics and/or specific projection of floccular Purkinje cell output to type II vestibular neurons.

ACKNOWLEDGEMENTS

Research sponsored by the Max-Planck-Gesellschaft and by the National Institutes of Health Grant EY00955-07. E. L. Keller was partially supported by a grant from the Alexander von Humboldt Foundation.

REFERENCES

Anderson, J.H., Blanks, R.H.I. and Precht, W. (1978) Response characteristics of semicircular canal and otolith systems in cat. I. Dynamic responses of primary vestibular fibers. *Exp. Brain Res.*, 32: 491–507.

Baker, R.G., Precht, W. and Llinás, R. (1972) Cerebellar modulatory action on the vestibulo-trochlear pathway in the cat. *Exp. Brain Res.*, 15: 364–385.

Duensing, F. and Schaefer, K.-P. (1958) Die Aktivität einzelner Neurone im Bereich der Vestibulariskerne bei horizontal Beschleunigungen unter besonderer Berüksichtigung des vestibulären Nystagmus. *Arch. Psychiat. Nervenkr.*, 198: 225–252.

Fukuda, J., Highstein, S.M. and Ito, M. (1972) Cerebellar inhibitory control of the vestibulo-ocular reflex investigated in rabbit IIIrd nucleus. *Exp. Brain Res.*, 14: 511–526.

Gauthier, G.M. and Robinson, D.A. (1975) Adaptation of the human vestibulo-ocular reflex to magnifying lenses. *Brain Res.*, 92: 331–335.

Ghelarducci, B., Ito, M. and Yagi, N. (1975) Impulse discharges from flocculus Purkinje cells of alert rabbits during visual stimulation combined with horizontal head rotation. *Brain Res.* 87: 66–72.

Gonshor, A. and Melvill Jones, G. (1971) Plasticity in the adult human vestibulo-ocular reflex arc. *Proc. Canad. Fed. Biol. Sci.*, 14: 11.

Gonshor, A. and Melvill Jones, G. (1976a) Short-term adaptive changes in the human vestibulo-ocular reflex arc. *J. Physiol. (Lond.)*, 256: 361–379.

Gonshor, A. and Melvill Jones, G. (1976b) Extreme vestibulo-ocular adaptation induced by prolonged optical reversal of vision. *J. Physiol., (Lond.)*, 256: 381–414.

Ito, M. (1970) Neurophysiological aspects of the cerebellar motor control system. *Int. J. Neurol.*, 7: 162–176.

Ito, M. (1977) Neuronal events in the cerebellar flocculus associated with an adaptive modification of the vestibulo-ocular reflex of the rabbit. In *Control of Gaze by Brain Stem Neurons*, R. Baker and A. Berthoz (Eds.), Elsevier, Amsterdam, pp. 391–398.

Ito, M., Nisimaru, N. and Yamamoto, M. (1977) Specific patterns of neuronal connections involved in the control of the rabbit's vestibulo-ocular reflexes by the cerebellar flocculus. *J. Physiol., (Lond.)*, 265: 833–854.

Ito, M., Shiida, T., Yagi, N. and Yamamoto, M. (1974) The cerebellar modification of rabbit's horizontal vestibulo-ocular reflex induced by sustained head rotation combined with visual stimulation. *Proc. Jap. Acad.*, 50: 85–89.

Keller, E.L. and Precht, W. (1978) Persistence of visual response in vestibular nucleus neurons in cerebellectomized cat. *Exp. Brain Res.*, 32: 591–594.

Melvill Jones, G. and Davies, P. (1976) Adaptation of cat vestibulo-ocular reflex to 200 days of optically reversed vision. *Brain Res.*, 103: 551–554.

Miles, F.A. and Fuller, J.H. (1974) Adaptive plasticity in the vestibulo-ocular responses of the rhesus monkey. *Brain Res.*, 80: 512–516.

Robinson, D.A. (1976) Adaptive gain control of vestibulo-ocular reflex by the cerebellum. *J. Neurophysiol.*, 39: 954–969.

Shimazu, H. and Precht, W. (1966) Inhibition of central vestibular neurons from the contralateral labyrinth and its mediating pathway. *J. Neurophysiol.*, 29: 467–492.

Shinoda, Y. and Yoshida, K. (1974) Dynamic characteristics of responses to horizontal head angular acceleration in vestibulo-ocular pathway in the cat. *J. Neurophysiol.*, 37: 653–673.

Adaptation of Optokinetic and Vestibulo-Ocular Reflexes to Modified Visual Input in the Rabbit

H. COLLEWIJN and A. F. GROOTENDORST

Department of Physiology, Faculty of Medicine, Erasmus University, Rotterdam, (The Netherlands)

Passive rotations of the head induce opposite, compensatory rotations of the eye which tend to maintain a relative (although not complete) stability of gaze. This reflex control of eye position is mainly achieved by the combined action of the vestibulo-ocular (VOR) and optokinetic (OKN) systems. In the rabbit, it has been shown (Baarsma and Collewijn, 1974) that the VOR is especially effective for the higher frequencies and the OKN for the lower ones, while the two systems in combination produce a rather constant gain (about 0.8) within the frequency range tested (0.05–2 Hz). In addition to this immediate interaction between visual and vestibular inputs, a long-term effect of retinal image motion on the VOR has been postulated (Ito, 1972). Since the VOR is a feedforward system, an indirect, adaptive gain control would maintain an optimal amplitude of the VOR throughout life. Ito (1972) proposed a vestibulocerebellar circuit with retinal image slip as the controlling input for such long-term adaptation. However, since the VOR and OKN function as an integrated system, changes in the VOR may be only one aspect of adaptation. The optokinetic loop alone may also show adaptability. The effect of long-term visual stimulation alone on OKN and/or VOR has been little investigated. Furthermore, it is insufficiently known whether adaptation is general or specific for the conditioned motion stimulus.

CONDITIONS FOR ADAPTATION

To examine some important factors in adaptation, let us consider the normal condition, in which the head is moving in a stationary visual world (Fig. 1A). If the compensatory eye movements are too small for total stability (gain < 1, as usually found), gaze will be displaced in the same direction as the head but at a much smaller velocity. The slip (movement of visual surroundings with respect to gaze) will be in the same direction as the movement of the eye in the head. Functionally meaningful adaptation in this condition would require an increase of the amplitude of the eye movements, to reduce slip. On the other hand, if compensatory eye movements are larger than the head movements (gain > 1), slip will be opposite in sign to the eye movements in the head (Fig. 1B). Of course, this condition would require a decrease of the eye movements. Thus, it can be argued that visually detected slip should be compared to the combined output of the VOR and OKN systems to determine the desirable direction of adaptation.

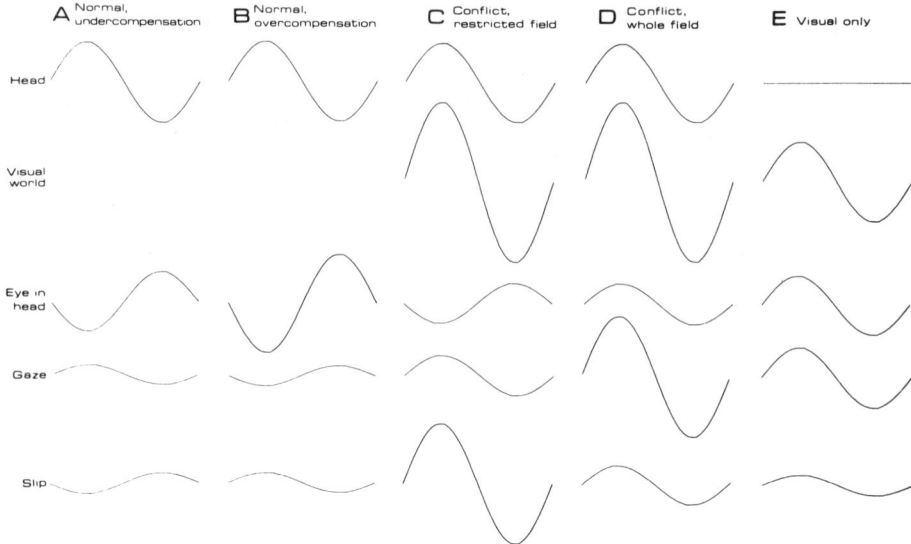

Fig. 1. Idealized, schematic stimulus-response relations of OKN and VOR under various experimental conditions. (A) Normal head movement in a stationary visual world. Compensatory eye movements are too small for complete stability. The gaze wanders with the head and the visual world slips with respect to the eye in the same direction as the eye moves in the head. This situation would require increase of the amplitude of the eye movements. (B) Normal situation as in A, but with too large an amplitude of the eye movements. Slip and eye movement have now opposite directions. (C) The visual world is moved in phase with the head, at twice the amplitude (inverted motion vision); the visual field is restricted to a sector of 90°. The eye movement is still in the direction of the VOR; the slip is in opposite direction as in B, which results in a decrease of the gain of the VOR. (D) Inverted motion vision as in C, but in the whole visual field. This visual input is so powerful that the VOR is overruled and eye movement has the same direction as the slip, as in A, which results in an increase of the gain of the VOR. (E) Rotation of the visual world alone, with the head stationary. Optokinetic eye movements are too small for complete stability. Slip is in the same direction as the eye movement (as in A) and the result is an increase of gain of OKN and VOR.
Gaze = head + eye in head; Slip = visual world − gaze.

The most physiological test of this system would be to change the magnitude of the slip signal. This has been done in monkey (Miles and Fuller, 1974) and man (Gauthier and Robinson, 1975) by fitting subjects with telescopic magnifying or reducing spectacles. Substantial increase and decrease, respectively, of the VOR were indeed induced by these conditions, which simulate under- and overcompensation.

In most other experiments on adaptation, visual motion information has been altered more drastically by right-left inversion. Although such conditions may seem quite artificial they have resulted in a strong decrease of the magnitude and sometimes even inversion of the VOR in man (Gonshor and Melvill Jones, 1976a,b) and cat (Robinson, 1976; Melvill Jones and Davies, 1976).

The species mentioned so far possess central or foveal vision, which may account for the effectiveness of the used optical devices, notwithstanding the small field of vision that they allow. In the rabbit, which has a panoramic and visual-streak type of visual field, strong restriction of the visual field seems undesirable and other strategies to modify visual slip have been designed. Ito et al. (1974) used a single vertical light slit which was moved in phase with, but at twice the amplitude of the head to simulate inverted vision and obtained reliable changes of the VOR in albino rabbits. Collewijn and Grootendorst (1978) found the light slit not to be very effective, particularly in pigmented Dutch belted rabbits, but obtained a consistent decrease of the VOR by inverting a 90° sector of the visual field with a mirror, fixed with respect to the head. On the other hand, in previous experiments Collewijn and Kleinschmidt (1975) had been unable to obtain a decrease of the VOR's amplitude when a whole striped drum was oscillated around the rabbit in phase with the head. The present experiments were performed to resolve these partly contradictory results, to further examine visuovestibular interaction in adaptation and to search for a possible pattern-specificity.

Methods were largely as described before (Collewijn and Grootendorst, 1978). Eye movements were measured in young adult Dutch belted rabbit with permanently implanted scleral search coils. The alert animals were fixed on a platform by screws implanted in the skull and a hammock, with the eyes close to the center of rotation. The platform was oscillated sinusoidally around a vertical axis at frequencies from 1/30 to 10/6 Hz and amplitudes of 2.5°–10° peak-to-peak. The platform was surrounded by a drum (diameter 140 cm, height 125 cm) lined with a random dot pattern (elements 1°). The drum could be moved continuously or sinusoidally, either independently or with some relation to the motion of the platform. The VOR was tested with the platform moving in darkness; OKN was tested with the drum alone moving in the light. Several combinations of stimuli were used to test the adaptability of the compensatory eye movements.

INVERTED VISION, RESTRICTED VISUAL FIELD

In a first type of experiment (7 rabbits) the animals were oscillated at 1/6 Hz, 10° peak-to-peak (p.p.). The drum was oscillated in phase with the head but at the double amplitude (20° p.p.). With a black box around the rabbit, the visual field was restricted to 90° anterior. Optically, this condition was equivalent to that with the mirror used by Collewijn and Grootendorst (1978), except that the visual pattern consisted now of the random dot pattern instead of the room. This condition was continued for 4 h. VOR, OKN (stimulus 1/6 Hz, 10° p.p.) and the response to the conditioning stimulus were measured briefly at hourly intervals.

The results (average of 7 rabbits) are shown in Fig. 2B. As expected, the amplitude of the VOR and the response to the conflict condition decreased, as in the previous experiments (Collewijn and Grootendorst, 1978). With the visual stimulus present, the gain of compensatory eye movements decreased from 0.60 ± 0.11 (SD) at 0 h to 0.29 ± 0.10 (SD) at 4 h. The gain of the VOR in the dark decreased from 0.74 ± 0.07 (SD) to 0.56 ± 0.07 (SD). Both of these changes were highly significant (t-test: $P < 0.0005$). The gain for the conflict condition is expressed in terms of VOR gain, thus the decrease indicates that the eye movements were progressively less determined by the vestibular, and more by the visual input. However, the sign of the responses remained positive, thus no inversion occurred even in the presence of the inverted visual stimulus. Quite remarkably, the responses to visual stimulation alone (OKN) significantly increased ($P < 0.025$) from 0.42 ± 0.13 to 0.54 ± 0.16 (SD) while the VOR decreased. Both trends would of course favour the stability of the retinal image and the decrease of slip.

The relations are further clarified in Fig. 1C. The eye movement is out of phase with the head, but the slip signal (difference between visual surroundings and gaze) is in phase with the head. Thus, slip is out of phase with the eye movement. This situation

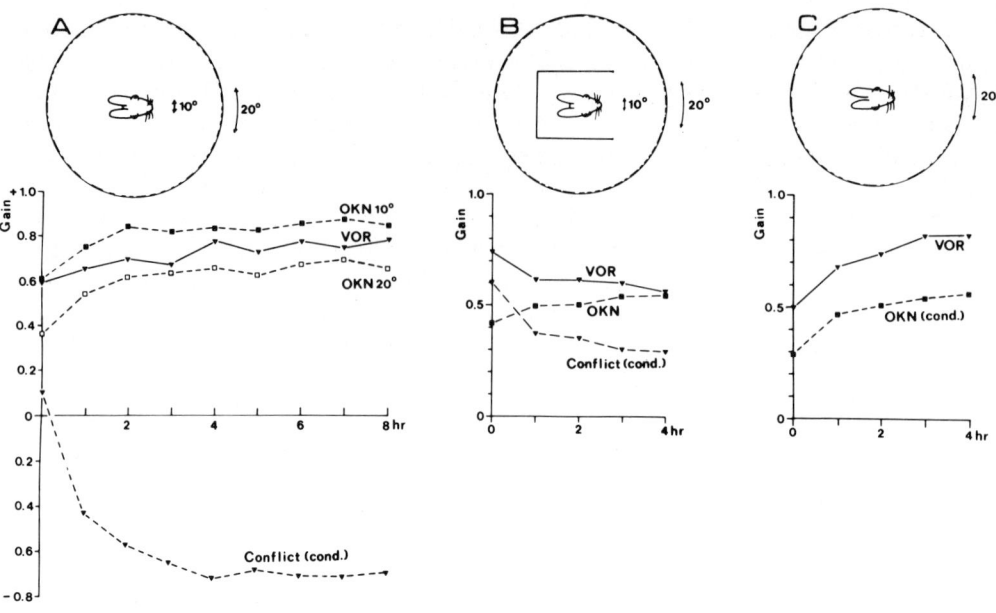

Fig. 2. Gain as a function of time for OKN, VOR and combined stimulation for various conditions. (A) Visual world oscillated in phase with the head at twice the amplitude, whole visual field exposed (average of 6 rabbits). (B) As A, but visual field restricted to anterior 90° (average of 7 rabbits). (C) Visual stimulation alone (average of 6 rabbits).

would normally occur if the compensatory eye movements were too large, and therefore the decrease of the VOR is physiologically meaningful. Since the OKN is a negative feedback system, an increase in gain is always a meaningful response to increased slip. During the adaptation of the VOR, a directional preponderance often occurred, which could result in strongly asymmetrical reactions. Gonshor and Melvill Jones (1976b) made a similar observation in man and attributed it to a temporary inequality of the adapting processes on both sides of the brain.

INVERTED VISION, WHOLE FIELD

Since the inverted vision in a sector of 90° was much more effective than a single light slit in modifying the VOR, it seemed worthwhile to repeat the experiments using the whole visual field. One would expect an even stronger effect of this stimulus, although in an earlier attempt (at an amplitude of 2° p.p.) it had been unproductive (Collewijn and Kleinschmidt, 1975). Six rabbits were oscillated at 1/6 Hz, 10° p.p., and the drum was moved in phase at 20° p.p. for a period of 8 h. The whole drum proved to be a much more powerful visual stimulus than the 90° field, and the response to the conflicting stimulus showed a strong adaptation towards visual stability, as shown in Fig. 2A.

At the start of the conditioning, the eye movements were still in the direction of a VOR (in counterphase with head and drum), but within 1 h the sign of the responses was inverted and the eye was tracking the visual pattern with a gain that gradually increased. The total change of the gain with the visual stimulus present was from 0.10 ± 0.21 (SD) at 0 h to -0.70 ± 0.24 (SD) at 8 h (t-test: $P < 0.0005$). This improvement in visual tracking was also reflected in the optokinetic gain. This went up from 0.60 ± 0.23 (SD) to 0.85 ± 0.13 (SD) for a stimulus of 1/6 Hz, 10° p.p. and from 0.37 ± 0.17 (SD) to 0.66 ± 0.19 (SD) for a stimulus of 1/6 Hz, 20° p.p. (t-tests: $P < 0.01$).

Paradoxically, however, the VOR gain showed no decrease, but an increase from 0.59 ± 0.07 (SD) to 0.79 ± 0.18 (t-test: $P < 0.025$). Actually, the changes in the VOR gain showed two varieties in the individual animals. In four out of the six rabbits, the VOR decreased in the first hour and increased later. In these same animals, the eye moved in the sense of a vestibular response in the beginning of the conditioning and in the direction of the optokinetic stimulus after 1 h. Representative recordings from such an animal are shown in Fig. 3. In the two other animals the response to the conflict stimulus was in the direction of the optokinetic stimulus immediately from the start of the experiment. In these same rabbits the VOR gain showed an increase from the very beginning. These findings offer a clue to the paradoxical increase to the VOR: the direction of change of the VOR appears to be related to the actual direction of the compensatory eye movements. This relation is further clarified in Fig. 1D, which shows idealized responses for whole field conflict stimulation at a time when the eye is following the visual, and not the vestibular stimulus. In this condition, the slip

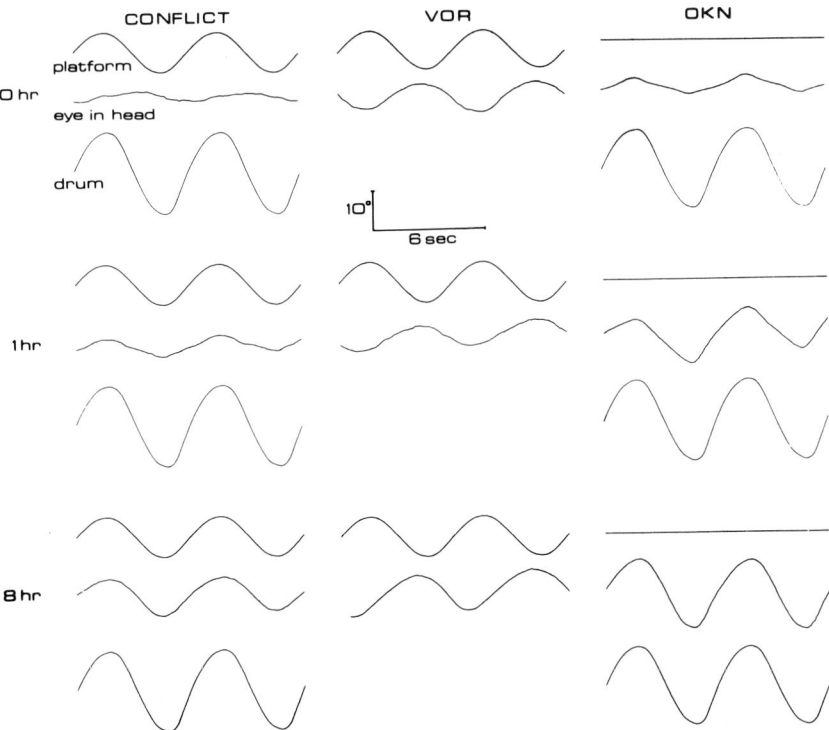

Fig. 3. Representative recordings from one rabbit conditioned with inverted motion vision in the whole visual field. At the start of the experiment (0 h) VOR (in darkness) and OKN are normal; the response to the combined stimulus (conflict) is still dominated by the VOR. After 1 h the eye is tracking the drum (eye movements in phase with the head) under the combined stimulus, as in Fig. 1D. The VOR has slightly decreased, the OKN is increased. After 8 h, the eye is tracking the drum almost perfectly in conflict and OKN conditions; the VOR gain has been paradoxically increased.

and the eye movement in the head have the same direction, which is normally the case when compensatory eye movements are too small (Fig. 1A). Thus, our findings suggest that the adaptation of the VOR is a function of slip and eye movement, not of slip and head movement.

ADAPTATION BY VISUAL STIMULATION ONLY

In a third experiment, six rabbits were mounted on the stationary platform and only the drum was oscillated at 1/6 Hz, 20° p.p. during a period of 4 h. This movement is rather fast for the rabbit's optokinetic system and the gain of the compensatory eye movements was substantially below 1.0.

The VOR was tested briefly at hourly intervals in complete darkness: the head was never moved with the lights on. As illustrated in Fig. 1E, the slip of the surroundings will be in phase with the eye movements. Head movement is of course absent. The averaged results are shown in Fig. 2C. The OKN gain increased from 0.29 ± 0.18 (SD) at 0 h to 0.56 ± 0.28 (t-test: $P < 0.001$). At the same time, the gain of the VOR increased from 0.50 ± 0.17 (SD) to 0.82 ± 0.42 (t-test: $P = 0.025$). Both trends consistently occurred in all rabbits tested. Thus, visual slip alone can modify both OKN and VOR, in the absence of any combined visuovestibular stimulation.

In none of the three described conditioning regimes were the changes in gain accompanied by any significant changes of phase relations. The VOR was always nearly in counterphase with the head movement, and optokinetic pursuit was nearly in phase with the drum.

FREQUENCY SPECIFICITY OF ADAPTATION

To investigate whether the adaptations described were generalized for all movements or somehow specific for the conditioned frequency (which was always 1/6 Hz), the VOR and OKN were tested in a frequency range around 1/6 Hz immediately before and after the total period of conditioning (4 or 8 h). Some evidence for specificity was indeed found. Fig. 4 illustrates this for a rabbit conditioned with visual stimulation only. Graphs a, b and c show the VOR before and after conditioning. The increase at 1/6 Hz is remarkable, but the responses at 1/12 and 1/3 Hz show only little changes. Fig. 4d, e and f show the same for the OKN responses. The enhancement at 1/6 Hz is very large; the reactions at the other stimulus frequencies were unchanged (1/3 Hz) or even clearly smaller (1/12 Hz).

The averaged frequency responses before and after the three types of conditioning used are shown in Fig. 5. Under all conditions, the changes in VOR as well as OKN were largest at 1/6 Hz, the conditioned frequency. For the conflict stimulus with the whole visual field (Fig. 5A) the VOR was markedly increased at 1/6 Hz, less so at 1/3 Hz and not (at the average) at 1/12, 5/6 and 10/6 Hz. The OKN was also mostly enhanced at 1/6 Hz, less at 1/3 and 1/12 Hz and even decreased at 1/30 Hz.

With the conflict stimulus limited to a 90° sector of the visual field (Fig. 5B) the VOR decreased most strongly at 1/6 Hz, less so at the other frequencies and even increased at 10/6 Hz. The OKN gain showed its largest increase at 1/6 Hz, with an average decrease at 1/30 Hz.

With visual stimulation only (Fig. 5C), the frequency specificity of the changes in VOR and OKN gain was even more distinct than in the other conditions.

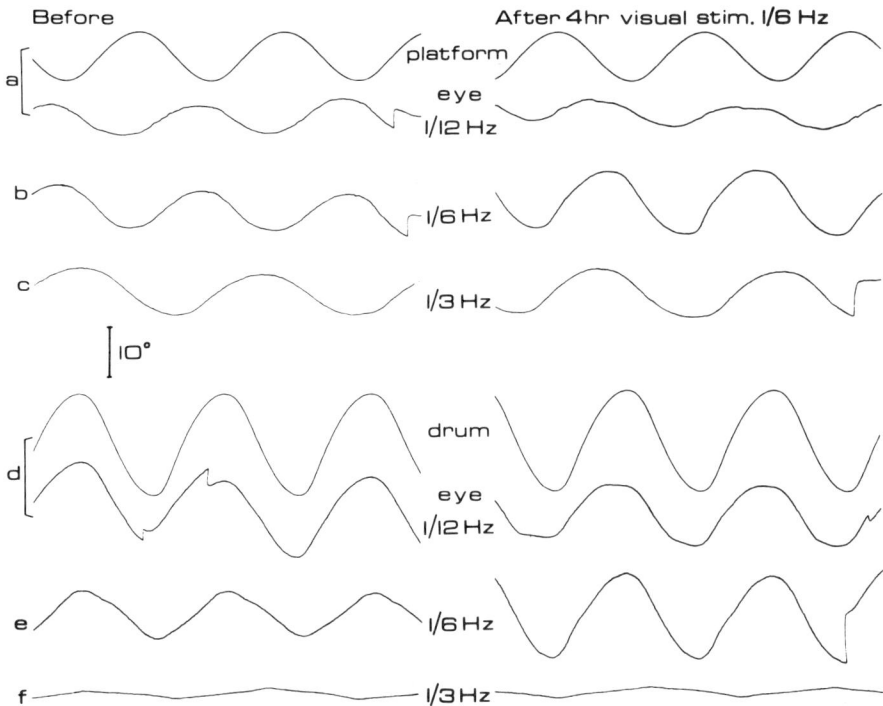

Fig. 4. Response of OKN and VOR to different frequencies before and after conditioning with visual motion alone (1/6 Hz, 20° p.p., 4 h). a, b, c) VOR tested in the dark. The platform motion (stimulus) is shown only for a and gives no phase information for b and c. Optokinetic responses shown in d (with the stimulus motion), e and f. Both VOR and OKN show an increased amplitude at 1/6 Hz but little changes at 1/12 and 1/3 Hz.

A further argument for the specific adaptation to the conditioned stimulus was found in the occasional occurrence of after-effects. Fig. 6A illustrates such a phenomenon in a rabbit that had been conditioned for 6 h with visual stimulation (1/6 Hz, 20° p.p.) only. When the VOR was tested in darkness at 1/12 Hz, the responses were strongly interfered with by the spontaneous generation of sinusoidal eye movements at 1/6 Hz. The platform was stopped, and in the absence of any stimulation the eyes continued to oscillate at exactly 1/6 Hz for several minutes. Another such event is shown in Fig. 6B. The rabbit had been conditioned for 2 h with the drum moving in phase with the head (conflict, whole visual field) at 1/6 Hz. The VOR was tested in darkness at this moment and when the platform was stopped the eyes continued to move spontaneously with a nearly perfect sine wave at exactly 1/6 Hz. These oscillations damped out gradually, but were reinstated several times after the occurrence of spontaneous saccades (Fig. 6B). These findings strongly suggest that the conditioned slip pattern was stored in the nervous system, although it should be emphasized that this spontaneous reproduction was seen only in a few rabbits.

A further test for the generality or specificity of adaptation was formed by the optokinetic responses to steady drum rotation at velocities from 1–100 degrees/sec, which were recorded before and after the conditioning with sinusoidal motion of the drum alone (1/6 Hz, 20° p.p.). Responses were measured after sufficient time for the

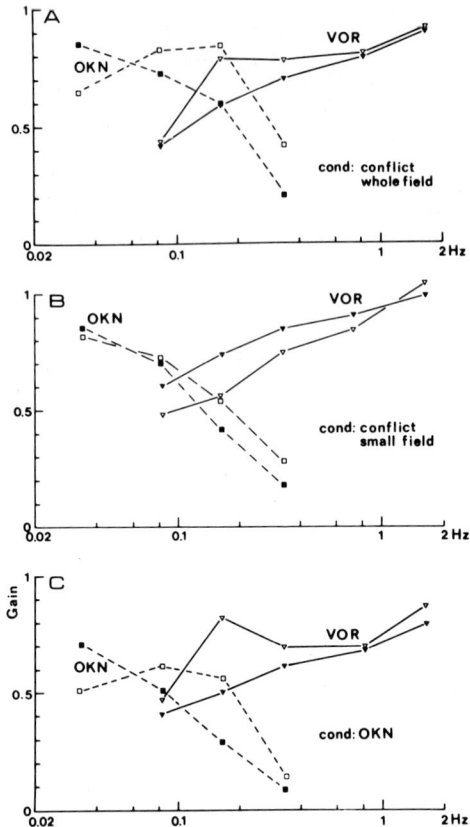

Fig. 5. Frequency specificity of adaptation of OKN (■, □) and VOR (▼, ▽). ■, ▼, before conditioning; □, ▽, after conditioning. Stimulus amplitudes were 10° p.p., except at 5/6 Hz (5° p.p.) and 10/6 Hz (2.5° p.p.). A) Inverted motion vision, whole field (average of 6 rabbits). B) Inverted motion vision, visual field restricted to 90° (average of 7 rabbits). C) Visual stimulation only (average of 6 rabbits).

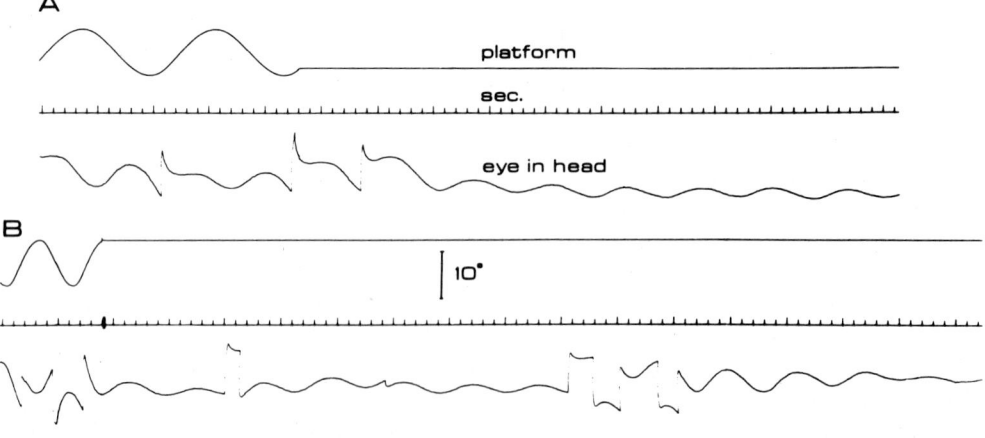

Fig. 6. Spontaneous postoscillatory eye movements as after-effect of conditioning. A) Spontaneous oscillation of the eye at 1/6 Hz during and after test of the VOR at 1/12 Hz. The animal had been conditioned 6 h with visual motion only at 1/6 Hz, 20° p.p. B) Similar effect after test of the VOR at 1/6 Hz. The animal had been conditioned during 2 h in a conflict (whole field) situation.

TABLE I

GAIN OF OPTOKINETIC RESPONSES BEFORE AND AFTER 4 H OF SINUSOIDAL VISUAL STIMULATION ALONE

(1/6 Hz, 20° p.p.) Average values of 6 rabbits

Conditioned stimulus (16 Hz)	Unconditioned constant drum velocities						
	1°/sec	2°/sec	5°/sec	10°/sec	20°/sec	50°/sec	100°/sec
0 h	0.29	0.79	0.75	0.77	0.75	0.69	0.23	0.049
4 h	0.56	0.73	0.68	0.61	0.51	0.41	0.067	0.019

pursuit velocities to become maximal. Although the responses to the conditioned frequency had markedly increased, the gain of the OKN for all steady velocities had decreased, particularly for the higher velocity range (Table I). This demonstrates again that the adaptive changes are not general but rather specific for the motion pattern that was conditioned.

DISCUSSION

The present experiments show that adaptive plasticity in the rabbit's compensatory eye movements is not limited to the VOR, but is also clearly present in the OKN. They also show that the VOR can be adapted, without being activated, by visual stimulation alone. Both of these facts were already noticed by Collewijn and Kleinschmidt (1975), who also failed to obtain a decrease of VOR gain by inverted whole field optokinetic stimulation. The present findings suggest that this apparent controversy with the earlier results of Ito et al. (1974) and the recent ones by Collewijn and Grootendorst (1978) can be solved if the actual relationship between visual slip and eye movements is taken into account. It appears now that visual slip in the same direction as the eye movements in the head will increase the gain of both the VOR and the OKN, while slip in the direction opposite to that of the eye movements will also increase the gain of the OKN, but decrease that of the VOR. Thus, slip and eye movements seem to be the relevant signals for the adaptation of the rabbit's visuovestibular oculomotor reflexes, as illustrated in Fig. 1.

A further finding is the specificity of the adaptation as documented in the frequency response relations of OKN and VOR (Fig. 5) and the occasional manifestation of highly specific after-effects (Fig. 6). A similar after-effect has been recorded by Kleinschmidt (1974) in conflict experiments of the same type. Apparently, the oculomotor control system has the possibility to store a consistent slip pattern. It appears legitimate, then, to postulate that the adaptation of compensatory eye movements involves the storage and retrieval of a specific movement pattern, which is derived from the visual slip.

Adaptation of compensatory eye movements could then roughly work as follows. A consistent pattern of slip occurring over a period of hours is stored in the nervous system. Also the relation of the slip to the eye movements is somehow recorded. The stored pattern interacts with the eye movements whenever the same motion pattern occurs. In the integrated visuovestibular response the interaction is always such to decrease visual slip. When the OKN alone is elicited the pattern interacts with the

OKN in a positive way, so that OKN gain for the particular motion pattern is increased. When the VOR alone is elicited, the pattern interacts in an enhancing way if during the previous storage period slip and eye movements had been in phase; if they had been in counterphase (opposite direction) then the pattern interacts with the VOR to decrease its gain.

The resultant compensatory eye movements then are a function of visual and vestibular inputs and of any stored patterns that are similar to the stimulus patterns. The retrieval of the pattern may occur by an input or output sufficiently similar to the one used to form it. Precise terms with regard to the nature of the interactions are avoided, since they have not been investigated and may be non-linear. It is apparent that any interaction by stored patterns will make it impossible to predict the integrated visuovestibular response from the visual and vestibular components alone. Baarsma and Collewijn (1974) and Collewijn and Kleinschmidt (1975) already noticed discrepancies between the VOR and OKN tested alone, and their integrated response in a synergistic or conflicting condition.

The current hypothesis obviously differs from earlier theories, which mainly assume a teaching function of the visual slip on the gain of the feedforward VOR loop (Ito, 1972; Robinson, 1976; Gonshor and Melvill Jones, 1976b). Although it may have some validity for other species, the hypothesis is only founded on the rabbit data at present. The relations in foveate species might be more complex, for instance due to the possibility of foveate animals to suppress the VOR during visual pursuit.

The marked, learning effect in optokinetic tracking of periodic stimuli has not been noted so far in other species; Gonshor and Melvill Jones (1976a) deny its occurrence in the human experiments with reversed vision. An improvement of optokinetic responses by repeated stimulation with a moving stripe pattern was described by Miyoshi et al. (1973).

In another type of experiment on visuovestibular interaction (Young and Henn, 1974, 1976), a period of OKN elicited in one direction was found to have a decreasing effect on the VOR with slow phases in the same direction, elicited by velocity steps. Although these results were obtained with different stimuli and species (man and monkey), their tendency is essentially opposite to that of the present rabbit data. Thus, care in extrapolating the latter is due.

The existence of a storage system for patterns of motion (a "pattern centre") has been postulated before, notably in connection with the occurrence of motion sickness, habituation and after-effects due to unfamiliar patterns of motion (Groen, 1957). An important question in this respect is, whether long-term stimulation of the VOR alone leads to habituation or adaptation effects. Such effects are well known for postrotatory nystagmus and have been also described for the rabbit (Hood and Pfaltz, 1954). However, it has been found that the VOR elicited by a sinusoidal stimulus is not in any consistent way affected by long-term stimulation (Collewijn and Kleinschmidt, 1975; Kleinschmidt and Collewijn, 1975; Gonshor and Melvill Jones, 1976a). In fact, this evidence for stability of the VOR in itself was specifically sought to justify the experiments on visual modification of the VOR. Also the after-effects of such a stimulus seem to be weak. Some spontaneous postoscillatory sinusoidal eye movements have been described in man (Festen and Clemens, 1970). In all rabbits oscillated for 24 h (1/6 Hz, 20° p.p.) in the dark, Kleinschmidt and Collewijn (1975) found some spontaneous oscillatory eye movements as after-effects, but the shape was quite irregular and the frequency was variable and unrelated to the stimulus frequency.

SUMMARY

Long-term (4–8 h) sinusoidal stimulation (1/6 Hz, amplitude 10°–20°) of the vestibulo-ocular (VOR) and/or optokinetic (OKN) reflex caused changes in the gain of both VOR and OKN. Visual slip always resulted in an improvement of visual tracking. The amplitude of VOR increased if slip was in phase with the eye movements in the head, but decreased if these signals were in counterphase. The present results resolve previous contradictory findings on adaptation in the rabbit. Visual stimulation alone was sufficient to increase the gain of both OKN and VOR, without any coincidence of vestibular and visual stimulation. The adaptations showed considerable frequency specificity and occasionally spontaneous eye movements at exactly the conditioned frequency were produced as an after-effect. It is hypothesized that during adaptation a copy of the slip pattern is accumulated in the nervous system and that this pattern is reproduced and interferes with OKN and VOR when a similar pattern of motion returns.

REFERENCES

Baarsma, E.A. and Collewijn, H. (1974) Vestibulo-ocular and optokinetic reactions to rotation and their interaction in the rabbit. *J. Physiol. (Lond.)*, 238: 603–625.

Collewijn, H. and Kleinschmidt, H.J. (1975) Vestibulo-ocular and optokinetic reactions in the rabbit: changes during 24 hours of normal and abnormal interaction. In *Basic Mechanisms of Ocular Motility and their Clinical Implications*, G. Lennerstrand and P. Bach-y-Rita (Eds.), Pergamon Press, Oxford, pp. 477–483.

Collewijn, H. and Grootendorst, A.F. (1978) Adaptation of the rabbit's vestibulo-ocular reflex to modified visual input: importance of stimulus conditions. *Arch. ital. Biol.*, 116: 273–280.

Festen, H. and Clemens, A. (1970) Pattern centre. *Adv. oto-rhino-laryngol.*, 17: 100–106.

Gauthier, G.M. and Robinson, D.A. (1975) Adaptation of the human vestibulo-ocular reflex to magnifying lenses. *Brain Res.*, 92: 331–335.

Gonshor, A. and Melvill Jones, G. (1976a) Short-term adaptive changes in the human vestibulo-ocular reflex arc. *J. Physiol. (Lond.)*, 256: 361–379.

Gonshor, A. and Melvill Jones, G. (1976b) Extreme vestibulo-ocular adaptation induced by prolonged optical reversal of vision. *J. Physiol. (Lond.)*, 256: 381–414.

Groen, J.J. (1957) Adaptation. *Pract. oto-rhino-laryngol. (Basel)*, 19: 524–530.

Hood, J.D. and Pfaltz, C.R. (1954) Observations upon the effects of repeated stimulation upon rotational and caloric nystagmus. *J. Physiol. (Lond.)*, 124: 130–144.

Ito, M. (1972) Neural design of the cerebellar motor control system. *Brain Res.*, 40: 81–84.

Ito, M., Shiida, T., Yagi, N. and Yamamoto, M. (1974) The cerebellar modification of rabbit's horizontal vestibulo-ocular reflex induced by sustained head rotation combined with visual stimulation. *Proc. Jap. Acad.*, 50: 85–89.

Kleinschmidt, H.J. (1974) *Effecten van Langdurige Prikkeling op de Vestibulo-Oculaire Reflexen van het Konijn*. Thesis, Erasmus University, Rotterdam.

Kleinschmidt, H.J. and Collewijn, H. (1975) A search for habituation of vestibulo-ocular reactions to rotatory and linear sinusoidal accelerations in the rabbit. *Exp. Neurol.*, 47: 257–267.

Melvill Jones, G. and Davies, P. (1976) Adaptation of cat vestibulo-ocular reflex to 200 days of optically reversed vision. *Brain Res.*, 103: 551–554.

Miles, F.A. and Fuller, J.H. (1974) Adaptive plasticity in the vestibulo-ocular responses of the rhesus monkey. *Brain Res.*, 80: 512–516.

Miyoshi, T., Pfaltz, C.R. and Piffko, P. (1973) Effect of repetitive optokinetic stimulation upon optokinetic and vestibular responses. *Acta oto-laryngol. (Stockh.)*, 75: 259–265.

Robinson, D.A. (1976) Adaptive gain control of vestibulo-ocular reflex by the cerebellum. *J. Neurophysiol.*, 39: 954–969.

Young, L.R. and Henn, V.S. (1974) Selective habituation of vestibular nystagmus by visual stimulation. *Acta oto-laryngol. (Stockh.)*, 77: 159–166.

Young, L.R. and Henn, V.S. (1976) Selective habituation of vestibular nystagmus by visual stimulation in the monkey. *Acta oto-laryngol. (Stockh.)*, 82: 165–171.

Visual Substitution of Labyrinthine Defects

J. H. COURJON and M. JEANNEROD

Laboratoire de Neuropsychologie Expérimentale, I.N.S.E.R.M. Unité 94, 69500 Bron (France)

INTRODUCTION

Labyrinthine and visual inputs contribute in maintaining the original angle of the retina in space when the body is tilted laterally. In the static conditions, counterrolling of the eyes compensates only for a small amount of the tilt (Krejcova et al., 1971) and preservation of the retinal angle is mostly achieved by a righting of the head. Magnus (1924, 1926) in his extensive studies in cats and dogs had identified labyrinthine and optical head righting reflexes, which he considered as additive components of the same predetermined reaction. However, a more modern conception of visual and vestibular interactions tends to admit that postural adjustments could be achieved through either input, each one being able to substitute for the other. Adaptation of postural mechanisms to natural, experimental or pathological situations, which is now well documented (see Dichgans et al., 1973a; Talbott, 1974) in fact requires a large amount of flexibility, where any available input may prevail in each particular case in achieving the correct adjustment.

Accordingly, the aim of our study was to evaluate the role of vision in reestablishing a normal head posture after the input from one of the two labyrinths has been suppressed. Hemilabyrinthectomy is known to produce a strong but transient postural asymmetry, which is easily recognizable in all studied vertebrates since the classical work of Flourens (1842), Ewald (1892) and Magnus (1924). In the cat, the postoperative syndrome is first marked by a critical stage lasting more than 24 and less than 48 h. The animal cannot stand up and lies on the side corresponding to the lesioned labyrinth. With a right hemilabyrinthectomy its body axis will be tonically bent to the left and twisted in the clockwise direction, so that its head will be tilted right side down. Within about 2 days, the animal is able to walk though its attempts are limited to inept leftward circling. After the third day, its head-body axis tends to become rectilinear. After 5–8 days gait is almost normal except for a hypotonia of the right limbs, and for occasional falls on the right side during spontaneous head-shakings. Finally, lateral head tilt may persist for several weeks or months.

A study of spontaneous head posture and of righting of the head after hemilabyrinthectomy thus appears to be a good way to describe the time course of postural compensation. We have used this index in animals recovering in normal conditions, and in animals submitted to a temporary exclusion of visual input during the postoperative stage. By varying the duration of the visual deprivation period, and its position in time with respect to the operation, it becomes possible to alter the normal process of recovery and consequently to assess the role of vision in re-acquisition and/or in maintenance of the new postural compensation (see also Putkonen et al., 1977).

METHODS

Experiments were conducted in adult cats. Destruction of the right labyrinth was performed under Nembutal anaesthesia. The bulla was opened through a ventral approach, and the bony labyrinth was destroyed under visual control with a dissecting microscope.

After the operation, the animals were directed to one out of two experimental groups. In the L (light) group (3 cats), hemilabyrinthectomized animals were kept in cages in a normal laboratory environment. Normal illumination was provided during day hours. In the D (dark) group (3 cats), animals were put in a light proof room immediately after surgery. Total darkness was severely controlled, including during feeding and cleaning procedures. Duration of the postoperative dark period was varied from 10 days to 6 months according to different animals.

The same dark room was used when animals from either group were put in the dark for short periods (2–15 days) during the late course of recovery.

Head posture was measured in standard conditions by using serial photography of cats' heads. The animals were placed in a box (45 × 14 × 18 cm) leaving the head free to move. The box, at a fixed distance from the camera, could be placed either in the upright position (0°), or tilted 45° to the right (+45°) or to the left (−45°). For each session, its position was alternated in a fixed sequence (0°, +45°, −45°) repeated 3 times. The cat's head was photographed in each position of the box (9 photographs per session). Sessions were repeated at fixed intervals of time (see below).

In the case of animals from the D group, the same procedure was used in the dark, except that single brief flashes of light were given to take the photographs. However, to ensure a further control of the lack of visual input during the photography sessions, two animals from this group had the lids sutured on the same day as the hemilabyrinthectomy.

Finally, unlesioned animals were also photographed in the box, in order to obtain control values of head posture and righting reflexes.

Each photograph was then analysed by tracing the interocular axis on the cat's face, and by measuring its angulation with respect to the physical horizontal. Values obtained for a given position (e.g., upright 0°) were averaged for each session.

RESULTS

Head posture and righting reflexes in normal cats

Normal cats tend to keep a symmetrical head posture when they have their body in the upright position. However, during static lateral body tilt, the head does not fully compensate for the inclination, and remains underrightened. Fig. 1A shows data from four intact animals observed under normal illumination. For a 45° body tilt to the right or to the left, the head rightens on both sides by about 25° on the average, which represents only a 50% compensation of body tilt. Underrighting of the head might be partly explained by the existence of a counter rolling of the eyes, thus compensating for

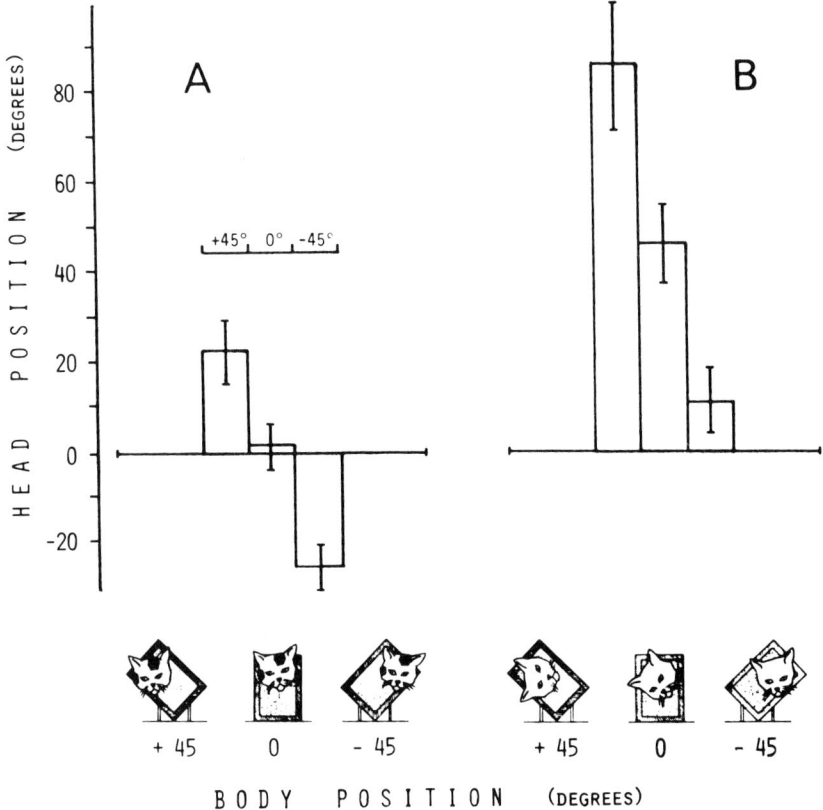

Fig. 1. Head position re body position in normal and hemilabyrinthectomized cats. A) Mean head position and SD in 4 normal cats tested in the box as indicated by the drawings in the lower row (redrawn from original photographs). +45, 45° body tilt to the right; 0, upright body position; −45, 45° tilt to the left. B) Same data from 2 right hemilabyrinthectomized cats, tested on the second postoperative day, i.e., the day with the maximum head tilt.

the residual head tilt, and bringing the retinal coordinates back to their original position. Though we have not measured ocular torsion in our animals, we know from experiments in monkeys by Krejcova et al., (1971) that under static conditions it should not exceed 10% of the amount of head tilt.

In the dark also, righting of the head in response to body tilt is incomplete. This is less surprising, however, since the need for a preservation of retinal position in space no longer exists.

Postoperative evolution of head posture in cats from the L group

After hemilabyrinthectomy, the normal pattern of static head posture in relation to body position was dramatically altered. When the cats had the body in the upright position, the head appeared permanently tilted to the right (lesioned) side. The amount of spontaneous head tilt was usually maximum on the second postoperative day. Concomitantly, the head righting was abolished when the body was tilted to the right. In this case, the resulting head posture was a passive addition of the spontaneous head tilt, plus the body tilt. By contrast, the righting reflex was preserved, at least partly, when the

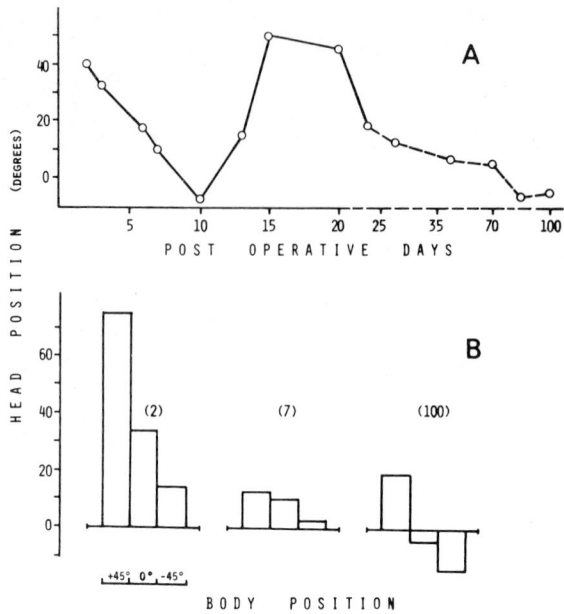

Fig. 2. Postoperative evolution of head posture in one hemilabyrinthectomized cat, recovering in normal conditions (L group). A) Postoperative evolution of head tilt with the body in the upright position. B) Head position re body position in postoperative days 2, 7 and 100.

body was tilted to the left. Fig. 1B shows average values of head tilt in the three positions of the box, obtained in two animals from the L group on their second postoperative day.

Recovery from this postural deficit was first marked by a steep decrease in spontaneous head tilt during the first 7–10 postoperative days, down to close to normal values of static head position. Righting reflex partly reappeared in response to body tilt toward the lesioned side. This postural improvement was only temporary, however. As exemplified by the animal shown in Fig. 2, static head-tilt reincreased during the third postoperative week, up to the values of the first days, and righting reflexes deteriorated. From this point, a long-lasting process took place, leading progressively to a stable compensation within about 3 months (Fig. 2). Indication of such a two-stage recovery process was also found in the other animals of the L group.

Examination of hemilabyrinthectomized cats at a very late postoperative stage (1 year) revealed a small residual head tilt to the right (average value in 4 animals: 9.4°, see Fig. 5B).

Postoperative evolution of head position in cats from the D group

Lack of visual input during the postoperative stage, resulted in "freezing" postural recovery at the level of the first or second postoperative day. In the two lid-sutured cats, the postlabyrinthectomy syndrome could be observed as described in the Introduction, throughout the whole period of light deprivation. Once vision was restored (on the 16th and 28th days, respectively), compensation began immediately and proceeded with an accelerated time course.

Head posture was examined in these two animals during the dark period. It was found that the initial head tilt in the upright body position was less pronounced than in animals using visual cues during recovery. However a progressive deterioration occurred over

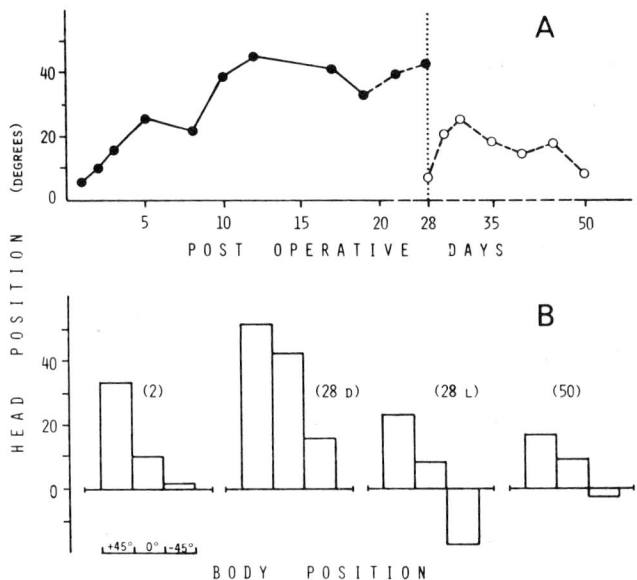

Fig. 3. Postoperative evolution of head posture in one hemilabyrinthectomized cat recovering in the dark (D group). A) Post-operative evolution of head tilt with the body in the upright position. Dashed line on day 28 indicates the return to the light condition. B) Head position re body position in postoperative days 2,28,50. On day 28, compare the data obtained in the dark (28 D) with those obtained in the light, a few min later (28 L).

time, so that, finally head tilt peaked at 42.7° on the 28th day in one animal, and at 112.5° on the 16th day in the other. This evolution is shown for the first animal in Fig. 3A.

Head righting in response to rightward body tilt was absent during the first days. It tended to improve, however, in that the resulting value of head tilt in that position was less than the summation of spontaneous head tilt and of body tilt (Fig. 3B). Thus, the pattern of head posture in animals of the D group at the end of the dark period differed somewhat from that of animals of the L group during the critical stage. Though they still had their head strongly tilted in the upright body position, the righting in response to lateral body tilt appeared to be normal in amplitude, in both directions.

Restoration of normal visual conditions (which had been preceded two days before by a reopening of the lids under a small dose of ketamine) resulted in an immediate decrease in head asymmetry in the upright body position. In the animal shown in Fig. 3, a few min spent in the light were sufficient to restore a quasi-normal pattern. However, this improved level of postural symmetry was not stable and could not be maintained. It deteriorated within 1 or 2 days and finally improved progressively over about 2 weeks, before compensated values of head posture could be reached and maintained. This temporal pattern was very reminiscent, though with a shorter time course, of the two stage recovery process observed in the animals from the L group.

In a different animal, the duration of the dark period was prolonged for 6 months. Spontaneous head tilt in the upright body position was about 50° after 3 months and about 55° after 6 months (Fig. 4), a value which would never be observed at this stage in animals recovering in normal conditions. Righting reflexes were also very poor and asymmetrical. Restoration of visual input produced an immediate improvement in head posture, to a lesser extent, however, than after shorter visual deprivation periods.

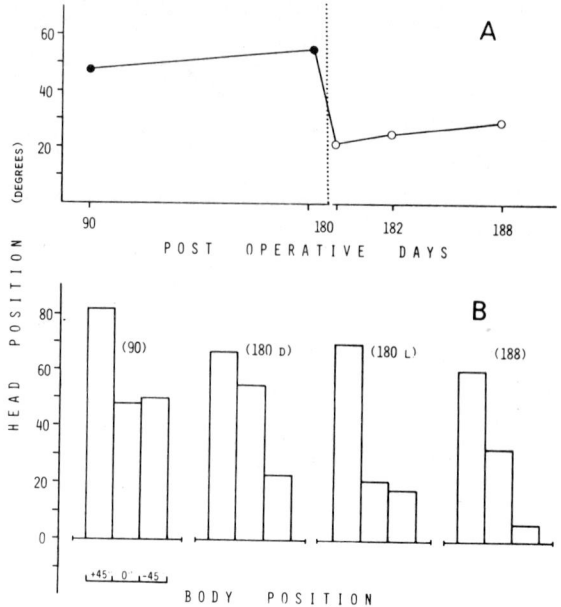

Fig. 4. Postoperative evolution of head posture in one hemilabyrinthectomized cat left in the dark for 6 months. A) Postoperative evolution of head tilt with the body in the upright position. Dashed line on day 180 indicates the return to the light condition. B) Head position re body position on postoperative days 90, 180, 188. On day 180 compare the data obtained in the dark (180 D) with those obtained in the light (180 L).

Role of visual input in maintenance of a symmetrical head posture in compensated cats

Animals from either group which had reached a sufficient level of postural recovery, were submitted to short (2–15 days) periods of visual deprivation. As a rule, head posture was examined in the dark on the last day of the period, and in the light on the first and the following days after restoration of normal vision.

In all cases, a deterioration of postural symmetry was observed at the end of the dark period. This effect occurred irrespectively of the duration of the visual deprivation, or of its position in postoperative time. For example, Fig. 5A shows the effect of a 15 day dark period in one animal, at the end of the second postoperative month. Head tilt had increased by a factor of 2 or 3 at the end of the period. An effect of a similar amplitude could be obtained in another animal after a dark period of only 2 days.

Four of these animals underwent another light deprivation period (duration, 10 days) more than 1 year after hemilabyrinthectomy. Mean angular position of the head in the four animals (as measured in the upright body position) increased from 9.4° prior to the deprivation period up to 24° after 10 days (Fig. 5B).

The specificity of the role of vision on this effect was ascertained by the immediate return to predeprivation values of head posture, when normal vision was restored.

DISCUSSION

Postural asymmetry following hemilabyrinthectomy reflects the imbalance between the activity of the vestibular nuclei on the two sides. According to Precht et al. (1966) and McCabe et al. (1972), the activity of the nuclei on the deafferented side is strongly

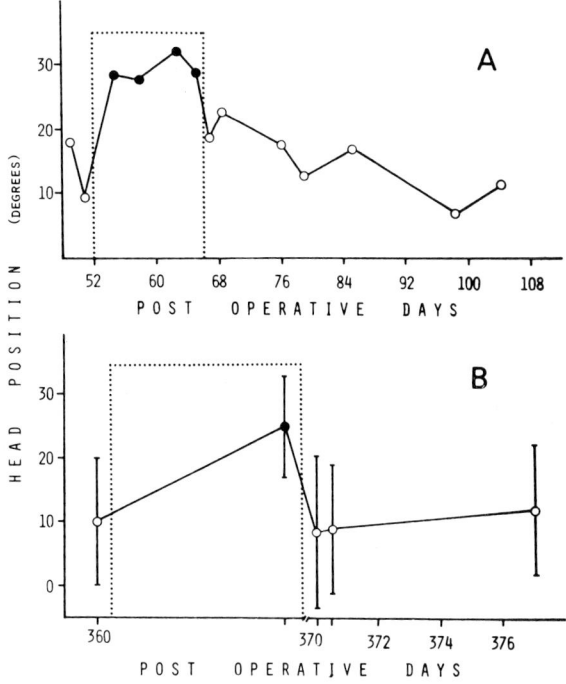

Fig. 5. Effect of late dark exposure on the postoperative evolution of head posture in hemilabyrinthectomized cats. A) Effect of a 15 day dark exposure in one animal between postoperative days 52 and 67. Compare the 2 values of head tilt obtained on day 67, in the dark (●) and in the light (○). B) Effects of a 10 days dark exposure in 4 animals 1 year after the hemilabyrinthectomy. ○, average value of head tilt (and SD) in the light. ●, in the dark. Compare the two values obtained on day 370.

depressed after the operation, while in the contralateral nuclei the resting discharge increases as a result of a reduced contralateral inhibition. Hence, the hyperactivity of neck muscles on the side of the lesion, leading to the head tilt, may be explained by the increased discharge from the intact vestibular nuclei, via crossed excitatory vestibulospinal projections onto neck motoneurons (Wilson and Yoshida, 1969; Wilson and Peterson, 1978).

Compensation in normal postoperative conditions begins during the first week with a strong inhibition of the resting activity in the vestibular nuclei on both sides. The evolution is then marked by a progressive regeneration of activity in the vestibular nuclei on the deafferented side (presumably via an increase in excitatory synaptic efficacy of the commissural system) (Dieringer and Precht, 1977), and later by a disinhibition on the normal side (Precht et al., 1966; Precht, 1974).

These neural changes may represent the substrate for the 2-stage recovery pattern observed in cats from the L group. The first stage would correspond to a rapid motor learning, tending to reduce a critical dysfunction through a massive inhibition of postural reactions. The second, long-term stage would correspond to the learning and to the fixation of new motor sets, based on the reorganization of activity within the vestibular system (Llinas et al., 1975).

Postoperative evolution of animals from the D group seems to indicate that normal recovery heavily relies upon visual cues. First, the fact that postural asymmetry is less marked during the early postoperative stage in animals kept in the dark shows that the visual factor aggravates the motor imbalance produced by the lesion. One possible explanation for this difference is that hemilabyrinthectomy would also bias central mechanisms responsible for the detection of visual coordinates (Bisti et al., 1972), thus resulting in an increased head tilt. If this hypothesis were correct, then it would become questionable whether the rapid decrease in postural asymmetry observed in the L group is due to motor learning per se, or to some adaptation within the visual system.

Second, we know from the animals of the D group that the motor imbalance cannot be compensated for in the absence of vision even if visual deprivation has been prolonged for as long as 6 months. This fact shows that the long-term learning which occurs during the second stage of normal recovery also requires static visual cues in order to re-equilibrate and to stabilize the postural system.

Finally, whether postural symmetry is recovered normally or whether it is "frozen" during visual deprivation, we know from our experiments that the visual vestibular interaction responsible for postural compensation remains fully flexible. Restoration of visual input at any postoperative stage in animals from the D group is always followed by recovery. We have no reason to believe that a visual deprivation longer than 6 months would not be followed also by some degree of recovery. On the other hand, deprivation of visual input in already compensated animals invariably produces a deterioration of postural symmetry, even 1 year after the operation. This flexibility is a well known fact from the literature in animals (Ewald, 1892; Magnus, 1924; Dow, 1938; Schaefer and Meyer, 1974), as well as in man (André-Thomas et al., 1941), showing that blind-folding abolishes temporarily postural compensation after labyrinthine lesions.

These experimental data lead to the conclusion that vision is just another input channel feeding into the vestibular system. In normal animals, electrophysiological evidence has been found that vision can influence vestibular neurons (Dichgans et al., 1973b; Azzena et al., 1974; Henn et al., 1974). An anatomical pathway has been identified, from the accessory optic tract and via the inferior olive and the cerebellar flocculus (Maekawa and Simpson, 1973) or more directly via the flocculus only (Brauth and Karten, 1977; Winfield et al., 1978), which may account for this influence.

Our suggestion is that the gain of this pathway increases under pathological conditions, where vision becomes predominant and is able to fully substitute for the labyrinthine input. In fact, totally delabyrinthed animals may have normal righting reflexes when allowed to use their vision (Magnus, 1926). This would mean either that fibers carrying visual input to the brain stem form new synapses with vestibular neurons responsible for postural control, or alternatively that already existing synapses are derepressed by the destruction of labyrinthine afferents. The latter hypothesis, postulated by Merrill and Wall (1972) for the somesthetic system would account for the rapid recovery when visual input is available to the animal.

SUMMARY

The evolution of lateral head-tilt following hemilabyrinthectomy has been studied in adult cats. Animals were maintained postoperatively in normally lit conditions (L group) or in total darkness (D group).

In cats from the L group, the head-tilt peaked at 45° (with the lesioned side down) on the second postoperative day, and decreased to about 0° within about 10 days. This evolution was followed by rebounds of head-tilt to larger angles before a stable compensated head position could be maintained (approximately at the end of the third postoperative month).

In cats from the D group, the head remained tilted by a large angle throughout the duration of the dark period up to 6 months. Re-exposure to light was followed by a rapid decrease of head-tilt.

Finally, when put back in the dark at a late postoperative stage (up to 1 year), already compensated animals were found to lose their symmetrical head position, and to re-acquire a strong head-tilt. This effect resumed on re-exposure to light.

It is inferred that static visual input is a necessary condition for compensation of the postural deficits of hemilabyrinthectomy in the cat. Maintenance of a stable head posture also depends upon continuous availability of visual input.

REFERENCES

André-Thomas, Sorrel, E. and Sorrel-Dejerine, M. (1941) Fracture du rocher, troubles vestibulaires, attitude de la tête, réactions méningées. *Rev. Neurol.*, 64–73.

Azzena, G. B., Azzena, M. T. and Marini, R. (1974) Optokinetic nystagmus and the vestibular nuclei. *Exp. Neurol.*, 42: 158–168.

Bisti, S., Maffei, L. and Piccolino, M. (1972) Variations of the visual responses of the superior colliculus in relation to body roll. *Science*, 175: 456–457.

Brauth, S. E. and Karten, M. J. (1977) Direct accessory optic projections to the vestibulo-cerebellum: a possible channel for oculomotor control systems. *Exp. Brain Res.*, 28: 73–84.

Dichgans, J., Bizzi, E., Morasso, P. and Tagliasco, V. (1973a) Mechanisms underlying recovery of eye-head coordination following bilateral labyrinthectomy in monkeys. *Exp. Brain Res.*, 18: 548–562.

Dichgans, J., Schmidt, C. L. and Graf, W. (1973b) Visual input improves the speedometer function of the vestibular nuclei in the goldfish. *Exp. Brain Res.*, 18: 319–322.

Dieringer, N. and Precht, W. (1977) Modification of synaptic input following unilateral labyrinthectomy. *Nature (Lond.)*, 269: 431–433.

Dow, R. S. (1938) The effects of bilateral and unilateral labyrinthectomy in monkey, baboon and chimpanzee. *Amer. J. Physiol.*, 121: 392–399.

Ewald, J. R. (1892) *Physiologische Untersuchungen über das Endorgan des N. Oktavus.* Wiesbaden, Bergmann.

Flourens, J.P.N. (1842) *Recherches Expérimentales sur les Propriétés et les Fonctions du Système Nerveux, dans les Animaux Vertébrés.* Baillère, Paris.

Henn, V., Young, L. R. and Finley, C. (1974) Vestibular nucleus units in alert monkeys are also influenced by moving visual fields. *Brain Res.*, 71: 144–149.

Krejcova, H., Highstein, S. and Cohen, B. (1971) Labyrinthine and extralabyrinthine effects on ocular counter-rolling. *Acta oto-laryngol. (Stockh.)*, 72: 165–171.

Llinás, R., Walton, R., Hillman, D. E. and Sotelo, C. (1975) Inferior olive: its role in motor learning. *Science*, 190: 1230–1231.

Maekawa, K. and Simpson, J. L. (1973) Climbing fiber response evoked in vestibulocerebellum of rabbit from visual system. *J. Neurophysiol.*, 36: 649–666.

Magnus, R. (1924) *Körperstellung.* J. Springer, Berlin.

Magnus, R. (1926) Some results of studies in the physiology of posture. *Lancet*, 211: 531–536 and 585–588.

McCabe, B. F., Ryu, J. H. and Sekitani, T. (1972) Further experiments on vestibular compensation. *Laryngoscope*, 82: 381.

Merill, E. G. and Wall, P. D. (1972) Factors forming the edge of a receptive field: the presence of relatively ineffective afferent terminals. *J. Physiol. (Lond.)*, 226: 825.

Precht, W. (1974) Characteristics of vestibular neurons after acute and chronic labyrinthine destruction. In *Handbook of Sensory Physiology, Vol. VI/1, Vestibular System, Part 2, Psychophysics, Applied Aspects and General Interpretations.* H. H. Kornhuber (Ed.), Springer-Verlag, Berlin, pp. 451–462.

Precht, W., Shimazu, H. and Markham, C. H. (1966) A mechanism of central compensation of vestibular function following hemilabyrinthectomy *J. Neurophysiol.*, 29: 996–1010.

Putkonen, P. T. S., Courjon, J. H. and Jeannerod, M. (1977) Compensation for postural effects of hemilabyrinthectomy in the cat. A sensory substitution process? *Exp. Brain Res.*, 28: 249–257.

Schäfer, K. P. and Meyer, D. L. (1974) Compensation of vestibular lesions. In *Handbook of Sensory Physiology, Vol. VI/1. Vestibular System, Part 2, Psychophysics, Applied Aspects and General Interpretations,* H. H. Kornhuber (Ed.), Springer-Verlag, Berlin, pp. 463–490.

Talbott, R. E. (1974) Modification of the postural response of the normal dog by blindfolding. *J. Physiol., (Lond.),* 243: 309–320.

Wilson, V. J. and Yoshida, M. (1969) Bilateral connections between labyrinths and neck motoneurons. *Brain Res.,* 13: 603–607.

Wilson, V. J. and Peterson, B. W. (1978) Peripheral and central substrates of vestibulo spinal reflexes. *Physiol. Rev.,* 58: 80–105.

Winfield, J. A., Hendrickson, A. and Kimm, J. (1978) Anatomical evidence that the medial terminal nucleus of the accessory optic tract in mammals provides a visual mossy fiber input to the flocculus. *Brain Res.,* 151: 175–182.

SECTION VI

COORDINATION OF EYE-HEAD MOVEMENTS

Strategies of Eye-Head Coordination

E. BIZZI

Department of Psychology, Massachusetts Institute of Technology, Cambridge, MA 02139 (U.S.A.)

Two modes of eye-head coordination in monkeys will be compared here. First, we will describe those eye-head movements following the appearance of an unexpected and stationary target in space and second, the coordination of the eyes with head observed during smooth tracking.

EYE-HEAD COORDINATION FOLLOWING THE APPEARANCE OF AN UNEXPECTED AND STATIONARY TARGET IN SPACE IN HIGH VERTEBRATES

To direct head and eyes toward a target, and ultimately fixate it with his fovea, an animal must solve three problems. First, he must compute the angular distance between his foveal lines of sight and the position of the target which is to be acquired. I shall call this angular distance a retinal error; its absolute magnitude will determine, to a first approximation, the amplitude of the saccadic eye movement which will be produced. Second, the animal must initiate a head movement that will be compatible in amplitude with the saccadic eye movement. Third, since the eyes usually move first and with higher velocity than the head their lines of sight will reach and fixate the target while the head is still moving. The stabilization of the eyes with respect to a stationary target during head movement is accomplished by performing a rotational movement that, by being counter to the movement of the head, allows the fovea to remain constantly on the target it has just acquired. This movement is termed compensatory.

In the next three sections (A, B and C), we will consider these three problems separately.

A. Saccades during head movement

Since in monkeys, on average, the head moves only 10–25 msec after the initiation of the saccade, it follows that relatively large saccades of more than 10–25 msec duration will often take place while the head is moving. A case in point is shown in Fig. 1b which illustrates coordinated eye head movements in response to a 30° target together with the sum of these movements, the "gaze". Fig. 1a displays the record of a saccadic eye movement made by the same animal to the same target while its head was artificially restrained. The comparison between the saccade in (a) and the gaze in (b) clearly shows that target acquisition is achieved with the same precision in both instances (Morasso et

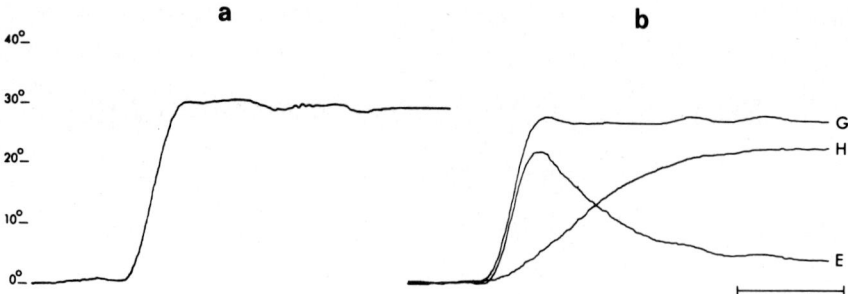

Fig. 1. Comparison of eye saccades and gaze. (a) Eye saccade to a suddenly appearing target with head fixed. (b) Coordinated eye saccade (E) and head movement (H) to the same target with head free. The gaze movement (G) represents the sum of E and H. Note the remarkable similarity of eye saccade in (a) and gaze trajectory in (b) as well as reduced saccade amplitude in (b). Time calibration: 100 msec. (From Morasso et al., 1973.)

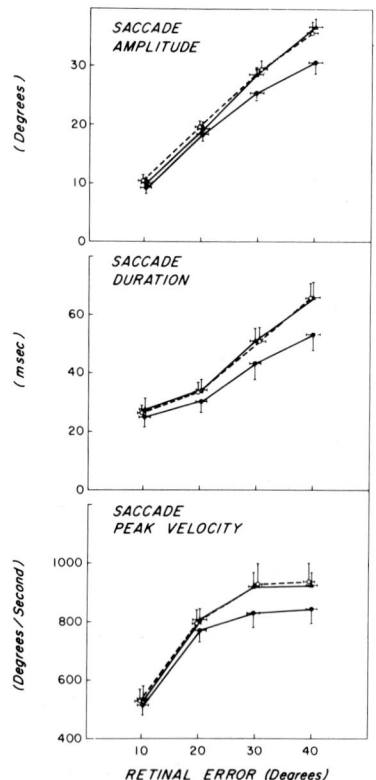

Fig. 2. Amplitude, duration and peak velocity of eye saccades and gaze. The abscissa represents the angular distance between starting position of the eyes and target position. ▲—▲, saccades during continuous immobilization of the head; ○—○, gaze movements; ●—●, saccades during head turning. Each point represents the mean of 20 measurements from three adult monkeys studied with SDs. (From Morasso et al., 1973.)

al., 1973). This identity is shown further in Fig. 2 which gives saccade duration, amplitude and peak velocity at various target angles. The upper two lines in each graph give, respectively, values for gaze movements and for saccades made with head restrained (broken line). The clear superimposition of these two sets of data demonstrates the fact that the task of bringing the fovea to the image of the target either with eyes alone or with eyes plus head is accomplished with the same duration and speed.

Fig. 2 also shows an additional curve below the two which are superimposed. This third curve describes the values for saccades made during head turning. Although it is clear that relatively small saccades (10°–20°) made in the coordinated eye-head mode do not differ significantly from those made with head fixed, it should be pointed out that these saccades are generally almost over at the time when the head begins to move. For larger responses (30°–40°) the curve for saccades with head movement clearly diverges from the curve for saccades with head fixed.

The findings illustrated in Figs. 1 and 2 raise the question as to whether in the monkey the decrease in saccade amplitude, duration and peak velocity with the head movement is the result of an adjustment of the central eye movement program occurring in conjunction with head movement, or instead is a correction mediated by reflex activity originating from structures which are excited by head turning. The results of Morasso et al. (1973) suggested that the central mechanisms responsible for programming saccades act only according to target position and that there is no information fed forward from a head programming mechanism to the oculomotor system. From these results it was argued that the decrease in saccade amplitude, duration and peak velocity must be due to a reflex mechanism activated by the movement of the head. The most likely candidates for this reflex action are the vestibular and neck afferents (Dichgans et al., 1973, 1974). As for the visual loop, its role was discounted by showing that saccade characteristics were not changed by turning off the target lights just before the saccade movement was initiated (Dichgans et al., 1973). There is evidence indicating that in the monkey the vestibular impulses are responsible for modulating the saccadic eye movements. Positive evidence of the crucial role of the vestibular afferent signals was demonstrated in monkeys by surgically interrupting the pathway linking the vestibular receptors to the vestibular nuclei (Dichgans et al., 1973, 1974). For several weeks after the operation (before the monkey had learned to compensate) the saccade amplitude during head-turning was identical with the saccade amplitude in the absence of head movement. This resulted in a remarkable overshooting of the target because the unmodulated eye movement was simply added to the head movement. Clearly the reflex mechanism is more advantageous to the animal than one based on a centrally preprogrammed modification of saccadic parameters. The reflex mode of organization greatly simplifies the task of the motor-programming systems required for eye-head coordination. The eye and head movements can be programmed independently, since the vestibular system "automatically" nullifies any displacements of the fovea from the target as a result of head movement. Furthermore, by relying on vestibular reflexes that monitor the actual movement of the head, the resulting adjustment of eye movements will be able to compensate for all the unpredictable peripheral loads and resistances that might change the course of the centrally initiated (intended) head movement.

B. Compensatory eye movements

The modification of saccade characteristics is one aspect of the interaction of central programming and reflex activities. Although this interaction plays a decisive part in the process of target acquisition by a combined eye-head movement, the role of feedback from peripheral sensory organs (vestibular and neck afferents) extends beyond saccadic modulation to control and generate compensatory eye movements. Such spatial stabilization, within certain frequency limits, is accomplished by a movement of the eyes which is counter to that of the head but of equal amplitude and velocity, thereby compensating for the head movement. Compensatory eye movements of this sort are observed in practically every species that possesses moving eyes.

It has long been known that compensatory eye movements are critically influenced by vestibular (Ewald, 1892; Szentágothai, 1950; Philipszoon, 1962) and neck proprioceptive inputs (Bárány, 1906; De Kleijn, 1918). In monkeys, however, ocular stabilization during active or passive head movements is largely due to the vestibulo-ocular reflex, as shown by the experiments of Bizzi et al. (1971, 1972) and Dichgans et al. (1973, 1974). In addition, results obtained from monkeys tested within the first 24 h after bilateral labyrinthectomy showed complete lack of ocular stabilization during head movements (Dichgans et al., 1973, 1974).

While in the monkey, ocular stabilization during active or passive head movements is essentially due to an unmodulated vestibulo-ocular reflex, data obtained in intact humans by Meiry (1971) and Sugie and Melvill Jones (1971) indicate that the gain of the vestibulo-ocular loop ranges around 0.6. Also, the gain of the neck loop in humans was shown to range between 30% at frequencies below 0.01 Hz and 8% at frequencies exceeding 0.4 Hz (Meiry, 1971). Hence, it appears that for ocular compensation, humans and perhaps some animal species (Suzuki, 1972) rely on the vestibular system plus the neck loop while monkeys achieve ocular stability through the vestibular system alone. However, before concluding that there is a species difference between man and monkey, it would be necessary to determine whether these differences might at least in part result from an artifact introduced by the way vestibular functions were tested in human subjects (sinusoidal rotation in the dark). Since, at least in humans, the vestibulo-ocular loop is extremely dependent upon attentiveness, it is possible that this mode of stimulation might induce an abnormally low vestibulo-ocular gain. The relative contribution of neck afferents was also examined by stimulating the neck to eye reflex through rotation of the body in the dark while the head was kept stationary in space. Only occasionally was there a slight eye deviation in the direction of body movement, but this never exceeded two or three times the body movement amplitude (Dichgans et al., 1973, 1974).

The question of whether visual input is a necessary factor in eye stabilization during head turning was investigated by observing compensatory eye movements during active and passive head movements executed with and without a visible target light. A comparison of compensatory eye movements and the gaze movements recorded under these conditions indicated that in the absence of visual control ocular stabilization is completely adequate within the limits of electro-oculographic measurements. Within the range of velocities displayed in active head movements (Dichgans et al., 1973, 1974) this was also true for passive head movements. Hence, these findings extend our previous conclusion that for the range of movements we tested, ocular stabilization is entirely achieved by afferents from the labyrinth.

C. Head movement

During eye-head movements, the saccadic eye movement occurred first (about 200 msec after stimulus presentation) and then after a 20–30 msec delay the head begins to move in the same direction. However, the electromyographic (EMG) records showed that the neck muscles are activated first and then after 20 msec the eye muscles begin to contract (Bizzi et al., 1971). Simultaneous recordings from several neck muscles during horizontal head rotation demonstrated that all of the agonists were activated synchronously (Bizzi et al., 1971). Concurrently, activity was suppressed in all of the antagonists. It is remarkable that the synchronous activation of agonist muscles occurred regardless of initial head position; however, the amplitude and duration of initial bursts of neck muscle activity were related to the starting positions and amplitudes of the head movement. These findings indicate that the order of neural commands is not reflected in the resulting overt sequence of eye-head movement.

Summary and schematic outline of eye-head coordination following the appearance of an unexpected and stationary target in space.

We are now in a position to outline a realistic scheme for how movements of the eye and head are coordinated in the very simple case in which a monkey is looking straight ahead and a single target is flashed in his visual field. The sequence begins with the detection of a target somewhere in the visual field. Motor programs involving the head and the eyes are activated and respond by sending impulses to eye and neck muscles. This results in a saccadic eye movement and a head movement that activates vestibular receptors, which in turn generate saccadic modulation and a compensatory eye movement. The compensatory eye movement allows the fovea to remain fixed in relation to a point in visual space during head-turning. The fixation allows a second visual sampling, then a third and so on, with opportunities for correcting errors at each sampling.

If our hypothesized closed loop correctly described the coordination of eye-head movements, it is clear that the role of the motor program stored in the central nervous system is simply to initiate, in an impulsive manner, movements of the eyes and head. Since there is no central programming of saccadic adjustment and of compensatory eye movement, it follows that the functional, or behavioral, coordination of head and eyes is the joint result of a central initiation (following a stored program) modified by the crucial intervention of modulating signals triggered by receptors in the vestibule of the inner ear. This conclusion somewhat simplifies our views of the neural mechanisms underlying motor coordination insofar as, contrary to common assumptions, we find no need to postulate a special central population of "executive" neurons with exclusive responsibility for coordinating the eyes and the head.

THE COORDINATION OF EYE AND HEAD MOVEMENT DURING SMOOTH PURSUIT

Man, monkeys and cats use a combination of eye and head movements to track a moving visual stimulus. The question of how these two movements are coordinated, that is, how centrally generated commands to the motor system of the eye and the head are integrated with afferent activity originating from visual, vestibular, as well as from neck proprioceptors, has recently been investigated in monkeys. It was found that the gaze (i.e., the sum of eye and head movements) remains on target just as accurately when the head is free as it does when the head is fixed and only the eyes pursue the target (Lanman et al., 1978). Although the total gaze is on target with or without head movement, the eye movements are very much different. The primary effect of freeing the head was that

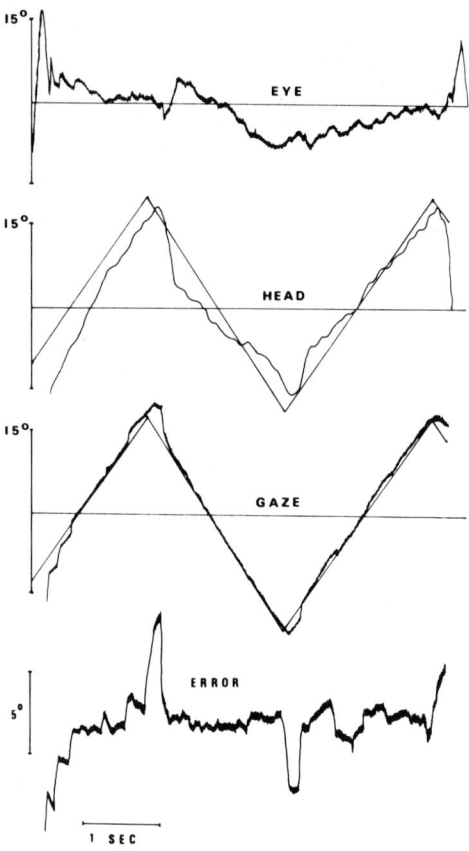

Fig. 3. Pursuit with combined eye and head movement. Pursuit movements of eye and head are shown together with the computed gaze and retinal error. The target, which periodically reverses direction, is superimposed on both head and gaze tracings. The head pursued the target, while the eyes remained relatively close to the center of the orbit. There is no obvious difference between the retinal error pattern recorded here and that observed in monkeys whose heads were restrained. (From Lanman et al., 1978.)

movements of smooth pursuit were almost completely accomplished by the head movement system, while the eyes tended to remain relatively stationary in the center of the orbit (Fig. 3). A direct comparison of the magnitude of retinal errors during pursuit with and without head movement indicated that there was little difference in accuracy of tracking between the two paradigms. To gain some insight in the coordinating mechanism Lanman et al. (1978) provoked a sudden and unexpected arrest of head movements during tracking. The result of this maneuver is shown in Fig. 4. The left part of the record shows the head tracking with eyes relatively steady in the orbit. The arrow indicates where the brake was applied to stop the head. The eyes responded with an acceleration within 15 msec. This acceleration was so fast and so accurate that the gaze continued almost uninflected, with a barely detectable retinal error. This quick change in eye velocity after head arrest could not be accomplished by either the visual loop, whose latency is too long, or neck afferents, which not only are too slow (70–80 msec), but have a very low gain in monkeys. Consequently, it must be due to the "release" of a signal representing target velocity from the opposing action of vestibular input. In other words

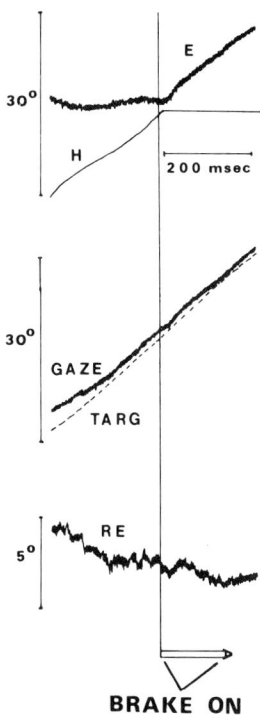

Fig. 4. The brake experiment. The result of suddenly and unexpectedly stopping head movement (H) during ongoing pursuit in the normal monkey. Before application of the brake, the head tracks the target (T) and the eyes (E) remain relatively stationary in the orbit. When the head is braked, the eyes start to move so that there is no detectable change in gaze velocity or retinal error (RE). (Modified from Lanman et al., 1978.)

there is a signal within the central nervous system that, to a first approximation, represents target velocity in space. Presumably this signal drives both the eye and head movement centers. During normal smooth pursuit with the head free, the head follows this hypothesized command with a lag which depends on the activation time of the neck musculature and on the amount of prediction involved in the pursuit strategy. The eyes, however, appear to receive, in addition to the postulated smooth pursuit signal, a signal which results from the activation of the vestibular system. This second signal specifies movements proportional to and opposite in direction from the head movement. When the two signals are combined in some part of the oculomotor system, the result is an eye movement that has an amplitude approximately equal to the difference between target and head amplitudes. Since this difference is small, eye-head smooth pursuit consists mainly of head tracking (Lanman et al., 1978).

Although these studies of eye-head coordination have led to postulate a signal representing either target velocity in space or gaze velocity, there is little evidence concerning its derivation. Visual information certainly plays an important role – a role recognized by the many investigators who have considered a retinal slip servo model for smooth pursuit (Puckett and Steinman, 1969; St-Cyr and Fender, 1969). But there is a growing body of information suggesting that retinal slip is at best only one of several inputs driving smooth pursuit eye movements (Rashbass, 1961; Kommerell and Täumer, 1972; Young, 1977).

Possible single-unit correlates to the postulated central representation of target velocity have been found. Miles and Fuller (1975) recorded from Purkinje cells in the monkey flocculus during smooth pursuit and found cells that fired at a rate proportional to the target's velocity in space whether or not the head was moving. Since the gaze is very nearly on target during smooth pursuit, these cells might code either target velocity in space or gaze velocity (Lisberger and Fuchs, 1974; Miles and Fuller, 1975). Although it remains for future investigations to show that these cells do drive eye, and possibly head systems, it is interesting that a representation of target velocity in space or of gaze velocity has been found to be encoded in the discharge of cerebellar cells. These physiological findings are complemented by the results of psychophysical investigations. Yasui and Young (1975), for example, recently proposed that a central process, identified as "perceived target velocity", is the stimulus for smooth pursuit.

During tracking with head free, eye movements appear to be driven by a combination of an internal command representing the target's motion in space and the vestibular feedback from head movement. Since head movements tend to follow the target, it is reasonable to hypothesize that they, too, are driven by this target movement signal. This hypothesis, however, must be qualified. Firstly, since the head movement is frequently smaller than the target movement, there must be some attenuation of the target movement command as it is forwarded to the head movement centers. Secondly, since the latency of head movement can vary considerably without affecting gaze, the head must be able to follow the target velocity command with a variable degree of latency or anticipation. Finally, although on the average the head follows the target, the head movements vary considerably in amplitude, indicating that they are not determined by target motion in space alone.

In conclusion, there are interesting similarities and contrasts between the "coordination" of visual triggered eye and head movements and the coordination of smooth tracking eye and head movements. It is remarkable not only that the vestibular apparatus plays a crucial role in both strategies, but that both strategies depend on a "signal" delivered in parallel and approximately simultaneously to both eye and head motor systems. These similarities indicate that the two strategies have a *common structural organization*. But while in the visually triggered mode, eye and head movements depend critically on the retinal error signal, during smooth pursuit they depend on a central representation of target motion in space. These different commands for triggered and smooth pursuit are presumably produced by different calculations and perhaps are produced in different CNS regions.

SUMMARY

In this article the similarities and contrasts between eye-head movements which are elicited by a stationary target and those observed during smooth pursuit are compared. It was found that the vestibular apparatus not only plays a crucial role in both strategies, but also both strategies depend on a "signal" delivered in parallel and approximately simultaneously to both eye and head motor systems. These similarities indicate that the two strategies have a *common structural organization*. But while in the visually triggered mode, eye and head movements depend critically on the retinal error signal, during smooth pursuit the coordination depends on a central representation of target motion in space. These different commands for triggered and smooth pursuit are presumably produced by different calculations and perhaps are produced in different CNS regions.

ACKNOWLEDGEMENT

This research was supported by National Institute of Neurological Diseases and Stroke Research Grant NS09343, and National Aeronautics and Space Administration Grant NGR 22-009-798.

REFERENCES

Bárány, R. (1906) Augenbewegungen durch Thoraxbewegungen ausgelöst. *Zbl. Physiol.*, 20: 298–302.
Bizzi, E., Kalil, R.E. and Tagliasco, V. (1971). Eye-head coordination in monkeys: evidence for centrally patterned organization. *Science*, 173: 452–454.
Bizzi, E., Kalil, R.E., Morasso, P. and Tagliasco, V. (1972) Central programming and peripheral feedback during eye-head coordination in monkeys. In *Cerebral Control of Eye Movements and Motion Perception*, J. Dichgans and E. Bizzi (Eds.), Karger, Basel, 220–232.
De Kleijn, A. (1918) Action réflexes du labyrinthe et du cou sur les muscles de l'oeil. *Arch. néerl. Physiol.*, 2: 644–649.
Dichgans, J., Bizzi, E., Morasso, P. and Tagliasco, V. (1973) Mechanisms underlying recovery of eye-head coordination following bilateral labyrinthectomy in monkeys. *Exp. Brain Res.*, 18: 548–562.
Dichgans, J., Bizzi, E., Morasso, P. and Tagliasco, V. (1974) The role of vestibular and neck afferents during eye-head coordination in the monkey. *Brain Res.*, 71: 225–232.
Ewald, J.R. (1892) *Physiologische Untersuchungen über das Endorgan des Nervus Octavus*. Bergmann, Wiesbaden.
Kommerell, G. and Täumer, R. (1972) Investigations of the eye tracking system through stabilized retinal images. In *Cerebral Control of Eye Movements and Motion Perception*, J. Dichgans and E. Bizzi (Eds.), Karger, Basel, pp. 288–297.
Lanman, J., Bizzi, E. and Allum, J.H.J. (1978) The coordination of eye and head movement during smooth pursuit. *Brain Res.*, 153: 39–53.
Lisberger, S. and Fuchs, A. (1974) Response of flocculus Purkinje cells to adequate vestibular stimulation in the alert monkey: fixation vs. compensatory eye movements. *Brain Res.*, 69: 347–353.
Meiry, J.L. (1971) Vestibular and proprioceptive stabilization of eye movement. In *The Control of Eye Movements*, P. Bach-y-Rita, C.C. Collins and J.E. Hyde (Eds.), Academic Press, New York, pp. 483–496.
Miles, F.A. and Fuller, J.H. (1975) Visual tracking and the primate flocculus. *Science*, 189: 1000–1002.
Morasso, P., Bizzi, E. and Dichgans, J. (1973) Adjustment of saccade characteristics during head movements. *Exp. Brain Res.*, 16: 492–500.
Philipszoon, A.J. (1962) Compensatory eye movements and nystagmus provoked by stimulation of the vestibular organ and the cervical nerve roots. *Pract. oto-rhino-laryngol.*, 24: 193–202.
Puckett, J. and Steinman, R.M. (1969) Tracking eye movements with and without saccadic correction. *Vision Res.*, 9: 695–703.
Rashbass, C. (1961) The relationship between saccadic and smooth tracking eye movements. *J. Physiol. (Lond.)*, 159: 326–338.
St-Cyr, G.J. and Fender, D.H. (1969) Nonlinearities of the human oculomotor system: gain. *Vision Res.*, 9: 1235–1246.
Sugie, N. and Melvill Jones, G. (1971) A model of eye movements induced by head rotation. *IEEE Trans. Syst. Man, Cybern.*, SMC-1: 251–260.
Suzuki, J.-L. (1972) Vestibular and spinal control of eye movements. In *Cerebral Control of Eye Movements and Motion Perception*, J. Dichgans and E. Bizzi (Eds.), Karger, Basel, pp. 109–115.
Szentágothai, J. (1950) The elementary vestibulo-ocular reflex arc. *J. Neurophysiol.*, 13: 395–407.
Yasui, S. and Young, L. (1975) Perceived visual motion as effective stimulus to pursuit eye movement system. *Science*, 190: 906–908.
Young, L.R. (1977) Pursuit eye movements. What is being pursued. In *Control of Gaze by Brain Stem Neurons*, R. Baker and A. Berthoz (Eds.). Elsevier, Amsterdam, pp. 29–36.

Neural Activity Pattern in Different Brain Stem Structures during Eye-Head Movements*

K.-P. SCHAEFER, D.L. MEYER and H. ZIERAU

Neurobiology Unit, Department of Psychiatry, University of Göttingen, 3400 Göttingen (F.R.G.)

INTRODUCTION

If an animal experiences a passive change of posture the vestibular system can only initiate appropriate body postural reflexes if the relative position between the labyrinths and the body is known. A continuous flow of information from neck-proprioceptors and other position receptors of the vertebral column is thus essential for a functioning of the vestibular system. Such afferents interact with postural control circuits of the brain stem at various levels including the vestibular nuclei (Pompeiano and Brodal, 1957; Fredrickson et al., 1965; Ebbesson, 1969; Brodal, 1974; Schaefer et al., 1977; and others); thereby not only gaining access to neuronal substrates which guide postural reflexes of the body but also to circuits mediating such reflexes of the oculomotor system. On the oculomotor system at least two types of influence are exerted by afferents from spinal position receptors: firstly, there are tonic changes of ocular position in dependence of twisting or bending the vertebral column (Lyon, 1900/1901; Bárány, 1918; De Kleyn, 1921; Grahe, 1922; Magnus, 1924; Harris, 1965; Takemori, 1969, 1971; Takemori and Suzuki, 1971; Suzuki and Takemori, 1971; Suzuki, 1972, Graf and Meyer, 1978; and others) and secondly, such stimuli result in modifications of oculomotor behavior during nystagmus (Schaefer et al., 1975).

Apart from conditions under which spinal afferences influence the oculomotor system, oculomotor and neck- as well as trunk-motor systems can also act in coordinated ways due to central initiation (Bizzi et al., 1972). This paper attempts to discuss both aspects of eye-head coordination.

METHODS

Our experiments were carried out on rabbits.
Optokinetic stimulation was performed by turning the animal at constant velocities in front of a striped drum. Vestibular stimulation was applied by accelerating or decelerating an animal on an electronically controlled turntable.
Nystagmogram recordings were obtained by inserting Grass platinum needle electrodes nasally and caudally to the eye bulb.

---

*Supported by the "Deutsche Forschungsgemeinschaft" (SFB 33, Schaefer).

Microrecordings were taken from animals that were neither anesthetized nor curarized. Glass insulated platinum wires and INSL X coated tungsten needles served as electrodes.

Locations of recording sites were histologically determined.

RESULTS

Effects of head-position on ocular nystagmus

EOG-recordings. Recording EOGs the number of nystagmus beats per 360° of optokinetic stimulation (15 degrees/sec) has been determined in different head-positions varying from 90° right to 90° left (11 rabbits). With the head in the normal position 360° of optokinetic stimulation induced about 22 nystagmus beats. If optokinetic stimulation was towards one side and the head was fixated in a position towards the other side the number of nystagmus beats per 360° of stimulation was increased (Fig. 1), whereas the amplitude of each beat was decreased. A slight decrease in number of nystagmus beats per 360° occurred when optokinetic stimulation and head deviation were towards the same side.

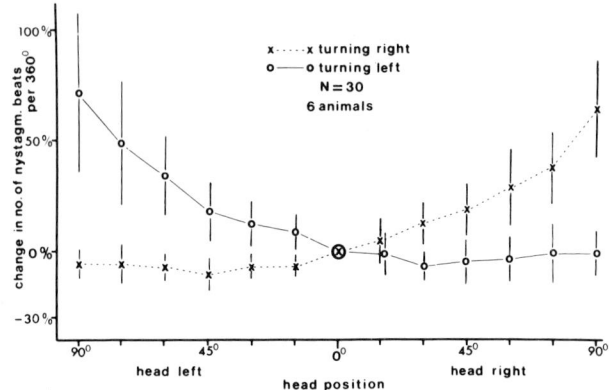

Fig. 1. Influence of head-position on optokinetic nystagmus in rabbits. Plotted is the change in number of nystagmus beats obtained per 360° of optokinetic stimulation at 10 degrees/sec in different head-positions. 0% is equivalent to a number of 19–24 beats per 360°. Optokinetic stimulation is applied by turning the animal at constant velocity. During turning in one direction there is thus an optokinetic stimulation towards the opposite side.

Recordings from oculomotor neurons. Recordings from oculomotor neurons (nucleus of the rectus internus) revealed neural phenomena that corresponded to the observations made by EOG-recordings. During head-positions to the right tonic increases of discharge rate were seen in cells of the N. rectus internus on the right side whereas decreases of firing frequencies were present on the contralateral side. Neck-proprioceptive reflexes obviously tended to maintain an ocular position left of the midposition. If nystagmus (vestibular and optokinetic) was induced while the head was fixated in left or right positions nystagmic discharge patterns of oculomotor neurons occurred on top of the changed basic firing rate (Fig. 2). Corresponding to the findings made during EOG-recordings the length of slow phase activations was changed (Fig. 2).

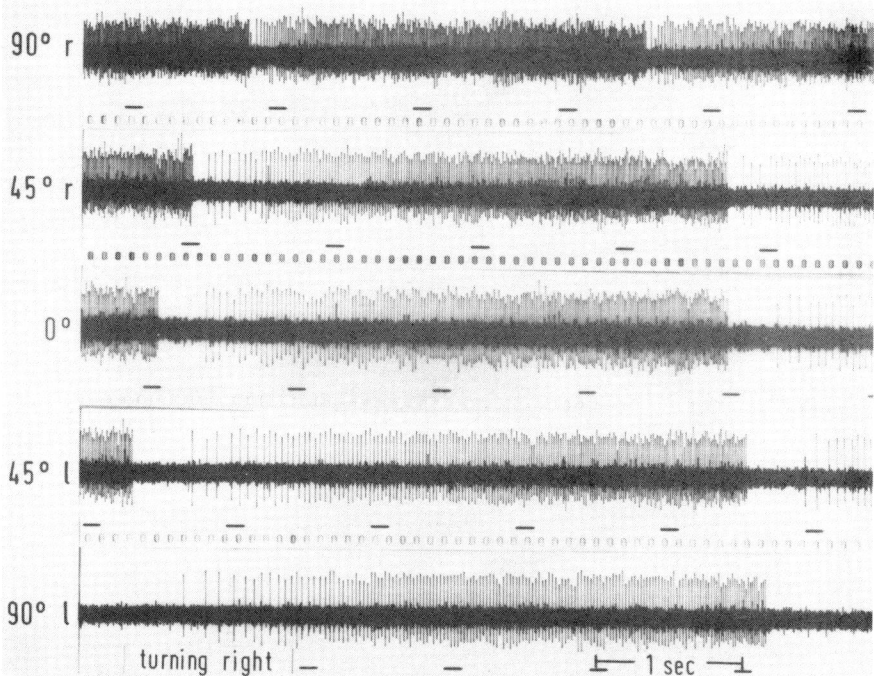

Fig. 2. Influence of head-position on an oculomotor neuron in the nucleus of the internal rectus on the right side during optokinetic nystagmus (stimulation at 5 degree/sec). A change in basic firing frequency and in length of slow phase activation depending on head-position can be noted. Turning right is equivalent to optokinetic stimulation towards the left.

Recordings from reticular formation neurons. According to Peterson (1977) the dorsal part of the nucleus reticularis gigantocellularis and the nucleus reticularis pontis caudalis are involved in eye-head coordination. Duensing and Schaefer (1960) also found neurons in these areas that discharged during eye and head movements. Recently, we recorded more than 50 neurons in these structures, the adjacent part of the FLP, and in the nucleus praepositus hypoglossi which responded directionally specific to vestibular and optokinetic stimulation. 80% of those neurons were H-neurons (main afferent from a horizontal semicircular canal) and 20% were A- or P-neurons (main afferent from a vertical canal). The H-neurons were mostly of the type II (main afferent from contralateral labyrinth). Of these, about 50% displayed a nystagmic modulation of discharge whereas the other half only increased or decreased the firing rate in a tonic fashion during optokinetic and vestibular nystagmus.

Practically all H-neurons were influenced by head-deviations to the left and to the right. Most type II neurons of the right reticular formation discharged at a higher rate when the head was in a left position and decreased their firing level during head-positions to the right. A few type II neurons were either activated or inhibited in both head-positions (Fig. 3). The rarely encountered type I neurons in this area were activated during head-positions towards the ipsilateral side.

The tonic influences of neck-proprioceptive afferents were also present during vestibular and optokinetic nystagmus. They added their effects to the ones seen during ocular nystagmus with the head in the midposition.

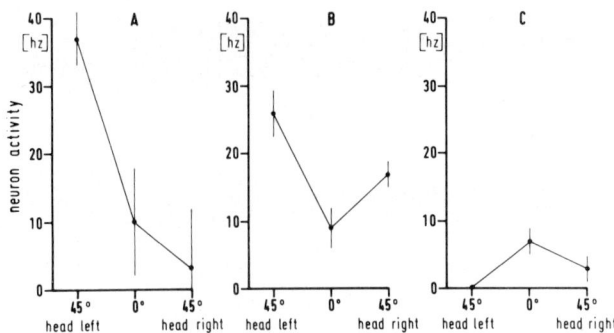

Fig. 3. Discharge rates of 3 reticular formation neurons at 3 head-positions. 10 measurements were taken at each head-position. SDs are indicated. (A) Type II neuron in the paramedian region of the ventral reticular formation right. The cell increases its firing rate when the head is in a position towards the contralateral side and decreases the discharge frequency when the head is in an ipsilateral position. (B) Neuron in the paramedian area of the dorsal reticular formation showing increases of discharge rate in both laterally deviated head-positions. (C) Neuron in the ventral reticular formation displaying a decrease of firing rate in right and left positions of the head.

Recordings from vestibular nuclei neurons. With certain modifications the basic types of discharge patterns found in the reticular formation were also encountered in the vestibular nuclei. In both brain stem structures neurons were present that responded to vestibular as well as to optokinetic stimulation and neurons that discharged in synchrony with nystagmus as well as neurons lacking such a correlation. More distinct differences between reticular and vestibular neurons were seen with respect to the influence of different head-positions on discharge rate and pattern.

The type H Ib neurons (excited during acceleration to ipsilateral side, inhibited during fast phase of nystagmus) of the rostral part of the nucleus vestibularis medialis (NVM) were not significantly influenced by the head-position whereas the corresponding neuron type in the reticular formation was clearly affected. Only some H Ib neurons of the NVM displayed slight influences which did never modify the spontaneous discharge level by more than 30%. These changes due to variations of head-position were somewhat easier to detect during vestibular and optokinetic stimulation. Type H IIa neurons (excited by acceleration to contralateral side, additional activation during fast phase of nystagmus) of the NVM which discharged somewhat more irregularly than the above were completely uninfluenced by the different positions of the head.

Neurons in the nucleus vestibularis lateralis (NVL) were not usually nystagmus-modulated. The influence of head-position on such cells varied. The irregularly and with low frequency discharging neurons of the dorsocaudal part of the NVL (Deiters' gamma, hindlimb region) were hardly ever sensitive to the head-position. Neurons of the rostromedial part (Deiters' alpha, forelimb region) that are more sensitive to optokinetic and vestibular stimulation could more clearly be affected by the head-position. In the NVL several A- and P-neurons were encountered. Some of them also were influenced by head-position when this stimulation occurred in the appropriate spatial plane.

Our sample of neurons from the nucleus vestibularis superior is too small to allow any conclusions.

All neurons of the vestibular nuclei behaved according to the rule that also holds for the reticular cells: if the basic frequency or the firing level during the slow phase of ocular nystagmus was, for example, increased by acceleration to the right a head-position towards the right also induced a higher discharge rate.

Neuronal discharge pattern during head movements

If an animal's head is not fixated as in the above experiments, but free to move one can compare the effects of active and passive movements on neuronal behavior. Speed and amplitude of such head movements have a significant influence on the eye movements occurring. A head nystagmus, for example, causes the ocular nystagmus to become more irregular than seen with fixated head (Fig. 4). As such conditions are relevant for an understanding of naturally occurring phenomena, we studied the discharge patterns of neurons located in the structures also investigated while the head was fixated under such circumstances. The observations made could partly be predicted from the findings obtained with fixated head.

Recordings from oculomotor neurons. Passive and active head movements to the left induced an activation of neurons in the left nucleus rectus medialis. The degree of activation was dependent on the speed of head movement. Initially there was an inhibition of discharges during active left movements of the head which correlated with an eye movement towards the right ("triggered" eye-head coordination according to Bizzi). The activation only occurred with the onset of the compensatory backward movement. This activation started out with a high rate of discharges and was lacking position coding.

Depending on the level of vigilance of the animal various modifications were observed.

Recordings from reticular formation neurons. Duensing and Schaefer (1960) obtained first insights into neuronal activity distribution in the reticular formation during active

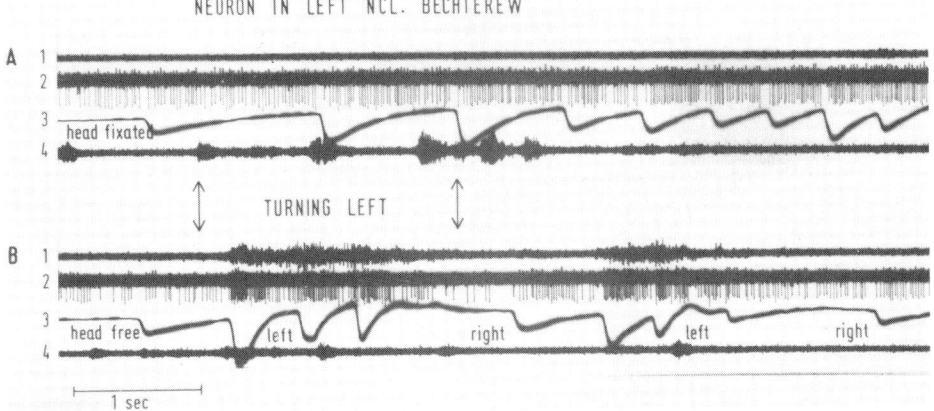

Fig. 4. Neuron in left nucleus vestibularis superior during acceleration of 6 degrees/sec² towards the left with the eyes open. 1, Myogram of neck-musculature; 2, discharge of neuron; 3, nystagmogram; 4, voice channel. A) Turning with the head fixated. The neuron displays an increase of discharge rate that is interrupted slightly before and during fast phases of ocular nystagmus. B) Turning with the head free. Now ahead nystagmus to the left can be seen in the myogram. The additional fast left movement of the head results in amplitude modulations of the ocular nystagmus. During fast phases of head nystagmus the speed of the slow phase of ocular nystagmus is increased. This may be considered a "return eye movement" (Cohen et al., 1967).

head movements. Those investigations were carried on and 30 new neurons have been recorded from. These are in part identical to the ones mentioned in the first section of this paper.

During passive head movements some neurons were excited or inhibited during movements in all directions. Most cells responded directionally specific. They were activated by head movements towards the contralateral side (type II). The response to active movements varied. This is understandable from the fact that agonists and antagonists of the neck muscles do not always act in a reciprocal way. Nevertheless, 60% of the neurons that were excited during passive head movements displayed an activation with active head movements towards the ipsilateral side and an inhibition of firing rate during such movements to the side contralateral to their location in the brain stem. The more rarely found opposite type of neuron (type I, afferents from ipsilateral labyrinth) behaved in an opposite way to the above described one during active as well as during passive head movements.

Thus our sample of reticular formation cells was reciprocally activated during active and passive head movements in the same direction.

Recordings from vestibular nuclei neurons. The discharge patterns of nystagmus-correlated neurons in the NVM were somewhat similar to the ones of the ipsilateral nucleus rectus internus. The vestibular neurons were less regular and modulation during head nystagmus could only be observed in some cases. Neurons located on the left side were mostly activated by passive head movements to the left and displayed an inhibition during the fast phase of ocular nystagmus. During active head movements to the left, which were accompanied by an initial eye movement towards the same side, activation only began with the onset of the compensatory backward movement of the eyes. Fast and abruptly occurring head movements caused the discharge rate to be very high from the beginning. Inhibitory phases could not be distinguished anymore under such conditions. With compensatory head movements a strong inhibition was observed. Position coding which was more or less clearly seen with slow nystagmus phases was not found then.

Neurons of the NVL and the nucleus vestibularis superior were less correlated with eye movements and discharged in a more continuous way during head movements. These clearly demonstrated that vestibular nuclei neurons responded similarly during active and passive head movements in the same direction. Such cells thus differed from reticular formation units. Apart from some rare exceptions type I neurons in the left vestibular nuclei were excited during passive and active head movements towards the left. Units of the type II behaved reciprocally to type I cells in both situations (see Fig. 4).

DISCUSSION

Considering the complexity of motor phenomena involved in eye-head coordination it has to be expected that discharge patterns of brain stem neurons mediating such mechanisms cannot be uniform. Nevertheless it is obvious that neurons of the left nucleus rectus medialis, units of the type II in the right reticular formation, and left vestibular nucleus cells of type I have several properties in common. Apart from their responsiveness to vestibular stimulation also the influence of neck-proprioception is qualitatively similar. Neck-proprioception exerts more distinct influences on reticular formation cells than on neurons of the vestibular nuclei; It can therefore be assumed that reticular formation neurons are responsible for mediating tonic neck reflexes on the eyes as well as for the change in the beating field of nystagmus during different head-positions.

Cells of the structures mentioned are also excited with directionally specific characteristics during active head movements. Excitations and inhibitions of discharge occur in particularly clear ways during compensatory backward movements of the eyes. This movement serves to stabilize the eyes in space and has the speed of the head movement. In contrast to slow phases of nystagmus position coding of excitations is often absent. The discharge rate is usually very high from the onset of the return movement on. One can therefore assume a speed-code to be present. This is already visible in neurons of the nucleus vestibularis medialis which project to the oculomotor nuclei via the MLF. According to Lorente de Nó (1933) and Szentágothai (1964) such cells and connections represent a direct reflex arc for the mediation of eye movements during head movements.

Neurons of the reticular formation and of vestibular nuclei which discharge independently of ocular nystagmus differ in one important aspect. During both active and passive head movements vestibular nucleus cells respond with either an increase or a decrease in firing rate. That is: such cells do not distinguish between active and passive movements; they are just driven by labyrinthine afferents. From the vestibular nuclei the information is conducted to the reticular formation as can be seen during passive head movements. But during active head movements labyrinthine drives are not noted on reticular neurons. Their activity pattern is reciprocal to the one present with a passive head-movement. This suggests the reticular formation neurons to be the site at which voluntary commands are fed into postural control circuits.

SUMMARY

Neurons of the oculomotor nucleus, of several areas of the reticular formation, and of the vestibular nuclei were investigated during ocular nystagmus while the head was fixated in different positions and during active and passive head movements. The findings allow the following statements that hold for the vast majority of units encountered:

1. If a cell receives vestibular and neck-proprioceptive afferents, its firing rate will be increased when the head is fixated in a position towards the same side as to which accelerative stimulation causes a raise in discharge frequency.

2. Neck-proprioceptive influences on vestibular neurons are significantly weaker than those on reticular formation and oculomotor neurons.

3. Vestibular nucleus cells respond in the same manner during active and passive head movements.

4. Reticular formation neurons respond in reciprocal ways during active and passive head movements, suggesting this brain stem area to be the location at which commands from higher centers interact with postural control circuits.

5. Oculomotor neurons behave alike to vestibular nuclei neurons in many respects.

REFERENCES

Bárány, R. (1918) Uber einige Augen- und Halsmuskelreflexe bei Neugeborenen. *Acta oto-laryngol (Stockh.)*, 1: 97–102.

Bizzi, E., Kalil, R.E., Morasso, P. and Tagliasso, V. (1972) Central programming and peripheral feedback during eye-head coordination in monkeys. *Bibl. ophthal. (Basel)*, 82: 220–232.

Brodal, A. (1974) Anatomy of vestibular nuclei and their connections. In *Handbook of Sensory Physiology, Vol. VI/1, Vestibular System, Part 1, Basic Mechanisms,* H.H. Kornhuber (Ed.), Springer-Verlag, Berlin-Heidelberg-New York, pp. 239–352.

Cohen, B., Goto, K. and Tokumasu, K. (1967) Return eye movements, an ocular compensatory reflex in the alert cat and monkey. *Exp. Neurol.,* 17: 172–185.

De Kleijn, A. (1921) Tonische Labyrinth- und Halsreflexe auf die Augen. *Pflügers Arch.,* 186: 82–97.

Duensing, F. and Schaefer, K.-P. (1960) Die Aktivität einzelner Neurone der Formatio reticularis des nicht gefesselten Kaninchens bei Kopfwendungen und vestibulären Reizen. *Arch. Psychiat. Nervenkr.,* 201: 97–122.

Ebbesson, S.O.E. (1969) Brain stem afferents from the spinal cord in a sample of reptilian and amphibian species. *Ann. N.Y. Acad. Sci.,* 167: 60–101.

Fredrickson, J.M., Schwartz, D. and Kornhuber, H.H. (1965) Convergence and interaction of vestibular and deep somatic afferents upon neurons in the vestibular nuclei of the cat. *Acta oto-laryngol. (Stockh.),* 61: 168–188.

Graf, W. and Meyer, D.L. (1978) Eye positions in fishes suggest different modes of interaction between commands and reflexes. *J. comp. Physiol.,* 128: 241–250.

Grahe, K. (1922) Über Halsreflexe und Vestibularisreaktionen beim Menschen. *Z. Hals-, Nas.-u. Ohrenheilk.* 3: 550–558.

Harris, A.J. (1965) Eye movements of the dogfish *Squalus acanthias* L. *J. exp. Biol.,* 43: 107–130.

Lorente de Nó, R. (1933) Vestibulo-ocular reflex arc. *Arch. Neurol. Psychiat. (Chic.),* 30: 245–291.

Lyon, E.P. (1900/1901) Compensatory motions in fishes. *Amer. J. Physiol.,* 4: 77–82.

Magnus, R. (1924) *Körperstellung.* Springer-Verlag, Berlin.

Peterson, B.W. (1977) Identification of reticulospinal projections that may participate in gaze control. In *Control of Gaze by Brain Stem Neurons,* R. Baker and A. Berthoz (Eds.), Elsevier, Amsterdam-New York, pp. 143–152.

Pompeiano, O. and Brodal, A. (1957) Spino-vestibular fibres in the cat. An experimental study. *J. comp. Neurol.,* 108: 353–378.

Schaefer, K.-P., Zierau, H. and Süß K.J. (1977) Differentiation of neuronal activity in the vestibular nuclei of rabbits. In *Control of Gaze by Brain Stem Neurons,* R. Baker and A. Berthoz (Eds.), Elsevier, Amsterdam-New York, pp. 257–260.

Schaefer, K.-P., Meyer, D.L., Büttner, U. and Schott, D. (1975) The effect of head position on oculomotor discharge patterns in rabbits. In *Basic Mechanisms of Ocular Motility and their Clinical Implications,* G. Lennerstrand and P. Bach-y-Rita (Eds.), Pergamon Press, Oxford-New York, pp. 457–459.

Suzuki, J. (1972) Vestibular and spinal control of eye movements. *Bibl. ophthal. (Basel),* 82: 109–115.

Suzuki, J. and Takemori, S. (1971) Eye movements induced from spinal nerves. *Equilibr. Res.,* Suppl. 2: 33–40.

Szentágothai, J. (1964) Pathways and synaptic articulation patterns connecting vestibular receptors and oculomotor nuclei. In *The Oculomotor System,* M. Bender (Ed.), Hoeber, New York-Evanston-London, pp. 205–223.

Takemori, S. (1969) A study on eye deviations from neck torsion in humans. *Jap. J. Otol. (Tokyo),* 72: 75–83.

Takemori, S. (1971) A study on eye deviations from neck torsion. *Pract. Otol. (Kyoto),* 64, 1361–1368.

Takemori, S. and Suzuki, J. (1971) Eye deviations from neck torsion in humans. *Ann. oto-rhin.-laryngol. (St. Louis),* 80: 439–444.

Subject Index

Abducens internuclear neurons
  connections, afferent
    from ipsilateral pontine reticular formation, 621
  connections, efferent
    to contralateral medial rectus division of oculomotor nucleus, 621
  functional aspects, responses to vestibular stimulation, 436
Abducens nucleus motoneurons
  connections, afferent
    from contralateral medullary reticular formation, 626
    from ipsilateral pontine reticular formation, 620
    from perihypoglossal nuclei, 640, 657
  functional aspects
    responses to both neck and vestibular (semicircular canal) stimulation, 561
    responses to neck stimulation, 552, 563
    responses to vestibular (semicircular canal and otolith) stimulation, 460, 551
Accessory oculomotor areas (rostral interstitial nucleus of the MLF and interstitial nucleus of Cajal)
  connections, afferent
    from brain stem reticular formation, 621
    from contralateral accessory oculomotor areas, 625
    from perihypoglossal nuclei, 660
    from vestibular nuclei, 621
  connections, efferent
    to contralateral accessory oculomotor areas, 625
    to extraocular motor nuclei, 131, 624
  functional aspects, influences on eye movements, 131, 660
Accessory optic system
  anatomy, 705, 715
  functional aspects
    effects of lesion of accessory optic system on optokinetic nystagmus, 707
    relation to vestibulocerebellum, 719
    responses to visual stimulation, 705, 717
Acoustic system
  functional aspects, influences on cerebellar cortex, 316
α-Motoneurons

α-Motoneurons–*cont.*
  contribution of different size motoneurons to Renshaw cell discharge, 52
  excitatory and/or inhibitory responses to stretch of agonistic muscles, 32, 37, 41
  influences from brain stem reticular formation, 128
  influences from interstitial nucleus of Cajal, 130
  influences from vestibular nuclei, 123
  inhibitory responses to stretch of antagonistic muscles, 40, 41
  reciprocal group Ia inhibition, 11
  recruitment of motoneurons according to presynaptic organization of synaptic input, 61
  recruitment of motoneurons according to size, xix, 61
  recurrent inhibition by Renshaw cell discharge, 46
  responses of different size motoneurons to muscle vibration, 45, 47
  responses of different size motoneurons to static muscle stretch, 50
  responses of homonymous motoneurons to stimulation of both group Ia and Ib afferents, 32
  responses of homonymous motoneurons to stimulation of group Ia afferents, 32
  responses of homonymous motoneurons to stimulation of group Ib afferents, 32
  sensitivity of different size motoneurons to Renshaw inhibition, 54
α–γ Linkage
  during locomotion, 148
  during the vestibulo-collic reflex, 546
Autogenetic excitation
  autogenetic excitation of motoneurons during muscle stretch, 37, 39, 41, 45, 50
    during muscle vibration, 45, 47
  autogenetic excitation of motoneurons from group Ia afferents, 32
Autogenetic inhibition
  autogenetic postsynaptic inhibition of motoneurons from group Ia afferents, 32, 40, 41
  autogenetic postsynaptic inhibition of motoneurons from group Ib afferents, 32

Autogenetic inhibition—*cont.*
   autogenetic presynaptic inhibition of group Ia afferents, 34, 36
   autogenetic recurrent inhibition of motoneurons, 46, 54
3-Acetylpyridine
   destruction of inferior olive by 3-acetylpyridine, 600, 759
   effects of 3-acetylpyridine on cerebellar corticovestibular inhibition 759
   effects of 3-acetylpyridine on discharge of corticocerebellar and vestibular neurons in compensated hemilabyrinthectomized animals, 601

Basal ganglia
   influences on locomotion, 148
   influences on posture, xx
β-Motoneurons
   β-dynamic motoneurons, 7
   co-excitation of muscle spindle and tendon organ afferents by β-motoneurons, 34
Body rotation
   influences on brain stem reticular formation, 385 (in pigeon)
   influences on vestibular nuclei, 385 (in pigeon)
Brain stem reticular formation, *see* Reticular formation

Caudal vestibulospinal tract
   origin and distribution within the cord, 123
Central pattern generator, *see* Locomotion
Cerebellum
   connections, afferent
     from lateral reticular nucleus, 80, 501
     from perihypoglossal nuclei, 657
   connections, efferent
     to perihypoglossal nuclei, 656
     to spinal cord, from cerebellar nuclei, 269
     to vestibular nuclei, from cerebellar cortex, 453, 517, 524
   functional aspects
     activity of cerebellar cortex neurons
       during saccadic eye movements, 316, 385, 645
       during smooth pursuit eye movements, 385
     activity of cerebellar nuclei neurons during saccadic eye movements, 645
     effects of destruction of inferior olive on cerebellar corticovestibular inhibition, 758, 759
     functional organization in sagittal zones, 83, 517
     influences from acoustic afferents, 316
     influences from both neck and vestibular

Cerebellum—*cont.*
     (otolith) afferents, 522
     influences from extraocular muscle afferents, 296, 317, 346
     influences from neck afferents
       on cerebellar vermis, 518, 522
       on vestibulocerebellum, 555
     influences from vestibular (otolith) afferents, 520, 522
     influences from vestibular (semicircular canal) afferents, 385 (in pigeon)
     influences from visual afferents, 316, 385
     influences on eye movements, 315, 453
     influences on vestibular nuclei,
       cerebellar corticovestibular, 82, 83, 757
       fastigiovestibular, 123
Cerebral cortex
   connections, afferent
     thalamocortical, 116, 575, 583
   connections, efferent
     corticoreticular, 80
     corticospinal to upper cervical cord, 264
   functional aspects
     activity of cortical neurons during movements, 141, 142
     influences on dorsal column nuclei, 165
     influences on dorsal spinocerebellar tract, 165
     influences on primary afferents (PAD), 165
     influences on spinocervical tract, 165
     influences on spinoolivocerebellar tracts, 86
     influences on spinoreticular tract (bVFRT), 82, 165
     influences on ventral spinocerebellar tract, 79, 165
     muscle fields of cortical neurons during movements, 138
     responses to both neck and vestibular stimulation, 570
     responses to both visual and vestibular stimulation, 584
     responses to electrical vestibular stimulation, 567
     responses to natural vestibular (semicircular canal and otolith) stimulation, 567, 569, 570, 583
     responses to neck stimulation, 570
     responses to stimulation of group Ia muscle and low threshold cutaneous afferents via thalamus, 116
Cervical spinocerebellar tract of central cervical nucleus
   influences from both neck and vestibular (semicircular canal) afferents, 532
   influences from neck afferents, 531, 532
   influences from vestibular (semicircular canal)

Cervical spinocerebellar tract of central cervical nucleus—*cont.*
    afferents, 532
Cervical spinoreticular tract, 501
    influences from both neck and vestibular (otolith) afferents, 504
    influences from lateral vestibulospinal tract, 81, 503
    influences from neck afferents, 504
    influences from vestibular (otolith) afferents, 169, 502, 503
Cervico-abducens reflex, *see* Neck receptors
Colliculus, superior
    connections, efferent
        tectoreticular, 80, 729, 739
        tectospinal, 269, 270, 739
        tectotectal, 740
        to perihypoglossal nuclei, 657
    connections, afferent
        from brain stem reticular formation, 740
        from contralateral superior colliculus, 740
        from perihypoglossal nuclei, 660
        from sensory trigeminal nuclei, 740
        from substantia nigra, 740
    functional aspects
        eye movements produced by stimulation of superior colliculus 729, 745
        head movements produced by stimulation of superior colliculus, 745
        influences from both extraocular muscle and visual afferents, 335, 338–340
        influences from both visual and vestibular (semicircular canal) afferents, 735, 738
        influences from extraocular muscle afferents, 296, 325, 337, 346
        influences from vestibular afferents
            on tectoreticular neurons, 737
            on tectospinal neurons, 736
        influences from vestibular (semicircular canal) afferents, 735–738
        influences from visual afferents, 335, 337–340, 735–738, 745
        influences from visual afferents
            on tectoreticular neurons, 737
            on tectospinal neurons, 736
        influences on neck muscle motoneurons, 258
        influences on perihypoglossal nuclei, 654
        unit activity during saccadic eye movements, 725
Commissural pathways, *see* Nuclei, vestibular
Convergence of extraocular muscle and visual afferents
    on superior colliculus, 335, 338–340
Convergence of neck and vestibular (semicircular canal and otolith) afferents
    influences on abducens motoneurons, 553, 561

Convergence of neck and vestibular (semicircular canal and otolith) afferents—*cont.*
    influences on cerebral cortex, 570
    influences on cervical spinocerebellar tract of central cervical nucleus, 532
    influences on cervical spinoreticular tract, 504
    influences on lateral reticular nucleus, 504
    influences on posture, 494, 525
    influences on Purkinje cells of cerebellar vermis, 522
    influences on spinal reflexes, 525
Convergence of vestibular (semicircular canal and otolith) afferents
    on cerebral cortex, 567, 569, 570, 583
    on extraocular motoneurons, 443–445, 459, 460, 551
    on vestibular nuclei, 378
Convergence of visual and vestibular (semicircular canal) afferents
    on cerebral cortex, 584
    on eye movements
        before flocculectomy, 695
        after flocculectomy, 697
    on Purkinje cells of cerebellar cortex, 385, 698, 757
    on superior colliculus, 735, 738
    on thalamic nuclei, 584
    on vestibular nuclei, 687, 688, 710
    on vestibulo-ocular reflexes
        before flocculectomy, 695, 703, 757, 771
        after flocculectomy, 757
Coordination of eye-head movements
    produced by a stationary target, 795
    produced by stimulation of superior colliculus, 745
    produced by stimulation of vestibular afferents, 795, 799
    produced by stimulation of visual afferents, 795, 799
    produced during smooth pursuit eye movements, 797
    unit activity in brain stem during eye-head movements, 805
Corticospinal tract
    origin and distribution to spinal cord, 264
    functional aspects
        activity of corticospinal tract neurons during movements, 141, 142
        influences on dorsal column nuclei, 165
        influences on dorsal spinocerebellar tract, 165
        influences on spinocervical tract, 165
        influences on spino-olivocerebellar tracts, 86
        influences on spinoreticular tract (bVFRT), 82, 165
        influences on spinothalamic tract, 163

influences on ventral spinocerebellar tract, 79, 165
   muscle fields of corticospinal tract neurons during movements, 138
Cristae ampullares, see Receptors, vestibular
Cuneocerebellar tract, 79
Cutaneous afferents
   influences on motor cortex via area 3A, 116
   influences on motor cortex via thalamus, 116
   influences on thalamic nucleus ventralis posterolateralis, 116

Descending propriospinal projections, 269
Descending supraspinal projections
   origin and distribution to upper cervical cord
      from brain stem reticular formation, 266, 272
      from cerebellar nuclei, 269
      from cerebral cortex, 264
      from dorsal column nuclei, 266
      from interstitial nucleus of Cajal, 266
      from locus coeruleus region, 269
      from mesodiencephalic structures, 269
      from nucleus of solitary tract, 266
      from raphe nuclei, 266
      from red nucleus, 266
      from superior colliculus, 269, 270
      from vestibular nuclei, 266
   functional aspects
      influences on dorsal column nuclei, 165
      influences on dorsal spinocerebellar tract, 165
      influences on FRA interneurons, 21
      influences on γ-motoneurons, 86, 147
      influences on primary afferents, 165
      influences on spinocervical tract, 165, 166
      influences on spinoreticular tract (bVFRT), 165, 166
      influences on spinothalamic tract, 163, 166
      influences on ventral spinocerebellar tract, 165, 167
Descending supraspinal projections, monoaminergic origin
      from raphe nuclei, 165, 266
      from locus coeruleus region, 269
   functional aspects
      influences on FRA interneurons, 82
      influences on spinoreticular tract (bVFRT), 82, 165
      influences on spinothalamic tract, 166
Development of visual processes
   by extraocular muscle afferents, 347
DOPA
   influences on γ-static motoneurons, 148
   influences on locomotion, 227
   influences on spinal reflexes, 14
Dorsal column nuclei

Dorsal column nuclei–*cont.*
   connections, efferent
      to upper cervical cord, 266
   functional aspects
      influences from brain stem reticular formation, 168
      influences from cerebral cortex, 165

Efferent vestibular system, see Nerve, vestibular
Extraocular motor nuclei and neighbouring regions,
   connections, afferent
      from internuclear neurons, 621
      from medullary reticular formation, 626
      from mesencephalic reticular formation, 621
      from paramedian pontine reticular formation, 620
      from perihypoglossal nuclei, 640, 657
      from vestibular nuclei, 431
   connections, efferent
      to perihypoglossal nuclei, 654
   functional aspects
      characteristics of extraocular motoneurons, 646
      effects of head movements on discharge of extraocular motoneurons, 809
      effects of head position on discharge of extraocular motoneurons during nystagmus, 806
      responses during vestibular or optokinetic nystagmus, 806
      responses to stimulation of the brain stem reticular formation, see Paramedian pontine reticular formation, Reticular formation, Saccadic eye movements
      responses to stimulation of the interstitial nucleus of Cajal, 131
      responses to vestibular (otolith) stimulation, 434, 443–445, 459, 460, 551
      responses to vestibular (semicircular canal) stimulation, 431, 433, 436, 443–445, 451, 452, 459, 460, 551, 553, 561, 562
      responses to visual stimulation, see Optokinetic reflex
      saccadic related discharge, 645
Extraocular muscle afferents
   connections, afferent
      central course in ungulata, 291
      peripheral course in ungulata, 291
   functional aspects
      influences on cerebellar cortex, 296, 317, 346
      influences on development of visual processes, 347
      influences on mesencephalic tegmentum, 296, 325, 337, 346

Extraocular muscle afferents—*cont.*
    influences on semilunar trigeminal ganglion in ungulata, 291
    influences on sensory trigeminal nuclei in ungulata, 294
    influences on superior colliculus, 296, 325, 335, 337–340, 346
    influences on ventrobasal complex of thalamus, 296
    influences on visual cortex, 342, 347
    properties, 306
Eye movements
    induced by accessory oculomotor areas, 660
    induced by cerebellum, 315, 453
    induced by combined visual and vestibular (semicircular canal) stimulation, 695, 703, 771
    induced by flocculus, 453
    induced by perihypoglossal nuclei, 660
    induced by stimulation of the brain stem reticular formation, *see* Paramedian pontine reticular formation, Reticular formation
    induced by vestibular (semicircular canal) stimulation, 695, 703, 771
    induced by visual stimulation (optokinetic response), 605, 703, 771
    premotor areas controlling horizontal eye movements, 620, 626
    premotor areas controlling vertical eye movements, 621
Eye muscle receptors
    muscle spindles
        morphological properties, 301
        physiological properties, 308
    tendon organs
        morphological properties, 303
        physiological properties, 310

Fall
    afferent responses
        from primary and secondary endings of muscle spindles, 425
    efferent motor responses
        after blindfolding, 424
        after labyrinthectomy, 205
        hemispheric dominance during fall, 237
        in darkness, 205
        in normal visual conditions, 204, 424
        reflex compensation of asymmetric motor responses during fall, 237
Fixation
    coordination eye-head movements induced by a stationary target, 795
    unit activity in perihypoglossal nuclei during fixation, 670
Flexion reflex afferents (FRA)
    influences from descending monoaminergic projections on FRA interneurons, 82
    influences from descending supraspinal projections on FRA interneurons, 21
    influences on dorsal spinocerebellar tract, 13
    influences on spinoolivocerebellar tracts, 13
    influences on spinoreticulocerebellar tract, 13, 80, 501
    influences on spinothalamic tract, 105, 106
    influences on ventral spinocerebellar tract, 13
    interneurons in the FRA pathway, 13
    reflex pathways from FRA to motoneurons, 13, 18, 86
    reflex pathways from FRA to primary afferents, 86
Flocculonodular lobe, *see* Flocculus
Flocculus
    connections, efferent
        to perihypoglossal nuclei, 656
        to vestibular nuclei, 453
    connections, afferent
        from perihypoglossal nuclei, 657
        from visual system, 705, 715
    functional aspects
        effects of flocculectomy on adaptive changes of the vestibulo-ocular reflex during combined visual and vestibular stimulation, 757
        effects of flocculectomy on dynamic characteristics of the responses of eye movements to combined visual and vestibular (semicircular canal) stimulation, 696
        influences on eye movements, 453
        influences on eye movements after destruction of inferior olive, 758
        influences on oculomotor responses to visual and vestibular (semicircular canal) stimulation, 696–698, 704
        influences on vestibulo-ocular reflex arcs, 453, 454
        relation of vestibulocerebellum to accessory optic system, 719
        responses to neck stimulation, 555
        responses to optokinetic stimulation, 698, 699, 757
        responses to vestibular stimulation, 453, 698
        responses to vestibular (semicircular canal) stimulation, 384, 698, 757

γ-Motoneurons
    γ-dynamic motoneurons, 3
        responses of γ-dynamic motoneurons to stimulation of rubrobulbospinal tract, 86
    γ-static motoneurons, 3
        effects of DOPA on γ-static motoneurons, 148

γ-Motoneurons—*cont.*
  recurrent inhibition of γ-static motoneurons, 55
  activity of γ-motoneurons during locomotion, 147
  co-excitation of muscle spindle and tendon organ afferents by γ-motoneurons, 34
  distribution of fusimotor axons to intrafusal muscle fibers, 3
  responses of primary endings to stimulation of γ-motoneurons, 5, 7
  supraspinal influences on γ-motoneurons, 147
Globus pallidus
  influences on locomotion, 148
  influences on posture, xx
Golgi tendon organs
  convergence of primary muscle spindle and Golgi tendon organ afferents on spinal motoneurons, 32
  influences on homonymous motoneurons, 32
  reflex pathways from tendon organ afferents, 29
  responses to fusimotor stimulation, 2, 34
  responses to muscle contraction, 160

Head movements
  induced by stimulation of superior colliculus, 745
  influences on brain stem reticular neurons during vestibular and optokinetic nystagmus, 809
  influences on extraocular motoneurons during vestibular and optokinetic nystagmus, 809
  influences on optokinetic nystagmus, 809
  influences on vestibular nuclei neurons during vestibular and optokinetic nystagmus, 810
Head position
  influences on brain stem reticular neurons during vestibular and optokinetic nystagmus, 807
  influences on extraocular motoneurons during vestibular and optokinetic nystagmus, 806
  influences on optokinetic nystagmus, 806
  influences on vestibular nuclei neurons during vestibular and optokinetic nystagmus, 808
Head posture
  after hemilabyrinthectomy in darkness, 783
  after hemilabyrinthectomy in light conditions, 783
  in intact animals, 783
Hemispheric dominance
  asymmetric motor responses during fall, 237
  reflex compensation of asymmetric motor responses during fall, 237

Inferior olive
  connections, afferent

Inferior olive—*cont.*
  rubroolivary tract, 86
  spinoolivary tract, 83
  connections, efferent
    olivocerebellar tract, 83
  functional aspects
    destruction of inferior olive by 3-acetylpyridine, 600, 759
    effects of destruction of inferior olive on cerebellar corticovestibular inhibition, 758, 759
    effects of destruction of inferior olive on eye movements induced by stimulation of flocculus, 758
    influences from visual afferents, 455
    influences on the vestibulo-ocular reflex, 455
Interaction, *see* Convergence
Interneurons
  convergence of group Ia and Ib afferents on spinal interneurons, 29
  effects of limb position on FRA interneurons, 18
  influences from descending monoaminergic pathways on FRA motoneurons, 82
  influences from descending pathways on FRA interneurons, 21
  inhibitory interneurons in autogenetic group Ia pathway, 29
  inhibitory interneurons in autogenetic group Ib pathway, 29
  inhibitory interneurons in group Ia pathway to motoneurons of the antagonistic muscles, 11
  interneurons in the FRA pathway, 13
  interneurons involved in locomotion, 16
  mutual inhibition between interneurons exciting flexors and extensors, 15, 20
  recurrent inhibition of group Ia inhibitory interneurons, 55
Interstitiospinal tract
  origin and distribution within spinal cord, 130, 266
  influences on back motoneurons, 130, 131
  influences on limb motoneurons, 130, 131
  influences on neck muscle motoneurons, 130, 131

Labyrinth
  macular input
    influences on cerebral cortex, 569, 583
    influences on extraocular motoneurons, 434, 443–445, 459, 460, 551
    influences on forelimb extensors, 413
    influences on lateral reticular nucleus, 502, 504
    influences on lateral vestibulospinal tract

Labyrinth—*cont.*
  neurons, 122
    influences on medial vestibulospinal tract neurons, 123
    influences on neck muscle motoneurons, 123, 124, 409, 410
    influences on posture, 204, 493, 515, 525
    influences on primary vestibular afferents, 359
    influences on Purkinje cells of the cerebellar vermis, 520, 522
    influences on reticulospinal tracts neurons, 130
    influences on spinal reflexes, 212, 214, 215
    influences on spinoreticulocerebellar tract neurons, 169, 502-504
    influences on vestibular nuclei, 122, 123, 376
  semicircular canal input
    influences on brain stem reticular formation, 285 (in pigeon)
    influences on cerebellar cortex (flocculus), 384, 698, 757
    influences on cerebellar nuclei, 385 (in pigeon)
    influences on cerebral cortex, 567, 569, 570, 583
    influences on cervical spinocerebellar tract neurons on the central cervical nucleus, 532
    influences on extraocular motoneurons, 431, 433, 436, 443-445, 451, 459, 460, 551, 553, 561, 462
    influences on eye movements, 695, 703 771
    influences on fastigial nucleus, 384
    influences on lateral vestibulospinal tract neurons, 122, 407
    influences on limb extensors, 122, 124, 126, 413
    influences on medial vestibulospinal tract neurons, 123, 407
    influences on neck muscle motoneurons, 122–124, 405, 407
    influences on neck muscles, 126, 537, 543, 546
    influences on perihypoglossal nuclei, 640, 654, 673
    influences on primary vestibular afferents, 357
    influences on reticulospinal tract neurons, 130
    influences on superior colliculus, 735, 738
    influences on thalamus, 583
    influences on vestibular nuclei, 122, 123, 372, 384, 684, 710, 763, 385 (in pigeon)

Labyrinthectomy
  unilateral
    compensation of labyrinthine deficits
      by somatosensory input, 591
      by visual input, 783
    decompensation by cerebellectomy, 593, 601
      by transverse section of the spinal cord, 601
      by i.p. administration of 3-acetylpyridine, 601
    influences on head posture in darkness, 783
    influences on head posture in light conditions, 783
    influences on postural asymmetry produced by
      unilateral deafferentation of the neck, 508
      unilateral lesion of lateral reticular nucleus, 508
    labyrinthine deficits, 783, 591
    synaptic mechanisms involved in compensation of labyrinthine deficits, 607
    unit activity of lateral vestibular neurons in compensated and decompensated animals, 599
  bilateral
    influences on motor responses to fall, 205
Landing
  responses of primary and secondary endings of muscle spindles, 425
Lateral reticular nucleus
  connections, afferent
    corticoreticular, 80
    rubroreticular, 80
    spinoreticular (bVFRT), 80, 501
    tectoreticular, 80
    vestibuloreticular, 503
  connections, efferent
    reticulocerebellar, 80, 501
  functional aspects
    convergence of neck and vestibular (otolith) afferents, 504
    effects of lesion of lateral reticular nucleus on placing reaction, 506, 511
    effects of lesion of lateral reticular nucleus on posture, 506, 508
    influences from neck afferents, 504
    influences from vestibular (otolith) afferents, 502, 504
Lateral vestibulospinal tract, *see* Nuclei, vestibular, lateral
Linear motion, *see* Fall
Locomotion
  α–γ linkage during locomotion, 148
  central control of reflex transmission during locomotion, 230
  central pattern generator, 15, 148, 227

Locomotion—*cont.*
  depression of peripheral effects on spino-
    reticular tract (bVFRT) neurons during
    locomotion, 82
  effects of DOPA on locomotion, 227
  influences from cerebellum, 82
  influences from globus pallidus, 148
  influences from mesencephalon, 148
  influences on γ-motoneurons, 147
  influences on ventral spinocerebellar tract
    neurons, 79
  interneurons in spinal cord involved in
    locomotion, 16
  peripheral control of locomotion, 228
  reflex gain of force during locomotion, 238
Locus coeruleus region
  connections, efferent
    to upper cervical cord, 269

Macular of utricle and saccule, *see* Receptors,
    vestibular
Medial vestibulospinal tract, *see* Nuclei, vestibular,
    medial
Monosynaptic spinal reflexes
  influences from labyrinth (otolith) stimulation,
    212, 214, 215, 525
  influences from neck and labyrinth (otolith)
    stimulation, 525
  influences from neck stimulation, 281–283,
    515, 525
  influences from visual (optokinetic)
    stimulation, 213–215
Movement
  activity of corticospinal tract neurons during
    movement, 141, 142
  muscle fields of corticospinal tract neurons
    during movement, 138
  postural responses during movement, xix, 219
Muscle contraction
  responses of primary and secondary endings of
    muscle spindles, 155
Muscle afferents
  convergence of group Ia and Ib afferents on
    spinal interneurons, 29, 32
  influences from group I muscle afferents on
    motor cortex via area 3A, 113
  influences from group II muscle afferents on
    motor cortex via thalamus, 116
  influences of group Ia afferents on homo-
    nymous motoneurons, 32, 37, 41
  influences of group Ia afferents on moto-
    neurons to antagonistic muscles, 11, 40, 41
  influences of group Ib afferents on homo-
    nymous motoneurons, 32
  influences on nucleus ventralis postero-
    lateralis of thalamus, 116
  reflex pathways from group Ia muscle afferents,

Muscle afferents—*cont.*
    11, 29
  reflex pathways from group Ib muscle
    afferents, 29
Muscle field
  of corticospinal tract neurons, 138
Muscle spindles
  fusimotor innervation of intrafusal muscle
    fibers, 3
  glycogen-depletion method, 3
  muscle spindle apparatus, xix
  nuclear-chain fibers, 3
  type 1 nuclear-bag fibers, 3
  type 2 nuclear-bag fibers, 3
Muscle spindle receptors
  influence on dorsal spinocerebellar tract
    neurons, 98
  responses of primary and secondary endings to
    fall and landing, 425
  responses of primary and secondary endings to
    muscle contractions of different speed, 155
  responses of primary and secondary endings to
    muscle stretch, 45
  responses of primary endings to fusimotor
    stimulation, 5, 7, 34
  responses of primary endings to muscle
    vibration, 45
Muscle stretch
  recruitment of different size motoneurons, 50
  responses of dorsal spinocerebellar tract
    neurons, 93
  responses of motoneurons to stretch of
    agonistic muscles, 32, 37, 41, 45
  responses of motoneurons to stretch of
    antagonistic muscles, 11, 40, 41
  responses of primary and secondary endings of
    muscle spindles, 3, 45
  responses of Renshaw cells, 52
Muscle units
  amplitude of EMG potential, 61
  amplitude of group Ia EPSP in corresponding
    motoneuron, 61
  conduction velocity of corresponding motor
    axon, 61
  fatigue resistance, 61
  force output, 61
  recruitment according to motoneuronal size, 61
  recruitment according to presynaptic
    organization of synaptic input, 61
  relation between group Ia EPSP amplitudes and
    other characteristics of motor units, 63
  size of corresponding motoneuron, 61
  threshold gradation, 61
  twitch contraction time, 61
Muscle vibration
  autogenetic excitation of motoneurons, 45, 47
  responses of different size motoneurons, 45, 47

Muscle vibration—*cont.*
  responses of dorsal spinocerebellar tract neurons, 95
  responses of primary endings of muscle spindles, 45
  responses of Renshaw cells during vibration, 52
  responses of vibrated muscle in man, xx, 150
Myotatic feedback
  adaptive properties, 69
  linear systems analysis, 64
  position control, 70

Neck muscle motoneurons
  cutaneous afferent projections to neck muscle motoneurons, 260
  influences from interstitiospinal tract, 131
  influences from lateral vestibulospinal tract, 122, 123, 407
  influences from medial vestibulospinal tract, 123, 124, 407
  influences from reticulospinal tracts, 128, 258
  influences from vestibular (otolith) afferents, 122–124, 409, 410
  influences from vestibular (semicircular canal) afferents, 123, 124, 126, 405, 407, 537, 543, 546
  influences from tectospinal tract, 258
  localization in upper cervical cord, 255
  muscle afferent projections to neck muscle motoneurons, 258, 545
  trigeminal afferent projections to neck muscle motoneurons, 259
Neck proprioceptors
  Golgi tendon organs, 249
  muscle spindles, 245, 258
  muscle spindle receptors
    responses during labyrinthine (semicircular canal) stimulation in the vestibulo-collic reflex, 543, 546
  functional aspects
    influences on abducens motoneurons, 552, 553, 561, 563
    influences on cerebral cortex, 570
    influences on cervical spinocerebellar tract neurons of the central cervical nucleus, 531, 532
    influences on fore- and hindlimb mono- synaptic reflexes, 281-283, 515, 525
    influences on group x, 535, 557
    influences on lateral reticular nucleus, 504
    influences on neck muscle motoneurons, 258, 545
    influences on placing reaction by unilateral deafferentation of the neck, 508, 511
    influences on posture by stimulation of neck receptors, 493, 494, 515, 525
    influences on posture by unilateral

Neck proprioceptors—*cont.*
    deafferentation of the neck, 506, 508
    influences on Purkinje cells of the cerebellar vermis, 518, 522
    influences on spinoreticulocerebellar tract neurons, 504
    influences on vestibular nuclei, 501, 557
    influences on vestibulocerebellum, 555
Neck reflexes, *see* Neck proprioceptors
Nerve vestibular
  afferent fibers
    diameter, 356
    otolith afferents, 359
    resting discharge, 356
    semicircular canal afferents, 357
    functional aspects, 356, 357, 359
  efferent fibers
    peripheral action, 361
    functional aspects, 363
Nociceptive afferents, 23
Nuclear-bag fibers, *see* Muscle spindles
Nuclear-chain fibers, *see* Muscle spindles
Nucleus (Nuclei)
  dentatus
    connections, afferent
      from perihypoglossal nuclei, 657
    functional aspects
      influences on red nucleus, 87
  fastigii
    connections, afferent
      from perihypoglossal nuclei, 657
    connections, efferent
      to brain stem reticular formation, 130
      to perihypoglossal nuclei, 656
    functional aspects
      influences from semicircular canal receptors, 384, 385
  intercalatus of Staderini, *see* Perihypoglossal nuclei
  interpositus
    connections, afferent
      from perihypoglossal nuclei, 657
  praepositus hypoglossi, *see* Perihypoglossal nuclei
  of the solitary tract
    descending projections to upper cervical cord, 266
  ventralis posterolateralis of thalamus
    connections, efferent
      to motor cortex, 116
    functional aspects
      influences from cutaneous and muscle afferents, 116
  vestibular
    connections, afferent
      corticocerebello-vestibular, 453, 517, 524

Nucleus (Nuclei)—*cont.*
    from contralateral vestibular nuclei, 641
    from periphypoglossal nuclei, 576, 632, 657
    primary vestibular, 123, 383, 407
    reticulovestibular, 631
  connections, efferent
    to accessory oculomotor areas, 621
    to cerebellum, 384
    to contralateral vestibular nuclei, 641
    to extraocular motor nuclei, 431
    to lateral reticular nucleus, 503
    to perihypoglossal nuclei, 637, 654
    to spinal cord, 122, 123, 266
    to superior colliculus, 740
    to thalamus, 574, 576, 582
  functional aspects
    effects of head movements on unit activity of vestibular nuclear neurons during nystagmus, 810
    effects of head position on unit activity of vestibular nuclear neurons during nystagmus, 808
    influences from cerebellar cortex, 81, 82, 123, 126, 759
    influences from contralateral vestibular system, 371, 452, 510
    influences from ipsilateral labyrinth, 122, 123, 369
    influences from perihypoglossal nuclei, 640
    influences from reticular formation, 632, 638
    influences from spinal cord, 123
    influences on abducens internuclear neurons, 436
    influences on cerebellar cortex, 384, 453, 698, 757
    influences on cerebral cortex, 567, 569, 583
    influences on extraocular motoneurons, 469, 472
    influences on perihypoglossal nuclei, 637, 640, 654, 660
    influences on spinal cord
      limb motoneurons, 123, 124
      neck muscle motoneurons, 123, 124, 407
    influences on superior colliculus, 735–738
    influences on tectoreticular neurons, 737
    influences on tectospinal neurons, 736
    influences on thalamus, 583
    responses to vestibular (otolith) stimulation, 376
    responses to vestibular (otolith and semicircular canal) stimulation, 378

Nucleus (Nuclei)—*cont.*
    responses to vestibular (semicircular canal) stimulation 372, 377, 384, 385, 684, 710, 763
    responses to visual (optokinetic) stimulation 685, 687, 688, 710
    responses to visual and vestibular (semicircular canal) stimulation, 687, 688, 710
    unit activity of vestibular nuclear neurons during vestibular and optokinetic nystagmus, 469, 472, 808
  vestibular, descending
    connections, efferent, 123, 266, 576
    functional aspects
      influences from neck afferents, 557
      influences from somatic afferents, 123
      influences from vestibular afferents, 123
  vestibular
    group f, 123
    group x, 529
      connections, afferent, 535, 557
      connections, efferent, 557
      functional aspects, 535, 557
    group y, 434
      connections, afferent, 434, 632
      connections, efferent, 436, 576
    group z, 576
    interstitial of vestibular nerve, 576
  vestibular, lateral
    connections, afferent, *see* Nuclei, vestibular
    connections, efferent, *see* Nuclei, vestibular
    functional aspects
      influence from brain stem reticular formation, 123
      influences from cerebellar cortex, 82, 123, 759
      influences from cerebellar cortex after destruction of inferior olive, 759
      influences from contralateral vestibular system, 122
      influences from fastigial nucleus, 123
      influences from ipsilateral labyrinth, 122, 407
      influences from spinal cord, 123
      influences on contralateral vestibular system, 122
      influences on crossed spinoreticular tract (bVFRT), 82, 169, 503
      influences on spinal cord
        limb motoneurons, 123, 125
        neck muscle motoneurons, 123, 407
      influences on ventral spinocerebellar tract, 79
      unit activity after hemilabyrinthectomy, 599
  vestibular, medial

Nucleus (Nuclei)—*cont.*
    connections, afferent, *see* Nuclei, vestibular
    connections, efferent, *see* Nuclei, vestibular
    functional aspects
        influences from ipsilateral labyrinth, 123, 407
        influences from spinal cord, 123
        influences on extraocular motoneurons, 472
        influences on spinal cord
            limb motoneurons, 124, 125
            neck motoneurons, 124, 407
        unit activity during vestibular and optokinetic nystagmus, 472, 808

Nystagmus
    vestibular, 469
        unit activity of medial vestibular nucleus neurons, 472
        unit activity of presynaptic vestibular axons within the abducens nucleus, 469
    optokinetic, *see also* Optokinetic reflex, Optokinetic stimulation
        effects of head movements, 809
        effects of head position, 806

Oculomotor nucleus neurons, *see* Extraocular motor nuclei
Optokinetic reflex, 695, 703, 771
    adaptive modification of the optokinetic reflex, 771
    effects of lesion of accessory optic nuclei on optokinetic reflex, 707
Optokinetic stimulation
    influences on cerebellar flocculus, 698, 699, 757
    influences on cerebral cortex neurons, 584
    influences on eye movements, 695, 703, 771
    influences on monosynaptic spinal reflexes, 213, 214, 215
    influences on vestibular nuclei neurons, 685 710
    influences on thalamic neurons, 584
Overbalancing
    motor reactions to overbalancing, 397

Parabigeminal nucleus
    connections, afferent
        from perihypoglossal nuclei, 660
    connections, efferent
        to superior colliculus, 740
Paramedian pontine reticular formation (PPRF)
    connection, efferent
        to ipsilateral abducens nucleus, 620
        to ipsilateral medullary reticular formation, 626, 639
        to ipsilateral perihypoglossal nuclei, 640, 654

Paramedian pontine reticular formation (PPRF) —*cont.*
        to ipsilateral vestibular nuclei, 631
    functional aspects, *see* Reticular formation
Periaqueductal gray
    influences on spinothalamic tract neurons, 169
Perihypoglossal nuclei
    connections, afferent
        from brain stem reticular formation, 637, 640, 654
        from cerebellar cortex, 656
        from cervical spinal cord, 657
        from fastigial nuclei, 656
        from flocculus, 656
        from ipsilateral and contralateral perihypoglossal nuclei, 637, 640, 656, 657
        from oculomotor nuclei and neighbouring regions, 654
        from sensory trigeminal nuclei, 657
        from spinal trigeminal nucleus, 657
        from superior colliculus, 657
        from vestibular nuclei, 637, 654
    connections, efferent
        to accessory oculomotor areas, 660
        to cerebellar cortex, 657
        to cerebellar nuclei, 657
        to extraocular motor nuclei, 640, 657
        to flocculus, 657
        to ipsilateral and contralateral perihypoglossal nuclei, 637, 640, 656, 657
        to medullary and pontine reticular formation, 657
        to superior colliculus – parabigeminal nucleus, 660
        to thalamus, 576
        to vestibular nuclei, 576, 632, 657
    functional aspects
        discharge pattern of perihypoglossal nuclei neurons
            during saccadic eye movements, 672
            during fixation, 670
        influences from brain stem reticular formation, 654
        influences from superior colliculus, 654
        influences from vestibular (semicircular canal) afferents, 637, 640, 654, 660, 673
        influences from visual cortex, 654
        influences on eye movements, 660
        influences on vestibular nuclei, 640
Placing reactions
    effects of unilateral deafferentation of neck, 508, 511
    effects of unilateral lesion of lateral reticular nucleus, 506, 511
Plasticity
    adaptive changes of the vestibulo-ocular reflex, 480, 757, 763, 771

Postsynaptic inhibition
　autogenetic postsynaptic inhibition of motoneurons from group Ia afferents, 32
　autogenetic postsynaptic inhibition of motoneurons from group Ib afferents, 32
　inhibitory responses to motoneurons to stretch of agonistic muscles, 32, 37, 41
　inhibitory responses to stretch of antagonistic muscles, 40, 41
　reciprocal group Ia inhibition, 11
　recurrent inhibition by Renshaw cell discharge, 46, 54
Posture
　general concepts, xix
　motor reactions to overbalancing, 397
　motor reactions to postural perturbations, 178, 185, 201
　postural changes produced by unilateral labyrinthectomy, 591, 783
　postural responses during limb movement, 219
　postural responses following unilateral deafferentation of the neck, 506, 508
　postural responses following unilateral lesion of lateral reticular nucleus, 506, 508
　postural responses to neck stimulation, 493, 515
　postural responses to neck and vestibular (otolith) stimulation, 494, 525
　postural responses to somatosensory stimulation, 177, 185
　postural responses to vestibular (otolith) stimulation, 204, 493, 494, 515, 525
　postural responses to visual stimulation, 197, 204, 783
Preprogrammed movements
　reflex interaction with preprogramming during fall, 237
　reflex interaction with preprogramming during locomotion, 228, 238
Presynaptic inhibition
　presynaptic inhibition in the group Ia pathway, 34, 46
　presynaptic inhibition in the FRA pathway, 86
Pretectal area
　influences from visual input, 455
Primary afferent depolarization (PAD)
　induced by muscle stretch and vibration, 34, 46
　induced by supraspinal descending pathways, 165
　PAD in the FRA pathway, 86
　PAD in the group Ia pathway, 34, 46
Propriospinal neurons
　afferent projections, 80
　efferent projections, 80

Raphe nuclei
　connections, efferent

Raphe nuclei—cont.
　raphespinal, 165, 266
　functional aspects
　　influences from spinoreticular tract, 166
　　influences on spinothalamic tract, 166
Rapid eye movements (REM)
　activity of vestibular nuclei neurons, 469
Receptors, vestibular
　morphology
　　polarization of the hair cells, 355
　　sensory epithelium, 355
　functional aspects
　　fiber caliber and afferent response, 356
　　otolith afferents, 359
　　resting discharge, 356
　　semicircular canal afferents, 357
Reciprocal inhibition
　reciprocal group Ia inhibition of motoneurons, 11, 40, 41
Recruitment of motoneurons
　according to size, xix, 61
　during muscle vibration, 47
　during static muscle stretch, 50
Recurrent inhibition
　recurrent influences on ventral spinocerebellar tract neurons, 79
　recurrent inhibition of γ-static motoneurons, 55
　recurrent inhibition of group Ia inhibitory interneurons, 55
　recurrent inhibition of motoneurons, 46
　sensitivity of different size motoneurons to recurrent inhibition, 54
Red nucleus
　connections, efferent
　　rubrobulbospinal, 86
　　rubro-olivary, 86
　　rubroreticular, 80
　　rubrospinal (to upper cervical cord), 266
　functional aspects
　　influences from dentate nucleus, 87
　　influences on γ-dynamic motoneurons, 86
　　influences on spinoolivocerebellar tracts, 86
　　influences on ventral spinocerebellar tract, 79
　　inhibition of effects from FRA to motoneurons and primary afferents, 86
Renshaw cells
　contribution of different size motoneurons to Renshaw cell discharge, 52
　recurrent influences on ventral spinocerebellar tract neurons, 79
　recurrent inhibition of group Ia inhibitory interneurons, 55
　recurrent inhibition of motoneurons by Renshaw cell discharge, 46
　recurrent inhibition of γ-static motoneurons, 55

Renshaw cells—*cont.*
  responses to muscle vibration, 52
  responses to static muscle stretch, 52
  sensitivity of different size motoneurons to Renshaw inhibition, 54
Reticular formation
  connections, afferent
    from brain stem reticular formation, 626
    from cerebellar nuclei, 130
    from cerebral cortex, 80, 129
    from perihypoglossal nuclei, 657
    from red nucleus, 80
    from spinal cord (bVFRT), 80, 129, 501
    from superior colliculus, 80, 129, 729, 739
    from vestibular nuclei, 129, 503
  connections, efferent
    to abducens motoneurons, 620, 626
    to accessory oculomotor areas, 621
    to cerebellar cortex, 80
    to extraocular motoneurons, 619, 629, 626
    to medullary reticular formation from ipsilateral paramedian pontine reticular formation, 626, 639
    to perihypoglossal nuclei, 637, 640, 654
    to spinal cord, 127, 266, 272
    to superior colliculus, 740
    to vestibular nuclei, 631
  functional aspects
    effects of head movements on reticular neurons during vestibular and optokinetic nystagmus, 809
    effects of head position on reticular neurons during vestibular and optokinetic nystagmus, 807
    influences from extraocular muscle afferents on mesencephalic tegmentum, 296, 325, 335, 346
    influences from FRA, 13
    influences from pontine to ipsilateral medullary reticular formation, 626
    influences on abducens internuclear neurons, 621
    influences on dorsal column nuclei, 168
    influences on ipsilateral abducens motoneurons, 620
    influences on locomotion, 148
    influences on neck muscle motoneurons, 128, 158
    influences on perihypoglossal nuclei, 654
    influences on spinal motoneurons, 128
    influences on spinocervical tract, 166, 169
    influences on spinoreticular tract (bVFRT), 82, 166, 169
    influences on spinothalamic tract, 166, 167, 169
    influences on ventral spinocerebellar tract, 167

Reticular formation—*cont.*
    influences on vestibular nuclei, 123, 632, 638
    responses to body rotation, 385 (in pigeon)
    responses to vestibular (otolith) stimulation, 130, 502, 504
    responses to vestibular (semicircular canal) stimulation, 130, 385 (in pigeon)
    saccadic related discharge in brain stem reticular formation, 645, 727
    unit activity during eye-head movements, 805
    unit activity during saccadic eye movements, 645, 727
    unit activity during vestibular and optokinetic nystagmus, 807
Reticulospinal tracts
  influences on dorsal column nuclei, 168
  influences on neck muscle motoneurons, 128, 258
  influences on spinal motoneurons, 128
  influences on spinocervical tract, 166, 169
  influences on spinoreticular tract (bVFRT), 82, 166, 169
  influences on spinothalamic tract, 166, 167, 169
  influences on ventral spinocerebellar tract, 167
  origin and distribution within the spinal cord, 127
  responses to vestibular (semicircular canal and otolith) stimulation, 130

Saccadic eye movements
  discharge patterns of saccadic eye movement related neurons, 646, 725
  following stimulation of cerebellar cortex, 315
  following stimulation of superior colliculus, 729, 745
  model, 647, 725
  saccadic related discharge in brain stem reticular formation, 645, 727
  saccadic related discharge in cerebellar nuclei, 645
  saccadic related discharge in extraocular motoneurons, 645
  saccadic related discharge in perihypoglossal nuclei, 672
  saccadic related discharge in superior colliculus, 725
  saccadic related discharge of Purkinje cells of the cerebellar cortex, 316, 385, 645
Scratching
  activity of ventral spinocerebellar tract neurons, 79
  depression of peripheral effects on spinoreticular (bVFRT) neurons, 82
Size principle

Size principle—*cont.*
  contribution of different size motoneurons to Renshaw cell discharge, 52
  recruitment of motoneurons according to size, xix, 61
  responses of different size motoneurons to muscle vibration, 45, 47
  responses of different size motoneurons to static muscle stretch, 50
  sensitivity of different size motoneurons to recurrent inhibition, 54
Smooth pursuit eye movements
  coordination of eye-head movements, 799
  unit activity of Purkinje cells of cerebellar cortex, 385
Spinal reflexes
  effects of DOPA, 14
  influences from labyrinth (otolith) stimulation, 212, 214, 215, 525
  influences from neck stimulation, 281–283, 515, 525
Spinocerebellar tract, dorsal, 79, 91
  influences from cerebral cortex, 165
  influences from FRA, 13
  influences from muscle spindle afferents, 98
  responses to sinusoidal muscle stretch, 95
  responses to static muscle stretch, 93
Spinocerebellar tract, ventral, 79
  influences from cerebral cortex, 79, 165
  influences from FRA, 13
  influences from lateral vestibulospinal tract, 79
  influences from recurrent discharge of Renshaw cells, 79
  influences from red nucleus, 79
  influences from tonic descending inhibitory system, 167
  unit activity during locomotion, 79
  unit activity during scratching, 79
Spinocervical tract
  influences from brain stem reticular formation, 169
  influences from cerebral cortex, 165
  influences from tonic descending inhibitory system, 166
Spinoolivocerebellar tracts, 83
  influences from cerebral cortex, 86
  influences from FRA, 13
  influences from red nucleus, 86
Spinoreticular tract (bVFRT), 80, 501
  depression of peripheral effects on bVFRT during locomotion and scratching, 82
  influences from corticospinal tract, 82, 165
  influences from descending monoaminergic projections and raphe nuclei, 82, 165
  influences from FRA, 13, 80, 501
  influences from lateral vestibulospinal tract, 81, 82, 169, 503

Spinoreticular tract (bVFRT)—*cont.*
  influences from both neck and vestibular (otolith) afferents, 504
  influences from neck muscle afferents, 504
  influences from reticulospinal tracts, 82, 169
  influences from tonic descending inhibitory system, 82, 166
  influences from vestibular (otolith) afferents, 169, 502–504
Spinothalamic tract, 105
  influences from cerebral cortex, 163
  influences from FRA, 105, 106
  influences from mesencephalic reticular formation, 169
  influences from periaqueductal gray, 169
  influences from pontomedullary reticular formation, 167
  influences from raphe nuclei, 165, 166
  influences from tonic descending inhibitory system, 166
Stretch reflex
  active mechanisms in stretch reflex, 185, 240
  coupled stretch reflexes, 185
  functional stretch reflex, 151, 178
  passive mechanisms in stretch reflex, 185, 240
  responses of different size motoneurons during muscle stretch, 50
Substantia nigra
  connections, efferent
    to superior colliculus, 740

Thalamus
  connections, afferent
    from group y, 576
    from interstitial nucleus of vestibular nerve, 576
    from nucleus intercalatus of Staderini, 576
    from vestibular nuclei, 574, 582
  connections, efferent
    thalamocortical, 116, 575, 583
  functional aspects
    influences from cutaneous and muscle afferents, 116
    influences from extraocular muscle afferents, 296
    responses to both visual and vestibular stimulation, 584
    responses to vestibular (semicircular canal) stimulation, 583
    responses to visual (optokinetic) stimulation, 584
Tonic descending inhibitory system
  influences on spinocervical tract, 166
  influences on spinoreticulocerebellar tract, 166
  influences on spinothalamic tract, 166
  influences on ventral spinocerebellar tract, 167

Tonic labyrinth reflexes, *see* Labyrinth macular input
Tonic vibration reflex, *see* Muscle vibration
Trigemino-collic reflex, *see* Neck muscle motoneurons
Trigeminal system
  connections, afferent
    peripheral course of extraocular muscle afferents in ungulata, 291
    somatotopic organization of extraocular muscle afferents within semilunar trigeminal ganglion in ungulata, 291
    somatotopic organization of extraocular muscle afferents within sensory trigeminal nuclei in ungulata, 294
  connections, efferent
    from sensory trigeminal nuclei to perihypoglossal nuclei, 657
    from sensory trigeminal nuclei to superior colliculus, 740
  functional aspects
    influences on neck muscle motoneurons, 259
    influences on semilunar trigeminal ganglion in ungulata, 291
    influences on sensory trigeminal nuclei in ungulata, 294

Vestibulo-collic reflex, 126
  α–γ linkage during the vestibulo-collic reflex, 546
  responses during vestibular (semicircular canal) stimulation, 537
    with cervical dorsal roots cut, 543
    with cervical dorsal roots intact, 538
Vestibulo-ocular reflexes
  adaptive modifications and plasticity of the vestibulo-ocular reflex, 480, 757, 763, 771
  horizontal vestibulo-ocular reflex by stimulation of horizontal semicircular canals, 436, 451, 452
  influences of inferior olive and preolivary structures on the vestibulo-ocular reflex, 455
  influences of visual system on the vestibulo-ocular reflex, 695, 703, 771

Vestibular-ocular reflexes–*cont.*
  mathematical models of the vestibulo-ocular reflex, 478, 483
  representation of the vestibulo-ocular reflex in flocculus, 453
  sacculo-ocular reflex, 434
  vertical vestibulo-ocular reflex
    by stimulation of anterior semicircular canals, 431, 451
    by stimulation of posterior semicircular canals, 433, 451, 452
    by stimulation of semicircular canal and otolith receptors, 443
  vestibulo-ocular reflex in darkness, 695, 703
Vibration reflex, *see* Muscle vibration
Visual cortex
  influences from extraocular muscle afferents, 342, 347
  influences on perihypoglossal nuclei, 654
Visual fixation reflex, 745
Visual system
  coordination of eye-head movements by visual input, 795
  influences on accessory optic nuclei, 705,·, 717
  influences on cerebellar cortex, 316
  influences on extraocular motoneurons, *see* Optokinetic reflex
  influences on flocculus, 698, 699, 704, 719, 757
  influences on inferior olive and preolivary structures, 455
  influences on motor responses to fall, 204, 205, 424
  influences on postural changes following hemilabyrinthectomy, 783
  influences on posture, 197, 204, 783
  influences on pretectal area, 455
  influences on superior colliculus, 335–340, 735–738, 745
  influences on tectoreticular neurons, 737
  influences on tectospinal neurons, 736
  influences on vestibular nuclei, 685, 710
  influences on vestibulo-ocular reflex, 695, 703, 771